Transcendental Representations with Applications to Solids and Fluids

Mathematics and Physics for Science and Technology

Series Editor

L.M.B.C. Campos

Director of the Center for Aeronautical
and Space Science and Technology
Lisbon Technical University

Complex Analysis with Applications to Flows and Fields
L.M.B.C. Campos

Transcendental Representations with Applications to Solids and Fluids
L.M.B.C. Campos

MATHEMATICS AND PHYSICS FOR SCIENCE AND TECHNOLOGY

Transcendental Representations with Applications to Solids and Fluids

L.M.B.C. Campos

Director of the Center for Aeronautical and Space Science and Technology
Lisbon Technical University

CRC Press is an imprint of the
Taylor & Francis Group, an **informa** business

CRC Press
Taylor & Francis Group
6000 Broken Sound Parkway NW, Suite 300
Boca Raton, FL 33487-2742

First issued in paperback 2019

ISBN-13: 978-1-4398-3431-2 (hbk)
ISBN-13: 978-0-367-38152-3 (pbk)

Library of Congress Cataloging-in-Publication Data

Campos, Luis Manuel Braga da Costa.
Transcendental representations with applications to solids and fluids / Luis Manuel Braga da Costa Campos.
p. cm. -- (Mathematics and physics for science and technology)
Includes bibliographical references and index.
ISBN 978-1-4398-3431-2 (hardback)
1. Transcendental functions. 2. Mathematical physics. 3. Physics--Mathematical models. 4. Mechanical engineering--Mathematical models. I. Title.

QC20.7.F87C36 2012
515'.22--dc23 2011043043

Visit the Taylor & Francis Web site at
http://www.taylorandfrancis.com

and the CRC Press Web site at
http://www.crcpress.com

to the memory of António Gouvea Portela

Contents

List of Tables, Notes, Diagrams, Classifications, Lists, and Examples

Tables

Notes

Diagrams

Classifications

Lists

Examples

Series Preface

The aim of the Mathematics and Physics for Science and Technology series is to describe the mathematical methods as they are applied to model natural physical phenomena and solve scientific and technological problems. The primary emphasis is on the application, including formulation of the problem, detailed solution, and interpretation of results. The mathematical methods are presented in sufficient detail to justify every step of solution and avoid superfluous assumptions.

The main areas of physics are covered, namely,

1. Mechanics of particles, rigid bodies, deformable solids, and fluids
2. Electromagnetism, thermodynamics, and statistical physics as well as their classical, relativistic, and quantum formulations
3. Interactions and combined effects (e.g., thermal stresses, magnetohydrodynamics, plasmas, piezoelectricity, and chemically reacting and radiating flows)

The examples and problems chosen include natural phenomena in our environment, geophysics, and astrophysics; the technological implications in various branches of engineering; and other mathematical models in biological, economic, and social sciences.

The coverage of areas of mathematics and branches of physics is sufficient to lay the foundations of all branches of engineering, namely,

1. Mechanical—including machines, engines, structures, and vehicles
2. Civil—including structures and hydraulics
3. Electrical—including circuits, waves, and quantum effects
4. Chemical—including transport phenomena and multiphase media
5. Computer—including analytical and numerical methods and associated algorithms

Particular emphasis is given to interdisciplinary areas such as electromechanics and aerospace engineering. These require a combined knowledge of several subjects and have an increasing importance in modern technology.

Analogies are applied in an efficient and concise way across distinct disciplines and also stress the differences and aspects specific to each area, for example,

1. Potential flow, electrostatics, magnetostatics, gravity field, steady heat conduction, and plane elasticity and viscous flow
2. Acoustic, elastic, electromagnetic, internal, and surface waves
3. Diffusion of mass, electricity and momentum

In each case the analogies are exploited by common mathematical methods with distinct interpretations in each context.

The series is organized as a sequence of mathematical methods, each with a variety of applications. As the mathematical methods progress, the range of applications widens. For example, complex functions are used to study potential flows and electrostatics in the plane. The three-dimensional extension uses generalized functions. The latter are used with differential equations to describe vibrations and waves. The series and integral transforms are applied to initial and boundary-value problems. Tensor calculus is used for elasticity, viscous fluids, and relativity. Special functions are used in quantum mechanics, acoustics and other waves in fluids, solids, and plasmas. High-order, nonlinear, and coupled systems are considered using analytical and numerical methods. Thus each method is consolidated with diverse applications before proceeding to the next.

The presentation of the material is intended to remain accessible to the university student. The subjects are introduced at a basic undergraduate level. The deductions and intermediate steps are detailed. Extensive illustrations and detailed legends promote visual and intuitive memory and understanding. The material is presented like a sequence of lectures and can be used to construct the subjects or disciplines of a university curriculum. It is possible to adjust the level of the curriculum by retaining the basic theory and simpler examples and using the rest as background material for further reading. Some topics chosen for greater insight may be included according to the motivation. The bibliography gives a choice of approaches to the subject and the possibility to focus more in specific subareas. The presentation follows a logical rather than historical sequence; some references to the original sources are used to give a historical perspective. The notes at the ends of chapters hint at the broader scope of the subject. The contents of each chapter are previewed in an introduction.

The present series embodies a concept of "interdisciplinary education in science and technology." The traditional approach is to study each area of mathematics separately (analysis, geometry, differential equations, etc.) as well as each branch of physics (mechanics, heat and thermodynamics, electromagnetism, etc.). The student is then expected to "merge" all these sources of information, for example, to know all that is needed about partial differential equations for the Maxwell theory of electromagnetism, calculus of variations for the minimum energy methods of elasticity, thermodynamics for the dynamics of compressible fluids, and so on. The time gaps and disjoint nature of this teaching implies a careful sequence of subjects to ensure each subject starts with the required background. Also there is considerable duplication in that similar methods and analogous problems recur in different contexts. Furthermore, the student discovers the utility of most of the mathematics much later, when it is applied to physical and engineering problems. The combined interdisciplinary study aims to resolve these issues.

Although the course starts at the undergraduate level it gradually proceeds to the research level and to the frontiers of current knowledge. The presentation of each subject takes into account from the very beginning not only the fundamentals but also the major topics of subsequent use. For example, the treatment of complex functions lays the basis for differential equations, integral transforms, asymptotics, and special functions. Linear algebra and analytic geometry lead to tensor analysis, differential geometry, variational calculus, and relativity. An introduction to fluid mechanics via the potential flow is followed by vortical, compressible, viscous, and multiphase flows. Electrostatics and magnetostatics are followed by unsteady electromagnetic fields and waves, magnetohydrodynamics, and plasmas. The basic subjects are treated at an early stage, analogies are presented, and at subsequent stages they are combined into multidisciplinary applications.

For example, a fluid may be subject to the four restoring forces associated with pressure, gravity, rotation, and magnetic fields. The corresponding wave motions are respectively acoustic, internal, inertial, and magnetic waves. They appear combined as magneto-, acoustic-gravity-inertial (MAGI) waves in a compressible, ionized, stratified rotating fluid. The simplest exact solutions of the MAGI wave equation require special functions. Thus the topic of MAGI waves combines gravity field, fluid mechanics, and electromagnetism and uses complex analyses, differential equations, and special functions. This is not such a remote subject, since many astrophysical phenomena involve the combination of several of these effects, as does the technology of controlled nuclear fusion. The latter is the main source of energy in the stars and in the universe; if harnessed, it would provide a clean and inexhaustible source of energy on earth. Closer to our everyday experience there is a variety of electromechanical and control systems that use modern interdisciplinary technology. The ultimate aim of the present series is to build up knowledge from the undergraduate to the research level across a range of subjects to cover contemporary and promising interdisciplinary needs. This requires a consistent treatment of all subjects so that their combination fits together as a whole.

The way these objectives are pursued in the present series is that each book contains (i) a related set of mathematical methods; (ii) its application to model an ensemble of analogous physical phenomena, highlighting both similarities and differences; and (iii) a number of practical utilizations in current technology that are thereby enabled. Each book presents its subject matter and lays the background for subsequent books. Successive books evolve on the same three complementary directions: (i) toward the next set of more general or powerful mathematical methods; (ii) enlarging the range and scope of physical phenomena that can be modeled, continuing to exploit further analogies and highlight limitations and differences; and (iii) enabling a wider variety of technological and scientific utilizations. The evolution of the content of successive books respects a consistency in terminology and simbology across the various branches of mathematics, areas of physics, and technological motivations. The new concepts are highlighted in bold, and the main conclusions appear in italics, integrated as much as possible in a continuous process of reasoning that starts from some fundamental idea in each chapter, with complementary aspects at the section and subsection levels. The uniform style and consistent notation are intended to facilitate the use of the series as a ready reference, connecting seamlessly the constituent books.

The approach followed in the present series is a combined study of mathematics, physics, and engineering so that the practical motivation develops side by side with the theoretical concepts: the mathematical methods are applied without delay to "real" problems, not just to exercises. The electromechanical and other analogies simulate the ability to combine different disciplines, which is the basis of much modern interdisciplinary science and technology. Starting with the simpler mathematical methods and consolidating them with the detailed solutions of physical and engineering problems gradually widens the range of topics that can be covered. The traditional method of separate monodisciplinary study remains possible, selecting mathematical disciplines (e.g., complex functions) or sets of applications (e.g., fluid mechanics). The combined multidisciplinary study has the advantage of connecting mathematics, physics, and technology at an earlier stage. Moreover, preserving that link provides a broader view of the subject and the ability to innovate. Innovation requires understanding the technical aims, the physical phenomena that can implement them, and the mathematical methods that quantify the expected results. The combined interdisciplinary approach to the study of mathematics, physics, and engineering is thus a direct introduction to a professional experience in scientific discovery and technological innovation.

Preface

Book 1 (*Complex Analysis with Applications to Flows and Fields*) in this series was concerned with complex analysis, including (Part 1) the representation of complex functions on the plane and Riemann surfaces; (Part 2) differential and integral calculus including residues; (Part 3) power series expansions in the neighborhood of regular and singular points; and (Part 4) conformal mapping, including bilinear, Schwartz–Christoffel, and other transformations. Book 2 (*Transcendental Representations with Applications to Solids and Fluids*) adds to the (a) series expansions three other infinite representations; (b) series of fractions for meromorphic functions, that is, functions with an infinite number of poles; (c) infinite products for functions with infinitely many zeros; and (d) continued fractions as alternative representation of items a–c. These four infinite representations (a–d) together with the main topics of complex analysis (1–4) have their simplest illustration in the elementary transcendental functions (and their inverses), namely, (1) exponential (logarithm) and (2) circular and hyperbolic (cyclometric) functions. Some higher transcendental functions such as the confluent and Gaussian hypergeometric functions are mentioned when necessary. Some related aspects of convergence (such as summation) and divergence (such as transfinite numbers) are also considered.

Book 1 applied the theory of complex functions mainly, but not exclusively, to five classes of two-dimensional potential fields: (1) steady irrotational and incompressible flow, (2) gravity field, (3 and 4) electrostatics/magnetostatics, and (5) steady heat conduction. These fields satisfy the Laplace (Poisson) equation in the absence (presence) of forcing terms such as multipoles. Book 2 continues the application of complex functions, including the elementary transcendental functions, to eight more classes of potential fields: (6) incompressible rotational flows, (7) some compressible irrotational flows, (8) some unsteady flows, (9) rotating flows, (10 and 11) surface tension and capillarity, (12) deflection of membranes under load, and (13) torsion of rods by torques. Besides these potential fields (1–13) specified by harmonic functions, that is, solutions of the Laplace equation, two biharmonic fields that are solutions of higher-order multiharmonic operators are considered: (14) plane elasticity and (15) plane viscous flows. Thus, Books 1 and 2 together cover complex analysis (including the elementary transcendental functions) and applications to fluid and solid media and force fields mostly in two dimensions.

Organization of the Book

This book is organized in a similar way to Book 1 of the Mathematics and Physics for Science and Technology series. This book consists of 10 chapters: (1) the odd-numbered chapters present mathematical developments, (2) the even-numbered chapters contain physical applications, and (3) the last chapter is a set of 20 detailed examples of items 1 and 2. The chapters are divided into sections and subsections, for example, Chapter 1, Section 1.1, Subsection 1.1.1. The formulas are numbered by chapters in curved brackets, for example, Equation 1.2 is Equation 2 of Chapter 1. When referring to Book 1, the symbol I is inserted at the beginning, for example, Chapter I.35, Section I.35.1, Subsection

I.35.1.1, Equation I.35.33a. The final part of each chapter includes (1) a conclusion referring to the figures as a kind of visual summary and (2) note(s), list(s), table(s), diagram(s), and classification(s) as additional support. The latter apply at the end of each chapter and are numbered within the chapter (e.g., Note 8.1, List 6.1, Table 5.1, Classification 8.1, Diagram 6.1); if there is more than one, they are numbered sequentially (e.g., Notes 4.1 through 4.11). The chapter starts with an introductory preview, and related topics may be mentioned in the notes at the end. The lists of mathematical symbols and physical quantities appear before the main text, and the bibliography and index of subjects are at the end of the book.

Acknowledgments

This book would justify a similar list of acknowledgments to the first book in the series, which is not repeated here. A few acknowledgments are renewed to those who contributed more directly to the final form of the present book: Mrs. L. Sousa and S. Pernadas for help with the manuscripts, Mr. J. Coelho for the drawings, Professor J.M.C.S. André for very helpful criticisms, Emeritus Professor A.G. Portela to whom this book is dedicated for several pages of written general and specific comments and suggestions, and last, but not least, to my wife as the companion of the author in preparing this work.

Author

Luis Manuel Braga da Costa Campos was born on 28 March 1950 in Lisbon, Portugal. He graduated in 1972 as a mechanical engineer from the Instituto Superior Técnico (IST) of Lisbon Technical University. The tutorials as student (1970) were followed by a career at the same institution (IST) through all levels: assistant on probation (1972), assistant (1974), auxiliary professor (1978), assistant professor (1982), and Chair de Applied Mathematics and Mechanics (1985). His present positions include coordinator of undergraduate and postgraduate degrees in aerospace engineering since their creation in 1991, coordinator of the applied and aerospace mechanics group in the Department of Mechanical Engineering, and director and founder of the Center for Aeronautical and Space Science and Technology.

Campos received his doctorate on "Waves in Fluids" at the Engineering Department of Cambridge University, England (1977). It was followed by a Senior Rouse Ball Scholarship at Trinity College, Cambridge, while also on leave from IST. The first sabbatical was as a senior visitor at the Department of Applied Mathematics and Theoretical Physics of Cambridge University, England (1984). The second sabbatical (1991) was as an Alexander von Humboldt scholar at the Max-Planck Institut für Aeronomie in Katlenburg-Lindau, Germany. Further sabbaticals abroad were excluded by major commitments at the home institution. The latter were always compatible with extensive professional travel related to participation in scientific meetings, individual or national representation in international institutions, and collaborative research projects.

Campos received a von Karman medal from the Advisory Group for Aerospace Research and Development (AGARD) and Research and Technology Organization (RTO). Participation in AGARD/RTO included the role of vice chairman of the System Concepts and Integration Panel and chairman of the Flight Mechanics Panel and of the Flight Vehicle Integration Panel. He was also a member of the Flight Test Techniques Working Group; the latter related to the creation of an independent flight test capability active in Portugal in the last 20 years, which has been used in national and international projects including Eurocontrol and the European Space Agency (ESA). Participation in the ESA has included roles on various committees from being a national representative up to the Council and Council of Minister levels.

His participation in activities sponsored by the European Union (EU) included (1) 27 research projects with industry, research, and academic institutions; (2) membership in various committees, including vice chairman of the Aeronautical Science and Technology Advisory Committee; and (3) participation in the Space Advisory Panel on the future role of the EU in space. The author has been a member of the Space Science Committee of the European Science Foundation and concerned with liaison with the Space Science Board of the National Science Foundation of the United States. He has also been a member of the Committee for Peaceful Uses of Outer Space of the United Nations. Various individual consultant and advisory missions were performed on behalf of these and other institutions. His participation in professional societies includes roles as a member and vice chairman of the Portuguese Academy of Engineering; fellow of the Royal Aeronautical Society, Astronomical Society, and Cambridge Philosophical Society; and associate fellow of the American Institute of Aeronautics and Astronautics.

His publications include 8 books, 126 papers in 55 journals, and 199 communications to symposia. His areas of research focus on four topics: acoustics, magnetohydrodynamics, special functions, and flight dynamics. His work on acoustics has concerned the generation, propagation, and refraction of sound in flows with mostly aeronautical applications. His work on magnetohydrodynamics has concerned magneto-acoustic-gravity-inertial waves in solar–terrestrial and stellar physics. Developments on special functions have been mostly based on differintegration operators, generalizing the ordinary derivative and primitive to complex order. His work on flight dynamics has concerned aircraft and rockets, including trajectory optimization, performance, stability, control, and atmospheric disturbances.

Campos's interests in topics from mathematics to physics and engineering fits with the aims and contents of the present series, and his university teaching and scientific and industrial research relates to the build-up of the series from the undergraduate to the research level. His professional activities on the technical side are balanced by other cultural and humanistic interests; they are not reflected in publications, except for one literary work. His complementary nontechnical interests include classical music (mostly orchestral and choral), plastic arts (painting, sculpture, and architecture), social sciences (psychology and biography), history (classical, renaissance, and overseas expansion), and technology (automotive, photo, and audio). Campos is listed in various biographical publications, including *Who's Who in the World* since 1986.

Mathematical Symbols

These mathematical symbols are commonly used in the context of (1) sets, quantifiers, and logic; (2) numbers, ordering, and bounds; (3) operations, limits, and convergence; and (4) vectors, functions, and the calculus. This section concludes with a list of functional spaces, most but not all of which appear in the present book. The book, chapter, and section where the symbol first appears may be indicated after a colon, for example, "I.2.3" means book 1, chapter 2, section 3, and II.5.8 means book 2, chapter 5, section 8.

1. Sets, Quantifiers, and Logic

1.1 Sets

$A \equiv \{x:\ldots\}$—set a whose elements x have the property...
$A \cup B$—union of sets A and B
$A \cap B$—intersection of sets A and B
$A \supset B$—set A contains set B
$A \subset B$—set A is contained in set B

1.2 Quantifiers

$\forall_{x \in A}$—for all x belonging to A holds...
$\exists_{x \in A}$—there exists at least one x belonging to A such that...
$\exists^1{}_{x \in A}$—there exists one and only one x belonging to A such that...
$\exists^\infty_{x \in A}$—there exist infinitely many x belonging to A such that...

1.3 Logic

$a \wedge b$—a and b
$a \vee b$—or (inclusive): a or b or both
$a \dot{\vee} b$—or (exclusive): a or b but not both
$a \Rightarrow b$—implication: a implies b
$a \Leftrightarrow b$—equivalence: a implies b and b implies a
$a \,\bar{\Rightarrow}\,$ b—nonimplication: a may not imply b

1.4 Constants

$e = 2.7182\ 81828\ 45904\ 52353\ 60287$
$\pi = 3.1415\ 92653\ 58979\ 32384\ 62643$
$\gamma = 0.5772\ 15664\ 90153\ 28606\ 06512$
$\log 10 = 2.3025\ 85092\ 99404\ 56840\ 179915$

2. Numbers, Ordering, and Bounds

2.1 Types of Numbers

$|C$—complex numbers: I.1.2
$|C^n$—ordered sets of n complex numbers
$|F$—transfinite numbers: 9.7–9.9
$|H$—hypercomplex
$|I$—irrational numbers, real nonrational numbers: I.1.2, II.9.7.2
$|L$—rational numbers, ratios of integers: I.1.1, II.9.7.1
$|N$—natural numbers, positive integers: I.1.1
$|N_0$—nonnegative integers, zero plus natural numbers: I.1.1
$|P$—prime numbers, numbers without divisors
$|Q$—quaternions: I.1.9
$|R$—real numbers: I.1.2, II.9.7.3
$|R^n$—ordered sets of n real numbers: II.9.7.3
$|Z$—integer numbers: I.1.1, II.9.7.1

2.2 Complex Numbers

$|\ldots|$—modulus of complex number…: I.1.4
arg (…)—argument of complex number…: I.1.4
Re (…)—real part of complex number…: I.1.3
Im (…)—imaginary part of complex number…: I.1.3
… *—conjugate of complex number…: I.1.6

2.3 Relations between Real Numbers

$a > b$—a greater than b
$a \geq b$—a greater or equal to b
$a = b$—a equal to b
$a \leq b$—a smaller or equal to b
$a < b$—a smaller than b

2.4 Intervals and Ordering of Numbers

(a,b)—closed interval: $a \leq x \leq b$
)a,b(—open interval: $a < x < b$
)a,b)—interval open (closed) at left (right): $a < x \leq b$
(a,b(—interval closed (open) at left (right): $a \leq x < b$

3. Operations, Limits, and Convergence

3.1 Operations between Numbers

$a + b$—sum: a plus b

$a - b$—difference: a minus b
$a \times b$—product: a times b
a/b—ratio: a divided by b (alternative a:b)
a^b—power: a to the power b
$\sqrt[b]{a}$ —root: root b of a

3.2 Iterated Sums and Products

$\sum_a$—sum over a set

$\sum_{n=a}^{b}$—sum from $n = a$ to $n = b$

$\sum_{n,m=a}^{b}$—double sum over $n,m = a,\ldots,b$

$\prod_a$—product over a set

$\prod_{n=a}^{b}$—product from $n = a$ to $n = b$

$[f,g]$—inner product of functions: 5.7.1
$\|f\|$—norm of a function: 5.7.1

3.3 Limits

lim—limit when x tends to a, $x \to a$
l.i.m.—limit in the mean
$a \sim O(b)$—a is of order b, $\lim b/a \neq 0,\infty$
$a \sim o(b)$—b is of lower order than a, $\lim b/a = 0$

3.4 Convergence

A.C.—absolutely convergent: I.21.2
A.D.—absolutely divergent: I.21.2
C.—convergent: I.21.2, II.9.2.1
C.C.—conditionally convergent: I.21.2, II.9.4
Cn—converges to class n, $C0 \equiv C$: II.9.6
D.—divergent: I.21.1, II.9.2.1
N.C.—nonconvergent, divergent or oscillatory: I.21.1
O.—oscillatory: I.21.1, II.9.2.1
T.C.—totally convergent: I.21.7
U.C.—uniformly convergent: I.21.5
applies to
—power series: I.21.1, II.1

—series of fractions: I.27.9, II.1.2
—infinite products: I.27.9, II.1.4
—continued fractions: I.1.6

4. Vectors, Functions, and the Calculus

4.1 Vectors

$\vec{A}.\vec{B}$—inner product

$\vec{A}\wedge\vec{B}$—outer product

$\vec{A}.(\vec{B}\wedge\vec{C})$—mixed product

$\vec{A}\wedge(\vec{B}\wedge\vec{C})$—double outer product

$|\vec{A}|$—modulus

ang $(\vec{A},\vec{B})$—angle of vector $\vec{B}$ with vector $\vec{A}$

4.2 Invariant Operators

$\nabla\Phi$—gradient of a scalar: I.11.7, II.4.4.5
$\nabla\vec{A}$—divergence of a vector $\vec{A}$: I.11.7, II.4.4.5
$\nabla\wedge\vec{A}$—curl of a vector $\vec{A}$: I.11.7, II.4.4.5
$\nabla^2\Phi$—Laplacian of a scalar Φ: I.11.6, II.4.4.5
$\nabla^2\vec{A}$—Laplacian of a vector $\vec{A}$: II.4.4.5

4.3 Values of Functions

$f(a)$—value of function f at point a
$f(a + 0)$—right-hand limit at a
$f(a - 0)$—left-hand limit at a
$f_{(n)}(a)$—residue at pole of order n at a: I.15.8, II.1.1.4
$\overline{B}$ or M—upper bound, $|f(z)|\le\overline{B}$ for z in …
$\underline{B}$ or m—lower bound, $|f(z)|\ge\underline{B}$ for z in …
$f\circ g$—composition of functions f and g

4.4 Integrals

$\int\ldots dx$—primitive of…with regard to x: I.13.1

$\int^{y}\ldots dx$—indefinite integral of…at y: I.13.1

$\int_{a}^{b}\ldots dx$—definite integral of…between a and b: I.13.2

$\int_{a}^{b}\ldots dx$—principal value of integral: I.17.8

$\int^{(z+)}$—integral along a loop around z in the positive (counterclockwise) direction: I.13.5

$\int^{(z-)}$—idem in the negative (clockwise) direction: I.13.5

$\int_L$—integral along a path L: I.13.2

$\int_C^{(+)}$—integral along a closed path or loop C in the positive direction: I.13.5

$\int_C^{(-)}$—integral along a closed path or loop C in the negative direction: I.13.5

[f,g]—inner product of functions: II.5.7.1

‖f‖—norm of a function: II.5.7.1

5. Functional Spaces

The sets of numbers and spaces of functions are denoted by calligraphic letters, in alphabetical order:
…(a, b)—set of functions over interval from a to b
omission of interval: set of function over real line)−∞,+∞(
$\mathcal{A}$ (…)—analytic functions in…: I.27.1
$\bar{\mathcal{A}}$ (…)—monogenic functions in…: I.31.1
$\mathcal{B}$ (…)—bounded functions in…: $\mathcal{B} \equiv \mathcal{B}^0$: I.13.3
$\mathcal{B}^n$ (…)—functions with bounded nth derivative in….
$\mathcal{C}$ (…)—continuous functions in…: $C \equiv C^\circ$: I.12.2
$\mathcal{C}^n$ (…)—functions with continuous nth derivative in…
$\bar{\mathcal{C}}$ (…)—piecewise continuous functions in…: $C \equiv C^\circ$
$\bar{\mathcal{C}}^n$ (…)—functions with piecewise continuous nth derivative in…
$\tilde{\mathcal{C}}$ (…)—uniformly continuous function in…: I.13.4
$\tilde{\mathcal{C}}^n$ (…)—function with uniformly continuous nth derivative in…
$\mathcal{D}$ (…)—differentiable functions in…: $\mathcal{D} \equiv \mathcal{D}^0$: I.11.2
$\mathcal{D}^n$ (…)—n-times differentiable functions in…
$\mathcal{D}^\infty$ (…)—infinitely differentiable or smooth functions in…: I.27.1
$\bar{\mathcal{D}}$ (…)—piecewise differentiable functions in…: $\bar{\mathcal{D}} \equiv \bar{\mathcal{D}}^\circ$
$\bar{\mathcal{D}}^n$ (…)—functions with piecewise continuous nth derivative in…
$\mathcal{E}$ (…)—Riemann integrable functions in…: I.13.2
$\bar{\mathcal{E}}$ (…)—Lebesgue integrable functions in…
$\mathcal{F}$ (…)—functions of bounded oscillation (or bounded fluctuation or bounded variation) in…; $\mathcal{F} \equiv \mathcal{E} \equiv \mathcal{F}^0$: 5.7.5
$\mathcal{F}^n$ (…)—functions with nth derivative of bounded oscillation (or fluctuation or variation) in….
$\mathcal{G}$ (…)—generalized functions (or distributions) in…
$\mathcal{H}$ (…)—harmonic functions in…: I.11.4, II.4.6.4
$\mathcal{H}_2$ (…)—biharmonic functions in…: II.4.6.4

$\mathcal{H}_n$ (…)—multiharmonic functions of order n in…: II.4.6.6
$\mathcal{I}$ (…)—integral functions in…: I.27.9, II.1.1.7
$\mathcal{I}_m$ (…)—rational–integral functions of degree m in…: $\mathcal{I} \equiv \mathcal{I}_0$: I.27.9, II.1.1.9
$\mathcal{J}$ (…)—square integrable functions with a complete orthogonal set of functions, *Hilbert space*
$\bar{\mathcal{K}}$ (…)—Lipshitz functions in…
$\mathcal{K}^n$ (…)—homogeneous functions of degree n in…
$\mathcal{L}^1$ (…)—absolutely integrable functions in…
$\mathcal{L}^2$ (…)—square integrable functions in…
$\mathcal{L}^p$ (…)—functions with power p of modulus integrable in…—*normed* space: $\mathcal{L}^p \equiv \mathcal{W}_0^p$
$\mathcal{M}^+$ (…)—monotonic increasing functions in…: 9.1.1
$\mathcal{M}_0^+$ (…)—monotonic nondecreasing functions in…: 9.1.1
$\mathcal{M}_0^-$ (…)—monotonic nonincreasing functions in…: 9.1.2
$\mathcal{M}^-$ (…)—monotonic decreasing functions in…: 9.1.2
$\mathcal{N}$ (…)—null functions in…
$\mathcal{O}$ (…)—orthogonal systems of functions in…: 5.7.2
$\bar{\mathcal{O}}$ (…)—orthonormal systems of functions in…: 5.7.2
$\tilde{\mathcal{O}}$ (…)—complete orthogonal systems of functions in…: 5.7.5
$\mathcal{P}$ (…)—polynomials in…: I.27.7, II.1.1.6
$\mathcal{P}_n$ (…)—polynomials of degree n in…: I.27.7, II.1.1.6
$\mathcal{Q}$ (…)—rational functions in…: I.27.7, II.1.1.6
$\mathcal{Q}_n^m$ (…)—rational functions of degrees n, m in…: I.27.7, II.1.1.6
$\mathcal{R}$ (…)—real functions, that is, with the real line as range
$\mathcal{S}$ (…)—complex functions, that is, with the complex plane as range
$\mathcal{T}$ (…)—functions with compact support, that is, which vanish outside a finite interval
$\mathcal{T}^n$ (…)—temperate functions of order n: n-times differentiable functions with first $(n–1)$ derivatives with compact support
$\mathcal{T}^\infty$ (…)—temperate functions: smooth or infinitely differentiable functions with compact support
$\mathcal{U}$ (…)—single-valued functions in…: I.9.1
$\tilde{\mathcal{U}}$ —injective functions in…: I.9.1
$\underline{\mathcal{U}}$ —surjective functions in…: I.9.1
$\tilde{\underline{\mathcal{U}}}$ —bijective functions in…: I.9.1
$\mathcal{U}_n$ (…)—multivalued functions with n branches in…: I.6.1
$\mathcal{U}_\infty$ (…)—many-valued functions in…: I.6.2
$\mathcal{U}^1$ (…)—univalent functions in…: I.37.4
$\mathcal{U}^m$ (…)—multivalent functions taking m values in…: I.37.4
$\mathcal{U}^\infty$ (…)—many-valent functions in…: note I.37.4
$\mathcal{U}_n^m$ (…)—multivalued multivalent functions with n branches and m values in…: note I.37.4
$\mathcal{V}$ (…)—good functions, that is, with decay at infinity faster than some power
$\mathcal{V}^N$ (…)—good functions of degree N, that is, with decay at infinity faster than the inverse of a polynomial of degree N
$\bar{\mathcal{V}}$ (…)—fairly good functions, that is, with growth at infinity slower than some power
$\tilde{\mathcal{V}}^N$ (…)—fairly good functions of degree N, that is, with growth at infinity slower than a polynomial of degree N
$\bar{\mathcal{V}}$ (…)—very good or fast decay functions, that is, with faster decay at infinity than any power
$\mathcal{W}_q^p$ (…)—functions with generalized derivatives of orders up to q such that all the power p of the modulus is integrable…—*Sobolev* space

$\mathcal{X}$ (...)—automorphic functions in...
$\mathcal{X}_0$ (...)—self-inverse linear functions in....: I.37.5
X_1 (...)—linear functions in...: I.35.2
$\mathcal{X}_2$ (...)—bilinear, homographic, or Mobius functions in...: I.35.4
$\mathcal{X}_3$ (...)—self-inverse bilinear functions in...: I.37.5
$\mathcal{X}_a$ (...)—automorphic functions in...: I.37.6
$\mathcal{X}_m$ (...)—isometric mappings in...: I.35.1
$\mathcal{X}_r$ (...)—rotation mappings in...: I.35.1
$\mathcal{X}_t$ (...)—translation mappings in...: I.35.1
$\mathcal{Y}$ (...)—meromorphic functions in...: I.37.9, II.1.1.10
$\mathcal{Z}$ (...)—polymorphic functions in...: I.37.9, I.1.1.8

6. Ordinary and Special Functions

cos(z)—circular cosine: II.5.1.1
cosh(z)—hyperbolic cosine: II.5.1.1
cot(z)—circular cotangent: II.5.2.1
coth(z)—circular hyperbolic cotangent: II.5.2.1
csc(z)—circular cosecant: II.5.2.1
csch(z)—hyperbolic cosecant: II.5.2.1
exp(z)—exponential: II.3.1
log(z)—logarithm of base *e*: II.3.5
$\log_a(z)$—logarithm of base *a*: II.3.7
sec(z)—circular secant: II.5.2.1
sech(z)—hyperbolic secant: II.5.2.1
sgn(z)—sign function: I.36.4.1
sin(z)—circular sine: II.5.1.1
sinh(z)—hyperbolic sine: II.5.1.1
tan(z)—circular tangent: II.5.2
tanh(z)—hyperbolic tangent: II.5.2
B_n—Bernoulli number: II.7.1.8
E_n—Euler number: II.7.1.7
$F(;c;z)$—hypogeometric function: II.1.9.1
$F(a;c;z)$—confluent hypergeometric function: II.3.9.8
$F(a,b;c;z)$—Gaussian hypergeometric function: II.3.9.7
${}_pF_q(a_1,..,a_p;c_1..,c_q;z)$—generalized hypergeometric function: Note 3.3

Physical Quantities

Suffixes may be suppressed in the text when no confusion can arise.

1. Small Arabic Letters

a—core radius of vortex: 2.2.1
—radius of circle/cylinder: 2.5.3
—height of step: 8.5.8
—half-thickness of plate: 8.5.8
—chord of flap: 8.7.2
$\vec{a}$—acceleration: 2.1.1
b—radius of cylindrical cavity: 8.1.1
—radius of second cylinder: 8.2.1
—width of channel upstream: 8.5.2
—width of slot between airfoil and flap: 8.7.2
c—capillary constant: 6.4.2
—distance between the centers of two cylinders: 8.2.1
—chord of an airfoil: 8.6.1
—speed of light in vacuo: E10.11.10
d—width of channel downstream: 8.5.2
e—irrational number: 3.1
f—complex potential, for the velocity of a potential flow f_v: 2.4.6
$\vec{f}$—force per unit area or volume: 2.3.1
f_d—complex potential for plane elasticity: 4.8.1
f_e—complex potential of electrostatic field: E10.11.6
f_m—complex potential of magnetostatic field: E10.11.10
f_t—complex potential for torsion: 6.5.3
g—Sedov function: 8.7.1
g_1—separation function: 8.6.4
g_2—slotted airfoil function: 8.7.2
g_3—airfoil function: 8.8.1
g_4—circulation function: 8.8.7
g_5—tandem airfoil function: 8.8.9
g_6—tandem circulation function: 8.8.10
g_7—cascade function: 8.8.12
i—imaginary unit $i \equiv \sqrt{-1}$
$\vec{j}$—electric current density, per unit volume: N.6.6, E10.11.10
k—adiabatic factor: 2.1.4
—curvature: 6.4.1
—thermal conductivity: N.6.6
—wavenumber: 8.9.1

$\vec{k}$—wavevector: 8.9.1
m—added mass of fluid entrained by a body: 8.2.6
m_0—mass of a body: 8.2.6
$\bar{m}$—total mass of a body, including entrainment of fluid: 8.2.6
n—coordinate normal to a curve: 2.3.1
$\vec{n}$—unit vector normal to a curve: 4.2.1
p—pressure: 2.1.1
p_0—stagnation pressure: 2.1.2
p_∞—pressure at infinity: 2.2.3
r—polar coordinate, distance from origin: 2.2.1
—cylindrical coordinate, distance from axis: 2.6.2
$\vec{r}$—position vector of observer: 4.1.1
s—coordinate tangent to a curve: 2.3.1
—arc length: 4.1.3
—fraction of upper surface of airfoil with attached flow: 8.6.1
$\vec{t}$—unit vector tangent to a curve: 4.2.1
$\vec{u}$—displacement vector: 4.1.1
v—complex velocity: 2.4.6
v^*—complex conjugate velocity: 2.4.6
$\vec{v}$—velocity vector: 2.1.1
w—density per unit volume of heat source/sink: N.6.6
x—Cartesian coordinate: 1.1.1
y—Cartesian coordinate: 1.1.1
z—complex number: 1.1.1

2. Capital Arabic Letters

A—homothety factor in the Schwartz–Christoffel transformation: 8.4.1
$\vec{B}$—magnetic induction vector: E10.11.10
C—torsional stiffness of a rod: 6.6.1
C_v—specific heat at constant volume: 2.1.4
C_p—specific heat at constant pressure: 2.1.4
C_D—drag coefficient: 8.4.7
C_L—lift coefficient: 2.6.3, 8.6.6
C_S—shear coefficient: 2.6.3
D—drag force: 8.4.7
D_2—relative area change: 4.1.2
$\dot{D}_2$—two-dimensional dilatation: 4.1.2
$\vec{E}_2$—two-dimensional rotation vector: 4.1.2
E—Young's modulus of elastic material: 4.3.5
E_0—uniform electric field: E10.11.6
E_2—two-dimensional rotation: 4.1.2
E_d—elastic energy: 4.3.8
E_v—kinetic energy: 2.3.4, 8.2.4
E^*—complex conjugate electric field: E10.11.6

$\vec{E}$—electric field vector: N.6.6
$G(z{:}\zeta)$—Green or influence function: 2.9.2
G—gravitational constant: N.6.6
$\vec{G}$—heat flux: N.6.6
$\vec{H}$—magnetic field vector: N.6.6
L—lift: 2.6.3
$\vec{L}$—angular momentum: 6.9.2
$\vec{M}$—moment of forces: 2.5.1, 2.6.3
—non-unit normal to a surface: 6.1.1
$\vec{N}$—unit normal to a surface: 6.1.1
P_n—moment of 2^n: multipole, such as monopole P_0, dipole P_1, quadrupole P_2
Q—volume flow rate: 2.4.11
R—radius of curvature of a curve: 6.4.1
R_1,R_2—principal radii of curvature: 6.3.1
S—entropy: 2.1.4
dS—area element of a surface: 2.1.1
d$\vec{S}$—vector area element of a surface: 2.1.1
S_{ij}—strain tensor: 4.1.2
$\dot{S}_{ij}$—rate-of-strain tensor: N.4.4
$\bar{S}_{ij}$—sliding tensor: 4.3.2
$\dot{\bar{S}}_{ij}$—rate-of-sliding tensor: N.4.4
T—temperature: N.6.6
$\vec{T}$—stress vector: 4.2.1
T_{ij}—stress tensor: 4.2.1
U—velocity of uniform stream: 2.4.7
V—flow velocity in a duct: 8.5.3
W_d—work of deformation: 4.3.7, 6.1.1

3. Lowercase Greek Letters

α—angle-of-attack of a uniform flow: 2.4.7
β—angle-of-attack of a unidirectional shear flow: 2.6.1
—internal angle in a corner: 8.6.2
β_{n}—angular position on a circle: E10.18
ε—dielectric permittivity: N.6.6
γ—adiabatic exponent: 2.1.4
—circulation density: N.6.6
—external angle in a corner: 8.6.2
η—shear viscosity N.4.4
—coaxial coordinate: 8.3.1
φ—polar angle: 2.1.8, 2.4.8
—cylindrical coordinate, azimuth: 2.4.9
μ—shear modulus: 4.3.2
—magnetic permeability: N.6.6, E10.11.10
υ—volume modulus of elasticity: 4.3.1

ω—angular frequency: 6.9.3
ρ—mass density per unit volume: 2.1.1
σ—Poisson ratio: 4.3.6
—surface electric charge density: N.6.6
σ_{ij}—viscous stress tensor: N.4.5
ϑ—surface electric current density: N.6.6
$\vec{\varpi}$—vorticity: 2.2.1
ζ—bulk viscosity: N.4.4
—displacement of membrane: 6.1.1
—complex confocal coordinate: 8.3.1
ξ—coaxial coordinate: 8.3.1

4. Capital Greek Letters and Others

Φ—scalar velocity potential: 2.4.1
Φ_t—warping function: 6.5.1
Φ_v—velocity potential: N.6.6
Γ—circulation: 2.4.11
Λ—path function: 6.9.1
$\vec{\Omega}$—angular velocity: 6.9.1
Θ—stress function: 4.2.3; 6.5.2
$\vec{\Theta}$—biharmonic vector: 4.6.8
Ξ—multiharmonic function: 4.6.5
Ψ—stream function: 2.4.1
Ψ_0—path function: 2.9.1
Ψ_m—field function of magnetic field: N.6.6
Ψ_t—torsion function: 6.5.3

5. Dimensionless Numbers

Re—Reynolds number: N.4.10

Introduction

There are seven types of general representations of functions, namely, by (1) limits involving the variable as a parameter (Section I.13.4*) for uniformly continuous functions; (2) integrals involving the variable as a parameter (Sections I.13.8 and 13.9), which applies to differentiable, that is, holomorphic, functions; (3) power series applying to functions either analytic (Chapter I.23) or with singularities (Chapter I.25) in a region; (4) series of fractions, applying to functions with an infinite number of poles (Sections I.1.2 and 1.3), such as meromorphic functions; (5) infinite products applying to functions with an infinite number of zeros (Sections 1.4 and 1.5); (6) continued fractions (Sections 1.6 through 1.9), which, for a given level of complexity, may give a better approximation than other infinite processes; and (7) other functional series such as Fourier series (Note 1.2; Section 5.7). All these infinite processes raise questions of convergence (Chapter 9), and they can be converted into each other in some cases. The simplest examples of these processes are the elementary transcendental functions, namely, (1) the exponential that can be defined in at least six distinct but equivalent ways (Chapter 3); (2) the exponential's even and odd parts are, respectively, the hyperbolic cosine and sine (Chapter 5); (3) the real and imaginary parts of the exponential of an imaginary variable define, respectively, the circular cosine and sine (Chapter 5); (4) the direct functions cosine and sine specify the secant and cosecant by inversion, and the tangent and cotangent by their ratio; (5) the function inverse to the exponential is the logarithm, and it is related to six inverse circular and hyperbolic functions, which are known collectively as the 12 cyclometric functions (Chapter 7). These 26 functions provide the simplest examples of the theory of complex functions, including power series, series of fractions, infinite products, and continued fractions. These methods apply not only (1) to the potential flow (Chapters I.12, 14, 34, 36, and 38) but also to vortical (Chapter 2) and unsteady (Chapter 8) flows; (2) to gravity (Chapter I.18), electric (Chapter I.24), and magnetic (Chapter I.26) fields and to plane elasticity (Chapter 4); and (3) to steady heat conduction (Chapter I.32) and to capillarity, membranes, and torsion of rods (Chapter 6). This set of 225 mathematical topics (150 physical problems) is addressed in Classification 9.1 (8.1), covering both Books 1 and 2 of the course of Mathematics and Physics for Science and Technology.

* Roman numeral I refers to Book 1 in this series, *Complex Analysis with Applications to Flows and Fields.*

1

Sequences of Fractions or Products

The extension of the concept of polynomial to infinite order (Chapter 1.1) is the power series (Chapters I.23 and I.25); it gives a convenient representation of integral functions (Chapter I.27) because, in that case, the radius of convergence is infinity, that is, the series applies in the whole plane, for example, for sin z. For a meromorphic function, for example, tan z, with an infinite number of poles without an accumulation point, a power series representation has a radius of convergence limited by the nearest pole, that is, an infinite number of analytic continuations (Chapter I.31) are needed to cover the whole complex plane. Because a meromorphic function is the ratio of two integral functions, it is the extension of the rational function to infinite degrees of the numerator and denominator, and may have a representation as a series of fractions (Section 1.2), one (a set) for each simple (multiple) pole, with such a series converging at all points of the complex plane other than the poles (Section 1.3). By means of logarithmic differentiation, a zero of any order becomes a simple pole, with residue equal to the order; the inverse process of integration of series of fractions follows by taking the exponential and leads to an infinite product (Section 1.4). The infinite product represents a function with an infinite number of zeros, for example, the circular sine (Section 1.5), that is one zero associated with the vanishing of each factor and generalizes the theorem on the factorization of polynomials. Both the infinite products (the series of powers or of fractions) have products (sums) as terms that may satisfy recurrence relations. The latter may be solved in terms of continued fractions (Section 1.6). Both series (Sections 1.1 through 1.3) and infinite products (Sections 1.4 and 1.5) can be transformed into continued fractions (Section 1.8); for example, the nonterminating continued fraction for the exponential proves that e is an irrational number. The truncation of a continued fraction may provide, for a given complexity, the best rational approximation, as well as an upper bound for the error of truncation (Section 1.7). Besides transformation of series and products (Section 1.8), another method to obtain continued fractions is to start from triple recurrence relations (Section 1.5) for series. This specifies the ratio of two convergent series as a continued fraction (Section 1.9).

1.1 Power Series, Singularities, and Functions

As a preliminary to the consideration of series of fractions (Sections 1.2 and 1.3), infinite products (Sections 1.4 and 1.5), and continued fractions (Sections 1.6 through 1.9), some properties of complex functions (Chapters I.21, 23, 25, 27, and 29) are briefly reviewed concerning: (1) the Taylor (Laurent) series [Subsection 1.1.1 (Subsection 1.1.2)] around a regular point (isolated singularity); (2) the classification of singularities [Subsection 1.1.3 (Subsection 1.1.5)] in the finite plane (at infinity); (3) the classification of functions from their singularities (Subsection 1.1.6); (4) the calculation of the coefficients of the principal part near a pole including the residue (Subsection 1.1.4); (5) the integral (meromorphic) functions [Subsection 1.1.7 (Subsection 1.1.10)] as extensions of polynomials (rational functions), with the primary (secondary) circular functions as examples [Subsection 1.1.8 (Subsection 1.1.11)]; and (6) the rational–integral and polymorphic functions (Subsection 1.1.9) to complete the classification of functions. The subject matter of Volume II thus starts with Section 1.2.

1.1.1 Taylor Series for Analytic Functions

A complex function (Equation 1.1b) of a complex variable (Equation 1.1a) corresponds (Equation 1.1c) to two real functions of two real variables:

$$z = x + iy, \quad f = \Phi + i\Psi: \quad f(x + iy) = \Phi(x,y) + i\Psi(x,y), \tag{1.1a–1.1c}$$

where $\Phi(\Psi)$ is the real (imaginary) part. A complex function is **differentiable** (Equation 1.2a) at a point (Section I.11.2) if the incremental ratio (Equation 1.2b) has a unique limit independent of the path taken from z to a:

$$f(a) \in \mathcal{D}: \qquad f'(a) \equiv \frac{df}{da} = \lim_{z \to a} \frac{f(z) - f(a)}{z - a}. \tag{1.2a, 1.2b}$$

In contrast with real functions (Section I.27.1), the existence of first-order derivative (Equation 1.2b) for complex functions (Section I.27.2) implies the existence of derivatives of all orders (Section I.15.4), so that (1) the function is **smooth**:

$$\mathcal{D}^\infty \equiv \left\{ f(z): \quad \forall_{n \in |N} \quad \exists f^{(n)}(z) \right\}; \tag{1.3}$$

and (2) *the function is* **analytic** *because (Section I.23.7) the* **Taylor series**

$$f \in \mathcal{A}: f(z) = \sum_{n=0}^{\infty} f^{(n)}(a) \frac{(z-a)^n}{n!} \quad \begin{cases} \text{A.C.} \quad \text{if} \quad |z-a| < R \equiv |z - z_v|, & (1.4a) \\ \text{T.C.} \quad \text{if} \quad |z-a| \le R - \varepsilon, & (1.4b) \end{cases}$$

converges (Chapter I.21) (1) absolutely (Equation 1.4a) in a circle with center at $z = a$ and radius R determined by the nearest singularity and (2) uniformly, and hence totally (Equation 1.4b), in a closed subcircle (Equation 1.4b) with $0 < \varepsilon < R$.

1.1.2 Laurent Series near an Isolated Singularity

A **singularity** is a point where the function is not differentiable, that is, the incremental ratio (Equation 1.2b) has no limit or has more than one limit (Example I.20.1). A singularity $z = a$ is **isolated** if a **neighborhood**

$$V_\varepsilon(a) = \{z: \quad |z-a| < \varepsilon\} \tag{1.5}$$

exists for some $\varepsilon > 0$ such that it contains no other singularity. *A function that is analytic in a region, except for an isolated singularity at $z = a$, has a* **Laurent series** *(Section I.25.6) that*

$$f(z) = \sum_{n=-\infty}^{+\infty} A_n (z-a)^n \begin{cases} \text{A.C.} & \text{if} \quad 0 < |z-a| < R \equiv |z - z_*|, & (1.6a) \\ \text{T.C.} & \text{if} \quad \delta \le |z-a| \le R - \varepsilon, & (1.6b) \end{cases}$$

converges (1) absolutely (Equation 1.6a) in a circular annulus with inner boundary excluding the singularity, and the outer boundary of radius R determined by the nearest singularity $z = z_*$; and (2) totally (Equation 1.6b) in a closed subannulus with ε, $\delta > 0$ and $R > \varepsilon + \delta$. The Taylor (Equations 1.4a and 1.4b) [Laurent (Equations 1.6a and 1.6b)] series near a regular point (singularity) consists of ascending (ascending and descending) powers. If the function is analytic at $z = a$, the Laurent (Equations 1.6a and 1.6b) reduces to the Taylor (Equations 1.4a and 1.4b) series, implying

$$n = 1,2,\ldots \in |N: \quad A_{-n} = 0, \quad A_0 = f(a), \quad n!\, A_n = f^{(n)}(a) \tag{1.7a–1.7c}$$

that (1) there are no powers with negative exponents (Equation 1.7a); and (2 and 3) the constant term (powers with positive exponents) has, for coefficient, the value of the function at $z = a$ in Equation 1.7b (its derivatives of all orders n divided by $n!$ in Equation 1.7c).

1.1.3 Classification of Singularities from the Principal Part

The Laurent series (Equations 1.6a and 1.6b) near a singularity consists of a **principal part** containing the terms with powers with negative exponents (Equation 1.8a) that are singular at $z = a$:

$$P\{f(z)\} \equiv \sum_{n=-\infty}^{1} A_n (z-a)^n; \qquad f(z) - P\{f(z)\} = \sum_{n=-\infty}^{1} A_n (z-a)^n = O(1) \in \mathcal{A}; \tag{1.8a and 1.8b}$$

and the remaining terms, that is, the difference between the singular function and its principal part, are nonsingular at $z = a$ and form an analytic function (Equation 1.8b). *The principal part of a function near a singularity can be used to* **classify the singularity**, *leading (Sections I.27.3 and 27.4) to three cases: (1) if there is a single inverse power*

$$\text{simple pole:} \quad f(z) = \frac{A_{-1}}{z-a} + O(1), \tag{1.9}$$

the singularity is a **simple pole** *with residue A_{-1}; (2) if the highest inverse power is N, the singularity is a* **pole of order** *N:*

$$\text{pole of order } N: \quad f(z) = \sum_{n=1}^{N} \frac{A_{-n}}{(z-a)^n} + O(1), \tag{1.10}$$

and the principal part involves $N-1$ *coefficients* $A_{-N}, A_{1-N}, A_{2-N}, \ldots$ *besides the residue* A_{-1}*; and (3) the remaining case is an* **essential singularity**, *implying that there is an infinite sequence of inverse powers:*

$$\text{essential singularity:} \quad \forall_{n \in |N} \quad \exists_{m \in |N}: \quad m > n \wedge A_{-m} \neq 0 \tag{1.11}$$

so that for any positive integer n, there is at least one $A_{-m} \neq 0$ *with* $m > n$ *also a positive integer.* The simple pole (Equation 1.9) is a pole of order $N = 1$ in Equation 1.10, and thus, the main distinction is between (1) a pole of order N, that is, a singularity that can be eliminated by multiplying by $(z-a)^N$, which leads to an analytic function:

$$\text{pole of order } N: \quad (z-a)^N f(z) = \sum_{n=1}^{N} A_{-n}(z-a)^{n-N} + O\left((z-a)^N\right) = O(1)A; \tag{1.12}$$

and (2) essential singularity that cannot be eliminated by multiplication by $(z-a)^N$ for any value of N. This distinction is to be used next to calculate the coefficients A_{-n} of the principal part at a pole (Equation 1.10).

1.1.4 Coefficients of the Principal Part at a Pole

A method similar to the partial fraction decomposition of a rational function (Section I.31.9) can be used to calculate the coefficients of the principal part (Equation 1.10) at a pole of order N in three steps: (1) multiplication by $(z-a)^N$, that leads to an analytic function (Equation 1.12 ≡ Equation 1.13):

$$g(z) \equiv (z-a)^N f(z) = \sum_{n=1}^{N} A_{-n}(z-a)^{N-n} + O\left((z-a)^N\right); \tag{1.13}$$

(2) the coefficient A_{-k} is brought to the leading position, differentiating $N-k$ times with regard to z:

$$\begin{aligned} g^{(N-k)}(z) &= \sum_{n=1}^{k} A_{-n}(z-a)^{k-n}(N-n)(N-n-1)\ldots(k-n+1) + O\left((z-a)^k\right) \\ &= (N-k)!\,A_{-k} + O(z-a); \end{aligned} \tag{1.14}$$

and (3) only the leading term remains because the remaining terms vanish as z tends to a:

$$g^{(N-k)}(a) = \lim_{z \to a} \frac{\mathrm{d}^{N-k} g}{\mathrm{d}z^{N-k}} = (N-k)!\,A_{-k}. \tag{1.15}$$

Substituting Equation 1.13 in Equation 1.15 follows the **theorem on pole coefficients**: *the principal part of a function (Equation 1.10) near a pole of order N has coefficients given by*

$$A_{-k} = \frac{1}{(N-k)!}\lim_{z\to a}\frac{\mathrm{d}^{N-k}}{\mathrm{d}z^{N-k}}\left[(z-a)^N f(z)\right], \tag{1.16}$$

of which the first is the **residue**:

$$A_{-1} = \frac{1}{(N-1)!}\lim_{z\to a}\frac{\mathrm{d}^{N-1}}{\mathrm{d}z^{N-1}}\left[(z-a)^N f(z)\right] \equiv f_{(N)}(a); \tag{1.17}$$

in the case of a simple pole (Equation 1.18a), the only coefficient is the residue (Equation 1.18):

$$N=1: \qquad A_{-1} = \lim_{z\to a}(z-a)f(z) \equiv f_{(1)}(a). \tag{1.18}$$

The residue at a simple (multiple) pole can be calculated from (1) the Laurent series [Equation 1.9 (Equation 1.10)], (2) the limit [Equation 1.18 (Equation 1.17)], and (3) the use of L'Hôspital's rule [Subsection I.19.9.1 (Subsection I.19.9.2)].

1.1.5 Pole or Essential Singularity at Infinity

The **inversion** (Equation 1.19a) maps the origin (Equation 1.19b) to infinity (Equation 1.19c) and is used to study the behavior of a function at infinity (Equation 1.19d):

$$\zeta = 1/z: \qquad z\to 0, \qquad \zeta\to\infty, \qquad f(\infty) = \lim_{\zeta\to\infty} f(\zeta) = \lim_{z\to 0} f\left(\frac{1}{z}\right). \tag{1.19a–1.19d}$$

As in the finite plane (Equations 1.9 through 1.11), *there are three types of* **singularities at infinity:** *(1) a simple pole if the function diverges linearly*:

$$\text{simple pole at infinity:} \quad f(z) = A_1 z + O(1); \tag{1.20}$$

(2) a pole of order N if the function diverges like a polynomial of degree N:

$$\text{pole of order } N \text{ at infinity:} \quad f(z) = \sum_{n=1}^{N} A_n z^n + O(1) \equiv P_N(z) + O(1); \tag{1.21}$$

and (3) the remaining case is an essential singularity at infinity, when there is an unending sequence of ascending powers:

$$\text{essential singularity at infinity:} \quad \forall_{n\in|N}; \quad m > n \wedge \mathrm{A}_n \neq 0. \tag{1.22}$$

The distinction is that a pole at infinity can be eliminated by dividing by z^N for some N, whereas an essential singularity cannot be eliminated in this way for any N.

1.1.6 Classification of Functions from Their Singularities

The singularities of a function at infinity (Subsection 1.1.5) and in the finite part of the plane (Subsection 1.1.3) can be used to classify functions (Chapter I.27) into seven classes (Tables I.27.1 and I.27.2). The simplest is a function without singularities: (1) at infinity (Equations 1.21 and 1.22), so that $0 = A_1 = A_2 = \ldots$; and (2) in the finite part of the plane (Equations 1.9 through 1.11), so that $0 = A_{-1} = A_{-2} = \ldots$ Thus, the Laurent series (Equations 1.6a and 1.6b) reduces to a constant $f(z) = A_0$, and thus, *a function without singularities, neither in the finite plane nor at infinity, is a* **constant**. This first case is the Liouville theorem, of which several extensions exist (Sections I.27.6 and I.39.3 through I.39.6).

The second case is a function analytic in the plane except for a pole of order N at infinity (Equation 1.21), which corresponds to a **polynomial** *of degree N. For the third case: (1) it retains the pole of order N at infinity, corresponding to a polynomial of degree N in the numerator; (2) it adds a finite number of poles in the finite plane with orders adding to M, corresponding to division by a polynomial of degree M; and (3) the ratio of polynomials of degrees N and M without common roots is* **a rational function** *of degrees (N, M)*:

$$Q_{N,M}(z) = \frac{P_N(z)}{Q_M(z)}. \tag{1.23}$$

These three cases exhaust the Laurent series with a finite number of terms.

1.1.7 Integral Functions as Extensions of Polynomials

In all other cases, that is, the four remaining cases of transcendental functions, the Laurent series has an infinite number of terms, raising convergence issues (Chapters I.21 and I.29). The simplest case is no singularities in the finite plane and an essential singularity at infinity, so that the Taylor series (Equations 1.4a and 1.4b) has an infinite radius of convergence; because the origin is a regular point, it may be used for the expansion in the **MacLaurin series**:

$$f \in \mathcal{J}: \qquad f(z) = \sum_{n=0}^{\infty} f^{(n)}(0)\frac{z^n}{n!} \quad \begin{cases} \text{A.C.} \quad \text{if} \quad |z| < \infty; & (1.24\text{a}) \\ \text{T.C.} \quad \text{if} \quad |z| \le M < \infty. & (1.24\text{b}) \end{cases}$$

Thus, *the fourth case is an* **integral function** *specified by a MacLaurin series, that converges in an open circle (Equation 1.24a) [closed subcircle (Equation 1.24b)] excluding the essential singularity at infinity*. The integral function may be seen as an extension of the polynomial to "infinite" degree as a convergent power series.

1.1.8 Exponential and Sine/Cosine as Integral Functions

The **exponential** is defined as a function that is equal to its derivative (Equation 1.25a) and is unity at the origin (Equation 1.25b):

$$\frac{d(e^z)}{dz} \equiv e^z, \qquad e^0 \equiv 1. \qquad \text{(1.25a and 1.25b)}$$

Six other equivalent definitions of the exponential are given in the Chapter 3. The definition (Equations 1.25a and 1.25b) implies that the derivatives of all orders of the exponential at the origin are unity (Equation 1.26a):

$$\lim_{z\to 0} \frac{d^n}{dz^n}(e^z) = \lim_{z\to 0} e^z = e^0 = 1: \qquad e^z = \sum_{n=0}^{\infty} \frac{z^n}{n!}, \qquad \text{(1.26a and 1.26b)}$$

and thus its MacLaurin series (Equations 1.24a and 1.24b) is Equation 1.26b. The general term (Equation 1.27b) of the series (Equation 1.26b ≡ Equation 1.27a):

$$e^z = \sum_{n=0}^{\infty} e_n(z), \qquad e_n(z) \equiv \frac{z^n}{n!} \qquad \text{(1.27a and 1.27b)}$$

satisfies

$$|z| \le M < \infty: \qquad \lim_{n\to\infty} \left| \frac{e_{n+1}(z)}{e_n(z)} \right| = \lim_{n\to\infty} \frac{|z|}{n+1} \le \lim_{n\to\infty} \frac{M}{n+1} = 0. \qquad (1.28)$$

Thus, by the D'Alembert ratio test (Subsection I.29.3.2), the series (Equation 1.26b) converges the finite z-plane (Equation 1.24b), proving that the exponential is an integral function.

If z is real, the real (imaginary) parts of the exponential of iz define the **circular cosine (sine)**:

$$e^{iz} \equiv \cos z + i \sin z. \qquad (1.29)$$

Using the series (Equation 1.26b) in Equation 1.29:

$$e^{iz} = \sum_{n=0}^{\infty} \frac{(iz)^n}{n!} = \sum_{n=0}^{\infty} \frac{(iz)^{2n}}{(2n)!} + i \sum_{n=0}^{\infty} \frac{i^{2n} z^{2n+1}}{(2n+1)!}, \qquad (1.30)$$

it follows that the circular cosine (sine) is specified by a series of even (Equation 1.31a) [odd (Equation 1.31b)] powers:

$$\cos z = \sum_{n=0}^{\infty} \frac{(-)^n}{(2n)!} z^{2n} = \cos(-z), \qquad \sin z = \sum_{n=0}^{\infty} \frac{(-)^n}{(2n+1)!} z^{2n+1} = -\sin(-z), \qquad \text{(1.31a and 1.31b)}$$

and hence is an even (odd) function. This implies

$$e^{iz} = \cos(-z) + i \sin(-z) = \cos z - i \sin z. \qquad (1.32)$$

Elimination between Equations 1.29 and 1.32 shows that the circular cosine (Equation 1.33a) and sine (Equation 1.33b) are linear combinations of exponentials:

$$\cos z = \frac{e^{iz} + e^{-iz}}{2}, \qquad \sin z = \frac{e^{iz} - e^{-iz}}{2i}, \tag{1.33a and 1.33b}$$

and hence are also integral functions. Other properties of circular and hyperbolic functions appear in the Chapters 5 and 7.

1.1.9 Rational–Integral and Polymorphic Functions

The fifth (second) case of complex (transcendental) functions are the **rational–integral functions** *that have (1) an essential singularity at infinity and (2) a finite number of poles in the finite plane.* Thus, the rational–integral functions of degree N (Equation 1.34a) are the ratio of an integral function (Equation 1.34b) by a polynomial of degree N without common roots (Equation 1.34c):

$$f \in \mathcal{P}_N, \; g \in \mathcal{I}: \qquad f(z) = \frac{g(z)}{P_N(z)}. \tag{1.34a–1.34c}$$

The remaining cases must involve either (1) an infinite number of poles or (2) at least one essential singularity in the finite plane. The seventh (fourth) and last case of a complex (transcendental) function is a **polymorphic function** *that has at least one essential singularity in the finite plane (2); it may or may not have other singularities in the finite plane and/or at infinity.*

1.1.10 Meromorphic Functions as Extensions of the Rational Functions

The remaining case to consider is a function with an infinite number of poles; two cases arise. First, if the poles lie in the finite plane: (1) they must have at least one accumulation point; (2) the accumulation point is a nonisolated singularity; (3) hence, it cannot be a pole; (4) thus, it must be an essential singularity; and (5) because it lies in the finite plane, the function is polymorphic. The other possibility is an infinite number of poles that when ordered by modulus (Equation 1.35a) tends to infinity (Equation 1.35b):

$$|z_1| \le |z_2| \le \cdots \le |z_n| \le |z_{n+1}| \le \cdots : \qquad \lim_{n\to\infty} z_n = \infty. \tag{1.35a and 1.35b}$$

In this case, the accumulation point is at infinity, and no essential singularity needs to exist in the finite plane. Thus, the remaining sixth (third) case of a complex (transcendental) function is the **meromorphic function** *that has an infinite number of poles accumulating at infinity (Equations 1.35a and 1.35b) and no essential singularity in the finite plane.* A meromorphic function is the ratio of two integral functions (Section 1.3), whereas a rational function is the ratio of two polynomials (Equation 1.23). Thus, integral (meromorphic) functions may be considered as the extension of polynomials (rational functions) to infinite degree, which leads to power series (Section 1.1) [series of fractions (Sections 1.2 and 1.3)] that involve convergence issues.

1.1.11 Tangent/Cotangent and Secant/Cosecant as Meromorphic Functions

The primary circular (and hyperbolic) functions, the cosine and sine (Subsection 1.1.8), are integral functions. The secondary circular (and hyperbolic) functions, defined by their ratios (Equations 1.36a and 1.36b) and inversion (Equations 1.36c and 1.36d), are meromorphic functions:

$$\tan z \equiv \frac{\sin z}{\cos z} \equiv \frac{1}{\cot z}, \qquad \csc z \equiv \frac{1}{\sin z}, \qquad \sec z \equiv \frac{1}{\cos z}, \qquad \text{(1.36a–1.36d)}$$

because they have an infinite number of poles accumulating at infinity, namely, the roots of the denominator. Thus, the circular cosecant (Equation 1.36c) and cotangent (Equation 1.36b) [secant (Equation 1.36d) and tangent (Equation 1.36a)] have simple poles at the roots of the sine (Equation 1.37a) [cosine (Equation 1.37b)]:

$$m \in Z: \qquad z_m^{(1)} = m\pi, \qquad z_m^{(2)} = m\pi + \frac{\pi}{2}, \qquad \text{(1.37a and 1.37b)}$$

as shown in Figure 1.1a (Figure 1.1b).

As an example, the residue (Equation 1.18) of the cosecant at its poles:

$$A_{-1}^{(m)} = \csc_{(1)}(m\pi) = \lim_{z \to m\pi} (z - m\pi)\csc z = (-)^m \qquad \text{(1.38)}$$

is calculated (Equation 1.39b) using the change of variable (Equation 1.39a):

$$u \equiv z - m\pi: \qquad A_{-1}^{(m)} = \lim_{u \to 0} \frac{u}{\sin(u + m\pi)} = (-)^m \lim_{u \to 0} \frac{u}{\sin u} = (-)^m, \qquad \text{(1.39a and 1.39b)}$$

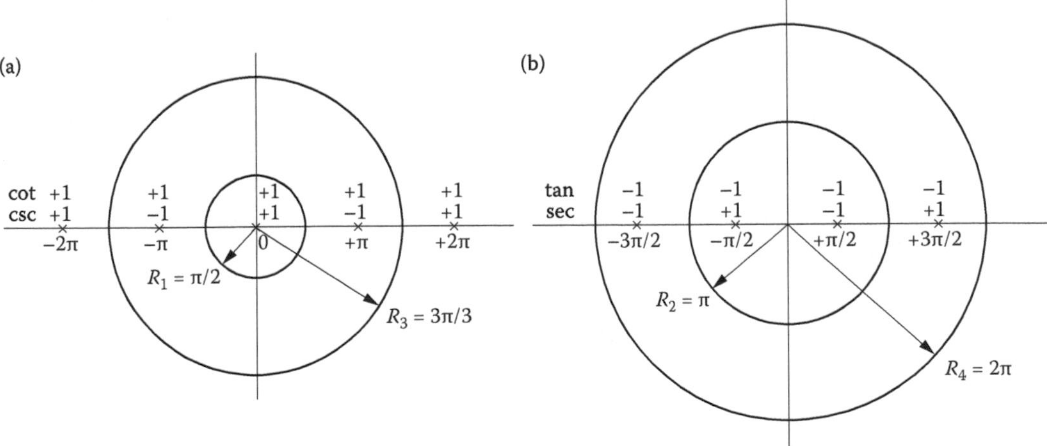

FIGURE 1.1
Meromorphic functions circular secant and tangent (cosecant and cotangent) have simple poles at the simple zeros of the circular cosine (sine) in the Equation 1.37b (Equation 1.37a) for integer m in Figure 1.1b (Figure 1.1a) with the residues indicated [Equations 1.52 and 1.79 (Equations 1.82 and 1.55)]. Thus, the circles with center at the origin and radius [Equation 1.41c (Equation 1.41b)] contain $2N(2N-1)$ poles with N a positive integer.

as well as Equation I.5.14b ≡ Equation I.19.37a in agreement with Equation I.19.39a ≡ Equation 1.31b. The residue could also be calculated using L'Hôspital's rule (Subsection I.19.9.1), differentiating once the numerator and denominator that are first-order infinitesimals, before applying the limit:

$$A_{-1}^{(m)} = \lim_{z\to m\pi}\frac{z-m\pi}{\sin z} = \lim_{z\to m\pi}\frac{(z-m\pi)'}{(\sin z)'} = \lim_{z\to m\pi}\frac{1}{\cos z} = \frac{1}{\cos(m\pi)} = (-)^m, \tag{1.40}$$

in agreement with Equation 1.38 ≡ Equation 1.39b ≡ Equation 1.40. This brief summary on (1) power series for integral functions (Section 1.1) based on the results proven in Volume I serves as a start to Volume II concerning (2) series of fractions for meromorphic functions (Sections 1.2 and 1.3); (3) infinite products for functions with an infinite number of zeros (Sections 1.4 and 1.5); and (4) continued fractions (Sections 1.6 through 1.9) as an alternative to (1) to (3) and optimal rational approximation.

1.2 Series of Fractions for Meromorphic Functions (Mittag-Leffler 1876, 1884)

A meromorphic function by definition (Subsection 1.1.10) has an infinite number of poles $z_1, \ldots, z_n \ldots$ and no essential singularity, so that the poles have no accumulation point in the finite plane, and tend to infinity (Equation 1.35b) if ordered by modulus (Equation 1.35a) [e.g., circular cosecant and tangent (secant and cotangent) in the Figure 1.1a (Figure1.1b)]. These examples are cases where a sequence of circles (Equation 1.41a) with center at the origin and radii [Equation 1.41b (Equation 1.41c)], each containing 2N–1 (2N) simple poles [Equation 1.37a (Equation 1.37b)], may be considered:

$$|\zeta| = R_N: \qquad R_N^{(1)} = N\pi - \frac{\pi}{2}, \qquad R_N^{(2)} = N\pi. \tag{1.41a–1.41c}$$

More generally, the poles may be of any order, possibly not the same for all of them, and some circles coincide with nondecreasing radii $R_{N+1} \ge R_N$. This leads to the consideration of a sequence of regions D_N, each containing the first N poles, assuming that the boundary ∂D_N lies outside a circle of radius R_N containing the same poles (Figure 1.2), and such that $R_N \to \infty$ as $N \to \infty$. Let the principal part of the function (Equation 1.8a) be specified at the nth pole z_n of order α_n, by

$$f(z) = \sum_{k=1}^{\alpha_n} A_{-k}^{(n)}(z-z_n)^{-k} + O(1), \tag{1.42}$$

with given coefficients, $A_{-k}^{(n)}$, of which $A_{-1}^{(n)}$ is the residue. The auxiliary function:

$$f_N(z) \equiv f(z) - \sum_{n=1}^{N}\sum_{k=1}^{\alpha_n} A_{-k}^{(n)}(z-z_n)^{-k} \tag{1.43}$$

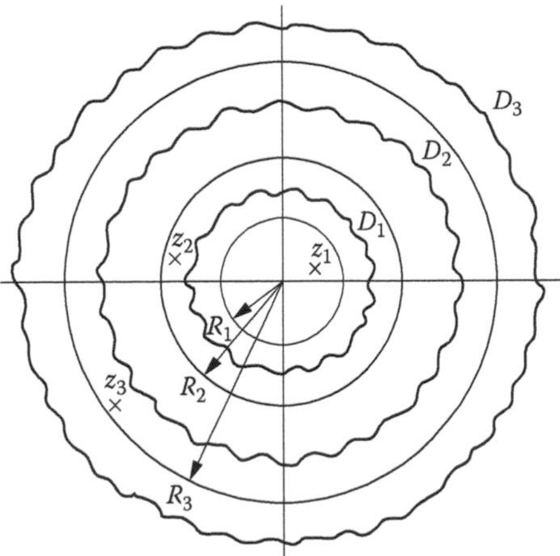

FIGURE 1.2
Figure 1.1a and 1.1b are examples of meromorphic functions such that (1) the first N poles ordered by modulus (Equations 1.35a and 1.35b) are contained in a finite region D_N; (2) the boundary of the region contains a circle of radius R_N; (3) the function is analytic within the circle of radius R_N except for the poles $z_1, \ldots, z_N$; (4) hence, it is bounded on the circles; (5) the radii diverge $R_N \to \infty$ as $N \to \infty$. These conditions allow the proof of the Mittag-Leffler (1876, 1884) theorems related to infinite products (series of fractions) in Sections 1.4 and 1.5 (Sections 1.2 and 1.3).

has no poles in the region D_N, and thus is analytic. The second Cauchy theorem (Equation I.15.8) can be applied, leading to

$$\begin{aligned}
\left|f_N(z) - f_N(a)\right| &= \left|\frac{1}{2\pi i}\oint_{\partial D_N} f(\zeta)\left(\frac{1}{\zeta - z} - \frac{1}{\zeta - a}\right)\mathrm{d}\zeta\right| \le \frac{1}{2\pi}\int_{\partial D_N}\left|f(\zeta)\right|\left|\frac{1}{\zeta - z} - \frac{1}{\zeta - a}\right|\mathrm{d}\zeta \\
&= \frac{\left|z - a\right|}{2\pi}\int_{|\zeta| = R_N}\frac{\left|f(z)\right|}{\left|\zeta - z\right|\left|\zeta - a\right|}\mathrm{d}\zeta \\
&= O\left(\frac{\left|z - a\right|}{2\pi}\frac{M}{\left(R_N\right)^2}2\pi R_N\right) = O\left(\left|z - a\right|\frac{M}{R_N}\right),
\end{aligned} \tag{1.44}$$

where (1) the boundary ∂D_N can be contracted to the circle because the integrand is analytic in between (Section I.15.2); (2) a is any point in the circle of radius $R_N > |a|$ not coinciding with a pole, and thus, $|f_N(\zeta)| \le M < \infty$ is bounded on $|\zeta| = R_N$; and (3) as $N \to \infty$, $R_N \to 0$ and $|\zeta - a| \sim R_N \sim |\zeta - z|$ and the perimeter of the circle $|\zeta| = R_N$ is $2\pi R_N$. It follows that taking the linct $R_N \to \infty$, then Equation 1.44 vanishes, so that the constant (Equation 1.43) can be evaluated by replacing z by a.

Thus, has been proven by the **theorem on series of fractions** *(Mittag-Leffler 1876, 1884): let f(z) be a meromorphic function with principal parts (Equation 1.42) at its poles z_n of orders α_n*

with coefficients $A_{-k}^{(n)}$, *with* $k = 1, \ldots, \alpha_n$; *then the poles tend to the point at infinity (Equations 1.35a and 1.35b) as* $n \to \infty$, *and the function* $f(z)$ *can be represented by a series of fractions:*

$$f(z) = f(a) + \sum_{n=1}^{\infty} \sum_{k=1}^{\alpha_n} A_{-k}^{(n)} \left\{ \left(z - z_n\right)^{-k} - \left(a - z_n\right)^{-k} \right\}, \tag{1.45}$$

where a is any point distinct from the poles; if the poles are all simple, $\alpha_n = 1$, *with residues (Equation 1.46a), then Equation 1.45 reduces to Equation 1.46b:*

$$B_n \equiv A_{-1}^{(n)}: \qquad f(z) = f(a) + \sum_{n=1}^{\infty} B_n \left\{ \left(z - z_n\right)^{-1} - \left(a - z_n\right)^{-1} \right\}; \tag{1.46a and 1.46b}$$

if the origin is not a pole setting, $a = 0$ *in Equation 1.5 (Equation 1.6b) leads to*

$$a = 0: \qquad f(z) = f(0) + \sum_{n=1}^{\infty} \sum_{k=1}^{\alpha_n} A_{-k}^{(n)} \left\{ \left(z - z_n\right)^{-k} - \left(-z_n\right)^{-k} \right\}, \tag{1.47}$$

$$f(z) = f(0) + \sum_{n=1}^{\infty} B_n \left\{ \left(z - z_n\right)^{-1} + z_n^{-1} \right\}, \tag{1.48}$$

for multiple (Equation 1.47) [simple (Equation 1.48)] poles. The series (Equations 1.45 through 1.48) converge at all points of the complex plane other than the poles $|z - z_n| > 0$ *for all n, and the convergence is uniform, hence, total, for* $|z - z_n| \geq \varepsilon$ *with* $\varepsilon > 0$.

1.3 Meromorphic Function as a Ratio of Two Integral Functions

The theorem on series of fractions (Section 1.2) can be used to prove (Subsection 1.3.1) that a meromorphic function is the ratio of two integral functions, as stated before (Subsection 1.1.10). The theorem (Equations 1.45 and 1.46) takes a simpler form (Equations 1.47 and 1.48) if the origin is not a pole; the latter restriction is not essential, as shown by the extended theorem of series of fractions (Subsection 1.3.4). Both the original (extended) theorems on series of fractions [Section 1.2 (Subsection 1.3.4)] are illustrated using as example the circular secant (tangent) in Subsection 1.3.2 (Subsection 1.3.3) and Figure 1.1b (Figure 1.1a).

1.3.1 Two Cases of Ratios of Integral Functions

All the series (Equations 1.45 through 1.48) are particular cases of Equation 1.45, that can be restated as

$$f(z) = \lim_{N \to \infty} \sum_{n=1}^{N} \sum_{k=1}^{\alpha_n} A_{-k}^{(n)} \left(z - z_n\right)^{-k} + A_0, \tag{1.49}$$

where A_0 is a constant. Multiplying both sides by $(z-z_n)^{\alpha_n}$ eliminates all poles from the right-hand side (r.h.s.) of Equation 1.49, leading to an integral function:

$$f(z)\left\{\lim_{N\to\infty}\prod_{n-1}^{N}\left(z-z_n\right)^{\alpha_n}\right\}=g(z)\in\mathcal{J}. \tag{1.50}$$

The term in curly brackets on the r.h.s. is an infinite product that can be made to converge by inserting exponential factors (Section 9.2), thus specifying an integral function $Q(z)$. The exponential factors also lead to an integral function $P(z)$ when multiplied by the integral function $g(z)$ on the r.h.s. It follows that *a meromorphic function $f(z) = P(z)/Q(z)$ is the ratio of two integral functions $P(z)$ and $Q(z)$*. This result is analogous to the statement that Equation 1.23, a rational function, is the ratio of two polynomials. In the case $N \to \infty$ of meromorphic functions, there is an infinite product in Equation 1.50, and the same result can be proven (Subsection 9.2.4) by a different method.

The converse theorem concerns the ratio of two integral functions $P(z)$ and $Q(z)$. The second Picard theorem (Subsection I.39.2.2) leaves *two possibilities for the ratio $P(z)/Q(z)$ of two integral functions without common roots: (1) if $c = 0$ is not an exceptional point of $Q(z)$, then the equation $Q(z) = 0$ has an infinite number of roots, and the ratio of two integral functions $f(z) = P(z)/Q(z)$ has an infinite number of poles and is a meromorphic function; (2) there is at most one exceptional point of the integral function $Q(z)$, and if it is $c = 0$, then $Q(z) = 0$ may have no zeros (only a finite number of zeros), and $f(z) = P(z)/Q(z)$ is not a meromorphic function, but rather an integral (rational–integral) function*. An example of 1 (2) is Equation 1.51a (Equation 1.51b):

$$f(z)=\frac{P(z)}{\sin z}=P(z)\csc z,\qquad f(z)=\frac{P(z)}{e^z}=e^{-z}P(z), \tag{1.51a and 1.51b}$$

because csc z (exp z) has an infinite number of zeros (no zeros), implying that Equation 1.51a (Equation 1.51b) is a meromorphic (integral) function.

1.3.2 Series of Fractions for the Circular Secant

As a first example of the derivation of a series of fractions, consider (Figure 1.1b) the function (Equation 1.36d) that is analytic in the circles (Equations 1.41a and 1.41c) except for poles; each circle contains $2N$ simple poles at Equation 1.37b; these are the simple zeros of the cosine, corresponding to simple poles of the secant with residues:

$$B_m\equiv\sec_{(1)}(m\pi+\pi/2)=\left(\frac{z-m\pi-\pi/2}{\cos z}\right)_{z=m\pi+\pi/2}=-\left\{\frac{1}{\sin z}\right\}_{z=m\pi+\pi/2}=(-)^{m+1}, \tag{1.52}$$

calculated by the L'Hôspital's rule (Equation I.19.41), as in Equation 1.40 for the cosecant. Because all poles are simple, and the origin is not pole, Equation 1.48 may be used:

$$\begin{aligned}\sec z &= \sec 0 + \sum_{m=-\infty}^{+\infty} (-)^{m+1} \left\{ (z - m\pi - \pi/2)^{-1} + (m\pi + \pi/2)^{-1} \right\} \\ &= 1 + \sum_{n=0}^{\infty} (-)^{n+1} \left\{ \left(z - (n+1/2)\pi\right)^{-1} + \left(n\pi + \pi/2\right)^{-1} - \left(z + (n+1/2)\pi\right)^{-1} - \left(n\pi + \pi/2\right)^{-1} \right\};\end{aligned} \tag{1.53}$$

the terms $m = 0,1,2,\ldots = n$ were paired with $-m = -1, -2,\ldots = -1 - n$ to simplify Equation 1.53 to

$$\sec z = 1 + \pi \sum_{n=0}^{\infty} \frac{(-)^{n+1}(2n+1)}{z^2 - (n+1/2)^2 \pi^2}. \tag{1.54}$$

This is *the series of fractions for the secant (Equation 1.54), that is, conditionally convergent,* because (Equation I.29.67c) the general term is $O(n^{-1})$.

1.3.3 Series of Fractions for the Circular Cotangent

As a further example, consider the cotangent (Equation 1.36b), which is (Figure 1.1a) analytic in the circles (Equations 1.41a and 1.41b) containing $2N + 1$ simple poles at Equation 1.37b that are the zeros of the sine. The residues of the cotangent at its simple poles:

$$B_m \equiv \cot_1(m\pi) = \left\{ \frac{\cos z}{(\sin z)'} \right\}_{z=m\pi} = 1 \tag{1.55}$$

are all unity. Because the origin is a pole with residue unity, its principal part is subtracted from the function:

$$\cot z - \frac{1}{z} = \frac{\cos z}{\sin z} - \frac{1}{z} = \frac{1 + O\left(z^2\right)}{z} - \frac{1}{z} = O(z). \tag{1.56}$$

This leads to no pole, but rather, a simple zero at the origin. Thus, Equation 1.48 can be used for the function Equation 1.56 that has the same poles outside the origin, with the same residues (Equation 1.55) as the cotangent:

$$\cot z - \frac{1}{z} = \sum_{\substack{m=-\infty \\ m \neq 0}}^{+\infty} \left\{ (z - m\pi)^{-1} + (m\pi)^{-1} \right\} = \sum_{n=1}^{\infty} \left\{ (z - n\pi)^{-1} + (z + n\pi)^{-1} + (n\pi)^{-1} - (n\pi)^{-1} \right\}. \tag{1.57}$$

This simplifies to a series general term $O(n^{-2})$, which is (Equation I.29.76d) absolutely convergent:

$$\cot z = \frac{1}{z} + 2z \sum_{n=1}^{\infty} \frac{1}{z^2 - n^2 \pi^2}. \tag{1.58}$$

The series of fractions for the cotangent (Equation 1.58) is absolutely convergent for $|z - n\pi| > 0$, *and totally convergent for* $|z - n\pi| \geq \varepsilon$ *with* $\varepsilon > 0$. The series of fractions for the cotangent (Equation 1.57 ≡ Equation 1.58) can be used to sum series of rational functions (Section 9.5; Example 10.1).

1.3.4 Extended Theorem on Series of Fractions

The preceding example shows that the series (Equations 1.47 and 1.48) can be used in all cases because (1) they hold for $f(z)$ if it is holomorphic at the origin; (2) if $f(z)$ has a pole of order α_0 at the origin, with principal part:

$$f(z) = \sum_{k=1}^{\alpha_0} A_{-k}^{(0)} z^{-k} + A_0 + O(z), \tag{1.59}$$

then the function Equation 1.60a:

$$F(z) \equiv f(z) - \sum_{k=1}^{\alpha_0} A_{-k}^{(0)} z^{-k}, \qquad F(0) = A_0 \qquad \text{(1.60a and 1.60b)}$$

is holomorphic at the origin (Equation 1.60b) and has the same poles (Equations 1.35a and 1.35b) and coefficients (Equation 1.42) as $f(z)$ elsewhere, so that Equations 1.47 and 1.48 apply to $F(z)$. Thus, the **extended theorem on series of fractions** *can be stated: let* $f(z)$ *be a meromorphic function with principal parts (Equation 1.42) at the poles (Equations 1.35a and 1.35b) outside the origin* $z_1 \neq 0$ *of orders* α_n *with coefficients* $A_{-k}^{(n)}$ *with* $k = 1, n, \ldots, \alpha_n$*; the origin may be a pole of order* α_0 *with principal part (Equation 1.59) with coefficients* $A_{-k}^{(0)}$ *with* $k = 1, \ldots, \alpha_0$ *then the function can be represented by the series of fractions:*

$$f(z) = A_0 + \sum_{k=1}^{\alpha_0} A_{-k}^{(0)} z^{-k} + \sum_{n=-1}^{\infty} \sum_{k=1}^{\alpha_n} A_{-k}^{n} \left\{ (z - z_n)^{-k} - (-z_n)^{-k} \right\}, \tag{1.61}$$

which simplifies to Equation 1.62e in the case of simple poles (Equations 1.62a and 1.62b) with residues (Equations 1.62c and 1.62d):

$$\alpha_n = 1 = \alpha_0, \quad A_{-1}^{(n)} \equiv B_n, A_{-1}^{(0)} \equiv B_0: \quad f(z) = A_0 + B_0 z^{-1} + \sum_{n=1}^{\infty} B_n \left\{ (z - z_n)^{-1} + z_n^{-1} \right\}.$$

(1.62a–1.62e)

The series (Equations 1.61 and 1.62e) converge for $|z| > 0$, $|z - z_n| > 0$ *for all n and converge uniformly for* $|z| \geq \varepsilon$, $|z - z_n| \geq \delta$ *with* $\varepsilon, \delta > 0$.

1.4 Factorization with Infinite Number of Zeros

The series [Equation 1.61 (Equation 1.62e)] are an alternative to Equation 1.45 (Equation 1.46b) in that instead of evaluating the sum (Equation 1.43) as $N \to \infty$ outside the origin a $\neq 0$ they evaluate it at the origin a = 0 after removing the pole (Equation 1.59) there. The process opposite to removing poles is creating them (Subsection 1.4.1), and leads to the Mittag-Leffler theorem on infinite products for a function with an infinite number of zeros, not including the origin (Subsection 1.4.2); a simple extension (Subsection 1.4.3) applies if the origin is a zero, as before for series of fractions (Section 1.2 and Subsection 1.3.4).

1.4.1 Removal of Poles and Relation with Zeros

Following the removal of poles (Subsection 1.3.4), the relation with zeros is addressed next. Starting with the zero z_n of order β_n of a function:

$$G(z_n) \neq 0: \qquad g(z) = (z - z_n)^{\beta_n} G(z), \tag{1.63a and 1.63b}$$

and performing a logarithmic differentiation:

$$\frac{d}{dz}\{\log[g(z)]\} = \beta_n \left\{\frac{d}{dz}\log(z - z_n)\right\} + \frac{d}{dz}\{\log[G(z)]\} = \frac{\beta_n}{z - z_n} + \frac{G'(z)}{G(z)} \tag{1.64}$$

leads to Equation 1.64, which consists of the following: (1) a simple pole with residue β_n for the first term; (2) the second term is analytic at z_n because G is analytic and nonzero. If the function $g(z)$ has an infinite number of zeros z_n of order β_n, its logarithmic derivative:

$$\frac{g'(z)}{g(z)} = \frac{d}{dz}\{\log[g(z)]\} = \frac{g'(0)}{g(0)} + \sum_{n=1}^{\infty} \beta_n \left\{(z - z_n)^{-1} + z_n^{-1}\right\} \tag{1.65}$$

has a series of fractions (Equation 1.46b), with simple poles at z_n and residues $B_n = \beta_n$. In deducing Equation 1.65, it was assumed that (1) the origin is not a zero of $g(z)$, that is, $g(0) \neq 0$; and (2) the function $g'(z)/g(z)$ is analytic except for poles in a sequence of circles $|\zeta| = R_N$ containing the first N zeros. The former restriction (1) will be removed next (Subsection 1.4.3) as was done before for series of fractions (Subsection 1.3.4); the latter restriction (2) is not essential and will be lifted subsequently (Section 9.2) for infinite products.

1.4.2 Theorem on Infinite Products

Formula 1.65 may be integrated from 0 to z:

$$\log\left\{\frac{g(z)}{g(0)}\right\} = z\frac{g'(0)}{g(0)} + \sum_{n=1}^{\infty} \frac{\beta_n}{z_n} z + \sum_{n=1}^{\infty} \beta_n \log\left(\frac{z - z_n}{-z_n}\right). \tag{1.66}$$

The exponential of a convergent series is an infinite product having the same convergence properties:

$$g(z) = g(0)\exp\left\{z\frac{g'(0)}{g(0)}\right\}\prod_{n=1}^{\infty}\left(1-\frac{z}{z_n}\right)^{\beta_n}\exp\left(\frac{\beta_n}{z_n}z\right), \tag{1.67}$$

leading to the **theorem on infinite products:** *let the function $g(z)$ have an infinite number of zeros z_n of order β_n in Equations 1.63a and 1.63b; suppose that $g'(z)/g(z)$ is analytic except for poles at the zeros of $g(z)$ in a sequence of circles $|\zeta| = R_N$ containing the first N zeros and such that $R_N \to \infty$ as $N \to \infty$. Then the function can be represented by the infinite product (Equation 1.67) that converges in the whole complex plane.* It is assumed that the origin is not a zero. The last restriction can be readily removed as shown next.

1.4.3 Extended Theorem on Infinite Products

Suppose that $g(z)$ has a zero of order β at the origin:

$$g(z) = Az^{\beta}\left\{1 + Bz + O\left(z^2\right)\right\} \equiv z^{\beta}G(z), \tag{1.68}$$

where the coefficients A, B are specified by the auxiliary function $G(z)$ and its logarithmic derivative:

$$G(0) = A \neq 0, \qquad G'(0) = AB = BG(0). \tag{1.69a and 1.69b}$$

Outside the origin $z \neq 0$, the auxiliary function $G(z)$ has the same zeros with the same order as $g(z)$, so that Equation 1.67 can be applied to $G(z)$:

$$z^{-\beta}g(z) = G(z) = G(0)\exp\left\{z\frac{G'(0)}{G(0)}\right\}\prod_{n=1}^{\infty}\left(1-\frac{z}{z_n}\right)^{\beta_n}\exp\left(\frac{\beta_n}{z_n}z\right). \tag{1.70}$$

Substituting Equations 1.69a and 1.69b leads to a slightly extended form of the original Mittag-Leffler's theorem, that is, the **extended theorem on infinite products:** *let the function $g(z)$ have a zero of order β at the origin (Equations 1.68, 1.69a and 1.69b) and an infinite number of other zeros $z_n \neq 0$ of orders β_n in Equations 1.63a and 1.63b; suppose that $g'(z)/g(z)$ is analytic except for poles at the zeros of $g(z)$ in a sequence of circles $|\zeta| = R_N$ containing the first N zeros and such that $R_N \to \infty$ as $N \to \infty$. Then, the function can be represented by the infinite product:*

$$g(z) = Az^{\beta z}e^{Bz}\prod_{n=1}^{\infty}\left(1-\frac{z}{z_n}\right)^{\beta_n}\exp\left(\frac{\beta_n z}{z_n}\right), \tag{1.71}$$

that converges in the whole z-plane. The exponential factors in Equations 1.67 and 1.71 do not affect the location z_n and order β_n of the zeros, but may be necessary to guarantee convergence of the infinite product (Section 9.2). Five more instances of infinite products and series of fractions are given in Example 10.1.

1.5 Infinite Products for Circular Functions

The original (extended) theorem on infinite products [Subsection 1.4.2 (Subsection 1.4.3)] is applied to the circular cosine (sine) as an example [Subsection 1.5.1 (Subsection 1.5.2)]. The series of fractions for the tangent and cosecant [Subsection 1.5.3 (Subsection 1.5.4)] lead to a discussion of zeros and poles of the tangent (Section 1.5.5).

1.5.1 Infinite Product for the Circular Cosine

As a first example of calculation of infinite product, consider (Figure 1.1b) the function cos z, which is analytic and has an infinity of simple zeros $\beta_n = 1$ at Equation 1.37b. The circles (Equations 1.41a and 1.41c) contain $2N$ zeros, and in them, the logarithmic derivative of cos z namely, cot z, is analytic except at the poles. Thus, Equation 1.67 may be used with cos 0 = 1, sin 0 = 0, namely:

$$\begin{aligned}\cos z &= \prod_{m=-\infty}^{+\infty}\left\{1-\frac{z}{m\pi+\pi/2}\right\}\exp\left\{\frac{z}{m\pi+\pi/2}\right\}\\ &= \prod_{n=0}^{\infty}\left\{1-\frac{z}{n\pi+\pi/2}\right\}\left\{1+\frac{z}{n\pi+\pi/2}\right\}\exp\left\{\frac{z}{n\pi+\pi/2}\right\}\exp\left\{-\frac{z}{n\pi+\pi/2}\right\},\end{aligned} \tag{1.72}$$

that simplifies to

$$\cos z = \prod_{n=0}^{\infty}\left\{1-\frac{z^2}{(n+1/2)^2\pi^2}\right\}. \tag{1.73}$$

This is *the infinite product for the cosine (Equation 1.73), that is absolutely convergent,* because the general term $1 + O(n^{-2})$ implies (Subsection 9.2.1) that it behaves like the series $O(n^{-2})$, which is uniformly convergent (Equation I.29.40d). The exponential factors in Equation 1.72 cancel out and are not needed to ensure the convergence of Equation 1.73 because the general term is $O(n^{-2})$.

1.5.2 Infinite Product for the Circular Sine

If instead of the cosine (Equation 1.73) the sine is considered, then Equation 1.71 should be used instead of Equation 1.67 because the sine has a simple zero at the origin (Equation 1.74a ≡ Equation 1.31b ≡ Equation I.19.39a):

$$\sin z = z + O(z^3), \quad A = 1 \quad B = 0, \tag{1.74a–1.74c}$$

corresponding to Equation 1.68 ≡ Equations 1.74b and 1.74c. All zeros of the sine are simple $\beta = 1 = \beta_n$, and $2N-1$ of (Figure 1.1a) them (Equation 1.37a) lie within the circle (Equations 1.41a and 1.41b), in which the logarithmic derivative cot z is analytic except for the poles. Thus, Equation 1.71 may be applied:

$$\sin z = z\prod_{\substack{m=-\infty \\ m\neq 0}}^{+\infty}\left(1-\frac{z}{m\pi}\right)\exp\left(\frac{z}{m\pi}\right) = z\prod_{n=1}^{\infty}\left(1-\frac{z^2}{n^2\pi^2}\right), \tag{1.75}$$

leading to *the infinite product for the sine (Equation 1.75), that is absolutely convergent,* like the series for the cotangent (Equation 1.58), which can be obtained by logarithmic differentiation of Equation 1.75.

$$\begin{aligned}\cot z = \frac{\cos z}{\sin z} &= \frac{d}{dz}[\log(\sin z)] = \frac{d}{dz}\left[\log z + \sum_{n=1}^{\infty}\log\left(1-\frac{z^2}{n^2\pi^2}\right)\right] \\ &= \frac{1}{z} + \sum_{n=1}^{\infty}\left(-\frac{2z}{n^2\pi^2}\right)\left(1-\frac{z^2}{n^2\pi^2}\right)^{-1} = \frac{1}{z} + 2z\sum_{n=1}^{\infty}\frac{1}{z^2-n^2\pi^2},\end{aligned} \tag{1.76a}$$

that is justified [Section I.21.6 (Section 9.2)] by the uniform convergence of the series (infinite product). The simplification of equation (1.75) used the pairs of terms m = ±1, ±2,..., ±n in:

$$\sin z = z\prod_{n=1}^{\infty}\left(1-\frac{z}{n\pi}\right)\exp\left(\frac{z}{n\pi}\right)\left(1+\frac{z}{n\pi}\right)\exp\left(-\frac{z}{n\pi}\right) = z\prod_{n=1}^{\infty}\left(1-\frac{z^2}{n^2\pi^2}\right). \tag{1.76b}$$

The first expression (Equation 1.75) needs (Section 9.2) exponential factors for convergence:

$$\left(1-\frac{z}{m\pi}\right)e^{z/m\pi} = \left(1-\frac{z}{m\pi}\right)\left\{1+\frac{z}{m\pi}+O\left(\frac{z^2}{m^2\pi^2}\right)\right\} = 1+O\left(\frac{z^2}{m^2\pi^2}\right) \tag{1.76c}$$

in the first but not in the second form because in both cases, the general term is $O(n^{-2})$, which is consistent with absolute convergence of Equations 1.75 and 1.58. The series of fractions for the cotangent (Equation 1.58 ≡ Equation I.36.142 with $b = \pi$) and infinite product for the sine (Equation 1.75 ≡ Equation I.36.141 with $b = \pi$) was obtained before (Section I.36.6) by a totally different approach, comparing the solution of the same problem, namely, a flow source/sink between parallel walls, by two distinct methods, namely, (1) conformal mapping and (2) the method of images.

1.5.3 Series of Fractions for Circular Tangent

The process (Section 1.4) of integration and exponentiation was used to deduce the infinite product (Equation 1.67) from the series of fractions (Equation 1.62e). Conversely, logarithmic differentiation transforms an infinite product into a series of fractions (Equation 1.76a). For example, taking the logarithm of Equation 1.73 and differentiating:

$$-\frac{\sin z}{\cos z} = \frac{d}{dz}\left[\log(\cos z)\right] = \sum_{n=0}^{\infty}\frac{d}{dz}\left\{\log\left[1-\frac{z^2}{(n\pi+\pi/2)^2}\right]\right\} \tag{1.77}$$

leads to

$$\tan z = 2z\sum_{n=0}^{\infty}\frac{1}{(n+1/2)^2\pi^2 - z^2}. \tag{1.78}$$

This is the series of fractions for the tangent (Equation 1.78), which is absolutely convergent for $|z - n\pi - \pi/2| > 0$ *and totally convergent for* $|z - n\pi - \pi/2| \geq \varepsilon > 0$. It could also be derived directly from Equation 1.48, bearing in mind that (1) the circular tangent has simple poles at the zeros (Equation 1.37b) of the cosine, with residues:

$$B_m \equiv \tan_{(1)}(m\pi/2) = \lim_{z\to m\pi+\pi/2}(z - m\pi - \pi/2)\frac{\sin z}{\cos z} = \lim_{z\to m\pi+\pi/2}\frac{\sin z}{(\cos z)'} = -1; \tag{1.79}$$

(2) tan 0 = 0, and thus Equation 1.48 with Equation 1.79 simplifies to

$$\begin{aligned}\tan z &= -\sum_{m=-\infty}^{\infty}\left(\frac{1}{z - m\pi - \pi/2} + \frac{1}{m\pi + \pi/2}\right)\\ &= -\sum_{n=0}^{\infty}\left(\frac{1}{z - n\pi - \pi/2} + \frac{1}{n\pi + \pi/2} + \frac{1}{z + n\pi + \pi/2} - \frac{1}{n\pi + \pi/2}\right)\\ &= -2z\sum_{n=-0}^{\infty}\frac{1}{z^2 - (n\pi + \pi/2)^2},\end{aligned} \tag{1.80}$$

which coincides with Equation 1.78 ≡ Equation 1.80.

1.5.4 Series of Fractions for the Circular Cosecant

The circular tangent also has an infinite number of zeros at Equation 1.37a but is not an integral function unlike the sine. Its logarithmic derivative:

$$\frac{d}{dz}\left\{\log\left[\tan\left(\frac{z}{2}\right)\right]\right\} = \frac{\sec^2(z/2)}{\tan(z/2)} = \frac{1}{2\cos(z/2)\sin(z/2)} = \frac{1}{\sin z} = \csc z \tag{1.81}$$

is a meromorphic function with simple poles at Equation 1.37a. The residues at the poles are given by (Equations 1.38, 1.39a and 1.39b ≡ Equation 1.82):

$$\begin{aligned} u = z - n\pi: \qquad B_n &= \csc_{(1)}(z) = \lim_{z\to n\pi/2}\frac{z - n\pi}{\sin z}\\ &= \lim_{u\to 0}\frac{u}{\sin(u + n\pi)} = (-)^n\lim_{u\to 0}\frac{u}{\sin u} = (-)^n.\end{aligned} \tag{1.82}$$

Near the pole at the origin, the function (Equation 1.81) has Laurent series (Equation 1.83a):

$$\frac{1}{\sin z} = \frac{1}{z\left[1 + O(z^2)\right]} = \frac{1}{z}\left[1 + O(z^2)\right] = \frac{1}{z} + O(z): \quad A_0 = 0, \ B_0 = 1, \tag{1.83a–1.83c}$$

implying Equations 1.83b and 1.83c in Equation 1.59. Substituting Equations 1.82, 1.83b, and 1.83c in Equation 1.62e leads to the series of fractions:

$$\csc(z) = \frac{1}{z} + \sum_{\substack{m=-\infty \\ m\neq 0}}^{+\infty} (-)^m \left(\frac{1}{z - m\pi} + \frac{1}{m\pi} \right)$$
$$= \frac{1}{z} + \sum_{n=1}^{\infty} (-)^n \left(\frac{1}{z - n\pi} + \frac{1}{n\pi} + \frac{1}{z + n\pi} - \frac{1}{n\pi} \right), \tag{1.84}$$

that simplifies to

$$\csc z = \frac{1}{z} + 2z \sum_{n=1}^{\infty} \frac{(-)^n}{z^2 - n^2\pi^2}, \tag{1.85}$$

which is *the series of fractions for the circular cosecant and converges absolutely for* $|z - n\pi| > 0$ *and uniformly for* $|z - n\pi| \geq \varepsilon > 0$.

1.5.5 Zeros and Poles of the Circular Tangent

So far, have been obtained (1) the series of fractions for the four secondary circular functions that are all meromorphic, namely, the tangent (Equation 1.36a ≡ Equation 1.78), cotangent (Equation 1.36b ≡ Equation 1.58), cosecant (Equation 1.36c ≡ Equation 1.85), and secant (Equation 1.36d ≡ Equation 1.54); and (2) the infinite products for the primary circular functions that are integral with an infinite number of zeros, namely, the sine (Equations 1.37a and 1.75) and cosine (Equations 1.37b and 1.73); logarithmic differentiation of the cosine (sine) leads to minus the tangent (plus the cotangent), transforming the infinite product into a series of fractions, namely, Equations 1.73, 1.77, and 1.78 (Equations 1.75 and 1.76a). The tangent (cotangent) also has an infinite number of zeros [Equation 1.37a (Equation 1.37b)], and although they are meromorphic rather than an integral functions, the question of the existence of infinite product is considered next for the circular tangent using (Equations 1.81 and 1.85):

$$\frac{d}{dz}\left\{\log\left[\tan\left(\frac{z}{2}\right)\right]\right\} = \frac{d}{dz}(\log z) + \sum_{n=1}^{\infty} (-)^n \frac{d}{dz}\left[\log\left(z^2 - n^2\pi^2\right)\right]$$
$$= \frac{d}{dz}\left\{\log\left[z\prod_{n=1}^{\infty}(-)^n\left(1 - \frac{z^2}{n^2\pi^2}\right)\right]\right\}, \tag{1.86}$$

where are omitted constant terms. The exchange of the derivative with the infinite sum is permissible (Section I.21.6) for a uniformly convergent series, which excludes the neighborhoods of the poles. Substituting $z/2$ by z leads to

$$\left|z - n\pi/2\right| \geq \varepsilon > 0: \qquad \tan z \neq 2z \prod_{n=1}^{\infty} (-)^n \left(1 - \frac{4z^2}{n^2\pi^2}\right). \tag{1.87}$$

For $z = m\pi$ the r.h.s. of (1.87) vanishes for $n = 2m$, and the l.h.s. also vanishes because it is a zero of the sine (Equation 1.37a) also hence of the tangent (Equation 1.36a); therefore, Equation 1.87

actually holds for Equation 1.36a. For $z = m\pi + \pi/2$ the r.h.s. of Equation 1.87 vanishes for $n = 2m + 1$, whereas the r.h.s. diverges, because (Equation 1.37b) is a zero of the cosine and hence a pole of the tangent (Equation 1.36a); therefore the quality in Equation 1.87 cannot possibly hold for Equation 1.37b. A similar conclusion was obtained by a different approach, namely, comparison of conformal mapping with the method of images (Subsection I.36.7.1). Concerning the conditions of validity of the theorem on infinite products (Subsection 1.4.2) and its extension (Subsection 1.4.3), it is the former that is relevant to the circular tangent because it has no pole at the origin: (1) the theorem allows poles of the logarithmic derivative (Equation 1.81) due to zeros (Equation 1.37a) of the denominator sin z where Equation 1.87 holds; and (2) the theorem requires the logarithmic derivative (Equation 1.81) to be analytic at all other points, and that is not the case at the poles (Equation 1.37b) of the numerator $1/(\cos z) = \sec z$, where Equation 1.87 does not hold. Thus, the theorem on infinite products cannot be used to justify Equation 1.87 for $|z - n\pi/2| > 0$, and indeed, it does not hold with this condition.

1.6 Recurrence Formulas and Continued Fractions (Wallis 1656; Euler 1737)

Both the series of functions (Sections 1.1 through 1.3) and infinite products (Sections 1.4 and 1.5) can be transformed into continued fractions (Section 1.8) whose truncation specifies the continuants. The properties of continued fractions are considered first (Sections 1.6 and 1.7), starting with recurrence formulas (Subsection 1.6.1) and leading to the convergence of continuants (Subsection 1.6.2). The representation as a terminating (nonterminating) continued fraction can be used (Subsection 1.6.3) to distinguish between rational (irrational) numbers and functions. The truncation of a continued fraction at any order leads to a rational number or function whose numerator and denominator can be determined by recurrence (Subsection 1.6.4). The exchange of the numerators and denominators specifies the continued fraction for the algebraic inverse (Subsection 1.6.5).

1.6.1 Multiple Backward Linear Recurrence Formulas

The general term f_n of a sequence may satisfy a **backward linear recurrence formula** of order N:

$$f_{n-1} = \sum_{k=0}^{N-1} A_{k,n} f_{n+k}, \tag{1.88}$$

with coefficients $A_{k,n}$ The simplest case is (Equation 1.89a) a recurrence formula of order 1 (Equation 1.89b) whose solution (Equation 1.89c) is immediate:

$$A_{0,n} \equiv a_n \neq 0: \qquad f_{n-1} = a_n f_n, \qquad f_n = \frac{f_{n-1}}{a_n} = \frac{f_0}{a_n \ldots a_1} \tag{1.89a–1.89c}$$

if the coefficients are nonzero (Equation 1.89a). A recurrence formula of order 2 (Equations 1.89a and 1.90a) has two terms (Equation 1.90b):

$$A_{2,n} \equiv b_{n+1}: \quad f_{n-1} = a_n f_n + b_{n+1} f_{n+1}, \tag{1.90a and 1.90b}$$

and determines f_{n-1} from f_n, f_{n+1} with given coefficients (a_n, b_{n+1}). It can be rewritten in the form:

$$f_n \neq 0: \qquad \frac{f_{n-1}}{f_n} = a_n + b_{n+1}\frac{f_{n+1}}{f_n} = a_n + \frac{b_n}{f_n/f_{n+1}}, \qquad \text{(1.91a and 1.91b)}$$

that can be applied recursively:

$$0 \neq f_n, f_{n+1}, \ldots: \qquad \frac{f_{n-1}}{f_n} = a_n + \frac{b_{n+1}}{a_{n+1} + b_{n+1}/\left(f_{n+1}/f_{n+2}\right)} = \ldots, \qquad \text{(1.92a and 1.92b)}$$

leading to the **continued fraction:**

$$\frac{f_{n+1}}{f_n} = a_n + \cfrac{b_{n+1}}{a_{n+1} + \cfrac{b_{n+2}}{a_{n+2} + \cfrac{b_{n+3}}{a_{n+3} + \cdots}}} \qquad \text{(1.93a)}$$

$$\equiv a_n + \frac{b_{n+1}}{a_{n+1} +}\,\frac{b_{n+2}}{a_{n+2} +}\,\frac{b_{n+3}}{a_{n+3} +}\cdots. \qquad \text{(1.93b)}$$

The **denominators** $a_{n+1}, \ldots$ *and the* **numerators** $b_{n+1}, \ldots$ *of the continued fraction were the coefficients in the backward linear recurrence formula of the second order (Equation 1.90b).* The notation (Equation 1.93b) should not be confused with

$$\frac{f_{n+1}}{f_n} \neq a_n + \frac{b_{n+1}}{a_{n+1}} + \frac{b_{n+2}}{a_{n+2}} + \frac{b_{n+3}}{a_{n+3}} + \cdots. \qquad \text{(1.94)}$$

It has been shown that *the solution of the two-point backward recurrence formula (Equation 1.90b) with the conditions (Equation 1.92a) is the continued fraction (Equation 1.93a ≡ Equation 1.93b).*

1.6.2 Continuants, Termination, Periodicity, and Convergence

Like the other infinite processes such as limits, integrals, series, and products, the continued fractions raise questions of convergence. Consider the continued fraction (Equations 1.93a and 1.93b) from the very beginning $n = 1$:

$$X = a_1 + \frac{b_2}{a_2 +}\,\frac{b_3}{a_3 +}\cdots\frac{b_n}{a_n +}\cdots. \qquad \text{(1.95)}$$

The nth **continuant** is defined as the result of truncating the continued fraction at the nth term:

$$X_n \equiv a_1 + \frac{b_2}{a_2 +}\,\frac{b_3}{a_3 +}\cdots\frac{b_{n-1}}{a_{n-1} +}\,\frac{b_n}{a_n} \equiv \frac{p_n}{q_n}, \qquad \text{(1.96)}$$

where $p_n(q_n)$ are the **numerators (denominators) of the continuants**, not to be confused with the numerators and denominators of the continued fraction (Equation 1.95). A continued fraction is **terminating (infinite)** if the continuants are equal beyond a certain order (if above any order, there are always distinct continuants). A terminating continued fraction specifies a rational number; thus, an irrational number can be specified only by a convergent nonterminating continued fraction (Subsection 1.7.4). A nonterminating fraction is **periodic (nonperiodic)** if the same set of terms repeats itself beyond a certain order (there are no sequences that repeat indefinitely). A nonterminating continued fraction raises the issue of convergence. As for series (Section I.21.1), a nonterminating continued fraction is (1) **convergent** if its continuants have a single finite limit (Equation 1.97a):

$$\lim_{n\to\infty} X_n = \begin{cases} X < \infty & C, \\ \infty & D, \\ \text{otherwise} & O; \end{cases} \tag{1.97a--1.97c}$$

(2) **divergent** if the continuants are unbounded (Equation 1.97b); and (3) **oscillatory** in the remaining cases, that is, the continuants are bounded but have no limit (Equation 1.97c).

1.6.3 Recurrence Formula for Numerators and Denominators of Convergents (Wallis 1656; Euler 1737)

The first three continuants are the fractions:

$$\frac{p_1}{q_1} = a_1, \qquad \frac{p_2}{q_2} = a_1 + \frac{b_2}{a_2} = \frac{a_1 a_2 + b_2}{a_2}, \tag{1.98a and 1.98b}$$

$$\frac{p_3}{q_3} = a_1 + \frac{b_2}{a_2 + b_3/a_3} = \frac{a_1 a_2 a_3 + a_1 b_3 + b_2 a_3}{a_2 a_3 + b_3}. \tag{1.98c}$$

The numerators p_n and denominators p_n satisfy:

$$p_1 = a_1, \quad q_1 = 1; \quad p_2 = a_1 a_2 + b_2, \quad q_2 = a_2; \tag{1.99a–1.99d}$$

$$p_3 = (a_1\, a_2 + b_2)\, a_3 + a_1 b_3, \quad q_3 = a_2\, a_3 + b_3. \tag{1.100a and 1.100b}$$

The last two formulas can be rewritten in the form:

$$p_3 = p_2\, a_3 + p_1\, b_3, \quad q_3 = q_2\, a_3 + q_1 b_3. \tag{1.101a and 1.101b}$$

The latter suggests that

$$p_n = a_n\, p_{n-1} + b_n\, p_{n-2}, \quad q_n = a_n\, q_{n-1} + b_n\, q_{n-2}. \tag{1.102a and 1.102b}$$

The **value of continuants** *(Wallis 1656; Euler 1737), the numerators p_n and denominators q_n of the continuants (Equation 1.96) of the continued fraction (Equation 1.95), can be calculated recursively* *by Equations 1.102a and 1.102b, using only the coefficients* a_n, b_n. Applying Equations 1.102a and 1.102b with n = 2 leads to

$$p_2 = a_2\, p_1 + b_2\, p_0, \quad q_2 = a_2\, q_1 + b_2\, q_0. \qquad \text{(1.103a and 1.103b)}$$

Bearing in mind the Equations 1.99c and 1.99d, the choices:

$$p_0 = 1, \quad q_0 = 0 \qquad \text{(1.104a and 1.104b)}$$

make consistent the following: Equations 1.99c and 1.99d ≡ Equations 1.103a and 1.103b.

1.6.4 Differences of Continuants and Odd/Even Commutators

To prove Equations 1.102a and 1.102b formally by induction (Section I.1.1), note that (1) they hold for $n = 3$ when they reduce to Equations 1.101a and 1.101b; and (2) and thus it is sufficient to show that if they hold for n, then they also hold for $n + 1$. To demonstrate the latter point, remark that the $(n + 1)$ th continuant p_{n+1}/q_{n+1} is obtained from the preceding p_n/q_n by replacing a_n by Equation 1.105a, leading from Equations 1.102a and 1.102b with n replaced by $n + 1$ to Equation 1.105b:

$$a_n \to a_n + \frac{b_{n+1}}{a_{n+1}}: \quad \frac{p_{n+1}}{q_{n+1}} = \frac{\left(a_n + b_{n+1}/a_{n+1}\right)p_{n-1} + b_n p_{n-2}}{\left(a_n + b_{n+1}/a_{n+1}\right)q_{n-1} + b_n q_{n-2}}$$

$$= \frac{a_{n+1}\left(a_n p_{n-1} + b_n p_{n-2}\right) + b_{n+1}p_{n-1}}{a_{n+1}\left(a_n q_{n-1} + b_n q_{n-2}\right) + b_{n+1}q_{n-1}} = \frac{a_{n+1}p_n + b_{n+1}p_{n-1}}{a_{n+1}q_n + b_{n+1}q_{n-1}}.$$

(1.105a and 1.105b)

Formulas 1.102a and 1.102b for n were used to prove Equation 1.105b, whose numerator and denominator coincide, respectively, with Equations 1.102a and 1.102b for $n + 1$.

The first (Equation 1.106a) and second (Equation 1.106b) commutators for the numerators p_n *and denominators* q_n *of the continuants* also follow from Equations 1.102a and 1.102b:

$$p_n q_{n-1} - p_{n-1} q_n = -b_n(p_{n-1}q_{n-2} - p_{n-2}q_{n-1}) = (-)^{n-1} b_n \ldots b_2 (p_1 q_0 - p_0 q_1) = (-)^n b_n \ldots b_2 \qquad \text{(1.106a)}$$

$$p_n q_{n-2} - p_{n-2} q_n = a_n(p_{n-1}q_{n-2} - p_{n-2}q_{n-1}) = (-)^{n-1} a_n b_{n-1} \ldots b_2 \qquad \text{(1.106b)}$$

using Equations 1.99a 1.99b, 1.104a and 1.104b. The results (Equations 1.106a and 1.106b) can be rewritten in the form

$$X_n - X_{n-1} = \frac{p_n}{q_n} - \frac{p_{n-1}}{q_{n-1}} = (-)^n \frac{b_n \ldots b_2}{q_n q_{n-1}}, \qquad \text{(1.107a)}$$

$$X_n - X_{n-2} = \frac{p_n}{q_n} - \frac{p_{n-2}}{q_{n-2}} = (-)^{n-1} \frac{a_n b_{n-1} \ldots b_2}{q_n q_{n-2}}, \qquad \text{(1.107b)}$$

as the differences between the first (Equation 1.107a) and second (Equation 1.107b) successive continuants.

1.6.5 Continued Fraction for the Algebraic Inverse

The preceding recurrence formulas can be used to prove the **lemma of inversion:** *the algebraic inverse of the continued fraction (Equation 1.95) is*

$$\frac{1}{X} = \frac{1}{a_1 +} \frac{b_2}{a_2 +} \frac{b_3}{a_3 +} \cdots \frac{b_n}{a_2 +} \cdots. \tag{1.108}$$

The proof is made by showing that for all continuants (Equation 1.109a):

$$\frac{1}{X_n} = \frac{1}{a_1 +} \frac{b_2}{a_2 +} \frac{b_3}{a_3 +} \cdots \frac{b_n}{a_n} = \frac{P_n}{Q_n}: \qquad P_n = q_n, \qquad Q_n = P_n, \tag{1.109a–1.109c}$$

the numerator and denominator are interchanged (Equations 1.109b and 1.109c) with Equation 1.96. This involves three steps: (1) the first continuants (Equations 1.110a and 1.110b) satisfy Equations 1.110c and 1.110d that prove Equations 1.109b and 1.109c for $n = 1$:

$$X_1 = a_1 = \frac{p_1}{q_1}, \qquad \frac{1}{X_1} = \frac{1}{a_1} = \frac{P_1}{Q_1}: \qquad P_1 = 1 = q_1, \qquad Q_1 = a = P_1; \tag{1.110a–1.110d}$$

(2) the second continuants (Equations 1.111a and 1.111b) satisfy Equations 1.111c and 1.111d that prove Equations 1.109b and 1.109c for $n = 2$:

$$X_2 = a_1 + \frac{b_2}{a_2} = \frac{b_2 + a_1 a_2}{a_2} = \frac{p_2}{q_2}, \qquad \frac{1}{X_2} = \frac{1}{a_1 + b_2/a_2} = \frac{a_2}{b_2 + a_1 a_2} = \frac{P_2}{Q_2}, \tag{1.111a and 1.111b}$$

$$P_2 = a_2 = q_2, \quad Q_2 = b_2 + a_1\, a_2 = p_2; \tag{1.111c and 1.111d}$$

(3) for all higher orders (Equation 1.112a), the recurrence relations Equations 1.102a and 1.102b lead to Equations 1.112b and 1.112c:

$$n = 3,\ldots; \quad P_n = a_n P_{n-1} + b_n P_{n-2} = a_n q_{n-1} + b_n q_{n-2} = q_n, \tag{1.112a and 1.112b}$$

$$Q_n = a_n Q_{n-1} + b_n Q_{n-2} = a_n p_{n-1} + b_n p_{n-2} = p_n, \tag{1.112c}$$

which complete the proof of Equations 1.109b and 1.109c.

1.7 Optimal and Doubly Bounded Sharpening Approximations

The continuants X_n of a convergent continued fraction provide rational approximations to the limit X in Equation 1.97a. It is useful to (a) have an upper bound for the error between the nth estimate and the ultimate value $|X_n - X|$; and (b) know whether, for a given level of complexity, the approximation X_n is the closest possible. The question (a) of accuracy is considered first (Section 1.7.1) followed by that (b) of optimality (Section 1.7.3). The estimate of the accuracy of the truncation of a continued fraction of the first kind (1) also proves (Subsection 1.7.2) that it cannot diverge, that is, it can only oscillate or converge. The simple continued fraction (2) provides a unique representation (Subsection 1.7.4) of a real number, that is, rational (irrational) if the associated simple continued fraction is terminating (nonterminating).

1.7.1 Error of Truncation for a Continued Fraction of the First Kind

Concerning the matter (a) of accuracy, consider (Equation 1.95) a **continued fraction of the first kind**, for which all terms are positive (Equation 1.113a):

$$a_n, b_n > 0 \quad X_{2n-1} < X_{2n} < X_{2n-2}, \quad X_{2n-1} < X_{2n+1} < X_{2n}, \tag{1.113a–1.113d}$$

then Equations 1.113c and 1.113d follow from Equations 1.107a and 1.107b with $2n$, $2n + 1$. Thus, it is concluded that

$$\ldots X_{2n} > X_{2n+2} > \ldots > X > \ldots X_{2n+1} > X_{2n-1} > \ldots. \tag{1.114}$$

For a continued fraction (Equation 1.95) of the first kind, Equations 1.113a and 1.113b ≡ Equations 1.115a and 1.115b holds (Figure 1.3) that (1) the even (odd) continuants are larger (smaller) than the limit, that is, provide an **upper (lower) bound** *for the ultimate value; (2) the even (odd) continuants form a monotonic decreasing (increasing) sequence, all terms of the former being larger than those of the latter, which leads to a* **sharpening approximation;** *and (3) the difference between successive continuants is an upper bound on the error between the last continuant and the limit (Equation 1.115c):*

$$a_n, b_n > 0: \quad |X_n - X| < |X_n - X_{n-1}|. \tag{1.115a–1.115c}$$

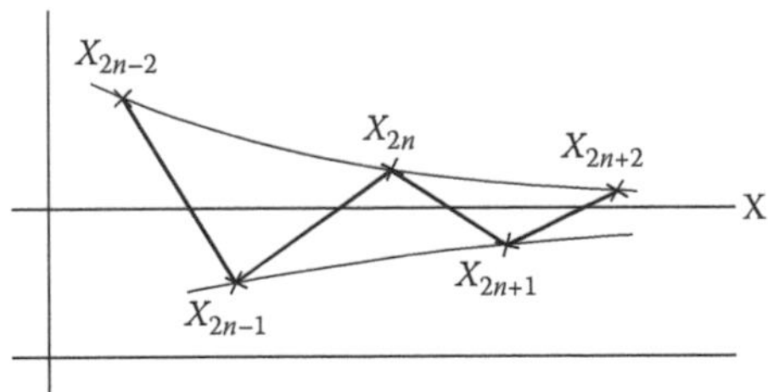

FIGURE 1.3
Continued fraction (Equation 1.95) of the first kind, that is, with all numerators and denominators positive (Equations 1.113a and 1.113b ≡ Equations 1.115a and 1.115b), provides through its continuants (Equation 1.96) a sharpening approximation (Equation 1.115c) to the limit, in the sense that (1) each continuant is closer to the limit than all the preceding; (2) successive continuants alternate above and below the limit; and (3) their difference is an upper bound for the error of truncation.

It has been shown that the continued fractions of the first kind meet the requirement (a) stated above of providing an upper bound for the error in the estimate of the limit arising from truncation at any order.

1.7.2 Convergence/Oscillation of a Continued Fraction of the First Kind

For a continued fraction of the first kind (Equation 1.113a), the even (odd) continuants form an increasing (decreasing) sequence (Equation 1.114). Because the latter are larger than the former, both must have a limit:

$$\lim_{n\to\infty} X_{2n} = X^+ \geq \lim_{n\to\infty} \frac{p_n}{q_n} \geq X^- = \lim_{n\to\infty} X_{2n-1}, \tag{1.116}$$

Because the limit of the continuants must lie below (above) the even (odd) continuants, *a simple fraction (Equation 1.113a ≡ Equations 1.117a and 1.117b) cannot diverge, leaving two possibilities: (1) if the limits of the even (odd) continuants (Equation 1.116) converge to distinct values (Equation 1.117b), it is oscillatory between them:*

$$a_n, b_n > 0: \qquad \lim_{n\to\infty} \frac{p_n}{q_n} \begin{cases} \text{oscillates in } \left(X^-, X^+\right) & \text{if} \quad X^+ > X^-, \\ = X & \text{if} \quad X^+ = X^- \equiv X; \end{cases} \tag{1.117a–1.117d}$$

(2) if the limits coincide, it converges to their common value (Equation 1.117d). The convergence of continued fractions is considered in Subsections 9.3.1 to 9.3.3.

1.7.3 Optimal Rational Approximation for a Simple Continued Fraction

Concerning requirement (b), consider a **simple continued fraction** (Equations 1.93a and 1.93b), defined (Equation 1.118b) as a convergent continued fraction, with numerators equal to unity (Equation 1.118a):

$$b_n = 1: \qquad Y \equiv a_1 + \frac{1}{a_2 +}\frac{1}{a_3 +}\cdots\frac{1}{a_n +}\cdots. \tag{1.118a and 1.118b}$$

A continued fraction that is simple (Equation 1.118a) and of the first kind (Equations 1.113a and 1.113b) is a **simple positive continued fraction** (Equation 1.119b) that by definition has positive (Equation 1.119a) denominators:

$$a_n > 0: \qquad Y = a_1 + \frac{1}{a_2 +}\frac{1}{a_3 +}\cdots\frac{1}{a_n +}\cdots. \tag{1.119a and 1.119b}$$

Next is to prove the **optimal property**: *any rational fraction p/q that is closer to the limit Y than the continuant p_n/q_n of a simple positive continued fraction has larger numerator and denominator:*

$$\left|\frac{p}{q} - Y\right| < \left|\frac{p_n}{q_n} - Y\right| \Rightarrow p > p_n \wedge q > q_n. \tag{1.120}$$

To prove the theorem (Equation 1.120), suppose that p/q is closer to Y than p_n/q_n; then it is also closer to Y than p_{n-1}/q_{n-1} and must lie between these two continuants (Equation 1.121a):

$$\left|\frac{p_{n-1}}{q_{n-1}}-\frac{p}{q}\right|<\left|\frac{p_{n-1}}{q_{n-1}}-\frac{p_n}{q_n}\right|, \qquad \frac{p_{n-1}}{q_{n-1}}><\frac{p}{q}><\frac{p_n}{q_n}, \tag{1.121a and 1.121b}$$

and the three form either an increasing or a decreasing sequence (Equation 1.121b). From Equations 1.106a and 1.106b, it follows that for a simple positive continued fraction (Equation 1.118a), the commutators (Equations 1.122a and 1.122b) hold:

$$p_n q_{n-1} - p_{n-1} q_n = (-)^n, \quad p_n q_{n-2} - p_{n-2} q_n = (-)^{n-1} a_n. \tag{1.122a and 1.122b}$$

From Equation 1.121a follows Equation 1.123b:

$$N \equiv \left|p_{n-1} q - q_{n-1} p\right|: \qquad \frac{N}{q_{n-1} q} < \frac{\left|p_{n-1} q_n - p_n q_{n-1}\right|}{q_{n-1} q_n} \le \frac{1}{q_{n-1} q_n}, \tag{1.123a and 1.123b}$$

where N is an integer (Equation 1.123a); thus, Equation 1.123b implies $q > N q_n \ge q_n$ and proves the second inequality in Equation 1.120. From Equation 1.121b, if the sequence is decreasing $p > p_n\, q/q_n \ge p_n$ (increasing $p > p_{n-1} q/q_{n-1} \ge p_{n-1}$), so that in both cases, the first inequality in Equation 1.120 is proven. The optimal property may be restated: *of all fractions p/q with numerator p and denominator q not exceeding $p \le p_n$ and $q \le q_n$, the n-th continuant p_n / q_n of a simple positive (Equation 1.119a) continued fraction (Equation 1.119b) is the closest to the limit of p_n/q_n as $n \to \infty$, and thus is the best approximation for a given level of complexity.*

1.7.4 Unique Simple Terminating (Nonterminating) Continued Fraction for a Rational (Irrational) Number

A simple continued fraction has the **property of unicity**: a given real number (Equation 1.124a) has a unique representation as a simple fraction (Equation 1.124c) whose denominators are all positive integers (Equation 1.124b):

$$c \in |R: \quad \exists^1 \{a_2, \ldots, a_n, \ldots\} \in |N: \qquad c = a_1 + \frac{1}{a_2+}\frac{1}{a_3+}\cdots\frac{1}{a_n+}\cdots. \tag{1.124a–1.124c}$$

The unique simple fraction for a real number is obtained by choosing before each fraction the largest integer that does not exceed the continuant, for example, the unique simple continued fraction for the number 11/3 is

$$\frac{11}{3} = 3 + \frac{2}{3} = 3 + \frac{1}{3/2} = 3 + \frac{1}{1+1/2} = 3 + \frac{1}{1+}\frac{1}{2}. \tag{1.124d}$$

The method to obtain a unique simple fraction for a real number c is described as follows: (1) take as the leading term a_1 of the simple continued fraction (Equation 1.119b) the integer part (Equation 1.124f) of c, that is, the largest integer (Equation 1.124e) not exceeding c; (2) if c equals its integer part, it is an integer; (3) if it is not equal its integer part (Equation 1.124g), it must satisfy Equation 1.124h with c_1 positive (Equation 1.124i):

$$a_1 \in |Z: \quad a_1 \le c < a_1 + 1; \qquad c \ne a_1: \quad c = a_1 + \frac{1}{c_1}, \quad c_1 > 0; \qquad (1.124e\text{–}1.124i)$$

(4) if c_1 is a positive integer, then Equation 1.124h is a terminating simple fraction (Equations 1.119a and 1.119b) specifying a rational number c; (5) if c_1 is not an integer, the process is repeated to find $c_2,\ldots, c_n,\ldots$, whose integer parts $a_3,\ldots, a_{n+1},\ldots$ are all positive (Equation 1.119a); (6) if at any stage c_n becomes an integer, the continued fraction terminates and specifies a rational number c; (7) if c_n is never an integer, the simple continued fraction does not terminate and it specifies an irrational number c, provided that it converges; and (8) the simple fraction has positive integer denominators a_n; thus, the series (Equation 9.53c) diverges and the simple continued fraction converges. It has been proven that *the unique simple continued fraction for a real number is terminating (nonterminating) if it is a rational (irrational) number.* This provides a method to prove that a number is irrational (Subsection 9.3.3), for example, e (π) in Subsection 1.8.5 (Subsection 1.8.7).

1.8 Transformation of Series and Products into Fractions (Euler 1785)

In order to use the optimal numerical approximation properties of continued fractions, it may be appropriate to convert to the latter other infinite processes, such as sums (Subsection 1.8.1) [products (Subsection 1.8.3)], either finite or infinite; the latter include convergent infinite series (products). An intermediate result (Subsection 1.8.2) is the continued fraction with given continuants. Examples of the conversion of an infinite series (product) to a continued fraction include [Subsection 1.8.4 (Subsection 1.8.5)] the exponential function (circular sine), that leads [Subsection 1.8.6 (Subsection 1.8.7)] to the proof that $e(\pi)$ is an irrational number and enables its computation.

1.8.1 Continued Fraction for a Series (Euler 1748)

In order that a series of general term f_n coincides with a continued fraction (Equations 1.93a and 1.93b), the sum of the first n terms may be equated (Equation 1.125a) to the nth continuant (Equation 1.96):

$$f_1 + \cdots + f_n = p_n = \frac{q_n}{p_n}; \qquad q_n \equiv 1. \qquad (1.125a \text{ and } 1.125b)$$

Equation 1.125a, like Equation 1.96, is unchanged, that is, multiplying p_n, q_n by the same factor λ leads to $(\lambda p_n, \lambda q_n)$; the choice (Equation 1.125b) is made to ensure unicity. Using the recurrence formula 1.102a leads to

$$\begin{aligned}0 &= \left(f_1 + \cdots + f_{n-1}\right)a_n + \left(f_1 + \cdots + f_{n-2}\right)b_n - \left(f_1 + \cdots + f_n\right) \\ &= \left(f_1 + \cdots + f_{n-2}\right)\left(a_n + b_n - 1\right) + f_{n-1}\left(a_n - 1\right) - f_n.\end{aligned} \tag{1.126}$$

From Equations 1.102b and 1.125b follows Equation 1.127a, that simplifies Equation 1.126 to Equations 1.127b and 1.127c:

$$a_n + b_n = 1, \quad a_n = 1 + \frac{f_n}{f_{n-1}} = \frac{f_{n-1} + f_n}{f_{n-1}}, \quad b_n = 1 - a_n = -\frac{f_n}{f_{n-1}}; \quad a_1 = p_1 = f_1, \qquad \text{(1.127a–1.127d)}$$

noting Equation 1.99a, that implies Equation 1.127d leads to the continued fractions (Equations 1.128a and 1.128b):

$$\sum_{k=1}^{n} f_k = \frac{f_1}{1-}\,\frac{f_2/f_1}{\left(f_2 + f_1\right)/f_1 -}\,\frac{f_3/f_2}{\left(f_3 + f_2\right)/f_2 -}\cdots\frac{f_n/f_{n-1}}{\left(f_n + f_{n-1}\right)/f_{n-1}}, \tag{1.128a}$$

$$= \frac{f_1}{1-}\,\frac{f_2}{f_2 + f_1 -}\,\frac{f_1 f_3}{f_2 + f_3 -}\,\frac{f_2 f_4}{f_3 + f_4 -}\cdots\frac{f_{n-2} f_n}{f_{n-1} + f_n} = -f_0 + \sum_{k=0}^{\infty} f_k, \qquad \text{(1.128b and 1.128c)}$$

where an independent first term f_0 may be inserted (Equation 1.128c) as in Equation 1.95. *This is the transformation (Equations 1.128a through 1.128c) of a sum (series $n \to \infty$) into the associated continued fraction.* Three more instances of transformation of series to continued fractions are given in the Example 10.2.

1.8.2 Continued Fraction with Given Continuants

The nth term of the series (Equation 1.125a) is the difference of two successive continuants:

$$f_n = \sum_{k=1}^{n} f_k - \sum_{k=1}^{n-1} f_k = X_n - X_{n-1}, \quad f_n + f_{n-1} = X_n - X_{n-2}, \quad f_1 = X_1, \quad f_1 + f_2 = X_2, \qquad \text{(1.129a–1.129d)}$$

which, on substitution into Equation 1.128b, leads to the continued fraction (Equation 1.130):

$$\frac{X_1}{1-}\,\frac{X_2 - X_1}{X_2 -}\,\frac{X_1\left(X_3 - X_2\right)}{X_3 - X_1 -}\,\frac{\left(X_2 - X_1\right)\left(X_4 - X_3\right)}{X_4 - X_2 -}\cdots\frac{\left(X_{n-2} - X_{n-3}\right)}{\left(X_n - X_{n-2}\right)}\left(X_n - X_{n-1}\right) = X_n. \tag{1.130}$$

Thus, *the continued fraction whose continuants are the quantities X_n given a priori is Equation 1.130.*

1.8.3 Continued Fraction for a Product

Concerning the product:

$$X_n = g_1 \ldots g_n, \quad X_n - X_{n-1} = g_1 \ldots g_{n-1}(g_n - 1), \quad X_n - X_{n-2} = g_1 \ldots g_{n-2}(g_{n-1}g_n - 1), \tag{1.131a–1.131c}$$

the substitution in Equation 1.130 leads to

$$\prod_{k=1}^{n} g_k = \frac{g_1}{1-}\,\frac{g_1(g_2-1)}{g_1g_2-}\,\frac{g_1g_1g_2(g_3-1)}{g_1(g_2g_3-1)-}\,\frac{g_1(g_2-1)g_1g_2g_3(g_4-1)}{g_1g_2(g_3g_4-1)-}\cdots\frac{g_1\cdots g_{n-3}(g_{n-2}-1)g_1\cdots g_{n-1}(g_n-1)}{g_1\cdots g_{n-2}(g_{n-1}g_n-1)}. \tag{1.132}$$

This simplifies to the continued fraction:

$$\prod_{k=1}^{n} g_k = \frac{g_1}{1-}\,\frac{g_2-1}{g_2-}\,\frac{g_2(g_3-1)}{g_2g_3-1-}\,\frac{g_3(g_2-1)(g_4-1)}{g_3g_4-1-}\cdots\frac{g_{n-1}(g_{n-2}-1)(g_n-1)}{g_{n-1}g_n-1}. \tag{1.133}$$

This is *the transformation of a product (infinite product $n \to \infty$) into an associated continued fraction. The limit $n \to \infty$ proves that the nonterminating continued fraction has the same convergence properties as the infinite series (Equations 1.128a through 1.128c) [product (Equation 2.133)].* The continued fraction may terminate representing a rational number or fraction. The nonterminating continued fractions raise issues of convergence (Subsections 1.6.2, 1.7.2, and Section 9.3) as do infinite series (Chapters I.21, 23, 25, and 29) and infinite products (Section 9.2).

1.8.4 Continued Fraction for the Exponential

The series (Equation 1.26b ≡ 1.134a) for the exponential has first (Equation 1.134b) and general (Equation 1.134c) terms and ratio of successive terms (Equation 1.134d):

$$e^z = \sum_{n=1}^{\infty} f_n, \qquad f_1 = 1, \qquad f_n = \frac{z^{n-1}}{(n-1)!}, \qquad \frac{f_{n+1}}{f_n} = \frac{z}{n}. \tag{1.134a–1.134d}$$

Substitution in Equation 1.128a leads to the continued fraction for the exponential:

$$e^z = \frac{1}{1-}\,\frac{z}{1+z-}\,\frac{z/2}{1+z/2-}\,\frac{z/3}{1+z/3-}\cdots\frac{z/n}{1+z/n-}\cdots, \tag{1.135a}$$

$$= \frac{1}{1-}\,\frac{z}{1+z-}\,\frac{z}{2+z-}\,\frac{2z}{3+z-}\,\frac{3z}{4+z-}\cdots\frac{nz}{n+1+z-}\cdots. \tag{1.135b}$$

An alternative is to write the exponential series (Equation 1.26b ≡ Equation 1.134a ≡ Equation 1.136a):

$$e^z = \sum_{n=0}^{\infty} f_n, \qquad f_0 = 1, \qquad f_n = \frac{z^n}{n!}, \qquad \frac{f_{n+1}}{f_n} = \frac{z}{n+1}, \tag{1.136a–1.136d}$$

leading (Equation 1.128c) to

$$e^z = 1 + \frac{z}{1-}\frac{z/2}{1+z/2-}\frac{z/3}{1+z/3-}\cdots\frac{z/n}{1+z/n-}\cdots \tag{1.137a}$$

$$= 1 + \frac{z}{1-}\frac{z}{2+z-}\frac{2z}{3+z-}\cdots\frac{nz}{n+1+z-}\cdots \tag{1.137b}$$

as another continued fraction for the exponential.

Applying the algebraic inversion Equations 1.95 and 1.108 to Equation 1.135b (Equation 1.137b) leads to Equation 1.138 (Equation 1.139):

$$e^{-z} = \frac{1}{e^z} = 1 - \frac{z}{1+z-}\frac{z}{2+z-}\frac{2z}{3+z-}\frac{3z}{4+z-}\cdots\frac{nz}{n+1+z-}\cdots, \tag{1.138}$$

$$e^{-z} = \frac{1}{e^z} = \frac{1}{1+}\frac{z}{1-}\frac{z}{2+z-}\frac{2z}{3+z-}\frac{3z}{4+z-}\cdots\frac{nz}{n+1+z-}. \tag{1.139}$$

The change of variable (Equation 1.140a) transforms Equation 1.138 (Equation 1.139) to the continued fraction [Equation 1.140b (Equation 1.141b)], which is valid like the preceding in the finite complex plane (Equation 1.141a):

$$z \to -z: \qquad e^z = 1 + \frac{1}{1-z+}\frac{z}{2-z+}\frac{2z}{3-z+}\frac{3z}{4-z+}\cdots\frac{nz}{n+1-z+}\cdots, \tag{1.140a and 1.140b}$$

$$|z| < \infty: \qquad e^z = \frac{1}{1-}\frac{z}{1+}\frac{z}{2-z+}\frac{2z}{3-z+}\frac{3z}{4-z+}\cdots\frac{nz}{n+1-z+}\cdots. \tag{1.141a and 1.141b}$$

The exponential is represented in the finite complex plane (Equation 1.141a) by the four continued fractions (Equations 1.135a, 1.135b, 1.137a, 1.137b, 1.140b, and 1.141b). These are reviewed in Section 3.2 together with two additional continued fractions (Subsection 3.9.9) obtained by a different method (Section 1.9).

1.8.5 Calculation of the Napier Irrational Number e

Setting Equation 1.142a in Equations 1.135b, 1.137b, 1.140b, and 1.141b leads, respectively, to Equations 1.142b through 1.142e. The continued fractions for the exponential [Equation

1.135b (Equation 1.137)] specify [Equation 1.138b (Equation 1.138c)] for (Equation 1.138a) the number e:

$$z=1: \quad 2.718=e=\frac{1}{1-}\frac{1}{2-}\frac{1}{3-}\frac{2}{4-}\frac{3}{5-}\cdots\frac{n}{n+2-}\cdots, \tag{1.142a and 1.142b}$$

$$=1+\frac{1}{1-}\frac{1}{3-}\frac{2}{4-}\frac{2}{5-}\cdots\frac{n}{n+2-}\cdots, \tag{1.142c}$$

$$=1+\frac{1}{0+}\frac{1}{1+}\frac{2}{2+}\frac{3}{3+}\cdots\frac{n}{n+}\cdots, \tag{1.142d}$$

$$=\frac{1}{1-}\frac{1}{1+}\frac{1}{1+}\frac{2}{2+}\frac{3}{3+}\cdots\frac{n}{n+}\cdots. \tag{1.142e}$$

The continued fractions (Equations 1.142b through 1.142e) are nonterminating, proving that e is an irrational number, and can be used to calculate its value. The second continuant of the continued fractions (Equations 1.142d and 1.142e) is infinite $X_2=\infty$; the following continuants X_n with $n \geq 3$ are finite and converge to the number e as $n \to \infty$. The preceding circumstance implies that the continued fractions (Equations 1.142d and 1.142e) converge more slowly than Equation 1.142b and 1.142c, in the sense that the same accuracy is attained by the latter with earlier continuants of lower order. Thus, Equations 1.142b and 1.142c are used to obtain rational approximations to the irrational number e using the first four continuants:

$$e-e_1=0.718: \quad e_1=\frac{1}{1-1/2}=2.000, \tag{1.143a and 1.143b}$$

$$e-e_2=0.218: \quad e_2=\frac{1}{1-1/(2-1/3)}=\frac{1}{1-3/5}=\frac{5}{2}=2.500, \tag{1.144a and 1.144b}$$

$$e-e_3=0.051: \quad e_3=\frac{1}{1-1/\left[2-1/(3-2/4)\right]}=\frac{1}{1-1/(2-2/5)}=\frac{1}{1-5/8}=\frac{8}{3}=2.667, \tag{1.145a and 1.145b}$$

$$e-e_4=0.010: \quad e_4=\frac{1}{1-1/\{2-1/[3-2/(4-3/5)]\}}=\frac{1}{1-1/[2-1/(3-10/17)]}$$
$$=\frac{1}{1-1/(2-17/41)}=\frac{1}{1-41/65}=\frac{65}{24}=2.708. \tag{1.146a and 1.146b}$$

The continuants form an increasing sequence (Equations 1.143b through 1.146b) approaching the exact value (Equation 1.142a) with decreasing error (Equations 1.143a through 1.146a), namely, 0.37% for the fourth continuant. The continued fraction (Equation 1.95 ≡ Equation 1.142b) has denominators (Equation 1.147a) and numerators (Equation 1.147b):

$$a_n = n + 2, \quad b_n = -n. \tag{1.147a and 1.147b}$$

Thus, it is neither of the first kind (Equations 1.113a and 1.113b) nor simple (Equation 1.118a); therefore, the theorems on sharpening oscillatory (optimal rational) approximation in Subsection 1.7.1 (Subsection 1.7.3) do not apply. Comparing with the former theorem (Section 1.7.1), the successive continuants come closer to the ultimate value as an increasing (Equations 1.143b through 1.146b) instead of an oscillating (Figure 1.3) sequence. Several methods of calculation of the irrational number e are indicated in Section 3.2, including two other continued fractions giving nine accurate digits (Table 3.2). The number e is given (Note 3.2) with 110 decimals in Equation 3.7. The example of derivation of a continued fraction from an infinite series is followed next by an infinite product.

1.8.6 Continued Fraction (Euler 1739) for the Sine

The change of variable (Equation 1.148a) in the infinite product for the circular sine (Equation 1.75) leads to (Equation 1.148b):

$$z \to \pi z: \qquad \frac{\sin(\pi z)}{\pi z} = \prod_{n=1}^{\infty}\left(1 - \frac{z^2}{n^2}\right) = \prod_{n=1}^{\infty}\left(1 - \frac{z}{n}\right)\left(1 + \frac{z}{n}\right)$$

$$= (1-z)(1+z)\left(1 - \frac{z}{2}\right)\left(1 + \frac{z}{2}\right)\left(1 - \frac{z}{3}\right)\left(1 + \frac{z}{3}\right)\cdots = \prod_{n=1}^{\infty} g_n. \tag{1.148a and 1.148b}$$

The coefficients in the infinite product satisfy

$$n \in| N: \qquad 1 - g_{2n-1} = \frac{z}{n} = g_{2n} - 1, \qquad g_{2n-1}g_{2n} - 1 = -\frac{z^2}{n^2},$$

$$g_{2n}g_{2n+1} - 1 = \left(1 + \frac{z}{n}\right)\left(1 - \frac{z}{n+1}\right) - 1 = \frac{z(1-z)}{n(n+1)}. \tag{1.149a–1.149e}$$

Substitution of Equations 1.149b through 1.149e in Equation 1.133 leads to the continued fraction for the circular sine:

$$\frac{\sin(\pi z)}{\pi z} = \frac{1-z}{1-}\,\frac{z}{1+z-}\,\frac{(1+z)(-z/2)}{z(1-z)/2-}\,\frac{(1-z/2)z^2/2}{-z^2/4-}\,\frac{(1+z/2)z^2/6}{z(1-z)/6-}\cdots$$

$$\frac{(1-z/n)z^2/[n(n-1)]}{-z^2/n^2-}\,\frac{(1+z/n)z^2/[n(n+1)]}{z(1-z)/[n(n+1)]-}\cdots, \tag{1.150}$$

that simplifies to

$$|z| < \infty: \qquad \frac{\sin(\pi z)}{\pi z} = \frac{1-z}{1-}\,\frac{z}{1+z+}\,\frac{1+z}{1-z+}\,\frac{2(2-z)}{z+}\,\frac{2(2+z)}{1-z+}\cdots\frac{n(n-z)}{z+}\,\frac{n(n+z)}{1-z+}\cdots. \tag{1.151a and 1.151b}$$

Algebraic inversion (Equations 1.95 and 1.108) leads to the continued fraction for the circular cosecant:

$$z \neq m \in |Z: \qquad \pi z \csc(\pi z) = \frac{1}{1-z} - \frac{z}{1+z+}\,\frac{1+z}{1-z+}\,\frac{2(2-z)}{z+}\,\frac{2(2+z)}{1-z+}\cdots\frac{n(n-z)}{z+}\,\frac{n(n+z)}{1-z+}\cdots. \tag{1.152a and 1.152b}$$

Thus, *the circular sine (cosecant) is specified by the continued fraction [Equation 1.151 (Equation 1.152b)] converging: (1) absolutely for* $|z| < \infty$ *(z is not an integer to exclude the poles); and (2) totally for* $|z| \le M < \infty$ *(*$|z - m\pi| \ge \varepsilon > 0$*) to exclude a neighborhood of the poles.* A suitable choice of variable leads to the calculation of the irrational number π and the quadrature of the circle.

1.8.7 Quadrature of the Circle and the Irrational Number π

One of the classical problems dating from the Greek civilization is the **quadrature of the circle**: find the length L of the side of the square with the same area $L^2 = \pi R^2$ as the circle of radius R. The solution amounts to the determination of the number π. The latter is specified, setting Equation 1.153a in Equation 1.151b, which leads to Equations 1.153b and 1.153c:

$$z = \frac{1}{2}: \qquad \frac{2}{\pi} = \frac{1/2}{1-}\,\frac{1/2}{3/2+}\,\frac{3/2}{1/2+}\,\frac{3}{1/2+}\,\frac{5}{1/2+}\cdots\frac{n(n-1/2)}{1/2+}\,\frac{n(n+1/2)}{1/2+}\cdots$$

$$= \frac{1}{2-}\,\frac{2}{3+}\,\frac{2.3}{1+}\,\frac{2^2.3}{1+}\,\frac{2^2.5}{1+}\cdots\frac{2^2(2n-1)}{1+}\,\frac{2^2(2n+1)}{1+}\cdots = 3.142. \tag{1.153a–1.153c}$$

Algebraic inversion Equations 1.95 and 1.108 of Equation 1.153c leads to

$$\frac{\pi}{2} = 2 - \frac{2}{3+}\,\frac{2.3}{1+}\,\frac{2^2.3}{1+}\,\frac{2^2.5}{1+}\cdots\frac{2^2(2n-1)}{1+}\,\frac{2^2(2n+1)}{1+}\cdots. \tag{1.154}$$

Thus, *the continued fractions (Equations 1.153b, 1.153c, and 1.154) prove (Euler 1739) that* π *is an irrational number, and can be used to compute its value.* The first four continuants of Equation 1.154 are

$$\pi - \pi_1 = 0.475: \qquad \pi_1 = 2\left(2 - \frac{2}{3}\right) = \frac{8}{3} = 2.667, \tag{1.155a and 1.155b}$$

$$\pi - \pi_2 = -0.414: \qquad \pi_2 = 2\left(2 - \frac{2}{3+2.3}\right) = 4\left(1 - \frac{1}{9}\right) = \frac{32}{9} = 3.556, \qquad \text{(1.156a and 1.156b)}$$

$$\pi - \pi_3 = +0.297: \qquad \pi_3 = 2\left[2 - \frac{2}{3+2.3/\left(1+2^2.3\right)}\right]$$
$$= 4\left(1 - \frac{1}{3+6/13}\right) = 4\left(1 - \frac{13}{45}\right) = \frac{128}{45} = 2.844, \qquad \text{(1.157a and 1.157b)}$$

$$\pi - \pi_4 = -0.272: \qquad \pi_4 = 2\left\{2 - \frac{2}{3+2.3/\left[1+2^2.3/\left(1+2^2.5\right)\right]}\right\}$$
$$= 4\left[1 - \frac{1}{3+6/\left(1+12/21\right)}\right] = 4\left(1 - \frac{1}{3+126/33}\right)$$
$$= 4\left(1 - \frac{33}{225}\right) = \frac{768}{225} = 3.413, \qquad \text{(1.158a and 1.158b)}$$

and show that (1) the successive approximations oscillate above and below the final value coming progressively closer (Figure 1.3); and (2) the fourth continuant (Equation 1.158b) of π gives a relative error of 8.6%, which is larger than the error of 0.37% for the fourth continuant (Equation 1.146b) of *e*. More rapidly convergent solutions to the quadrature of the circle are given in Section 7.9, and the value of π with 110 decimals (Note 3.2) is given in Equation 7.307. Other continued fractions are obtained for the exponential (Section 3.2), logarithm (Section 3.6), circular and hyperbolic, and inverse or cyclometric (Section 7.8) functions. Several of these are obtained by the Lambert method (Section 1.9), which uses a recurrence formula of order 2 (Subsection 1.6.1) to obtain a continued fraction for the ratio of two series, such as hypergeometric (hypogeometric) function [Section 1.9 (Section 3.9)].

1.9 Continued Fraction for the Ratio of Two Series (Lambert 1770)

The transformation of series into continued fractions was made before (Subsection 1.8.1) based on the condition that the sum of the first N terms of the series coincides with the Nth continuant of the continued fraction. Another approach is to represent the ratio of two series as a continued fraction (Lambert 1861); in this case, the continued fraction represents the ratio of the series with all terms, and there is no required relation between their truncated forms. The Lambert method has three stages, best understood in reverse order: (1) final stage III is a continued fraction (Subsection 1.9.4) arising from a recurrence formula of order 2 (Subsection 1.4.1); (2) intermediate stage II is to establish (Subsection

1.9.3) the recurrence formula for example the hypogeometric function $F(;z;c)$, relating the functions of successive parameters $(c, c \pm 1)$ for the same variable; and (3) first stage I is to define the hypogeometric function and show that it is an integral (meromorphic) function [Subsection 1.9.1 (Subsection 1.9.2)] of the variable z (parameter c). The combination of stages I to III leads to the continued fraction for the ratio of two hypogeometric series with (Subsection 1.9.3) the same variable z and parameters differing by unity $(c, c + 1)$, as well as the restrictions on its validity (Subsection 1.9.4). As an example, the circular sine (cosine) can be expressed as hypogeometric functions with parameters (1/2, 3/2) differing by unity and leading (Subsection 1.9.5) to the continued fraction for their ratio, which is the circular tangent.

1.9.1 Convergence of the Hypogeometric Series

Consider the **hypogeometric series** with variable z and parameter c defined by

$$\begin{aligned} {}_0F_1(c;z) &\equiv 1+\frac{z}{1!}\frac{1}{c}+\frac{z^2}{z!}\frac{1}{c(c+1)}+\cdots+\frac{z^n}{n!}\frac{1}{c(c+1)\cdots(c+n-1)}+\cdots \\ &= \sum_{n=0}^{\infty}\frac{z^n}{n!}\frac{1}{c(c+1)\cdots(c+n-1)}=\sum_{n=0}^{\infty}\frac{z^n}{n!}\frac{1}{(c)_n}\equiv F(;c;z), \end{aligned} \tag{1.159}$$

where the **Pochhammer symbol** is used (Equation I.29.79a ≡ Equation 1.160):

$$(c)_n \equiv c(c+1)\cdots(c+n-1)=\prod_{m=0}^{n-1}(c+m). \tag{1.160}$$

The series (Equation 1.159b) resembles the exponential series (Equation 1.26b) that has $n! = (1)_n$ in the denominator, inserting as an extra factor the Pochhammer symbol (Equation 1.160) also in the denominator, as a lower parameter. The hypogeometric series is the particular case of the generalized hypergeometric series (Equation 3.155) with zero upper $p = 0$ and one lower $q = 1$ parameters, hence the notation $F(;c;z)$. Other particular cases $p = 1 = q$ $(p = 1, q = 2)$ are the confluent (Gaussian) hypergeometric functions [Section I.29.9 (Example I.30.20)] to which the Lambert method that follows can also be applied (Section 3.9). The series (Equation 1.159 ≡ Equation 1.161b) has a general term (Equation 1.161c) and ratio of successive terms (Equation 1.161d):

$$|z|<\infty: \quad F(;c;z)=\sum_{n=0}^{\infty}f_n, \qquad f_n=\frac{z^n}{n!}\frac{1}{(c)_n}, \qquad \frac{f_{n+1}}{f_n}=\frac{z}{n+1}\frac{1}{c+n}\sim O\left(\frac{1}{n^2}\right). \tag{1.161a–1.161d}$$

The latter shows that the series is absolutely and uniformly convergent (Equation I.29.40d) in the finite z-plane (Equation 1.161a), and thus specifies an integral function (Subsection

1.1.7) of the variable z. It is shown next that it is a meromorphic function of the parameter c and its residues are calculated (Subsection 1.9.2).

1.9.2 Integral (Meromorphic) Function of the Variable (Parameter)

The hypogeometric function defined by the series (Equation 1.159) is (is not) an integral function of the variable z (parameter c). For zero or negative values of the parameter (Equation 1.162a), the denominators in Equation 1.159 vanish for higher order terms; the simple zeros in the denominator correspond to simple poles (Equations 1.9 and 1.18) of the hypogeometric function, with residues (Equation 1.162b):

$$c = 0, 1, -2, \ldots, -k: \qquad F_{(1)}(;-k;z) = \lim_{c \to -k} (c+k) F(;c;z) = \lim_{c \to -k} (c+k) \sum_{n=0}^{\infty} \frac{z^n}{n!} \frac{1}{(c)_n}. \tag{1.162a and 1.162b}$$

The calculation of the residues of the hypogeometric (hypergeometric) function [Equation 1.159 (Equation I.29.74)] at their simple poles [Equation 1.163a ≡ (Equation I.29.75a)] is similar, which leads to Equation 1.168b (Equation I.29.75b). The method of calculation of residues at the simple poles for zero or negative integer values of the lower parameter c is detailed next for the hypogeometric function (Equation 1.159) in five steps: (1) the series (Equation 1.159) has the Pochhammer symbol (Equation 1.160 ≡ Equation 1.163b) in the denominator:

$$c = -k: \quad (c)_n = c\,(c+1)\ldots(c+n-1) = (-k)(-k+1)\ldots(n-k-1), \tag{1.163a and 1.163b}$$

that first vanishes (Equation 1.163a) for $n = k + 1$; (2) thus, the product (Equation 1.163b) is nonzero for $n = 0, \ldots, k$, and the corresponding $k + 1$ first terms of the series (Equation 1.162b) are finite and vanish when multiplied by $c + k$ in the limit (Equation 1.164a):

$$\lim_{c \to -k} (c+k) \sum_{n=0}^{k} \frac{z^n}{n!} \frac{1}{(c)_n} = 0; \qquad F_{(1)}(;-k;z) = \lim_{c \to -k} (c+k) \sum_{n=k+1}^{\infty} \frac{z^n}{n!} \frac{1}{(c)_n}; \tag{1.164a and 1.164b}$$

(3) it follows that in Equation 1.162b, only the terms of the sum (Equation 1.164b) remain with $n \geq k + 1$, and in Equation 1.164b, the ratios (Equation 1.165b) appear for Equation 1.165a:

$$n = k+1, \ldots, k+m+1: \qquad \frac{c+k}{(c)_n} = \frac{c+k}{c(c+1)\cdots c(c+n-1)}$$
$$= \frac{1}{c(c+1)\cdots(c+k-1)(c+k+1)\cdots(c+n-1)}; \tag{1.165a and 1.165b}$$

(4) the numerator in Equation 1.165b cancels with one of the factors in the denominator, which leads to the finite limit:

$$\lim_{c\to-k}\frac{c+k}{(c)_n}=\frac{1}{(-k)(-k+1)\cdots(-1)(+1)\cdots(n-k-1)}=\frac{(-)^k}{k!(n-k-1)!}; \tag{1.166}$$

(5) the limit (Equation 1.166) substituted in Equation 1.164b specifies the residue (Equation 1.167b):

$$n=k+1+m: \qquad F_{(1)}(;-k;z)=\sum_{n=k+1}^{\infty}\frac{z^n}{n!}\frac{(-)^k}{k!(n-k-1)!}=\frac{(-)^k}{k!}z^{k+1}\sum_{m=0}^{\infty}\frac{z^m}{m!}\frac{1}{(k+1+m)!}$$

$$=\frac{(-)^k}{k!}\frac{z^{k+1}}{(k+1)!}\sum_{m=0}^{\infty}\frac{z^m}{m!}\frac{1}{(k+2)_m}=\frac{(-)^k}{k!}\frac{z^{k+1}}{(k+1)!}F(;k+2;z),$$

(1.167a and 1.167b)

where Equations 1.167a and 1.167c were used.

$$(k+1+m)!=1.2\ldots(k+1)\,(k+2)\ldots(k+2+m-1)=(k+1)!\,(k+2)_m. \tag{1.167c}$$

Thus, the **hypogeometric function** *defined by the hypogeometric series (Equation 1.159) is (1) an integral function (Equation 1.161a) of the variable z and (2) a meromorphic function of the parameter c with simple poles at the origin and negative integer values (Equation 1.168a ≡ Equation 1.162a):*

$$-c\in|N_0: \qquad F_{(1)}(;-k;z)=\lim_{c\to-k}(c+k)F(;z;c)=\frac{(-)^k}{k!}\frac{z^{k+1}}{(c+k)!}F(;k+2;z),$$

(1.168a and 1.168b)

with residue specified by (Equation 1.168b) hypogeometric functions with parameter k + 2. The hypogeometric functions, with parameters (c, $c \pm 1$) differing by an integer, satisfy a recurrence relation (Subsection 1.9.3) that leads to a continued fraction.

1.9.3 Contiguity Relations and Continued Fractions

The **contiguity relation** for the hypogeometric function follows from the series (Equation 1.159) for the same value of the variable and parameters differing by unity (Equation 1.169b):

$$n=m+1: \qquad F(;c+1;z)-F(;c;z)=\sum_{n=1}^{\infty}\frac{z^n}{n!}\left[\frac{1}{(c+1)_n}-\frac{1}{(c)_n}\right]$$

$$=\sum_{n=1}^{\infty}\frac{z^n}{n!}\frac{1}{(c+1)\ldots(c+n)}\left(1-\frac{c+n}{c}\right)$$

$$=-\frac{1}{c(c+1)}\sum_{n=1}^{\infty}\frac{z^n}{(n-1)!}\frac{1}{(c+2)\ldots(c+n)} \qquad \text{(1.169a and 1.169b)}$$

$$=-\frac{z}{c(c+1)}\sum_{m=0}^{\infty}\frac{z^m}{m!}\frac{1}{(c+2)\ldots(c+m+1)}$$

$$=-\frac{z}{c(c+1)}\sum_{m=0}^{\infty}\frac{z^m}{m!}\frac{1}{(c+2)_m}=-\frac{z}{c(c+1)}F(;c+2;z).$$

The hypogeometric functions (Equation 1.159), with the same variable z and three successive parameters (c, c + 1, c + 2) of which the first two are nonzero (Equation 1.170a), satisfy the contiguity relation (Equation 1.170b):

$$c\neq 0,-1: \qquad F(;c+1;z)=F(;c;z)-\frac{z}{c(c+1)}F(;c+2;z); \qquad \text{(1.170a and 1.170b)}$$

that is, a recurrence formula of degree 2 (Equation 1.90b). If in addition to Equation 1.170a the hypogeometric function of parameter c + 1 does not vanish (Equation 1.171a), its ratio (Equation 1.171b) by the hypogeometric function of parameter c satisfies (Equation 1.171c):

$$F(;c+1;z)\neq 0: \qquad F_+(;c;z)\equiv\frac{F(;c+1;z)}{F_+(;c;z)}, \qquad \frac{1}{F_+(;c;z)}=1+\frac{z}{c(c+1)}F_+(;c+1;z).$$

(1.171a–1.171c)

The latter relation (Equations 1.170a, 1.171a, and 1.171b ≡ Equations 1.172a through 1.172c):

$$c\neq 0,-1; \quad 0\neq F(;c+1;z): \qquad F_+(;c;z)=\frac{1}{1+\dfrac{z/[c(c+1)]}{F_+(;c+1;z)}}, \qquad \text{(1.172a–1.172c)}$$

when applied repeatedly leads to the continued fractions (Equations 1.173c and 1.173d) for the ratio of two contiguous hypergeometric functions:

$$c\neq 0,-1,\ldots,-m,\ldots; \quad 0\neq F(;c+1;z),\ldots,F(;c+n;z),\ldots:$$

$$F_+(;c;z)=\frac{1}{1+}\frac{z/[c(c+1)]}{1+}\frac{z/[(c+1)(c+2)]}{1+}\ldots\frac{z/\left[(c+n)(c+n+1)\right]}{1+}\ldots \qquad \text{(1.173a–1.173c)}$$

$$=\frac{c}{c+}\frac{z}{c+1+}\frac{z}{c+2+}\ldots\frac{z}{c+n+}\ldots=\frac{F(;c+1;z)}{F(;c;z)},$$

$$\frac{F(;c;z)}{F(;c+1;z)} = \frac{1}{c} + \frac{z}{c+1+} \frac{z}{c+2+} \cdots \frac{z}{c+n+} \cdots, \tag{1.173d}$$

which are subject to the restrictions (Equations 1.173a and 1.173b). The continued fractions (Equations 1.173c and 1.173d) are related by algebraic inversion (Equations 1.96 and 1.108). Before giving an example (Subsection 1.9.5), some general properties of the *Lambert method (1770)* are discussed next (Subsection 1.9.4).

1.9.4 Terminating and Nonterminating Continued Fractions

The two restrictions (Equations 1.173a and 1.173b) are considered next. The first restriction (Equation 1.173a) excludes the poles (Equation 1.168a) of the hypogeometric function, ensuring that it is an integral function of the variable z and the parameter c. This applies both for c and $c + 1$, which is the ratio (Equation 1.171b) of the meromorphic function (Subsection 1.3.1) to which the continued fraction applies (Equation 1.173c). The restriction (Equation 1.173b), if removed, allows (Equation 1.174a) a hypogeometric function to vanish for one value $c+n=\bar{c}$ of the parameter. If for the same value of the variable z the hypogeometric function with parameter $\bar{c}+1$ would also vanish (Equation 1.174b), then the contiguity relation (Equation 1.170b) with Equation 1.174a,b implies that all hypogeometric functions with parameters $c + n$ with n as a positive integer vanish (Equation 1.174c):

$$F(;c;z) = 0 = F(;c+1;z) \iff 0 = F(;c+n;z) = \lim_{n\to\infty} F(;c+n;z) = F(;\infty;z) = 1, \tag{1.174a–1.174d}$$

contradicting the limit (Equation 1.174d) of Equation 1.159 as $n \to \infty$ that is unity. Thus, *the hypogeometric functions with contiguous parameters c + 1, neither of which is zero (Equations 1.175a and 1.175b), cannot (Equation 1.175c) both vanish for the same values of the variable z:*

$$c \neq 0, -1: \qquad F(;c;z) = 0 \iff F(;c+1;z) \neq 0. \tag{1.175a–1.175c}$$

Thus, two cases can arise in the Lambert method (Equations 1.170a, 1.170b, 1.171a through 1.171c, 1.172a through 1.172c, and 1.173a through 1.173d): (1) if the hypogeometric function vanishes for the value c + n + 1 of the parameter (Equation 1.176a), the continued fraction (Equation 1.172c) for the ratio of two contiguous functions terminates at the nth continuant (Equation 1.176b):

$$F(;c+n+1;z) = 0: \qquad F_+(;c;z) = \frac{c}{c+} \frac{z}{c+1+} \cdots \frac{z}{c+n-2} \frac{z}{c+n-1} = \frac{F(;c+1;z)}{F(;c;z)}; \tag{1.176a and 1.176b}$$

and (2) if none of the hypogeometric functions with parameters c + 1, . . ., c + n, . . . vanish (Equation 1.173b), the continued fraction (Equation 1.173c) for the ratio of contiguous hypergeometric functions is nonterminating. In the latter case, the issue of convergence of nonterminating continued fractions arises, leading to two cases: (1) the operations on infinite series (products) may justify the convergence of a nonterminating continued fraction [Subsection 1.8.1 (Subsection 1.8.3)]; and (2) regardless of the manner it is obtained, there are convergence tests for continued fractions (Subsection 1.7.2 and Section 9.3) as for infinite series (Chapters I.21, 23, 25, and 29) and infinite products (Section 9.2).

1.9.5 Continued Fraction for the Circular Tangent

As an example, the Lambert method (Subsections 1.9.3 and 1.9.4) may be applied to the circular cosine (sine) that are [Equation 1.31a (Equation 1.31b)] particular cases [Equation 1.177a (Equation 1.177b)] of the hypogeometric function (Equation 1.159) with the same variable (Equation 1.180a) and contiguous parameters (Equation 1.180b):

$$\cos z = \sum_{n=0}^{\infty} \frac{(-)^n}{(2n)!} z^{2n} = 1 + \sum_{n=1}^{\infty} \left(-\frac{z^2}{4}\right)^n \frac{2^n}{(2n)!!} \frac{2^n}{(2n-1)!!}$$

$$= 1 + \sum_{n=0}^{\infty} \frac{(-z^2/4)^n}{n!} \frac{1}{(1/2)_n} = F\left(;\frac{1}{2};-\frac{z^2}{4}\right), \tag{1.177a}$$

$$\sin z = \sum_{n=0}^{\infty} \frac{(-)^n}{(2n+1)!} z^{2n+1} = z\left\{1 + \sum_{n=1}^{\infty} \left(-\frac{z^2}{4}\right)^n \frac{2^n}{(2n)!!} \frac{2^n}{(2n+1)!!}\right\}$$

$$= z\left\{1 + \sum_{n=0}^{\infty} \frac{\left(-z^2/4\right)^n}{n!} \frac{1}{(3/2)_n}\right\} = zF\left(;\frac{3}{2};-\frac{z^2}{4}\right), \tag{1.177b}$$

where four properties of the Pochhammer symbol (Equation 1.160) and the double factorial were used:

$$(2n)! = 2n(2n-2)\ldots 4.2(2n-1)(2n-3)\ldots 3.1 = (2n)!!(2n-1)!!, \tag{1.178a}$$

$$2^{-n}(2n)!! = 2^{-n}\, 2n(2n-2)\ldots 4.2 = n(n-1)\ldots 2.1 = n!, \tag{1.178b}$$

$$2^{-n}(2n-1)!! = 2^{-n}(2n-1)(2n-3)\ldots 3.1 = \left(n-\frac{1}{2}\right)\left(n-\frac{3}{2}\right)\ldots\frac{3}{2}\cdot\frac{1}{2} = \left(\frac{1}{2}\right)_n, \tag{1.178c}$$

$$2^{-n}(2n+1)!! = 2^{-n}(2n+1)(2n-1)\ldots 3.1 = \left(n+\frac{1}{2}\right)\left(n-\frac{1}{2}\right)\ldots\frac{5}{2}\cdot\frac{3}{2} = \left(\frac{3}{2}\right)_n, \tag{1.178d}$$

namely, Equations 1.178a and 1.178b and Equation 1.178c (Equation 1.178d) in Equation 1.177a (Equation 1.177b).

Thus, *the circular cosine (sine) are particular cases [Equation 1.177a ≡ Equation 1.179a (Equation 1.177b ≡ Equation 1.179b)] of the hypogeometric function*:

$$\cos z = F\left(;\frac{1}{2};-\frac{z^2}{4}\right), \qquad \sin z = zF\left(;\frac{3}{2};-\frac{z^2}{4}\right), \tag{1.179a and 1.179b}$$

with the same variable (Equation 1.180a) and contiguous parameters (Equation 1.180b):

$$\zeta = -\frac{z^2}{4}; \qquad c = \frac{1}{2}, \frac{3}{2}: \qquad \frac{\tan z}{z} = \frac{1}{z}\frac{\sin z}{\cos z} = \frac{F(;1/2;\zeta)}{F(;3/2;\zeta)} = \frac{1/2}{1/2+}\frac{\zeta}{3/2+}\frac{\zeta}{5/2+}\cdots\frac{\zeta}{n+1/2+}\cdots$$

$$= \frac{1}{1+}\frac{4\zeta}{3+}\frac{4\zeta}{5+}\cdots\frac{4\zeta}{2n+1+}\cdots = \frac{1}{1-}\frac{z^2}{3-}\frac{z^2}{5-}\cdots\frac{z^2}{2n+1-}\cdots; \tag{1.180a–1.180c}$$

applying the Lambert theorem (Equation 1.173c) leads to (Equation 1.180c ≡ Equation 1.181b) *the continued fraction for the circular tangent:*

$$z \neq n\pi - \frac{\pi}{2}: \qquad \tan z = \frac{z}{1-}\frac{z^2}{3-}\frac{z^2}{5-}\cdots\frac{z^2}{2n+1-}\cdots, \tag{1.181a and 1.181b}$$

that is valid at all points of the complex plane except (Equation 1.181a) those that are its simple poles (Figure 1.1b). The circular cosine (sine) is the particular case [Equation 1.177a (Equation 1.177b)] of the hypogeometric function (Equation 1.159); the latter is the particular case $p = 0$, $q = 1$ of the (Equation 3.155) generalized hypergeometric function (p, q), which includes the Gaussian (confluent) hypergeometric function [Section I.29.9 and Subsection 3.9.2 (Subsection 3.9.7)] for $(p =, q = 2)$ $[(p = 1 = q)]$. The Lambert method also applies to the Gaussian and confluent hypergeometric functions. Because these functions have as particular cases the most elementary functions (exponential, logarithm, binomial, circular, hyperbolic, and inverse), the Lambert method is a powerful tool to obtain their representations as continued fractions (Sections 3.9 and 7.8).

NOTE 1.1 FINITE (INFINITE) REPRESENTATIONS FOR RATIONAL (TRANSCENDENTAL) FUNCTIONS

The Table 1.1 compares the rational (transcendental) functions and their corresponding finite (infinite) representations: (1) the power series representation of a holomorphic (singular) function, that is [Subsection 1.1.1 (Subsection 1.1.2)], the Taylor (Laurent) series of a function near a regular point (isolated singularity) corresponds to a polynomial in z (in z and $1/z$) of infinite degree; (2) the factorization of a polynomial of degree N into N factors

TABLE 1.1

Comparison of Rational and Transcendental Functions

Function	Rational (I.31; II.1.1.6)	Transcendental (I.39.7 through I.39.9; II.3, 5, 7)
Power expansions	Ratio of polynomials (I.31.6; II.1.1.6)	Power series (I.23 and I.25; II.1.1)
Factorization	Finite (I.31.5 through I.31.7)	Infinite product[a] (I.27.9.2; II.1.4 and II.1.5)
Decomposition into fractions	Finite (I.31.8 and 31.9)	Series of fractions[a] (I.27.9.2; II.1.2 and 1.3)
Continued fractions	Terminating (II.1.6 through II.1.9)	Nonterminating[a] (II.1.6 through II.1.9)
Convergence issues	No	Yes (I.21 and I.29; II.9)

Note: Volume, chapter, section, and subsection are in parentheses.

[a] These representations do not exist always but only in some cases.

(Section I.31.6) for each root (repeated for multiple roots) corresponds to the infinite product (Sections 1.4 and 1.5) for a function with an infinite number of zeros; (3) the partial fraction decomposition of rational function (Sections I.31.8 and I.3.19), that is, the ratio of two polynomials, corresponds to the series of fractions for a meromorphic function (Section 1.2), that is the ratio of two integral functions (Subsection 1.3.1); (4) a continued fraction that is terminating (nonterminating) represents a rational (transcendental) function and does not raise [raises (Chapters I.21, I.29, and II.9)] convergence issues; (5) the functional series (Note 1.2) using infinite linear combinations of base functions, such as orthogonal systems of functions; and (6) the integral transforms (Note 1.3) with choice of kernel and path of integration. The use of each of these requires a study of its properties, starting with existence and convergence. Some series of fractions (infinite products) can be obtained by two distinct approaches: (1) the theorems of Mittag-Leffler (Sections 1.2 through 1.5) [Weierstrass (Section 9.2)] provide the general mathematical method and (2) an alternative physical approach in specific cases of multipoles (Sections I.36.6 and 36.7) is to compare the solution of the same problem using (a) conformal mapping into a region and (b) infinite images on its boundaries.

NOTE 1.2 TEIXEIRA, LAGRANGE–BURMANN, FOURIER, AND FUNCTIONAL SERIES

The first general representation of a singular function as a Laurent series (Equations 1.6a and 1.6b) is a particular case of the Teixeira series (Chapter I.25) that expands in powers of an auxiliary function:

$$F(z) = \sum_{n=-\infty}^{+\infty} B_n \left[f(z) \right]^n \begin{cases} \text{A.C.} & \text{if} \quad r < |f(z)| < R, & (1.182a) \\ \text{T.C.} & \text{if} \quad r + \varepsilon \le |f(z)| \le R - \delta, & (1.182b) \end{cases}$$

with $\varepsilon > 0$, $\delta > R - \varepsilon - r > 0$; in the case of a regular function, only ascending powers appear in the Lagrange series (Chapter I.23):

$$F(z) = \sum_{n=0}^{\infty} \frac{\mathrm{d}^n F}{\mathrm{d}f^n} \frac{[f(z)]^n}{n!} \begin{cases} \text{A.C.} & \text{if} \quad |f(z)| < R, & (1.183a) \\ \text{T.C.} & \text{if} \quad |f(z)| \le R - \delta, & (1.183b) \end{cases}$$

with $0 < \delta < R$. The particular choice of the auxiliary function (Equation 1.184a), with a simple zero (Equations 1.184b and 1.184c) at z = a:

$$f(z) = z - a \qquad f(a) = 0 \ne 1 = f'(a) \tag{1.184a–1.184c}$$

leads (Diagram I.25.1) from the Teixeira (Equations 1.182a and 1.182b) [Lagrange (Equations 1.183a and 1.183b)] to the Laurent (Equations 1.6a and 1.6b) [Taylor (Equations 1.1a and 1.1b)] series. Another choice of auxiliary function (Equation 1.185a) for a power series (Equation 1.185b) leads from the Teixeira (Equations 1.183a and 1.183b) to the Fourier (Equation 1.185c) series:

$$f(z) = e^{iz}, \qquad [f(z)]^n = e^{inz}: \qquad F(z) = \sum_{n=-\infty}^{+\infty} C_n e^{inz}. \tag{1.185a–1.185c}$$

The latter is a particular case of functional series (Equation 1.186a) using the base functions (Equation 1.186b):

$$F(z)=\sum_{n=-\infty}^{+\infty} C_n\Phi_n(z), \qquad \Phi_n(z)=e^{inz}, \tag{1.186a and 1.186b}$$

that are orthogonal with unit weight in the interval $(-\pi, +\pi)$:

$$(\Phi_n,\Phi_m)\equiv\int_{-\pi}^{+\pi}\Phi_n(z)\Phi_m^*(z)\,dz=\int_{-\pi}^{+\pi}e^{i(n-m)z}\,dz=\begin{cases}2\pi & \text{if} \quad n=m, \quad (1.187a)\\ 0 & \text{if} \quad n\neq m. \quad (1.187b)\end{cases}$$

The system of functions Ψ_n (z) is orthogonal with weight $w(z)$ in the interval (a, b) if (Subsection 5.7.2) it satisfies:

$$(\Psi_n,\Psi_m)\equiv\int_a^b w(z)\,\Phi_n(z)\,\Phi_m^*(z)\,dz=\begin{cases}0 & \text{if} \quad n\neq m, \quad (1.188a)\\ \|\Psi_n\|=|\Psi_n|^2>0 & \text{if} \quad n=m, \quad (1.188b)\end{cases}$$

where $\|\Psi_n\|$ is the norm ($|\Psi_n|$ the modulus) that must be positive (real). The functional series apply to functions of bounded fluctuation (Subsection I.27.9.3) or oscillation (Subsection 5.7.5). All these series raise issues of existence, convergence, accuracy, differentiability, integrability, and other properties, as do as the integral transforms that follow (Note 1.3).

NOTE 1.3 FOURIER, LAPLACE, AND OTHER INTEGRAL TRANSFORMS

The continuous (Equation 1.189a) analog (Equation 1.189b) of the Fourier series (Equation 1.185c) is the Fourier transform (Equation 1.189c):

$$\sum_{n=-\infty}^{+\infty}\to\int_{-\infty}^{+\infty}d\omega, \qquad C_n\to\tilde{F}(\omega): \qquad F(z)=\int_{-\infty}^{+\infty}\tilde{F}(\omega)e^{i\omega z}\,d\omega. \tag{1.189a–1.189c}$$

Changing to a negative exponential (Equation 1.190a) and integrating over the positive, instead of the whole real axis for convergence, lead to the Laplace transform (Equation 1.190b):

$$i\omega=-s: \qquad F(z)=\int_0^{\infty}\bar{F}(s)e^{-sz}\,ds. \tag{1.190a and 1.190b}$$

The Fourier (Equation 1.189) [Laplace (Equation 1.190b)] transform is a particular case of the general integral transform:

$$F(z)=\int_{\Gamma}G(w)K(z,w)\,dw, \tag{1.191}$$

where (1) the path of integration Γ in the complex plane is chosen to be the whole (the positive) real axis (−∞, +∞) [0, ∞] and (2) the kernel is an exponential imaginary (Equation 1.192a) [real negative (Equation 1.192b)] argument:

$$K(z, w) = e^{izw}, \ e^{-zw}. \qquad (1.192a \text{ and } 1.192b)$$

The kernels (Equations 1.192a and 1.192b) are both of product type (Equation 1.193a) that is symmetric:

$$K(z, w) = g(zw), g(z - w), \qquad (1.193a \text{ and } 1.193b)$$

but other forms are possible, such as Equation 1.193b; the latter reduces (Equation 1.191) to

$$F(z) = \int G(w)\, K(z - w)\, dw = G^*K(z), \qquad (1.194)$$

which is a convolution integral (Section 17.4). The integral transform besides the questions of existence and convergence raises the issue of inversion: how to determine the integral transform $G(w)$ from the original function $F(z)$. An integral transform (Equation 1.191) is a parametric integral (Sections I.13.8 and I.13.9) from which differentiability and other properties can be proven.

NOTE 1.4 SIX GENERAL REPRESENTATIONS OF FUNCTIONS

Table 1.2 summarizes six general representations of functions: (1) the power series for complex functions analytic except at isolated singularities (Section 1.1); (2) the series of fractions for complex meromorphic functions with an infinite number of poles accumulating at infinity (Sections 1.2 and 1.3); (3) the infinite products for integral functions with an infinite number of zeros whose logarithmic derivative is a meromorphic function (Sections

TABLE 1.2

Representations of Transcendental Functions

Number	I	II	III	IV	V	VI
Representation by	*Power series*	*Series of fractions*	*Infinite products*	*Continued fractions*	*Functional series*	*Parametric integrals*
Volume I	21, 23, 25, 27, and 29	27.9.2	27.9.2	–	Notes 25.2 and 25.3	13.8 and 13.9
Volume II	1.1	1.2 and 1.3	1.4 and 1.5; 9.2	1.6 through 1.9, 9.3	5.7; Note 1.2	Note 1.3
Type of function	*With isolated singularities, such as integral*	*With infinite number of poles, such as meromorphic*	*With infinite number of zeros, such as periodic*	*Double recurrence formula*	*Bounded fluctuation (I.27.9.5)*	–
General case	Teixeira series	Multiple poles	Origin is a zero		Orthogonal series	Hilbert transform
Particular case A	Laurent series	Simple poles	Origin is not a zero	Continued fraction of first kind	Fourier series	Laplace transform
Particular case B	Fourier series	Pole at the origin	–	Simple continued fraction	–	Fourier transform

1.4 and 1.5); (4) the continued fractions (Sections 1.6 through 1.9) that provide an alternative to (1)–(3); (5) the functional series (Note 1.2) using infinite linear combinations of base functions, for example, an orthogonal system of functions; and (6) the integral transforms (Note 1.3) using parametric integrals with choice of kernel and path of integration. The use of each of these requires a study of its properties, starting with existence and convergence.

1.10 Conclusion

A meromorphic function (Figure 1.2) has an infinite number of poles z_n, without accumulation point in the finite plane; it may be possible to select regions D_N, or circles $|\zeta| = R_N$, containing the first n poles, and such that $R_n \to \infty$ as $n \to \infty$. For example, for the secant (cotangent) [Figure 1.1a (Figure 1.1b)], the poles are at $z_n = n\pi + \pi/2$ ($z_n = n\pi$) and on the circles of radius $R_{2N} = N\pi$ ($R_{2N+1} = N\pi + \pi$) containing $2N$ ($2N + 1$) poles in which the function is bounded. The secondary circular functions, namely, the tangent (cotangent), have series of fractions that are related to the infinite products for the primary circular functions, namely, the cosine (sine). Both series and products can be transformed into continued fractions that, if they are convergent of the first kind, provide rational approximations X_n to the limit X, which are (Figure 1.3) doubly bounded, that is, (1) alternating from above and below and (2) sharpening, that is, with increasing accuracy in modulus.

2

Compressible and Rotational Flows

The previous consideration of flows (Chapters 12, 14, 16, 28, 34, 36, and 38 in volume I) has been mostly restricted to plane, steady, irrotational, and incompressible conditions. The compressible flow in homentropic conditions was considered in connection with the sound speed (Subsection I.14.6.1) and the measurement of velocity using Pitot (Venturi) tubes [Subsection I.14.6.2 (Subsection I.14.7.3)]. The homentropic compressible flow limits the velocity and hence sets a minimum radius (Section 2.1) for the flow due to a source, sink, vortex, or their combination in a spiral flow. Another way to avoid the singular velocity at the center of a potential vortex is to match the outer irrotational flow to a rotational core with zero velocity at the center (Section 2.2), for example, rigidly rotating or some other angular or smooth radial profile of the tangential velocity. Unlike the line vortex, for which vorticity is concentrated at the center, in the case of a nonpotential core, the vortical flow occupies a finite or infinite domain. For the same boundary conditions, an irrotational flow has less kinetic energy than a rotational flow (Section 2.3); the equations of motion in intrinsic coordinates, that is, parallel and tangent to the velocity, show that, in a rotational flow, the stagnation pressure is conserved only along streamlines but varies between streamlines due to the vorticity, and thus, Bernoulli's equation does not apply; although there is no scalar potential, for an incompressible rotational flow, there is a stream function satisfying a Poisson equation (Section 2.4). The Blasius theorem, specifying the lift and drag forces and pitching moment on a body, can be expressed in terms of the stream function alone (Section 2.5); likewise, the effect of inserting a cylinder in a rotational flow is specified by the second circle theorem, in terms of the stream function alone. The combination allows the consideration of a vortical flow past a cylinder (Section 2.6), showing that the mean flow vorticity can add to or subtract from the lift due to the circulation. Returning to assemblies of line monopoles, it can be shown that their centroid moves uniformly (Section 2.7), even though the monopoles move relative to each other; in particular, two monopoles move relative to each other, but can be in static equilibrium in specific conditions, for example, in a duct with parallel walls (Section 2.7) or behind a cylinder in a stream (Section 2.8). A monopole starting outside an equilibrium position follows a path specified by a path function (Section 2.9) that involves the Green's or influence function.

2.1 Source, Sink, and Vortex in a Compressible Flow

Considering an inviscid fluid (Subsection 2.1.1), the simplest case is a potential flow (Subsection 2.1.2) that is both irrotational and incompressible. The two simplest extensions are (1) compressible irrotational flow (Section 2.1) and (2) incompressible rotational flow (Sections 2.2 through 2.9). The simplest compressible irrotational flow is that of a barotropic fluid (Subsection 2.1.3), of which the homentropic flow is a particular case (Subsection 2.1.4). In the latter case, all flow variables can be expressed in terms of one variable and a constant stagnation value (Subsection 2.1.6). The flow variables include the velocity, pressure and mass density (Subsections 2.1.1 through 2.1.4), the sound speed (Subsection 2.1.5), and the Mach number (Subsection 2.1.7). A source/sink (vortex) in a potential flow [Section I.12.4 (Section I.12.5)] has a singularity of the radial (azimuthal) velocity at the center in two dimensions, corresponding to the axis in three dimensions. The flow is incompressible only at a low Mach number, that is, for flow velocity that is small compared with the sound speed (Subsection 2.1.7). For a compressible flow, for example, homentropic in the absence of heat exchanges, there is an upper limit to the flow velocity, corresponding to the critical flow condition (Subsection 2.1.6). This sets a lower limit to the radius of a source/sink (vortex) in a homentropic flow [Subsection 2.1.8 (Subsection 2.1.9)]; this compressibility limit extends to the spiral flow due to a line monopole, that is, a superposition of a source/sink and a vortex (Subsection 2.1.10).

2.1.1 Momentum Equation in an Inviscid Fluid (Euler 1755, 1759)

In the absence of viscosity, the only internal stresses in an inviscid fluid are due to an isotropic pressure, which acts inward orthogonally to a boundary surface ∂D, that is, opposite to the normal vector and area element (Equation 2.1a):

$$d\vec{S} = \vec{N}\,dS: \qquad -\int_{\partial D} p\,d\vec{S} = \int_D \vec{a}\,dm. \tag{2.1a and 2.1b}$$

The resultant force due to the pressure distribution on the boundary is balanced by the inertia force in the interior of the domain D, equal to the acceleration $\vec{a}$ times the mass element. The mass per unit volume defines the mass density Equation 2.2a, leading to Equation 2.2b:

$$\rho \equiv \frac{dm}{dV}: \qquad 0 = \int_{\partial D} p\,d\vec{S} + \int_D \vec{a}\,\rho\,dV. \tag{2.2a and 2.2b}$$

Assume that the boundary surface ∂D is closed and regular, that is, has a unique unit normal vector everywhere (Equation 2.3a), thus excluding edges or cusps; then, the gradient theorem holds (Equation 2.3b):

$$\left|\vec{N}\right|^2 \equiv \vec{N}.\vec{N} = 1: \qquad \int_{\partial D} p\,d\vec{S} = \int_D \nabla p\,dV. \tag{2.3a and 2.3b}$$

Substituting Equation 2.3b in Equation 2.2b yields Equation 2.4a:

$$\int_D \left(\nabla p + \rho\,\vec{a}\right) dV = 0: \qquad \rho\vec{a} + \nabla p = 0, \tag{2.4a and 2.4b}$$

and because the domain D is arbitrary, the integrand must vanish (Equation 2.4b). The latter is *(Euler 1755, 1759) the* **momentum equation** *for an inviscid fluid:*

$$a_i \equiv \frac{dv_i}{dt}: \qquad \rho\frac{dv_i}{dt} = -\frac{\partial p}{\partial x_i}, \tag{2.5a and 2.5b}$$

stating that the inertia force per unit volume, equal to the product of the mass density (Equation 2.2a) by the acceleration (Equation 2.5a), is balanced by minus the pressure gradient (Equation 2.5b); the minus sign implies that the motion is from the high to the low pressures.

2.1.2 Potential Flow: Irrotational and Incompressible (Bernoulli 1693)

The velocity is the derivative of the position vector with regard to time (Equation 2.6a) and generally depends on both:

$$v_i = \frac{dx_i}{dt}: \qquad a_i = \frac{dv_i}{dt} = \frac{dv_i}{\partial t} + \frac{dv_i}{\partial x_j}\frac{dx_j}{dt} = \left(\frac{\partial v_i}{\partial t} + v_j \frac{\partial}{\partial x_j}\right) v_i. \qquad \text{(2.6a and 2.6b)}$$

The **acceleration** *(Equation 2.5a) is the total or* **material derivative** *of the velocity with regard to time and consists of two parts: (1) the local acceleration that is linear and is specified by the partial derivative of the velocity with regard to time at a fixed position; and (2) the convective acceleration that is nonlinear and is specified by the partial derivative of the velocity with regard to the spatial coordinates, bearing in mind that the latter depend on time (Equation 2.6a).* In the last term on the right-hand side (r.h.s.), the repeated index j = 1,2,3 implies a summation over the three spatial coordinates. Substituting Equation 2.6b, *the momentum equation (Equation 2.5b) can be written:*

$$\frac{\partial v_i}{\partial t} + v_j \frac{dv_i}{\partial x_j} = -\frac{1}{\rho}\frac{\partial p}{\partial x_i}, \qquad (2.7)$$

for an inviscid fluid.

A flow is **irrotational** *iff the curl of the velocity is zero (Equation 2.8a):*

$$\text{irrotational flow:} \quad \frac{\partial v_i}{\partial x_j} = \frac{\partial v_j}{\partial x_i} \quad \Leftrightarrow \quad v_j = \frac{\partial \Phi}{\partial x_i}. \qquad \text{(2.8a and 2.8b)}$$

This is equivalent to the existence of a **scalar velocity potential** *Φ whose gradient is the velocity vector (Equation 2.8b).* Substituting Equation 2.8b in Equation 2.7 follows Equation 2.9b:

$$v^2 = \left|\vec{v}\right|^2 = v_j v_j: \qquad -\frac{1}{\rho}\frac{\partial p}{\partial x_i} = \frac{\partial^2 \Phi}{\partial x_i \partial t} + v_j \frac{dv_j}{\partial x_i} = \frac{\partial}{\partial x_i}\left(\frac{\partial \Phi}{\partial t} + \frac{1}{2}v^2\right), \qquad \text{(2.9a and 2.9b)}$$

where Equation 2.9a is the square of the modulus of the velocity vector. An incompressible flow has constant mass density (Equation 2.10a), implying Equation 2.10b:

$$\text{incompressible:} \quad \rho = \text{const}; \qquad \frac{\partial}{\partial x_i}\left(\frac{p}{\rho} + \frac{1}{2}v^2 + \frac{\partial \Phi}{\partial t}\right) = 0. \qquad \text{(2.10a and 2.10b)}$$

The latter (Equation 2.10b) states that the term in curved brackets does not depend on the position, and thus:

$$\frac{p}{\rho}+\frac{1}{2}v^2+\frac{\partial \Phi}{\partial t}=f(t) \tag{2.11}$$

can be at most a function of time.

If the flow is steady, that is, time independent (Equation 2.12a), the function reduces to a constant (Equation 2.12b):

$$\text{steady:} \quad \partial/\partial t=0; \qquad p+\frac{1}{2}\rho v^2=\text{const}\equiv p_0. \qquad \text{(2.12a and 2.12b)}$$

The constant in Equation 2.12b is the **stagnation pressure**, that is, pressure at a point where the velocity is zero. Thus, *in a steady irrotational incompressible flow of an inviscid fluid, stagnation pressure is constant, so that changes in pressure are balanced by changes in kinetic energy per unit volume.* For example, in the flow past an edge, the velocity is higher near the edge, and the pressure is lower; thus, there is a pressure gradient toward the edge, which turns the fluid around it. This tendency of a fluid to remain attached to a wall is known as the **Coanda effect** and can be demonstrated using the hodograph method (Section I.38.7).

2.1.3 Irrotational Flow of a Barotropic Fluid

In an incompressible fluid (Equation 2.10a), flow pressure is independent of the mass density of the fluid. In a general compressible fluid, pressure can depend on mass density and another thermodynamic variable such as temperature or entropy. The simplest compressible fluid is a **barotropic fluid** for which the pressure depends only on the mass density (Equation 2.13a):

$$\text{barotropic:} \quad p=p(\rho): \qquad \frac{\partial}{\partial x_i}\left(\int\frac{dp}{\rho}+\frac{1}{2}v^2+\frac{\partial \Phi}{\partial t}\right)=0. \qquad \text{(2.13a and 2.13b)}$$

In this case, the momentum or Euler equation (Equation 2.7) in Bernoulli form (Equation 2.9b) leads to Equation 2.13b. For a steady flow (Equation 2.12a), this simplifies to Equation 2.14b:

$$c^2\equiv\frac{dp}{d\rho}: \qquad \text{const}=\frac{1}{2}v^2+\int\frac{dp}{\rho}=\frac{1}{2}v^2+\int\frac{c^2}{\rho}d\rho, \qquad \text{(2.14a–2.14c)}$$

where the derivative of the pressure with regard to the mass density specifies the **sound speed** squared (Equation 2.14a). Thus, *the Bernoulli equation for a steady irrotational flow of an inviscid fluid takes the form Equation 2.12b in the incompressible case and Equation 2.14b ≡ Equation 2.14c in the compressible barotropic case.*

2.1.4 Bernoulli Equation for a Homentropic Flow

A particular case of barotropic fluid is the **homentropic** case that excludes heat exchanges by keeping the entropy constant (Equation 2.15a) and leads to the **adiabatic relation** (Equation 2.15b) between the pressure and the mass density:

$$\text{homentropic:} \quad S = \text{const}; \qquad p = k\rho^{\gamma}, \qquad \gamma \equiv \frac{C_p}{C_v} > 1. \tag{2.15a–2.15c}$$

The **adiabatic exponent** (Equation 2.15c) is the ratio of **specific heat** at constant pressure C_p and volume C_v; because the latter is smaller than the former, the adiabatic exponent exceeds unity. The constant k does not appear in the adiabatic (Equation 2.16) sound speed (Equation 2.14a):

$$c^2 = \left(\frac{dp}{d\rho}\right)_s = \frac{d\left(k\rho^{\gamma}\right)}{d\rho} = k\gamma\rho^{\gamma-1} = \gamma\frac{p}{\rho}. \tag{2.16}$$

It does not appear either in the last term on the r.h.s. of the Bernoulli equation (Equation 2.14b):

$$\gamma = \text{const:} \quad \int \frac{dp}{\rho} = \int \frac{d\left(k\rho^{\gamma}\right)}{\rho} = \int k\gamma\rho^{\gamma-2}\, d\rho = \frac{k\gamma}{\gamma-1}\rho^{\gamma-1} = \frac{\gamma}{\gamma-1}\frac{p}{\rho} = \frac{c^2}{\gamma-1}, \tag{2.17a and 2.17b}$$

which simplifies to Equation 2.17b for constant adiabatic exponent (Equation 2.17a). Substituting Equation 2.17b in Equation 2.14b follows *the Bernoulli equation (Equations 2.18a and 2.18b) for the steady irrotational homentropic flow of an inviscid fluid with constant adiabatic exponent:*

$$\frac{\gamma-1}{2}v^2 + \gamma\frac{p}{\rho} = \frac{\gamma-1}{2}v^2 + c^2 = \text{const} \equiv c_0^2. \tag{2.18a and 2.18b}$$

The constant in Equations 2.18a and 2.18b is the stagnation sound speed c_0, that is, the sound speed at a point where the velocity is zero. The Bernoulli equation relating the pressure and the velocity in the steady irrotational flow of an inviscid fluid involves the mass density in the incompressible case (Equations 2.12a and 2.12b); in the compressible homentropic case, the mass density in Equation 2.18a may be replaced by the sound speed in Equation 2.18b, and the adiabatic exponent (Equation 2.15b) appears in both forms (Equations 2.18a and 2.18b).

2.1.5 Sound Speed as a Flow Variable

From the adiabatic relation (Equation 2.15b ≡ Equation 2.19a) and sound speed (Equation 2.16 ≡ Equation 2.19b) follow the ratios of free to stagnation values for the pressure, mass density, and sound speed by

$$\frac{p}{p_0} = \left(\frac{\rho}{\rho_0}\right)^{\gamma}: \quad \left(\frac{c}{c_0}\right)^2 = \frac{p}{p_0}\frac{\rho_0}{\rho} = \left(\frac{\rho}{\rho_0}\right)^{\gamma-1} = \left(\frac{p}{p_0}\right)^{1-\frac{1}{\gamma}}; \quad \left(\frac{c}{c_0}\right)^2 = 1 - \frac{\gamma-1}{2}\frac{v^2}{c_0^2}, \tag{2.19a–2.19c}$$

using Equation 2.18b ≡ Equation 2.19c; these lead to *the ratios of the sound speed (Equation 2.20b), mass density (Equation 2.20c), and pressure (Equation 2.20d) to their stagnation values:*

$$M_0 \equiv \frac{v}{c_0}: \qquad 1 - \frac{\gamma-1}{2} M_0^2 = \left(\frac{c}{c_0}\right)^2 = \left(\frac{\rho}{\rho_0}\right)^{\gamma-1} = \left(\frac{p}{p_0}\right)^{1-\frac{1}{\gamma}}, \tag{2.20a–2.20d}$$

in terms of the **reference Mach number** *(Equation 2.20a) defined by the ratio of the flow velocity to the stagnation sound speed.*

It follows from Equation 2.20b that *a flow is incompressible (Equation 2.10a ≡ Equation 2.21a) if the Mach number is low (Equation 2.21b), that is, the flow velocity is small compared with the stagnation sound speed (Equation 2.21c):*

$$\text{incompressible:} \quad \rho = \rho_0 \quad \Leftrightarrow \quad (M_0)^2 \ll 1 \quad \Leftrightarrow \quad v^2 \ll (c_0)^2. \tag{2.21a–2.21c}$$

Using the binomial theorem (Equations I.25.37a through I.25.37c) in Equation 2.20d specifies the pressure:

$$\begin{aligned} p &= p_0\left(1 - \frac{\gamma-1}{2}M_0^2\right)^{\frac{\gamma}{\gamma-1}} \\ &= p_0\left[1 - \frac{\gamma}{\gamma-1}\frac{\gamma-1}{2}M_0^2 + \frac{1}{2}\frac{\gamma}{\gamma-1}\left(\frac{\gamma}{\gamma-1}-1\right)\left(\frac{\gamma-1}{2}M_0^2\right)^2 + O\left(M_0^6\right)\right] \\ &= p_0\left[1 - \frac{\gamma}{2}M_0^2 + \frac{\gamma}{8}M_0^4 + O\left(M_0^6\right)\right] = p_0 - \frac{\gamma}{2}p_0\frac{v^2}{c_0^2}\left[1 - \frac{1}{4}M_0^2 + O\left(M_0^4\right)\right] \\ &= p_0 - \frac{1}{2}\rho_0 v^2\left[1 - \left(\frac{v}{2c_0}\right)^2 + O\left(\left(\frac{v}{c_0}\right)^4\right)\right], \end{aligned} \tag{2.22}$$

where the sound speed (Equation 2.16) was used in stagnation conditions. *At low Mach number, Equation 2.22 reduces to the incompressible Bernoulli equation (Equation 2.12b), corresponding to the first two terms on the r.h.s.; the third term is the lowest order compressibility correction.* The sound speed c is constant at a low Mach number and becomes a flow variable like the velocity v, pressure p, and density ρ in a compressible flow.

2.1.6 Critical Flow at the Sonic Condition

A **sonic flow condition** corresponds to the flow velocity equal to the sound speed (Equation 2.23a) and leads from Equations 2.18b to 2.23b:

$$v_* = c_*: \qquad c_0^2 = \frac{\gamma-1}{2}v_*^2 + v_*^2 = \frac{\gamma+1}{2}v_*^2 = \frac{\gamma+1}{2}c_*^2. \tag{2.23a and 2.23b}$$

The **critical Mach number** (Equation 2.24a) is defined by the ratio of velocity to the critical velocity (Equation 2.23b) and is related to the reference Mach number (Equation 2.20a):

$$M_* = \frac{v}{c_*} = \sqrt{\frac{2}{\gamma+1}}\frac{v}{c_0} = M_0\sqrt{\frac{2}{\gamma+1}}: \qquad 1-\frac{\gamma-1}{\gamma+1}M_*^2 = \left(\frac{c}{c_0}\right)^2 = \left(\frac{\rho}{\rho_0}\right)^{\gamma-1} = \left(\frac{p}{p_0}\right)^{1-\frac{1}{\gamma}}. \tag{2.24a–2.24d}$$

The relation between the reference (Equation 2.20a) and critical (Equation 2.24a) Mach numbers can be used in Equations 2.20b through 2.20d to express in terms of the latter the ratio of sound speed (Equation 2.24b), mass density (Equation 2.24c), and pressure (Equation 2.24d) to their stagnation values. At the sonic condition (Equation 2.25a):

$$M_* = 1 \quad \Leftrightarrow \quad v_* = c_*: \qquad \frac{2}{\gamma+1} = \left(\frac{c_*}{c_0}\right)^2 = \left(\frac{\rho_*}{\rho_0}\right)^{\gamma-1} = \left(\frac{p_*}{p_0}\right)^{1-\frac{1}{\gamma}}, \tag{2.25a–2.25d}$$

Equations 2.24a through 2.24d lead to the ratios of the critical sound speed c_*, mass density ρ_*, and pressure p_* to their stagnation values (Equations 2.25b through 2.25d).

2.1.7 Local, Reference, and Critical Mach Numbers

The local c, stagnation c_0, and critical c_* sound speeds are related to the velocity by Equations 2.18b and 2.23b:

$$c^2 = c_0^2 - \frac{\gamma-1}{2}v^2 = \frac{\gamma+1}{2}c_*^2 - \frac{\gamma-1}{2}v^2. \tag{2.26}$$

Thus, *the* **local Mach number** *(Equation 2.27a) defined as the ratio of the velocity to the local sound speed is related to the reference (Equation 2.20b) [critical (Equation 2.24a)] Mach numbers by*

$$M \equiv \frac{v}{c}: \qquad M^2 = \frac{2}{1-\gamma+(\gamma+1)/M_*^2} = \frac{2}{1-\gamma+2/M_0^2}. \tag{2.27a–2.27c}$$

The Equation 2.27b follows from Equation 2.26 in the form

$$\frac{\gamma-1}{2}M_*^2 = \frac{\gamma-1}{2}\left(\frac{v}{c_*}\right)^2 = \frac{\gamma+1}{2} - \left(\frac{c}{c_*}\right)^2 = \frac{\gamma+1}{2} - \left(\frac{c}{v}\frac{v}{c_*}\right)^2 = \frac{\gamma+1}{2} - \left(\frac{M_*}{M}\right)^2, \tag{2.28}$$

solving for M^2; substituting Equation 2.24a in Equation 2.27b leads to Equation 2.27c. *The sonic condition Equation 2.23a ≡ Equation 2.29a corresponds to (1 and 2) unit critical (Equation 2.29b) and local (Equation 2.29c) Mach numbers:*

$$v_* = c_*: \qquad M_* = 1 = M, \qquad M_0 = \sqrt{\frac{2}{\gamma+1}} = \frac{v_*}{c_0}; \tag{2.29a–2.29d}$$

(3) the reference Mach number (Equation 2.29d), as follows from Equations 2.27b and 2.29b for Equation 2.29c; (2) Equations 2.27c and 2.29c for Equation 2.29d, where was also used Equation 2.23b.

Because the sound speed must be real and positive (Equation 2.30a), it follows from Equation 2.24b that the left-hand side (l.h.s.) must be positive, setting *an upper bound on the range of variation of (1) the critical Mach number (Equation 2.30b); (2) also on the reference Mach number (Equation 2.30c) from Equation 2.24a:*

$$c \geq 0: \qquad 0 \leq M_* \leq \sqrt{\frac{\gamma+1}{\gamma-1}} \quad \Leftrightarrow \quad 0 \leq M_0 \leq \sqrt{\frac{2}{\gamma-1}} \quad \Leftrightarrow \quad 0 \leq M \leq \infty; \tag{2.30a–2.30d}$$

And (3) the corresponding (Equations 2.27b and 2.27c) local Mach number is unlimited (Equation 2.30d). The ratio sound speed, pressure, and mass density to the stagnation values can be expressed in terms of the three Mach numbers: (1) reference (Equations 2.20a and 2.30c) by Equations 2.20b through 2.20d; (2) critical (Equations 2.24a and 2.30b) by Equations 2.24b through 2.24d; and (3) local (Equations 2.27a and 2.30d) by Equations 2.31a through 2.31c:

$$1 + \frac{\gamma-1}{2} M^2 = \left(\frac{c_0}{c}\right)^2 = \left(\frac{\rho_0}{\rho}\right)^{\gamma-1} = \left(\frac{p_0}{p}\right)^{1-\frac{1}{\gamma}}. \tag{2.31a–2.31c}$$

All these results follow from the Bernoulli equation (Equations 2.18a and 2.18b) for a steady irrotational homentropic flow of an inviscid fluid with a constant adiabatic exponent. In the same conditions, the velocity has the upper limit (Equation 2.32b):

$$c \geq 0: \qquad v \leq c_0 \sqrt{\frac{\gamma-1}{2}} = c_* \sqrt{\frac{\gamma+1}{\gamma-1}} = v_* \sqrt{\frac{\gamma+1}{\gamma-1}}, \tag{2.32a–2.32d}$$

which is related by Equation 2.32c (Equation 2.32d) to the coincident (Equation 2.23a) [critical (Equation 2.23b)] sound speed c_* (velocity v_*). Higher velocities in an inviscid irrotational steady flow exclude homentropic conditions and must involve heat exchanges.

2.1.8 Minimum/Critical Radii for a Vortex in a Homentropic Flow

In the case (Figure I.12.2b) of a vortex of circulation Γ, the velocity (Equation 2.6a ≡ Equation I.12.33b) is azimuthal (Equation 2.33b):

$$M_\varphi \equiv \frac{v_\varphi}{c}: \qquad \Gamma = 2\pi v_\varphi r, \quad r = \frac{\Gamma}{2\pi v_\varphi} = \frac{\Gamma}{2\pi c M_\varphi} = \frac{\Gamma}{2\pi c_0} \frac{1}{M_\varphi} \frac{c_0}{c}, \tag{2.33a–2.33c}$$

and leads to the radius (Equation 2.33c) as a function of the local azimuthal Mach number (Equation 2.33a); the latter appears in the sound speed (Equation 2.31a) in a homentropic flow:

$$r = \frac{\Gamma}{2\pi c_0} \frac{1}{M_\varphi} \left| 1 + \frac{\gamma - 1}{2} M_\varphi^2 \right|^{1/2} = \frac{\Gamma}{2\pi c_0} \left| \frac{1}{M_\varphi^2} + \frac{\gamma - 1}{2} \right|^{1/2}. \quad (2.34)$$

For a vortex of circulation Γ in a homentropic flow with stagnation sound speed c_0, the radius is a function of the Mach number (Equation 2.34) and is minimum for $M_{\varphi *} = \infty$ in Equation 2.35a:

$$r \geq r_1 = r\left(M_{\varphi 1} = \infty\right) = \frac{\Gamma}{2\pi c_0} \sqrt{\frac{\gamma - 1}{2}}; \qquad \frac{r}{r_1} = \left| 1 + \frac{2}{(\gamma - 1) M_\varphi^2} \right|^{1/2}. \quad (2.35\text{a and } 2.35\text{b})$$

The radius (Equation 2.34) divided by the minimum radius (Equation 2.35a) is given by Equation 2.35b. The sonic condition specifies the critical radius:

$$r_a \equiv r\left(M_{\varphi *} = 1\right) = \frac{\Gamma}{2\pi c_0} \sqrt{\frac{\gamma + 1}{2}} = \frac{\Gamma}{2\pi c_*} = r_1 \sqrt{\frac{\gamma + 1}{\gamma - 1}} > r_1, \quad (2.36\text{a and } 2.36\text{b})$$

that (1) involves (Equation 2.36a) the critical sound speed (Equation 2.25b); (2) is larger (Equation 2.36b) than the minimum radius (Equation 2.35a); and (3) separates the supersonic (subsonic) flow for smaller (larger) radii $r_1 < r < r_*(r_* < r < \infty)$, as shown in Figure 2.1a.

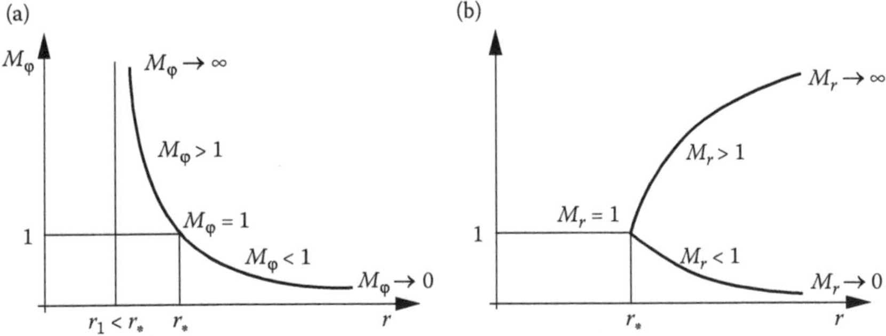

FIGURE 2.1
(a) A potential vortex with constant circulation in a homentropic flow has a minimum radius r_1 smaller than the critical radius r_* for the sonic condition. Thus, the local Mach number M_φ corresponding to the tangential velocity decays to zero (diverges toward infinity) away from the radius and toward the infinity (the minimum radius). (b) A source/sink with constant mass rate in a homentropic flow has a minimum radius corresponding to the sonic condition; for all other larger values of the radius, there are two branches for the local radial Mach number M_r, one supersonic (subsonic) tending to infinity (zero). In the case of a spiral or monopole flow, combining (a) vortex with constant circulation and (b) source/sink with constant mass, the latter predominates in the sense that there are subsonic and supersonic branches issuing from the minimum radius at the sonic condition: the minimum radius r_3 generalizes both r_1 and r_2 because both circulation and mass rate are present.

The preceding results can be confirmed from the first two derivatives of the radius (Equation 2.34) with regard to the azimuthal Mach number:

$$\frac{2\pi c_0}{\Gamma}\frac{dr}{dM_\varphi} = -\frac{1}{M_\varphi^3}\left|\frac{1}{M_\varphi^2}+\frac{\gamma-1}{2}\right|^{-1/2}, \tag{2.37}$$

$$\begin{aligned}\frac{2\pi c_0}{\Gamma}\frac{d^2r}{dM_\varphi^2} &= \left|\frac{1}{M_\varphi^2}+\frac{\gamma-1}{2}\right|^{-3/2}\left[-\frac{1}{M_\varphi^6}+\frac{3}{M_\varphi^4}\left(\frac{1}{M_\varphi^2}+\frac{\gamma-1}{2}\right)\right]\\ &= \frac{1}{M_\varphi^4}\left(\frac{2}{M_\varphi^2}+3\frac{\gamma-1}{2}\right)\left|\frac{1}{M_\varphi^2}+\frac{\gamma-1}{2}\right|^{-3/2}.\end{aligned} \tag{2.38}$$

The stationary value (Equation 2.39a) of Equation 2.37 specifies the Mach number $M_{\varphi 1} = \infty$ corresponding to the minimum radius (Equation 2.35a):

$$\lim_{M_\varphi\to\infty}\frac{dr}{dM_\varphi} = \lim_{M_\varphi\to\infty} O\left(\frac{1}{M_\varphi^2}\right) = 0; \qquad \lim_{M_\varphi\to\infty}\frac{d^2r}{dM_\varphi^2} = \lim_{M_\varphi\to\infty} O\left(\frac{1}{M_\varphi^3}\right) = 0. \tag{2.39a and 2.39b}$$

The second-order derivative (Equation 2.38) is also zero (Equation 2.39b) at the minimum radius, as are all derivatives (Equation 2.40b) of higher order (Equation 2.40a):

$$n = 1,2,\ldots \in |N: \qquad \lim_{M_\varphi\to\infty}\frac{d^n r}{dM_\varphi^n} = \lim_{M_\varphi\to\infty} O\left(M_\varphi^{-1-n}\right) = 0. \tag{2.40a and 2.40b}$$

The conclusion that $M_{\varphi 1} = \infty$ is a minimum follows most simply from the algebraic relation (Equation 2.34).

2.1.9 Conservation of the Mass Rate for a Source/Sink

A source (sink) with radial positive $v_r > 0$ (negative $v_r < 0$) velocity can be considered in a compressible flow, provided that the mass flux (Equation 2.41a) is conserved:

$$\dot{m} = \rho Q = \rho 2\pi r v_r = 2\pi r M_r c\rho, \tag{2.41a and 2.41b}$$

where (1) the mass flux $\dot{m}$ is (Equations 2.41a and 2.41b) the product of the volume flux (Equation I.12.25a) by the mass density; (2) in an incompressible flow, the mass density is constant, and the conservation of the mass flux is equivalent to the conservation of the volume flux; (3) in a compressible flow, such as homentropic, the mass density ρ is related to the stagnation mass density ρ_0 by the compressible Bernoulli equation (Equation 2.31b);

and (4) using also the radial local Mach number (Equation 2.42a) introduces the sound speed (Equation 2.31a) and leads to Equation 2.42b:

$$M_r \equiv \frac{v_r}{c}: \qquad r = \frac{\dot{m}}{2\pi\rho_0 c_0}\frac{1}{M_r}\frac{\rho_0 c_0}{\rho c} = \frac{\dot{m}}{2\pi\rho_0 c_0}\frac{1}{M_r}\left(1+\frac{\gamma-1}{2}M_r^2\right)^{\frac{1}{\gamma-1}+\frac{1}{2}}$$
$$= \frac{\dot{m}}{2\pi\rho_0 c_0}\frac{1}{M_r}\left(1+\frac{\gamma-1}{2}M_r^2\right)^{\frac{\gamma+1}{2(\gamma-1)}}. \qquad \text{(2.42a and 2.42b)}$$

The Mach number (Equation 2.42a) of the homentropic flow due to a source/sink of mass rate (Equation 2.41a) is related to the radius by Equation 2.42b, where $\rho_0(c_0)$ is the stagnation mass density (sound speed). The derivative of the radius (Equation 2.42b) with regard to the Mach number:

$$\frac{dr}{dM_r} = \frac{\dot{m}}{2\pi\rho_0 c_0}\left(1-\frac{1}{M_r^2}\right)\left(1+\frac{\gamma-1}{2}M_r^2\right)^{\frac{3-\gamma}{2(\gamma-1)}} \qquad (2.43)$$

shows that the stationary value (Equation 2.44a) occurs at (Equation 2.43) the sonic condition, that specifies the minimum radius (Equation 2.44b):

$$\lim_{M_r\to 1}\frac{dr_2}{dM_r} = 0: \quad r \ge r_2 = r(M_{r*}=1) = r(M_{r2}=1) = \frac{\dot{m}}{2\pi\rho_0 c_0}\left(\frac{\gamma+1}{2}\right)^{\frac{\gamma+1}{2(\gamma-1)}} = \frac{\dot{m}}{2\pi\rho_* c_*}. \qquad \text{(2.44a and 2.44b)}$$

The minimum radius depends on the critical mass density (Equation 2.25c) and sound speed (Equation 2.25b) at the sonic condition. For a supersonic $M_r > 1$ (subsonic $M_r < 1$) flow, Equation 2.43 implies that $dr/dM_r > 0$ ($dr/dM_r < 0$), and hence, the local Mach number increases (decreases) away from the sonic condition at the stationary radius (Figure 2.1b) along two branches issuing from the sonic condition.

The derivative (Equation 2.43) can be obtained from Equation 2.42b:

$$\frac{2\pi\rho_0 c_0}{\dot{m}}\left(1+\frac{\gamma-1}{2}M_r^2\right)^{1-\frac{\gamma+1}{2(\gamma-1)}}\frac{dr}{dM_r}$$
$$= -\frac{1}{M_r^2}\left(1+\frac{\gamma-1}{2}M_r^2\right) + \frac{1}{M_r}\frac{\gamma+1}{2(\gamma-1)}\frac{\gamma-1}{2}2M_r = 1-\frac{1}{M_r^2}. \qquad (2.45)$$

Because $\gamma > 1$ in Equation 2.15c, the exponent in Equation 2.43 is positive for $\gamma < 3$, and the only root of Equation 2.44a is $M_r = 1$. From Equation 2.42b, it follows that

$$\lim_{M_r\to 0} r = \lim_{M_r\to 0} O\left(\frac{1}{M_r}\right) = \infty = \lim_{M_r\to\infty} O\left(M_r^{\frac{3-\gamma}{2(\gamma-1)}}\right) = \lim_{M_r\to\infty} O\left(M_r^{-1+\frac{\gamma+1}{2(\gamma-1)}}\right) = \lim_{M_r\to\infty} r, \qquad \text{(2.46a and 2.46b)}$$

showing that the infinite radius corresponds to two branches (Figure 2.1b) issuing from $r(M_{r*} = r_2)$, namely, the subsonic (supersonic) branch for Equation 2.46a (Equation 2.46b). *The radius (Equation 2.42b) divided by the minimum radius (Equation 2.44b) is given by*

$$\frac{r}{r_2} = \frac{1}{M_r}\left(\frac{2}{\gamma+1} + \frac{\gamma-1}{\gamma+1}M_r\right)^{\frac{\gamma+1}{2(\gamma-1)}}, \tag{2.47}$$

with the subsonic $M_r < 1$ and supersonic $M_r > 1$ branches both starting at $r = r_1$ for $M_r = 1$, as shown in Figure 2.1b for the source or sink. This contrasts with the vortex in Figure 2.1a, for which there is a single curve. Thus, *a given radius corresponds to (1) a single azimuthal Mach number for a vortex, that is, supersonic (subsonic) before $v_1 < v < r_*$ (after $v_* < r < \infty$) the sonic condition; and (2) two radial Mach numbers for a source or sink, that is, one on the supersonic $M_r > 1$ and another on the subsonic $M_r < 1$ branch.*

2.1.10 Combination in a Compressible Spiral Monopole

The main difference between the source and the sink (Subsection 2.1.9) [vortex (Subsection 2.1.8)] in a compressible homentropic flow is that because the velocity has nonzero (zero) divergence, the mass density does (does not) appear, as seen in the superposition of the radial (Equation 2.41a) [azimuthal (Equation 2.33b)] velocity components in the total velocity (Equation 2.48b):

$$M \equiv \frac{v}{c}: \qquad c^2M^2 = v^2 = (v_r)^2 + (v_\varphi)^2 = \frac{\Gamma^2 + \dot{m}^2/\rho^2}{(2\pi r)^2}. \tag{2.48a and 2.48b}$$

The total Mach number (Equation 2.48a) appears explicitly and also implicitly in the sound speed (Equation 2.31a) and mass density (Equation 2.31b):

$$r = \frac{\Gamma}{2\pi c M}\left|1 + \frac{\dot{m}^2}{\rho^2\Gamma^2}\right|^{1/2} = \frac{\Gamma}{2\pi c_0}\frac{1}{M}\frac{c_0}{c}\left|1 + \frac{\dot{m}^2}{\rho_0^2\Gamma^2}\left(\frac{\rho_0}{\rho}\right)^2\right|^{1/2}. \tag{2.49}$$

Substituting Equations 2.31a and 2.31b in Equation 2.46 leads to

$$r \equiv \frac{\Gamma}{2\pi c_0}\frac{1}{M}\left|1 + \frac{\gamma-1}{2}M^2\right|^{1/2}\left|1 + \left(\frac{\dot{m}}{\rho_0\Gamma}\right)^2\left(1 + \frac{\gamma-1}{2}M^2\right)^{\frac{2}{\gamma-1}}\right|^{1/2}, \tag{2.50}$$

specifying the radius as a function of the Mach number for a monopole with mass rate $\dot{m}$ and circulation Γ in a flow with stagnation sound speed c_0 and mass density ρ_0. Because Equation 2.50 implies

$$\lim_{M\to 0} r = \lim_{M\to 0} O\left(\frac{1}{M}\right) = \infty = \lim_{M\to\infty} O\left(M^{\frac{2}{\gamma-1}}\right) = \lim_{M\to\infty} O\left(M^{-1+1+\frac{2}{\gamma-1}}\right) = \lim_{M\to\infty} r, \tag{2.51a and 2.51b}$$

there must be a minimum radius, that reduces to Equation 2.43b (Equation 2.35a) for a source/sink (vortex).

Using the notation (Equation 2.52a) in Equation 2.50 ≡ Equation 2.52b:

$$X \equiv \left(\frac{\dot{m}}{\rho_0 \Gamma}\right)^2 \left(1+\frac{\gamma-1}{2}M^2\right)^{\frac{2}{\gamma-1}}: \qquad \frac{2\pi c_0}{\Gamma} r = \frac{1}{M} \left|1+\frac{\gamma-1}{2}M^2\right|^{1/2} \left|1+X\right|^{1/2} \quad \text{(2.52a and 2.52b)}$$

follows:

$$\frac{dX}{dM} = \left(\frac{\dot{m}}{\rho_0 \Gamma}\right)^2 \left(1+\frac{\gamma-1}{2}M^2\right)^{\frac{2}{\gamma-1}-1} \frac{2}{\gamma-1}\frac{\gamma-1}{2}2M = 2MX\left(1+\frac{\gamma-1}{2}M^2\right)^{-1}, \tag{2.52c}$$

and hence the derivative of the radius with regard to the local Mach number:

$$\begin{aligned} &\frac{2\pi c_0}{\Gamma}\left|1+\frac{\gamma-1}{2}M^2\right|^{1/2} \left|1+X\right|^{1/2} \frac{dr}{dM} \\ &= \left[-\frac{1}{M^2}\left(1+\frac{\gamma-1}{2}M^2\right)+\frac{1}{2M}\frac{\gamma-1}{2}2M\right](1+X)+\frac{1}{2M}\left(1+\frac{\gamma-1}{2}M^2\right)\frac{dX}{dM} \\ &= -\frac{1}{M^2}(1+X)+X = -\frac{1}{M^2}\left[1-(M^2-1)X\right] = -\frac{1}{M^2}\left[1-(M^2-1)\left(\frac{\dot{m}}{\rho_0\Gamma}\right)^2\left(1+\frac{\gamma-1}{2}M^2\right)^{\frac{2}{\gamma-1}}\right]. \end{aligned} \tag{2.52d}$$

For monopole (Equation 2.48b) with circulation Γ and mass rate $\dot{m}$, the stationary radius (Equation 2.53a) corresponds to a Mach number M_3 satisfying Equation 2.53b:

$$\lim_{M\to M_3}\frac{dr}{dM} = 0: \qquad \left(M_3^2-1\right)\left(1+\frac{\gamma-1}{2}M_3^2\right)^{\frac{2}{\gamma-1}} = \left(\frac{\rho_0\Gamma}{\dot{m}}\right)^2. \quad \text{(2.53a and 2.53b)}$$

The critical Mach number (Equation 2.53b) leads by Equation 2.50 to

$$\begin{aligned} r_3 \equiv r(M_3) &= \frac{\Gamma}{2\pi c_0}\frac{1}{M_3}\left|1+\frac{\gamma-1}{2}M_3^2\right|^{1/2}\left|1+\frac{1}{M_3^2-1}\right|^{1/2} \\ &= \frac{\Gamma}{2\pi c_0}\left|M_3^2-1\right|^{-1/2}\left|1+\frac{\gamma-1}{2}M_3^2\right|^{1/2}, \end{aligned} \tag{2.54}$$

as the minimum radius (Equation 2.54).

For a source/sink $\dot{m} \neq 0 = \Gamma$ (vortex $\dot{m} = 0 \neq \Gamma$), the Mach number satisfying Equation 2.53b is $M_2 = 1 = M_{r*} = M_{r2}(M_1 = \infty = M_{\varphi 1})$ in agreement with Equation 2.44a (Equation 2.35a). The limit $M \to \infty$ in Equation 2.54 is Equation 2.35a for the vortex. For the source/sink $\Gamma = 0$, the limit $M = 1$ in Equation 2.50 leads to

$$\begin{aligned} \lim_{M\to 1,\Gamma\to 0} r &= \lim_{M\to 1}\frac{1}{2\pi c_0}\frac{1}{M}\left|1+\frac{\gamma-1}{2}M^2\right|^{1/2}\frac{\dot{m}}{\rho_0}\left(1+\frac{\gamma-1}{2}M^2\right)^{\frac{1}{\gamma-1}} \\ &= \frac{\dot{m}}{2\pi\rho_0 c_0}\left(\frac{\gamma+1}{2}\right)^{\frac{1}{2}+\frac{1}{\gamma-1}} = \frac{\dot{m}}{2\pi\rho_0 c_0}\left(\frac{\gamma+1}{2}\right)^{\frac{\gamma+1}{2(\gamma-1)}} = \frac{\dot{m}}{2\pi\rho_* c_*} = r_2, \end{aligned} \tag{2.55}$$

in agreement with Equation 2.44b. This confirms that the results for the spiral flow (Subsection 2.1.10) reduce to those for the vortex (source/sink) in the absence [Subsection 2.1.8 (Subsection 2.1.9)] of mass flow (circulation). The coincidence of the limits [Equations 2.51a and 2.51b (Equations 2.46a and 2.46b)] shows that in the monopole, the source/sink predominates over the vortex in that there are two branches as in Figure 2.1b, starting at Equation 2.54 instead of Equation 2.44b. For all radii beyond the minimum $r > r_3$, there are two possible flows: one supersonic (subsonic) with local Mach number decaying (diverging) with the radius.

2.2 Potential Vortex with Rotational Core (Rankine; Hallock and Burnham 1997)

The singularity at the center of a potential flow is excluded by compressibility effects that limit the minimum radius for (1) a source/sink (Section I.12.4 and Subsection 2.1.9); (2) vortex (Section I.12.5 and Subsection 2.1.8); and (3) their superposition as a monopole or spiral flow (Section I.12.6 and Subsection 2.1.10). Considering the vortex, another way to avoid the singularity at the center is to match the azimuthal velocity of the outer potential flow (Equation 2.33b) to a rotational core; this serves as the simplest case (Section 2.2) of incompressible rotational flow (Sections 2.3 through 2.9). A vortex core with azimuthal velocity proportional to a power of the radius is the generalized Rankine vortex that leads (Subsection 2.2.1) to a zero velocity at the center; the core radius can be chosen to have a continuous tangential velocity where it is matched to the outer potential vortex with given circulation; this leads (Subsection 2.2.1) to a discontinuous vorticity at the core radius because the vorticity is zero outside and finite and nonzero inside. A continuous radial vorticity profile is obtained for a smooth azimuthal velocity such as the generalized Hallock–Burnham vortex (Subsection 2.2.2). In this case, both the azimuthal velocity and vorticity are continuous functions for all values of the radius. The azimuthal velocity perturbation is related to the pressure (Subsection 2.2.3) by the inviscid momentum equation (Subsection 2.1.1); for a rotational flow, the latter does not reduce to the Bernoulli equation (Subsection 2.2.4), and thus, the stagnation pressure is not generally constant in a vortex, except outside the vortex core. The pressure distribution must be nonnegative in all spaces, leading to bounds on the core radius of the generalized Rankine (Hallock–Burnham) vortex [Subsection 2.2.5 (Subsection 2.2.6)]. The stagnation pressure varies radially (Subsection 2.2.7) and its radial gradient (Subsection 2.2.8) equals minus the vortical force (Subsection 2.2.9). Vortices occur in the atmosphere as hurricanes, in the wakes of aircraft, and as depressions of the free surface of oceans (Subsection 2.2.10).

2.2.1 Vorticity Concentrated in a Finite Core (Rankine)

The singularity of the azimuthal velocity (Equation 2.33b) of a potential vortex can be avoided (Kelvin 1880) in an incompressible flow by (1) restricting it to the outside of a vortex core (Equation 2.56b); and (2) imposing inside the vortex core another azimuthal velocity profile, such as (Rankine) a rigid body rotation (Equation 2.56c) with angular velocity Ω:

$$n = 0: \qquad v_{1\varphi}(r \geq a) = \frac{\Gamma}{2\pi r}, \qquad v_{1\varphi}(r \leq a) = \Omega r. \qquad (2.56a\text{–}2.56c)$$

The continuity of the velocity at the vortex core radius (Equation 2.57a) relates (Equation 2.57b) the angular velocity of the core (Equation 2.56c) to the circulation of the outer potential vortex (Equation 2.56b):

$$\frac{\Gamma}{2\pi r} = v_{1\varphi}(a) = \Omega a: \qquad \Gamma = 2\pi\Omega a^2. \tag{2.57a and 2.57b}$$

The circulation (Equation 2.57b) is the product of the perimeter of the core radius $2\pi a$ by the tangential velocity Ωa there. The tangential velocity of the original Rankine vortex is thus given by

$$n = 0: \qquad v_{1\varphi}(r) = \begin{cases} \dfrac{\Gamma}{2\pi r} = \dfrac{\Omega a^2}{r} & \text{if} \quad r \geq a, \qquad (2.58\text{a}) \\[2ex] \Omega r = \dfrac{\Gamma r}{2\pi a^2} & \text{if} \quad r \leq a, \qquad (2.58\text{b}) \end{cases}$$

and it vanishes both at the infinity and at the center. The latter property is retained in the generalized Rankine vortex family, allowing a power law radial dependence of the azimuthal velocity in the core:

$$v_{1\varphi}(r \leq a) = \Omega r \left(\frac{r}{a}\right)^n = \Omega \frac{r^{n+1}}{a^n}, \tag{2.59}$$

with the velocity reduced as n increases (Figure 2.1a), starting at zero on axis and reaching Ωa at the core radius, where it matches in the same way to the outer potential vortex. The original Rankine vortex (Equation 2.56c) is the case $n = 0$ of the generalization (Equation 2.59).

Thus, the **generalized Rankine vortex** *is specified by the azimuthal velocity (1) of a potential vortex (Equation 2.33b ≡ Equation 2.56b ≡ Equation 2.60a) outside the core and (2) a power law (Equation 2.59 ≡ Equation 2.60b) inside the core:*

$$v_{1\varphi}(r) = \begin{cases} \dfrac{\Gamma}{2\pi r} = \Omega \dfrac{a^2}{r}, & \text{if} \quad a \leq r < \infty, \qquad (2.60\text{a}) \\[2ex] \dfrac{\Omega r^{n+1}}{a^n} = \dfrac{\Gamma r^{n+1}}{2\pi a^{n+2}}, & \text{if} \quad 0 \leq r \leq a, \qquad (2.60\text{b}) \end{cases}$$

where the circulation Γ and angular velocity Ω are related by the continuity of the tangential velocity at the core radius (Equation 2.61a):

$$r_4 = a: \qquad v_{1\max} = \frac{\Gamma}{2\pi a} = v_{1\varphi}(r_4) = \Omega a; \qquad \varpi_{1\max} = \varpi_1(r_4) = (n+2)\Omega, \tag{2.61a–2.61c}$$

where is the maximum of (1) the (Equation 2.61a) tangential velocity (Equations 2.60a and 2.60b); and (2) also (Equation 2.61c) of the vorticity (Equation 2.62a):

$$\vec{\varpi}_1(r<a) \equiv \nabla \wedge \vec{v}_1 = \vec{e}_3 \frac{\vec{e}_3}{r}\frac{d}{dr}\left(v_{1\varphi}r\right) = (n+2)\Omega\left(\frac{r}{a}\right)^n, \vec{\varpi}_1(r>a) = 0, \qquad \text{(2.62a and 2.62b)}$$

which vanishes for the potential flow outside the core (Equation 2.62b); the curl in polar coordinates (Equation I.11.35b) was used in Equation 2.62a, where $\vec{e}_3$ is the unit vector orthogonal to the polar plane (r,φ). The case $n = 0$ corresponds to *the Rankine vortex for which (1) the core (Equation 2.56a) is in* **rigid body rotation** $v_\varphi = \Omega r$ *with angular velocity* Ω *equal to one-half of the vorticity* $\varpi = 2\Omega$ *in Equation 2.61c; and (2) outside the core is a potential vortex (Equation 2.60b) with zero vorticity (Equation 2.62b).* In all cases $n = 0$ or $n \neq 0$, the vorticity is discontinuous at $r = a$, because (1) it vanishes for the potential vortex outside as $r \to a + 0$; and (2) it has the finite limit $(n + 2)\Omega$ on the inside (Equation 2.61c) as $r \to a - 0$. *The velocity (Equations 2.60a and 2.60b ≡ Equations 2.63f and 2.63g) [vorticity (Equations 2.62a and 2.62b ≡ Equations 2.63d and 2.63e)] of a generalized Rankine vortex is given in dimensionless [Equation 2.63a (Equation 2.63b)] form versus (Equation 2.63c) the radial distance divided by the core radius:*

$$\bar{v} \equiv \frac{v_\varphi}{\Omega a}, \bar{\varpi} \equiv \frac{\varpi}{\Omega}, s \equiv \frac{r}{a}: \quad \bar{\varpi}_1(s) = \begin{cases} 0 & \text{if} \quad s>1, \\ (n+2)s^n, & \text{if} \quad s<1, \end{cases} \qquad \bar{v}_1(s) = \begin{cases} 1/s & \text{if} \quad s \geq 1, \\ s^{n+1} & \text{if} \quad s \leq 1, \end{cases} \qquad \text{(2.63a–2.63g)}$$

showing for increasing exponent n = 0,1,2,3,4 at the core radius (1) for the velocity (Equation 2.63e), a sharper angular point (Panel 2.1a); and (2) for the vorticity (Equation 2.63d), a larger peak value (Panel 2.2a). The pressure associated with the generalized Rankine vortex is calculated in Subsections 2.2.4 and 2.2.5.

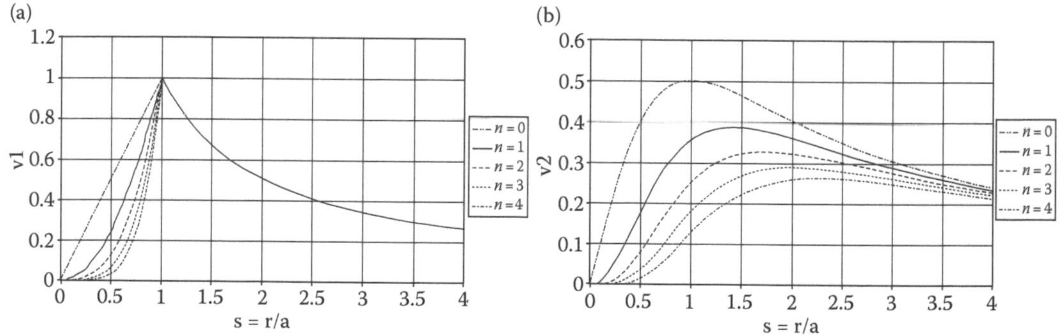

PANEL 2.1
Dimensionless tangential velocity (Equation 2.63a) versus radial distance divided by the core radius (Equation 2.63c) for the generalized (a) Rankine [(b) Hallock–Burnham] vortex [Equations 2.63f and 2.63g (Equation 2.76a)]. The flow outside (inside) the core of the generalized (a) Rankine vortex is irrotational (rotational), with a single profile (several profiles) of the tangential velocity. The increasing exponent of the tangential velocity inside the core leads to a slower rise from zero at the center to the same value at the core radius. For the tangential velocity of the generalized (b) Hallock–Burnham vortex, an increasing exponent implies a slower rise to a lower peak farther from the center. The tangential velocity vanishes at the center and infinity for all values of the exponent of the generalized (a) Rankine [(b) Hallock–Burnham] vortex.

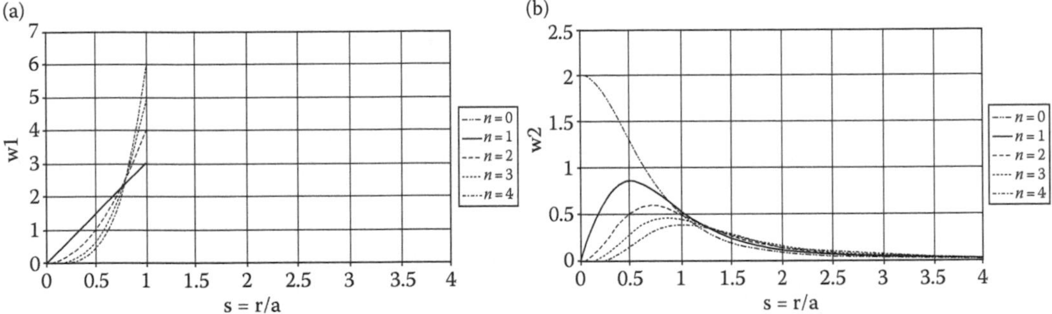

PANEL 2.2
Dimensionless vorticity (Equation 2.63b) versus radial distance divided by the core radius (Equation 2.63c) for the generalized (a) Rankine [(b) Hallock–Burnham] vortex [Equations 2.63d and 2.63e (Equation 2.76b)]. The generalized (a) Rankine vortex has a potential flow, hence zero vorticity outside the core. Because the core is rotational, there is a discontinuity of the vorticity at the core radius (a) corresponding to the angular point of the tangential velocity (Panel 2.1a) at the same point. Because the tangential velocity in the core is proportional to a power of the radius (Panel 2.1a), the peak vorticity (a) increases with the exponent. In the case of the generalized Hallock–Burnham vortex, both the tangential velocity (Panel 2.1b) and vorticity (b) are smooth functions of the radial distance, and as the exponent increases, the peak is lower and farther from the center. The tangential velocity and vorticity vanish at the center and infinity for all values of the exponent n of the generalized Rankine (Hallock–Burnham) vortex, with two exceptions: (1) the original Rankine vortex $n = 0$ has constant vorticity (a) equal to twice the angular velocity in the core because the latter is in rigid body rotation; (2) the original Hallock–Burnham vortex $n = 0$ has the same value of the vorticity at the center (b) because the flow nearby is also in rigid body rotation.

2.2.2 Smooth Radial Variation of the Vorticity (Hallock and Burnham 1997)

The discontinuity of the vorticity at the core radius can be avoided by a smooth tangential velocity profile, valid for all radii, such as the Hallock–Burnham (1962) vortex:

$$n = 0: \qquad v_{2\varphi} = \frac{\Omega a^2 r}{r^2 + a^2} \sim \begin{cases} \dfrac{\Omega a^2}{r} & \text{if} \quad r^2 >> a^2, \quad (2.64a) \\ \Omega r & \text{if} \quad r^2 << a^2, \quad (2.64b) \end{cases}$$

that (1) scales like the Rankine vortex for radius both large (Equation 2.64a ≡ Equation 2.58a) and small (Equation 2.64b ≡ Equation 2.58b) compared with the core radius; and (2) matches the two smoothly with a tangential velocity $v_{2\varphi}(a) = \Omega a/2$ at the core radius $r = a$. This is the particular case $n = 0$ of the **generalized Hallock–Burnham vortex** *with tangential velocity profile (Equation 2.65a):*

$$v_{2\varphi}(r) = \frac{\Omega a^2 r^{n+1}}{(a^2 + r^2)^{1+n/2}} \sim \begin{cases} \Omega \dfrac{a^2}{r} & \text{if} \quad r^2 >> a^2, \\ \Omega \dfrac{r^{n+1}}{a^n} & \text{if} \quad r^2 << a^2, \end{cases} \qquad (2.65\text{a}–2.65\text{c})$$

which is smooth and (1) scales like the generalized Rankine vortex both for large (Equation 2.65b ≡ Equation 2.60a) and small (Equation 2.65c ≡ Equation 2.60b) radius compared with the core radius; (2) has a vorticity that is a smooth function of the radius:

$$\vec{\varpi}_2(r) = \frac{\vec{e}_3}{r}\left[\frac{\mathrm{d}}{\mathrm{d}r}\left(v_{2\varphi} r\right)\right] = \vec{e}_3 (n+2) r^n \frac{\Omega a^4}{(a^2+r^2)^{2+n/2}}; \tag{2.66}$$

and (3) both the tangential velocity and vorticity vanish only at infinity and at the center (the exception is the finite vorticity 2Ω at the center for n = 0). The value of the velocity (Equation 2.65a) [vorticity (Equation 2.66)] at the core radius is given by Equation 2.67a (Equation 2.67b):

$$v_{2\varphi}(a) = 2^{-1-n/2}\,\Omega a, \quad \varpi_2(a) = 2^{-2-n/2}\,\Omega(n+2). \tag{2.67a and 2.67b}$$

The vorticity (Equation 2.66) is calculated from the tangential velocity (Equation 2.65a), leading to

$$\begin{aligned}\vec{\varpi}_2(r) &= \frac{\vec{e}_3}{r}\frac{\mathrm{d}}{\mathrm{d}r}\left[\Omega a^2 r^{n+2}(a^2+r^2)^{-1-n/2}\right] \\ &= \vec{e}_3 \frac{\Omega a^2}{r}(a^2+r^2)^{-2-n/2} r^{n+1}\left[\left(-1-\frac{n}{2}\right)2r^2 + (n+2)(a^2+r^2)\right] \\ &= \vec{e}_3 (n+2)\Omega a^4 r^n (a^2+r^2)^{-2-n/2},\end{aligned} \tag{2.68}$$

for the generalized Hallock–Burnham vortex, in agreement with Equation 2.66 ≡ Equation 2.68.

The location of the maxima of the tangential velocity (Equation 2.65a) [vorticity (Equation 2.66)] is specified by the vanishing of their radial derivatives [Equation 2.69 (Equation 2.70)]:

$$\begin{aligned}\frac{\mathrm{d}v_{2\varphi}}{\mathrm{d}r} &= \Omega a^2 \frac{\mathrm{d}}{\mathrm{d}r}\left[r^{n+1}(a^2+r^2)^{-1-n/2}\right] \\ &= \Omega a^2 (a^2+r^2)^{-2-n/2} r^n\left[\left(-1-\frac{n}{2}\right)2r^2 + (n+1)(a^2+r^2)\right] \\ &= \Omega a^2 r^n (a^2+r^2)^{-2-n/2}\left[(n+1)a^2 - r^2\right],\end{aligned} \tag{2.69}$$

$$\begin{aligned}\frac{\mathrm{d}\varpi_2}{\mathrm{d}r} &= \vec{e}_3 \Omega a^4 (n+2)\frac{\mathrm{d}}{\mathrm{d}r}\left[r^n (a^2+r^2)^{-2-n/2}\right] \\ &= \vec{e}_3 \Omega a^4 (n+2)(a^2+r^2)^{-3-n/2} r^{n-1}\left[\left(-2-\frac{n}{2}\right)2r^2 + n(r^2+a^2)\right] \\ &= \vec{e}_3 \Omega a^4 (n+2) r^{n-1}(a^2+r^2)^{-3-n/2}(na^2-4r^2).\end{aligned} \tag{2.70}$$

Thus, *the derivatives [Equation 2.69 ≡ Equation 2.71a (Equation 2.70 ≡ Equation 2.71b)] of the velocity (Equation 2.65a) [vorticity (Equation 2.66)] of a generalized Hallock–Burnham vortex:*

$$\frac{dv_{2\varphi}}{dr} = \Omega^2 a^2 r^n \frac{(n+1)a^2 - r^2}{(a^2 + r^2)^{2+n/2}}, \tag{2.71a}$$

$$\frac{d\varpi_2}{dr} = \Omega(n+2)a^4 r^{n-1} \frac{na^2 - 4r^2}{(a^2 + r^2)^{3+n/2}}, \tag{2.71b}$$

show that (1) the velocity Equation 2.65a peaks (Equation 2.72a) at the radius Equation 2.72b with the value Equation 2.72c:

$$\frac{dv_{2\varphi}}{dr_5} = 0: \qquad r_5 = a\sqrt{n+1}, \quad v_{2\max} = v_{2\varphi}(r_5) = \Omega a(n+1)^{(n+1)/2}(n+2)^{-1-n/2}; \tag{2.72a–2.72c}$$

(2) the vorticity (Equation 2.66) peaks (Equation 2.73a) at the radius Equation 2.73b with the value Equation 2.73c:

$$\frac{d\varpi}{dr} = 0: \qquad r_6 = \frac{a\sqrt{n}}{2} < r_5, \quad \varpi_{2\max} = \varpi(r_6) = \Omega 2^{1-n} n^{n/2}(1+n/2)(1+n/4)^{-2-n/2}. \tag{2.73a–2.73c}$$

The peak (Equation 2.72b) of the tangential velocity (Equation 2.65a) always lies outside $r_5 < r_6$ the peak (Equation 2.73b) of the vorticity (Equation 2.66). In the case of the original Hallock–Burnham vortex (Equation 2.74a), the vorticity (Equation 2.66) decays monotonically (Equation 2.74b) from the maximum value (Equation 2.74c) on the axis:

$$n = 0: \qquad \varpi_2(r) = \frac{2\Omega a^4}{(a^2 + r^2)^2}, \qquad \varpi_{\max} = \varpi(0) = 2\Omega. \tag{2.74a–2.74c}$$

In this case, Equation 2.73c does not hold (or holds with $0^0 = 1$) and is replaced by Equation 2.74c.

The derivation of Equation 2.72c (Equation 2.73c) from Equation 2.69 (Equation 2.70) uses Equation 2.72b (Equation 2.73b):

$$a^2 + (r_5)^2 = a^2(n+2), \qquad a^2 + (r_5)^2 = a^2\left(1 + \frac{n}{4}\right). \tag{2.75a and 2.75b}$$

The original Hallock–Burnham (Equations 2.64a and 2.64b) vortex contrasts with the original Rankine vortex (Equations 2.58a and 2.58b) due to its smooth velocity and vorticity distributions. *The same contrast applies to the generalized Rankine (Hallock–Burnham) vortex, in particular, at the core radius, where (1) the velocity has an angular point (is smooth) in Panel 2.1a (Panel 2.1b); and (2) the vorticity is discontinuous (continuous) in Panel 2.2a (Panel 2.2b). The velocity (Equation 2.65a ≡ Equation 2.76a) [vorticity (Equation 2.66 ≡ Equation 2.76b)] of the generalized Hallock–Burnham vortex appears in Panel 2.1b (Equation 2.2b) in dimensionless form (Equations 2.63a through 2.63c), namely:*

$$\bar{v}_2(s) = \frac{s^{n+1}}{(1+s^2)^{1+n/2}}, \qquad \bar{\varpi}_2(s) = (n+2)\frac{s^n}{(1+s^2)^{2+n/2}}, \qquad \text{(2.76a and 2.76b)}$$

showing for increasing exponent n = 0,1,2,3,4 that (1) the peak velocity (vorticity) is farther from axis; (2) the maximum has a lower value; and (3) the peak is broader around the maximum. The generalized Rankine (Hallock–Burnham) vortex [Subsection 2.2.1 (Subsection 2.2.2)] has the following: (1) a velocity profile with an angular point (Equations 2.60a and 2.60b) [a smooth velocity profile (Equation 2.65a)]; (2) a discontinuous (Equations 2.62a and 2.62b) [smooth (Equation 2.66)] vorticity; and (3) both have the same limits, for example, the scaling of the velocity at the infinity (Equation 2.60a ≡ Equation 2.65b) and near the center (Equation 2.60b ≡ Equation 2.65c). *The Table 2.1 lists for the generalized Rankine (Hallock–Burnham) vortex with exponents n = 0,1,2,3,4: (1) the velocity at r = a the core radius [Equation 2.61b (Equation 2.67a)] divided by Ωa; (2 and 3) the peak velocity [Equation 2.61b (Equation 2.72c)] and radius for which the maximum occurs [Equation 2.61a (Equation 2.72b)]; (4) the vorticity at the core radius [Equation 2.61c (Equation 2.67b)] divided by Ω; (5 and 6) the peak vorticity [Equation 2.61c (Equation 2.73c)] and the radius for which the maximum occurs [Equation 2.61a (Equation 2.73b)].* The pressure due to the generalized Rankine (Hallock–Burnham) vortex is considered [Subsection 2.2.5 (Subsection 2.2.6)], distinguishing rotational (irrotational) flows for which the Bernoulli equation does not (does) apply [Subsection 2.2.3 (Subsection 2.2.4)]; the condition that the pressure is positive everywhere sets a lower limit on the core radius.

2.2.3 Momentum Equation for an Inviscid Vortex

The inviscid momentum equation (Equation 2.4b) in polar coordinates:

$$-\frac{1}{\rho}\frac{\partial p}{\partial r} = a_r = \frac{dv_r}{dt} - \frac{1}{r}v_\varphi^2, \qquad -\frac{1}{\rho r}\frac{\partial p}{\partial \varphi} = a_\varphi = \frac{dv_\varphi}{dt} + \frac{1}{r}v_r v_\varphi, \qquad \text{(2.77a and 2.77b)}$$

uses (1) the gradient operator (Equation I.11.31b) and (2) the polar components of the acceleration (Equations I.16.41a and I.16.41b). The velocity field of the vortex is azimuthal in direction with modulus depending only on the radius (Equation 2.78a), simplifying Equations 2.77a and 2.77b to Equations 2.78b and 2.78c:

$$\vec{v} = \vec{e}_\varphi v_\varphi(r): \quad -\frac{1}{\rho}\frac{\partial p}{\partial r} = -\frac{1}{r}[v_\varphi(r)]^2, \qquad \frac{\partial p}{\partial \varphi} = 0. \qquad \text{(2.78a–2.78c)}$$

Thus, *for an inviscid vortex (Equation 2.78a), the pressure gradient is radial and equal to the* **centrifugal force** *that equals the mass density times the square of the linear v_φ [angular ω (Equation 2.79a)] velocity divided (multiplied) by the radius [Equation 2.79b (Equation 2.79c)]:*

$$\omega(r) = \frac{1}{r}v_\varphi(r): \qquad \frac{dp}{dr} = \frac{\rho}{r}[v_\varphi(r)]^2 = \rho r[\omega(r)]^2. \qquad \text{(2.79a–2.79c)}$$

TABLE 2.1

Vortices with Smooth and Discontinuous Vorticity

Quantity		Velocity			Vorticity		
Dimensionless		$\bar{v}_\varphi = \frac{v_\varphi}{\Omega a}$			$\bar{\varpi} \equiv \frac{\varpi}{\Omega}$		
Radial Location		At the Core Radius	Peak Value	Location $\frac{r}{a} = \cdots$	At the Core Radius $r = a$	Peak Value	Location $\frac{r}{a} = \cdots$
Discontinuous[a]		1	1	1	$n + 2$	$n + 2$	1
Smooth[b]	n	$2^{-1-n/2}$	$(n+1)^{(n+1)/2}(n+2)^{-1-n/2}$	$\sqrt{n+1}$	$2^{-2-n/2}(n+2)$	$2^{1-n}n^{n/2}(1+n/2)(1+n/4)^{-2-n/2}$	$\frac{\sqrt{n}}{2}$
	$n = 0$	$\frac{1}{2} = 0.500$	$\frac{1}{2} = 0.5$	1	$\frac{1}{2} = 0.500$	2	0
	$n = 1$	$\frac{1}{2\sqrt{2}} = 0.354$	$\frac{2}{3\sqrt{3}} = 0.385$	$\sqrt{2} = 1.414$	$\frac{3}{4\sqrt{2}} = 0.530$	$\frac{48}{25\sqrt{5}} = 0.859$	$\frac{1}{2} = 0.500$
	$n = 2$	$\frac{1}{4} = 0.250$	$\frac{3\sqrt{3}}{16} = 0.325$	$\sqrt{3} = 1.732$	$\frac{4}{8} = 0.500$	$\frac{16}{27} = 0.593$	$\frac{1}{\sqrt{2}} = 0.707$
	$n = 3$	$\frac{1}{4\sqrt{2}} = 0.177$	$\frac{16}{25\sqrt{5}} = 0.286$	2	$\frac{5}{8\sqrt{2}} = 0.442$	$\frac{240}{343}\sqrt{\frac{3}{7}} = 0.458$	$\frac{\sqrt{3}}{2} 0.866$
	$n = 4$	$\frac{1}{8} = 0.125$	$\frac{25\sqrt{5}}{216} = 0.259$	$\sqrt{5} = 2.236$	$\frac{3}{8} = 0.375$	$\frac{3}{8} = 0.375$	1
	$n \to \infty$	0	0	∞	0	0	∞

Note: Comparison of the generalized Rankine (Hallock–Burnham) vortices that have discontinuous (smooth) vorticity distribution [Subsection 2.2.1 (Subsection 2.2.2)] with regard to the dimensionless velocity (Equation 2.63a) and vorticity (Equation 2.63b): (1) at the core radius $r = a$; (2) their peak value and the value of the dimensionless radius (Equation 2.63c) where it occurs. In the case of the generalized Hallock–Burnham vortex, explicit values are given for six values of the exponent; in the case of the generalized Rankine vortex, this is not needed because the value for arbitrary n is very simple.

[a] Subsection 2.2.1: generalized Rankine vortex.

[b] Subsection 2.2.2: generalized Hallock–Burnham vortex.

If the fluid is at rest at infinity with static pressure p_∞, the integration of Equation 2.79b specifies the pressure at all points:

$$p(r) - p_\infty = \int_\infty^r \frac{\rho}{\xi} [v_\varphi(\xi)]^2 \, d\xi. \tag{2.80}$$

For an incompressible fluid (Equation 2.81a), the mass density is constant and may be taken out of the integral (Equation 2.81b):

$$\text{rotational}; \quad \rho = \text{const:} \;\; p(r) = p_\infty + \rho \int_\infty^r \xi^{-1} [v_\varphi(\xi)]^2 \, d\xi. \tag{2.81a and 2.81b}$$

In the case of an irrotational flow, the pressure is given without integrations by the constancy of the stagnation pressure (Equation 2.82b), where in this case, Equation 2.82a is the static pressure for the fluid at rest at infinity:

$$\text{irrotational}; \qquad p_0 = p_\infty: \qquad p_0(r) = p_\infty - \frac{1}{2} \rho [v_\varphi(r)]^2. \tag{2.82a and 2.82b}$$

From Equations 2.81b and 2.82b, it follows that *the stagnation pressure in the vortex core:*

$$p_0(r) = p(r) - \frac{1}{2} \rho [v_\varphi(r)]^2 - \rho \int_\infty^r \xi^{-1} [v_\varphi(\xi)]^{-1} \, d\xi, \tag{2.83}$$

is generally a function of the radius, implying that it is constant azimuthally (varies radially), that is, along (across) the streamlines that are circles. These conclusions will be confirmed in Subsections 2.2.5 and 2.2.6 for vortex cores and subsequently in Subsection 2.3.3 for general plane vortical incompressible flow.

2.2.4 Pressure and Core Radius for a Vortex

In the case of a potential vortex with circulation Γ, the azimuthal velocity is given by Equation 2.33b ≡ Equation 2.84a, leading (Equation 2.81b) to the pressure (Equation 2.84b):

$$v_\varphi(r) = \frac{\Gamma}{2\pi r}: \qquad p_1(r) - p_\infty = \rho \left(\frac{\Gamma}{2\pi} \right)^2 \int_\infty^r \xi^{-3} \, d\xi = -\frac{\rho \Gamma^2}{8\pi r^2}, \tag{2.84a and 2.84b}$$

that corresponds (Equation 2.84a) to the conservation of the stagnation pressure (Equation 2.85):

$$p_\infty = p_1(r) + \frac{\rho \Gamma^2}{8\pi r^2} = p(r) + \frac{\rho}{2} [v_\varphi(r)]^2 = p_0. \tag{2.85}$$

The potential vortex flow applies outside the core of the generalized Rankine vortex (Equation 2.60a) specifying the pressure (Equation 2.86b) in terms of the angular velocity Ω instead (Equation 2.85 ≡ Equation 2.84b) of the circulation Γ:

$$p_1(r \geq a) = p_\infty - \frac{\rho\Gamma^2}{8\pi r^2} = p_\infty - \frac{\rho\Omega^2 a^4}{2r^2}. \qquad \text{(2.86a and 2.86b)}$$

The pressure (Equations 2.86a and 2.86b) in the generalized Rankine vortex decreases from infinity as the velocity increases:

$$p_1(r \geq a) \geq p_{1\min}(r \geq a) = p_1(a) = p_\infty - \frac{\rho\Gamma^2}{8\pi a^2} = p_\infty - \frac{\rho}{2}\Omega^2 a^2 \equiv p_{11}, \qquad \text{(2.87a and 2.87b)}$$

and takes the minimum value at the vortex core radius equal to the pressure at infinity minus the kinetic energy of rotation $(\rho/2)[v_\varphi(a)]^2$ at the core radius $v_\varphi(a) = \Omega a$. The condition of nonnegative pressure at the core radius (Equation 2.88a):

$$p_1(a) \geq 0: \quad a \leq \frac{1}{\Omega}\sqrt{\frac{2p_\infty}{\rho}} \equiv a_1 \qquad \text{(2.88a and 2.88b)}$$

sets a lower limit (Equation 2.88b) on the core radius for an external potential vortex flow. The constraint on the vortex core radius (Equation 2.88b) is expressed in terms of the angular velocity Ω, so that it can be compared with the cores of the generalized Rankine (Hallock–Burnham) vortices [Subsections 2.2.1, 2.2.4, and 2.2.5 (Subsections 2.2.2 and 2.2.6)]. The condition of near zero pressure in a liquid corresponds to the onset of cavitation (Equations I.28.6.2, I.28.7.2, and I.34.2.3), such as the mixing of bubbles and water, generating noise in the flow and causing stresses in nearby bodies.

2.2.5 Pressure in the Core of the Generalized Rankine Vortex

The pressure distribution for an incompressible rotational flow (Equation 2.81b) applies in the core of the generalized Rankine vortex (Equation 2.60b), integrating from the vortex radius:

$$p_1(r \leq a) - p_{11} = \rho\int_a^r \xi^{-1}[v_{1\varphi}(\xi)]^2\, d\xi = \rho\Omega^2 a^{-2n}\int_a^r \xi^{2n+1}\, d\xi, \qquad (2.89)$$

where the pressure is given by (Equation 2.87b):

$$p_1(r \leq a) = p_{11} - \frac{\rho}{2}\frac{\Omega^2 a^2}{n+1}\left[1 - \left(\frac{r}{a}\right)^{2n+2}\right] = p_\infty - \frac{\rho}{2}\Omega^2 a^2\left[1 + \frac{1-(r/a)^{2n+2}}{n+1}\right]$$

$$= p_\infty - \frac{\rho}{2}\Omega^2 a^2\frac{n+2}{n+1} + \frac{\rho}{2}\frac{\Omega^2 a^2}{n+1}\left(\frac{r}{a}\right)^{2n+2}. \qquad \text{(2.90a–2.90c)}$$

It follows that *the pressure in the generalized Rankine vortex is given by Equations 2.60a, 2.86a, and 2.86b outside [Equations 2.60b and 2.90a through 2.90c inside] the core radius. The Panel 2.3a for the exponents $n = 0,1,2,3,4$ shows that the pressure is almost constant inside the vortex for lower values of the exponent n. The pressure is minimum at the center:*

$$p_{10} = p_1(0) = p_{11} - \frac{\rho\Omega^2 a^2}{2n+2} = p_\infty - \frac{\rho}{2}\Omega^2 a^2 \frac{n+2}{n+1} = p_\infty - \frac{\rho\Gamma^2}{8\pi a^2}\frac{n+2}{n+1}, \tag{2.91}$$

where it is less than the pressure at infinity. The condition that the pressure is nonnegative at the center (Equation 2.92a) sets a minimum value (Equation 2.92b) on the core radius:

$$p_{10} \equiv p_1(0) \geq 0: \qquad a \geq \frac{1}{\Omega}\sqrt{\frac{2p_\infty}{\rho}}\sqrt{\frac{n+1}{n+2}} \equiv a_2 = a_1\sqrt{\frac{n+1}{n+2}} < a_1. \tag{2.92a and 2.92b}$$

This value is smaller than Equation 2.88b, and thus, $a \geq a_2$ is the dominant constraint for the core radius of the generalized Rankine vortex. The pressure field in the generalized Rankine vortex is given in dimensionless forms as a function of the radial distance (Equation 2.93a) divided by the core radius by

$$s \equiv \frac{r}{a}: \qquad 2\frac{p_\infty - p_1(r)}{\rho\Omega^2 a^2} \equiv \bar{p}_1(s) = \begin{cases} \dfrac{1}{s^2} & \text{if} \quad s \geq 1, \\ 1 + \dfrac{1 - s^{2n+2}}{n+1} & \text{if} \quad s \leq 1, \end{cases} \tag{2.93a–2.93c}$$

outside (Equation 2.86b ≡ Equation 2.93b) [inside (Equation 2.90b ≡ Equation 2.93c)] the core radius.

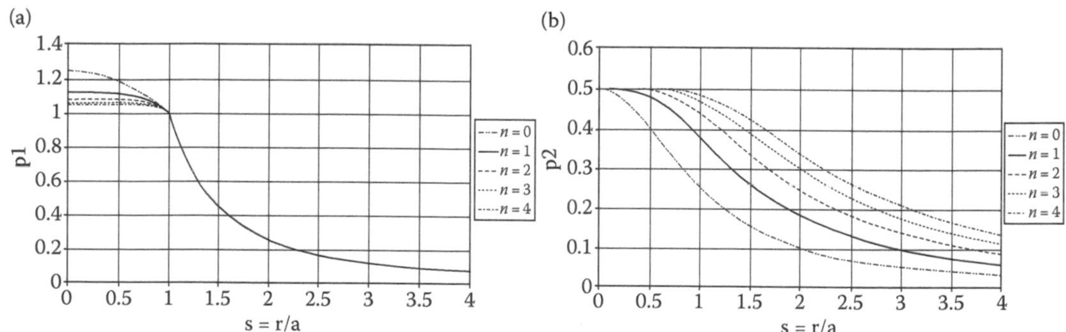

PANEL 2.3
Dimensionless pressure in the generalized (a) Rankine [(b) Hallock–Burnham] vortex [Equations 2.93b and 2.93c (Equation 2.100)] as a function of radial distance divided by the core radius (Equation 2.93a). In the case of the generalized Rankine vortex (a), the potential flow outside the core leads to the conservation of the stagnation pressure, and a rapid reduction of pressure toward the static pressure at infinity; in the core, the flow is rotational, the stagnation pressure is not conserved, and the pressure varies less, being almost constant for larger values of the exponent. In the case of the generalized Hallock–Burnham vortex (b), the pressure has a slower decay and forms a plateau for a small radius for smaller values of the exponent; the larger plateau corresponds to a slower decay of the pressure toward infinity. The curves in the Panel 2.3a (Panel 2.3b) with the vertical axis reversed to form a depression corresponding to the shape of the free surface of an incompressible fluid in a uniform gravity field containing a columnar generalized Rankine (Hallock–Burnham) vortex with vertical axis.

2.2.6 Pressure for the Generalized Hallock–Burnham Vortex

The tangential velocity profile is smooth (Equation 2.65a) for the generalized Hallock–Burnham vortex, leading to a single expression for the pressure (Equation 2.81b) at all radial distances:

$$p_2(r) - p_\infty = \rho\,\Omega^2 a^4 \int_\infty^r \xi^{2n+1}\left(a^2+\xi^2\right)^{-2-n} d\xi = \rho\,\Omega^2 a^2 I_n, \tag{2.94}$$

in terms of the dimensionless (Equation 2.95a) integral (Equation 2.95c) with upper limit Equation 2.95b:

$$\psi \equiv \frac{\xi}{a}, \psi_0 = \frac{r}{a}: \qquad I_n = \int_\infty^{\psi_0} \psi^{2n+1}(1+\psi^2)^{-2-n} d\psi. \tag{2.95a–2.95c}$$

A further change of variable (Equation 2.96a) simplifies the integral to Equation 2.96c with upper limit Equation 2.96b:

$$\begin{aligned} \psi \equiv \sinh\varphi, \quad \varphi_0 \equiv \arg\sinh\psi_0: \qquad I_n &= \int_\infty^{\varphi_0} \tanh^{2n+1}\varphi\,\mathrm{sech}^2\varphi\, d\varphi \\ &= \frac{\left[\tanh^{2n+2}\varphi\right]_\infty^{\varphi_0}}{(2n+2)} = \frac{\tanh^{2n+2}\varphi_0 - 1}{(2n+2)}. \end{aligned} \tag{2.96a–2.96c}$$

In the integrand of Equation 2.95c, the following were used:

$$\begin{aligned} \psi^{2n+1}(1+\psi^2)^{-2-n} d\psi &= \sinh^{2n+1}\varphi(\cosh^2\varphi)^{-2-n}\cosh\varphi\, d\varphi = \sinh^{2n+1}\varphi\cosh^{-2n-3}\varphi\, d\varphi \\ &= \tanh^{2n+1}\varphi\,\mathrm{sech}^2\varphi\, d\varphi = \frac{d(\tanh^{2n+2}\varphi)}{2n+2}. \end{aligned} \tag{2.97}$$

Substituting back the changes of variable (Equations 2.96b and 2.95b) evaluates the integral (Equation 2.96c as Equation 2.97 ≡ Equation 2.98):

$$(2n+2)I_n = \left(\frac{\psi_0^2}{1+\psi_0^2}\right)^{n+1} - 1 = \left(\frac{r^2}{r^2+a^2}\right)^{n+1} - 1. \tag{2.98}$$

It can be checked by direct differentiation:

$$\begin{aligned} \frac{d}{dr}[r^{2n+2}(r^2+a^2)^{-n-1}] &= r^{2n+1}(r^2+a^2)^{-n-2}\left[(2n+2)(r^2+a^2)+2(-n-1)r^2\right] \\ &= 2(n+1)a^2 r^{2n+1}(r^2+a^2)^{-2-n}, \end{aligned} \tag{2.99}$$

that Equation 2.99 is the primitive of Equation 2.94.

Substitution of Equation 2.98 in Equation 2.94 specifies *the pressure in the generalized Hallock–Burnham vortex at all radial distances:*

$$p_2(r) = p_\infty - \frac{\rho}{2}\frac{\Omega^2 a^2}{n+1}\left[1 - \left(\frac{r^2}{r^2 + a^2}\right)^{n+1}\right]. \tag{2.100}$$

The pressure is minimum at the center (Equation 2.102a) of the generalized Hallock–Burnham vortex:

$$p_{2\min} = p_2(0) = p_\infty - \frac{\rho}{2}\frac{\Omega^2 a^2}{n+1} \geq 0. \tag{2.101}$$

The condition that it be nonnegative (Equation 2.103a) specifies a minimum value for the core radius (Equation 2.103b).

$$p_2(0) \geq 0: \quad a \leq \sqrt{\frac{2p_\infty}{\rho}}\frac{\sqrt{n+1}}{\Omega} \equiv a_3 = a_2\sqrt{n+2} = a_1\sqrt{n+1} \geq a_2\sqrt{2}, a_1. \tag{2.102a and 2.102b}$$

The minimum core radius of the generalized Hallock–Burnham vortex (Equation 2.102b) is larger than both the smaller (Equation 2.92b) and larger (Equation 2.88b) minimum core radii for the generalized Rankine vortex; in the case $n = 0$ of the original Hallock–Burnham vortex, the minimum core radius (1) exceeds the lowest core radius of the original Rankine vortex $a_3 = a_2\sqrt{2}$, that is, the dominant constraint; and (2) coincides with the larger minimum core radius of the Rankine vortex $a_3 = a_1$, that is, the lesser constraint. The values of the three radii are compared in Table 2.2 for exponents $n = 0,1,2,3,4$ as complement to Table 2.1 for the velocity and vorticity of the same generalized Rankine and Hallock–Burnham vortices. The pressure (Equation 2.100) in the generalized Hallock–Burnham vortex is given by Equation 2.103b in dimensionless form (Equation 2.103a):

$$s \equiv \frac{r}{a}: \qquad \bar{p}_2(s) = 2\frac{p_\infty - p_2(r)}{\rho\Omega^2 a^2} = \frac{1}{n+1}\left[1 - \left(\frac{s^2}{1+s^2}\right)^{n+1}\right] = \frac{1}{n+1}\left[1 - \left(\frac{1}{1+1/s^2}\right)^{n+1}\right]. \tag{2.103a and 2.103b}$$

The Panel 2.3b for the exponents $n = 0,1,2,3,4$ shows that the pressure drop with distance to the vortex center is more rapid for lower n values.

2.2.7 Radial Variation of the Stagnation Pressure

For the generalized Rankine vortex, velocity is zero both at the center and at infinity, but the pressure (Equation 2.91) is not equal: this would not be possible in an irrotational homentropic flow, which satisfies the Bernoulli equation requiring that the stagnation

TABLE 2.2
Minimum Core Radius for Vortices

Vortex		**Generalized**		
		Rankine		**Hallock–Burnham**
		Subsections 2.2.1, 2.2.4, and 2.2.5		**Subsections 2.2.2, 2.2.6 and 2.2.7**
Section		**Inside the Core Radius**	**Outside the Core Radius**	**Everywhere**
Profiles	Velocity	Equation 2.60a	Equation 2.60b	Equation 2.65a
	Vorticity	Equation 2.62a	Equation 2.62b	Equation 2.66
	Pressure	Equations 2.90a–2.90c	Equation 2.86a	Equation 2.100
Dimensionless profiles	Velocity	Equation 2.63g	Equation 2.63f	Equation 2.76a
	Vorticity	Equation 2.63e	Equation 2.63d	Equation 2.76b
	Pressure	Equation 2.93c	Equation 2.93b	Equation 2.103b
Minimum core radii	General	a_1	a_2	a_3
	$\frac{1}{\Omega}\sqrt{\frac{2p_\infty}{\rho}}\times\cdots$	1	$\sqrt{\frac{n+1}{n+2}}$	$\sqrt{n+1}$
	$n=0$	1	$\frac{1}{\sqrt{2}}=0.707$	1
	$n=1$	1	$\sqrt{\frac{2}{3}}=0.816$	$\sqrt{2}=1.414$
	$n=2$	1	$\sqrt{\frac{3}{4}}=0.866$	$\sqrt{3}=1.732$
	$n=3$	1	$\sqrt{\frac{4}{5}}=0.894$	2.000
	$n=4$	1	$\sqrt{\frac{5}{6}}=0.913$	$\sqrt{5}=2.236$
	$n\to\infty$	1	1	∞
Stagnation pressure		Equation 2.104	p_∞	Equation 2.106

Note: Minimum core radii for (1 and 2) the generalized Rankine vortex, so that the pressure is nonnegative both in the core (a_2) and in the outer potential flow (a_1), the former being the dominant constraint $a_1 < a_1$ for all finite exponents, and $a_2 \to 1 = a_1$ as $n\to\infty$; (3) the generalized Hallock–Burnham vortex; it is a single smaller value $a_3 < a_2$. The core radii are given generally for any exponent n and in particular for five values.

pressure be the same everywhere. *The stagnation pressure inside the core of the generalized Rankine vortex is given (Equations 2.60b and 2.90b) by*

$$p_{01}(r\le a)=p_1(r\le a)+\frac{1}{2}\rho[v_{1\varphi}(r)]^2=p_\infty-\frac{\rho}{2}\Omega^2a^2\left[1+\frac{1-(r/a)^{2n+2}}{n+1}-\left(\frac{r}{a}\right)^{2n+2}\right]$$
$$=p_\infty-\frac{\rho}{2}\Omega^2a^2\frac{n+2}{n+1}\left[1-\left(\frac{r}{a}\right)^{2n+2}\right], \tag{2.104}$$

and is not constant: (1) it simplifies to Equation 2.91 ≡ Equation 2.105a on the axis:

$$p_{01}(0) = p_1(0) = p_\infty - \frac{\rho}{2}\Omega^2 a^2 \frac{n+2}{n+1} = p_{10}; \qquad p_{01}(a) = p_\infty; \tag{2.105a and 2.105b}$$

(2) at the core radius, it takes the same value (Equation 2.105b) as at infinity (Equation 2.105b) because the flow outside the core is potential.

The stagnation pressure in the generalized Hallock–Burnham vortex is given (Equations 2.65a and 2.100) by

$$\begin{aligned} p_{02}(r) &= p_2(r) + \frac{1}{2}\rho[v_{2\varphi}(r)]^2 \\ &= p_\infty - \frac{\rho}{2}\frac{\Omega^2 a^2}{n+1}\left[1 - \left(\frac{r^2}{r^2+a^2}\right)^{n+1}\right] + \frac{\rho}{2}\frac{\Omega^2 a^4 r^{2n+2}}{(r^2+a^2)^{n+2}} \\ &= p_\infty - \frac{\rho}{2}\frac{\Omega^2 a^2}{n+1} + \frac{\rho}{2}\Omega^2 a^2 \left(\frac{r^2}{r^2+a^2}\right)^{n+1}\left(\frac{1}{n+1} + \frac{a^2}{r^2+a^2}\right). \end{aligned} \tag{2.106}$$

The stagnation pressure coincides with the static pressure at infinity (Equation 2.107a):

$$p_{20}(\infty) = p_\infty, \qquad p_{20}(0) = p_\infty - \frac{\rho}{2}\frac{\Omega^2 a^2}{n+1}, \tag{2.107a and 2.107b}$$

and on axis is given by Equation 2.107b. The stagnation pressure at the center for the generalized Hallock–Burnham (Equations 2.102b and 2.107b) [Rankine (Equations 2.92b and 2.95a)] vortex vanishes for the minimum core radius [Equations 2.108a and 2.108b (Equations 2.108a and 2.108d)]:

$$a = a_3: \quad p_{20}(0) = 0; \quad a = a_2: \quad p_{10}(0) = 0, \tag{2.108a–2.108d}$$

because (1) the flow velocity is zero at the center and (2) the minimum core radius corresponds to zero pressure at the center.

2.2.8 Vorticity and the Gradient of the Stagnation Pressure

The stagnation pressure in the generalized Rankine vortex (Equation 2.104) has a radial gradient (Equation 2.109b) inside the core (Equation 1.109a):

$$r \le a: \qquad \frac{dp_{01}}{dr} = \rho\Omega^2(n+2)\frac{r^{2n+1}}{a^{2n}} = \rho v_{1\varphi}(\mathrm{r})\varpi_1(r), \tag{2.109a and 2.109b}$$

where tangential velocity (Equation 2.60b) and vorticity (Equation 2.62a) were used; likewise, the stagnation pressure in the generalized Hallock–Burnham vortex (Equation 2.106) has a radial gradient:

$$\begin{aligned}\frac{dp_{02}}{dr} &= \frac{\rho}{2}\Omega^2 a^2 \frac{d}{dr}\left\{ r^{2n+2}\left[\frac{(r^2+a^2)^{-n-1}}{n+1} + a^2(r^2+a^2)^{-n-2}\right]\right\} \\ &= \rho\Omega^2 a^2 r^{2n+1}\left\{(r^2+a^2)^{-n-1} + (n+1)a^2(r^2+a^2)^{-n-2}\right. \\ &\quad \left. - r^2(r^2+a^2)^{-n-2} - (n+2)a^2r^2(r^2+a^2)^{-n-3}\right\} \\ &= \rho\Omega^2 a^2 r^{2n+1}(r^2+a^2)^{-n-3}\left\{(r^2+a^2)^2 + [(n+1)a^2 - r^2](r^2+a^2) - (n+2)a^2r^2\right\} \\ &= \rho\Omega^2 a^6 r^{2n+1}(r^2+a^2)^{-n-3}(n+2) = \rho v_{2\varphi}(r)\varpi_2(r),\end{aligned} \tag{2.110}$$

using the tangential velocity (Equation 2.65a) and vorticity (Equation 2.66). Thus, *for the core (the whole flow region) of the generalized Rankine (Equations 2.109a and 2.109b) [Hallock–Burnham (Equation 2.110)] vortex, the gradient of the stagnation pressure is radial and equals the product of the mass density by the tangential velocity and vorticity:*

$$\nabla p_0 = \vec{e}_r \rho v_\varphi(r)\varpi(r), \tag{2.111}$$

corresponding to minus the vortical force, as shown next.

2.2.9 Vortical and Local Inertia Forces

Using the identity:

$$\frac{\partial}{\partial x_i}\left(\frac{v^2}{2}\right) = \frac{\partial}{\partial x_i}\left(\frac{1}{2}v_j v_j\right) = v_j\frac{\partial v_j}{\partial x_i} = v_j\frac{\partial v_i}{\partial x_j} + v_j\left(\frac{\partial v_j}{\partial x_i} - \frac{\partial v_i}{\partial x_j}\right), \tag{2.112}$$

in vector form:

$$\nabla\left(\frac{v^2}{2}\right) = (\vec{v}.\nabla)\vec{v} + \vec{v}\wedge(\nabla\wedge\vec{v}), \tag{2.113}$$

the inviscid momentum Equation 2.7 becomes

$$-\nabla p = \rho\vec{a} = \rho\left[\frac{\partial\vec{v}}{\partial t} + (\vec{v}.\nabla)\vec{v}\right] = \rho\left[\frac{\partial\vec{v}}{\partial t} + \nabla\left(\frac{v^2}{2}\right) + (\nabla\wedge\vec{v})\wedge\vec{v}\right]. \tag{2.114}$$

In the case of an incompressible fluid (Equation 2.10a ≡ Equation 2.115a), introducing the stagnation pressure (Equation 2.12b ≡ Equation 2.115b) and the vorticity (Equation 2.115c):

$$\rho = \text{const}, \qquad p_0 = p + \frac{1}{2}\rho v^2, \qquad \vec{\varpi} \equiv \nabla \wedge \vec{v}, \tag{2.115a–2.115c}$$

it follows from Equation 2.114 that

$$-\nabla p_0 = -\nabla p - \nabla\left(\frac{1}{2}\rho v^2\right) = \rho\left(\frac{\partial \vec{v}}{\partial t} + \vec{\varpi} \wedge \vec{v}\right). \tag{2.116}$$

It has been shown that *in the incompressible (Equation 2.115a) flow of an inviscid fluid, the momentum equation (Equation 2.5b ≡ Equation 2.7 ≡ Equation 2.114) implies that minus the gradient of the stagnation pressure (Equation 2.115b ≡ Equation 2.12b) is equal to the sum (Equation 2.116 ≡ Equation 2.117a) of (1) the* **local inertia force** *density, that is, the product of the mass density by the local acceleration or partial derivative of the velocity with regard to time (Equation 2.117b); (2) the* **vortical force** *density (Equation 2.117c ≡ Equation I.28.50b), that is, the outer vector product of the vorticity by the velocity multiplied by the mass density:*

$$-\nabla p_0 = \vec{f}_a + \vec{f}_\ell: \qquad \vec{f}_a \equiv \rho \frac{\partial \vec{v}}{\partial t}, \quad \vec{f}_\ell \equiv \rho \vec{\varpi} \wedge \vec{v}. \tag{2.117a–2.117c}$$

In the case of a steady flow (Equation 2.118a), only the vortical force remains (Equation 2.118b):

$$\partial \vec{v}/\partial t = 0: \qquad \nabla p_0 = -\vec{f}_\ell = -\rho \vec{\varpi} \wedge \vec{v}, \quad \vec{v}.\nabla p_0 = 0 = \vec{\varpi}.\nabla p_0; \tag{2.118a–2.118d}$$

and thus *(3) the stagnation pressure is constant along the streamlines (Equation 2.118c) and vortex lines (Equation 2.118d); and (4) it can vary only in the direction normal to both (Equation 2.118b).* In a plane flow, the vortex lines are orthogonal to the plane. For a plane vortex, the velocity is tangential (Equation 2.119a) and the vorticity is normal to the plane (Equation 2.119b), leading to the gradient of the stagnation pressure (Equation 2.119c):

$$\vec{v} = \vec{e}_\varphi v_\varphi, \ \vec{\varpi} = \varpi \vec{e}_3: \qquad \nabla p_0 = -\rho v_\varphi \varpi \vec{e}_3 \wedge \vec{e}_\varphi = \rho v_\varphi \varpi \vec{e}_r, \tag{2.119a–2.119c}$$

in agreement with Equation 2.111 ≡ Equation 2.119c.

2.2.10 Columnar Vortex in the Gravity Field

The vortices in the atmosphere and ocean are affected by the gravity force (Equation 2.120a), that appears in the inviscid momentum equation (Equation 2.114 ≡ Equation 2.120b) together with minus the pressure gradient balancing the inertia force:

$$\vec{f}_g = \rho \vec{g}, \qquad \vec{f}_g - \nabla p = \rho \vec{a}. \tag{2.120a and 2.120b}$$

Taking uniform gravity vertical downward (Equation 2.120c), the corresponding gravity potential is Equation 2.120d:

$$\vec{f}_g = -\rho g \vec{e}_z = -\frac{d\phi}{dz}\vec{e}_z, \quad \phi = \rho g z. \tag{2.120c and 2.120d}$$

The latter appears in the vertical component of the momentum (Equation 2.120b ≡ Equation 2.121a) adding to the pressure (Equation 2.121b):

$$-\rho a_z = \frac{\partial p}{\partial z} - f_{gz} = \frac{\partial(p+\phi)}{\partial z} = \frac{\partial \bar{p}}{\partial z}, \qquad \bar{p} = p + \phi = p + \rho g z. \tag{2.121a and 2.121b}$$

Thus *the incompressible Bernoulli equation (Equation 2.12b) in the presence of a uniform gravity field (Equations 2.120c and 2.120d) with acceleration g becomes 2.121d where* p_0 *is the stagnation pressure:*

$$p \to p + \rho g z: \quad p_0 = \bar{p} + \frac{1}{2}\rho r^2 = p + \rho g z + \frac{1}{2}\rho v^2; \tag{2.121c and 2.121d}$$

the effect of uniform gravity (Equations 2.120c and 2.120d) is to add to the pressure the weight of column of fluid of density p and height z corresponding to the transformation (Equation 2.121b ≡ Equation 2.121c).

2.2.11 Shape of the Free Surface of a Swirling Liquid

In the case of *an incompressible fluid in a uniform gravity field containing a generalized Hallock–Burnham vortex (Equation 2.100) with vertical axis the transformation (Equation 2.121c) leads to the pressure (Equation 2.122a):*

$$p(r,z) = p_\infty - \rho g z - \frac{\rho}{2}\frac{\Omega^2 a^2}{n+1}\left[1 - \left(\frac{r^2}{r^2+a^2}\right)^{n+1}\right]. \tag{2.122a}$$

The pressure equals the pressure at infinity (Equation 2.122b) at the free surface (Equation 2.122c):

$$p(r,z) = p_\infty: \quad z(r) = -\frac{\rho}{2}\frac{\Omega^2 a^2}{n+1}\left[1 - \left(\frac{r^2}{r^2+a^2}\right)^{n+1}\right]; \quad z_{\min} = z(0) = -\frac{\rho\Omega^2 a^2}{2n+2}, \tag{2.122b–2.122d}$$

the free surface is a depression (Equation 2.122c) with lowest point on the axis at depth given by Equation 2.122d.

In the case of *an incompressible fluid in a uniform gravity field containing a generalized Rankine vortex with vertical axis the transformation (Equation 2.121c) leads to the pressure [Equation 2.90c (Equation 2.86b)] inside (outside) the vortex core [Equation 2.123a (Equation 2.123b)]:*

$$p(r,z) = p_\infty - \rho g z - \frac{\rho}{2}\Omega^2 a^2 \times \begin{cases} \dfrac{n+2}{n+1} - \dfrac{(r/a)^{2n+2}}{n+1} & \text{if} \quad r \le a, \qquad (2.123a) \\ \dfrac{a^2}{r^2} & \text{if} \quad r \ge a. \qquad (2.123b) \end{cases}$$

The pressure equals the pressure at infinity on the free surface (Equation 2.122b), leading to its shape inside (outside) the vortex core radius [Equation 2.123c (Equation 2.123d)]:

$$z(r) = -\frac{\Omega^2 a^2}{2g} \times \begin{cases} \dfrac{n+2}{n+1} - \dfrac{(r/a)^{2n+2}}{n+1} & \text{if} \quad r \le a, \qquad (2.123c) \\ (a/r)^2 & \text{if} \quad r \ge a. \qquad (2.123d) \end{cases}$$

The shape of the free surface is a depression with lowest point on the axis at the depth (Equation 2.123e) that is lower than the depth (Equation 2.123f) at the vortex core radius:

$$z_{\min} = z(0) = -\frac{\Omega^2 a^2}{2g}\frac{n+2}{n+1} < -\frac{\Omega^2 a^2}{2g} = z(a). \qquad (2.123e \text{ and } 2.123f)$$

The shape of the free surface corresponding to the generalized Rankine (Equations 2.123c and 2.123d) [Hallock–Burnham (Equation 2.122c)] corresponds to the curves in the Panel 2.4a (Panel 2.4b) with the vertical axis reversed to represent a depression.

2.2.12 Free Surfaces, Tornadoes, and Wake Vortices

Vortices form naturally in the atmosphere and oceans due to the rotation of the earth. For example a warm column of moist air rising from the surface of the ocean due to evaporation of sea water under the tropical sun will tend to rotate entraining dry ambient air. This leads to the formation of a columnar vortex that becomes an hurricane if the circulation is large enough. The depression on the surface of the ocean due to a strong vortex may generate large waves that can propagate long distances. To the surface storm are added the winds in the atmosphere associated with the swirling motion around the vortex core. The vorticity is dissipated only by the shear viscosity (Note 4.6) that is small for air, and thus strength of the hurricane decays slowly when traveling across the ocean, since the surface friction is also small. Thus hurricane can travel long distances over the ocean, and only starts to decay more significantly after landfall, due to surface friction on the ground. The destructive effect of the hurricane is due to (1) the suction force associated with the low pressure in the core that pulls up roofs and uproots trees; (2) the large tangential velocities around the core that spins debris until they move sufficiently far to fall under gravity. The path of the hurricane is determined by the vortical force (Equation 2.117c) that depends on wind direction, and can be difficult to predict in variable conditions.

The vortices issuing from the wing tips of an aircraft (Figure I.34.7) are a man-made phenomenon that can affect following aircraft by causing unwanted motion, such as rolling around the longitudinal axis. The wake vortex strength or circulation of the leading aircraft is proportional to the lift (Equation I.28.29b ≡ Equation 2.233a) and thus is larger for a heavy aircraft flying at larger angle-of-attack, for example during the climb after take-off or the descent on the approach to land. The wake vortices decay slowly due to the small viscosity of air and can persist in the atmosphere for some time. The path of the wake vortices is determined by the vortical force (Equation 2.117c) and thus is affected by: (1) the prevailing winds; (2) the vortex images on the ground and obstacles (Sections I.16.2, I.16.6 and Subsection I.36.3.1). The wake vortices can sink to the ground (Subsection 2.7.4) and rebound. If a following aircraft on the take-off or landing sequence encounters the wake vortices of the leading aircraft it may or may not have sufficient control power to compensate for the unwanted motion. Thus the safe separation distance due to wake vortices is larger for a heavier leading aircraft and lighter following aircraft, and is strongly dependent on local conditions like wind direction and ground topography. Besides the generalized Hallock–Burnham (Rankine) vortices (Section 2.2) another vortex with continuous (discontinuous) vorticy profile is considered in the Example 10.3 (Example 10.4).

2.3 Minimum Energy (Thomson 1849) and Intrinsic Equations of Motion

In the particular case of vortices (Section 2.2), it was shown that in the plane, steady, incompressible flow of an inviscid fluid, the stagnation pressure (1) is constant along streamlines and (2) varies across the streamlines in proportion to minus the vortical force, that is, as the product of the mass density by the vorticity by the velocity. This is the particular two-dimensional case of a more general three-dimensional theorem (Subsection 2.2.9); it can also be proved (Subsection 2.3.3) using the momentum equation (Subsection 2.3.1) involving the vorticity (Subsection 2.3.2) in intrinsic coordinates, that is, in curvilinear coordinates locally tangent and normal to the streamlines. Thus, one distinction between the plane steady incompressible flow of an inviscid fluid in the irrotational (rotational) case is that the stagnation pressure is conserved everywhere (only along streamlines). Another distinction between incompressible rotational and irrotational flows with the same normal velocity at the boundary of a closed region is that the total kinetic energy in the interior is smaller in the latter case (Subsection 2.3.4); thus, adding vorticity always increases the kinetic energy; this result is valid for an inviscid or viscous fluid in the plane or space.

2.3.1 Momentum Equation in Intrinsic Coordinates

The streamlines are the curves to which the velocity is tangent and, for a steady flow, correspond to the paths of fluid particles. The **intrinsic coordinates** along the streamlines lie (Figure 2.2) (1) along the arc length s and (2) in the inward normal direction n. The inviscid momentum equation (Equation 2.4b), in the presence of an external force density per unit volume such as gravity (Equations 2.120a and 2.120b), is given by Equation 2.124a:

$$\vec{f} - \nabla p = \rho\vec{a}: \qquad f_s - \frac{\partial p}{\partial s} = \rho a_s \qquad f_n - \frac{\partial p}{\partial s} = \rho a_n. \qquad (2.124\text{a–}2.124\text{c})$$

The external $\vec{f}$ and inertia $\rho\vec{a}$ forces can be written in intrinsic coordinates (Equations 2.124a through 2.124c) in terms of the tangential a_s and normal a_n components of the acceleration. The latter can be determined starting with the velocity: (1) the path tangent to the velocity is locally equivalent to a circle whose radius $r = R$ is the **radius of curvature**

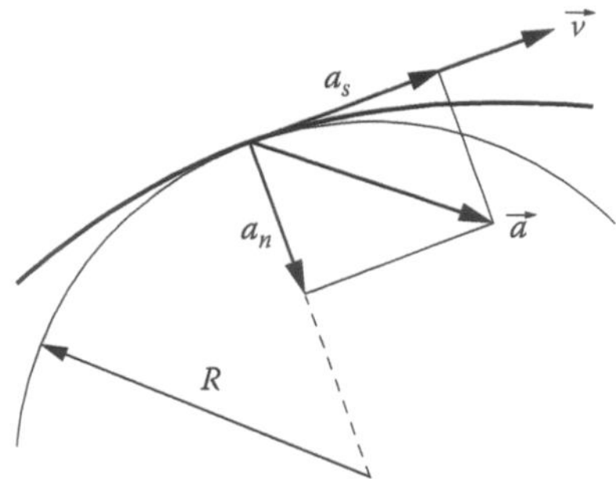

FIGURE 2.2
Curve such as the path of a fluid particle or a streamline of a flow can be approximated locally by a circle with a radius equal to the radius of curvature. Choosing intrinsic coordinates along the tangent and normal to the curve, the velocity has only a tangential component. Acceleration has two components: (1) tangential equal to the rate of change with regard to the arc length of half the square of the modulus of the velocity; and (2) normal equal to the square of the velocity divided by the radius of curvature, with a centripetal direction, that is, toward the center. The tangential acceleration is zero only if the modulus of velocity is constant (2.2), that is, for uniform motion; the normal acceleration is zero only if the radius of curvature is infinite, that is, for a straight line.

(Equation 2.125a); (2) the velocity is tangential (Equation 2.125b) and equal to the arc length covered to per unit time (Equation 2.125c); and (3) hence, it has only one component (Equation 2.125d) in intrinsic coordinates:

$$r = R, \quad -v_n = v_r = 0, \quad -v_\varphi = v_s = \frac{\mathrm{d}s}{\mathrm{d}t} = v, \quad \vec{v} = \vec{e}_s v = \vec{e}_s \frac{\mathrm{d}s}{\mathrm{d}t}. \tag{2.125a–2.125d}$$

Note that the radial (polar) component of a vector is opposite to the inward normal (Equation 2.125b) [tangent in the direction of motion (Equation 2.125c)]. The polar components of the acceleration (Equations 2.77a and 2.77b ≡ Equations I.16.41a and I.16.41b) applied to Equations 2.125a through 2.125c show that (1) the tangential acceleration is the rate of change of the tangential velocity with time, that in the steady case is given (Equation 2.126a) by the derivative with regard to the arc length of half the square of the modulus of the velocity; and (2) the normal acceleration is purely centripetal (Equation 2.126b) and equal to the square of the modulus of the velocity divided by the radius of curvature:

$$a_s \equiv -a_\varphi = -\frac{\mathrm{d}v_\varphi}{\mathrm{d}t} - \frac{v_r v_\varphi}{r} = \frac{\mathrm{d}v_s}{\mathrm{d}t} = \frac{\mathrm{d}v}{\mathrm{d}t} = \frac{\mathrm{d}v}{\mathrm{d}s}\frac{\mathrm{d}s}{\mathrm{d}t} = v\frac{\mathrm{d}v}{\mathrm{d}s} = \frac{\mathrm{d}}{\mathrm{d}s}\left(\frac{v^2}{2}\right), \tag{2.126a}$$

$$a_n \equiv -a_r = -\frac{\mathrm{d}v_r}{\mathrm{d}t} - \frac{v_\varphi^2}{R} = -\frac{v^2}{R}. \tag{2.126b}$$

Substituting Equations 2.127a and 2.127b in Equations 2.125b and 2.125c leads to

$$f_s - \frac{\partial p}{\partial s} = \rho a_s = \rho v \frac{\mathrm{d}v}{\mathrm{d}s}, \qquad f_n - \frac{\partial p}{\partial n} = \rho \frac{v^2}{R}, \tag{2.127a and 2.127b}$$

the inviscid momentum equation in intrinsic coordinates, involving the components of the external force density per unit volume $\vec{f}$, normal f_n, and parallel f_s to the velocity (Figure 2.2), minus the pressure gradient, balanced against the inertia force, that is, mass density ρ times acceleration $\vec{a}$, with tangential (Equation 2.126a) and normal (Equation 2.126b) components.

2.3.2 Vorticity in Terms of Intrinsic Coordinates

The acceleration is related to the vorticity (Equation 2.114 ≡ Equation I.14.19) in a steady flow (Equation 2.128a) by Equation 2.128b:

$$\frac{\partial \vec{v}}{\partial t} = 0: \quad \vec{a} - \nabla\left(\frac{v^2}{2}\right) = \vec{\varpi} \wedge \vec{v} = \varpi\, \vec{e}_3 \wedge v\vec{e}_s = \varpi\, v\, \vec{e}_n, \qquad \text{(2.128a and 2.128b)}$$

where (1) the velocity is tangent to trajectory with unit vector $\vec{e}_s$; (2) the vorticity lies along the binormal vector that is orthogonal to the osculating plane (s,n) of the trajectory with unit vector $\vec{e}_3$; and (3) their outer product lies along the normal with unit vector $\vec{e}_n$. Substituting in Equation 2.128b the intrinsic components of the acceleration (Equations 2.126a and 2.126b) leads to

$$a_s = \frac{1}{2}\frac{\partial(v^2)}{\partial s} = v\frac{\partial v}{\partial s}, \quad a_n = \frac{1}{2}\frac{\partial(v^2)}{\partial n} + \varpi v = v\left(\frac{\partial v}{\partial n} + \varpi\right), \qquad \text{(2.129a and 2.129b)}$$

where (1) there is coincidence of the tangential components (Equation 2.129a ≡ Equation 2.126a); and (2) comparison of the normal components (Equation 2.129b) with Equation 2.126b specifies the vorticity:

$$\varpi = \frac{a_n}{v} - \frac{\partial v}{\partial n} = \frac{v}{R} - \frac{\partial v}{\partial n}, \qquad \vec{\varpi} = \varpi \vec{e}_3. \qquad \text{(2.130a and 2.130b)}$$

The vorticity is given in intrinsic coordinates by Equation 2.130a in terms of the tangential velocity v, its normal derivative $\partial v/\partial n$, and the radius of curvature R; the vorticity is normal (Equation 2.130b) to the osculating plane of the trajectory. The first term in Equation 2.130a is the angular velocity (Equation 2.79a) for a circular trajectory, that is, with constant radius of curvature; the effect of variable radius of curvature along a noncircular trajectory appears in the second term on the r.h.s. of Equation 2.130a.

2.3.3 Variation of the Stagnation Pressure across Streamlines due to the Vorticity

The inviscid momentum equation in intrinsic coordinates (Equations 2.127a and 2.127b) for a steady (Equation 2.131a) incompressible (Equation 2.131b) flow and conservative external forces that derive from a potential (Equation 2.131c) is given by Equations 2.131d and 2.131e:

$$\frac{\partial \vec{v}}{\partial t} = 0, \qquad \rho = \text{const}, \qquad \vec{f} = -\nabla\phi: \qquad \frac{\partial(\rho v^2 + p + \phi)}{\partial s} = 0, \qquad \text{(2.131a–2.131d)}$$

$$\frac{\partial(\rho v^2 + p + \phi)}{\partial n} = -\rho\frac{v^2}{R} + \rho v\frac{\partial v}{\partial n} = -\rho v \varpi, \qquad \text{(2.131e)}$$

where the vorticity (Equation 2.130a) was introduced, showing that the stagnation pressure (Equation 2.132a) (1) is conserved along the streamlines (Equation 2.131d ≡ Equation 2.132b):

$$p_0 \equiv p + \frac{1}{2}\rho v^2 + \phi: \qquad \frac{\partial p_0}{\partial s} = 0; \qquad \frac{\partial p_0}{\partial n} = -\rho v \varpi = f_e; \tag{2.132a–2.132c}$$

and (2) varies across streamlines in proportion to the vorticity, velocity, and mass density (Equation 2.132c corresponding to Equations 2.119a through 2.119c) to the vortical force (Equation 2.117c). In an irrotational flow $\varpi = 0$, the stagnation pressure (Equation 2.132a) is conserved everywhere $p_0 =$ const, in agreement with the Bernoulli theorem [(Equation 2.12b) without external forces $\phi = 0$]. The derivative of the stagnation pressure (Equation 2.132c) across streamlines is nonvanishing due to the vorticity and corresponds to (Equation 2.133a) the Kutta–Joukowski theorem (Equation I.28.29b) on the lift (Equation 2.133b) due to the circulation (Equation 2.133c).

$$\frac{F_y}{\Gamma} = -\rho v = \frac{1}{\varpi}\frac{\partial p_0}{\partial n}, \qquad F_y \leftrightarrow \frac{\partial p_0}{\partial n}, \qquad \Gamma \leftrightarrow \varpi. \tag{2.133a–2.133c}$$

Another distinction between rotational and irrotational flow relates to the kinetic energy and is considered next.

2.3.4 Potential Flow as the Minimum of the Kinetic Energy (Kelvin 1849)

Consider an incompressible rotational flow, whose velocity (Equation 2.134a) consists of a potential and rotational part:

$$\vec{v} = \nabla\Phi + \vec{u}, \qquad \nabla.\vec{u} = 0, \qquad 0 = \left[v_N - \frac{\partial \Phi}{\partial N} \right] = u_N \Big|_{\partial D}, \tag{2.134a–2.134c}$$

where the rotational part is incompressible (Equation 2.134b) and does not change the normal velocity on the closed boundary (Equation 2.134c) or vessel containing the fluid. The total kinetic energy, the domain D, in the interior of the closed boundary surface ∂D:

$$E = \frac{1}{2}\rho \int_D \left|\vec{v}\right|^2 dS = \frac{1}{2}\rho \int (\nabla\Phi + \vec{u}).(\nabla\Phi + \vec{u}) dS = E_1 + E_2 + E_{12} \tag{2.135}$$

is the sum of three parts: (1) the interaction energy (Equation 2.136a) that will be shown to vanish:

$$E_{12} \equiv \rho \int_D (\vec{u}.\nabla\Phi) dS = 0; \quad E_2 \equiv \frac{1}{2}\rho \int_D \left|\vec{u}\right|^2 dS > 0, \quad E_1 \equiv \frac{1}{2}\rho \int (\nabla\phi)^2 ds = E - E_2 \le E; \tag{2.136a–2.136c}$$

(2) the kinetic energy of the rotational velocity perturbation (Equation 2.136b) that is positive; and (3) the kinetic energy of the potential flow (Equation 2.136c). The latter is thus the smallest possible value of the total kinetic energy, obtained when $E_2 = 0$, that is, $\vec{u} = 0$,

implying that the flow is potential. This corresponds to **Kelvin theorem (1849):** *of all incompressible (Equation 2.135b) flows (Equation 2.135a) with a given normal velocity on the boundary (Equation 2.135c), the potential flow has less kinetic energy (Equation 2.136c) than any rotational flow (Equation 2.136b).*

The proof of the theorem rests on the interaction energy (Equation 2.136a) being zero, as will be shown next. The integrand is (Equation 2.137a) in vector notation:

$$\vec{u}.\nabla\Phi = \nabla.(\vec{u}\Phi) - \Phi(\nabla.\vec{u}), \tag{2.137a}$$

that is equivalent to

$$u_x \frac{\partial \Phi}{\partial x} + u_y \frac{\partial \Phi}{\partial y} = \frac{\partial(u_x\Phi)}{\partial x} + \frac{\partial(u_y\Phi)}{\partial y} - \Phi\left(\frac{\partial u_x}{\partial x} + \frac{\partial u_y}{\partial y}\right). \tag{2.137b}$$

Concerning the former (Equation 2.137a): (1) the second term on the r.h.s. vanishes on account of Equation 2.134b ≡ Equation 2.138a:

$$\nabla.\vec{u} = 0: \qquad \int_D (\vec{u}.\nabla\Phi)\mathrm{d}S = \int_D \nabla.(\vec{u}\Phi)\mathrm{d}S = \int_D \Phi(\vec{u}.\vec{N})\mathrm{d}s = \int_{\partial D} \Phi u_N \,\mathrm{d}s = 0;$$

(2.138a and 2.138b)

(2) the divergence theorem (Equation I.28.1b) can be applied to the first term on the r.h.s. of Equation 2.138b, transforming the integral over the domain D of area $\mathrm{d}S$ and normal $\vec{N}$ into an integral along the boundary ∂D with arc length ds; and (3) the latter (Equation 2.138b) vanishes because the rotational velocity perturbation has zero normal component (Equation 2.134c) on the boundary. Thus, adding vorticity to a potential flow (Equation 2.134a) without changing the normal velocity on the boundary (Equation 2.134b) increases the kinetic energy (Equations 2.135, 2.136b, and 2.136c) in incompressible conditions; the result holds for a viscous or inviscid fluid because the momentum equation was not used at all.

2.4 Laplace/Poisson Equations in Complex Conjugate Coordinates

Continuing with the distinctions between plane flows (Subsection 2.4.1): (1) an incompressible (irrotational) flow has a stream function (potential) that satisfies a Poisson equation forced by the vorticity (dilatation); and (2) a potential flow is both incompressible and irrotational, and hence has both a stream function and a potential satisfying the Laplace equation, leading to a complex potential. The complex conjugate coordinates (Subsection 2.4.2) lead readily to the general (complete) integral of the Laplace (Poisson) equation [Subsection 2.4.3 (Subsection 2.4.4)]. The Cartesian (polar) components of the flow velocity, or their combination as the complex conjugate velocity, can be expressed [Subsection 2.4.6 (Subsection 2.4.9)] in terms of the following: (1) the complex potential, real potential, or stream function for a potential flow; (2 and 3) only the real potential (stream function) exists for a compressible irrotational (incompressible rotational) flow; and (4) for a compressible rotational flow, both the potential and stream function are needed. The simplest incompressible rotational (irrotational) flow is [Subsection 2.4.5 (Subsection 2.4.7)] the unidirectional shear flow (uniform flow with angle of attack). For the flow due to a source/sink, vortex, or their combination in a monopole (Subsection 2.4.11), the relation between polar and complex conjugate coordinates is used (Subsection 2.4.8), specifying also the Laplace operator in polar coordinates (Subsection 2.4.9).

2.4.1 Incompressible, Irrotational, and Potential Flows

In a plane irrotational flow, the vorticity vanishes (Equation 2.8a ≡ Equation 2.139a) and the velocity is the gradient of the potential (Equation 2.8b ≡ Equations 2.139b and 2.139c):

$$\text{irrotational:} \qquad 0 = \varpi = \frac{\partial v_y}{\partial x} - \frac{\partial v_x}{\partial y}, \quad \Leftrightarrow \quad v_x = \frac{\partial \Phi}{\partial x}, \quad v_y = \frac{\partial \Phi}{\partial y}; \qquad (2.139a\text{–}2.139c)$$

the potential satisfies the Poisson equation forced by the dilatation:

$$\dot{D}_2 \equiv \nabla.\vec{u} = \frac{\partial u_x}{\partial x} + \frac{\partial u_y}{\partial y} = \frac{\partial^2 \Phi}{\partial x^2} + \frac{\partial^2 \Phi}{\partial y^2} \equiv \nabla^2 \Phi. \qquad (2.140)$$

In a plane incompressible flow (Equation 2.10a), the dilatation is zero (Equation 2.141a), implying the existence of a stream function (Equations 2.141b and 2.141c):

$$\text{incompressible:} \quad 0 = \dot{D}_2 = \frac{\partial u_x}{\partial x} + \frac{\partial u_y}{\partial y} = 0 \quad \Leftrightarrow \quad v_x = \frac{\partial \Psi}{\partial y}, \; u_y = -\frac{\partial \Psi}{\partial x}; \qquad (2.141a\text{–}2.141c)$$

the stream function satisfies a Poisson equation forced by minus the vorticity:

$$-\varpi = \frac{\partial v_x}{\partial y} - \frac{\partial v_y}{\partial x} = \frac{\partial^2 \Psi}{\partial y^2} + \frac{\partial^2 \Psi}{\partial x^2} \equiv \nabla^2 \Psi. \qquad (2.142)$$

A potential flow is irrotational (Equation 2.143a) [incompressible (Equation 2.143b)] and thus has a potential (Equation 2.143c) [stream function (Equation 2.143d)] satisfying the Laplace equation:

$$\text{potential:} \quad \varpi = 0 = \dot{D}_2 \quad \Leftrightarrow \quad \nabla^2 \Phi = 0 = \nabla^2 \Psi. \qquad (2.143a\text{–}2.143d)$$

The solution of the Laplace (Equations 2.143c and 2.143d) [Poison (Equations 2.140 and 2.142)] equations is conveniently expressed [Subsection 2.4.3 (Subsection 2.4.4)] in terms of complex conjugate coordinates (Subsection 2.4.2).

2.4.2 Complex Conjugate Coordinates

The **complex conjugate coordinates** are the pair formed by a complex number (Equation 2.144a) and its conjugate (Equation 2.144b) can be used as an alternative to the Cartesian coordinates (Equations 2.144c and 2.144d ≡ Equations I.3.17a and I.3.17b):

$$z = x + iy, \quad z^* = x - iy: \qquad x = \frac{z+z^*}{2}, \quad y = \frac{z-z^*}{2i}. \tag{2.144a–2.144d}$$

The relation from Cartesian (x, y) to complex conjugate (z, z^*) coordinates leads to the **partial derivatives of the coordinate transformation:**

$$\frac{\partial z}{\partial x} = 1 = \frac{\partial z^*}{\partial y}, \qquad \frac{\partial z}{\partial y} = i = -\frac{\partial z^*}{\partial y}. \tag{2.145a–2.145d}$$

These appear in the **Jacobian** or determinant of the direct transformation matrix:

$$X \equiv \frac{\partial(z,z^*)}{\partial(x,y)} \equiv \begin{vmatrix} \dfrac{\partial z}{\partial x} & \dfrac{\partial z^*}{\partial x} \\ \dfrac{\partial z}{\partial y} & \dfrac{\partial z^*}{\partial y} \end{vmatrix} = \begin{vmatrix} 1 & 1 \\ i & -i \end{vmatrix} = -2i \neq 0, \infty; \tag{2.146}$$

because it is nonzero and finite, the linear transformation (Equations 2.144a and 2.144b) is invertible everywhere. The inverse transformation (Equations 2.144c and 2.144d) from complex conjugate (z, z^*) to Cartesian (x, y) coordinates leads to the partial derivatives of the inverse coordinate transformation:

$$\frac{\partial x}{\partial z} = \frac{1}{2} = \frac{\partial x}{\partial z^*}, \qquad \frac{\partial y}{\partial z} = \frac{1}{2i} = -\frac{i}{2} = -\frac{\partial y}{\partial z^*}. \tag{2.147a–2.147d}$$

The determinant of the inverse transformation matrix:

$$\frac{\partial(x,y)}{\partial(z,z^*)} = \begin{vmatrix} \dfrac{\partial x}{\partial z} & \dfrac{\partial y}{\partial z} \\ \dfrac{\partial x}{\partial z^*} & \dfrac{\partial y}{\partial z^*} \end{vmatrix} = \begin{vmatrix} \dfrac{1}{2} & -\dfrac{i}{2} \\ \dfrac{1}{2} & \dfrac{i}{2} \end{vmatrix} = \frac{i}{2} = \frac{1}{-2i} = \frac{1}{X} \neq 0, \infty \tag{2.148}$$

is the inverse of Equation 2.147, showing that the inverse transformation is also invertible everywhere.

The partial derivatives [Equations 2.145a through 2.145d (Equation 2.147a through 2.147d)] can be used to calculate the **compound derivatives of the coordinate transformation:**

$$\frac{\partial}{\partial z} = \frac{\partial x}{\partial z}\frac{\partial}{\partial x} + \frac{\partial y}{\partial z}\frac{\partial}{\partial y} = \frac{1}{2}\left(\frac{\partial}{\partial x} - i\frac{\partial}{\partial y}\right), \tag{2.149a}$$

$$\frac{\partial}{\partial z^*} = \frac{\partial x}{\partial z^*}\frac{\partial}{\partial x} + \frac{\partial y}{\partial z^*}\frac{\partial}{\partial y} = \frac{1}{2}\left(\frac{\partial}{\partial x} + i\frac{\partial}{\partial y}\right), \tag{2.149b}$$

for the complex conjugate (Equations 2.149a and 2.149b) [Cartesian (Equations 2.150a and 2.150b)] coordinates:

$$\frac{\partial}{\partial x} = \frac{\partial z}{\partial x}\frac{\partial}{\partial z} + \frac{\partial z^*}{\partial x}\frac{\partial}{\partial z^*} = \frac{\partial}{\partial z} + \frac{\partial}{\partial z^*}, \tag{2.150a}$$

$$\frac{\partial}{\partial y} = \frac{\partial z}{\partial y}\frac{\partial}{\partial z} + \frac{\partial z^*}{\partial y}\frac{\partial}{\partial z^*} = i\left(\frac{\partial}{\partial z} - \frac{\partial}{\partial z^*}\right). \tag{2.150b}$$

Either set [Equations 2.149a and 2.149b (Equations 2.150a and 2.150b)] may be used in the Laplace operator [Equation 2.151a (Equation 2.151b)]:

$$\nabla^2 \equiv \frac{\partial^2}{\partial x^2} + \frac{\partial^2}{\partial y^2} = \left(\frac{\partial}{\partial x} + i\frac{\partial}{\partial y}\right)\left(\frac{\partial}{\partial x} - i\frac{\partial}{\partial y}\right) = 4\frac{\partial^2}{\partial z^* \partial z}, \tag{2.151a}$$

$$\nabla^2 \equiv \frac{\partial^2}{\partial x^2} + \frac{\partial^2}{\partial y^2} = \left(\frac{\partial}{\partial z} + \frac{\partial}{\partial z^*}\right)^2 - \left(\frac{\partial}{\partial z} - \frac{\partial}{\partial z^*}\right)^2 = 4\frac{\partial^2}{\partial z^* \partial z}, \tag{2.151b}$$

leading to the same *simple form of the Laplace operator in complex conjugate coordinates (Equation 2.151a ≡ Equation 2.151b ≡ Equation 2.152c)*. The solution of the Laplace (Poisson) equation [Subsection 2.4.3 (Subsection 2.4.4)] will be applied (Equation 2.142) to the stream function in complex conjugate coordinates $\Psi(z, z^*)$ and would be simplest for any other function such as (Equation 2.140) the potential forced by the dilatation.

2.4.3 General Integral of the Laplace Equation

In the absence of vorticity (Equation 2.152a), the stream function of an incompressible (Equation 2.152b) flow (Equation 2.143d) satisfies (Equation 2.151a ≡ Equation 2.151b) the Laplace equation (Equation 2.152c):

$$\varpi = 0 = \dot{D}_2: \quad \frac{1}{4}\nabla^2\Psi_0 \equiv \frac{\partial^2\Psi_0}{\partial z^* \partial z} = 0, \quad \Psi_0(z, z^*) = f(z) + g(z^*), \tag{2.152a–2.152d}$$

whose **general integral** *(Equation 2.152d) involves two arbitrary differentiable functions, as it should be for a second-order partial differential equation. In this case, the potential (Equation 2.143c) also satisfies the Laplace equation.* In the case of the stream function, a real solution is obtained by taking the second function in Equation 2.152d to be the complex conjugate of the first (Equation 2.153a), leading to Equation 2.153b:

$$g(z) = f^*(z): \quad \Psi_0(z, z^*) = f(z) + f^*(z^*) = 2\text{Re}[f(z)], \quad f \in \mathcal{D}(|C), \tag{2.153a–2.153c}$$

where $f^*(z^*)$ is analytic if $f(z)$ is (Equation 2.153c) analytic (Equation I.31.19). Thus, *the Laplace equation (Equation 2.152c) has real solution (Equation 2.153b) where Equation 2.153c is a complex analytic function.*

The integration of the Laplace equation in complex conjugate coordinates (Equation 2.152a) can be made in two steps: (1) from Equation 2.152c ≡ Equation 2.154a, it follows that the term in brackets does not depend on z^* and hence can be an arbitrary function $h(z)$ only of z in Equation 2.154b:

$$0 = \frac{\partial}{\partial z^*}\left(\frac{\partial \Psi_0}{\partial z}\right) \quad \Rightarrow \quad \frac{\partial \Psi_0}{\partial z} = h(z); \tag{2.154a and 2.154b}$$

and (2) the latter (Equation 2.154b) can be integrated again leading to Equation 2.155a:

$$\Psi_0(z) = f(z) + g(z^*), \quad f(z) \equiv \int h(z)\,\mathrm{d}z, \tag{2.155a and 2.155b}$$

where g and f are also arbitrary differentiable functions. *The general integral (Equation 2.152d ≡ Equation 2.156b) of the Laplace equation in complex conjugate coordinates (Equation 2.154c) corresponds to*

$$\frac{\partial^2 \Psi_0}{\partial x^2} + \frac{\partial^2 \Psi_0}{\partial y^2} = 0: \quad \Psi_0(x, y) = f(x + iy) + g(x - iy), \; f, g \in \mathcal{D}^2\left(|R^2\right) \tag{2.156a and 2.156b}$$

in Cartesian coordinates (Equation 2.156a), where f and g are twice differentiable functions in the plane (Equation 2.156c).

2.4.4 Complete Integral of the Poisson Equation

In the presence of vorticity, the stream function of an incompressible flow satisfies the Poisson equation (Equations 2.142, 2.151a, and 2.151b ≡ Equation 2.157a) that has a particular integral (Equation 2.157b):

$$4\frac{\partial^2 \Psi_1}{\partial z^* \partial z} = -\varpi(z, z^*), \quad \Psi_1(z, z^*) = -\frac{1}{4}\int \mathrm{d}z^* \int \mathrm{d}z\, \varpi(z, z^*). \tag{2.157a and 2.157b}$$

The most general solution of the second-order partial differential equation (Equation 2.157a) must involve two arbitrary functions and thus cannot reduce to Equation 2.157b. Subtracting the particular solution Ψ_1 in Equation 2.157b from the complete solution Ψ in Equation 2.158a leads to Equation 2.158b:

$$4\frac{\partial^2\Psi}{\partial z^*\partial z} = -\varpi, \qquad 0 = \nabla^2\Psi - \nabla^2\Psi_1 = \nabla^2\Psi_0. \qquad \text{(2.158a and 2.158b)}$$

The latter is the Laplace equation (Equation 2.152c) that has solution Equation 2.152d. Thus, Equations 2.157a and 2.158b imply

$$\Psi(z,z^*) = \Psi_0(z,z^*) + \Psi_1(z,z^*) = f(z) + g(z^*) - \frac{1}{4}\int dz^* \int dz\varpi(z,z^*), \qquad (2.159)$$

showing that *the* **complete integral** *of the Poisson equation (Equation 2.142 ≡ Equation 2.158a) involving two arbitrary functions is (Equation 2.159) the sum of (1) the general integral (Equation 2.152d) of the Laplace equation (Equation 2.152c) and (2) the particular integral (Equation 2.157b) of the Poisson equation (Equation 2.157a).* In the Equation 2.159, the particular integral of the Poisson or forced Laplace equation (Equation 2.157a) may be taken as simple as possible (Equation 2.157b) because the general integral (Equation 2.152d) of the Laplace equation (Equation 2.152c) already involves two arbitrary functions; this is sufficient because the solution of a second-order partial differential equation involves two arbitrary functions, no more and no less. If a flow is specified by the vorticity (Equations 2.157a and 2.157b), it is possible to superimpose any potential flow (Equations 2.152c and 2.152d) without changing (Equation 2.159) the vorticity in the Poisson equation (Equation 2.158a). The potential flow will be omitted next, when considering the simplest rotational flow.

2.4.5 General/Simple Unidirectional Shear Flow

The simplest rotational flow in the whole plane is the **general linear unidirectional shear flow** that has constant vorticity (Equation 2.160a):

$$\varpi = \text{const} \equiv \varpi_0: \qquad \Psi = -\frac{\varpi_0}{4}z^*z = -\frac{\varpi_0}{4}r^2 = -\frac{\varpi_0}{4}(x^2+y^2), \qquad \text{(2.160a and 2.160b)}$$

leading (Equation 2.157b) to the stream function (Equation 2.160b); the latter specifies by (Equations 2.141b and 2.141c) the Cartesian components of the velocity (Equations 2.161a and 2.161b):

$$v_x = \frac{\partial\Psi}{\partial y} = -\frac{\varpi_0}{2}y, \quad v_y = -\frac{\partial\Psi}{\partial x} = \frac{\varpi_0}{2}x, \quad \frac{\partial v_y}{\partial x} - \frac{\partial v_x}{\partial y} = \varpi_0, \qquad \text{(2.161a–2.161c)}$$

and confirms (Equation 2.161c) that the total vorticity is ϖ_0.

The general unidirectional shear flow (Figure 2.3) consists of the superposition of two **simple linear unidirectional shear flows**, *with equal vorticities: (1) the first (Equation 2.162a)*

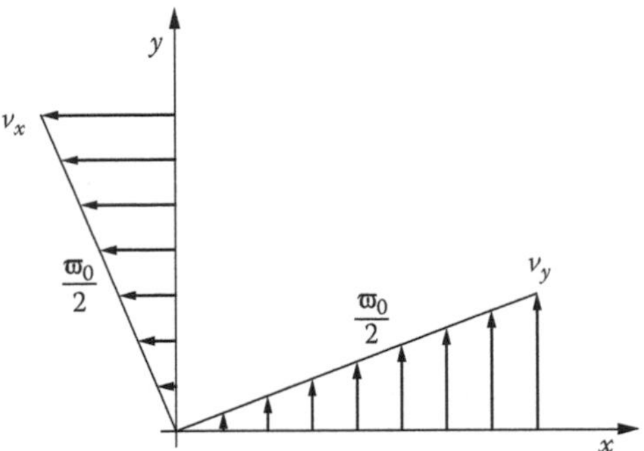

FIGURE 2.3
Simplest incompressible rotational flow has constant vorticity in the whole plane. It corresponds to the general linear unidirectional shear flow that consists of the superposition of two simple unidirectional shear flows each with one-half the vorticity: (1) one with velocity in the y-direction whose magnitude varies linearly in the transverse x-direction; (2) the other in the negative x-direction varying linearly in the transverse y-direction. Because each velocity varies only transversely, the dilatation is zero, and the shear flow is incompressible; because the velocity varies linearly with position, the rotational flow has constant vorticity.

has velocity in the x-direction (Equation 2.162c), varying linearly in the transverse y-direction (Equation 2.162b) corresponding to half the vorticity (Equation 2.162d):

$$\Psi^+ = -\frac{\varpi_0}{4}y^2, \quad v_x^+ = \frac{\partial\Psi^+}{\partial y} = -\frac{\varpi_0}{2}y, \quad v_y^+ = -\frac{\partial\Psi^+}{\partial x} = 0, \varpi_+ = -\frac{\partial v_x^+}{\partial y} = \frac{\varpi_0}{2}, \tag{2.162a–2.162d}$$

$$\Psi^- = -\frac{\varpi_0}{4}x^2, \quad v_x^- = \frac{\partial\Psi^-}{\partial y} = 0, \quad v_y^- = -\frac{\partial\Psi^-}{\partial x} = \frac{\varpi_0}{2}x, \quad \varpi_- = \frac{\partial v_y^-}{\partial x} = \frac{\varpi_0}{2}; \tag{2.163a–2.163d}$$

and (2) the second (Equation 2.163a) has velocity in the y-direction (Equation 2.163b), varying linearly in the transverse x-direction (Equation 2.163c) corresponding to the other half (Equation 2.163d) of the total vorticity:

$$\Psi = \Psi^+ + \Psi^-, \quad v_x = v_x^+ + v_x^-, \quad v_y = v_y^+ + v_y^-, \quad \varpi^+ + \varpi^- = \varpi_0 = 2\varpi_\pm. \tag{2.164a–2.164d}$$

A constant total vorticity would also result from two simple unidirectional flows with distinct vorticities $\varpi^+ \neq \varpi^-$ instead of $\varpi^+ \neq \varpi_0/2 = \varpi^-$ in Equation 2.164d.

A more general unidirectional shear flow is Equation 2.165a, consisting also of two simple unidirectional shear flows:

$$\vec{v}(x,y) = \vec{e}_x v_x(y) + \vec{e}_y v_y(x): \quad \varpi = \frac{dv_x}{dy} - \frac{dv_y}{dx}, \quad D_2 = \frac{\partial v_x}{\partial y} + \frac{\partial v_y}{\partial x} = 0, \tag{2.165a–2.165c}$$

where (1) the velocity profile (Equation 2.165a) need not be linear; (2) vorticity need not be constant (Equation 2.165b); and (3) the dilatation remains zero (Equation 2.165c). The general linear unidirectional shear flow is the particular case for a stream function (Equation 2.165d) bilinear in the two variables (z^*, z):

$$\Psi(z^* z) = A + Bz + Cz^* + Dz^* z: \quad A = 0 = B = C, \quad D = -\frac{\varpi_0}{4}, \tag{2.165d–2.165h}$$

where (1) the constant term can be omitted (Equation 2.165e) because it does not affect the velocity; (2 and 3) the next two terms correspond to uniform flows (Subsection 2.4.7) that are potential and may be omitted (Equations 2.165f and 2.165g) to concentrate on the rotational part; and (4) the latter is specified by the last term on the r.h.s. of Equation 2.165d whose coefficient is determined by the constant vorticity (Equation 2.165h). The simplest incompressible rotational (irrotational) flow is the general unidirectional shear flow (uniform flow with angle of attack) in Subsection 2.4.5 (Subsection 2.4.7). The consideration of the latter is preceded by the complex conjugate velocity (Subsection 2.4.6).

2.4.6 Complex Conjugate Velocity and Cartesian Components

For a potential flow that is both irrotational (Equation 2.139a ≡ Equation 2.166a) and incompressible (Equation 2.141a ≡ Equation 2.166b), the velocity can be calculated from the potential (Equations 2.139b and 2.139c) [stream function (Equations 2.141a and 2.141c)], leading to [Equations 2.166a and 2.166e (Equations 2.166d and 2.166f)]:

$$\varpi = 0 = \dot{D}_2: \qquad v_x = \frac{\partial \Phi}{\partial x} = \frac{\partial \Psi}{\partial y}, \qquad v_y = \frac{\partial \Phi}{\partial y} = -\frac{\partial \Psi}{\partial x}. \tag{2.166a–2.166f}$$

The relations (Equation 2.166b through 2.166f) are the Cauchy–Riemann conditions (Section I.12.3) that ensure the existence of a complex potential (Equation 2.167a), that is, an analytic function whose real (imaginary) part is the potential (stream function):

$$f(x+iy) = \Phi(x,y) + i\Psi(x,y) \in \mathcal{D}\left(|C\right): \quad \Phi, \Psi \in C^1\left(|R^2\right) \quad \Leftrightarrow \quad \vec{v} \in C\left(|R^2\right), \tag{2.167a–2.167c}$$

provided that they have continuous first-order derivatives (Equation 2.167b). This is equivalent to the velocity being continuous (Equation 2.167c); discontinuous flows such as shock waves or shear layers (Section 10.9) are, in general, not potential.

The derivative of the complex potential specifies the complex conjugate velocity:

$$v^*(z) = \frac{df}{dz} = \frac{\partial(\Phi + i\Psi)}{\partial x} = \frac{\partial \Phi}{\partial x} + i\frac{\partial \Psi}{\partial x} = v_x - iv_y, \tag{2.168a}$$

$$= \frac{\partial(\Phi + i\Psi)}{i\partial y} = \frac{\partial \Psi}{\partial x} - i\frac{\partial \Phi}{\partial y} = v_x - iv_y, \tag{2.168b}$$

where, because the derivative is independent of direction, it can be calculated along the x-axis in Equation 2.168a or y-axis in Equation 2.168b. Equating the real and imaginary parts of Equations 2.168a and 2.168b leads to the Cauchy–Riemann conditions (Equation 2.166c and 2.166d) that are necessary for the complex potential (Equation 2.167a) to be a differentiable function f of its complex variable z; these Cauchy–Riemann conditions arise from an equality in two directions (x and y) and are not sufficient to ensure that a different result would arise in another direction. The latter possibility is excluded if the first-order partial derivatives are continuous (Equation 2.167b), leading to the sufficient conditions that $f(z)$ be differentiable (Section I.11.3), namely, the Cauchy–Riemann conditions plus Equation 2.167b. *The Cartesian components of the continuous velocity (Equation 2.167c) in a potential flow can be combined (Equations 2.166c through 2.166f) in the* **complex conjugate velocity** *(Equation 2.169a):*

$$v^*(z) = \frac{df}{dz} = v_x - iv_y = \frac{\partial\Phi}{\partial x} + i\frac{\partial\Psi}{\partial x} = \frac{\partial\Psi}{\partial y} - i\frac{\partial\Phi}{\partial y} = \frac{\partial\Phi}{\partial x} - i\frac{\partial\Phi}{\partial y} = \frac{\partial\Psi}{\partial y} + i\frac{\partial\Psi}{\partial x}, \tag{2.169a–2.169e}$$

that is given (1 and 2) using only derivatives with regard to x(y) by Equation 2.169b (Equation 2.169c); and (3 and 4) using only the potential (stream function) by Equation 2.169d (Equation 2.169e). If the flow is irrotational compressible (incompressible rotational), only the potential (stream function) can be used in Equation 2.169d (Equation 2.169e). The simplest potential flow is uniform with angle of attack and is considered next.

2.4.7 Uniform Flow with Angle of Attack

A uniform flow (Subsection I.14.8.1) with angle-of-attack α has Cartesian velocity components (Equations 2.170a and 2.170b) leading (Equation 2.169a) to the complex conjugate velocity (Equation 2.170c):

$$\{v_x, v_y\} = U\{\cos\alpha, \sin\alpha\}: \quad v^* = v_x - iv_y = U(\cos\alpha - i\sin\alpha) = Ue^{-i\alpha} = \frac{df}{dz}. \tag{2.170a–2.170c}$$

Because the latter depends only on z and not on z^*, the flow has complex potential (Equation 2.171a):

$$f(z) = Ue^{-i\alpha}z: \quad f(z) = A + Bz,\ A = 0,\ |B| = U,\ \arg(B) = -\alpha. \tag{2.171a–2.171e}$$

The latter corresponds to the simplest complex potential, namely, a linear function (Equation 2.171b) where (1) the constant term can be omitted (Equation 2.171c) because it leads to zero velocity, that is, the complex potential (potential and stream function) is defined to within an added arbitrary complex (real) constant; and (2) the coefficient of the linear term specifies the velocity (angle of attack) as its modulus (Equation 2.171d) [minus its argument (Equation 2.171e)]. Separating real and imaginary parts in Equation 2.171a ≡ Equation 2.172:

$$\Phi(x, y) + i\Psi(x, y) = f(z) = Ue^{-i\alpha}z = U(x + iy)(\cos\alpha - i\sin\alpha) \tag{2.172}$$

specifies the potential (Equation 2.173a) [stream function (Equation 2.173b)]:

$$\Phi(x, y) = U(x \cos\alpha + y \sin\alpha), \quad \Psi(x, y) = U(y \cos\alpha - x \sin\alpha) \qquad \text{(2.173a and 2.173b)}$$

in Cartesian coordinates.

Using Equations 2.144c and 2.144d, the potential (Equation 2.173a ≡ Equation 2.174a) [stream function (Equation 2.173b ≡ Equation 2.174b)] can be expressed in complex conjugate coordinates:

$$\begin{aligned}\Phi(z,z^*) &= \frac{U}{2}\left[(z+z^*)\cos\alpha - i(z-z^*)\sin\alpha\right]\\ &= \frac{U}{2}\left[z(\cos\alpha - i\sin\alpha) + z^*(\cos\alpha + i\sin\alpha)\right]\\ &= \frac{U}{2}(ze^{-i\alpha} + z^*e^{i\alpha}) = \frac{f(z)+f^*(z)}{2},\end{aligned} \qquad (2.174a)$$

$$\begin{aligned}\Psi(z,z^*) &= \frac{U}{2}\left[-i(z-z^*)\cos\alpha - (z+z^*)\sin\alpha\right]\\ &= \frac{U}{2}\left[-iz(\cos\alpha - i\sin\alpha) + iz^*(\cos\alpha + i\sin\alpha)\right]\\ &= \frac{U}{2i}(ze^{-i\alpha} - z^*e^{i\alpha}) = \frac{f(z)-f^*(z)}{2i},\end{aligned} \qquad (2.174b)$$

confirming that (1) it coincides with the real (imaginary) part of the complex potential (Equation 2.171a); and (2) both are the sum of a function of z and a function of z^* and thus Equation 2.152d satisfies the Laplace equation (Equation 2.152c) that applies to a potential flow [Equation 2.143c (Equation 2.143d)]. The vortex (source/sink) [Subsection 2.1.8 (Subsection 2.1.9)] and their combination as a monopole (Subsection 2.1.10) in the case of a potential flow are discussed more simply in polar coordinates (Subsection 2.4.11); these are related next (Subsection 2.4.8) to complex conjugate coordinates (Subsection 2.4.2) and used to derive in polar coordinates (1) the Laplace operator (Subsection 2.4.9) and (2) the complex conjugate velocity (Subsection 2.4.10).

2.4.8 Relation between Polar and Complex Conjugate Coordinates

The complex conjugate coordinates (z, z^*) may be related to the Cartesian coordinates (x, y) by Equations 2.144a and 2.144b, or to the polar coordinates (r, φ) by (Equations 2.175a and 2.175b):

$$z = re^{i\varphi}, \quad z^* = re^{-i\varphi}: \qquad r = \sqrt{z^*z} \equiv (z^*z)^{1/2} \quad \varphi = -\frac{i}{2}\log\left(\frac{z}{z^*}\right), \qquad \text{(2.175a–2.175d)}$$

whose inverses are Equations 2.175c and 2.175d; the latter (Equation 2.175d) follows from

$$\varphi = \frac{\log(e^{i2\varphi})}{2i} = -\frac{i}{2}\log\left(\frac{z}{z^*}\right) = \frac{i}{2}(\log z^* - \log z). \tag{2.176}$$

The direct transformation from polar (r, φ) to complex conjugate (z, z^*) coordinates has (Equations 2.173a and 2.173b) partial derivatives:

$$\frac{\partial z}{\partial r} = e^{i\varphi}, \frac{\partial z^*}{\partial r} = e^{-i\varphi} = \left(\frac{\partial z}{\partial r}\right)^{-1}, \frac{\partial z}{\partial \varphi} = ire^{i\varphi}, \quad \frac{\partial z^*}{\partial \varphi} = -ire^{-i\varphi} = \left(\frac{\partial z}{\partial \varphi}\right)^*. \tag{2.177a–2.177d}$$

The Jacobian or determinant of the direct transformation matrix:

$$Y = \frac{\partial(z,z^*)}{\partial(r,\varphi)} = \begin{vmatrix} \dfrac{\partial z}{\partial r} & \dfrac{\partial z^*}{\partial r} \\ \dfrac{\partial z}{\partial \varphi} & \dfrac{\partial z^*}{\partial \varphi} \end{vmatrix} = \begin{vmatrix} e^{i\varphi} & e^{-i\varphi} \\ ire^{i\varphi} & -ire^{-i\varphi} \end{vmatrix} = -2ir \neq 0, \infty \tag{2.178}$$

is nonzero in the whole plane except at the origin $r = 0$ and infinity $r = \infty$; at these two points, the polar angle φ is undetermined, and everywhere else, the transformation is invertible. The inverse transformation from complex conjugate (z, z^*) to polar (r, φ) coordinates has (Equations 2.175c and 2.176) partial derivatives:

$$\frac{\partial r}{\partial z} = \frac{1}{2}\sqrt{\frac{z^*}{z}}, \frac{\partial r}{\partial z^*} = \frac{1}{2}\sqrt{\frac{z}{z^*}}, = \frac{1}{4}\left(\frac{\partial r}{\partial z}\right)^{-1}, \frac{\partial \varphi}{\partial z} = -\frac{i}{2z}, \frac{\partial \varphi}{\partial z^*} = \frac{i}{2z^*} = \left(\frac{\partial \varphi}{\partial z}\right)^*. \tag{2.179a–2.179d}$$

The determinant of the inverse transformation matrix:

$$\frac{\partial(r,\varphi)}{\partial(z,z^*)} = \begin{vmatrix} \dfrac{\partial r}{\partial z} & \dfrac{\partial \varphi}{\partial z} \\ \dfrac{\partial r}{\partial z^*} & \dfrac{\partial \varphi}{\partial z^*} \end{vmatrix} = \begin{vmatrix} \dfrac{1}{2}\sqrt{\dfrac{z^*}{z}} & -\dfrac{i}{2z} \\ \dfrac{1}{2}\sqrt{\dfrac{z}{z^*}} & \dfrac{i}{2z^*} \end{vmatrix} = \frac{i}{2\sqrt{z^* z}} = \frac{i}{2r} = \frac{1}{Y} \tag{2.180a}$$

is the inverse of Equation 2.178 and leads to the same conditions $r \neq 0,\infty$ for the coordinate transformation to be invertible.

2.4.9 Compound Derivates and Laplacian in Polar Coordinates

The partial derivatives [Equations 2.179a through 2.179d (Equations 2.177a through 2.177d)] can be used to calculate the compound derivatives:

$$\frac{\partial}{\partial z}=\frac{\partial r}{\partial z}\frac{\partial}{\partial r}+\frac{\partial \varphi}{\partial z}\frac{\partial}{\partial \varphi}=\frac{1}{2}\left(\sqrt{\frac{z^*}{z}}\frac{\partial}{\partial r}-\frac{i}{z}\frac{\partial}{\partial \varphi}\right)=\frac{e^{-i\varphi}}{2}\left(\frac{\partial}{\partial r}-\frac{i}{r}\frac{\partial}{\partial \varphi}\right), \tag{2.181a}$$

$$\frac{\partial}{\partial z^*}=\frac{\partial r}{\partial z^*}\frac{\partial}{\partial r}+\frac{\partial \varphi}{\partial z^*}\frac{\partial}{\partial \varphi}=\frac{1}{2}\left(\sqrt{\frac{z}{z^*}}\frac{\partial}{\partial r}+\frac{i}{z^*}\frac{\partial}{\partial \varphi}\right)=\frac{e^{i\varphi}}{2}\left(\frac{\partial}{\partial r}+\frac{i}{r}\frac{\partial}{\partial \varphi}\right), \tag{2.181b}$$

for the complex conjugate (Equations 2.181a and 2.181b) [polar (Equations 2.182a and 2.182b)] coordinates:

$$\frac{\partial}{\partial r}=\frac{\partial z}{\partial r}\frac{\partial}{\partial z}+\frac{\partial z^*}{\partial r}\frac{\partial}{\partial z^*}=e^{i\varphi}\frac{\partial}{\partial z}+e^{-i\varphi}\frac{\partial}{\partial z^*}=\sqrt{\frac{z}{z^*}}\frac{\partial}{\partial z}+\sqrt{\frac{z^*}{z}}\frac{\partial}{\partial z^*}, \tag{2.182a}$$

$$\frac{\partial}{\partial \varphi}=\frac{\partial z}{\partial \varphi}\frac{\partial}{\partial z}+\frac{\partial z^*}{\partial \varphi}\frac{\partial}{\partial z^*}=ir\left(e^{i\varphi}\frac{\partial}{\partial z}-e^{-i\varphi}\frac{\partial}{\partial z^*}\right)=i\left(z\frac{\partial}{\partial z}-z^*\frac{\partial}{\partial z^*}\right). \tag{2.182b}$$

Using Equations 2.181a and 2.181b in the Laplacian in complex conjugate coordinates (Equation 2.151a ≡ Equation 2.151b) leads to

$$\begin{aligned}\nabla^2 &= 4\frac{\partial^2}{\partial z^*\partial z}=e^{i\varphi}\left(\frac{\partial}{\partial r}+\frac{i}{r}\frac{\partial}{\partial \varphi}\right)e^{-i\varphi}\left(\frac{\partial}{\partial r}-\frac{i}{r}\frac{\partial}{\partial \varphi}\right)\\ &=\frac{\partial^2}{\partial r^2}+\frac{i}{r^2}\frac{\partial}{\partial \varphi}-\frac{i}{r}\frac{\partial^2}{\partial r\partial \varphi}+\frac{i}{r}\frac{\partial^2}{\partial \varphi\partial r}+\frac{1}{r^2}\frac{\partial^2}{\partial \varphi^2}+\frac{1}{r}\left(\frac{\partial}{\partial r}-\frac{i}{r}\frac{\partial}{\partial \varphi}\right)\\ &=\frac{\partial^2}{\partial r^2}+\frac{1}{r}\frac{\partial}{\partial r}+\frac{1}{r^2}\frac{\partial^2}{\partial \varphi^2}=\frac{1}{r}\frac{\partial}{\partial r}\left(r\frac{\partial}{\partial r}\right)+\frac{1}{r^2}\frac{\partial^2}{\partial \varphi^2};\end{aligned} \tag{2.183a}$$

$$\begin{aligned}\nabla^2 &= 4\frac{\partial^2}{\partial z\partial z^*}=e^{-i\varphi}\left(\frac{\partial}{\partial r}-\frac{i}{r}\frac{\partial}{\partial \varphi}\right)e^{i\varphi}\left(\frac{\partial}{\partial r}+\frac{i}{r}\frac{\partial}{\partial \varphi}\right)\\ &=\frac{\partial^2}{\partial r^2}-\frac{i}{r^2}\frac{\partial}{\partial \varphi}+\frac{i}{r}\frac{\partial^2}{\partial r\partial \varphi}-\frac{i}{r}\frac{\partial^2}{\partial \varphi\partial r}+\frac{1}{r^2}\frac{\partial^2}{\partial \varphi^2}+\frac{1}{r}\left(\frac{\partial}{\partial r}+\frac{i}{r}\frac{\partial}{\partial \varphi}\right)\\ &=\frac{\partial^2}{\partial r^2}+\frac{1}{r}\frac{\partial}{\partial r}+\frac{1}{r^2}\frac{\partial^2}{\partial \varphi^2}=\frac{1}{r}\frac{\partial}{\partial r}\left(r\frac{\partial}{\partial r}\right)+\left(\frac{1}{r}\frac{\partial}{\partial \varphi}\right)^2,\end{aligned} \tag{2.183b}$$

as the **Laplace operator in polar coordinates:**

$$\nabla^2 = \frac{\partial^2}{\partial r^2} + \frac{1}{r}\frac{\partial}{\partial r} + \frac{1}{r^2}\frac{\partial^2}{\partial \varphi^2} = \frac{1}{r}\frac{\partial}{\partial r}\left(r\frac{\partial}{\partial r}\right) + \left(\frac{1}{r}\frac{\partial}{\partial \varphi}\right)^2, \tag{2.184a–2.184c}$$

in agreement with Equations 2.184a and 2.184b ≡ Equations I.11.28b and I.11.28c.

2.4.10 Polar Coordinates and Components of the Velocity

Using Equation 2.169a (Equation 2.175a), *the complex conjugate velocity is given in polar coordinates by*

$$v^*(z) = \frac{df}{dz} = \frac{\partial(\Phi + i\Psi)}{e^{i\varphi}\partial r} = e^{-i\varphi}\left(\frac{\partial \Phi}{\partial r} + i\frac{\partial \Psi}{\partial r}\right), \tag{2.185a}$$

$$= \frac{\partial(\Phi + i\Psi)}{r\partial(e^{i\varphi})} = \frac{1}{ire^{i\varphi}}\frac{\partial(\Phi + i\Psi)}{\partial \varphi} = \frac{e^{-i\varphi}}{r}\left(\frac{\partial \Psi}{\partial \varphi} - i\frac{\partial \Phi}{\partial \varphi}\right), \tag{2.185b}$$

by differentiation with regard to $r(\varphi)$ at constant $\varphi(r)$ in Equation 2.185a (Equation 2.185b). For an irrotational flow (Equation 2.8a), velocity is the gradient of the potential (Equation 2.8b); using the latter in polar coordinates (Equation I.11.31b) leads to the polar components of the velocity in Equations 2.186a and 2.186c; (1) equating Equations 2.185a and 2.185b leads to the Cauchy–Riemann conditions in polar coordinates (Section I.11.4) and specifies the polar components of the velocity in terms of the stream function (Equations 2.186b and 2.186d). The comparison of Equations 2.185a and 2.185b is consistent with (2) the components of the velocity in polar coordinates:

$$v_r = \frac{\partial \Phi}{\partial r} = \frac{1}{r}\frac{\partial \Psi}{\partial \varphi}, \qquad v_\varphi = \frac{1}{r}\frac{\partial \Phi}{\partial \varphi} = -\frac{\partial \Psi}{\partial r}. \tag{2.186a–2.186d}$$

Substituting Equations 2.186a through 2.186d in Equations 2.185a and 2.185b specifies *the complex conjugate velocity in polar coordinates* (Equation 2.187a):

$$\begin{aligned} v^* = e^{-i\varphi}(v_r - iv_\varphi) &= e^{-i\varphi}\left(\frac{\partial \Phi}{\partial r} + i\frac{\partial \Psi}{\partial r}\right) = \frac{e^{-i\varphi}}{r}\left(\frac{\partial \Psi}{\partial \varphi} - i\frac{\partial \Phi}{\partial \varphi}\right) \\ &= e^{-i\varphi}\left(\frac{\partial \Phi}{\partial r} - \frac{i}{r}\frac{\partial \Phi}{\partial \varphi}\right) = e^{-i\varphi}\left(\frac{1}{r}\frac{\partial \Psi}{\partial \varphi} + i\frac{\partial \Psi}{\partial r}\right), \end{aligned} \tag{2.187a–2.187e}$$

that can be calculated using (1 and 2) only derivatives with regard to the radius (azimuthal angle) in Equation 2.187b (Equation 2.187c); and (3 and 4) only the potential (Equation 2.187d) [stream function (Equation 2.187e)]. For a plane compressible irrotational (incompressible rotational) flow, only the potential (stream function) [Equation 2.187d (Equation 2.187c)] can be used.

The complex conjugate velocity in Cartesian coordinates (Equation 2.169a) can be calculated from the potential (Equation 2.169d) [stream function (Equation 2.169e)] by [Equation 2.188a (Equation 2.188b)]:

$$v^* = v_x - iv_y = \frac{\partial\Phi}{\partial x} - i\frac{\partial\Phi}{\partial y} = 2\frac{\partial\Phi}{\partial z}, \tag{2.188a}$$

$$v^* = v_x - iv_y = \frac{\partial\Psi}{\partial y} + i\frac{\partial\Psi}{\partial x} = i\left(\frac{\partial\Psi}{\partial x} - i\frac{\partial\Psi}{\partial y}\right) = 2i\frac{\partial\Psi}{\partial z}, \tag{2.188b}$$

where the compound derivative (Equation 2.149a) was used. The same result (Equations 2.189a and 2.189b ≡ Equations 2.188a and 2.188b) can be obtained using the complete conjugate velocity (Equation 2.187a) in polar coordinates [Equation 2.187d (Equation 2.187e)]:

$$v^* = e^{-i\varphi}(v_r - iv_\varphi) = e^{-i\varphi}\left(\frac{\partial\Phi}{\partial r} - \frac{i}{r}\frac{\partial\Phi}{\partial\varphi}\right) = 2\frac{\partial\Phi}{\partial z}, \tag{2.189a}$$

$$v^* = e^{-i\varphi}(v_r - iv_\varphi) = e^{-i\varphi}\left(\frac{1}{r}\frac{\partial\Psi}{\partial\varphi} + i\frac{\partial\Psi}{\partial r}\right) = ie^{-i\varphi}\left(\frac{\partial\Psi}{\partial r} - \frac{i}{r}\frac{\partial\Psi}{\partial\varphi}\right) = 2i\frac{\partial\Psi}{\partial z}, \tag{2.189b}$$

from the compound derivative (Equation 2.181a). Thus, *the complex conjugate velocity can be calculated from (1 and 2) the Cartesian (Equation 2.190a) [polar (Equation 2.190b)] components of the velocity by Equations 2.169a through 2.169e (Equations 2.187a through 2.187e); (3) from the complex potential (Equation 2.190c) in a potential flow; and (4 and 5) in complex conjugate coordinates from the potential (Equation 2.190d) [stream function (Equation 2.190e)] in a compressible irrotational (incompressible rotational) flow, that includes the potential flow.*

$$v^* = v_x - iv_y = e^{-i\varphi}(v_r - iv_\varphi) = \frac{\mathrm{d}f}{\mathrm{d}z} = 2\frac{\partial\Phi}{\partial z} = 2i\frac{\partial\Psi}{\partial z}. \tag{2.190a–2.190e}$$

Several of these relations in polar coordinates are applied next to the vortex and source/sink combined in a monopole.

2.4.11 Potential Vortex, Source/Sink, and Monopole

The monopole for a potential flow has tangential (Equation 2.33b ≡ Equation 2.191a) [radial (Equation 2.41b ≡ Equation 2.191b)] velocity associated with the circulation Γ (flow rate Q), leading (Equation 2.190b) to the complex conjugate velocity (Equation 2.191c):

$$v_\varphi = \frac{\Gamma}{2\pi r}, v_r = \frac{Q}{2\pi r}: \qquad v^* = e^{-i\varphi}(v_r - iv_\varphi) = \frac{Q - i\Gamma}{2\pi re^{i\varphi}} = \frac{Q - i\Gamma}{2\pi z} = \frac{\mathrm{d}f}{\mathrm{d}z}. \tag{2.191a–2.191c}$$

Because the complex conjugate velocity (Equation 2.191c) depends only on z and not on z^*, the flow is potential except for the singularity at the origin that is a simple pole (Equation 1.9) and corresponds to the location of the line-monopole; since the monopole consists of a source/sink (plus a vortex) the flow is not incompressible (irrotational) at the singularity at the origin. Integration of Equation 2.191a specifies the complex potential (Equation 2.192a):

$$f(z) = \frac{Q - i\Gamma}{2\pi}\log z: \qquad 2\pi(\Phi + i\Psi) = \left(Q - i\Gamma\right)\log\left(re^{i\varphi}\right) = (Q - i\Gamma)(\log r + i\varphi), \tag{2.192a and 2.192b}$$

and equating real (imaginary) parts in Equation 2.192b specifies the potential (Equation 2.193a) [stream function (Equation 2.193b)]:

$$2\pi\Phi(r,\varphi) = Q\log r + \Gamma\varphi, \quad 2\pi\Psi(r,\varphi) = Q\varphi - \Gamma\log r \tag{2.193a and 2.193b}$$

in polar coordinates.

Using (Equations 2.175c and 2.175d) complex conjugate coordinates:

$$\Phi(z,z^*) = \frac{Q}{2\pi}\log\left(\sqrt{z^* z}\right) - \frac{i\Gamma}{4\pi}\log\left(\frac{z}{z^*}\right) = \frac{Q}{4\pi}\log(z^* z) - \frac{i\Gamma}{4\pi}\log\left(\frac{z}{z^*}\right)$$
$$= \frac{Q - i\Gamma}{4\pi}\log z + \frac{Q + i\Gamma}{4\pi}\log z^* = \frac{f(z) + f^*(z)}{2} = \mathrm{Re}\{f(z)\}, \tag{2.194a}$$

$$\Psi(z,z^*) = -\frac{iQ}{4\pi}\log\left(\frac{z}{z^*}\right) - \frac{\Gamma}{2\pi}\log\left(\sqrt{z^* z}\right) = -i\left[\frac{Q}{4\pi}\log\left(\frac{z}{z^*}\right) - i\frac{\Gamma}{4\pi}\log(z^* z)\right]$$
$$= \frac{1}{2i}\left(\frac{Q - i\Gamma}{2\pi}\log z - \frac{Q - i\Gamma}{2\pi}\log z^*\right) = \frac{f(z) - f^*(z)}{2i} = Im\{f(z)\} \tag{2.194b}$$

confirms that (1) the potential (Equation 2.194a) [stream function (Equation 2.194b)] is the real (imaginary) part of the complex potential (Equation 2.192a); (2) both satisfy the Laplace equation (Equation 2.152c) because Equations 2.194a and 2.194b are a sum (Equation 2.152d) of functions of z and z^*; and (3) the complex conjugate velocity is given by Equation 2.195a (Equation 2.195b) using Equations 2.190d and 2.194a (Equations 2.190e and 2.194b):

$$v^*(z) = 2\frac{\partial\Phi}{\partial z} = \frac{Q - i\Gamma}{2\pi z} = 2i\frac{Q - i\Gamma}{4\pi i z} = 2i\frac{\partial\Psi}{\partial z}, \tag{2.195a and 2.195b}$$

in agreement with Equation 2.191c ≡ Equation 2.195a ≡ Equation 2.195b.

2.5 Second Forces/Moment and Circle Theorems

The theorems on an incompressible flow may be extended from the irrotational to the rotational case, replacing the complex potential $f(z)$ in Equation 2.190c by the stream function $\Psi(z, z^*)$ in Equation 2.190e. This leads in a plane incompressible irrotational (rotational) flow to (1) the first (second) Blasius theorem specifying the components of the force and of the pitching moment on a body [Section I.28.2 (Subsection 2.5.2)]; and (2) the first (second) circle theorem [Section 24.7 (Subsection 2.5.3)] concerning the introduction of a cylinder in a potential (vortical) flow. The first and second Blasius theorems are based on the forces and moments due to the pressure in an inviscid fluid (Subsection 2.5.1).

2.5.1 Forces and Pitching Moment due to the Pressure

Let L be a closed boundary curve representing in two-dimensional form a rigid impermeable body that corresponds to the directrix of a cylinder in three dimensions with generators orthogonal to the plane. If it is a regular curve, there are at each point a unique tangent (Equation 2.196b) and normal (Equation 2.196c) vector:

$$(\mathrm{d}s)^2 = (\mathrm{d}x)^2 + (\mathrm{d}y)^2: \quad \vec{t} = \left(\frac{\mathrm{d}x}{\mathrm{d}s}, \frac{\mathrm{d}y}{\mathrm{d}s}\right), \quad \vec{n} = \left(\frac{\mathrm{d}y}{\mathrm{d}s}, -\frac{\mathrm{d}x}{\mathrm{d}s}\right), \tag{2.196a–2.196c}$$

where Equation 2.196a is the arc length in Cartesian coordinates. The tangent (Equation 2.196b) [normal (Equation 2.196c)] is a unit vector [Equation 2.197a (Equation 2.197b)]:

$$\left|\vec{t}\right|^2 = \left(\frac{\mathrm{d}x}{\mathrm{d}s}\right)^2 + \left(\frac{\mathrm{d}y}{\mathrm{d}s}\right)^2 = \left|\vec{n}\right|^2 = \frac{(\mathrm{d}x)^2 + (\mathrm{d}y)^2}{(\mathrm{d}s)^2} = 1, \tag{2.197a and 2.197b}$$

$$\vec{t}.\vec{n} = \frac{\mathrm{d}x}{\mathrm{d}s}\frac{\mathrm{d}y}{\mathrm{d}s} + \frac{\mathrm{d}y}{\mathrm{d}s}\left(-\frac{\mathrm{d}x}{\mathrm{d}s}\right) = 0, \quad \vec{n} \wedge \vec{t} = \vec{e}_3 \begin{vmatrix} \dfrac{\mathrm{d}y}{\mathrm{d}s} & -\dfrac{\mathrm{d}x}{\mathrm{d}s} \\ \dfrac{\mathrm{d}x}{\mathrm{d}s} & \dfrac{\mathrm{d}y}{\mathrm{d}s} \end{vmatrix} = \vec{e}_3\left[\left(\frac{\mathrm{d}x}{\mathrm{d}s}\right)^2 + \left(\frac{\mathrm{d}y}{\mathrm{d}s}\right)^2\right] = \vec{e}_3, \tag{2.197c and 2.197d}$$

and they are orthogonal because (1) their inner product is zero (Equation 2.197c), and (2) their outer product is the unit vector (Equation 2.197d) orthogonal to the (x, y)-plane. In an inviscid fluid, the only internal stress is an isotropic pressure p acting along the inward normal (Equation 2.1b) and leading in the plane case to the Cartesian force components:

$$\vec{F} = -\int_L p\vec{n}\,\mathrm{d}s = -\int_L p\left(\frac{\mathrm{d}y}{\mathrm{d}s}, -\frac{\mathrm{d}x}{\mathrm{d}s}\right)\mathrm{d}s = \int_L p(-\mathrm{d}y, \mathrm{d}x). \tag{2.198a and 2.198b}$$

The horizontal F_x and vertical F_y force components may be combined in the **complex conjugate force:**

$$F^* = F_x - iF_y = \int_L p(-dy - i dx) = -i\int_L p(dx - i dy) = -i\int p dz^*, \tag{2.199}$$

using Equation 2.144b. The forces in the (x, y)-plane correspond to a **pitching moment** in the direction of the unit normal to the plane:

$$\vec{M} = \vec{e}_3 \int \begin{vmatrix} x & y \\ dF_x & dF_y \end{vmatrix} = \vec{e}_3 \int_L (x dF_y - y dF_x) = \vec{e}_3 \int_L p(x dx + y dy). \tag{2.200}$$

The pitching moment is positive (negative) in the counterclockwise (clockwise) direction corresponding to a nose-down (nose-up). Using Equations 2.144a and 2.144b in the form:

$$z dz^* = (x + iy)(dx - idy) = x dx + y dy + i(y dx - x dy), \tag{2.201}$$

it follows that the pitching moment (Equation 2.200) is given by

$$M = \int_L p(x dx + y dy) = \frac{1}{2}\int_L p d(x^2 + y^2) = \text{Re}\left\{\int_L p z dz^*\right\}. \tag{2.202}$$

Thus, *a body with closed regular boundary L in an inviscid fluid experiences a pitching moment (Equation 2.202) and a horizontal (Equation 2.198a), vertical (Equation 2.198b), and a complex conjugate force (Equation 2.198) due to the pressure on the surface.*

2.5.2 Forces and Moment in an Incompressible Rotational Flow

In a steady incompressible flow, the stagnation pressure (Equation 2.12b) is constant: (1) in all space for an irrotational flow; and (2) along streamlines (Equation 2.132b) for a rotational flow. The surface of a rigid impermeable body is a streamline in the steady flow of an inviscid fluid because the velocity is tangent to it. Thus, for a rigid impermeable body in the steady incompressible flow of an inviscid fluid, the Bernoulli equation (Equation 2.12b) may be substituted in the complex conjugate force (Equation 2.199) [the pitching moment (Equation 2.202)]:

$$F^* = -i\int_L \left(p_0 - \frac{\rho}{2}v^2\right) dz^*, M = \text{Re}\left\{\int_L \left(p_0 - \frac{\rho}{2}v^2\right) z dz^2\right\}, \tag{2.203a and 2.203b}$$

leading to Equation 2.203a (Equation 2.203b) regardless of whether the flow is rotational or irrotational [Equation 2.203a ≡ Equation I.28.18a (Equation 2.203b ≡ Equation I.28.18b)]. The stagnation pressure is constant (Equation 2.204a), and thus, the integrals (Equations 2.205b and 2.205c) around a closed loop vanish:

$$p_0 = \text{const:} \qquad \int_L p_0 dz^* = 0 = \int_L p_0 z dz^*. \tag{2.204a–2.204c}$$

The assumption of incompressible flow made in the Bernoulli equation (Equation 2.12b) implies constant mass density (Equation 2.205a) that may be taken out of the integrals for the complex conjugate force (Equation 2.203a) [pitching moment (Equation 2.203b)], leading to Equation 2.205b (Equation 2.205c) on account of Equation 2.204b (Equation 2.204c):

$$p_0 = \text{const:} \qquad F^* = \frac{i}{2}\rho\int_L v^2\,\mathrm{d}z^*, \quad M = \mathrm{Re}\left\{-\frac{\rho}{2}\int_L v^2 z\mathrm{d}z^*\right\}. \tag{2.205a–2.205c}$$

The expressions for *the complex conjugate force (Equation 2.205b) and pitching moment (Equation 2.205c) apply to a rigid impermeable body with a closed regular boundary L in the plane steady incompressible flow of an inviscid fluid, regardless of whether the flow is irrotational or rotational.*

The difference lies in the next step: (1) in the irrotational case, the flow is potential and the complex potential (Equation 2.190c) can be used (Subsection I.28.2.2); and (2) in the rotational case, the complex potential does not exist and must be replaced by the stream function (Equation 2.190e) that specifies the square of the modulus of the velocity:

$$v^2 \equiv \left|\vec{v}\right|^2 = v^* v = \left(2i\frac{\partial\Psi}{\partial z}\right)\left(2i\frac{\partial\Psi}{\partial z}\right)^* = \left(2i\frac{\partial\Psi}{\partial z}\right)\left(-2i\frac{\partial\Psi}{\partial z^*}\right) = 4\frac{\partial\Psi}{\partial z}\frac{\partial\Psi}{\partial z^*}. \tag{2.206}$$

Substituting Equation 2.206 in Equation 2.205b (Equation 2.205c) leads to the conjugate force (Equation 2.207a) [pitching moment (Equation Equation 2.207b)]:

$$F^* = 2i\rho\int_L \frac{\partial\Psi}{\partial z}\frac{\partial\Psi}{\partial z^*}\mathrm{d}z^*, \quad M = \mathrm{Re}\left\{-2\rho\int_L \frac{\partial\Psi}{\partial z}\frac{\partial\Psi}{\partial z^*} z\mathrm{d}z^*\right\}. \tag{2.207a and 2.207b}$$

The stream function is constant along the surface L of the body that is a streamline (Equation 2.208a), implying (Equation 2.208b)

$$0 = \mathrm{d}\Psi = \frac{\partial\Psi}{\partial z}\mathrm{d}z + \frac{\partial\Psi}{\partial z^*}\mathrm{d}z^*: \qquad \frac{\partial\Psi}{\partial z}\frac{\partial\Psi}{\partial z^*}\mathrm{d}z^* = -\left(\frac{\partial\Psi}{\partial z}\right)^2\mathrm{d}z. \tag{2.208a and 2.208b}$$

Substituting Equation 2.208b in Equations 2.207a and 2.207b leads to

$$F^* \equiv F_x - iF_y = -2i\rho\int_L\left(\frac{\partial\Psi}{\partial z}\right)^2\mathrm{d}z, \quad M = \mathrm{Re}\left\{2\rho\int_L\left(\frac{\partial\Psi}{\partial z}\right)^2 z\mathrm{d}z\right\}. \tag{2.209a and 2.209b}$$

The **second Blasius theorem** *states that, for a rotational flow, a rigid impermeable body with a closed regular boundary L in a plane steady flow of an inviscid with mass density ρ and stream function Ψ is subject to (1) a horizontal F_x, vertical F_y, and complex conjugate F^* force specified by Equation 2.209a; and (2) a pitching moment specified by Equation 2.209b.* The stream function also replaces the complex potential when passing from the first circle theorem (Sections I.26.7 and I.28.6) in an irrotational flow to the second circle theorem in a rotational flow (Subsection 2.4.3); both are incompressible.

2.5.3 Second Circle Theorem in a Rotational Flow

The real integral of the complex Poisson equation (Equation 2.158a) for the stream function consists (Equation 2.210b) of (1) the general real integral (Equation 2.153a ≡ Equation 2.210a) of the Laplace equation (Equation 2.152b) corresponding to a potential flow, (2) plus a particular integral (Equation 2.157b) of the Poisson equation (Equation 2.157a) with constant vorticity (Equations 2.160a and 2.160b) that appears as the second term on the r.h.s. of Equation 2.210b; and (3) a vortex (Equation 2.194b) with circulation Γ may be added as the third term on the r.h.s. of Equation 2.210b because its streamlines are circular and so the cylinder remains a streamline:

$$\Psi_0(z_1 z^*) = f(z) + f^*(z^*), \quad \Psi_2(z,z^*) = \Psi_0(z,z^*) - \frac{\varpi_0}{4} z^* z - \frac{\Gamma}{4\pi}\log(z^* z). \tag{2.210a and 2.210b}$$

An arbitrary point in the complex plane (Equation 2.211a) has, for **reciprocal point** (Sections I.24.7, I.26.7, I.28.6, and I.35.7) relative to the circle of radius a and center at the origin, the point Equation 2.211b

$$z = re^{i\varphi}, \zeta = \frac{a^2}{z^*} = \frac{a}{r}e^{i\varphi}: \qquad \arg(z) = \varphi = \arg(\zeta), \quad |z||\zeta| = a^2, \tag{2.211a–2.211d}$$

(1) that lies on the same radius through the center (Equation 2.211c); and (2) the product of the distances of the original (Equation 2.211a) and reciprocal (Equation 2.211b) points to the center of the circle is the square of the radius (Equation 2.211d). The insertion of a cylinder of radius a corresponds, by analogy with the first circle theorem (Equation I.24.47), to subtracting Equation 2.210a evaluated at the conjugate point (Equation 2.211b), leading to the stream function:

$$\Psi(z,z^*) = \Psi_2(z,z^*) - \Psi_0\left(\frac{a^2}{z^*}, \frac{a^2}{z}\right) = \Psi_0(z,z^*) - \Psi_0\left(\frac{a^2}{z^*}, \frac{a^2}{z}\right) - \frac{\varpi_0}{4} z^* z - \frac{\Gamma}{4\pi}\log(z^*, z), \tag{2.212}$$

where Equation 2.210b was used.

Substitution of Equation 2.210a in Equation 2.212 yields the **second circle theorem:** *the effect on the plane steady incompressible rotational flow of an inviscid fluid of the insertion of rigid impermeable cylinder with radius a and center at the origin is specified by the stream function:*

$$\Psi(z,z^*) = f(z) - f\left(\frac{a^2}{z^*}\right) + f^*(z^*) - f^*\left(\frac{a^2}{z}\right) - \frac{\varpi_0}{4} z^* z - \frac{\Gamma}{4\pi}\log(z^* z), \tag{2.213}$$

where (1) there are no boundaries other than the cylinder; (2) the singularities of the function f(z) lie inside (outside) the cylinder for (Equation 2.213) the stream function outside (inside) the cylinder; and (3) the result is unaffected by the presence of a linear unidirectional shear flow (Equation

2.160b) with constant vorticity ϖ_0 and a vortex (Equation 2.194b) with circulation Γ. The proof of the second circle theorem comprises the three statements about the stream function Ψ in Equation 2.213: (1) it does not change the flow (Equations 2.210a and 2.210b) at infinity; (2) it does not introduce any additional singularities outside (inside) the cylinder relative to Equations 2.210a and 2.210b; and (3) it is a constant on the cylinder, that is, the latter is a streamline.

Concerning the first statement, the perturbation terms due to the cylinder are constant at infinity (Equation 2.214):

$$\lim_{z\to\infty} \Psi(z,z^*) - \Psi_2(z,z^*) = \lim_{z\to\infty}\left[-f\left(\frac{a^2}{z^*}\right) - f^*\left(\frac{a^2}{z}\right)\right] - \tag{2.214}$$

$$= -f(0) - f^*(0) = -2\,\mathrm{Re}[f(0)] = \text{const},$$

and do not change the velocity in the far field. Concerning statement 2, there are two subcases; (2a) for the flow outside the cylinder (Equation 2.215a), the reciprocal point (Equation 2.211b) lies inside the cylinder (Equation 2.215b), and the functions (Equation 2.215c):

$$|z| > a \Leftrightarrow |\zeta| = \frac{a^2}{|z|} < a: \qquad \Psi(z,z^*) - \Psi_2(z,z^*) = -f\left(\frac{a^2}{z^*}\right) - f^*\left(\frac{a^2}{z}\right) \tag{2.215a–2.215c}$$

may have singularities inside the cylinder but do not introduce any new singularities outside, that is, in the region of interest; (2b) concerning the flow inside the cylinder (Equation 2.216a), the reciprocal point (Equation 2.211b) is outside the cylinder (Equation 2.216b), and the functions (Equation 2.216c):

$$|z| < a \Leftrightarrow |\zeta| = \frac{a^2}{|z|} > a: \qquad -f(z) - f^*(z) = -2\,\mathrm{Re}\{f(z)\} \tag{2.216a–2.216c}$$

may have singularities outside the cylinder, but do not introduce any singularities inside the cylinder. Concerning statement 3 on the cylinder (Equation 2.217a), the stream function (Equation 2.213) is real and constant (Equation 2.217b):

$$z = ae^{i\varphi}: \quad \Psi\left(ae^{i\varphi}, ae^{-i\varphi}\right) = f\left(ae^{i\varphi}\right) - f\left(ae^{i\varphi}\right) + f^*\left(ae^{-i\varphi}\right) - f^*\left(ae^{-i\varphi}\right)$$

$$-\frac{\varpi_0}{4}a^2 - \frac{\Gamma}{4\pi}\log\left(a^2\right) = -\frac{\varpi}{4}a^2 - \frac{\Gamma}{2\pi}\log a, \tag{2.217a and 2.217b}$$

and thus the cylinder is a streamline.

2.6 Cylinder in a Unidirectional Shear Flow

The simplest case of a body in a stream is the cylinder in a uniform flow at angle of attack (Section I.28.6), to which circulation may be added to produce lift or downforce (Section

I.28.7); the problem is solved using the first force and moment (Section I.28.2) and circle (Section I.24.7) theorems for a potential flow. The second forces and moment (Subsection 2.5.2) and circle (Subsection 2.5.3) theorems allow the addition of a linear unidirectional shear flow with a possibly distinct angle of attack (Subsection 2.6.1). The resulting stream function (Subsection 2.6.2) represents a cylinder in a flow at angle of attack accounting for the relative motion, plus a unidirectional shear flow. This corresponds to a body moving relative to a fluid and allows for both the motion of the body relative to an external reference such as the ground and the presence of a uniform wind; a sheared wind causes a rotational flow, for which the simplest example is the linear unidirectional shear flow (Subsection 2.4.5). The vorticity in the incident stream leads to a shear force that can add or subtract (Subsection 2.6.3) to the lift due to the circulation; thus, a sheared wind can affect the lift on an aircraft, with possibly serious consequences in critical flight phases such as takeoff and landing (Subsection 2.6.4).

2.6.1 Unidirectional Shear Flow with Angle of Attack

The linear unidirectional shear flow with velocity in the x-direction sheared in the y-direction (Figure 2.3) is specified (Equation 2.162a) by the stream function (Equation 2.218b):

$$\varpi = 2\varpi_0: \qquad \Psi_2 = \Psi^+ = -\frac{\varpi_0}{4}y^2 = -\frac{\varpi}{2}y^2 = -\frac{\varpi}{2}\left(\frac{z-z^*}{2i}\right)^2 = \frac{\varpi}{8}(z-z^*)^2, \tag{2.218a and 2.218b}$$

for a constant vorticity (Equation 2.218a); a rotation (Equation 2.219a) by an angle β in the positive or counterclockwise direction, such as Equation 2.171b by α, leads to Equation 2.219b:

$$z \to ze^{-i\beta}: \quad \Psi_3 = \frac{\varpi}{8}(ze^{-i\beta} - z^*e^{i\beta})^2 = -\frac{\varpi}{4}z^*z + \frac{\varpi}{8}(z^2e^{-2i\beta} + z^{*2}e^{2i\beta}), \tag{2.219a and 2.219b}$$

that is, *the stream function (Equation 2.219b) for a linear unidirectional shear flow with vorticity* ϖ *and angle of attack* β. Adding (Equation 2.174b) a uniform flow of velocity U and angle of attack α leads to the stream function:

$$\Psi_4 = \Psi_3 + \frac{U}{2i}(ze^{-i\alpha} - z^*e^{i\alpha}) - \frac{\Gamma}{2\pi}\log\left(\sqrt{z^*z}\right), \tag{2.220}$$

where a vortex with circulation Γ was included as the last term (Equation 2.213) from Equation 2.194b. The insertion of a cylinder of radius a by the circle theorem (Equation 2.213) corresponds in Equations 2.220 and 2.219b ≡ Equation 2.221:

$$\Psi_4 = -\frac{\varpi}{4}z^*z - \frac{\Gamma}{4\pi}\log(z^*z) - i\frac{U}{2}(ze^{-i\alpha} - z^*e^{i\alpha}) + \frac{\varpi}{8}(z^2e^{-2i\beta} + z^{*2}e^{2i\beta}), \tag{2.221}$$

to the following: (1) leaving unchanged the first two terms on the r.h.s. of Equation 2.221, that coincide (Equation 2.220) with the last two terms of Equation 2.213, corresponding to the vortex and the first term on the r.h.s. of Equation 2.219b; and (2) to the remaining two terms, which are last on the r.h.s. of Equation 2.221, the same expression at the reciprocal point (Equation 2.211b) is subtracted (Equation 2.213), leading to

$$\Psi(z,z^*) = -\frac{\varpi}{4}z^*z - \frac{\Gamma}{4\pi}\log(z^*z) - i\frac{U}{2}\left[\left(z-\frac{a^2}{z^*}\right)e^{-i\alpha} - \left(z^*-\frac{a^2}{z}\right)e^{i\alpha}\right]$$
$$+\frac{\varpi}{8}\left[\left(z^2-\frac{a^4}{z^{*2}}\right)e^{-2i\beta}+\left(z^{*2}-\frac{a^4}{z^2}\right)e^{2i\beta}\right]. \tag{2.222}$$

The stream function (Equation 2.222) applies to (Figure 2.4) a rigid impermeable cylinder of radius a with center at the origin in the presence of (1) uniform flow with velocity U and angle-of-attack α corresponding to the third term on the r.h.s.; (2) a linear unidirectional shear flow with vorticity ϖ and angle-of-attack β corresponding to the first and fourth terms on the r.h.s.; and (3) a line vortex with circulation Γ per unit length and center on the axis of the cylinder corresponding to the second term on the r.h.s.

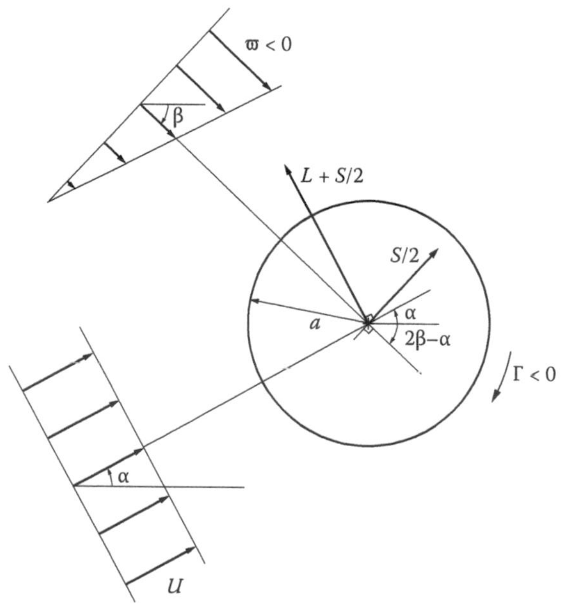

FIGURE 2.4
Rigid impermeable cylinder in a uniform stream of velocity U at angle-of-attack α is acted upon by a lift force L orthogonal to the free stream in the presence of negative circulation Γ. An incident linear unidirectional shear flow with vorticity ϖ and angle-of-attack β causes a shear force S with two components: (1) one-half parallel to the lift and orthogonal to the angle-of-attack α of the free stream; (2) the other half is orthogonal to the direction $2\beta - \alpha$, involving also the angle of attack of the incident linear unidirectional shear flow. If the incident uniform flow and linear unidirectional shear flow have the same angle-of-attack $\alpha = \beta$, then (1) and (2) have the same direction $\alpha = 2\beta - \alpha$, and the shear force adds to the lift (Figures 2.5 and 2.6).

2.6.2 Free Stream, Vorticity, and Circulation Effects

The stream function (Equation 2.222) is specified in polar coordinates (Equation 2.223a) by Equation 2.223b:

$$z = re^{i\varphi}: \qquad \Psi(r,\varphi) = -\frac{\varpi}{4}r^2 - \frac{\Gamma}{2\pi}\log r + U\left(r - \frac{a^2}{r}\right)\sin(\varphi-\alpha) + \frac{\varpi}{4}\left(r^2 - \frac{a^4}{r^2}\right)\cos[2(\varphi-\beta)] \tag{2.223a and 2.223b}$$

where in the passage from Equations 2.222 through 2.223b, the following were used:

$$-i\frac{U}{2}\left[\left(z - \frac{a^2}{z^*}\right)e^{-i\alpha} - \left(z^* - \frac{a^2}{z}\right)e^{i\alpha}\right] = \frac{U}{2i}\left(r - \frac{a^2}{r}\right)\left[e^{i(\varphi-\alpha)} - e^{-i(\varphi-\alpha)}\right] = U\left(r - \frac{a^2}{r}\right)\sin(\varphi-\alpha), \tag{2.224a}$$

$$\frac{\varpi}{8}\left[\left(z^2 - \frac{a^4}{z^{*2}}\right)e^{-2i\beta} + \left(z^{*2} - \frac{a^4}{z^4}\right)e^{2i\beta}\right] = \frac{\varpi}{8}\left(r^2 - \frac{a^4}{r^2}\right)\left[e^{2i(\varphi-\beta)} + e^{-2i(\varphi-\beta)}\right] = \frac{\varpi}{4}\left(r^2 - \frac{a^4}{r^2}\right)\cos[2(\varphi-\beta)], \tag{2.224b}$$

confirming that the stream function consists of real terms for (1) vortex with circulation Γ; (2 and 3) the uniform stream (linear unidirectional shear flow) with velocity U (vorticity ϖ) and angle-of-attack α (β). *The stream function (Equation 2.223b) specifies the polar components of the velocity (Equations 2.186b and 2.186d)*

$$v_r(r,\varphi) = \frac{1}{r}\frac{\partial\Psi}{\partial\varphi} = U\left(1 - \frac{a^2}{r^2}\right)\cos(\varphi-\alpha) - \frac{\varpi}{2}\left(r - \frac{a^4}{r^3}\right)\sin[2(\varphi-\beta)], \tag{2.225a}$$

$$v_\phi(r,\varphi) = -\frac{\partial\Psi}{\partial r} = \frac{\varpi}{2}r + \frac{\Gamma}{2\pi r} - U\left(1 + \frac{a^2}{r^2}\right)\sin(\varphi-\alpha) - \frac{\varpi}{2}\left(r + \frac{a^4}{r^3}\right)\cos[2(\varphi-\beta)]. \tag{2.225b}$$

The stream function (Equation 2.223b) is constant on the cylinder (Equation 2.226a):

$$\Psi(a,\varphi) = -\frac{\varpi}{4}a^2 - \frac{\Gamma}{2\pi}\log a, \quad v_r(a,\varphi) = 0, \qquad p(a,\varphi) = p_0 - \rho[v_\varphi(a,\varphi)]^2, \tag{2.226a–2.226c}$$

$$v_\varphi(a,\varphi) = -2U\sin(\varphi-\alpha) + \frac{\Gamma}{2\pi a} + \varpi a\left\{\frac{1}{2} - \cos[2(\varphi-\beta)]\right\}, \tag{2.226d}$$

where the velocity (Equation 2.226b) is tangential (Equation 2.226d), leading to the pressure distribution (Equation 2.226c). The tangential velocity (Equation 2.226d) consists of three terms due to the (1) uniform free stream, (2) vortex, and (3) the linear unidirectional shear flow. Concerning the uniform stream (1), the velocity (Equation I.28.94b) vanishes at the stagnation points $\varphi = \alpha$, $\alpha + \pi$ in the flow direction, is maximum in modulus $|v_\varphi| = 2U$ at the transverse direction $\varphi = \alpha \pm \pi/2$, and takes intermediate values in between $0 \le |v_\varphi| \le 2U$. Concerning the circulation Γ, it adds a tangential velocity (Equation 2.33b) at the radius $r = a$, leading to the first two terms on the r.h.s. of Equation 2.226d, which coincide with Equation I.28.118b, with $\varphi \to \varphi - \alpha$. The angle–of-attack β also applies to the contribution of the unidirectional shear flow that vanishes for $\cos[2(\varphi-\beta)] = 1/2$, that is, the directions $2(\varphi-\beta) = \pm\pi/3$ or $\varphi_\pm = \beta + \pi/6$.

2.6.3 Lift and Shear Forces and Coefficients

The components of the force and pitching moment on the cylinder can be obtained by either integrating Equations 2.199 and 2.202 with the pressure distribution (Equations 2.226c and 2.226d) or using the stream function (Equation 2.222) in the second Blasius theorem (Equations 2.209a and 2.209b). The second approach is simpler, starting from the stream function (Equation 2.222) and calculating (1) the complex conjugate velocity (Equation 2.190e):

$$-\frac{i}{2}v^* = \frac{\partial\Psi}{\partial z} = -\frac{\varpi}{4}z^* - \frac{\Gamma}{4\pi z} - i\frac{U}{2}\left(e^{-i\alpha} - \frac{a^2}{z^2}e^{i\alpha}\right) + \frac{\varpi}{4}\left(ze^{-2i\beta} + \frac{a^4}{z^3}e^{2i\beta}\right); \tag{2.227}$$

(2) on the cylinder (Equations 2.228a and 2.228b), the complex conjugate velocity (Equation 2.227) simplifies to Equation 2.228c:

$$z = ae^{i\varphi}, \quad z^* = ae^{-i\varphi} = \frac{a^2}{z}: \qquad -\frac{i}{2}v_0^* = \frac{\varpi}{4}ze^{-2i\beta} - i\frac{U}{2}e^{-i\alpha} - \left(\frac{\varpi}{4}a^2 + \frac{\Gamma}{4\pi}\right)\frac{1}{z} + i\frac{U}{2}a^2e^{i\alpha}\frac{1}{z^2} + \frac{\varpi}{4}a^4e^{2i\beta}\frac{1}{z^3}; \tag{2.228a–2.228c}$$

(3) the complex conjugate velocity on the cylinder has been written (Equation 2.228c) as a descending sequence of powers of z, leading to its square (Equation 2.229a):

$$(v_0^*)^2 = O(1) + \frac{A_{-2}}{z^2} + \frac{A_{-1}}{z} + O\left(\frac{1}{z^3}\right), \tag{2.229a}$$

$$A_{-1} = \frac{i}{4}\varpi U a^2 e^{i\alpha - 2i\beta} + iU\left(\frac{\varpi}{4}a^2 + \frac{\Gamma}{4\pi}\right)e^{-i\alpha}, \tag{2.229b}$$

$$A_{-2} = \frac{\varpi}{8}a^4 + \frac{1}{2}U^2 a^2 + \left(\frac{\varpi}{4}a^2 + \frac{\Gamma}{4\pi}\right)^2, \tag{2.229c}$$

where only the coefficients A_{-1} (A_{-2}) of the first (second) inverse power z^{-1} (z^{-2}) have been written explicitly [Equation 2.229b (Equation 2.229c)]; (4) the integrals over the surface of the body L in the second forces and moment theorem (Equations 2.209a and 2.209b) may be shrunk to a small circle of radius ε around the origin (Equation 2.231a) where the singularities of Equation 2.229a lie, leading to (Equation 2.209b):

$$z = \varepsilon e^{i\varphi}: \qquad \int_{|z|=\varepsilon} z^n \, dz = i\varepsilon^{1+n}\int_0^{2\pi} e^{i(1+n)\varphi}\, d\varphi = \begin{cases} 0 & \text{if } n \neq -1, \\ 2\pi i & \text{if } n = -1, \end{cases} \tag{2.230a and 2.230b}$$

in agreement with Equations I.28.24a through I.28.24c ≡ Equation 2.230b; (5) thus, when substituting in the conjugate force (Equation 2.209a) [pitching moment (Equation 2.209b)], only the term $O(z^{-1})$ [$O(z^{-2})$] in Equation 2.229a gives a nonzero contribution:

$$F^* = F_x - iF_y = -2i\rho \int_L \frac{A_{-1}}{z}\, dz = -2i\rho\, 2\pi i\, A_{-1} = 4\pi\rho\, A_{-1}, \tag{2.231a}$$

$$M = \operatorname{Re}\left\{2\rho \int_L \frac{A_{-2}}{z}\, dz\right\} = \operatorname{Re}\left(2\rho\, 2\pi i\, A_{-2}\right) = \operatorname{Re}\left(i4\pi\rho\, A_{-2}\right), \tag{2.231b}$$

involving the coefficient [Equation 2.229b (Equation 2.229c)] in Equation 2.231a (Equation 2.231b).

Because A_{-2} is real (Equation 2.229c), the pitching moment (Equation 2.231b) is zero (Equation 2.232a); the coefficient A_{-1} in Equation 2.229b specifies (Equation 2.231a) the complex conjugate force (Equation 2.232b). Thus, *a rigid impermeable cylinder of radius a with circulation Γ in the plane steady incompressible flow of an inviscid fluid with density ρ in a uniform stream (unidirectional shear flow) of velocity U (vorticity ϖ) with angle-of-attack α(β) experiences no pitching moment (Equation 2.232a) and a complex conjugate force (Equation 2.232b):*

$$M = 0, \quad F^* = F_x - iF_y = i\rho U(\Gamma + \pi\varpi a^2)e^{-i\alpha} + i\rho\pi U\varpi a^2\, e^{i\alpha - 2i\beta}. \tag{2.232a and 2.232b}$$

The pitching moment is zero (Equation 2.232a) because (1) in a potential flow, the pitching moment is due to dipole or spiral terms (Equation I.28.31b) both absent here; and (2) the shear flow causes no pitching moment. The complex conjugate force (Equation 2.232b) involves the **lift (shear) force** *[Equation 2.233a (Equation 2.233c)]:*

$$L = -\rho U\Gamma = C_L \rho U^2 a, \quad C_L \equiv -\frac{\Gamma}{Ua}; \quad S \equiv -\rho U 2\pi\varpi a^2, \quad C_S \equiv -\frac{2\pi a\varpi}{U}. \tag{2.233a–2.233d}$$

The lift (Equation 2.233a) [shear (Equation 2.233c)] force scales (Equations 2.234b and 2.234c) on the diameter 2a of the circle multiplied by the dynamic pressure (Equation 2.234a), that is, the kinetic energy per unit volume, and coincides with the difference of the stagnation p_0 and free stream p pressures in the Bernoulli equation (Equation 2.12b ≡ Equation 2.234a):

$$q \equiv \frac{1}{2}\rho U^2 = p_0 - p: \qquad \{L,S\} = 2qa\{C_L,C_S\} = \rho U^2 a\{C_L,C_S\}. \qquad (2.234a–2.234c)$$

The dimensionless factor in Equation 2.233a ≡ Equation 2.234b (Equation 2.233c ≡ Equation 2.234c) is the **lift (shear) coefficient** *that is specified [Equation 2.233b (Equation 2.233d)] by the circulation Γ (vorticity ϖ).* The analogy $C_L = C_S$ leads to the correspondence (Equation 2.240b) between circulation and vorticity that will be established next.

2.6.4 Correspondence between the Circulation and the Vorticity

Substituting (Equations 2.233a and 2.233c) *the complex conjugate force (Equation 2.232b):*

$$F^* = F_x - iF_y = -i\left(L + \frac{S}{2}\right)e^{-i\alpha} - i\frac{S}{2}e^{i\alpha - 2i\beta} \qquad (2.235)$$

specifies the horizontal (Equation 2.236a) and vertical (Equation 2.236b) force components:

$$F_x = -\left(L + \frac{S}{2}\right)\sin\alpha + \frac{S}{2}\sin(\alpha - 2\beta), \quad F_y = \left(L + \frac{S}{2}\right)\cos\alpha + \frac{S}{2}\cos(\alpha - 2\beta), \qquad (2.236a \text{ and } 2.236b)$$

showing (Figure 2.4) that (1) the lift (Equations 2.233a and 2.233b) is orthogonal (Equation 2.239b) to the free stream direction (Equation 2.237a) and to it adds half of the shear force (Equations 2.233c and 2.233d):

$$\left(L + \frac{S}{2}\right)\{-\sin\alpha, \cos\alpha\} = \left(L + \frac{S}{2}\right)\left\{\cos\left(\alpha + \frac{\pi}{2}\right), \ \sin\left(\alpha + \frac{\pi}{2}\right)\right\}; \qquad (2.237a)$$

$$\frac{S}{2}\{-\sin(2\beta - \alpha), \cos(2\beta - \alpha)\} = \frac{S}{2}\left\{\cos\left(2\beta - \alpha + \frac{\pi}{2}\right), \ \sin(2\beta - \alpha + \frac{\pi}{2}\right); \qquad (2.237b)$$

(2) the other half of the shear force acts orthogonally (Equation 2.237b) to the direction $2\beta - \alpha$. The two directions (1) and (2) coincide if the angle of attack is the same (Equation 2.238a) for the free stream and unidirectional shear flow, leading (Equation 2.235) to the complex conjugate force (Equation 2.238b):

$$\beta = \alpha: \quad F_x - F_y = F^* = i\rho U(\Gamma + \pi a^2 \varpi)e^{-i\alpha} = -i(L + S)e^{-i\alpha}. \qquad (2.238a \text{ and } 2.238b)$$

In this case, the total force (Equation 2.238b):

$$F = L + S: \qquad F_x = -F\sin\alpha = F\cos\left(\alpha + \frac{\pi}{2}\right), \quad F_y = F\sin\left(\alpha + \frac{\pi}{2}\right) \qquad \text{(2.239a–2.239c)}$$

adds the lift and shear forces (Equation 2.239a) in the direction (Equations 2.239b and 2.239c) orthogonal to the common angle of attack for the free stream and unidirectional shear flow.

From Equations 2.239a, 2.233a, and 2.233c ≡ Equation 2.240a follows the extended **Kutta–Joukowski theorem:** *for the lift in a potential flow due to the circulation is equivalent to the shear force due to the vorticity (Equation 2.240b):*

$$F = -\rho U(\Gamma + 2\pi a^2\varpi): \qquad \Gamma = 2\pi a^2\varpi, \qquad v_\varphi = \frac{\Gamma}{2\pi a} = \varpi a, \qquad \text{(2.240a–2.240d)}$$

corresponding to a tangential velocity equal to (1) that (Equation 2.33b ≡ Equation 2.240c) of a vortex of circulation Γ at radius a; (2) the product of the vorticity by the radius of the cylinder (Equation 2.242d). Both the lift and shear force (Equation 2.240b) relate in the same way to the Lamb's vector or the vortical force (Equation 2.117c ≡ Equation 2.241c):

$$\vec{v} = \vec{e}_x U, \quad \vec{\varpi} = -\vec{e}_3\varpi: \qquad \vec{F}_v = \rho\vec{\varpi} \wedge \vec{v} = -\vec{e}_y \rho U \varpi = \vec{e}_y \frac{S}{2\pi a^2} = -\vec{e}_y \frac{\rho U \Gamma}{2\pi a^2} = \vec{e}_y \frac{L}{2\pi a^2}, \qquad \text{(2.241a–2.241c)}$$

choosing the free stream velocity in the x-direction (Equation 2.241a) and noting that the vorticity is orthogonal to the plane of the flow (Equation 2.241b). Thus, (1) a negative circulation (vorticity) produces a lift (Figure 2.5a) because it increases (decreases) the flow

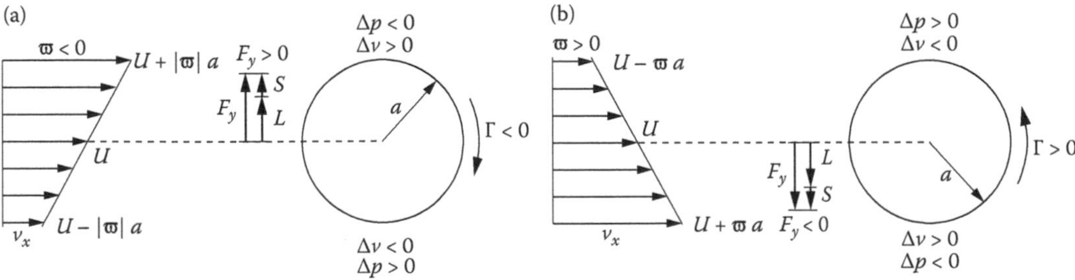

FIGURE 2.5
Negative circulation (vorticity) around the cylinder (in the incident free stream) with the same angle of attack (a) in a uniform stream both: (1) increase the velocity and decrease the pressure above the cylinder; (2) decrease the velocity and increase the pressure below the cylinder. Thus, they are parallel and contribute together to the lift on the cylinder. Conversely, a positive circulation (vorticity) around the cylinder (in the incident free stream) with the same angle of attack (b) both (1) reduce the velocity and increase the pressure above the cylinder; (2) increase the velocity and reduce the pressure below the cylinder. Thus, they are parallel and both contribute to the downforce on the cylinder.

velocity above (below), thus reducing (increasing) the pressure and resulting in a net upward force or lift; (2) conversely, a positive circulation (vorticity) causes (Figure 2.5b) a downforce. When the vorticity and circulation are of the same (opposite) sign, they reinforce (cancel) each other. In particular, a positive vorticity counters the lift of a negative circulation and decreases (increases) the velocity above (below), thus increasing (decreasing) the pressure and causing (Figure 2.6a) reduced vertical upward force, which can be a dangerous shear flow when landing an aircraft; conversely, a negative vorticity decreases the downforce due to a positive vortex (Figure 2.6b). The vorticity effect is important only when the shear force (Equation 2.233a) becomes comparable to the lift (Equation 2.233c). The wind near the ground has a negative vorticity (Figure 2.7a) because the velocity must be zero at the surface; thus, it increases (decreases) the lift on approach to land (climb after takeoff). A wind inversion may occur at low altitude (Figure 2.7b), and if an aircraft crosses at this level, the effect of the shear flow is to reduce (increase) the lift on approach to land (climb after takeoff). The reduction in the lift at takeoff implies (case I) that the aircraft will require a higher speed and a longer runway to lift off; this may be unsafe for takeoff from short runways. A lift increase (case II) is no problem at takeoff because it leads to a shorter run on the ground and earlier liftoff. An increase in lift at landing (case III) will lead to an overshoot, that is, landing farther from the threshold of the runway; this leaves less distance in the ground roll to brake the aircraft to a halt. A decrease in lift at landing (case IV) will lead to an aircraft crash short of the runway unless lift is restored. The lift (Equation 2.233a) can be restored by (1) increasing the velocity U if there is enough thrust to overcome the higher drag (Subsection I.34.9) at the higher speed; and (2) increasing the lift coefficient by flying at a higher angle of attack below stall, that is, below the maximum angle of attack for which the flow separates and there is a lift loss (Subsection I.28.7.7). Thus, the possibility to recover lift is limited by (1) the extra thrust available and the time delay in the response of the engine to throttling to a higher thrust level; and (2) the margin between the current and maximum angle of attack and the ability to avoid stall or other

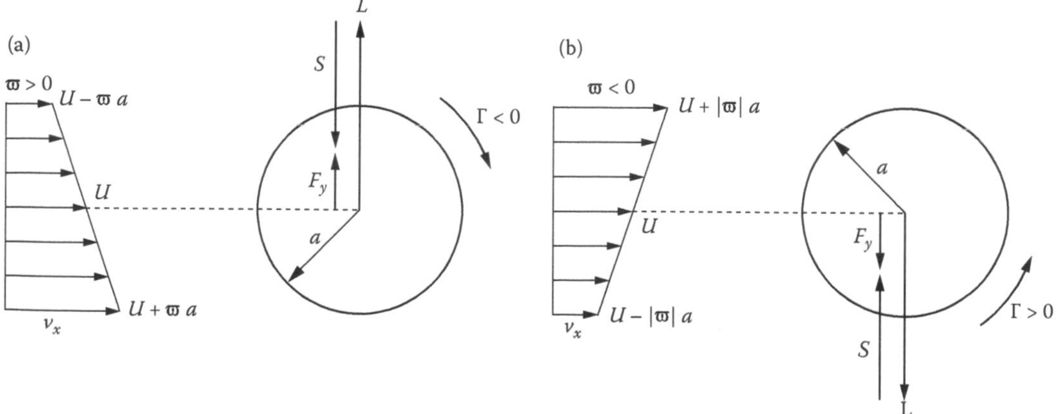

FIGURE 2.6
Lift on a cylinder in a uniform stream due to a negative circulation around the cylinder may be countered (a) by an opposite shear force due to a positive vorticity in the free stream, such as the wind profile in Figure 2.7b; the result is a vortical force smaller than the lift, that is, a lift loss. Conversely, (b) positive circulation around the cylinder would lead to a downforce, which would be reduced by a negative vorticity in the free stream, such as the wind profile in Figure 2.7a.

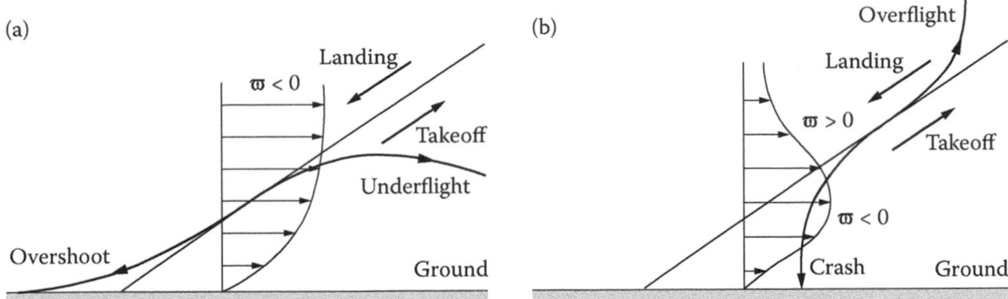

FIGURE 2.7
Wind speed must be zero on the ground and usually increases with altitude leading to the wind profile (a) with negative vorticity. For an aircraft on approach to land, this increases lift (Figure 2.5a), leading if uncompensated to an overshoot, that is, touchdown farther from the edge of the runway, leaving less distance to brake to halt in the ground roll. For an aircraft in the initial climb after takeoff, the vorticity is positive and causes a lift loss, which if uncompensated causes an underflight, that is, a slower rate of climb or a descent toward the ground. In some weather conditions, there is a wind inversion (b) leading to a positive vorticity at some altitude above the ground. An aircraft crossing this altitude on approach to land will experience a lift loss, which if uncompensated, can lead to a crash short of the threshold of the runway. The same positive vorticity encountered on a climb after takeoff increases the lift and leads to an overflight, in which the climb rate is increased, leading to higher flight path. The compensation of a lift gain can be made by reducing the velocity using (1) aerodynamic brakes to increase drag; and (2) throttling back the engine to reduce thrust. Compensation of a lift loss may be more problematic because it reacquires (1) higher velocity if the thrust is below maximum, and a thrust increase is available with a short response time; (2) an increase in angle of attack if below stall, and there is a thrust margin to keep velocity constant and no control problems arise. To try to compensate the lift loss by diving to gain speed involves an altitude loss, and near the ground may lead to a more serious crash with larger vertical velocity. The lift is maximized by flying at the highest possible angle of attack, at the highest possible velocity with maximum thrust. In the case of a crash, this will minimize the vertical impact speed, and if the vorticity is not too strong gives the best chance of flying away from the ground.

forms of loss of control causing departure from stable flight. Of the four cases above, three (cases I, III, and IV) thus raise safety issues related to flying in a sheared wind.

2.7 Monopole Interactions and Equilibrium Positions

The vorticity may be continuously distributed (Sections 2.2 through 2.6) or concentrated into line vortices (Sections 2.7 through 2.9). The line vortices (Subsection 2.1.8) may be combined with sources/sinks (Subsection 2.1.9) to form monopoles (Subsection 2.1.10). An assembly of mutually interacting monopoles is a potential flow that has a centroid at rest (Subsection 2.7.1); for two monopoles, this condition must be satisfied by their relative motion, that is, radial (azimuthal) [Subsection 2.7.2 (Subsection 2.7.3)] for two sources/sinks (vortices). The cases of constant (zero) velocity represent [Subsection 2.7.5 (Subsection 2.7.6)] equilibrium in convection (at a fixed position); the uniform (nonuniform convection) leads to the paths of monopoles [Subsection 2.7.4 (Subsection 2.7.7)]. The "vortex shooting" methods place line vortices and trace their evolution in the flow; in the following cases (Sections 2.7 through 2.9), the problem is generalized, combining the vortices with sources/sinks to form interacting monopoles and find their paths, equilibrium positions, and stability or instability of the latter.

2.7.1 Centroid of an Ensemble of Line Monopoles

An assembly of N line monopoles with moments (Equation 2.242b), consisting of sources/sinks with flow rate Q_n and vortices with circulation Γ_n, has the complex potential (Equation 2.242c) [conjugate velocity (Equation 2.242d)]:

$$z \neq z_n, \quad P_{0n} \equiv Q_n + i\Gamma_n: \quad 2\pi f(z) = \sum_{n=1}^{N} P_{0n} \log\left(z - z_n\right), \quad 2\pi v^*(z) = \sum_{n=1}^{N} \frac{P_{0n}}{z - z_n}, \tag{2.242a–2.242d}$$

that is, the sum of the contributions [Equation 2.192a (Equation 2.191c)] of each monopole at an external flow point is not coincident with any monopole (Equation 2.242a). Because a monopole does not act upon itself, the monopole self-effect must be subtracted out when considering the complex potential (Equation 2.243b) [conjugate velocity (Equation 2.243c)] induced at (Equation 2.243a) monopole m by all others, that is, $n \neq m$ in Equations 2.243b and 2.243c:

$$z \neq z_m: \qquad 2\pi\left\{f_m, v_m^*\right\} = \sum_{\substack{n=1 \\ n\neq m}}^{N} P_{0m} \,\{\log(z_m - z_n), (z_m - z_n)^{-1}\}. \tag{2.243a–2.243c}$$

The **centroid** of an ensemble of monopoles is defined (Equation 2.244a) as the average or mean position weighted by the moments (Equation 2.242b):

$$\bar{x} + i\bar{y} \equiv \bar{z} = \frac{\sum_{n=1}^{N} P_{0n} z_n}{\sum_{n=1}^{N} P_{0n}}: \qquad \bar{v}_x + i\bar{v}_y = \bar{v} = \frac{d\bar{z}}{dt} = \frac{\sum_{n=1}^{N} P_{0n} v_n}{\sum_{n=1}^{N} P_{0n}}. \tag{2.244a and 2.244b}$$

It leads to the **centroid velocity** (Equation 2.244b) by differentiation with regard to time, bearing in mind that the monopole moments (Equation 2.242b) are constant (Equation 2.246a). The numerator in Equation 2.244b is zero because (Equation 2.243c) the induced conjugate velocities of each pair of monopoles cancel:

$$2\pi \sum_{n=1}^{N} P_{0n} v_n = \sum_{n=1}^{N} \sum_{\substack{m=1 \\ m\neq n}}^{N} \frac{P_{0n} P_{0m}}{z_n - z_m} = \sum_{n=1}^{N} \sum_{m=1}^{n-1} P_{0n} P_{0m} \left(\frac{1}{z_n - z_m} + \frac{1}{z_m - z_n}\right) = 0. \tag{2.245}$$

Thus, *the centroid of an assembly of N monopoles is at rest (Equations 2.246c and 2.246d):*

$$P_{0n} \equiv \frac{Q_n - i\Gamma_n}{2\pi} = \text{const}; \qquad \sum_{n=1}^{N} P_{0n} \neq 0: \qquad \bar{v} = 0, \qquad \bar{z} = \text{const}, \tag{2.246a–2.246d}$$

for any values of the constant monopole moments (Equation 2.246a) whose sum is not zero (Equation 2.246b).

2.7.2 Radial Motion of a Pair of Sources/Sinks

Two sources/sinks of flow rates $Q_1(Q_2)$ at positions $z_1(z_2)$ induce on each other velocities along the relative position vector (Equations 2.247a and 2.247b):

$$2\pi v_1^* = \frac{Q_2}{z_1 - z_2}, \qquad 2\pi v_2^* = \frac{Q_1}{z_2 - z_1}. \qquad \text{(2.247a and 2.247b)}$$

From Equations 2.247a and 2.247b follows Equation 2.248a, that is, a particular case of Equation 2.245:

$$Q_1 v_1^* + Q_2 v_2^* = \frac{Q_1 Q_2}{2\pi}\left(\frac{1}{z_1 - z_2} + \frac{1}{z_2 - z_1}\right) = 0; \qquad Q_1 z_1 + Q_2 z_2 = \text{const} = 0. \qquad \text{(2.248a and 2.248b)}$$

The latter (Equation 2.248a) is integrated (Equation 2.248b), where the constant is set to zero by suitable choice of the origin of the coordinate system. If the sources/sinks do not cancel (Equation 2.249a), the centroid is at the origin (Equation 2.249c) for all time because its velocity is zero (Equation 2.249b):

$$Q_1 + Q_2 \neq 0: \qquad \bar{v} = \frac{Q_1 v_1^* + Q_2 v_2^*}{Q_1 + Q_2} = 0, \qquad \bar{z} = \frac{Q_1 z_1 + Q_2 z_2}{Q_1 + Q_2} = 0. \qquad \text{(2.249a–2.249c)}$$

The result (Equations 2.249a through 2.249c) is a particular case of Equations 2.246a through 2.246d. From Equation 2.248b, it follows that the two sources/sinks lie on the same line through the origin, leading to two cases: (I) two sources (Equation 2.250a) or two sinks (Equation 2.250b) are (Equation 2.250c) on opposite sides (Equation 2.250d) of the centroid (Figure 2.8a and b):

$$Q_1 > 0 < Q_2, \quad Q_1 < 0 > Q_2: \quad Q_1 Q_2 > 0, \quad |Q_1| z_1 = -|Q_2| z_2; \qquad \text{(2.250a–2.250d)}$$

$$Q_1 > 0 > Q_2, \quad Q_1 Q_2 < 0: \quad |Q_1| z_1 = |Q_2| z_2; \qquad \text{(2.251a–2.251c)}$$

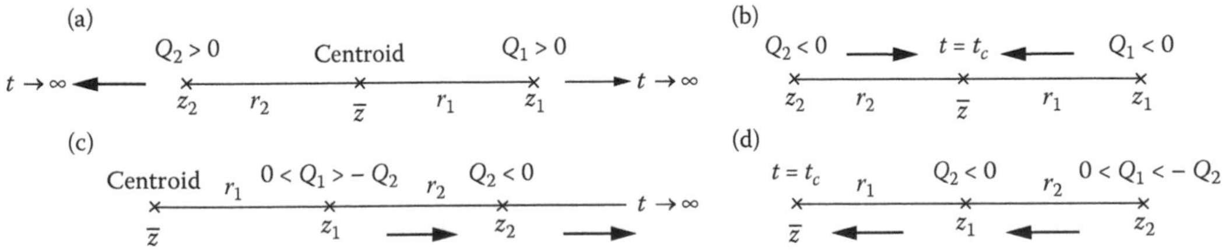

FIGURE 2.8
Two sources or sinks cause upon each other induced velocities corresponding to a radial motion passing through a fixed point, namely, the centroid: (a) two sources repel each other, taking an infinite time to reach infinity because the finite-induced velocities reduce as the distance increases; (b) two sinks attract each other and collide at the centroid in a finite time because the induced velocity increases as they approach each other; (c) a source with a larger flow rate than the modulus of the flow rate of the sink lies closer to the centroid, and thus, the source pushes the sink to infinity, and the sink pulls the source to infinity, that is, reached by both in an infinite time as in case (a); (d) conversely, if the source has a smaller flow rate than the modulus of the flow rate of the sink, then the sink is closer to the centroid and pulls the source to the centroid, and the source pushes the sink to the centroid, both colliding there after the same finite time as in case (b).

(II) a source and a sink [Equation 2.251a (Equation 2.251b)] are on the same side (Equation 2.151c) of the centroid (Figure 2.8c and d).

In case I of two sources or sinks (Equations 2.250a through 2.250d), because they lie on the opposite sides of the centroid: (1) the azimuthal angles φ_1 and φ_2 are constant and equal to the initial value (Equations 2.252a and 2.252b) with a difference of π between the two; and (2) the positions (Equations 2.252c and 2.252d) lead Equation 2.248b to Equation 2.252e:

$$\varphi_1(t) = \varphi_0 = \text{const} = \varphi_2(t) - \pi: \qquad z_1 = r_1 e^{i\varphi_1} = r_1 e^{i\varphi_0}, \qquad z_2 = r_1 e^{i\varphi_2} = -r_2 e^{i\varphi_0}, \quad r_1 Q_1 = r_2 Q_2. \tag{2.252a–2.252e}$$

The latter (Equation 2.252e) may be used in the radial velocities (Equations 2.247a and 2.247b):

$$2\pi v_{1r} = 2\pi \frac{dr_1}{dt} = \frac{Q_2}{r_1 + r_2} = \frac{1}{r_1} \frac{Q_2}{1 + Q_1/Q_2} = \frac{Q_2}{r_1} \frac{Q_2}{Q_1 + Q_2}, \tag{2.253a}$$

$$2\pi v_{2r} = 2\pi \frac{dr_2}{dt} = \frac{Q_1}{r_1 + r_2} = \frac{1}{r_2} \frac{Q_1}{1 + Q_2/Q_1} = \frac{Q_1}{r_2} \frac{Q_1}{Q_1 + Q_2}. \tag{2.253b}$$

Thus, *the relative motion of two sources (Equation 2.250a) or sinks (Equation 2.250b) is along the same radius (Equations 2.252a and 2.252b) in opposite directions (Equations 2.252c and 2.252d) with velocities (Equations 2.253a and 2.253b ≡ Equations 2.254a and 2.254b) specified by the* **effective flow rate** *(Equations 2.254c and 2.254d):*

$$\{v_1, v_2\} = \left\{\frac{dr_1}{dt}, \frac{dr_2}{dt}\right\} = \frac{1}{2\pi}\left\{\frac{\bar{Q}_2}{r_1}, \frac{\bar{Q}_1}{r_2}\right\}, \quad \{\bar{Q}_1, \bar{Q}_2\} = \frac{\left\{(Q_1)^2, (Q_2)^2\right\}}{Q_1 + Q_2}, \tag{2.254a–2.254d}$$

that are (1 and 2) positive for two sources (Equation 2.250a) or a source stronger than the sink (Equation 2.264a); and (3 and 4) negative for two sinks (Equation 2.250b) or a sink stronger than the source (Equation 2.265a).

The integration of Equations 2.254a and 2.254b from the initial radial positions (r_{10}, r_{20}) at time $t = 0$ leads at time $t \neq 0$ to the radial positions $\{r_1, r_2\}$:

$$\begin{aligned} \frac{\{\bar{Q}_2, \bar{Q}_1\}}{\pi} t &= \frac{\left\{(Q_2)^2, (Q_1)^2\right\}}{Q_1 + Q_2} \frac{1}{\pi} \int_0^t d\tau = 2\int_{\{r_{10}, r_{20}\}}^{\{r_1, r_2\}} \xi \, d\xi \\ &= [r^2]_{\{r_{10}, r_{20}\}}^{\{r_1, r_2\}} = \left\{(r_1)^2 - (r_{10})^2, (r_2)^2 - (r_{20})^2\right\}. \end{aligned} \tag{2.255}$$

Thus, *the trajectories of two sources or two sinks are given by*

$$r_1(t) = \left|(r_{10})^2 + \frac{\bar{Q}_1}{\pi} t\right|^{1/2} = \left|(r_{10})^2 + \frac{(Q_2)^2}{Q_1 + Q_2} \frac{t}{\pi}\right|^{1/2}, \tag{2.256a}$$

$$r_2(t) = \left| (r_{20})^2 + \frac{\bar{Q}_2}{\pi} t \right|^{1/2} = \left| (r_{20})^2 + \frac{(Q_1)^2}{Q_1 + Q_2} \frac{t}{\pi} \right|^{1/2}, \tag{2.256b}$$

leading to two subcases: (I-A) two sources (Equation 2.250a ≡ Equation 2.257a) have a positive effective flow rate (Equations 2.254c and 2.254d) and repel each other (Figure 2.8a) toward infinity (Equations 2.257c and 2.257d) because their induced velocities point outward:

$$Q_1 > 0 < Q_2: \qquad \bar{Q}_1 > 0 < \bar{Q}_2, \quad \lim_{t\to\infty} r_1(t) = \infty = \lim_{t\to\infty} r_2(t), \tag{2.257a–2.257d}$$

$$Q_1 < 0 > Q_2: \qquad \bar{Q}_1 > 0 < \bar{Q}_2, \quad r_1(t_c) = 0 = r_2(t_c); \tag{2.258a–2.258d}$$

(I-B) two sinks (Equation 2.250b ≡ Equation 2.258a) have negative (Equation 2.258b) effective flow rates (Equations 2.254c and 2.254d) and attract each other toward the centroid that they reach together (Equations 2.258c and 2.258d) at the same finite time.

$$\begin{aligned} t_c &= -\pi(Q_1 + Q_2)\left(\frac{r_{10}}{Q_2}\right)^2 = -\pi(Q_1 + Q_2)\left(\frac{r_{20}}{Q_1}\right)^2 \\ &= \pi|Q_1 + Q_2|\left(\frac{r_{10}}{Q_2}\right)^2 = \pi|Q_1 + Q_2|\left(\frac{r_{20}}{Q_1}\right)^2. \end{aligned} \tag{2.259a–2.259d}$$

The two times (Equations 2.259a and 2.259b) coincide because Equation 2.248b implies (Equation 2.260a) in all cases of sources or sinks:

$$(Q_1 r_1)^2 = (Q_2 r_2)^2: \qquad \left(\frac{r_{10}}{Q_2}\right)^2 = \left(\frac{r_{20}}{Q_1}\right)^2, \tag{2.260a and 2.260b}$$

since (Equation 2.260a) must hold all the time and it also holds for the initial time (Equation 2.260b), confirming that the two expressions (Equation 2.259a ≡ Equation 2.259b) coincide, that is, the monopoles must reach the centroid at the same time. The two sources (Equation 2.257a) take an infinite time to reach infinity (Equations 2.257c and 2.257d) because the distance is infinite and the velocity (Equations 2.253a and 2.253b) is finite and decays to zero with distance; the two sinks (Equation 2.258a) take a finite time (Equations 2.259a through 2.259d) to reach the origin (Equations 2.258c and 2.258d) because the distance is finite and the velocities (Equations 2.253a and 2.253b) increase without bound as the centroid is approached.

In the case II of a source and sink (Equation 2.251a), index 1 (2) will be used for the source (sink). They lie (Equation 2.251c) along the same line through the center (Equations 2.261a and 2.261b) on the same side (Equations 2.263c and 2.263d), leading to the radial velocities (Equations 2.261e and 2.261f):

$$\varphi_1(t) = \varphi_0 = \text{const} = \varphi_2(t), \quad \{z_1, z_2\} = \{r_1, r_2\} e^{i\varphi_0}: \quad \{v_1, v_2\} = \left\{\frac{dr_1}{dt}, \frac{dr_2}{dt}\right\} = \frac{\{Q_2, Q_1\}}{r_1 - r_2}. \tag{2.261a–2.261f}$$

Using Equation 2.251a ≡ Equations 2.262a and 2,262b and Equations 2.251c, 2.261c, and 2.261d ≡ Equation 2.262c:

$$0 < Q_1 = |Q_1|, 0 > Q_1 = -|Q_2|: \quad Q_1 r_1 = |Q_1||z_1| = |Q_2||z_1| = -Q_2 r_2 \tag{2.262a–2.262c}$$

leads to the radial velocities

$$2\pi v_1 = 2\pi \frac{dr_1}{dt} = \frac{Q_2}{r_1 - r_2} = \frac{1}{r_1} \frac{Q_2}{1 + Q_1/Q_2} = \frac{(Q_2)^2}{Q_1 + Q_2} \frac{1}{r_1} = \frac{\bar{Q}_2}{r_1}, \tag{2.263a}$$

$$2\pi v_2 = 2\pi \frac{dr_2}{dt} = \frac{Q_2}{r_1 - r_2} = \frac{1}{r_2} \frac{Q_1}{1 + Q_2/Q_1} = \frac{(Q_1)^2}{Q_1 + Q_2} \frac{1}{r_1} = \frac{\bar{Q}_1}{r_1}, \tag{2.263b}$$

that coincide with Equations 2.254a and 2.254b ≡ Equations 2.263a and 2.263b. Hence (Equation 2.255), *the integrals (Equations 2.256a and 2.256b) are the same for a source and sink (Equation 2.251a), leading to two subcases: (II-A) if the source is stronger than the sink (Equation 2.264a), it lies closer to the centroid (Equation 2.264b) and both move away from the centroid toward infinity (Equations 2.264c and 2.264d):*

$$Q_1 = |Q_1| > |Q_2| = -Q_2; \quad r_1 = \left|\frac{Q_2}{Q_1}\right| r_2 < r_2: \qquad \lim_{t\to\infty} r_1(t) = \infty = \lim_{t\to\infty} r_2(t), \tag{2.264a–2.264d}$$

because (Figure 2.8c) the source pushes (sink pulls) the sink (source) outward toward infinity; and (II-B) if the sink is stronger than the source (Equation 2.265a), then it lies closer to the centroid (Equation 2.265b) and both fall to the centroid (Equations 2.265c and 2.265d) in a finite $t = t_c$ *in Equations 2.259c and 2.259d:*

$$-Q_2 = |Q_2| > |Q_1| = Q_1; \qquad r_2 = \left|\frac{Q_1}{Q_2}\right| r_2 < r_1: \qquad r_1(t_c) = 0 = r_2(t_c), \tag{2.265a–2.265d}$$

because (Figure 2.8d) the sink pulls (source pushes) the source (sink) toward the centroid. The time of fall to the centroid (Equations 2.259a through 2.259d) is the same in both cases of (1) two sinks (Equations 2.258c and 2.258d) and (2) a sink stronger than a source (Equations 2.265a through 2.265d).

2.7.3 Azimuthal Motion of Two Vortices

Two vortices with circulation Γ_1, Γ_2 have induced velocities orthogonal to the relative position vector (Figure 2.9a and 2.9b):

$$v_1^* = -\frac{i}{2\pi} \frac{\Gamma_2}{z_1 - z_2}, \qquad v_2^* = -\frac{i}{2\pi} \frac{\Gamma_1}{z_2 - z_1}. \tag{2.266a and 2.266b}$$

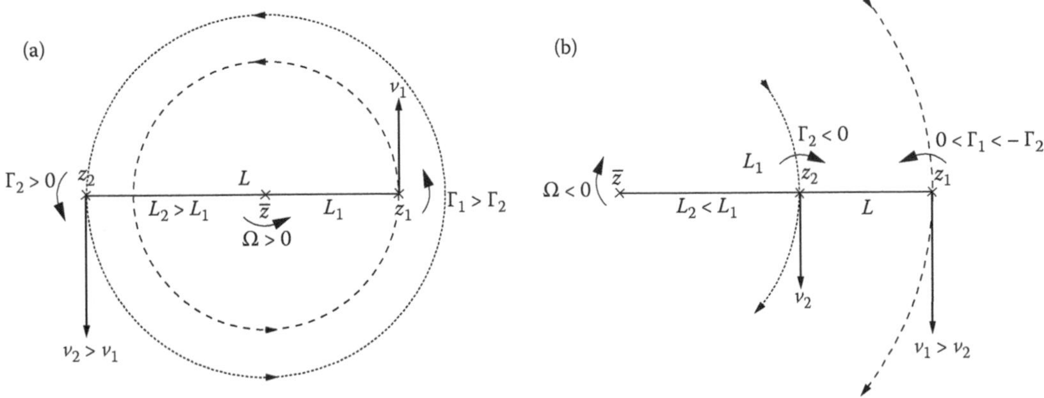

FIGURE 2.9
Two vortices induce on each other an azimuthal velocity and thus rotate around the centroid with the same angular velocity, staying on the same radial line through the centroid. If the vortices have circulations with the same (opposite) sign, they lie on (a) opposite sides [(b) the same side] of the centroid. The rotation is (a) counterclockwise (clockwise) for two vortices with positive (negative) circulation. If the vortices have circulations with opposite signs, the circulation with a larger modulus determines the direction of rotation, that is, a dominant positive (negative) circulation leads to a counterclockwise [(b) clockwise] rotation. In all cases, the vortices stay on the same radial line through the centroid, implying that the angular velocity is the same in modulus and direction. All parameters of the two vortices, such as azimuthal velocities, distance from the centroid, and trajectories, can be determined from three parameters, for example, the two circulations and the mutual distance.

From Equations 2.266a and 2.266b follows Equation 2.267a:

$$v_1^*\Gamma_1 + v_2^*\Gamma_2 = -i\frac{\Gamma_1\Gamma_2}{2\pi}\left(\frac{1}{z_1 - z_2} + \frac{1}{z_2 - z_1}\right) = 0; \quad z_1\Gamma_1 + z_2\Gamma_2 = \bar{z} = 0. \qquad \text{(2.267a and 2.267b)}$$

Integration of Equation 2.267a leads to Equation 2.267b, where the constant vanishes by a suitable choice of origin. If the vortices do not balance (Equation 2.168a), then Equations 2.267a and 2.267b imply that the centroid is at rest (Equation 2.268b) at the origin (Equation 2.268c):

$$\Gamma_1 + \Gamma_2 \neq 0: \qquad \bar{v} = \frac{\Gamma_1 v_1 + \Gamma_2 v_2}{\Gamma_1 + \Gamma_2} = 0, \qquad \bar{z} = \frac{\Gamma_1 z_1 + \Gamma_2 z_2}{\Gamma_1 + \Gamma_2} = 0. \qquad \text{(2.268a–2.268c)}$$

This is a particular case (Equations 2.268a through 2.268c) of the general theorem (Equations 2.246a through 2.246d). The vortices have velocities orthogonal to the line joining them that passes through the centroid, leading to two cases: (1) if the circulations have the same sign (Equation 2.269a or 2.269b), implying (Equation 2.169c) that the vortices are on the opposite sides of the centroid (Figure 2.9a):

$$\Gamma_1 > 0 < \Gamma_2 \text{ or } \Gamma_1 < 0 > \Gamma_2: \quad \Gamma_1\Gamma_2 > 0, \quad |\Gamma_1|z_1 = -|\Gamma_2|z_2; \qquad \text{(2.269a–2.269c)}$$

$$\Gamma_1 > 0 > \Gamma_2: \quad \Gamma_1\Gamma_2 < 0, \quad |\Gamma_1|z_1 = |\Gamma_2|z_2; \qquad \text{(2.270a–2.270c)}$$

(2) if the vortices have circulation with opposite signs (Equation 2.270a), then (Equation 2.270b) they lie on the same side (Equation 2.270c) of the centroid (Figure 2.9b). In both cases, the vortices move in circle around the centroid with a radius equal to the initial distance (Equations 2.271a and 2.271b):

$$r_1(t) = |z_1| = |z_{10}| = L_1, \quad r_2(t) = |z_2| = |z_{20}| = L_2, \quad L = |z_2 - z_2| = L_1 + L_2, \tag{2.271a–2.271c}$$

and because they lie on the same line through the centroid, their mutual distance (Equation 2.271c) is also constant.

The velocities (Equations 2.266a and 2.266b) of the two vortices are azimuthal (Equations 2.272a and 2.272b) and have moduli whose ratio (Equation 2.272c) is the inverse of the ratio of moduli of the circulations:

$$\left|v_1^*\right| = \left|v_{\varphi 1}\right| = \frac{\left|\Gamma_2\right|}{2\pi L}, \quad \left|v_2^*\right| = \left|v_{\varphi 2}\right| = \frac{\left|\Gamma_1\right|}{2\pi L}: \qquad \frac{\left|v_1\right|}{\left|v_2\right|} = \frac{\left|v_{\varphi 1}\right|}{\left|v_{\varphi 2}\right|} = \frac{\left|\Gamma_2\right|}{\left|\Gamma_1\right|}. \tag{2.272a–2.272c}$$

Because the vortices rotate around the centroid while staying on the same radial line through the centroid (Equation 2.267b), they have the same angular velocity Equation 2.273a:

$$\frac{\left|v_{\varphi 1}\right|}{L_1} = \frac{\left|v_1^*\right|}{\left|z_1\right|} = \frac{\left|v_2^*\right|}{\left|z_2\right|} = \frac{\left|v_{\varphi 2}\right|}{L_2} \equiv \Omega, \quad L = L_1 + L_2 = \frac{\left|v_1\right| + \left|v_2\right|}{\Omega} = \frac{\left|\Gamma_1\right| + \left|\Gamma_2\right|}{2\pi L \Omega}, \tag{2.273a and 2.273b}$$

that can be calculated from Equation 2.273b.

Thus, *all parameters of the relative motion of two vortices can be expressed in terms of the circulations* (Γ_1, Γ_2) *and mutual distance* L, *namely, (1) the angular velocity (Equation 2.273b ≡ Equation 2.274b) that involves the* **total circulation**, *defined (Equation 2.274a) as the sum of the moduli of the circulations:*

$$\Gamma_0 \equiv \left|\Gamma_1\right| + \left|\Gamma_2\right|: \qquad \Omega = \frac{\Gamma_0}{2\pi L^2}; \qquad \frac{\{L_1, L_2\}}{L} = \frac{\left\{\left|\Gamma_2\right|, \left|\Gamma_1\right|\right\}}{2\pi \Omega L^2} = \frac{\left\{\left|\Gamma_2\right|, \left|\Gamma_1\right|\right\}}{\Gamma_0}; \tag{2.274a–2.274d}$$

(2) the distances of the vortices from the centroid (Equations 2.274c and 2.274d); (3) the velocities (Equations 2.275a and 2.275b):

$$\left\{\left|v_1\right|, \left|v_2\right|\right\} = \Omega\{L_1, L_2\} = \frac{\Gamma_0}{2\pi} \frac{\{L_1, L_2\}}{L^2} = \frac{\left\{\left|\Gamma_2\right|, \left|\Gamma_1\right|\right\}}{2\pi L}; \qquad \frac{\left|v_1\right|}{\left|v_2\right|} = \frac{\left|\Gamma_2\right|}{\left|\Gamma_1\right|} = \frac{L_1}{L_2}; \tag{2.275a–2.275c}$$

(4) the **lever rule** *(Equation 2.275c) implies that the vortex with larger circulation* $|\Gamma_2| > |\Gamma_1|$ *has lower velocity* $|v_2| < |v_1|$ *and smaller distance to the centroid* $L_2 < L_1$.

The preceding results can be obtained as follows: (1) solving Equation 2.273b yields the angular velocity (Equation 2.274b) in terms of the total circulation (Equation 2.274a); (2) substituting Equation 2.274b in Equations 2.272a and 2.272b and using Equations 2.273a and 2.273b leads to

$$\{|\Gamma_1|, |\Gamma_2|\} = 2\pi L\{|v_2|, |v_1|\} = 2\pi \Omega L\{L_2, L_1\}, \tag{2.276}$$

that can be solved for the distances (Equations 2.271a and 2.271b) of the vortices from the centroid (Equations 2.274c and 2.274d); (3) substituting Equations 2.274c and 2.274d in Equations 2.273a and 2.273b yields the azimuthal velocities (Equations 2.275a and 2.275b) in agreement with Equations 2.266a, 2.266b, and 2.271c; and (4) the lever rule (Equation 2.275c) follows from Equations 2.266a and 2.266b or Equations 2.275b and 2.275c or Equation 2.273a. *The trajectories of the two vortices are circles with a radius equal to the initial distance from the centroid (Equations 2.271a and 2.271b); the angular velocity is the same for the two vortices (Equations 2.277a and 2.277b) both in modulus (Equations 2.274a and 2.274b) and direction (Equation 2.277c):*

$$\varphi_1(t) = \varphi_{10} + \Omega t, \qquad \varphi_2(t) = \varphi_{20} + \Omega t, \qquad \frac{\varphi_1(t) - \varphi_{10}}{\varphi_2(t) - \varphi_{20}} = 1. \qquad (2.277a\text{–}2.277c)$$

The vortices start and remain in the same radial line through the centroid either on the same side (Equations 2.278a and 2.278b) or on opposite sides (Equations 2.278c and 2.278d) of the centroid:

$$\varphi_{20} = \varphi_{10}\text{:} \quad \varphi_2(t) = \varphi_1(t); \quad \varphi_{20} = \varphi_{10} + \pi\text{:} \quad \varphi_2(t) = \varphi_1(t) + \pi. \qquad (2.278a\text{–}2.278d)$$

Thus, not only the modulus of the angular velocity but also the direction of rotation must be the same: (1) clockwise rotation (Equation 2.779a) applies to two vortices in Figure 2.9a with positive circulation (Equation 2.279b), or a counterclockwise vortex with larger circulation in modulus than the clockwise vortex (Equation 2.279c):

$$\Omega = \frac{d\varphi}{dt} > 0\text{:} \qquad \Gamma_1 > 0 < \Gamma_2 \qquad \text{or} \qquad \Gamma_1 > 0 > -\Gamma_2; \qquad (2.279a\text{–}2.279c)$$

$$\Omega < 0\text{:} \quad \Gamma_1 < 0 > \Gamma_2 \quad \text{or} \quad 0 < \Gamma_1 < = \Gamma_2; \qquad (2.280a\text{–}2.280c)$$

(2) counterclockwise rotation (Equation 2.280d) for two vortices with a negative circulation (Equation 2.280b), or a counterclockwise vortex with (Figure 2.9b) smaller circulation in modulus than the clockwise vortex (Equation 2.278c).

The motion of two vortices (Subsection 2.7.3) is specified by the circulations through their ratios (sum), for example, in Equation 2.275c (Equation 2.274a). The motion of two sources/sinks (Subsection 2.7.2) involves *the effective flow rates of the sources/sinks that relate (Equations 2.254c and 2.254d ≡ Equations 2.281b and 2.281c) to a single combined* **reduced flow rate** *(Equation 2.281a):*

$$\bar{Q} = \frac{Q_1 Q_2}{Q_1 + Q_2}\text{:} \qquad \{\bar{Q}_1, \bar{Q}_2\} = \bar{Q}\left\{\frac{Q_1}{Q_2}, \frac{Q_2}{Q_1}\right\}, \qquad (2.281a\text{–}2.281c)$$

through inverse factors. The inverse of the reduced flow rate is the sum of the inverses of the individual flow rates (Equation 2.282a), such as the law of association (Equation I.4.27a) of mechanical (electrical) circuits in series (parallel):

$$\frac{1}{\bar{Q}} = \frac{Q_1 + Q_2}{Q_1 Q_2} = \frac{1}{Q_1} + \frac{1}{Q_2}; \qquad Q_1 = Q_2 \equiv Q_0\text{:} \quad \bar{Q} = \frac{Q_0}{2}; \qquad (2.282a\text{–}2.282c)$$

in particular, for equal sources/sinks (Equation 2.182b), the reduced flow rate is one-half of the individual flow rate (Equation 2.182c). If one source/sink has a much larger flow rate in modulus (Equation 2.283a), the reduced flow rate (Equation 2.281a) corresponds to the weakest source/sink in the lowest order approximation:

$$|Q_1| >> |Q_2|: \qquad \bar{Q} = \frac{Q_2}{1+Q_2/Q_1} = Q_2 \sum_{n=0}^{\infty} \left(-\frac{Q_2}{Q_1} \right)^n = Q_2 - \frac{(Q_2)^2}{Q_1} + \cdots + (-)^n \frac{(Q_2)^{n+1}}{(Q_1)^n},$$

(2.283a and 2.283b)

with the higher order approximations (Equation 2.283b) specified by the geometric series (Equation I.21.62c). Given two sources (Equation 2.284a) or sinks (Equation 2.284b) implying (Equation 2.284c)

$$Q_1 > 0 < Q_2, \quad Q_1 < 0 > Q_2: \qquad Q_1, Q_2 > 0: \qquad |\bar{Q}| < |Q_1|, \quad |Q_2|,$$

(2.284a–2.284d)

the reduced flow rate cannot exceed the smallest flow rate, all taken in modulus (Equation 2.284d). The lower limit in Equation 2.284d is set by Equations 2.283a and 2.283b, with one source/sink much weaker than the other.

2.7.4 Source/Sink, Vortex, or Monopole near a Wall

The influence of a wall on the flow due to a monopole is specified by its image(s) and thus corresponds to a system of monopoles. The simplest case is a monopole at a distance a from a rigid impermeable wall (Figure 2.10): the velocity must be tangent to the wall, implying

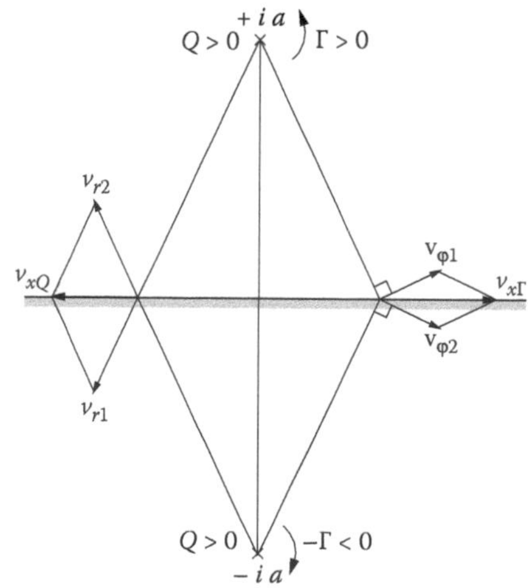

FIGURE 2.10
Monopole consisting of a source/sink and a vortex at a distance a from a rigid impermeable straight infinite wall must lead to a tangential velocity corresponding to an image at the same distance from the wall and complex conjugate monopole moment, that is, (1) a source/sink with the same flow rate, so that the radial velocities add to a tangential velocity at the wall; (2) a vortex with opposite circulation, so that the azimuthal velocities also add to a tangential velocity at the wall.

that the image of the monopole at $z_0 = ia$ is located at $z_0^* = -ia$ with the same flow rate (opposite circulation), so that the sum of the radial (azimuthal) velocities is parallel to the wall. Thus (Equation 2.192a), *the flow for a monopole of moment (Equation 2.285a) at a distance a from a wall along the x-axis has complex potential Equation 2.285b:*

$$P_0 = Q - i\Gamma: \qquad f(z) = P_0 \log(z - ia) + P_0^* \log(z - ia)$$
$$= \frac{Q - i\Gamma}{2\pi} \log(z - ia) + \frac{Q + i\Gamma}{2\pi} \log(z + ia) \qquad \text{(2.285a and 2.285b)}$$
$$= \frac{1}{2\pi}\left[Q \log(z^2 + a^2) + i\Gamma \log\left(\frac{z + ia}{z - ia}\right)\right],$$

corresponding to the complex conjugate velocity Equation 2.285d:

$$z \neq \pm ia: \qquad v^*(z) = \frac{df}{dz} = \frac{P_0}{z - ia} + \frac{P_0^*}{z + ia} = \frac{1}{2\pi}\left(\frac{Q - i\Gamma}{z - ia} + \frac{Q + i\Gamma}{z + ia}\right) = \frac{1}{\pi}\frac{Qz + \Gamma a}{z^2 + a^2},$$
$$\text{(2.285c and 2.285d)}$$

valid at any point excluding (Equation 2.285c) the monopole and its image. On the wall, $z = x$ is real, and thus, the complex conjugate velocity (Equation 2.285d) is real ($v^* = v_x$), showing that the velocity is tangential; because the normal velocity is zero, the boundary condition for a rigid impermeable wall is met.

In Equation 2.286a, the monopole (its image) in the induced velocity (Equation 2.286b) is due only to its image (original monopole), that is, the second (first) term on the r.h.s. of Equation 2.285d with $z = ia$:

$$z = \pm ia: \qquad v^*(\pm ia) = \frac{Q \pm i\Gamma}{2\pi 2ia} = \pm\frac{1}{4\pi a}(\Gamma \mp iQ) = v_x - iv_y. \qquad \text{(2.286a and 2.286b)}$$

Thus, *in the presence of a rigid impermeable flat infinite wall (Figure 2.11): (1) a source (sink) is repelled from (attracted to) the wall with vertical velocity (Equation 2.287a); and (2) a vortex moves parallel to the wall (Equation 2.287c) with constant velocity (Equation 2.287b) to the right (left) for positive (negative) circulation:*

$$v_y = \frac{Q}{4\pi a}, \quad v_x = \frac{\Gamma}{4\pi a}, \quad y = a; \qquad \frac{dy}{dt} = \frac{Q}{4\pi y}, \qquad \frac{dx}{dt} = \frac{\Gamma}{4\pi y}. \qquad \text{(2.287a–2.287e)}$$

For a monopole, both Equations 2.287a and 2.287b apply, corresponding (Equation 2.287c) to the coupled system of two ordinary differential equations (Equations 2.287d and 2.287e) for the trajectory. The vertical velocity (Equation 2.287a ≡ Equation 2.287d) depends (Equation 2.287c) on the distance from the wall:

$$\frac{Q}{2\pi}t = \frac{Q}{2\pi}\int_0^t d\tau = 2\int_{y_0}^{y} \xi\, d\xi = \left[\xi^2\right]_{y_0}^{y} = y^2 - y_0^2, \qquad \text{(2.288a and 2.288b)}$$

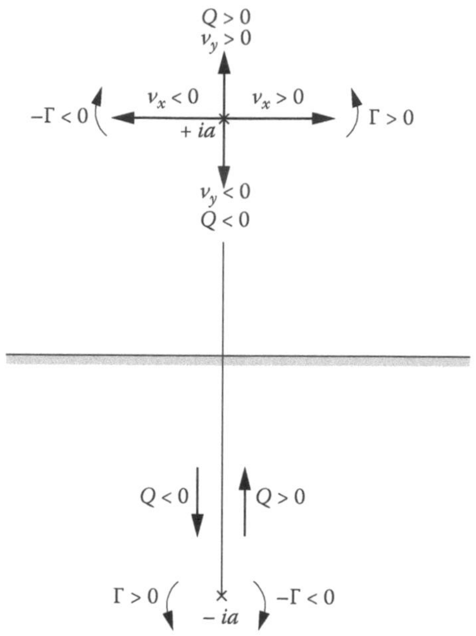

FIGURE 2.11
Monopole at a distance a from a rigid impermeable infinite flat wall is acted upon by an induced velocity due to its image (Figure 2.10), consisting of (1) a vertical velocity orthogonal to the wall, away from (toward) the wall for a source (sink); (2) a horizontal velocity parallel to the wall with direction to the right (left) for positive (negative) circulation.

leading to (Equation 2.288b ≡ Equation 2.289a):

$$y(t) = \left| y_0^2 + \frac{Qt}{2\pi} \right|^{1/2}; \qquad x(t) = x_0 + \frac{\Gamma t}{4\pi y} = x_0 + \frac{\Gamma t}{4\pi} \left| y_0^2 + \frac{Qt}{2\pi} \right|^{-1/2}. \qquad (2.289a–2.289c)$$

From the horizontal velocity (Equation 2.287e) follows Equation 2.289b; substitution of Equation 2.289a in Equation 2.289b yields Equation 2.289c.

Thus, *a monopole of moment (Equation 2.285a) at an initial position (x_0, y_0) in the presence of (Figures 2.10 and 2.11) a straight infinite rigid impermeable wall at the real axis has a trajectory specified by the coordinates (Equations 2.289a and 2.289c) as a function of time. Elimination of the time (Equation 2.289a ≡ Equation 2.290a) leads to the equation for the path (Equation 2.290b):*

$$\frac{t}{y} = \frac{2\pi}{Q}\left(y - \frac{y_0^2}{y} \right): \qquad x = x_0 + \frac{\Gamma}{2Q}\left(y - \frac{y_0^2}{y} \right). \qquad (2.290a and 2.290b)$$

Thus, four cases arise (Figure 2.12): (1 and 2) for a source (case I), the monopole (Equation 2.291a) moves away from the wall (Equation 2.291b) to the right (left) for the subcase I-A (I-B) of positive (negative) circulation, diverging to infinity (Equation 2.291c):

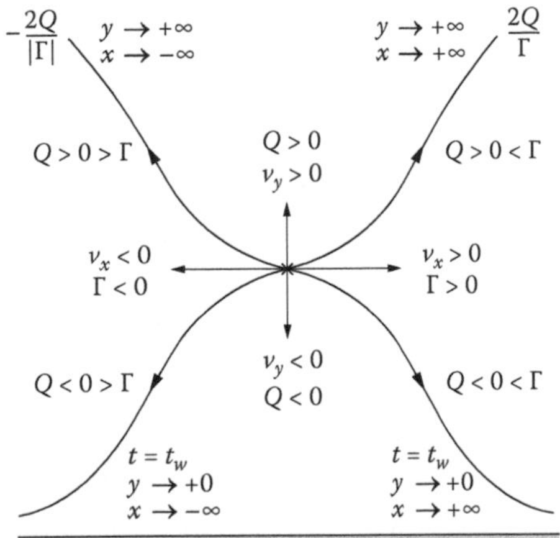

FIGURE 2.12
Induced velocities (Figure 2.11) on a monopole near a rigid impermeable flat infinite wall (Figure 2.10) lead to a path that (1 and 2) for a source, it implies a motion away from the wall toward infinity, to the right (left) for positive (negative) circulation, the time taken to reach infinity being infinite, because the velocity decreases as the distance from the wall and the image increases; (3 and 4) for a sink, it implies a collision with the wall in a finite time, to the right (left) for a positive (negative) circulation, at an infinite distance because the induced velocity increases as the distance from the wall and the image decreases. In cases 1 and 2, the path is asymptotically a straight line with a slope equal to the ratio of the double of the flow rate to the circulation; in cases 3 and 4, the path is tangent to the wall at infinity after a finite time when the monopole and its image meet and cancel each other at the wall.

$$Q > 0: \quad \lim_{t\to\infty} y(t) = \infty, \quad \lim_{t\to\infty} x(t) = \infty\,\text{sgn}(\Gamma) = \begin{cases} +\infty & \text{if} \quad \Gamma > 0, \\ -\infty & \text{if} \quad \Gamma < 0; \end{cases} \tag{2.291a–2.291c}$$

$$Q < 0: \quad y(t_w) = 0, \quad t_w = -\frac{2\pi y_0^2}{Q} = \frac{2\pi y_0^2}{|Q|}, \quad \lim_{t\to t_w} x(t) = \infty\,\text{sgn}(\Gamma); \tag{2.292a–2.292d}$$

(3 and 4) for a sink (case II), the monopole (Equation 2.292a) is attracted toward the wall that it reaches (Equation 2.292b) after a finite time (Equation 2.292c), at a position (Equation 2.292d ≡ Equation 2.291c) at infinity to the right (left) for positive (negative) circulation. The infinite distance in Equation 2.292d arises because the horizontal velocity diverges (Equation 2.287e) as the monopole approaches the wall and its image $y \to 0$. The finite time (Equation 2.292c) corresponds to the vanishing (Equation 2.292b) of the argument of the square root in Equation 2.289a for a sink (Equation 2.292a).

The asymptotic shape of the path (Figure 2.12) is as follows: (1) for a source (Equation 2.293a) at large time (Equations 2.289a and 2.289c), a straight line with slope (Equation 2.293b) equal to the ratio of the double of the flow rate to the circulation:

$$Q > 0: \quad \lim_{t\to\infty} \frac{y(t)}{x(t)} = \frac{2Q}{\Gamma}; \qquad Q < 0: \quad \lim_{t\to t_2} \frac{y(t)}{x(t)} = \frac{y(t_w)}{x(t_w)} = 0; \tag{2.293a–2.293d}$$

(2) for a sink (Equation 2.293c), the path is tangent (Equation 2.293d) to the wall (Equation 2.293b) after a finite time (Equation 2.293c). The asymptotic direction of the path (Equation 2.293b) for a source for large time follows from (Equations 2.289a and 2.289c):

$$\lim_{t\to\infty}\frac{y(t)}{x(t)}=\lim_{t\to\infty}\sqrt{\frac{Qt}{2\pi}}\frac{4\pi}{\Gamma t}\sqrt{\frac{Qt}{2\pi}}=\frac{2Q}{\Gamma}. \tag{2.294}$$

In conclusion (1) a monopole combining a source/sink and a vortex is never at rest near a rigid impermeable wall; (2) a source (sink) alone moves (Equation 2.287a) nonuniformly (Section I.16.1) away from (toward the) wall, tending to infinity (hitting the wall) at infinite time $t \to \infty$ [finite time (Equation 2.292c)]; and (3) a vortex alone moves (Equation 2.287b) uniformly (Section I.16.2) parallel to the wall to the right (left) for positive negative circulation. Superposing a uniform stream with horizontal velocity (Equation 2.295c) opposite to Equation 2.287b on the complex potential (Equation 2.285b) [conjugate velocity (Equation 2.285d)] due to a vortex only (Equation 2.295a) leads to

$$\Gamma\neq 0=Q: \qquad F(z)=f(z)+Uz=\frac{\Gamma}{2\pi}\left[i\log\left(\frac{z+ia}{z-ia}\right)-\frac{z}{2a}\right] \tag{2.295a and 2.295b}$$

$$U=-\frac{\Gamma}{4\pi a}: \qquad V^*(z)=v^*(z)-U=\frac{\Gamma}{\pi}\left(\frac{a}{z^2+a^2}-\frac{1}{4a}\right), \tag{2.295c and 2.295d}$$

the complex potential (Equation 2.295b) [conjugate velocity (Equation 2.295d) of a vortex (Equation 2.295a) of circulation Γ at rest at a distance a from a rigid impermeable infinite flat wall in a uniform flow with velocity (Equation 2.295c) parallel to the wall.

2.7.5 Pair of Conjugate Monopoles in a Parallel-Sided Duct

Consider again the pair of conjugate monopoles, such as the original monopole and its image on a wall (Figure 2.10), but remove the wall, leaving a pair of conjugate monopoles in free space, specifying the same potential flow (Subsection 2.7.4). Next, place the pair inside a duct with parallel infinite flat rigid impermeable walls at $\mathrm{Im}(z) = \pm b/2$, the whole being symmetric (Figure 2.13). This can be done in two ways: (1) by a conformal mapping from the ζ-plane into a horizontal strip $|\mathrm{Im}(z)| \le b/2$ in the z-plane (Figure 2.13) and (2) by considering the doubly infinite system of images on two walls (Figure 2.14). The same result is obtained by the two methods of images (conformal mapping), as shown before for a single source/sink [Section I.36.1 (Section I.36.2)] or vortex (Section I.37.1) between walls. For the present case of a pair of conjugate monopoles symmetrically placed between parallel walls (Figure 2.13), the conformal mapping from free space is used to obtain the complex potential and conjugate velocity (Subsection 2.7.5), that proves that the boundary condition of zero normal velocity at the rigid impermeable walls is met (Subsection 2.7.6); they also specify the velocity of the monopoles (Subsection 2.7.7) leading to the positions for equilibrium at rest (Subsection 2.7.8) and otherwise to the paths in Subsection 2.7.9. The potential flow due to two monopoles with complex conjugate moments symmetrically placed between parallel walls (Figure 2.15) is obtained in three steps. First, consider two monopoles (Equation 2.192a) with complex

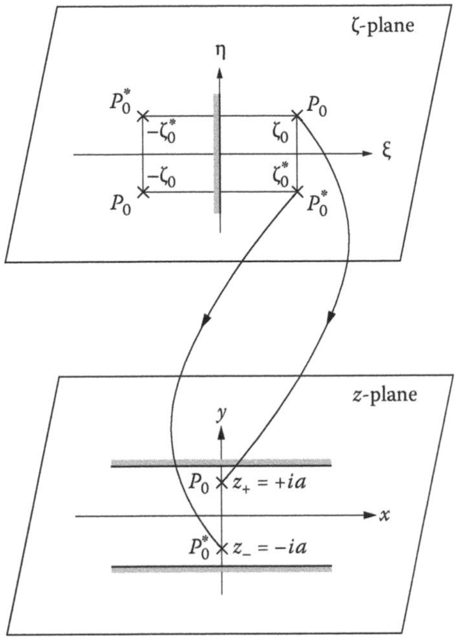

FIGURE 2.13
Conformal mapping of the r.h.s. half-ζ-plane into a horizontal strip of the z-plane transforms the potential flow due to a pair of monopoles with complex conjugate moments near an infinite wall into the potential flow due to the same monopoles in a parallel-sided duct with rigid impermeable flat infinite walls. The conformal mapping is equivalent to the consideration of the doubly infinite system of images on the two walls, shown in Figure 2.14 for monopoles at equal distance from the walls and axis of the duct. These are the positions of equilibrium at rest (Figure 2.15) that are unstable because small deviations exist for which the resulting motion is further away from the position of rest.

conjugate moments (Equations 2.301a and 2.301b) symmetrically placed (Figure 2.10) relative to the x-axis at $\left(\zeta_0, \zeta_0^*\right)$ in the ζ-plane (Equation 2.296):

$$2\pi f_0(\zeta) = P_0 \log(\zeta - \zeta_0) + P_0^* \log\left(\zeta - \zeta_0^*\right). \tag{2.296}$$

Second, insert a vertical wall along the y-axis (Figure 2.13 top) corresponding (Figure I.16.4) to

$$\begin{aligned} 2\pi f(\zeta) &= 2\pi f_0(\zeta) + P_0^* \log\left(\zeta + \zeta_0^*\right) + P_0 \log(\zeta + \zeta_0) \\ &= P_0\left[\log(\zeta - \zeta_0) + \log(\zeta + \zeta_0)\right] + P_0^*\left[\log\left(\zeta - \zeta_0^*\right) + \log\left(\zeta + \zeta_0^*\right)\right] \\ &= P_0 \log\left(\zeta^2 - \zeta_0^2\right) + P_0^* \log\left(\zeta_0^2 - \zeta_0^{*2}\right), \end{aligned} \tag{2.297}$$

adding the image monopoles of $\zeta_0(\zeta_0^*)$ at $-\zeta_0^*(-\zeta_0)$ with moments $P_0^*(P_0)$.

Third, map the right-hand half-ζ-plane into a horizontal strip in the z-plane (Figure 2.13, bottom) by a conformal mapping $\zeta(z)$ specified by an analytic function. The function

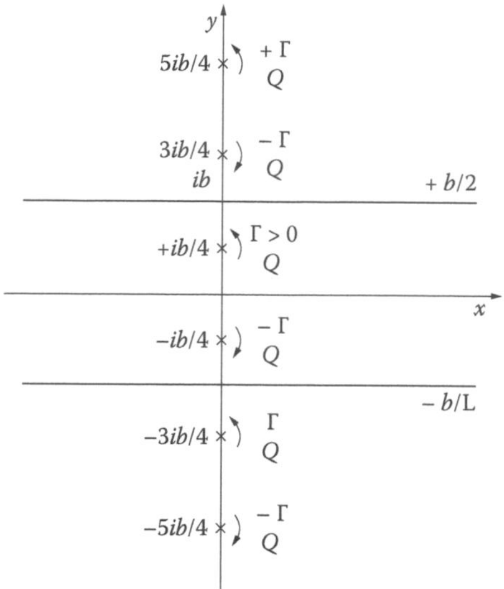

FIGURE 2.14
Considering a pair of monopoles with complex conjugate moments symmetrically placed relative to the axis of a parallel-sided duct (Figure 2.13), they can be in equilibrium at rest only if (1) for opposite vortices they are at equal distance from the axis and the walls, that is, at positions $y = a = \pm b/4$; (2) for sources (sinks), a uniform flow with velocity $U = |Q|/b$ must be added in the direction to the right (left). The equilibrium at rest is possible because for a vortex (source/sink), the horizontal (vertical)-induced velocities due to the other vortex (source/sink) and all the images cancel.

(Equation I.36.145b) maps the right-hand half-ζ-plane into a vertical strip, and for a horizontal strip, the factor i is omitted, leading to Equation 2.298a whose inverse is Equation 2.298b:

$$\xi + i\eta = \zeta = e^{\pi z/b}: \qquad x + iy = z = \frac{b}{\pi}\log\zeta = \frac{b}{\pi}\log|\zeta| + i\frac{b}{\pi}\arg(\zeta). \qquad \text{(2.298a and 2.298b)}$$

It can be confirmed that the right-hand half-ζ-plane (Equations 2.299a and 2.299c) is projected by the inverse (Equation 2.298b) conformal mapping (Equation 2.298a) into a horizontal strip (Equations 2.299b and 2.299d) of the z-plane:

$$0 \le |\zeta| < \infty: \qquad -\infty < x = \frac{b}{\pi}\log|\zeta| = \frac{b}{\pi} < +\infty; \quad -\frac{\pi}{2} < \arg(\zeta) < \frac{\pi}{2}: \quad -\frac{b}{2} < y = \frac{b}{\pi}\arg(\zeta) \le \frac{b}{2}.$$
(2.299a–2.299d)

Besides (1) the rest of the z-plane is divided into horizontal strips of width $2b$ (b), where the conformal mapping (Equation 2.298a), that is periodic, has the same value (Equation 2.300a) [with alternating sign (Equation 2.300b)]:

$$n \in |Z: \qquad e^{\pi(z+2nb)/b} = e^{\pi z/b + 2n\pi} = e^{\pi z/b}, \quad e^{\pi(z+nb)/b} = e^{\pi z/b + n\pi} = (-)^n e^{\pi z/b};$$
(2.300a and 2.300b)

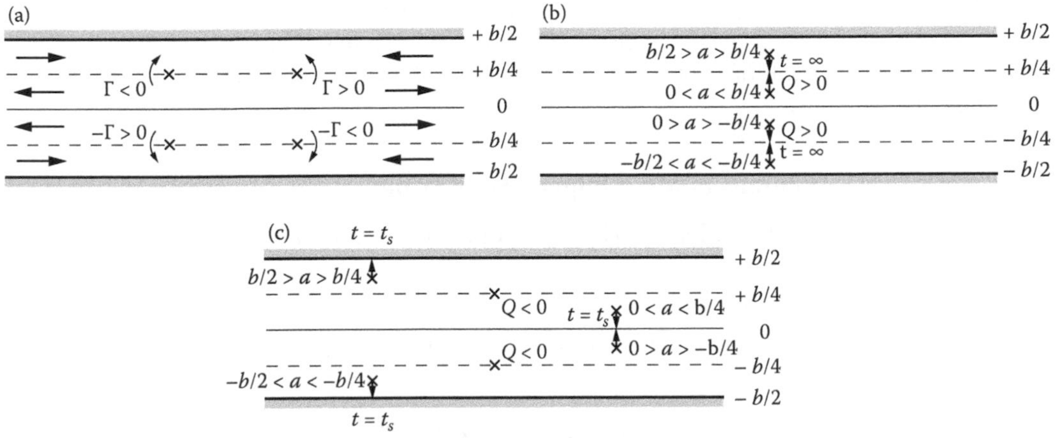

FIGURE 2.15
Considering a pair of monopoles with complex conjugate moments placed in a parallel-sided duct (Figure 2.13), the positions of equilibrium are at equal distance from the axis and the walls (Figure 2.14), and are unstable (stable) for vortices and sinks (sources) as shown by the following deviations (Figure 2.15) that lead to motion further away from (back toward) equilibrium. (a) Two vortices with opposite circulations are at rest at $a = b/4$ equal distance from the axis and walls of the duct. If the vortices are displaced vertically toward the wall $|a| > b/4$ (axis $|a| < b/4$), the other vortex and all the images induce a net horizontal velocity to the left (right), if the upper vortex has positive circulation; the opposite for negative circulation. (b) Two sources are at rest in a duct at equal distance from the walls and axis $|a| = b/4$ in the presence of a uniform flow with horizontal velocity $U = -Q/b$ to the left; if displaced vertically to any position above or below the velocities induced by the other source, and all the images are dominated by the nearest source, this causes a motion back toward the position of stable equilibrium, that is, reached in an infinite time, because the vertical velocity vanishes as it is approached. (c) Two sinks are at rest at equal distance from the walls and axis of the duct $|a| = b/4$ in the presence of a uniform flow with horizontal velocity $U = |Q|/b$ to the right. If displaced vertically toward the wall $|a| > b/4$ (axis $|a| < b/4$), the velocity is dominated by the nearest image (real) sink, which causes an increasing attraction as it is approached, so the motion continues away from the unstable equilibrium with the wall (axis) being reached in a finite time.

(2) the left-hand half-ζ-plane is not mapped into the z-plane at all. In order that the monopoles with complex conjugate moments (Equations 2.301a and 2.301b) be placed at Equations 2.301c and 2.301d in the z-plane, they must come from the points (Equations 2.301e and 2.301f) in the ζ-plane (Equation 2.298a):

$$P_0 = Q - i\Gamma,\ P_0^* = Q - i\Gamma: \qquad z_\pm = \pm ia, \quad \zeta_0 = \zeta(+ia) = e^{i\pi a/b}, \zeta_0^* = \zeta(-ia) = e^{-i\pi a/b}. \tag{2.301a–2.301f}$$

Substituting Equations 2.301a, 2.301b, 2.301e, and 2.301f in Equation 2.297 specifies

$$\begin{aligned} 2\pi f(z) &= (Q - i\Gamma)\left[\log(\mathrm{e}^{\pi z/b} - \mathrm{e}^{i\pi a/b}) + \log(\mathrm{e}^{\pi z/b} + \mathrm{e}^{i\pi a/b})\right] \\ &\quad + (Q + i\Gamma)\left[\log(\mathrm{e}^{\pi z/b} - \mathrm{e}^{-i\pi a/b}) + \log(\mathrm{e}^{\pi z/b} + \mathrm{e}^{-i\pi a/b})\right] \\ &= (Q + i\Gamma)\log(\mathrm{e}^{2\pi z/b} - \mathrm{e}^{i2\pi a/b}) + (Q + i\Gamma)\log(\mathrm{e}^{2\pi z/b} - \mathrm{e}^{-i2\pi a/b}), \end{aligned} \tag{2.302}$$

with the complex potential (Equation 2.302) [conjugate velocity (Equation 2.303)] due to two monopoles with complex conjugate moments (Equations 2.301a and 2.301b) symmetrically placed (Equations 2.301c and 2.301d) in a parallel-sided channel of width b with (Figure 2.15) flat infinite rigid impermeable walls.

2.7.6 Boundary Conditions at Rigid Impermeable Walls

The complex potential (Equation 2.302) corresponds to the conjugate velocity:

$$\begin{aligned} v^*(z) = \frac{df}{dz} &= \frac{Q-i\Gamma}{2b}\left[\frac{1}{1-e^{\pi(ia-z)/b}}+\frac{1}{1+e^{\pi(ia-z)/b}}\right] \\ &+\frac{Q+i\Gamma}{2b}\left[\frac{1}{1-e^{-\pi(ia+z)/b}}+\frac{1}{1+e^{-\pi(ia+z)/b}}\right] \\ &=\frac{1}{b}\left[\frac{Q-i\Gamma}{1-e^{2\pi(ia-z)/b}}+\frac{Q+i\Gamma}{1+e^{-2\pi(ia+z)/b}}\right], \end{aligned} \tag{2.303}$$

where the following was used:

$$\frac{d}{dz}\left[\log(e^{\pi z/b}\pm e^{\pm i\pi a/b})\right]=\frac{\pi}{b}\frac{e^{\pi z/b}}{e^{\pi z/b}\pm e^{\pm i\pi a/b}}=\frac{\pi/b}{1\pm e^{\pi(\pm ia-z)/b}}. \tag{2.304}$$

The complex potential (Equation 2.302) and conjugate velocity (Equation 2.303) are both real (Equations 2.305b and 2.306a, respectively) on the walls (Equation 2.305a):

$$\begin{aligned} z=x\pm i\frac{b}{2}:\quad 2\pi f(z) &= (Q-i\Gamma)\log(-e^{2\pi x/b}-e^{i2\pi a/b}) \\ &+(Q+i\Gamma)\log(-e^{2\pi x/b}-e^{-i2\pi a/b}) \\ &=2\,\mathrm{Re}\left\{(Q-i\Gamma)\log(-e^{2\pi x/b}-e^{i2\pi a/b})\right\}=\Phi,\ \Psi=0, \end{aligned} \tag{2.305a–2.305c}$$

$$bv^*(z)=\frac{Q-i\Gamma}{1+e^{2\pi(ia-x)/b}}+\frac{Q+i\Gamma}{1+e^{-2\pi(ia-x)/b}}=\mathrm{Re}\left\{\frac{Q-i\Gamma}{1+e^{\pm 2\pi(a-2)/b}}\right\}=v_x,\ v_y=0, \tag{2.306a and 2.306b}$$

showing that the duct walls are streamlines (Equation 2.305c) [the velocity is horizontal (Equation 2.306b), hence tangent to the walls (Equation 2.306a)]. Thus, the rigid impermeable wall boundary conditions are met.

2.7.7 Velocity of the Monopoles between Parallel Walls

The complex potential (Equation 2.302) [conjugate velocity (Equation 2.303)] is valid only outside the monopoles. The induced velocity at a monopole is due to the other monopoles plus the images, that is, it can be obtained by subtracting out of Equation 2.303 the self-induced velocity to eliminate the singularity at $z = a$, so that the limit $z \to a$ can be taken, leading to

$$v_0^* = \lim_{z \to ia}\left[v^*(z) - \frac{Q - i\Gamma}{2\pi}\frac{1}{z - ia}\right]. \tag{2.307}$$

Only the first term of the r.h.s. of Equation 2.303 is singular at $z = ia$, as can be seen from its Laurent series expansion to $O(z - ia)$, to be evaluated next. The first three terms of (Equation 1.26b) the MacLaurin series for the exponential:

$$1 - e^{\pi(ia - z)/b} = -\pi\frac{ia - z}{b}\left[1 + \pi\frac{ia - z}{2b} + O\left(\left(\frac{ia - z}{b}\right)^2\right)\right] \tag{2.308}$$

lead to the first two terms of the Laurent series expansion:

$$\begin{aligned}\frac{Q - i\Gamma}{2b}\frac{1}{1 - e^{\pi(ia - z)/b}} &= \frac{Q - i\Gamma}{2\pi}\frac{1}{(z - ia)\left[1 - \pi\dfrac{z - ia}{2b} + O\left(\left(\dfrac{z - ia}{b}\right)^2\right)\right]}\\ &= \frac{Q - i\Gamma}{2\pi}\frac{1 + \pi(z - ia)/(2b) + O\left(\left(z - ia\right)/b\right)^2}{z - ia}.\end{aligned} \tag{2.309}$$

The terms singular at $z = ia$ on the l.h.s. and r.h.s. of Equation 2.309 differ by

$$\frac{Q - i\Gamma}{2b}\frac{1}{1 - e^{\pi(ia - z)/b}} - \frac{Q - i\Gamma}{2\pi}\frac{1}{z - ia} = \frac{Q - i\Gamma}{4b} + O\left(\frac{z - ia}{b}\right). \tag{2.310}$$

In the limit $z \to ia$, only the constant term, that is, the first on the r.h.s. of Equation 2.310, appears in Equation 2.307, and hence in the velocity (Equation 2.303) of the monopole:

$$\begin{aligned}2bv_0^* &= \lim_{z \to ia}\left\{\frac{Q - i\Gamma}{1 - e^{\pi(ia - z)/b}} - \frac{Q - i\Gamma}{\pi/b}\frac{1}{z - ia}\right\}\\ &\quad + \left\{\frac{Q - i\Gamma}{1 + e^{\pi(ia - z)/b}} + (Q + i\Gamma)\left[\frac{1}{1 + e^{-\pi(ia + z)/b}} + \frac{1}{1 - e^{-\pi(ia + z)/b}}\right]\right\}_{z = ia}\\ &= \frac{Q - i\Gamma}{2} + \frac{Q - i\Gamma}{2} + (Q + i\Gamma)\left(\frac{1}{1 + e^{-i2\pi a/b}} + \frac{1}{1 - e^{-i2\pi a/b}}\right)\\ &= Q - i\Gamma + (Q + i\Gamma)\frac{2}{1 - e^{-i4\pi a/b}} = Q - i\Gamma + (Q + i\Gamma)\frac{2e^{i2\pi a/b}}{e^{i2\pi a/b} - e^{i2\pi a/b}}\\ &= Q - i\Gamma + (Q + i\Gamma)\frac{\cos(2\pi a/b) + i\sin(2\pi a/b)}{i\sin(2\pi a/b)}\\ &= Q - i\Gamma + (\Gamma - iQ)\left[\cot\left(\frac{2\pi a}{b}\right) + i\right] = 2Q + (\Gamma - iQ)\cot\left(\frac{2\pi a}{b}\right).\end{aligned} \tag{2.311}$$

A steady motion is possible only parallel to the walls, implying that the velocity (Equation 2.311 ≡ Equation 2.312):

$$v_1^* = v_0^* - \frac{Q}{b} = \frac{\Gamma - iQ}{2b}\cot\left(\frac{2\pi a}{b}\right) \tag{2.312}$$

must be real for convective equilibrium. Its vanishing specifies the conditions of equilibrium at rest.

2.7.8 Uniform Convection and Equilibrium at Rest

For a pair of monopoles with complex conjugate moments (Equations 2.301a and 2.301b) symmetrically placed (Equations 2.301c and 2.301d) between parallel walls (Equation 2.299d), the velocity is given by Equation 2.312, implying that (1) for a vortex (Equation 2.313a), the velocity (Equation 2.313b) is parallel to the walls and coincides with Equation I.36.214a, where the parallel walls (Figure I.36.14c) are vertical instead of horizontal, hence the factor $i = e^{i\pi/2}$:

$$Q = 0 \neq \Gamma: \qquad v_1^* = v_0^* = \frac{\Gamma}{2b}\cot\left(\frac{2\pi a}{b}\right) = \frac{i\Gamma}{2b}\coth\left(i\frac{2\pi a}{b}\right); \qquad v_0^*\left(\frac{b}{4}\right) = 0; \tag{2.313a–2.313c}$$

(2) the velocity of the vortex vanishes when (Equation 2.313c) the distance from the wall is one quarter the width of the channel (Figure 2.14); (3) for the same position (Equation 2.314c ≡ Equation 2.313c), the velocity (Equation 2.314b) of the source (Equation 2.314a) reduces:

$$Q \neq 0 = \Gamma: \qquad v_0^* - \frac{Q}{b} = Q\cot\left(\frac{2\pi a}{b}\right) = v_1^*, \quad v_1^*\left(a = \frac{b}{4}\right) = \frac{Q}{b} \tag{2.314a–2.314c}$$

to Q/b; (4) this velocity is independent of the positions of the sources/sinks; (5) the sources/sinks produce a net volume flux into the channel, which is not the case for vortices; (6) in the case of a pair of sources/sinks with flow rate Q, inside a channel of width b, the total flow rate 2Q is split in two directions, leading in each direction to a flow rate Q and the velocity Q/b that appears in Equation 2.314c; (7) if this velocity is subtracted out of the complex potential (Equation 2.302) [conjugate velocity (Equation 2.303)] in the resulting flows [Equation 2.315a (Equation 2.315b)]:

$$U = \frac{Q}{b}: \quad F(z) = f(z) - Uz = f(z) - Q\frac{z}{b}, \quad V^*(z) = \frac{dF}{dz} = \frac{df}{dz} - Uz = v^*(z) - \frac{Q}{b}, \tag{2.315a and 2.315b}$$

the sources are also at rest like the vortices; (8) this corresponds (Equation 2.315a) to the suppression of the motion of a vortex near a wall by superimposing a uniform flow (Equations 2.295a through 2.295d); (9) both the velocities $v_0^*(v_1^*)$ *of the vortex (Equation 2.313b) [source (Equation 2.314b)] are positive for monopole closer to the centerline* $0 < a < b/4$ *than to the wall, when the images below the channel predominate; (10) the reverse for monopole closer to the wall than to* *the centerline $b/4 < a < b/2$ because the images on the other side of the wall predominate; (11) the reverse of (9 and 10) for the other monopole with conjugate moment located below the centerline $a < 0$; (12) the reverse of (9–11) for the sink instead of the source.* The preceding remarks suggest (Figure 2.14) the following predictions concerning the stability of the position of equilibrium: (1) if a vortex is displaced from the equilibrium position towards the nearest wall (axis), the velocities induced by the other vortex and all the images of both vortices are out of balance, and the resultant motion parallel to the walls shows that the equilibrium position is unstable; (2) if a sink is displaced from the equilibrium position towards the nearest wall (axis), the dominant attraction by the nearest image (other sink) causes the motion to continue towards the wall (axis), showing that the equilibrium position is again unstable; (3) if a source is displaced from the equilibrium position towards the nearest wall (axis), the dominant repulsion from the nearest image (other source), causes a return to the equilibrium position, that thus is stable. The prediction that *the equilibrium position is stable (unstable) for the source (sink and vortex) pair between parallel walls* needs to be substantiated by a formal stability analysis. A formal proof of the instability of a row of equally spaced alternating conjugate monopoles would require (1) decomposing the position of a monopole $z = z_0 + \varepsilon$ into a position of rest z_0 plus a perturbation $\varepsilon(t)$ and (2) determining the temporal evolution of the perturbation. There are three cases: (1) if the perturbation always decays to zero $\varepsilon \to 0$ as $t \to \infty$, the equilibrium is **stable**; (2) if it can diverge $\varepsilon \to \infty$ as $t \to \infty$, it is **unstable**; and (3) if it remains in the vicinity without decaying or diverging, the equilibrium is **indifferent**. The stability analysis is simpler for linear perturbations $|\varepsilon|^2 << |z_0|^2$. A linear stability analysis will be made for the next configuration that also allows equilibrium positions at rest: two complex conjugate monopoles symmetrically placed relative to a cylinder in a uniform stream (Section 2.8). If the equilibrium position is unstable, a deviation will cause the monopole to follow a path away from it. The path of a monopole can be calculated from any position of nonequilibrium as shown next (Subsection 2.7.9); this determines the stability of the stability of the equilibrium positions, for arbitrary large perturbations, with no need for linearizations, that is not restricted to small perturbations (Figure 2.15).

2.7.9 Paths of Vortices, Sources, and Sinks in a Duct

A pair of monopoles, with conjugate moments (Equations 2.301a and 2.301b), that is, the same flow rate Q for the source/sink and opposite circulations $\pm\Gamma$ for the vortices, symmetrically placed at the positions (Equations 2.301c and 2.301d) in a parallel-sided duct (Figures 2.13 and 2.14) of width b in Equation 2.299d have a velocity (Equation 2.312) orthogonal (parallel) to the wall that vanishes if $2\pi a/b = \pi/2$, that is, $a = b/4$, implying that the monopoles are at equilibrium at a distance from the walls equal to a quarter of the width of the parallel-sided channel (Figure 2.14) for the vortices; for the sources/sinks, the flow velocity they produce in the channel (Equation 2.315a) must be subtracted out (Equations 2.315b and 2.315c). In the general case of arbitrary positions, a vortex (Equation 2.316a) moves uniformly (Equation 2.313b ≡ Equation 2.316c) down the duct parallel to the walls (Equation 2.316b), to (Figure 2.15a) the right (left) for positive (negative) circulation and $a < b/4$:

$$Q = 0 \neq \Gamma: \qquad y(t) = a, \qquad x(t) = \frac{\Gamma t}{2b}\cot\left(\frac{2\pi a}{b}\right), \tag{2.316a–2.316c}$$

and vice-versa for a > b/4. In the case of a source/sink (Equation 2.317a), the induced velocity (Equation 2.312) is (Equation 2.317b) perpendicular to the walls:

$$Q \neq 0 = \Gamma: \qquad -i\frac{dy}{dt} = -iv_y = v_0^* = -i\frac{Q}{2b}\cot\left(\frac{2\pi y}{b}\right), \tag{2.317a and 2.317b}$$

leading

$$\frac{Qt}{2b} = \int_a^y \tan\left(\frac{2\pi\xi}{b}\right)d\xi = -\frac{b}{2\pi}\left\{\log\left[\cos\left(\frac{2\pi\xi}{b}\right)\right]\right\}_a^y = -\frac{b}{2\pi}\log\left[\frac{\cos(2\pi y/b)}{\cos(2\pi a/b)}\right] \tag{2.317c}$$

to the trajectory (Equation 2.317b ≡ Equation 2.318b):

$$x(t) = a: \qquad \cos\left[\frac{2\pi}{b}y(t)\right] = \cos\left(\frac{2\pi a}{b}\right)\exp\left(-\frac{\pi Qt}{b^2}\right), \tag{2.318a and 2.318b}$$

that is discussed next separately for sources and sinks.

In contrast with the longitudinal motion of vortices, both the sources and sinks move transversely across the channel normal to the walls but in opposite directions. The source (Equation 2.319a) is (1) repelled both by the other source in the duct and by the equal image source on the other side of the wall; (2) thus, it moves toward (Figure 2.15b) the equilibrium position (Equation 2.319b) that it takes an infinite time to reach (Equation 2.319c) because the velocity is reduced as it is approached:

$$Q > 0: \qquad \lim_{t\to\infty}\cos\left[\frac{2\pi}{b}y(t)\right] = 0, \qquad \lim_{t\to\infty} y(t) = \pm\frac{b}{4}. \tag{2.319a–2.319c}$$

In the case (Equation 2.320a) of a sink, (1) the motion (Equation 2.320b) lasts a finite time (Equation 2.320c):

$$Q < 0: \qquad \lim_{t\to t_s}\left|\cos\left[\frac{2\pi}{b}y(t)\right]\right| = 1, \quad t_s = \frac{b^2}{\pi|Q|}\log\left|\sec\left(\frac{2\pi a}{b}\right)\right|; \tag{2.320a–2.320c}$$

(2) if the sink is perturbed starting below (Equation 2.321b) [above (Equation 2.322b)] the equilibrium position, it leads to a positive (negative) factor [Equation 2.321c (Equation 2.322c)], implying by Equation 2.318b that [Equation 2.321d (Equation 2.322d)] it reaches the centerline (Equation 2.321e) [wall (Equation 2.322e)] in the finite time (Equation 2.320c):

$$Q < 0, 0 < a < \frac{b}{4}: \quad \cot\left(\frac{2\pi a}{b}\right) > 0, \cos\left[\frac{2\pi}{a}y(t_s)\right] = +1, y(t_s) = 0, \tag{2.321a–2.321e}$$

$$Q > 0, \frac{b}{4} < a < \frac{b}{2}: \quad \cot\left(\frac{2\pi a}{b}\right) < 0, \cos\left[\frac{2\pi}{a}y(t_s)\right] = -1, y(t_s) = \pm\frac{b}{2}; \tag{2.322a–2.322d}$$

(3) in the lower half duct, the sink also reaches the centerline (Equation 2.321e) [the lower wall (Equation 2.321e) with – sign] after the same time (Equation 2.320c) if it starts closer to it; (4) in cases 2 and 3, the equilibrium position is unstable for the sink. If the sink starts closer to the axis, it is attracted toward the axis by the other sink in the duct (Figure 2.15c); if the sink starts closer to the wall, it moves toward the wall because the strongest attraction is from the closest "image" sink on the other side of the wall. In contrast, the equilibrium position is always stable for the source because if it moves away, the closest real (image) source in the duct (on the other side of the wall) pushes it back. The motion of the source toward the stable equilibrium position is monotonic rather than oscillatory because the velocity tends to zero as the equilibrium position is approached, and thus, there is no overshoot to the other side; the motion of the sink away from the unstable equilibrium position is also monotonic with increasing velocity and finite time to reach the axis (wall) and merge with the other sink (the image on the wall). The monotonic motion in all cases of source or sink corresponds to the fixed sign of Equation 2.318b, that is, positive (negative) for $|a| < b/4$ ($|a| > b/4$) regardless of the sign of the flow rate; the modulus of Equation 2.318b increases (decreases) with time for a sink $Q < 0$ (source $Q > 0$), corresponding to motion away from (toward the) equilibrium position. The stability (instability) of the equilibrium position also applies to a source (sink) near a semi-infinite plate (Subsection I.36.3.2); in that case, a vortex also has no position of equilibrium at rest (Subsection I.36.3.1). The consideration of two monopoles with complex conjugate moments symmetrically placed relative to a cylinder in a stream also leads to positions of equilibrium, whose stability is analyzed (Section 2.8); a deviation from an unsteady equilibrium position or a start from any nonequilibrium position leads to a path (Section 2.9).

2.8 Cylinder in a Stream with Two Trailing Monopoles (Föppl 1913)

For a cylinder in a uniform stream (von Karmann 1911) without (with) superimposed [Figure I.28.4 (Figure 2.4)] shear flow [Sections I.28.6 and I.28.7 (Section 2.6)]: (1) on approach to the cylinder, the streamlines converge, the velocity increases, the pressure decreases, and the favorable pressure gradient keeps the flow attached; (2) moving away from the cylinder, the streamlines diverge, the velocity reduces, the pressure increases, and the adverse pressure gradient can lead to flow separation (Subsection I.28.7.7); (3) as the velocity of the free stream is increased, the first stage of flow separation is the appearance of two vortices with opposite circulations symmetrically placed downstream (Figure 2.16); (4) as the velocity of the free stream is increased further, the wake of the cylinder develops (von Karmann 1912) into a double vortex stream consisting of two parallel rows of vortices with opposite circulations (Figure I.36.17) with possible stagger if $d \neq 0$. The configuration (3) was shown to be unstable (Föppl 1913). In the present section, the problem (3) is generalized to two monopoles with complex conjugate moments symmetrically placed near a cylinder in a stream (Figure 2.17); this includes as a particular case both two opposite vortices (Figures 2.16 and 2.18) and two equal sources/sinks (Figure 2.19).

The potential flow due to two monopoles with complex conjugate moments in a uniform stream is modified by the presence of cylinder (Subsection 2.8.1) as indicated by the first circle theorem (Section I.24.7). The corresponding conjugate velocity shows that the radial velocity vanishes on the cylinder (Subsection 2.8.2). The velocity of the monopoles is obtained by subtraction of the singular self-induced velocity of the monopole from the complex conjugate velocity (Subsection 2.8.3). The condition of zero monopole velocity specifies the position of

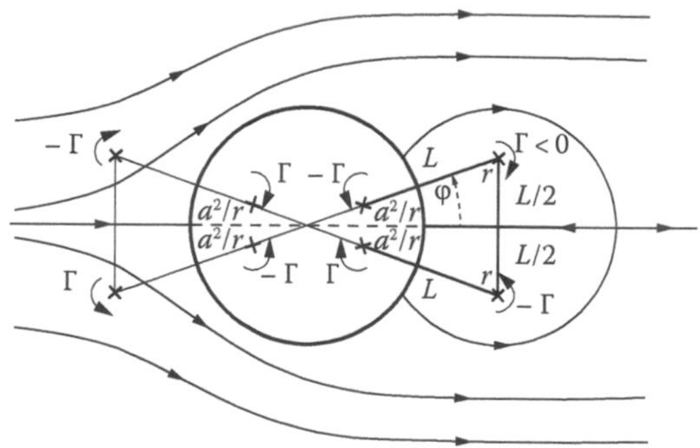

FIGURE 2.16
Two monopoles with conjugate moments, that is, equal flow rate and opposite circulations, can be at rest near a rigid cylinder in a uniform stream at any pair of symmetric positions for a suitable choice of flow rate and circulation. The equilibrium at rest results from adding to zero the velocities induced by (1) the conjugate monopole in the stream; (2) the two image monopoles in the cylinder; and (3) the uniform stream perturbed by the cylinder. The equilibrium configuration is possible with conjugate monopoles either forward or aft of the cylinder; the aft case is of practical relevance as it includes the simplest configuration of flow separation behind a cylinder with two symmetric trailing vortices with opposite circulations.

equilibrium at rest, leading to the "direct" and "inverse" problems. The "direct problem" shows that for any symmetric position of the monopoles in a stream given *a priori*, there is a unique flow rate and circulation that will enforce it as a position of rest; one example is the monopole strength that will ensure that two complex conjugate monopoles remain at rest symmetrically placed (Figure 2.18) on the line through the axis of the cylinder perpendicular to the mean flow (Subsection 2.8.4) only if they are vortices with opposite circulations. The "inverse problem" specifies the flow rate and circulation *a priori* and seeks the corresponding

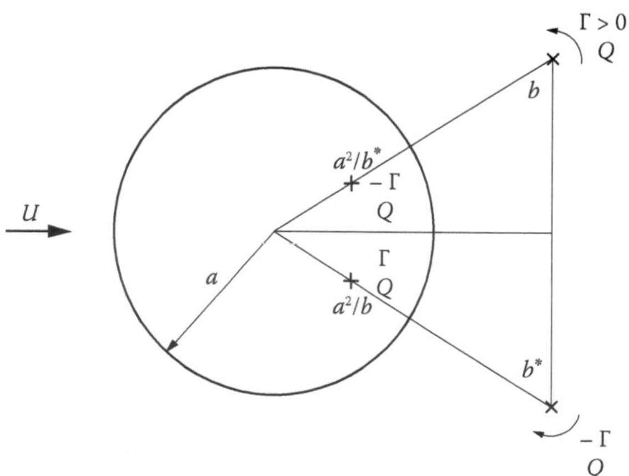

FIGURE 2.17
Consider two conjugate monopoles in a uniform stream near a cylinder. The cylinder (1) perturbs the uniform flow (Figure I.28.4a); (2) each monopole has an image with conjugate moment at the reciprocal point relative to the circle. Thus, a monopole with the same moment at the cross-reciprocal point a^2/b is added to the original monopole at b with flow rate Q and circulation Γ; the conjugate monopole at b^* has the same flow rate and opposite circulation, which also apply to the image at the cross-reciprocal point a^2/b^*. The remarkable feature of this combination is that it is possible to cancel the contributions to the induced velocity at both monopoles, leading to static equilibrium for a suitable choice of flow rate circulation and incident stream velocity; one of them can be chosen at will, with the other two specified by the conditions of equilibrium at rest.

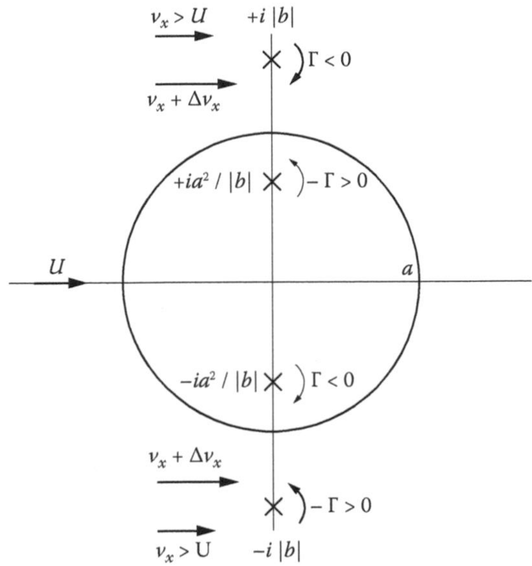

FIGURE 2.18
Last statement in the legend of Figure 2.17 corresponds to the "direct problem": given two symmetric positions of conjugate monopoles near a cylinder in a stream with specified velocity, find the flow rate and circulation that ensure equilibrium at rest. As an example, equilibrium at rest is sought along the line perpendicular to the flow direction passing through the axis of the cylinder. It is found that equilibrium is possible only in the absence of flow rate, that is, for vortices. Furthermore, the circulation increases in modulus along the loci of vortices at rest, from zero at the surface of the cylinder. Because the velocity of the flow around the cylinder is larger closer to the wall, the equilibrium at rest requires that this be countered by a clockwise (counterclockwise) or negative (positive) circulation above (below) the cylinder, not just for this position (Figure 2.18) but also for any other (Figure 2.19).

positions for which the monopoles are at rest; two examples are a pair of opposite vortices (equal sources/sinks) at rest symmetrically placed [Figures 2.17 and 2.18 (Figure 2.19)] relative to the cylinder [Subsection 2.8.5 (Subsection 2.8.6)] that may be compared (Subsection 2.8.7). The linear stability of these equilibrium positions can be established by analyzing the motion following a small perturbation from the position at rest (Subsection 2.8.8). This leads (Subsection 2.8.9) to two stability conditions: (1) the first is violated by a sink with vortex or a vortex alone, so that neither can be stable (Subsection 2.8.10); (2) the first stability condition is met by a source (with vortex or not) that can be stable (Subsection 2.8.11) if it meets the second condition. The stability conditions (Subsection 2.8.12) are independent of the free stream velocity (Subsection 2.8.13) and thus depend only on the monopole position (Subsections 2.8.14 and 2.8.15). The monopole combines a vortex (source/sink) and thus can represent (Subsection I.29.2.4) a combined lift (and propulsion) system near a body.

2.8.1 Pair of Conjugate Monopoles as the Wake of a Cylinder

Considering two monopoles (Equation 2.192a) with complex conjugate moments (Equations 2.301a and 2.301b ≡ Equations 2.323a and 2.323b) symmetrically placed at $z = b, b^*$ in a uniform stream (Equation 2.171a) of velocity U without angle of attack leads to the complex potential (Equation 2.323c):

$$P_0 \equiv Q - i\Gamma,\ P_0^* = Q - i\Gamma: \quad f_0(z) = Uz + P_0 \log(z-b) + P_0^* \log(z-b^*). \qquad (2.323a–2.323c)$$

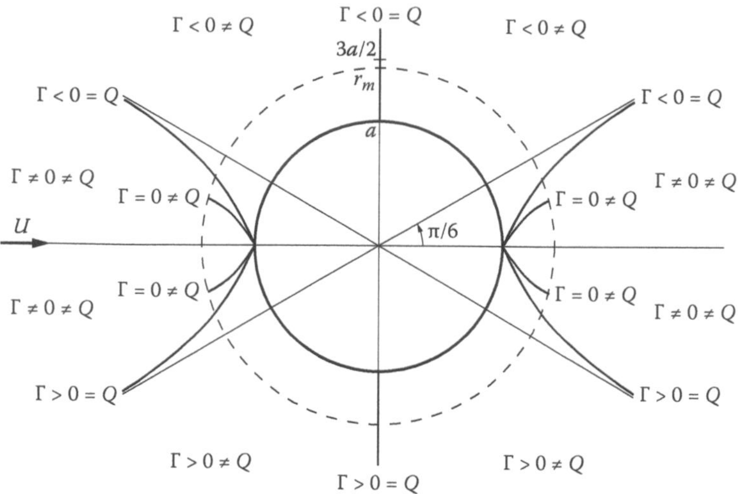

FIGURE 2.19
"Inverse problem" is to specify the flow rate and circulation and determine the corresponding positions of equilibrium at rest. The two simplest "inverse problems" are zero flow rate (circulation), that is, the symmetric positions of rest of vortices (sources/sinks) with opposite circulations (equal flow rate). The "first inverse problem" of positions of rest of opposite vortices leads to six loci: (1) the two side lines in Figure 2.18 in the direct problem; (2) four curves issuing from the two stagnation points in the inverse problem. Because the direct problem is linear, it has only one solution in the upper half-plane, that is, two by symmetry in the whole flow: the inverse problem being nonlinear can have zero, one, or multiple solutions. In the case of the opposite vortices, the inverse problem has three pairs of solutions (Figure 2.19), adding two pairs to the direct problem (Figure 2.18). The "second inverse problem" of the positions of rest of two sources/sinks with the same flow rate has four pairs of "solutions" of which (1) two pairs are complex and only two are real for a "small" radius, that is, below the maximum radius r_m; (2) above the maximum radius, all four pairs are complex, and there is no real solution. The two equal sources/sinks can be at rest only sufficiently close to the cylinder, whereas two opposite vortices can be at rest at any distance. For opposite vortices, there are six directions of rest at any distance; for equal sources/sinks, there are four directions of rest only below the maximum radius. Thus, four loci of equal sources/sinks at rest start at the stagnation points and lie closer to the axis than the four loci of rest of opposite vortices that start at the same two stagnation points. The opposite vortices at rest have two more loci on the line perpendicular to the free stream through the axis of the cylinder.

The insertion of a cylinder of radius and center at the origin (Figure 2.17) is specified by the first circle theorem (Equation I.24.47) using

$$f_0^*\left(\frac{a^2}{z^*}\right) - U\frac{a^2}{z} = U\left[P_0 \log\left(\frac{a^2}{z^*} - b\right) + P_0^* \log\left(\frac{a^2}{z^*} - b^*\right)\right]^* = P_0^* \log\left(\frac{a^2}{z} - b^*\right) + P_0 \log\left(\frac{a^2}{z} - b\right), \tag{2.323d}$$

in the complex potential:

$$\begin{aligned} f(z) = f_0(z) + f_0^*\left(\frac{a^2}{z^*}\right) = U\left(z + \frac{a^2}{z}\right) + P_0 \log(z-b) + P_0^* \log(z-b^*) \\ + P_0^* \log\left(\frac{a^2}{z} - b^*\right) + P_0 \log\left(\frac{a^2}{z} - b\right). \end{aligned} \tag{2.324}$$

Substituting Equations 2.323a and 2.323b in Equation 2.324 leads to *the complex potential (Equation 2.325) consisting of the following: (1) a uniform stream of velocity U without angle of attack past a cylinder of radius a and center at the origin (Equation I.28.91a); (2) a pair of sources/sinks (Equation I.12.26a) of flow rate Q at symmetric positions b, b*; (3) a pair of vortices (Equation I.12.32a) with opposite circulations ± Γ co-located at b, b* to form conjugate monopoles (Equation 2.192a); (4) the images of all monopoles according to the first circle theorem (Equation I.24.47) representing a sphere of radius a with center at the origin:*

$$
\begin{aligned}
2\pi f(z) = 2\pi U\left(z+\frac{a^2}{z}\right) + (Q-i\Gamma)\log(z-b) + (Q+i\Gamma)\log(z-b^*) \\
+(Q+i\Gamma)\log\left(\frac{a^2}{z}-b^*\right) + (Q-i\Gamma)\log\left(\frac{a^2}{z}-b\right);
\end{aligned}
\tag{2.325}
$$

(5) the image of the monopole $P_0(P_0^*)$ at $b(b^*)$ is located (Figure 2.17) at the reciprocal point (Equation 2.211b), that is, at $a^2/b^*(a^2/b)$ with moment $P_0^*(P_0)$; and (6) the circle theorem also applies to the free stream. The complex potential can be split into (1) the direct potential due to the free stream and monopoles (Equation 2.326a):

$$
f(z) = U\left(z+\frac{a^2}{z}\right) + \frac{Q}{2\pi}\log\left[(z-b)(z-b^*)\right] - \frac{i\Gamma}{2\pi}\log\left[\left(\frac{z-b}{z-b^*}\right)\right] + h(z;b,b^*), \tag{2.326a}
$$

$$
h(z;b,b^*) \equiv \frac{Q}{2\pi}\log\left[\left(\frac{a^2}{z}-b\right)\left(\frac{a^2}{z}-b^*\right)\right] - \frac{i\Gamma}{2\pi}\log\left[\left(\frac{a^2/z-b}{a^2/z-b^*}\right)\right]; \tag{2.326b}
$$

and (2) the **image function** *(Equation 2.326b) representing the images of the monopoles on the cylinder. The corresponding complex conjugate velocity is Equation 2.327a:*

$$
v^*(z) = \frac{df}{dz} = U\left(1-\frac{a^2}{z^2}\right) + \frac{Q}{2\pi}\left(\frac{1}{z-b}+\frac{1}{z-b^*}\right) - \frac{i\Gamma}{2\pi}\left(\frac{1}{z-b}-\frac{1}{z-b^*}\right) + g(z;b,b^*), \tag{2.327a}
$$

$$
\begin{aligned}
g(z;b,b^*) \equiv \frac{dh(z;b,b^*)}{dz} = \frac{Q}{2\pi z}\left(\frac{1}{bz/a^2-1}-\frac{1}{b^*z/a^2-1}\right) \\
-\frac{i\Gamma}{2\pi z}\left(\frac{1}{bz/a^2-1}-\frac{1}{b^*z/a^2-1}\right),
\end{aligned}
\tag{2.327b}
$$

where Equation 2.327b is the **image velocity**.

2.8.2 Rigid Wall Boundary Condition on the Impermeable Cylinder

On the cylinder (Equation 2.328a), the complex conjugate (Equation 2.327a) velocity including the image velocity (Equation 2.237b) is given by Equation 2.328b:

$$
\begin{aligned}
z = ae^{i\varphi}: \quad & v_r(a,\varphi) - i v_\varphi(a,\varphi) = e^{i\varphi} v^*(ae^{i\varphi}) = U(e^{i\varphi} - e^{-i\varphi}) \\
& + \frac{Q}{2\pi}\left[\frac{1}{a - be^{-i\varphi}} + \frac{1}{a - b^* e^{-i\varphi}}\right] - \frac{i\Gamma}{2\pi}\left[\frac{1}{a - be^{-i\varphi}} - \frac{1}{a - b^* e^{-i\varphi}}\right] \\
& + \frac{Q}{2\pi}\left[\frac{1}{be^{i\varphi} - a} + \frac{1}{b^* e^{i\varphi} - a}\right] - \frac{i\Gamma}{2\pi}\left[\frac{1}{be^{i\varphi} - a} - \frac{1}{b^* e^{i\varphi} - a}\right] \\
& = 2iU \sin\varphi + \frac{1}{2\pi}\frac{2Qa - Qe^{-i\varphi}(b + b^*) - i\Gamma e^{-i\varphi}(b - b^*)}{a^2 + b^* b e^{-2i\varphi} - ae^{-i\varphi}(b + b^*)} \\
& - \frac{1}{2\pi}\frac{2Qa - Qe^{i\varphi}(b + b^*) - i\Gamma e^{i\varphi}(b - b^*)}{a^2 + b^* b e^{i2\varphi} - ae^{i\varphi}(b + b^*)}.
\end{aligned}
\tag{2.328a and 2.328b}
$$

The last two terms on the r.h.s. of Equation 2.328b are complex conjugate, so their difference is $2i$ times the imaginary part; thus, the whole expression is imaginary, implying that the radial velocity is zero on the cylinder (Equation 2.329a) and the tangential velocity is given by Equation 2.329b:

$$
v_r(a,\varphi) = 0, \qquad v_\varphi(a,\varphi) = -2U \sin\varphi - \frac{2}{\pi}\,\mathrm{Im}\left\{\frac{Qa + e^{-i\varphi}\left[\Gamma\,\mathrm{Im}(b) - Q\,\mathrm{Re}(b)\right]}{a^2 + |b|^2 e^{-2i\varphi} - 2ae^{-i\varphi}\,\mathrm{Re}(b)}\right\}.
\tag{2.329a and 2.329b}
$$

The pressure on the cylinder (Equation 2.330a) follows from the Bernoulli equation (Equation 2.12b):

$$
p(a,\varphi) = p_0 - \frac{\rho}{2}[v_\varphi(a,\varphi)]^2; \qquad \{F_x, F_y\} = \frac{\rho}{2} a \int_0^{2\pi} p(a,\varphi)\{\cos\varphi, \sin\varphi\}\, d\varphi.
\tag{2.330a and 2.330b}
$$

Because the pressure is radial inward, it specifies the horizontal F_x (vertical F_y) force by integration (Equation 2.330b) over the cylinder. The velocity (Equation 2.329b) changes sign, replacing φ by $-\varphi$ so that it is symmetric relative to the x-axis, implying that the pressure is also symmetric and there is no lift on the cylinder because the opposite circulations of the vortices cancel (Subsections 2.8.4 and 2.8.5). The change of variable $\varphi \rightarrow \pi - \varphi$ does change the modulus of the velocity of the cylinder, that can be subject to a thrust (drag) if the two sources lie in the rear (forward) arc and *vice versa* for sinks.

2.8.3 Circulation and Flow Rate for Static Equilibrium

The velocity induced at each monopole is specified by subtracting its own field, that is, Equation 2.331a (Equation 2.331b) for the monopole (conjugate monopole) at $z = b(z = b^*)$:

$$\dot{b}^* \equiv \frac{db}{dt} = \lim_{z \to b}\left[v^*(z) - \frac{Q - i\Gamma}{2\pi}\frac{1}{z - b}\right], \tag{2.331a}$$

$$\dot{b} \equiv \frac{db^*}{dt} = \lim_{z \to b^*}\left[v^*(z) - \frac{Q + i\Gamma}{2\pi}\frac{1}{z - b^*}\right]. \tag{2.331b}$$

Substituting Equations 2.327a and 2.327b in Equation 2.331a (Equation 2.331b) leads to the velocity [Equation 2.332a (Equation 2.332b)] of the monopole (conjugate monopole) at $z = b(z = b^*)$:

$$\dot{b}^* \equiv \frac{db^*}{dt} = U\left(1 - \frac{a^2}{b^2}\right) + \frac{Q + i\Gamma}{2\pi}\frac{1}{b - b^*} + g(b;b,b^*), \tag{2.332a}$$

$$\dot{b} \equiv \frac{db}{dt} = U\left(1 - \frac{a^2}{b^{*2}}\right) + \frac{Q - i\Gamma}{2\pi}\frac{1}{b^* - b} + g(b^*;b,b^*), \tag{2.332b}$$

where the image velocities (Equation 2.327b) are given by [Equation 2.333a (Equation 2.333b)]:

$$2\pi b g(b;b,b^*) = \frac{Q - i\Gamma}{b^2/a^2 - 1} + \frac{Q + i\Gamma}{b^* b/a^2 - 1}, \tag{2.333a}$$

$$2\pi b^* g(b^*;b,b^*) = \frac{Q + i\Gamma}{b^{*2}/a^2 - 1} + \frac{Q - i\Gamma}{b^* b/a^2 - 1} = 2\pi\left[b g(b;b,b^*)\right]^*, \tag{2.333b and 2.333c}$$

and are complex conjugates (Equation 2.333c). The monopoles will be at rest if the induced velocities (Equations 2.332a and 2.332b) vanish (Equations 2.334a and 2.334b), leading by subtraction (Equation 2.334c) [sum (Equation 2.334d)] to

$$\dot{b}^* = 0 = \dot{b}: \quad \pi U a^2\left(\frac{1}{b^{*2}} - \frac{1}{b^2}\right) + \frac{Q}{b - b^*} + \pi\left[g(b;b,b^*) - g(b^*;b,b^*)\right] = 0, \tag{2.334a–2.334c}$$

$$\pi U\left[2-a^2\left(\frac{1}{b^{*2}}+\frac{1}{b^2}\right)\right]+\frac{i\Gamma}{b-b^*}+\pi\left[g(b;b,b^*)+g(b^*;b,b^*)\right]=0. \tag{2.334d}$$

Although Equations 2.334c and 2.334d have complex terms, bearing in mind Equation 2.333c, they are actually two real equations:

$$4\pi U\,\mathrm{Im}\left(\frac{a^2}{b^2}\right)+\frac{Q}{\mathrm{Im}(b)}-4\pi\,\mathrm{Im}\left[g(b;b,b^*)\right]=0, \tag{2.335a}$$

$$4\pi U\left[1-\mathrm{Re}\left(\frac{a^2}{b^2}\right)\right]+\frac{\Gamma}{\mathrm{Im}(b)}+4\pi\,\mathrm{Re}\left[g(b;b,b^*)\right]=0. \tag{2.335b}$$

Because Equations 2.335a and 2.335b are a linear algebraic system of equations in (U, Q, Γ), they can be solved for two of them given b, unless the system is redundant, if the corresponding determinant vanishes, in which case only one is independent. The case $\mathrm{Im}(b) = 0$ is excluded because it would correspond to the coincidence on the real axis $b = 0$ for symmetric monopoles at (b, b^*).

Substituting Equation 2.333a in Equations 2.335a and 2.335b, it follows that

$$\begin{aligned}4\pi U\,\mathrm{Im}\left(\frac{a^2}{b^2}\right)+\frac{Q}{\mathrm{Im}(b)}&=2Q\left\{\left(\frac{|b|^2}{a^2}-1\right)^{-1}\mathrm{Im}(b^{-1})+\mathrm{Im}\left[b^{-1}\left(\frac{b^2}{a^2}-1\right)^{-1}\right]\right\}\\&\quad+2\Gamma\left\{\left(\frac{|b|^2}{a^2}-1\right)^{-1}\mathrm{Re}(b^{-1})-\mathrm{Re}\left[b^{-1}\left(\frac{b^2}{a^2}-1\right)^{-1}\right]\right\},\end{aligned} \tag{2.336a}$$

$$\begin{aligned}4\pi U\left[1-\mathrm{Re}\left(\frac{a^2}{b^2}\right)\right]+\frac{\Gamma}{\mathrm{Im}(b)}&=-2Q\left\{\left(\frac{|b|^2}{a^2}-1\right)^{-1}\mathrm{Re}(b^{-1})+\mathrm{Re}\left[b^{-1}\left(\frac{b^2}{a^2}-1\right)^{-1}\right]\right\}\\&\quad+2\Gamma\left\{\left[\left(\frac{|b|^2}{a^2}-1\right)^{-1}\mathrm{Im}(b^{-1})\right]-\mathrm{Im}\left[b^{-1}\left(\frac{b^2}{a^2}-1\right)^{-1}\right]\right\}\end{aligned} \tag{2.336b}$$

relate the flow rate Q, circulation Γ, and free stream velocity U linearly, and the monopole positions (b, b^*) nonlinearly. This proves that *two monopoles with complex conjugate moments (Equations 2.323a and 2.323b) at symmetric positions (b, b*) near a cylinder of radius a in a stream* *of velocity U are (1) at rest* **(direct problem)** *for a flow rate Q and circulations ±Γ uniquely specified by the solution of the linear algebraic system of equations (Equations 2.336a and 2.336b); (2) if the flow rate Q and circulation ±Γ are specified* **(inverse problem)**, *then symmetric positions of rest exist if Equations 2.336a and 2.336b have solutions for complex b; (3) instead of determining (specifying) {Q, Γ} in the direct (inverse) problem, the pairs {Q, U} or {Γ, U} involving the mean flow velocity can be used as alternatives; (4) because Equations 2.336a and 2.336b are linear in {Q, Γ, U}, there is also a unique solution [as in the direct problem] if one pair is chosen together with the position of the monopole.* In case 2 of the inverse problem, the solution may not be unique, that is, there could be none or several. As an example, it will be shown [Subsection 2.8.5 (Subsection 2.8.6)] (Figure 2.19) that in the absence of flow rate (circulation), that is, for vortices with opposite circulations (identical sources/sinks), there are six (four) curves with positions of rest, that is, six (four) solutions of the inverse problem for all (some) values of the radius (Subsection 2.8.7).

2.8.4 Direct Equilibrium Problem and Downforce Coefficient

As an example of the direct problem, assume that the monopoles are (Figure 2.18) at the sides (Equation 2.337a) of the cylinder, that is, on the line through the center of the cylinder orthogonal to the free stream velocity. In this case, Equations 2.336a and 2.336b simplify to Equations 2.337b and 2.337c:

$$b = ic = i|b|: \quad 0 = \frac{Q}{c}\left(\frac{1}{2} + \frac{1}{c^2/a^2 - 1} - \frac{1}{c^2/a^2 + 1}\right) = \frac{Q}{c}\left(\frac{1}{2} + \frac{2}{c^4/a^4 - 1}\right), \tag{2.337a and 2.337b}$$

$$2\pi Uc\left(1 + \frac{a^2}{c^2}\right) + \Gamma\left(\frac{1}{2} + \frac{1}{c^2/a^2 - 1} + \frac{1}{c^2/a^2 + 1}\right) = 0. \tag{2.337c}$$

Because the monopoles are outside the cylinder $c > a$, the term in the brackets in Equation 2.337b cannot vanish; it follows (Equation 2.238a) that the monopoles are at rest (Equation 2.337a) if (1) there is no source/sink (Equation 2.337b ≡ Equation 2.338b) and (2) the vortices have circulation (Equation 2.337c ≡ Equation 2.338b) involving the **downforce parameter** (Equation 2.338c):

$$Q = 0, \qquad \Gamma = -4\pi Uch, \qquad h \equiv \left(1 + \frac{a^2}{c^2}\right)\frac{c^4/a^4 - 1}{c^4/a^4 + 4c^2/a^2 - 1}, \tag{2.338a–2.338c}$$

given by

$$\frac{1 + a^2/c^2}{h} = -\frac{4\pi Uc}{\Gamma}\left(1 + \frac{a^2}{c^2}\right) = 1 + \frac{2}{c^2/a^2 - 1} + \frac{2}{c^2/a^2 + 1} = 1 + \frac{4c^2/a^2}{c^4/a^4 - 1}, \tag{2.339}$$

that coincides with Equation 2.339 ≡ Equation 2.338c.

The absence (presence) of a source/sink (vortex) at the side (Equation 2.337a) of the cylinder implies that the images have opposite circulations at the reciprocal points (Figure 2.18), leading

[Equation I.28.29a (Equation 28.29b)] to no thrust or drag (Equation 2.340a) [a **downforce** *(Equation 2.340b)]:*

$$D = 0, \qquad L = -\rho U(-\Gamma) = \rho U\Gamma = -4\pi\rho U^2 ch = -\rho U^2 a C_F, \qquad C_F = 4\pi h \frac{c}{a}, \tag{2.340a–2.340c}$$

where the **downforce coefficient** *(Equation 2.340c) was calculated for the cylinder diameter 2a. In particular, if the monopoles are at the sides of the cylinder at a distance (Equation 2.341a) from the center, the downforce parameter (Equation 2.338c) is given by Equation 2.341b:*

$$c = \frac{a}{3}\sqrt{10}: \qquad h = \frac{19^2}{3790} = 0.09525, \qquad C_L = \frac{4}{3}\pi h\sqrt{10} = 1.2617, \tag{2.341a–2.341c}$$

whereas the downforce coefficient is given by Equation 2.341c. Because there are two vortices with opposite circulations, the downforce cancels. The upper vortex in Figure 2.18 reduces the velocity above the cylinder, thus increases the pressure and causes a downforce; the lower vortex also decreases the velocity and increases the pressure by the same amount, causing the opposite lift that cancels the downforce. Furthermore, the insertion of the cylinder in a uniform stream causes a nonuniform velocity that is higher and closer to the surface (Figure 2.18). In order to be in equilibrium at rest, a vortex must counter the higher incident flow velocity near the wall; this implies a negative (positive) circulation for a monopole above (below) the real axis (Figures 2.18 and 2.19).

2.8.5 First Inverse Problem: Pair of Opposite Vortices at Rest (Föppl 1913)

As a first inverse problem, the case when there is no source/sink (Equation 2.342a) is considered, and thus (Equations 2.332a and 2.333a), the second vortex (Equation 2.332a) is at rest if Equation 2.342b is met:

$$\begin{aligned} Q = 0 \neq \Gamma: \qquad i\frac{2\pi U}{\Gamma}\frac{b^2 - a^2}{b^2} &= \frac{1}{b - b^*} + \frac{1}{b}\left(\frac{a^2}{b^* b - a^2} - \frac{a^2}{b^2 - a^2}\right) \\ &= \frac{1}{b - b^*} + \frac{a^2(b - b^*)}{(b^* b - a^2)(b^2 - a^2)} \\ &= \frac{(b^2 - a^2)(b^* b - a^2) + a^2(b - b^*)^2}{(b - b^*)(b^2 - a^2)(b^* b - a^2)}. \end{aligned} \tag{2.342a and 2.342b}$$

The first vortex is at rest (Equations 2.332b and 2.333b) if the complex conjugate of Equation 2.342b holds. The ratio of the two expressions can be used to eliminate the mean flow velocity U and circulation Γ:

$$\frac{b^2 - a^2}{b^{*2} - a^2}\frac{b^{*2}}{b^2} = \frac{(b^2 - a^2)(b^* b - a^2) + a^2(b - b^*)^2}{(b^{*2} - a^2)(b^* b - a^2) + a^2(b^* - b)^2}\frac{b^{*2} - a^2}{b^2 - a^2}, \tag{2.343a}$$

leading to an equation for the vortex positions (b,b^*):

$$\begin{aligned}&(b^2-a^2)^2b^{*2}\left[(b^{*2}-a^2)(b^*b-a^2)+a^2(b-b^*)^2\right]\\&=(b^{*2}-a^2)^2b^2\left[(b^2-a^2)(b^*b-a^2)+a^2(b-b^*)^2\right].\end{aligned}\tag{2.343b}$$

The r.h.s. and l.h.s. of Equation 2.343b have common factors:

$$\begin{aligned}0&=(b^2-a^2)(b^{*2}-a^2)(b^*b-a^2)\left[b^{*2}(b^2-a^2)-b^2(b^{*2}-a^2)\right]\\&+a^2(b-b^*)^2\left[b^{*2}(b^2-a^2)^2-b^2(b^{*2}-a^2)^2\right]\\&=a^2(b^2-b^{*2})(b^*b-a^2)\left[(b^2-a^2)(b^{*2}-a^2)+(b-b^*)^2(b^*b+a^2)\right].\end{aligned}\tag{2.344a}$$

In Equation 2.344a, the following was used:

$$b^{*2}(b^2-a^2)^2-b^2(b^{*2}-a^2)^2=(b^{*2}b^2-a^4)(b^2-b^{*2})=(b^*b-a^2)(b^*b+a^2)(b^2-b^{*2})\tag{2.344b}$$

in the second term.

The first factor in Equation 2.344a cannot vanish because the vortex lies outside the cylinder $b^*b=|b|^2>a^2$. The case $b^2=b^{*2}$ allows two possibilities: (1) real $b=b^*$, in which case the vortices coincide on the real axis, and their opposite circulations cancel; (2) $b=i|b|$, that is, the case (Equation 2.337a) of vortices on (Figure 2.18) the side of the cylinder (Subsection 2.8.4). That is, for the unique solution of the "direct problem" (Subsection 2.8.4) that is linear: given a position (the sides of the cylinder), there is only one flow rate (zero) and circulation (nonzero) for that position to be at rest. The "inverse problem" (Subsection 2.8.5) specifies vortices only, so flow rate is zero and thus must include the side position. However, because the inverse problem is nonlinear, it could have several solutions. Thus, any new (Equation 2.345a) equilibrium position for symmetric vortices with opposite circulations must correspond to the roots of the term (Equation 2.345b) in square brackets in Equation 2.344a:

$$\begin{aligned}b^2\neq b^{*2}:\qquad 0&=(b^2-a^2)(b^{*2}-a^2)+(b-b^*)^2(b^*b+a^2)\\&=a^4-a^2\left[b^2+b^{*2}-(b-b^*)^2\right]+(b^*b)^2+b^*b(b-b^*)^2\\&=a^4-2b^*ba^2+(b^*b)^2+b^*b(b-b^*)^2.\end{aligned}\tag{2.345a and 2.345b}$$

In terms of the radial distance r and polar angle φ of the vortex position (Equation 2.346a), Equation 2.345b reduces to Equation (2.346b):

$$b=re^{i\varphi}:\quad O=(b^*b-a^2)^2+b^*b(b-b^*)^2=(r^2-a^2)^2-4r^4\sin^2\varphi,\tag{2.346a and 2.346b}$$

that simplifies further to Equation 2.347a:

$$r - \frac{a^2}{r} = \pm 2r \sin\varphi; \qquad \Gamma = \mp \frac{2\pi U}{r^5}(r^2 - a^2)^2(r^2 + a^2). \qquad \text{(2.347a and 2.347b)}$$

Substitution of Equation 2.344a in Equation 2.342b specifies the circulation (Equation 3.347b).
The latter result (Equation 2.347b) is obtained (Equation 2.342b) using Equations 2.345b and 2.346a:

$$i\frac{2\pi U}{\Gamma}\left(1 - \frac{a^2}{r^2}e^{-2i\varphi}\right) = i\frac{2\pi U}{\Gamma}\left(1 - \frac{a^2}{b^2}\right) = \frac{b^*b - a^2 + a^2(b - b^*)^2/(b^2 - a^2)}{(b - b^*)(b^*b - a^2)}$$

$$= \frac{b^*b - a^2 - a^2(b^{*2} - a^2)/(b^*b + a^2)}{(b - b^*)(b^*b - a^2)} = \frac{(b^{*2}b^2 - a^4) - a^2(b^{*2} - a^2)}{(b - b^*)(b^{*2}b^2 - a^4)}$$

$$= \frac{r^4 - a^4 - a^2(r^2e^{-2i\varphi} - a^2)}{r(e^{i\varphi} - e^{-i\varphi})(r^4 - a^4)} = \frac{r^4 - a^2r^2e^{-2i\varphi}}{2ir\sin\varphi(r^4 - a^4)} = \frac{1 - (a/r)^2e^{-2i\varphi}}{\left[1 - (a/r)^4\right]2ir\sin\varphi}. \qquad \text{(2.348)}$$

Solving for the circulation and using Equation 3.347a lead to

$$\Gamma = -4\pi U\, r\sin\varphi\left(1 - \frac{a^4}{r^4}\right) = \mp\frac{2\pi U}{r^5}(r^2 - a^2)(r^4 - a^4), \qquad \text{(2.349)}$$

that coincides with Equation 2.347b ≡ Equation 2.349. Thus, *a pair of opposite vortices in symmetric positions behind a cylinder of radius a in a uniform flow of velocity U (1) are at rest (Equation 2.347a) if (Figure 2.16) their mutual distance equals the distance from the respective reciprocal point on the circle; (2) there are four positions of rest of vortices satisfying condition 1; (3) the two rear positions correspond to the detached flow (Föppl 1913); (4) the two forward positions, although mathematically possible, do not occur in practice due to the attached flow; (5) there are in addition the two side positions (Figure 2.18) for a total of six positions of rest of vortices (Figure 2.19); (6) the vortices with opposite circulations all have negative (positive) circulation above (below) the real axis (Subsection 2.8.4). In conclusion, for each radial distance, there are six angular directions for which a vortex can be at rest. The image at the reciprocal point inside the cylinder has opposite circulation corresponding to a downforce or negative lift (Equation 2.350a):*

$$F_y = -\rho U(-\Gamma) = \rho U\Gamma = -\rho U^2 a C_F, \qquad C_F \equiv -\frac{\Gamma}{Ua} = \frac{2\pi}{r^5 a}(r^2 - a^2)^2(r^2 + a^2), \qquad \text{(2.350a and 2.350b)}$$

and downforce coefficient (Equation 2.350b); for a vortex at a distance, Equation 2.351a leads to the downforce coefficient Equation 2.351b:

$$r = a\sqrt{\frac{3}{2}}, \qquad C_F = \frac{5\pi}{9}\sqrt{\frac{2}{3}} = 1.4251, \qquad \text{(2.351a and 2.351b)}$$

which is comparable in magnitude to Equation 2.341c for two vortices at the sides (Equation 2.337a) of the cylinder (Figure 2.18). The preceding calculation concerns one vortex only; because there are two opposite vortices, there is no net downforce. It was shown (Subsection 2.8.4) that the "direct problem" specifying the position (Equation 2.337a) corresponds to a vortex (Equations 2.338a and b); direct substitution of Equation 2.337a shows that it satisfies Equation 2.342b in the form (2.337c). The case (Equation 2.337a) was eliminated (Equation 2.244a) in the simplification from Equations 2.243b through 2.245b because it corresponds to the factor $b^2 - b^{*2} = 0$ omitted (Equation 2.345a) when passing from Equations 2.343b through 2.345b. Thus, a pair of opposite vortices in a uniform stream outside a cylinder are [Figure 2.18 (Figure 2.17)] in equilibrium in two cases [Equation 2.337a (Equation 2.347a)] corresponding to the circulation [Equations 2.338b and 2.338c (Equation 2.347b)]. The equilibrium of a pair of vortices inside a cylinder will be considered subsequently (Subsection 2.9.4); next, symmetric conjugate monopoles in a free stream outside a cylinder are considered, in the case of source/sink(s) (Subsection 2.8.6) instead of vortices (Subsection 2.8.5).

2.8.6 Second Inverse Problem: Pair of Equal Sources/Sinks at Rest

The second inverse problem omits the vortices (Equation 2.352a), so that (Equations 2.332a and 2.333a) the second source/sink is at rest at $z = b^*$ if Equation 2.352b is met:

$$\Gamma = 0 \neq Q: \qquad \frac{2\pi U}{Q}\frac{a^2 - b^2}{b^2} = \frac{1}{b-b^*} + \frac{a^2}{b}\left(\frac{1}{b^2-a^2} + \frac{1}{b^* b - a^2}\right)$$

$$= \frac{1}{b-b^*} + \frac{a^2}{b}\frac{b^2 + b^* b - 2a^2}{(b^2-a^2)(b^* b - a^2)}$$

$$= \frac{b(b^2-a^2)(b^* b - a^2) + a^2(b-b^*)(b^2 + b^* b - 2a^2)}{b(b-b^*)(b^2-a^2)(b^* b - a^2)}. \qquad \text{(2.352a and 2.352b)}$$

The first source/sink is at rest if the complex conjugate of Equation 2.352b holds. The free stream velocity and flow rate are eliminated by their ratio:

$$-\left(\frac{b^2-a^2}{b^{*2}-a^2}\right)^2 \frac{b^*}{b} = \frac{b(b^2-a^2)(b^* b - a^2) + a^2(b-b^*)(b^2 + b^* b - 2a^2)}{b^*(b^{*2}-a^2)(b^* b - a^2) - a^2(b-b^*)(b^{*2} + b^* b - 2a^2)}, \tag{2.353a}$$

specifying an equation for the positions $\{b, b^*\}$:

$$0 = (b^2-a^2)^2 b^*[b^*(b^{*2}-a^2)(b^* b - a^2) - a^2(b-b^*)(b^{*2} + b^* b - 2a^2)]$$
$$+(b^{*2}-a^2)^2 b[b(b^2-a^2)(b^* b - a^2) + a^2(b-b^*)(b^2 + b^* b - 2a^2)]. \tag{2.353b}$$

The latter has common factors:

$$\begin{aligned}0 &= (b^2-a^2)(b^{*2}-a^2)(b^*b-a^2)[b^{*2}(b^2-a^2)+b^2(b^{*2}-a^2)]\\&\quad+a^2(b-b^*)\{b(b^{*2}-a^2)^2[b(b+b^*)-2a^2]-b^*(b^2-a^2)^2[b^*(b+b^*)-2a^2]\}\\&=(b^2-a^2)(b^{*2}-a^2)(b^*b-a^2)[2b^{*2}b^2-a^2(b^2+b^{*2})]\\&\quad+a^2(b-b^*)\{(b+b^*)[b^2(b^{*2}-a^2)^2-b^{*2}(b^2-a^2)^2]-2a^2[b(b^{*2}-a^2)^2-b^*(b^2-a^2)^2]\}.\end{aligned} \tag{2.354a}$$

This equality can be written:

$$\begin{aligned}0 &= (b^2-a^2)(b^{*2}-a^2)(b^*b-a^2)[2b^{*2}b^2-a^2(b^2+b^{*2})]\\&\quad+a^2(b^2-b^{*2})^2(a^4-b^{*2}b^2)-2a^4(b-b^*)\{a^2(a^2+2b^*b)(b-b^*)+b^*b(b^{*3}-b^3)\},\end{aligned} \tag{2.354b}$$

using Equation 2.344b (Equation 2.355a) in the second (third) term:

$$b(b^{*2}-a^2)^2-b^*(b^2-a^2)^2=(b-b^*)a^2(a^2+2b^*b)+b^*b(b^{*3}-b^3). \tag{2.355a}$$

The cubic polynomial (Equation 2.355b) has the root $b = b^*$ and thus can be factorized:

$$b^{*3}-b^3=(b^*-b)(b^{*2}+b^*b+b^2). \tag{2.355b}$$

Substituting Equation 2.355b in Equation 2.354b yields

$$\begin{aligned}0 &= [b^{*2}b^2+a^4-a^2(b^2+b^{*2})](b^*b-a^2)[2b^{*2}b^2-a^2(b^2+b^{*2})]\\&\quad-a^2(b^*-b)^2(b^*b+a^2)[(b+b^*)^2(b^*b-a^2)+2a^4]-2a^4b^*b(b^*-b)^2(a^2+b^2+b^{*2}+b^*b),\end{aligned} \tag{2.356}$$

as the equation whose roots are the positions of static equilibrium of sources/sinks with equal flow rate.

Next, polar coordinates (Equation 2.346a ≡ Equation 2.357a) are used in Equations 2.357b through 2.357e:

$$b=re^{i\varphi}:\qquad \begin{aligned}(b-b^*)^2 &= r^2(e^{i\varphi}-e^{-i\varphi})^2=(2i\sin\varphi)^2\\&=-4r^2\sin^2\varphi=-2r^2[1-\cos(2\varphi)],\end{aligned} \tag{2.357a and 2.357b}$$

$$(b+b^*)^2=r^2(e^{i\varphi}+e^{-i\varphi})^2=(2r\cos\varphi)^2=4r^2\cos^2\varphi=2r^2[1+\cos(2\varphi)], \tag{2.357c}$$

$$b^*b=r^2,\quad b^2+b^{*2}=r^2(e^{2i\varphi}+e^{-2i\varphi})=2r^2\cos(2\varphi), \tag{2.357d and 2.357e}$$

for substitution in Equation 2.356 for the source/sink positions of rest:

$$\begin{aligned}0 = &\left[r^4 + a^4 - 2a^2r^2\cos(2\varphi)\right](r^2 - a^2)r^2[r^2 - a^2\cos(2\varphi)] \\ &+ 2a^2r^2(r^2 + a^2)[1 - \cos(2\varphi)]\{(r^2 - a^2)r^2[1 + 2\cos(2\varphi)] + a^4\} \\ &+ 2a^4r^4[1 - \cos(2\varphi)]\{a^2 + r^2[1 + 2\cos(2\varphi)]\}.\end{aligned} \tag{2.358}$$

The latter (Equation 2.358 ≡ Equation 2.359a) is a binomial in cos(2φ) whose roots specify two pairs of angles (Equation 2.359b) corresponding to positions of rest of the sources/sinks:

$$A\cos^2(2\varphi) - B\cos(2\varphi) + C = 0: \quad \pm\varphi_{\pm} = \frac{1}{2}\arg\cos\left\{\frac{B \pm \sqrt{B^2 - 4AC}}{2A}\right\}. \tag{2.359a and 2.359b}$$

Comparing Equations 2.358 and 2.359a specifies the coefficients:

$$\begin{aligned}\{A, B, C\} = &\left\{-2a^2r^2(2r^6 + a^2r^4 - a^4r^2),\right. \\ &\left. a^2r^2(r^6 - 5a^2r^4 + 7a^4r^2 + a^6), \quad r^2(r^8 + a^2r^6 + 3a^4r^4 + a^6r^2 + 2a^8)\right\}.\end{aligned} \tag{2.360a–2.360c}$$

The loci of position of rest of two symmetrically placed sources/sinks with the same flow rates are at most eight curves, that is, the four angles (Equation 2.359b) plus $\pi \pm \varphi_{\pm}$ that does not change cos(2φ). The number of curves could be reduced to four (zero) if one (both) of the arguments in Equation 2.359b is either (1) not real or (2) real and greater than unity in modulus. This is analyzed next in comparison with the case of two vortices with opposite circulations.

2.8.7 Comparison of Equilibria for Vortices and Sources/Sinks

Considering a cylinder of radius a with center at the origin in a uniform stream of velocity U along the real axis with two symmetrically placed monopoles with conjugate moments (Figure 2.17), two extremes are compared: (1) the cases of zero flow rate (circulation), that is, two vortices with opposite circulations (Subsections 2.8.4 and 2.8.5; Figures 2.16 and 2.18) [two equal sources/sinks with the same flow rate (Subsection 2.8.6; Figure 2.19)]; (2) there are, for a given free stream velocity, four (six) loci along which the equal sources/sinks (opposite vortices) are at rest, with the circulation (flow rate) varying with position along each locus; (3) the vortices are at rest along the line orthogonal to free stream passing through the center, that is, the "side" positions (Figure 2.18); (4) all four (the remaining four) loci for equal sources/sinks (opposite vortices) are (Figure 2.19) the curves [Equation 2.358 ≡ Equations 2.359b and 2.360a through 2.360c (Equation 2.347a)]; (5) these curves touch the cylinder [Equation 2.361a (Equation 2.362a)] at the stagnation points [Equations 2.361b through 2.361d (Equations 2.362b through 2.362d), also at the side positions].

$$r = a: \quad \Gamma = 0 \neq Q, \quad \cos(2\varphi) = 1, \quad \varphi = 0, \pi, \qquad (2.361a\text{–}2.361d)$$

$$r = a: \qquad \Gamma \neq 0 = Q, \qquad \sin\varphi = 0, \pm 1, \quad \varphi = 0, \pi, \pm\frac{\pi}{2}; \qquad (2.362a\text{–}2.362d)$$

(6) the curves do not reach infinity (Equations 2.363a through 2.363c) [go to infinity (Equation 2.362a) along directions making a 30° angle with the stream direction besides the side position (Equations 2.364b through 2.364d)]:

$$r \to \infty: \quad \Gamma = 0 \neq Q, \quad r \leq r_m < \infty. \qquad (2.363a\text{–}2.363c)$$

$$r \to \infty: \qquad \Gamma \neq 0 = Q, \qquad \sin\varphi = \frac{1}{2}, 1, \qquad \varphi = \pm\frac{\pi}{6}, \pm, \frac{\pi}{2}, \pm\frac{5\pi}{6}. \qquad (2.364a\text{–}2.364d)$$

The latter aspect (6) is further analyzed next.

A clear difference is for a vortex for every value for the radius (Equation 2.365a), there are four real values of the angle φ in Equations 2.347a and 2.347b not more than $\pi/6$ from the flow direction (Equations 2.365b through 2.365d):

$$0 < r < \infty: \qquad \Gamma \neq 0 = Q, \qquad |\sin\varphi| = \frac{1}{2}\left(1 - \frac{a^2}{r^2}\right) \leq \frac{1}{2}, \qquad |\varphi| < \frac{\pi}{6} > |\pi - \varphi|. \qquad (2.365a\text{–}2.365d)$$

For a source/sink (Equation 2.366a), real values of the angle (Equations 2.359a and 2.359b) exist only if two constraints on the radius hold for the coefficients (Equations 2.360a through 2.360c), namely, the first (Equation 2.366b) and second (Equation 2.366c), so that the argument of arc cos is real (does not exceed unity in modulus):

$$Q \neq 0 = \Gamma, \qquad B^2 \geq 4AC, \qquad \left|B \pm \sqrt{B^2 - 4AC}\right| < 2|A|. \qquad (2.366a\text{–}2.366c)$$

For example, exceeding the radius of the cylinder by one half (Equation 2.367a ≡ Equation 2.368a): (1) the vortices (Equation 2.367b) lie in the directions Equations 2.367c and 2.367d:

$$r = \frac{3}{2}a: \qquad \Gamma \neq 0 = Q, \quad \sin\varphi = \pm\frac{5}{18}, 1, \quad \varphi = \pm 16.13°, \pm 90°, \pm 163.87°; \qquad (2.367a\text{–}2.367d)$$

$$r = \frac{3}{2}a: \quad \Gamma = 0 \neq Q, \{A, B, C\} = \left\{-\frac{7371}{64}, \frac{1629}{256}, \frac{130077}{1024}\right\}, \cos(2\varphi) = +1.078, -1.023;$$

(2.368a–2.368d)

(2) for sources/sinks (Equation 2.368b), the parameters (Equations 2.360a through 2.360c) for the radius (Equation 2.368a) take the values (Equation 2.368c) corresponding to imaginary angles (Equation 2.368d). Thus, the positions of rest of vortices (sources/sinks) extend to infinity [are limited to a maximum radius (Equation 2.363c) that $r_m < 3a/2$ is less than Equation 2.368a]. The implication is that (1) the rotation around the centroid of two vortices with opposite circulations (Subsection 2.7.3) can be canceled by the mean flow and images on the cylinder, for any distance in six directions; and (2) the repulsion (attraction) of two sources (sinks) with equal flow rates (Subsection 2.7.2) can be canceled by the mean flow and images on the cylinder in four directions, but only sufficiently close to the cylinder, that is, at less than the maximum radius (Figure 2.19).

For a large radius (Equation 2.369a), the coefficients (Equations 2.360a through 2.360c) scale as Equation 2.369b leading (Equation 2.369c) to the cosines greater than unity (Equation 2.369d), thus excluding real angles:

$$r \to \infty: \quad \{A, B, C\} \sim \{-4\,a^2\,r^8,\ a^2\,r^8,\ r^{10}\},$$
$$B^2 - 4\,A\,C \sim 16\,a^2\,r^{18}, \quad \cos(2\varphi) \sim \pm\frac{4\,a\,r^9}{8\,a^2\,r^8} \sim \pm\frac{1}{2}\frac{r}{a}. \tag{2.369a–2.369d}$$

In contrast, at the surface of the cylinder (Equation 2.369e), the parameters (Equations 2.360a through 2.360c) lead to Equations 2.369f and 2.369g:

$$r = a: \quad \{A, B, C\} = 4a^{10}\{-1, 1, 2\} \quad \cos(2\varphi) = -2, 1, \tag{2.369e–2.369g}$$

showing that the loci of constant flow rate touch the cylinder only at the stagnation points (Equations 2.361c and 2.361d). Consider an intermediate radius that is larger between that of the cylinder (Equation 2.369a) but smaller than Equation 2.368a, which exceeds the maximum value, for example, Equation 2.370a:

$$r = \frac{5}{4}a: \qquad \{A, B, C\} = \left\{-\frac{1742500}{4096}, \frac{363025}{65536}, \frac{32214025}{1048576}\right\},$$
$$\cos(2\varphi) = 1.184, -0.976, \quad \varphi = \pm 12.63°, \pm 167.37°. \tag{2.370a–2.370d}$$

The parameters (Equations 2.360a through 2.360c) take the values Equation 2.370b leading (Equation 2.370c) to four real angles (Equation 2.370d) on the loci of equal sources/sinks at rest. For the same radius (Equation 2.370a ≡ Equation 2.371a), the loci of opposite vortices at rest (Equation 2.347a) are (Equations 2.371b and 2.371c) farther from the flow direction:

$$r = \frac{5}{4}a: \qquad \sin\varphi = \pm\frac{9}{50}, \quad \varphi = \pm 10.37°, \pm 169.63°. \tag{2.371a–2.371c}$$

In conclusion *(Figure 2.19): (1) inside the maximum radius $r < r_m$, there are six (four) directions with positions of rest for opposite vortices (equal sources/sinks); (2) for larger radii $r > r_m$, six positions of rest exist for opposite vortices (none for equal sources/sinks); (3) the four loci of equal sources/sinks are closer to the flow direction than the four loci of opposite vortices; (4) excluding the two side loci for opposite vortices, the other four loci for opposite vortices issue from the stagnation points* *together with the four loci of equal sources/sinks; (5) thus, at the stagnation points, the flow rate and circulation both vanish, and there are no monopoles to perturb the flow; and (6) all points outside the 10 loci in Figure 2.19 are equilibrium positions for monopoles whose flow rate and circulation are both nonzero (7) the maximum radius for the existence of equilibrium positions for sources/sinks lie $5a/4 < r_m < 3a/2$ between the limits set by Equations 2.370a (2.368a), since equilibrium positions exist (Equations 2.370b through 2.370d) [do not exist (Equations 2.368b through 2.368d)] in the former (latter) case.* The linear stability of the equilibrium positions is analyzed next (Subsections 2.8.8 through 2.8.14) for any combination of circulation and flow rate.

2.8.8 Small Perturbations from Positions of Rest

The velocity of the monopole combining vortices and sources/sinks is specified by (Equations 2.332b and 2.333b):

$$\frac{db}{dt} = U\left(1-\frac{a^2}{b^{*2}}\right)+\frac{Q-i\Gamma}{2\pi}\left(\frac{1}{b^*-b}+\frac{1}{b^*}\frac{a^2}{b^*b-a^2}\right)+\frac{Q+i\Gamma}{2\pi}\frac{1}{b^*}\frac{a^2}{b^{*2}-a^2} \equiv F(b,b^*) \qquad (2.372)$$

and its conjugate. The equilibrium position corresponds to zero velocity (Equations 2.373a and 2.373b):

$$F(b_0,b_0^*) = \frac{db_0}{dt} = 0 = \frac{db_0^*}{dt} = F(b_0^*,b_0): \qquad \varepsilon = b-b_0, \quad \varepsilon^* \equiv b^*-b_0^*, \qquad (2.373\text{a–}2.373\text{d})$$

and the deviation from static equilibrium (Equations 2.373c and 2.373d) satisfies

$$\begin{aligned}\frac{d\varepsilon}{dt} = \frac{db}{dt} &= F(b,b^*) = F(b,b_0^*) + \frac{\partial F}{\partial b_0}(b-b_0) + \frac{\partial F}{\partial b_0^*}(b^*-b_0)\\ &\quad + O((b-b_0)^2,(b^*-b_0)^2,(b-b_0)(b^*-b_0)) \qquad (2.374)\\ &= \varepsilon\frac{\partial F}{\partial b_0} + \varepsilon^*\frac{\partial F}{\partial b_0^*} + O(\varepsilon^2,\varepsilon^{*2},\varepsilon^*\varepsilon),\end{aligned}$$

(1) using only the first three terms of the Taylor series (Equations 1.4a and 1.4b) in two variables (b, b^*) around (b, b_0^*); (2) the latter are positions of equilibrium at rest (Equations 2.373a and 2.373b), so that the constant term on the r.h.s. of Equation 2.374 vanishes; (3) the nonlinear terms of order 2 or higher may be neglected for small perturbations; and (4) this leaves only the linear terms in Equation 2.374 ≡ Equation 2.375a and its complex conjugate Equation 2.375d:

$$\lambda \equiv \frac{\partial F}{\partial b_0}, \mu \equiv \frac{\partial F}{\partial b_0^*}: \qquad \frac{d\varepsilon}{dt} = \lambda\varepsilon + \mu\varepsilon^*, \qquad \frac{d\varepsilon^*}{dt} = \mu^*\varepsilon + \lambda^*\varepsilon^*, \qquad (2.375\text{a–}2.375\text{d})$$

involving as coefficients in the system of two complex coupled first-order ordinary differential equations (Equations 2.375c and 2.375d) the constant **stability parameters**

(Equations 2.375a and 2.375b). *The stability of the equilibrium positions of the monopoles for small deviations is specified by the temporal evolution of the linearized perturbations (Equations 2.375a through 2.375d),* which is considered next (Subsection 2.8.9).

The solution of the system of two linear coupled first-order ordinary differential equations (Equations 2.375c and 2.375d) is sought in the form Equation 2.376a leading to the linear algebraic system Equation 2.376b:

$$\{\varepsilon(t), \varepsilon^*(t)\} = \{\varepsilon_0, \varepsilon_0^*\}e^{\vartheta t}: \quad \begin{bmatrix} \lambda - \vartheta & \mu \\ \mu^* & \lambda^* - \vartheta \end{bmatrix} \begin{bmatrix} \varepsilon_0 \\ \varepsilon_0^* \end{bmatrix} = 0. \qquad \text{(2.376a and 2.376b)}$$

A nontrivial solution with initial perturbations not both zero (Equation 2.377a) is possible only if the determinant in Equation 2.376b vanishes (Equation 2.377b):

$$\{\varepsilon_0, \varepsilon_0^*\} \neq (0,0): \quad 0 = (\lambda - \vartheta)(\lambda^* - \vartheta) - \mu^*\mu = \vartheta^2 - (\lambda + \lambda^*)\vartheta + \lambda^*\lambda - \mu^*\mu$$
$$= \vartheta^2 - 2\,\mathrm{Re}(\lambda)\vartheta + |\lambda|^2 - |\mu|^2 = (\vartheta - \vartheta_+)(\vartheta - \vartheta_-). \qquad \text{(2.377a and 2.377b)}$$

The binomial (Equation 2.377b) for ϑ has real coefficients and its roots specify the **eigenvalues** (Equation 2.378):

$$\vartheta_\pm = \mathrm{Re}(\lambda) \pm \{[\mathrm{Re}(\lambda)]^2 + |\mu|^2 - |\lambda|^2\}^{1/2}, \qquad (2.378)$$

that (1) depend on the stability parameters (Equations 2.375a and 2.375b) and (2) appear in the temporal evolution (Equation 2.376a) of the perturbations.

The latter (Equation 2.376a ≡ Equation 2.379):

$$\{\varepsilon(t), \varepsilon^*(t)\} = \{\varepsilon_0, \varepsilon_0^*\}\exp[t\,\mathrm{Re}(\vartheta)]\exp[it\,\mathrm{Im}(\vartheta)] \qquad (2.379)$$

has modulus:

$$\left\{|\varepsilon(t)|, |\varepsilon^*(t)|\right\} = \left\{|\varepsilon_0|, |\varepsilon_0^*|\right\}\exp[t\,\mathrm{Re}(\vartheta)]. \qquad (2.380)$$

It follows that for a long time, the perturbation decays to zero (Equation 2.381a) [diverges to infinity (Equation 2.381b)] if the eigenvalues have negative (positive) real part, implying stability (instability):

$$\lim_{t\to\infty}\left\{|\varepsilon(t)|, |\varepsilon^*(t)|\right\} = \begin{cases} 0 & \text{if} \quad \mathrm{Re}(\vartheta) < 0: \text{ stable;} & \text{(2.381a)} \\ \infty & \text{if} \quad \mathrm{Re}(\vartheta) > 0: \text{ unstable.} & \text{(2.381b)} \end{cases}$$

The intermediate case of eigenvalue with zero real part (Equation 2.382a) leads to neutral stability:

$$\mathrm{Re}(\vartheta)=0;\omega\equiv\mathrm{Im}(\vartheta): \qquad \{\varepsilon(t),\varepsilon^*(t)\}=\{\varepsilon_0\varepsilon_0^*\}\times\begin{cases} e^{i\omega t} & \text{if} \quad \omega\neq 0, \\ 1 & \text{if} \quad \omega=0, \end{cases} \tag{2.382a–2.382c}$$

(1) if the imaginary part is also zero (Equation 2.382b), it is a static equilibrium at the displaced position; and (2) if the imaginary part is nonzero (Equation 2.382c), it is the frequency of an oscillation around the equilibrium positions with amplitude equal to the initial perturbations.

2.8.9 Necessary and Sufficient Conditions for Stability

The general solution of Equations 2.375c and 2.375d consists of a linear (Equation 2.383b) combination of Equation 2.376a for the two eigenvalues if (1) they are distinct (Equation 2.383a):

$$\vartheta_+\neq\vartheta_-: \qquad \varepsilon(\mathrm{t})=A_+e^{\vartheta_+t}+A_-e^{\vartheta_+t}, \tag{2.383a and 2.383b}$$

$$\vartheta_+=\vartheta_-\neq\vartheta_0: \qquad \varepsilon(\mathrm{t})=(A_++A_-t)e^{\vartheta_+t}; \tag{2.384a and 2.384b}$$

(2) for coincident eigenvalues (Equation 2.384a), the general displacement is specified by Equation 2.384b as a function of time; and (3) in both cases, the constants $A_\pm$ are determined by the initial conditions (Equation 2.377a). Combining Equations 2.381a through 2.381c and 2.382a through 2.382c with Equations 2.383a and 2.383b, it follows that for distinct eigenvalues (Equation 2.388a ≡ Equation 2.385a): (1) stability requires both eigenvalues to have negative real parts (Equation 2.385b); (2) for instability, it suffices that one eigenvalue has a positive real part (Equation 2.385c); (3) for neutral stability, at least one eigenvalue must have zero real part and the other must have nonpositive real part (Equation 2.385d):

$$\vartheta_+\neq\vartheta_-; \qquad \mathrm{Re}(\vartheta_+)\geq\mathrm{Re}(\vartheta_-): \qquad \begin{cases} \mathrm{Re}(\vartheta_+)<0: & \text{stable}, \\ \mathrm{Re}(\vartheta_+)>0: & \text{unstable}, \\ \mathrm{Re}(\vartheta_+)=0: & \text{neutral}. \end{cases} \tag{2.385a–2.385d}$$

Combining Equations 2.381a through 2.381c and 2.382a through 2.382c with Equations 2.384a and 2.384b, it follows that if the eigenvalues coincide (Equation 2.386a): (1) a negative real part ensures stability (Equation 2.386b); (2) neutral stability requires a zero real part and (Equation 2.386c) one zero coefficient, implying particular initial conditions:

$$\vartheta_+=\vartheta_-\equiv\vartheta_0: \qquad \begin{cases} \mathrm{Re}(\vartheta_0)<0: & \text{stable}, \\ \mathrm{Re}(\vartheta_0)=0=\mathrm{A}_-: & \text{neutral}, \\ \text{otherwise}: & \text{unstable}; \end{cases} \tag{2.386a–2.386d}$$

and (3) all other cases are unstable, that is, positive real part $\text{Re}(\vartheta_0) > 0$ or zero real part general initial conditions $\text{Re}(\vartheta_0) = 0 \neq A_-$.

Applying Equations 2.385a through 2.385d and 2.386a through 2.386d to Equation 2.378 specifies the stability conditions for conjugate monopoles near a cylinder in a stream. Starting with distinct eigenvalues (Equation 2.385a), stability requires both (Equation 2.385b) to have negative real parts, leading to three cases: (1) the eigenvalues (Equation 2.378) are complex conjugate (Equation 2.387b) if Equation 2.387a is met, in which case Equation 2.387c ensures stability:

$$[\text{Re}(\lambda)]^2 < |\lambda|^2 - |\mu|^2: \quad \vartheta_+ = \vartheta_-^*, \quad 0 > \text{Re}(\vartheta_+) = \text{Re}(\vartheta_-) = \text{Re}(\lambda); \tag{2.387a–2.387d}$$

(2) the eigenvalues are real and distinct (Equation 2.388b) if Equation 2.388a is met, requiring that the first term be negative (Equation 2.388c) and larger in modulus than the second (Equation 2.388d):

$$[\text{Re}(\lambda)]^2 > |\lambda|^2 - |\mu|^2: \quad \vartheta_+ > \vartheta_-, \quad \text{Re}(\lambda) < 0, \quad |\mu| < |\lambda|; \tag{2.388a–2.388d}$$

and (3) in the intermediate case (Equation 2.389a) of equal eigenvalues (Equation 2.389b), stability is ensured by the negative real part for Equation 2.389c:

$$[\text{Re}(\lambda)]^2 = |\lambda|^2 - |\mu|^2: \quad \vartheta_+ > \vartheta_- = \vartheta_0 = \text{Re}(\lambda) < 0. \tag{2.389a–2.389c}$$

Comparing cases 1–3, it follows that (1) in all cases, Equation 2.387c ≡ Equation 2.388c ≡ Equation 2.389c is common (Equation 2.390a):

$$\text{Re}(\lambda) < 0; \quad |\lambda| > |\mu|; \tag{2.390a and 2.390b}$$

(2) the second stability condition (Equation 2.388d ≡ Equation 2.390b) required in case 2 is met by Equation 2.387a in case 1 and also (Equation 2.389a) in case 3, so it is also general. Thus, *the conditions (Equations 2.390a and 2.390b) are necessary and sufficient to ensure the stability of the equilibrium positions (Equations 2.372, 2.373a, and 2.373b) of the monopoles and are specified by the eigenvalues (Equation 2.378) in terms of the constant stability parameters (Equations 2.375a and 2.375b).* The latter are calculated next (Subsections 2.8.10 and 2.8.11) to apply the stability conditions.

2.8.10 Sufficient (Necessary) Condition for Instability (Stability)

In order to apply the stability conditions (Equations 2.390a and 2.390b), the stability parameters (λ, μ) must be determined. They follow from Equation 2.372, substituting b by $b_0 + \varepsilon$, expanding in powers of $(\varepsilon, \varepsilon^*)$ to first order (Equation 2.374), and picking (λ, μ) as the coefficient of $(\varepsilon, \varepsilon^*)$. The first stability parameter Equation 2.375a follows (Equation 2.391) from Equation 2.372:

$$\lambda = \lim_{b \to b_0} \frac{\partial F}{\partial b} = \frac{Q - i\Gamma}{2\pi}\left[\frac{1}{(b_0^* - b_0)^2} - \frac{a^2}{(b_0^* b_0 - a)^2}\right]. \tag{2.391}$$

Using the monopole position in polar coordinates (Equation 2.392a), the stability parameter (Equation 2.391) is given (Equations 2.357b and 2.357d) by Equation 2.392b:

$$b_0 = re^{i\varphi}: \qquad \lambda = -\frac{Q - i\Gamma}{2\pi}\left[\frac{1}{4r^2 \sin^2\varphi} + \frac{a^2}{(r^2 - a^2)^2}\right]. \qquad \text{(2.392a and 2.392b)}$$

The first stability condition (Equation 2.390a) is Equation 2.393b:

$$Q > 0: \quad 0 > 8\pi r^2 \sin^2\varphi (r^2 - a^2)^2 \operatorname{Re}(\lambda) = -Q[(r^2 - a^2)^2 + 4a^2 r^2 \sin^2\varphi]. \qquad \text{(2.393a and 2.393b)}$$

Because both the coefficient of Re(λ) on the l.h.s. and the term in square brackets on the r.h.s. are positive, the first stability condition (Equation 2.390a) is met (violated) by a positive (negative) flow rate.

It follows that *for a pair of monopoles with complex conjugate moments (Equations 2.323a and 2.323b) symmetrically placed (Equation 2.392a) near a cylinder of radius a in a uniform stream (Equation 2.393b): (1) they are unstable for a negative flow rate (Equation 2.394a) because the first stability condition (Equation 2.390a) is violated; (2) stability is possible only for a positive flow rate (Equation 2.393a) if the second stability condition (Equation 2.390b) is also met (Subsections 2.8.11 and 2.8.12):*

$$Q \neq \begin{cases} \text{I}: & Q < 0: \quad \text{unstable}, & \text{(2.394a)} \\ \text{II}: & Q > 0: \quad \text{stable if } |\lambda| > |\mu|. & \text{(2.394b)} \end{cases}$$

In cases 1 and 2, the monopole may or may not include vortices. In the case of zero flow rate (Equation 2.395a), then Equation 2.393b implies Equation 2.395b, and the eigenvalues (Equation 2.378) are given by Equation 2.395c:

$$Q = 0: \quad \operatorname{Re}(\lambda) = 0, \quad \vartheta_\pm = \pm\{|\mu|^2 - |\lambda|^2\}^{1/2}, \qquad \text{(2.395a–2.395c)}$$

leading to three subcases for zero flow rate: (1) for conjugate imaginary roots (Equation 2.396a), there is neutral stability; (2) for real symmetric roots (Equation 2.96b), there is instability;

$$Q = 0: \begin{cases} |\lambda| > |\mu|: & \text{neutral}, & \text{(2.396a)} \\ |\lambda| < |\mu|: & \text{unstable}, & \text{(2.396b)} \\ |\lambda| - |\mu| = 0 \neq A_-: & \text{unstable}, & \text{(2.396c)} \\ |\lambda| - |\mu| = 0 = A_-: & \text{neutral}; & \text{(2.396d)} \end{cases}$$

and (3) for coincident zero root (Equation 2.396c), there is also instability except for peculiar initial conditions (Equation 2.396d). Violating one of the two stability conditions (Equations 2.380a and 2.380b) is sufficient to prove instability. To prove stability, it is necessary (sufficient) to prove one (two) stability condition. Thus, stability is ensured (Equation 2.390a) for positive flow rate (Equation 2.393a) if Equation 2.390b also holds; the latter is considered next (Subsection 2.8.11).

2.8.11 Sufficient Conditions for Stability

The second stability condition (Equation 2.390b) involves the second stability parameter (Equations 2.372 and 2.375b):

$$\mu = \lim_{b \to b_0} \frac{\partial F}{\partial b^*} = \frac{2Ua^2}{b_0^{*3}} - \frac{Q - i\Gamma}{2\pi} \left[\frac{1}{(b_0^* - b_0)^2} + \frac{a^2}{b_0^{*2}} \frac{2b_0^* b_0 - a^2}{(b_0^* b_0 - a^2)^2} \right] - \frac{Q + i\Gamma}{2\pi} \frac{a^2}{b_0^{*2}} \frac{3b_0^{*2} - a^2}{(b_0^{*2} - a^2)^2}, \tag{2.397}$$

that may be expressed (Equations 2.357b and 2.357d) in terms of the position of the monopole (Equation 2.392a) by

$$\mu = \frac{2Ua^2}{r^3} e^{i3\varphi} + \frac{Q - i\Gamma}{2\pi} \left[\frac{1}{4r^2 \sin^2 \varphi} - e^{i2\varphi} \frac{a^2}{r^2} \frac{2r^2 - a^2}{(r^2 - a^2)^2} \right] - \frac{Q + i\Gamma}{2\pi} \frac{a^2}{r^2} \frac{3r^2 - a^2 e^{i2\varphi}}{(r^2 e^{-i2\varphi} - a^2)^2}. \tag{2.398}$$

In conclusion, *consider a cylinder of radius a and center at the origin in a uniform stream of velocity U along the x-axis (Figure 2.17) in the presence of two symmetrically placed* (b_0, b_0^*) *monopoles (Equation 2.392a) with complex conjugate moments. Table 2.3 shows that (case I) a negative flow rate (with or without circulation) leads to instability (Equation 2.394a) because the first stability condition (Equation 2.390a) is violated Equation (2.392b) ≡ Equation (2.393b); (case II) a zero flow*

TABLE 2.3

Static Stability of Conjugate Monopoles near a Cylinder in a Stream

Case	Monopole	Stability	Condition	Subcase
I	Sink (with or without vortex)	Unstable	$\text{Re}(\lambda) > 0$	–
II	Vortex (zero flow rate)	Neutral or unstable	$\text{Re}(\lambda) = 0$	–
III	Source (with or without vortex)	Stable	$\text{Re}(\lambda) < 0 > \|\lambda\| - \|\mu\|$	III-A
		Neutral or unstable	$\text{Re}(\lambda) = 0 = \|\lambda\| - \|\mu\|$	III-B
		Unstable	$\text{Re}(\lambda) < 0 < \|\lambda\| - \|\mu\|$	III-C

Note: $\lambda(\mu)$: First (second) stability parameter [Equations 2.391 and 2.392b (Equation 2.397 ≡ Equation 2.398)] with Equation 2.392a. Stability of monopoles with complex conjugate moments, that is, equal flow rate and opposite circulations at symmetric positions near a cylinder in a uniform flow. Any position is a position of rest for a suitable choice of flow rate and circulation. The equilibrium is always unstable for negative flow rate, with or without circulation; for a zero flow rate, that is, circulation alone, the equilibrium is at best indifferent. Stability is possible only for a positive flow rate, with or without circulation, and a range of free stream velocities.

rate for two vortices with opposite circulations leads to neutral stability (Equations 2.396a and 2.396d) or instability (Equations 2.396b and 2.396c) because the first stability condition (Equation 2.390a) is replaced by an equality; (case 3) a positive flow rate (Equation 2.393a) satisfies (Equation 2.393b) the first stability condition (Equation 2.390a), leading (sub-cases III A/B/C) to instability/indifferent stability/stability if the second stability condition (Equation 2.390b) is not met/met with equality/met with inequality. The stability parameters $\lambda(\mu)$ are given by Equations 2.391, 2.392a, and 2.392b (Equations 2.392a, 2.397, and 2.398). The second stability condition is shown next to depend on the free stream velocity as well as the positions of the monopoles:

2.8.12 Circulation and Flow Rate for Equilibrium

An arbitrary position of the monopole b is an equilibrium position at rest provided that the flow rate Q, circulation Γ and free stream velocity U satisfy the system of Equations 2.336a and 2.336b, that is equivalent to:

$$A_{11}Q + A_{12}\Gamma = 4\pi U \operatorname{Im}\left(\frac{a^2}{b_0^2}\right), \quad A_{12}Q + A_{22}\Gamma = 4\pi U\left[1 - \operatorname{Re}\left(\frac{a^2}{b_0^2}\right)\right], \qquad (2.399\text{a and } 2.399\text{b})$$

involving the **stability matrix** (Equations 2.400a through 2.400d):

$$A_{11} = -\frac{1}{\operatorname{Im}(b_0)} + \frac{a^2}{|b_0|^2 - a^2}\operatorname{Im}\left(\frac{2}{b_0}\right) + \operatorname{Im}\left(\frac{2}{b_0}\frac{a^2}{b_0^2 - a^2}\right), \qquad (2.400\text{a})$$

$$A_{12} = \frac{a^2}{|b_0|^2 - a^2}\operatorname{Re}\left(\frac{2}{b_0}\right) - \operatorname{Re}\left(\frac{2}{b_0}\frac{a^2}{b_0^2 - a^2}\right), \qquad (2.400\text{b})$$

$$A_{21} = -\frac{a^2}{|b_0|^2 - a^2}\operatorname{Re}\left(\frac{2}{b_0}\right) - \operatorname{Re}\left(\frac{2}{b_0}\frac{a^2}{b_0^2 - a^2}\right), \qquad (2.400\text{c})$$

$$A_{22} = -\frac{1}{\operatorname{Im}(b_0)} + \frac{a^2}{|b_0|^2 - a^2}\operatorname{Im}\left(\frac{2}{b_0}\right) - \operatorname{Im}\left(\frac{2}{b_0}\frac{a^2}{b_0^2 - a^2}\right), \qquad (2.400\text{d})$$

For all positions such that the determinant of the stability matrix (Equations 2.400a through 2.400d) does not vanish (Equation 2.401a) the equilibrium at rest is ensured Equations 2.399a and 2.399b by the flow rate and circulation given by Equations 2.401b and 2.401c:

$$A \equiv A_{11}A_{22} - A_{12}A_{21} \neq 0: \quad \frac{A}{4\pi U}Q = A_{22}\operatorname{Im}\left(\frac{a^2}{b_0^2}\right) - A_{12}\left[1 - \operatorname{Re}\left(\frac{a^2}{b_0^2}\right)\right]$$

$$\frac{A}{4\pi U}\Gamma = -A_{21}\operatorname{Im}\left(\frac{a^2}{b_0^2}\right) + A_{11}\left[1 - \operatorname{Re}\left(\frac{a^2}{b_0^2}\right)\right], \qquad (2.401\text{a-}2.401\text{c})$$

as a linear function of the free stream velocity U and nonlinear function of positions (b,b^*) of the monopoles. Thus *two monopoles with complex conjugate moments Equations 2.323a and 2.323b at symmetric positions* (b_0, b_0^*) *relative to a cylinder of radius a in a free stream of velocity U are in equilibrium at rest if the flow rate (circulation) is given by Equation 2.401b (Equation 2.401c) involving Equations 2.401a and 2.400a through 2.400d.* The stability of the equilibrium positions is specified by the two stability parameters considered next.

2.8.13 Stability Independent of the Free Stream Velocity

Substituting Equations 2.401b and 2.401c in the first (Equation 2.391) [second (Equation 2.397)] stability parameters leads to Equation 2.402a (Equation 2.402b):

$$\frac{\lambda A}{2U} = \left\{ \left(A_{22} + iA_{21}\right) \mathrm{Im}\left(\frac{a^2}{b_0^2}\right) - \left(A_{12} + iA_{11}\right) \left[1 - \mathrm{Re}\left(\frac{a^2}{b_0^2}\right)\right] \right\} \times \left[\frac{1}{\left(b_0^* - b_0\right)^2} - \frac{a^2}{\left(b_0^* b_0 - a^2\right)^2} \right] \equiv B_1\left(b_0, b_0^*, a\right), \tag{2.402a}$$

$$\begin{aligned} \frac{\mu A}{2U} = \frac{a^2 A}{b_0^{*3}} - &\left\{ \left(A_{22} + iA_{12}\right) \mathrm{Im}\left(\frac{a^2}{b_0^2}\right) - \left(A_{12} + iA_{11}\right) \left[1 - \mathrm{Re}\left(\frac{a^2}{b_0^2}\right)\right] \right\} \\ &\times \left[\frac{1}{\left(b_0^* - b_0\right)^2} + \frac{a^2}{b_0^{*2}} \frac{2b_0^* b_0 - a^2}{\left(b_0^* - b_0 - a^2\right)^2} \right] \\ &- \left\{ \left(A_{22} - iA_{21}\right) \mathrm{Im}\left(\frac{a^2}{b_0^2}\right) - \left(A_{12} - iA_{11}\right) \left[1 - \mathrm{Re}\left(\frac{a^2}{b_0^2}\right)\right] \right\} \\ &\frac{a^2}{b_0^2} \frac{3b_0^{*2} - a^2}{\left(b_0^{*2} - a^2\right)^2} \equiv B_2\left(b_0, b_0^*, a\right). \end{aligned} \tag{2.402b}$$

Since both stability parameters Equation 2.402a and 2.402b are linear on the free stream velocity the latter drops out of the stability conditions (Equations 2.390a and 2.390b).

2.8.14 Application of the Two Stability Conditions

The coefficients in Equations 2.402a and 2.402b may be simplified using Equations 2.400a through 2.400d:

$$A_{22} + iA_{21} = -\frac{1}{\mathrm{Im}\left(b_0\right)} - \frac{2i}{b_0} \frac{a^2}{\left|b_0\right|^2 - a^2} - \frac{2i}{b_0^*} \frac{a^2}{b_0^{*2} - a^2}, \tag{2.403a}$$

$$A_{12} + iA_{11} = -\frac{i}{\operatorname{Im}(b_0)} + \frac{2}{b_0}\frac{a^2}{|b_0|^2 - a^2} - \frac{2}{b_0^*}\frac{a^2}{b_0^{*2} - a^2}, \tag{2.403b}$$

$$A_{22} - iA_{21} = -\frac{1}{\operatorname{Im}(b_0)} + \frac{2i}{b_0^*}\frac{a^2}{|b_0|^2 - a^2} + \frac{2i}{b_0}\frac{a^2}{b_0^2 - a^2}, \tag{2.403c}$$

$$A_{12} - iA_{11} = \frac{i}{\operatorname{Im}(b_0)} + \frac{2}{b_0^*}\frac{1}{|b_0|^2 - a^2} - \frac{2}{b_0}\frac{a^2}{b_0^2 - a^2}, \tag{2.403d}$$

where were used the identities:

$$z \equiv \frac{2}{b_0}\frac{a^2}{b_0^2 - a^2}: \quad \operatorname{Re}(z) + i\operatorname{Im}(z) = z, \quad \operatorname{Re}(z) - i\operatorname{Im}(z) = z^*,$$
$$z \equiv \frac{2}{b_0}: \quad \operatorname{Im}(z) + i\operatorname{Re}(z) = iz^*, \quad \operatorname{Im}(z) - i\operatorname{Re}(z) = -iz. \tag{2.404a–2.404g}$$

Thus in all cases *the stability of equilibrium at rest of two monopoles with complex conjugate moments Equations 2.323a and 2.323b at symmetric positions* (b_0, b_0^*) *relative to a cylinder of radius a is independent of the free stream velocity: (1) the equilibrium is stable if the conditions (Equations 2.390a and 2.390b; 2.402a and 2.402b) are met Equations 2.405a and 2.405b:*

$$\operatorname{Re}\left\{A^{-1}B_1\left(b_0, b_0^*; a\right)\right\} < 0, \quad \left|B_1\left(b_0, b_0^*; a\right)\right| > \left|B_2\left(b_0, b_0^*; a\right)\right|; \tag{2.405a and 2.405b}$$

(2) if one or two of the inequalities (Equations 2.405a and 2.405b) is replaced by an equality the equilibrium is indifferent; (3) in all other cases, that is when at least one of the inequalities (Equations 2.405a and 2.405b) is reversed, the equilibrium is unstable. The preceding stability conditions involve (Equations 2.402a and 2.402b) the stability matrix (Equations 2.400a through 2.400d) through its determinant (Equation 2.401a) and the combinations of terms in Equations 2.403a through 2.403d.

2.8.15 Opposite Vortices at Sideline Positions

The simplest application of the two stability criteria (Equations 2.390a and 2.390b) is to the direct equilibrium problem (Subsection 2.8.4) of two vortices (Equation 2.406a) with opposite circulations at the sideline position (Equation 2.406b) that is (Figure 2.18) on the line orthogonal to the free stream velocity passing through the center of the cylinder. In this case the first stability parameter (Equation 2.391) simplifies to Equation 2.406c:

$$\Gamma \neq 0 = Q, \quad b_0 = ic = i|b_0|: \quad \lambda = \frac{i\Gamma}{8\pi c^2}\left[1 + \frac{4a^2c^2}{\left(c^2 - a^2\right)^2}\right], \tag{2.406a–2.406c}$$

and the second stability parameter (Equation 2.397) is given by (Equation 2.406d):

$$\mu = \frac{i2Ua^2}{c^3} - \frac{i\Gamma}{8\pi c^2}\left\{1 + 4a^2\left[\frac{2c^2 - a^2}{\left(c^2 - a^2\right)^2} + \frac{3c^2 + a^2}{\left(c^2 + a^2\right)^2}\right]\right\}. \tag{2.406d}$$

The vortices are at rest at the sideline position only if the circulation is given by Equations 2.338b and 2.338c, implying (Equations 2.406c and 2.406d) that the stability parameters are given by:

$$\left\{\lambda, \mu - i\frac{2Ua^2}{c^3}\right\} = \frac{iU}{2c}\left(1 + \frac{a^2}{c^2}\right)\frac{c^4 - a^4}{c^4 + 4a^2c^2 - a^4}$$
$$\times\left\{1 + \frac{4a^2c^2}{\left(c^4 - a^4\right)^2}, -1 - 4a^2\frac{5\left(c^4 + a^4\right) - 2a^2c^2\left(c^2 + 2a^2\right)}{\left(c^4 - a^4\right)^2}\right\}, \tag{2.407a and 2.407b}$$

where was made the simplification:

$$\frac{2c^2 - a^2}{\left(c^2 - a^2\right)^2} + \frac{3c^2 + a^2}{\left(c^2 - a^2\right)^2} = \frac{\left(2c^2 - ac^2\right)\left(c^2 + a^2\right)^2 + \left(3c^2 + a^2\right)\left(c^2 - a^2\right)^2}{\left(c^4 - a^4\right)^2}$$
$$= \frac{5c^2\left(c^4 - a^4\right) - 2a^2c^2\left(c^2 + 2a^2\right)}{\left(c^4 - a^4\right)^2}. \tag{2.407c}$$

The term in curly brackets in Equation 2.406d is larger than the term in square brackets in Equation 2.406c, implying that (Equation 2.407a and 2.407b) the second stability parameter is larger in modulus than the first:

$$\frac{2c}{U}|\mu| = 4\frac{a^2}{c^2} + \left(1 + \frac{a^2}{c^2}\right)\frac{c^4 - a^4}{c^4 + 4a^2c^2 - a^4}\left\{1 + 4a^2\left[\frac{2c^2 - a^2}{\left(c^2 - a^2\right)^2} + \frac{3c^2 - a^2}{\left(c^2 - a^2\right)^2}\right]\right\}$$
$$> \left(1 + \frac{a^2}{c^2}\right)\frac{c^4 - a^4}{c^4 + 4a^2c^2 - a^4}\left[1 + \frac{4a^2c^2}{\left(c^2 - a^2\right)^2}\right] = \frac{2c}{U}|\lambda|; \tag{2.407d}$$

this proves (Equation 2.396b) instability (Equation 2.407d ≡ Equation 2.408d):

$$\Gamma \neq 0 = Q, \quad b_0 = ic = i|b_0|: \quad \mathrm{Re}(\lambda) = 0, \quad |\mu| > |\lambda|. \tag{2.408a–2.408d}$$

The second instability condition dominates the first that would suggest (Equations 2.406c and 2.408c) indifferent equilibrium (Equation 2.390a). Thus *two vortices (Equation 2.408a) with opposite circulations (Figure 2.18) are at rest at the sideline positions (Equation 2.408b) if the circulation is specified by Equations 2.388b and 2.388c, but the equilibrium is unstable.*

2.9 Reciprocity Theorem (Green 1828) and Path Function (Routh 1881)

Outside the preceding curve (Equation 2.347a), a pair of vortices will not be at rest. Thus, a perturbation may lead to the start of a motion along a path. Starting at any nonequilibrium position will also lead to a path that can be calculated from the Routh path or stream function (Subsection 2.9.1) using the Green's or influence function (Subsection 2.9.2). A simple example is one vortex near a cylinder (Subsection 2.9.3). In the preceding case (Section 2.8) of two vortices with opposite circulations near a cylinder without mean flow, the relevant Green's function (Subsection 2.9.4) leads to the path function and to the paths of a pair of opposite vortices outside or inside a cylinder (Subsection 2.9.5); the latter allows indifferent equilibrium at rest at two points only (Subsection 2.9.6).

2.9.1 Induced Velocity and Path Function (Routh 1881)

The path of a singularity in a potential flow is not a streamline of the flow because the **induced velocity** on the singularity differs from the flow velocity by subtracting the self-effect, that is, singular; for example:

$$\dot{b}^* \equiv \frac{db^*}{dt} = v_0^*(b) = \lim_{z \to b} v_0^*(z) = \lim_{z \to b} \left[v^*(z;b) - \frac{Q - i\Gamma}{2\pi} \frac{1}{z-b} \right]\Bigg\} \tag{2.409}$$

is the induced velocity on a monopole of flow rate Q and circulation Γ at position $z = b$ in a potential flow of complex conjugate velocity $v^*(z)$. This method was used to calculate the paths of monopoles, that is, sources, sinks, and vortices: (1) in a rectangular corner (Section I.16.6); (2) past a semi-infinite plate (Section I.36.3); (3) for a single (Section I.36.8) [double (Section I.36.9)] row; (4) a monopole near a wall (Section I.16.2; Subsection 2.7.4); and (5) for a pair between parallel walls (Subsections 2.7.5 through 2.7.9). In the cases 2 and 5 the velocity was uniform, whereas in the cases 2 and 4 (case 1) it was not, and the trajectory (path) was found by integration of the induced velocity.

In order to avoid the latter integration, Equation 2.409 could be written in the form

$$\frac{df_0}{dz} = \lim_{z \to b} \frac{d}{dz} \left\{ f(z;b) - \frac{Q - i\Gamma}{2\pi} \log(z-b) \right\}. \tag{2.410}$$

If it is permissible to exchange the derivative with the limit in Equation 2.410, it follows that

$$f_0(b) = \lim_{z \to b} \left\{ f(z;b) - \frac{Q - i\Gamma}{2\pi} \log(z-b) \right\} + \text{const}; \tag{2.411}$$

Then, taking the imaginary part:

$$\Psi_0(r,\varphi) = \lim_{z \to b = re^{i\varphi}} \left\{ \Psi(r,\varphi) - \frac{Q}{2\pi}\arg(z-b) + \frac{\Gamma}{2\pi}\log|z-b| \right\} \tag{2.412}$$

relates the stream function of the flow Ψ to the **Routh or path function** Ψ_0; the latter is the analog of the stream function for the singularity because its path is specified by $\Psi_0 = \text{const}$.

The analogy is that the fluid particles (monopoles) in a steady potential flow with stream (path) function $\Psi(\Psi_0)$ follow the generally distinct lines $\Psi = const$ ($\Psi_0 = const$). This leads to the **first path theorem:** *the path of a singularity such as a monopole of flow rate Q and circulation Γ in a potential flow of stream function Ψ is specified by the constancy of the path function (Equation 2.412), on the assumption that the exchange of limit with derivative in the passage from Equations 2.410 to 2.411 is valid.* The exchange of a limit with a derivative is valid in the case of uniform convergence (Section I.13.6); the convergence in Equations 2.410 and 2.411 may be nonuniform because both expressions are the difference of singular functions. Thus, it is necessary to find some general conditions under which the passage is legitimate.

2.9.2 Influence Function (Green 1828) and Reciprocity Principle

Suppose a function $h(b)$ can be found such that Equation 2.413a has imaginary part (Equation 2.413b):

$$F(z;b) \equiv f(z;b) + h(b), \quad G(z;b) \equiv \mathrm{Im}\{F(z;b)\} = G(b;z), \tag{2.413a and 2.413b}$$

with the reciprocal property (Equation 2.413b). Then, Equation 2.413b is the **Green's or influence function** *because (1) it coincides with the stream function of a potential flow, satisfying all boundary conditions to within an irrelevant constant h(b); and (2) it satisfies the* **reciprocity principle** *that the stream function at z due to a singularity at b is equal to the stream function at b due to a singularity at z.* The Green's function can be introduced in general as the fundamental solution of a differential equation forced by the generalized function Dirac unit impulse: in this case, with the Laplace operator for the potential flow; like any linear self-adjoint operator, the Laplacian leads to the reciprocity principle (Equation 2.413b). Property 1 (Equation 2.413a) implies Equation 2.414a:

$$G(z;b) = \Psi(r,\varphi) + \mathrm{Im}\{h(b)\}; \qquad \frac{\partial G(z;b)}{\partial z} = \frac{\partial G(z;b)}{\partial b}. \tag{2.414a and 2.414b}$$

Property 2 implies that Equation 2.414b can be differentiated equivalently with regard to b or z. The condition (Equation 2.414b) allows exchanging $\mathrm{d}/\mathrm{d}z$ by $\mathrm{d}/\mathrm{d}b$ in Equation 2.410, justifying the passage to Equations 2.411 and 2.412. This proves the **second path theorem**: *(1) let f(z;b) be the complex potential of a flow with a singularity at b, such as a monopole with flow rate Q and circulation Γ; (2) if a function h(b) can be found such that Equation 2.413a has imaginary part with the reciprocal property (Equation 2.413b), then it specifies the Green's or influence function; (3) then the Routh or path function (Equation 2.412) is constant along the path of the singularity. This specifies the path of the singularity without integrations.*

2.9.3 Vortex near a Rigid Impermeable Cylinder

As an example, consider a vortex of circulation Γ at a position b near a cylinder of radius a (Section I.28.8) for which the complex potential (Equation 2.192a) specifies the first term on the r.h.s. of Equation 2.415a and the circle theorem (Equation I.24.47) adds the second term:

$$2\pi f(z;b) = -i\Gamma\log(z-b) + i\Gamma\log\left(\frac{a^2}{z} - b^*\right), \quad 2\pi h(b) = -i\Gamma\log b^*. \qquad \text{(2.415a and 2.415b)}$$

The imaginary part of Equation 2.415a is not reciprocal in (z, b). The additional term (Equation 2.415b) on the r.h.s. of Equation 2.416 is a constant that does not change the flow:

$$2\pi F(z;b) = 2\pi[f(z;b) + h(b)] = -i\Gamma\log(z-b) + i\Gamma\log\left(\frac{a^2}{b^* z} - 1\right). \qquad \text{(2.416)}$$

The imaginary part of Equation 2.416 has the reciprocal property:

$$2\pi G(z;b) \equiv 2\pi\,\mathrm{Im}\{F(z;b)\} = -\Gamma\log|z-b| + \Gamma\log\left|\frac{a^2}{b^* z} - 1\right| = 2\pi G(b;z), \qquad \text{(2.417)}$$

and hence is the Green's or influence function. Thus, the second path theorem applies, and the path of the vortex near a cylinder is specified (Equation 2.412) by

$$\begin{aligned} \text{const} = \Psi_0(r,\varphi) &= \lim_{z\to b}\left\{\mathrm{Im}[F(z,b)] + \frac{\Gamma}{2\pi}\log|z-b|\right\} \\ &= \lim_{z\to b}\left[\frac{\Gamma}{2\pi}\log\left|\frac{a^2}{b^* z} - 1\right|\right] = \frac{\Gamma}{2\pi}\log\left|\frac{a^2}{|b|^2} - 1\right|, \end{aligned} \qquad \text{(2.418)}$$

implying (Equation 2.419a) that the path is a circle. Thus, *the path of a vortex near a cylinder is a circle (Equation 2.419a) with radius equal to its initial distance from the center. The velocity of the vortex is given by Equation 2.409, leading to Equation 2.419b:*

$$\begin{aligned} \text{const} = b^* b = |b|^2: \quad v_0^* &= \lim_{z\to b}\left(\frac{df}{dz} + \frac{i\Gamma}{2\pi}\frac{1}{z-b}\right) = \lim_{z\to b}\frac{d}{dz}\left[\frac{i\Gamma}{2\pi}\log\left(\frac{a^2}{z} - b^*\right)\right] \\ &= -\lim_{z\to b}\frac{i\Gamma}{2\pi}\frac{a^2/z^2}{a^2/z - b^*} = \lim_{z\to b}\frac{i\Gamma a^2}{2\pi z}\frac{1}{b^* z - a} \\ &= \frac{i\Gamma a^2}{2\pi b}\frac{1}{b^* b - a} = \frac{i\Gamma a^2}{2\pi b}\frac{1}{|b|^2 - a^2}, \end{aligned} \qquad \text{(2.419a and 2.419b)}$$

in agreement with Equation 2.419b [≡ (Equation I.28.148b) for $Q = 0$]. The second path theorem applies as well to one vortex near a cylinder as to any number; for example, two opposite vortices at symmetric positions are considered next, that is, as in Section 2.8, but without mean flow.

2.9.4 Influence/Path Function for a Pair of Vortices near a Cylinder

The second path theorem is applied as follows: (1) the complex potential for a pair of vortices with opposite circulations $\pm\Gamma$ at symmetric positions $\{b, b^*\}$ is given by Equation 2.325 ≡ Equation 2.420c, omitting the free stream (Equation 2.420a) and source/sink (Equation 2.420b):

$$U = 0 = Q: \qquad f(z;b) = -\frac{i\Gamma}{2\pi}\log\left(\frac{z-b}{z-b^*}\,\frac{a^2/z-b}{a^2/z-b^*}\right); \tag{2.420a–2.420c}$$

(2) subtracting the self-effect of the vortex at b specifies the complex potential due to the vortex at b^* and the images of both vortices:

$$f_0(z;b) = f(z;b) + \frac{i\Gamma}{2\pi}\log(z-b) = \frac{i\Gamma}{2\pi}\log\left[(z-b^*)\frac{b^*z-a^2}{bz-a^2}\right]; \tag{2.421}$$

(3) its imaginary part is reciprocal:

$$G(z;b) = \operatorname{Im}\{f_0(z)\} = \frac{i\Gamma}{2\pi}\log\left[|z-b^*|\frac{|b^*z-a^2|}{|bz-a^2|}\right] = G(b;z), \tag{2.422}$$

and hence coincides with the Green's function; (4) hence, the auxiliary function $h(b) = 0$ is not needed in Equation 2.413a, and the value of Equation 2.411 at the vortex is

$$f_0(b;b) = \frac{i\Gamma}{2\pi}\log\left[(b-b^*)\frac{b^*b-a}{b^2-a^2}\right]; \tag{2.423}$$

(5) the imaginary part of Equation 2.422 specifies the path function:

$$\Psi_0 = \operatorname{Im}\{f_0(b;b)\} = \frac{\Gamma}{2\pi}\log\left[|b-b^*|\frac{|b^*b-a^2|}{|b^2-a^2|}\right], \tag{2.424}$$

that is constant along the path; (6) the path is specified by

$$|b-b^*|\frac{|b^*b-a^2|}{|b^2-a^2|} = \exp\left(\frac{2\pi\Psi_0}{\Gamma}\right) = \text{const} \equiv 2k, \tag{2.425}$$

Equivalently, Equation 2.425 ≡ Equation 2.424.

2.9.5 Path of a Pair of Vortices Outside/Inside a Cylinder

The path of the vortex (Equation 2.425 ≡ Equation 2.426):

$$|b - b^*|^2 \, |b^*b - a^2|^2 = 4k^2|b^2 - a^2|^2 \tag{2.426}$$

can be written in Cartesian (Equations 2.427a and 2.427b) [polar (Equations 2.428a and 2.428b)] coordinates:

$$b = x + iy: \quad y^2(x^2 + y^2 - a^2)^2 = k^2[(x^2 - y^2 - a^2)^2 + 4x^2y^2], \qquad \text{(2.427a and 2.427b)}$$

$$b = re^{i\varphi}: \quad r^2(r^2 - a^2)^2 \sin^2\varphi = k^2[r^4 + a^4 - 2a^2r^2\cos(2\varphi)], \qquad \text{(2.428a and 2.428b)}$$

where were used for the same three factors in Equation 2.426: (1) in Cartesian (Equation 2.427a) coordinates:

$$|b^* - b|^2 = |i2y|^2 = 4y^2, \; |b^*b - a^*|^2 = (|b|^2 = a^2)^2 = (x^2 + y^2 - a^2)^2, \qquad \text{(2.429a and 2.429b)}$$

$$|b^2 - a^2|^2 = |(x + iy)^2 - a^2|^2 = |x^2 - y^2 - a^2 + 2\,ixy|^2 = (x^2 - y^2 - a^2)^2 + 4x^2y^2, \tag{2.429c}$$

that were substituted in Equation 2.426, leading to Equation 2.427b; (2) in polar coordinates (Equation 2.428a):

$$|b^* - b|^2 = |r(e^{-i\varphi} - e^{i\varphi})|^2 = r^2|-2i \sin \varphi|^2 = 4r^2 \sin^2 \varphi, \quad (b^*b - a)^2 = (r^2 - a^2)^2, \qquad \text{(2.430a and 2.430b)}$$

$$\begin{aligned} \left|a^2 - b^2\right|^2 &= \left|a^2 - r^2e^{i2\varphi}\right|^2 = (a^2 - r^2e^{i2\varphi})(a^2 - r^2e^{-i2\varphi}) \\ &= a^4 - r^4 - 2a^2r^2(e^{i2\varphi} + e^{-i2\varphi}) = a^4 + r^4 - 2a^2r^2\cos(2\varphi), \end{aligned} \tag{2.430c}$$

that were substituted in Equation 2.426, yielding Equation 2.428b.

From Equation 2.427b (Equation 2.428b), the constant (Equation 2.431a) corresponds to the separating streamline (Equations 2.431b and 2.431c) and to the cylinder (Equations 2.431d and 2.431e):

$$k = 0: \quad \mathrm{y} = 0 = \varphi \quad \text{or} \quad \mathrm{r} - a = 0 = x^2 + y^2 - a^2. \qquad \text{(2.431a–2.431e)}$$

The paths of two vortices with opposite circulations near a cylinder with radius a (Figure 2.20a) are (1) two open curves outside the cylinder for which $y \to \pm k$ *as* $x \to \infty$*, so that k is the* **aiming distance** *(Equation 2.440a), that is, the distance of the path of the vortex from the axis at a large distance; and (2) two closed curves inside the cylinder. The distance of closest approach to the cylinder (Equation 2.432a ≡ Equation 2.432b) satisfies Equation 2.432c:*

$$\bar{r} \equiv r_{\min} = r\left(\varphi = \frac{\pi}{2}\right) = \left|y(x = 0)\right| = \bar{y}: \qquad \bar{r}^3 - k\bar{r}^2 - a^2\bar{r} - ka^2 = 0. \qquad \text{(2.432a–2.432c)}$$

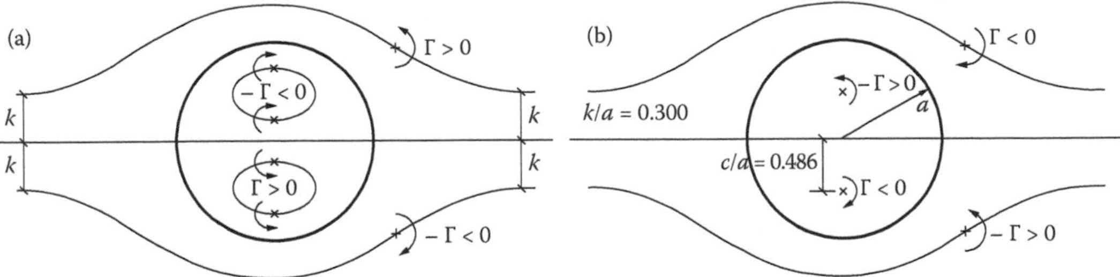

FIGURE 2.20
Problem in Figures 2.17 through 2.19 may be reconsidered for (1) two opposite vortices, that is, zero flow rate; (2) symmetric positions relative to a cylinder without mean flow. The vortices follow symmetric paths that (a) are (1) open curves outside the cylinder, specified by the aiming distance, that is, distance from the axis at infinity, implying that there are no positions of rest; (2) closed nested paths to the cylinder with opposite circulations to the open "semi-parallel" paths outside. The smallest closed paths inside the cylinder are (b) a pair of points where the opposite vortices are at rest. The equilibrium is indifferent because if displaced, the vortices will move to the neighboring closed path, neither returning to the position of rest nor moving further away from it.

The condition [Equation 2.432a (Equation 2.432b)] substituted in Equation 2.428b (Equation 2.429b) yields

$$\bar{r}(\bar{r}^2 - a^2) = k\left|\bar{r}^4 + a^4 + 2a^2\bar{r}\right|^{1/2} = k(\bar{r}^2 + a^2), \tag{2.433}$$

that is equivalent to Equation 2.432c ≡ Equation 2.433. The paths of the vortices inside the cylinder (Figure 2.20a) are closed curves; the curves do not intersect, that is, they are nested inside each other. The innermost curve reduces to a point where the vortex must be at rest. Thus, *the closed paths inside the cylinder have a pair of limit points, corresponding to the positions of rest of the vortices, as proved next (Subsection 2.9.6).*

2.9.6 Indifferent Equilibrium of Vortices Inside the Cylinder

The vortices will be at rest inside the cylinder if the closed paths reduce to two points (Figure 2.20b), which must lie on the transverse axis (Equation 2.436a) at the same distance c from the center. From Equation 2.421 ≡ Equation 2.434:

$$f_0(z;b) = \frac{i\Gamma}{2\pi}\log\left[(z - b^*)\frac{a^2/z - b^*}{a^2/z - b}\right], \tag{2.434}$$

follows the induced velocity at a vortex:

$$\begin{aligned} v_0^*(b) &= \lim_{z\to b}\frac{df_0(z;b)}{dz} = \lim_{z\to b}\frac{i\Gamma}{2\pi z}\left(\frac{1}{1 - b^*/z} + \frac{1}{b^* z/a^2 - 1} - \frac{1}{bz/a^2 - 1}\right) \\ &= \frac{i\Gamma}{2\pi b}\left(\frac{1}{1 - b^*/b} + \frac{1}{b^* b/a^2 - 1} - \frac{1}{b^2/a^2 - 1}\right). \end{aligned} \tag{2.435}$$

For the positions on the imaginary axis (Equation 2.436a), the condition of zero-induced velocity (Equation 2.435) is (Equation 2.246b) the vanishing of the term in brackets on the r.h.s. of Equation 2.435, leading to Equation 2.436c

$$b = \pm ic = \pm i|b|: \quad v_0^*(b) = 0 = \frac{1}{2} + \frac{1}{c^2/a^2 - 1} + \frac{1}{c^2/a^2 + 1} = \frac{1}{2} + \frac{2c^2/a^2}{c^4/a^4 - 1} = \frac{1}{2}\frac{c^4/a^4 + 4c^2/a^2 - 1}{c^4/a^4 - 1}. \tag{2.436a–2.436c}$$

The latter agrees (Equation 2.436c ≡ Equation 2.337c) for zero mean flow velocity $U = 0$. The equilibrium positions correspond to the real positive roots (Equation 2.437b) of the numerator in Equation 2.436c ≡ Equation 2.437a, leading to real Equation 2.437c:

$$\frac{c^4}{a^4} + 4\frac{c^2}{a^2} - 1 = 0: \quad \frac{c^2}{a^2} = -2 + \sqrt{5} = 0.236068; \quad \gamma \equiv \frac{c}{a} = \pm 0.48568. \tag{2.437a–2.437c}$$

The corresponding value of the path parameter from Equation 2.427b (Equation 2.428b) are Equations 2.438a through 2.438c (Equations 2.439a through 2.439c):

$$x = 0, y = \pm c: \quad \left(\frac{k_*}{a}\right)^2 = \lim_{\substack{z\to 0\\ y\to\pm c}} \frac{y^2}{a^2}\frac{(x^2+y^2-a^2)^2}{(x^2-y^2-a^2)^2+4x^2y^2} = \frac{c^2}{a^2}\frac{(a^2-c^2)^2}{(c^2+a^2)^2} = \gamma^2\left(\frac{\gamma^2-1}{\gamma^2+1}\right)^2. \tag{2.438a–2.438c}$$

$$r = c, \varphi = \pm\frac{\pi}{2}: \quad \left(\frac{k_*}{a}\right)^2 = \lim_{\substack{r\to c\\ \varphi\to\pm\pi/2}} \frac{r^2}{a^2}\frac{(r^2-a^2)^2\sin^2\varphi}{r^4+a^4-2a^2r^2\cos(2\varphi)} = \frac{c^2}{a^2}\frac{(c^2-a^2)^2}{(c^2+a^2)^2} = \left(\gamma\frac{\gamma^2-1}{\gamma^2+1}\right)^2. \tag{2.439a–2.439c}$$

These coincide (Equation 2.338c ≡ Equation 2.339c) and specify the value of the path parameter corresponding to the positions of rest inside the cylinder:

$$\pm k = \lim_{r\to\infty} r\sin\varphi = \lim_{r\to\infty} y: \quad \frac{k_*}{a} = \pm\gamma\frac{1-\gamma^2}{1+\gamma^2} = \pm\frac{c}{a}\frac{a^2-c^2}{a^2+c^2} = \pm 0.300283. \tag{2.440a and 2.440b}$$

Two vortices with opposite circulations inside a cylinder of radius a are at rest (Figure 2.20b) on the transverse axis (Equation 2.436a) at a distance (Equation 2.347c) from the center, that is, a little less than half the radius; the same vortices outside the cylinder would follow the path (Equation 2.427b ≡ Equation 2.428b) corresponding to the path parameter (Equation 2.440b) in Equation 2.425 ≡ Equation 2.426 with (Figure 2.20a) an aiming distance (Equation 2.440a). The positions of rest correspond to **neutral equilibrium** *because the displacement of the vortices (Figure 2.20b) leads to nearby closed paths inside the cylinder (Figure 2.20a) neither returning to the position of rest nor moving farther from it.* The distinct motions and positions of equilibrium of vortices, sources, and sinks in potential flows considered in the present chapter and in volume I are listed in Table 2.3.

NOTE 2.1 DISTRIBUTIONS OF VORTICITY AND FLOW STABILITY

It may be stated in a somewhat simplistic but generally accurate way that irrotational (rotational) flows tend to be stable (unstable); a possible explanation is Kelvin's theorem (Subsection 2.3.3) that an irrotational flow has the minimum kinetic energy relative to rotational flows with the same rigid wall boundary conditions. Flow singularities such as sources, sinks, and vortices, and their combinations into monopoles, dipoles, quadrupoles, and multipoles (Chapter I.12), are generally not at rest in the presence of obstacles due to the velocities induced by their images. Some instances of equilibrium were found, for example, (1) a monopole near a wall (Subsection 2.7.4); (2 and 3) a pair of conjugate monopoles [Subsection 2.7.5 (Section 2.8)] in a parallel-sided duct (symmetrically placed near a cylinder in a uniform stream); (4) two opposite vortices inside a cylinder (Subsection 2.9.6); (5 and 6) a single row (Section I.36.8) or parallel or staggered double row (Section I.36.9) of vortices; and (7) the combination of the preceding, namely, a single or double row of vortices between parallel walls should also be in equilibrium. The existence of a static (dynamic) equilibrium, that is, at rest (in motion), leads to (1) linear ordinary differential equations (Subsections 2.8.8 through 2.8.12) for small perturbations; and (2) the consideration of the motion from arbitrary positions for large or nonlinear perturbations (Subsections 2.7.5, 2.7.6, and Section 2.9). Instability corresponds to a growth in time from a small (large) perturbation in the linear (nonlinear) case; for example, for a vortex sheet (Section 8.9) separating two tangential flows with distinct velocity, the growth of a perturbation is exponential in time. The question of stability is thus relevant to rotational flows with either concentrated vorticity, that is, line vortices (Subsections 2.7.3 and 2.9.6), or distributed vorticity such as vortex sheets (Section 8.9) or spherical vortices and general vortical flow. The question of stability applies not only to rotational but also to heated and viscous flows subject to external force fields; the transition to turbulence is a manifestation of the instability of laminar flows and may follow flow separation. Stability issues arise not only in fluid mechanics but also in the mechanics of particles and rigid bodies, deformable solids, electromagnetism, and nearly all areas of physics and engineering. Table 2.4 indicates the cases considered so far of equilibrium of monopoles, that is, sources, sinks, and vortices, in potential flows, distinguishing three situations: (1) static for equilibrium at rest; (2) convective for cases of motion. The latter distinguish (3a) the path, that is, the curve described, and (3b) the trajectory, that is, the position as a function of time.

NOTE 2.2 EQUILIBRIUM, STABILITY, AND TRAJECTORIES/PATHS FOR MULTIPOLES IN POTENTIAL FLOWS

The calculation of paths and trajectories of multipoles can be used to identify (1) particular positions of rest because, for static equilibrium, the path reduces to a point unchanging in time; and (2) the stability of positions of rest by tracing the motion from a nearby position toward/away/neither for stable/neutral/unstable equilibrium. Table 2.4 lists 11 cases with 25 subcases. Case I is a source/sink/vortex/monopole near a wall (Section I.16.1/I.16.2/2.7.4)

TABLE 2.4

Equilibria, Stability, and Trajectories for Multipoles in Potential Flows

Case	Subcase	Potential Flow	Multipole	Section	Figure	Equilibrium at Rest?	Stability	Stability Analysis	Path	Trajectory
I	1–3	Near an infinite wall	Source/sink	I.16.1	I.16.1 and 16.2	No	–	–	Yes	Yes
			Vortex	I.16.2	I.16.3	Convective	Unstable	Nonlinear	Yes	Yes
			Monopole	2.7.4	2.10 through 2.12	No	–	–	Yes	Yes
II	4–5	In a square corner	Source/sink	I.16.1	I.16.5	No	–	–	Yes	No
			Vortex	I.16.2	I.16.5	No	–	–	Yes	No
III	6–7	Near cylinder	Source/sink	I.28.8	I.28.13	No	–	–	Yes	Yes
			Vortex	I.28.8 and 2.9.3	I.28.13	No	–	–	Yes	Yes
IV	8–10	Near a semi-infinite plate	Vortex	I.36.3.1	I.36.6a	No	Unstable	Nonlinear	Yes	Yes
			Source	I.36.3.2	I.36.6b and I.36.7a	Static	Stable	Nonlinear	Yes	Yes
			Sink	I.36.3.2	I.36.6b and I.36.7b	Static	Unstable	Nonlinear	Yes	Yes
V	11–12	Between parallel walls	Source/sink	I.36.6	I.36.14a and b	Convective	Unstable	Nonlinear	Yes	Yes
			Vortex	I.36.7	I.36.14c and d	Static	Unstable	Nonlinear	Yes	Yes
VI	13–15	In a square well	Source/sink	I.36.8	I.36.16a and b	No	–	–	Yes	Yes
			Vortex	I.36.9	I.36.16c and d	Convective	Unstable	Nonlinear	Yes	Yes
			Staggered vortex	I.36.9	I.36.17	Convective	(1)	Nonlinear	Yes	Yes

VII	16–18	Pair between parallel walls	Equal sources/sinks	2.7.5 through 2.7.9	2.15b and c	Convective	Unstable	Nonlinear	Yes	Yes
			Opposite vortices	2.7.5 through 2.7.9	2.15a	Static	Unstable	Nonlinear	Yes	Yes
			Conjugate monopoles	2.7.5 through 2.7.9	2.13 and 2.14	Convective	Unstable	Nonlinear	Yes	Yes
VIII	19–21	Pair outside cylinder	Equal sources/sinks	2.8.6	2.19	Static	(2)	Linear	No	No
			Oppositive vortices	2.8.4 and 2.8.5; 2.9.4 and 2.9.5	2.17 and 2.20a	Static	Unstable	Linear	Yes	No
			Conjugate monopoles	2.8.3	2.16 and 2.18	Static	(2)	Linear	No	No
IX	22	Conjugate pair inside cylinder	Opposite vortices	2.9.5 and 2.9.6	2.20b	Static	Indifferent	Nonlinear	Yes	No
X	23	Uniform with cylinder	Cylinder plus with circulation plus vortex	I.28.6	I.28.6	No	–	–	Yes	No
XI	24–25	Cylinder near a wall	Parallel motion	Ex. 10.11	10.3	No	–	–	Yes	Yes
			Orthogonal motion	8.2	8.3	No	–	–	Yes	Yes

Note: (1) Stability possible for suitable separation of two rows. (2) Stability possible for positive flow rate in a range of free stream velocities. The cases of equilibrium at rest and trajectories of multipoles in potential flows are listed. The multipole could be a source, sink, vortex, or a monopole combination, or a dipole representing a cylinder. An equilibrium position of rest may or may not exist. If an equilibrium position exists, it may be (1) stable if there is a return to equilibrium for any perturbation; (2) unstable if there are some perturbations that lead further away from equilibrium; (3) neutral if a perturbation neither grows nor decays. A perturbation of unstable equilibrium leads to a trajectory. Starting from a nonequilibrium position also leads to a trajectory. Linear stability analysis considers only small perturbations from equilibrium. If the trajectories are obtained for any initial position, and there are positions of rest, then the nonlinear stability with large perturbations is included.

with one image leading to a static equilibrium only for the vortex with a cross flow that is an unstable configuration. In case II of a source/sink (vortex) in a rectangular corner [Section I.16.1 (Section I.16.2)], the three images allow no position of rest. In case IV of a source/sink/vortex near a semi-infinite plate (Subsection I.36.3.1/2/2), only the source (sink) can be at rest and aligned with the edge, with stable (unstable) static equilibrium. A source/sink/vortex between (case V) parallel walls (Section I.36.6/I.36.7/Subsections 2.7.5 through 2.7.9) can be in connective equilibrium, that is, uniform motion that can be canceled by a counter flow. The equilibrium is unstable in case V (VI) of source/sink/vortex in a parallel-sided duct (square well) corresponding [Sections I.36.6–3 (Sections I.36.8 and I.36.9)] to a single (double) row of images; the only stable case are two staggered rows of opposite vortices for a specific spacing (the von Karman vortex street).

For a conjugate pair of monopoles, including equal sources/sinks and opposite vortices, stable (unstable) static equilibrium is possible in a parallel-sided duct (case VII) at equal distance from the walls and axis (Subsections 2.7.5 through 2.7.9) for a source (sink or vortex). A source/sink/vortex near a cylinder (case III) are never at rest (Section I.28.8) due to the image at the reciprocal point. A conjugate pair symmetrically placed near a cylinder in a uniform stream (case VIII) can be at rest in any position for suitable flow rate and circulation (Subsections 2.8.1 through 2.8.7); stability applies (Section 2.8.8 through 2.8.12) only for positive flow rate and a range of free stream velocities. In the absence of a free stream, two opposite vortices symmetrically placed near a cylinder cannot be at rest outside without a mean flow (Subsection 2.9.5); inside, they have a single pair of positions (case IX) of neutral equilibrium (Subsection 2.9.6). A cylinder in a uniform stream is at rest (case X) only in the absence of circulation (Section I.28.6); otherwise, the cylinder has a cycloidal or trochoidal motion (Section I.28.7). A cylinder near a wall (case XI) is also not at rest, the simplest trajectories being motion parallel (orthogonal) to the wall [Example 10.11 (Section 8.2)].

2.10 Conclusion

An example of compressible flows is a vortex (source/sink) with azimuthal (radial positive/negative) velocity and constant circulation (flow rate), leading to a minimum radius in Figure 2.1a (Figure 2.1b). The singularity at the center for a vortex in an incompressible flow can be avoided by matching to a rotational core, for example, for the generalized Rankine (Hallock–Burnham) vortex, with azimuthal velocity that has an angular point (is smooth) at the matching point [Panel 2.1a (Panel 2.1b)], corresponding to a discontinuous (smooth) vorticity [Panel 2.2a (Panel 2.2b)] and specifying the pressure [Panel 2.3a (Panel 2.3b)] as a function of the radius. The condition of nonnegative pressure specifies the minimum core radius of vortices (Table 2.2) and the values of the velocity and vorticity at the core radius (Table 2.1). These are examples of steady incompressible vortices, for which the use of intrinsic coordinates (Figure 2.2) shows that the stagnation pressure is conserved (varies) along the tangent (normal) to the streamlines. The vorticity uniformly distributed in all space corresponds to a general unidirectional shear flow that consists of (Figure 2.3) two simple unidirectional shear flows, each with the velocity varying linearly with the transverse coordinate. A cylinder with circulation in a uniform stream plus a unidirectional shear flow (Figure 2.4) both with distinct angles of attack is subject to a lift (downforce) due to [Figure 2.5a (Figure 2.5b)] both the negative (positive) vorticity and circulation, which have equivalent effects. If they have opposite signs, a positive vorticity reduces the lift due to a negative circulation (Figure 2.6a), and a negative vorticity reduces the downforce due to a positive circulation (Figure 2.6b). These effects occur when an aircraft flies through a negatively (positively) sheared wind [Figure 2.7a (Figure 2.7b)] on approach to land or climb after takeoff.

The centroid is static for (1) a pair of sources and sinks that move either toward the centroid in a finite time (Figure 2.8b and 2.8d) or to infinity in an infinity time (Figure 2.8a and 2.8c); and (2) two vortices that rotate around the centroid between (outside) them [Figure 2.9a (Figure 2.9b)] while staying on the same radial line. A monopole consisting of a source/sink and a vortex has an image near an infinite wall (Figure 2.10) that determines the induced velocity (Figure 2.11) and trajectory (Figure 2.12). A pair of conjugate monopoles symmetrically placed in a parallel-sided duct (Figure 2.15) can be obtained: (1) taking the doubly infinite set of images on the walls (Figure 2.14); (2) mapping conformally into a strip (Figure 2.12) from the pair of conjugate monopoles near a wall. Two conjugate monopoles, that is, sources/sinks with the same flow rate Q plus vortices with opposite circulations $\pm\Gamma$, can be (Figure 2.17) at rest at any symmetric positions near a circular cylinder in a stream of velocity U if one of the three $\{Q, \Gamma, U\}$ is chosen arbitrarily and the other two are determined uniquely. In particular, two opposite vortices ($Q = 0$) are at rest if (1) they lie on the two lines (Figure 2.18) through the center of the cylinder orthogonal to the mean flow; (2) at four positions for which their mutual distance equals the distance from the corresponding reciprocal point in the cylinder (Figure 2.16). Whereas two opposite vortices are at rest (Figures 2.16 and 2.18) at six directions for all radii (Figure 2.19), two equal sources/sinks can be at rest only in four directions for radii not exceeding a maximum value. Although all the positions are of static equilibrium for a suitable choice of flow rate and circulation, the equilibrium is stable only (Table 2.3) for positive flow rate and a range of free stream velocities. A perturbation from an unstable equilibrium position or a start away from an equilibrium position leads to the trajectory of a multipole in a potential flow; 11 cases with 25 sub-cases are indicated in Table 2.4. One of the examples concerns the open (closed) paths (Figure 2.20a) of two opposite vortices outside (inside) a cylinder, leading to zero (two) positions of rest (Figure 2.20b).

3

Exponential and Logarithmic Functions

The exponential is the simplest integral function that can be defined (Subsection 3.3.3) in six equivalent ways: (D1) by being equal to its own derivative (Sections 1.1.8 and 3.1); (D2) by a power series (Sections 1.1.8 and 3.1); (D3) by the limit of a binomial (Section 3.1); (D4) by one of a number of equivalent continued fractions (Section 3.2); (D5) by the property of changing sums to products (Subsection 3.3.1); and (D6) by the property of changing products to powers (Subsection 3.3.2). For unicity, the definition D1 (D5 and D6) requires that the function (its derivative) be unity at the origin. Of the representations (D2, D3, and D4), those that lead to numerical computation with faster convergence are the series (D2) and the continued fraction (D3); the latter also proves that the number e is irrational. Any of these definitions leads to (Section 3.4) the other properties of the exponential: (1) it has an essential singularity at infinity, and so grows faster than any power; (2) it has no zeros (poles) and hence no representation as infinite product (series of fractions); (3) the latter property can be extended to show that any integral function without zeros is the exponential of another integral function. The exponential is periodic, hence many-valent, and thus, its inverse (Section 3.5), the logarithm, is many-valued and becomes single-valued in the complex plane with a branch-cut (Sections I.7.2 and I.7.3). The existence of a branch-point limits the region of convergence of the power series for the logarithm, that can also be represented by continued fractions (Section 3.6). The logarithm transforms products to sums, and powers to products, thus owing its usefulness to this simplification of operations. The exponential and its inverse, the natural logarithm, allow the definition of powers and logarithms with arbitrary base (Section 3.7), that is, other than the number e. All these functions can be calculated from logarithmic tables (Section 3.8), whose computation can be performed using series expansions (Section 3.6) combined with elementary properties. Some of the continued fractions for elementary transcendental functions, such as the exponential (logarithm) in spite of their apparent simplicity [Section 3.2 (Section 3.6)], require for proof the Lambert method (Section 1.9), involving higher transcendental functions (Section 3.9) such as the confluent (Gaussian) hypergeometric series.

3.1 Derivation Property, Series, and Rational Limit

Perhaps the simplest of the six equivalent definitions of the exponential is that it equals its derivative (Equation 1.25a); this definition is not unique because Equation 1.25a also holds if multiplied by an arbitrary constant C. The constant C is specified by prescribing the value of the exponential at one point: the simplest is to impose (Equation 1.25b) the value unity at the origin. This leads to the first definition (D1): *the exponential is the only function that equals its derivative (Equation 1.25a) and is unity at the origin (Equation 1.25b).* This is the first (D1) of six equivalent definitions (D1–D6) of the exponential. A consequence of Equation 1.25a is that the exponential has MacLaurin series absolutely (totally) convergent in the finite complex plane (Equation 1.24a) [a disk of finite radius (Equation 1.24b)], and hence is an integral function (Section 1.17). From Equations 1.25a and 1.25b also follow the coefficients (Equation 1.26a) of the MacLaurin series (Equation 1.26b), whose convergence in the finite plane can be confirmed (Equations 1.27a, 1.27b, and 1.28) by the D'Alembert ratio test (Section I.29.3.2). This provides the second definition (D2): *the exponential is the integral function defined by the series (Equation 1.26b ≡ Equation 3.1):*

$$|z| < \infty: \qquad e^z = \exp(z) = \sum_{n=0}^{\infty} \frac{z^n}{n!} = 1 + z + \frac{z^2}{2} + \frac{z^3}{6} + \frac{z^4}{24} + \frac{z^5}{120} + \frac{z^6}{720} + \cdots, \tag{3.1}$$

which has one essential singularity at infinity (Section 1.1.5). This complex series is equivalent to two real series (Subsection 3.6.4). Complex generalizations of the series (Equation 3.1) are considered in Example 10.5.

The series (Equation 3.1) can be obtained from the binomial expansion (Equation I.25.38):

$$\left(1+\frac{z}{N}\right)^N = \sum_{n=0}^{N} \binom{N}{n} \left(\frac{z}{N}\right)^n = \sum_{n=0}^{\infty} \frac{z^n}{n!} \frac{N!}{(N-n)!N^n} \sum_{n=0}^{\infty} \frac{z^n}{n!} \frac{N(N-1)\cdots(N-n+1)}{N^n} \tag{3.2}$$

by taking the limit as $N \to \infty$:

$$\lim_{N\to\infty} \sum_{n=0}^{N} \left\{1\left(1-\frac{1}{N}\right)\cdots\left(1-\frac{n-1}{N}\right)\right\} \frac{z^n}{n!} = \sum_{n=0}^{\infty} \frac{z^n}{n!}, \tag{3.3}$$

thus follows:

$$\exp(z) = \lim_{N\to\infty} \left(1+\frac{z}{N}\right)^N \tag{3.4}$$

the third definition (D3): *the exponential is defined by the binomial limit (Equation 3.4).*

3.2 Continued Fractions and Computation of the Number *e*

The terms [Equations 1.134a through 1.134d (Equations 1.136a through 1.136d)] of the power series (Equation 1.26b ≡ Equation 3.1) for the exponential lead to the continued fraction [Equation 1.135a (Equation 1.135b)]. Two alternative (Section 3.9.9) continued fractions are Equations 3.5a ≡ Equation 3.136c and Equation 3.5b ≡ Equation 3.138b:

$$e^z = \frac{1}{1-}\,\frac{z}{1+}\,\frac{z}{2-}\,\frac{z}{3+}\,\frac{z}{2-}\,\frac{z}{5+}\cdots\frac{z}{2n+1+}\,\frac{z}{2-}\cdots, \tag{3.5a}$$

$$= 1 + \frac{z}{1-}\,\frac{z}{2+}\,\frac{z}{3-}\,\frac{z}{2+}\,\frac{z}{5-}\cdots\frac{z}{2n+1-}\,\frac{z}{2+}\cdots. \tag{3.5b}$$

This provides a fourth definition (D4): *the exponential is specified by any of the continued fractions [Equations 1.135a and 1.135b (Equations 3.5a and 3.5b)].*

If z is an integer, then Equations 1.135a, 1.135b, 3.5a, and 3.5b are nonterminating continued fractions, showing that *e^n is an irrational number for all integer values n*. In particular, for $n = 1$, it follows that setting $z = 1$ in Equations 1.26b, 3.4, 3.5a, and 3.5b:

$$e \equiv \exp(1) = \sum_{n=0}^{\infty} \frac{1}{n!} = \lim_{n\to\infty}\left(1+\frac{1}{n}\right)^n = 1+\frac{1}{1-}\frac{1}{2+}\frac{1}{3-}\frac{1}{2+}\frac{1}{5-}\cdots\frac{1}{2+}\frac{1}{2n+1-}. \quad (3.6a\text{–}3.6c)$$

The **irrational number** ***e*** *can be calculated by the series (Equation 3.6a), the limit (Equation 3.6b), or the continued fraction (Equation 3.6c) and is given by*

$$\begin{aligned} e = 2.71828\ &18284\ 59045\ 23536\ 02874\ 71352\ 66249\ 77572\ 47093\ 69995\ 95749 \\ &66987\ 62772\ 40766\ 30353\ 54759\ 45713\ 82178\ 52516\ 64274\ 27466\ 39193 \end{aligned} \quad (3.7)$$

with 110 decimals (Note 3.2). The first 10 approximations to the irrational number e in Table 3.1 show that (1) the limit of the binomial (Equation 3.6b) converges slowly, giving four accurate digits for $N = 10^4$; and (2) the continued fraction (Equation 3.6c) and the series converge much faster, giving four accurate digits for $N = 6$ and seven accurate digits for $N = 9$. The calculation of the continued fraction (Equation 3.6c), using Equations 1.99a, 1.99b, 1.102a, 1.102b, 1.104a, and 1.104b for the continuants (Equation 1.96), is indicated in Table 3.2, that shows the successive rational approximations to number e.

TABLE 3.1

Approximations to the Irrational Number e

N	$\sum_{n=0}^{N}\frac{1}{n!}$	N	$\left(1+\frac{1}{N}\right)^N$	N	$\frac{1}{1-}\frac{1}{2+}\frac{1}{3-}\frac{1}{2+}\frac{1}{5+}\cdots\frac{1}{N}$
1	2.	1	2.	1	1.
2	2.5	2	2.	2	2.
3	2.7	3	2.	3	3.
4	2.7	6	2.	4	2.7
5	2.72	10	3.	5	2.7
6	2.718	10^2	2.7	6	2.718
7	2.7183	10^3	2.72	7	2.7183
8	2.71828	10^4	2.718	8	2.71828
9	2.718282	10^5	2.7183	9	2.718282
10	2.7182818	10^6	2.71828	10	2.71828183

Note: Comparison of the computation of the irrational number e by three methods. The limit of the binomial (Equation 3.6b) gives limited accuracy even for large values of N, for example, four digits for $N = 10^4$. Furthermore, large values of N can lead to truncation errors that limit the accuracy achievable in practice. The MacLaurin series (Equation 3.6a) converges rapidly because the general term is $O\left((n!)^{-1}\right)$ and gives four (seven) accurate digits with six (nine) terms. The continued fraction (Equation 3.6c) converges even faster than the MacLaurin series, giving nine instead of eight accurate digits for $n = 10$; the computation of the continued fraction is detailed in Table 3.2.

TABLE 3.2

Continuants of the Continued Fraction (Equation 3.6c) for e

n	a_n	b_n	p_n	q_n	p_n/q_n
0	–	–	1	0	–
1	1	–	1	1	1
2	1	1	2	1	2
3	2	–1	3	1	3
4	3	1	11	4	2.7
5	2	–1	19	7	2.7
6	5	1	106	39	2.718
7	2	–1	193	71	2.7183
8	7	1	1457	536	2.71828
9	2	–1	2721	1001	2.718282
10	9	1	25946	9545	2.71828183

Note: Computation of the first four continuants of the continued fraction (Equation 3.6b) is made: (1) starting with the numerators b_n and denominators a_n of the continued fraction (Equation 1.95); (2) using the recurrence formulas (Equations 1.102a and 1.102b) to obtain the numerators p_n and denominators q_n of the continuants (Equation 1.96); and (3) the continuants X_n of the convergent continued fraction (Equation 3.6b) provide successive closer approximations to the irrational number e as shown in Table 3.2.

3.3 Transformation of Sums to Products and Powers

Besides the first three (D1–D3) [fourth (D4)] definitions of exponential [Section 3.1 (Section 3.2)], two more (Section 3.3) are provided by the property D5 (D6) of Subsection 3.3.1 (Subsection 3.3.2), transforming sums (products) to products (powers). Because all these definitions are equivalent, any of them can be taken as the starting point to prove all the others (Subsection 3.3.3).

3.3.1 Transformation of Sums to Products

Besides Equations 1.25a, 1.25b, and 1.26b ≡ Equations 3.1 and 3.4, another (fifth) equivalent definition of the exponential is (D5) that *a differentiable function transforms sums to products iff it coincides with the exponential:*

$$f \in \mathcal{D}: \quad f(u+v) = f(u)\, f(u) \Leftrightarrow f(z) = \exp(az) \tag{3.8}$$

to within a constant factor a in the argument. The latter can be omitted by requiring the function to have derivative unity at the origin:

$$1 = f'(0) = \lim_{z\to 0} \frac{\mathrm{d}\left(e^{az}\right)}{\mathrm{d}z} = \lim_{z\to 0} e^{az} = a. \tag{3.9}$$

To prove the sufficient condition, that is, the exponential transforms sums to products, the exponential of the sum is the expanded MacLaurin series of v:

$$e^{u+v} = \sum_{n=0}^{\infty} \frac{v^n}{n!} \lim_{v\to 0} \frac{\mathrm{d}^n}{\mathrm{d}v^n}\left(e^{u+v}\right) = \sum_{n=0}^{\infty} \frac{v^n}{n!} \lim_{v\to 0} e^{u+v} = e^u \sum_{n=0}^{\infty} \frac{v^n}{n!} = e^u e^v. \tag{3.10}$$

To prove the necessary condition, assume that the analytic function transforms sums to products (Equation 3.8) and hence takes the same value when differentiated with regard to u or v:

$$f(v)\frac{\mathrm{d}f}{\mathrm{d}u} = \frac{\mathrm{d}[f(u+v)]}{\mathrm{d}u} = \frac{\mathrm{d}[f(u+v)]}{\mathrm{d}v} = f(u)\frac{\mathrm{d}f}{\mathrm{d}v}. \tag{3.11}$$

This identity can be rewritten in the form

$$\frac{f'(u)}{f(u)} = \frac{f'(v)}{f(v)} = a; \qquad \frac{f'(z)}{f(z)} = a = \text{const.} \tag{3.12a and 3.12b}$$

Because the first (second) term in Equation 3.12a is a function of u (v) alone, they can be equal only if they reduce to a constant a in Equation 3.12b. From Equation 3.12b ≡ Equation 3.13a follows Equation 3.13b:

$$a = \frac{f'(z)}{f(z)} = \frac{\mathrm{d}}{\mathrm{d}z}\{\log[f(z)]\}, \quad az + \log A = \log[f(z)], \quad f(z) = A\exp(az), \tag{3.13a–3.13c}$$

where A in Equation 3.13c is another arbitrary constant. Setting $v = 0$ in Equation 3.8 shows (Equation 3.14a) that the constant is unity (Equation 3.14b):

$$f(u) = f(0)\, f(u), \qquad 1 = f(0) = \mathrm{A}. \tag{3.14a and 3.14b}$$

Thus, the constant A can be omitted from Equation 3.13c, which leads to Equation 3.8. Applying the property (Equation 3.8) of the sum $N - 1$ times leads to Equation 3.15a. Then, the change of variable (Equation 3.15b) leads to Equation 3.15c:

$$\left(e^z\right)^N = e^{Nz}: \qquad z = \frac{w}{M}, \qquad e^{(N/M)w} = \sqrt[M]{\left(e^z\right)^N}. \tag{3.15a–3.15c}$$

It has been shown that the exponential transforms the product to a power (Equation 3.15a), and for rational numbers N/M, it leads to the Mth root of the Nth power (Equation 3.15c). This property extends to complex exponents and leads to the sixth definition of exponential as shown next (Section 3.3.2).

3.3.2 Transformation of Products into Powers

This corresponds to the sixth definition of exponential (D6): *a differentiable function transforms the product into power iff it is the exponential*

$$\mathrm{f} \in \mathcal{D}: \quad f(uv) = \{f(u)\}^v \Leftrightarrow f(z) = \exp(az) \tag{3.16}$$

to within a constant a, determined as in Equation 3.9. The sufficient condition that the exponential transforms products to powers can be proven (Equation 3.17b) using the limit (Equation 3.4):

$$N = \frac{n}{v}: \qquad \left(e^u\right)^v = \lim_{N\to\infty}\left(1+\frac{u}{N}\right)^{Nv} = \lim_{n\equiv Nv\to\infty}\left(1+\frac{uv}{n}\right)^n = e^{uv}, \quad \text{(3.17a and 3.17b)}$$

where the substitution (Equation 3.17a) was made. To prove the necessary condition, assume that the analytic function transforms products to powers (Equation 3.16), which leads to two identities:

$$\frac{1}{v}\frac{df(uv)}{du} = \frac{1}{v}\frac{d(uv)}{du}\frac{df(uv)}{d(uv)} = \frac{df(uv)}{d(uv)} = f'(uv), \tag{3.18a}$$

$$\frac{1}{v}\frac{df(uv)}{du} = \frac{1}{v}\frac{d}{du}\left\{\left[f(u)\right]^v\right\} = \left\{f(u)\right\}^{v-1}\frac{df}{du} = f(uv)\frac{f'(u)}{f(u)}. \tag{3.18b}$$

Their equality (Equation 3.18a ≡ Equation 3.18b) implies (Equation 3.19a) for arbitrary (u, v):

$$\frac{f'(uv)}{f(uv)} = \frac{f'(u)}{f(u)} = a; \qquad \frac{f'(z)}{f(z)} = a. \quad \text{(3.19a and 3.19b)}$$

It follows from Equation 3.19a that Equation 3.19b is a constant, and hence Equations 3.13a through 3.13c. Setting $u = 0$ in Equation 3.16 gives Equation 3.20a:

$$f(0) = \{f(0)\}^v, \qquad 1 = f(0) = A, \quad \text{(3.20a and 3.20b)}$$

implying (Equation 3.20b) that the constant A can be omitted in Equation 3.13c, which leads to Equation 3.16.

3.3.3 Equivalence of the Six Definitions of Exponential

The exponential can be defined in six equivalent ways: (D1) it equals its derivative (Equation 1.25a) and is unity at the origin (Equation 1.25b); (D2) it is the integral function defined by the MacLaurin series (Equation 1.26b ≡ Equation 3.1); (D3) it is specified by the limit (Equation 3.4) of a binomial; (D4) it is specified by any of the continued fractions (Equations 1.135a, 1.135b, 3.5a, and 3.5b); (D5 and D6) it is the differentiable function that transforms sums to products (Equation 3.8) [products to powers (Equation 3.16)] and has a unit derivative at the origin (Equation 3.9). Because all these definitions are equivalent, any of them can be taken as the starting point to establish all the others. The preceding properties were proved from Equations 1.25a and 1.25b, but Equation 1.26b ≡ Equation 3.1 could equally well be taken as a starting point, bearing in mind that the exponential series is a series of analytic functions z^n totally convergent for $|z| \le M$. This allows the operations and transformations (Table I.21.1) that prove from (D2) the MacLaurin series the other five properties: (D1) it is unity at the origin (Equation 1.25b) and is unchanged by derivation:

$$n = m+1: \qquad \frac{d\left(e^z\right)}{dz} = \frac{d}{dz}\left\{\sum_{n=0}^{\infty}\frac{z^n}{n!}\right\} = \sum_{n=1}^{\infty}\frac{z^{n-1}}{(n-1)!} = \sum_{m=0}^{\infty}\frac{z^m}{m!} = e^z, \qquad \text{(3.21a and 3.21b)}$$

in agreement with Equation 1.25a; (D3) the binomial limit (Equation 3.4) also follows Equations 3.2 through 3.3 from the series (Equation 3.1); (D4) the continued fractions (Equations 1.135a, 1.135b, 3.5a, and 3.5b) can also be obtained from the series; (D5) the Cauchy rule for the product of series (Equation I.21.25c), replacing the double sum by rows and columns by sums along the diagonals, when applied to Equation 3.1 leads to

$$e^u e^v = \sum_{n=0}^{\infty}\frac{u^n}{n!}\sum_{n=0}^{\infty}\frac{v^m}{m!} = \sum_{n=0}^{\infty}\sum_{k=1}^{n}\frac{u^k v^{n-k}}{k!(n-k)!} = \sum_{n=0}^{\infty}\frac{1}{n!}\sum_{k=0}^{n}\binom{n}{k}u^k v^{n-k} = \sum_{n=0}^{\infty}\frac{(u+v)^n}{n!} = e^{u+v}, \qquad (3.22)$$

which proves the property of the product of exponentials; (D6) the property of the power follows from the preceding D5 for rational exponent (Equations 3.15a through 3.15c) and from D3 for complex exponent (Equations 3.17a and 3.17b).

3.4 Limits, Period, and Absence of Zeros

Separating real and imaginary parts in the exponential:

$$z = x + iy: \qquad \exp(z) = e^{x+iy} = e^x\, e^{iy} = e^x\,(\cos y + i \sin y) \qquad \text{(3.23a and 3.23b)}$$

leads to *the modulus and argument of the exponential:*

$$|\exp(z)| = \exp\{\mathrm{Re}(z)\}, \qquad \arg\{\exp(z)\} = \mathrm{Im}(z). \qquad \text{(3.24a and 3.24b)}$$

This shows that the modulus (argument) of the complex exponential is constant on vertical (horizontal) lines (Figure 3.1) and leads to the carpet plot (Figure 3.2) shaped like a curved ramp. The divergence of the ramp corresponds to the essential singularity at infinity, implying that the *exponential dominates all powers*:

$$|a| < \infty: \qquad 0 = \lim_{z\to\infty}\begin{cases} z^a e^z & \text{if } \mathrm{Re}(z) < 0, & \text{(3.25a)} \\ z^a e^{-z} & \text{if } \mathrm{Re}(z) > 0; & \text{(3.25b)} \end{cases}$$

in particular

$$x, a \in |R: \qquad \lim_{x\to+\infty} x^a e^{-x} = 0, \qquad \lim_{x\to+\infty} x^a e^x = +\infty, \qquad \text{(3.26a–3.26d)}$$

for real variable and exponent (Equations 3.26a and 3.26b), the exponential dominates all powers (Equation 3.26c and 3.26d) the ends of the real line (Figure 3.3). From Equation 3.23b, it follows that

$$m \in |Z: \qquad \exp(z) = \exp(z + 2\pi i) = \exp(z + m2\pi i). \qquad \text{(3.27a and 3.27b)}$$

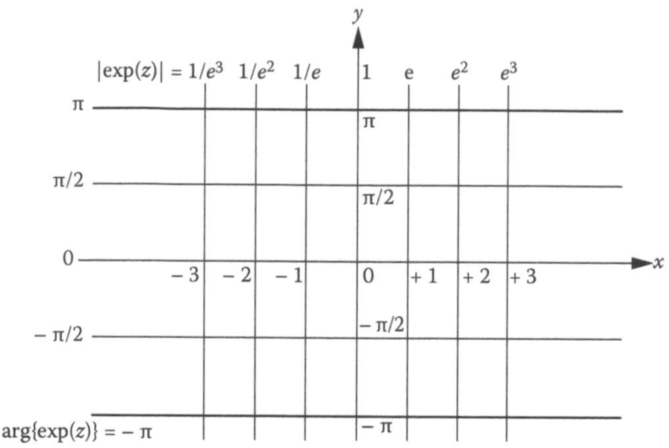

FIGURE 3.1
Level curves for which the modulus and argument of the complex exponential are constant are drawn over the complex plane. The modulus of the complex exponential is the exponential of the real part and is constant along vertical lines; when these are equally spaced to the right (left) of the imaginary axis, the constant value on each of them grows (decays) exponentially (Figure 3.2). The argument of the complex exponential is the imaginary part of the variable and is constant over horizontal lines; if the latter are equally spaced, so are the constant values on each of them. The two sets of level curves for the modulus and argument of the complex exponential (Figure 3.1) correspond to the carpet plot in three-dimensional perspective in Figure 3.2.

The exponential (Equations 3.23a and 3.23b) has the fundamental period $2\pi i$ (Equation 3.27b), and its multiples (Equation 3.27a) are also periods; hence, its fundamental region (Figure 3.4) is the horizontal strip (Equation 3.28a) because

$$-\pi < \operatorname{Im}(z) \le \pi: \quad m \in |Z, \qquad (2m-1)\pi < \operatorname{Im}(z) \le (2m+1)\pi. \tag{3.28a–3.28c}$$

If e^z is known in this region (Equation 3.28a), then Equation 3.27b can be used to calculate e^z for any z in the whole z-plane by choosing (Equation 3.28b) an integer n such that Equation 3.28c holds.

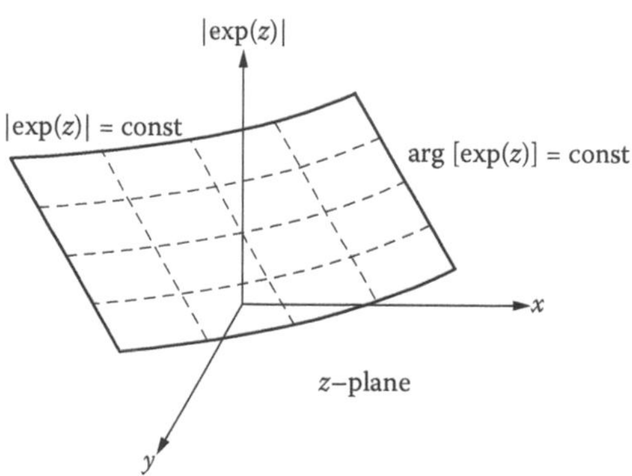

FIGURE 3.2
Carpet plot of the modulus of the complex exponential showing exponential growth (decay) as $x \to +\infty$ ($x \to -\infty$) parallel to the real axis (Figure 3.3) and constant argument parallel to the imaginary axis. The surface $|\exp(z)|$ can be obtained by translation of the curve e^x in Figure 3.3 normal to the plane of the paper.

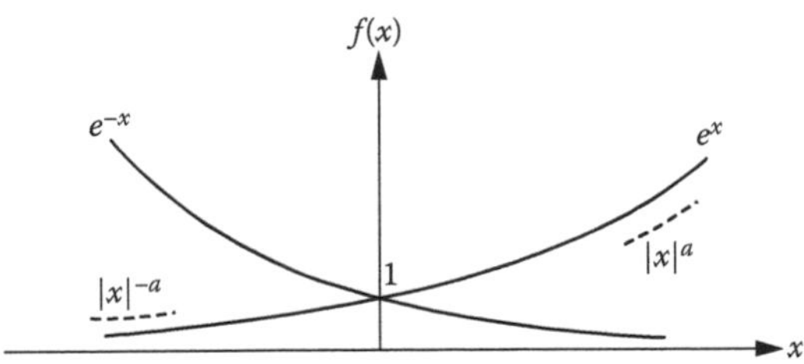

FIGURE 3.3
Exponential of real variable (Figure 3.3) corresponds to a section of the carpet plot (Figure 3.2) by the (x, z)-plane and projection of the exponential with complex variable (Figure 3.1) on the real axis. The exponential of real variable grows (decays) faster than any power or polynomial toward infinity along the positive (negative) real axis.

The property of the product (Equation 3.22 ≡ 3.29a) implies Equation 3.29c for Equation 3.29b:

$$e^u\, e^v = e^{u+v}; \qquad u = -v \equiv z: \qquad e^z\, e^{-z} = e^0 = 1. \tag{3.29a–3.29c}$$

Because e^z is analytic in the finite z-plane (Equation 3.30a), it is bounded in modulus (Equation 3.30b) and the other factor in Equation 3.29 cannot vanish (Equation 3.30c):

$$|z| \le R < \infty: \qquad |e^z| < B < \infty \Rightarrow e^{-z} \ne 0. \tag{3.30a–3.30c}$$

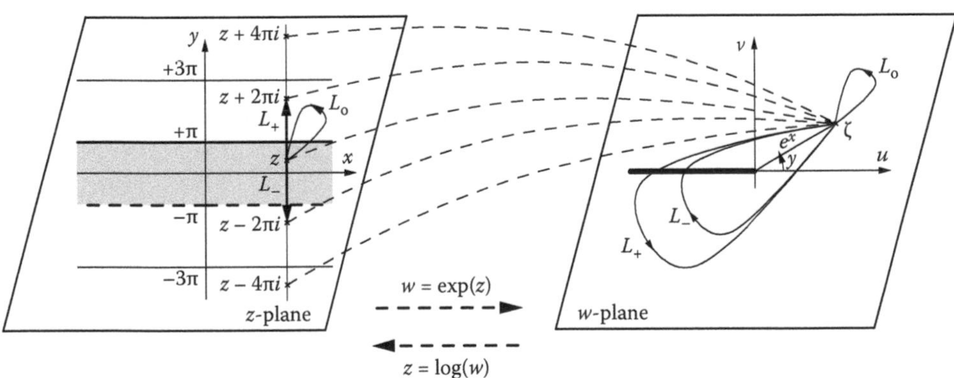

FIGURE 3.4
Complex exponential is a periodic function with period $2\pi i$. It is defined in the horizontal strip (Equation 3.28a) in the z-plane that includes (excludes) the upper (lower) boundary [solid (dotted)] line. It is repeated in parallel horizontal strips of width 2π. Thus, all congruent points spaced $2\pi i$ in the z-plane mapped to the same point ζ-plane, that is, the complex exponential is a single-valued many-valent function; its inverse, the logarithm, is a univalent many-valued function because each value of ζ corresponds to an infinite number of values of z, one for each branch. The principal branch corresponds to the fundamental strip in the z-plane. A loop L_0 in the ζ-plane that does not enclose the origin keeps the same branch. A loop $L_+(L_-)$ in the positive (negative) or counterclockwise (clockwise) direction around the origin in the z-plane passes the next branch up (down) the next strip in the ζ-plane. This is prevented by a branch-cut along the negative real axis in the z-plane. This is equivalent to folding the fundamental strip in the ζ-plane into a cylinder with the upper boundary coincident with the lower boundary. This keeps the principal branch of the logarithm and causes a jump across the branch-cut. Thus, there are two alternatives: (1) to keep continuity across an infinite number of branches of the algorithm, horizontal strips in the ζ-plane, or sheets of the Riemann surface (Section I.7.3); and (2) to remain in the principal branch of the logarithm, and fundamental strip in the range and a single sheet of the Riemann surface, by introducing in the domain a branch-cut joining the branch-point at the origin to infinity, causing a jump across the branch-cut.

This gives an alternative proof not using the Cauchy fourth integral theorem (Section I.31.4) that *the exponential has no zeros (Equation 3.31b) in the finite plane (Equation 3.31a):*

$$|z| < \infty: \qquad e^z \neq 0; \quad \lim_{x \to +\infty} e^x = \infty, \quad \lim_{x \to -\infty} e^{-x} = 0. \qquad \text{(3.31a–3.31d)}$$

Because the exponential has an essential singularity at infinity, the series (Equation 3.1) implies that it diverges along the positive real line (Equation 3.31c), and by Equation 3.29c, it can vanish only along the negative real line (Equation 3.31d). The result (Equations 3.31c and 3.31d) is a particular case $a = 0$ of Equations 3.26a through 3.26d, which, in turn, is a particular case of Equations 3.25a and 3.25b. The result (Equations 3.31a and 3.31b) can be generalized as the **theorem of absence of zeros:** *an integral function (Equation 3.32a) has no zeros (Equation 3.32b) in the finite plane iff it is the exponential (Equation 3.32c) of another integral function (Equation 3.32d):*

$$f \in \mathcal{D}\,(|z| \le R < \infty) \wedge f(z) \neq 0 \Leftrightarrow f(z) = \exp[g(z)] \wedge g \in \mathcal{D}\,(|z| \le R < \infty). \qquad \text{(3.32a–3.32d)}$$

The proof of the sufficient condition is immediate: if g is analytic, $e^g \neq 0$ by Equations 3.31a and 3.31b. The proof of the sufficient condition relies on Equation 3.33a being an integral function (Equation 3.32d) because Equation 3.32a is an integral function without zeros (Equation 3.32b):

$$g(z) = \frac{f'(z)}{f(z)} = \frac{\mathrm{d}}{\mathrm{d}z}\{\log[f(z)]\}; \quad f(z) = \exp[c + g(z)]. \qquad \text{(3.33a and 3.33b)}$$

The primitive of Equation 3.33a is Equation 3.33b, where the arbitrary constant c can be incorporated in the function $g(z)$, which leads to Equation 3.32c.

3.5 Logarithm as the Function Inverse to Exponential

The **natural logarithm** is defined as the function inverse of the exponential:

$$w \equiv \log z \Leftrightarrow z = \exp(w). \qquad \text{(3.34)}$$

The derivative of the logarithm is calculated by the rule of the inverse function:

$$\frac{\mathrm{d}(\log z)}{\mathrm{d}z} = \frac{\mathrm{d}w}{\mathrm{d}z} = \left(\frac{\mathrm{d}z}{\mathrm{d}w}\right)^{-1} = \left[\frac{\mathrm{d}(e^w)}{\mathrm{d}w}\right]^{-1} = \frac{1}{e^w} = \frac{1}{z}, \qquad \text{(3.35)}$$

which can be generalized to the *nth order derivate of the logarithm:*

$$\frac{\mathrm{d}^n(\log z)}{\mathrm{d}z^n} = \frac{\mathrm{d}^{n-1}}{\mathrm{d}z^{n-1}}\left(\frac{1}{z}\right) = \frac{(-1)(-2)\cdots(1-n)}{z^n} = (n-1)!\frac{(-)^n}{z^n}. \qquad \text{(3.36)}$$

The derivates (Equations 3.35 and 3.36) are singular at the origin, which is a branch-point (Sections I.7.2 and I.7.3) of the logarithm, that is, because the exponential is a periodic function (Equations 3.27a and 3.27b), its inverse, *the logarithm (Equation 3.37d), is a many-valued function, that becomes single-valued, choosing the principal branch (Equation 3.37b) in the complex z-plane with a semi-infinite cut (Equation 3.37a) joining the branch-point at the origin to infinity, along the negative real axis (Figure 3.4):*

$$z \notin \,)-\infty, 0\,(: \log z = \log|z| + i\arg(z); \qquad m \in |N: (\log z)_m = \log z + m\,2\pi i. \quad (3.37a\text{–}3.37d)$$

The branch-cut prevents counterclockwise (clockwise) loops (Figure 3.3) around the branch-point at the origin, which add (subtract) multiples of 2π of the argument and lead to the other branches (Equations 3.37c and 3.37d); the branches differ (Equation 3.38a) by multiples of $2\pi i$:

$$(\log z)_n - (\log z)_m = i2\pi(n-m); \qquad x > 0: \log(-x \pm i0) = \log|x| \pm i\pi. \qquad (3.38a\text{–}3.38c)$$

In order to prevent passage to the next jump, *the principal branch (Equation 3.37b) of the logarithm takes different values (Equation 3.38c) above and below the branch-cut (Equations 3.37a and 3.38b), and thus*

$$x > 0: \quad \log(-x \pm i0) - \log(-x - i0) = 2\pi i \qquad (3.38d \text{ and } 3.38e)$$

has a jump (Equation 3.38e) across the branch-cut (Equation 3.38d).

From Equation 3.37b follow *the real and imaginary parts of the logarithm:*

$$\operatorname{Re}\{\log z\} = \log|z|, \quad \operatorname{Im}\{\log z\} = \arg z, \qquad (3.39a \text{ and } 3.39b)$$

which show that the real (imaginary) part of the logarithm is constant on circles with center (radial lines passing through) at the origin (Figure 3.5), which leads to a carpet plot (Figure 3.6) like a

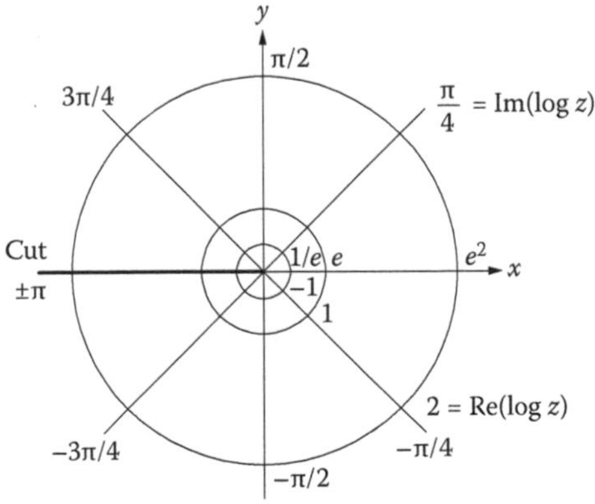

FIGURE 3.5
Level curves for which the real and imaginary parts of the logarithm are projected on the complex plane. The real (imaginary) part of the logarithm is constant on circles with center at (radial lines through) the origin; for equally spaced values, the radial lines are equally spaced and the circles become progressively farther outside (closer inside) the unit circle toward the origin. This corresponds to the carpet plot in three-dimensional perspective in Figure 3.6.

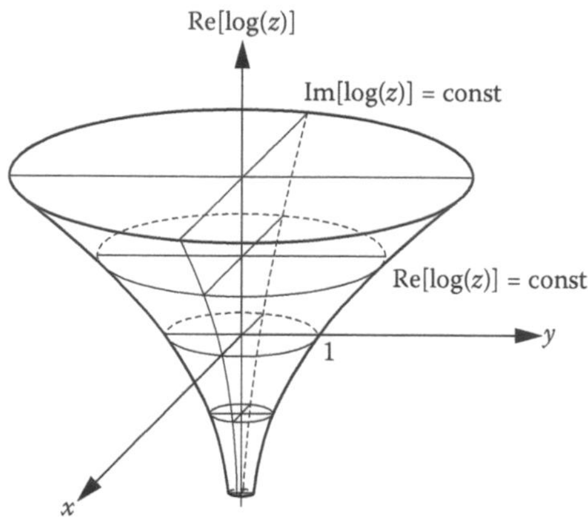

FIGURE 3.6
Carpet plot of the real part of the complex logarithm showing the funnel shape due to the increasing (decreasing) radius of the circles outside (inside) the unit circle for positive (negative) equally spaced values. The lines of constant argument of the logarithm are the generators of the axisymmetric surface and correspond to the curve $\log x$ in Figure 3.7.

curved funnel. The particular values (Equations 1.25b, 3.31c, and 3.31d) of the exponential imply the following *values of the logarithm with real variable:*

$$\log 1 = 0 \qquad \log \infty = \infty, \qquad \log 0 = -\infty. \tag{3.40a–3.40c}$$

The analog of Equations 3.26a through 3.26d states that *the logarithm diverges at the origin and infinity slower than any power (Figure 3.7):*

$$\text{Re}(a) > 0: \qquad \lim_{u \to 0} \; u^a \log u = 0 = \lim_{v \to \infty} v^{-a} \log v. \tag{3.41a–3.41c}$$

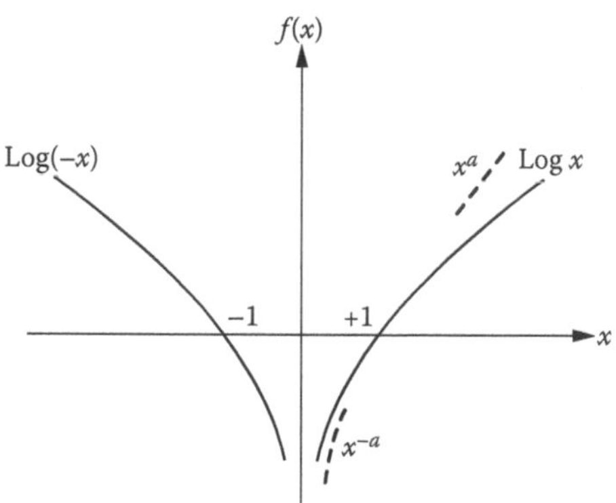

FIGURE 3.7
Logarithm of real variable (Figure 3.7) rotated around the *f*-axis generates the axisymmetric surface of the constant real part of the logarithm of complex variable (Figure 3.6). For real variable, the logarithm exists only for positive real values. The negative real values correspond to the branch-cut (Figure 3.5) joining the branch-point at the origin to infinity, so that only the complex algorithm exists above and below the branch-cut with a discontinuity across it. Because the logarithm is the inverse of the exponential, interchanging the axis in Figure 3.3 follows Figure 3.7, showing that the logarithm diverges to $+\infty$ ($-\infty$) as $x \to \infty$ ($x \to 0$), slower than any power or polynomial, that is, any power or polynomial dominates the logarithm.

The proof of Equations 3.41a through 3.41c follows in two stages: (1) the property (Equation 3.41b) at the origin can be reduced (Equation 3.42b) by inversion (Equation 3.42a) to the property at infinity (Equation 3.41c):

$$v = \frac{1}{u}: \qquad \lim_{u\to 0} u^a \log u = -\lim_{v\to\infty} v^{-a} \log v, \qquad \text{(3.42a and 3.42b)}$$

so that only the latter needs to be proven; (2) the latter follows from Equation 3.26c:

$$v = e^x: \qquad \lim_{v\to\infty} \; v^{-a} \log v = \lim_{z\to\infty} xe^{-ax} = 0, \qquad \text{(3.43a and 3.43b)}$$

after making in Equation 3.41c a substitution (Equation 3.43a) to eliminate the logarithm in Equation 3.43b.

The properties inverse to Equations 3.8 and 3.16:

$$f(uv) = f(u) + f(u) \Leftrightarrow f(z) = A \log z \Leftrightarrow f(u^v) = vf(u) \qquad \text{(3.44a and 3.44b)}$$

state that *a function transforms products to sums (powers to products) iff it is the logarithm to within a multiplying constant A. The constant can be omitted:*

$$1 = \lim_{z\to 1} \frac{\mathrm{d}}{\mathrm{d}z}(A \log z) = \lim_{z\to 1} \frac{A}{z} = A \qquad \text{(3.44c)}$$

by specifying a unit derivative at the point unity. The logarithm is defined with the following limit:

$$\lim_{N\to\infty} N\left(w^{1/N} - 1\right) = \log w, \qquad \text{(3.45a)}$$

as follows from the change of variable (Equation 3.45b):

$$w = e^z: \qquad w = \left(1 + \frac{\log w}{N}\right)^N, \qquad w^{1/N} - 1 = \frac{\log w}{N}, \qquad \text{(3.45b–3.45d)}$$

in the definition (D3) of the exponential (Equation 3.4 ≡ Equation 3.45c ≡ Equation 3.45d) as the binomial limit $N \to \infty$.

3.6 Series Expansions and Continued Fractions

The branch-cut of the logarithm limits the radius of convergence of series expansions for the logarithm; for example, convergence in the unit disk is possible if the branch-points are shifted to a point with modulus unity (Section 3.6.1). The series expansions lead to some continued fractions for the logarithm (Section 3.6.2); other continued fractions are obtained by applying the Lambert method to the Gaussian hypergeometric function (Section 3.9.6). The series expansions in the unit disk (Section 3.6.1) lead via changes of variable to series expansions valid in other regions, such as offset circles and half-planes (Section 3.6.3). The real and imaginary parts of complex series specify two real series, for example, starting with the complex exponential and logarithm in polar coordinates (Section 3.6.4).

3.6.1 Series for the Logarithm in Unit Disk

In order to apply MacLaurin's theorem to the logarithm, the branch-point may be shifted to $z = -1$ by choosing log $(1 + z)$. The latter series was obtained (Equation I.21.64b) by term-by-term integration of the geometric series. Its coefficients can also be reduced from (Equations 3.24a and 3.24b) MacLaurin's theorem:

$$\log(1+z) = \log 1 + \sum_{n=1}^{\infty} \frac{z^n}{n!} \lim_{a\to 0} \frac{d^n}{da^n}\left[\log(1+a)\right]$$

$$= \sum_{n=1}^{\infty} \frac{z^n}{n!} \lim_{a\to 0} (-)^{n-1} \frac{(n-1)!}{(1+a)^n} = \sum_{n=1}^{\infty} \frac{(-)^{n-1}}{n} z^n, \tag{3.46}$$

where Equation 3.36 was used. The radius of convergence of the series (Equation 3.46 ≡ Equation 3.47b) is unity because of the presence of the branch-point (Equation 3.47a):

$$z \notin)-\infty, -1): \quad \log(1+z) = \sum_{n=1}^{\infty} (-)^{n-1} \frac{z^n}{n} = z - \frac{z^2}{2} + \frac{z^3}{3} - \frac{z^4}{4} + \frac{z^5}{5} - \frac{z^6}{6} \cdots, \quad \text{(3.47a and 3.47b)}$$

$$z \notin (+1, +\infty(: \quad \log(1-z) = -\sum_{n=1}^{\infty} \frac{z^n}{n} = -z - \frac{z^2}{2} - \frac{z^3}{3} - \frac{z^4}{4} - \frac{z^5}{5} - \frac{z^6}{6} \cdots. \quad \text{(3.48a and 3.48b)}$$

The series (Equations 3.48a and 3.48b) follows from Equations 3.47a and 3.47b by replacing with $-z$. Furthermore:

$$z \notin)-\infty, -1) \cup (+1, \infty(: \quad \log\left(1-z^2\right) = -\sum_{n=1}^{\infty} \frac{z^{2n}}{n} = -z^2 - \frac{z^4}{2} - \frac{z^6}{3} - \frac{z^8}{4} - \cdots, \quad \text{(3.49a and 3.49b)}$$

$$z \notin)-\infty, -1) \cup (+1, \infty(: \quad \log\left(\frac{1+z}{1-z}\right) = \sum_{n=0}^{\infty} \frac{z^{2n+1}}{n+1/2} = 2z + \frac{2}{3}z^3 + \frac{2}{5}z^5 + \frac{2}{7}z^7 + \cdots$$

(3.50a and 3.50b)

are obtained [Equation 3.49b (Equation 3.50b)] from the sum (difference) of Equations 3.47b and 3.48b.

The branch-cuts exclude (Equations 3.51b through 3.51e) negative real values (Figure 3.8) of the arguments of the logarithms in Equations 3.47b, 3.48b, 3.49b, and 3.50b:

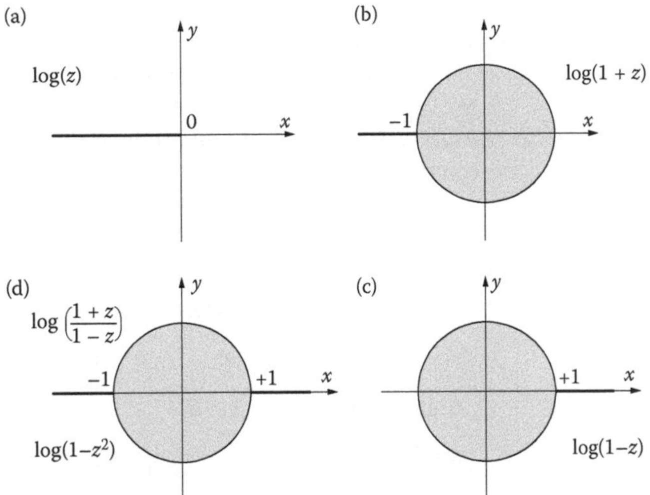

FIGURE 3.8
Logarithm is single-valued in the complex plane with a branch-cut from the branch-point at the origin to infinity along the negative real axis, as seen in Figures 3.4, 3.5, and (a). Changing the variable to $1 + z$ $(1 - z)$ moves the branch-cut (b and c) to the real axis to the left (right) of the branch-point at $z = -1$ $(z = +1)$, so that the argument of the logarithm is not a real negative number. Changing the variable to $1 - z^2$ or $(1 + z)/(1 - z)$ leads to two branch-points at $z = \pm 1$ and "two" branch-cuts joining them to the point-at-infinity. Because there is only one point at infinity (Section I.92), this is actually only one branch-cut joining the branch-points $z = \pm 1$ through the point at infinity. In all cases (b–d), the branch-points lie on the unit circle and limit the radius of convergence of the corresponding power series for the logarithm. The series diverge at the branch-points, which are included in the branch-cuts, because the logarithm is singular there.

$$x > 0: \qquad -x \neq 1+z,\, 1-z,\, 1-z^2,\, \frac{1+z}{1-z}; \qquad z \neq -1-x,\, 1+x,\, \sqrt{1+x},\, \frac{x+1}{x-1}. \qquad \text{(3.51a–3.51i)}$$

Solving (Equations 3.51b through 3.51e) for z specifies (Equations 3.51f through 3.51i) for Equation 3.51a the branch-cuts (Equations 3.47a, 3.48a, 3.49a, and 3.50a) (Figure 3.8b through d). The branch-cuts include the branch-points (Equations 3.52a through 3.52d) where the logarithm is singular, that limit the radius of convergence to unity (Equation 3.52e):

$$z_0 \neq -1, +1, \pm 1, \pm 1: \qquad |z| < |z_0| = 1: \qquad |a_n| = \left\{\frac{1}{n}, \frac{1}{n+1/2}\right\}. \qquad \text{(3.52a–3.52g)}$$

This can be confirmed from the coefficients [Equation 3.52f (Equation 3.52g)] of the series [Equations 3.47b, 3.48b, and 3.49b (Equation 3.50b)], whose ratio is the same:

$$\begin{aligned} \left|\frac{a_{n+1}}{a_n}\right| &= \left\{\frac{n}{n+1}, \frac{n+3/2}{n+1/2}\right\} = \left\{\frac{1}{1+1/n}, \frac{1}{1+1/(n+1/2)}\right\} \\ &= 1 - \left\{\frac{1}{n}, \frac{1}{n+1/2}\right\} + O\left(\frac{1}{n^2}\right) = 1 - \frac{1}{n} + O\left(\frac{1}{n^2}\right), \end{aligned} \qquad \text{(3.53)}$$

with regard to the combined convergence test (Section I.29.1.1).

Thus, *the logarithmic series [Equations 3.47b, 3.48b, and 3.49b (Equation 3.50b)] are (1) single-valued [Figure 3.8b through d (Figure 3.8d)] in the complex z-plane with branch-cuts [Equations 3.47a, 3.48a, and 3.49a (Equation 3.50a)] including the branch-points [Equations 3.52a through 3.52c (Equation 3.52d)]; and (2) the branch-points specify the radius of convergence to unity (Equation 3.52e), and the ratio (Equation 3.53) of coefficients [Equation 3.52f (Equation 3.52g)] specifies convergence in the whole complex z-plane:*

$$\delta > 0,\ 0 < \varepsilon < 1: \quad \begin{cases} \text{divergence:} & |z| > 1 \text{ or } z = z_0, & (3.54a) \\ \text{conditional convergence:} & |z| = 1 \text{ and } z \neq z_0, & (3.54b) \\ \text{absolute convergence:} & |z| < 1 & (3.54c) \\ \text{uniform convergence:} & |z| \leq 1 \text{ and } |z - z_0| > \delta, & (3.54d) \\ \text{total convergence:} & |z| \leq 1 - \varepsilon. & (3.54e) \end{cases}$$

namely, there is (1) divergence (Equation 3.54a) outside the unit disk and at branch-point(s); (2) conditional convergence, that is, convergence provided that the terms of the series are not deranged (Section I.21.3), on the unit circle excluding the branch-point(s) (Equation 3.54b); (3) absolute convergence, that is, convergence of the series of moduli, allowing derangement of the series (Section I.21.3) in the open unit disk (Equation 3.54c); (4) uniform convergence allowing differentiation, integration, and taking limits term-by-term (Section I.21.6) in the closed unit disk excluding a neighborhood of the branch-points (Equation 3.54d); and (5) total, that is, absolute and uniform, convergence (Section I.21.7) in a closed subdisk (Equation 3.54e). The complex series (Equations 3.47a, 3.47b, 3.48a, 3.48b, 3.49a, 3.49b, 3.50a, and 3.50b) are equivalent to two real series (Subsection 3.6.4); complex extensions of the series (Equations 3.48a and 3.48b) are considered in Example 10.6.

3.6.2 Continued Fractions for Logarithm

The preceding series can be transformed into continued fractions (Section 1.8.1). For example, the series [Equation 3.47b (Equation 3.50b)] has first term [Equation 3.55a (Equation 3.55b)], general term [Equation 3.55c (Equation 3.55d)], and ratio of successive terms [Equation 3.55e (Equation 3.55f)]:

$$f_0 = z,\ 2z, \qquad f_n = \frac{(-)^{n-1}}{n} z^n, \frac{z^{2n+1}}{n+1/2}, \qquad \frac{f_{n+1}}{f_n} = -z\frac{n}{n+1}, z^2\frac{2n+1}{2n+3}. \qquad (3.55a\text{–}3.55f)$$

Substitution in Equation 1.128a leads to the continued fraction [Equation 3.56 (Equation 3.57)]:

$$\log(1+z) = \frac{z}{1-}\,\frac{-z/2}{1-z/2-}\,\frac{-2z/3}{1-2z/3-}\,\frac{-3z/4}{1-3z/4-}\cdots\frac{-(nz)/(n+1)}{1-(nz)/(n+1)-}\cdots, \qquad (3.56)$$

$$\log\left(\frac{1+z}{1-z}\right) = \frac{2z}{1-}\,\frac{z^2/3}{1+z^2/3-}\,\frac{3z^2/5}{1+3z^2/5-}\cdots\frac{(2n+1)z^2/(2n+3)}{1+(2n+1)z^2/(2n+3)-}-\cdots, \qquad (3.57)$$

which can be simplified [Equation 3.56 ≡ Equation 3.58b (Equation 3.57≡ Equation 3.59b)]:

$$z \notin)-\infty, -1): \quad \log(1+z) = \frac{z}{1+} \frac{z}{2-z+} \frac{4z}{3-2z+} \frac{9z}{4-3z+} \cdots \frac{n^2 z}{n+1-nz+} \cdots, \quad \text{(3.58a and 3.58b)}$$

$$z \notin)-\infty,-1) \cup (+1,+\infty: \quad \log\left(\frac{1+z}{1-z}\right) = \frac{2z}{1-} \frac{z^2}{3+z^2-} \frac{9z^2}{5+3z^2-} \frac{25z^2}{7+5z^2-} \cdots \frac{(n+1)^2 z^2}{2n+3+(2n+1)z-} \cdots. \quad \text{(3.59a and 3.59b)}$$

The continued fractions [Equation 3.58b (Equation 3.59b)] are valid outside the branch-cuts for negative real argument [Equation 3.58a ≡ Equation 3.47a ≡ Equation 3.60a (Equation 3.59a ≡ Equation 3.50a ≡ Equation 3.61a)]. The same branch-cuts apply to the alternative continued fractions [Equation 3.60b ≡ Equation 3.120 (Equation 3.61b ≡ Equation 3.122e)] proven in Section 3.9.5:

$$z \neq)-\infty, -1): \quad \log(1+z) = \frac{z}{1+} \frac{z}{2+} \frac{z}{3+} \frac{4z}{4+} \frac{4z}{5+} \frac{9z}{6+} \frac{9z}{7+} \cdots \frac{n^2 z}{2n+} \frac{n^2 z}{2n+1+} \cdots, \quad \text{(3.60a and 3.60b)}$$

$$z \notin)-\infty,-1) \cup (+1,+\infty(: \quad \log\left(\frac{1+z}{1-z}\right) = \frac{2z}{1-} \frac{z^2}{3-} \frac{4z^2}{5-} \frac{9z^2}{7-} \frac{16z^2}{9-} \frac{25z^2}{11-} \cdots \frac{n^2 z^2}{2n+1-} \cdots. \quad \text{(3.61a and 3.61b)}$$

The logarithm has continued fractions (Equations 3.58a, 3.58b, 3.59a, 3.59b, 3.60a, 3.60b, 3.61a, and 3.61b). These show that the logarithm of a positive integer log N is an irrational number, for example, the natural logarithm of the decimal base:

$$K = \log 10 = 2.30258\ 50929\ 94045\ 68401\ 79914\ 54684\ 36420\ 76011\ 01488\ 62877$$
$$29760\ 33327\ 90096\ 75726\ 09677\ 35248\ 02359\ 97205\ 08959\ 82983\ 41967\ 78404, \quad (3.62)$$

with 110 decimals (Note 3.2).

3.6.3 Series in Offset Circles and Half-Planes

The series (Equations 3.47a, 3.47b, 3.48a, 3.48b, 3.49a, 3.49b, 3.59a, and 3.59b) are valid in the unit disk (Equations 3.54a through 3.54e) with distinct branch-points (Equations 3.52a through 3.52d) and branch-cuts (Figure 3.8b through d). An alternative is to retain the branch-cut of the logarithm (Figure 3.8a) and change the region of convergence (Figure 3.9a through d) via changes of variable. The transformation (Figure 3.9a)

$$\zeta = 1 - z^2, \qquad z = \sqrt{1-\zeta} \qquad \text{(3.63a and 3.63b)}$$

offsets (Equation 3.64a) the center $z = 0$ of the circle to $\zeta = 1$ and transforms the series Equation 3.49b to Equation 3.64c:

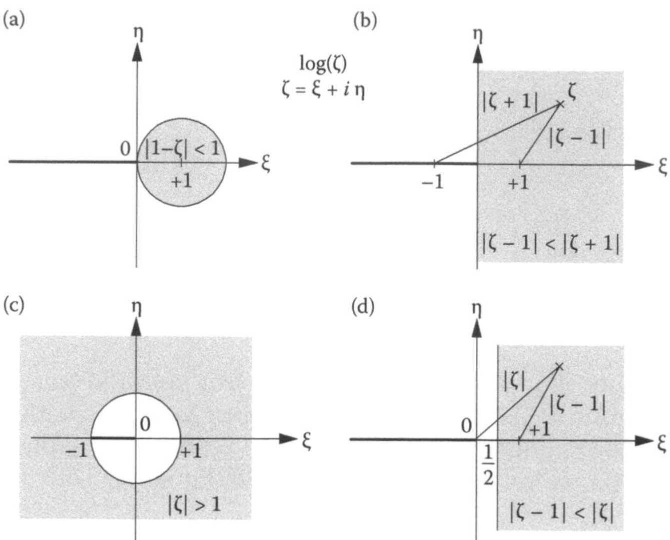

FIGURE 3.9
Logarithm of ζ is single-valued in the complex ζ-plane with a branch-cut along the negative real axis joining the branch-point at the origin to infinity, as shown in Figures 3.4, 3.5, 3.8a, and (a). If log is expanded in a power series of $1 - \zeta$, the series converges (a) for $|\zeta - 1| < 1$, that is, in a circle with center at $\zeta = 1$ and radius unity, to exclude the branch-point at the origin. If $\log \zeta$ is expanded in power series of $(\zeta - 1)/\zeta + 1)$, the series converges (b) for $|\zeta - 1| < |\zeta + 1|$, that is, at the points ζ closer to +1 than to −1, corresponding to the right-hand ζ-half plane, which lies to the right of the vertical line limited by the branch-point at the origin. If $\log \zeta$ is expanded in power series of $1 + 1/\zeta$, the series converges (c) for $|\zeta| > 1$ outside the unit disk excluding the branch-point on the unit circle and leaving the branch-cut inside. If $\log \zeta$ is expanded in power series of $1 - 1/\zeta$, the series converges (d) for $|\zeta -1| < |\zeta|$, that is, at points closer to 1 than to the origin, corresponding to the region to the right of $\mathrm{Re}(z) = 1/2$, which does not touch the branch-cut at all. The convergence of these four series at all points of the complex ζ-plane (Table 3.4) corresponds in every case to the following five types: (1) divergence outside the convergence boundary and at the branch-point(s) on the convergence boundary; (2) conditional convergence on the convergence boundary except the branch-points; (3) absolute convergence inside the convergence boundary; (4) uniform convergence inside and on the convergence boundary except in a neighborhood of the branch-point(s); and (5) total convergence in a closed subregion of the convergence region.

$$|1-\zeta| \le 1, \zeta \ne 0: \qquad \log\zeta = -\sum_{n=1}^{\infty} \frac{(1-\zeta)^n}{n} = \zeta - 1 - \frac{(\zeta-1)^2}{2} + \frac{(\zeta-1)^3}{3} - \frac{(\zeta-1)^4}{4} + \cdots, \quad (3.64\text{a}–3.64\text{c})$$

excluding the branch-point (Equation 3.64b). The bilinear (Section I.35.5) transformation

$$\zeta = \frac{1+z}{1-z}, \qquad z = \frac{\zeta-1}{\zeta+1} \qquad (3.65\text{a and } 3.65\text{b})$$

leads from Equation 3.50b to the series Equation 3.66c:

$$\mathrm{Re}(\zeta) \ge 0, \zeta \ne 0: \ \log\zeta = \sum_{n=0}^{\infty} \frac{1}{n+1/2}\left(\frac{\zeta-1}{\zeta+1}\right)^{2n+1} = 2\frac{\zeta-1}{\zeta+1} + \frac{2}{3}\left(\frac{\zeta-1}{\zeta+1}\right)^3 + \frac{2}{5}\left(\frac{\zeta-1}{\zeta+1}\right)^5 + \cdots, \quad (3.66\text{a}–3.66\text{c})$$

holding in the region $|\zeta - 1| \le |\zeta + 1|$, that is, the set of points closer to +1 than to −1 (Figure 3.9b), corresponding to the right-hand half-plane (Equation 3.66a) and excluding the branch-point (Equation 3.66b).

The change of variable (Equations 3.68a and 3.68b) in Equation 3.47b leads to the series

$$|\zeta| \ge 1, \quad \zeta \ne -1: \quad \log\left(1+\frac{1}{\zeta}\right) = \sum_{n=1}^{\infty} \frac{(-)^{n-1}}{n} \frac{1}{\zeta^n} = \frac{1}{\zeta} - \frac{1}{2\zeta^2} + \frac{1}{3\zeta^3} - \frac{1}{4\zeta^4} + \cdots, \qquad (3.67a\text{–}3.67c)$$

valid (Figure 3.9c) outside the unit circle (Equation 3.67a), excluding the branch-point (Equation 3.67b). The branch-cut (Figure 3.9c) for which the argument of the logarithm in Equation 3.67c is negative (Equations 3.67d and 3.67e) lies between the branch-point (Equation 3.67f) and the origin (Equation 3.67g):

$$x > 0: \qquad -x \ne 1 + \frac{1}{\zeta}, \qquad \zeta \ne -\frac{1}{1+x}, \qquad \zeta \notin (-1, 0). \qquad (3.67d\text{–}3.67g)$$

The transformation (Equations 3.68c and 3.68d):

$$\zeta = \frac{1}{z}, \qquad z = \frac{1}{\zeta}; \qquad \zeta = \frac{1}{1-z}, \qquad z = \frac{\zeta - 1}{\zeta} = 1 - \frac{1}{\zeta} \qquad (3.68a\text{–}3.68d)$$

leads from Equation 3.48b to the series

$$\mathrm{Re}(\zeta) \ge \frac{1}{2}, \ \zeta \ne \infty: \qquad \log\zeta = \sum_{n=1}^{\infty} \frac{1}{n}\left(1 - \frac{1}{\zeta}\right)^n = 1 - \frac{1}{\zeta} + \frac{1}{2}\left(1-\frac{1}{\zeta}\right)^2 + \frac{1}{3}\left(1-\frac{1}{\zeta}\right)^3 + \cdots, \quad (3.69a\text{–}3.69c)$$

which converges for $|\zeta - 1| \le |\zeta|$, that is, for points closer to +1 than to the origin (Figure 3.9d), corresponding to the right of $x = ½$ and excluding the branch-point (Equation 3.69b). The branch-points (Equations 3.52a through 3.52d ≡ Equations 3.70a through 3.70d) for the series (Equations 3.47a, 3.47b, 3.48a, 3.48b, 3.49a, 3.49b, 3.50a, and 3.50b) specify the excluded points (Equations 3.70e through 3.70h)

$$z \ne z_0 = -1, \ +1, \ \pm 1, \ \pm 1: \qquad \zeta \ne \zeta_0 = \frac{1}{z_0}, \frac{1}{1-z_0}, 1 - z_0^2, \frac{1+z_0}{1-z_0} = -1, \infty, 0, \infty \qquad (3.70a\text{–}3.70h)$$

of the series (Equations 3.64a through 3.64c, 3.66a through 3.66c, 3.67a through 3.67c, and 3.69a through 3.69d). The second branch-point $\zeta_0 = 0$ in Equation 3.70h was omitted (Equation 3.69b) because it lies outside the region of convergence (Equation 3.69a), leaving only the first branch-point (Equation 3.69b) that lies inside. The branch-cut lies along the negative real axis (Equation 3.37a) in the cases (Equations 3.64a through 3.64c and 3.69a through 3.69c) in Figure 3.9a, b, and d, and only within the circle (Equation 3.67g) in the case (Equations 3.67a through 3.67c) in Figure 3.9c. Recalling the convergence conditions (Equations 3.54a through 3.54c) for the series (Equations 3.47a, 3.47b, 3.48a, 3.48b, 3.49a, 3.49b, 3.50a, and 3.50b) stated before (Section 3.6.1), it follows that *the logarithmic series (Equations 3.64a through 3.64c, 3.66a through 3.66c, 3.67a through 3.67c, and 3.69a through 3.69c) have the*

TABLE 3.3

Convergence of Series for the Logarithm

Series	Equation 3.64c	Equation 3.66c	Equation 3.67c	Equation 3.69c
Variable	$1-\zeta$	$\frac{\zeta-1}{\zeta+1}$	$\frac{1}{\zeta}$	$1-\frac{1}{\zeta}$
Divergent	$\lvert\zeta-1\rvert>1$ or $\zeta=0$	$\text{Re}(\zeta)<0$ or $\zeta=0$	$\lvert\zeta\rvert<1$ or $\zeta=-1$	$\text{Re}(\zeta)<1/2$ and $\zeta\neq 0\ \infty$
Convergent: conditionally	$\lvert\zeta-1\rvert=1$ and $\zeta\neq 0$	$\text{Re}(\zeta)=0$ and $\zeta\neq 0$	$\lvert\zeta\rvert=1$ and $\zeta\neq -1$	$\text{Re}(\zeta)=1/2$
Absolutely	$\lvert\zeta-1\rvert<1$	$\text{Re}(\zeta)>0$	$\lvert\zeta\rvert>1$	$\text{Re}(\zeta)>1/2$
Uniformly	$\lvert\zeta-1\rvert\le 1$ and $\lvert\zeta\rvert>\varepsilon$	$\text{Re}(\zeta)>0$ and $\lvert\zeta\rvert>\delta$	$\lvert\zeta\rvert>1$ and $\lvert\zeta+1\rvert>\delta$	$\text{Re}(\zeta)>1/2$
Totally	$\lvert\zeta-1\rvert\le 1-\varepsilon$	$\text{Re}(\zeta)\ge\delta$	$\lvert\zeta\rvert\ge 1+\delta$	$\text{Re}(\zeta)\ge 1/2+\delta$

Note: $\delta>0$ and $1>\varepsilon>0$. The convergence of the four series for the logarithm in Subsection 3.6.3 is established at all points of the complex plane (Figure 3.9a through d) either by using directly the combined convergence test (Chapter I.29) or the latter is used to establish the convergence (Figure 3.8b through d) of the series in Subsection 3.6.1 that lead via changes of variable to the series in Subsection 3.6.3.

convergence properties in the complex z-plane indicated in Table 3.3, where $\delta > 0$ and $1 > \varepsilon > 0$. The Stirling–MacLaurin series for the exponential (Equation 3.1) [logarithm (Equations 3.47a and 3.47b)] (1) can be written in complex form in polar coordinates (Subsection 3.6.4); and (2) lead to rational exponential (generalized logarithmic) series [Example 10.5 (Example 10.6)] with general term involving rational functions.

3.6.4 Real Series in Polar Coordinates

The preceding complex series lead to two real series when separated into real and imaginary parts. For example, starting with the series for the complex exponential (Equation 3.1 ≡ Equation 3.71b) [logarithm (Equations 3.47a and 3.47b ≡ Equation 3.71c)] in polar coordinates (Equation 3.71a):

$$z = re^{i\varphi}: \qquad \exp\left(re^{i\varphi}\right) = \sum_{n=0}^{\infty} \frac{\left(re^{i\varphi}\right)^n}{n!} \qquad \log\left(1+re^{i\varphi}\right) = \sum_{n=1}^{\infty} \frac{(-)^n}{n}\left(re^{i\varphi}\right)^n \quad \text{(3.71a–3.71c)}$$

leads to Equations 3.72a and 3.72b (Equations 3.73a and 3.73b):

$$r<\infty: \qquad \exp(r\cos\varphi)\exp(ir\sin\varphi) = \sum_{n=0}^{\infty} \frac{r^n}{n!} e^{in\varphi}, \qquad \text{(3.72a and 3.72b)}$$

$$r<1: \qquad \log\,(1 + r\cos\varphi + ir\sin\varphi) = \sum_{n=1}^{\infty} \frac{(-)^n}{n!} r^n e^{in\varphi}. \qquad \text{(3.73a and 3.73b)}$$

Separating real and imaginary parts in Equations 3.72a and 3.72b (Equations 3.73a and 3.73b) yields Equations 3.74a through 3.74d (Equations 3.75a through 3.75d)]:

$$r < \infty: \qquad \exp(r\cos\varphi)\cos(r\sin\varphi) = \sum_{n=0}^{\infty} \frac{r^n}{n!}\cos(n\varphi), \qquad \text{(3.74a and 3.74b)}$$

$$r < \infty: \qquad \exp(r\cos\varphi)\sin(r\sin\varphi) = \sum_{n=0}^{\infty} \frac{r^n}{n!}\sin(n\varphi), \qquad \text{(3.74c and 3.74d)}$$

$$r < 1: \qquad \log\left(1 + r^2 + 2r\cos\varphi\right) = 2\sum_{n=1}^{\infty} \frac{(-)^n}{n}\, r^n \cos(n\varphi), \qquad \text{(3.75a and 3.75b)}$$

$$r < 1: \qquad \operatorname{arc\,cot}\left(\cot\varphi + r^{-1}\csc\varphi\right)\} = \sum_{n=1}^{\infty} \frac{(-)^n}{n!} r^n \sin(n\varphi), \qquad \text{(3.75c and 3.75d)}$$

with radius of convergence infinity for Equations 3.74b and 3.74d (unity for Equations 3.75b and 3.75d). Thus, *the exponential (logarithm) and circular functions satisfy the pair of real series [Equations 3.74b and 3.74d (Equations 3.75b and 3.75d)] in polar coordinates (Equation 3.71a), with radius of convergence infinity for Equations 3.74b and 3.74d (unity for Equations 3.75b and 3.75d).*

3.7 Exponential and Logarithm with Complex Base

The logarithm is used (Section I.5.7) to define the **power with complex exponent:**

$$z \notin) - \infty, 0): \qquad z^a \equiv \exp(\log z^a) = \exp(a \log z) = \exp\{a\,[\log|z| + i\arg(z)]\}, \qquad \text{(3.76a–3.76c)}$$

with the same branch-point, branch-cut (Equation 3.76a ≡ Equation 3.37a), and principal branch (Equation 3.76c ≡ Equation 3.37b) corresponding to that (Equation 3.37b) of the logarithm. If a is exchanged with z, that is, the exponent is taken in the place of the base as the variable, this leads to (Equations 3.77a and 3.77b), **the exponential with complex base a:**

$$a \neq) -\infty, 0): \qquad a^z \equiv \exp(z \log a); \qquad m \in Z|: \qquad a^z = a^{z + i2\pi m/(\log a)}. \qquad \text{(3.77a–3.77d)}$$

This is an integral function with period (Equations 3.77c and 3.77d) whose inverse is a many-valued function (Equation 3.73b), namely, the **logarithm of base** *a that is single-valued in the cut-plane (Equation 3.78a):*

$$w \neq) -\infty, 0): \qquad z \equiv \log_a w; \qquad \log \equiv \log_e. \qquad \text{(3.78a–3.78c)}$$

It includes the ordinary logarithm (Equation 3.78c) as the particular case with the base $a = e$. The rules of derivation of the exponential (Equation 3.79a) [logarithm (Equation 3.79b)] with base a imply (Equation 3.79c)

$$\frac{d(a^z)}{dz} = a^z \log a, \qquad \frac{d(\log_a w)}{dw} = \frac{1}{w \log a}, \qquad \frac{d(z^z)}{dz} = (1 + \log z) z^z. \quad (3.79a\text{–}3.79c)$$

These differentiation rules are proved from the definition of exponential (Equation 3.77b) [logarithm (Equation 3.78b)] of base a as follows:

$$\frac{d(a^z)}{dz} \equiv \frac{d[\exp(z \log a)]}{dz} = (\log a)\exp(z \log a) = a^z \log a, \qquad (3.80a)$$

$$\frac{d(\log_a w)}{dw} \equiv \frac{dz}{dw} = \left(\frac{dw}{dz}\right)^{-1} = \frac{1}{a^z \log a} = \frac{1}{w \log a}, \qquad (3.80b)$$

$$\frac{d(z^z)}{dz} = \frac{d[\exp(z \log z)]}{dz} = \exp(z \log z)\frac{d}{dz}(z \log z) = (1 + \log z) z^z. \qquad (3.80c)$$

Taking as base the irrational number e, namely, $a = e = 1$, then Equation 3.80a (Equation 3.80b) reduces to Equation 1.25a (Equation 3.35).

The most important properties of the exponential and logarithm are not affected by the base, in that *considering the operations sum–subtraction, product–ratio, and power–root, the exponential (logarithm) of arbitrary base transforms toward the more complicated (simpler):*

$$a^{u+v} = a^u a^v, \qquad a^{u-v} = \frac{a^u}{a^v}, \qquad a^{uv} = (a^u)^v, \qquad a^{u/v} = \sqrt[v]{a^u}, \qquad (3.81a\text{–}3.81d)$$

$$\log_a(uv) = \log_a u + \log_a v, \qquad \log_a\left(\frac{u}{v}\right) = \log_a u - \log_a v, \quad (3.82a \text{ and } 3.82b)$$

$$\log_a(u^v) = v \log_a u, \qquad \log_a\left(\sqrt[v]{u}\right) = \frac{1}{v}\log_a u. \qquad (3.83a \text{ and } 3.83b)$$

Performing the changes of variable (Equations 3.84a and 3.84b) in Equation 3.84c:

$$u \equiv \log_a b, \qquad v \equiv \log_b z: \qquad z = b^v = (a^u)^v = a^{uv}, \qquad (3.84a\text{–}3.84c)$$

and applying logarithms lead to Equation 3.85a:

$$\log_a z = (\log_a b)\,(\log_b z); \qquad z = a: \qquad (\log_a b)\,(\log_b a) = 1, \qquad (3.85a\text{–}3.85c)$$

and, in particular, to Equation 3.85c for Equation 3.85b. These are *the formulas (Equations 3.85a and 3.85c) for change of base of the logarithm; for example, the* **natural** *$\log \equiv \log_e$ and* **decimal** *$\log_{10}$ logarithms are related by*

$$\log z = K \log_{10} z, \qquad \log_{10} z = \frac{1}{K} \log z, \qquad \text{(3.86a and 3.86b)}$$

with K given by Equation 3.62. Other bases for the logarithm are considered in Example 10.7.

3.8 Tables of Natural (Napier 1614) and Decimal (Briggs 1624) Logarithms

The exponential and logarithm with any base a can be calculated from a table of logarithms in any other given base, the two most useful being: (1) the natural logarithms (Napier 1614), with the irrational number e as base, for which analytic properties such as Equations 3.79a and 3.79b become simpler (Equations 1.25a and 3.35); and (2) the decimal logarithms (Briggs 1624), which are well suited to the decimal system because given any number (Equation 3.87a):

$$10^n \le x < 10^n + 1, \qquad n \le \log_{10} x < n + 1, \qquad 0 \le y = \log_{10} x - n < 1, \qquad \text{(3.87a–3.87c)}$$

the **integer part** *n is immediately determined (Equation 3.87b), and only the* **mantissa** *or decimal part (Equation 3.87c) needs to be tabled over the unit interval.* The series Equations 3.47a, 3.47b, 3.48a, 3.48b, 3.49a, 3.49b, 3.50a, and 3.50b and Equations 3.64a through 3.64c, 3.66a through 3.66c, 3.67a through 3.67c, and 3.69a through 3.69c have a general term $O\left(\frac{1}{n}\right)$, that is, they are conditionally convergent and give poor accuracy, unless the variable is small. The series Equations 3.66a through 3.66c is valid over positive real axis and can be chosen to calculate logarithmic tables. Setting Equation 3.88a in Equation 3.66c leads to Equation 3.83b:

$$\zeta = \frac{u}{v}: \qquad \log u - \log v = \sum_{n=0}^{\infty} \frac{2}{2+n} \left(\frac{u-v}{u+v} \right)^{2n+1}$$
$$= 2\frac{u-v}{u+v} + \frac{2}{3}\left(\frac{u-v}{u+v}\right)^3 + \frac{2}{5}\left(\frac{u-v}{u+v}\right)^5 + \frac{2}{7}\left(\frac{u-v}{u+v}\right)^7 + \ldots, \qquad \text{(3.88a–3.88c)}$$

which is a series for the logarithm that converges rapidly if the numbers u, v are close and not small $u - v \ll u + v$.

For example, the first four terms of the series that are written explicitly in Equation 3.88b allow the calculation with four accurate decimal places of the logarithms of all prime numbers up to 10:

$$u = 2, v = 1: \qquad \frac{u-v}{u+v} = \frac{1}{3}, \qquad \log 2 = \frac{2}{3} + \frac{2}{3^4} + \frac{2}{5.3^5} + \frac{2}{7.3^7} = 0.6931, \qquad \text{(3.89a–3.89d)}$$

$$u = 3^2, v = 2^3 \qquad \frac{u-v}{u+v} = \frac{1}{17}, \qquad \log 3 = \frac{3}{2}\log 2 + \frac{1}{17} + \frac{1}{3.17^3} = 1.0986, \qquad \text{(3.90a–3.90d)}$$

TABLE 3.4

Logarithms of the First 10 Integers

x	2	3	4	5
$\log x$	0.6931	1.0986	1.3863	1.6094
$\log_{10} x$	0.3010	0.4771	0.6021	0.6990
x	6	7	8	9
log	1.7918	1.9459	2.0794	2.1972
$\log_{10} x$	0.7782	0.8451	0.9031	0.9542

Note: Example of computation (Section 3.8) of small sample table of logarithms of the first 10 integers with five accurate digits. The same methods can be used to compute more extensive tables of logarithms with higher accuracy, such as the tables of logarithms with 110 decimals (Note 3.2) of Mansell (1929).

$$u = 5^2, v = 3^3: \quad \frac{u-v}{u+v} = -\frac{1}{26}, \quad \log 5 = \frac{3}{2}\log 3 - \frac{1}{26} - \frac{1}{3.26^3} = 1.6094, \quad (3.91a\text{–}3.91d)$$

$$u = 7^2, v = 2.5^2: \quad \frac{u-v}{u+v} = -\frac{1}{99}, \quad \log 7 = \frac{1}{2}\log 2 + \log 5 - \frac{1}{99} - \frac{1}{3.99^3} = 1.9459. \quad (3.92a\text{–}3.92d)$$

The logarithms of nonprime numbers are readily calculated:

$$\log 4 = 2\log 2 = 1.3863, \quad \log 6 = \log 2 + \log 3 = 1.7918, \quad \log 8 = 3\log 2 = 2.0794, \quad (3.93a\text{–}3.93c)$$

$$\log 9 = 2\log 3 = 2.1972, \quad \log 10 = \log 2 + \log 5 = 2.3026, \quad (3.93d \text{ and } 3.93e)$$

as the sum of the logarithms of their prime factors; the value (Equation 3.93e) agrees with Equation 3.62 with five accurate digits. All the values (Equations 3.89d, 3.90d, 3.91d, 3.92d, and 3.93a through 3.93e) should be calculated with one extra decimal (i.e., five decimals) to be sure of rounding off correctly (i.e., to four decimals). This example has led to Table 3.4, a table of logarithms, with four decimals, for the first 10 integers, where the conversion from natural to decimal was made using Equations 3.62 and 3.86b.

3.9 Gaussian and Related Hypergeometric Functions

The only formulas that remain to be proven in the present chapter are the continued fractions [Equations 3.5a and 3.5b (Equations 3.60a, 3.60b, 3.61a, and 3.61b)] for the exponential (logarithm). Although they may appear simpler than the continued fractions [Equations 1.135a and 1.135b (Equations 3.58a, 3.58b, 3.59a and 3.59b)] for the exponential (logarithm), the proofs are less elementary and have been deferred to the present section. The material in the present section will not be needed in the sequel, except to prove some more continued fractions in Section 7.8. The present section and Section 7.8 show that the proof of some properties of elementary transcendental functions requires recourse to higher transcendental functions. The continued fractions for transcendental functions have been obtained by three methods: first, by transformation from series (Section 1.8.1), which leads to the continued fractions for the exponential (Equations 1.135a and 1.135b, 1.137a and 1.137b, 1.140a and 1.140b and 1.141a and 1.141b) and logarithm (Equations 3.58a, 3.58b, 3.59a, and 3.59b); the latter is a particular case of a more general continued fraction (Section 3.9.1); second, by transformation from a product (Section 1.8.3), which leads to the continued fraction for the circular sine (Equation 1.151a and 1.151b) and cosecant (Equation 3.152a and 3.152b); and third, by the Lambert method (Section 1.9) that was applied to the hypogeometric function, which leads to the continued fraction for the circular tangent (Equations 1.181a and 1.181b). The hypogeometric function (Sections 1.9.1 and 1.9.2) is a particular case of the Gaussian hypergeometric function (Sections 3.9.2 and 3.9.3) to which the Lambert method also applies: (1) it starts from a contiguity relation for the Gaussian hypergeometric function (Section 3.9.4); and (2) it leads to a continued fraction for the ratio of two Gaussian hypergeometric functions (Section 3.9.5). It proves as a particular case the continued fractions (Section 3.9.6) for the logarithm. The Gaussian hypergeometric function leads by suppression of one (two) upper parameter to the confluent hypergeometric (Section 3.9.5) [the hypogeometric (Sections 1.9.1 and 1.9.2)] function. This specifies a continued fraction for the ratio of confluent hypergeometric functions (Section 3.9.8) that proves the continued fraction (Equation 3.127c ≡ Equation 3.5a and Equation 3.129 ≡ Equation 3.5b) for the exponential, as well as related Padé approximants (Section 3.9.9). Suppressing one upper and one lower parameters in the Gaussian hypergeometric function leads to the binomial series and its continued fractions (Section 3.9.10).

3.9.1 Extension of Continued Fraction for the Logarithm

The function (Equation 3.94b) is defined for Equation 3.94a:

$$\operatorname{Re}(a) > -1: \qquad f(a;z) = \int_0^z \frac{\zeta^a}{1+\zeta}\, d\zeta, \tag{3.94a and 3.94b}$$

including two cases: (1) if $\operatorname{Re}(a) \geq 0$, the integral is a proper integral, that is, the integrand is finite at the origin $\zeta = 0$; and (2) if $0 > \operatorname{Re}(a) > -1$, the integrand is infinite at the origin, but the improper unilateral integral of the second kind converges. Using the geometric series (Equation I.21.36a ≡ Equation 3.95c ≡ Equation 3.147b) imposes the restriction (Equations 3.95a and 3.95b) for uniform convergence (Section I.21.9):

$$0 < \varepsilon < 1, \ |\zeta| \leq 1-\varepsilon: \qquad \frac{1}{1+\zeta} = \sum_{m=0}^{\infty} (-\zeta)^m = 1 - \zeta + \zeta^2 - \zeta^3 + \zeta^4 - \zeta^5 + \cdots, \tag{3.95a–3.95c}$$

and allows (Section I.21.6) term-by-term integration of Equation 3.95c, which leads to Equations 3.96a and 3.96b:

$$n = m+1: \qquad f(a;z) = \sum_{m=0}^{\infty} (-)^m \int_0^z \zeta^{a+m}\, d\zeta = \sum_{m=0}^{\infty} \frac{(-)^m}{a+m+1} z^{a+m+1} = \sum_{n=1}^{\infty} \frac{(-)^{n-1}}{a+n} z^{a+n}. \tag{3.96a and 3.96b}$$

It follows that

$$\mathrm{Re}(a) < -1,\ |z| < 1: \qquad f(a;z) \equiv \int_0^z \frac{\zeta^a}{1+\zeta}\,d\zeta = \sum_{m=1}^{\infty} \frac{(-)^{n-1}}{a+n} z^{a+n}. \qquad (3.97a\text{–}3.97c)$$

The function (Equation 3.94b) with parameter a satisfying Equation 3.94a ≡ Equation 3.97a has power series expansion (Equation 3.97c) in the unit disk (Equation 3.97b).

The series Equation 3.96b has first term (Equation 3.98a), general term (Equation 3.98b), and ratio of successive terms (Equation 3.98c):

$$f_1 = \frac{z^{a+1}}{a+1}, \qquad f_n = \frac{(-)^{n-1}}{a+n} z^{a+n}, \qquad \frac{f_{n+1}}{f_n} = -z\frac{a+n}{a+n+1}, \qquad (3.98a\text{–}3.98c)$$

and leads by Equation 1.128a to the continued fraction

$$z^{-a} f(a;z) = \frac{z/(a+1)}{1-}\ \frac{-z(a+1)/(a+2)}{1-z(a+1)/(a+2)-} \cdots \frac{-z(a+n)/(a+n+1)}{1-z(a+n)/(a+2)+} \cdots. \qquad (3.98d)$$

This can be simplified (Equation 3.98d ≡ Equation 3.99c):

$$\mathrm{Re}(a) > -1,\ |z| < 1: \quad z^{-a}\int_0^z \frac{\zeta^a}{1+\zeta}\,d\zeta = \frac{z}{a+1+}\ \frac{(a+1)^2 z}{a+2-z(a+1)+} \cdots \frac{(a+n)^2 z}{a+n+1-z(a+n)+} \cdots, \qquad (3.99a\text{–}3.99c)$$

recalling the restriction (Equation 3.94a ≡ Equation 3.99a) and relaxing Equations 3.95a and 3.95b to Equation 3.99b for absolute convergence. Thus, *the function (Equations 3.94a and 3.94b) has the series expansion (Equations 3.97a through 3.97c) and associated continued fraction (Equations 3.99a through 3.99c).* The particular case (Equation 3.100a):

$$a = 0,\ |z| < 1: \quad \log(1+z) = \int_0^z \frac{d\zeta}{1+\zeta} = \frac{z}{1+}\ \frac{z}{2-z+}\ \frac{4z}{3-2z+}\ \frac{9z}{4-3z+}\ \frac{16z}{5-4z+} \cdots \frac{n^2 z}{n+1-nz+} \cdots \qquad (3.100a\text{–}3.100c)$$

is the continued fraction for the logarithm (Equation 3.100c ≡ Equation 3.58b) valid outside the branch-cut (Equation 3.58a).

3.9.2 Gaussian Hypergeometric Function (Gauss 1812)

The Gaussian hypergeometric function with variable z, upper parameters a,b, and lower parameter c is defined by the series (Equation 3.155 ≡ Equation 3.101b):

$$|z| < 1: \qquad {}_2F_1(a,b;c;z) = \sum_{n=0}^{\infty} \frac{z^n}{n!} \frac{(a)_n (b)_n}{(c)_n} \equiv F(a,b;c;z)$$
$$= 1 + \sum_{n=1}^{\infty} \frac{z^n}{n!} \frac{ab}{c} \cdots \frac{(a+n-1)(b+n-1)}{c+n-1}, \qquad (3.101a \text{ and } 3.101b)$$

using the Pochhammer symbol (Equation 1.160). The Gaussian hypergeometric series (Equation 3.101b ≡ Equation 3.102a) has general term (Equation 3.102b):

$$F(a, b; c; z) = \sum_{n=0}^{\infty} f_n: \qquad f_n = \frac{z^n}{n!} \frac{(a)_n (b)_n (b)_n}{(c)_n} = \frac{z^n}{n!} \frac{a \cdots (a+n-1) b \cdots (b+n-1)}{c \cdots (c+n-1)}, \tag{3.102a and 3.102b}$$

and ratio of two successive terms (Equation 3.102b):

$$\frac{f_{n+1}}{f_n} = \frac{z}{n+1} \frac{(a+n)(b+n)}{(c+n)} = z \frac{(1+a/n)(1+b/n)}{(1+1/n)(1+c/n)} = z\left[1 - \frac{c+1-a-b}{n} + O\left(\frac{1}{n^2}\right)\right], \tag{3.102c}$$

proves that (1) the **Gaussian hypergeometric function** *(Equation 3.101b) is an analytic function of the variable z in the unit circle (Equation 3.101a); (2) the convergence at all points of the z-plane including the boundary of convergence is given in Section I.29.9 and Table I.29.2; (3) it is an integral function of the upper parameters (a,b), that reduces to a polynomial when one (Equations 3.103a through 3.103d) [both (Equations 3.104a through 3.104d)] is a negative integer:*

$$p, q \in| N; b \neq -q, a = -p: \quad F(-p, b; c; z) = \sum_{n=0}^{p} \frac{z^n}{n!} \frac{(-p)_n (b)_n}{(c)_n} = 1 + \sum_{n=0}^{p} (-z)^n \frac{(b)_n}{(c)_n} \binom{p}{n}, \tag{3.103a–3.103d}$$

$$q > p, a = -p, b = -q: \quad F(-p, -q; c; z) = \sum_{n=0}^{q} \frac{z^n}{n!} \frac{(-p)_n (-q)_n}{(c)_n} = 1 + \sum_{n=0}^{p} \frac{z^n}{(c)_n} \frac{q!}{(q-n)!} \binom{p}{n}; \tag{3.104a–3.104d}$$

and (4) it is a meromorphic function of the lower parameter c, with simple poles for zero or negative integer values (Equations 3.105a and 3.105b):

$$k \in| N_0; c = -k = 0, -1, \ldots: \qquad F_{(1)}(a, b; -k; z) = \lim_{c \to -k} (c+k) F(a, b; c; z), \tag{3.105a–3.105c}$$

with residues (Equation 3.105c). The result (Equation 3.103d) follows from

$$(-p)_n = (-p)(-p+1) \cdots (-p+n-1) = (-)^n p(p-1) \cdots (p-n+1) = (-)^n \frac{p!}{(p-n)!}, \tag{3.106a}$$

$$\binom{p}{n} \equiv \frac{p!}{n!(p-n)!}: \qquad \frac{z^n}{n!} (-p)_n = \frac{(-z)^n}{n!} \frac{p!}{(p-n)!} = (-z)^n \binom{p}{n}, \tag{3.106b and 3.106c}$$

using the definitions of Pochhammer symbol (Equation 1.160) [arrangements (Equation 3.101b) of p undistinguishable objects in sets of n]. It implies

$$\frac{z^n}{n!}(-p)_n(-q)_n = \frac{z^n}{n!}\frac{p!q!}{(p-n)!(q-n)!} = z^n\binom{p}{n}\frac{q!}{(q-n)!}, \tag{3.107}$$

which is used in Equation 3.104d.

3.9.3 Residues at Poles of the Lower Parameter

The calculation of the residues is similar (Section 1.9.2) for the hypogeometric (Equations 1.168a and 1.168b) and hypergeometric (Equation 3.96b) functions because the upper parameters (a,b) play no role:

$$\begin{aligned}
n = k+m+1:\quad F_{(1)}(a,b;c;z) &= \lim_{c\to -k}(c+k)F(a,b;c;z) = \lim_{c\to -k}(c+k)\sum_{n=k+1}^{\infty}\frac{z^n}{n!}\frac{(a)_n(b)_n}{c(c+1)\ldots(c+n-1)}\\
&= \lim_{c\to -k}\sum_{n=k+1}^{\infty}\frac{z^n}{n!}\frac{(a)_n(b)_n}{c\ldots(c+k-1)(c+k+1)\ldots(c+n-1)} = \sum_{n=k+1}^{\infty}\frac{z^n}{n!}\frac{a\ldots(a+n-1)b\ldots(b+n-1)}{(-)^k k!(n-k-1)!}\\
&= \frac{(-)^k}{k!}z^{k+1}\sum_{m=0}^{\infty}\frac{z^m}{m!}\frac{a\ldots(a+k+m)b\ldots(b+k+m)}{(k+m+1)!}\\
&= \frac{(-)^k}{k!}z^{k+1}\frac{a\ldots(a+k)b\ldots(b+k)}{1\ldots(k+1)}\sum_{m=0}^{\infty}\frac{z^m}{m!}\frac{(a+k+1)_m(b+k+1)_m}{(k+2)_m}\\
&= \frac{(-)^k}{k!}\frac{z^{k+1}}{(k+1)!}(a)_{k+1}(b)_{k+1}F(a+k+1,b+k+1;k+2;z),
\end{aligned}$$

(3.108a and 3.108b)

so that the same five steps apply: (1) the first k terms of the series (Equation 3.108b) are finite and vanish when multiplied by $c + k$ in the limit $c \to -k$; (2) the terms of the series starting with $k + 1$ have the factor $c + k$ in the denominator that cancels with the same factor in the numerator; (3) thus, the limit $c \to -k$ leads to a finite expression; (4) the change of summation variable (Equation 3.108a) leads to a split into factors; and (5) the series summed over $m = 0,1, \ldots,$ is another Gaussian hypergeometric function with parameters $(a + k + 1, b + k + 1; k + 2)$, and the terms independent of m can be treated as a constant factor. *The residues (Equation 3.108b) of the Gaussian hypergeometric function (Equation 3.101a and 3.101b) at its poles (Equations 3.105a through 3.105c) can be written*

$$F_{(1)}(a,b;-k;z) = \frac{z^{k+1}}{(k+1)!}F(a+k+1,b+k+1;k+2;z)(a+k)(b+k)\prod_{s=0}^{k-1}\frac{(a+s)(b+s)}{s-k}, \tag{3.108c}$$

in agreement with Equation I.29.75b ≡ Equation 3.108c. As with the hypogeometric function (Section 1.9.3), the hypergeometric function satisfies a contiguity formula, that leads to the continued fraction for the ratio of two functions.

3.9.4 Contiguity Relations for Three Gaussian Hypergeometric Functions (Gauss 1812)

The Gaussian hypergeometric function (Equation 3.101b) satisfies 15 contiguity relations with functions whose parameters differ by plus or minus unity, namely, (1) one contiguity relating the function with parameters $(a,b;c)$ to $(a \pm 1, b; c)$; (2) four contiguity relations for $(a,b \pm 1; c \pm 1)$; and (3) the five contiguity relations 1 + 2 are multiplied by 3, replacing a by b or c. Any of the contiguity relations is a recurrence formula of order 2 (Equation 1.90b) and leads to a continued fraction for the ratio of two functions (Section 1.6.1). An example is the difference of Gaussian hypergeometric functions:

$$n = m+1: \qquad F(a, b; c; z) - F(a, b+1; c+1; z) = \sum_{n=1}^{\infty} \frac{z^n}{n!}(a)_n \frac{(b+1)\dots(b+n-1)}{(c+1)\dots(c+n-1)}\left(\frac{b}{c} - \frac{b+n}{c+n}\right)$$

$$= \frac{b-c}{c}\sum_{n=1}^{\infty} \frac{z^n}{(n-1)!}(a)_n \frac{(b+1)\dots(b+n-1)}{(c+1)\dots(c+n)} = \frac{b-c}{c}\sum_{m=0}^{\infty} \frac{z^{m+1}}{m!}(a)_{m+1} \frac{(b+1)\dots(b+m)}{(c+1)\dots(c+m+1)}$$

$$= \frac{b-c}{c}\frac{a}{c+1} z \sum_{m=0}^{\infty} \frac{z^m}{m!}\frac{(a+1)_m (b+1)_m}{(c+2)_m}, \qquad \text{(3.109a and 3.109b)}$$

where the property (Equation 3.109c) of the Pochhammer (Equation 1.160) symbol is used:

$$(a)_{m+1} = a(a+1)\dots(a+m) = a(a+1)\dots(a+1+m-1) = (a+1)_m\, a. \qquad \text{(3.109c)}$$

This leads to *the contiguity relations (Equations 3.110a and 3.110b) for the Gaussian hypergeometric function:*

$$F(a, b; c; z) - F(a, b+1; c+1; z) = \frac{a}{c}\frac{b-c}{c+1} zF(a+1, b+1; c+2; z), \qquad \text{(3.110a)}$$

$$F(a, b; c; z) - F(a+1, b; c+1; z) = \frac{b}{c}\frac{a-c}{c+1} zF(a+1, b+1; c+2; z), \qquad \text{(3.110b)}$$

that are equivalent due to the symmetry in the two upper parameters (Equation 3.111a):

$$F(a, b; c; z) = F(a, b; c; z); \qquad (a, b, c) \to (a+n, b+n, c+2n). \qquad \text{(3.111a and 3.111b)}$$

The change of parameters (Equation 3.111b) substituted in Equation 3.110a (Equation 3.110b) leads to the system:

$$F_n^+(a,b;c;z) \equiv \frac{F(a+n, b+n; c+2n; z)}{F(a+n, b+n+1; c+2n+1; z)} = 1 + \frac{\alpha_{2n+1} z}{F_n^-(a,b;c;z)}, \qquad \text{(3.112a)}$$

$$F_n^-(a,b;c;z) \equiv \frac{F(a+n, b+n+1; c+2n+1; z)}{F(a+n+1, b+n+1; c+2n+2; z)} = 1 + \frac{\alpha_{2n+2} z}{F_{n+1}^+(a,b;c;z)}, \qquad \text{(3.112b)}$$

with coefficients

$$\alpha_{2n+1} = -\frac{a+n}{c+2n}\frac{c-b+n}{c+2n+1}, \qquad \alpha_{2n+2} = -\frac{b+n+1}{c+2n+1}\frac{c-a+n+1}{c+2n+2}. \quad \text{(3.113a and 3.113b)}$$

This system is reducible to a continued fraction, as shown next (Subsection 3.9.5).

3.9.5 Continued Fraction for Ratio of Two Gaussian Hypergeometric Functions (Gauss 1812)

Substitution of Equation 3.112b in Equation 3.112a leads to

$$F_n^+(a,b;c;z) = 1 + \frac{\alpha_{2n+1}z}{1+}\frac{\alpha_{2n+2}z}{F_{n+1}^+(a,b;c;z)}, \quad (3.114)$$

that can be continued as Equation 3.115c:

$$|z|<1; \ -c \notin |N_0: \ F_0^+(a,b;c;z) \equiv \frac{F(a,b;c;z)}{F(a,b+1;c+1;z)} = 1 + \frac{\alpha_1 z}{1+}\frac{\alpha_2 z}{1+}\dots\frac{\alpha_{2n+1}z}{1+}\frac{\alpha_{2n+2}z}{1+}\dots, \quad \text{(3.115a–3.115c)}$$

$$\frac{F(a,b+1;c+1;z)}{F(a,b;c;z)} = \frac{1}{1+}\frac{\alpha_1 z}{1+}\frac{\alpha_2 z}{1+}\dots\frac{\alpha_{2n+1}z}{1+}\frac{\alpha_{2n+2}z}{1+}\dots, \quad (3.115d)$$

provided that the lower parameter c is not a negative integer (Equation 3.115b) to avoid poles (Equations 3.108a through 3.108c). The Gaussian hypergeometric series converges (Equation 3.152b) in the unit disk (Equation 3.101a ≡ Equation 3.115a). The continued fraction (Equation 3.115d) follows from Equation 3.115c by algebraic inversion (Equations 1.96 and 1.108). Thus, *the ratio of two Gaussian hypergeometric functions (Equation 3.101b) in the unit disk (Equation 3.101a ≡ Equation 3.115a), with one contiguous upper and lower parameter and excluding poles (Equation 3.115b) of the latter, has the continued fraction (Equation 3.115c) with coefficients Equations 3.113a and 3.113b.* Choosing an upper parameter to be zero (Equation 3.116a) (1) reduces the Gaussian hypergeometric function (Equation 3.101b) to unity (Equation 3.116b) as a particular case (Equations 3.103a through 3.103d) of polynomial of degree zero $p = 0$; and (2) together with the change of parameter (Equation 3.116c) simplifies the coefficients Equations 3.113a and 3.113b to Equations 3.116d and 3.116e:

$$b=0: \quad F(a,1;c;z)=1; \quad c \to c-1:$$

$$\{\alpha_{2n+1},\alpha_{2n+2}\} = -\left\{\frac{a+n}{c+2n-1}\frac{c+n-1}{c+2n}, \frac{n+1}{c+2n}\frac{c-a+n}{c+2n+1}\right\} \equiv \{\beta_{2n+1},\beta_{2n+2}\}. \quad \text{(3.116a–3.116e)}$$

Substituting Equations 3.116a through 3.116e in Equations 3.115a through 3.115c leads to Equations 3.117a through 3.117c:

$$|z|<1;\, c\neq 0,\,-1,\,-2,\cdots:\quad \frac{1}{F(a,1;c;z)}=1+\frac{\beta_1 z}{1+}\frac{\beta_2 z}{1+}\cdots\frac{\beta_{2n+1} z}{1+}\frac{\beta_{2n+2} z}{1+}\cdots, \tag{3.117a–3.117c}$$

$$\frac{1}{1+}\frac{\beta_1 z}{1+}\frac{\beta_2 z}{1+}\cdots\frac{\beta_{2n+1} z}{1+}\frac{\beta_{2n+2} z}{1+}\cdots = F(a,1;c;z)=\sum_{n=0}^{\infty} z^n\frac{(a)_n}{(c)_n}=1+\sum_{n=0}^{\infty} z^n\frac{a\cdots(a+n-1)}{c\cdots(c+n-1)}. \tag{3.117d}$$

The algebraic inversion (Equations 1.95 and 1.108) of the continued fraction (Equation 3.117c) leads to the continued Equation 3.117d for the Gaussian hypergeometric function (Equation 3.101b) with one upper parameter unity. Thus, *the Gaussian hypergeometric function (Equation 3.101b) in the unit disk (Equation 3.101a ≡ Equation 3.117a), with one upper parameter unity, and outside (Equation 3.117b) the poles of the lower parameter, has the representations by algebraically inverse continued fractions (Equations 3.117c and 3.117d) with coefficients Equations 3.116d and 3.116e.*

3.9.6 Two Continued Fractions for Logarithm (Jacobi 1827; Rouché 1858)

The simplest case of the Gaussian hypergeometric function (Equation 3.117d) is equal upper and lower parameter:

$$|z|<1:\qquad F(1,c;c;z)=\sum_{n=0}^{\infty} z^n=\frac{1}{1-z}, \tag{3.118a and 3.118b}$$

which leads to the series (Equations 3.118a and 3.118b ≡ Equations I.21.62a and I.21.62b) that is a particular case $b=1$ of the geometric series (Equation 3.148b). The next simplest case is the logarithm (Equation 3.46 ≡ Equation 3.114a) that is a particular case of the Gaussian hypergeometric function (Equation 3.101b):

$$\log(1+z)=\sum_{n=1}^{\infty}\frac{(-)^{n-1}}{n}z^n=z\sum_{m=0}^{\infty}\frac{(-)^m}{m+1}z^m=z\sum_{m=0}^{\infty}\frac{(-z)^m}{m!}\frac{(1)_m(1)_m}{(2)_m}=zF(1,1;2;-z). \tag{3.119a}$$

The coefficients (Equations 3.116d and 3.116e ≡ Equations 3.119c and 3.119d) in this case (Equations 3.119a and 3.119b):

$$a=1,\, c=2:\qquad \{\beta_{2n+1},\beta_{2n+2}\}=-\left\{\frac{n+1}{4n+2},\frac{n+1}{4n+6}\right\} \tag{3.119b–3.119e}$$

lead to the continued fraction:

$$\begin{aligned}\log(1+z)&=\frac{z}{1+}\frac{z/2}{1+}\frac{z/6}{1+}\frac{z/3}{1+}\frac{z/5}{1+}\cdots\frac{(n+1)z/(4n+2)}{1+}\frac{(n+1)z/(4n+6)}{1+}\cdots\\&=\frac{z}{1+}\frac{z}{2+}\frac{z}{3+}\frac{2^2z}{4+}\frac{2^2z}{5+}\frac{3^2z}{6+}\frac{3^2z}{7+}\cdots\frac{n^2z}{2n+}\frac{n^2z}{2n+1+}\cdots.\end{aligned} \tag{3.120}$$

This proves the continued fraction for the logarithm (Equation 3.120 ≡ Equation 3.60b), which is valid outside the branch-cut (Equation 3.60a). The series (Equation 3.50b) for the logarithm is also a particular case of the Gaussian hypergeometric function (Equation 3.101b):

$$\log\left(\frac{1+z}{1-z}\right) = 2z\sum_{n=0}^{\infty}\frac{z^{2n}}{2n+1} = 2z\sum_{n=0}^{\infty}\frac{(z^2)^n}{n!}\frac{(1)_n\,(1/2)_n}{(3/2)_n} = 2zF\left(\frac{1}{2},1;\frac{3}{2};z^2\right), \tag{3.121a}$$

bearing in mind that

$$\frac{(1/2)_n}{(3/2)_n} = \frac{(1/2)(3/2)\cdots(n-1/2)}{(3/2)(5/2)\cdots(n+1/2)} = \frac{1/2}{n+1/2} = \frac{1}{2n+1}. \tag{3.121b}$$

The parameters Equations 3.122a and 3.122b lead (Equations 3.116d and 3.116e) to the coefficients Equation 3.122c and 3.122d:

$$a = \frac{1}{2},\; c = \frac{3}{2}: \qquad \{\beta_{2n+1},\beta_{2n+2}\} = -\left\{\frac{n+1/2}{2n+1/2}\frac{n+1/2}{2n+3/2},\; \frac{n+1}{2n+3/2}\frac{n+1}{2n+5/2}\right\}, \tag{3.122a–3.122d}$$

and by Equation 3.117d to the continued fraction:

$$\log\left(\frac{1+z}{1-z}\right) = \frac{2z}{1+}\frac{-z^2/3}{1+}\frac{-4z^2/15}{1+}\frac{-9z^2/35}{1+}\frac{-16z^2/63}{1+}$$

$$\cdots\frac{-(2n+1)^2 z/[(4n+1)/4n+3]}{1+}\frac{-(2n+2)^2 z^2/[(4n+3)(4n+5)]}{1+} \tag{3.122e}$$

$$= \frac{2z}{1-}\frac{z^2}{3-}\frac{4z^2}{5-}\frac{9z^2}{7-}\frac{16z^2}{9-}\cdots\frac{n^2z^2}{2n+1-}\cdots.$$

This proves the continued fraction (Equation 3.122e ≡ Equation 3.61b) with branch-cut (Equation 3.61a). A change of variable $z \to 1/z$ leads to the branch-cut (Equation 3.125a) and continued fraction (Equation 3.125b):

$$z \notin (-1,\,+1): \qquad \log\left(\frac{z+1}{z-1}\right) = \log\left(\frac{1+1/z}{1-1/z}\right) = \frac{2/z}{1-}\frac{1/z^2}{3-}\frac{4/z^2}{5-}\frac{9/z^2}{7-}\frac{16/z^2}{9-}\cdots\frac{n^2/z^2}{2n+1-}\cdots$$

$$= \frac{2}{z-}\frac{1}{3z-}\frac{4}{5z-}\frac{9}{7z-}\frac{16}{9z-}\cdots\frac{n^2}{(2n+1)z-}\cdots. \tag{3.123a and 3.123b}$$

Thus, *the logarithms [Equation 3.119a (Equation 3.121a)] are particular cases [Equations 3.119b and 3.119c (Equations 3.122a and 3.122b)] of the Gaussian hypergeometric function with one upper parameter unity (Equations 3.117a through 3.117d), which by Equations 3.116d and 3.116e leads to the continued fraction [Equation 3.120 ≡ Equation 3.60b (Equation 3.122e ≡ Equation 3.61b)] valid in the complex z-plane with branch-cut [Equation 3.60a (Equation 3.61a)]. The latter (Jacobi 1827; Rouché 1858) leads to the continued fraction (Equation 3.123b) valid with the branch-cut (Equation 3.123a), which is the complement of Equation 3.61a relative to the real axis.*

3.9.7 Confluent Hypergeometric Function (Kummer 1836)

The generalized hypergeometric function (Equation 3.155) with one upper a and one lower c parameter specifies the **confluent hypergeometric function** (Kummer 1836):

$$|z| < \infty: \quad {}_1F_1(b;c;z) \equiv \sum_{n=0}^{\infty} \frac{z^n}{n!}\frac{(b)_n}{(c)_n} = 1 + \sum_{n=1}^{\infty} \frac{z^n}{n!}\frac{b}{c} \cdots \frac{b+n-1}{c+n-1} \equiv F(b;c;z). \qquad \text{(3.124a and 3.124b)}$$

The designation "confluent" arises because it is obtained from the "Gaussian" suppressing one upper parameter, which is equivalent (Equation 3.125c) to the transformation (Equation 3.125a) followed by the limit (Equation 3.125b):

$$z \to \frac{z}{a};\ a \to \infty: \qquad \lim_{a\to\infty} F\left(a,b;c;\frac{z}{a}\right) = \sum_{n=0}^{\infty} \frac{z^n}{n!}\frac{(b)_n}{(c)_n} \lim_{a\to\infty} \frac{(a)_n}{a^n} = \sum_{n=0}^{\infty} \frac{z^n}{n!}\frac{(b)_n}{(c)_n} = F(b;c;z), \qquad \text{(3.125a–3.125c)}$$

where the property (Equation 3.126) of the Pochhammer symbol (Equation 1.160) was used:

$$\lim_{a\to\infty} \frac{(a)_n}{a^n} = \lim_{a\to\infty} \frac{a(a+1)\cdots(a+n-1)}{aa\ldots a} = \lim_{a\to\infty} 1\left(1+\frac{1}{a}\right)\cdots\left(1+\frac{n-1}{a}\right) = 1. \qquad \text{(3.126)}$$

The transformation $z \to bz$ inverse to Equation 3.125a expands the unit disk $|z| < 1$ to the finite plane $|z| < \infty$ as $b \to \infty$, and thus, the confluent hypergeometric function (Equation 3.124b) is an integral function of the variable (Equation 3.124a). This can be confirmed from the confluent hypergeometric series (Equation 3.124b ≡ Equation 3.127a) that has general term (Equation 3.127b) and ratio of successive terms (Equation 3.127c):

$$F(a;c;z) = \sum_{n=0}^{\infty} f_n, \qquad f_n = \frac{z^n}{n!}\frac{(b)_n}{(c)_n}, \qquad \frac{f_{n+1}}{f_n} = \frac{z}{n+1}\frac{b+n}{c+n} \sim O\left(\frac{z}{n}\right), \qquad |z| < \infty: \qquad \lim_{x\to\infty} \frac{f_{n+1}}{f_n} = 0, \qquad \text{(3.127a–3.127e)}$$

implying (Equation 3.127e) convergence in the finite z-plane (Equation 3.127d). From Equation 3.127c, it also follows that the confluent hypergeometric function is an integral function of the upper parameter, which reduces to a polynomial (Equation 3.128b) when it is zero or a negative integer (Equation 3.128a):

$$b=-p, \quad p\in|N_0: \qquad F(-p;c;z)=\sum_{n=0}^{p}\frac{z^n}{n!}\frac{(-p)_n}{(c)_n}=\sum_{n=0}^{p}\frac{(-z)^n}{(c)_n}\binom{p}{n}, \qquad (3.128a\text{–}3.128c)$$

where Equations 3.106a through 3.106c were used. The confluent hypergeometric function is a meromorphic function of the lower parameter, with simple poles for zero or negative integer values (Equation 3.129a) with residues (Equation 3.129b):

$$c=0,-1,\ldots,-k\ldots: \qquad F_{(1)}(b;-k;z)=\lim_{c\to-k}(c+k)F(a;c;z) \;=\frac{(-)^k}{k!}\frac{z^{k+1}}{(k+1)!}(b)_{k+1}F(b+k+1;k+2;z),$$

(3.129a and 3.129b)

which follow from those of the Gaussian hypergeometric function (Equation 3.108b) using the limit (Equations 3.125a and 3.125b) to suppress the first upper parameter. It has been proved that *the confluent hypergeometric function (Equation 3.129b) is (1) an integral function (Equations 3.127a through 3.127e) of the variable (Equation 3.124a); (2) an integral function of the upper parameter a that for zero or negative integer values b = −p reduces Equations 3.128a and 3.128b to a polynomial (Equation 3.128c) of degree p; and (3) a meromorphic function of the lower parameter c with simple poles for zero or negative integer values (Equation 3.129a) with residues (Equation 3.129b).*

3.9.8 Continued Fractions for Confluent Hypergeometric Function

The transformation (Equations 3.125a and 3.125b) from the Gaussian hypergeometric function (Equation 3.125c) can be used to (1) prove the properties stated before (Section 3.9.7); and (2) determine the corresponding (Equations 3.130a and 3.130b) coefficients (Equations 3.113a and 3.113b) in the continued fraction (Equations 3.115a through 3.115d):

$$\lim_{a\to\infty}\frac{z}{a}\{\alpha_{2n+1},\alpha_{2n+2}\}=\lim_{a\to\infty}-\frac{z}{a}\left\{\frac{a+n}{c+2n}\frac{c-b+n}{c+2n+1},\;\frac{b+n+1}{c+2n+1}\frac{c-a+n+1}{c+2n+2}\right\}$$

$$=\frac{z}{c+2n+1}\left\{-\frac{c-b+n}{c+2n},\frac{b+n+1}{c+2n+2}\right\}\equiv z\{\gamma_{2n+1},\gamma_{2n+2}\}. \qquad (3.130a and 3.130b)$$

Using Equations 3.115a through 3.115c, 3.130a, and 3.130b in Equations 3.125a through 3.125c, it follows that *the ratio of two confluent hypergeometric functions in the finite z-plane (Equation 3.124a ≡ Equation 3.131a), with contiguous lower parameter that is not zero or a negative integer (Equation 3.131b), has the continued fraction (Equation 3.131c) with coefficients Equations 3.130a and 3.130b:*

$$|z|<\infty;\, c\neq 0,-1,\ldots: \qquad \frac{F(b;c;z)}{F(b;c+1;z)}=1+\frac{\gamma_1 z}{1+}\frac{\gamma_2 z}{1+}\cdots\frac{\gamma_{2n+1}z}{1+}\frac{\gamma_{2n+1}z}{1+}\cdots; \qquad (3.131a\text{–}3.131c)$$

$$\frac{F(b;c+1;z)}{F(b;c;z)}=\frac{1}{1+}\frac{\gamma_1 z}{1+}\frac{\gamma_2 z}{1+}\cdots\frac{\gamma_{2n+1}z}{1+}\frac{\gamma_{2n+2}z}{1+}\cdots. \qquad (3.131d)$$

and also the continued fraction (Equation 3.131d) that follows from Equation 3.131c by algebraic inversion (Equations 1.96 and 1.108). Choosing a zero upper parameter (Equation 3.132a) reduces the confluent hypergeometric function to unity (Equation 3.132b) as a particular case of the polynomial (Equations 3.128a and 3.128b) of degree zero $p = 0$:

$$b=0: \quad F(0;c;z)=1; \quad c \to c-1: \quad \lim_{b\to 0}\left\{\gamma_{2n+1}, \gamma_{2n+2}\right\} = \frac{1}{c+2n}\left\{-\frac{c+n-1}{c+2n-1}, \frac{n+1}{c+2n+1}\right\}. \tag{3.132a–3.132e}$$

Furthermore, the substitution (Equation 3.132c) simplifies the parameters (Equations 3.130a and 3.130b) to Equations 3.132d and 3.132e. Alternatively, the limit (Equations 3.125a and 3.125b) applied to Equation 3.116e leads to Equations 3.132d and 3.132e. Substitution of Equations 3.132d and 3.132e in Equation 3.131c leads to the continued fraction:

$$\begin{aligned}\frac{1}{F(1;c;z)} &= 1+\frac{-z/c}{1+}\,\frac{z/[c(c+1)]}{1+}\,\frac{-cz/[(c+1)(c+2)]}{1+}\,\frac{2z/[(c+2)(c+3)]}{1+}\cdots \\ &\quad \frac{-z(c+n-1)/[(c+2n)(c+2n-1)]}{1+}\,\frac{(n+1)z/[(c+2n)(c+2n+1)]}{1-}\cdots \\ &= 1-\frac{z}{c+}\,\frac{z}{c+1-}\,\frac{cz}{c+2+}\,\frac{2z}{c+3-}\cdots\frac{(c+n-1)z}{c+2n+}\,\frac{(n+1)z}{c+2n+1-}\cdots.\end{aligned} \tag{3.133}$$

Thus, the confluent hypergeometric function (Equation 3.124b) in the finite *z*-plane (Equation 3.124a ≡ Equation 3.134a), with upper parameter unity and lower parameter that is not zero or a negative integer (Equation 3.134b), has the continued fraction (Equation 3.133 ≡ Equation 3.134c), whose algebraic inverse is Equation 3.134d:

$$\begin{aligned}|z|<1; \quad c \neq 0,-1, \quad \frac{1}{F(1;c;z)} &= 1-\frac{z}{c+}\,\frac{z}{c+1-}\,\frac{cz}{c+2+}\,\frac{2z}{c+3-}\,\frac{(c+1)z}{c+4+} \\ &\quad \cdots\frac{(c+n-1)z}{c+2n+}\,\frac{(n+1)z}{c+2n+1-}\,\frac{(c+n)z}{c+2n+2+}\cdots,\end{aligned} \tag{3.134a–3.134c}$$

$$\begin{aligned}F(1;c;z) &= \frac{1}{1-}\,\frac{z}{c+}\,\frac{z}{c+1-}\,\frac{cz}{c+2+}\,\frac{2z}{c+3-}\cdots\frac{(c+n-1)z}{c+2n+}\,\frac{(n+1)z}{c+2n+1-}\cdots \\ &= \sum_{n=0}^{\infty}\frac{z^n}{n!}\frac{(1)_n}{(c)_n}\sum_{n=0}^{\infty}\frac{z^n}{(c)_n} = 1+\sum_{n=1}^{\infty}\frac{z^n}{c\cdots(c+n-1)}.\end{aligned} \tag{3.134d}$$

The passage from Equations 3.133 through 3.134c uses Equations 1.95 and 1.108.

3.9.9 Continued Fraction for Exponential (Lagrange 1776) and Approximants (Padé 1892, 1899)

The exponential Equation 1.26b is a particular case (Equation 3.130b) of the confluent hypergeometric function (Equation 3.124b):

$$|z| < \infty: \qquad e^z = \sum_{n=0}^{\infty} \frac{z^n}{n!} = \sum_{n=0}^{\infty} \frac{z^n}{n!} \frac{(c)_n}{(c)_n} = F(c;c;z) = F(1;1;z), \qquad \text{(3.135a and 3.135b)}$$

with the same region of validity, that is, the finite complex z-plane (Equation 3.124a ≡ Equation 3.135a); the same upper and lower parameters (Equation 3.136a) that can be given the value unity (Equation 3.136b) to specify (Equation 3.134d) the continued fraction for the exponential (Equation 3.125c):

$$c = b = 1: \qquad e^z = \frac{1}{1-}\frac{z}{1+}\frac{z}{2-}\frac{z}{3+}\frac{2z}{4-}\frac{2z}{5+}\cdots\frac{nz}{2n+1+}\frac{(n+1)z}{2(n+1)-}\frac{(n+1)z}{2n+3+}\cdots$$
$$= \frac{1}{1-}\frac{z}{1+}\frac{z}{2-}\frac{z}{3+}\frac{z}{2-}\frac{z}{5+}\cdots\frac{z}{2n+1+}\frac{z}{2-}\frac{z}{2n+3+}\cdots \qquad \text{(3.136a–3.136c)}$$

This proves (Lagrange 1776) the continued fraction (Equation 3.136c ≡ Equation 3.5a). Using Equations 1.95 and 1.108, the algebraic inverse of Equation 3.136c is

$$e^{-z} = \frac{1}{e^z} = 1 - \frac{z}{1+}\frac{z}{2-}\frac{z}{3+}\frac{z}{2-}\cdots\frac{z}{2n+1+}\frac{z}{2-}\cdots. \qquad \text{(3.137)}$$

The change of variable (Equation 3.138a) leads to Equation 3.138b:

$$z \to -z: \qquad e^z = 1 + \frac{z}{1-}\frac{z}{2+}\frac{z}{3-}\frac{z}{2+}\cdots\frac{z}{2n+1-}\frac{z}{2+}\cdots, \qquad \text{(3.138a and 3.138b)}$$

that proves the continued fraction (Equation 3.138b ≡ Equation 3.5b). In the confluent hypergeometric function (Equation 3.124b) with upper parameter unity (Equation 3.139a) and lower parameter, a positive integer (Equation 3.139b) may be inserted in the exponential series (Equation 3.124):

$$b = 1, c = m \in |N: \quad F(1;m;z) = 1 + \frac{z}{m} + \frac{z^2}{m(m+1)} + \cdots$$
$$= 1 + \frac{(m-1)!}{z^{m-1}}\left[\frac{z^m}{m!} + \frac{z^{m+1}}{(m+1)!} + \cdots\right] \qquad \text{(3.139a–3.139c)}$$
$$= 1 + \frac{(m-1)!}{z^{m-1}}\left[e^z - 1 - z - \frac{z^2}{2!} - \cdots \frac{z^{m-1}}{(m-1)!}\right].$$

Solving for the exponential leads to

$$e^z = 1 + z + \frac{z^2}{2} + \cdots + \frac{z^{m-1}}{(m-1)!} + \frac{z^{m-1}}{(m-1)!}[F(1;m;z) - 1], \tag{3.140}$$

where the continued fraction (Equation 3.134d) may be substituted:

$$|z| < 1: \; e^z = 1 + z + \frac{z^2}{2} + \cdots + \frac{m^{m-2}}{(m-2)!} + \frac{z^{m-1}}{(m-1)!}\left[\frac{1}{1+}\frac{z}{m+}\frac{z}{m+1-}\frac{2z}{m+3+}\cdots\frac{z(m+n-1)}{m+2n+}\frac{(n+1)z}{m+2n+1-}\right].$$

(3.141a and 3.141b)

The **Padé approximant** *for the exponential (Equation 3.141b) in the finite plane (Equation 3.141a) relates to the power series (Equation 3.135b) [continued fraction (Equation 3.136c)] in isolation for* $m \to \infty$ $(m = 1)$ *and combines them in varying proportions for* $2 \le m < \infty$.

3.9.10 Continued Fraction for Binomial

Suppressing Equation 3.142a in the Gaussian hypergeometric function (Equation 3.101b) besides the first upper parameter as in the confluent hypergeometric function (Equations 3.125a through 3.125c), also the second upper parameter (Equations 3.142b and 3.142c):

$$z \to \frac{z}{ab};\, a, b \to \infty: \qquad \lim_{a,b\to\infty} F(a,b;c;z) = \sum_{n=0}^{\infty} \frac{z^n}{n!}\frac{1}{(c)_n} \lim_{a\to\infty}\frac{(a)_n}{a^n} \lim_{b\to\infty}\frac{(b)_n}{b^n}$$

$$= \sum_{n=0}^{\infty} \frac{z^n}{n!}\frac{1}{(a)_n} = F(;c;z), \equiv {}_0F_1(;c;z)$$

(3.142a–3.142d)

also leads to the hypogeometric function (Equation 3.142d ≡ Equation 1.159), that is an integral (meromorphic) function of the variable z (parameter c). The zero or negative integer values of the lower parameter correspond to simple poles in all three cases of the Gaussian (Equations 3.101a and 3.101b) [confluent (Equations 3.124a and 3.124b)] hypergeometric (Equation 1.164) functions and the hypogeometric functions (Equation 1.159), with residues [Equations 3.108a through 3.108c (Equations 3.129a and 3.129b)/(Equations 1.168a and 1.168b)] obtained by suppressing one (two) upper parameter(s). The suppression of one upper and one lower parameter (Equation 3.143a) requires no limit and leads to Equation 3.143b:

$$b = c: \quad \mathrm{F(a,c;c;z)} = \sum_{n=0}^{\infty} \frac{z^n}{n!}(a)_n = \sum_{n=0}^{\infty} \frac{(-z)^n}{n!}(-a)(-a+1)\cdots(-a+n-1) = (1-z)^{-a},$$

(3.143a and 3.143b)

corresponding to a particular case (Equation 3.143b) of the binomial series:

$$|z| < |b|: \qquad (z+b)^v = \sum_{n=0}^{\infty} \binom{v}{n} b^{v-n} z^n, \qquad \binom{v}{n} \equiv \frac{v\,(v-1)\cdots(v-n+1)}{n!}. \tag{3.144a–3.144c}$$

This coincides with Equations 3.144a through 3.144c ≡ Equations I.25.37a through I.25.37c, apart from a misprint in b^{v-n}.

The binomial series corresponds to the MacLaurin series (Subsections I.23.7 and 1.1.7) for the function (Equation 3.145a) that has derivatives (Equation 3.145b) at the origin (Equations 3.145c and 3.145d):

$$f(z) = (z+b)^v, \qquad f^{(n)}(z) = (z+b)^{v-n}\, v(v-1)\ldots v\,(v-n+1), \qquad \text{(3.145a and 3.145b)}$$

$$f^{(n)}(0) = b^{v-n} n! \binom{v}{n}, \quad \binom{v}{n} \equiv \frac{v(v-1)(v-1)\cdots(v-n+1)}{n!} = \frac{(v-n+1)_n}{n!}. \qquad \text{(3.145c and 3.145d)}$$

The binomial series (Equation 3.144b) has (1) coefficients specified by (Equation 3.144c ≡ Equation 3.145d) using the Pochhammer symbol (Equation 1.160); and (2) radius of convergence (Equation 3.144a) limited by the nearest singularity z = −b of the function (Equation 3.145a), which is a branch-point (pole of order N) for v not an integer (a negative integer v = −N). In the case of v = N, which is an integer (Equation 3.146a), the series terminates and leads to a polynomial (Equation 3.146b) of degree N with coefficients Equation 3.146c:

$$v = N \in| N: \qquad (z+b)^N = \sum_{n=0}^{N} \binom{N}{n} b^{N-n} z^n, \qquad \text{(3.146a and 3.146b)}$$

$$\binom{N}{n} = \frac{N(N-1)\cdots(N-n+1)}{n!} = \frac{N!}{(N-n)!\,n!}. \qquad \text{(3.146c)}$$

The convergence of the binomial series (Equations 3.144b and 3.144c) for v not a positive integer, that is, excluding the finite polynomial case (Equations 3.146a through 3.146c), is indicated at all points of the complex plane in Section I.29.1.3 and Table I.29.1. The particular case (Equation 3.147a) of the binomial series (Equation 3.144b) is the geometric series (Equation 3.147b) that has (Equation 3.144c) coefficients (Equation 3.147c) with alternating sign:

$$v = -1: \qquad \frac{1}{b+z} = \frac{1}{b}\sum_{n=0}^{\infty} (-)^n \left(\frac{z}{b}\right)^n, \quad \binom{-1}{n} = \frac{(-1)(-2)\cdots(-n)}{n!} = (-)^n; \qquad \text{(3.147a–3.147c)}$$

$$z \to -z: \qquad \frac{1}{b-z} = \frac{1}{b}\sum_{n=0}^{\infty} \left(\frac{z}{b}\right)^n = \sum_{n=0}^{\infty} z^n b^{1-n}. \qquad \text{(3.148a and 3.148b)}$$

The alternating coefficients in Equation 3.147b are suppressed in Equation 3.148b, which changes the sign of the variable (Equation 3.148a). The convergence of the geometric series (Equations 3.147b and 3.148b) at all points of the z-plane (also the remainders after N terms and related series) is discussed in Section I.21.9 (Subsection I.21.8).

Thus, *the binomial series is a hypergeometric function (Equation 3.149a):*

$$(1+z)^a = \sum_{n=0}^{\infty}\binom{a}{n} z^n = \sum_{n=0}^{\infty}\frac{z^n}{n!}a\cdots(a-n+1) = \frac{1}{F(a,1;1;-z)} \tag{3.149a and 3.149b}$$

$$= 1+\frac{az}{1+}\frac{(1-a)z}{2+}\frac{(1+a)z}{3+}\frac{2(2-a)z}{4+}\frac{2(2+a)}{5+}\cdots\frac{n(n-a)z}{2n+1}\frac{n(n+a)}{2n+1+}\cdots,$$

$$z\notin)-\infty,-1):\quad (1+z)^a = \frac{1}{1-}\frac{az}{1+}\frac{(1+a)}{2+}\frac{(1-a)z}{3+}\frac{2(2+a)z}{4+}\frac{2(2-a)z}{5+}\cdots\frac{n(n+a)z}{2n+}\frac{n(n-a)z}{2n+1+'}\cdots, \tag{3.149c and 3.149d}$$

leading to the continued fraction (Equation 3.149b), whose algebraic inverse is Equation 3.149d, both valid in the complex z-plane with a branch-cut (Equation 3.149c) to the left of –1 along the negative real axis. The branch-cut (Equation 3.149c ≡ Equation 3.37a) is the same as for the logarithm, and the continued fractions are obtained as follows: (1) the parameters (Equations 3.150a and 3.150b) of the hypergeometric function (Equation 3.149a) specify (Equations 3.116d and 3.116e) the coefficients (Equations 3.150c and 3.150d):

$$b=1=c:\quad \beta_1=-a,\quad \{\beta_{2n+1},\beta_{2n+2}\} = -\left\{\frac{a+n}{4n+2},\frac{n+1-a}{4n+2}\right\}, \tag{3.150a–3.150d}$$

leading (Equation 3.117c) to the continued fraction:

$$\begin{aligned}(1+z)^a &= \frac{1}{F(a,1;1;-z)} = 1-\frac{\beta_1 z}{1-}\frac{\beta_2 z}{1-}\cdots\frac{\beta_{2n+1}z}{1-}\frac{\beta_{2n+2}z}{1-}\\ &= 1-\frac{-az}{1+}\frac{(1-a)z/2}{1+}\frac{(1+a)z/6}{1+}\frac{(2-a)z/6}{1+}\frac{(2+a)z/10}{1+}\\ &\quad\cdots\frac{(n+1-a)z/(4n+2)}{1+}\frac{(n+1+a)z/(4n+6)}{1+}\cdots\\ &= 1+\frac{az}{1+}\frac{(1-a)z}{2+}\frac{(1+a)z}{3+}\frac{2(2-a)z}{4+}\frac{2(2+a)z}{5+}\cdots\frac{n(n-a)z}{2n+}\frac{n(n+a)z}{2n+1+}\cdots,\end{aligned} \tag{3.151}$$

that proves Equation 3.149b ≡ Equation 3.151; (2) the algebraic inverse (Equations 1.95 and 1.108) of Equation 3.151 is

$$\begin{aligned}(1+z)^{-a} = F(a,1;1;-z) &= \frac{1}{1+}\frac{az}{1+}\frac{(1-a)z}{2+}\frac{(1+a)z}{3+}\frac{2(2-a)z}{4+}\frac{2(2+a)z}{5+}\\ &\quad\cdots\frac{n(n-a)z}{2n+}\frac{n(n+a)z}{2n+1+}\cdots;\end{aligned} \tag{3.152}$$

and (3) the change of parameter (Equation 3.153a) leads to

$$a \to -a: \qquad (1+z)^a = \frac{1}{1-}\frac{az}{1+}\frac{(1+a)z}{2+}\frac{(1-a)z}{3+}\frac{2(2+a)z}{4+}\frac{2(2-a)z}{5+}\cdots\frac{n(n+a)z}{2n+}\frac{n(n-a)z}{2n+1-}\cdots, \tag{3.153a and 3.153b}$$

proving Equation 3.153b ≡ Equation 3.144d.

NOTE 3.1 SIX EQUIVALENT DEFINITIONS OF THE EXPONENTIAL

Although the six definitions of the exponential are equivalent, some lead more readily than others to certain properties, for example: (1) defining the exponential to be equal to its derivative proves that it is analytic, has no zeros, and leads to the power series; (2) defining the exponential by its power series leads to the same properties in reverse order; (3) the rational limit is the least useful for accurate calculation of exponentials due to its slow convergence; (4) conversely, the continued fraction and power series lead to the most accurate estimate of the exponential for a given level of complexity; (5 and 6) the property of changing sums to products (products to powers) can be used as the definition of exponential, instead of being derived from one of the other definitions 1–4.

NOTE 3.2 TABLES OF LOGARITHMS WITH 110 DECIMALS (MANSELL 1929)

The tables of logarithms with 110 decimals were calculated for the first 1000 integers before the computer age (Mansell 1929). The method was that indicated in Section 3.8 and the values were written in long strips of paper. It is remarkable that numerous tests made using modern digital computers failed to detect any error in these manually computed tables. One test was to sum all 1000 values in the table:

$$\sum_{n=1}^{1000} \log n = \log\left(\prod_{n=1}^{1000} n\right) = 1000! \tag{3.154}$$

to compare this very large number with the asymptotic approximation for the factorial $N!$ for large N. The values Equations 3.7 and 3.62 come from these remarkable tables.

NOTE 3.3 GENERALIZED HYPERGEOMETRIC FUNCTION AND PARTICULAR CASES

The series (Equation 3.155 ≡ Equation I.30.112) using the Pochhammer symbol (Equation 1.160) specifies

$$\begin{aligned} {}_pF_q\left(a_1,\ldots,a_p;b_1,\ldots,b_q;z\right) &= F\left(a_1,\ldots,b_1;b_q,\ldots,z\right) \\ &= \sum_{n=0}^{\infty}\frac{z^n}{n!}\frac{\left(a_1\right)_n\cdots\left(a_p\right)_n}{\left(b_1\right)_n\cdots\left(b_p\right)_n} = 1+\sum_{n=1}^{\infty}\frac{z^n}{n!}\frac{a_1\ldots a_p}{b_n\ldots b_q}\cdots\frac{\left(a_1+n-1\right)\cdots\left(a_p+n-1\right)}{\left(b_1+n-1\right)\cdots\left(b_q+n-1\right)_n}. \end{aligned} \tag{3.155}$$

the **generalized hypergeometric function** with (1) variable z; (2) p upper parameters $a_1,\ldots, a_p$; and (3) q lower parameters $b_1,\ldots, b_q$. The particular cases include (1) the binomial (Equation 3.143b) with one upper parameter (Subsection 3.9.10); (2) the hypogeometric function (Equation 1.159) with one lower parameter (Section 1.9); (3) the confluent hypergeometric function (Equation 3.124b) with one upper and one lower parameter (Subsections

3.9.7 through 3.9.9); and (4) the Gaussian hypergeometric function (Equation 3.101b) with two upper and one lower parameters (Subsections 3.9.2 through 3.9.6). The generalized hypergeometric function is symmetric in the upper (lower) parametric $a_1,\ldots, a_p$ ($b_1,\ldots, b_q$), for example, (a,b) in the Gaussian case (Equation 3.111a). From Equation 3.156a ≡ Equation 3.155, the general term (Equation 1.156b):

$$F(a_1,\ldots,a_p;b_1,\ldots,b_q;z)=\sum_{n=0}^{\infty} f_n, \qquad f_n=\frac{z^n}{n!}\frac{(a_1)_n\cdots(a_p)_n}{(b_1)_n\cdots(b_q)_n}, \tag{3.156a and 3.156b}$$

and ratio of successive terms:

$$\frac{f_{n+1}}{f_n}=\frac{z}{n+1}\frac{(a_1+n)\cdots(a_p+n)}{(b_1+n)\cdots(b_q+n)}\sim zO(n^{p-q-1}), \tag{3.156c}$$

it follows that *the generalized hypergeometric function is an analytic function of the variable z: (1) in the finite plane (Equation 3.157a) for $q \geq p$, for example, $p = 1 = q$ for the confluent hypergeometric function (3.124b), and $1 = q > p = 0$ for the hypogeometric function (Equation 1.159); (2) the unit disk (Equation 3.157b) if $p = q + 1$, for example, ($p = 2$, $q = 1$) for the Gaussian hypergeometric function (Equation 3.101a) and ($p = 1$, $q = 0$) for the binomial (Equation 3.143b). For $q > p + 1$, the generalized hypergeometric function diverges (Equation 3.157c) everywhere except at the origin:*

$$F(a_1,\ldots,a_p;b_1,\ldots,b_q;z)\begin{cases}\in \mathcal{D}(|z|<\infty) & \text{if} \quad q\geq p, & (3.157a)\\ \in \mathcal{D}(|z|<1) & \text{if} \quad p=q+1, & (3.157b)\\ \text{diverges for } z\neq 0 & \text{if} \quad p>q+1. & (3.157c)\end{cases}$$

In the case 2 of convergence in the unit disk (Equation 3.157b ≡ Equation 3.158a), the ratio of successive terms (Equation 3.156c) leads to Equation 3.158b:

$$p=q+1: \qquad \frac{f_{n+1}}{f_n}=z\left[1+\frac{a_1+\cdots+a_p-b_1-\cdots-b_q-1}{n}+O\left(\frac{1}{n^2}\right)\right], \tag{3.158a and 3.158b}$$

specifies the convergence on the boundary (Example I.30.20), in particular, ($p = 2$, $q = 1$) for the Gaussian hypergeometric function (Subsections I.29.9 and Table I.29.2).

From Equation 3.155, it follows that *the generalized hypergeometric function is an integral function of all upper parameters and reduces to a polynomial when any of them is a negative integer, for example:*

$$-p\in |N: \qquad F(-p,a_2,\ldots,a_p;b_1,\ldots,b_q;z)=\sum_{n=0}^{p}(-z)^n\binom{n}{p}\frac{(a_2)_n\cdots(a_p)_n}{(b_1)_n\cdots(b_q)_n}; \tag{3.159a and 3.159b}$$

particular cases include Equations 3.103a through 3.103d (Equations 3.128a through 3.128c) for the Gaussian (confluent) hypergeometric function. The generalized hypergeometric function is a meromorphic function of all lower parameters, with simple poles at their zero or negative integer values, with residues given, for example, by

$$k \in |N_0: \quad F\left(a_1,\ldots,a_p;-k,b_2,\ldots,b_q;z\right) = \lim_{b\to -k}(c+k)F\left(a_1,\ldots,a_p;b_1,b_2,\ldots,b_q;z\right)$$

$$= \frac{z^{k+1}}{(k+1)!}\frac{(-)^k}{k!}\cdots\frac{\left(a_1\right)_k\cdots\left(a_p\right)_k}{\left(b_2\right)_k\cdots\left(b_q\right)_k}F\left(a_1+k+1,\ldots,a_p+k+1;k+2,b_2+k+1,\ldots,b_q+k+1;z\right);$$

(3.160a and 3.160b)

particular cases include the hypogeometric ($p = 0$, $q = 1$) function (Equations 1.168a and 1.168b), and $p = 1 = q$ ($p = 2$, $q = 1$) for the confluent (Gaussian) hypergeometric functions [Equations 3.108a and 3.108b (Equations 3.129a and 3.129b)]. The proof of Equations 3.160a and 3.160b is similar.

3.10 Conclusion

The surface of constant modulus of the exponential is a curved ramp above the (x, y)-plane (Figure 3.2), with (1) the argument of the exponential constant on horizontal lines (Figure 3.1); and (2) the profile of the ramp specified by the real exponential (Figure 3.3). The complex exponential is a single-valued many-valent function with period $2\pi i$, whose inverse (Figure 3.4) is the many-valued univalent function logarithm. The logarithm becomes single-valued, choosing the principal branch in the cut-plane (Figure 3.5), on which the axisymmetric funnel-shaped surface of the constant real part of the logarithm is projected (Figure 3.6), whose generators are the logarithm of a real variable (Figure 3.7). The logarithm can be expanded in power series in two ways: (1) changing the variable in the logarithm that displaces the branch-cuts but preserves the unit circle of convergence (Figure 3.8a through d); and (2) changing the variable in the power series, that keeps the branch-cut but changes the region of convergence (Figure 3.9a through d). The irrational number e, like the exponential, is best calculated from the MacLaurin series (Table 3.1) or continued fraction (Table 3.2) rather than from the binomial limit. The series for the logarithm (Table 3.3) can be chosen with variables allowing accurate computation of tables of logarithms (Table 3.4) using just a few terms of the series.

4

Plane Elasticity and Multiharmonic Functions

Whereas a fluid can have unlimited displacements resulting from integration of velocity over time, an elastic, nonplastic solid has finite displacements. A translation and a rotation are displacements that change the position but not the shape of a body. The remaining part of the displacement represents the deformation of the body, specified by a strain tensor (Section 4.1). The deformations are associated with stresses (Section 4.2), that is, the surface forces that together with volume forces balance the inertia force in the momentum equation. The stresses and strains are related by an elastic constitutive relation (Section 4.3) that depends on the properties of a material. Using the preceding relations, it is found that the displacement vector satisfies a second-order differential equation (Section 4.4); in the case of bounded media, the differential equation has a unique solution satisfying boundary conditions such as given (or zero) stresses/forces/displacements at a boundary. There are three main methods to approach problems of steady plane elasticity. Method I is based on the solution of the momentum equation (Section 4.4) for the displacement vector; for example, a radial (azimuthal) displacement such as the velocity of a source/sink (vortex) leads to the stresses due to a line pressure (torque) in a massive or hollow cylinder or cylindrical cavity or shell (Section 4.5). Method II relies on all components of the stress tensor deriving from a scalar stress function that is biharmonic, so that its double Laplacian is zero (Section 4.6); for example, this method can be used to determine the displacements, stresses, and strains in a wedge, with applied or concentrated force or torque or distributed fluid loading as in a dam (Section 4.7). Method III uses the divergence (curl) of the displacement vector, that is, the linearized relative area change (rotation), with which can be formed a complex elastic potential that is an analytic function, whose real and imaginary parts satisfy the Laplace equation; an example is a monopole that corresponds to a line force in an unbounded medium (Section 4.8). Because the equations of elasticity are linear, the principle of superposition applies. It can be combined with method II (method III) to specify the displacements, strains, and stress in a semi-infinite (infinite) medium [Section 4.7 (Section 4.8)] due to an arbitrary load distribution on the boundary (in the interior). An example of multiple loads is the case of the displacements, stresses, and strains on a wheel (Section 4.9) due to the superposition of (1) its own weight; (2) a traction/braking force associated with forward motion/retardation; and (3) a torque applied by a driving engine.

4.1 Displacement Vector and Deformation and Strain Tensors

The deformation of a medium is the part of the displacement that is neither a translation (Subsection 4.1.1) nor a rotation (Subsection 4.1.2). It can be specified in polar (Subsection 4.1.3) as well as in Cartesian coordinates, and specifies the (1) scalar volume change; (2) displacement and rotation vectors; and (3 and 4) displacement and strain tensors (Subsection

4.1.4). The strain tensor derives from the displacement vector, and its components cannot be chosen arbitrarily, that is, they must satisfy a set of compatibility relations (Subsection 4.1.5).

4.1.1 Translation and Displacement Vector and Tensor

A continuous medium, such as a rigid body, a deformable solid or a fluid, or a material with similar or hybrid properties, is specified by the position vector $\vec{r}$ of its points. For a plane medium, the position vector may be identified with a complex number (Equation 4.1a) and specified by its Cartesian coordinates (Equation 4.1b):

$$\vec{r} \equiv z = x + iy: \qquad \vec{r} = \vec{e}_x x + \vec{e}_y y. \tag{4.1a and 4.1b}$$

The real coordinate normal to the (x, y)-plane is designated x_3 to distinguish from z, which is the complex coordinate (Equation 4.1a) in the (x, y)-plane. A displacement of the medium implies that the point of position $\vec{r}$ goes to another position $\vec{r} + \vec{u}$ in Equation 4.2a, where the **displacement vector:**

$$\vec{r} \to \vec{r} + \vec{u}: \qquad \vec{u} = u_x(x,y,t)\vec{e}_x + u_y(x,y,t)\vec{e}_y \tag{4.2a and 4.2b}$$

may depend on position in the plane (x, y) and on time t. The simplest displacement is a **translation**, which moves all points by the same amount. Thus, *a translation corresponds to a constant displacement vector and is eliminated by taking the differential of the displacement, that specifies the* **relative displacement** *(Equation 4.3c):*

$$\mathrm{d}\vec{e}_x = 0 = \mathrm{d}\vec{e}_y: \qquad \mathrm{d}\vec{u} = \vec{e}_x \mathrm{d}u_x + \vec{e}_y \mathrm{d}u_y, \tag{4.3a–4.3c}$$

where the Cartesian unit base vectors (Equations 4.3a and 4.3b) are constant. *Assuming that the components of the displacement vector are differentiable functions of position (Equation 4.4a), their differentials (Equation 4.4b):*

$$\vec{u}(\vec{r}) \in D(|R^2): \qquad \mathrm{d}u_x = D_{xx}\mathrm{d}x + D_{xy}\mathrm{d}y, \quad \mathrm{d}u_y = D_{yx}\mathrm{d}x + D_{yy}\mathrm{d}y, \tag{4.4a and 4.4b}$$

$$D_{xx} = \partial_x u_x \equiv \frac{\partial u_x}{\partial x}, \qquad D_{yy} = \partial_y u_y, \qquad D_{xy} = \partial_y u_x, \qquad D_{yx} = \partial_x u_y, \tag{4.5a–4.5d}$$

specify a **displacement tensor** *(Equations 4.5a through 4.5d), that appears in*

$$\mathrm{d}\vec{u} = D_{xx}\vec{e}_x\mathrm{d}x + D_{xy}\vec{e}_x\mathrm{d}y + D_{yx}\vec{e}_y\mathrm{d}x + D_{yy}\vec{e}_y\mathrm{d}y \tag{4.6}$$

the relative displacement (Equation 4.6). The latter is unaffected by translations.

4.1.2 Rotation, Area Change, and Strain Tensor

The rotation, like the translation, is a displacement that does not cause a deformation. Unlike the translation, the rotation causes a change in relative position. A rotation in the

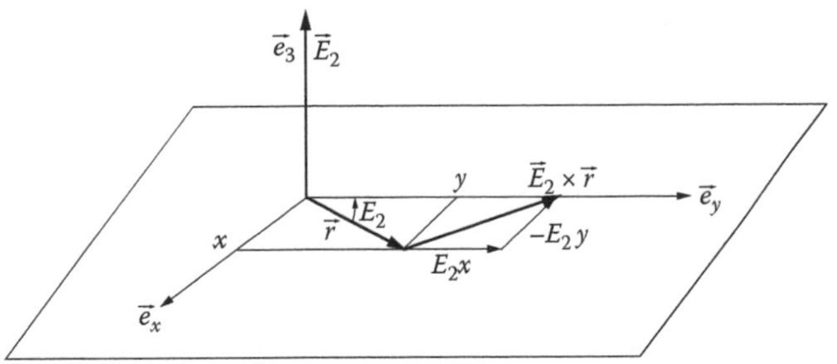

FIGURE 4.1
Rigid body displacement consists of a translation and a rotation, which cause no deformation, because the distance between two arbitrary points is unchanged. A rotation in the plane is specified by a rotation vector normal to the plane, whose modulus is one-half of the modulus of the curl of the displacement vector.

plane (x, y) is specified by a vector orthogonal to the plane, that is, in the x_3-direction. A *rigid plane rotation* by an angle E_2 (Figure 4.1) leads to the displacement (Equation 4.7):

$$\vec{u}_R = \vec{E}_2 \wedge \vec{r} = E_2 \vec{e}_3 \wedge \left(\vec{e}_x x + \vec{e}_y y\right) = E_2 \left(\vec{e}_y x - \vec{e}_x y\right) \tag{4.7}$$

which has components (Equations 4.8a and 4.8b):

$$u_{Rx} = -E_2 y, \qquad u_{Ry} = E_2 x: \qquad \nabla \wedge \vec{u}_R = \vec{e}_3 \left(\partial_x u_{Ry} - \partial_y u_{Rx}\right) = \vec{e}_3 2E_2 = 2\vec{E}_2, \quad \text{(4.8a–4.8c)}$$

whose curl (Equation I.11.35a) is twice the rotation (Equation 4.8c). Thus, *a rotation is specified by one-half of the curl of the displacement (Equation 4.9a). The latter coincides with the skew-symmetric part of the displacement tensor (Equation 4.9b):*

$$2E_2 = \partial_x u_y - \partial_y u_x = D_{xy} - D_{yx}, \qquad d\vec{u}_R = E_2 \left(-\vec{e}_x dy + \vec{e}_y dx\right) \quad \text{(4.9a–4.9c)}$$

and is represented locally (Equation 4.9c) by the displacement (Equation 4.7). A **deformation** *is the part of the displacement due neither to a translation nor to a rotation:*

$$d\vec{u} - d\vec{u}_R = D_{xx} \vec{e}_x dx + \left(D_{xy} + E_2\right) \vec{e}_x dy + \left(D_{yx} - E_2\right) \vec{e}_y dx + D_{yy} \vec{e}_y dy. \tag{4.10}$$

It specifies a linear **strain tensor:**

$$d\vec{u} - d\vec{u}_R \equiv S_{xx} \vec{e}_x dx + S_{xy} \vec{e}_x dy + S_{yx} \vec{e}_y dx + S_{yy} \vec{e}_y dy, \tag{4.11}$$

whose (1) diagonal components coincide with the displacement tensor (Equations 4.12a and 4.12b):

$$S_{xx} = D_{xx} = \partial_x u_x, \quad S_{yy} = D_{yy} = \partial_y u_y; \quad \text{(4.12a and 4.12b)}$$

$$2S_{xy} = 2D_{xy} + 2E_2 = 2\partial_y u_x + \partial_x u_y - \partial_y u_x = \partial_x u_y + \partial_y u_x = D_{yx} + D_{xy}; \tag{4.12c}$$

$$2S_{yx} = 2D_{yx} - 2E_2 = 2\partial_x u_y - \partial_x u_y + \partial_y u_x = \partial_x u_y + \partial_y u_x = 2S_{xy}; \tag{4.12d}$$

(2) off-diagonal components are equal (Equation 4.12c = Equation 4.12d); and (3) thus the strain tensor is symmetric.

The strain tensor (Equations 4.12a through 4.12d) [rotation scalar (Equations 4.9a and 4.9b)] is specified by the symmetric (skew-symmetric) part of the displacement tensor:

$$\begin{bmatrix} D_{xx} & D_{xy} \\ D_{yx} & D_{yy} \end{bmatrix} = \begin{bmatrix} D_{xy} & \dfrac{D_{xy}+D_{yx}}{2} \\ \dfrac{D_{xy}+D_{yx}}{2} & D_{yy} \end{bmatrix} + \begin{bmatrix} 0 & \dfrac{D_{xy}-D_{yx}}{2} \\ \dfrac{D_{yx}-D_{xy}}{2} & 0 \end{bmatrix} \tag{4.13}$$

$$= \begin{bmatrix} S_{xx} & S_{xy} \\ S_{xy} & S_{yy} \end{bmatrix} + \begin{bmatrix} 0 & +E_2 \\ -E_2 & 0 \end{bmatrix}.$$

The diagonal strains (Equations 4.12a and 4.12b) are an **extension (contraction)** *in the x/y-direction if positive (negative) (Figure 4.2); the off-diagonal components (Equation 4.12c ≡ Equation 4.12d) are a* **distortion** *counterclockwise (clockwise) if positive (negative). The* **relative area change** *is given (Equation 4.14a):*

$$D_2\,\mathrm{d}x\mathrm{d}y \equiv (\mathrm{d}x + D_{xx}\mathrm{d}x)(\mathrm{d}y + D_{yy}\mathrm{d}y) - \mathrm{d}x\mathrm{d}y = (D_{xx} + D_{yy} + D_{xx}D_{yy})\mathrm{d}x\mathrm{d}y \tag{4.14a}$$

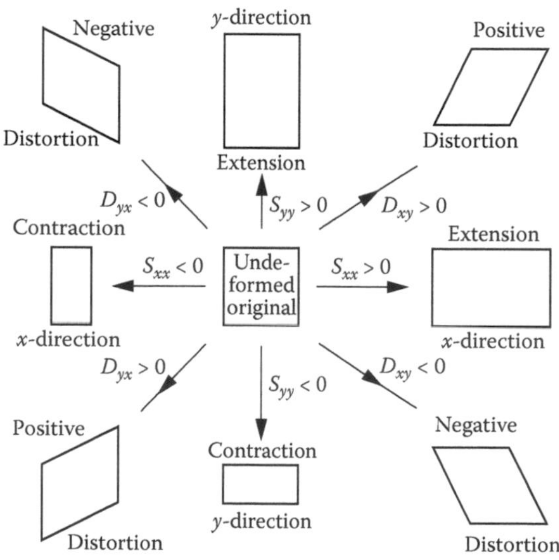

FIGURE 4.2
Spatial derivatives of the displacement vector form the displacement tensor, whose symmetric (skew-symmetric) part specifies the strain tensor (rotation vector). The deformation is completely specified by the strain tensor: (1) the diagonal components S_{xx} (S_{yy}) specify an extension in the $x(y)$-direction if $S_{xx} > 0$ ($S_{yy} > 0$) and a contraction in the opposite case; (2) the symmetric nondiagonal components $S_{xy} = S_{yx}$ specify a distortion in the positive (negative) or counterclockwise (clockwise) direction if $S_{xy} > 0$ ($S_{xy} < 0$). The linearized relative area change is unaffected by the distortion and is specified by the sum of the diagonal strains, that is, the trace of the strain tensor that equals the divergence of the displacement vector. The curl specifies twice the rotation vector in Figure 4.1.

exactly by Equation 4.14b, that includes terms that are both linear and nonlinear in the displacement tensor:

$$D_2 = D_{xx} + D_{yy} + D_{xx}D_{yy}. \tag{4.14b}$$

Neglecting the second-order terms by Equation 4.15a leads to the linearized relative area change (Equation 4.15b):

$$D_{xx}D_{yy} << 1: \qquad D_2 \sim D_{xx} + D_{yy} = S_{xx} + S_{yy} = \partial_x u_x + \partial_y u_y = \nabla \cdot \vec{u}, \qquad \text{(4.15a and 4.15b)}$$

which is specified by the divergence of the displacement (Equation I.11.33a). Thus, the curl (divergence) of the displacement vector (Equation 4.2b) specifies twice the rotation (Equation 4.9a) [linearized (Equation 4.15a) relative area change (Equation 4.15b)]. The dependence of the displacement vector (Equation 4.2b) on position (time) concerns **nonuniform (unsteady)** displacements [Equations 4.3 through 4.15 (Equations 4.16 through 4.17)]. Concerning the unsteady displacements, *the derivative with regard to time leads from (1) the displacement (Equation 4.2b) to the velocity (Equation 4.16a); (2) the rotation (Equation 4.8c) to the vorticity (Equation 4.16b); and (3) the linearized relative area change (Equation 4.15b) to the dilatation (Equation 4.16c):*

$$\vec{v} \equiv \dot{\vec{u}} \equiv \frac{\partial \vec{u}}{\partial t}, \quad \vec{\varpi} \equiv \nabla \wedge \vec{v} = \nabla \wedge \dot{\vec{u}} = 2\dot{\vec{\Omega}} = \dot{\vec{\Omega}} = 2\dot{\vec{E}}_2, \quad \nabla.\vec{v} = \nabla.\dot{\vec{u}} = \dot{D}_2; \qquad \text{(4.16a–4.16c)}$$

$$i, j = 1, 2: \qquad 2\dot{S}_{ij} = \partial_i \dot{u}_j + \partial_j \dot{u}_i = \partial_i v_j + \partial_j v_i; \qquad \text{(4.17a and 4.17b)}$$

the rate-of-strain tensor has the same dependence on the velocity vector (Equations 4.17a and 4.17b) as the strain tensor on the displacement vector (Equations 4.12a through 4.12d). The velocity, vorticity, and dilatation (Equations 4.16a through 4.16c) are relevant to inviscid fluids (Chapter 2), and the rates of strain (Equations 4.17a and 4.17b) to viscous fluids (Note 4.4).

4.1.3 Relative Displacement in Polar Coordinates

The preceding results can be expressed in three dimensions instead of two and in non-Cartesian coordinate systems, such as orthogonal curvilinear coordinates (Subsection I.11.9), of which polar coordinates are the examples considered next. The polar representation (Subsection I.1.4) of a complex number (Equation 4.18b) with unit modulus (Equation 4.18a) and its rotation (Equation 4.18c) by $\pi/2$:

$$|z| = 1: \qquad z = e^{i\varphi} = \cos\varphi + i\sin\varphi, \qquad iz = i\,e^{i\varphi} = -\sin\varphi + i\cos\varphi \qquad \text{(4.18a–4.18c)}$$

leads to *the unit base radial (Equation 4.19a) [azimuthal (Equation 4.19b)] vectors in polar coordinates (Figure 4.3a):*

$$\vec{e}_r = (\cos\varphi,\ \sin\varphi), \qquad \vec{e}_\varphi = (-\sin\varphi,\ \cos\varphi). \qquad \text{(4.19a and 4.19b)}$$

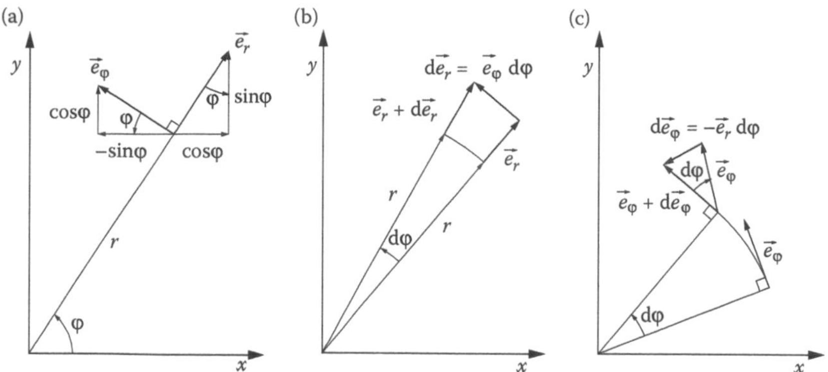

FIGURE 4.3
Unit base vectors in polar coordinates (a) are radial outward $\vec{e}_r$ and azimuthal counterclockwise $\vec{e}_\varphi$. Because their modulus is constant, their differential is orthogonal: (b) differential of the unit radial vector lies in the counterclockwise azimuthal direction and thus is parallel to the unit azimuthal base vector; (c) differential of the unit azimuthal base vector lies in the inward radial direction and thus is antiparallel to the unit radial vector.

Their differentials

$$d\vec{e}_r = (-\sin\varphi,\ \cos\varphi)\, d\varphi = \vec{e}_\varphi\, d\varphi, \quad d\vec{e}_\varphi = (-\cos\varphi,\ -\sin\varphi)\, d\varphi = -\vec{e}_r d\varphi \qquad \text{(4.20a and 4.20b)}$$

show that the radial (azimuthal) unit vector in Figure 4.3b (Figure 4.3c) varies in the positive (negative) azimuthal (radial) direction. From the differential of a complex number in the polar representation (Equation 4.21a) follows (1) by comparison with Equation 4.18b ≡ Equation 4.19a and Equation 4.18c ≡ Equation 4.19b the differential of the position vector (Equation 4.21b):

$$dz = d(re^{i\varphi}) = e^{i\varphi}(dr + i d\varphi), \qquad d\vec{r} = \vec{e}_r\, dr + \vec{e}_\varphi r\, d\varphi, \quad (ds)^2 = \left|dz\right|^2 = (dr)^2 + r^2(d\varphi)^2; \qquad \text{(4.21a–4.21c)}$$

(2) the arc length in polar coordinates (Equation 4.21c ≡ Equation 4.21d) is the modulus of the differential of a complex number (Equation 4.21a) [differential of the position vector (Equation 4.21b)]:

$$\begin{aligned}(ds)^2 = (dx)^2 + (dy)^2 = \left|dz\right|^2 = \left|d(re^{i\varphi})\right|^2 = \left|e^{i\varphi}(dr + i d\varphi)\right|^2 \\ = \left|dr + i d\varphi\right|^2 = (dr)^2 + r^2(d\varphi)^2 = \left|\vec{e}_r dr + \vec{e}_\varphi r d\varphi\right|^2\end{aligned} \qquad (4.21d)$$

because both lead to the same result. The displacement vector in polar coordinates (Equation 4.22):

$$\vec{u} = u_r(r,\varphi,t)\ \vec{e}_r + u_\varphi(r,\varphi,t)\vec{e}_\varphi \qquad (4.22)$$

leads to the relative infinitesimal displacement:

$$\mathrm{d}\vec{u} = \mathrm{d}u_r\vec{e}_r + \mathrm{d}u_\varphi\vec{e}_\varphi + u_r\mathrm{d}\vec{e}_r + u_\varphi\mathrm{d}\vec{e}_\varphi = \left(\mathrm{d}u_r - u_\varphi\mathrm{d}\varphi\right)\vec{e}_r + \left(\mathrm{d}u_\varphi + u_r\mathrm{d}\varphi\right)\vec{e}_\varphi, \tag{4.23}$$

where the unit base polar vectors (Equations 4.20a and 4.20b) are not constant, unlike their Cartesian counterparts in Equations 4.3a and 4.3b.

4.1.4 Displacement and Strain Tensors in Polar Coordinates

Using the differential of the polar components, the displacement vector:

$$\mathrm{d}u_r = (\partial_r u_r)\,\mathrm{d}r + (\partial_\varphi u_r)\,\mathrm{d}\varphi, \qquad \mathrm{d}u_\varphi = (\partial_r u_\varphi)\,\mathrm{d}r + (\partial_\varphi u_\varphi)\,\mathrm{d}\varphi \tag{4.24a and 4.24b}$$

in the infinitesimal relative displacement (Equation 4.24) leads to

$$\mathrm{d}\vec{u} = \left(\partial_r u_r\right)\vec{e}_r\,\mathrm{d}r + \left(\partial_\varphi u_r - u_\varphi\right)\vec{e}_r\,\mathrm{d}\varphi + \left(\partial_r u_\varphi\right)\vec{e}_\varphi\,\mathrm{d}r + \left(\partial_\varphi u_\varphi + u_r\right)\vec{e}_\varphi\,\mathrm{d}\varphi. \tag{4.25}$$

The displacement tensor in polar coordinates is defined as in Cartesian coordinates (Equation 4.26), with the differentials of the position vector (Equation 4.21b) in polar form:

$$\mathrm{d}\vec{u} = D_{rr}\vec{e}_r\,\mathrm{d}r + D_{r\varphi}\vec{e}_r r\,\mathrm{d}\varphi + D_{\varphi r}\vec{e}_\varphi\mathrm{d}r + D_{\varphi\varphi}\vec{e}_\varphi r\,\mathrm{d}\varphi. \tag{4.26}$$

Comparing Equations 4.25 and 4.26 *specifies in polar coordinates the following: (1) the components of the displacement tensor:*

$$D_{rr} = \partial_r u_r = S_{rr}, \qquad D_{\varphi\varphi} = r^{-1}(\partial_\varphi u_\varphi + u_r) = S_{\varphi\varphi}, \tag{4.27a and 4.27b}$$

$$D_{r\varphi} = r^{-1}(\partial_\varphi u_r - u_\varphi), \qquad D_{\varphi r} = \partial_r u_\varphi; \tag{4.27c and 4.27d}$$

(2) the diagonal components of the displacement and strain tensors coincide (Equations 4.27a and 4.27b); (3) the off-diagonal components of the strain tensor (Equation 4.28) are the arithmetic mean of the off-diagonal components of the displacement tensor (Equations 4.27c and 4.27d), and hence are symmetric:

$$2S_{r\varphi} = D_{r\varphi} + D_{\varphi r} = \partial_r u_\varphi + r^{-1}\,(\partial_\varphi u_r - u_\varphi) = 2S_{\varphi r}; \tag{4.28}$$

(4) the difference of the off-diagonal components of the displacement tensor (Equations 4.27c and 4.27d) specifies the curl of the displacement vector (Equation 4.29), which coincides (Equation 4.8c) with twice the rotation vector:

$$2\vec{E}_2 = \vec{e}_3\left(D_{\varphi r} - D_{r\varphi}\right) = \partial_r u_\varphi + r^{-1}u_\varphi - r^{-1}\partial_\varphi u_r = r^{-1}\partial_r\left(u_\varphi r\right) - r^{-1}\partial_\varphi u_r = \nabla\wedge\vec{u}; \tag{4.29}$$

and (5) the sum of the diagonal components of the displacement or strain tensors (Equations 4.27a and 4.27b) specifies (Equation 4.15a) the linearized relative area change or divergence of the displacement vector:

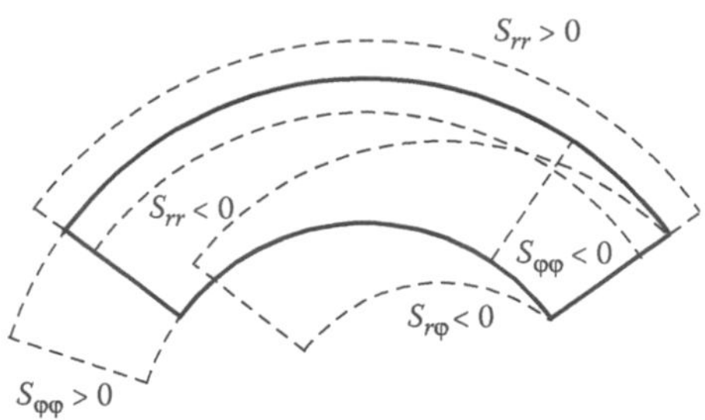

FIGURE 4.4
Cartesian (polar) components of the strain tensor are visualized best [Figure 4.2 (Figure 4.4)] for a rectangle or square (angular sector of a circle between two radii). The radial diagonal strain increases (decreases) the radius for $S_{rr} > 0$ ($S_{rr} < 0$). The azimuthal diagonal strain stretches (contracts) the circular sector if $S_{\varphi\varphi} > 0$ ($S_{\varphi\varphi} < 0$). The symmetric nondiagonal component $S_{r\varphi} = S_{\varphi r}$ distorts the angular sector outward (inward) if $S_{r\varphi} > 0$ ($S_{r\varphi} < 0$).

$$D_2 \sim D_{rr} + D_{\varphi\varphi} = S_{rr} + S_{\varphi\varphi} = \partial_r u_r + r^{-1} u_r + r^{-1}\partial_\varphi u_\varphi = r^{-1}\partial_r\left(u_r r\right) + r^{-1}\partial_\varphi u_\varphi = \nabla \cdot \vec{u}. \quad (4.30)$$

The curl (Equation 4.29) [divergence (Equation 4.30)] in polar coordinates coincides with Equation I.11.35b (Equation I.11.33b) obtained before from the Cauchy–Riemann conditions for a complex holomorphic function (Subsection I.11.7). They are a particular application of invariant differential operators in the tensor calculus. The latter also specifies the displacement and strain tensors using the covariant derivative. The strain tensor has its simplest presentations in Cartesian (Figure 4.2) and polar (Figure 4.4) coordinates. In Cartesian (polar) coordinates, the strain tensor [Equations 4.12a through 4.12d (Equations 4.27a, 4.27b, and 4.28)] involves only derivatives of displacements (also some displacements). Thus, in Cartesian coordinates, a constant displacement causes no strain. In polar coordinates, a constant displacement causes strains (Figure 4.4) because (1) a radial outward $u_r > 0$ (inward $u_r < 0$) displacement causes (Equation 4.27b) a tangential extension $S_{\varphi\varphi} > 0$ (contraction $S_{\varphi\varphi} < 0$); and (2) a positive (negative), that is, counterclockwise $u_\varphi > 0$ (clockwise $u_\varphi < 0$), displacement causes (Equation 4.28) a negative $S_{r\varphi} < 0$ (positive $S_{r\varphi} > 0$) distortion.

4.1.5 Compatibility Relation for the Strains

The three components of the strain tensor are obtained from the two components of the displacement vector by differentiation, for example, in Cartesian (Equations 4.12a through 4.12d) [polar (Equations 4.27a, 4.27b, and 4.28)] components and coordinates. The inverse process involves integrations and may have no solution for three arbitrary strains because there are only two displacements to satisfy three differential equations; a compatibility relation is needed to ensure that a solution exists. In other words, because the three components (S_{xx}, S_{yy}, S_{xy}) of the strain tensor derive (Equations 4.12a through 4.12d) from two components (u_x, u_y) of the displacement vector, they cannot be given independently and must be connected by a compatibility relation. To obtain the latter, consider the four differential coefficients (Equations 4.5a through 4.5d and 4.12a through 4.12d) of the displacement vector:

$$\partial_x u_x \equiv D_{xx} = S_{xx}, \quad \partial_y u_x \equiv D_{xy} = S_{xy} - E_2, \quad \partial_x u_y \equiv D_{yx} = S_{yx} + E_2, \quad \partial_y u_y \equiv D_{yy} = S_{yy}. \quad (4.31a\text{–}4.31d)$$

The compatibility conditions for the displacement vector that (Equations 4.32a and 4.32b) are exact differentials:

$$du_x = (\partial_x u_x)\,dx + (\partial_y u_x)\,dy = S_{xx}\,dx + (S_{xy} - E_2)\,dy, \tag{4.32a}$$

$$du_y = (\partial_x u_y)\,dx + (\partial_y u_y)\,dy = (S_{xy} + E_2)\,dx + S_{yy}\,dy \tag{4.32b}$$

are the equality of the cross-second-order derivatives:

$$0 = \partial_{yx}u_x - \partial_{xy}u_x = \partial_y S_{xx} - \partial_x S_{xy} + \partial_x E_2, \tag{4.33a}$$

$$0 = \partial_{yx}u_y - \partial_{xy}u_y = \partial_y S_{xy} + \partial_y E_2 - \partial_x S_{yy}. \tag{4.33b}$$

The conditions (Equations 4.33a and 4.33b) involve the strain tensor and the rotation; eliminating the rotation leads to a compatibility relation involving the strain tensor only.

To obtain the latter, Equations 4.33a and 4.33b are solved for the partial derivatives of the rotation scalar:

$$\partial_x E_2 = \partial_x S_{xy} - \partial_y S_{xx}, \qquad \partial_y E_2 = \partial_x S_{yy} - \partial_y S_{xy}. \tag{4.34a and 4.34b}$$

These must also form an exact differential:

$$dE_2 = (\partial_x E_2)\,dx + (\partial_y E_2)\,dy = (\partial_x S_{xy} - \partial_y S_{xx})\,dx + (\partial_x S_{yy} - \partial_y S_{xy})\,dy; \tag{4.34c}$$

therefore, the cross-derivatives must be equal:

$$0 = \partial_{yx}E_2 - \partial_{xy}E_2 = \partial_{yx}S_{xy} - \partial_{yy}S_{xx} - \partial_{xx}S_{yy} + \partial_{xy}S_{xy}. \tag{4.35}$$

This is the **compatibility relation** *linking the second-order derivatives of the plane strain tensor:*

$$\partial_{xx}S_{yy} + \partial_{yy}S_{xx} - 2\partial_{xy}S_{xy} = 0. \tag{4.36}$$

The linear relation (Equation 4.36) is both (1) sufficient, that is, if it is met, then Equations 4.32a, 4.32b, and 4.34c are all exact differentials, and a rotation scalar E_2 (displacement vector $\vec{u}$) exists, satisfying Equations 4.34a and 4.34b (Equations 4.31a through 4.31d); and (2) necessary, that is, the strain tensor defined by Equations 4.31a through 4.31d must satisfy Equation 4.36. The proof of sufficiency 1 was made before Equations 4.31 through 4.35 to deduce the compatibility relation (Equation 4.36); the proof of necessity 2 follows, substituting Equations 4.31a through 4.31d in Equation 4.36, which leads to an identity:

$$\partial_{xx}S_{yy} + \partial_{yy}S_{xx} - 2\partial_{xy}S_{xy} = \partial_{xxy}u_y + \partial_{yyx}u_x - \partial_{xyx}u_y - \partial_{xyy}u_x = 0. \tag{4.37}$$

The compatibility relation for the strain tensor in polar coordinates is given in Example 10.8.

4.2 Stress Vector, Tensor, and Function

The inclusion of the surface forces or the stress vector in the balance of inertia and volume forces (Subsection 4.2.1) leads to the stress tensor. In the absence of external forces, the three components of the stress tensor can be derived from a single scalar stress function (Subsection 4.2.2), either in Cartesian or in polar coordinates (Subsection 4.2.3).

4.2.1 Balance of Inertia, Volume, and Surface Forces

Considering an infinitesimal square (Figure 4.5), *the* **surface forces** *are (1) a* **traction (compression)** *in the x/y-direction if the force acts along the normal outward (inward); and (2) a* **shear stress** *if the force is tangential, that is, positive (negative) counterclockwise (clockwise). The* **momentum equation** *balances (1) the inertial force per unit volume, equal to mass density* ρ *times acceleration* $\vec{a}$*; (2) the density of force per unit volume* $\vec{f}$*; and (3) the variation of surface forces projected in the x-direction (Equation 4.38a) [y-direction (Equation 4.38b)]:*

$$\rho a_x = f_x + \partial_x T_{xx} + \partial_y T_{xy}, \qquad \rho a_y = f_y + \partial_x T_{yx} + \partial_y T_{yy}. \qquad \text{(4.38a and 4.38b)}$$

In a medium at rest or in uniform motion, the inertial force vanishes (Equation 4.39a), and if the external forces $\vec{f}$ *are uniform (Equation 4.39b) or absent, the momentum equation (Equations 4.38a and 4.38b) simplifies to Equations 4.39c and 4.39d:*

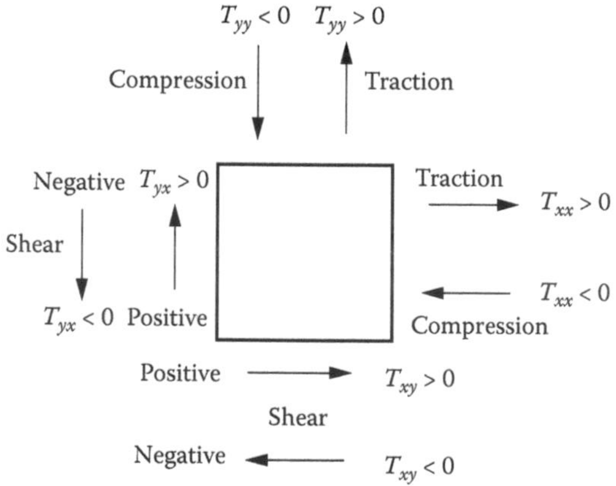

FIGURE 4.5
Cartesian components of the stress tensor consist (Figure 4.5) of (1) the diagonal components specifying a traction $T_{xx} > 0$ ($T_{yy} > 0$) in the $x(y)$-direction if positive and a compression if negative $T_{xx} < 0$ ($T_{yy} < 0$); (2) the balance of moments requires that the nondiagonal components be symmetric $T_{xy} = T_{yx}$ because they represent tangential forces, that is, the $x(y)$-direction for a surface with unit normal in the $y(x)$-direction. The similarity of Cartesian strains (stresses) suggests [Figure 4.2 (Figure 4.5)] that they are proportional for an isotropic elastic material. In addition, a normal stress causes an area change and affects the normal strains both in the same direction and the orthogonal direction, and vice versa for a normal strain. This is the reason for the existence of two elastic moduli, specifying the compression/expansion and shear properties of the elastic material.

$$\vec{a} = 0, \vec{f} = \text{const}: \qquad f_x + \partial_x T_{xx} + \partial_y T_{xy} = 0 = f_y + \partial_x T_{yx} + \partial_y T_{yy}. \qquad \text{(4.39a–4.39d)}$$

Equations 4.39c and 4.39d are satisfied by a **stress function**:

$$\partial_{xx} \equiv \frac{\partial^2}{\partial x^2}: \quad \left\{T_{xx}, T_{yy}, T_{xy} \equiv T_{yx}\right\} = \left\{\partial_{yy}\Theta, \partial_{xx}\Theta, -\partial_{xy}\Theta - yf_x - xf_y\right\}, \qquad \text{(4.40a–4.40c)}$$

as follows from

$$f_x + \partial_x T_{xx} + \partial_y T_{xy} = f_x + \partial_{xyy}\Theta - \partial_{yxy}\Theta - f_x = 0, \qquad \text{(4.41a)}$$

$$f_y + \partial_x T_{yx} + \partial_y T_{yy} = f_y - \partial_{xyx}\Theta - f_y + \partial_{yxx}\Theta = 0. \qquad \text{(4.41b)}$$

The stress tensor is symmetric (Equation 4.40c) and has only three independent components, as a consequence of the balance of moments counterclockwise T_{xy} and clockwise T_{yx} in Figure 4.5.

For an oblique facet with inclination θ in Figure 4.6 with unit normal (Equation 4.42a), the surface force or **stress vector** *is specified by Equations 4.42b and 4.42c:*

$$\vec{n} = (\sin\theta, \cos\theta): \qquad \begin{aligned} T_x &= T_{xx}\sin\theta + T_{xy}\cos\theta = T_{xx}n_x + T_{xy}n_y \\ T_y &= T_{yx}\sin\theta + T_{yy}\cos\theta = T_{yx}n_x + T_{yy}n_y, \end{aligned} \qquad \text{(4.42a–4.42c)}$$

in terms of the stress tensor. At a regular point of a boundary curve, there exists a unique unit tangent (Equation 4.43a) [normal (Equation 4.43b)] vector that is orthogonal (Equation 4.43c) where the arc length is used (Equation 4.21d):

$$\vec{t} = \left\{\frac{dx}{ds}, \frac{dy}{ds}\right\}, \quad \vec{n} = \left\{\frac{dy}{ds}, -\frac{dx}{ds}\right\}, \quad \vec{n}\cdot\vec{t} = 0. \qquad \text{(4.43a–4.43c)}$$

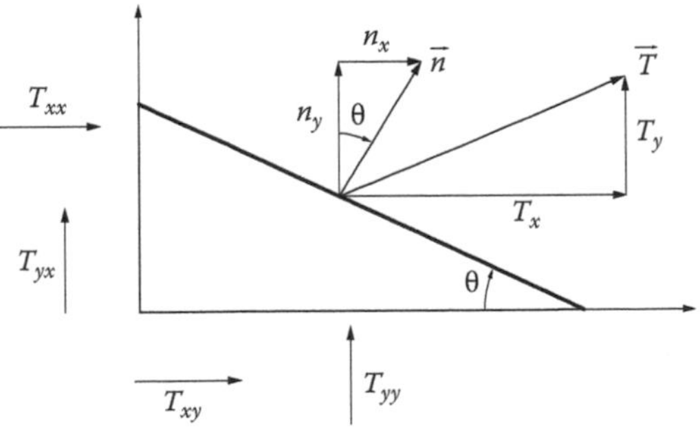

FIGURE 4.6
In the case of a facet or element of a surface at an arbitrary position relative to the coordinate axis, the force per unit area or stress vector need not lie along the unit outward normal vector. The linear relation between these two vectors specifies the stress tensor, whose components appear separately when the facets are parallel or orthogonal to the coordinate axis (Figure 4.5).

The components of the stress vector (Equations 4.42b and 4.42c) are given by

$$T_x = T_{xx}\frac{dy}{ds} - T_{xy}\frac{dx}{ds} = \frac{dy}{ds}\partial_{yy}\Theta + \frac{dx}{ds}\partial_{xy}\Theta = \frac{d}{ds}\left(\partial_y\Theta\right), \tag{4.44a}$$

$$T_y = T_{xy}\frac{dy}{ds} - T_{yy}\frac{dx}{ds} = -\frac{dy}{ds}\partial_{xy}\Theta - \frac{dx}{ds}\partial_{xx}\Theta = -\frac{d}{ds}\left(\partial_x\Theta\right) \tag{4.44b}$$

in terms of the stress function (Equations 4.40a through 4.40c). In deriving Equations 4.44a and 4.44b, it was taken into account that the volume forces are negligible on the boundary.

4.2.2 Stress Tensor and Momentum Balance in Polar Coordinates

The relation (Equations 4.42a through 4.42c) between the stress tensor and vector is an orthogonal projection, which holds equally in Cartesian (Figure 4.6) or orthogonal curvilinear and polar coordinates (e.g., Figure 4.7). The force balance equations (Equations 4.38a and 4.38b) are changed in polar coordinates and also the relation (Equations 4.40a through 4.40c) between the stress tensor and stress function. The right-hand side (r.h.s.) of Equations 4.38a and 4.38b is the divergence of the stress tensor, which can be interpreted as

$$\rho a_i - f_i = \sum_{j=1}^{2}\vec{e}_i\left(\nabla_j T_{ij}\right) = \sum_{j=1}^{2}\nabla_j\left(T_{ij}\vec{e}_i\right) - \sum_{j=1}^{2}T_{ij}\left(\nabla_j\vec{e}_i\right), \tag{4.45}$$

where $i, j = 1, 2 = r, \varphi$. In the second term on the r.h.s. of Equation 4.45, the gradient (Equation 4.46a) of the unit vectors is needed:

$$\nabla \equiv \vec{e}_r\frac{\partial}{\partial r} + \vec{e}_\varphi\frac{1}{r}\frac{\partial}{\partial\varphi}: \qquad \nabla\vec{e}_r \equiv \left(\vec{e}_r\partial_r + r^{-1}\vec{e}_\varphi\partial_\varphi\right)\vec{e}_r = r^{-1}\vec{e}_\varphi\partial_\varphi\vec{e}_r = r^{-1}\vec{e}_\varphi\vec{e}_\varphi, \tag{4.46a and 4.46b}$$

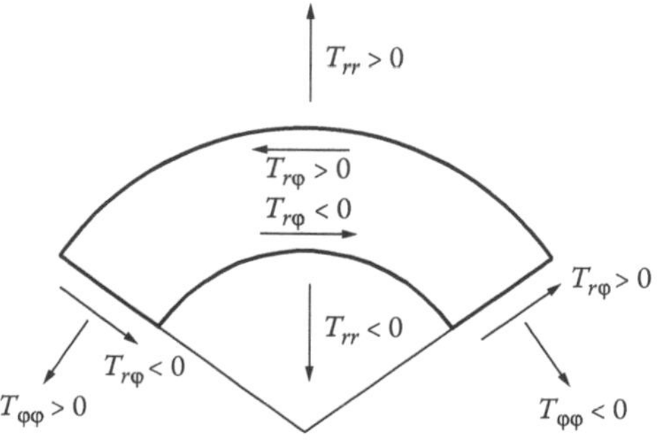

FIGURE 4.7
Cartesian (polar) components can be used both for the strain [Figure 4.2 (Figure 4.4)] and stress [Figure 4.6 (Figure 4.7)] tensors. As for the strains (Figure 4.4), the polar components of the stress tensor are best visualized for the angular sector of a circle between two radii. The radial diagonal stress expands (contracts) the "tube" if positive $T_{rr} > 0$ (negative $T_{rr} < 0$). The azimuthal diagonal stress causes azimuthal tractions (compressions) for $T_{\varphi\varphi} > 0$ ($T_{\varphi\varphi} < 0$). The balance of moments implies that the nondiagonal shear stress is symmetric $T_{r\varphi} = T_{\varphi r}$ because it represents the tangential force on a radial or azimuthal face T.

$$\nabla \vec{e}_\varphi = \left(\vec{e}_r \partial_r + r^{-1}\vec{e}_\varphi \partial_\varphi\right)\vec{e}_\varphi = r^{-1}\vec{e}_\varphi \partial_\varphi \vec{e}_\varphi = -r^{-1}\vec{e}_\varphi \vec{e}_r, \tag{4.46c}$$

where Equations 4.20a and 4.20b were used in Equations 4.46b and 4.46c in the form:

$$\partial_r \vec{e}_r = 0 = \partial_r \vec{e}_\varphi, \qquad \partial_\varphi \vec{e}_r = \vec{e}_\varphi, \qquad \partial_\varphi \vec{e}_\varphi = -\vec{e}_r. \tag{4.47a–4.47d}$$

Substituting Equations 4.47a through 4.47d in the second term on the r.h.s. of Equation 4.45 and interpreting the first term as the divergence (Equation 4.30) lead to Equations 4.48a and 4.48b:

$$\rho a_r - f_r = \nabla \cdot \left(T_{rr}\vec{e}_r + T_{r\varphi}\vec{e}_\varphi\right) - T_{\varphi\varphi}\left(\nabla \vec{e}_r\right)_{\varphi\varphi} = r^{-1}\partial_r\left(T_{rr} r\right) + r^{-1}\partial_\varphi T_{r\varphi} - r^{-1}T_{\varphi\varphi}, \tag{4.48a}$$

$$\rho a_\varphi - f_\varphi = \nabla \cdot \left(T_{\varphi r}\vec{e}_r + T_{\varphi\varphi}\vec{e}_\varphi\right) - T_{r\varphi}\left(\nabla \vec{e}_\varphi\right)_{\varphi r} = r^{-1}\partial_r\left(T_{\varphi r} r\right) + r^{-1}\partial_\varphi T_{\varphi\varphi} + r^{-1}T_{r\varphi}, \tag{4.48b}$$

where the last term on the left-hand side (l.h.s.) of Equation 4.48a (Equation 4.48b) involves only the nonvanishing components of Equations 4.46a through 4.46c. Simplifying Equations 4.48a and 4.48b leads to

$$\rho a_r = f_r + \partial_r T_{rr} + r^{-1}\,\partial_\varphi T_{r\varphi} + r^{-1}\,(T_{rr} - T_{\varphi\varphi}), \tag{4.49a}$$

$$\rho a_\varphi = f_\varphi + \partial_r T_{\varphi r} + r^{-1}\,\partial_\varphi T_{\varphi\varphi} + r^{-1}\,(T_{\varphi r} + T_{r\varphi}) \tag{4.49b}$$

the momentum equation balancing the inertia force against external forces and internal stresses in polar coordinates. In Cartesian (Equations 4.38a and 4.38b) [polar (Equations 4.49a and 4.49b)] coordinates, the volume forces can arise only from derivatives of stresses (also from the stresses). In particular, volume forces arise (1) radially if the difference of normal stresses per unit length of a circle is not zero and (2) tangentially if the shear stress is not zero.

4.2.3 Stress Function in Polar Coordinates (Michell 1900a)

In the absence of inertia and external forces (Equations 4.50a and 4.50b), the momentum equation (Equations 4.49a and 4.49b) involves only the polar components of the stress tensor (Equations 4.50c and 4.50e):

$$\vec{a} = 0 = \vec{f}: \qquad r\partial_r T_{rr} + \partial_\varphi T_{r\varphi} + T_{rr} - T_{\varphi\varphi} = 0, \tag{4.50a–4.50c}$$

$$T_{r\varphi} = T_{\varphi r}: \qquad r\partial_r T_{r\varphi} + \partial_\varphi T_{\varphi\varphi} + 2T_{r\varphi} = 0, \tag{4.50d and 4.50e}$$

assuming (Equation 4.50d) the balance of moments, as in Equation 4.40c. The stress function (Equations 4.40a through 4.40c) satisfies the stress balance (Equations 4.41a and 4.41b) in Cartesian coordinates in the absence of inertia (Equation 4.51a) and external (Equation 4.51b) forces by virtue of the identities (Equations 4.51c and 4.51d):

$$\vec{f} = 0 = \vec{a}: \qquad 0 = \partial_x T_{xx} + \partial_y T_{xy} = \partial_{xyy}\Theta - \partial_{yxy}\Theta, \qquad 0 = \partial_x T_{yx} + \partial_y T_{yy} = -\partial_{xyx}\Theta + \partial_{yxx}\Theta, \tag{4.51a–4.51d}$$

resulting from the symmetry of derivatives of the second or higher order, that is, third order in the present case. To obtain the stress function in polar coordinates, similar identities must be applied to Equations 4.50c and 4.50e, namely, (1) assume Equation 4.52a; (2) then Equation 4.50e can be written in the form (Equation 4.52b):

$$T_{\varphi\varphi} = \partial_{rr}\,\Theta: \quad -\partial_{\varphi rr}\Theta = -\partial_\varphi T_{\varphi\varphi} = r\partial_r T_{r\varphi} + 2T_{r\varphi} = r^{-1}\,\partial_r\,(r^2\,T_{r\varphi}); \qquad \text{(4.52a and 4.52b)}$$

(3) the latter (Equation 4.52b ≡ Equation 4.53a) is satisfied by Equation 4.53b:

$$-\partial_{rr\varphi}\Theta = r^{-1}\partial_r\,(r^2 T_{r\varphi}), \qquad T_{r\varphi} = -\partial_r\,(r^{-1}\partial_\varphi\,\Theta), \qquad \text{(4.53a and 4.53b)}$$

as follows from

$$r^{-1}\partial_r(r^2 T_{r\varphi}) = -\,r^{-1}\partial_r[r^2\partial_r(r^{-1}\partial_\varphi\,\Theta)] = -\,r^{-1}\partial_r[r\,\partial_{r\varphi}\,\Theta - \partial_\varphi\,\Theta] = -\,\partial_{rr\varphi}\,\Theta; \qquad (4.54)$$

(4) it remains to satisfy the first stress balance equation (Equation 4.50c), where Equations 4.52a and 4.53b can be substituted:

$$r\,\partial_r T_{rr} + T_{rr} = T_{\varphi\varphi} - \partial_\varphi T_{r\varphi} = \partial_{rr}\,\Theta + \partial_r\,(r^{-1}\partial_{\varphi\varphi}\,\Theta); \qquad (4.55)$$

(5) the latter (Equation 4.55 ≡ Equation 4.56a) leads to Equation 4.56b:

$$\partial_r\,(r\,T_{rr}) = \partial_r\,(\partial_r\,\Theta + r^{-1}\partial_{\varphi\varphi}\,\Theta), \qquad T_{rr} = r^{-1}\partial_r\,\Theta + r^{-2}\partial_{\varphi\varphi}\,\Theta. \qquad \text{(4.56a and 4.56b)}$$

Thus, *if the inertia or volume forces balance (Equation 4.57a), the polar components of the stress tensor are specified by the stress function (Equations 4.57b through 4.57d):*

$$\vec{f} = \rho\vec{a}: \quad \left\{T_{rr}, T_{\varphi\varphi}, T_{r\varphi} = T_{\varphi r}\right\} = \left\{r^{-1}\partial_r\Theta + r^{-2}\partial_{\varphi\varphi}\Theta,\ \partial_{rr}\Theta, -\partial_r(r^{-1}\partial_\varphi\Theta)\right\}. \qquad \text{(4.57a–4.57d)}$$

These relations can be compared with the Cartesian case (Equations 4.40a through 4.40c ≡ Equations 4.58b through 4.58d):

$$\vec{f} = \rho\vec{a}: \qquad \left\{T_{xx}, T_{yy}, T_{xy} = T_{yx}\right\} = \left\{\partial_{yy}\Theta, \partial_{xx}\Theta, -\partial_{xy}\Theta\right\}, \qquad \text{(4.58a–4.58d)}$$

when the inertia and external forces balance (Equation 4.57a ≡Equation 4.58a) or, in particular, are both zero (Equations 4.51a and 4.51b).

4.3 Elastic Energy and Moduli of a Material (Hooke 1678; Poisson 1829a; Lamé 1852)

The stresses (Section 4.2) cause deformations and strains (Section 4.1), where the stress–strain relations depend on the constitutive properties of the material (Subsections I.4.1 and I.4.2). In the case of elasticity, the stress–strain relations are specified by the Hooke law (Subsection 4.3.3) that is linear and homogeneous, and involves two elastic moduli.

Three alternative choices for the pair of elastic moduli are (1) the moduli of compression and shear (Subsections 4.3.1 and 4.3.2); (2) the Lamé moduli (Subsection 4.3.4); and (3) the Young's modulus and Poisson ratio (Subsection 4.3.5). The latter are most readily measured experimentally in a uniaxial traction (compression) test. The elastic moduli satisfy inequalities (Subsection 4.3.6) that ensure that the elastic energy is positive (Subsection 4.3.8); the elastic energy is defined as the work performed by the stresses in a deformation (Subsection 4.3.7). The compatibility condition can be expressed [Subsection 4.1.5 (Subsection 4.3.9)] in terms of strains (stresses for an isotropic elastic material).

4.3.1 Pressure, Volume Change, and Volume Modulus

The simplest **state of stress** *is isotropic and corresponds to an equal* **pressure** *in all directions acting inward:*

$$T_{xx} = T_{yy} = T_{33} = -p, \qquad T_{xy} = T_{y3} = T_{x3} = 0, \tag{4.59a and 4.59b}$$

where the stresses are considered in three dimensions, including the direction x_3 normal to the plane (x, y) consisting of three normal stresses (Equation 4.59a), that is, compressions/tractions in the $x,y,3$-directions; the three shear stresses in the planes (x, y), $(x,3)$, and $(y,3)$ vanish (Equation 4.59b). *The* **pressure** *may be defined more generally than in Equation 4.59a, but consistently with it, as minus the average of normal stresses:*

$$\nu \equiv -\frac{\partial p}{\partial D_3} > 0: \qquad -p = \frac{\tilde{T}_{xx} + \tilde{T}_{yy} + \tilde{T}_{33}}{3} = \nu D_3 = \nu\left(S_{xx} + S_{yy} + S_{33}\right), \tag{4.60a and 4.60b}$$

in which case it is proportional to the linearized relative volume change (Equation 4.60b) through the **volume modulus** *(Equation 4.60a). The latter must be positive so that a compression $p > 0$ (expansion $p < 0$) causes a reduction $D_3 < 0$ (increase $D_3 > 0$) in volume.* The linearized relative volume (Equation 4.61b) [area (Equation 4.15b ≡ Equation 4.61a)] change is the three- (two-) dimensional divergence of the displacement:

$$D_2 \sim S_{xx} + S_{yy}, \qquad D_3 \sim S_{xx} + S_{yy} + S_{33} = D_2 + S_{33} \tag{4.61a and 4.61b}$$

and involves the **trace** of the strain tensor, that is, the sum of its two (three) diagonal components.

4.3.2 Sliding Tensor and Modulus

The strain tensor in three dimensions has six components, that is, Equations 4.12a through 4.12d plus another three, namely, (1) an extension or contraction normal to (x, y)-plane (Equation 4.62a):

$$S_{33} = \partial_3 u_3, \qquad 2S_{x3} = \partial_3 u_x + \partial_x u_3, \qquad 2S_{y3} = \partial_3 u_y + \partial_y u_3; \tag{4.62a–4.62c}$$

and (2) two shear strains [Equation 4.62b (Equation 4.62c)] between the x-direction (y-direction) in the (x, y)-plane and the normal x_3-direction. The **sliding tensor** is the strain tensor without linearized volume change, so that (1) the shear strains are unaffected (Equation 4.63a):

$$\left\{\bar{S}_{xy}, \bar{S}_{x3}, \bar{S}_{y3}\right\} \equiv \left\{S_{xy}, S_{x3}, S_{y3}\right\}; \qquad \left\{\bar{S}_{xx}, \bar{S}_{yy}, \bar{S}_{33}\right\} \equiv \left\{S_{xx}, S_{yy}, S_{33}\right\} - \frac{1}{3}D_3; \tag{4.63a and 4.63b}$$

and (3) the normal strains (Equation 4.63b) subtract one-third of the linearized relative volume change (Equation 4.61b), so that *the sliding strain (Equation 4.63b ≡ Equation 4.64a) involves no volume change (Equation 4.64b):*

$$\left\{\bar{S}_{xx}, \bar{S}_{yy}, \bar{S}_{33}\right\} = \frac{1}{3}\left\{2S_{xx} - S_{yy} - S_{33}, 2S_{yy} - S_{xx} - S_{33}, 2S_{33} - S_{xx} - S_{yy}\right\}, \tag{4.64a}$$

$$\bar{D}_3 \equiv \bar{S}_{xx} + \bar{S}_{yy} + \bar{S}_{33} = S_{xx} + S_{yy} + S_{33} - D_3 = 0. \tag{4.64b}$$

This suggests that *each component of the stress, that is, traction/compression/shear stress, is proportional to the corresponding component of sliding tensor, namely (Equations 4.65b through 4.65g), extension/contraction/distortion:*

$$\mu > 0: \quad \left\{\bar{T}_{xx}, \bar{T}_{yy}, \bar{T}_{33}, \bar{T}_{xy}, \bar{T}_{x3}, \bar{T}_{y3}\right\} = 2\mu\left\{\bar{S}_{xx}, \bar{S}_{yy}, \bar{S}_{xy}, \bar{S}_{33}, \bar{S}_{x3}, \bar{S}_{y3}\right\} \tag{4.65a and 4.65b}$$

through the **shear modulus** *(Equation 4.65a) that is positive.*

4.3.3 Direct and Inverse Hooke Law

The Hooke law or stress–strain relation for an elastic medium states that the total stress is the sum of (1) those associated (Equation 4.65b) with the sliding tensor (Equations 4.63a and 4.63b); and (2) those due to volume changes (Equation 4.60b). It follows that (1) the shear stresses are proportional to the corresponding distortions (Equation 4.66b ≡ Equation 4.65b) through the shear modulus that is positive (Equation 4.65a ≡ Equation 4.66a):

$$\mu > 0: \quad \left\{T_{xy}, T_{x3}, T_{y3}\right\} = \left\{\bar{T}_{xy}, \bar{T}_{x3}, \bar{T}_{y3}\right\} = 2\mu\left\{S_{xy}, S_{x3}, S_{y3}\right\}; \tag{4.66a and 4.66b}$$

$$\begin{aligned}
\nu > 0: \quad \left\{T_{xx}, T_{yy}, T_{33}\right\} &= \left\{\bar{T}_{xx} + \tilde{T}_{xx}, \bar{T}_{yy} + \tilde{T}_{yy}, \bar{T}_{33} + \tilde{T}_{33}\right\} \\
&= 2\mu\left\{\bar{S}_{xx}, \bar{S}_{yy}, \bar{S}_{33}\right\} + \nu\left(S_{xx} + S_{yy} + S_{33}\right) \\
&= 2\mu\left\{S_{xx}, S_{yy}, S_{33}\right\} + \left(\nu - \frac{2\mu}{3}\right)\left(S_{xx} + S_{yy} + S_{33}\right) \\
&= \left(\nu + \frac{4\mu}{3}\right)\left\{S_{xx}, S_{yy}, S_{33}\right\} + \left(\nu - \frac{2\mu}{3}\right)\left\{S_{yy} + S_{33}, S_{xx} + S_{33}, S_{xx} + S_{yy}\right\};
\end{aligned} \tag{4.66c and 4.66d}$$

and (2) the normal stresses (tractions/compressions) are due (Equation 4.66d) to the corresponding (Equation 4.65b) normal strains (extensions/contractions) through the modulus of sliding (Equation 4.65a) plus the volume changes (Equation 4.62b) through the volume modulus, which is also positive (Equation 4.62b ≡ Equation 4.66c). The **direct (inverse) Hooke law** *specifies stresses from strains (Equations 4.66a through 4.66d) [vice versa, i.e., the strains from the stresses (Equations 4.67a through 4.67c)]:*

$$2\mu\,\{S_{xy}, S_{x3}, S_{y3}\} = \{T_{xy}, T_{x3}, T_{y3}\}, \tag{4.67a}$$

$$2\mu\left\{S_{xx}, S_{yy}, S_{33}\right\} = \left\{T_{xx}, T_{yy}, T_{33}\right\} - \left(1 - \frac{2\mu}{3\nu}\right)\frac{T_{xx} + T_{yy} + T_{33}}{3}, \tag{4.67b}$$

$$6\mu\left\{S_{xx}, S_{yy}, S_{33}\right\} = 2\left(1 + \frac{\mu}{3\nu}\right)\left\{T_{xx}, T_{yy}, T_{33}\right\} - \left(1 - \frac{2\mu}{3\nu}\right)\left\{T_{yy} + T_{33}, T_{xx} + T_{33}, +T_{xx} + S_{yy}\right\}. \tag{4.67c}$$

Concerning the derivation of Equations 4.67a through 4.67c from Equations 4.66b and 4.66d: (1) the identity (Equation 4.66b ≡ Equation 4.67a) is immediate; (2) from the direct Hooke law (Equation 4.66d) follows the relation between the trace or sum of diagonal components of the stress and strain tensors:

$$T_{xx} + T_{yy} + T_{33} = \left[2\mu + 3\left(\nu - \frac{2\mu}{3}\right)\right]\left(S_{xx} + S_{yy} + S_{33}\right) = 3\nu\left(S_{xx} + S_{yy} + S_{33}\right), \tag{4.68a}$$

that is, the relation (Equation 4.68a ≡ Equation 4.60b) between the pressure and linearized relative volume change (Equation 4.61b); (3) solving Equation 4.66d for the strains and using Equation 4.68a leads to

$$\begin{aligned} &2\mu\left\{S_{xx}, S_{yy}, S_{33}\right\} - \left\{T_{xx}, T_{yy}, T_{33}\right\} \\ &= -\left(\nu - \frac{2\mu}{3}\right)\left(S_{xx} + S_{yy} + S_{33}\right) = -\left(1 - \frac{2\mu}{3\nu}\right)\frac{T_{xx}, T_{yy} + T_{33}}{3}, \end{aligned} \tag{4.68b}$$

which coincides with Equation 4.68b ≡ Equation 4.67b; and (4) a rearrangement of the latter (Equation 4.67b) leads to Equation 4.67c.

4.3.4 First and Second Lamé Moduli

The direct (Equations 4.66b and 4.66d) and inverse (Equations 4.67a and 4.67b) Hooke laws are equivalent and involve the same five implicit assumptions: (1) the stresses vanish for vanishing strain, that is, there are no **residual stresses** remaining if deformed material is returned to the undeformed state; (2) the stress–strain relations are **linear**, thus excluding large deformations and nonlinear materials such as rubber; (3) the material is **isotropic**, that is, its properties do not depend on the direction and are specified by two scalars; (4) the two elastic moduli do not depend on position, so that the material is **homogeneous**; and (5) the two elastic moduli do not depend on time, so that the material is steady or has no **memory**. Assumptions 1 and 2 imply that the relation between stresses and strains is linear without a constant term; assumptions 3–5 reduce the elastic properties of the material to two scalars that depend neither on position nor on time, and thus are two elastic constants. The three most common choices of scalars are (1) the moduli of volume (Equation 4.60a ≡ Equation 4.66c) and sliding (Equation 4.65a ≡ Equation 4.66a), which are both positive; (2) the Young's modulus and Poisson ratio (Subsection 4.3.5); (and 3) *the* **Lamé moduli** *that simplify the direct Hooke law (Equation 4.66d) to Equations 4.69b and 4.69c:*

$$\lambda \equiv \nu - \frac{2\mu}{3}: \quad \left\{T_{xx}, T_{yy}, T_{33}\right\} = 2\mu\left\{S_{xx}, S_{yy}, S_{33}\right\} + \lambda\left(S_{xx} + S_{yy} + S_{33}\right),$$
$$= (\lambda + 2\mu)\left\{S_{xx}, S_{yy}, S_{33}\right\} + \lambda\left\{S_{yy} + S_{33}, S_{xx} + S_{33}, S_{xx} + S_{yy}\right\}, \tag{4.69a–4.69c}$$

by (1) introducing the first Lamé modulus (Equation 4.69a), and (2) taking the shear modulus as the second Lamé modulus (Equation 4.65a ≡ Equation 4.66a) leaves the rest of the direct Hooke law unchanged (Equation 4.66b). The inverse Hooke law becomes (Equations 4.67a and 4.70):

$$(2\mu + 3\lambda)2\mu\left\{S_{xx}, S_{yy}, S_{33}\right\} = (2\mu + 3\lambda)\left\{T_{xx}, T_{yy}, T_{33}\right\} - \lambda\left(T_{xx} + T_{yy} + T_{33}\right),$$
$$= 2(\mu + \lambda)\left\{T_{xx}, T_{yy}, T_{33}\right\} - \lambda\left\{T_{yy} + T_{33}, T_{xx} + T_{33}, T_{zz} + T_{yy}\right\},$$
$$\text{(4.70a and 4.70b)}$$

in terms of the Lamé moduli. The latter result is obtained as for Equations 4.67c and 4.67d, that is, (1) the traces of the stress and strain tensors are related (Equation 4.69b) by Equation 4.71a:

$$T_{xx} + T_{yy} + T_{33} = (2\mu + 3\lambda)\,(S_{xx} + S_{yy} + S_{33}); \tag{4.71a}$$

$$2\mu\left\{S_{xx}, S_{yy}, S_{zz}\right\} - \left\{T_{xx}, T_{yy}, T_{33}\right\} = -\lambda\left(S_{xx} + S_{yy} + S_{33}\right)$$
$$= -\frac{\lambda}{2\mu + 3\lambda}\left(T_{xx} + T_{yy} + T_{33}\right); \tag{4.71b}$$

(2) substitution of Equation 4.71a in Equation 4.69b leads to Equation 4.71b, which coincides with Equation 4.70a ≡ Equation 4.71b; and (3) from Equation 4.69b follows Equation 4.69c by a rearrangement of terms.

4.3.5 Uniaxial Traction/Compression Test

A pair of elastic moduli alternative to the (1) moduli of volume and sliding and (2) Lamé moduli, and (3) more readily measurable by experiment, are associated with *the* **uniaxial traction (compression) test** *in Figure 4.8a (Figure 4.8b). A long thin rod is subject to a traction (compression) as the only stress (Equation 4.72a), resulting in (1) a longitudinal extension (contraction), whose ratio to the stress specifies the* **Young's modulus** *(Equation 4.72b):*

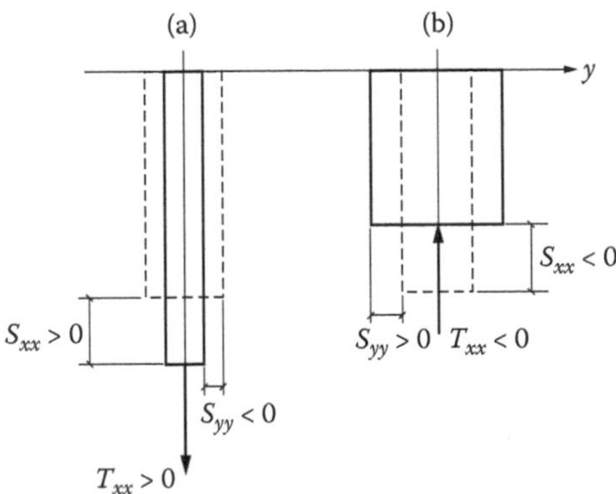

FIGURE 4.8
Two elastic moduli of a material can both be determined in a single uniaxial traction (a) [compression (b)] test: (1) the only applied force is an axial traction (compression), hence the name of the test; (2) the ratio to the longitudinal strain specifies the Young's modulus E, which is positive, because a traction (compression) causes an extension (contraction); (3) the ratio of the transversal to the longitudinal strain with minus sign specifies the Poisson ratio, which is positive if an axial extension (contraction) causes a transversal contraction (extension). The latter (item 3) is the case for most natural materials, although some man-made structures can behave otherwise, for example, expanding transversely under an axial traction.

$$T_{33} \neq 0: \qquad E \equiv \frac{T_{33}}{S_{33}}; \qquad \sigma = -\frac{S_{xx}}{S_{33}} = -\frac{S_{yy}}{S_{33}}; \tag{4.72a–4.72c}$$

and (2) a transverse contraction (extension), whose ratio specifies the **Poisson ratio** *(Equation 4.72c).* The Poisson ratio (Equation 4.72c ≡ Equation 4.73c) follows from Equation 4.67c with Equations 4.73a and 4.73b:

$$T_{xx} = 0 = T_{yy}: \qquad \sigma = -\frac{S_{xx}}{S_{33}} = \frac{1-2\mu/3\nu}{2(1+\mu/3\nu)} = \frac{3\nu-2\mu}{6\nu+2\mu}, \tag{4.73a–4.73c}$$

$$E = \frac{T_{33}}{S_{33}} = 3\nu\frac{S_{xx}+S_{yy}+S_{33}}{S_{33}} = 3\nu\left(1+2\frac{S_{xx}}{S_{33}}\right) = 3\nu(1-2\sigma), \tag{4.73d}$$

and from Equation 4.68a the Young's modulus (Equation 4.72b ≡ Equation 4.73d). Because the volume (Equation 4.60a ≡ Equation 4.66c ≡ Equation 4.74a) and sliding (Equation 4.65a ≡ Equation 4.66a ≡ Equation 4.74b) moduli are both positive, it follows from Equations 4.73c and 4.73d that the Young's modulus (Poisson ratio) satisfies the inequalities [Equation 4.74c (Equation 4.74d)]:

$$\nu > 0 < \mu: \qquad -1 \le \sigma = \frac{3\nu-2\mu}{6\nu+2\mu} \le \frac{1}{2}, \qquad E = 3\nu(1-2\sigma) \ge 0. \tag{4.74a–4.74d}$$

The Young's modulus (Equation 4.74d) is positive, implying that all materials extend [contract under traction (compression)]. The positive range $0 \le \sigma \le 1/2$ of the Poisson ratio

(Equation 4.74c) implies that a longitudinal extension (contraction) corresponds to a transversal (contraction) extension up to one-half. The negative range $-1 \le \sigma < 0$ of the Poisson ratio would imply that a longitudinal extension (contraction) would correspond to a transversal extension (contraction) up to the same amount. Most natural materials do not behave this way in the uniaxial traction/compression test, although artificial structures can be conceived with this property. With the latter exception, the Poisson ratio varies in the range $0 \le \sigma \le 1/2$.

4.3.6 Young's Modulus (1678) and Poisson Ratio (1829a)

Solving Equations 4.74c and 4.74d *specifies the volume (Equation 4.75a) and shear (Equation 4.75b) moduli in terms of the Young's modulus (1878) and Poisson ratio (1829a):*

$$3\nu = \frac{E}{1-2\sigma}, \qquad 2\mu = \frac{E}{1+\sigma}; \qquad \lambda = \frac{E}{1+\sigma}\frac{\sigma}{1-2\sigma}; \tag{4.75a–4.75c}$$

the first Lamé parameter (Equation 4.69a) then follows (Equation 4.75c). These relations obtained the following: (1) the identity (Equation 4.73d ≡ Equation 4.75a) is immediate; (2) solving Equation 4.73c for μ and substituting ν from Equation 4.75a leads to Equation 4.75d ≡ Equation 4.75b:

$$2\mu\,(1+\sigma) = 3\nu - 6\nu\sigma = 3(1-2\sigma)\nu = E; \tag{4.75d}$$

$$\lambda = \nu - \frac{2\mu}{3} = \frac{E}{3}\left(\frac{1}{1-2\sigma} - \frac{1}{1+\sigma}\right) = \frac{E}{3}\frac{3\sigma}{(1+\sigma)(1-2\sigma)}; \tag{4.75e}$$

and (3) substituting Equations 4.75a and 4.75b in Equation 4.69a yields Equation 4.75a ≡ Equation 4.75c. Using Equation 4.75a (Equations 4.75a and 4.75c) in the direct Hooke law [Equation 4.66b (Equation 4.69b)] leads to Equation 4.76a (Equation 4.76b ≡ Equation 4.76c):

$$\left\{T_{xy}, T_{x3}, T_{y3}\right\} = \frac{E}{1+\sigma}\left\{S_{xy}, S_{x3}, S_{y3}\right\}, \tag{4.76a}$$

$$\frac{1+\sigma}{E}\left\{T_{xx}, T_{yy}, T_{33}\right\} = \left\{S_{xx}, S_{yy}, S_{33}\right\} + \frac{\sigma}{1-2\sigma}\left(S_{xx} + S_{yy} + S_{zz}\right), \tag{4.76b}$$

$$\frac{(1+\sigma)(1-2\sigma)}{E}\left\{T_{xx}, T_{yy}, T_{33}\right\} = (1-\sigma)\left\{S_{xx}, S_{yy}, S_{33}\right\} + \sigma\left\{S_{yy} + S_{33}, S_{xx} + S_{33}, S_{xx} + S_{yy}\right\}, \tag{4.76c}$$

with the Young's modulus E and Poisson ratio σ as material constitutive properties. From Equation 4.76b follows

$$-3p = T_{xx} + T_{yy} + T_{33} = \frac{E}{1+\sigma}\left(1 + \frac{3\sigma}{1-2\sigma}\right)\left(S_{xx} + S_{yy} + S_{33}\right) = \frac{E}{1-2\sigma}\left(S_{xx} + S_{yy} + S_{33}\right), \tag{4.77}$$

that can be used in the direct Hooke law (Equations 4.76a through 4.76c), which then leads to the inverse Hooke law (Equations 4.78a through 4.78c):

$$E\{S_{xy}, S_{x3}, S_{y3}\} = (1+\sigma)\{T_{xy}, T_{x3}, T_{y3}\}, \tag{4.78a}$$

$$\begin{aligned} E\{S_{xx}, S_{yy}, S_{33}\} &= (1+\sigma)\{T_{xx}, T_{yy}, T_{33}\} - \sigma(T_{xx} + T_{yy} + T_{33}) \\ &= \{T_{xx}, T_{yy}, T_{33}\} - \sigma\{T_{yy} + T_{33}, T_{xx} + T_{33}, T_{xx} + T_{yy}\}. \end{aligned} \qquad \text{(4.78b and 4.78c)}$$

The latter is the simplest form of the Hooke law (Equations 4.78a and 4.78c). The value of the Poisson ratio $\sigma = 1/2$ corresponds to incompressibility, that is, an infinite pressure is needed (Equation 4.77) to effect a volume change. Thus, *the direct (inverse) Hooke law relating strains to stresses (vice versa) in terms of a pair of elastic moduli has been obtained (1) using [Equations 4.76a through 4.76c (Equations 4.78a through 4.78c)] the Young's modulus (Equation 4.72b) and Poisson ratio (Equation 4.72c) measured experimentally in a uniaxial traction (compression) test [Figure 4.8a (Figure 4.8b)]; (2) using [Equations 4.66b and 4.66d (Equations 4.67a through 4.67c)] the shear (Equation 4.65a ≡ Equation 4.66a ≡ Equation 4.75b) and volume (Equation 4.60a ≡ Equation 4.66c ≡ Equation 4.75a) moduli; and (3) using [Equations 4.66a, 4.69b, and 4.69c (Equations 4.67a, 4.71a, and 4.71b)] the first (Equation 4.69a ≡ Equation 4.75c) and second (Equation 4.65a ≡ Equation 4.66a ≡ Equation 4.75b) Lamé moduli. These also appear in the relation (Equation 4.60b ≡ Equation 4.68a ≡ Equation 4.71a ≡ Equation 4.77) between the pressure and the linearized relative volume change.*

4.3.7 Work of Deformation Associated with Displacement

The **work of deformation** is the work performed by the applied minus the inertia forces in a displacement (Equation 4.79a):

$$\delta W_d = (\vec{f} - \rho\vec{a})\cdot\delta\vec{u} = -(\partial_x T_{xx} + \partial_y T_{xy})\delta u_x - (\partial_x T_{yx} + \partial_{yy} T_y)\delta u_y, \qquad \text{(4.79a and 4.79b)}$$

which can be expressed in terms of the stresses (Equation 4.79b) using the momentum equation (Equations 4.38a and 4.38b). Integrating by parts the first term on the r.h.s. of Equation 4.79b leads to

$$-(\partial_x T_{xx} + \partial_y T_{xy})\,\delta u_x = -\partial_x(T_{xx}\,\delta u_x) - \partial_y(T_{xy}\,\delta u_y) + T_{xx}\,\partial_x(\delta u_x) + T_{xy}\,\partial_y(\delta u_x); \tag{4.80}$$

proceeding likewise for the second term leads to

$$\delta W_d = T_{xx}\,\partial_x(\delta u_x) + T_{yy}\,\partial_y(\delta u_y) + T_{xy}\,[\partial_x(\delta u_y) + \partial_y(\delta u_x)] + \delta W_0 \tag{4.81}$$

where the last term on the r.h.s. of Equation 4.81 may be integrated over a domain:

$$\begin{aligned} -\int_D \delta W_0\,dx\,dy &\equiv \int_D \left[\partial_x(T_{xx}\delta u_x + T_{xy}\delta u_y) + \partial_y(T_{xy}\delta u_x + T_{yy}\delta u_y)\right]dx\,dy \\ &= \int_{\partial D} \left\{(T_{xx}\delta u_x + T_{xy}\delta u_y)dx + (T_{xy}\delta u_x + T_{yy}\delta u_y)dy\right\} \end{aligned} \tag{4.82}$$

using the divergence theorem (Equation I.28.1b) to transform into an integral over the regular boundary. Because the displacements are fixed on the boundary, their variation vanishes (Equation 4.83a) there and so does the term (Equation 4.82 ≡ Equation 4.83b) in Equation 4.81. The strains (Equations 4.12a through 4.12d) may be substituted in the remaining terms of Equation 4.81, which leads to Equation 4.83c:

$$\delta\vec{u}\,|_{\partial D}= 0: \qquad \delta W_0 = 0, \qquad \delta W_d = T_{xx}\delta S_{xx} + T_{yy}\delta S_{yy} + 2T_{xy}\delta S_{xy}, \qquad \text{(4.83a–4.83c)}$$

which specifies the work of deformation in terms of stresses and strains. Thus, *the work of the applied minus inertia forces in a displacement (Equation 4.79a) in a domain with fixed boundaries (Equation 4.83a) equals the sum of the stresses multiplied by the corresponding strains (Equation 4.83c). The stresses appear in symmetric form like the strain (Equation 4.12c ≡ Equation 4.12d) and specify the work of deformation of the material (Equation 4.83c). In the case of an isotropic pressure (Equations 4.59a and 4.59b) corresponding to Equations 4.84a through 4.84c in two dimensions, the work (Equation 4.84d) is the product of minus the pressure times the linearized relative area change (Equation 4.15b):*

$$T_{xx} = -p = T_{yy}, T_{xy} = 0: \qquad \delta W_d = -p\,(S_{xx} + S_{yy}) = -p\,\mathrm{d}D_2. \qquad \text{(4.84a–4.84d)}$$

In three dimensions (Equations 4.85a through 4.85c), the linearized relative area change (Equation 4.15b ≡ Equation 4.61a) would be replaced in the work (Equation 4.85d):

$$-p = T_{33}\, T_{x3} = T_{y3} = 0: \qquad -p\,(S_{xx} + S_{yy} + S_{33}) = -p\,\mathrm{d}D_3. \qquad \text{(4.85a–4.85d)}$$

by the linearized relative volume change (Equation 4.61b). For an elastic material, the stresses and strains are related linearly, which leads to a quadratic work of deformation. The latter specifies the elastic energy, which is positive, as shown next (Subsection 4.3.8) in the two-dimensional case.

4.3.8 Elastic Energy in Terms of Strains and Stresses

The work of deformation is the exact differential of the **elastic energy** (Equation 4.86a):

$$\delta W_d = \mathrm{d}E_d: \qquad 2E_d = T_{xx}\, S_{xx} + T_{yy}\, S_{yy} + T_{xy}\, S_{xy} + T_{yx}\, S_{yx}; \qquad \text{(4.86a and 4.86b)}$$

for example, in the two-dimensional case: (1) it equals one-half the sum of the products of the corresponding strains and stresses (Equation 4.86b); (2 and 3) the factor one-half appears because it is a quadratic function [Equation 4.87a (Equation 4.87b)] of the strains (stresses), using the direct (Equations 4.76a and 4.76c) [inverse (Equations 4.78a and 4.78c)] Hooke law:

$$2\frac{1+\sigma}{E}E_d = \left(S_{xy}\right)^2 + \left(S_{yx}\right)^2 + \frac{1}{1-2\sigma}\left\{(1-\sigma)\left[\left(S_{xx}\right)^2 + \left(S_{yy}\right)^2\right] + 2\sigma S_{xx}S_{yy}\right\}, \qquad \text{(4.87a)}$$

$$2E\,E_d = (1+\sigma)\,[(T_{xy})^2 + (T_{yx})^2] + (T_{xx})^2 + (T_{yy})^2 - 2\sigma\, T_{xx}\, T_{yy}; \qquad \text{(4.87b)}$$

(4) the derivatives of the elastic energy (Equation 4.87a) with regard to the strains are (Equations 4.76a and 4.76c) the stresses (Equations 4.88a through 4.88c):

$$\begin{aligned}\left\{T_{xx}, T_{yy}, T_{xy} = T_{yx}\right\} &= \left\{\frac{\partial E_d}{\partial S_{xx}}, \frac{\partial E_d}{S_{yy}}, \frac{\partial E_d}{\partial S_{xy}} = \frac{\partial E_d}{\partial S_{yx}}\right\} \\ &= \frac{E}{1+\sigma}\left\{\frac{(1-\sigma)S_{xx}+\sigma S_{yy}}{1-2\sigma}, \frac{(1-\sigma)S_{yy}+\sigma S_{xx}}{1-2\sigma}, S_{xy} = S_{yx}\right\};\end{aligned} \tag{4.88a–4.88c}$$

(5) vice versa, the derivatives of the elastic energy (Equation 4.87b) with regard to the stresses are (Equations 4.78a and 4.78c) the strains (Equations 4.89a through 4.89c):

$$\begin{aligned}\left\{S_{xx}, S_{yy}, S_{xy} = S_{yx}\right\} &= \left\{\frac{\partial E_d}{\partial T_{xx}}, \frac{\partial E_d}{T_{yy}}, \frac{\partial E_d}{\partial T_{xy}} = \frac{\partial E_d}{\partial T_{yx}}\right\} \\ &= \frac{1}{E}\left\{T_{xx} - \sigma T_{yy}, T_{yy} - \sigma T_{xx}, (1+\sigma)T_{xy}\right\};\end{aligned} \tag{4.89a–4.89c}$$

and (6) the elastic energy is always positive:

$$2\frac{1+\sigma}{E}E_d = 2\left(S_{xy}\right)^2 + \left(S_{xx}\right)^2 + \left(S_{yy}\right)^2 + \frac{\sigma}{1-2\sigma}\left(S_{xx}+S_{yy}\right)^2 \ge 0, \tag{4.90a}$$

$$2EE_d = 2(1+\sigma)T_{xy}^2 + \left(T_{xx} - \sigma T_{yy}\right)^2 + T_{yy}^2\left(1-\sigma^2\right) \ge 0 \tag{4.90b}$$

for nonzero strains (Equation 4.90a) [stresses (Equation 4.90b)], bearing in mind the inequalities (Equation 4.74c) satisfied by the Poisson ratio.

The result [Equation 4.90a (Equation 4.90b)] follows from Equation 4.87a (Equation 4.87b) via a rearrangement of terms [Equation 4.90c ≡ Equation 4.90a (Equation 4.90d ≡ Equation 4.90b)]:

$$\begin{aligned}&2\frac{1+\sigma}{E}E_d - 2\left(S_{xy}\right)^2 - \left(S_{xx}\right)^2 - \left(S_{yy}\right)^2 \\ &= \left[\left(S_{xx}\right)^2 + \left(S_{yy}\right)^2\right]\left(\frac{1-\sigma}{1-2\sigma} - 1\right) + \frac{2\sigma}{1-2\sigma}S_{xx}S_{yy} \\ &= \frac{\sigma}{1-2\sigma}\left[\left(S_{xx}\right)^2 + \left(S_{yy}\right)^2 + 2S_{xx}S_{yy}\right] = \frac{\sigma}{1-2\sigma}\left(S_{xx}+S_{yy}\right)^2,\end{aligned} \tag{4.90c}$$

$$\begin{aligned}2EE_d - 2(1+\sigma)\left(T_{xy}\right)^2 &= \left(T_{xx}\right)^2 + \left(T_{yy}\right)^2 - 2\sigma T_{xx}T_{yy} \\ &= \left(T_{xx} - \sigma T_{yy}\right)^2 - \sigma^2\left(T_{yy}\right)^2 + \left(T_{yy}\right)^2.\end{aligned} \tag{4.90d}$$

As an alternative to Equation 4.87a, the elastic energy (Equation 4.86b) can be calculated from Equations 4.66b and 4.66d as follows:

$$\begin{aligned}2E_d &= \bar{T}_{xx}S_{xx} + \bar{T}_{yy}S_{xy} + 2\bar{T}_{xy}S_{xy} + \tilde{T}_{xx}S_{xx} + \tilde{T}_{yy}S_{yy} \\ &= 2\mu\left[S_{xx}\bar{S}_{xx} + S_{yy}\bar{S}_{yy} + \left(S_{xy}\right)^2\right] + \nu\left(S_{xx} + S_{yy}\right)^2\end{aligned} \tag{4.91a}$$

using the two-dimensional (Equation 4.91b) analog of Equation 4.63b, that is,

$$\left\{S_{xx}, S_{yy}\right\} = \left\{\bar{S}_{xx}, \bar{S}_{yy}\right\} + \frac{1}{2}D_2 = \left\{S_{xx}, S_{yy}\right\} + \frac{1}{2}\left(S_{xx} + S_{yy}\right). \tag{4.91b}$$

From Equation 4.91a, it also follows that the elastic energy is nonnegative:

$$2E_d = 2\mu\left[\left(S_{xx}\right)^2 + \left(S_{yy}\right)^2 + \left(S_{xy}\right)^2\right] + (\nu + \mu)\left(S_{xx} + S_{yy}\right)^2 \geq 0, \tag{4.91c}$$

bearing in mind the inequalities for the shear (Equation 4.65a) and volume (Equation 4.60a) moduli of elasticity. Thus, *the elastic energy (Equation 4.91c) is quadratic in the sliding tensor* $\bar{S}_{ij}$ *in Equation 4.91b and in the linearized relative area change (Equation 4.15b ≡ Equation 4.61a) with the modulus of shear (Equation 4.66a) [volume (Equation 4.66c)] as a coefficient. The condition that the latter is positive is equivalent to the condition that the elastic energy is positive for nonzero stress.* This is the same result as Equation 4.90a ≡ Equation 4.90b, which was obtained via the Young's modulus and Poisson ratio. This justifies the assumptions [Equation 4.60a ≡ Equation 4.66a (Equation 4.65a ≡ Equation 4.66c)] made [Subsection 4.3.1 (Subsection 4.3.2)] about the constitutive properties of an elastic material, that is, (1) an isotropic compression (traction) decreases (increases) the area; and (2) a positive (negative) shear stress causes a positive (negative) shear strain.

4.3.9 Compatibility Equation for Isotropic Elastic Stress

Using the inverse Hooke law (Equations 4.78a through 4.78c) in the compatibility relation for the strain tensor (Equation 4.36):

$$\begin{aligned}0 &= \frac{1}{E}\left(\partial_{xx}S_{yy} + \partial_{yy}S_{xx} - 2\partial_{xy}S_{xy}\right) \\ &= (1+\sigma)\left(\partial_{xx}T_{yy} + \partial_{yy}T_{xx} - 2\partial_{xy}T_{xy}\right) - \sigma\left(\partial_{xx} + \partial_{yy}\right)\left(T_{xx} + T_{yy}\right)\end{aligned} \tag{4.92a}$$

leads to

$$\partial_{xx}T_{yy} + \partial_{yy}T_{xx} - 2\,\partial_{xy}T_{xy} = \sigma\,(\partial_{xx}T_{xx} + \partial_{yy}T_{yy} + 2\,\partial_{xy}T_{xy}). \tag{4.92b}$$

Thus, *the compatibility relation for the stress tensor of an isotropic elastic material satisfying the Hooke law (Equation 4.92b) involves only the Poisson ratio and not the Young's modulus.* The compatibility relation for the stress tensor in polar coordinates appears in Example 10.8.

Substituting in Equation 4.92b the stress tensor in terms of the stress function (Equations 4.40a through 4.40c) yields

$$\begin{aligned}\nabla^4\Theta &\equiv \left(\partial_{xx}+\partial_{yy}\right)^2\Theta = \partial_{xxxx}\Theta+\partial_{yyyy}\Theta+2\partial_{xxyy}\Theta = \partial_{xx}T_{yy}+\partial_{yy}T_{xx}-2\partial_{xy}T_{xy}\\ &= \sigma\left(\partial_{xx}T_{xx}+\partial_{yy}T_{yy}+2\partial_{xy}T_{xy}\right) = \sigma\left(\partial_{xxyy}\Theta+\partial_{yyxx}\Theta-2\partial_{xyxy}\Theta\right)=0.\end{aligned} \tag{4.92c}$$

Thus, the compatibility relation for the stress tensor is automatically satisfied if the stress function satisfies a biharmonic equation (Equation 4.92c ≡ Equation 4.181b). An independent proof of the latter will be given in the sequel (Equation 4.106e). In an inviscid fluid, the stresses reduce to a pressure (Equations 4.84a through 4.84d and 4.85a through 4.85d); the viscous stresses are added in a viscous fluid. The viscous stresses are related to the rates of strain (Equations 4.17a and 4.17b) in a Newtonian fluid (Note 4.4), in the same way as the stresses are related to strains by the Hooke law (Equations 4.66b and 4.66d) for an elastic material. The modulus of shear (Equation 4.66a) corresponds to the shear viscosity (Equation 4.333a), and the modulus of volume (Equation 4.66c) to the bulk viscosity (Equation 4.333b). Because the inertia and external forces generally do not balance in a fluid, there is no stress function for a plane flow. The compatibility relations apply both to (1) the strain (rate-of-strain) tensor (Equation 4.36) and (2) the stress (viscous stress) tensor (Equation 4.92b). These and other fundamental equations of solids and fluids are compared in Note 4.2, in particular, for elastic solids (Newtonian viscous fluids) in Note 4.3. The starting point is the momentum equation, considered next for a solid (Section 4.4).

4.4 Momentum Equation for Isotropic Elasticity

The consideration of three-dimensional stress–strain relations (Section 4.3) is necessary to perform the simplification to plane elasticity (Subsection 4.4.1) of the momentum balance equations involving the (1 and 2) stress and strain tensors (Subsection 4.4.2); (3 and 4) displacement vector and tensor; (5 and 6) rotation and linearized relative volume change; and (7) stress function (Subsection 4.4.3). The balance of stresses and inertia and external forces leads to the momentum equation in Cartesian (polar) coordinates [Subsection 4.4.2 (Subsection 4.4.4)]. Comparison of the latter specifies in polar coordinates (Subsection 4.4.5) the invariant differential operators, namely, (1) the gradient and Laplacian of a scalar; and (2) the divergence, curl, and Laplacian of a vector. These results (Subsections 4.4.1 through 4.4.5) suggest three main methods of solution (Subsection 4.4.6) of problems in elasticity (Diagram 4.1).

4.4.1 Plane Elasticity and Normal Out-of-Plane Stress

The displacement vector in plane elasticity (Equation 4.2b) implies that (1) only the three components of the strain tensor in the plane (Equations 4.12a through 4.12d) are generally nonzero; (2) the out-of-plane distortions (Equations 4.63b and 4.63c) as well as the out-of-plane extensions and contractions (Equation 4.63a) all vanish (Equations 4.93a through 4.93c):

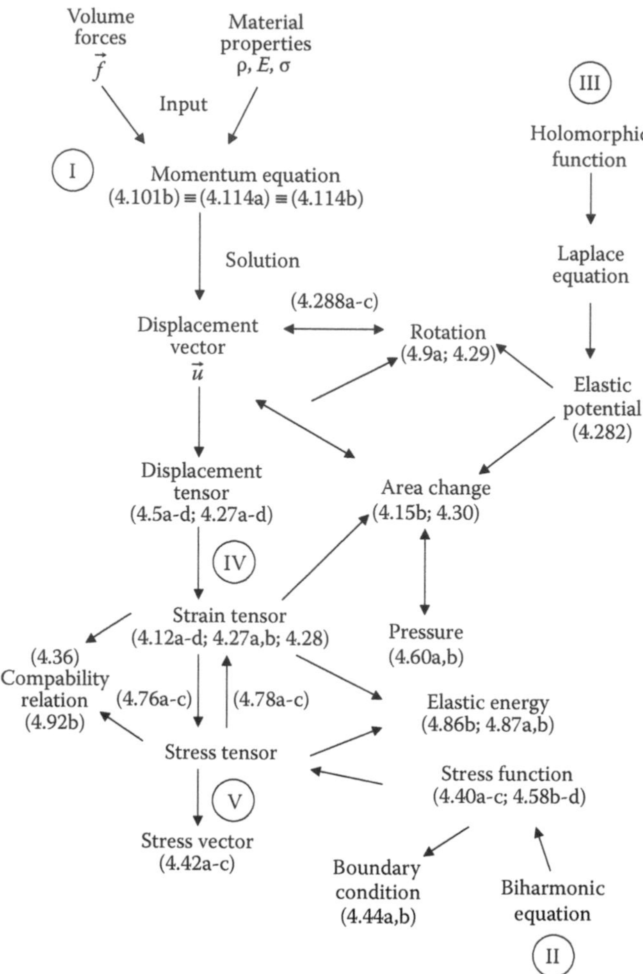

DIAGRAM 4.1
There are three methods of solution of elasticity problems starting from: (I) the momentum equation for the displacement vector; (II) a biharmonic stress function that specifies the stress tensor; (III) the complex elastic potential that is an analytic function, whose real (imaginary) part is related to the scalar linearized relative area change (rotation scalar). Methods II and III apply only to plane elasticity and assume that the inertia and external forces balance, both vanish, or the external force is constant or irrotational. Method I is general and applies in two or three dimensions in the presence of the inertia force and arbitrary external forces. All three methods specify the displacements, stresses, and strains, in a different sequence, as shown by 24 elasticity problems in List 4.1.

$$S_{x3} = S_{y3} = S_{33} = 0; \qquad T_{x3} = T_{y3} = 0; \tag{4.93a–4.93e}$$

and (3) Hooke law (Equation 4.76a) implies that the out-of-plane shear stresses also vanish (Equations 4.93d and 4.93e), but the normal out-of-plane stress (Equation 4.76c) does not vanish (Equation 4.93f) if there is a linearized relative area change:

$$\frac{1+\sigma}{E}\frac{1-2\sigma}{\sigma}T_{33} = S_{xx} + S_{yy} = \nabla\cdot\vec{u} = D_2. \tag{4.93f}$$

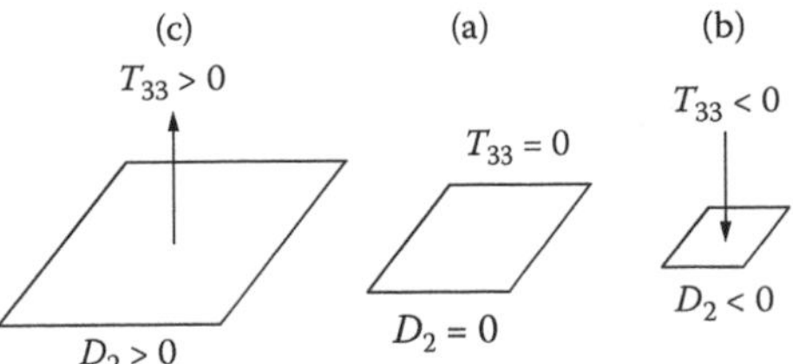

FIGURE 4.9
Plane elasticity is specified by a displacement vector (u_x, u_y) in the (x, y) plane, implying that the strain tensor has nonzero components only in the plane (S_{xx}, S_{yy}, S_{xy}), with corresponding stresses (T_{xx}, T_{yy}, T_{xy}). Because there are no out-of-plane distortions $S_{3x} = 0 = S_{3y}$, there also no out-of-plane shears $T_{3x} = 0 = T_{3y}$. In order to have no extension or contraction normal to the plane $S_{33} = 0$, two cases arise: (1) if there is no linearized area change $D_2 = S_{xx} + S_{yy} = 0$, no out-of-plane stress is needed $T_{33} = 0$; (2) if the linearized area in the plane increases $D_2 > 0$ (decreases $D_2 < 0$), the contraction (extension) in the normal direction must be countered by an out-of-plane traction $T_{33} > 0$ (compression $T_{33} < 0$), so that there is no extension or contraction along the normal. Thus, a plane deformation (1) does not require an out-of-plane stress $T_{33} = 0$ if there is no area change (a); (2 and 3) if the area increases (c) [decreases (b)], an out-of-plane traction $T_{33} > 0$ (compression $T_{33} < 0$) is needed to prevent an out-of-plane normal strain.

Thus, *an increase (decrease) in area in plane elasticity is associated with a traction (compression) normal to the plane (Figure 4.9) to ensure that there is no out-of-plane deformation. Thus, a two-dimensional state-of-strain and displacement require a three-dimensional stress, without cross-shears between the plane and the normal. The in-plane stresses are specified by the Hooke law (Equations 4.76a and 4.76c) in terms of the Young's modulus and Poisson ratio:*

$$\frac{1+\sigma}{E}T_{xy} = S_{xy}, \frac{(1+\sigma)(1-2\sigma)}{E}\{T_{xx}, T_{yy}\} = \{(1-\sigma)S_{xx} + \sigma S_{yy}, (1-\sigma)S_{yy} + \sigma S_{xx}\}. \quad (4.94a\text{–}4.94c)$$

The in-plane and out-of-plane stresses are related in the absence of out-of-plane strain (Equation 4.95a) by

$$S_{33} = 0: \qquad T_{33} = \sigma(T_{xx} + T_{yy}) = \frac{E}{1+\sigma}\frac{\sigma}{1-2\sigma}(S_{xx} + S_{yy}) = \frac{E}{1+\sigma}\frac{\sigma}{1-2\sigma}D_2 \qquad (4.95a\text{–}4.95d)$$

to (1) the in-plane normal stresses (Equation 4.78c ≡ Equation 4.95b); (2) the in-plane normal strains (Equations 4.94b and 4.94c ≡ Equation 4.95c); (3) the linearized relative area change (Equation 4.15b ≡ Equation 4.61a ≡ Equation 4.95d); and (4) the Laplacian of the stress function (Equations 4.58b and 4.58a ≡ Equation 4.96a):

$$S_{33} = 0 = \vec{f} = -\rho\vec{a}: \qquad T_{33} = \sigma(T_{xx} + T_{yy}) = \sigma(\partial_{yy}\Theta + \partial_{xx}\Theta) = \sigma\nabla^2\Theta, \quad (4.96a\text{–}4.96c)$$

if the external and inertia forces balance (Equation 4.96b ≡ Equation 4.58a). Substituting Equation 4.95b in the inverse Hooke law (Equation 4.78a) specifies

$$ES_{xx} = T_{xx} - \sigma[T_{yy} + \sigma(T_{xx} + T_{yy})] \equiv (1-\sigma^2)T_{xx} - \sigma(1+\sigma)T_{yy}, \qquad (4.96d)$$

$$ES_{yy} = T_{yy} - \sigma[T_{xx} + \sigma(T_{xx} + T_{yy})] \equiv (1-\sigma^2)T_{yy} - \sigma(1+\sigma)T_{xx}, \qquad (4.96e)$$

The normal in-plane strains in plane elasticity (Equations 4.96d and 4.96e)

4.4.2 Balance of Inertia and In-Plane Stresses in Cartesian Coordinates

The balance of inertia force plus external forces with internal stresses in the plane can be written (1) in terms of the stress tensor (Equations 4.38a and 4.38b) for any material; (2) for an elastic material using the Hooke law (Equations 4.76a and 4.76c) in terms of the strain tensor:

$$\frac{1+\sigma}{E}\left(\rho a_x - f_x\right) = \partial_y S_{xy} + \frac{1-\sigma}{1-2\sigma}\partial_x S_{xx} + \frac{\sigma}{1-2\sigma}\partial_x S_{yy}, \tag{4.97a}$$

$$\frac{1+\sigma}{E}\left(\rho a_y - f_y\right) = \partial_x S_{yx} + \frac{1-\sigma}{1-2\sigma}\partial_y S_{yy} + \frac{\sigma}{1-2\sigma}\partial_y S_{xx}; \tag{4.97b}$$

(3) using Equations 4.12a through 4.12d in terms of displacement tensor:

$$\frac{1+\sigma}{E}\left(\rho a_x - f_x\right) = \frac{1}{2}\partial_y\left(D_{xy} + D_{yx}\right) + \frac{1-\sigma}{1-2\sigma}\partial_x D_{xx} + \frac{\sigma}{1-2\sigma}\partial_x D_{yy}, \tag{4.98a}$$

$$\frac{1+\sigma}{E}\left(\rho a_y - f_y\right) = \frac{1}{2}\partial_x\left(D_{xy} + D_{yx}\right) + \frac{1-\sigma}{1-2\sigma}\partial_y D_{yy} + \frac{\sigma}{1-2\sigma}\partial_y D_{xx}; \tag{4.98b}$$

(4) using Equations 4.5a through 4.5d in terms of the displacement vector:

$$2\frac{1+\sigma}{E}\left(\rho a_x - f_x\right) = \partial_{yy} u_x + \frac{1}{1-2\sigma}\partial_{xy} u_y + \frac{2-2\sigma}{1-2\sigma}\partial_{xx} u_x, \tag{4.99a}$$

$$2\frac{1+\sigma}{E}\left(\rho a_x - f_x\right) = \partial_{xx} u_y + \frac{1}{1-2\sigma}\partial_{xy} u_x + \frac{2-2\sigma}{1-2\sigma}\partial_{yy} u_y; \tag{4.99b}$$

(5) in terms of the Laplacian, and successive application of the gradient and divergence of the displacement vector:

$$2\frac{1+\sigma}{E}\left(\rho a_x - f_x\right) = \left(\partial_{xx} + \partial_{yy}\right)u_x + \frac{1}{1-2\sigma}\partial_x\left(\partial_x u_x + \partial_y u_y\right); \tag{4.100a}$$

$$2\frac{1+\sigma}{E}\left(\rho a_x - f_x\right) = \left(\partial_{xx} + \partial_{yy}\right)u_x + \frac{1}{1-2\sigma}\partial_y\left(\partial_x u_x + \partial_y u_y\right); \tag{4.100b}$$

corresponding to

$$\vec{a} \equiv \ddot{\vec{u}} = \frac{\partial^2 \vec{u}}{\partial t^2}: \qquad 2\frac{1+\sigma}{E}\left(\rho\ddot{\vec{u}} - \vec{f}\right) = \nabla^2\vec{u} + \frac{1}{1-2\sigma}\nabla\left(\nabla\cdot\vec{u}\right) \qquad \text{(4.101a and 4.101b)}$$

in vector form using the operators Laplacian, gradient, and divergence (Subsection I.11.7).

4.4.3 External Forces in an Elastic Medium in Uniform Motion

In a medium at rest or in uniform motion (Equation 4.102a), the **elastic momentum balance equation** *(Equation 4.101b) simplifies to Equation 4.102b:*

$$\ddot{\vec{u}} = 0: \qquad -2\frac{1+\sigma}{E}\vec{f} = \nabla^2\vec{u} + \frac{1}{1-2\sigma}\nabla(\nabla\cdot\vec{u}). \qquad \text{(4.102a and 4.102b)}$$

Taking the divergence (curl) implies that (1) the divergence (curl) of the displacement vector, that is, the linearized relative area change (Equation 4.15b) [rotation scalar (Equation 4.9a)], satisfies a Poisson equation forced by the divergence (Equation 4.103b) [curl (Equation 4.103d)] of the external forces:

$$\ddot{\vec{u}} = 0: \qquad \nabla^2 D_2 \equiv \nabla^2(\nabla\cdot\vec{u}) = \nabla\cdot(\nabla^2\vec{u}) = -\frac{1+\sigma}{E}\frac{1-2\sigma}{1-\sigma}(\nabla\cdot\vec{f}), \qquad \text{(4.103a and 4.103b)}$$

$$\ddot{\vec{u}} = 0: \qquad \nabla^2\vec{E}_2 = \nabla^2(\nabla\wedge\vec{u}) = \nabla\wedge(\nabla^2\vec{u}) = -2\frac{1+\sigma}{E}(\nabla\wedge\vec{f}); \qquad \text{(4.103c and 4.103d)}$$

and (2) if the external force field is solenoidal (Equation 4.104b) [irrotational (Equation 4.104e)], the linearized relative area change (Equation 4.104c) [rotation (Equation 4.104f)] is a harmonic function:

$$\ddot{\vec{u}} = 0 = \nabla\cdot\vec{f}: \quad \nabla^2 D_2 = 0; \qquad \ddot{\vec{u}} = 0 = \nabla\wedge\vec{f}: \quad \nabla^2\vec{\Omega} = 0. \qquad \text{(4.104a–4.104d)}$$

In the case of a medium at rest or in uniform motion (Equation 4.102a ≡ Equation 4.103a ≡ Equation 4.103c ≡ Equation 4.104a ≡ Equation 4.104d ≡ Equation 4.105a ≡ Equation 4.105c), the divergence of the external forces appears as the forcing term in the Poisson equation for the linearized relative area change (Equation 4.105b) and for (Equations 4.95b through 4.95d) the out-of-plane stress (Equation 4.105d):

$$\ddot{\vec{u}} = 0: \qquad \nabla^2 T_{33} = \frac{E}{1+\sigma}\frac{\sigma}{1-2\sigma}\nabla^2(\nabla\cdot\vec{u}) = -\frac{\sigma}{1-\sigma}(\nabla\cdot\vec{f}), \qquad \text{(4.105a and 4.105b)}$$

$$\ddot{\vec{u}} = 0: \qquad \nabla^4\Theta = \frac{1}{\sigma}\nabla^2 T_{33} = -\frac{1}{1-\sigma}(\nabla\cdot\vec{f}); \qquad \text{(4.105c and 4.105d)}$$

the latter is a forced biharmonic equation (Equation 4.105d) for the stress function (Equation 4.96c). For an elastic medium at rest or in uniform motion (Equation 4.106a) under solenoidal external forces (Equation 4.106b):

$$\vec{a} = 0 = \nabla\cdot\vec{f}: \qquad \nabla^2 D_2 = 0 = \nabla^2 T_{33}, \qquad \nabla^4\Theta = 0, \qquad \text{(4.106a–4.106e)}$$

linearized relative to the area change (Equation 4.106c) and out-of-plane stress (Equation 4.106d) [stress function (Equation 4.106e)] are harmonic functions (is a **biharmonic function***), that is, solutions (solution) of the Laplace (biharmonic) equation.* The existence of a stress function was proven (Equations 4.40a and 4.40b) for constant external volume force (Equation 4.39b) and no acceleration (Equation 4.39a), when Equation 4.105d reduces to a biharmonic equation (Equation 4.106e). The balance of inertia and external forces (Equation 4.58a) also leads (Equations 4.58b through 4.58d) to the existence of a stress function. From the compatibility relation for the strains (Equation 4.36) and elastic stresses (Equation 4.92b), the stress function satisfies a biharmonic equation (Equation 4.92c). The forcing of the biharmonic equation for the stress function thus corresponds to violation of the preceding conditions, for example, the presence of volume forces with non-zero divergence (Equation 4.105d) in an elastic medium at rest or in uniform motion (Equation 4.105c), such as a point force (Section 4.8).

4.4.4 Momentum Equation for the Displacement Vector in Polar Coordinates

The momentum balance equation in polar coordinates can be written (1) in terms of the stress tensor (Equations 4.49a and 4.49b) for any material and (2) for an isotropic elastic solid using the Hooke law (Equations 3.76a and 3.76c) in terms of the strain tensor:

$$\frac{1+\sigma}{E}\left(\rho a_r - f_r\right) = r^{-1}\partial_\varphi S_{r\varphi} + r^{-1}\left(S_{rr} - S_{\varphi\varphi}\right) + (1-2\sigma)^{-1}\partial_r\left[(1-\sigma)S_{rr} + \sigma S_{\varphi\varphi}\right], \tag{4.107a}$$

$$\frac{1+\sigma}{E}\left(\rho a_\varphi - f_\varphi\right) = \partial_r S_{\varphi r} + 2r^{-1}S_{\varphi r} + (1-2\sigma)^{-1}r^{-1}\partial_\varphi\left[(1-\sigma)S_{\varphi\varphi} + \sigma S_{rr}\right]; \tag{4.107b}$$

expressing the strains (Equations 4.27a, 4.27b, and 4.28) in terms of the displacement vector and separating the terms independent of the Poisson ratio σ from those with the factor $(1-\sigma)^{-1}$ leads to

$$\begin{aligned} 2\frac{1+\sigma}{E}\left(\rho a_r - f_r\right) &= r^{-1}\partial_r\left(r\partial_r u_r\right) + r^{-2}\partial_{\varphi\varphi}u_r - 2r^{-2}\partial_\varphi u_\varphi - r^{-2}u_r \\ &\quad + (1-2\sigma)^{-1}\partial_r\left[r^{-1}\partial_r\left(ru_r\right) + r^{-1}\partial_\varphi u_\varphi\right], \end{aligned} \tag{4.108a}$$

$$\begin{aligned} 2\frac{1+\sigma}{E}\left(\rho a_\varphi - f_\varphi\right) &= r^{-1}\partial_r\left(r\partial_r u_\varphi\right) + r^{-2}\partial_{\varphi\varphi}u_\varphi + 2r^{-2}\partial_\varphi u_r - r^{-2}u_\varphi \\ &\quad + (1-2\sigma)^{-1}r^{-1}\partial_\varphi\left[r^{-1}\partial_r\left(ru_r\right) + r^{-1}\partial_\varphi u_\varphi\right]. \end{aligned} \tag{4.108b}$$

The comparison of the momentum equation for the displacement vector in the plane in Cartesian (polar) coordinates (Subsection 4.4.2 (Subsection 4.4.4) leads to the invariant differential operators in polar coordinates (Subsection 4.4.5). The passage from Equations 4.107a and 4.107b to Equations 4.108a and 4.108b is justified first.

Substitution of the strain tensor in polar coordinates (Equations 4.27a, 4.27b, and 4.28) in Equation 4.107a and separation of the terms without (with) the factor $(1-2\sigma)^{-1}$ lead to

$$
\begin{aligned}
&2\frac{1+\sigma}{E}\left(\rho a_r - f_r\right)\\
&= \frac{1}{r}\partial_{\varphi r}u_\varphi + \frac{1}{r^2}\partial_{\varphi\varphi}u_r - \frac{3}{r^2}\partial_\varphi u_\varphi + \frac{2}{r}\partial_r u_r - \frac{2}{r^2}u_r + \frac{2}{1-2\sigma}\partial_r\left[(1-\sigma)\partial_r u_r + \frac{\sigma}{r}\left(u_r + \partial_\varphi u_\varphi\right)\right]\\
&= \frac{1}{r^2}\partial_{\varphi\varphi}u_r + \frac{2-2\sigma}{1-2\sigma}\partial_{rr}u_r + \left(1+\frac{2\sigma}{1-2\sigma}\right)\frac{1}{r}\partial_{\varphi r}u_\varphi - \left(3+\frac{2\sigma}{1-2\sigma}\right)\frac{1}{r^2}\partial_\varphi u_\varphi + \left(2+\frac{2\sigma}{1-2\sigma}\right)\left[\frac{1}{r}\partial_r u_r - \frac{1}{r^2}u_r\right]\\
&= \frac{1}{r^2}\partial_{\varphi\varphi}u_r + \partial_{rr}u_r + \frac{1}{r}\partial_r u_r - \frac{2}{r^2}\partial_\varphi u_\varphi - \frac{1}{r^2}u_r + \frac{1}{1-2\sigma}\left(\partial_{rr}u_r + \frac{1}{r}\partial_r u_r - \frac{1}{r^2}u_r + \frac{1}{r}\partial_{\varphi r}u_\varphi - \frac{1}{r^2}\partial_\varphi u_\varphi\right)\\
&= \frac{1}{r}\partial_r\left(r\partial_r u_r\right) + \frac{1}{r^2}\partial_{\varphi\varphi}u_r - \frac{2}{r^2}\partial_\varphi u_\varphi - \frac{1}{r^2}u_r + \frac{1}{1-2\sigma}\partial_r\left[\frac{1}{r}\partial_r\left(ru_r\right) + \frac{1}{r}\partial_\varphi u_\varphi\right],
\end{aligned}
\tag{4.109a}
$$

that coincides with Equation 4.108a ≡ Equation 4.109a. A similar substitution in Equation 4.107b leads to

$$
\begin{aligned}
&2\frac{1+\sigma}{E}\left(\rho a_\varphi - f_\varphi\right)\\
&= \partial_{rr}u_\varphi + \frac{1}{r}\partial_{r\varphi}u_r + \frac{1}{r}\partial_r u_\varphi + \frac{1}{r^2}\partial_\varphi u_r - \frac{1}{r^2}u_\varphi + \frac{2}{1-2\sigma}\frac{1}{r}\partial_\varphi\left[\frac{1-\sigma}{r}\left(u_r + \partial_\varphi u_\varphi\right) + \sigma\partial_r u_r\right]\\
&= \partial_{rr}u_\varphi + \frac{2-2\sigma}{1-2\sigma}\frac{1}{r^2}\partial_{\varphi\varphi}u_\varphi + \left(1+\frac{2\sigma}{1-2\sigma}\right)\frac{1}{r}\partial_{r\varphi}u_r + \left(1+\frac{2-2\sigma}{1-2\sigma}\right)\frac{1}{r^2}\partial_\varphi u_r + \frac{1}{r}\partial_r u_\varphi - \frac{1}{r^2}u_\varphi\\
&= \partial_{rr}u_\varphi + \frac{1}{r}\partial_r u_\varphi + \frac{1}{r^2}\partial_{\varphi\varphi}u_\varphi + \frac{2}{r^2}\partial_\varphi u_r - \frac{1}{r^2}u_\varphi + \frac{1}{1-2\sigma}\left(\frac{1}{r}\partial_{r\varphi}u_r + \frac{1}{r^2}\partial_\varphi u_r + \frac{1}{r^2}\partial_{\varphi\varphi}u_\varphi\right)\\
&= \frac{1}{r}\partial_r\left(r\partial_r u_\varphi\right) + \frac{1}{r^2}\partial_{\varphi\varphi}u_\varphi + \frac{2}{r^2}\partial_\varphi u_r - \frac{1}{r^2}u_\varphi + \frac{1}{1-2\sigma}\frac{1}{r}\partial_\varphi\left[\frac{1}{r}\partial_r\left(ru_r\right) + \frac{1}{r}\partial_\varphi u_\varphi\right],
\end{aligned}
\tag{4.109b}
$$

that coincides with Equation 4.108b ≡ Equation 4.109b.

4.4.5 Invariant Differential Operators in Polar Coordinates

Comparing Equations 4.108a and 4.108b ≡ Equation 4.101b, it follows that the coincidence of (1) the second term on the r.h.s. leads to the gradient (Equation 4.46a) [divergence (Equation 4.30)] in polar coordinates; (2) the first term on the r.h.s. leads to the **vector Laplacian operator** *(Equation 4.110a):*

$$
\begin{aligned}
\nabla^2\vec{B} &= \vec{e}_r\left[r^{-1}\partial_r\left(r\partial_r B_r\right) + r^{-2}\partial_{\varphi\varphi}B_r - 2r^{-2}\partial_\varphi B_\varphi - r^{-2}B_r\right]\\
&\quad + \vec{e}_\varphi\left[r^{-1}\partial_r\left(r\partial_r B_r\right) + r^{-2}\partial_{\varphi\varphi}B_\varphi + 2r^{-2}\partial_\varphi B_r - r^{-2}B_\varphi\right]\\
&= \vec{e}_r\left(\nabla^2 B_r - 2r^{-2}\partial_\varphi B_\varphi - r^{-2}B_r\right) + \vec{e}_\varphi\left[\nabla^2 B_\varphi + 2r^{-2}\partial_\varphi B_r - r^{-2}B_\varphi\right],
\end{aligned}
\tag{4.110a and 4.110b}
$$

that has additional terms (Equation 4.110b) relative to the scalar Laplacian operator (Equation I.11.28b ≡ Equation 4.111):

$$\nabla^2 A = r^{-1}\,\partial_r(r\,\partial_r A) + r^{-2}\,\partial_{\varphi\varphi}\,A. \tag{4.111}$$

The complete set of invariant differential operators in polar coordinates consists of (1) the gradient (Equation 4.46a ≡ Equation I.11.31b); (2) the curl (Equation 4.29 ≡ Equation I.11.35b); (3) the divergence (Equation 4.30 ≡ Equation I.11.33b); (4) the scalar Laplacian (Equation 4.111 ≡ Equation I.11.28b); and (5) the vector Laplacian (Equations 4.110a and 4.110b). Of these five invariant differential operators, four were obtained from Cauchy–Riemann conditions for holomorphic functions (Subsections I.11.6 and I.11.7) and three from the momentum equation in plane elasticity (Subsection 4.4.4).

The vector Laplacian satisfies the identity

$$\nabla^2\vec{B} = \nabla\left(\nabla\cdot\vec{B}\right) - \nabla\wedge\left(\nabla\wedge\vec{B}\right), \tag{4.112}$$

that can be proven for each component in Cartesian coordinates, for example, for the x-component:

$$\begin{aligned}\left[\nabla\wedge\left(\nabla\wedge\vec{B}\right)\right]_1 &= \partial_2\left(\nabla\wedge\vec{B}\right)_3 - \partial_3\left(\nabla\wedge\vec{B}\right)_2 = \partial_2\left(\partial_1 B_2 - \partial_2 B_1\right) - \partial_3\left(\partial_3 B_1 - \partial_1 B_3\right)\\ &= -\partial_{22}B_1 - \partial_{33}B_1 + \partial_{12}B_2 + \partial_{13}B_3\\ &= \partial_1\left(\partial_1 B_1 + \partial_2 B_2 + \partial_3 B_3\right) - \left(\partial_{11} + \partial_{22} + \partial_{33}\right)B_1\\ &= \partial_1\left(\nabla\cdot\vec{B}\right) - \nabla^2 B_1.\end{aligned} \tag{4.113}$$

Cyclic permutation of (1, 2, 3) ≡ (x, y, z) in Equation 4.113 proves the (y, z) components of Equation 4.112. Substituting Equation 4.112 in Equation 4.101b leads to

$$2\frac{1+\sigma}{E}\left(\rho\ddot{\vec{u}} - \vec{f}\right) = -\nabla\wedge\left(\nabla\wedge\vec{u}\right) + 2\frac{1-\sigma}{1-2\sigma}\nabla\left(\nabla\cdot\vec{u}\right), \tag{4.114a}$$

$$2(1-2\sigma)\frac{1+\sigma}{E}\left(\rho\ddot{\vec{u}} - \vec{f}\right) = \nabla\wedge\left(\nabla\wedge\vec{u}\right) + 2(1-\sigma)\nabla^2\vec{u}. \tag{4.114b}$$

Thus, *there are three alternate forms (Equation 4.101b ≡ Equation 4.114a ≡ Equation 4.114b) of the momentum equation in an elastic medium. It is a second-order linear vector differential equation satisfied by the displacement vector $\vec{u}$, forced by the external forces per unit volume $\vec{f}$, and involves the properties of the material through the mass density ρ, Young's modulus E, and Poisson ratio σ.*

4.4.6 Three Methods of Solution of Elasticity Problems

There are three main methods (Diagram 4.1) of solution of elasticity problems, starting from (I) the momentum equation in three or two (Section 4.5) dimensions; (II) the stress function (Sections 4.6, 4.7, and 4.9) only in two dimensions; and (III) the complex potential (Section 4.8), also only in two dimensions. Starting with the general method I at the top left of Diagram 4.1, there are 12 steps: (1) the volume forces and material properties are the input to the momentum equation (Equation 4.101b ≡ Equation 4.114a ≡ Equation 4.114b); (2) its solution specifies the displacement vector to within two arbitrary vector constants (functions) in the steady (unsteady) case, that is, the integration constants (functions); (3) the displacement tensor (Equations 4.5a through 4.5d and 4.27a through 4.27d) followed by differentiation of the displacement vector; (4 and 5) the strain tensor (Equations 4.12a through 4.12d, 4.27a, 4.27b, and 4.28) and rotation (Equations 4.9a and 4.29) follow from the displacement tensor or vector; (6) the linearized relative area change (Equations 4.15b and 4.30) follows from the displacement vector or displacement tensor or strain tensor; (7) the stress (strain) tensor is specified by the strain (stress) tensor via the direct (inverse) Hooke law [Equations 4.76a through 4.76c (Equations 4.78a through 4.78c)]; (8) the strain (stress) tensor automatically satisfies the compatibility relation [Equation 4.36 (Equation 4.92a)]; (9) the stress tensor specifies (Equations 4.42a through 4.42c) the stress vector; (10) the stress vector or strains or displacements on the boundaries lead to the boundary conditions (such as Equations 4.44a and 4.44b) that determine the arbitrary constants (functions) in the steady (unsteady) solution of the momentum equation; (11) the mean of the normal stresses specifies the pressure (Equation 4.60b); and (12) the elastic energy is specified by the stress (strain) tensor alone [Equation 4.87b (Equation 4.87a)] for an isotropic elastic material or by both together (Equation 4.86b) for an arbitrary material.

The other two methods of solution apply only to plane problems, when (II) the external forces are constant (Equation 4.39b) and the acceleration is zero (Equation 4.39a), so that a stress function (Equations 4.40a through 4.40c) exists; (III) the inertia and external forces balance [Equation 4.57a ≡ Equation 4.58a (Equation 4.276a)], so that a stress function (Equations 4.57b through 4.57d and 4.58b through 4.58d) [complex potential (Equation 4.282 ≡ Equations 4.281a through 4.281c)] exists. Concerning method II, the steps as distinct from method I are as follows: (1) the solution of a biharmonic equation (Equation 4.106e) specifies the stress function; (2) the latter leads to the stress tensor (Equations 4.57b through 4.57d and 4.58b through 4.58d); (3) the inverse Hooke law (Equations 4.78a through 4.78c) specifies the strain tensor; (4) the compatibility relation (Equation 4.36) is automatically satisfied; (5–8) this ensures the existence of a displacement vector (Equations 4.32a and 4.32b) and tensor (Equations 4.5a through 4.5d), rotation (Equation 4.34c), and dilatation (Equation 4.15b); (9) the stress tensor specifies (Equations 4.42a through 4.42c) the stress vector; and (10) the boundary conditions such as Equations 4.44a and 4.44b are applied as before. Method III relies on the existence in plane elasticity of (1) a complex elastic potential (Section 4.7) that specifies (Equation 4.281c) the rotation (Equation 4.281b) and linearized relative area change (Equation 4.281a), as well as the potential and field function (Equation 4.282); (2) the latter determine (Equations 4.288a through 4.288c) the displacement vector; and (3–10) the remaining steps follow from the displacement vector as in method I. It is possible to conceive other methods, for example, method IV (V) specifying *a priori* the strain (stress) tensor; in this case: (1) the compatibility relation [Equation 4.36 (Equation 4.92b)] must be satisfied; and (3–10) the remaining steps follow as in method II.

4.5 Cavities and Static and Rotating Cylinders

The first essential steps of method I of solution of problems in plane elasticity consist of (1) solving the momentum equation (Equation 4.101b ≡ Equation 4.114a ≡ Equation 4.114b) for the displacement vector, for example, in Cartesian (Equations 4.100a and 4.100b) [polar (Equations 4.108a and 4.108b)] coordinates; (2) determining the displacement and strain tensors [Equations 4.12a through 4.12d (Equations 4.27a through 4.27d and 4.28)]; (3) the Hooke law (Equations 4.76a through 4.76c) specifies the stresses; and (4) the constants of integration can be determined from boundary conditions, for example, on the displacement or stress such as Equations 4.44a and 4.44b. As examples, the stresses with cylindrical symmetry are determined for (1 and 2) a solid (Subsection 4.5.1) or hollow (Subsection 4.5.3) cylinder; (3 and 4) a cylindrical cavity (Subsection 4.5.2) or thin shell (Subsection 4.5.4), for example, with different inner and outer pressures; and (5–7) two concentric tubes made of different materials (Subsection 4.5.5), in particular, one material is much stiffer than the other (Subsection 4.5.6), and one pressure much larger than the other (Subsection 4.5.7). In the preceding cases (1–7), there are no external forces, whereas a solid rotating cylinder is subject to stresses due to the centrifugal force (Subsection 4.5.8).

4.5.1 Displacement, Strains, and Stresses (Method I) with Cylindrical Symmetry

Assuming that the external and inertia forces balance (Equation 4.115a), **cylindrical symmetry** requires that both the radial (Equation 4.115b) [azimuthal (Equation 4.115c)] displacements depend only on the radius:

$$\vec{f} = \rho\vec{a}, \vec{u}(r) \equiv \{u_r(r), u_\varphi(r)\}: \; 0 = r(ru')' - u = r^2u'' + ru' - u = r^2\left[r^{-1}(ru)'\right]'. \quad (4.115a\text{–}4.115d)$$

Both satisfy (Equations 4.108a and 4.108b) the same equation (Equation 4.115d), where prime denotes (Equation 4.116a) a derivative with regard to r, that is, the azimuthal (radial) displacement satisfies [Equation 4.108b ≡ Equation 4.116b (Equation 4.108a ≡ Equation 4.116c)]:

$$u' = \frac{du}{dr}: \qquad 0 = \frac{1}{r}\frac{d}{dr}\left(r\frac{du_\varphi}{dr}\right) - \frac{u_\varphi}{r^2} = \frac{1}{r^2}\left(r^2u''_\varphi + ru'_\varphi - u_\varphi\right), \quad (4.116a \text{ and } 4.116b)$$

$$0 = \left(1 + \frac{1}{1-2\sigma}\right)\left[\frac{1}{r}\frac{d}{dr}\left(r\frac{du_r}{dr}\right) - \frac{u_r}{r^2}\right] = \frac{2-2\sigma}{1-2\sigma}\left(u''_r + \frac{1}{r}u'_r - \frac{1}{r^2}u_r\right)$$
$$= 2\frac{1-\sigma}{1-2\sigma}\frac{d}{dr}\left[\frac{1}{r}\frac{d}{dr}(u_r r)\right] = \frac{1-\sigma}{1-2\sigma}\frac{2}{r^2}\left(r^2u''_r + ru'_r - u_r\right). \quad (4.116c)$$

For the azimuthal displacement (Equation 4.115c), there is coincidence of Equation 4.116b ≡ Equation 4.115d. The same applies (Equation 4.116c ≡ Equation 4.115d) for the radial displacement (Equation 4.115c), bearing in mind that the factor in Equation 4.116c involving the Poisson ratio σ cannot vanish or diverge by Equation 4.74c. The ordinary differential equation (Equation 4.115d) is linear with variable coefficients of the Euler type that has power solutions (Equation 4.117a):

$$u \sim r^n: \qquad 0 = r^2u'' + r^2u' - u = r^n\,[n\,(n-1) + n - 1] = u(n^2 - 1), \quad (4.117a \text{ and } 4.117b)$$

where the exponent satisfies Equation 4.117b. Because $u(r)$ is not zero everywhere, the term in square brackets must vanish: it is a polynomial of the second degree with roots (Equation 4.118a). Each root is a particular solution $r^{\pm 1}$ of the differential equation (Equation 4.115d), and because the latter is linear, a linear combination is also a solution. The two arbitrary constants are generally distinct for the radial (Equation 4.118b) and azimuthal (Equation 4.118c) displacements:

$$n = \pm 1: \qquad u_r(r) = Ar + \frac{B}{r}, \qquad u_\varphi(r) = Cr + \frac{D}{r}, \qquad \text{(4.118a–4.118c)}$$

thus leading to four constants of integration.

For a solid cylinder (Figure 4.10a), two constants of integration must vanish (Equations 4.119c and 4.119d) for the displacement to be finite at the center (Equations 4.119a and 4.119b):

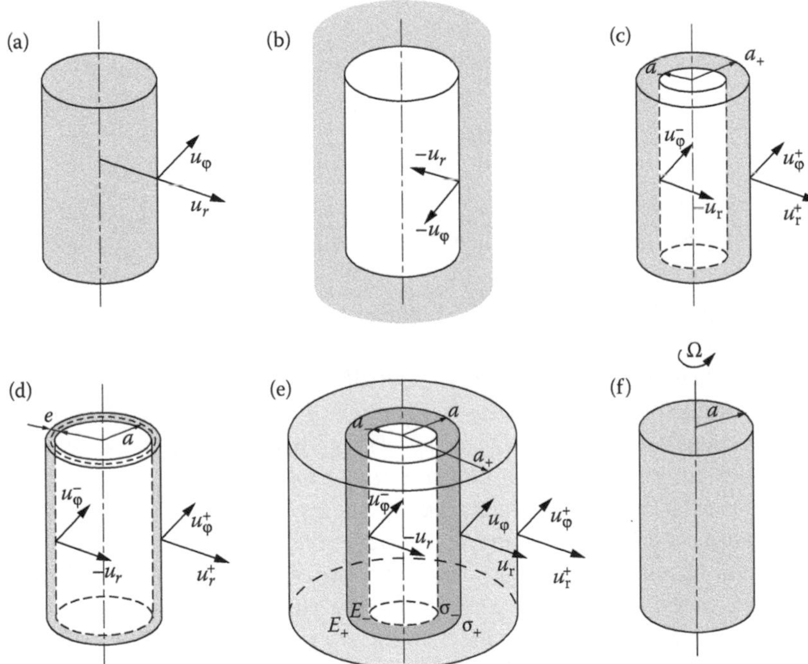

FIGURE 4.10
Elastic deformation with axial or cylindrical symmetry consists of a displacement vector that depends only on the radius or distance from the axis and lies on the plane orthogonal to the axis, allowing radial u_r and azimuthal u_ϕ components. The particular cases include (1 and 2) a solid cylinder (a) [cylindrical cavity in an infinite medium (b)] for which the elastic material lies inside (outside) the boundary cylindrical surface, which has opposite outward unit normal vectors that is radial outward (inward); (3 and 4) the elastic material between two coaxial cylindrical surfaces corresponds to a cylindrical tube (c) that in the limit of small thickness compared to the average radius leads to a cylindrical shell (d); (5) three (or more) coaxial cylindrical surfaces lead to a compound tube with elastic materials with possibly different properties between each pair of successive surfaces. The tubes (a through e) may be subject to external forces, for example, the centrifugal force on a rotating cylindrical shaft (f).

$$u_r(0), u_\varphi(0) < \infty: \qquad B = 0 = D, \qquad u_{r0} \equiv u_r(a) = Aa, \qquad u_{\varphi 0} \equiv u_\varphi(a) = Ca, \tag{4.119a–4.119f}$$

and the other two (Equations 4.119e and 4.119f) are determined (Equation 4.120a) by the displacement at the surface (Equations 4.120b and 4.120c):

$$0 \le r \le a: \qquad u_r(r) = u_{r0}\frac{r}{a}, \qquad u_\varphi(r) = u_{\varphi 0}\frac{r}{a}. \tag{4.120a–4.120c}$$

The corresponding strains (Equations 4.27a, 4.27b, and 4.28) are constant (Equations 4.121a through 4.121c):

$$S_{rr} = \frac{u_{r0}}{a} = S_{\varphi\varphi} = \frac{1}{2}\left(\nabla \cdot \vec{u}\right) \equiv \frac{1}{2}D_2, \qquad S_{r\varphi} = 0; \tag{4.121a–4.121c}$$

$$T_{rr} = \frac{E}{1+\sigma}\frac{1}{1-2\sigma}\frac{u_{r0}}{a} = T_{\varphi\varphi}, \quad T_{r\varphi} = 0, \quad T_{33} = \frac{E}{1+\sigma}\frac{2\sigma}{1-2\sigma}\frac{u_{r0}}{a}; \tag{4.122a–4.122c}$$

the strains imply by the Hooke law [Equations 4.76a and 4.76c (by Equation 4.95d)] also the in-plane (Equations 4.120a through 4.120c) [out-of-plane (Equation 4.120d)] stresses *and lead (Equation 4.86b) to the elastic energy density:*

$$E_d = \frac{1}{2}\left(T_{rr}S_{rr} + T_{\varphi\varphi}S_{\varphi\varphi}\right) = T_{rr}S_{rr} = T_{\varphi\varphi}S_{\varphi\varphi} = \frac{E}{1+\sigma}\frac{1}{1-2\sigma}\left(\frac{u_{r0}}{a}\right)^2. \tag{4.122d}$$

Thus, *the deformation (Figure 4.10a) with cylindrical symmetry of a solid cylinder (Equation 4.120a) with radius a, made of a material with Young's modulus E and Poisson ratio σ, has linear radial (Equation 4.120b) [azimuthal (Equation 4.120c)] displacements and hence constant strains (Equations 4.121a through 4.121c) and stresses (Equations 4.122a through 4.122c). The constants of integration in Equations 4.118b and 4.118c are specified (Equations 4.119a through 4.119f) by two displacements at the surface, or two independent stresses, for example: (1) the radial displacement (Equation 4.120b) is associated with equal normal in-plane (Equations 4.122a and 4.122b) stresses and no shear (Equation 4.122c) stress; and (2) the azimuthal displacement in the form (Equation 4.120c) preserves the radius of a circle* $u_\varphi r = \text{const}$ *and causes no stress or strain. The equal normal stresses imply equal normal strains, each contributing one-half of the linearized relative area change (Equations 4.121a through 4.121c) and to the elastic energy density (Equation 4.122d) and leading to the out-of-plane stress (Equation 4.121d).*

4.5.2 Solid Cylinder versus Cylindrical Cavity in an Unbounded Medium

In the case of a cylindrical cavity in an unbounded medium (Figure 4.10b), the condition of finite displacement at infinity (Equations 4.123a and 4.123b) implies that the other two constants of integration vanish (Equations 4.123c and 4.123d) in Equations 4.118b and 4.118c:

$$u_r(\infty), u_\varphi(\infty) < \infty: \qquad A = 0 = C, \quad u_{r0} \equiv u_r(a) = \frac{B}{a}, \quad u_{\varphi 0} \equiv u_\varphi(a) \equiv \frac{D}{a}, \tag{4.123a–4.123f}$$

and the remaining two constants are determined from the displacement at the surface (Equations 4.123e and 4.123f). The displacement vector instead of varying linearly (Equations 4.120b and 4.120c) for the solid cylinder (Equation 4.120a) varies inversely with distance from the axis (Equations 4.124b and 4.124c) for a cylindrical cavity (Equation 4.124a):

$$a \le r < \infty: \qquad u_r(r) = u_{r0}\frac{a}{r}, \qquad u_\varphi(r) = u_{\varphi 0}\frac{a}{r}. \tag{4.124a–4.124c}$$

This leads to nonconstant strains (Equations 4.27a, 4.27b, and 4.28 ≡ Equations 4.125a through 4.125c) with zero linearized relative area change (Equation 4.30 ≡ Equation 4.125d) and hence (Equation 4.95d) no out-of-plane stress (Equation 4.125e):

$$S_{rr}(r) = -u_{r0}\frac{a}{r^2} = -S_{\varphi\varphi}, \qquad S_{r\varphi}(r) = -u_{\varphi 0}\frac{a}{r^2}, \qquad D_2 \equiv \nabla\cdot\vec{u} = 0 = T_{33}; \tag{4.125a–4.125e}$$

$$T_{rr}(r) = -\frac{E}{1+\sigma}u_{r0}\frac{a}{r^2} = -T_{\varphi\varphi}, \qquad T_{r\varphi}(r) = -\frac{E}{1+\sigma}u_{\varphi 0}\frac{a}{r^2}. \tag{4.126a–4.126c}$$

By the Hooke law (Equations 4.76a and 4.76b), there are opposite normal stresses (Equations 4.126a and 4.126b) and a shear stress (Equation 4.126c) all in the same ratio $E/(1 + \sigma)$ to the strains. The elastic energy density (Equation 4.86b) has an equal contribution from the radial and tangential displacements:

$$E_d(r) = \frac{1}{2}\left(T_{rr}S_{rr} + T_{\varphi\varphi}S_{\varphi\varphi}\right) + T_{r\varphi}S_{r\varphi} = \frac{E}{1+\sigma}\frac{a^2}{r^4}\left[\left(u_{r0}\right)^2 + \left(u_{\varphi 0}\right)^2\right] = \frac{E}{1+\sigma}\frac{a^2}{r^4}\left|u_0\right|^2. \tag{4.126d}$$

The stresses at the surface are related to the displacements by

$$T_{rr}(a) = -\frac{E}{1+\sigma}\frac{u_{r0}}{a} = -T_{\varphi\varphi}(a), \quad T_{r\varphi}(a) = -\frac{E}{1+\sigma}\frac{u_{\varphi 0}}{a} \tag{4.127a–4.127c}$$

and can be used to determine the constants of integration in Equations 4.123e and 4.123f. Thus, *a cylindrical cavity (Figure 4.10b) with radius a in an unbounded medium (Equation 4.124a) with Young's modulus E and Poisson ratio σ has displacement vector (Equations 4.124b and 4.124c), strain tensor (Equations 4.125a through 4.125d), and stress tensor (Equations 4.126a through 4.126c) and elastic energy density (Equation 4.126d) with cylindrical symmetry. The surface radial (tangential) displacement [Equation 4.123e (Equation 4.123f)] is related to the surface normal (Equations 4.127a and 4.127b) [(shear (Equation 4.127c)] stresses. The solid cylinder (Equation 4.128a) [cylindrical cavity (Equation 4.128c)] needs (Equation 4.128b) [does not need (Equation 4.128d)]:*

$$0 \le r \le a: \qquad T_{33} = \frac{E}{1+\sigma}\frac{2\sigma}{1-2\sigma}\frac{u_{r0}}{a}; \qquad a \le r < \infty: \quad T_{33} = 0 \tag{4.128a–4.128d}$$

an axial stress (Equation 4.95b) to prevent an axial displacement because there is [Equation 4.121b (there is no (Equation 4.125d)] linearized relative area change. An outward radial displacement causes a radial traction (Equation 4.122a) [compression (Equation 4.126a)] in the solid cylinder (cylindrical cavity) and vice versa for an inward radial displacement. A tangential displacement causes no distortion (Equation 4.121c) [a distortion (Equation 4.125c)] and hence no shear stress (Equation 4.122c) [a shear stress (Equation 4.126c)] both of opposite signs. The elastic energy density is constant (Equation 4.122d) [decays with the radial distance (Equation 4.126d)] for the solid cylinder (cylindrical cavity).

4.5.3 Hollow Cylinder with Surface Displacements or Stresses

In the case of a hollow cylinder (Figure 4.10c) with inner (outer) radius a_- (a_+) in Equation 4.131a, none of the four constants of integration in Equations 4.118b and 4.118c is zero, that is, they can be determined from the surface displacements:

$$u_r^+ \equiv u_r\left(a_+\right) = Aa_+ + \frac{B}{a_+}, \qquad u_r^- \equiv u_r\left(a_-\right) = Aa_- + \frac{B}{a_-}; \quad \text{(4.129a and 4.129b)}$$

for example, for the radial displacement:

$$A\left(a_+^2 - a_-^2\right) = u_r^+ a_+ - u_r^- a_-, \qquad B\left(\frac{1}{a_+^2} - \frac{1}{a_-^2}\right) = \frac{u_r^+}{a_+} - \frac{u_r^-}{a_-}; \quad \text{(4.130a and 4.130b)}$$

this leads to the radial displacement (Equation 4.131b):

$$a_- \le r \le a_+: \qquad \left(a_+^2 - a_-^2\right)u_r(r) = \left(u_r^+ a_+ - u_r^- a_-\right)r - \left(u_r^+ a_- - u_r^- a_+\right)\frac{a_+ a_-}{r}, \quad \text{(4.131a and 4.131b)}$$

likewise for the azimuthal displacement. The strains (Equations 4.27a, 4.27b, and 4.28) corresponding to the displacements (Equations 4.118b and 4.118c) are given by Equations 4.132a through 4.132c, and the linearized relative area change is Equation 4.132d:

$$S_{rr}(r) = A - \frac{B}{r^2}, \qquad S_{\varphi\varphi}(r) = A + \frac{B}{r^2}, \qquad S_{r\varphi}(r) = -\frac{D}{r^2}, \qquad \nabla \cdot \vec{u} = S_{rr} + S_{\varphi\varphi} = 2A, \quad \text{(4.132a–4.132d)}$$

$$\frac{1+\sigma}{E}\left\{T_{rr}(r), T_{\varphi\varphi}(r)\right\} = \frac{A}{1-2\sigma} \mp \frac{B}{r^2}, \qquad \frac{1+\sigma}{E}T_{r\varphi}(r) = -\frac{D}{r^2}, \qquad T_{33} = \frac{E}{1+\sigma}\frac{2\sigma}{1-2\sigma}A. \quad \text{(4.133a–4.133d)}$$

The corresponding in-plane (out-of-plane) stresses specified by Hooke law (Equations 4.76a through 4.76c) [by (Equation 4.95d)] are Equations 4.133a through 4.133c (Equation 4.133d). The elastic energy density (Equation 4.86b) corresponding to the strains (Equations 4.125a through 4.125c) and stresses (Equation 4.126a through 4.126c):

$$E_d(r) = \frac{1}{2}\left(T_{rr}S_{rr} + T_{\varphi\varphi}S_{\varphi\varphi}\right) + T_{r\varphi}S_{r\varphi}$$

$$= \frac{E}{1+\sigma}\left[\frac{1}{2}\left(A - \frac{B}{r^2}\right)\left(\frac{A}{1-2\sigma} - \frac{B}{r^2}\right) + \frac{1}{2}\left(A + \frac{B}{r^2}\right)\left(\frac{A}{1-2\sigma} + \frac{B}{r^2}\right) + \frac{D^2}{r^4}\right] \quad (4.133e)$$

$$= \frac{E}{1+\sigma}\left(\frac{A^2}{1-2\sigma} + \frac{B^2 + D^2}{r^4}\right),$$

simplifies to (Equation 4.133e).

Several combinations of boundary conditions are possible; for example, the radial and azimuthal normal stresses at one surface (Equations 4.133a and 4.133b) can determine the constants of integration A and B, and the constants of integration C and D can be determined by the shear stress (Equation 4.132c) and tangential displacement (Equation 4.118c) at one surface. Another case is a purely radial displacement (Equations 4.134a and 4.134b) due to pressures p_+ (p_-) at the outer (inner) surface [Equation 4.134c (Equation 4.134d)]:

$$C = 0 = D: -p_+ = T_{rr}\left(a_+\right) = \frac{E}{1+\sigma}\left(\frac{A}{1-2\sigma} - \frac{B}{a_+^2}\right), \qquad p_- = T_{rr}\left(a_-\right) = \frac{E}{1+\sigma}\left(\frac{A}{1-2\sigma} - \frac{B}{a_-^2}\right),$$

(4.134a–4.134d)

where the pressure acts inward (Equation 4.134c) [outward (Equation 4.134d)] at the outer (inner) surface. Solving Equations 4.134c and 4.134d specifies the constants of integration:

$$\frac{E}{1+\sigma}\left(a_-^2 - a_+^2\right)\{A, B\} = \left\{(1-2\sigma)\left(p_+a_+^2 + p_-a_-^2\right), \left(p_+ + p_-\right)a_+^2a_-^2\right\}. \quad (4.135a \text{ and } 4.135b)$$

Thus, *a hollow cylinder (Figure 4.10c) with (Equation 4.131a) inner (outer) radius a_- (a_+) made of a material with Young's modulus E and Poisson ratio σ has displacements (Equations 4.118b and 4.118c), strains (Equations 4.132a through 4.132c), and stresses (Equations 4.133a through 4.133c) and elastic energy density (Equation 4.133e) with cylindrical symmetry. The four constants of integration can be determined by (1) the displacements (Equation 4.131b) at the surfaces and (2) any other compatible set of stresses, for example, no surface shears and pressures p_+ (p_-) at the inner (outer) surface (Equations 4.134a through 4.134d). Using Equation 4.130a (Equation 4.135a), the linearized relative area change (Equation 4.132d):*

$$D_2 \equiv \nabla \cdot \vec{u} = S_{rr} + S_{\varphi\varphi} = 2A = \frac{1+\sigma}{E}\frac{1-2\sigma}{\sigma}T_{33}$$

$$= \left\{2\frac{u_r^+a_+ - u_r^-a_-}{a_+^2 - a_-^2}, \frac{1+\sigma}{E}\frac{1-2\sigma}{a_-^2 - a_+^2}\left(p_+a_+^2 + p_-a_-^2\right)\right\} \quad (4.136a \text{ and } 4.136b)$$

is given by Equation 4.136a (Equation 4.136b) in case 1 (case 2) and also specifies (Equation 4.95b) the out-of-plane stress.

4.5.4 Thin Cylindrical Shell with Distinct Inner and Outer Pressures

In the case of a hollow cylinder (Subsection 4.5.3), the thickness (Equation 4.137a) [average radius (Equation 4.137b)] is the difference (arithmetic mean) of the outer a_+ and inner a_- radii (Equation 4.137c):

$$e \equiv a_+ - a_-, \qquad 2a \equiv a_+ + a_-: \qquad a_\pm = a \pm \frac{e}{2}, \qquad \frac{a}{e} = \frac{1}{2}\frac{a_+ + a_-}{a_+ - a_-}. \quad (4.137a–4.137d)$$

For a **thin cylindrical shell** (Figure 4.10d), the thickness is small compared with the average radius (Equation 4.138a), which leads to the first-order approximations (Equations 4.138b and 4.138c):

$$e^2 << a^2: \qquad p_+a_+^2 + p_-a_-^2 = a^2\left(p_+ + p_-\right), \quad (4.138a and 4.138b)$$

$$a_+^2a_-^2\left(p_+ + p_-\right) = a^4\left(p_+ + p_-\right), \quad a_-^2 - a_+^2 = \left(a - \frac{e}{2}\right)^2 - \left(a + \frac{e}{2}\right)^2 = -2ae \quad (4.138c and 4.138d)$$

and exact relation (Equation 4.138d). Using Equations 4.138b through 4.138d specifies the constants of integration (Equations 4.134a, 4.134b, 4.135a, and 4.135b) for the case (Equations 4.139b through 4.139e) of a thin cylindrical shell (Equation 4.139a) with unequal pressures and no shears at the surfaces:

$$e^2 << a^2: \quad C = 0 = D; \quad \{A, B\} = -\frac{1+\sigma}{E}\bar{p}\left\{(1-2\sigma)\frac{a}{e}, \frac{a^3}{e}\right\}, \quad \bar{p} = \frac{p_+ + p_-}{2} \quad (4.139a–4.139f)$$

in terms of the arithmetic mean of the pressures (Equation 4.139f). The case of a hollow cylinder (cylindrical shell) represents a thick (thin) tube separating two fluids at different pressures [Subsection 4.5.3 (Subsection 4.5.4)]: (1) at rest and (2) axial in motion either inviscid or viscous with no azimuthal stresses. Because the problem is two-dimensional, it assumes an infinitely long tube in the axial direction, so that no boundary conditions are needed at the ends.

The constants (Equations 4.139b through 4.139e) specify (Equations 4.118b and 4.118c) (1) a radial displacement (Equations 4.140a and 4.140b):

$$u_\varphi(r) = 0, \qquad u_r(r) = -\frac{1+\sigma}{E}\bar{p}\frac{a}{e}\left[(1-2\sigma)r + \frac{a^2}{r}\right]; \quad (4.140a and 4.140b)$$

(2) the associated (Equations 4.27a, 4.27b, and 4.28) strains:

$$\left\{S_{rr}(r), S_{\varphi\varphi}(r), S_{r\varphi}(r)\right\} = -\frac{1+\sigma}{E}\bar{p}\frac{a}{e}\left\{1 - 2\sigma \mp \frac{a^2}{r^2}, 0\right\}; \quad (4.141a–4.141c)$$

(3) the linearized relative area change (Equation 4.30 ≡ Equation 4.142a) and axial stress (Equation 4.95d ≡ Equation 4.142b):

$$D_2 \equiv \nabla \cdot \vec{u} = S_{rr} + S_{\varphi\varphi} = -\frac{1+\sigma}{E}(1-2\sigma)2\bar{p}\frac{a}{e} = -\frac{1+\sigma}{E}(1-2\sigma)(p_+ + p_-)\frac{a}{e}, \tag{4.142a}$$

$$T_{33} = \frac{E}{1+\sigma}\frac{\sigma}{1-2\sigma}(\nabla \cdot \vec{u}) = -2\sigma\bar{p}\frac{a}{e} = -\sigma(p_+ + p_-)\frac{a}{e} \tag{4.142b}$$

which both involve the total pressure $p_+ + p_-$ and the ratio of the average radius (Equation 4.137b) to thickness (Equation 4.137a); (4) by the Hooke law (Equations 4.76a and 4.76c), the elastic stresses

$$T_{r\varphi}(r) = 0: \quad \{T_{rr}(r), T_{\varphi\varphi}(r)\} = \frac{E}{1+\sigma}\left[\{S_{rr}, S_{\varphi\varphi}\} + \frac{\sigma}{1-2\sigma}(\nabla \cdot \vec{u})\right] = -\bar{p}\frac{a}{e}\left(1 \mp \frac{a^2}{r^2}\right); \tag{4.143a–4.143c}$$

(5) by the stresses (Equations 4.143a through 4.143c) and strains (Equations 4.141a through 4.141c) the elastic energy density (Equation 4.86b):

$$\begin{aligned} E_d(r) &= \frac{1}{2}(T_{rr}S_{rr} + T_{\varphi\varphi}S_{\varphi\varphi}) \\ &= \frac{1+\sigma}{2E}\left(\bar{p}\frac{a}{e}\right)^2\left[\left(1-2\sigma-\frac{a^2}{r^2}\right)\left(1-\frac{a^2}{r^2}\right) + \left(1-2\sigma+\frac{a^2}{r^2}\right)\left(1+\frac{a^2}{r^2}\right)\right] \\ &= \frac{1+\sigma}{E}\left(\bar{p}\frac{a}{e}\right)^2\left(1-2\sigma+\frac{a^4}{r^4}\right); \end{aligned} \tag{4.143d}$$

and (6) at the average radius (Equation 4.144a), the stresses

$$r = a: \qquad T_{r\varphi}(a) = 0 = T_{rr}(a), \qquad T_{\varphi\varphi}(a) = -2\bar{p}\frac{a}{e} = -(p_+ + p_-)\frac{a}{e} \tag{4.144a–4.144d}$$

are tangential (Equations 4.144b through 4.144d).

At the surfaces (Equation 4.145a), the approximation to the lowest order (Equation 4.145b):

$$r = a_\pm = a \pm \frac{e}{a}: \qquad \frac{a^2}{a_\pm^2} = \left(1 \pm \frac{e}{2a}\right)^{-2} = 1 \mp \frac{e}{a} + O\left(\frac{e^2}{a^2}\right) \tag{4.145a and 4.145b}$$

leads to the stresses:

$$T_{r\varphi}(a_\pm) = 0, T_{rr}(a_\pm) = \pm\bar{p} = \pm\frac{p_+ + p_-}{2}, T_{\varphi\varphi}(a_\pm) = -2\bar{p}\frac{a}{e} = -(p_+ + p_-)\frac{a}{e}. \tag{4.146a–4.146c}$$

Thus, *a thin cylindrical shell (Equation 4.138a) with average radius (Equation 4.137b) and thickness (Equation 4.137a), hence (Equation 4.137c) inner (outer) radii a_- (a_+) subject to distinct* *pressures p_- (p_+) and no shears at the surfaces, has (1) a purely radial displacement (Equations 4.140a and 4.140b); (2 and 3) only normal strains (Equations 4.141a through 4.141c) and stresses (Equations 4.143a through 4.143c); and (4 and 5) linearized relative area change (Equation 4.142a) and out-of-plane stress (Equation 4.142b). All items 1–5 are proportional to (1) the arithmetic mean of the pressure at the two surfaces (Equation 4.139f) and (2) the ratio of the average radius to the thickness (Equation 4.137d). The same factors to the square appear in the elastic energy density (Equation 4.143d). Concerning the stresses: (1) the shear stress is zero (Equations 4.143a, 4.144a, and 4.146a) at all radii; (2) the azimuthal stress (Equation 4.143c) is nearly constant to the lowest order (Equation 4.138a ≡ Equation 4.147a) and hence equal at the mean radius (Equation 4.144d ≡ Equation 4.146c) and free surfaces (Equation 4.147b):*

$$(a_+ + a_-)^2 \ll 4\,(a_+ + a_-)^2: \qquad T_{\varphi\varphi}(a_\pm) = T_{\varphi\varphi}(a), \qquad 2T_{rr}(a) = T_{rr}(a_+) + T_{rr}(a_-) = 0; \tag{4.147a–4.147c}$$

and (3) the radial stress (Equation 4.143b) varies most rapidly, with opposite signs at the surfaces (Equation 4.146b) and their arithmetic mean (Equation 4.147c) that is zero (Equation 4.144c) at the mean radius. The boundary conditions for the normal stress are satisfied exactly (Equations 4.139c and 4.139d) [only approximately (Equation 4.146b)] for the thick (thin) hollow cylinder.

4.5.5 Concentric Tubes of Different Materials

The deformation of a cylindrical tube at the inner (outer) surface may be reduced by inserting inside (outside) a cylinder (sleeve) of a harder material. This leads to the problem of stresses in concentric cylindrical tubes made of different materials (Figure 4.10e). In the case when the inner and outer surfaces are subject to a pressure and no shear stress, the displacement is radial (Equation 4.148a) and given by Equation 4.118b with distinct constants of integration for the inner (Equations 4.148b and 4.148c) [outer (Equations 4.148d and 4.148e)] tube:

$$u_\varphi(r) = 0; \quad a_- \le r \le a: u_r(r) = A_- r + \frac{B_-}{r}; \quad a \le r \le a_+: u_r(r) = A_+ r + \frac{B_+}{r}. \tag{4.148a–4.148e}$$

The corresponding radial stress (Equation 4.133a) is in the inner (outer) tube made of an elastic material of Young's modulus E_- (E_+) and Poisson ratio σ_- (σ_+) [Equations 4.149a and 4.149b (Equations 4.149c and 4.149d)]:

$$a_- \le r \le a: \qquad T_{rr}(r) = \frac{E_-}{1+\sigma_-}\left(\frac{A_-}{1-2\sigma_-} - \frac{B_-}{r^2}\right), \tag{4.149a and 4.149b}$$

$$a \le r \le a_+: \qquad T_{rr}(r) = \frac{E_+}{1+\sigma_+}\left(\frac{A_+}{1-2\sigma_+} - \frac{B_+}{r^2}\right). \tag{4.149c and 4.149d}$$

The continuity of the displacement (Equation 4.150a) [radial stress (Equation 4.150b)]:

$$A_+ + \frac{B_+}{a^2} = A_- + \frac{B_-}{a^2}: \qquad \frac{1+\sigma_+}{1+\sigma_-}\frac{E_-}{E_+}\left(\frac{A_-}{1-2\sigma_-} - \frac{B_-}{a^2}\right) = \frac{A_+}{1-2\sigma_+} - \frac{B_+}{a^2} \tag{4.150a and 4.150b}$$

relates the four constants of integration in Equations 4.148b and 4.148d (Equations 4.149b and 4.149d).

Solving Equation 4.150a for B_+ and substituting in Equation 4.150b leads to

$$\begin{aligned}\frac{1+\sigma_+}{1+\sigma_-}\left(\frac{A_-}{1-2\sigma_-}-\frac{B_-}{a^2}\right)\frac{E_-}{E_+}&=\frac{A_+}{1-2\sigma_+}+A_+-A_--\frac{B_-}{a^2}\\&=2\frac{1-\sigma_+}{1-2\sigma_+}A_+-A_--\frac{B_-}{a^2},\end{aligned}\tag{4.151a}$$

which specifies the two constants of integration of the outer tube (Equation 4.148e) terms and those of the inner tube (Equation 4.148c):

$$2\frac{1-\sigma_+}{1-2\sigma_+}A_+=A_-+\frac{B_-}{a^2}+\frac{1+\sigma_+}{1+\sigma_-}\frac{E_-}{E_+}\left(\frac{A_-}{1-2\sigma_-}-\frac{B_-}{a^2}\right),\tag{4.151b}$$

$$2B_+=2B_-+2a^2A_--2a^2A_+=\frac{B_-+a^2A_-}{1-\sigma_+}-a^2\frac{1-2\sigma_+}{1-\sigma_+}\frac{1+\sigma_+}{1+\sigma_-}\frac{E_-}{E_+}\left(\frac{A_-}{1-2\sigma_-}-\frac{B_-}{a^2}\right).\tag{4.151c}$$

The two remaining constants of integration are determined by the boundary conditions specifying the pressure at the inner (Equation 4.152a) [outer (Equation 4.152b)] surface:

$$p_-=T_{rr}(a_-)=\frac{E_-}{1+\sigma_-}\left(\frac{A_-}{1-2\sigma_-}-\frac{B_-}{a_-^2}\right),\quad p_+=-T_{rr}(a_+)=-\frac{E_+}{1+\sigma_+}\left(\frac{A_+}{1-2\sigma_+}-\frac{B_+}{a_+^2}\right).$$

(4.152a and 4.152b)

Substituting Equations 4.151b and 4.151c in Equation 4.152b specifies the outer boundary condition in terms of the constants of integration of the inner cylinder:

$$-2p_+=\frac{E_+}{1-\sigma_+^2}\left(A_-+\frac{B_-}{a^2}\right)\left(1-\frac{a^2}{a_+^2}\right)+\frac{E_-}{1+\sigma_-}\left(\frac{A_-}{1-2\sigma_-}-\frac{B_-}{a^2}\right)\left(1+\frac{a^2}{a_+^2}\frac{1-2\sigma_+}{1-\sigma_+}\right).\tag{4.152c}$$

The preceding results may be summarized as follows: *consider two coaxial cylinders (Figure 4.10e) with inner (outer) radii a_- (a_+) and an interface at $r = a$ separating the inner (outer) elastic material with Young's modulus E_- (E_+) and Poisson ratio σ_- (σ_+); the inner (outer) surface is subject to a pressure p_- (p_+) and no shear stress. The displacements are radial and are specified in the inner cylinder (Equation 4.148b) by Equation 4.148c with the constants of integration (A_-, B_-) given by the solution of the linear system (Equations 4.152a and 4.152c); this specifies (Equations 4.151b and 4.151c) constants of integration (A_+, B_+) in the radial displacement (Equation 4.148e) in the outer cylinder (Equation 4.148d). The radial stresses are given by Equation 4.149b (Equation 4.149d) in the inner (Equation 4.149a) [outer (Equation 4.149c)] cylinder. The nonzero normal stresses (Equations 4.133a and 4.133b), strains (Equations 4.132a and 4.132c), linearized relative area change (Equation*

4.132d), out-of-plane stress (Equation 4.133d) and elastic energy density (Equation 4.133e) are given replacing (A,B) [(C,D)] by $(A_-,B_-)[(A_+, B_+)]$ in the inner (outer) cylinder.

4.5.6 Materials with Very Different Stiffness

As a particular case, consider an outer material much harder than the inner material (Equation 4.153a), so that Equations 4.151b and 4.151c simplify to Equations 4.153b and 4.153c:

$$E_+ \gg E_-: \qquad 2\frac{1-\sigma_+}{1-2\sigma_+}A_+ = A_- + \frac{B_-}{a^2}, \qquad 2B_+ = \frac{B_- + a^2 A_-}{1-\sigma_+}. \qquad \text{(4.153a–4.153c)}$$

These (Equations 4.153a and 4.153b) simplify the outer boundary condition (Equation 4.152b) to Equation 4.154a:

$$-2\frac{1+\sigma_+}{E_+}p_+ = \left(A_- + \frac{B_-}{a^2}\right)\frac{1}{1-\sigma_+}\left(1-\frac{a^2}{a_+^2}\right); \qquad \frac{1+\sigma_-}{E_-}p_- = \frac{A_-}{1-2\sigma_-} - \frac{B_-}{a_-^2}. \qquad \text{(4.154a and 4.154b)}$$

The inner boundary condition (Equation 4.152a ≡ Equation 4.154b) together with the outer (Equation 4.154a) determines the two constants of integration in the inner cylinder:

$$\left(1+\frac{1}{1-2\sigma_-}\frac{a_-^2}{a^2}\right)A_- = \frac{1+\sigma_-}{E_-}p_-\frac{a_-^2}{a^2} - 2\frac{1-\sigma_+^2}{E_+}\frac{p_+a_+^2}{a_+^2-a^2}, \qquad \text{(4.155a)}$$

$$-\left(1+\frac{1}{1-2\sigma}\frac{a_-^2}{a^2}\right)B_- = \frac{1+\sigma_-}{E_-}p_-a_-^2 + 2\frac{1-\sigma_+^2}{E_+}\frac{a_-^2}{1-2\sigma_-}\frac{p_+a_+^2}{a_+^2-a^2}, \qquad \text{(4.155b)}$$

which is equivalent to

$$\left(1+\frac{1}{1-2\sigma}\frac{a_-^2}{a^2}\right)\left\{A_-, -\frac{B_-}{a^2}\right\} = \frac{1+\sigma_-}{E_-}p_-\frac{a_-^2}{a^2} - 2\frac{1-\sigma_+^2}{E_+}\frac{p_+a_+^2}{a_+^2-a^2}\left\{1, -\frac{1}{1-2\sigma_-}\frac{a_-^2}{a^2}\right\}. \qquad \text{(4.155c and 4.155d)}$$

The approximation (Equation 4.153a ≡ Equation 4.156a) simplifies Equations 4.155a and 4.155b to Equations 4.156b and 4.156c:

$$E_+ \gg E_-: \qquad \left(\frac{a^2}{a_-^2} + \frac{1}{1-2\sigma_-}\right)A_- = \frac{1+\sigma}{E_-}p_- = -B_-\left(\frac{1}{a_-^2} + \frac{1}{1-2\sigma_-}\frac{1}{a^2}\right). \qquad \text{(4.156a–4.156c)}$$

The latter (Equations 4.156b and 4.156c) implies Equation 4.157b. This (Equation 4.157b) also follows from Equations 4.155c and 4.155d by neglecting the second term on the r.h.s. on account of Equation 4.157a. From Equation 4.157b, it follows (Equations 4.153b and

4.153c) that the displacements (Equations 4.157c and 4.157d) are negligible in the outer cylinder (Equations 4.157e and 4.157f) in the limit (Equation 4.157a):

$$E_+ \gg E_-: \qquad a^2 A_- + B_- = 0, \qquad A_+ = 0 = B_+ \qquad a \le r \le a_+: \qquad u_r(r) = 0. \qquad (4.157a–4.157f)$$

Thus, the outer cylinder causes zero displacement (Equation 4.157b) at the interface, and the radial displacement in the inner cylinder (Equation 4.148b ≡ Equation 4.158b) is given (Equation 4.148c) by Equation 4.158c:

$$E_+ \gg E_-: \;\; a_- \le r \le a: \;\; u_r(r) = A_-\left(r - \frac{a^2}{r}\right), \quad \left(\frac{a^2}{a_-^2} + \frac{1}{1-2\sigma_-}\right)A_- = \frac{1+\sigma_-}{E_-}p_-, \quad (4.158a–4.158d)$$

where the constant A_- is determined (Equation 4.158d) by the inner pressure (Equation 4.152a). Thus, *if a cylinder with inner (outer) radius a (a_-) is surrounded by another cylinder with inner (outer) radius a (a_+) made of a much stiffer material (Equation 4.158a): (1) there is no deformation of the outer cylinder (Equations 4.157e and 4.157f); and (2) the displacement (Equation 4.158c) in the inner cylinder (Equation 4.158b) is determined (Equation 4.158d) by the inner pressure p_- and the condition of zero displacement at the interface. The strains (Equations 4.159c through 4.159e), linearized relative area change (Equation 4.160a), and out-of-plane (in-plane) stresses [Equation 4.160b (Equations 4.161a through 4.161c)] and elastic energy density (Equation 4.161d):*

$$E_+ \gg E_-; \;\; a_- \le r \le a: \;\; \left\{S_{rr}(r), S_{\varphi\varphi}(r), S_{r\varphi}(r)\right\} = A_-\left\{1 + \frac{a^2}{r^2}, 1 - \frac{a^2}{r^2}, 0\right\}, \qquad (4.159a–4.159e)$$

$$D_2 = S_{rr} + S_{\varphi\varphi} = 2A_-, \qquad T_{33} = \frac{E_-}{1+\sigma_-}\frac{\sigma_-}{1-2\sigma_-}D_2 = \frac{E_-}{1+\sigma_-}\frac{2\sigma_-}{1-2\sigma_-}A_-, \qquad (4.160a \text{ and } 4.160b)$$

$$\left\{T_{rr}(r), T_{\varphi\varphi}(r), T_{r\varphi}(r)\right\} = \frac{E_- A_-}{1+\sigma_-}\left\{1 \pm \frac{a^2}{r^2} + \frac{2\sigma_-}{1-2\sigma_-}, 0\right\}, \qquad (4.161a–4.161c)$$

$$E_d(r) = \frac{1}{2}\left(T_{rr}S_{rr} + T_{\varphi\varphi}S_{\varphi\varphi}\right) = \frac{E_-}{1+\sigma_-}(A_-)^2\left(1 + \frac{a^4}{r^4} + \frac{2\sigma_-}{1+2\sigma_-}\right), \qquad (4.161d)$$

follow, respectively, from Equations 4.27a, 4.27b, 4.28, 4.30, 4.95d, and 4.76a through 4.76c, and 4.76b. In Equation 4.161d were used the stresses (Equations 4.161a through 4.161c) and strains (Equations 4.159a through 4.159e)

$$\begin{aligned} E_d(r) &= \frac{E_-}{1+\sigma_-}\frac{(A_-)^2}{2}\left[\left(1 + \frac{a^2}{r^2} + \frac{2\sigma_-}{1-2\sigma_-}\right)\left(1 + \frac{a^2}{r^2}\right) + \left(1 - \frac{a^2}{r^2} + \frac{2\sigma_-}{1-2\sigma_-}\right)\left(1 - \frac{a^2}{r^2}\right)\right] \\ &= \frac{E_-}{1+\sigma_-}\frac{(A_-)^2}{2}\left[\left(1 + \frac{a^2}{r^2}\right)^2 + \left(1 - \frac{a^2}{r^2}\right)^2 + \frac{4\sigma_-}{1-2\sigma_-}\right], \end{aligned} \qquad (4.161e)$$

and Equation 4.161e simplifies to Equation 4.161d.

4.5.7 Concentric Tubes with Very Different Inner/Outer Pressures

Consider next the case of very different inner and outer pressures, for example, the firing of a cylindrical gun or a tube containing steam at high pressure. In this case, the inner pressure is much larger than the outer pressure (Equation 4.162a) that may be neglected (Equation 4.162b) in the outer boundary condition (Equation 4.152b):

$$p_- \gg p_+ \sim 0: \qquad A_+ = B_+ \frac{1-2\sigma_+}{a_+^2}; \qquad a \le r \le a_+: \qquad u_r(r) = A_+ r\left(1 + \frac{1}{1-2\sigma_+}\frac{a_+^2}{r^2}\right). \tag{4.162a–4.162d}$$

This leads (Equation 4.148e) to the displacement (Equation 4.162d) in the outer cylinder (Equation 4.148d ≡ Equation 4.162c). Substituting Equation 4.162b in the system (Equations 4.150a and 4.150b) leads to

$$A_- a^2 + B_- = A_+ a^2 + B_+ = A_+\left(a^2 + \frac{a_+^2}{1-2\sigma_+}\right), \tag{4.163a}$$

$$\frac{A_-}{1-2\sigma_-} - \frac{B_-}{a^2} = \frac{1+\sigma_-}{1+\sigma_+}\frac{E_+}{E_-}\left(\frac{A_+}{1-2\sigma_+} - \frac{B_+}{a^2}\right) = \frac{1+\sigma_-}{1+\sigma_+}\frac{E_+}{E_-}\frac{A_+}{1-2\sigma_+}\left(1 - \frac{a_+^2}{a^2}\right). \tag{4.163b}$$

Together with the inner pressure (Equation 4.152a ≡ Equation 4.154b), the system (Equations 4.163a and 4.163b) specifies the three constants (A_-, B_-, A_+).

In the case when the material of the inner tube is much stiffer (Equation 4.164a), then Equation 4.163b simplifies to Equation 4.164b:

$$E_- \gg E_+: \qquad B_- = A_- \frac{a^2}{1-2\sigma_-}; \qquad \frac{1+\sigma_-}{E_-} p_- = \frac{A_-}{1-2\sigma_-}\left(1 - \frac{a^2}{a_-^2}\right). \tag{4.164a–4.164c}$$

The use of Equation 4.164b in the inner boundary condition (Equation 4.154b) determines the constant (Equation 4.164c). Thus, the displacement (Equation 4.148c) in the inner tube (Equation 4.148b ≡ Equation 4.165a) is given (Equations 4.164b and 4.164c) by Equation 4.165b:

$$a_- \le r \le a: \qquad u_r(r) = A_- r\left(1 + \frac{1}{1-2\sigma_-}\frac{a^2}{r^2}\right) = \frac{1+\sigma_-}{E_-}\frac{p_- a_-^2}{a_-^2 - a^2}\left[(1-2r_-)\sigma + \frac{a^2}{r}\right]. \tag{4.165a and 4.165b}$$

Summarizing, *consider two coaxial cylinders (Figure 4.10e) with an interface at r = a, with (1) the inner cylinder (Equation 4.148b) being made of a much stiffer material (Equation 4.164a) than the outer cylinder (Equation 4.148d); and (2) the pressure at the inner surface is much larger than the pressure at the outer surface (Equation 4.162a), so that the latter may be neglected. There is no displacement in the outer cylinder (Equations 4.157a through 4.157f) in approximation 1; the displacement in the inner (Equation 4.165a) cylinder is given by Equation 4.165b. The corresponding strains (Equations 4.27a, 4.27b, and 4.28) are*

$$a_- \le r \le a: \quad \left\{S_{rr}(r), S_{\varphi\varphi}(r), S_{r\varphi}(r)\right\} = \frac{1+\sigma_-}{E_-}\frac{p_- a_-^2}{a_-^2 - a^2}\left\{1-2\sigma_- \mp \frac{a^2}{r^2}, 0\right\}. \quad (4.166a\text{–}4.166d)$$

The linearized relative area change (Equation 4.30) [out-of-plane stress (Equation 4.95d)] is given by [Equation 4.167b (Equation 4.167c)]:

$$a_- \le a \le a: \quad D_2 = S_{rr} + S_{\varphi\varphi} = 2\frac{1+\sigma_-}{E_-}\frac{p_- a_-^2}{a_-^2 - a^2} p_-(1-2\sigma_-), T_{33} = 2p_-\sigma_- \frac{a_-^2}{a_-^2 - a^2}. \quad (4.167a\text{–}4.167d)$$

The Hooke law (Equation 4.76a through 4.76c) specifies the stresses (Equations 4.168b through 4.168d):

$$a_- \le r \le a: \quad \left\{T_{rr}(r), T_{\varphi\varphi}(r), T_{r\varphi}(r)\right\} = \frac{p_- a_-^2}{a_-^2 - a^2}\left\{1 \mp \frac{a^2}{r^2}, 0\right\} \quad (4.168a\text{–}4.168d)$$

$$E_d(r) = \frac{1}{2}\left(T_{rr}S_{rr} + T_{\varphi\varphi}S_{\varphi\varphi}\right) = \frac{1+\sigma_-}{E_-}\left(\frac{p - a_-^2}{a_-^2 - a^2}\right)^2\left(1-2\sigma_- + \frac{a^4}{r^4}\right). \quad (4.168e)$$

The elastic energy density (Equation 4.168e) follows using also the strains (Equations 4.166b through 4.166d) in:

$$\begin{aligned} E_d(r) &= \frac{1+\sigma_-}{2E_-}\left(\frac{p - a_-^2}{a_-^2 - a^2}\right)^2\left[\left(1-\frac{a^2}{r^2}\right)\left(1-2\sigma_- - \frac{a^2}{r^2}\right) + \left(1+\frac{a^2}{r^2}\right)\left(1-2\sigma_- + \frac{a^2}{r^2}\right)\right] \\ &= \frac{1+\sigma_-}{2E_-}\left(\frac{p - a_-^2}{a_-^2 - a^2}\right)^2\left[\left(1-\frac{a^2}{r^2}\right)^2 + \left(1+\frac{a^2}{r^2}\right)^2 - 4\sigma_-\right], \end{aligned} \quad (4.168f)$$

that simplifies to Equation 4.168e ≡ Equation 4.168f.

4.5.8 Stresses due to the Centrifugal Force in a Rotating Cylinder

In the presence of external forces, the forced displacements must be added to the free displacements due to the boundary conditions (Subsections 4.5.1 through 4.5.7); for example, for a rotating solid cylinder, (1) the free displacements are linear and lead to constant stresses (Subsection 4.5.1); and (2) the centrifugal force is proportional to the mass density ρ and radial acceleration, and leads to the differential equation (Equations 4.169b and 4.169c) for the radial displacement (Equation 4.169a):

$$\vec{u} = u(r)\vec{e}_r; \qquad u' \equiv \frac{du}{dr}: \qquad u'' + \frac{u'}{r} - \frac{u}{r^2} = \frac{1+\sigma}{E}\frac{1-2\sigma}{1-\sigma}\rho a_r, \qquad (4.169a–4.169c)$$

that can be obtained from Equation 4.108a using Equation 4.116c. Alternatively, Equation 4.169c can be derived from Equation 4.114b noting that (1) the displacement (Equations 4.169a and 4.169b) is solenoidal $\nabla \wedge \vec{u} = 0$ (Equation 4.29); and (2) the operator $\nabla(\nabla \cdot \vec{u})$ is the divergence (Equation 4.30) followed by the gradient (Equation 4.46a) for a radial (Equation 4.169a) vector:

$$\nabla\left[\nabla \cdot \left(\vec{e}_r u\right)\right] = \vec{e}_r \frac{\mathrm{d}}{\mathrm{d}r}\left[\frac{1}{r}\frac{\mathrm{d}}{\mathrm{d}r}(ru)\right] = \vec{e}_r\left(u'' + \frac{u'}{r} - \frac{u}{r^2}\right). \qquad (4.169d)$$

For a circular motion (Figure 4.10f) with constant angular (linear azimuthal) velocity [Equation 4.170a (Equation 4.170b)], the acceleration is centripetal (Equation 2.126b), that is, radial inward (Equation 4.170c):

$$\Omega = \dot{E}_2 = \dot{\varphi} = \frac{d\varphi}{dt} = \text{const}; \qquad v_\varphi = \Omega r: \qquad a_r = -\Omega^2 r; \qquad k \equiv \rho\Omega^2 \frac{1+\sigma}{E}\frac{1-2\sigma}{1-\sigma}; \qquad (4.170a–4.170d)$$

thus, only one parameter (Equation 4.170d) appears in the second-order ordinary differential equation (Equation 4.179c ≡ Equation 4.171a) for the displacement:

$$-kr = u'' + \frac{u'}{r} - \frac{u}{r^2} = \left[\frac{(ur)'}{r}\right]'; \qquad (ur)' = -\frac{1}{2}kr^3 + 2Ar. \qquad (4.171a \text{ and } 4.171b)$$

A first integration leads to Equation 4.171b, and the second integration to Equation 4.172a. One constant must be zero (Equation 4.172c) for the displacement to be finite at the center (Equation 4.172b):

$$u(r) = -\frac{1}{8}kr^3 + Ar + \frac{B}{r}; \qquad u(\infty) < 0: \qquad B = 0. \qquad (4.172a–4.172c)$$

The remaining constant of integration is determined by the condition of zero radial stress at the surface of the cylinder of radius a:

$$\begin{aligned} 0 &= \frac{1+\sigma}{E}(1-2\sigma)T_{rr}(a) = (1-\sigma)S_{rr}(a) + \sigma S_{\varphi\varphi} = (1-\sigma)u'(a) + \frac{\sigma}{a}u(a) \\ &= (1-\sigma)\left(A - \frac{3}{8}ka^2\right) + \sigma\left(A - \frac{1}{8}ka^2\right) = A - \frac{3-2\sigma}{8}ka^2, \end{aligned} \qquad (4.173)$$

where the Hooke law (Equation 4.76c) and the strain tensor (Equation 4.27a and b) were used.

Substituting A from Equation 4.173, B from Equation 4.172c, and k from Equation 4.170d in Equation 4.172a specifies:

$$0 \le r \le a: \qquad u(r) = \rho \frac{\Omega^2}{8} \frac{1+\sigma}{E} \frac{1-2\sigma}{1-\sigma} r\left[(3-2\sigma)a^2 - r^2\right] \qquad \text{(4.174a and 4.174b)}$$

the radial displacement (Equation 4.174b) in a cylinder (Equation 4.174a) of radius a, rotating (Figure 4.10f) with constant angular velocity (Equation 4.170a), made of a material with mass density ρ, Young's modulus E, and Poisson ratio σ. The corresponding strains (Equations 4.27a, 4.27b, and 4.28) are Equations 4.175a through 4.175c and lead to Equation 4.30 and linearized relative area change (Equation 4.175d):

$$\left\{S_{rr}(r),\, S_{\varphi\varphi}(r),\, S_{r\varphi}(r),\, D_2(r)\right\}$$
$$= \rho \frac{\Omega^2}{8} \frac{1+\sigma}{E} \frac{1-2\sigma}{1-\sigma} \left\{(3-2\sigma)a^2 - 3r^2,\, (3-2\sigma)a^2 - r^2,\, 0,\, 2\left[(3-2\sigma)a^2 - 2r^2\right]\right\}. \qquad \text{(4.175a–4.175d)}$$

The in-plane stresses (Equations 4.76a through 4.76c) are normal (Equations 4.176a through 4.176c), and there is also (Equation 4.95c) an out-of-plane normal stress (Equation 4.176d):

$$\left\{T_{rr}(r),\, T_{\varphi\varphi}(r),\, T_{r\varphi}(r),\, T_{33}(r)\right\} = \frac{\rho}{8} \frac{\Omega^2}{1-\sigma} \left\{(3-2\sigma)\left(a^2 - r^2\right),\right.$$
$$\left.(3-2\sigma)a^2 - (1+2\sigma)r^2, 0, 2\sigma\left[(3-2\sigma)a^2 - 2r^2\right]\right\}. \qquad \text{(4.176a–4.176d)}$$

The elastic energy density (Equation 4.86b) is given by:

$$E_d(r) = \frac{1+\sigma}{E}\left(\frac{\rho}{8}\frac{\Omega^2}{1-\sigma}\right)^2 (1-2\sigma)\left[(3-2\sigma)^2 a^4 + (5-2\sigma)r^4 - 4(3-2\sigma)a^2 r^2\right]. \qquad \text{(4.176e)}$$

The displacement (Equation 4.174) at the surface is Equation 4.177:

$$u_0 \equiv u(a) = \rho \frac{\Omega^2}{4} \frac{1+\sigma}{E} a^2 (1-2\sigma), \qquad \text{(4.177)}$$

$$\left\{T_{rr}(a),\, T_{\varphi\varphi}(a),\, T_{z\varphi},\, T_{rr}(a)\right\} = \rho \frac{\Omega^2}{4} \frac{1-2\sigma}{1-\sigma} a^2 \{0,1,0,\sigma\} = \frac{E}{1-\sigma^2} \frac{u_0}{a} \{0,1,0,\sigma\}, \qquad \text{(4.178a–4.178d)}$$

and the surface stresses are (1) tangential (Equations 4.178a through 4.178c) in agreement with the boundary condition (Equation 4.173 ≡ Equation 4.178b) of zero radial stress; and (2) the in-plane (Equation 4.178b) and out-of-plane (Equation 4.178d) stresses are in the ratio σ, as required (Equation 4.95b) by the absence of deformation (Equation 4.95a) along the axis of the cylinder.

The passage from the strains (Equations 4.175a and 4.175b) [linearized relative area change (Equation 4.175c)] to the normal in-plane (Equations 4.176a and 4.176b) [out-of-plane (Equation 4.176d)] stresses uses Equations 4.76a through 4.76c (Equation 4.95d):

$$\begin{aligned}T_{rr}(r) &= \rho\frac{\Omega^2}{8}\frac{1-2\sigma}{1-\sigma}\left\{(3-2\sigma)a^2-3r^2+\frac{2\sigma}{1-2\sigma}\left[(3-2\sigma)a^2-2r^2\right]\right\}\\ &= \rho\frac{\Omega^2}{8}\frac{1-2\sigma}{1-\sigma}\left[(3-2\sigma)\left(1+\frac{2\sigma}{1-2\sigma}\right)a^2-r^2\left(3+\frac{4\sigma}{1-2\sigma}\right)\right]\\ &= \frac{\rho}{8}\frac{\Omega^2}{1-\sigma}(3-2\sigma)\left(a^2-r^2\right),\end{aligned} \tag{4.179a}$$

$$\begin{aligned}T_{\varphi\varphi}(r) &= \rho\frac{\Omega^2}{8}\frac{1-2\sigma}{1-\sigma}\left\{(3-2\sigma)a^2-r^2+\frac{2\sigma}{1-2\sigma}\left[(3-2\sigma)a^2-2r^2\right]\right\}\\ &= \rho\frac{\Omega^2}{8}\frac{1-2\sigma}{1-\sigma}\left[(3-2\sigma)a^2\left(1+\frac{2\sigma}{1-2\sigma}\right)-r^2\left(1+\frac{4\sigma}{1-2\sigma}\right)\right]\\ &= \frac{\rho}{8}\frac{\Omega^2}{1-\sigma}\left[(3-2\sigma)a^2-(1+2\sigma)r^2\right],\end{aligned} \tag{4.179b}$$

$$T_{33}(r) = \frac{E}{1+\sigma}\frac{\sigma}{1-2\sigma}D_2(r) = \rho\frac{\Omega^2}{4}\frac{\sigma}{1-\sigma}\left[(3-2\sigma)a^2-2r^2\right] \tag{4.179c}$$

in agreement with Equations 4.176a, 4.176b, and 4.176d ≡ Equations 4.179a through 4.179c. The elastic energy density (Equation 4.76b) is specified by the stresses (Equations 4.176b through 4.176d) and strains (Equations 4.175a through 4.175c)

$$\begin{aligned}E_d &= \frac{1}{2}\left(T_{rr}S_{rr}+T_{\varphi\varphi}S_{\varphi\varphi}\right)\\ &= \frac{1+\sigma}{E}\left(\frac{\rho}{8}\frac{\Omega^2}{1-\sigma}\right)^2\frac{1-2\sigma}{2}\\ &\times\left\{(3-2\sigma)(a^2-r^2)\left[(3-2\sigma)a^2-3r^2\right]+\left[(3-2\sigma)a^2-(1+2\sigma)r^2\right]\left[(3-2\sigma)a^2-r^2\right]\right\}\\ &= \frac{1+\sigma}{E}\left(\frac{\rho}{8}\frac{\Omega^2}{1-\sigma}\right)^2\frac{1-2\sigma}{2}\\ &\times\left\{2(3-2\sigma)^2a^4+\left[3(3-2\sigma)+1+2\sigma\right]r^4-(3-2\sigma)a^2r^2(3-2\sigma+5+2\sigma)\right\},\end{aligned} \tag{4.179d}$$

that simplifies to Equation 4.179d ≡ Equation 4.176e.

4.6 Multiharmonic Equation and Fluid Loading on a Dam

Instead of solving the second-order vector differential equation for the displacement (Equation 4.101b ≡ Equation 4.114a ≡ Equation 4.114b) in method I, the fourth-order scalar differential equation (Equation 4.106e) for the stress function is solved in method II (Subsection 4.6.1). This specifies stresses, for example, in Cartesian (Equations 4.40a through 4.40c) [polar (Equations 4.58b through 4.58d)] coordinates and also the displacement vector (Subsection 4.6.8). The stress function is a solution of a biharmonic equation with (without) forcing [Subsection 4.6.1 (Subsections 4.6.2 and 4.6.3)] if the external and inertia forces do not (do) balance. The biharmonic functions are the harmonic functions plus others (Subsection 4.6.3): the biharmonic functions (Subsections 4.6.4 and 4.6.5) in turn are a subset of the multiharmonic functions of higher order (Subsections 4.6.6 and 4.6.7). Method II is applied to (1) the calculation of the stresses (displacements) in a dam [Subsection 4.6.9 (Subsection 4.6.10)] with a wedge shape; (2) the forces and moments applied at the tip of a wedge or at the boundary of a semi-infinite medium (Section 4.7); and (3) the loads and deformations of a rolling wheel (Section 4.9).

4.6.1 Complete Integral of the Forced Biharmonic Equation

The stress function satisfies (Equation 4.105d) in a medium at rest (Equation 4.180a ≡ Equation 4.105e) the biharmonic equation (Equation 4.180b) with forcing (Equation 4.180c):

$$\vec{a}=0: \qquad \nabla^4\Theta = Q, \qquad Q = -\frac{\sigma}{1-\sigma}\left(\nabla\cdot\vec{f}\right). \tag{4.180a–4.180c}$$

The existence of the stress function was proven in the case when (1) the external and inertia forces balance (Equation 4.58a), for example (Equation 4.57a), in polar coordinates (Equations 4.57b through 4.57d); and (2) in Cartesian coordinates (Equations 4.40a through 4.40c) when the inertia force is zero (Equation 4.39a) and the external forces are constant (Equation 4.39b). In both cases, the forcing term (Equation 4.180c) is zero (Equation 4.181a ≡ Equation 4.92c), which leads to a biharmonic equation (Equation 4.181b ≡ Equation 4.92c) for the stress function:

$$Q=0: \qquad 0 = \nabla^4\Theta = \left(\frac{\partial^2}{\partial x^2}+\frac{\partial^2}{\partial y^2}\right)^2\Theta(x,y) = \left(\frac{\partial^4}{\partial x^4}+\frac{\partial^4}{\partial y^4}+2\frac{\partial^4}{\partial x^2\partial y^2}\right)\Theta(x,y). \tag{4.181a and 4.181b}$$

If the forcing is not zero at some isolated points, then the stress function exists elsewhere, and Equation 4.180c acts as a singularity, such as source, sink, or vortex, in a potential flow. The biharmonic equation (Equation 4.181b) can be written in complex conjugate coordinates (Equations 2.144a and 2.144b), which leads Equations 2.151a and 2.151b to Equation 4.182a:

$$Q\left(z,z^*\right) = \nabla^4\Theta = \left(4\frac{\partial^4}{\partial z^*\partial z}\right)^2\Theta = 16\frac{\partial^4\Theta}{\partial z^{*2}\partial z^2}. \tag{4.182a}$$

A particular integral is obtained (Equation 4.182b) by four integrations:

$$\Theta_1\left(z, z^*\right) = \frac{1}{16}\int^z d\xi \int^n d\eta \int^{z*} du \int^u dv\; Q(\eta, v). \tag{4.182b}$$

The difference (Equation 4.183a) between the general (Equation 4.182a) and the particular (Equation 4.182b) integrals satisfies an unforced biharmonic equation (Equation 4.183b):

$$\Theta_0\,(z, z^*) = \Theta\,(z, z^*) - \Theta_1\,(z, z^*): \qquad \nabla^4\,\Theta_0 = \nabla^4\,\Theta - \nabla^4\,\Theta_1 = 0. \quad \text{(4.183a and 4.183b)}$$

The latter (Equation 4.183b ≡ Equation 4.184b) is integrated (Equation 4.184c) in terms of four arbitrary twice differentiable functions (Equation 4.184a):

$$f, g, h, j \in \mathcal{D}^2(|C): \qquad \frac{\partial^4\Theta_0}{\partial z^2 \partial z^{*2}} = 0 \Leftrightarrow \Theta_0\left(z, z^*\right) = f(z) + z^* h(z) + g\left(z^*\right) + zj\left(z^*\right).$$

(4.184a–4.184c)

The result (Equation 4.184c) follows, noting that Equation 4.184b is satisfied by a polynomial of the first degree in z (z^*), whose coefficients can be functions of z^* (z).

The preceding results can be summarized in the statement: *the forced biharmonic equation (Equation 4.180b) in complex conjugate coordinates (Equation 4.182a) has* **complete integral:**

$$\Theta\left(z, z^*\right) = f(z) + z^* h(z) + g\left(z^*\right) + zj\left(z^*\right) + \frac{1}{16}\int^z d\xi \int^\xi d\eta \int^{z*} du \int^u dv\, Q(\eta, v) \tag{4.185}$$

consisting of (1) the **general integral** *(Equation 4.184c) of the unforced equation (Equation 4.184b), involving four arbitrary twice differentiable functions (Equation 4.184a); and (2) a* **particular integral** *(Equation 4.182b) of the forced equation (Equation 4.182a), which simplifies in the case of constant forcing term (Equation 4.186a) to Equation 4.186b:*

$$Q\left(z, z^*\right) = \text{const:} \quad \Theta\left(z, z^*\right) - f(z) - z^* g(z) - h\left(z^*\right) - zj\left(z^*\right) = Q\frac{z^{*2} z^2}{64} = Q\frac{|z|^4}{64}.$$

(4.186a and 4.186b)

A real solution (Equation 4.187c) is obtained by using only two (Equations 4.187a and 4.187b) arbitrary functions:

$$f = g, h = j: \qquad \Theta\left(z, z^*\right) = f(z) + f\left(z^*\right) + z^* g(z) + zg\left(z^*\right) + Q\frac{|z|^2}{64}, \quad \text{(4.187a–4.187c)}$$

since analytic functions (Equations 4.188a and 4.188b) satisfy (Equation I.31.18 ≡ Equations 4.188c and 4.188d):

$$f, g \in \mathcal{A}\,(|C): \qquad f(z^*) = f^*(z), \qquad g(z^*) = g^*(z)|, \quad \text{(4.188a–4.188d)}$$

$$\Theta\,(z, z^*) = f(z) + f^*(z) + z^*g(z) + [z^*g(z)]^* = 2\mathrm{Re}\,[f(z)] + 2\mathrm{Re}\,[z^*g(z)], \tag{4.188e}$$

that lead to Equation 4.188e.

4.6.2 Biharmonic Functions in Cartesian and Polar Coordinates

If the inertia and external forces balance, the stress function exists and satisfies a biharmonic equation, which can be written in Cartesian coordinates (Equation 4.181b ≡ Equation 4.189 ≡ Equation 4.92c):

$$0 = \nabla^2\,\Theta = (\partial_{xx} + \partial_{yy})^2\,\Theta = \partial_{xxxx}\,\Theta + 2\partial_{xxyy}\,\Theta + \partial_{yyyy}\,\Theta. \tag{4.189}$$

A solution is a homogeneous polynomial of third degree:

$$\Theta(x,y) = Ax^3 + Bx^2\,y + Cxy^2 + Dy^3, \tag{4.190}$$

where A, B, C, and D are four arbitrary constants. Note that (1) a polynomial of the second degree would lead to constant stresses by Equations 4.58b through 4.58d; and (2) a polynomial of fourth degree or higher would not satisfy the biharmonic equation if the coefficients were left arbitrary.

Using the scalar Laplacian in polar coordinates (Equation 4.111 ≡ Equation I.11.28c), the biharmonic equation becomes

$$0 = \nabla^4\,\Theta = (r^{-1}\,\partial_r\,r\partial_r + r^{-2}\partial_{\varphi\varphi})^2\,\Theta = (\partial_{rr} + r^{-1}\partial_r + r^{-2}\partial_{\varphi\varphi})^2\,\Theta. \tag{4.191}$$

The solution depending only on the azimuthal angle (Equation 4.192a) [two-dimensional radius (Equation 4.193b)] is Equation 4.192b (Equation 4.193b):

$$\frac{\mathrm{d}^4\Theta}{\mathrm{d}\varphi^4} = 0, \qquad \Theta(\varphi) = B + A\varphi + C\varphi^2 + D\varphi^3, \tag{4.192a and 4.192b}$$

$$\frac{1}{r}\frac{\mathrm{d}}{\mathrm{d}r}\left\{r\frac{\mathrm{d}}{\mathrm{d}r}\left[\frac{1}{r}\frac{\mathrm{d}}{\mathrm{d}r}\left(r\frac{\mathrm{d}\Theta}{\mathrm{d}r}\right)\right]\right\} = 0: \qquad \Theta(r) = r^2(A\log r + B) + C\log r + D, \tag{4.193a and 4.193b}$$

where (A, B, C, D) are arbitrary constants.

The former result (Equation 4.192b) is immediate. The latter result (Equation 4.193b) is obtained next: (1) since $\nabla^2\,\Theta$ is a harmonic function, a double integration shows that its axially symmetric solution (Equation 4.194b) is a logarithmic potential (Equation 4.194c), where $\left(\bar{A}, \bar{B}\right)$ are arbitrary constants:

$$0 = \nabla^4\Theta = \frac{1}{r}\frac{\mathrm{d}}{\mathrm{d}r}\left[r\frac{\mathrm{d}}{\mathrm{d}r}\left(\nabla^2\Theta\right)\right]: \quad r\frac{\mathrm{d}}{\mathrm{d}r}\left(\nabla^2\,\Theta\right) = \bar{A}, \;\; \nabla^2\Theta = \bar{A}\log r + \bar{B}; \tag{4.194a–4.194c}$$

(2) rewriting Equation 4.194c ≡ Equation 4.195a leads to Equation 4.195b after a third integration:

$$\frac{1}{r}\frac{d}{dr}\left(r\frac{d\Theta}{dr}\right) = \bar{A}\log r + \bar{B}, \qquad r\frac{d\Theta}{dr} = \frac{r^2}{2}\left[\bar{A}\left(\log r - \frac{1}{2}\right) + \bar{B}\right] + C; \quad \text{(4.195a and 4.195b)}$$

(3) it can be checked that differentiating Equation 4.195b with regard to r and dividing by r leads back to Equation 4.195a:

$$\begin{aligned} &\frac{1}{r}\frac{d}{dr}\left[\bar{A}\frac{r^2}{2}\left(\log r - \frac{1}{2}\right) + \bar{B}\frac{r^2}{2} + C\right] \\ &\qquad = \frac{1}{r}\left\{\bar{A}\left[r\left(\log r - \frac{1}{2}\right) + \frac{r^2}{2}\cdot\frac{1}{r}\right] + \bar{B}r\right\} = \bar{A}\log r + \bar{B}; \end{aligned} \tag{4.195c}$$

(4) a fourth and final integration leads from Equation 4.195b to Equation 4.196a:

$$\Theta(r) = \frac{\bar{A}}{4}r^2(\log r - 1) + \frac{\bar{B}}{4}r^2 + C\log r + D; \quad A \equiv \frac{\bar{A}}{4}, \quad B \equiv \frac{\bar{B} - \bar{A}}{4}; \quad \text{(4.196a–4.196c)}$$

(5) it can be checked that differentiating Equation 4.196a with regard to r and multiplying by r leads back to Equation 4.195b:

$$\begin{aligned} &r\frac{d}{dr}\left[\frac{\bar{A}}{4}r^2(\log r - 1) + \frac{\bar{B}}{4}r^2 + C\log r + D\right] \\ &= r\left\{\frac{\bar{A}}{4}\left[2r(\log r - 1) + r^2\frac{1}{r}\right] + \frac{\bar{B}}{2}r + \frac{C}{r}\right\} \\ &= \frac{\bar{A}}{2}r^2\left(\log r - \frac{1}{2}\right) + \frac{\bar{B}}{2}r^2 + C; \end{aligned} \tag{4.196d}$$

and (6) a modified choice of arbitrary constants of integration (Equation 4.196b and 4.196c) leads from Equation 4.196a to Equation 4.193b, proving that it is the general integral of Equation 4.193a. Thus, *solutions of the biharmonic equation have been obtained (1) in Cartesian coordinates (Equation 4.189) as a homogeneous polynomial of degree 3 (Equation 4.190); and (2) in polar coordinates (Equation 4.191) with purely azimuthal (Equations 4.192a and 4.192b) [radial (Equations 4.193a and 4.193b)] dependence.*

4.6.3 Biharmonic Derived from Harmonic Functions

If Φ is a harmonic function (Equation 4.197a), then (1) it is also a biharmonic function (4.197b); (2) it specifies other (Equation 4.197c) biharmonic functions:

$$\nabla^2\Phi = 0, \qquad \nabla^4\Theta = 0: \qquad \Theta = \Phi\, r\,(Cr + B\cos\varphi + A\sin\varphi); \qquad \text{(4.197a–4.197c)}$$

for example, the harmonic function (Equation 4.198a) leads to the biharmonic functions (Equation 4.198b):

$$\Phi = \varphi, \qquad \Theta = r\,\varphi\,(Cr + B\cos\varphi + A\sin\varphi). \qquad \text{(4.198a and 4.198b)}$$

The proof of Equations 4.197a through 4.197c follows from Θ in Equation 4.197b, having a factor z or z^*z relative to Φ in Equation 4.197a, that is, if Φ satisfies a Laplace equation (Equation 4.199a):

$$\Phi\,(z, z^*) = f(z) + g(z^*), \qquad \Theta\,(z, z^*) = z\,\Phi\,(z, z^*),\ z^*z\,\Phi\,(z, z^*), \qquad \text{(4.199a and 4.199b)}$$

then Equation 4.199b satisfies Equation 4.182a, a biharmonic equation (Equation 4.199a ≡ Equation 4.197b):

$$\frac{\partial^4\Theta}{\partial z^2\partial z^{*2}} = \frac{\partial^4}{\partial z^2\partial z^{*2}}\left\{zf(z), zg\left(z^*\right), z^*zf(z), z^*zg\left(z^*\right)\right\} = 0. \qquad \text{(4.199c)}$$

The harmonic functions (Equations 4.197a and 4.198a) are also biharmonic (Equation 4.197b), but the biharmonic functions (Equations 4.197c and 4.198b) are not harmonic as shown next.

The second (third) term on the r.h.s. of Equation 4.197a corresponds to the real (imaginary) part of Equation 4.200a:

$$z = re^{i\varphi} = r\cos\varphi + ir\sin\varphi: \qquad r\{A\cos\varphi, B\sin\varphi\} = \{A\,\mathrm{Re}\,(z), B\,\mathrm{Im}\,(z)\}, \qquad \text{(4.200a–4.200c)}$$

multiplied by a constant A (B) in Equation 4.200b (Equation 4.200c), and thus are harmonic functions. The first term on the r.h.s. of Equation 4.197c is Cr^2 and is not a harmonic function because it does not satisfy (Equation 4.111) the Laplace equation (Equation 4.200d):

$$0 = \nabla^2\left(Cr^2\right) = \frac{1}{r}\frac{d}{dr}\left[r\frac{d}{dr}\left(Cr^2\right)\right] = \frac{1}{r}\frac{d}{dr}\left(2Cr^2\right) = 4C, \qquad \text{(4.200d)}$$

unless C = 0. It is only when this term is omitted that the biharmonic function (Equation 4.197a) is also harmonic. Concerning (Equation 4.198b) none of the terms is a harmonic function:

$$\nabla^2\left(r^2\varphi\right) = \varphi\frac{1}{r}\frac{d}{dr}\left[r\frac{d}{dr}\left(r^2\right)\right] = 4\varphi$$

$$\nabla^2\left(r\varphi\,{}^{\cos}_{\sin}\varphi\right) = \varphi\,{}^{\cos}_{\sin}\varphi\,\frac{1}{r}\frac{d}{dr}\left[r\frac{d}{dr}(r)\right] + \frac{1}{r}\frac{d^2}{d\varphi^2}\left(\varphi\,{}^{\cos}_{\sin}\varphi\right) \qquad \text{(4.201a)}$$

$$= \frac{\varphi}{r}\,{}^{\cos}_{\sin}\varphi - \frac{\varphi}{r}\,{}^{\cos}_{\sin}\varphi \mp \frac{2}{r}\,{}^{\sin}_{\cos}\varphi = \mp\frac{2}{r}\,{}^{\sin}_{\cos}\varphi.$$

Thus, the classes of biharmonic functions are wider than the class of harmonic functions, as shown also in the next example.

4.6.4 Harmonic as Subset of the Biharmonic Functions

The Laplace equation (Equation 2.152c ≡ Equation 4.202a) has a general integral (Equation 2.152d), which includes as particular cases (Equations 4.202b through 4.202d):

$$\nabla^2 \Phi = 0; \quad a \in| R: \quad \Phi_+ = z^a + z^{*a} = r^a \left(e^{ia\varphi} + e^{-ia\varphi}\right) = 2r^a \cos(a\varphi), \tag{4.202a–4.202c}$$

$$\Phi_- = -i\left(z^a + z^{*a}\right) = -ir^a\left(e^{ia\varphi} + e^{-ia\varphi}\right) = 2r^a \sin(a\varphi). \tag{4.202d}$$

The biharmonic equation (Equation 4.184b ≡ Equation 4.201a) has a general integral (Equation 4.184c) that includes as particular integrals not only Equations 4.202b through 4.202d but also Equations 4.203b through 4.203d:

$$\nabla^4 \Theta = 0; a \in| R: \quad \Theta_+ = z^{a+1} z^* + z^{*a+1} z = z^*z\left(z^a + z^{*a}\right) = 2r^{a+2}\cos(a\varphi), \tag{4.203a–4.203c}$$

$$\Phi_- = -i\left(z^* z^{a+1} + z^{*a+1} z\right) = -iz^*z\left(z^a + z^{*a}\right) = 2r^{a+2}\sin(a\varphi). \tag{4.203d}$$

This can be checked by using the Laplacian in polar coordinates (Equation 4.111) as follows: (1) the functions (Equations 4.202c and 4.202d) are of the form Equation 4.204a; (2) they satisfy the Laplace equation (Equation 4.204b) if Equation 4.204c is met:

$$\Phi(r,\varphi) = r^b \begin{matrix}\cos\\ \sin\end{matrix}(a\varphi): \quad 0 = \nabla^2\Phi = \left(\frac{1}{r}\frac{\partial}{\partial r} r \frac{\partial}{\partial r} + \frac{1}{r^2}\frac{\partial^2}{\partial\varphi^2}\right)\Phi$$

$$= r^{b-2}\left(b^2 - a^2\right)\begin{matrix}\cos\\ \sin\end{matrix}(a\varphi), \quad b = \pm a; \tag{4.204a–4.204c}$$

and (3) substitution of Equation 4.204c in Equation 4.204a yields Equations 4.202c and 4.202d. Thus, *the Laplace equation (Equation 4.202a ≡ Equation 4.205c) has solutions (Equation 4.205b) with a real Equation 4.205a:*

$$a \in| R: \quad \Phi(r,\varphi) = r^{\pm a}\left[A\cos(a\varphi) + B\sin(a\varphi)\right] \Rightarrow \nabla^2\Phi = 0. \tag{4.205a–4.205c}$$

If $a = +n$ ($a = -n$) with n a positive integer $n \in| N$, the solution for a potential flow corresponds [Chapter I.14 (Chapter I.12)] to a corner of angle $\beta = \pi/n$ (2^n– multiple of order n).

The general form (Equation 4.204a ≡ Equation 4.206a) also includes Equations 4.203b and 4.203d, and is a biharmonic function (Equation 4.206b):

$$\Theta(r,\varphi) = r^b \begin{matrix}\cos\\ \sin\end{matrix}(a\varphi) \qquad 0 = \nabla^4\Theta = \left(\frac{1}{r}\frac{\partial}{\partial r} r \frac{\partial}{\partial r} + \frac{1}{r^2}\frac{\partial^2}{\partial\varphi^2}\right)^2 r^b \begin{matrix}\cos\\ \sin\end{matrix}(a\varphi)$$

$$= \left(\frac{1}{r}\frac{\partial}{\partial r} r \frac{\partial}{\partial r} + \frac{1}{r^2}\frac{\partial^2}{\partial\varphi^2}\right) r^{b-2}\left(b^2 - a^2\right)\begin{matrix}\cos\\ \sin\end{matrix}(a\varphi) = r^{b-4}\left[(b-2)^2 - a^2\right]\left(b^2 - a^2\right)\begin{matrix}\cos\\ \sin\end{matrix}(a\varphi) \tag{4.206a and 4.206b}$$

if one of Equations 4.206c and 4.206d is met:

$$b = \pm a, 2 \pm a. \tag{4.206c and 4.206d}$$

Substitution of Equation 4.206c (Equation 4.206d) in Equation 4.206a yields Equations 4.202c and 4.202d (Equations 4.203c and 4.203d). The solution [Equation 4.207b ≡ Equations 4.205b and 4.206c (Equation 4.207c ≡ Equations 4.206a and 4.206d)] corresponds to Equations 4.202c and 4.202d (Equations 4.203c and 4.203d), replacing a by $\pm a$. From Equations 4.206a through 4.206d, it follows that *the biharmonic equation (Equation 4.207d):*

$$a \in| R: \Theta_\pm (r,\varphi) = r^{\pm a} \{1, r^2\} [A \cos (a\varphi) + B \sin (a\varphi)] \Rightarrow \nabla^4\Theta = 0 \quad (4.207a–4.207d)$$

has (1) the solution (Equation 4.207b ≡ Equations 4.202c and 4.202d) that is both a harmonic and a biharmonic function; and (2) the solution (Equation 4.207c) that is a biharmonic but not a harmonic function. The class of biharmonic functions contains the class of harmonic functions:

$$\mathcal{H} \equiv \{\Phi: \nabla^2\Phi = 0\} \subset \{\Theta: \nabla^4\Theta = 0\} \equiv \mathcal{H}_2 \quad (4.208)$$

because every harmonic is biharmonic, but not the reverse. This follows from the general biharmonic function (Equation 4.184c) that is harmonic (Equation 2.152d) iff (h, j) are both zero (or constant, in which case, they can be incorporated in other terms).

4.6.5 Multiharmonic Operator, Equation, and Functions

The reasoning from Equations 4.204a through 4.204c and 4.205a through 4.205c to Equations 4.206a through 4.206d and 4.207a through 4.207d may be extended (Equation 4.209a) by induction (Equation 4.207b) to Equation 4.207c:

$$\begin{aligned} a \in| R,\ n \in| N: \quad 0 &= \left\{ \frac{1}{r}\frac{\partial}{\partial r} r \frac{\partial}{\partial r} + \frac{1}{r^2}\frac{\partial^2}{\partial \varphi^2} \right\}^n r^b \begin{matrix}\cos\\ \sin\end{matrix}(a\varphi) \\ &= r^{b-2n} \begin{matrix}\cos\\ \sin\end{matrix}(a\varphi) \prod_{m=0}^{n-1} \left[(b-2m)^2 - a^2 \right], \end{aligned} \quad (4.209a–4.209c)$$

which vanishes for

$$b = \pm a, 2 \pm a, 4 \pm a, \ldots, 2m - 2 \pm a. \quad (4.209d)$$

Thus, *the* **multiharmonic equation** *(Equation 4.210e) of order n (Equation 4.210a) has 2n solutions (Equations 4.210b through 4.210d) in polar coordinates:*

$$n \in| N; \qquad a \in| R; m = 0, 1, \ldots, n-1:$$

$$\Xi(r,\varphi) = r^{2m\pm a} \begin{matrix}\cos\\ \sin\end{matrix}(a\varphi) \quad \Rightarrow \quad 0 = \nabla^{2n}\Xi = \left(\partial_{rr} + r^{-1}\partial_r + r^{-2}\partial_{\varphi\varphi}\right)^n \Xi. \quad (4.210a–4.210e)$$

The **multiharmonic operator** *(Equation 4.210e) of order n (Equation 4.211a) is written (Equation 4.211b) in complex conjugate coordinates:*

$$n \in| N: \qquad \nabla^{2n}\Xi = 2^{2n} \frac{\partial^{2n}\Xi}{\partial z^n \partial z^{*n}}, \quad (4.211a \text{ and } 4.211b)$$

as follows from Equations 2.151a and 2.151b. The particular solutions (Equations 4.210a through 4.210e) corresponding to Equation 4.212a (Equation 4.212b) in complex conjugate (polar coordinates) are

$$\begin{aligned}\Xi(z,z^*) &= (z^*z)^m\left(z^a \pm z^{*a}\right) = z^{m+a}z^{*a} \pm z^m z^{*m+a} = r^{2m+a}\left(e^{ia\varphi} \pm e^{-ia\varphi}\right)\\ &= 2r^{a+2m}\{\cos(a\varphi), i\sin(a\varphi)\}.\end{aligned} \tag{4.212a and 4.212b}$$

These are particular cases of the general solution obtained next.

The multiharmonic operator in complex conjugate coordinates (Equations 4.211a and 4.211b) implies that (1) the unforced multiharmonic equation (Equation 4.213b) of order n (Equation 4.213a) has a general integral (Equation 4.213d) consisting of the sum of two polynomials in z (z) with degree n − 1, whose coefficients (Equation 4.213c) are arbitrary n-times differentiable functions of the other complex conjugate variable z*(z):*

$$n \in |N; \qquad \nabla^{2n}\Xi_0 = 0 \Leftrightarrow \left\{A_0,\ldots,A_{N-1},B_0,\ldots,B_{n-1}\right\} \in \mathcal{D}^n\left(|C\right):$$

$$\Xi_0\left(z,z^*\right) = \sum_{m=0}^{n-1}\left\{z^m A_m\left(z^*\right) + z^{*m}B_m(z)\right\}; \tag{4.213a–4.213d}$$

(2) the complete integral (Equation 4.214c) of the forced multiharmonic equation (Equation 4.214b) of order n (Equation 4.214a) adds to the general integral (Equations 4.213c and 4.213d) of the unforced equation (Equations 4.213a and 4.213b):

$$n \in |N; \quad \nabla^{2n}\Xi = f\left(z,z^*\right): \quad \Xi f\left(z,z^*\right) - \Xi_0\left(z,z^*\right)$$

$$= 2^{-2n}\int^{z} dz_1 \int^{z_1} dz_2 \cdots \int^{z_{n-1}} dz_n \int^{z^*} dz_1^* \int^{z_1^*} dz_2^* \cdots \int^{z_{n-1}^*} dz_n^* f\left(z_n, z_n^*\right), \tag{4.214a–4.214c}$$

the forcing term integrated n times with regard to z(z) with (Equation 4.214c) a constant factor from Equation 4.211b.* The preceding results (Equations 4.213a through 4.213d and 4.214a through 4.214c) are proven for any n positive integer starting from the particular cases $n = 1$ ($n = 2$) of the harmonic (Equations 2.157a, 2.159, and 2.156b) [biharmonic (Equations 4.182a, 4.185, and 4.184a)] functions [Subsection 2.4.4 (Subsection 4.6.1)].

4.6.6 Chain of Inclusion of Multiharmonic Functions

The set of multiharmonic functions of order n contains all of lower order, in particular, the biharmonic and harmonic functions:

$$n \in |N: \qquad \mathcal{H}_n \equiv \{\Xi: \nabla^{2n}\,\Xi = 0\} \supset \mathcal{H}_{n-1} \ldots \supset \mathcal{H}_2 \supset \mathcal{H}. \tag{4.215}$$

As an example, the function (Equation 4.216b) is neither harmonic (Equation 4.216c) nor biharmonic (Equation 4.216d), but is multiharmonic (Equation 4.216e) for all higher orders (Equation 4.216a):

$$m = 2,3,\ldots: \qquad \Phi(r,\varphi) = r^{4\pm a}\,{\cos \atop \sin}(a\varphi) \;\Rightarrow\; \nabla^2\Phi \neq 0 \neq \nabla^4\Phi, \quad \nabla^{2m}\Phi = 0. \tag{4.216a–4.216e}$$

The result [Equation 4.216c (Equation 4.216d)] follows from Equation 4.217a (Equation 4.217b):

$$\nabla^2\Phi = \left[(4 \pm a)^2 - a^2\right] r^{2\pm a} \begin{matrix}\cos\\ \sin\end{matrix}(a\varphi) = 8(2 \pm a) r^{2\pm a} \begin{matrix}\cos\\ \sin\end{matrix}(a\varphi), \tag{4.217a}$$

$$\nabla^4\Phi = \left[(4 \pm a)^2 - a^2\right]\left[(2 \pm a)^2 - a^2\right] r^{\pm a} \begin{matrix}\cos\\ \sin\end{matrix}(a\varphi) = 16(2 \pm a)^2 r^{\pm a} \begin{matrix}\cos\\ \sin\end{matrix}(a\varphi), \tag{4.217b}$$

and Equation 4.216e follows from (Equation 4.217c):

$$\nabla^6\Phi = \left[(4 \pm a)^2 - a^2\right]\left[(2 \pm a)^2 - a^2\right]\left(a^2 - a^2\right) r^{\pm a-2} \begin{matrix}\cos\\ \sin\end{matrix}(a\varphi) = 0, \tag{4.217c}$$

for $m = 3$ and hence also for $m = 4, \ldots$. Choosing Equation 4.218a in Equations 4.210a through 4.210d, it follows that (1) the Laplace equation (Equation 4.218d) has (Equation 4.218b) solutions (Equation 4.218c):

$$a = 2; m = 0: \quad \Phi(r,\varphi) = \left(Ar^2 + \frac{B}{r^2}\right)\cos(2\varphi) + \left(Ar^2 + \frac{D}{r^2}\right)\sin(2\varphi) \quad \Rightarrow \quad \nabla^2\Phi = 0; \tag{4.218a–4.218d}$$

(2) the biharmonic equation (Equation 4.219c) has (Equation 4.219a) in addition to Equation 4.218c the solution (Equation 4.219b):

$$m = 1: \quad \Theta(r, \varphi) = (E + Fr^4)\cos(2\varphi) + (G + Hr^4)\sin(2\varphi) \Rightarrow \nabla^4\Theta = 0; \tag{4.219a–4.219c}$$

(3) the triharmonic equation (Equation 4.220c) has (Equation 4.220a) in addition to Equations 4.218c and 4.219b the solution (Equation 4.220b):

$$m = 2: \quad \Xi(r, \varphi) = r^6\left[I\cos(6\varphi) + J\sin(6\varphi)\right] \Rightarrow \nabla^6\Xi = 0; \tag{4.220a–4.220c}$$

and (4) the multiharmonic equation (Equation 4.221c) of order n (Equation 4.221a) has (in addition to all the preceding Equations 4.218c, 4.219b, 4.220b,…) the solution (Equation 4.221b):

$$n \in |N: \quad \Xi(r,\varphi) = r^n\left[\left(Kr^2 + \frac{L}{r^2}\right)\cos(2\varphi) + \left(Mr^2 + \frac{N}{r^2}\right)\sin(2\varphi)\right] \Rightarrow \nabla^{2n}\Xi = 0. \tag{4.221a–4.221c}$$

In Equations 4.218 through 4.221, all coefficients (A–N) are arbitrary constants. *The solutions of the biharmonic equation include (1) the particular integral (Equation 4.190) in Cartesian coordinates (Equation 4.189); (2) the particular integrals (Equations 4.198a and 4.198b) [Equations*

4.207b and 4.207c including Equation 4.218c] in polar coordinates (Equation 4.191); (3 and 4) the particular integral [Equation 4.192b (Equation 4.193b)] depending only on the azimuthal angle (Equation 4.192a) [radius (Equation 4.193a)]; and (5) whereas the particular solutions 1–4 involve arbitrary constants, the general integral (Equation 4.184c) in complex conjugate coordinates (Equation 4.184b) involves four arbitrary twice differentiable functions (Equation 4.184a). The solutions in Cartesian (polar) coordinates are used to determine the stresses in a wedge-shaped dam (Subsections 4.6.9 and 4.6.10) [stresses in a wedge with a concentrated force and/or moment at the tip (Subsections 4.7.2 and 4.7.3)]. The particular and general solutions of the biharmonic equation can be generalized to the scalar multiharmonic equation (Subsections 4.6.5 through 4.6.8).

4.6.7 Multiharmonic Functions in Cartesian, Conjugate, and Polar Coordinates

The multiharmonic operator in Cartesian coordinates (Equation 4.222) may be expanded by the binomial theorem (Equations 3.146a through 3.146c ≡ Equation I.25.38):

$$\nabla^{2n}\Xi = \left(\frac{\partial^2}{\partial x^2} + \frac{\partial^2}{\partial y^2}\right)^n \Xi = \sum_{m=0}^{n} \binom{n}{m} \left(\frac{\partial}{\partial x}\right)^{2m} \left(\frac{\partial}{\partial y}\right)^{2n-2m} \Xi = \sum_{m=0}^{n} \frac{n!}{m!(n-m)!} \frac{\partial^{2n}\Xi}{\partial^{2m}x\partial^{2n-2m}y}. \quad (4.222)$$

The particular case $n = 2$ ($n = 3$) is [Equation 4.92c ≡ Equation 4.181b ≡ Equation 4.189 (Equation 4.223)]:

$$\nabla^6\Xi = \left(\frac{\partial^2}{\partial x^2} + \frac{\partial^2}{\partial y^2}\right)^3 \Xi = \frac{\partial^6\Xi}{\partial x^6} + \frac{\partial^6\Xi}{\partial y^6} + 3\frac{\partial^6\Xi}{\partial x^2\partial y^4} + 3\frac{\partial^6\Xi}{\partial x^4\partial y^2}. \quad (4.223)$$

Because the multiharmonic equation (Equation 4.224c) of order n (Equation 4.224a) in Cartesian coordinates (Equation 4.222) involves derivatives with regard to x (y) of order at least $2n$, a homogeneous polynomial of degree $2n - 1$ is a solution (Equation 4.224b):

$$n \in| N: \quad \Xi_0(z,y) = \sum_{m=0}^{2n-1} A_m x^m y^{2n-m-1} \quad \Rightarrow \quad \nabla^{2n}\Xi_0 = 0. \quad (4.224a–4.224c)$$

The particular case $n = 2$ ($n = 3$) is [Equation 4.190 (Equations 4.225a and 4.225b)]:

$$n = 3: \quad \Xi_0(z, y) = A_0x^5 + A_1x^4y + A_2x^3y^2 + A_3x^2y^3 + A_yxy^4 + A_5 y. \quad (4.225a \text{ and } 4.225b)$$

The solution (Equation 4.224b) remains valid:

$$\Xi_0(z,y) = \sum_{k=0}^{2n} \sum_{m=0}^{k} A_{m,k} x^m y^{2k-m-1}, \quad (4.226)$$

adding polynomials of lower degree.

The multiharmonic equation (Equation 4.210e) of order n (Equation 4.227a) depending only on the azimuthal angle (Equation 4.227c) has solution as a polynomial (Equation 4.227b) of degree $2n - 1$:

$$n \in| N: \quad \Xi(\varphi) = \sum_{m=0}^{2n-1} A_m \varphi^m \quad \Rightarrow \quad 0 = \nabla^{2n}\Xi = r^{-2n}\frac{d^{2n}\Xi}{d\varphi^{2n}}. \quad (4.227a–4.227c)$$

The multiharmonic equation (Equation 4.210e) of order n (Equation 4.228a) depending only on the radius (Equation 4.228c) has solution (Equation 4.228b):

$$n \in| N: \quad \Xi(r) = \sum_{m=0}^{2n-1} r^{2m}\left(A_m \log r + B_m\right) \Rightarrow 0 = \nabla^{2n}\Xi = \left[\frac{1}{r}\frac{d}{dr}\left(r\frac{d}{dr}\right)\right]^n \Xi. \quad (4.228a–4.228c)$$

The latter result is proven by induction: (1) starting for $n = 1$ with the Laplace equation (Equation 4.229b) that has radial solution (Equation 4.229a) as shown in Equations 4.194a through 4.194c:

$$\Psi(r) = A + B\log r \Rightarrow 0 = \nabla^2\Phi = \frac{1}{r}\frac{d}{dr}\left(r\frac{d\Phi}{dr}\right); \quad (4.229a \text{ and } 4.229b)$$

(2) continuing for $n = 2$ with the biharmonic equation (Equation 4.193a) that has radial solution (Equation 4.193b); (3) it follows by induction that the multiharmonic equation (Equation 4.228c) has radial solution (Equation 4.228b). Thus, *the unforced multiharmonic equation has (1) particular solutions (Equations 4.224b and 4.226) in Cartesian coordinates (Equation 4.222); (2) particular solutions (Equation 4.210a through d) including Equations 4.218a through 4.218d, 4.219a through 4.219c, 4.220a through 4.220c, and 4.221a through 4.221c in polar coordinates (Equation 4.210e); (3 and 4) particular solution [Equation 4.227b (Equation 4.228b)] depending only on the azimuthal angle (Equation 4.227c) [distance from the cylindrical axis (Equation 4.228c)]; and (5) whereas the particular solutions 1–4 involve arbitrary constants, the general solution (Equations 4.214a through 4.214c) in complex conjugate coordinates (Equations 4.211a and 4.211b) involves 2n arbitrary n-times differentiable functions (Equations 4.213a through 4.213d).* The scalar biharmonic functions (Subsections 4.6.3 and 4.6.4) may be extended to (1) scalar multiharmonic functions (Subsections 4.6.5 through 4.6.7) and (2) vector biharmonic functions (Section 4.6.8). The double extension to vector biharmonic functions is also possible.

4.6.8 Elastic Displacement Specified by a Biharmonic Vector (Galerkin 1930, 1931)

Method II of solution of problems of plane elasticity starts with the calculation of the stress function as the solution of a biharmonic equation (Subsections 4.6.1 through 4.6.4). The latter specifies the stress tensor (Equations 4.40a through 4.40c and 4.57b through 4.57d) by its second-order derivatives, and hence the strains by the inverse Hooke law (Equations 4.78a through 4.78c); the displacement vector involves one integration of the strains (Equations 4.12a through 4.12d) to within a rigid body displacement, that is, translation plus rotation (Equations 4.32a and 4.32b). This suggests that the displacement vector should be related to the derivatives of the biharmonic vector, providing a general integral of the momentum equation without forcing. In order to obtain this relation, consider the momentum equation for the displacement vector (Equation 4.101b ≡ Equation 4.230b) with balance of inertia and external forces (Equation 4.230a):

$$\rho\vec{a} = \vec{f}: \quad 0 = (1-2\sigma)\nabla^2\vec{u} + \nabla\left(\nabla\cdot\vec{u}\right); \quad A\vec{u} = \nabla^2\vec{\Theta} + B\nabla(\nabla\cdot\vec{\Theta}). \quad (4.230a–4.230c)$$

Its solution is sought in the form of Equation 4.230c, where A and B are constants and the vector $\vec{\Theta}$ satisfies Equation 4.231a:

$$A\left(\nabla\cdot\vec{u}\right)=(1+B)\nabla\cdot\left(\nabla^2\vec{\Theta}\right); \qquad 0=(1-2\sigma)\nabla^4\vec{\Theta}+[1+B+(1-2\sigma)B]\nabla\left[\nabla\cdot\left(\nabla^2\vec{\Theta}\right)\right]; \tag{4.231a and 4.231b}$$

substituting Equations 4.230c and 4.231a in Equation 4.230b yields Equation 4.231b. The vector $\vec{\Theta}$ is biharmonic (Equation 4.232a) if the constant B is chosen in Equation 4.231b to satisfy Equation 4.232b ≡ Equation 4.232c:

$$\nabla^4\vec{\Theta}=0: \qquad 1+B+(1-2\sigma)B=0, \qquad 2B=-\frac{1}{1-\sigma}. \tag{4.232a and 4.232b}$$

Substituting Equation 4.232b in Equation 4.230c, it follows that *the displacement vector in a plane elastic medium (Equation 4.320b) with balance of inertia and external forces (Equation 4.230a ≡ Equation 4.233a) is specified by Equation 4.233b in terms of a biharmonic vector (Equation 4.233c):*

$$\rho\vec{a}=\vec{f}: \qquad A\vec{u}=\left\{\nabla^2\vec{\Theta}-\frac{1}{2(1-\sigma)}\nabla\left(\nabla\cdot\vec{\Theta}\right)\right\}, \qquad \nabla^4\vec{\Theta}=0. \tag{4.233a–4.233c}$$

In particular, in Cartesian coordinates, the displacement vector is given by Equations 4.234d and 4.234e:

$$\vec{\Theta}=\Theta_1\vec{e}_x+\Theta_2\vec{e}_y; \quad \nabla^4\Theta_1=0=\nabla^4\Theta_2:$$

$$A\left\{u_x,u_y\right\}=\nabla^2\left\{\Theta_1,\Theta_2\right\}-\frac{1}{2(1-\sigma)}\left\{\partial_{xx}\Theta_1+\partial_{xy}\Theta_2,\ \partial_{xy}\Theta_1+\partial_{yy}\Theta_2\right\}, \tag{4.234a–4.234e}$$

where each of the components of the biharmonic vector (Equation 4.234a) is a scalar biharmonic function (Equations 4.234b and 4.234c).

4.6.9 Stresses in a Wedge-Shaped Dam (Levy 1898)

As a first example of method II of solution of problems of plane elasticity using the stress function (Figure 4.11), consider a wedge-shaped dam with (1) one face at an angle β in Equation 4.235a, where unit outward normal (Equation 4.235b) is free from loads (Equations 4.235c and 4.235d):

$$y=x\cot\beta:$$

$$\vec{n}_\beta=\left[\cos\beta,-\sin\beta\right]; \quad 0=\begin{bmatrix} T_{xx} & T_{xy} \\ T_{xy} & T_{yy} \end{bmatrix}\begin{bmatrix} \cos\beta \\ -\sin\beta \end{bmatrix}=\begin{bmatrix} T_{xx}\cos\beta-T_{xy}\sin\beta \\ T_{xy}\cos\beta-T_{yy}\sin\beta \end{bmatrix}; \tag{4.235a–4.235d}$$

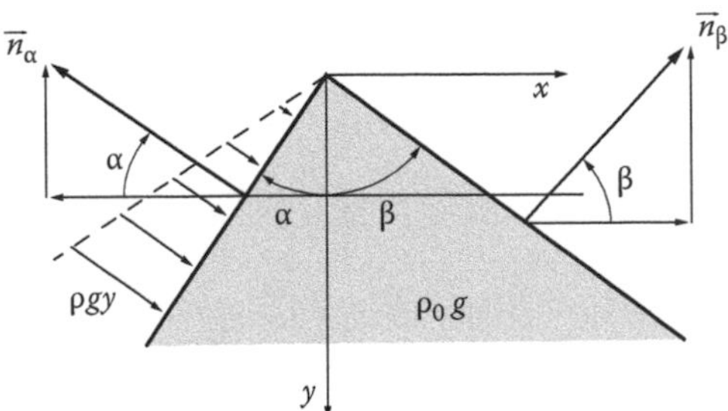

FIGURE 4.11
Simple model of a hydraulic dam is a wedge with the two flat sides making different angles with the vertical direction: (1) the wetted side is subject to a normal pressure proportional to the depth of the water; (2) the stresses on the dry side are small by comparison and may be neglected. The gravity forces in the dam (water) have the same direction but generally distinct modulus because the mass densities are different.

and (2) the other face at angle α in Equation 4.236a with unit outward normal (Equation 4.236b) is subject to a normal hydrostatic pressure corresponding to the weight of a column of fluid of height y and density ρ in a uniform gravity field (Equations 4.236c and 4.236d) with acceleration g:

$$y = -x\cot\alpha:$$

$$\vec{n}_\alpha = [-\cos\alpha, -\sin\alpha]: \quad -\rho g y \begin{bmatrix} \cos\alpha \\ \sin\alpha \end{bmatrix} = \begin{bmatrix} T_{xx} & T_{xy} \\ T_{xy} & T_{yy} \end{bmatrix} \begin{bmatrix} \cos\alpha \\ \sin\alpha \end{bmatrix} = \begin{bmatrix} T_{xx}\cos\alpha + T_{xy}\sin\alpha \\ T_{xy}\cos\alpha + T_{yy}\sin\alpha \end{bmatrix}. \tag{4.236a–4.236d}$$

In the present case, the momentum equation at rest (Equations 4.39c and 4.39d) involves a constant vertical force density per unit volume, namely, the weight $\rho_0 g$ of the material of the dam (Equation 4.237a):

$$\vec{f} = \rho_0 g \vec{e}_y: \quad \{T_{xx}, T_{yy}, T_{xy}\} = \{\partial_{yy}\Theta,\ \partial_{xx}\Theta,\ -\partial_{xy}\Theta - \rho_0 g x\}, \tag{4.237a–4.237d}$$

and thus the stresses (Equations 4.40a through 4.40c) are specified by Equations 4.237b through 4.237d in terms of the stress function. The stress function is taken as a homogeneous polynomial of degree 3 (Equation 4.190), that leads to in-plane stresses (Equations 4.237b through 4.237d) that are linear functions of position (Equations 4.238a through 4.238c):

$$\{T_{xx}, T_{yy}, T_{xy}, T_{33}\} = \{6Dy + 2Cx, 6Ax + 2By, -2Bx - 2Cy - \rho_0 g x, 2\sigma[(C + 3A)x + (B + 3D)y]\}, \tag{4.238a–4.238d}$$

as well as (Equation 4.95b) the out-of-plane stress (Equation 4.238d).

Substituting the stresses (Equations 4.238a through 4.238c) in the boundary conditions [Equations 4.235c and 4.235d (Equations 4.236c and 4.236d)] leads to Equations 4.239a and 4.239b (Equations 4.240a and 4.240b):

$$2B\sin\beta + 4C\cos\beta + 6D\cos\beta\cot\beta = -\rho_0 g \sin\beta, \tag{4.239a}$$

$$6A \sin\beta + 4B \cos\beta + 2C \cos\beta \cot\beta = \rho_0 g \cos\beta, \tag{4.239b}$$

$$2B \sin\alpha - 4C \cos\alpha + 6D \cos\alpha \cot\alpha = -g\,(\rho_0 \sin\alpha + \rho \cos\alpha \cot\alpha), \tag{4.240a}$$

$$6A \sin\alpha - 4B \cos\alpha + 2C \cos\alpha \cot\alpha = (\rho_0 + \rho) g \cos\alpha. \tag{4.240b}$$

Solving Equations 4.239a, 4.239b, 4.240a, and 4.240b specifies

$$6A = g\,[\rho_0 (\tan\beta - \tan\alpha) - \rho\,(2 - 3\tan\alpha\tan\beta - \tan^2\alpha)]/(\tan\alpha + \tan\beta)^2, \tag{4.241a}$$

$$4B = -g\,[\rho_0 + \rho\tan\alpha/(\tan\alpha + \tan\beta)] + 6A\,(\tan\beta - \tan\alpha), \tag{4.241b}$$

$$2C = \tan\alpha\tan\beta\,[\rho g/(\tan\alpha + \tan\beta) - 6A], \tag{4.241c}$$

$$12D = -\tan^2\beta\,[\rho_0 g/3\,\rho g\tan\alpha/(\tan\alpha + \tan\beta) - 6A\,(3\tan\alpha + \tan\beta)]. \tag{4.241d}$$

The four constants (Equations 4.241a through 4.241d) appear in the stress function (Equation 4.190) and stress tensor (Equations 4.238a through 4.238d) for a wedge-shaped dam (Figure 4.11) with the face at an angle β free from loads and the face at an angle α wetted by a fluid of density ρ distinct from ρ_0 of the dam, in the uniform gravity field with acceleration g. The stresses depend only on the loads and the geometry, and not on the properties of the material, that is, the Young's modulus E and Poisson ratio σ appear only in the strains, rotation, and displacement (Subsection 4.6.10). The preceding is an example of method II of solution (Diagram 4.1) of problems in elasticity: (1) it applies to plane elasticity with zero inertia force (Equation 4.39a) and constant external force density (Equation 4.39b) [balance of inertia and external forces (Equation 4.57a ≡ Equation 4.58a)], so that a stress function exists [Equations 4.40a through 4.40c (Equations 4.57b through 4.57d and 4.58b through 4.58d)]; (2) in this case, it satisfies a biharmonic equation (Equation 4.106e), whose solution specifies the stress function; (3) the stress function specifies the stress tensor in Cartesian (Equations 4.40a through 4.40c) [polar (Equations 4.58b through 4.58d)] coordinates; (4) the strain tensor follows from the inverse Hooke law (Equations 4.78a through 4.78c); (5) the strain tensor specifies the linearized relative area change (Equation 4.15b); (6–8) because the strain (stress) tensor satisfies the compatibility relation [Equation 4.36 (Equation 4.92b)], and the rotation (Equation 4.33c) and displacement vector (Equations 4.32a and 4.32b) exist and specify the displacement tensor (Equations 4.12a through 4.12d); and (9) the boundary conditions can be expressed in terms of the displacement, strain or stress vector (Equations 4.42b and 4.42c), or stress function (Equations 4.44a and 4.44b). Steps 6–8 of integration for the rotation and displacement vector and tensor are illustrated next.

4.6.10 Solution of Biharmonic Equation for Stress Function (Method II)

As an example of method II, consider the stress function in the form of a homogeneous cubic (Equation 4.190), that satisfies the biharmonic equation (Equation 4.189). For zero inertia force (Equation 4.39a), that is, uniform motion or rest, and constant external forces (Equation 4.39b), the corresponding in-plane stresses (Equations 4.40a through 4.40c) are (Equations 4.242a through 4.242c):

$$\{T_{xx}, T_{yy}, T_{xy}, T_{33}\} = \{6Dy + 2Cx, 6Ax + 2By, -2Bx - 2Cy - f_x y - f_y x, 2\sigma\,[(3A + C)x + (3D + B)y]\}, \tag{4.242a–4.242d}$$

implying (Equation 4.95b) an out-of-plane normal stress (Equation 4.242d). The inverse Hooke law (Equations 4.78a through 4.78c) leads to the corresponding strains:

$$S_{xy} = \frac{1+\sigma}{E} T_{xy} \equiv ax + by, \tag{4.243a}$$

$$\left\{S_{xx}, S_{yy}\right\} = \frac{1}{E}\left\{T_{xx}, T_{yy}\right\} - \frac{\sigma}{E}\left\{T_{yy} + T_{33}, T_{xx} + T_{33}\right\} \equiv a_{\pm}x + b_{\pm}y, \quad \text{(4.243b and 4.243c)}$$

involving the six constants (a, $a\pm$; b, $b\pm$) specified by (Equations 4.244a through 4.244f):

$$E\,\{a, b\} \equiv -\,(1+\sigma)\,\{2B + f_y, 2C + f_x\}, \quad \text{(4.244a and 4.244b)}$$

$$E\,\{a_+, b_+\} = \{2C - 6A\sigma - 2\sigma^2\,(3A + C), 6D - 2B\sigma - 2\sigma^2\,(3D + B)\}, \quad \text{(4.244c and 4.244d)}$$

$$E\,\{a_-, b_-\} = \{6A - 2C\sigma - 2\sigma^2\,(3A + C), 2B - 6D\sigma - 2\sigma^2\,(3D + B)\}. \quad \text{(4.244e and 4.244f)}$$

The rotation (Equation 4.34c) satisfies (Equation 4.245a):

$$\mathrm{d}E_2 = (a - b_+)\,\mathrm{d}x + (a_- - b)\,\mathrm{d}y, \quad E_2 = E_{20} + (a - b_+)\,x + (a_- - b)\,y \quad \text{(4.245a and 4.245b)}$$

and is given by Equation 4.245b, where E_{20} is a constant rotation.

The displacement vector (Equations 4.32a and 4.32b) satisfies

$$\mathrm{d}u_x = (a_+\,x + b_+y)\,\mathrm{d}x + [b_+x + (2b - a_-)\,y - E_{20}]\mathrm{d}y, \tag{4.246a}$$

$$\mathrm{d}u_y = [(2a - b_+)\,x + a_-\,y + E_{20}]\,\mathrm{d}x + (a_-\,x + b_-\,y)\,\mathrm{d}y \tag{4.246b}$$

and is given by

$$u_x = u_{x0} - E_{20}y + \frac{1}{2}a_+x^2 + \left(b - \frac{1}{2}a_-\right)y^2 + b_+xy, \tag{4.247a}$$

$$u_y = u_{y0} + E_{20}x + \left(a - \frac{1}{2}b_+\right)x^2 + \frac{1}{2}b_-y^2 + a_-xy, \tag{4.247b}$$

where the constants (u_{x0}, u_{y0}) specify a translation. Thus, *the cubic homogeneous polynomial in two variables (Equation 4.190) is a biharmonic function (Equation 4.189) and hence a stress function specifying (1) stress tensor (Equations 4.242a through 4.242d); (2) strain tensor (Equations 4.243a through 4.243c) involving the constants (Equations 4.244a through 4.244f); (3) the scalar rotation (Equation 4.245b); and (4) the displacement vector (Equations 4.247a and 4.247b). The latter consists of five terms on the r.h.s. of Equations 4.247a and 4.247b: (1) the first is a translation* $\vec{u}_0 = \text{const}$; *(2) the second is (Equations 4.8a and 4.8b) a rigid body rotation in the (x,y)-plane* $E_{20} =$

const; and (3) the last three are deformations. The translation does not appear in the displacement tensor:

$$\{D_{xx}, D_{yx}, D_{xy}, D_{yy}\} = \{a_+x + b_+y, -E_{20} + (2b - a_-)y + b_+x, E_{20} + (2a - b_+)x + a_-y, b_-y + a_-x\}, \tag{4.248a–4.248d}$$

that follows from Equations 4.247a and 4.247b using Equations 4.5a through 4.5d.

4.7 Forces and Moments on a Wedge

Method II of the stress function applied to the wedge-shaped dam (Subsections 4.6.9 and 4.6.10) also specifies the stresses in a wedge due to a force (moment) applied at the tip [Subsection 4.7.2 (Subsection 4.7.3)]. The stress field is simplest in polar components (Subsection 4.7.1), that can be converted into Cartesian components, for example, for the wedge of half-angle $\beta = \pi/4$, which corresponds to a half-plane or semi-infinite medium (Subsection 4.7.4). This leads to the stresses in a semi-infinite medium due to an arbitrary distribution of forces and moments on the boundary (Subsection 4.7.5).

4.7.1 Relation between Polar and Cartesian Components of Tensors

The relation between the Cartesian and polar components of the stress tensor can be obtained by considering the corresponding [Equation 4.249b (Equation 4.250b)] scalar **stress quadric:**

$$d\vec{x} = \vec{e}_x dx + \vec{e}_y dy: \quad d^2Q = T_{xx}(dx)^2 + T_{yy}(dy)^2 + 2T_{xy}dxdy, \tag{4.249a and 4.249b}$$

$$d\vec{x} = \vec{e}_r dx + \vec{e}_\varphi r d\varphi: \quad d^2Q = T_{rr}(dr)^2 + T_{\varphi\varphi}r^2(d\varphi)^2 + 2T_{r\varphi}r\, dr\, d\varphi \tag{4.250a and 4.250b}$$

using Cartesian (Equation 4.1a ≡ Equation 4.249a) [polar (Equation 4.21b ≡ Equation 4.250a)] coordinates, that are related by

$$\{x,y\} = r\{\cos\varphi, \sin\varphi\}: \ dx = \cos\varphi\, dr - r\sin\varphi\, d\varphi, \ \ dy = \sin\varphi\, dr + r\cos\varphi\, d\varphi. \tag{4.251a–4.251c}$$

Substitution of Equations 4.251b and 4.251c in Equation 4.249b leads to

$$\begin{aligned} d^2Q &= \left(T_{xx}\cos^2\varphi + T_{yy}\sin^2\varphi + 2T_{xy}\cos\varphi\sin\varphi\right)(dr)^2 \\ &+ r^2\left(T_{xx}\sin^2\varphi + T_{yy}\cos^2\varphi - 2T_{xy}\cos\varphi\sin\varphi\right)(d\varphi)^2 \\ &+ 2r\left[\left(T_{yy} - T_{xx}\right)\cos\varphi\sin\varphi + T_{xy}\left(\cos^2\varphi - \sin^2\varphi\right)\right]dr\, d\varphi. \end{aligned} \tag{4.252}$$

Comparison of Equation 4.252 with Equation 4.250b specifies *the relation between the Cartesian and polar components of the stress tensor (Equations 4.253a through 4.253c):*

$$T_{rr} = T_{xx}\cos^2\varphi + T_{yy}\sin^2\varphi + T_{xy}\sin(2\varphi), \tag{4.253a}$$

$$T_{\varphi\varphi} = T_{xx}\sin^2\varphi + T_{yy}\cos^2\varphi - T_{xy}\sin(2\varphi), \tag{4.253b}$$

$$T_{r\varphi} = (T_{yy} - T_{xx})\cos\varphi\sin\varphi + T_{xy}\cos(2\varphi) \tag{4.253c}$$

and its inverse (Equations 4.254a through 4.254c), specifying the Cartesian in terms of the polar components:

$$T_{xx} = T_{rr}\cos^2\varphi + T_{\varphi\varphi}\sin^2\varphi - T_{r\varphi}\sin(2\varphi), \tag{4.254a}$$

$$T_{yy} = T_{rr}\sin^2\varphi + T_{\varphi\varphi}\cos^2\varphi + T_{r\varphi}\sin(2\varphi), \tag{4.254b}$$

$$T_{xy} = (T_{rr} - T_{\varphi\varphi})\cos\varphi\sin\varphi + T_{r\varphi}\cos(2\varphi), \tag{4.254c}$$

that follow from each other reversing the sign of φ. The result (Equations 4.254a through 4.254c) can be obtained by (1) performing the derivation (Equations 4.251a through 4.251c and 4.252) in the reverse direction (r, φ) to (x,y) and (2) algebraic inversion of Equations 4.253a through 4.253c. The consistency of Equations 4.253a through 4.253c and 4.154a through 4.154c can be checked by substituting one in the other, leading to an identity. Both can be derived from the transformation law of a tensor and apply as well to any tensor in two dimensions, for example, the inertia tensor.

4.7.2 Stresses due to a Force Applied at the Tip of a Wedge (Michell 1900b)

Consider an infinite wedge of angle 2β subject to a concentrated force at the tip (Figure 4.12), with arbitrary modulus and direction, specified by the horizontal and vertical components (Equation 4.255a):

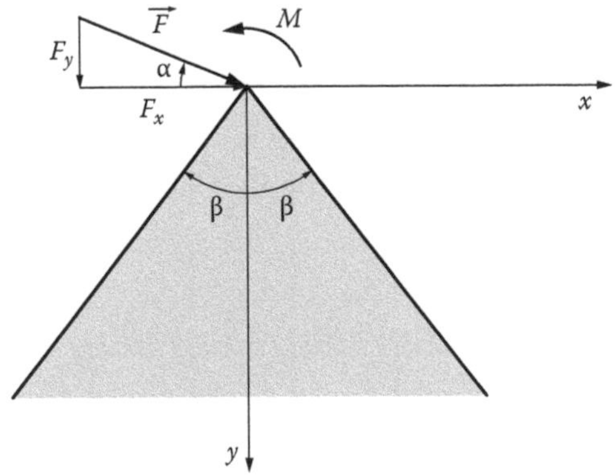

FIGURE 4.12
If a wedge is subject only to surface forces such as in the dam (Figure 4.11), the stress function exists in the interior. If a point force with an arbitrary direction and/or a moment is applied at the tip of the wedge (Figure 4.12), the wedge may be considered symmetric, and the stress function exists in the interior. The stress function specifies the stresses, which may have a singularity at the vertex. The singularity must be integrable to be compatible with a finite force and/or moment applied at the vertex.

$$\vec{F} = F_x\vec{e}_x + F_y\vec{e}_y = F\left(\vec{e}_x\cos\alpha + \vec{e}_y\sin\alpha\right): \quad \Theta_1(r,\varphi) = r\varphi(A\sin\varphi + B\cos\varphi), \tag{4.255a and 4.255b}$$

and assign the stress function (Equation 4.255b), that is, a solution (Equation 4.198b) with $C = 0$ of the biharmonic equation (Equation 4.197b). The corresponding stresses (Equations 4.57b through 4.57d) are radial:

$$T_{\varphi\varphi} = 0 = T_{r\varphi}: \qquad T_{rr} = \frac{2}{r}(A\cos\varphi - B\sin\varphi). \tag{4.256a–4.256c}$$

The stresses (Equations 4.256a and 4.256b) satisfy the boundary conditions (Equations 4.257a and 4.257b) at the surfaces where the y-axis is chosen (Figure 4.12) along the downward diagonal:

$$T_{\varphi\varphi}(r,\pm\beta) = 0 = T_{r\varphi}(r,\pm\beta): \qquad 0 = \left\{F_x, F_y\right\} + \int_{-\beta}^{+\beta} T_{rr} r\{\cos\varphi, \sin\varphi\}\, d\varphi. \tag{4.257a–4.257d}$$

The balance of forces at the tip (Equations 4.257c and 4.257d) specifies the two constants:

$$\begin{aligned}
\left\{F_x, F_y\right\} &= -2\int_{-\beta}^{+\beta} (A\cos\varphi - B\sin\varphi)\,\{\cos\varphi, \sin\varphi\}\, d\varphi \\
&= \int_{-\beta}^{+\beta} \left\{-A\left[1+\cos(2\varphi)\right] + B\sin(2\varphi), -A\sin(2\varphi) + B[1-\cos(2\varphi)]\right\} d\varphi \\
&= \left\{-A\left[\varphi + \frac{1}{2}\sin(2\varphi)\right] - \frac{B}{2}\cos(2\varphi), \frac{A}{2}\cos(2\varphi) + B\left[\varphi - \frac{1}{2}\sin(2\varphi)\right]\right\}_{-\beta}^{+\beta} \\
&= \left\{-A[2\beta + \sin(2\beta)],\ \ B[2\beta - \sin(2\beta)]\right\} \equiv \left\{-\frac{A}{C_+}, \frac{B}{C_-}\right\}.
\end{aligned} \tag{4.258a and 4.258b}$$

Substituting Equations 4.258a and 4.258b ≡ Equations 4.259a and 4.259b in Equation 4.255b (Equations 4.256a through 4.256c) specifies the stress function (Equation 4.259b) [in-plane stresses (Equation 4.259c through 4.259e)]:

$$\frac{1}{C_\pm} \equiv 2\beta \pm \sin(2\beta): \qquad \Theta_1(r,\varphi) = r\varphi\left(F_y C_- \cos\varphi - F_x C_+ \sin\varphi\right), \tag{4.259a and 4.259b}$$

$$T_{\varphi\varphi} = 0 = T_{r\varphi}: \qquad T_{rr}(r,\varphi) = -\frac{2}{r}\left(F_y C_- \sin\varphi + F_x C_+ \cos\varphi\right). \tag{4.259c–4.259e}$$

implying (Equation 4.96c) the out-of-plane stress (Equation 4.260a). The inverse Hooke law (Equations 4.78a through 4.78c) leads to the strains (Equations 4.260b through 4.260d) and hence (Equation 4.86b) to the elastic energy density (Equation 4.260e):

$$T_{33} = \sigma T_{rr} = -\frac{2\sigma}{r}\left(F_y C_- \sin\varphi + F_x C_+ \cos\varphi\right), \quad S_{r\varphi}(r,\varphi) = 0, \qquad \text{(4.260a and 4.260b)}$$

$$E S_{rr}(r,\varphi) = T_{rr} - \sigma T_{33} = \left(1-\sigma^2\right) T_{rr} = -2\frac{1-\sigma^2}{r}\left(F_y C_- \sin\varphi + F_x C_+ \cos\varphi\right), \qquad \text{(4.260c)}$$

$$E S_{\varphi\varphi}(r,\varphi) = -\sigma(T_{rr} + T_{\varphi\varphi}) = -\sigma\left(1+\sigma\right) T_{rr} = 2\frac{\sigma\left(1+\sigma\right)}{r}\left(F_y C_- \sin\varphi + F_x C_+ \cos\varphi\right), \quad \text{(4.260d)}$$

$$E_d(r,\varphi) = \frac{1}{2} T_{rr} S_{rr} = \frac{1-\sigma^2}{r}(T_{rr})^2 = \frac{1-\sigma^2}{r}\left(F_y C_- \sin\varphi + F_x C_+ \cos\varphi\right)^2. \qquad \text{(4.260e)}$$

Thus, *a force (Equation 4.255a) applied at the tip of a wedge (Figure 4.12) of angle 2β causes radial stresses (Equations 4.260a through 4.260c) associated with the stress function (Equations 4.259a through 4.259c), out-of-plane stress (Equation 4.260a), strains (Equations 4.260b through 4.260d) and elastic energy density (Equation 4.260e), with the constants $C_\pm$ given by Equation 4.259a.*

4.7.3 Stresses due to a Moment Applied at the Tip of a Wedge

In the case of a moment (Equation 4.261a) applied at the tip of the wedge (Figure 4.12):

$$\vec{M} = M\vec{e}_3: \qquad \Theta_2(r,\varphi) = A\varphi + B\sin(2\varphi), \qquad T_{\varphi\varphi}(r,\varphi) = 0, \qquad \text{(4.261a–4.261c)}$$

the stress function is sought in the form of Equation 4.261b, that is, a linear combination of two biharmonic functions, namely, the particular cases (1) of Equation 4.192b with $B = C = D = 0$ and (2) of Equation 4.219b with $E = F = H = 0 \neq G \equiv B$. The biharmonic function (Equation 4.261b) leads by Equations 4.57b through 4.57d to the absence of only azimuthal stresses (Equation 4.261c), that is, the nonzero stresses are (Equations 4.262a and 4.262b)

$$T_{r\varphi}(r,\varphi) = r^{-2}\left[A + 2B\cos(2\varphi)\right], \quad T_{rr}(r,\varphi) = -4Br^{-2}\sin(2\varphi). \qquad \text{(4.262a and 4.262b)}$$

The boundary condition (Equation 4.257a) is satisfied by Equation 4.261c. The boundary condition (Equation 4.257b) by Equation 4.262a relates the two constants of integration in (Equation 4.263a ≡ Equation 4.263b):

$$0 = T_{r\varphi}(r, \pm\beta) = \left[A + 2B\cos(2\beta)\right], \quad A = -2B\cos(2\beta). \qquad \text{(4.263a and 4.263b)}$$

The balance of moments at the tip:

$$M = \int_{-\beta}^{+\beta} r T_{r\varphi} r \, d\varphi = \int_{-\beta}^{+\beta}\left[A + 2B\cos(2\varphi)\right] d\varphi = \left[A\varphi + B\sin(2\varphi)\right]_{-\beta}^{+\beta} = 2A\beta + 2B\sin(2\beta) \qquad \text{(4.264)}$$

completes with (Equation 4.263b) the determination of the constants of integration:

$$M = 2B\,[\sin(2\beta) - 2\beta\cos(2\beta)] = A\,[2\beta - \tan(2\beta)] \qquad \text{(4.265a and 4.265b)}$$

in the alternate forms (Equations 4.266a and 4.266b ≡ Equations 4.266c and 4.266d):

$$A = \frac{M}{2\beta - \tan(2\beta)}, \qquad B = \frac{1}{2}\frac{M}{\sin(2\beta) - 2\beta\cos(2\beta)}, \qquad \text{(4.266a and 4.266b)}$$

$$\{A, B\} = \frac{M}{\sin(2\beta) - 2\beta\cos\left(2\beta\right)}\left\{-\cos(2\beta), \frac{1}{2}\right\}. \qquad \text{(4.266c and 4.266d)}$$

Substitution of Equations 4.266c and 4.226d in Equation 4.261b (Equations 4.261c, 4.262a, and 4.262b) specifies the stress function (Equation 4.267a) [tensor (Equations 4.267b through 4.267d)]:

$$\begin{aligned}&\left[\sin(2\beta) - 2\beta\cos(2\beta)\right]\left\{\Theta_2(r,\varphi), T_{\varphi\varphi}(r,\varphi), T_{rr}(r,\varphi), T_{r\varphi}(r,\varphi)\right\}\\ &= M\left\{\frac{1}{2}\sin(2\varphi) - \varphi\cos(2\beta), 0, -\frac{2}{r^2}\sin(2\varphi), \frac{1}{r^2}[\cos(2\varphi) - \cos(2\beta)]\right\}.\end{aligned} \qquad \text{(4.267a–4.267d)}$$

Together with (Equation 4.96b) the out-of-plane stress (Equation 4.268a), the inverse Hooke law (Equations 4.78a through 4.78c) leads to the strains (Equations 4.268b through 4.268d) and also (Equation 4.86b) to the elastic energy density (Equation 4.268e):

$$\begin{aligned}&\left\{T_{33}(r,\varphi), S_{\varphi\varphi}(r,\varphi), S_{rr}(r,\varphi), S_{r\varphi}(r,\varphi)\right\}\\ &= \left\{\sigma T_{rr}, -\frac{\sigma}{E}\left(T_{rr} + T_{33}\right), \frac{T_{rr} - \sigma T_{33}}{E}, \frac{1+\sigma}{E}T_{r\varphi}\right\}\\ &\left\{\sigma T_{rr}, \frac{\sigma(1+\sigma)}{E}T_{rr}, \frac{1-\sigma^2}{E}T_{rr}, \frac{1+\sigma}{E}T_{r\varphi}\right\}\\ &= \frac{M}{\sin(2\beta) - 2\beta\cos(2\beta)}\left\{-\frac{2\sigma}{r^2}\sin(2\varphi), -\frac{\sigma(1+\sigma)}{E}\frac{2}{r^2}\sin(2\varphi),\right.\\ &\left.-\frac{1-\sigma^2}{E}\frac{2}{r^2}\sin(2\varphi), \frac{1+\sigma}{Er^2}\left[\cos(2\varphi) - \cos(2\beta)\right]\right\},\end{aligned} \qquad \text{(4.268a–4.268d)}$$

$$E_d(r,\varphi)=\frac{1}{2}T_{rr}S_{rr}+T_{r\varphi}S_{r\varphi}=\frac{1-\sigma^2}{2E}\left(T_{rr}\right)^2+\frac{1+\sigma}{E}\left(T_{r\varphi}\right)^2$$
$$=\frac{1+\sigma}{E}\frac{M^2}{\left[\cos(2\beta)-\text{coss}(2\beta)\right]^2}\left\{(1-\sigma)\frac{2}{r^4}\sin^2(2\varphi)+\frac{1}{r^4}\left[\cos(2\varphi)-\cos(2\beta)\right]^2\right\}. \tag{4.268e}$$

Thus, *a moment (Equation 4.261a) applied at the edge of a wedge (Figure 4.12) of angle 2β causes the in-plane stresses (Equations 4.267b through 4.267d) associated with the stress function (Equation 4.267a), out-of-plane stress (Equation 4.268a), strains (Equations 4.268b through 4.268d) and elastic energy density (Equation 4.268e).*

4.7.4 Line Force and Moment at the Boundary of a Semi-Infinite Medium

The wedge (Figure 4.12) with angle shown in Equation 4.269a corresponds to a semi-infinite medium (Figure 4.13), and a line force (Equation 4.255a) [moment (Equation 4.261a)] at the origin $\xi = 0$ leads to the total stress function (Equation 4.269b), which is the sum of Equations 4.259a through 4.259c (Equation 4.267a):

$$\beta=\frac{\pi}{2}:\quad \Theta(r,\varphi)=\Theta_1+\Theta_2=\frac{r\varphi}{\pi}\left(F_y\cos\varphi-F_x\sin\varphi\right)+\frac{M}{\pi}\left[\varphi+\frac{1}{2}\sin(2\varphi)\right]. \tag{4.269a and 4.269b}$$

The associated stresses (Equations 4.270a through 4.270c) also add Equations 4.259c through 4.259e (Equations 4.267b through 4.267d):

$$T_{\varphi\varphi}(r,\varphi)=0,\qquad T_{r\varphi}(r,\varphi)=\frac{M}{\pi r^2}[1+\cos(2\varphi)]=\frac{2M}{\pi r^2}\cos^2\varphi \tag{4.270a and 4.270b}$$

$$T_{rr}(r,\varphi)=-\frac{2}{\pi r}\left(F_x\cos\varphi+F_y\sin\varphi\right)-\frac{2M}{\pi r^2}\sin(2\varphi)=\frac{1}{\sigma}T_{33}(r,\varphi); \tag{4.270c and 4.270d}$$

The out-of-plane stress (Equation 4.270d) follows from Equation 4.96b, and the inverse Hooke law (Equations 4.78a through 4.78c) leads to the strains (Equations 4.170e through 4.270g), and also (Equation 4.86b) to the elastic energy density (Equation 4.270b):

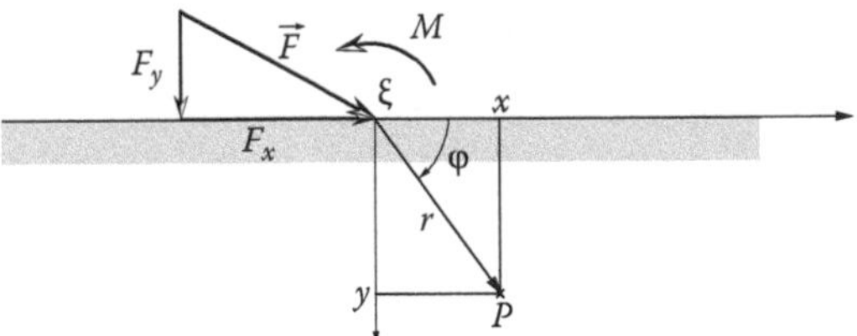

FIGURE 4.13
Wedge (Figure 4.12) with a right half-angle $\beta = \pi/2$ is a semi-infinite medium (Figure 4.13), that leads to the displacements, strains, and stresses at an arbitrary point in the interior due to normal or tangential forces and moments on the boundary, which may be concentrated at a point or continuously distributed over one or more segments. Because the momentum equation and boundary conditions of elasticity are all linear, the principle of superposition may be applied to add the effects of various loads.

$$\begin{aligned}
&E\left\{S_{\varphi\varphi}(r,\varphi),S_{rr}(r,\varphi),S_{r\varphi}(r,\varphi)\right\}\\
&=\left\{-\sigma(T_{rr}+T_{33}),T_{rr}-\sigma T_{33},(1+\sigma)T_{r\varphi}\right\}\\
&=\left\{-\sigma(1+\sigma)T_{rr},(1-\sigma^2)T_{rr},(1+\sigma)T_{r\varphi}\right\} \qquad (4.270\text{e}–4.270\text{g})\\
&=\left\{\frac{2\sigma(1+\sigma)}{\pi r}\left[F_x\cos\varphi+F_y\sin\varphi+\frac{M}{r}\sin(2\varphi)\right]\right.\\
&\left.-\frac{2}{\pi}\frac{1-\sigma^2}{r}\left[F_x\cos\varphi+F_y\sin\varphi+\frac{M}{r}\sin(2\varphi)\right],\frac{1+\sigma}{\pi}\frac{2M}{r^2}\cos^2\varphi\right\},
\end{aligned}$$

$$\begin{aligned}
E_d(r,\varphi)&=\frac{1}{2}T_{rr}S_{rr}+T_{r\varphi}S_{r\varphi}=\frac{1-\sigma^2}{2E}\left(T_{rr}\right)^2+\frac{1+\sigma}{E}\left(T_{r\varphi}\right)^2\\
&=\frac{1+\sigma}{E}\left\{\frac{2}{\pi^2}\frac{1-\sigma}{r^2}\left[F_x\cos\varphi+F_y\sin\varphi+\frac{M}{r}\sin(2\varphi)\right]^2+\frac{4M^2}{\pi^2 r^4}\cos^4\varphi\right\}
\end{aligned} \qquad (4.270\text{h})$$

The principle of superposition was used both in Equations 4.269b and 4.270a through 4.270g, which holds due to the linearity of both (1) the biharmonic differential equation satisfied by the stress function (Equation 4.191) and (2) the boundary conditions (Equations 4.257a and 4.257b) that involve only its first-order derivatives (Equations 4.57b through 4.57d). If the line force and moment are a position ξ along the boundary (Figure 4.13), and the stresses are determined at a position (x, y) in the semi-infinite medium, then the stress function (tensor) is specified by Equation 4.269a (Equations 4.270a through 4.270c), where

$$\varphi=\operatorname{arc\,cot}\left(\frac{x-\xi}{y}\right),r\equiv\left|\left(x-\xi\right)^2+y^2\right|^{1/2},\quad \left\{\cos\varphi,\sin\varphi\right\}=\frac{1}{r}\{x-\xi,y\} \qquad (4.271\text{a}–4.271\text{d})$$

specify the angle (Equation 4.271a) and distance (Equation 4.271b), in agreement with Equations 4.271c and 4.271d.

4.7.5 Arbitrary Distribution of Forces and Moments on the Boundary of a Semi-Infinite Medium

The linearity of the equations of plane elasticity and boundary conditions again allows the application of the principle of superposition by integrating Equation 4.269b with Equations 4.271a through 4.271d along the boundary, which leads to

$$\begin{aligned}
&\left\{m,h_x,h_y\right\}=\left[\frac{\mathrm{d}M}{\mathrm{d}\xi},\frac{\mathrm{d}F_x}{\mathrm{d}\xi},\frac{\mathrm{d}F_y}{\mathrm{d}\xi}\right]:\quad \pi\Theta(x,y)=\int_{-\infty}^{+\infty}m(\xi)\frac{y\left(x-\xi\right)}{\left(x-\xi\right)^2+y^2}\mathrm{d}\xi\\
&+\int_{-\infty}^{+\infty}\left[\operatorname{arc\,cot}\left(\frac{x-\xi}{y}\right)\right]\left[m(\xi)+(x-\xi)h_y(\xi)-yh_x(\xi)\right]\mathrm{d}\xi,
\end{aligned} \qquad (4.272\text{a}–4.272\text{d})$$

which determines the stress function (Equation 4.272d) in a semi-infinite medium (Figure 4.13) due to an arbitrary distribution of normal h_y and tangential h_x forces and moments m per unit length (Equations 4.274a through 4.274c) along the boundary. The stresses (Equations 4.273a through 4.273d) follow from Equations 4.270a through 4.270c and 4.271a through 4.271d:

$$y < 0: T_{\varphi\varphi}(x,y) = 0, \quad T_{r\varphi}(x,y) = \frac{2}{\pi}\int_{-\infty}^{+\infty} m(\xi)\frac{(x-\xi)^2}{\left[(x-\xi)^2+y^2\right]^2}\,d\xi, \qquad (4.273a–4.273c)$$

$$T_{rr}(x,y) = -\frac{2}{\pi}\int_{-\infty}^{+\infty}\left[(x-\xi)h_x(\xi) + yh_y(\xi) - 2\frac{(x-\xi)y}{(x-\xi)^2+y^2}m(\xi)\right]\frac{d\xi}{(x-\xi)^2+y^2}. \qquad (4.273d)$$

The out-of-plane stresses (Equation 4.274a), strains (Equations 4.274b) through 4.274d) and elastic energy density (Equation 4.274e) follow from Equations 4.273b through 4.273d as before (Equations 4.270e through 4.270h):

$$T_{33}(x,y) = \sigma T_{rr}(x,y), \quad E_d(x,y) = \frac{1+\sigma}{E}\left\{\left[T_{r\varphi}(x,y)\right]^2 + \frac{1-\sigma}{2}T_{rr}(x,y)^2\right\}, \qquad (4.274a \text{ and } 4.274e)$$

$$E\left\{S_{\varphi\varphi}(r,\varphi), S_{rr}(r,\varphi), S_{r\varphi}(r,\varphi)\right\}$$
$$= \left\{-\sigma(1+\sigma)T_{rr}(x,y), \left(1-\sigma^2\right)T_{rr}(x,y), (1+\sigma)T_{r\varphi}(x,y)\right\}. \qquad (4.274b–4.274d)$$

The stress function (Equation 4.272d) and in-plane components of the stress tensor (Equations 4.273b through 4.273d) in the semi-infinite medium (Equation 4.273a) are specified uniquely by the density of the moments (Equation 4.272a) and forces (Equations 4.272b and 4.272c) along the boundary and do not depend on the properties of the material. The material properties, namely, the Young's modulus and Poisson ratio, do appear in the strain tensor (Equations 4.274b through 4.274d), linearized relative area change (Equation 4.15b), out-of-plane stress (Equation 4.274a) and elastic energy density (Equation 4.274e), rotation (Equation 4.34c), and displacement vector (Equations 4.32a and 4.32b). A simple case of stresses and strains in a semi-infinite medium is due to a uniform force and moment on a strip on the boundary. If the forces (Equations 4.275b and 4.275c) [moments (Equation 4.275i)] are finite and applied to a set of segments along the boundary that is compact, in the sense that it does not extend to infinity [Equation 4.275a (Equation 4.275b)], then

$$|\xi| < \infty > |h_x|, |h_y|: \quad \Theta \sim O(r), \quad T_{rr} \sim O\left(\frac{1}{r}\right) \sim S_{rr}, \quad u_r \sim O(\log r) \qquad (4.275a–4.275g)$$

$$|\xi| < \infty > |m|: \quad \Theta \sim O(1), \quad T_{r\varphi} \sim O\left(\frac{1}{r^2}\right) \sim S_{r\varphi}, \quad u_\varphi \sim O\left(\frac{1}{r}\right), \qquad (4.275h–4.275m)$$

as $r \to \infty$ in Equation 4.271b: (1) the stress function is bounded (diverges like the distance r) for the moment (Equation 4.275j) [forces (Equation 4.275d)]; (2) the radial normal (Equations 4.275k and 4.275l) [shear (Equations 4.275e and 4.275f)] stresses and strains decay like the inverse (inverse square) of the distance; and (3) the radial (azimuthal) displacement may diverge (Equation 4.275g) [decay (Equation 4.275m)] like the logarithm (inverse) of the radial distance. The divergent cases may be prevented by a suitable dependence of the loads (Equations 4.272a through 4.272c) on the position ξ along the boundary.

4.8 Elastic Potential and Stresses in an Infinite Medium

The displacements, strain, and stresses in a plane semi-infinite (infinite) medium can be obtained [Section 4.7 (Section 4.8)] by method II (method III) of solution of elasticity problems (Subsection 4.4.6) based on the stress function (complex elastic potential). Method III applies when the inertia and external forces balance; in this case, there exists a complex elastic potential, whose derivative specifies the linearized relative area change (rotation) through its real (imaginary) part (Subsection 4.8.1). The real (imaginary) part of the complex potential is the real potential (field function) and specifies the displacement vector (Subsection 4.8.2), for example, a monopole (Subsection 4.8.3) in an infinite medium represents a line force (Subsections 4.8.4 and 4.8.5). Together with earlier results (Subsection 4.8.6), this leads to the displacements in an infinite medium due to an arbitrary distribution of forces and moments (Subsection 4.8.7).

4.8.1 Momentum Balance in Terms of Rotation and Dilatation

The momentum equation (Equations 4.100a and 4.100b) for a plane elastic medium simplifies when the inertia and external forces balance (Equation 4.276a) or vanish to Equations 4.276b and 4.276d:

$$\vec{f} = \rho\vec{a}: \quad (1-2\sigma)\left[\partial_x\left(\partial_x u_x\right)+\partial_y\left(\partial_y u_x\right)\right]+\partial_x D_2 = 0, \qquad \text{(4.276a and 4.276b)}$$

$$D_2 \equiv \partial_x u_x + \partial_y u_y \equiv \nabla\cdot\vec{u}: \quad (1-2\sigma)\left[\partial_y\left(\partial_y u_y\right)+\partial_x\left(\partial_x u_y\right)\right]+\partial_y D_2 = 0, \qquad \text{(4.276c and 4.276d)}$$

where the linearized relative area change (Equation 4.15b ≡ Equation 4.30 ≡ Equation 4.276c) was introduced. The rotation (Equation 4.9a) may also be introduced via Equations 4.277a and 4.277b:

$$\partial_y\,(\partial_y u_x) = \partial_y\,(\partial_y u_x - \partial_x u_y) + \partial_x\,(\partial_y u_y) = -\,2\partial_y E_2 + \partial_x\,(\partial_y u_y), \qquad \text{(4.277a)}$$

$$\partial_x\,(\partial_x u_y) = \partial_x\,(\partial_x u_y - \partial_y u_x) + \partial_y\,(\partial_x u_x) = 2\partial_x E_2 + \partial_y\,(\partial_x u_x). \qquad \text{(4.277b)}$$

Substitution of Equations 4.277a and 4.277b in Equations 4.276b and 4.276d leads to Equations 4.278a and 4.278b:

$$0 = (1 - 2\sigma)\,[\partial_x\,(\partial_x u_x + \partial_y u_y) - 2\partial_y E_2] + \partial_x D_2 = 2\,(1 - \sigma)\,\partial_x D_2 - 2\,(1 - 2\sigma)\,\partial_y E_2, \quad (4.278a)$$

$$0 = (1 - 2\sigma)\,[\partial_y\,(\partial_y u_y + \partial_x u_x) + 2\partial_x E_2] + \partial_y D_2 = 2\,(1 - \sigma)\,\partial_y D_2 + 2\,(1 - 2\sigma)\,\partial_x E_2. \quad (4.278b)$$

The r.h.s. of Equations 4.277a and 4.277b ≡ Equation 4.278b corresponds to the two-dimensional form of the momentum equation (Equation 4.114a ≡ Equation 4.279b) when the inertia and external forces balance (Equation 4.276a ≡ Equation 4.279a):

$$\vec{f} = \rho\vec{a}: \qquad 0 = (1-2\sigma)\nabla\wedge(\nabla\wedge\vec{u}) - 2(1-\sigma)\nabla(\nabla\cdot\vec{u}) = (1-2\sigma)\nabla\wedge\left(2E_2\vec{e}_3\right) - 2(1-\sigma)\nabla D_2$$
$$= \vec{e}_x 2\left[(1-2\sigma)\partial_y E_2 - (1-\sigma)\partial_x D_2\right] - \vec{e}_y 2\left[(1-2\sigma)\partial_x E_2 + (1-\sigma)\partial_y D_2\right]. \quad (4.279a \text{ and } 4.279b)$$

Thus, either Equations 4.278a and 4.278b or Equation 4.279b proves that *if the inertia and external forces balance or vanish in a plane elastic medium (Equation 4.276a ≡ Equation 4.279a ≡ Equation 4.280a), the linearized relative area change (Equation 4.15b ≡ Equation 4.30 ≡ Equation 4.276c ≡ Equation 4.281a) and rotation (Equation 4.9b ≡ 4.281b) satisfy the momentum equation in the form (Equations 4.280b and 4.280c):*

$$\rho\vec{a} = \vec{f}: \qquad (1-\sigma)\partial_x D_2 - (1-2\sigma)\partial_y E_2 = 0 = (1-\sigma)\partial_y D_2 + (1-2\sigma)\partial_x E_2. \quad (4.280a\text{–}4.280c)$$

These imply that the derivative (Equation 4.281c) of the complex **elastic potential** *is a holomorphic function:*

$$D_2 \equiv \partial_x u_x + \partial_y u_y, \qquad 2E_2 = \partial_x u_y - \partial_y u_x: \qquad \frac{df}{dz} = (1-\sigma)D_2 + i(1-2\sigma)E_2 \quad (4.281a\text{–}4.281c)$$

because the Cauchy–Riemann conditions (Equations 2.166c through 2.166f ≡ Equations I.11.10a and I.11.10b) for Equation 4.281c coincide with Equations 4.280b and 4.280c; hence, the elastic potential is also a holomorphic function (Equation 4.282). The real (imaginary) part of its derivative (Equation 4.281c) specifies directly the linearized relative area change (Equation 4.281a) [rotation (Equation 4.281b)] from which follows the displacement $\vec{u}$ by integration. This integration can be avoided by expressing the displacement vector in terms of the complex elastic potential. This completes the solution of the elasticity problem by method III since the displacement vector specifies the strain (Equations 4.12a through 4.12d) and stress (Equations 4.76a through 4.76c) tensors (Diagram 4.1).

4.8.2 Displacement Vector in Terms of the Complex Elastic Potential

The elastic potential satisfies (Equation 4.281c ≡ Equation 4.282)

$$\Phi + i\Psi \equiv f\,(x + iy) = \int [(1 - \sigma)\,D_2 + i\,(1 - 2\sigma)\,E_2]\,(dx + i\,dy), \quad (4.282)$$

that implies through the two relations (Equations 4.283a and 4.283b):

$$d\Phi = \int [(1 - \sigma)\,D_2\,dx - (1 - 2\sigma)\,E_2\,dy], \quad (4.283a)$$

$$d\Psi = \int [(1 - 2\sigma)\,E_2\,dx + (1 - \sigma)\,D_2\,dy] \quad (4.283b)$$

the four identities:

$$\partial_x \Phi = (1-\sigma)\, D_2 = \partial_y \Psi, \qquad -\partial_y \Phi = (1-2\sigma)\,\Omega = \partial_x \Psi. \tag{4.284a–4.284d}$$

These confirm the Cauchy–Riemann conditions (Equations 4.284a through 4.284d ≡ Equations 2.66c through 2.66f ≡ Equations I.11.10a and I.11.10b) for the complex elastic potential (Equation 4.282). The Cartesian components of the displacement vector are given by

$$\left\{\bar{u}_x, \bar{u}_y\right\} = \frac{1}{2(1-\sigma)}\left\{\partial_x, \partial_y\right\}(y\Psi) + \frac{1}{1-2\sigma}\left\{\partial_y, -\partial_x\right\}(y\Phi). \tag{4.285a and 4.285b}$$

This can be checked from

$$\nabla^2\Psi = 0: \qquad (1-\sigma)D_2 \equiv (1-\sigma)\left(\partial_x \bar{u}_x + \partial_y \bar{u}_y\right) = \frac{1}{2}\left(\partial_{xx} + \partial_{yy}\right)(y\Psi)$$

$$= \frac{y}{2}\nabla^2\Psi + \partial_y\Psi = \frac{\partial\Psi}{\partial y}, \tag{4.286a and 4.286b}$$

$$\nabla^2\Phi = 0: \qquad (1-2\sigma)E_2 \equiv \frac{1}{2}(1-2\sigma)\left(\partial_x \bar{u}_y - \partial_y \bar{u}_x\right) = -\frac{1}{2}\left(\partial_{xx} + \partial_{yy}\right)(y\Phi)$$

$$= -\frac{y}{2}\nabla^2\Phi - \partial_y\Phi = -\frac{\partial\Phi}{\partial y} \tag{4.287a and 4.287b}$$

using Equation 4.286a (Equation 4.287a) to prove the coincidence of Equation 4.286b ≡ Equation 4.284b (Equation 4.287b ≡ Equation 4.284c).

From Equations 4.285a and 4.285b, it follows that *in a plane elastic medium with balance of inertial and external forces (Equation 4.280a), the complex elastic potential (Equation 4.282), which involves the linearized relative area change (Equation 4.281a) and rotation (Equation 4.281b), specifies the displacement vector substituting Equations 4.284a and 4.284b in Equation 4.286b (Equation 4.287b):*

$$\nabla^2 g = 0: \qquad u_x = \frac{1}{1-2\sigma}\left\{\Phi + \frac{y}{2(1-\sigma)}\partial_y\Phi\right\} + \partial_x g, \tag{4.288a and 4.288b}$$

$$u_y = \frac{1}{2(1-\sigma)}\left\{\Psi - y\frac{1}{1-2\sigma}\partial_y\Psi\right\} + \partial_y g \tag{4.288c}$$

and adding the gradient of a harmonic function (Equation 4.288a). The passage from Equation 4.285a (Equation 4.285b) to Equation 4.288b (Equation 4.288c) uses Equations 4.284c and 4.284d (Equations 4.284a and 4.284b):

$$
\begin{aligned}
u_x - \partial_x g = \bar{u}_x &= \frac{1}{2(1-\sigma)}\partial_x\{y\Psi\} + \frac{1}{1-2\sigma}\partial_y\{y\Phi\} \\
&= \frac{1}{1-2\sigma}\left\{\Phi + y\partial_y\Phi\right\} + \frac{y}{2(1-\sigma)}\partial_x\Psi \\
&= \frac{1}{1-2\sigma}\Phi + y\partial_y\Phi\left[\frac{1}{1-2\sigma} - \frac{1}{2(1-\sigma)}\right] \\
&= \frac{1}{1-2\sigma}\left[\Phi + \frac{y}{2(1-\sigma)}\partial_y\Phi\right],
\end{aligned}
\tag{4.289a}
$$

$$
\begin{aligned}
u_y - \partial_y g = \bar{u}_y &= \frac{1}{2(1-\sigma)}\partial_y\{y\Psi\} - \frac{1}{1-2\sigma}\partial_x\{y\Phi\} \\
&= \frac{1}{2(1-\sigma)}\left(\Psi + y\partial_y\Psi\right) - \frac{y}{1-2\sigma}\partial_x\Phi \\
&= \frac{1}{2(1-\sigma)}\Psi + y\partial_y\Psi\left[\frac{1}{2(1-\sigma)} - \frac{1}{1-2\sigma}\right] \\
&= \frac{1}{2(1-\sigma)}\left(\Psi - \frac{y}{1-2\sigma}\partial_y\Psi\right).
\end{aligned}
\tag{4.289b}
$$

The difference between Equations 4.289a and 4.289b and Equations 4.286a and 4.286b concerns the addition of the gradient of an arbitrary harmonic vector (Equation 4.288a) that changes neither the linearized relative area change (Equation 4.290a) nor the rotation (Equation 4.290b):

$$
\partial_x\left(u_x - \bar{u}_x\right) + \partial_y\left(u_y - \bar{u}_y\right) = \partial_{xx}g + \partial_{yy}g = 0,
\tag{4.290a}
$$

$$
\partial_x\left(u_y - \bar{u}_y\right) - \partial_y\left(u_x - \bar{u}_x\right) = \partial_{xy}g - \partial_{yx}g = 0,
\tag{4.290b}
$$

and thus (1) does not affect the derivative (Equation 4.281c) of the complex elastic potential; and (2) it adds at most a constant to the complex elastic potential (Equation 4.282). The arbitrary harmonic function (Equation 4.288a) in Equations 4.288b and 4.288c can be used to ensure that the displacement is single-valued, as will be shown in the next example (Subsection 4.8.3).

4.8.3 Elastic Monopole as Line Force

The monopole (Equation 2.192a) corresponds to the elastic potential (Equation 4.291a) [its derivative (Equation 4.292a)]:

$$
f(z) = \frac{A}{2\pi}\log z: \qquad \Phi = \frac{A}{2\pi}\log r, \qquad \Psi = \frac{A\varphi}{2\pi},
\tag{4.291a–4.291c}
$$

$$\frac{df}{dz} = \frac{A}{2\pi z}: \qquad \{D_2, E_2\} = \frac{A}{2\pi r}\left\{\frac{\cos\varphi}{1-\sigma}, -\frac{\sin\varphi}{1-2\sigma}\right\}, \tag{4.292a–4.292c}$$

leading by Equation 4.282 (Equation 4.281c) to the real potential (Equation 4.291b) and field function (Equation 4.291c) [linearized relative area change (Equation 4.292b) and rotation (Equation 4.292c)]. The corresponding displacements (Equations 4.288b and 4.288c) are

$$\frac{2\pi}{A}u_x = \frac{1}{1-2\sigma}\left\{\log r + \frac{1}{2(1-\sigma)}\frac{y^2}{r^2}\right\} + \partial_x g, \tag{4.293a}$$

$$\frac{2\pi}{A}u_y = \frac{1}{2(1-\sigma)}\left\{\varphi - \frac{1}{1-2\sigma}\frac{xy}{r^2}\right\} + \partial_y g, \tag{4.293b}$$

where Equations 4.298a through 4.298c were used and

$$\partial_y(\log r) = r^{-1}\partial_y r = r^{-1}\partial_y\left\{\left|x^2+y^2\right|^{1/2}\right\} = \frac{y}{r^2}, \tag{4.294a}$$

$$\partial_y \varphi = \partial_y\left[\arctan\left(\frac{y}{x}\right)\right] = \frac{x^{-1}}{1+y^2/x^2} = \frac{x}{r^2}. \tag{4.294b}$$

The displacement (Equation 4.293b) is many-valued because of the cyclic term φ. It can be made single-valued by eliminating this term by choosing the arbitrary harmonic function to be Equation 4.295a:

$$\partial_y g = -\frac{\varphi}{2(1-\sigma)}, \quad \frac{dg}{dz} = \frac{1}{2(1-\sigma)}\log z, \quad \partial_x g = \frac{1}{2(1-\sigma)}\log r, \tag{4.295a–4.295c}$$

that implies Equations 4.295b and 4.295c. This simplifies Equations 4.293a and 4.293b to

$$\frac{4\pi}{A}(1-\sigma)(1-2\sigma)\{u_x, u_y\} = \left\{(3-4\sigma)\log r + \frac{y^2}{r^2}, -\frac{xy}{r^2}\right\}, \tag{4.296a and 4.296b}$$

which leads to a single-valued displacement. The passage from Equations 4.293a to 4.296a uses Equation 4.296c:

$$\begin{aligned}\frac{4\pi}{A}(1-\sigma)(1-2\sigma)u_x - \frac{y^2}{r^2} &= 2(1-\sigma)\left[\log r + (1-2\sigma)\partial_x g\right] \\ &= [2(1-\sigma)+1-2\sigma]\log r = (3-4\sigma)\log r.\end{aligned} \tag{4.296c}$$

Furthermore, Equation 4.296b follows from Equations 4.293b and 4.295a.

4.8.4 Stresses due to Line Force in an Infinite Medium

The displacements (Equations 4.296a and 4.296b) lead Equations 4.12a through 4.12c to the strains:

$$\frac{4\pi}{A}(1-\sigma)(1-2\sigma)\left\{S_{xx}, S_{yy}, S_{xy}\right\} = \left\{\left(3-4\sigma-\frac{2y^2}{r^2}\right)\frac{x}{r^2}, \left(\frac{2y^2}{r^2}-1\right)\frac{x}{r^2}, \left(1-2\sigma+\frac{2x^2}{r^2}\right)\frac{y}{r^2}\right\}, \tag{4.297a–4.297c}$$

where

$$r = \left|x^2+y^2\right|^{1/2}: \qquad \partial_x r = \frac{x}{r}, \qquad \partial_y r = \frac{y}{r} \tag{4.298a–4.298c}$$

were used to prove the coincidence of Equations 4.297a through 4.297c ≡ Equations 4.299a through 4.299c:

$$\frac{4\pi}{A}(1-\sigma)(1-2\sigma) \quad S_{xx} = \partial_x\left[(3-4\sigma)\log r + \frac{y^2}{r^2}\right] = (3-4\sigma)\frac{x}{r^2} - \frac{2xy^2}{r^4}, \tag{4.299a}$$

$$\frac{4\pi}{A}(1-\sigma)(1-2\sigma) \quad S_{yy} = -\partial_y\left(\frac{xy}{r^2}\right) = -\frac{x}{r^2} + \frac{2xy^2}{r^4}, \tag{4.299b}$$

$$\begin{aligned}\frac{4\pi}{A}(1-\sigma)(1-2\sigma) \quad S_{xy} &= \frac{1}{2}\partial_y\left[(3-4\sigma)\log r + \frac{y^2}{r^2}\right] - \frac{1}{2}\partial_x\left(\frac{xy}{r^2}\right)\\ &= (3-4\sigma)\frac{y}{2r^2} + \frac{y}{r^2} - \frac{y^3}{r^4} - \frac{y}{2r^2} + \frac{x^2y}{r^4}\\ &= \frac{y}{r^2}\left(2-2\sigma+\frac{x^2-y^2}{r^2}\right) = \frac{y}{r^2}\left(2-2\sigma+\frac{2x^2-x^2-y^2}{r^2}\right)\\ &= \frac{y}{r^2}\left(1-2\sigma+\frac{2x^2}{r^2}\right).\end{aligned} \tag{4.299c}$$

The linearized relative area change (Equation 4.15b) [out-of-plane stress (Equation 4.95d)] is given by Equation 4.300a (Equation 4.300b):

$$D_2 = S_{xx} + S_{yy} = \frac{A}{4\pi}\frac{x}{r^2}\frac{3-4\sigma-1}{(1-\sigma)(1-2\sigma)} = \frac{1}{2\pi}\frac{A}{1-\sigma}\frac{x}{r^2}; \tag{4.300a}$$

$$T_{33} = \frac{E}{1+\sigma}\frac{\sigma}{1-2\sigma}D_2 = \frac{1}{2\pi}\frac{A}{1-\sigma}\frac{E}{1+\sigma}\frac{\sigma}{1-2\sigma}\frac{x^2}{r}. \tag{4.300b}$$

The latter (Equation 4.300b ≡ Equation 4.301b) uses the parameter (Equation 4.301a):

$$F \equiv \frac{E}{1+\sigma}\frac{A}{1-2\sigma}: \qquad T_{33} = \frac{F}{2\pi}\frac{\sigma}{1-\sigma}\frac{x}{r^2}. \tag{4.301a and 4.301b}$$

The parameter (Equation 4.301a) also appears in the in-plane stresses:

$$\left\{T_{xx}, T_{yy}, T_{xy}\right\} = \frac{1}{2\pi}\frac{F}{1-\sigma}\left\{\frac{x}{r^2}\left(\frac{3}{2}-\sigma-\frac{y^2}{r^2}\right), \frac{x}{r^2}\left(-\frac{1}{2}+\sigma+\frac{y^2}{r^2}\right), \frac{y}{r^2}\left(\frac{1}{2}-\sigma+\frac{x^2}{r^2}\right)\right\} \tag{4.302a–4.302c}$$

obtained by substitution of Equations 4.297a through 4.297c and their trace (Equation 4.300a) in the Hooke law (Equations 4.76a and 4.76b):

$$T_{xy} = \frac{A}{4\pi}\frac{1}{1-\sigma}\frac{1}{1-2\sigma}\frac{E}{1+\sigma}\frac{y}{r^2}\left(1-2\sigma+\frac{2x^2}{r^2}\right) = \frac{1}{2\pi}\frac{F}{1-\sigma}\frac{y}{r^2}\left(\frac{1}{2}-\sigma+\frac{x^2}{r^2}\right), \tag{4.303a}$$

$$T_{xx} = \frac{1}{4\pi}\frac{F}{1-\sigma}\frac{x}{r^2}\left[3-4\sigma-\frac{2y^2}{r^2}+\frac{\sigma}{1-2\sigma}(3-4\sigma-1)\right] = \frac{1}{2\pi}\frac{F}{1-\sigma}\frac{x}{r^2}\left(\frac{3}{2}-\sigma-\frac{y^2}{r^2}\right), \tag{4.303b}$$

$$T_{yy} = \frac{1}{4\pi}\frac{F}{1-\sigma}\frac{x}{r^2}\left[\frac{2y^2}{r^2}-1+\frac{\sigma}{1-2\sigma}(3-4\sigma-1)\right] = \frac{1}{2\pi}\frac{F}{1-\sigma}\frac{x}{r^2}\left(-\frac{1}{2}+\sigma+\frac{y^2}{r^2}\right), \tag{4.303c}$$

with the coincidences Equation 4.303a ≡ Equation 4.302c, Equation 4.303b ≡ Equation 4.302a, and Equation 4.303c ≡ Equation 4.302b.

4.8.5 Moment and Horizontal and Vertical Force Components

The horizontal (Equation 4.304a) and vertical (Equation 4.304b) force and moment (Equation 4.304c) are given generally in terms of the stress vector and tensor (Equations 4.42b and 4.42c) by

$$\left\{F_x, F_y\right\} = \int_{\partial D}\left\{T_x, T_y\right\}\mathrm{d}s = \int_{\partial D}\left\{T_{xx}n_x + T_{xy}n_y, T_{xy}n_x + T_{yy}n_y\right\}\mathrm{d}s$$
$$= \int_{\partial D}\left\{T_{xx}\mathrm{d}y - T_{xy}\mathrm{d}x, T_{xy}\mathrm{d}y - T_{yy}\mathrm{d}x\right\}, \tag{4.304a and 4.304b}$$

$$\begin{aligned} M &= \int_{\partial D} \left(\vec{r} \wedge \vec{T}\right) r^{-1} \mathrm{d}s = \int_{\partial D} \left(T_y x - T_x y\right) r^{-1} \mathrm{d}s \\ &= \int_{\partial D} \left[x\left(T_{xy} n_x + T_{yy} n_y\right) - y\left(T_{xx} n_x + T_{xy} n_y\right)\right] r^{-1} \mathrm{d}s \\ &= \int_{\partial D} \left[T_{xy}\frac{y}{r} - T_{yy}\frac{x}{r}\right] \mathrm{d}x + \left[T_{xy}\frac{x}{r} - T_{xx}\frac{y}{r}\right] \mathrm{d}y \end{aligned} \tag{4.304c}$$

integration along the closed boundary ∂D of a domain D, with arc length (Equation 4.21d) and unit normal vector (Equation 4.43b). In the present case, the forces and moment at the origin are obtained by integration along a circle with coordinates (Equation 4.251a) with inward normal (Equation 4.305a), which leads from Equations 4.304a through 4.304c to 4.305b through 4.305d:

$$\begin{aligned} \vec{n} = -\{\cos\varphi, \sin\varphi\}: \quad \left\{F_x, F_y\right\} &= \int_0^{2\pi} \left\{T_x, T_y\right\} r \,\mathrm{d}\varphi \\ &= -\int_0^{2\pi} \left\{T_{xx}\cos\varphi + T_{xy}\sin\varphi,\ T_{xy}\cos\varphi + T_{yy}\sin\varphi\right\} r \,\mathrm{d}\varphi \\ &= -\int_0^{2\pi} \left\{T_{xx}x + T_{xy}y, T_{xy}x + T_{yy}y\right\} \mathrm{d}\varphi, \end{aligned} \tag{4.305a–4.305c}$$

$$\begin{aligned} M = \int_0^{2\pi} \left(T_y x - T_x y\right) r \,\mathrm{d}\varphi &= \int_0^{2\pi} \left[\left(T_{xx}\cos\varphi + T_{xy}\sin\varphi\right)\sin\varphi \right. \\ &\quad \left. - \left(T_{xy}\cos\varphi + T_{yy}\sin\varphi\right)\cos\varphi\right] r^2 \mathrm{d}\varphi, \\ &= \int_0^{2\pi} \left[\left(T_{xx} - T_{yy}\right)xy + T_{xy}\left(y^2 - x^2\right)\right] \mathrm{d}\varphi \end{aligned} \tag{4.305d}$$

that are all independent of the radius. From Equations 4.302a through 4.302c, it follows that the integrands in Equations 4.305b through 4.305d are

$$\begin{aligned} -2\pi\frac{1-\sigma}{F}\left(T_{xx}x + T_{xy}y\right) &= \frac{x^2}{r^2}\left(\frac{3}{2} - \sigma - \frac{y^2}{r^2}\right) + \frac{y^2}{r^2}\left(\frac{1}{2} - \sigma + \frac{x^2}{r^2}\right) \\ &= \left(\frac{1}{2} - \sigma\right)\frac{x^2 + y^2}{r^2} + \frac{x^2}{r^2} = \frac{1}{2} - \sigma + \frac{x^2}{r^2}, \end{aligned} \tag{4.306a}$$

$$\begin{aligned} -2\pi\frac{1-\sigma}{F}\left(T_{xy}x + T_{yy}y\right) &= \frac{xy}{r^2}\left(\frac{1}{2} - \sigma + \frac{x^2}{r^2}\right) + \frac{xy}{r^2}\left(-\frac{1}{2} + \sigma + \frac{y^2}{r^2}\right) \\ &= \frac{xy}{r^2}\frac{x^2 + y^2}{r^2} = \frac{xy}{r^2}, \end{aligned} \tag{4.306b}$$

$$\begin{aligned}
&\frac{2\pi}{F}(1-\sigma)\left[\left(T_{xx}-T_{yy}\right)xy+T_{xy}\left(y^2-x^2\right)\right] \\
&=\frac{x^2y}{r^2}\left(2-2\sigma-2\frac{y^2}{r^2}\right)+\left(y^2-x^2\right)\frac{y}{r^2}\left(\frac{1}{2}-\sigma+\frac{x^2}{r^2}\right) \\
&=\frac{x^2y}{r^2}\left(\frac{3}{2}-\sigma-\frac{y^2}{r^2}-\frac{x^2+y^2}{r^2}\right)+\frac{y^3}{r^2}\left(\frac{1}{2}-\sigma+\frac{x^2}{r^2}\right) \\
&=\left(\frac{1}{2}-\sigma\right)y\frac{x^2+y^2}{r^2}=\left(\frac{1}{2}-\sigma\right)y.
\end{aligned} \tag{4.306c}$$

The azimuthal integrals (Equations 4.306a through 4.306c) are evaluated by

$$\left\{F_x,F_y,M\right\}=\frac{1}{2\pi}\frac{F}{1-\sigma}\int_0^{2\pi}\left\{\frac{1}{2}-\sigma+\frac{x^2}{r^2},\frac{xy}{r^2},y\left(\frac{1}{2}-\sigma\right)\right\}d\varphi=\{F,0,0\}, \tag{4.307a–4.307c}$$

where some of the elementary integrals were used:

$$\int_0^{2\pi}\left\{\frac{x^2}{r^2},\frac{y^2}{r^2},\frac{xy}{r^2},\frac{x}{r},\frac{y}{r},1\right\}d\varphi=\int_0^{2\pi}\left\{\cos^2\varphi,\sin^2\varphi,\sin(2\varphi),\cos\varphi,\sin\varphi,1\right\}d\varphi=\{\pi,\pi,0,0,0,2\pi\}. \tag{4.308a–4.308f}$$

This confirms that F in Equation 4.301a is a line force in the horizontal direction. Thus, *a line force (Equation 4.301a) in the horizontal direction corresponds to the complex elastic potential (Equation 4.291a), real potential (Equation 4.291b) and field function (Equation 4.291c), linearized relative area change (Equation 4.292b ≡ Equation 4.300a), and rotation (Equation 4.292c). Together with the auxiliary harmonic function (Equations 4.295a through 4.295c), it leads by Equations 4.293a and 4.293b to the displacement vector (Equations 4.296a and 4.296b), strains (Equations 4.297a through 4.297c), and in-plane stresses (Equations 4.302a through 4.302c) [out-of-plane stress (Equation 4.301b)].*

4.8.6 Complex Elastic Potential (Method III)

For method III of solution of elasticity problems (Diagram 4.1): (1) it applies in the plane if the inertia and external forces balance (Equation 4.276a); (2) it starts with a holomorphic function (Equation 4.282) as the complex elastic potential; (3 and 4) its real (imaginary) part is the real potential (field function); (5) the latter specifies displacement vector by Equations 4.288b and 4.288c that may include the gradient of a harmonic function (Equation 4.288a), for example, to ensure unicity; (6–8) the derivative of the complex potential as a holomorphic function specifies (Equation 4.281c) the linearized relative area change (rotation) from its real (imaginary) part; (8 and 9) the strain tensor follows from the displacements (Equations 4.12a through 4.12d), and together with the rotation, it specifies displacement tensor; and (10 and 11) the direct Hooke law (Equations 4.76a through 4.76c) specifies the stress tensor and vector (Equations 4.42b and 4.42c). The same problem can be solved by more than one method, and distinct methods can be combined. For example, (1) the analog in potential flow (Section 2.4 and Chapter I.12) of a monopole

(Equation 2.192a) for the complex elastic potential (Subsections 4.8.2 through 4.8.5) is a line force (Equations 4.291a and 4.301a); and (2 and 3) a source/sink (Equation 2.41a) [vortex (Equation 2.33b)], in the sense that the displacement is radial (azimuthal) and decays like the inverse of the radius [Equation 4.124b (Equation 4.124c)], is a pressure (Equation 4.126a) [torque (Equation 4.126c)] in a cylindrical cavity (Subsection 4.5.2). Case 2 corresponds to an inner pressure and is distinct from (4) an outer pressure, which leads to a linear displacement (Equation 4.120b). The last three cases (cases 2–4) are as follows: *(4) a linear radial displacement (Equation 4.120b ≡ Equation 4.309c) that corresponds (Subsection 4.5.1) to an external line pressure (Figure 4.14) directed inward (Equations 4.122a and 4.122b ≡ Equation 4.309a):*

$$\frac{E}{1+\sigma}\frac{1}{1-2\sigma}\frac{u_{r0}}{a} = T_{rr} = T_{\varphi\varphi} = -p = \frac{T_{33}}{2\sigma}, \tag{4.309a–4.309b}$$

$$\vec{u} = \vec{e}_r u_{r0}\frac{r}{a} = -\vec{e}_r r p(1-2\sigma)\frac{1+\sigma}{E}, \quad S_{rr} = S_{\varphi\varphi} = \frac{u_{r0}}{a} = -p(1-2\sigma)\frac{1+\sigma}{E}, \tag{4.309c and 4.309d}$$

$$E_d = \frac{1}{2}\left(T_{rr}S_{rr} + T_{\varphi\varphi}S_{\varphi\varphi}\right) = T_{rr}S_{rr} = \frac{E}{(1+\sigma)(1-2\sigma)}\left(\frac{u_{r0}}{a}\right)^2 = \frac{(1+\sigma)(1-2\sigma)}{E}p^2. \tag{4.309e}$$

and nonzero strains (Equations 4.121a and 4.121b ≡ Equation 4.302d) and in-plane (Equations 4.122a through 4.122b ≡ Equation 4.309a) [out-of-plane (Equation 4.95b ≡ Equation 4.309b)] stress and elastic energy density (Equation 4.122d ≡ Equation 4.309e); (3) an azimuthal displacement (Equation 4.124c ≡ Equation 4.310b) corresponds to (Subsection 4.5.2) a line moment (Equation 4.126c ≡ Equation 4.310a):

$$M = 2\pi r^2 T_{r\varphi} = -\frac{2\pi E}{1+\sigma}u_{\varphi 0}a, \quad \vec{u} = \vec{e}_\varphi u_{\varphi 0} = -\vec{e}_\varphi \frac{1+\sigma}{E}\frac{M}{2\pi r}, \tag{4.310a and 4.310b}$$

$$S_{r\varphi} = -u_{\varphi 0}\frac{a}{r^2} = \frac{1+\sigma}{E}\frac{M}{2\pi r^2}, \qquad T_{r\varphi} = -\frac{E}{1+\sigma}u_{\varphi 0}\frac{a}{r^2} = \frac{M}{2\pi r^2} \tag{4.310c and 4.310d}$$

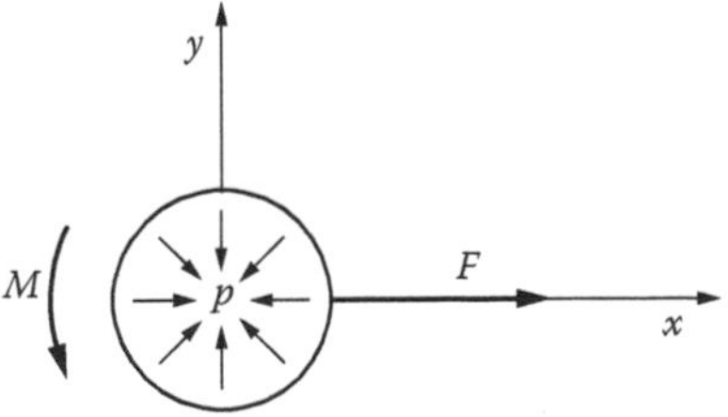

FIGURE 4.14
Replacing the semi-infinite medium (Figure 4.13) by an infinite medium (Figure 4.14), the forces, moments, and pressure lie in the interior and method II of the stress function is replaced by method III of the complex elastic potential. A point pressure (torque) causes a radial (tangential) displacement such as the velocity of a sink (vortex) in a potential flow. The elastic complex potential of a monopole corresponds to a force concentrated at a point, that is, a line force.

$$T_{33} = 0, \quad E_d(r) = T_{r\varphi} S_{r\varphi} = \frac{E}{1+\sigma} \frac{a^2}{r^4} \left(u_{\varphi 0}\right)^2 = \frac{1+\sigma}{E} \frac{M^2}{4\pi^2 r^4}, \qquad \text{(4.310e and 4.310f)}$$

and nonzero strains (Equation 4.125c ≡ Equation 4.310c) and stresses (Equation 4.126c ≡ Equation 4.310d and no out-of-plane stress (Equation 4.97b ≡ Equation 4.310e) and elastic energy density (Equation 4.86b ≡ Equation 4.310f); and (2) a radial displacement (Equation 4.124b ≡ Equation 4.311b):

$$p = -T_{rr}(a) = \frac{E}{1+\sigma} \frac{u_{r0}}{a}, \quad \vec{u} = \vec{e}_r u_{r0} \frac{a}{r} = \vec{e}_r \frac{1+\sigma}{E} p \frac{a^2}{r}, \qquad \text{(4.311a and 4.311b)}$$

$$T_{rr} = -\frac{E}{1+\sigma} u_{r0} \frac{a}{r^2} = -p \frac{a^2}{r^2} = -T_{\varphi\varphi}, \quad S_{rr} = -u_{r0} \frac{a}{r^2} = -\frac{1+\sigma}{E} p \frac{a^2}{r^2} = -S_{\varphi\varphi}, \qquad \text{(4.311c–4.311f)}$$

$$T_{33} = 0, \quad E_d(r) = \frac{1}{2}\left(T_{rr} S_{rr} + T_{\varphi\varphi} S_{\varphi\varphi}\right) = T_{rr} S_{rr} = \frac{E}{1+\sigma} \frac{a^2}{r^4} \left(u_{r0}\right)^2 = \frac{1+\sigma}{E} \frac{a^2}{r^4} p^2, \qquad \text{(4.311g and 4.311h)}$$

corresponds (Subsection 4.5.2) to the nonzero stresses (Equations 4.126a and 4.126b ≡ Equations 4.311c and 4.311d) and strains (Equations 4.125a and 4.125b ≡ Equations 4.311e and 4.311f) and no out-of-plane stress (Equation 4.97b ≡ Equation 4.311g) and elastic energy density (Equation 4.86b ≡ Equation 4.311h). The displacements [Equations 4.296a and 4.296b (Equation 4.310b], strains [Equations 4.297a through 4.297c (Equation 4.310c)], and stresses [Equations 4.302a through 4.302c (Equation 4.310d)] due to a line force (Equation 4.301a) [moment (Equation 4.310a)] can be superimposed (Figure 4.14) and integrated over an arbitrary two-dimensional domain in an unbounded medium, as was done (Subsection 4.7.5) for a one-dimensional distribution along the boundary of a semi-infinite medium.

4.8.7 Distribution of Line Forces and Moments in Infinite Medium

The radial (azimuthal) displacement due to an internal pressure (Equation 4.311b) [moment (Equation 4.310b)] leads by Equation 4.19a (Equation 4.19b) to the Cartesian components [Equations 4.312a and 4.312b (Equations 4.313a and 4.313b)]:

$$\left\{u_x, u_y\right\} = p \frac{1+\sigma}{E} \frac{a^2}{r} \{\cos\varphi, \sin\varphi\}, \qquad \text{(4.312a and 4.312b)}$$

$$\left\{u_x, u_y\right\} = \frac{1+\sigma}{E} \frac{M}{2\pi r} \left\{\sin\varphi, -\cos\varphi\right\}. \qquad \text{(4.313a and 4.313b)}$$

The displacements (Equations 4.296a and 4.296b) are due (Equations 4.307a through 4.307c) to a force $F = F_x$ in the x-direction (Equation 4.301a) and also apply to a force F_y in the y-direction exchanging x for y, so that both force components appear in

$$\{u_x, u_y\} = \frac{1}{4\pi E}\frac{1+\sigma}{1-\sigma}\left\{\left(F_x + F_y\right)(3-4\sigma)\log r + \frac{1}{r^2}(F_x y^2 + F_y x^2),\right.$$
$$\left. -\frac{xy}{r^2}\left(F_x + F_y\right)\right\} + \frac{1+\sigma}{E}\frac{M}{2\pi r^2}\{y,-x\}, \tag{4.314}$$

where the displacement due to torque or moment (Equations 4.313a and 4.313b) is also included.

If instead of being at the origin the forces and moments are located (Figure 4.15) at a position (ξ, η), displacements at a position (x,y) are specified by Equation 4.314 with the substitutions:

$$x \to x - \xi, \qquad y \to y - \eta, \qquad r = \left|(x-\xi)^2 + (y-\eta)^2\right|^{1/2}. \tag{4.315a–4.315c}$$

Using the principle of superposition, it follows that *an arbitrary distribution of line forces and torque or moments with densities (Equations 4.317b through 4.317d) per unit area (Equation 4.317a) in an unbounded medium (Figure 4.15), with Young's modulus E and Poisson ratio σ, leads to the displacements:*

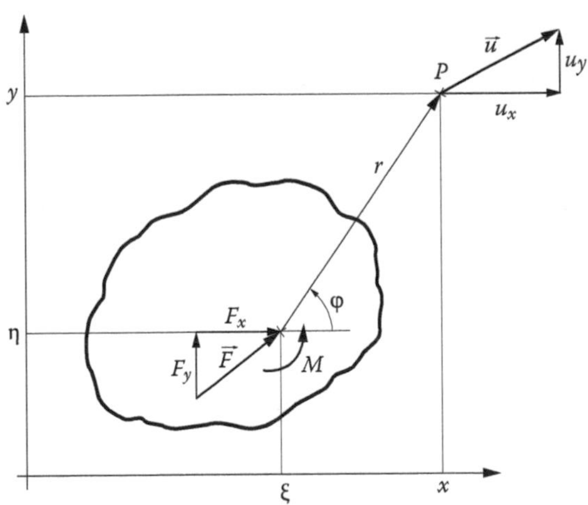

FIGURE 4.15
For the semi-infinite (infinite) medium [Figure 4.13 (Figure 4.15)], the principle of superposition specifies the displacements, strains, and stresses at all interior points due to an arbitrary distribution of forces and moments on the boundary (in any set of interior domains). The calculation of displacements, strains, and stresses at the interior of the force and/or moment distribution (Figure 4.15) may require Cauchy principal values of the integrals to avoid singularities due to coincident points of observation and application of the loads. Both for the semi-infinite (infinite) medium, the stresses in the interior depend only on the forces and moments on the boundary (interior domains) and not on the elastic properties of the material. The latter, that is, the Young's modulus and Poisson ratio, appear in the displacements and strains, out-of-plane stress and elastic energy density.

$$u_x(x,y) = \fint_{-\infty}^{+\infty} d\xi \fint_{-\infty}^{+\infty} d\eta \left\{ C_2\left(\frac{3}{4}-\sigma\right)\left[f_x(\xi,\eta)+f_y(\xi,\eta)\right]\log\left|(x-\xi)^2+(y-\eta)^2\right| \right.$$
$$\left. +\frac{1}{2}C_2\left|(x-\xi)^2+(y-\eta)^2\right|^{-1}\left[(x-\xi)^2 f_x(\xi,\eta)+(x-\xi)^2 f_x(\xi,\eta)+C_1(y-\eta)q(\xi,\eta)\right]\right\}, \tag{4.316a}$$

$$u_y(x,y) = -\fint_{-\infty}^{+\infty} d\xi \fint_{-\infty}^{+\infty} d\eta \left|(x-\xi)^2+(y-\eta)^2\right|^{-1}$$
$$\left\{C_2(x-\xi)(y-\eta)\left[f_x(\xi,\eta)+f_y(\xi,\eta)\right]+C_1(x-\xi)q(\xi,\eta)\right\}, \tag{4.316b}$$

where the material properties, that is, the Young's modulus and Poisson ratio, appear in the constants (Equations 4.317e and 4.317f):

$$dS = dx\,dy: \quad \left\{f_x, f_y, q\right\} \equiv \left\{\frac{dF_x}{dS}, \frac{dF_y}{dS}, \frac{dM}{dS}\right\}: \quad C_1 \equiv \frac{1+\sigma}{2\pi E}, C_2 = \frac{C_1}{1-\sigma}. \tag{4.317a–4.317f}$$

The strains follow from Equations 4.12a through 4.12d, the out-of-plane (in-plane) stress(es) from Equation 4.97b (Equation 4.76a through 4.76c) and the elastic energy density from Equation 4.86b.

If the displacements are determined within the force, pressure, and moment distribution, the integrals (Equation 4.316a and b) are taken as principal values (Subsection I.17.9). An example would be the strain and stresses in an infinite medium due to a uniform force and moment in a compact region, for example, square, rectangle, circle, or ellipse. In the case of a compact region [Equation 4.318a (Equation 4.319a)] with finite forces (Equation 4.318b) [moments (Equation 4.319b)]:

$$\xi^2+\eta^2 < \infty > \left(f_x\right)^2+\left(f_y\right)^2: \quad \vec{u} \sim O(\log r), O(1), \quad T_{ij} \sim O\left(\frac{1}{r}\right) \sim S_{ij}; \tag{4.318a–4.318d}$$

$$\xi^2+\eta^2 < \infty > |q|: \quad \vec{u} \sim O\left(\frac{1}{r}\right), \quad T_{ij} \sim O\left(\frac{1}{r^2}\right) \sim S_{ij}; \tag{4.319a–4.319d}$$

then (1) the displacement diverges logarithmically or is constant (Equation 4.318c) [decays like the inverse of distance (Equation 4.319c)]; (2) the stresses and strains decay like the inverse (Equation 4.318d) [inverse square (Equation 4.319d)] of distance; (3) the integral of item 2 over the boundary of a circle of a large radius is constant (decays like the inverse of distance).

4.9 Driven Loaded Wheel with Traction or Braking

As a final example of plane elasticity, the superposition of stress states is applied to a wheel on a rigid flat horizontal ground (Figure 4.16), subject to (1) its own weight and a fraction of that of the vehicle (Subsection 4.7.4); (2) a tangential tractive (braking) force (Subsection

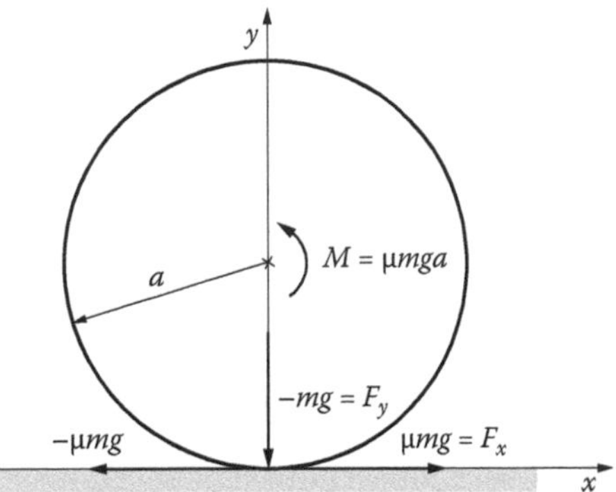

FIGURE 4.16
Rolling wheel, for example, of a car, is an example of loads applied at the boundary and interior of a finite body, consisting of (1) the weight of the wheel and part of the weight of the vehicle, that is, a vertical force normal to the surface; (2) the friction on the surface is associated with a tangential force, with opposite directions for traction (braking) in the direction of motion (opposite to it); (3) the wheel is driven by the engine by a moment applied at its axis. Considering the forces and moments as point loads and neglecting the effects of finite size of the wheel, the superposition principle applied to semi-infinite (infinite) media [Figure 4.13 (Figure 4.15)] specifies the displacements, strains, and stresses.

4.7.4); and (3) a driving (braking) moment on the axis (Subsection 4.5.2) besides the forces at the point of contact with the ground. The applied force and moments (Subsection 4.9.1) specify the stress function (Subsection 4.9.2) using method II.

4.9.1 Weight and Traction or Braking Force and Moment

A wheel of radius a supported on a flat rigid horizontal ground is subject to the following loads taken as static due to the balance of inertia and external forces: (1) its own weight and a fraction of the vehicle weight, equal to mass m times acceleration of gravity applied at the point of contact (Equation 4.320a); (2) a traction or (braking) force with sign + (−) in Equation 2.320b equal to the weight times a friction coefficient μ, also applied at the point of contact; and (3) a driving or braking torque at the center of the wheel equal to the force times (Equation 4.320b) the radius (Equation 4.320c):

$$F_y = -mg, \qquad F_x = \pm\,\mu mg, \qquad M = \pm\,\mu mga. \tag{4.320a–4.320c}$$

The vertical (Equation 4.320a) [horizontal (Equation 4.320b)] force, taken as a point force at the boundary of a semi-infinite medium, corresponds to the first two terms of the stress function (Equation 4.269b ≡ Equation 4.321):

$$\Theta(r,\varphi) = \frac{r\varphi}{\pi}\left(F_y \cos\varphi - F_x \sin\varphi\right) + \frac{M\varphi}{2\pi}, \tag{4.321}$$

and the moment is used in the last term of Equation 4.310b ≡ Equation 4.322b. The moment (Equation 4.310d ≡ Equation 4.322a) corresponds to the stress function (Equation 4.322b):

$$T_{r\varphi} = \frac{M}{2\pi r^2} = -\partial_r\left(r^{-1}\partial_\varphi \Theta\right), \qquad \Theta = \frac{M\varphi}{2\pi}. \tag{4.322a and 4.322b}$$

Substituting Equations 4.320a through 4.320c in Equation 4.321 leads to the stress function:

$$\pi\Theta = mg\varphi\left[r(-\cos\varphi \mp \mu\sin\varphi) \pm \frac{\mu a}{2}\right] = mg\left[\arc\tan\left(\frac{y}{x}\right)\right]\left\{-x \mp \mu\left(y - \frac{a}{2}\right)\right\} \quad \text{(4.323a and 4.323b)}$$

in polar (Equation 4.323a) [Cartesian (Equation 4.323b)] coordinates (Equation 4.248a).

4.9.2 Stress Function and Tensor in Polar and Cartesian Coordinates

Applying Equations 4.57b through 4.57d to Equation 4.323a specifies the polar components of the stress tensor:

$$T_{\varphi\varphi} = 0, \qquad \pi T_{r\varphi} = \pm\frac{\mu}{2}mg\frac{a}{r^2} = \pm\frac{\mu}{2}mg\frac{a}{x^2+y^2}, \quad \text{(4.324a–4.324c)}$$

$$\pi T_{rr} = \frac{2mg}{r}(\sin\varphi \mp \mu\cos\varphi) = 2mg\frac{y \mp \mu x}{x^2+y^2} \quad \text{(4.324d and 4.324e)}$$

in Cartesian (Equations 4.324b and 4.324d) and polar (Equations 4.324c and 4.324e) coordinates. The Cartesian components of the stress tensor can be obtained: (1) applying Equations 4.58b through 4.58d to the stress function (Equation 2.323b); and (2) converting (Equations 4.254a through 4.254c) polar (Equations 4.324a through 4.324e) to Cartesian components:

$$\pi\left\{T_{xx}, T_{yy}, T_{xy}\right\} = \frac{2mg}{r}(\sin\varphi \mp \mu\cos\varphi)\left\{\cos^2\varphi, \sin^2\varphi, \cos\varphi\sin\varphi\right\}$$
$$\pm\frac{\mu}{2}mg\frac{a}{r^2}\{-\sin(2\varphi), \sin(2\varphi), \cos(2\varphi)\}, \quad \text{(4.325a–4.325c)}$$

$$\pi\left|x^2+y^2\right|^2\left\{T_{xx}, T_{yy}, T_{xy}\right\} = 2mg(y \mp \mu x)\left\{x^2, y^2, xy\right\} \pm \frac{\mu}{a}mga\left\{-2xy, 2xy, x^2-y^2\right\}. \quad \text{(4.325d–4.325f)}$$

The Cartesian components of the stress tensor are given by Equations 4.325a through 4.325c (Equations 4.325d through 4.325f) in polar (Cartesian) coordinates. The preceding are the in-plane stresses (Equations 4.324a through 4.324e and 4.325a through 4.325f), and the out-of-plane stress is given (Equation 4.96b ≡ Equation 4.326a ≡ Equation 4.326b) in polar (Equation 4.326c) [Cartesian (Equation 4.326d)] coordinates:

$$T_{33} = \sigma T_{rr} = \sigma\left(T_{xx} + T_{yy}\right) = \frac{2mg\sigma}{\pi r}(\sin\varphi \mp \mu\cos\varphi) = \frac{2mg\sigma}{\pi}\frac{y \mp \mu x}{x^2+y^2}. \quad \text{(4.326a–4.326d)}$$

The polar components of the strain tensor follow from the stresses (Equations 4.324a through 4.324e and 4.326c and 4.326d) in polar (Equations 4.327c and 4.327f) [Cartesian (Equations 4.327d and 4.327g)] coordinates:

$$\left\{S_{rr}, S_{\varphi\varphi}\right\} = \frac{1}{E}\left\{T_{rr} - \sigma T_{33}, -\sigma\left(T_{rr} + T_{\varphi\varphi}\right)\right\} = \frac{1}{E}\left\{1-\sigma^2, -\sigma(1+\sigma)\right\}T_{rr}$$

$$= \frac{2mg}{\pi E}\frac{\sin\varphi \mp \mu\cos\varphi}{r}\left\{1-\sigma^2, -\sigma(1+\sigma)\right\} \tag{4.327a–4.327d}$$

$$= \frac{2mg}{\pi E}\frac{y \mp \mu x}{x^2+y^2}\left\{1-\sigma^2, -\sigma(1+\sigma)\right\},$$

$$S_{r\varphi} = \frac{1+\sigma}{E}T_{r\varphi} = \pm\frac{1+\sigma}{2\pi E}\frac{\mu mga}{r^2} = \pm\frac{1+\sigma}{2\pi E}\frac{\mu mga}{x^2+y^2}, \tag{4.327e–4.327g}$$

where was used the inverse Hooke law (Equations 4.76a through 4.76c).

4.9.3 Strains, Out-of (In)-Plane Stress(es) and Elastic Energy

The Cartesian components of the strain tensor may be obtained either (1) using the stress tensor (Equations 4.325a through 4.325f) in the inverse Hooke law (Equation 4.76a through 4.76c) or (2) using the polar components of the strain tensor (Equations 4.327c, 4.327d, 4.327f and 4.327g) in the transformation to Cartesian components (Equations 4.254a and 4.254c):

$$\left\{S_{xx}, S_{yy}, S_{xy}\right\} = S_{rr}\left\{\cos^2\varphi, \sin^2\varphi, \cos\varphi\right\} + S_{\varphi\varphi}\left\{\sin^2\varphi, \cos^2\varphi, -\cos\varphi\sin\varphi\right\}$$

$$+ S_{r\varphi}\left\{-2\cos\varphi\sin\varphi, 2\cos\varphi\sin\varphi, \cos^2\varphi - \sin^2\varphi\right\}$$

$$= \frac{mg}{\pi E r^2}\Big\{2r(\sin\varphi \mp \mu\cos\varphi)\left[\left(1-\sigma^2\right)\cos^2\varphi - \sigma(1+\sigma)\sin^2\varphi\right] \mp (1+\sigma)\mu\, a\sin\varphi\cos\varphi,$$

$$2r(\sin\varphi \mp \mu\cos\varphi)\left[\left(1-\sigma^2\right)\sin^2\varphi - \sigma(1+\sigma)\cos^2\varphi\right] \pm (1+\sigma)\mu\, a\sin\varphi\cos\varphi;$$

$$-4r(\sin\varphi \mp \mu\cos\varphi)\sin\varphi\cos\varphi\left[1-\sigma^2 - \sigma(1+\sigma)\right] + (1+\sigma)(\mu\, a/2)\left(\cos^2\varphi - \sin^2\varphi\right)\Big\}$$

$$= \frac{mg}{\pi E r^2}\Big\{2r(\sin\varphi \mp \mu\cos\varphi)\left(\cos^2\varphi - \varphi\sin^2\varphi - \sigma^2\right) \mp (1+\sigma)\mu\, a\sin\varphi\cos\varphi,$$

$$2r(\sin\varphi \mp \mu\cos\varphi)\left(\sin^2\varphi - \varphi\cos^2\varphi - \sigma^2\right) \mp (1+\sigma)\mu\, a\sin\varphi\cos\varphi,$$

$$-2r(\sin\varphi \mp \mu\cos\varphi)\left(1-\sigma-2\sigma^2\right)\sin\varphi\cos\varphi \mp (1+\sigma)(\mu\, a/2)\left(\cos^2\varphi - \sin^2\varphi\right)\Big\}$$

$$= \frac{mg}{\pi E}\frac{1}{\left(x^2+y^2\right)^2}\Big\{2(y \mp \mu x)\left(x^2 - \sigma y^2 - \sigma^2\right) \pm (1+\sigma)\mu\, axy,$$

$$2(y \mp \mu x)\left(x^2 - \sigma y^2 - \sigma^2\right) \mp \mu\, axy,$$

$$-2(y \mp \mu x)\left(1-\sigma-\sigma^2\right)xy \mp (1+\sigma)\mu\, a\left(x^2 - y^2\right)\Big\} \tag{4.328a–4.328d}$$

The relative area change (Equation 4.329a) follows from Equation 4.329c ≡ Equation 4.329d:

$$D_2 = S_{rr} + S_{\varphi\varphi} = S_{xx} + S_{yy} = \frac{2mg}{\pi r E}(\sin\varphi \mp \mu\cos\varphi)\left(1-\sigma-2\sigma^2\right)$$
$$= \frac{2mg}{\pi E}\frac{y \mp \mu x}{x^2+y^2}\left(1-\sigma-2\sigma^2\right) = \frac{T_{33}}{\sigma E}\left(1-\sigma-2\sigma^2\right) = \frac{T_{33}}{\sigma E}(1+\sigma)(1-2\sigma), \qquad (4.329\text{a}–4.329\text{d})$$

in agreement (Equation 4.326d) with Equation 4.95d ≡ Equation 4.329d. The elastic energy density (Equation 4.86b) is specified by the stresses (Equations 4.324a through 4.324e) and strains (Equations 4.327a through 4.327g):

$$E_d = \frac{1}{2}T_{rr}S_{rr} + T_{r\varphi}S_{r\varphi} = \frac{1-\sigma^2}{2E}(T_{rr})^2 + \frac{1+\sigma}{E}(T_{r\varphi})^2$$
$$= \frac{1+\sigma}{E}\left(\frac{mg}{\pi r}\right)^2\left[2(1-\sigma)(\sin\varphi \mp \mu\cos\varphi)^2 + \left(\frac{\mu a}{2r}\right)^2\right] \qquad (4.329\text{e}–4.329\text{h})$$
$$= \frac{1+\sigma}{\pi^2 E}\frac{m^2 g^2}{\left(x^2+y^2\right)^2}\left[2(1-\sigma)(y \mp \mu x) + \left(\frac{\mu a}{2}\right)^2\right].$$

A wheel with radius a (Figure 4.16) supporting a weight (Equation 4.320a), a tangential traction or braking force (Equation 4.320b) associated with a friction coefficient μ, and a traction/braking torque (Equation 4.320c) at the center has a plane elastic state, with balance of inertia and external forces, specified in polar (Cartesian) coordinates by (1 and 2) the stress function [Equation 4.323a (Equation 4.323b)] and by the polar [Equations 4.324b and 4.324d (Equations 4.324c and 4.324e)] and Cartesian [Equations 4.325a through 4.325c (Equations 4.325d through 4.325f) components of the in-plane stresses that depend only on the loads; (3 and 4) the properties of the material, namely the Young's modulus and Poisson ratio appear in the out-of-plane stress [Equation 4.326c (Equation 4.326d)] and relative area change [Equation 4.329c (Equation 4.329d)], and also (5 and 6) in the polar [Equations 4.327c and 4.327f) (Equation 4.328c)] and Cartesian [Equations 4.327d and 4.327g (Equation 4.328d)] components of the strains, and also (7) in the elastic energy density [Equation 4.329g (Equation 4.329h)]. The detailed set of data (Equations 4.320 through 4.329) for the wheel problem is a concluding example (Section 4.9) of the similar kinds of data that could be obtained for the other 24 problems of plane elasticity solved in the present chapter (see List 4.1, page 307).

NOTE 4.1 PLANE ELASTICITY VERSUS POTENTIAL FLOWS AND FIELDS

The problems in plane elasticity tend to be more involved than potential flows and fields (gravity, electric, and magnetic) and steady heat conduction for three reasons: (1) the existence of two material properties, namely, the Young's modulus (and Poisson ratio), which has dimensions (is dimensionless), and thus usually appears only when either multiplying or dividing (forms various fractions involving σ, 1 ± σ, 1 − 2σ, etc.); (2) besides the scalar (stress function) and vector (displacement) quantities, tensors are also involved (displacement, strain, and stress tensors); and (3) the solutions involve both harmonic (out-of-plane stress, dilatation) [biharmonic (stress function)] equations of degree 2 (4). On the one hand,

LIST 4.1

Twenty-Four Problems in Plane Elasticity

A. Displacements depending on radius (Subsection 4.5.1)
- 1. Radial
- 2. Azimuthal

B. Cylinders
- BA. Wheel (Section 4.9)
  - 3. Under its own weight
  - 4. With traction/braking
  - 5. Driven by torque
- BB. Solid
  - 6. Nonrotating (Subsection 4.5.1)
  - 7. Rotating (Subsection 4.5.9)
- BC. Cavity (Subsection 4.5.2)
  - 8. Surface pressure
  - 9. Surface shear
- BD. Hollow cylinder (Subsection 4.5.3)
  - 10. Inner and outer pressure
  - 11. Inner and outer displacement
- BE. Thin cylindrical shell (Subsection 4.5.4) inner and outer pressures
- BF. Two coaxial cylinders
  - 12. General (Subsection 4.5.5)
  - 13. Dissimilar materials (Subsection 4.5.6)
  - 14. Dissimilar pressures (Subsection 4.5.7)

C. Wedge
- 15. Horizontal force at the tip (Subsection 4.7.1)
- 16. Vertical force at the tip (Subsection 4.7.1)
- 17. Moment at the tip (Subsection 4.7.2)
- 18. Dam with hydraulic pressure on one side (Subsections 4.6.9 and 4.6.10)

D. Semi-infinite medium-combined loads on boundary
- 19. Point loads (Subsection 4.7.4)
- 20. Arbitrary distribution (Subsection 4.7.5)

E. Infinite medium
- 21. Point force (Subsections 4.8.3 through 4.8.5)
- 22. Point torque (Subsection 4.8.6)
- 23. Point pressure (Subsection 4.8.6)
- 24. Arbitrary combined distribution (Subsection 4.8.7).

---

special cases of elasticity are simpler in that they involve mainly scalar quantities, for example, (1) deflection of membranes (Sections 6.1 and 6.2); (2) surface tension and capillarity (Sections 6.3 and 6.4); and (3) torsion of beams (Sections 6.5 through 6.8). On the other hand, tensors, multiple material properties, and differential equations of order higher than the second are by no means unique to elasticity and occur in (1) viscous and turbulent fluids that have rates-of-strain and stresses (Notes 4.2 through 4.11); (2) the electric and magnetic fields have pressure and stresses; and (3) heat conduction and other physical processes in anisotropic materials involve multiple material properties.

NOTE 4.2 FUNDAMENTAL EQUATIONS FOR SOLIDS AND FLUIDS

The momentum equation (Equations 4.38a and 4.38b) applies equally well to a solid and a fluid, with the following differences: (1) in an inviscid fluid, the stresses reduce to a pressure (Equations 4.59a and 4.59b); (2) in a viscous fluid, the viscous stresses are added:

$$T_{xx} = -p + \sigma_{xx}, \qquad T_{yy} = -p + \sigma_{yy}, \qquad T_{xy} = \sigma_{xy}, \tag{4.330a–4.330c}$$

(3) the acceleration in a solid is the second-order local time derivative of the displacement, hence a linear operator (Equation 4.331a):

$$\vec{a} \equiv \vec{u}_x \equiv \frac{\partial^2 \vec{u}}{\partial t^2} = \vec{e}_x \frac{\partial_2 u_x}{\partial t^2} + \vec{e}_y \frac{\partial_2 u_y}{\partial t^2}, \tag{4.331a}$$

$$\vec{a} \equiv \frac{d\vec{v}}{dt} = \left(\frac{\partial}{\partial t} + \frac{dx_i}{dt}\frac{\partial}{\partial x_i}\right)\vec{v} = \frac{\partial \vec{v}}{\partial t} + (\vec{v}\cdot\nabla)\vec{v} = \left(\frac{\partial}{\partial t} + v_x\frac{\partial}{\partial x} + v_y\frac{\partial}{\partial y}\right)\left(\vec{e}_x v_x + \vec{e}_y v_y\right), \tag{4.331b}$$

whereas in a fluid, it is the first-order material derivative of the velocity (Equation 4.331b ≡ Equation 2.6b), which consists of a linear local time derivative plus a nonlinear convective term. In addition, (4) the pressure in a fluid is an independent variable satisfying the equation of state, relating it to the mass density and temperature; (5) in a solid, the mass density is an input, whereas in a fluid, it is a flow variable related to the velocity by the equation of continuity (Subsection I.13.1); and (6) the temperature satisfies the energy equation, which extends the heat equation (Chapter I.31), and applies both to solids and fluids, with several differences, as for the momentum equation. Thus, for a solid, there are two fundamental balance equations, namely, momentum and energy; for a compressible fluid, there are four, the other two being the equations of continuity and state.

NOTE 4.3 ELASTIC SOLID AND NEWTONIAN FLUID

The fundamental equations involve material properties that are specified by constitutive relations. There is an analogy between the constitutive relation for an elastic solid (Newtonian viscous fluid) (Note 4.4), relating the elastic stresses (viscous stresses) linearly to (1) the strains (rates-of-strain) by the shear modulus (shear viscosity); and (2) the linearized relative volume change (dilatation) by the volume modulus (bulk viscosity). Some viscous fluids are not Newtonian, such as the blood flow in arteries, as some solids are not elastic, for example, rubber or plastics. The Hooke law for an elastic solid applies to (1) small deformations of (2) an isotropic solid (3) without residual stresses; other constitutive relations apply when items 1–3 are violated, for example, for plastics or rubber (1), for crystals (2), or materials with memory (3). There exist intermediate constitutive relations such as viscoelasticity, specifying the stresses in terms of both strains and rates-of-strain. There are also very distinct phenomena, for example, (1) turbulence in a fluid is a random motion for which the viscous stresses are not sufficient as a model; and (2) the dislocations in solids can lead to creep and fracture, which are not modeled by the Hooke law. There are also broadly analogous phenomena such as instabilities, vibrations, and waves.

Focusing henceforth on the comparison between an elastic solid and a Newtonian viscous fluid, there is an initial analogy between the Hooke (Section 4.3) [Newton (Note 4.4)] law specifying the elastic (viscous) stresses in terms of the strains (Equations 4.12a through 4.12d) [rates-of-strain (Equations 4.17a and 4.17b)] through two constitutive parameters, namely, the elastic moduli and static viscosities. The momentum equation for a fluid is nonlinear due to the convective acceleration, even for the Euler equation in an inviscid fluid (Subsections 2.1.1 and 2.1.2). The Newtonian viscous stresses add linear terms, which leads to the Navier–Stokes (vorticity) equation [Note 4.5 (Note 4.6)]. This leads to a comparison of momentum and vorticity transport (Notes 4.7 and 4.8). The momentum and vorticity equations can be linearized, neglecting the convective terms compared with the viscous terms if the Reynolds number is small (Note 4.7), that is, for slow creeping flow. For a plane incompressible flow, there is a stream function that satisfies (1) the Laplace equation for an inviscid irrotational flow (Subsection 2.4.1); (2) a nonlinear fourth-order equation derived from the vorticity equation for the viscous flow of a Newtonian fluid; and (3) for a steady creeping flow at low Reynolds number, the latter reduces to a biharmonic equation (Note 4.11). The analogy of item 3 with the biharmonic equation for the stress function is incomplete because (a) the biharmonic equation for the stress (stream) function applies if the inertial and external forces balance in an elastic solid (for the creeping flow of a Newtonian viscous fluid); (b) the stress (stream) function specifies the stress tensor (velocity vector) through its second (first)-order partial derivatives. In the inviscid case, the biharmonic equation for the stream function is replaced by a nonlinear Poisson equation (Note 4.11) in the presence of vorticity, and a Laplace equation in an irrotational flow. The properties of Newtonian viscous fluids stated above are proved next (Notes 4.4 through 4.11).

NOTE 4.4 VISCOUS STRESSES IN A NEWTONIAN FLUID

As the displacement vector in elasticity (Equation 4.2b) specifies a strain tensor (Equations 4.12a and 4.12b), the velocity vector of a flow specifies the **rate-of-strain tensor** (Equations 4.17a and 4.17b ≡ Equations 4.332a through 4.332c).

$$\dot{S}_{xx} = \partial_x v_x, \qquad \dot{S}_{yy} = \partial_y v_y, \qquad 2\dot{S}_{xy} = \partial_x v_y + \partial_y v_x = 2\dot{S}_{yx}. \qquad (4.332\text{a}–4.332\text{c})$$

The normal rates of strain are time rates for extension (contraction), for example, $\dot{S}_{xx} > 0\left(\dot{S}_{xx} < 0\right)$, and the symmetric cross-term $\dot{S}_{xy}$ is a rate of distortion with regard to time. *The rate-of-strain tensor is related to the* **viscous stress tensor** *as in the Hooke law* (Equations 4.66b and 4.66d):

$$\sigma_{xy} = 2\eta\dot{S}_{xy}, \quad \left\{\sigma_{xx}, \sigma_{yy}\right\} = 2\eta\left\{\dot{\bar{S}}_{xx}, \dot{\bar{S}}_{yy}\right\} + \zeta\dot{D}_2, \qquad (4.333\text{a and } 4.333\text{b})$$

by (1) the **bulk viscosity** *ζ that specifies the normal viscous stresses due (Equation 4.333b) to the* **dilatation**, *that is, the linearized rate-of-change with time of the relative area (Equation 4.334), and is specified by the divergence of the velocity [compare with Equations 4.15a and 4.15b]:*

$$\dot{D}_2 \equiv \dot{S}_{xx} + \dot{S}_{yy} = \partial_x v_x + \partial_y v_y = \nabla \cdot \vec{v}; \qquad (4.334)$$

(2) the **shear viscosity** *η relates (Equations 4.333a and 4.333b) all viscous stresses to the* **rate-of-sliding tensor** *(Equations 4.335a and 4.335b):*

$$\left\{\dot{\bar{S}}_{xy}, \dot{\bar{S}}_{x3}\dot{\bar{S}}_{y3}\right\} = \left\{\dot{S}_{xy}, \dot{S}_{x3}, \dot{S}_{y3}\right\}, \quad \left\{\dot{\bar{S}}_{xx}, \dot{\bar{S}}_{yy}, \dot{\bar{S}}_{33}\right\} = \left\{\dot{S}_{xx}, \dot{S}_{yy}, \dot{S}_{33}\right\} - \frac{1}{3}\dot{D}_3, \qquad (4.335\text{a and } 4.335\text{b})$$

which has zero divergence (Equation 4.335d) in three dimensions (Equation 4.3235c):

$$\dot{D}_3 \equiv \dot{S}_{xx} + \dot{S}_{yy} + \dot{S}_{33}: \quad \dot{\bar{S}}_{xx} + \dot{\bar{S}}_{yy} + \dot{\bar{S}}_{33} = \dot{S}_{xx} + \dot{S}_{yy} + \dot{S}_{33} - \dot{D}_3 = 0. \qquad \text{(4.335c and 4.335d)}$$

The shear (bulk) viscosity for a Newtonian fluid replaces the shear (volume) modulus in the Hooke law for an elastic solid.

NOTE 4.5 NAVIER–STOKES EQUATION (NAVIER 1822; POISSON 1829b; SAINT-VENANT 1843; STOKES 1845)

The momentum equation for a viscous fluid adds the divergence of the viscous stresses to the terms in the Euler equation (Equation 2.7) for an inviscid fluid as in Equations 4.38a and 4.38b:

$$\rho\left(\dot{v}_x + v_x\partial_x v_x + v_y\partial_y v_x\right) - \partial_x p = \partial_x\sigma_{xx} + \partial_y\sigma_{xy}, \qquad (4.336\text{a})$$

$$\rho\left(\dot{v}_y + v_x\partial_x v_y + v_y\partial_y v_y\right) - \partial_x p = \partial_x\sigma_{yx} + \partial_y\sigma_{yy}. \qquad (4.336\text{b})$$

The viscous stress terms on the r.h.s of Equations 4.336a and 4.336b are specified for a Newtonian fluid (Equations 4.333a and 4.333b) by

$$\begin{aligned}\partial_x\sigma_{xx} + \partial_y\sigma_{xy} &= \zeta\partial_x\dot{D}_2 + 2\eta\left(\partial_x\dot{\bar{S}}_{xx} + \partial_y\dot{\bar{S}}_{xy}\right)\\ &= 2\eta\left(\partial_x\dot{S}_{xx} + \partial_y\dot{S}_{xy}\right) + (\zeta - 2\eta/3)\partial_x\dot{D}_2\\ &= \eta\left(2\partial_{xx}v_x + \partial_{yy}v_x + \partial_{xy}v_y\right) + (\zeta - 2\eta/3)\left(\partial_{xx}v_x + \partial_{xy}v_y\right)\\ &= \eta\left(\partial_{xx} + \partial_{yy}\right)v_x + (\zeta + \eta/3)\partial_x\left(\partial_x v_x + \partial_y v_y\right).\end{aligned} \qquad (4.337\text{a})$$

Together with the analogous term (Equation 4.337b) in the y-direction:

$$\partial_y\sigma_{xy} + \partial_y\sigma_{yy} = \eta\left(\partial_{xx} + \partial_{yy}\right)v_y + \left(\zeta + \frac{\eta}{3}\right)\partial_y\left(\partial_x v_x + \partial_y v_y\right), \qquad (4.337\text{b})$$

it can be written in the vector form:

$$\sum_{j=1}^{2}\partial_i\sigma_{ij} = \eta\nabla^2\vec{v} + \left(\zeta + \frac{\eta}{3}\right)\nabla(\nabla\cdot\vec{v}). \qquad (4.338)$$

The divergence of the viscous stresses for a Newtonian fluid is specified by Equation 4.338 in terms of the shear η and bulk ζ viscosities and the velocity $\vec{v}$. Adding the viscous stress term (Equation 4.338) to the external forces in the Euler inviscid momentum equation (Equation 2.7) leads to the **Navier–Stokes equation:**

$$\vec{f} - \nabla p + \eta\nabla^2\vec{v} + \left(\zeta + \frac{\eta}{3}\right)\nabla(\nabla\cdot\vec{v}) = \rho\vec{a}, \qquad (4.339)$$

or viscous momentum equation balancing (1) the volume density of the external forces; (2) minus the pressure gradient; (3) the viscous stresses; and (4) against the inertia force equal to the acceleration times mass density.

NOTE 4.6 NONLINEAR DIFFUSION OF VORTICITY BY THE SHEAR KINEMATIC VISCOSITY

The mass density is constant for an incompressible fluid (Equation 4.340a), and if the external forces are conservative [i.e., irrotational (Equation 4.340b)], the vorticity (Equation 4.340c) satisfies Equation 4.340d:

$$\rho = \text{const}, \qquad \nabla \wedge \vec{f} = 0: \qquad \vec{\varpi} = \nabla \wedge \vec{v}, \qquad \eta \nabla^2 \vec{\varpi} = \rho \nabla \wedge \vec{a}, \qquad (4.340\text{a}–4.340\text{d})$$

obtained by taking the curl of Equation 4.339 and noting that the curl of the gradient is zero, for example, in two dimensions:

$$\nabla \wedge (\nabla A) = \vec{e}_3 \left[\partial_x \left[(\nabla A)_y \right] - \partial_x (\nabla A)_x \right] = \vec{e}_3 \left[\partial_x \left(\partial_y A \right) - \partial_y \left(\partial_x A \right) \right] = 0. \qquad (4.341)$$

Decomposing the total acceleration into (Equation 2.114) local, rotational, and irrotational parts, the latter is eliminated by the curl from Equation 4.341.

$$\nabla \wedge \vec{a} = \frac{\partial (\nabla \wedge \vec{v})}{\partial t} + \nabla \wedge \left[\nabla \left(\frac{v^2}{2} \right) \right] + \nabla \wedge [(\nabla \wedge \vec{v}) \wedge \vec{v}] = \frac{\partial \vec{\varpi}}{\partial t} + \nabla \wedge (\vec{\varpi} \wedge \vec{v}). \qquad (4.342)$$

Substituting Equation 4.342 in Equation 4.340d leads to

$$\rho = \text{const}, \quad \nabla \wedge \vec{f} = 0: \quad \frac{\eta}{\rho} \nabla^2 \vec{\varpi} = \frac{\partial \vec{\varpi}}{\partial t} + \nabla \wedge (\vec{\varpi} \wedge \vec{v}) \qquad (4.343\text{a}–4.343\text{c})$$

the **vorticity equation** *(Equation 4.343c) for a viscous incompressible (Equation 4.343a) fluid under conservative (Equation 4.343b) external forces, where (1) the first term on the r.h.s. is linear and corresponds to a vector "heat" or diffusion equation with diffusivity of the medium specified by the* **kinematic shear viscosity** *(Equation 4.344a), that is, the ratio of the static shear viscosity η to the mass density per unit volume; (2) the second term on the r.h.s. is a nonlinear convective term, which when added to the first equals the transport derivative of the vorticity (Equation 4.344b):*

$$\bar{\eta} \equiv \frac{\eta}{\rho}; \qquad \frac{D\vec{\varpi}}{dt} \equiv \frac{\partial \vec{\varpi}}{\partial t} + \nabla \wedge (\vec{\varpi} \wedge \vec{v}): \qquad \frac{D\vec{\varpi}}{dt} = \bar{\eta} \nabla^2 \vec{\varpi} = \frac{\eta}{\rho} \nabla^2 \vec{\varpi}; \quad (4.344\text{a}–4.344\text{c})$$

and (3) thus the vorticity equation (Equation 4.343c ≡ Equation 4.344c) is a nonlinear (Equation 4.344b) diffusion equation (Equation 4.344c) with the shear kinematic viscosity (Equation 4.344a) as diffusivity.

NOTE 4.7 COMPARISON OF MOMENTUM AND VORTICITY TRANSPORT

Because the velocity (Equation 4.345a) [vorticity (Equation 4.340c)] is a polar (axial) vector, the **material** *(Equation 4.331b)* **[transport** *(Equation 4.344b)***] derivatives** *are distinct:*

$$\vec{v} = \frac{d\vec{x}}{dt}; \qquad \rho\frac{d\vec{v}}{dt} = -\nabla p + \eta\nabla^2\vec{v} + \left(\zeta + \frac{\eta}{3}\right)\nabla(\nabla\cdot\vec{v}) + \vec{f} \qquad \text{(4.345a and 4.345b)}$$

and appear in the momentum (Equation 4.339 ≡ Equation 4.345b) [vorticity (Equation 4.344c)] equation for a Newtonian viscous fluid [i.e., incompressible (Equation 4.343a) and subject to irrotational external forces (Equation 4.343b)]. The transport derivative of the vorticity (Equation 4.344b ≡ Equation 4.346a):

$$\frac{D\vec{\varpi}}{dt} = \frac{\partial\vec{\varpi}}{\partial t} + (\vec{v}\cdot\nabla)\vec{\varpi} - (\vec{\varpi}\cdot\nabla)\vec{v} + \vec{\varpi}(\nabla.\vec{v}) = \frac{d\vec{\varpi}}{dt} - (\vec{\varpi}\cdot\nabla)\vec{v} + \vec{\varpi}(\nabla\cdot\vec{v}) \qquad \text{(4.346a and 4.346b)}$$

differs (Equation 4.346b) from the material derivative (Equation 4.331b) due to (1) velocity transport by the vorticity and (2) the dilatation. In the incompressible flow (Equation 4.343a ≡ Equation 4.347a) of a fluid without shear viscosity (Equation 4.347b), the conservation of the vorticity (Equation 4.344c ≡ Equation 4.347c) implies that its material derivative is given by Equation 4.347d:

$$\nabla\cdot\vec{v} = 0 = \eta: \quad \frac{D\vec{\varpi}}{dt} = 0 \Leftrightarrow (\vec{\varpi}\cdot\nabla)\vec{v} = \frac{d\vec{\varpi}}{dt} = \frac{\partial\vec{\varpi}}{\partial t} + (\vec{v}.\nabla)\vec{\varpi}. \qquad \text{(4.347a–4.347d)}$$

Thus, the vorticity is not conserved along the streamlines due to the variation of the velocity along vorticity lines (Equation 4.347d). The passage from Equations 4.344b to 4.346a and 4.346b uses two identities involving the invariant differential operators that are proved next.

NOTE 4.8 SOME IDENTITIES FOR THE CURL, DIVERGENCE, AND GRADIENT OPERATORS

The passage from Equation 4.344b to 4.346a uses (1) the vector identity:

$$\nabla\wedge(\vec{A}\wedge\vec{B}) = (\vec{B}\cdot\nabla)\vec{A} - (\vec{A}\cdot\nabla)\vec{B} + \vec{A}(\nabla\cdot\vec{B}) - \vec{B}(\nabla\cdot\vec{A}); \qquad \text{(4.348)}$$

(2) this can be proven for the x-component:

$$\begin{aligned}\left[\nabla\wedge(\vec{A}\wedge\vec{B})\right]_1 &= \partial_2(\vec{A}\wedge\vec{B})_3 - \partial_3(\vec{A}\wedge\vec{B})_2 \\ &= \partial_2\left(A_1B_2 - A_2B_1\right) - \partial_3\left(A_3B_1 - A_1B_3\right) \\ &= \left(B_2\partial_2 + B_3\partial_3\right)A_1 - \left(A_2\partial_2 + A_3\partial_3\right)B_1 \\ &+ A_1\left(\partial_2B_2 + \partial_3B_3\right) - B_1\left(\partial_2A_2 + \partial_3A_3\right) \\ &= (\vec{B}\cdot\nabla)A_1 - (\vec{A}\cdot\nabla)B_1 + A_1(\nabla\cdot\vec{B}) - B_1(\nabla\cdot\vec{A});\end{aligned} \qquad \text{(4.349)}$$

(3) the y,z-components of Equation 4.348 follow by circular permutation of (1–3) in (Equation 4.349); and (4) substitution (Equation 4.348) in Equation 4.344b leads to Equation 4.346a, bearing in mind that (Equation 4.350a):

$$0 = \nabla \cdot \vec{\varpi} = \nabla \cdot (\nabla \wedge \vec{v}); \qquad \nabla \wedge (\nabla A) = 0. \tag{4.350a and 4.350b}$$

The result that the divergence of the curl of a vector is zero (Equation 4.350a) is analogous to the curl of the gradient of a scalar is zero (Equation 4.350b ≡ Equation 4.341). From

$$\begin{aligned}\nabla \cdot (\nabla \wedge \vec{B}) &= \partial_1 (\nabla \wedge \vec{B})_1 + \partial_2 (\nabla \wedge \vec{B})_2 + \partial_3 (\nabla \wedge \vec{B})_3 \\ &= \partial_1 \left(\partial_2 B_3 - \partial_3 B_2\right) + \partial_2 \left(\partial_3 B_1 - \partial_1 B_3\right) + \partial_3 \left(\partial_1 B_2 - \partial_2 B_1\right) \\ &= \left(\partial_2 \partial_3 - \partial_3 \partial_2\right) B_1 + \left(\partial_3 \partial_1 - \partial_1 \partial_3\right) B_2 + \left(\partial_1 \partial_2 - \partial_2 \partial_1\right) B_3 = 0\end{aligned} \tag{4.351}$$

follows the proof of the property Equation 4.350a ≡ Equation 4.351.

NOTE 4.9 PLANE INCOMPRESSIBLE UNSTEADY VISCOUS FLOW

The continuity equation (Subsection I.13.1) is unchanged by viscosity, and thus *an incompressible flow (Equation 4.352a) has a stream function (Equation 4.352b):*

$$0 = \nabla \cdot \vec{v} = \partial_x v_x + \partial_y v_y \quad \Leftrightarrow \quad \left\{v_x, v_y\right\} = \left\{\partial_y \Psi, -\partial_x \Psi\right\}, \tag{4.352a and 4.352b}$$

the Laplacian of the stream function is the minus vorticity (Equation 2.142 ≡ Equation 4.353):

$$-\varpi = \partial_y v_x - \partial_x v_y = \partial_{yy} \Psi + \partial_{xx} \Psi = \nabla^2 \Psi. \tag{4.353}$$

The vorticity equation (Equation 4.343c) is linear except for the convective term, that simplifies in two dimensions to

$$\vec{\varpi} \wedge \vec{v} = \varpi \vec{e}_3 \wedge \left(v_x \vec{e}_x + v_y \vec{e}_y\right) = \varpi \left(v_x \vec{e}_y - v_y \vec{e}_x\right), \tag{4.354a}$$

$$\nabla \wedge (\vec{\varpi} \wedge \vec{v}) = \vec{e}_3 \left[\partial_x \left(v_x \varpi\right) + \partial_y \left(v_y \varpi\right)\right] = \vec{e}_3 \left(v_x \partial_x \varpi + v_y \partial_y \varpi\right), \tag{4.354b}$$

where Equation 4.352a was used in Equation 4.354b. Substituting Equation 4.354b in the vorticity equation (Equation 4.343c) leads to

$$\frac{\eta}{\rho} \nabla^2 \varpi = \bar{\eta} \nabla^2 \varpi = \frac{\partial \varpi}{\partial t} + v_x \partial_x \varpi + v_y \partial_y \varpi. \tag{4.355}$$

This can be written entirely in terms of the stream function using Equations 4.352b and 4.353:

$$D: \quad \bar{\eta} \nabla^4 \Psi = \frac{\partial}{\partial t} \left(\nabla^2 \Psi\right) - \left(\partial_y \Psi\right) \partial_x \left(\nabla^2 \Psi\right) + \left(\partial_x \Psi\right) \partial_y \left(\nabla^2 \Psi\right). \tag{4.356a and 4.356b}$$

Thus, *the stream function of a two-dimensional (Equation 4.352b) incompressible (Equation 4.352a) viscous flow satisfies the nonlinear vorticity equation (Equation 4.356b) in a domain (Equation 4.356a). In all cases:*

$$\partial D: \qquad v_s = \frac{\partial \Psi}{\partial n} = 0 = \frac{\partial \Psi}{\partial s} = -v_n \qquad \text{(4.357a–4.357c)}$$

both components of the velocity (Equation 4.352b), that is, the normal (Equation 4.357b) and tangential (Equation 4.357c) derivatives of the stream function, must vanish at the boundary (Equation 4.357a).

NOTE 4.10 REYNOLDS (1883) NUMBER AND CREEPING OR INERTIAL FLOW

The Navier–Stokes (Equation 4.339) [vorticity (Equation 4.343c)] equation involves (1) a nonlinear term due to convection that scales like the variable, that is, the velocity (vorticity) times Equation 4.358a, where L is a length scale:

$$\vec{v}\cdot\nabla \sim \frac{v}{L}, \quad \bar{\eta}\nabla^2 \sim \frac{\bar{\eta}}{L^2} = \frac{\eta}{\rho L^2}: \qquad \frac{\vec{v}\cdot\nabla}{\bar{\eta}\nabla^2} \sim \frac{vL}{\bar{\eta}} = \frac{\rho v L}{\eta} \equiv \mathrm{Re}; \qquad \text{(4.358a–4.358c)}$$

(2) a viscous term that is linear and scales like Equation 4.358b; and (3) the ratio of the two terms specifies the dimensionless **Reynolds number** (Equation 4.358c). It follows that *(1) the convective inertia predominates over the viscous stresses for* **inertial flow** *at large Reynolds number (Equation 4.359a), leading to the Euler equation (Equation 4.359b ≡ Equation 2.7) [the equation of convection of vorticity (Equation 4.359c)] if the viscous stresses are omitted from the Navier–Stokes (Equation 4.339) [nonlinear vorticity equation (Equation 4.343c)]:*

$$\mathrm{Re} >> 1: \quad \vec{f} - \nabla p = \rho\left[\frac{\partial \vec{v}}{\partial t} + (\vec{v}\cdot\nabla)\vec{v}\right], \quad \frac{D\vec{\varpi}}{dt} \equiv \frac{\partial \vec{\varpi}}{\partial t} + \nabla\wedge(\vec{\varpi}\wedge\vec{v}) = 0; \qquad \text{(4.359a–4.359c)}$$

(2) the viscous stresses predominate over the convective inertia for the **creeping flow** *at low Reynolds number (Equation 4.360a), and the Navier–Stokes (Equation 4.339) [vorticity (Equation 4.343c)] equation becomes (Equation 4.360b) [a linear diffusion equation (Equation 4.360c)]:*

$$\mathrm{Re} << 1: \qquad \vec{F} - \nabla p = \eta\nabla^2\vec{v} + \left(\zeta + \frac{\eta}{3}\right)\nabla(\nabla\cdot\vec{v}), \qquad \frac{\partial \vec{\varpi}}{\partial t} = \bar{\eta}\nabla^2\vec{\varpi}; \qquad \text{(4.360a–4.360c)}$$

and (3) both terms are relevant for intermediate Reynolds number. The preceding cases (1–3) assume **laminar flow**, so that the Reynolds number does not exceed the critical value for which transition to turbulence occurs. For an incompressible fluid, the last term in Equation 4.360b can be omitted; in the two-dimensional case, the velocity vector can be replaced by the stream function, which leads to scalar instead of vector equation (Equation 4.360c), as shown next (Note 4.11).

NOTE 4.11 STREAM FUNCTION FOR PLANE INCOMPRESSIBLE VISCOUS FLOW

The stream function of a two-dimensional incompressible (Equations 4.352a and 4.352b) viscous flow under irrotational external force (Equation 4.343b) satisfies the vorticity equation (Equation 4.356b) that reduces (1) to Equation 4.361b in the linear case of creeping flow at low Reynolds number (Equation 4.361a):

$$\text{Re} << 1: \quad \bar{\eta}\nabla^4\Psi = \nabla^2\left(\frac{\partial \Psi}{\partial t}\right); \qquad \frac{\partial \Psi}{\partial t} = 0: \quad \nabla^4\Psi = 0; \qquad (4.361\text{a–}4.361\text{d})$$

(2) it simplifies further to a biharmonic equation (Equation 4.361d) for a steady flow (Equation 4.361c); (3) the vorticity equation (Equation 4.356b) to Equation 4.362b in the opposite limit of inertial flow at large Reynolds number (Equation 4.362a):

$$\text{Re} >> 1: \qquad \frac{\partial}{\partial t}\left(\nabla^2\Psi\right) = +\left(\partial_y\Psi\right)\partial_x\left(\nabla^2\Psi\right) - \left(\partial_x\Psi\right)\partial_y\left(\nabla^2\Psi\right); \quad (4.362\text{a and }4.362\text{b})$$

(4) in the case of steady (Equation 4.363a) inertial flow (Equation 4.363b), the stream function satisfies the **nonlinear Poisson equation** *(Equation 4.363c):*

$$\frac{\partial \Psi}{\partial t} = 0, \ \text{Re} >> 1: \quad \nabla^2\Psi = f(\Psi) \equiv -\varpi; \qquad \vec{\varpi} = \nabla \wedge \vec{v} = 0: \quad \nabla^2\Psi = 0; \qquad (4.363\text{a–}4.363\text{e})$$

and (5) the vorticity function (Equation 4.363 ≡ Equation 4.363c) is zero for irrotational flow (Equation 4.363d), which leads to Laplace equation (Equation 4.363e) for the stream function. Equation 4.363c follows from Equation 4.362b in the steady case (Equation 4.364a), when the inviscid form (Equation 4.364b) of Equation 4.355 leads to Equation 4.364c:

$$\frac{\partial}{\partial t} = 0 = \eta: \qquad \frac{\partial_x\varpi}{\partial_y\varpi} = -\frac{v_y}{v_x} = \frac{\partial_x\Psi}{\partial_y\Psi} \qquad (4.364\text{a–}4.364\text{c})$$

using Equation 4.352b. The latter (Equation 4.364c) is equivalent to Equation 4.365e:

$$\rho = \text{const}, \ \eta = 0 = \frac{\partial}{\partial t}, \nabla \wedge \vec{f} = 0: \ \left(\frac{dy}{dx}\right)_{\varpi} = -\frac{\partial_x\varpi}{\partial_y\varpi} = -\frac{\partial_x\Psi}{\partial_y\Psi} = \left(\frac{dy}{dx}\right)_{\Psi}. \qquad (4.365\text{a–}4.365\text{e})$$

From Equation 4.365e, it follows that *although the vorticity $\varpi(x,y)$ and stream function $\Psi(x,y)$ depend on two variables, they have the same curves Ψ = const and ϖ =* const, *and thus are a function of each other $-\varpi = f(\Psi)$ for a steady (Equation 4.365c), incompressible (Equation 4.365a) rotational flow of a fluid with zero shear viscosity (Equation 4.365b) under irrotational external forces (Equation 4.365d). Therefore, in the preceding conditions (Equations 4.365a through 4.365d), the vorticity is conserved along the streamlines of the flow.* The latter result also follows from Equation 4.347d, applied to a plane flow (Equation 4.366a) for which the vorticity is normal to the plane (Equation 4.366b), implying Equation 4.366c and hence Equation 4.366d:

$$\nabla = \vec{e}_x\frac{\partial}{\partial x} + \vec{e}_y\frac{\partial}{\partial y}, \vec{\varpi} = \vec{e}_3\varpi: \qquad \vec{\varpi}\cdot\nabla = 0, \qquad \frac{d\vec{\varpi}}{dt} = 0. \qquad (4.366\text{a–}4.366\text{d})$$

For an irrotational flow (Equation 4.363d), the vorticity function vanishes $f(\Psi) = 0$ and the nonlinear Poisson equation (Equation 4.363c) reduces to the Laplace equation (Equation 4.363e) for the stream function.

4.10 Conclusion

The general displacement consists of a translation, rotation (Figure 4.1), and deformation. The latter involves (Figure 4.2) extensions, contractions, and distortions. The displacements and deformations can be described in different coordinate systems, for example, Cartesian (polar) coordinates [Figures 4.1 and 4.2 (Figures 4.3 and 4.4)]. The deformations are due to stresses (Figure 4.5), that is, tractions, compressions, and shears. The stresses lead to surface forces that depend on the orientation of a plane facet (Figure 4.6); a circular element may also be considered [Figure 4.7 (Figure 4.4)] for the stresses (strains). The free traction (compression) [Figure 4.8a (Figure 4.8b)] test serves to determine the elastic properties of a material. When an in-plane deformation causes (Figure 4.9) a reduction (increase) in an area, an out-of-plane compression (traction) is needed to prevent out-of-plane deformations as assumed in plane elasticity. The radial and azimuthal deformations with axisymmetric (symmetric) part apply to the following: (1) a solid cylinder (Figure 4.10a); (2) a cylindrical cavity (Figure 4.10b) in an unbounded medium; (3) a hollow cylinder (Figure 4.10c) with surface displacements or stresses; (4) for small thickness, item 3 leads to a cylindrical shell (Figure 4.10d); (5) the coaxial cylindrical tubes may use different materials (Figure 4.10e); and (6) there is an external force for a rotating cylinder (Figure 4.10f), for example, a shaft subject to centrifugal forces. A wedge-shaped dam (Figure 4.11) is subject to a hydraulic pressure on one side. A wedge with applied forces and moment at the edge (Figure 4.12) includes, as a particular case of angle $2\beta = \pi$, a semi-infinite medium with a plane boundary (Figure 4.13). The applied forces, pressure, and moments may be distributed on the boundary or interior of a material, or both, for example, a loaded wheel subject to weight and traction or braking forces and moments (Figure 4.16). The displacements, strains, and stresses can be determined for arbitrary line force and moment distributions (Figure 4.14) on the boundary (in the interior) of a semi-infinite (infinite) medium [Figure 4.13 (Figure 4.15)]. These various problems (List 4.1) are solved by three methods (Diagram 4.1).

5

Circular and Hyperbolic Functions

The hyperbolic (circular) cosine and sine can be defined (Section 5.1), respectively, as the even and odd parts of the exponential with real (imaginary) argument. From these two primary functions, namely, the cosine and sine, it is possible to define four secondary functions, namely, the secant and cosecant by inversion, and the cotangent and tangent by their ratios (Section 5.2). All six primary plus secondary functions transform between circular and hyperbolic via an imaginary change of variable that corresponds to a rotation by $\pi/2$ in the complex plane. The circular (hyperbolic) functions satisfy a general complex identity, which in the case of real variable allows their geometrical representation (Sections 5.1 and 5.9) on an ellipse (hyperbola): this includes, in particular, the unit circle (equilateral hyperbola), and hence the designation of circular (hyperbolic) functions. The property of the exponential of transforming sums into products implies that the circular (hyperbolic) cosine, and sine of a sum of arguments, can be expressed (Section 5.3) as a sum of products of powers of circular (hyperbolic) sines and cosines of each of the arguments. This is the addition formula for arguments, which can be applied also to the circular (hyperbolic) tangent and cotangent. The particular case of equal arguments leads to the multiplication formula for arguments (Section 5.4). In the case of two arguments, these formulas can be inverted (Section 5.5) as formulas for the sum, subtraction, and product of circular (hyperbolic) functions. These can be extended to formulas for the powers (Section 5.5) of circular and hyperbolic functions expressed as sums of cosines and sines of multiple angles. The inverse formulas, for cosine and sine of multiple arguments as sums of powers of products of cosine and sines (Section 5.5), can be restated as sums of powers of the cosine or sine alone. This leads to two kinds of Chebychev polynomials (Section 5.6) that form orthogonal systems of functions related to the trigonometric orthogonal systems of functions (Section 5.7). This leads to the complex and real Fourier series, cosine and sine series, and Chebychev series of two kinds. The complex circular and hyperbolic functions can be calculated (Section 5.8) in terms of the functions of real variable. Because the functions are periodic, their values need only be calculated (Section 5.9) on a fundamental strip of width 2π; the width can be reduced to $\pi/12$ using symmetry properties. A number of particular values of circular (hyperbolic) functions of real variable can be calculated (Section 5.9) from their geometric representation (Section 5.2) on the unit circle (equilateral hyperbola).

5.1 Sine/Cosine Representations on the Ellipse/Hyperbola

The basic circular (hyperbolic) functions are the cosine and sine that can be defined (Subsection 5.1.1) from the exponential of imaginary (real) argument and are represented (Subsection 5.1.2) by the coordinates of points on an eclipse (hyperbola).

5.1.1 Cosine, Sine, and Imaginary Transformation

The function **hyperbolic cosine (sine)** is the even (odd) part of the exponential (Equation 5.1a):

$$\exp(\pm z) \equiv \cosh z \pm \sinh z, \qquad \exp(\pm iz) \equiv \cos z \pm i \sin z, \qquad (5.1a \text{ and } 5.1b)$$

and the **circular cosine (sine)** is defined as the even (odd) part of the exponential with argument multiplied by i (Equation 5.1b). The latter definition (Equation 5.1b) implies that for a real variable $z = u$, the circular cosine (sine) is the real (imaginary) part of the exponential of imaginary argument iu. Adding and subtracting the two equations [Equation 5.1a (Equation 5.1b)] lead to:

$$\cosh(z) = \frac{e^z + e^{-z}}{2} = \cosh(-z), \quad \sinh(z) = \frac{e^z - e^{-z}}{2} = -\sinh(-z), \quad (5.2a \text{ and } 5.2b)$$

$$\cos(z) = \frac{e^{iz} + e^{-iz}}{2} = \cos(-z), \quad \sin z = \frac{e^{iz} - e^{-iz}}{zi} = -\sin(z), \qquad (5.3a \text{ and } 5.3b)$$

that *relate the circular (Equations 5.3a and 5.3b) [hyperbolic (Equations 5.2a and 5.2b)] cosine (Equations 5.2a and 5.3a) [sine (Equations 5.2b and 5.3b)] to the exponential and confirm that it is an even (odd) function.*

Using the definitions [Equation 5.1a (Equation 5.1b)] in the form [Equation 5.4a (Equation 5.4b)]:

$$\cosh(iz) + \sinh(iz) \equiv e^{iz} \equiv \cos(z) + i \sin(z), \qquad (5.4a)$$

$$\cosh(z) + \sinh(z) \equiv e^{z} = e^{-iiz} \equiv \cos(iz) - i \sin(iz), \qquad (5.4b)$$

and equating even and odd parts in each equation [Equation 5.4a (Equation 5.4b)] follows [Equations 5.5a and 5.5b (Equations 5.6a and 5.6b)]:

$$\cosh(iz) = \cos(z), \qquad \sinh(iz) = i \sin(z), \qquad (5.5a \text{ and } 5.5b)$$

$$\cos(iz) = \cosh(z), \qquad \sin(iz) = i \sinh(z). \qquad (5.6a \text{ and } 5.6b)$$

Thus, *the circular and hyperbolic cosine (sine) transform into each other via an imaginary change of variable [Equations 5.5a and 5.6a (Equations 5.5b and 5.6b)].* The presence (absence) of the factor i for the sine (cosine) is related:

$$\cos(z) = \cos(-iiz) = \cos(iiz) = \cosh(iz) = \cos(z), \qquad (5.7a)$$

$$\sin(z) = \sin(-iiz) = -\sin(iiz) = -i \sinh(iz) = i^2 \sin(z) = \sin(z) \qquad (5.7b)$$

to the fact that it is an odd (even) function.

5.1.2 Representations on the Ellipse, Circle, and Hyperbola

From the definition [Equation 5.1a (Equation 5.1b)] follows [Equation 5.8a (Equation 5.8b)]:

$$1 = e^z e^{-z} = \{\cosh(z) + \sinh(z)\}\{\cosh(z) - \sinh(z)\}, \tag{5.8a}$$

$$1 = e^{iz} e^{-iz} = \{\cos(z) + i\sin(z)\}\{\cos(z) - i\sin(z)\} \tag{5.8b}$$

the **hyperbolic (circular) identity** *for the cosine and sine with complex variable:*

$$\cosh^2(z) - \sinh^2(z) = 1 = \cos^2(z) + \sin^2(z). \tag{5.9a and 5.9b}$$

The two identities are equivalent [Equation 5.10a (Equation 5.10b)] as follows from Equations 5.5a and 5.5b (Equations 5.6a and 5.6b):

$$1 = \cos^2 z + \sin^2 z = [\cosh(iz)]^2 + [-i\sinh(iz)]^2 = \cosh^2(iz) - \sinh^2(iz), \tag{5.10a}$$

$$1 = \cosh^2 z - \sinh^2 z = [\cos(iz)] - [-i\sin(iz)]^2 = \cos^2(iz) + \sin^2(iz). \tag{5.10b}$$

The **circular identity** *(Equation 5.9b) is interpreted geometrically, considering a real cyclic variable in the interval (Equation 5.11a):*

$$0 \le \varphi < 2\pi: \qquad x(\varphi) = a\cos\varphi, \qquad y(\varphi) = b\sin\varphi, \tag{5.11a–5.11c}$$

so that Equation 5.12b (Equation 5.12c) is the Cartesian horizontal (vertical) coordinate of a point (Figure 5.1a) on an **ellipse** *(Equation 5.12a) with the half-axis a (b) along the 0x (0y) coordinate axis:*

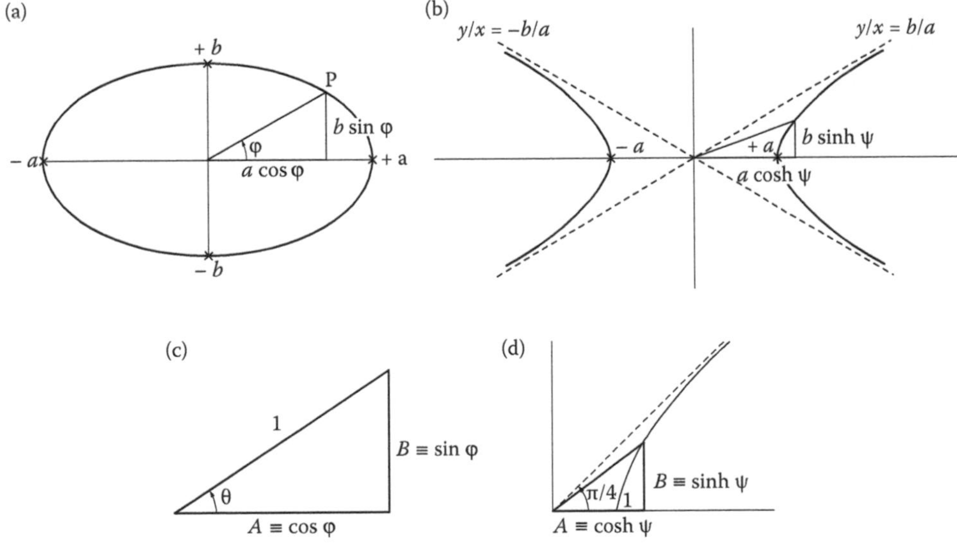

FIGURE 5.1
Horizontal (vertical) coordinate of a point on [a (b)] an ellipse with half-axis a (b) along the coordinate axis (hyperbola that cuts the real axis at $\pm a$ and has asymptotes with slopes $\pm b/a$) is given by a (b) times the circular (hyperbolic) cosine (sine). The particular case $a = b = R$ leads to the circle of radius R and center at the origin (the equilateral hyperbola that cuts the real axis at $\pm R$ and has the diagonals of quadrants as asymptotes). The simplest case $R = 1$ of the unit circle with center at the origin (unit equilateral hyperbola cutting the real axis at ± 1 and with the diagonal of quadrants as asymptotes) allows direct reading [c (d)] of (cos, sin) [(cosh, sinh)] from the (horizontal, vertical) coordinates of a point.

$$\frac{x^2}{a^2}+\frac{b^2}{b^2}=1; \quad a=b\equiv R: \quad x^2+y^2=R^2. \tag{5.12a–5.12c}$$

This becomes (Equation 5.12c) a **circle** *of radius R in the case of equal half-axis (Equation 5.12b). Taking the unit circle (Equations 5.13a through 5.13c):*

$$a=b\equiv R=1= x^2+y^2: \qquad x(\varphi)=\cos\varphi, \qquad y(\varphi)=\sin\varphi, \tag{5.13a–5.13e}$$

the cosine (sine) can be measured [Equation 5.13d (Equation 5.13e)] as the horizontal A (vertical B) coordinate of a point in Figure 5.1c.

The **hyperbolic identity** *(Equation 5.9a) can be interpreted geometrically, considering a real variable* $z=\psi$*, on the whole real line (Equation 5.14a):*

$$-\infty<\psi<+\infty: \qquad x(\psi)=a\cosh\psi, \qquad y(\psi)=b\sinh\psi, \tag{5.14a–5.14c}$$

so that Equation 5.14b (Equation 5.14c) is the Cartesian horizontal (vertical) coordinate of a point on (Equation 5.15a) a **hyperbola** *(Figure 5.1b): (1) with axis of symmetry coincident with the 0x coordinate axis; (2) cutting the 0x axis (Equation 5.15b) at Equation 5.15c; and (3) with asymptotes with slope (Equation 5.15d):*

$$\frac{x^2}{a^2}-\frac{y^2}{b^2}=1; \quad y=0: \quad x=\pm a; \quad \lim_{x\to\infty}\frac{y}{x}=\pm\lim_{x\to\infty} b\sqrt{\frac{1}{a^2}-\frac{1}{x^2}}=\pm\frac{b}{a}. \tag{5.15a–5.15d}$$

If the parameters are equal (Equation 5.16a), it becomes an **equilateral hyperbola** *(Equation 5.16b) that (1) cuts the real axis (Equation 5.16c) at Equation 5.16d and (2) has the diagonals of the quadrants (Equation 5.16e) for asymptotes:*

$$a=b\equiv R: \quad x^2-y^2=R^2; \quad y=0: \quad x=\pm R, \quad \lim_{x\to\infty}\frac{y}{x}=\pm 1. \tag{5.16a–5.16e}$$

If both parameters are unity (Equations 5.17a through 5.17c), the horizontal A (vertical B) coordinate [Equation 5.17d (Equation 5.17e)] of a point on the unit equilateral hyperbola

$$a=b=R=1\; y^2-x^2: \qquad x(\psi)=\cosh\psi, \qquad y(\varphi)=\sinh\psi, \tag{5.17a–5.17d}$$

specifies the hyperbolic cosine (sine), as indicated in Figure 5.1d.

5.2 Secant, Cosecant, Tangent, and Cotangent

Cosine and sine are the two primary functions (Subsection 5.1.1) from which the secondary functions secant and cosecant (cotangent and tangent) are defined (Section 5.2) by algebraic inversion (ratios). They are related (Subsection 5.3.1) to the exponential (Chapter 3). There is a relation between corresponding circular (hyperbolic) functions (Subsection 5.2.1) by an imaginary change of variable (Section 5.1), and they satisfy identities (Subsection 5.2.2)

TABLE 5.1

Relations between Circular Functions

	$a \equiv \sin z$	$b \equiv \cos z$	$c \equiv \tan z$	$d \equiv \cot z$	$e \equiv \sec z$	$f \equiv \csc z$
$a \equiv \sin z$	$a = a$	$b = \sqrt{1-a^2}$	$c = \dfrac{1}{\sqrt{\dfrac{1}{a^2}-1}}$	$d = \sqrt{\dfrac{1}{a^2}-1}$	$e = \dfrac{1}{\sqrt{1-a^2}}$	$f = \dfrac{1}{a}$
$b \equiv \cos z$	$a = \sqrt{1-b^2}$	$b = b$	$c = \sqrt{\dfrac{1}{b^2}-1}$	$d = \dfrac{1}{\sqrt{\dfrac{1}{b^2-1}}}$	$e = \dfrac{1}{b}$	$f = \dfrac{1}{\sqrt{1-b^2}}$
$c \equiv \tan z$	$a = \dfrac{1}{\sqrt{1+\dfrac{1}{c^2}}}$	$b = \dfrac{1}{\sqrt{1+c^2}}$	$c = c$	$d = \dfrac{1}{c}$	$e = \sqrt{1+c^2}$	$f = \sqrt{1+\dfrac{1}{c^2}}$
$d \equiv \cot z$	$a = \dfrac{1}{\sqrt{1+d^2}}$	$b = \dfrac{1}{\sqrt{1+\dfrac{1}{d^2}}}$	$c = \dfrac{1}{d}$	$d = d$	$e = \sqrt{1+\dfrac{1}{d^2}}$	$f = \sqrt{1+d^2}$
$e \equiv \sec z$	$a = \sqrt{1-\dfrac{1}{e^2}}$	$b = \dfrac{1}{e}$	$c = \sqrt{e^2-1}$	$d = \dfrac{1}{\sqrt{e^2-1}}$	$e = e$	$f = \dfrac{1}{\sqrt{1-\dfrac{1}{e^2}}}$
$f \equiv \csc z$	$a = \dfrac{1}{f}$	$b = \sqrt{1-\dfrac{1}{f^2}}$	$c = \dfrac{1}{\sqrt{f^2-1}}$	$d = \sqrt{f^2-1}$	$e = \dfrac{1}{\sqrt{1-\dfrac{1}{f^2}}}$	$f = f$

Note: Any five of the six circular functions can be expressed in terms of the sixth by a simple algebraic relation involving at most square roots.

arising from the circular (hyperbolic) identity (Subsection 5.1.2). All circular (hyperbolic) functions are algebraic functions of any one of them (Subsection 5.3.3) as shown in Table 5.1 (Table 5.2).

5.2.1 Relation between Exponential and Circular and Hyperbolic Functions

From the two **primary functions,** namely, the cosine (sine), the **secondary functions** are defined, namely, the **secant (cosecant)** by inversion:

$$\sec z \equiv \frac{1}{\cos z}, \quad \csc z \equiv \frac{1}{\cos z}, \tag{5.18a and 5.18b}$$

and the **tangent (cotangent)** by their ratio:

$$\tan z \equiv \frac{\sin z}{\cos z} = \sin z \sec z = \frac{1}{\cos z \csc z} = \frac{\sec z}{\csc z} \equiv \frac{1}{\cot z}. \tag{5.18c}$$

TABLE 5.2

Relations between Hyperbolic Functions

	$a \equiv \sinh z$	$b \equiv \cosh z$	$c \equiv \tanh z$	$d \equiv \coth z$	$e \equiv \text{sech}\, z$	$f \equiv \text{csch}\, z$
$a \equiv \sinh z$	$a = a$	$b = \sqrt{1+a^2}$	$c = \dfrac{1}{\sqrt{1+\dfrac{1}{a^2}}}$	$d = \sqrt{1+\dfrac{1}{a^2}}$	$e = \dfrac{1}{\sqrt{1+a^2}}$	$f = \dfrac{1}{a}$
$b \equiv \cosh z$	$a = \sqrt{b^2-1}$	$b = b$	$c = \sqrt{1-\dfrac{1}{b^2}}$	$d = \dfrac{1}{\sqrt{1-\dfrac{1}{b^2}}}$	$e = \dfrac{1}{b}$	$f = \dfrac{1}{\sqrt{b^2-1}}$
$c \equiv \tanh z$	$a = \dfrac{1}{\sqrt{\dfrac{1}{c^2}-1}}$	$b = \dfrac{1}{\sqrt{1-c^2}}$	$c = c$	$d = \dfrac{1}{c}$	$e = \sqrt{1-c^2}$	$f = \sqrt{\dfrac{1}{c^2}-1}$
$d \equiv \coth z$	$a = \dfrac{1}{\sqrt{d^2-1}}$	$b = \dfrac{1}{\sqrt{1-\dfrac{1}{d^2}}}$	$c = \dfrac{1}{d}$	$d = d$	$e = \sqrt{1-\dfrac{1}{d^2}}$	$f = \sqrt{d^2-1}$
$e \equiv \text{sech}\, z$	$a = \sqrt{\dfrac{1}{e^2}-1}$	$b = \dfrac{1}{e}$	$c = \sqrt{1-e^2}$	$d = \dfrac{1}{\sqrt{1-e^2}}$	$e = e$	$f = \dfrac{1}{\sqrt{\dfrac{1}{e^2}-1}}$
$f \equiv \text{csch}\, z$	$a = \dfrac{1}{f}$	$b = \sqrt{1+\dfrac{1}{f^2}}$	$c = \dfrac{1}{\sqrt{1+f^2}}$	$d = \sqrt{1+f^2}$	$e \equiv \dfrac{1}{\sqrt{1+\dfrac{1}{f^2}}}$	$f = f$

Note: Algebraic relations between the six hyperbolic functions are similar to the algebraic relations between the six circular functions in Table 5.1 except for some changes of sign.

Similar definitions apply to the secondary circular (Equations 5.18a through 5.18c) and hyperbolic (Equations 5.19a through 5.19c) functions:

$$\text{sech}\, z \equiv \frac{1}{\cosh z}, \quad \text{csch}\, z \equiv \frac{1}{\sinh z}, \quad \tanh z \equiv \frac{\sinh z}{\cosh z} \equiv \frac{1}{\coth z}. \tag{5.19a–5.19c}$$

From Equations 5.2a and 5.2b (Equations 5.3a and 5.3b) applied to Equations 5.19a through 5.19c (Equations 5.18a through 5.18c) follow Equations 5.20a through 5.20d (Equations 5.21a through 5.21d):

$$\text{sech}\,(z) = \frac{2}{e^z + e^{-z}} = \frac{2e^z}{e^{2z}+1} = \frac{2e^{-z}}{1+e^{-2z}} = \text{sech}(-z), \tag{5.20a}$$

$$\text{csch}\,(z) = \frac{2}{e^z - e^{-z}} = \frac{2e^z}{e^{2z}-1} = \frac{2e^{-z}}{1-e^{-2z}} = -\text{csch}\,(-z), \tag{5.20b}$$

$$\tanh\,(z) = \frac{e^z - e^{-z}}{e^z + e^{-z}} = \frac{e^{2z}-1}{e^{2z}+1} = \frac{1-e^{-2z}}{1+e^{-2z}} = -\tanh(-z), \tag{5.20c}$$

$$\coth(z) = \frac{e^z + e^{-z}}{e^z - e^{-z}} = \frac{e^{2z} + 1}{e^{2z} - 1} = \frac{1 + e^{-2z}}{1 - e^{-2z}} = -\coth(-z), \tag{5.20d}$$

$$\sec(z) = \frac{2}{e^{iz} + e^{-iz}} = \frac{2e^{iz}}{e^{i2z} + 1} = \frac{2e^{-iz}}{1 + e^{-2iz}} = \sec(-z), \tag{5.21a}$$

$$\csc(z) = \frac{2i}{e^{iz} - e^{-iz}} = \frac{2ie^{iz}}{e^{i2z} - 1} = \frac{2ie^{-iz}}{1 - e^{-2iz}} = -\csc(-z), \tag{5.21b}$$

$$i\tan(z) = \frac{e^{iz} - e^{-iz}}{e^{iz} + e^{-iz}} = \frac{e^{2iz} - 1}{e^{2iz} + 1} = \frac{1 - e^{-2iz}}{1 + e^{-2iz}} = -i\tan(-z), \tag{5.21c}$$

$$-i\cot(z) = \frac{e^{iz} + e^{-iz}}{e^{iz} - e^{-iz}} = \frac{e^{2iz} + 1}{e^{2iz} - 1} = \frac{1 + e^{-2iz}}{1 - e^{-2iz}} = -i\cot(-z), \tag{5.21d}$$

which *relate the secondary hyperbolic (Equations 5.20a through 5.20d) [circular (Equations 5.21a through 5.21d)] functions to the exponential and show that the secant is even, and the cosecant, tangent, and cotangent are odd functions.*

5.2.2 Imaginary Change of Variable between Circular and Hyperbolic Functions

The hyperbolic (circular) identity can be written in terms of primary [Equation 5.9a (Equation 5.9b)] or secondary [Equations 5.22a and 5.22b (Equations 5.23a and 5.23b)] functions:

$$\tanh^2(z) + \operatorname{sech}^2(z) = 1 = \coth^2(z) - \operatorname{csch}^2(z), \tag{5.22a and 5.22b}$$

$$\sec^2(z) - \tan^2(z) = 1 = \csc^2(z) - \cot^2(z). \tag{5.23a and 5.23b}$$

These agree with *the imaginary transformations between the circular and hyperbolic secondary functions secant (Equations 5.24a and 5.24b), cosecant (Equations 5.25a and 5.25b), tangent (Equations 5.26a and 5.26b), and cotangent (Equations 5.27a and 5.27b):*

$$\operatorname{sech}(iz) = \sec(z), \quad \sec(iz) = \operatorname{sech}(z), \tag{5.24a and 5.24b}$$

$$\operatorname{csch}(iz) = -i\csc(z), \quad \csc(iz) = -i\operatorname{csch}(z), \tag{5.25a and 5.25b}$$

$$\tanh(iz) = i\tan(z), \quad \tan(iz) = i\tanh(z), \tag{5.26a and 5.26b}$$

$$\coth(iz) = -i\cot(z), \quad \cot(iz) = -i\coth(z), \tag{5.27a and 5.27b}$$

that follow from those of the primary circular and hyperbolic functions cosine (Equations 5.5a and 5.6a) and sine (Equations 5.5b and 5.6b). Given the circular (hyperbolic) cosine A and sine B as, respectively, the horizontal and vertical Cartesian coordinates of a point on the unit circle (unit equilateral hyperbola) in Figure 5.1c (Figure 5.1d), the inversion leads to the secant 1/A and cosecant 1/B and the ratio to the tangent B/A and cotangent A/B; all six functions are plotted for real variable in Figure 5.2a through c (Figure 5.2d through f).

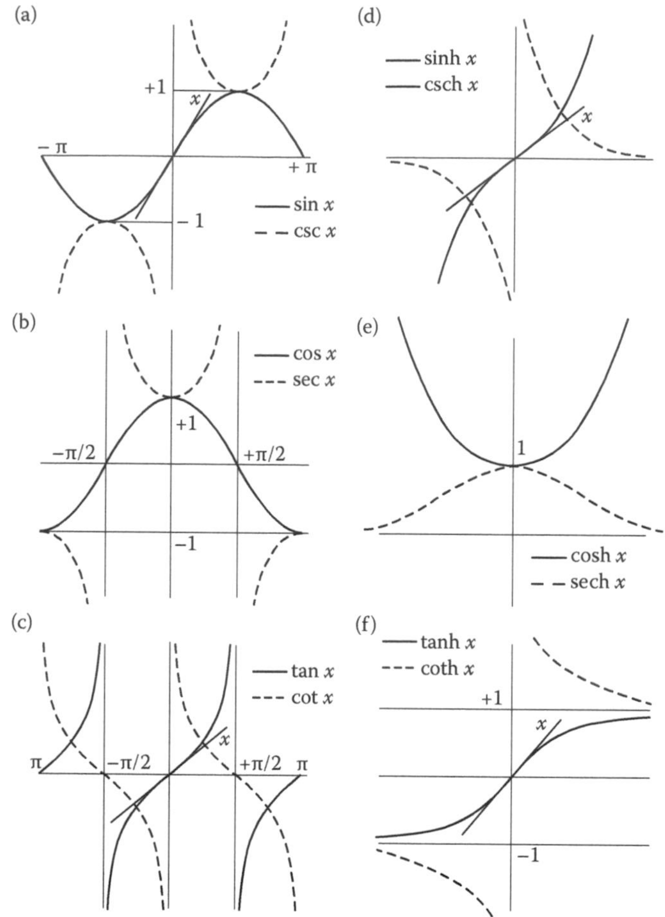

FIGURE 5.2
Because the circular (hyperbolic) functions have a real (imaginary) period [a (b)], they are periodic (are not periodic) as functions of a real variable [a–c (d–f)], as also implied by the representation [a and b (c and d)] on the eclipse (hyperbola). The six circular (hyperbolic) functions [a–c (d–f)] can be represented as three algebraically inverse pairs: (1) sine and cosecant [a (d)]; (2) cosine and secant [b (e)]; and (3) tangent and cotangent [c (f)].

5.2.3 All Circular (Hyperbolic) Functions as Algebraic Functions of Any One of Them

The algebraic relations [Equations 5.9a, 5.22a, and 5.22b (Equations 5.9b, 5.23a, and 5.23b)] show that all circular (hyperbolic) functions can be written as rational functions of any one of them in Table 5.1 (Table 5.2). For example, starting with the circular (hyperbolic) sine [Equation 5.28a (Equation 5.29a)], all other functions are specified by algebraic relations, namely, the cosecant [Equation 5.28b (Equation 5.29b)], the cosine and secant [Equations 5.28c and 5.28d (Equations 5.29c and 5.29d)], and the tangent and cotangent [Equations 5.28e and 5.28f (Equations 5.29e and 5.29f)]:

$$a \equiv \sin z: \quad \csc z = \frac{1}{a}, \quad \cos z = \sqrt{1-a^2} = \frac{1}{\sec z}, \quad \tan z = \frac{a}{\sqrt{1-a^2}} = \frac{1}{\sqrt{1/a^2-1}} = \frac{1}{\cot z}, \tag{5.28a–5.28f}$$

$$a \equiv \sinh z: \qquad \operatorname{csch} z = \frac{1}{a}, \quad \cosh z = \sqrt{1+a^2} = \frac{1}{\operatorname{sechz}}, \quad \tanh z = \frac{a}{\sqrt{1+a^2}} = \frac{1}{\sqrt{1/a^2+1}} = \frac{1}{\coth z}. \tag{5.29a–5.29f}$$

The relations [Equations 5.28a through 5.28f (Equations 5.29a through 5.29f)] follow from (1) the circular (hyperbolic) identity applying to the primary functions [Equation 5.9b (Equation 5.9a)]; and (2) the definitions of secondary functions [Equations 5.18a through 5.18c (Equations 5.19a through 5.19c)]. The preceding relations can be inverted, for example, Equation 5.28e (Equation 5.29e) leads to Equations 5.30a and 5.30b (Equations 5.31a and 5.31b), which specifies the circular (Equation 5.30c) [hyperbolic (Equation 5.31c)] sine in terms of the tangent or cotangent [Equations 5.30d and 5.30e (Equations 5.31d and 5.31e)]:

$$c \equiv \tan z: \quad c^2\left(1-a^2\right) = a^2, \quad a = \frac{c}{\sqrt{1+c^2}}, \quad \sin z = \frac{\tan z}{\sqrt{1+\tan^2 z}} = \frac{1}{\sqrt{\cot^2 z+1}}, \tag{5.30a–5.30e}$$

$$c \equiv \tanh z: \quad c^2\left(1+a^2\right) = a^2, \quad a = \frac{c}{\sqrt{1-c^2}}, \quad \sinh z = \frac{\tanh z}{\sqrt{1-\tanh^2 z}} = \frac{1}{\sqrt{\coth^2 z-1}}. \tag{5.31a–5.31e}$$

The relations [Equations 5.28a through 5.28f and 5.30a through 5.30e (Equations 5.29a through 5.29f and 5.31a through 5.31e)] are a subset of those appearing in Table 5.1 (Table 5.2), and the remaining can be obtained similarly. Table 5.1 (Table 5.2) expresses all circular (hyperbolic) functions in terms of any one of them.

5.3 Formulas of Addition of Several Variables

The property of the exponential of transforming sums to products (Subsection 3.3.1) leads to the addition formulas specifying the cosine and sine (tangent and cotangent) of the sum of several variables in terms of primary (secondary) functions of each argument (Subsection 5.3.1 (Subsection 5.3.2)]. These multiple addition formulas apply to circular (hyperbolic) functions and (1) have as simplest case the formulas for the addition of two variables (Subsection 5.3.3) and (2 and 3) the next two cases by order complexity are the triple [Subsection I.3.7 (Example 10.9)] and quadruple [Example I.10.7 (Example 10.9)] addition formulas.

5.3.1 Formulas for Addition of Variables of Primary Functions

The property of the exponential (Equation 3.22 ≡ Equation 5.32) of transforming sums into products:

$$\exp\left(i\sum_{n=1}^{N} z_n\right) = \prod_{n=1}^{N} \exp\left(iz_n\right) \tag{5.32}$$

implies for circular functions (Equation 5.lb):

$$\cos\left(\sum_{n=1}^{N} z_n\right) + i\sin\left(\sum_{n=1}^{n} z_n\right) = \prod_{n=1}^{N}\left\{\cos(z_n) + i\sin(z_n)\right\}; \tag{5.33}$$

equating even and odd parts leads to

$$\cos(z_1 + \cdots + z_N) = A_N - A_{N-2} + A_{N-4} - \cdots + (-)^n A_{N-2n} + \cdots = \sum_{n=0}^{\leq N/2} (-)^n A_{N-2n}, \tag{5.34a}$$

$$\sin(z_1 + \cdots + z_N) = A_{N-1} - A_{N-3} + A_{N-5} - \cdots + (-)^n A_{N-2n-1} + \cdots = \sum_{n=0}^{\leq (N-1)/2} (-)^n A_{N-2n-1}, \tag{5.34b}$$

$$A_n \equiv \Sigma' \cos(z_1) \ldots \cos(z_n) \sin(z_{n+1}) \ldots \sin(z_N), \tag{5.34c}$$

where Σ' means the sum over all possible distinct combinations (Equation 5.34c), in the formulas of addition of several variables for the circular cosine (Equation 5.34a) and sine (Equation 5.34b). These have analogs (Equation 5.35c) for the hyperbolic cosine (Equation 5.35a) and sine (Equation 5.35b):

$$\cosh(z_1 + \cdots + z_N) = B_N + B_{N-2} + B_{N-4} + \cdots + B_{N-2n} + \cdots = \sum_{n=0}^{\leq N/2} B_{N-2n}, \tag{5.35a}$$

$$\sinh(z_1 + \cdots + z_N) = B_{N-1} + B_{N-3} + A_{N-5} + \cdots + B_{N-2n-1} + \cdots \sum_{n=0}^{\leq (N-1)/2} B_{N-2n-1}, \tag{5.35b}$$

$$B_n \equiv \Sigma' \cosh(z1) \ldots \cosh(z_n) \sin(z_n+1) \ldots \sinh(z_N). \tag{5.35c}$$

The two differ in the suppression of alternating signs in the hyperbolic (Equations 5.35a through 5.35c) relative to the circular (Equations 5.34a through 5.34c ≡ Equations I.3.27a and I.3.27b) case, on account of the imaginary transformation (Equations 5.5a, 5.5b, 5.6a, and 5.6b). The sums up to $n \leq N/2$ mean for $N = 2m$ even ($N = 2m + 1$ odd) sums are up to $n = m$; the sums up to $n = (N - 1)/2$ mean for $N = 2m$ even ($N = 2m + 1$ odd) sums are up to $n = m - 1$ ($n = m$). The identity (Equation 5.36) analogous to Equations 5.32 and 5.33:

$$\cosh\left(\sum_{n=1}^{N} z_n\right) + \sinh\left(\sum_{n=1}^{N} z_n\right) = \exp\left(\sum_{n=1}^{N} z_n\right) = \prod_{n=1}^{N}\left[\cosh(z_n) + \sinh(z_n)\right] \tag{5.36}$$

proves directly Equations 5.35a through 5.35c). The passage from Equation 5.36 to 5.35a (Equation 5.35b) is justified, noting that the left-hand side (l.h.s.) is an even (odd) function of all variables $z_1, \ldots, z_N$, and the right-hand side (r.h.s.) is also even (odd) if the hyperbolic sine appears as a factor an even (odd) number of times. The passage from Equation 5.33 to

5.34a (Equation 5.34b) is similar for the circular cosine (sine) of the sum on the l.h.s., except that on the r.h.s., when the sine appears an even (odd) number $2n$ $(2n + 1)$ of times, there is a factor $i^{2n} = (-)^n$ $[i^{2n+1} = (-)^n\, i]$, hence the alternating signs.

5.3.2 Secondary Functions of the Sum of Several Variables

Rewriting Equation 5.33 in the form:

$$\cos\left(\sum_{n=1}^{N} z_n\right)\left\{1+i\tan\left(\sum_{n=1}^{N} z_n\right)\right\}=\prod_{n=1}^{N}\cos(z_n)\{1+i\tan(z_n)\} \tag{5.37}$$

and equating odd parts lead to

$$\tan(z_1+\cdots+z_N)=\left[\cos(z_1+\cdots+z_N)\right]^{-1}\sum_{n=1}^{\leq N/2}(-)^{n-1}C_{2n-1}\cos(z_{2n})\ldots\cos(z_N), \tag{5.38}$$

where Equation 5.39a appears:

$$C_n \equiv \Sigma' \tan(z_1)\ldots\tan(z_n), \qquad D_n \equiv \Sigma' \tanh(z_1)\ldots\tanh(z_n), \qquad \text{(5.39a and 5.39b)}$$

Substituting Equations 5.34a and 5.34c in Equation 5.38b leads to Equation 5.40:

$$\tan(z_1+\cdots+z_N)=\frac{C_1-C_3+C_5-\cdots+(-)^{n-1}C_{2n+1}+\cdots}{1-C_2+C_4-\cdots+(-)^{n-1}C_{2n}+\cdots}=\frac{\sum_{n=0}^{\leq(N-1)/2}(-)^nC_{2n+1}}{1+\sum_{n=1}^{\leq N/2}(-)^nC_{2n}}; \tag{5.40}$$

$$\tanh(z_1+\cdots+z_N)=\frac{D_1+D_3+D_5+\cdots+D_{2n-1}+\cdots}{1+D_2+D_4+\cdots+D_{2n}+\cdots}=\frac{\sum_{n=0}^{\leq(N-1)/2}D_{2n+1}}{1+\sum_{n=1}^{\leq N/2}D_{2n}}. \tag{5.41}$$

The imaginary change of variable (Equations 5.26a and 5.26b) leads from Equations 5.40 and 5.39a to Equations 5.41 and 5.39b. Alternatively, Equations 5.39a and 5.40 (Equations 5.39b and 5.341) also follow from the ratio of Equations 5.34a and 5.34b (Equations 5.35a and 5.35b), dividing all terms in Equation 5.34c (Equation 5.35c) by $\cos(z_1)\ldots\cos(z_N)$ [$\cosh(z_1)\ldots\cosh(z_n)$].

From Equation 5.42 ≡ Equation 5.37

$$i\sin\left(\sum_{n=1}^{N} z_n\right)\left\{1-i\cot\left(\sum_{n=1}^{N} z_n\right)\right\}=\prod_{n=1}^{N}i\sin(z_n)\{1-i\cot(z_n)\}; \tag{5.42}$$

equating odd parts leads to

$$\cot\left(z_1+\cdots+z_N\right)=\left[\sin\left(z_1+\cdots+z_n\right)\right]^{-1}\sum_{n=1}^{\le N/2}(-)^n E_{2n}\sin\left(z_{2n}\right)\ldots\sin\left(z_N\right), \tag{5.43}$$

where Equation 5.44a appears:

$$E_n \equiv \Sigma' \cot(z_1)\ldots\cot(z_n),\; F_n \equiv \Sigma' \coth(z_1)\ldots\coth(z_n). \tag{5.44a and 5.44b}$$

Using Equations 5.34b and 5.34c in Equation 5.43 leads to Equation 5.45:

$$\cot\left(z_1+\cdots+z_N\right)=\frac{-1+E_2-E_4+\cdots+(-)^{n-1}E_{2n}+\cdots}{E_1-E_3+E_3-\cdots+(-)^{n-1}E_{2n-1}+\cdots}=\frac{-1+\sum\limits_{n=1}^{\le N/2}\left(-\right)^{n-1}E_{2n}}{\sum\limits_{n=0}^{\le(N-1)/2}\left(-\right)^{n}E_{2n+1}}; \tag{5.45}$$

$$\coth\left(z_1+\cdots+z_N\right)=\frac{1+F_2+F_4+\cdots+F_{2n}+\cdots}{F_1+F_3+F_5+\cdots+F_{2n+1}+\cdots}=\frac{1+\sum\limits_{n=0}^{\le N/2}F_{2n+1}}{1+\sum\limits_{n=0}^{\le(N-1)/2}F_{2n+1}}. \tag{5.46}$$

The imaginary transformation (Equations 5.27a and 5.27b) leads from Equations 5.44a and 5.45 to Equations 5.44b and 5.46. The multiple addition formulas [Equation 5.45 (Equation 5.46)] can also be proved by the ratio of Equations 5.34a through 5.34c (Equations 5.35a through 5.35c). This completes the proofs of *the formulas for the addition of several variables for the circular (hyperbolic) tangent [Equations 5.39a and 5.40 (Equations 5.39b and 5.41)] and cotangent [Equations 5.44a and 5.45 (Equations 5.44b and 5.46)].*

5.3.3 Double/Triple/Quadruple Addition Formulas

The simplest instances of Equations 5.34a through 5.34c/5.40; 5.39a/5.45; 5.44a/5.35a through 5.35c/5.41; 5.39b/5.46; 5.44b are, respectively, Equations 5.47a and 5.47b; 5.48a/5.48b/5.49a and 5.49b/5.50a; 5.50b for $N = 2$, $z_1 \equiv u$, $z_2 \equiv v$, which yield *the addition and subtraction formulas for two variables:*

$$\cos(u \pm v) = \cos u \cos v \mp \sin u \sin v, \qquad \sin(u \pm v) = \sin u \cos v \pm \sin v \cos u, \tag{5.47a and 5.47b}$$

$$\tan\left(u\pm v\right)=\frac{\tan u\pm\tan v}{1\mp\tan u\tan v}, \quad \cot\left(u\pm v\right)=\frac{\cot u\cot v\mp 1}{\cot v\pm\cot u}, \tag{5.48a and 5.48b}$$

$$\cosh(u \pm v) = \cosh u \cosh v \pm \sinh u \sinh v, \quad \sinh(u \pm v) = \sinh u \cosh v \pm \sinh v \cosh u, \tag{5.49a and 5.49b}$$

$$\tanh(u \pm v) = \frac{\tanh u \pm \tanh v}{1 \pm \tanh u \tanh v}, \quad \coth(u \pm v) = \frac{\coth u \coth v \pm 1}{\coth v \pm \coth u}. \tag{5.50a and 5.50b}$$

The subtraction formulas with argument $u - v$ can be obtained from the addition formulas with argument $u + v$ by changing the sign of $v \to -v$ and using the even or odd properties. The addition formulas of next higher order of complexity, such as triplication (quadruplication) with three (four) variables, were given in Subsection I.3.7 (Example I.10.7) for the circular sine and cosine and are given in Example 10.9 for the hyperbolic sine and cosine.

5.4 Formulas for Multiple, Double, and Half Variable

Setting all variables equal in the formulas for the sum of several variables (Subsections 5.3.1 and 5.3.2) leads to the multiplication formulas for a multiple variable (Subsection 5.4.1). The simplest particular cases are the duplication/triplication/quadruplication formulas (Subsection 5.4.2) corresponding to the double/triple/quadruple addition formulas (Subsections 5.3.2 and 5.3.3) with equal variable, thus leading to the double, triple, and quadruple variable, respectively. The opposite results are the half-variable formulas (Subsection 5.5.3).

5.4.1 Circular/Hyperbolic Functions of a Multiple Variable

Setting all arguments equal in Equations 5.34a through 5.34c; 5.35a through 5.35c, Equations 5.40; 5.41; 5.39a and 5.39b, and Equations 5.45; 5.46; 5.44a and 5.44b (Equation 5.51a) leads to the formulas for multiplication of the variable (Equation 5.51b), which simplify the sums to a number (Equation 5.51c) of identical terms:

$$z_1 = z_2 = \cdots = z_N \equiv z; \quad z_1 + \cdots + z_N = Nz: \quad \begin{pmatrix} N \\ n \end{pmatrix} = \frac{N!}{n!(N-n)!}, \tag{5.51a–5.51c}$$

$$\{A_n, B_n, C_n, D_n, E_n, F_n\} \equiv \binom{N}{n}\{\cos^n z \sin^{N-n} z,\ \cosh^n z \sinh^{N-n} z,\ \tan^n z,\ \tanh^n z,\ \cot^n z,\ \coth^n z\}, \tag{5.52a–5.52f}$$

for Equation 5.34c ≡ Equation 5.52a, Equation 5.35c ≡ Equation 5.52b, Equations 5.39a and 5.39b ≡ Equations 5.52c and 5.52d, and Equations 5.44a and 5.44b ≡ Equations 5.52e and 5.52f. Substituting Equations 5.52a through 5.52f, respectively, in Equation 5.34a (Equation 5.35a)/Equation 5.34b (Equation 5.35b)/Equation 5.40 (Equation 5.45)/Equation 5.41 (Equation 5.46) leads to *the formulas of multiplication of the variable, for the circular (hyperbolic) cosine [Equation 5.53 (Equation 5.57)], sine [Equation 5.54 (Equation 5.58)], tangent [Equation 5.55 (Equation 5.59)], and cotangent [Equation 5.36 (Equation 5.60)]:*

$$\cos(Nz) = \sum_{n=0}^{\leq N/2} (-)^n \begin{pmatrix} N \\ 2n \end{pmatrix} \cos^{N-2n} z \sin^{2n} z, \tag{5.53}$$

$$\sin(Nz) = \sum_{n=0}^{\leq(N-1)/2} (-)^n \binom{N}{2n+1} \cos^{N-2n-1} z \sin^{2n+1} z, \tag{5.54}$$

$$\tan(Nz) = \frac{\displaystyle\sum_{n=0}^{\leq(N-1)/2} (-)^n \binom{N}{2n+1} \tan^{2n+1} z}{1 + \displaystyle\sum_{n=1}^{\leq N/2} (-)^n \binom{N}{2n} \tan^{2n} z}, \tag{5.55}$$

$$\cot(Nz) = \frac{-1 + \displaystyle\sum_{n=1}^{\leq N/2} (-)^n \binom{N}{2n} \cot^{2n} z}{\displaystyle\sum_{n=0}^{\leq(N-1)/2} (-)^n \binom{N}{2n+1} \cot^{2n+1} z}, \tag{5.56}$$

$$\cosh(Nz) = \sum_{n=1}^{\leq N/2} \binom{N}{n} \cosh^{N-2n} z \sinh^{2n} z, \tag{5.57}$$

$$\sinh(Nz) = \sum_{n=0}^{\leq(N-1)/2} \binom{N}{2n+1} \cos^{N-2n-1} z \sinh^{2n+1} z, \tag{5.58}$$

$$\tanh(Nz) = \frac{\displaystyle\sum_{n=0}^{\leq(N-1)/2} \binom{N}{2n+1} \tanh^{2n+1} z}{1 + \displaystyle\sum_{n=1}^{\leq N/2} \binom{N}{2n} \tanh^{2n} z}, \tag{5.59}$$

$$\coth(Nz) = \frac{1 + \displaystyle\sum_{n=1}^{\leq N/2} \binom{N}{2n} \coth^{2n} z}{\displaystyle\sum_{n=0}^{\leq(N-1)/2} \binom{N}{2n+1} \coth^{2n+1} z}, \tag{5.60}$$

where Equation 5.53 ≡ Equation I.3.28b and Equation 5.54 ≡ Equation I.3.28c. These formulas differ only in the suppression of alternating signs in the hyperbolic relative to the circular case.

5.4.2 Duplication/Triplication/Quadruplication Formulas for the Variable

The simplest particular instance $N = 2$ of Equations 5.53 through 5.60 leads to *the duplication formulas (Equations 5.61a through 5.61c; 5.62a through 5.62c; 5.63a and 5.63b; 5.64a through 5.64d; and 5.65a through 5.65d):*

$$\cos(2z) = \cos^2 z - \sin^2 z = 2\cos^2 z - 1 = 1 - 2\sin^2 z, \qquad (5.61a–5.61c)$$

$$\cosh(2z) = \cosh^2 z + \sinh^2 z = 2\cosh^2 z - 1 = 1 + 2\sinh^2 z, \qquad (5.62a–5.62c)$$

$$\sin(2z) = 2\sin z\cos z, \quad \sinh(2z) = 2\cosh z\sinh z, \qquad (5.63a\text{ and }5.63b)$$

$$\tan(2z) = \frac{2\tan z}{1-\tan^2 z} = \frac{2}{\cot z - \tan z} = \frac{2\cot z}{\cot^2 z - 1} = \frac{1}{\cot(2z)}, \qquad (5.64a–5.64d)$$

$$\tanh(2z) = \frac{2\tanh z}{1+\tanh^2 z} = \frac{2}{\coth z + \tanh z} = \frac{2\coth z}{1+\coth^2 z} = \frac{1}{\coth(2z)}. \qquad (5.65a–5.65d)$$

These can also be deduced from Equations 5.47a and 5.47b; 5.48a and 5.48b; 5.49a and 5.49b; and 5.50a and 5.50b with $u = v \equiv z$, that is, for $u + v = 2z$. Next in order of complexity are the triplicating $3z$ (quadruplicating $4z$) formulas given in Subsection I.3.7 (Example I.10.7) for the circular sine and cosine and in Example 10.9 for the hyperbolic sine and cosine.

5.4.3 Half-Variable Formulas for Cosine, Sine, and Tangent

Replacing z by $z/2$ and solving Equations 5.61b and 5.61c (Equations 5.62b and 5.62c) for the circular (hyperbolic) function with smaller variable lead to

$$\cos\left(\frac{z}{2}\right) = \pm\left|\frac{1+\cos z}{2}\right|^{1/2}, \quad \sin\left(\frac{z}{2}\right) = \pm\left|\frac{1-\cos z}{2}\right|^{1/2}, \qquad (5.66a\text{ and }5.66b)$$

$$\cosh\left(\frac{z}{2}\right) = \left|\frac{1+\cosh z}{2}\right|^{1/2}, \quad \sinh\left(\frac{z}{2}\right) = \pm\left|\frac{\cosh z - 1}{2}\right|^{1/2} \qquad (5.67a\text{ and }5.67b)$$

the half-variable formulas for the circular (hyperbolic) cosine [Equation 5.66a (Equation 5.67a)] and sine [Equation 5.66b (Equation 5.67b)]. The choice of sign is clear from the value of z; for example, for z real, it is always the + sign in Equation 5.67a in agreement with Figure 5.2e, and for real z, it is the sign of z in Equation 5.66b as follows from Figure 5.2a in the range $-\pi < z < +\pi$. The half-variable formulas for the circular (hyperbolic) tangent are obtained from the preceding [Equations 5.66a and 5.66b (Equations 5.67a and 5.67b)] together with the circular (Equation 5.9b) [hyperbolic (Equation 5.9a)] identity:

$$\tan\left(\frac{z}{2}\right) = \pm\left|\frac{1-\cos z}{1+\cos z}\right|^{1/2} = \pm\frac{\left|1-\cos^2 z\right|^{1/2}}{1+\cos z} = \frac{\sin z}{1+\cos z} = \pm\frac{1}{\csc z + \cot z}, \qquad (5.68a–5.68d)$$

$$= \pm\frac{1-\cos z}{\left|1-\cos^2 z\right|^{1/2}} = \frac{1-\cos z}{\sin z} = \csc z - \cot z = \left[\cot\left(\frac{z}{2}\right)\right]^{-1}, \qquad (5.68e–5.68h)$$

$$\tanh\left(\frac{z}{2}\right) = \pm\left|\frac{\cosh z - 1}{\cosh z + 1}\right|^{1/2} = \pm\frac{\left|\cosh^2 z - 1\right|^{1/2}}{\cosh z + 1} = \frac{\sinh z}{\cosh z + 1} = \frac{1}{\operatorname{csch} z + \coth z}, \quad (5.69a\text{–}5.69d)$$

$$= \pm\frac{\cosh z - 1}{\left|\cosh^2 z - 1\right|^{1/2}} = \frac{\cosh z - 1}{\sinh z} = \coth z - \operatorname{csch} z = \left[\coth\left(\frac{z}{2}\right)\right]^{-1}. \quad (5.69e\text{–}5.69h)$$

The ambiguity in sign in Equations 5.68a, 5.68b, 5.69a, and 5.69b (Equations 5.68e and 5.69e) is resolved in Equations 5.68c, 5.68d, 5.69c, and 5.69d (Equations 5.68f, 5.68g, 5.69f, and 5.69g) because for real z: (1) the denominator (numerator) in Equations 5.68c and 5.69c (Equations 5.68f and 5.69f) is positive; and (2) the sign is the same for the tangent, sine, and cotangent, both in the circular and hyperbolic cases. *The half-variable formulas for the circular (hyperbolic) functions are Equation 5.66a (Equation 5.67a) for the cosine, Equation 5.66b (Equation 5.67b) for the sine, and Equations 5.68a through 5.68h (Equations 5.69a through 5.69h) for the tangent and cotangent.*

5.5 Powers, Products, and Sums of the Functions

The cosine and sine of a multiple variable can be expressed (Subsection 5.4.1) as a sum of products of powers of the cosine and sine. The reverse formulas specify the powers of the cosine and sine as a sum of cosines or sines of a multiple variable (Subsections 5.5.1 and 5.5.2). The simplest particular cases are the squares (cubes) of sines and cosines expressed as sums of sines and cosines of double (triple) variable (Subsection 5.5.3). The squares of cosines and sines are also a particular case with equal variable of the product of cosines and sines with several unequal variables (Subsection 5.5.4); the latter can be transformed into formulas for the sum and difference of (of squares of) cosines and sines with different variables [Subsection 5.5.5 (Subsection 5.5.6)].

5.5.1 Even and Odd Powers of the Circular Cosine

The formulas for the circular (hyperbolic) cosine [Equation 5.53 (Equation 5.57)] and sine [Equation 5.54 (Equation 5.58)] of a multiple variable in terms of sums of products of powers of the sine and cosine of the variable have for "inverses" formulas giving the powers of a cosine or sine as a sum of cosine or sine multiples of the variable. As an example of deduction of the latter, consider the odd power of the circular cosine and apply the binomial formula (Equation I.25.37a ≡ Equations 3.146a through 3.146c):

$$\cos^{2N+1} z = \left(\frac{e^{iz} + e^{-iz}}{2}\right)^{2N+1} = 2^{-2N-1}\sum_{n=0}^{2N+1}\binom{2N+1}{n} e^{-inz} e^{i(2N+1-n)z}$$
$$= 2^{-2N-1}\left[\sum_{n=0}^{N} + \sum_{n=N+1}^{2N+1}\right]\binom{2N+1}{n} e^{i(2N-2n+1)z}. \quad (5.70)$$

The second sum on the r.h.s. of Equation 5.70 can be written in a similar form (Equation 5.71b) to the first using the change of dummy summation variable (Equation 5.71a):

$$m = 2N+1-n: \quad \sum_{n=N+1}^{2N+1} \binom{2N+1}{n} \exp\left[i(2N+1-2n)z\right]$$
$$= \sum_{m=0}^{N} \binom{2N+1}{m} \exp\left\{i\left[2N+1-2(2N+1-m)\right]z\right\} \qquad \text{(5.71a and 5.71b)}$$
$$= \sum_{m=0}^{N} \binom{2N+1}{m} \exp\left[-i(2N+1-2m)z\right],$$

where the following was used:

$$\binom{2N+1}{n} = \binom{2N+1}{2N+1-m} = \frac{(2N+1)!}{(2N+1-m)!m!} = \binom{2N+1}{m}. \tag{5.71c}$$

Substituting Equation 5.71b in Equation 5.70 leads to

$$\cos^{2N+1} z = 2^{-2N-1} \sum_{n=0}^{N} \binom{2N+1}{n} \left[e^{i(2N-2n+1)z} - e^{-i(2N-2n+1)z}\right]$$
$$= 2^{-2N} \sum_{n=0}^{N} \binom{2N+1}{n} \cos\left[(2N-2n+1)z\right], \tag{5.72}$$

that expresses the odd power of the cosine as a sum of cosines of odd multiples of the variable.

In the case of odd powers $(2N + 1)$ of the cosine, the sum (Equation 5.70) has an even number of terms $2N + 2 = 2(N + 1)$, which leads to $(N + 1)$ pairs in Equation 5.72. In the case of even powers $2N$ of the cosine, there are $2N + 1$ terms in

$$\cos^{2N} z = \left(\frac{e^{iz} + e^{-iz}}{2}\right)^{2N} = 2^{-2N} \sum_{n=0}^{N} \binom{2N}{n} e^{i(2N-n)z}$$
$$= 2^{-2N} \left\{\binom{2N}{N} + \left[\sum_{n=0}^{N-1} + \sum_{n=N+1}^{2N}\right] \binom{2N}{n} e^{i(2N-2n)z}, \tag{5.73}\right.$$

so that (1) the term $n = N$ is singled out because it is a constant independent of z; and (2) the remaining $2N$ terms that depend on z form N pairs as before in Equations 5.71a through c and 5.72. The second sum on the r.h.s. of Equation 5.73 can be written (Equation 5.74b) as the first using the change of dummy variable of summation (Equation 5.74a):

$$m = 2N - n: \quad \sum_{n=N+1}^{2N} \binom{2N}{n} e^{i(2N-2n)z} = \sum_{m=0}^{N-1} \binom{2N}{2N-m} \exp\{i[2N - 2(2N-m)]\} = \sum_{m=0}^{N-1} \binom{2N}{m} e^{-i(2N-2m)z}, \tag{5.74a and 5.74b}$$

where Equation 5.74c was used:

$$\binom{2N}{2N-m} = \frac{(2N)!}{(2N-m)!m!} = \binom{2N}{m}. \tag{5.74c}$$

Substituting Equation 5.74c in Equation 5.73 leads to

$$\begin{aligned} \cos^{2N} z &= 2^{-2N} \left\{ \binom{2N}{N} + \sum_{n=0}^{N-1} \binom{2N}{n} \left[e^{i2(N-n)z} + e^{-i2(N-n)z} \right] \right\} \\ &= 2^{-2N} \left\{ \binom{2N}{N} + 2 \sum_{n=0}^{N-1} \binom{2N}{n} \cos[(2N-2n)z], \right. \end{aligned} \tag{5.75}$$

specifying the even power of the cosine as a sum of cosine of even multiples of the variable. From the expressions for the even (odd) power of the circular cosine [Equation 5.75 (Equation 5.72)] follow by simple transformations those for the even (odd) power of the sine and for the corresponding hyperbolic functions (Subsection 5.5.2).

5.5.2 Powers of Circular/Hyperbolic Sine/Cosine

The formula for the even (odd) power [Equation 5.75 (Equation 5.72)] of the circular cosine with a translation by $\pi/2$ leads to [Equation 5.76a (Equation 5.76b)]

$$\begin{aligned} \sin^{2N} z &= (-)^{2N} \cos^{2N}\left(z + \frac{\pi}{2}\right) = 2^{-2N} \left\{ \binom{2N}{n} + 2 \sum_{n=0}^{N-1} \binom{2N}{n} \cos[(2N-2n)z + (N-n)\pi] \right\} \\ &= 2^{-2N} \left\{ \binom{2N}{N} + 2 \sum_{n=0}^{N-1} (-)^{N-n} \binom{2N}{n} \cos[(2N-2n)z] \right\}, \end{aligned} \tag{5.76a}$$

$$\begin{aligned}\sin^{2N+1} z &= (-)^{2N+1}\cos^{2N+1}\left(z+\frac{\pi}{2}\right)\\ &= -2^{-2N}\sum_{n=0}^{N}\binom{2N+1}{n}\cos\left[(2N+1-2n)z+(N-n)\pi+\pi/2\right] \\ &= 2^{-2N}\sum_{n=0}^{N}(-)^{N-n}\binom{2N+1}{n}\sin\left[(2N-2n+1)z\right]\end{aligned} \tag{5.76b}$$

the formulas for the even (Equation 5.76a) [odd (Equation 5.76b)] power of the circular sine: (1) the even power of the sine is an even function and appears as a sum of cosines of even multiples of the variable (Equation 5.76a); and (2) the odd power of the sine is an odd function and appears as a sum of sines of odd multiples of the variable (Equation 5.76b). The even (odd) power of the cosine is an even function and appears as a sum of cosines of even (odd) multiples of the variable [Equation 5.75 (Equation 5.72)]. Thus, (1) the sum of sines, which are odd functions, appears only in the odd power of the sine (Equation 5.76b), which is also an odd function; and (2) the even powers of the sine (Equation 5.76a) and the even (Equation 5.75) and odd (Equation 5.72) powers of the cosine are all even functions and appear as a sum of cosines, which are also even functions. Passing from the circular to the hyperbolic functions, the imaginary change of variable shows that (1) for the cosine (Equation 5.6a), the formulas are the same (Equation 5.77a):

$$\cosh^N z = \cos^N(iz), \qquad \sinh^{2N+1} z = [-i\sin(iz)]^{2N} = (-)^N \sin^{2N}(iz), \tag{5.77a and 5.77b}$$

$$\sinh^{2N+1} z = [-i\sin(iz)]^{2N+1} = -i\,(-)^N \sin^{2N+1}(iz); \tag{5.77c}$$

and (2) for the sine (Equation 5.6b), there are changes of sign (Equations 5.77b and 5.77c). The formulas are collected for the even (odd) powers of the circular cosine [Equation 5.75 ≡ Equation 5.78a (Equation 5.72 ≡ Equation 5.78b)] and sine [Equation 5.76a ≡ Equation 5.79a (Equation 5.76b ≡ Equation 5.79b)]:

$$\cos^{2N} z = 2^{-2N}\left\{\binom{2N}{N}+2\sum_{n=0}^{N-1}\binom{2N}{n}\cos\left[(2N-2n)z\right]\right\}, \tag{5.78a}$$

$$\cos^{2N+1} z = 2^{-2N}\sum_{n=0}^{N}\binom{2N+1}{n}\cos\left[(2N-2n+1)z\right], \tag{5.78b}$$

$$\sin^{2N} z = 2^{-2N}\left\{\binom{2N}{N}+2\sum_{n=0}^{N-1}(-)^{N-n}\binom{2N}{n}\cos\left[(2N-2n)z\right]\right\}, \tag{5.79a}$$

$$\sin^{2N+1} z = 2^{-2N}\sum_{n=0}^{N}(-)^{N-n}\binom{2N+1}{n}\sin\left[(2N-2n+1)z\right]. \tag{5.79b}$$

Applying the imaginary transformation [Equation 5.6a (Equation 5.6b)] to Equations 5.78a and 5.78b (Equations 5.79a and 5.79b) leads to the corresponding formulas for the even (odd) power of the hyperbolic cosine [Equation 5.80a (Equation 5.80b)] and sine [Equation 5.81a (Equation 5.81b)]:

$$\cosh^{2N}(z) = 2^{-2N}\left\{\binom{2N}{N} + 2\sum_{n=0}^{N-1}\binom{2N}{n}\cosh\left[(2N-2n)z\right]\right\}, \tag{5.80a}$$

$$\cosh^{2N+1}(z) = 2^{-2N}\sum_{n=0}^{N}\binom{2N+1}{n}\cosh\left[(2N-2n+1)z\right], \tag{5.80b}$$

$$\sinh^{2N}(z) = 2^{-2N}\left\{(-)^N\binom{2N}{N} + 2\sum_{n=0}^{N-1}(-)^n\binom{2N}{n}\cosh\left[(2N-2n)z\right]\right\}, \tag{5.81a}$$

$$\sinh^{2N+1}(z) = 2^{-2N}\sum_{n=0}^{N}(-)^n\binom{2N+1}{n}\sinh\left[(2N-2n+1)z\right], \tag{5.81b}$$

which (1) do not change the sign for the cosine (Equation 5.77a) in Equations 5.80a and 5.80b; and (2) introduce a factor $(-)^N$ for the odd (Equation 5.77c) [even (Equation 5.77b)] power of the sine in Equation 5.81b (Equation 5.81a). Thus, *the even (odd) powers of the circular cosine [Equation 5.77a (Equation 5.77b)] and sine [Equation 5.78a (Equation 5.78b)] and hyperbolic cosine [Equation 5.79a (Equation 5.79b)] and sine [Equation 5.80a (Equation 5.80b)] can be expressed in terms of multiples of the variable.*

5.5.3 Square and Cubes of Sines and Cosines

The simplest instance $N = 1$ of Equations 5.78a, 5.79a, 5.80a, and 5.81a (Equations 5.78b, 5.79b, 5.80b, and 5.81b) are, respectively *the formula for the square (cube) of the circular and hyperbolic cosine and sine [Equations 5.82a, 5.82b, 5.83a, and 5.83b (Equations 5.84a, 5.84b, 5.85a, and 5.85b]:*

$$\cos^2 z = \frac{1+\cos(2z)}{2}, \quad \sin^2 z = \frac{1-\cos(2z)}{2}, \tag{5.82a and 5.82b}$$

$$\cosh^2 z = \frac{1+\cosh(2z)}{2}, \quad \sinh^2 z = \frac{\cosh(2z)-1}{2}, \tag{5.83a and 5.83b}$$

$$\cos^3 z = \frac{1}{4}\cos(3z) + \frac{3}{4}\cos z, \quad \sin^3 z = -\frac{1}{4}\sin(3z) + \frac{3}{4}\sin z, \tag{5.84a and 5.84b}$$

$$\cosh^3 z = \frac{1}{4}\cosh(3z) + \frac{3}{4}\cosh z, \quad \sinh^3 z = \frac{1}{4}\sinh(3z) - \frac{3}{4}\sinh z, \tag{5.85a and 5.85b}$$

that can also be obtained from previous results: (1) the formulas for squares of the circular (hyperbolic) cosine and sine coincide [Equations 5.82a and 5.82b ≡ Equations 5.61b and 5.61c (Equations 5.83a and 5.83b ≡ Equations 5.62b and 5.62c)] with the circular (hyperbolic) cosine of double variable; and (2) the latter (Equations 5.61a through 5.61c and 5.62a through 5.62c) and the circular (hyperbolic) sine of double variable [Equation 5.63a (Equation 5.63b)] appear in the cubes of the circular (hyperbolic) cosine [Equation 5.86a (Equation 5.87a)] and sine [Equation 5.86b (Equation 5.87b)]:

$$\cos^3 z = \cos z \frac{1+\cos(2z)}{2} = \frac{\cos z}{2} + \frac{1}{2}\frac{\cos(3z)+\cos z}{2} = \frac{3}{4}\cos z + \frac{1}{4}\cos(3z), \tag{5.86a}$$

$$\sin^3 z = \sin z \frac{1-\cos(2z)}{2} = \frac{\sin z}{2} - \frac{1}{2}\frac{\sin(3z)-\sin z}{2} = \frac{3}{4}\sin z - \frac{1}{4}\sin(3z), \tag{5.86b}$$

$$\cosh^3 z = \cosh z \frac{\cosh(2z)+1}{2} = \frac{1}{2}\cosh z + \frac{1}{2}\frac{\cosh(3z)+\cosh z}{2} = \frac{3}{4}\cosh z + \frac{1}{4}\cosh(3z), \tag{5.87a}$$

$$\sinh^3 z = \sinh z \frac{\cosh(2z)-1}{2} = -\frac{1}{2}\sinh z + \frac{1}{2}\frac{\sinh(3z)-\sinh z}{2} = -\frac{3}{4}\sinh z + \frac{1}{4}\sinh(3z), \tag{5.87b}$$

in agreement with Equations 5.86a and 5.86b ≡ Equations 5.84a and 5.84b and Equations 5.87a and 5.87b ≡ Equations 5.86a and 5.86b. Formulas for the product of circular (hyperbolic) cosines and sines [Equations 5.88a through 5.88c (Equations 5.89a through 5.89c)] were used in Equations 5.87a and 5.87b (Equations 5.86a and 5.86b), and are proved next (Subsection 5.6.3). Combining the formulas for addition and subtraction of two variables for the circular and hyperbolic cosine and sine (Subsection 5.3.3) leads to formulas for their products (Subsection 5.5.4). From these, the formulas for the sum and difference of circular and hyperbolic cosines and sines of two variables (Subsection 5.5.5) follow by two changes of variables.

5.5.4 Products of Two Circular/Hyperbolic Sines or Cosines

Adding and subtracting the formulas [Equations 5.47a and 5.47b (Equations 5.49a and 5.49b)] for the circular (hyperbolic) cosine and sine of sum $u + v$ and difference $u - v$ of variables lead to *the formulas for the product of circular (Equations 5.88a through 5.88c) [hyperbolic (Equations 5.89a through 5.89c)] cosines and sines:*

$$2\cos u \cos v = \cos(u+v) + \cos(u-v), \tag{5.88a}$$

$$2\sin u \sin v = \cos(u-v) - \cos(u+v), \tag{5.88b}$$

$$2\sin u \cos v = \sin(u+v) + \sin(u-v), \tag{5.88c}$$

$$2\cosh u \cosh v = \cosh(u+v) + \cosh(u-v), \tag{5.89a}$$

$$2\sinh u \sinh v = \cosh(u+v) - \cosh(u-v), \tag{5.89b}$$

$$2\sinh u \cosh v = \sinh(u+v) + \sinh(u-v). \tag{5.89c}$$

The formulas [Equations 5.88a through 5.88c (Equations 5.89a through 5.89c)] reduce to the duplication formulas for the circular (Equations 5.61a through 5.61c and 5.63a) [hyperbolic (Equations 5.62a through 5.62c and 5.63b)] cosine and sine for $u = v \equiv z$.

5.5.5 Sums and Differences of Circular and Hyperbolic Sines and Cosines

Performing the changes of variable (Equations 5.90a and 5.90b ≡ Equations 5.90c and 5.90d):

$$a \equiv u + v, \quad b \equiv u - v: \quad u = \frac{a+b}{2}, \quad v = \frac{a-b}{2}, \tag{5.90a–5.90d}$$

the formulas Equations 5.88a through 5.88c (Equations 5.89a through 5.89c) can be rewritten [Equations 5.91a through 5.91d (Equations 5.92a through 5.92c)] and Equation 5.91d (Equation 5.92d) follows from Equation 5.91c (Equation 5.92c), exchanging the sign of b:

$$\cos a + \cos b = 2\cos\left(\frac{a+b}{2}\right)\cos\left(\frac{a-b}{2}\right), \tag{5.91a}$$

$$\cos a - \cos b = -2\sin\left(\frac{a+b}{2}\right)\sin\left(\frac{a-b}{2}\right), \tag{5.91b}$$

$$\sin a + \sin b = 2\sin\left(\frac{a+b}{2}\right)\cos\left(\frac{a-b}{2}\right), \tag{5.91c}$$

$$\sin a - \sin b = 2\cos\left(\frac{a+b}{2}\right)\sin\left(\frac{a-b}{2}\right), \tag{5.91d}$$

$$\cosh a + \cosh b = 2\cosh\left(\frac{a+b}{2}\right)\cosh\left(\frac{a-b}{2}\right), \tag{5.92a}$$

$$\cosh a - \cosh b = 2\sinh\left(\frac{a+b}{2}\right)\sinh\left(\frac{a-b}{2}\right), \tag{5.92b}$$

$$\sinh a + \sinh b = 2\sinh\left(\frac{a+b}{2}\right)\cosh\left(\frac{a-b}{2}\right), \tag{5.92c}$$

$$\sinh a - \sinh b = 2\cosh\left(\frac{a+b}{2}\right)\sinh\left(\frac{a-b}{2}\right). \tag{5.92d}$$

These are *the formulas of addition and subtraction of functions for the circular (Equations 5.91a through 5.91d) [hyperbolic (Equations 5.92a through 5.92d)] sine and cosine*. The differences of squares of cosines and sines are considered next (Subsection 5.5.6).

5.5.6 Differences of Squares of Cosines and Sines

The product of the circular (hyperbolic) cosine of the sum and difference [Equation 5.47a (Equation 5.49a)] leads to [Equation 5.93 (Equation 5.94)]:

$$\begin{aligned}\cos(u+v)\cos(u-v) &= (\cos u\cos v-\sin u\sin v)(\cos u\cos v+\sin u\sin v)\\ &= \cos^2 u\cos^2 v-\sin^2 u\sin^2 v\\ &= \cos^2 u(1-\sin^2 v)-\sin^2 u\sin^2 v\\ &= \cos^2 u-\sin^2 v,\end{aligned} \tag{5.93}$$

$$\begin{aligned}\cosh(u+v)\cosh(u-v) &= (\cosh u\cosh v-\sinh u\sinh v)(\cosh u\cosh v+\sinh u\sinh v)\\ &= \cosh^2 u\cosh^2 v-\sinh^2 u\sinh^2 v\\ &= \cosh^2 u(1+\sinh^2 v)-\sinh^2 u\sinh^2 v\\ &= \cosh^2 u+\sinh^2 v=\sinh^2 u+\cosh^2 v\end{aligned} \tag{5.94}$$

the sum (difference) of the squares of the circular (hyperbolic) cosine and sine.

The product of the circular (hyperbolic) sine of the sum and difference [Equation 5.47b (Equation 5.49b)] leads to [Equation 5.95 (Equation 5.96)]:

$$\begin{aligned}\sin(u+v)\sin(u-v) &= (\sin u\cos v+\sin v\cos u)(\sin u\cos v-\sin v\cos u)\\ &= \sin^2 u\cos^2 v-\sin^2 v\cos^2 u\\ &= \sin^2 u(1-\sin^2 v)-\sin^2 v\cos^2 u\\ &= \sin^2 u-\sin^2 v=\cos^2 v-\cos^2 u,\end{aligned} \tag{5.95}$$

$$\begin{aligned}\sinh(u+v)\sinh(u-v) &= (\sinh u\cosh v+\sinh v\cosh u)(\sinh u\cosh v-\sinh v\cosh u)\\ &= \sinh^2 u\cosh^2 v-\sinh^2 v\cosh^2 u\\ &= \sinh^2 u(1+\sinh^2 v)-\sinh^2 v\cosh^2\\ &= \sinh^2 u-\sinh^2 v=\cosh^2 u-\cosh^2 v\end{aligned} \tag{5.96}$$

the difference of squares of (1) sines and (2) cosines with the same (reversed) sign.

Thus:

$$\cos^2 u - \sin^2 v = \cos(u+v)\cos(u-v), \tag{5.97}$$

$$\sin^2 u - \sin^2 v = \sin(u+v)\sin(u-v) = \cos^2 v - \cos^2 u, \tag{5.98a and 5.98b}$$

$$\cosh^2 u + \sinh^2 v = \cosh(u+v)\cosh(u-v) = \cosh^2 v + \sinh^2 u, \tag{5.99a and 5.99b}$$

$$\cosh^2 u - \cosh^2 v = \sinh(u+v)\sinh(u-v) = \sinh^2 u - \sinh^2 v, \tag{5.100a and 5.100b}$$

specify the sums and/or differences of squares of circular (hyperbolic) cosines and sines as products of circular (hyperbolic) cosine and sine of the sum and difference of variables [Equations 5.97, 5.98a, and 5.98b (Equations 5.99a, 5.99b, 5.100a, and 5.100b)].

5.6 Chebychev (1859) Polynomials of Two Kinds

The circular cosine and sine of a multiple variable have been expressed as a sum of products of powers of the cosine and sine (Subsections 5.5.1 and 5.5.2). The powers of cosines and sines are related by the circular identity (Subsection 5.1.2), so it should be possible to use powers of the only one of the cosine (or sine) in the sum for the cosine and sine of a multiple variable (Subsection 5.6.2), distinguishing an odd or even multiple variable [Subsection 5.6.3 (5.6.4)]. The Chebychev polynomial of the first (second) kind is defined from the cosine (sine) of a multiple variable as a function of the cosine of that variable, and its coefficients follow from the preceding relations (Subsection 5.6.7). There are analogue formulas for the hyperbolic cosine and sine of a multiple variable as a sum of powers of the hyperbolic cosine and sine (Subsection 5.6.5). The simplest cases are duplication and triplication formulas for circular/hyperbolic cosines/sines (Subsection 5.6.6). All the polynomial expansions (Subsections 5.6.2 through 5.6.6) including the Chebychev polynomials of two kinds (Subsection 5.6.7) arise from two algebraic identities involving the sum and product of the roots of a binomial, which are proven first (Subsection 5.6.1). The algebraic relations derived in Section 5.6 are not essential in the sequel, and it is possible to proceed directly to Section 5.7 if the main interest concerns (1) the properties of orthogonal systems of functions in general and (2) in particular, the orthogonal systems of trigonometric functions and their relation with Chebychev polynomials of the first and second kinds.

5.6.1 Two Algebraic Identities between Sums, Products, and Powers

Given two numbers (a and b), their sum (Equation 5.101a) and product (Equation 5.101b) appear in Equations 5.101c and 5.101d:

$$p \equiv a+b,\ q \equiv ab: \quad \frac{1}{1-az} \pm \frac{1}{1-bz} = \frac{\{2-(a+b)z,\ (a-b)z\}}{1-(a+b)z+abz^2} = \frac{\{2-pz,\ (a-b)z\}}{1-pz+qz^2}. \qquad (5.101\text{a}–5.101\text{d})$$

For small enough z in modulus (Equations 5.102a and 5.102b), the l.h.s. of Equation 5.101c (Equation 5.101d) can be expanded as geometric series (Equation I.21.102a ≡ Equation 3.102b) of powers [Equation 5.102c (Equation 5.102d)] of z:

$$\frac{1}{|b|} > |z| < \frac{1}{|a|}: \quad \left(\frac{1}{1-az} \pm \frac{1}{1-bz}\right) = \sum_{N=0}^{\infty}\left[(az)^N \pm (bz)^N\right]$$
$$= \left\{2+\sum_{N=1}^{\infty} z^N\left(a^N+b^N\right),\ \sum_{N=1}^{\infty} z^N\left(a^N-b^N\right)\right\}, \qquad (5.102\text{a}–5.102\text{d})$$

whose coefficients are sums and differences of powers $a^N \pm b^N$. The latter are expressed in terms of sums (Equation 5.101a) and products (Equation 5.101b) of the two numbers (a and b) by expanding the coefficients of z^N. Therefore, again for small enough z in modulus (Equation 5.103a), the r.h.s. of Equation 5.101c (Equation 5.101d) can be expanded: (1) first in a geometric series (Equation I.21.62a ≡ Equation 3.148b) and (2) after in a binomial series (Equations I.25.37a and I.25.37b ≡ Equations 3.146a through 3.146c):

$$\left|pz - qz^2\right| < 1: \quad \frac{\{2 - pz,\, (a-b)z\}}{1 - pz + qz^2} = \{2 - pz,\, (a-b)z\} \sum_{m=0}^{\infty} z^m (p - qz)^m$$

$$N = n + m: \qquad = \{2 - pz,\, (a-b)z\} \sum_{m=0}^{\infty} z^m \sum_{n=0}^{m} (-)^n \binom{m}{n} p^{m-n} q^n z^n$$

$$= \{2 - pz,\, (a-b)z\} \sum_{N=0}^{\infty} z^N \sum_{n=0}^{\le N/2} (-)^n \binom{N-n}{n} p^{N-2n} q^n, \tag{5.103a–5.103d}$$

which lead to Equation 5.103c (Equation 5.103d), where Equation 5.103b was used, and only powers with positive exponents appear.

Equating the coefficients of z^N in the second relation (Equation 5.103d ≡ Equation 5.102d) leads to the identity:

$$\frac{a^N - b^N}{a - b} = \sum_{n=0}^{\le (N-1)/2} (-)^n \binom{N-n-1}{n} p^{N-2n-1} q^n = \sum_{n=0}^{\le N/2-1} \frac{(-)^n (N-n-1)!}{n!(N-2n-1)!} (a+b)^{N-2n-1} (ab)^n$$

$$= (a+b)^{N-1} - (N-2)(a+b)^{N-3} ab + \frac{(N-3)(N-4)}{2!} (a+b)^{N-5} (ab)^2 + \cdots \tag{5.104}$$

$$+ (-)^n \frac{(N-n-1)(N-n-2)\cdots(N-2n)}{n!} (a+b)^{N-2n-1} (ab)^n + \cdots,$$

where Equations 5.101a and 5.101b were substituted. Equating powers of z^N in Equation 5.102c ≡ Equation 5.103c leads to

$$a^N + b^N = \sum_{n=0}^{\le N/2} (-)^n p^{N-2n} q^n \left[2\binom{N-n}{n} - \binom{N-n-1}{n} \right], \tag{5.105}$$

where n was replaced by $n - 1$ in the second term because of the factor z. Using

$$2\binom{N-n}{n} - \binom{N-n-1}{n} = \frac{(N-n-1)\cdots(N-2n+1)}{n!} \left[2(N-n) - (N-2n)\right]$$

$$= \frac{N}{n!} (N-n-1)\cdots(N-2n+1) \tag{5.106}$$

and substituting Equations 5.106, 5.101a, and 5.101b in Equation 5.105 yield the identity:

$$a^N + b^N = \sum_{n=0}^{\le N/2} \frac{(-)^n}{n!} (a+b)^{N-2n} (ab)^n N(N-n-1)\cdots(N-2n+1)$$

$$= (a+b)^N - N(a+b)^{N-2} ab + \frac{N(N-3)}{2!} (a+b)^{N-4} (ab)^2 \tag{5.107}$$

$$- \frac{N(N-4)(N-5)}{3!} (a+b)^{N-6} (ab)^3 + \cdots.$$

Thus, *two algebraic identities expressing* $a^N \pm b^N$ *in terms [Equation 5.104 (5.107)] of powers of* $a + b$ *and* ab have been proved.

5.6.2 Cosine/Sine of a Multiple Variable as Power Series of Cosines

The choice (Equations 5.108a and 5.108b) leads to Equations 5.108c through 5.108f:

$$a = e^{iz},\ b = e^{-iz}: \qquad ab = 1,\ a + b = 2\cos z,\ a^N \pm b^N = e^{iNz} \pm e^{-iNz} = 2\{\cos(Nz), i\sin(Nz)\}. \tag{5.108a–5.108f}$$

Substituting Equations 5.108c through 5.108f in Equation 5.104 (Equation 5.107) gives the expansion of the sine (Equation 5.109) [cosine (Equation 5.110)] of a multiple variable in power series of the cosine of a variable:

$$\begin{aligned}\frac{\sin(Nz)}{\sin z} &= \sum_{n=0}^{\leq(N-1)/2} (-)^n \binom{N-n-1}{n} 2^{N-2n-1} \cos^{N-2n-1} z \\ &= 2^{N-1}\cos^{N-1} z - (N-2)\,2^{N-3}\cos^{N-3} z + \frac{(N-3)(N-4)}{2!} 2^{N-5}\cos^{N-5} z \\ &\quad - \cdots + \frac{(-)^n}{n!}(N-n-1)(N-n-2)\cdots(N-2n)\,2^{N-2n-1}\cos^{N-2n-1} z + \cdots,\end{aligned} \tag{5.109}$$

$$\begin{aligned}\cos(Nz) &= \sum_{n=0}^{\leq N/2} \frac{(-)^n}{n!} N(N-n-1)\cdots(N-2n+1)\,2^{N-2n-1}\cos^{N-2n} z \\ &= 2^{N-1}\cos^N z - 2^{N-3} N\cos^{N-2} z + 2^{N-5}\frac{N(N-3)}{2}\cos^{N-4} z + \cdots \\ &\quad + \frac{(-)^n}{n!} 2^{N-2n-1} N(N-n-1)\cdots(N-2n+1)\cos^{N-2n} z + \cdots.\end{aligned} \tag{5.110}$$

The expansions of the circular sine (cosine) of a multiple variable [Equation 5.109 (5.110)] in descending powers of the cosine of the variable hold for N a positive integer and contain only nonnegative powers, that is, stop at the lowest positive power or at a constant term.

5.6.3 Sine/Cosine of Even/Odd Multiple Arguments

This is made explicit replacing N by $2N$ ($2N+1$) in the formula for the multiple variable (Equation 5.109) and rearranging in ascending powers, which leads to Equation 5.111 (Equation 5.112) for the sine of even (odd) multiple variable in terms of a sum of powers of cosines:

$$\frac{\sin(2Nz)}{\sin z} = \sum_{n=0}^{N-1} (-)^n \binom{2N-n-1}{n} 2^{2N-2n-1} \cos^{2N-2n-1} z$$

$$= (-)^{N-1} \binom{N}{N-1} 2\cos z + (-)^{N-2} \binom{N+1}{N-2} 2^3 \cos^3 z + (-)^{N-3} \binom{N+2}{N-3} 2^5 \cos^5 z$$

$$+ \cdots + (-)^{N-1-m} \binom{N+m}{N-m-1} 2^{2m+1} \cos^{2m+1} z + \cdots + 2^{2N-1} \cos^{2N-1} z$$

$$= (-)^{N-1} \left[2N \cos z - \frac{8}{3!} N(N^2-1)\cos^3 z + \frac{32}{5!} N(N^2-1)(N^2-4)\cos^5 z \right. \quad (5.111)$$

$$+ \cdots + (-)^{N-1-m} \frac{2^{2m+1}}{(2m+1)!} N(N^2-1)(N^2-4) \quad (N^2-9)\ldots(N^2-m^2)\cos^{2m+1} z$$

$$\left. + \cdots + (-)^{N-1} 2^{2N-1} \cos^{2N-1} z \right],$$

$$\frac{\sin[(2N+1)z]}{\sin z} = \sum_{n=0}^{N} (-)^n \binom{2N-n}{N} 2^{2N-2n} \cos^{2N-2n} z$$

$$= (-)^N \binom{N}{N} + (-)^{N-1} \binom{N+1}{N-1} 4\cos^2 z + (-)^{N-2} \binom{N+2}{N-2} 16 \cos^4 z$$

$$+ \cdots + (-)^{N-m} 2^{2m} \binom{N+m}{N-m} \cos^{2m} z + \cdots + 2^{2N} \cos^{2N} z \quad (5.112)$$

$$= (-)^N \left\{ 1 - \frac{4}{2!} N(N+1)\cos^2 z + \frac{16}{4!} N(N+2)(N^2-1)\cos^4 z + \cdots \right.$$

$$+ (-)^m \frac{2^{2m}}{(2m)!} N(N+m) \left(N^2-1\right)\left(N^2-4\right)\cdots\left[N^2-(m-1)^2\right]\cos^{2m} z$$

$$\left. + \cdots + (-)^N 2^{2N} \cos^{2N} z \right\}.$$

Similarly, replacing N by $2N$ ($2N + 1$) in Equation 5.110 leads to the cosine of even (odd) multiple variable expanded as a sum of ascending powers of the cosine [Equation 5.113 (Equation 5.114)]:

$$
\begin{aligned}
\cos(2Nz) &= \sum_{n=0}^{N} \frac{(-)^n}{n!} N(2N-n-1)\cdots(2N-2n+1)2^{2N-2n}\cos^{2N-2n} z \\
&= \frac{(-)^N}{N!} N(N-1)\cdots 1 + \frac{(-)^{N-1}}{(N-1)!} 4N^2(N-1)\cdots 3\cos^2 z + \cdots \\
&+ \frac{(-)^{N-2}}{(N-2)!} 16N(N+1)N(N-1)\cdots 5\cos^4 z + \cdots \\
&+ \frac{(-)^{N-m}}{(N-m)!} 2^{2m} N(N+m-1)\,(N+m-2)\cdots(2m+1)\cos^{2m} z + \cdots + 2^{2N}\cos^{2N} z \\
&= (-)^N \left\{ 1 - \frac{4}{2!} N^2 \cos^2 z + \frac{16}{4!} N^2(N^2-1)\cos^4 z + \cdots \right. \\
&\left. + (-)^m \frac{2^{2m}}{(2m)!} N^2(N^2-1)\cdots\left[N^2-(m-1)^2\right]\cos^{2m} z + \cdots + (-)^N 2^{2N}\cos^{2N} z \right\},
\end{aligned}
\tag{5.113}
$$

$$
\begin{aligned}
\cos\left[(2N+1)z\right] &= \sum_{n=0}^{N} \frac{(-)^n}{n!} (2N+1)(2N-n)\cdots(2N-2n+2)2^{2N-2n}\cos^{2N-2n+1} z \\
&= \frac{(-)^N}{N!}(2N+1)N(N-1)\cdots 2\cos z + \frac{(-)^{N-1}}{(N-1)!}(2N+1)\,(N+1)N\cdots 5.4.4\cos^3 z \\
&+ \frac{(-)^{N-2}}{(N-2)!}(2N+1)N(N+2)(N+1)\ldots 7.6.16\cos^5 z + \cdots + \\
&+ \frac{(-)^{N-m}}{(N-m)!}(2N+1)(N+m)(N+m-1)\cdots(2m+3)\,(2m+2)2^{2m}\cos^{2m+1} z \\
&+ \cdots + 2^{2N}\cos^{2N+1} z \\
&= (-)^N(2N+1)\left\{ \cos z - \frac{4}{3!} N(N+1)\cos^3 z + \frac{16}{5!} N(N+2)\,(N^2-1)\cos^5 z \right. \\
&+ \cdots + \frac{2^{2m}}{(2m+1)!} N(N+m)(N^2-1)\cdots\left[N^2-(m-1)^2\right]\cos^{2m+1} z \\
&\left. + \cdots + \frac{(-)^N 2^{2N}}{2N+1}\cos^{2N+1} z \right\}.
\end{aligned}
\tag{5.114}
$$

The circular sine/cosine of an even (odd) multiple of a variable is expressed as a sum of ascending powers [Equation 5.111/5.113 (Equation 5.112/5.114)] of the circular cosine of the variable.

5.6.4 Cosine/Sine of Multiple Variable as Sum of Powers of the Cosine/Sine

The powers of cosine in the sine of even (odd) multiple arguments [Equation 5.111 (Equation 5.112)] can be replaced by powers of sine by substituting the variable z by $z + \pi/2$, which leads to Equation 5.115 (Equation 5.116):

$$\begin{aligned}
\frac{\sin(2Nz)}{\cos z} &= (-)^N \frac{\sin[2N(z+\pi/2)]}{\sin(z+\pi/2)} \\
&= \sum_{n=0}^{N-1} (-)^{N+n} \binom{2N-n-1}{n} 2^{2N-2n-1} \cos^{2N-2n-1}(z+\pi/2) \\
&= \sum_{n=0}^{N-1} (-)^{N+n+1} \binom{2N-n-1}{n} 2^{2N-2n-1} \sin^{2N-2n-1} z \\
&= 2N\sin z - \frac{8}{3!} N(N^2-1)\sin^3 z + \frac{32}{5!} N(N^2-1)(N^2-4)\sin^5 z + \cdots \\
&\quad + (-)^m \frac{2^{2m+1}}{(2m+1)!} N(N^2-1)(N^2-4)\cdots(N^2-m^2)\sin^{2m+1} z \\
&\quad + \cdots + (-)^{N+1} 2^{2N-1} \sin^{2N-1} z,
\end{aligned} \tag{5.115}$$

$$\begin{aligned}
\frac{\cos[(2N+1)z]}{\cos z} &= \frac{\sin[(2N+1)z+\pi/2]}{\sin(z+\pi/2)} = (-)^N \frac{\sin[(2N+1)(z+\pi/2)]}{\sin(z+\pi/2)} \\
&= \sum_{n=0}^{N} (-)^{N+n} \binom{2N-n}{n} 2^{2N-2n} \cos^{2N-2n}\left(z+\frac{\pi}{2}\right) \\
&= \sum_{n=0}^{N} (-)^{N+n} \binom{2N-n}{n} 2^{2N-2n} \sin^{2N-2n}(z) \\
&= 1 - \frac{4}{2!} N(N+1)\sin^2 z + \frac{16}{4!} N(N+2)(N^2-1)\sin^4 z + \cdots \\
&\quad + (-)^m \frac{2^{2m}}{(2m)!} N(N+m)(N^2-4)(N^2-2)\left[N^2-(m-1)^2\right]\sin^{2m} z \\
&\quad + \cdots + (-)^N 2^{2N} \sin^{2N} z.
\end{aligned} \tag{5.116}$$

The powers of the cosine in the cosine of even (odd) multiple variables [Equation 5.113 (Equation 5.114)] can be replaced by powers of the sine via the change of variable $z \to z + \pi/2$, which leads to Equation 5.117 (Equation 5.118):

$$\begin{aligned}
\cos(2Nz) &= (-)^N \cos\left[2N\left(z+\pi/2\right)\right] \\
&= \sum_{n=0}^{N} \frac{(-)^{N+n}}{n!} N(2N-n-1)\cdots(2N-2n+1)2^{2N-2n}\cos^{2N-2n}\left(z+\frac{\pi}{2}\right) \\
&= \sum_{n=0}^{N} \frac{(-)^{N+n}}{n!} N(2N-n-1)\cdots(2N-2n+1)2^{2N-2n}\sin^{2N-2n} z \qquad (5.117)\\
&= 1-\frac{4}{2!}N^2\sin^2 z+\frac{16}{4!}N^2(N^2-1)\sin^4 z+\cdots \\
&\quad +(-)^m\frac{2^{2m}}{(2m)!}N^2\left(N^2-1\right)\ldots\left[N^2-(m-1)^2\right]\sin^{2m} z+\cdots+(-)^N 2^{2N}\sin^{2N} z,
\end{aligned}$$

$$\begin{aligned}
\sin\left[(2N+1)z\right] &= -\cos\left[(2N+1)z+\pi/2\right] = (-)^{N+1}\cos\left[(2N+1)\left(z+\pi/2\right)\right] \\
&= \sum_{n=0}^{N} \frac{(-)^{n+N+1}}{n!}(2N+1)(2N-n)\cdots(2N-2n+2)2^{2N-2n}\cos^{2N-2n+1}\left(z+\frac{\pi}{2}\right) \\
&= \sum_{n=0}^{N} \frac{(-)^{N+n}}{n!}(2N+1)(2N-n)\cdots(2N-2n+2)2^{2N-2n}\sin^{2N-2n+1} z \qquad (5.118)\\
&= (2N+1)\left\{\sin z-\frac{4}{3!}N(N+1)\sin^3 z+\frac{16}{5!}N(N+2)\left(N^2-1\right)\sin^5 z+\cdots\right. \\
&\quad +(-)^m\frac{2^{2m}}{(2m+1)!}N(N+m)\left(N^2-1\right)\left(N^2-4\right)\ldots\left[N^2-(m-1)^2\right]\sin^{2m+1} z \\
&\quad \left.+\cdots+\frac{(-)^N 2^{2N}}{2N+1}\sin^{2N+1} z\right\}.
\end{aligned}$$

The circular cosine/sine of an even (odd) multiple of the variable is specified by a sum of powers of sines of the variable by Equation 5.117/5.115 (Equation 5.116/5.118).

5.6.5 Circular/Hyperbolic Sine/Cosine of Multiple Variable

The sine (cosine) of multiple [Equation 5.109 (Equation 5.110)], even multiple [Equation 5.111 (Equation 5.113)], and odd multiple [Equation 5.112 (Equation 5.114)] variable can be expressed as a sum of powers of the cosine by identical formulas for circular and hyperbolic functions, that is, (sin, cos) may be replaced by (sinh, cosh) in Equations 5.109 through 5.114. This result is an immediate consequence of the imaginary change of variable (Equations 5.6a and 5.6b). The same transformation shows that Equations 5.115 through 5.118 hold for hyperbolic instead of circular functions suppressing the alternating signs:

$$\frac{\sinh(2Nz)}{\cosh z} = \sum_{n=0}^{N-1} \binom{2N-n-1}{n} 2^{2N-2n-1} \sinh^{2N-2n-1} z$$

$$= 2N \sinh z + \frac{8}{3!} N(N^2-1)\sinh^3 z + \frac{32}{5!} N(N^2-1)(N^2-4)\sinh^5 z + \cdots \tag{5.119}$$

$$+ \frac{2^{2m+1}}{(2m+1)!} N(N^2-1)\cdots(N^2-m^2)\sinh^{2m+1} z + \cdots + 2^{2N-1}\sinh^{2N-1} z,$$

$$\frac{\cosh[(2N+1)z]}{\cosh z} = \sum_{n=0}^{N} \binom{2N-n}{n} 2^{2N-2n} \sinh^{2N-2n} z$$

$$= 1 + \frac{4}{2!} N(N+1)\sinh^2 z + \frac{16}{4!} N(N+2)^2 (N^2-1)\sinh^4 z + \cdots \tag{5.120}$$

$$+ \frac{2^{2m}}{(2m)!} N(N+m)(N^2-4)(N^2-2)\cdots\left[N^2-(m-1)^2\right]\sinh^{2m} z + \cdots$$

$$+ 2^{2N}\sinh^{2N} z,$$

$$\cosh(2Nz) = \sum_{n=0}^{N} \frac{N}{n!}(2N-n-1)\cdots(2N-2n+1)2^{2N-2n}\sinh^{2N-2n} z$$

$$= 1 + \frac{4}{2!} N^2 \sinh^2 z + \frac{16}{4!} N^2(N^2-1)\sinh^4 z + \cdots \tag{5.121}$$

$$+ \frac{2^{2m}}{(2m)!} N^2(N^2-1)\cdots\left[N^2-(m-1)^2\right]\sinh^{2m} z + \cdots + 2^{2N}\sinh^{2N} z,$$

$$\sinh[(2N+1)z] = \sum_{n=0}^{N} \frac{2N+1}{n!}(2N-n)\cdots(2N-2n+2)2^{2N-2n}\sinh^{2N-2n+1} z$$

$$= (2N+1)\left\{\sinh z + \frac{4}{3!} N(N+1)\sinh^3 z + \frac{16}{5!} N(N+2)(N^2-1)\sinh^5 z + \cdots\right.$$

$$+ \frac{2^{2m}}{(2m+1)!} N(N+m)(N^2-1)(N^2-4)\cdots\left[N^2-(m-1)^2\right]\sinh^{2m+1} z \tag{5.122}$$

$$\left. + \cdots + 2^{2N}\sinh^{2N+1} z\right\}.$$

The hyperbolic sine/cosine of even (odd) multiples of the variable are given by Equations 5.119/5.121 (Equations 5.120/5.122) as a sum of powers of the hyperbolic sine of the variable.

5.6.6 Circular/Hyperbolic Sines/Cosines of Double/Triple Variable

The simplest particular case $N = 1$ of expansions in powers of cosines (sines) are (1) from Equations 5.111 and 5.115 (Equations 5.111 and 5.119), respectively, Equation 5.123a (Equation 5.123b); (2) from Equations 5.113 and 5.117 (Equations 5.113 and 5.121), respectively, Equations 5.124a and 5.124b (Equations 5.124c and 5.124d); (3) from Equations 5.112 and 5.118 (Equations 5.112 and 5.122), respectively, Equations 5.125a and 5.125b (Equations 5.125c and 5.125d); (4) from Equations 5.114 and 5.116 (Equations 5.114 and 5.120), respectively, Equations 5.126a and 5.126b (Equations 5.126c and 5.126d):

$$\sin(2z) = 2\sin z\cos z, \quad \sinh(2z) = 2\sinh z\cosh z, \tag{5.123a and 5.123b}$$

$$\cos(2z) = 2\cos^2 z - 1 = 1 - 2\sin^2 z, \; \cosh(2z) = 2\cosh^2 z - 1 = 1 + 2\sinh^2 z, \tag{5.124a–5.124d}$$

$$\frac{\sin(3z)}{\sin z} = 4\cos^2 z - 1 = 3 - 4\sin^2 z, \quad \frac{\sinh(3z)}{\sinh z} = 4\cosh^2 z - 1 = 3 + 4\sinh^2 z, \tag{5.125a–5.125d}$$

$$\cos(3z) = 4\cos^3 z - 3\cos z = \cos z\,(1 - 4\sin^2 z), \tag{5.126a and 5.126b}$$

$$\cosh(3z) = 4\cosh^3 z - 3\cosh z = \cosh z\,(1 + 4\sinh^2 z); \tag{5.126c and 5.126d}$$

as a check, an alternative derivation follows.

The first set coincides with the formulas of duplication of the variable [Equations 5.123a and 5.123b ≡ Equations 5.63a and 5.63b (Equations 5.124a through 5.124d ≡ Equations 5.61b, 5.61c, 5.62b, and 5.62c)]: the second set can be obtained using the formulas for the sum and duplication of the variable:

$$\begin{aligned}\sin(3z) &= \sin z\cos(2z) + \sin(2z)\cos z\\ &= \sin z\left(2\cos^2 z - 1\right) + 2\sin z\cos^2 z = \sin z\left(4\cos^2 z - 1\right)\\ &= \sin z\left[4\left(1 - \sin^2 z\right) - 1\right] = \sin z\left(3 - 4\sin^2 z\right),\end{aligned} \tag{5.127a and 5.127b}$$

$$\begin{aligned}\sinh(3z) &= \sinh z\cosh(2z) + \sinh(2z)\cosh(z)\\ &= \sinh z\left(2\cosh^2 z - 1\right) + 2\sinh z\cosh^2 z = \sinh z\left(4\cosh^2 z - 1\right)\\ &= \sinh z\left[4\left(1 + \sinh^2 z\right) - 1\right] = \sinh z\left(4\sinh^2 z + 3\right),\end{aligned} \tag{5.127c and 5.127d}$$

$$\begin{aligned}\cos(3z) &= \cos z\cos(2z) - \sin z\sin(2z)\\ &= \cos z\left(2\cos^2 z - 1\right) - 2\sin^2 z\cos z\\ &= \cos z\left[2\cos^2 z - 1 - 2\left(1 - \cos^2 z\right)\right] = 4\cos^3 z - 3\cos z\\ &= \cos z\left[4\left(1 - \sin^2 z\right) - 3\right] = \cos z\left(1 - 4\sin^2 z\right),\end{aligned} \tag{5.128a and 5.128b}$$

$$\begin{aligned}\cosh(3z) &= \cosh z \cosh(2z) + \sinh z \sinh(2z) \\ &= \cosh z\left(2\cosh^2 z - 1\right) + 2\sinh^2 z \cosh z \\ &= \cosh z\left[2\cosh^2 z - 1 + 2\left(\cosh^2 z - 1\right)\right] = 4\cosh^3 z - 3\cosh z \\ &= \cosh z\left[4\left(1+\sin^2 z\right) - 3\right] = \cosh z\left(1 + 4\sinh^2 z\right),\end{aligned} \quad \text{(5.128c and 5.128d)}$$

which lead to the coincidences [Equations 5.125a through 5.125d ≡ Equations 5.127a through 5.127d (Equations 5.126a through 5.126d ≡ Equations 5.128a through 5.128d)]. This provides two alternative derivations of the **triplication formulas** *for the circular (hyperbolic) sine [Equations 5.125a and 5.125b (Equations 5.125c and 5.125d)] and cosine [Equations 5.126a and 5.126b (Equations 5.126c and 5.126d)].*

5.6.7 Chebychev (1859) Polynomials of the First/Second Kind

The cosine (sine) of a multiple variable [Equation 5.129a (Equation 5.129b)] as a function of the cosine of the variable (Equation 5.130a) defines the **Chebychev polynomials of the first (second) kind and degree** N:

$$T_N(\cos z) \equiv 2^{1-N}\cos(Nz), \quad U_N(\cos z) \equiv 2^{1-N}\frac{\sin[(N+1)z]}{\sin z}, \quad \text{(5.129a and 5.129b)}$$

that are even functions specified equivalently by Equation 5.130b (Equation 5.130c) in terms of the variable (Equation 5.130a):

$$\zeta = \cos z: \qquad T_N(\zeta) = 2^{1-N}\cos(N \arccos \zeta) = T_N(-\zeta), \quad \text{(5.130a and 5.130b)}$$

$$U_N(\zeta) = 2^{1-N}\,|1-\zeta^2|^{-1/2}\sin[(N+1)\arccos(\zeta)] = U_N(-\zeta) \quad \text{(5.130c)}$$

It follows that *the Chebychev polynomials of the first (Equation 5.129a ≡ Equation 5.130b) [second (Equation 5.129b ≡ Equation 5.130c)] kind and variable (Equation 5.130a) are given (1) in general for any degree by [Equation 5.110 ≡ Equation 5.131 (Equation 5.109 ≡ Equation 5.132)]:*

$$\begin{aligned}T_N(\zeta) &= \sum_{n=0}^{\le N/2}\frac{(-)^n}{n!}N(N-n-1)\cdots(N-2n+1)2^{-2n}\zeta^{N-2n} \\ &= \zeta^N - \frac{N}{4}\zeta^{N-2} + \frac{N(N-3)}{16.2}\zeta^{N-4} + \cdots \\ &+ \frac{(-)^n}{n!}\frac{N(N-n-1)\cdots(N-2n+1)}{2^{2n}}\zeta^{N-2n} + \cdots,\end{aligned} \quad \text{(5.131)}$$

$$\begin{aligned}U_N(\zeta) &= \sum_{n=0}^{\le N/2}(-)^n\binom{N-n}{n}2^{-2n}\zeta^{N-2n} \\ &= \zeta^N - \frac{N-1}{4}\zeta^{N-2} + \frac{(N-2)(N-3)}{16.2}\zeta^{N-4} + \cdots \\ &+ \frac{(-)^n}{n!}\frac{(N-n)\cdots(N-2n+1)}{2^{2n}}\zeta^{N-2n} + \cdots;\end{aligned} \quad \text{(5.132)}$$

(2) in particular, for even degree by [Equation 5.113 ≡ Equation 5.133 (Equation 5.112 ≡ Equation 5.134)]:

$$T_{2N}(\zeta)=\sum_{n=0}^{N}\frac{(-)^n}{n!}N(2N-n-1)\cdots(2N-2n+1)2^{1-2n}\zeta^{2N-2n}$$
$$=2^{1-2N}(-)^N\left\{1-\frac{4}{2!}N^2\zeta^2+\frac{16}{4!}N^2(N^2-1)\zeta^4+\cdots\right. \tag{5.133}$$
$$\left.+(-)^m\frac{2^{2m}}{(2m)!}N^2(N^2-1)\cdots\left[N^2-(m-1)^2\right]\zeta^{2m}+\cdots+(-)^N2^{2N}\zeta^{2N}\right\},$$

$$U_{2N}(\zeta)=\sum_{n=0}^{N}(-)^n\binom{2N-n}{n}2^{1-2n}\zeta^{2N-2n}$$
$$=2^{1-2N}(-)^N\left\{1-\frac{4}{2!}N(N+1)\zeta^2+\frac{16}{4!}N(N+2)(N^2-1)\zeta^4\right.$$
$$+\cdots+(-)^m\frac{2^{2m}}{(2m)!}N(N+m)(N^2-1)(N^2-4)\cdots\left[N^2-(m-1)^2\right]\zeta^{2m} \tag{5.134}$$
$$\left.+\cdots+(-)^N2^{2N}\zeta^{2N}\right\};$$

and (3) for odd degree by [Equation 5.114 ≡ Equation 5.135 (Equation 5.111 ≡ Equation 5.136)]:

$$T_{2N+1}(\zeta)=\sum_{n=0}^{N}\frac{(-)^n}{n!}(2N+1)(2N-n)\cdots(2N-2n+2)2^{-2n}\zeta^{2N-2n+1}$$
$$=(-)^N2^{-2N}(2N+1)\left\{\zeta-\frac{4}{3!}N(N+1)\zeta^3+\frac{16}{5!}N(N+2)(N^2-1)\zeta^5+\cdots\right.$$
$$+\frac{2^{2m}}{(2m+1)!}N(N+m)(N^2-1)(N^2-4)\cdots\left[N^2-(m-1)^2\right]\zeta^{2n+1} \tag{5.135}$$
$$\left.+\cdots+\frac{(-)^N2^{2N}}{2N+1}\zeta^{2N+1}\right\},$$

$$U_{2N-1}(\zeta)=\sum_{n=0}^{N-1}(-)^n\binom{2N-n-1}{n}2^{1-2n}\zeta^{2N-2n-1}$$
$$=(-)^N2^{2-2N}\left[2N\zeta-\frac{8}{3!}N(N^2-1)\zeta^3+\frac{32}{5!}N(N^2-1)(N^2-4)\zeta^5\right.$$
$$+\cdots+(-)^m\frac{2^{2m+1}}{(2m+1)!}N(N^2-1)(N^2-4)\cdots(N^2-m^2)\zeta^{2m+1} \tag{5.136}$$
$$\left.+\cdots+(-)^N2^{2N-1}\zeta^{2N-1}\right].$$

The Chebychev polynomials of each kind form an orthogonal system of functions related to the trigonometric functions (Section 5.7).

5.7 Orthogonal and Normalized Trigonometric Functions

The inner product and norm of complex functions (Subsection 5.7.1) lead to the definition of orthogonal systems of functions (Subsection 5.7.2). Three (two) examples of orthogonal systems of functions and their normalization are the imaginary exponentials, cosines, and sines of a multiple variable (Subsection 5.7.3) [Chebychev polynomials of first and second kinds (Subsection 5.7.4)]. The representation of a function with bounded oscillation by orthogonal series (Subsection 5.7.5) applies to the preceding five orthogonal systems, which lead to the Fourier, cosine, sine, and first/second-kind Chebychev series (Subsection 5.7.6). The Chebychev polynomials of the first kind have the unique property (Subsection 5.7.7) of having the smallest oscillation in an interval among all the polynomials with the same leading power. The expansion of even and odd powers of the circular cosine and sine as a sum of cosines (sines) of multiple angles (Subsection 5.5.1) is not a particular case of a terminating Fourier series (Subsection 5.7.8) because written in that form, the coefficients are not constant. However, because the odd powers of the sine (the even powers of the sine and the even and odd powers of the cosine) are even (odd) functions, they can be expanded in infinite sine (cosine) series (Subsection 5.7.9). The coefficients can be found by three methods: (1 and 2) use of trigonometric relations (Subsection 5.7.10) that are equivalent to orthogonality relations (Subsection 5.7.11); and (3) evaluation of the corresponding trigonometric integrals (Subsection I.17.2) by residues (Subsection 5.7.12), which in the present case correspond to the constant part of polynomials (Subsection 5.7.13).

5.7.1 Inner Product and Norm for Complex Functions

The **inner product** of two integrable complex functions (Equations 5.137a and 5.137b) with the integrable **weighting function** (Equation 5.137c) is defined by Equation 5.137d:

$$f, g, w \in \mathcal{E}(|C): \qquad [f, g] \equiv \int_a^b f(z)\, g^*(z)\, w(z)\, \mathrm{d}z. \qquad (5.137a\text{–}5.137d)$$

The complex conjugate appears in Equation 5.137d, so that for a real weighting function (Equation 5.138a) positive (Equation 5.138d) in the interval (Equation 5.138c), the **norm** of a complex function (Equation 5.138b), defined as the inner product by itself (Equation 5.138e), is nonnegative:

$$w \in \mathcal{E}(|R), \quad f \in \mathcal{E}(|C); \quad a \le x \le b; \quad w(x) > 0; \qquad (5.138a\text{–}5.138d)$$

$$\|f\| \equiv [f, f] = \int_a^b w(x) f(x) f^*(x)\, \mathrm{d}x = \int_a^b w(x) |f(x)|^2\, \mathrm{d}x \ge 0. \qquad (5.138e)$$

In particular, if the function is continuous (Equation 5.139a), its norm is zero (Equation 5.139b) only if it vanishes (Equation 5.139d) everywhere in the interval (Equation 5.139c):

$$f \in \mathcal{C}(a,b): \qquad \left\| f \right\| = 0 \qquad \Leftrightarrow \qquad a \le x \le b: \quad f(x) = 0. \tag{5.139a–5.139f}$$

An equivalent statement is

$$w \in \mathcal{E}(|R); \quad \forall_{a \le x \le b}: \quad w(x) > 0; \quad f \in \mathcal{C}(|C), \quad \exists_{a \le x \le b}: \quad f(x) \neq 0, \tag{5.140a–5.140d}$$

$$\left\| f \right\| \equiv [f, f] \equiv \int_a^b w(x) f(x) f^*(x) dx = \int_a^b w(x) \left| f(x) \right|^2 dx > 0 \tag{5.140e}$$

a complex function continuous (Equation 5.140c) and not identically zero (Equation 5.140d) in a real interval and has positive norm (Equation 5.140e) with regard to an integrable real (Equation 5.140a) positive (Equation 5.140b) weighting function in the same interval.

5.7.2 Orthogonal and Orthonormal Trigonometric Functions

A set of functions forms an **orthogonal system** of functions iff (1) the inner product (Equations 5.137a through 5.137d) is zero for distinct functions (Equation 5.141a); and (2) the inner product of a function by itself or norm (Equations 5.138a through 5.138e) is bounded (Equation 5.141b):

$$\Phi_n \text{ orthogonal:} \quad [\Phi_n, \Phi_m] = \begin{cases} 0 & \text{if} \quad n \neq m, & (5.141a) \\ \left\| \Phi_n \right\| < \infty & \text{if} \quad n = m. & (5.141b) \end{cases}$$

Defining the infinite **identity matrix** or **Kronecker delta** as zero for all terms (Equation 5.142a) except for unity along the diagonal (Equation 5.142b):

$$\delta_{nm} \equiv \begin{cases} 0 & \text{if} \quad n \neq m, & (5.142a) \\ 1 & \text{if} \quad n = m, & (5.142b) \end{cases}$$

it follows that a system of functions is orthogonal (Equations 5.141a and 5.141b) iff it satisfies

$$\Phi_n \text{ orthogonal:} \qquad [\Phi_n, \Phi_m] = \left\| \Phi \right\| \delta_{nm}. \tag{5.143}$$

An **orthonormal system** of functions is an orthogonal system with norm unity:

$$\Psi_n \text{ orthogonal:} \qquad \left\| \Psi_n \right\| = 1 \qquad \Leftrightarrow \qquad [\Psi_n, \Psi_m] = \delta_{nm}. \tag{5.144a and 5.144b}$$

An orthogonal system of functions (Equations 5.141a and 5.141b ≡ Equations 5.143, 5.142a, and 5.142b) can be **normalized**, *that is, transformed to an orthonormal system (Equations 5.144a and 5.144b):*

$$\Psi_n(z) = \left\| \Phi_n \right\|^{-1/2} \Phi_n(z) \tag{5.145}$$

by dividing each function by the square root of its norm (Equation 5.145).

5.7.3 Orthogonal System of Exponentials, Cosines, and Sines

The systems of functions formed by exponentials with imaginary argument (Equation 5.146a) [cosines (Equation 5.146b) and sines (Equation 5.146c)] of a multiple variable:

$$\Phi_n(\varphi) \equiv \{e^{in\varphi}, \cos(n\varphi), \sin(n\varphi)\} \tag{5.146a–5.146c}$$

are orthogonal with weighting function unity (Equation 5.147b) in the circular interval (Equation 5.147a):

$$0 \le \varphi \le 2\pi: \qquad w_n(\varphi) = 1: \qquad (\Phi_n, \Phi_m) = (2\pi, \pi, \pi)\,\delta_{nm}, \tag{5.147a–5.147e}$$

with norm 2π (π) in Equation 5.147c (Equation 5.147d ≡ Equation 5.147e). The corresponding orthonormal systems are Equation 5.148a (Equations 5.148b and 5.148c):

$$\Psi_n(\varphi) = \frac{e^{in\varphi}}{\sqrt{2\pi}}, \quad \frac{\cos(n\varphi)}{\sqrt{\pi}}, \quad \frac{\sin(n\varphi)}{\sqrt{\pi}} \tag{5.148a–5.148c}$$

using Equation 5.145. The proof of orthogonality follows for the imaginary exponential (Equations 5.146a, 5.147c, and 5.148a)/cosine (Equations 5.146b, 5.147d, and 5.148b)/sine (Equations 5.146c, 5.147e, and 5.148c), respectively, in Equations 5.149/5.150/5.151:

$$\left[e^{in\varphi}, e^{im\varphi}\right] = \int_0^{2\pi} e^{i(n-m)\varphi}\,d\varphi = 2\pi\delta_{nm}, \tag{5.149}$$

$$\begin{aligned}\left[\cos(n\varphi), \cos(m\varphi)\right] &= \int_0^{2\pi} \cos(n\varphi)\cos(m\varphi)\,d\varphi \\ &= \frac{1}{2}\int_0^{2\pi} \left\{\cos\left[(n+m)\varphi\right] + \cos\left[(n-m)\varphi\right]\right\} d\varphi \\ &= \frac{1}{2}\int_0^{2\pi} \cos\left[(n-m)\varphi\right] d\varphi = \pi\delta_{nm},\end{aligned} \tag{5.150}$$

$$\begin{aligned}\left[\sin(n\varphi), \sin(m\varphi)\right] &= \int_0^{2\pi} \sin(n\varphi)\sin(m\varphi)\,d\varphi \\ &= \frac{1}{2}\int_0^{2\pi} \left\{\cos\left[(n-m)\varphi\right] - \cos\left[(n+m)\varphi\right]\right\} d\varphi \\ &= \frac{1}{2}\int_0^{2\pi} \cos\left[(n-m)\varphi\right] d\varphi = \pi\delta_{nm}\end{aligned} \tag{5.151}$$

because (1) all integrals in Equations 5.149, 5.150, and 5.151 vanish if $n \neq m$; (2) if $n = m$, the integrals in Equation 5.149 (Equations 5.150 and 5.151) are equal to 2π (π), which leads to Equation 5.147c (Equations 5.147d and 5.147e). It is shown next (Subsection 5.7.4) that the trigonometric orthogonal system of cosines (sines) leads to the orthogonality of the Chebychev polynomials of the first (second) kind with a distinct weighting function.

5.7.4 Weighting Functions for Orthogonal Chebychev Polynomials

Considering the change (Equation 5.152a) of real variables (Equations 5.152b and 5.152c), the Chebychev polynomials of the first (Equation 5.130b) [second (Equation 5.130c)] kind are given by Equation 5.152d (Equation 5.152e):

$$\psi = \cos\varphi: \quad 0 \le \varphi \le 2\pi \quad \Leftrightarrow \quad -1 \le \psi \le 1:$$

$$\left\{T_n(\psi),\, S_n(\psi)\right\} = 2^{1-n}\left\{\cos(n\varphi),\, \frac{\sin\left[(n+1)\varphi\right]}{\sin\varphi}\right\}. \tag{5.152a–5.152e}$$

The transformation (Equation 5.153a) inverse to Equation 5.152a and their differentials (Equations 5.153b and 5.153c):

$$\varphi = \operatorname{arc}\cos\psi: \quad d\psi = -\sin\varphi\, d\varphi, \quad d\varphi = \frac{d\psi}{\sqrt{1-\psi^2}} \tag{5.153a–5.153c}$$

lead from the orthogonality relations for the cosines (Equation 5.150) [sines (Equation 5.151)] to Equation 5.154a (Equation 5.154b):

$$\begin{aligned}\pi\delta_{nm} &= \int_0^{2\pi} \cos(n\varphi)\cos(m\varphi)\, d\varphi \\ &= 2^{2n-2}\int_{-1}^{+1} T_n(\psi) T_m(\psi)\left|1-\psi^2\right|^{-1/2} d\psi,\end{aligned} \tag{5.154a}$$

$$\begin{aligned}\pi\delta_{nm} &= \int_0^{2\pi} \sin\left[(n+1)\varphi\right]\sin\left[(m+1)\varphi\right] d\varphi \\ &= 2^{2n-2}\int_0^{2\pi} U_n(\cos\varphi) U_m(\cos\varphi)\sin^2\varphi\, d\varphi \\ &= 2^{2n-2}\int_{-1}^{+1} U_n(\psi) U_m(\psi)\left|1-\psi^2\right|^{1/2} d\psi,\end{aligned} \tag{5.154b}$$

which are the orthogonality relations for the Chebychev polynomials of the first (Equation 5.152d) [second (Equation 5.152e)] kind. It has been shown that *the Chebychev polynomials of the first (Equation 5.152d ≡ Equation 5.155b) [second (Equation 5.152e ≡ Equation 5.155c)] kind of real variable in the interval (Equation 5.155a):*

$$-1 \le \psi \le +1: \qquad T_n(\psi) = 2^{1-n}\cos\left[(\operatorname{arc}\cos\psi)\right], \tag{5.155a and 5.155b}$$

$$U_n(\psi) = 2^{1-n}\csc(\psi)\sin\left[(n+1)\operatorname{arc}\cos\psi\right] \tag{5.155c}$$

are orthogonal with inverse (Equation 5.156c) weighting functions [Equation 5.156a (Equation 5.156b)]:

$$w_T(\psi) = \left|1-\psi^2\right|^{-1/2}, \qquad w_U(\psi) = \left|1-\psi^2\right|^{1/2}, \qquad w_T(\psi)\, w_U(\psi) = 1, \tag{5.156a–5.156c}$$

and norm [Equation 5.154a ≡ Equation 5.157 (Equation 5.156b ≡ Equation 5.158)]:

$$\left[T_n, T_m\right] = \int_{-1}^{+1} T_n(\psi) T_m(\psi) w_T(\psi) \mathrm{d}\psi = 2^{2-2N} \delta_{nm} \pi, \tag{5.157}$$

$$\left[U_n, U_m\right] = \int_{-1}^{+1} U_n(\psi) U_m(\psi) w_U(\psi) \mathrm{d}\psi = 2^{2-2N} \delta_{nm} \pi. \tag{5.158}$$

The corresponding orthonormal system (Equation 5.145 ≡ Equation 5.159a) is Equation 5.159b (Equation 5.159c):

$$\begin{aligned}\left[\Psi_n, \Psi_m\right] = \delta_{nm}: \quad \Psi_n(\psi) &= \frac{1}{\sqrt{\pi}} 2^{n-1} \left\{T_n(\psi), U_n(\psi)\right\} \\ &= \frac{1}{\sqrt{\pi}} \left\{\cos\left[n(\operatorname{arc}\cos\psi)\right], \csc\psi \sin\left[(n+1)\operatorname{arc}\cos\psi\right]\right\}.\end{aligned} \tag{5.159a–5.159c}$$

The analogy between orthogonal systems of vectors (functions) is discussed next (Subsections 5.7.5 and 5.7.6) before proving a property unique to the Chebychev polynomials of the first kind (Subsection 5.7.7).

5.7.5 Expansion in a Series of Orthogonal Functions

A system of N vectors (Equation 5.160a) forms a **base** of an N-dimensional space iff any other vector can be represented as a linear combination (Equation 5.160b):

$$n = 1, \ldots, N: \quad \vec{f} = \sum_{n=1}^{N} f_n \vec{e}_n; \quad f_n = \left\|\vec{e}_n\right\|^{-1} \left(\vec{f} \cdot \vec{e}_n\right). \tag{5.160a–5.160c}$$

The coefficients are given by Equation 5.160c ≡ Equation 5.161c if the base is orthogonal (Equation 5.161a) with nonzero norm (Equation 5.161b):

$$\left\|\vec{e}_n\right\| \neq 0: \quad \left(\vec{e}_n \cdot \vec{e}_m\right) = \left\|\vec{e}_n\right\| \delta_{nm}, \tag{5.161a and 5.161b}$$

as follows from

$$\left(\vec{f} \cdot \vec{e}_n\right) = \left(\sum_{m=1}^{N} f_m \vec{e}_m \cdot \vec{e}_n\right) = \sum_{m=1}^{N} f_m \left(\vec{e}_m \cdot \vec{e}_n\right) = \left\|\vec{e}_n\right\| \sum_{m=1}^{N} f_m \delta_{nm} = \left\|\vec{e}_n\right\| f_n. \tag{5.162}$$

The finite identity matrix (Equations 5.142a and 5.142b) was used in Equation 5.162. The passage from Equation 5.162 to Equation 5.160c assumes a nonzero norm (Equation 5.161a). A system of vectors forms a base iff it is **complete** in the sense that it spans the whole space. By analogy, a system of orthogonal functions (Equation 5.143) is complete if any function belonging to a certain class (Equation 5.161a) can be represented by a linear combination (Equation 5.161c) with coefficients (Equation 5.161d):

$$f(z) \in \mathcal{F}(a,b); \quad \left\|\Phi_n\right\| \neq 0: \quad f(z) = \sum_{n=1}^{N} C_n \Phi_n(z), \quad C_n = \left\|\Phi_n\right\|^{-1} \left[f \cdot \Phi_n\right]. \quad \text{(5.163a–5.163d)}$$

The coefficients (Equation 5.161c) are obtained assuming that the series (Equation 5.161c) is uniformly convergent and thus (Subsection I.21.5) can be integrated term-by-term, that is, the sum can be exchanged with the inner product (Equations 5.137a through 5.137d) in

$$\left[f, \Phi_n\right] = \left[\sum_{m=1}^{\infty} C_m \Phi_m, \Phi_n\right] = \sum_{m=1}^{\infty} C_m \left[\Phi_m, \Phi_n\right] = \left\|\Phi\right\|_n \sum_{m=1}^{\infty} C_m \delta_{nm} = \left\|\Phi_n\right\| C_n. \tag{5.164}$$

The passage from Equation 5.164 to 5.161d assumes (Equation 5.161b) that the norm is not zero (Equations 5.142a through 5.142e).

The orthogonal series (Equations 5.163a through 5.163c) involves, besides the calculation of coefficients (Equation 5.164), (1) the choice of an orthogonal system of functions (Equation 5.143) that is complete; and (2) the proof that the series (Equation 5.163c) converges for a suitable class of functions (Equation 5.163a). These statements (1 and 2) are proved by functional analysis, which shows that the preceding five orthogonal systems of functions are complete, namely, (1) the exponentials of imaginary variable (Equation 5.146a); and (2–5) the cosines (Equation 5.146b) and Chebychev polynomials of the first (Equation 5.159b) and second (Equation 5.159c) kinds [sines (Equation 5.146c)] are even (odd) functions and thus can represent only even (odd) functions in Equation 5.163c. A class of functions for which convergence holds in Equation 5.163c is the functions of bounded oscillation. In order to define **oscillation** (*or fluctuation or variation*) of a function in a real interval (Subsection I.27.9.5), a **partition** is first made (Equation 5.165a):

$$a = x_0 < x_1 < \cdots < x_n < \cdots < x_N \equiv b: \quad F\left(f(x); x_0, \ldots, x_N\right) = \sum_{n=1}^{N} \left|f\left(x_n\right) - f\left(x_{n-1}\right)\right|,$$

(5.165a and 5.165b)

and the fluctuation (Equation 5.165b) is the sum of moduli of the differences at successive points. The total fluctuation is the supremum (Equation 5.167) of Equation 5.165b for all possible partitions (Equation 5.165a) of the interval. The function has **bounded oscillation** (or fluctuation) if the total fluctuation is finite. A function of bounded fluctuation can be discontinuous but has right- and left-hand limits (Equations 5.166a and 5.166b):

$$\varepsilon > 0: \quad f\left(x \pm 0\right) = \lim_{\varepsilon \to 0} f\left(x \pm \varepsilon\right); \quad f\left(x\right) \equiv \frac{f\left(x+0\right) + f\left(x-0\right)}{2}. \quad \text{(5.166a–5.166c)}$$

At a point of discontinuity, the value of the function is defined (Equation 5.166c) as the arithmetic mean of the right- and left-hand limits (Equations 5.166a and 5.166b). It is to this value that the orthogonal series (Equation 5.163c) converges. The proof of these results is deferred to a subsequent volume, and the preceding statements may be stated as the

theorem of orthogonal series: *a function of bounded oscillation (Equations 5.165a and 5.165b) in an interval (Equation 5.163a ≡ Equation 5.167):*

$$F(a, b) \equiv \{f(x): \sup[F(f(x); x_0, \dots, x_N)] < \infty\} \tag{5.167}$$

can be expanded in an orthogonal series (Equation 5.163c) in terms of a complete orthogonal system (Equation 5.143) of functions with coefficients (Equation 5.163d) involving the nonzero norm (Equation 5.163b). The series (1) converges to the function at the points where it is continuous; and (2) converges to the arithmetic mean of left- and right-hand limits (Equation 5.166c) at the points of discontinuity (Equations 5.166a and 5.166b). The theorem applies to the five preceding systems of orthogonal functions that are reconsidered next.

5.7.6 Fourier (1818), Cosine/Sine, and Chebychev (1859) Series

Each of the five preceding orthogonal systems leads to an orthogonal series (Equations 5.163a through 5.163c), namely, (1) a function with bounded oscillation in the interval *(Equation 5.168a) has* **Fourier (1818) complex series** *(Equation 5.168b) with coefficients (Equation 5.168c):*

$$F \in \mathcal{F}(-\pi,+\pi): \ F(\varphi) = \sum_{n=-\infty}^{+\infty} C_n e^{in\varphi}, \ C_n = \frac{1}{2\pi}\left[F, e^{in\varphi}\right] = \frac{1}{2\pi}\int_{-\pi}^{+\pi} F(\varphi) e^{-in\varphi}\, d\varphi; \tag{5.168a–5.168c}$$

(2 and 3) an even (odd) function with bounded oscillation in the interval [Equation 5.169a (Equation 5.170a)] has **cosine (sine) series** *[Equation 5.169b (Equation 5.170b)]:*

$$E \in \mathcal{F}(-\pi, +\pi): \ E(\varphi) = \sum_{n=0}^{\infty} A_n \cos(n\varphi) = E(-\varphi), \tag{5.169a and 5.169b}$$

$$D \in \mathcal{F}(-\pi, +\pi): \ D(\varphi) = \sum_{n=1}^{\infty} B_n \sin(n\varphi) = -D(-\varphi), \tag{5.170a and 5.170b}$$

with coefficients [Equation 5.171a (Equation 5.171b)]:

$$\begin{aligned} \{A_n, B_n\} &= \frac{1}{\pi}\{(E(\varphi)\cos(n\varphi)), (D(\varphi), \sin(n\varphi))\} \\ &= \frac{1}{\pi}\int_{-\pi}^{+\pi} \{E(\varphi)\cos(n\varphi), D(\varphi)\sin(n\varphi)\}\, d\varphi \\ &= \frac{2}{\pi}\int_{0}^{\pi} \{E(\varphi)\cos(n\varphi), D(\varphi)\sin(n\varphi)\}\, d\varphi; \end{aligned} \tag{5.171a and 5.171b}$$

(4 and 5) an even function with bounded oscillation in the interval [Equation 5.172a (Equation 5.173a)] can be expanded in a **Chebychev (1859) series of the first (second) kind** *[Equation 5.172b (Equation 5.173b)]:*

$$E \in \mathcal{F}(-1,+1): \quad E(\Psi) = \sum_{n=0}^{\infty} E_n T_n(\psi) = E(-\psi), \qquad \text{(5.172a and 5.172b)}$$

$$E \in \mathcal{F}(-1,+1): \quad E(\psi) = \sum_{n=0}^{\infty} F_n U_n(\psi) = E(-\psi), \qquad \text{(5.173a and 5.173b)}$$

with coefficients [Equation 5.174b (Equation 5.174c)]:

$$\begin{aligned} \psi = \cos\varphi: \quad E_n &= \frac{1}{\pi}\int_{-1}^{+1} E(\psi) T_n(\psi) \left|1-\psi^2\right|^{-1/2} d\psi \\ &= \frac{1}{\pi}\int_0^{2\pi} E(\cos\varphi) T_n(\cos\varphi)\, d\varphi \\ &= \frac{2}{\pi}\int_0^{\pi} E(\cos\varphi)\cos(n\varphi)\, d\varphi, \end{aligned} \qquad \text{(5.174a and 5.174b)}$$

$$\begin{aligned} F_n &= \frac{1}{\pi}\int_{-1}^{+1} E(\psi) U_n(\psi) \left|1-\psi^2\right|^{1/2} d\psi \\ &= \frac{1}{\pi}\int_0^{2\pi} E(\cos\varphi) U_n(\cos\varphi)\sin^2\varphi\, d\varphi \\ &= \frac{2}{\pi}\int_0^{\pi} E(\cos\varphi)\sin\left[(n+1)\varphi\right]\sin\varphi\, d\varphi, \end{aligned} \qquad \text{(5.174c)}$$

where the following were used: (1) the change of variable (Equation 5.174a); (2) the weighting function [Equation 5.156a (Equation 5.156b)]; and (3) the definition of Chebychev polynomial of the first (Equation 5.155b) [second (Equation 5.155c)] kind. Next, a unique property of the Chebychev polynomials of the first kind is considered.

5.7.7 Polynomials with the Least Oscillation in an Interval

Any finite interval of the real axis (Equation 5.175b) can be mapped by the linear change of variable (Equation 5.175a) into the circular interval (Equation 5.175c):

$$\varphi = 2\pi\frac{n-a}{b-a}: \quad a \le x \le b \quad \Leftrightarrow \quad 0 \le \varphi \le 2\pi. \qquad \text{(5.175a–5.175c)}$$

The Chebychev polynomial (Equation 5.152d ≡ 5.176a) of the first kind and degree N has in the interval (Equation 5.175c) a number N of (Equation 5.176b) maxima (Equation 5.176c) [minima (Equation 5.176d)] all equal to +1 (−1):

$$T_N(\varphi) = \cos(N \arccos\varphi);\, n = 1,\ldots,N: \quad (T_N)_{\max} = T_N(\varphi_n^+) = 1 = -T_N(\varphi_n^-) = (T_N)_{\min}, \qquad \text{(5.176a–5.176d)}$$

corresponding to the points [Equations 5.177a and 5.177b (Equations 5.177c and 5.177d)]:

$$N\cos\varphi_n^+ = 2n\pi:\quad \varphi_n^+ = \arc\cos\left(\frac{2n\pi}{N}\right);\quad N\cos\varphi_n^- = (2n-1)\pi,\ \varphi_n^- = \arc\cos\left[(2n-1)\frac{\pi}{N}\right]. \tag{5.177a–5.177d}$$

This implies that *the oscillation (Equations I.27.79a and I.27.79b ≡ Equations 5.165a, 5.165b, and 5.167) of the Chebychev polynomial is of the first kind and the interval degree N (Equation 5.175c) is 2N:*

$$F\left(T_N(\varphi);\, 0,\, 2\pi\right) = \sum_{n=1}^{N}\left|T_N\left(\varphi_n^+\right) - T_N\left(\varphi_n^-\right)\right| = 2N \tag{5.178}$$

because it varies monotonically N times between the maxima (Equation 5.176c) and minima (Equation 5.176d).

The Chebychev polynomial of the first kind and degree N has a coefficient unity for the leading power (Equation 5.131 ≡ Equation 5.179a), and thus the difference from an arbitrary polynomial with the same leading power (Equation 5.179b) is a polynomial (Equation 5.179c) of degree at most $N - 1$:

$$T_N(\varphi) = \varphi^N + O(\varphi^{N-1}) = P_N(\varphi):\quad T_N(\varphi) - P_N(\varphi) = O(\varphi^{N-1}) \equiv Q_{N-1}(\varphi). \tag{5.179a–5.179c}$$

Suppose that the arbitrary polynomial (Equation 5.179b) varies less in the interval (Equation 5.175c) than the Chebychev polynomial of the first kind, that is, its maxima (Equation 5.180a) [minima (Equation 5.180b)] are less than +1 (more than –1) in Equation 5.176c (Equation 5.176d):

$$+1 = \left(T_N\right)_{\max} = T_N\left(\varphi_n^+\right) > P_N\left(\varphi_n^+\right),\quad -1 = \left(T_N\right)_{\min} = T_N\left(\varphi_n^-\right) < P_N\left(\varphi_n^-\right). \tag{5.180a and 5.180b}$$

Then, the difference of the two polynomials (Equation 5.179c) changes sign at least N times (Equation 5.181a):

$$Q_N\left(\varphi_n^+\right) > 0 > Q_N\left(\varphi_n^-\right) \ \Rightarrow\ Q_N(\varphi) \sim O\left(\varphi^N\right), \tag{5.181a and 5.181b}$$

implying that it must be a polynomial of degree (Equation 5.181b) at least N. This (Equation 5.181b) contradicts Equation 5.179c, so it cannot be true that the arbitrary polynomial fluctuates less than the Chebychev polynomial of the first kind. Thus, has been proven: the **property of minimum oscillation** *of all polynomials of degree N with highest power with coefficient unity (Equation 5.179b), that having the smallest possible oscillation interval (Equation 5.175c) is the Chebychev polynomial (Equation 5.176a) of the first kind and degree N that has (Equation 5.178) oscillation equal to 2N.*

5.7.8 Decomposition into Even Plus Odd Functions/Series

If a function is neither even (Equation 5.169a) [nor odd (Equation 5.170a)], it has neither cosine (Equation 5.169b) [nor sine (Equation 5.170b)] series, and the general Fourier series

(Equations 5.168a through 5.168c) must be used. For a real variable (Equation 5.182a), the Fourier series (Equation 5.168b) can be written in a real form Equation 5.182b:

$$-\pi < \varphi < \pi: \quad F(\varphi) = C_0 + \sum_{n=1}^{\infty} \left(C_n e^{in\varphi} + C_{-n} e^{-in\varphi} \right)$$
$$= C_0 + \sum_{n=1}^{\infty} \left[\left(C_n + C_{-n} \right) \cos(n\varphi) + i \left(C_n - C_{-n} \right) \sin(n\varphi) \right] \qquad \text{(5.182a and 5.182b)}$$

with real coefficients

$$C_n + C_{-n} = \frac{1}{2\pi} \int_{-\pi}^{+\pi} F(\varphi) \left(e^{-in\varphi} + e^{in\varphi} \right) d\varphi = \frac{1}{\pi} \int_{-\pi}^{+\pi} F(\varphi) \cos(n\varphi) d\varphi \equiv A_n, \qquad (5.183)$$

$$i\left(C_n - C_{-n} \right) = \frac{i}{2\pi} \int_{-\pi}^{+\pi} F(\varphi) \left(e^{-in\varphi} - e^{in\varphi} \right) d\varphi = \frac{1}{\pi} \int_{-\pi}^{+\pi} F(\varphi) \sin(n\varphi) d\varphi \equiv B_n. \qquad (5.184)$$

In the case of an even (odd) function [Equation 5.185a (Equation 5.186b)], the coefficients of (1) sine (Equation 5.184) [cosine (Equation 5.183)] vanish [Equation 5.185b (Equation 5.186b)]:

$$E(\varphi) = E(-\varphi): \quad B_n = 0, \quad A_n = \frac{2}{\pi} \int_0^{\pi} E(\varphi) \cos(n\varphi) d\varphi, \qquad \text{(5.185a–5.185c)}$$

$$D(\varphi) = -D(-\varphi): \quad \mathrm{A}_n = 0, \quad B_n = \frac{2}{\pi} \int_0^{\pi} D(\varphi) \sin(n\varphi) d\varphi; \qquad \text{(5.186a–5.186c)}$$

and (2) cosine (Equation 5.183) [sine (Equation 5.184)] are not zero and coincide with Equation 5.185c ≡ Equation 5.171a (Equation 5.186c ≡ Equation 5.171b). For $n = 0$, the coefficient (Equation 5.184) vanishes (Equation 5.187a), and Equation 5.183 specifies the **mean value** (Equation 5.187b) of the function in the interval:

$$B_0 = 0: \quad C_0 = \frac{1}{2\pi} \int_{-\pi}^{+\pi} F(\varphi) d\varphi = \frac{A_0}{2}. \qquad \text{(5.187a and 5.187b)}$$

This proves the **Fourier theorem (1818)** *in real form: a function with bounded oscillation (Equations 5.165a, 5.165b and 5.167 ≡ Equation 5.188a) in the interval (Equation 5.168a) has Fourier series (Equation 5.188b).*

$$F \in \mathcal{F})-\pi,+\pi): \quad F(\varphi) = \frac{A_0}{2} + \sum_{n=1}^{\infty} \left[A_n \cos(n\varphi) + B_n \sin(n\varphi) \right], \qquad \text{(5.188a and 5.188b)}$$

that converges to (1) the value of the function at points of continuity; and (2) the arithmetic mean (Equation 5.166c) of right- and left-hand limits (Equations 5.166a and 5.166b) at points of discontinuity; (3) at the boundaries of the interval the values of the function or limit of the Fourier series is:

$$f(\pi) \equiv \frac{1}{2}\left[f(\pi - 0) + f(-\pi + 0)\right] \equiv f(-\pi). \tag{5.188c}$$

The coefficients are given by Equations 5.183 and 5.187b ≡ Equations 5.189a and 5.189b (Equations 5.184 and 5.187a ≡ Equations 5.190a and 5.190b) and are for a real function:

$$n \notin |N_0 \equiv \{0,1,\ldots,\}: \quad A_n = \frac{1}{2\pi}\int_{-\pi}^{+\pi} F(\varphi)\cos(d\varphi)\,d\varphi, \quad \text{(5.189a and 5.189b)}$$

$$n \in |N \equiv \{1,2,\ldots,\}: \quad B_n = \frac{1}{2\pi}\int_{-\pi}^{+\pi} F(\varphi)\sin(d\varphi)\,d\varphi \quad \text{(5.190a and 5.190b)}$$

The coefficient (Equation 5.187b) specifies the mean value of the function in the interval (Equation 5.175c).

5.7.9 Cosine/Sine Series for Powers of Cosines and Sines

The expansion of the odd power of the circular sine (Equation 5.79b), that is, an odd function, as a finite sum of sines of odd multiples of the variable is not an example of a finite or terminating Fourier series (Equation 5.191):

$$\sin^{2N+1}\varphi = 2^{-2N}\sum_{n=0}^{N}(-)^{N-n}\binom{2N+1}{n}\{\sin[(2N-1)\varphi]\cos(2n\varphi) - \cos[(2N-1)\varphi]\sin(2n\varphi)\} \tag{5.191}$$

because the coefficients of cos, sin ($2n\varphi$) are not constant, but rather functions of φ. Likewise, the expansions of the even power of the sine (Equation 5.79a ≡ Equation 5.192) and even (odd) power of the cosine [Equation 5.78a ≡ Equation 5.193 (Equation 5.78b ≡ Equation 5.194)] that are even functions, as a finite sum of cosines of even multiples of the variable:

$$\sin^{2N}\varphi = 2^{-2N}\left\{\binom{2N}{N} + 2\sum_{n=0}^{N-1}(-)^{N-n}\binom{2N}{n}\left[\cos(2N\varphi)\cos(2n\varphi) + \sin(2N\varphi)\sin(2N\varphi)\right]\right\}, \tag{5.192}$$

$$\cos^{2N}\varphi = 2^{-2N}\left\{\binom{2N}{N} + 2\sum_{n=0}^{N-1}\binom{2N}{n}\left[\cos(2N\varphi)\cos(2n\varphi) + \sin(2N\varphi)\sin(2n\varphi)\right]\right\}, \tag{5.193}$$

$$\cos^{2N+1}\varphi = 2^{-2N}\sum_{n=0}^{N}\binom{2N+1}{n}\{\cos[(2N+1)\varphi]\cos(2n\varphi) + \sin[(2N+1)\sin(2n\varphi)]\} \tag{5.194}$$

are not terminating Fourier series because the coefficients of cos, sin ($2n\varphi$) are functions of φ rather than constants. *Because Equation 5.79b ≡ Equation 5.191 (Equations 5.79a, 5.78a, and 5.78b ≡ Equations 5.192, 5.193, and 5.194) are odd (even) real functions with period 2π, they can* *be represented not only in the interval (Equation 5.182a) but also on the whole real line (Equation 5.195a) by the sine (Equation 5.195b) [cosine (Equations 5.196, 5.197, and 5.198)] series:*

$$-\infty < \varphi < +\infty: \quad \sin^{2N+1}\varphi = 2^{-2N}\sum_{n=0}^{\infty}(-)^n\binom{2N+1}{N-n}\sin\left[(2n+1)\varphi\right] \tag{5.195a and 5.195b}$$

$$\sin^{2N}\varphi = 2^{-2N}\left\{\binom{2N}{N}+2\sum_{n=0}^{\infty}(-)^n\binom{2N}{N-n}\cos(2n\varphi)\right\}, \tag{5.196}$$

$$\cos^{2N}\varphi = 2^{-2N}\left\{\binom{2N}{N}+2\sum_{n=0}^{N-1}\binom{2N}{N-n}\cos(2n\varphi)\right\}, \tag{5.197}$$

$$\cos^{2N+1}\varphi = 2^{-2N}\sum_{n=0}^{N-1}\binom{2N+1}{N-m}\cos\left[(2n+1)\varphi\right]. \tag{5.198}$$

The coefficients (Equations 5.189a, 5.189b, 5.190a, and 5.190b) of these series are evaluated next by three methods using (1) formulas for the products of sines and cosines (Subsection 5.7.10); (2) the orthogonality relations (Subsection 5.7.11); and (3) the calculus of residues (Subsections 5.7.12 and 5.7.13).

5.7.10 Fourier Coefficients for the Powers of Cosine and Sine

The odd (even) powers of the circular sine [Equation 5.199a (Equation 5.200a)] and cosine [Equation 5.202a (Equation 5.201a)] have Fourier coefficients (Equations 5.189a, 5.189b, 5.190a, and 5.190b) specified for the sine (Equation 5.171b) [cosine (Equation 5.171a)] series by Equation 5.199b (Equations 5.200b, 5.202b, and 5.201b):

$$\sin^{2N+1}\varphi = \sum_{n=1}^{\infty}a_n\sin(n\varphi), \quad a_n = \frac{1}{\pi}\int_{-\pi}^{+\pi}\sin^{2N+1}\varphi\sin(n\varphi)\,d\varphi, \tag{5.199a and 5.199b}$$

$$\sin^{2N}\varphi = \frac{b_0}{2}+\sum_{n=1}^{\infty}b_n\cos(n\varphi), \quad b_n = \frac{1}{\pi}\int_{-\pi}^{+\pi}\sin^{2N}\varphi\cos(n\varphi)\,d\varphi, \tag{5.200a and 5.200b}$$

$$\cos^{2N}\varphi = \frac{c_0}{2}+\sum_{n=1}^{\infty}c_n\cos(n\varphi), \quad c_n = \frac{1}{\pi}\int_{-\pi}^{+\pi}\cos^{2N}\varphi\cos(n\varphi)\,d\varphi, \tag{5.201a and 5.201b}$$

$$\cos^{2N+1}\varphi = \frac{d_0}{2}+\sum_{n=1}^{\infty}d_n\cos(n\varphi), \quad d_n = \frac{1}{\pi}\int_{-\pi}^{+\pi}\cos^{2N+1}\varphi\cos(n\varphi)\,d\varphi. \tag{5.202a and 5.202b}$$

The first method of evaluation, using orthogonality relations, consists of six steps, illustrated first for Equation 5.199b: (1) substitution of the odd power of sine as a sum of sines of a multiple variable (Equation 5.79b) in Equation 5.199b yields

$$a_n = \frac{2^{-2N}}{\pi} \sum_{k=1}^{N} (-)^{N-k} \binom{2N+1}{k} \int_{-\pi}^{+\pi} \sin\left[(2N-2k+1)\varphi\right] \sin(n\varphi)\, d\varphi; \tag{5.203}$$

(2) using Equation 5.88b for the product of sines leads to

$$a_n = \frac{2^{-1-2N}}{\pi} \sum_{k=1}^{N} (-)^{N-k} \binom{2N+1}{k} \int_{-\pi}^{+\pi} \cos\left\{\left[(2N-2k-n+1)\varphi\right] - \cos\left[(2N-2k+n+1)\varphi\right] d\varphi\right\}; \tag{5.204}$$

(3) in the second term on the r.h.s. of Equation 5.204, the cosine has nonzero variable (Equation 5.205a), and hence its integral is zero (Equation 5.205b):

$$2N-2k+n+1 \in |N: \quad \int_{-\pi}^{+\pi} \cos\left[\left(2N-2k+n+1\right)\varphi\right] d\varphi = \left\{\frac{\sin\left[\left(2N-2k+n+1\right)\varphi\right]}{2N-2k+n+1}\right\}_{-\pi}^{+\pi} = 0; \tag{5.205a and 5.205b}$$

(4) in the first term on the r.h.s. of Equation 5.204, all integrals vanish as in Equations 5.205a and 5.205b, unless the variable of the first cosine is zero; and (5) for $n = 2m$, even the variable of the cosine is an odd multiple of φ, so it cannot be zero, and all integrals vanish, implying that the even coefficients are zero (Equation 5.206a):

$$a_{2m} = 0; \quad a_{2m+1} = \frac{2^{-1-2N}}{\pi} 2\pi \left[(-)^{N-k} \binom{2N+1}{k}\right]_{2N-2k-2m=0} = 2^{-2N} \left[(-)^{N-k} \binom{2N+1}{k}\right]_{k=N-m} = 2^{-2N} (-)^{m} \binom{2N+1}{N-\mathrm{m}}; \tag{5.206a and 5.206b}$$

and (6) for $n = 2m + 1$ odd, the first cosine has variable zero for $2N - 2k - (2m + 1) + 1 = 0$, that is, only the term with $k = N - m$ is present, and it is multiplied by the value 2π of the integral, which leads to Equation 5.206b. Substituting Equations 5.206a and 5.206b in Equation 5.199a proves Equations 5.195a and 5.195b.

5.7.11 Evaluation of Fourier Coefficients by Orthogonality Relations

The passage from Equations 5.203 to 5.206a and 5.206b is equivalent to the use of the orthogonality relation (Equation 5.151) for sines:

$$a_n = \frac{2^{-2N}}{\pi} \sum_{k=0}^{N} (-)^{N-k} \binom{2N+1}{k} \pi \delta_{n,2N-2k+1}. \tag{5.207}$$

The identity matrix in Equation 5.207 is zero only for all even coefficients (Equation 5.206a); it is unity only for one odd coefficient (Equation 5.209a), which leads to (Equation 5.209b ≡ Equation 5.206b):

$$2m+1 = 2N-2k+1: \quad a_{2m+1} = 2^{-2N} \left[(-)^{N-k} \binom{2N+1}{k} \right]_{k=N-m} = 2^{-2N} (-)^m \binom{2N+1}{N-m}. \tag{5.208a and 5.208b}$$

This provides an alternative derivation (Equations 5.208a and 5.208b ≡ Equations 5.206a and 5.206b) of the coefficients (Equation 5.199b) of the sine series (Equation 5.199a ≡ Equation 5.195b). The same method is used with Equations 5.79a, 5.78a, and 5.78b and the orthogonally relation (Equation 5.150) for the cosines, respectively, in the case of Equations 5.200b, 5.201b, and 5.202b, which lead, respectively, to Equations 5.209, 5.210, and 5.211:

$$\begin{aligned} b_n &= \frac{2^{-2N}}{\pi} \left\{ \binom{2N}{N} \int_{-\pi}^{+\pi} \cos(n\varphi) \right. \\ &\quad + 2 \sum_{k=0}^{N-1} (-)^{N-k} \binom{2N}{k} \int_{-\pi}^{\pi} \cos[(2N-2k)\varphi] \cos(n\varphi) \, d\varphi \\ &= 2^{1-2N} \left\{ \binom{2N}{N} \delta_{0,n} + \sum_{k=0}^{N-1} (-)^{N-k} \binom{2N}{k} \delta_{n,2N-2k} \right\}, \end{aligned} \tag{5.209}$$

$$\begin{aligned} c_n &= \frac{2^{-2N}}{\pi} \left\{ \binom{2N}{N} \int_{-\pi}^{+\pi} \cos(n\varphi) \right. \\ &\quad \left. + 2 \sum_{k=0}^{N-1} \binom{2N}{k} \int_{-\pi}^{+\pi} \cos[(2N-2k)\varphi] \cos(n\varphi) \, d\varphi \right\} \\ &= 2^{1-2N} \left\{ \binom{2N}{N} \delta_{0,n} + \sum_{k=0}^{N-1} \binom{2N}{k} \delta_{n,2N-2k} \right\}, \end{aligned} \tag{5.210}$$

$$\begin{aligned} d_n &= \frac{2^{-2N}}{\pi} \sum_{k=0}^{N} \left\{ \binom{2N+1}{k} \int_{-\pi}^{+\pi} \cos[(2N-2k+1)\varphi] \cos(n\varphi) \, d\varphi \right. \\ &= 2^{-2N} \sum_{k=0}^{N} \binom{2N+1}{k} \delta_{n,2N-2k+1}. \end{aligned} \tag{5.211}$$

The coefficients (Equation 5.211) are evaluated like Equation 5.207, which leads to Equations 5.212a and 5.212b instead of Equations 5.207a and 5.207b:

$$d_{2m} = 0, \quad d_{2m+1} = 2^{1-2N} \binom{2N+1}{N-m}; \qquad \text{(5.212a and 5.212b)}$$

substitution of Equations 5.212a and 5.212b in Equation 5.202a proves Equation 5.198. Concerning Equation 5.210 (Equation 5.209) for (1) $n = 0$, only the first term on the r.h.s. appears, which leads to Equation 5.213a (Equation 5.213b):

$$b_0 = 2^{1-2N} \binom{2N}{N} = c_0; \quad b_{2m+1} = 0 = c_{2m+1}; \qquad \text{(5.213a–5.213d)}$$

(2) for $n = 2m + 1$ odd, the identity matrix vanishes, so all odd coefficients are zero [Equation 5.213c (Equation 5.213d)]; and (3) for $n = 2m$, even the only nonzero term of the sum is Equation 5.214a, which leads to Equation 5.214c (Equation 5.214b):

$$2m = n = 2N - 2k: \quad c_{2m} = 2^{1-2N} \left[\binom{2N}{k} \right]_{k=N-m} = 2^{1-2N} \binom{2N}{N-m}, \qquad \text{(5.214a and 5.214b)}$$

$$b_{2m} = 2^{1-2N} \left[(-)^{N-k} \binom{2N}{k} \right]_{k=N-m} = 2^{1-2N} (-)^m \binom{2N}{N-m}. \qquad \text{(5.214c)}$$

Substitution of Equations 5.213a, 5.213c, and 5.214c (Equations 5.213b, 5.213d, and 5.214b) in Equation 5.200a (Equation 5.201a) proves Equation 5.195 (Equation 5.196).

5.7.12 Evaluation of Trigonometric Integrals by Residues

A third way to determine the Fourier coefficients (Equations 5.199b through 5.202b), quite distinct from the first two, is the general method of evaluation of trigonometric integrals by residues (Subsection I.17.2). It uses the change of variable (Equation 5.215a) that maps the interval (Equation 5.182a ≡ Equation 5.215b) to the unit circle (Equation 5.215c):

$$\zeta = e^{i\varphi}: \quad -\pi \le \varphi \le +\pi \quad \Leftrightarrow \quad |\zeta| = 1; \quad d\zeta = ie^{i\varphi}\, d\varphi = i\zeta\, d\varphi. \qquad \text{(5.215a–5.215d)}$$

The differential (Equation 5.215d) appears in the Fourier coefficients (Equations 5.199b through 5.202b ≡ Equations 5.216a through 5.216d):

$$\{a_n, b_n, c_n, d_n\} = \frac{1}{i\pi} \int_{|\zeta|=1} f(\zeta) \frac{d\zeta}{\zeta}. \qquad \text{(5.216a–5.216d)}$$

The integral over the unit circle has integrand involving cosines (sines) of single [Equation 5.218a (Equation 5.217b)] and multiple [Equation 5.218a (Equation 5.218b)] arguments:

$$\{2\cos\varphi, 2i\sin\varphi\} = e^{i\varphi} \pm e^{i\varphi} = \zeta \pm \zeta^{-1}, \qquad \text{(5.217a and 5.217b)}$$

$$\{2\cos(n\varphi), 2i\sin(n\varphi)\} = e^{in\varphi} \pm e^{-in\varphi} = \zeta^n \pm \zeta^{-n}, \qquad \text{(5.218a and 5.218b)}$$

in the function

$$\begin{aligned}&\left\{\sin^{2N+1}\varphi\sin(n\varphi),\ \sin^{2N}\varphi\cos(n\varphi),\ \cos^{2N}\varphi\cos(n\varphi),\ \cos^{2N+1}\varphi\cos(n\varphi)\right\}\\&=\left\{(-)^{N+1}2^{-2-2N}\left(\zeta-\zeta^{-1}\right)^{2N+1}\left(\zeta^n-\zeta^{-n}\right),(-)^N2^{-1-2N}\left(\zeta-\zeta^{-1}\right)^{2N}\left(\zeta^n+\zeta^{-n}\right),\right. \qquad \text{(5.219a–5.219d)}\\&\left.2^{-1-2N}\left(\zeta+\zeta^{-1}\right)^{2N}\left(\zeta^n+\zeta^{-n}\right),2^{-2-2N}\left(\zeta+\zeta^{-1}\right)^{2N+1}\left(\zeta^n+\zeta^{-n}\right)\right\}\equiv f(\zeta),\end{aligned}$$

which is a polynomial of ζ and $1/\zeta$.

The **constant part** of an expansion (Equation 5.220a) in powers of ζ and $1/\zeta$ is defined (Equation 5.220b) as the constant term independent of ζ:

$$N, M \in |N: \quad f(\zeta) = \sum_{n=-M}^{N} f_n\zeta^n: \quad Cp\{f(\zeta)\} = f_0. \qquad \text{(5.220a and 5.220b)}$$

The sum in Equation 5.220a may be (1) a polynomial of degree N (M) in ζ ($1/\zeta$) if N (M) is finite; and (2 and 3) if $M = 0$ ($M \geq 1$) and $N = \infty$, it is a MacLaurin (Laurent–MacLaurin) series [Equations 1.24a and 1.24b (Equations 1.6a and 1.6b)]. In all cases 1–3, the constant part (Equation 5.220b) of the function (Equation 5.220a) is the residue (Equation 5.221b) of Equation 5.221a:

$$\frac{f(\zeta)}{\zeta} = \sum_{n=-M-1}^{N-1} f_n\zeta^{n-1} = \sum_{n=-M}^{N} A_m\zeta^m: \quad f_0 = A_{-1}. \qquad \text{(5.221a and 5.221b)}$$

The substitution of Equations 5.219a through 5.219d in Equation 5.216d leads to integrals of the form (Equations 5.222a and 5.222b ≡ Equations I.28.24a through I.28.24c):

$$\int_{|\zeta|=1} \zeta^m \, d\zeta = i\int_0^{2\pi} e^{i(m+1)\varphi}\, d\varphi = \begin{cases} 2\pi i & \text{if} \quad m = -1, & \text{(5.222a)} \\ 0 & \text{if} \quad m \neq -1. & \text{(5.222b)} \end{cases}$$

Thus, substituting Equation 5.220a in Equation 5.216d yields

$$\{a_n, b_n, c_n, d_n\} = \frac{1}{i\pi}\sum_{n=-M}^{+N} f_n \int_{|\zeta|=1} \zeta^{n-1}\, d\zeta = 2f_0 = 2Cp\left[f(\zeta)\right] \qquad \text{(5.223)}$$

because the nonzero (Equation 5.222b) term of the sum in Equation 5.223 corresponds to $m = -1 = n - 1$, that is, $n = 0$ and is multiplied by $2\pi i$. Thus, *the evaluation of the coefficients [Equation 5.199b (Equations 5.200b through 5.202b)] of the sine (Equation 5.199a) [cosine (Equations 5.200a through 5.202a)] series reduces to the determination (Equation 5.223) of the constant part (Equations 5.220a and 5.220b) of the respective functions (Equations 5.219a through 5.219d).*

5.7.13 Determination of the Constant Part of an Expansion

The functions (Equations 5.219a through 5.219d) are polynomials in ζ and $1/\zeta$, whose coefficients are specified by the binomial theorem (Equation I.25.38 ≡ Equations 3.146a through 3.146c):

$$f(\zeta) = \begin{cases} (-)^{N+1} 2^{-2-2N} \left(\zeta^n - \zeta^{-n}\right) \sum_{k=0}^{2N+1} (-)^{2N+1-k} \binom{2N+1}{k} \zeta^{2k-2N-1}, & (5.224a) \\ (-)^N 2^{-1-2N} \left(\zeta^n + \zeta^{-n}\right) \sum_{k=0}^{2N} (-)^{2N-k} \binom{2N}{k} \zeta^{2k-2N}, & (5.224b) \\ 2^{-1-2N} \left(\zeta^n + \zeta^{-n}\right) \sum_{k=0}^{2N} \binom{2N}{k} \zeta^{2k-2N}, & (5.224c) \\ 2^{-2-2N} \left(\zeta^n + \zeta^{-n}\right) \sum_{k=0}^{2N+1} \binom{2N+1}{k} \zeta^{2k-2N-1}. & (5.224d) \end{cases}$$

The constant part (Equation 5.221b) of Equation 5.220a ≡ Equations 5.224a through 5.224d is the coefficient of the power with zero exponent, which leads to Equation 5.225b (Equation 5.225d) for Equation 5.225a (Equation 5.225c) in Equations 5.224a and 5.224d (Equations 5.224b and 5.224c):

$$a_n, d_n: \quad 0 = 2k - 2N - 1 \pm n; \qquad b_n, c_n: \quad 0 = 2k - 2N \pm n. \qquad (5.225a\text{–}5.225d)$$

Concerning Equation 5.225a (Equation 5.225b), k is not an integer if $n = 2m$ ($n = 2m + 1$) is even (odd), and thus the corresponding coefficients vanish [Equations 5.226a and 5.226b (Equations 5.226c and 5.226d)]:

$$a_{2m} = 0 = d_{2m}, \qquad b_{2m+1} = 0 = c_{2m+1}. \qquad (5.226a\text{–}5.226d)$$

This agrees with the results obtained before by the method of orthogonality relation (Equation 5.206a ≡ Equation 5.226a, Equation 5.212a ≡ Equation 5.226b, and Equations 5.213c and 5.213d ≡ Equations 5.226c and 5.226d).

The nonzero coefficients [Equation 5.227a (Equation 5.228a)] in Equations 5.199a and 5.202a (Equations 5.200a and 5.201a)] thus arise from the odd (Equation 5.227b) [even (Equation 5.228b)] coefficients in Equation 5.225b (Equation 5.225d), which lead to Equations 5.227c and 5.227d (Equations 5.228c and 5.228d):

$$a_n, d_n: \quad n = 2m + 1, \quad k = (2N + 1 \mp n)/2 = N - m, N + m + 1, \qquad (5.227a\text{–}5.227d)$$

$$b_n, c_n: \quad n = 2m, \quad k = (2N \mp n)/2 = N \mp m. \qquad (5.228a\text{–}5.228d)$$

The two values [Equations 5.227c and 5.227d (Equations 5.228c and 5.228d)] of k lead (Equation 5.223) in Equations 5.224a and 5.224d (Equations 5.224b and 5.224c) to Equations 5.229a and 5.229d (Equations 5.229b and 5.229c):

$$a_{2m+1} = (-)^{N+1} 2^{-1-2N} \left[(-)^{N+m+1} \binom{2N+1}{N-m} - (-)^{N-m} \binom{2N+1}{N+m+1} \right] = (-)^m 2^{-2N} \binom{2N+1}{N-m}, \tag{5.229a}$$

$$b_{2m} = (-)^N 2^{-2N} \left[(-)^{N+m} \binom{2N}{N-m} + (-)^{N-m} \binom{2N}{N+m} \right] = (-)^m 2^{1-2N} \binom{2N}{N-m}, \tag{5.229b}$$

$$c_{2m} = 2^{-2N} \left[\binom{2N}{N-m} + \binom{2N}{N+m} \right] = 2^{1-2N} \binom{2N}{N-m}, \tag{5.229c}$$

$$d_{2m+1} = 2^{-1-2N} \left[\binom{2N+1}{N-m} + \binom{2N+1}{N+m+1} \right] = 2^{-2N} \binom{2N+1}{N-m}. \tag{5.229d}$$

Substitution of Equations 5.226a and 5.229a/5.226c and 5.229b/5.226d and 5.229c/5.226b and 5.229d, respectively, in Equations 5.199a/5.200a/5.201a/5.202a proves Equations 5.195b/5.196/5.197/5.198.

In Equations 5.229a through 5.229d, there is a factor 2 coming from Equation 5.223. In Equations 5.229a and 5.229d (Equations 5.229b and 5.229c), the identities [Equation 5.230a (Equation 5.230b)] were used:

$$\binom{2N+1}{N-m} = \frac{(2N+1)!}{(N-m)!(N+m+1)!} = \binom{2N+1}{N+m+1}, \tag{5.230a}$$

$$\binom{2N}{N-m} = \frac{(2N)!}{(N+m)!(N-m)!} = \binom{2N}{N+m}. \tag{5.230b}$$

The case $n = 0$ is included in Equations 5.228a through 5.228d, and is evaluated distinctly from Equations 5.229b and 5.229c, because (Equation 5.231a) there is only one value (Equation 5.231b) of k leading to Equation 5.231c:

$$\frac{1}{2}b_0 = \frac{1}{2}c_0 = 2^{-2N}\binom{2N}{N} = 2^{-2N}\frac{(2N)!}{(N!)^2}$$
$$= \frac{1}{2\pi}\int_{-\pi}^{+\pi}\cos^{2N}\varphi\, d\varphi = \frac{1}{2\pi}\int_{-\pi}^{+\pi}\sin^{2N}\varphi\, d\varphi, \tag{5.231a–5.231c}$$

which (1) specifies the mean value of the even power of the circular cosine and sine in the interval (Equation 5.182a); (2) agrees with Equation I.20.69 ≡ Equation 5.231c; and (3) corresponds to the constant term in Equations 5.196 and 5.197. The Equation 2.531c agrees with Equations 5.229b and 5.229c for $n = 0$ in Equation 2.531a. The case $n = 0$ is not included in Equations 5.227a through 5.227d, implying that there is no constant term in Equations 5.195b and 5.198. There is no constant term in Equation 5.195b because the odd power of the sine is an odd function, and thus has zero mean value in the interval (Equation 5.182a). There is no constant term in Equation 5.198 because the mean value of the odd power of the cosine in the interval (Equation 5.182a) also has zero mean value, as can be checked from

$$\frac{d_0}{2} = \frac{1}{2\pi}\int_{-\pi}^{+\pi}\cos^{2N+1}\varphi\, d\varphi = \frac{1}{2\pi i}\int_{|\zeta|=1}\left(\frac{\zeta+\zeta^{-1}}{2}\right)^{2N+1}\frac{d\zeta}{\zeta}$$
$$= \frac{1}{2\pi i}2^{-2N-1}\sum_{k=0}^{2N+1}\binom{2N+1}{k}\int_{|\zeta|=1}\zeta^{2N-2k}\, d\zeta = 0, \tag{5.232}$$

where (1) the integral along the interval (Equation 5.182a) was transformed to loop integral along the unit circle via the change of variable (Equations 5.216a through 5.216d); (2) the binomial theorem (Equation I.25.38 ≡ Equations 3.146a and 3.146b) was used; and (3) all the integrals vanish by Equation 5.222b.

5.8 Relations between Complex and Real Functions

All primary and secondary circular (hyperbolic) functions of a complex variable can be expressed in terms of functions of a real variable [Subsection 5.8.1 (Subsection 5.8.2)] by using (Subsection 5.8.3) the addition formulas and imaginary change of variable. As a consequence, the moduli and arguments of the functions of a complex variable in terms of functions of real variable (Subsection 5.8.4) are obtained.

5.8.1 Real and Imaginary Parts of Complex Circular and Hyperbolic Functions

The real and imaginary parts of all complex circular functions are given in terms of the circular and hyperbolic cosine and sine of real variable by

$$\cos(x + iy) = \cos x \cosh y - i \sin x \sinh y, \tag{5.233}$$

$$\sin(x + iy) = \sin x \cosh y + i \cos x \sinh y, \tag{5.234}$$

$$\sec(x+iy) = 2\frac{\cos x \cosh y + i \sin x \sinh y}{\cosh(2y) + \cos(2x)}, \tag{5.235}$$

$$\csc(x+iy) = 2\frac{\sin x \cosh y - i \cos x \sinh y}{\cosh(2y) - \cos(2x)}, \tag{5.236}$$

$$\tan(x+iy) = \frac{\sin(2x) + i\sinh(2y)}{\cosh(2y) + \cos(2x)}, \tag{5.237}$$

$$\cot(x+iy) = \frac{\sin(2x) - i\sinh(2y)}{\cosh(2y) - \cos(2x)}, \tag{5.238}$$

namely, the circular cosine (Equation 5.233), sine (Equation 5.234), secant (Equation 5.235), cosecant (Equation 5.236), tangent (Equation 5.237), and cotangent (Equation 5.238).

5.8.2 Circular/Hyperbolic Functions of Complex/Real Variable

The imaginary change of variable for the primary (Equations 5.5a, 5.5b, 5.6a, and 5.6b) [secondary (Equations 5.24a, 5.24b, 5.25a, 5.25b, 5.26a, 5.26b, 5.27a, and 5.27b)] functions can be used to transform circular (Equations 5.233 through 5.238) [hyperbolic (Equations 5.239 through 5.244)] functions of complex variable. This leads to *the real and imaginary parts of the hyperbolic functions in terms of the circular and hyperbolic cosine and sine of real variable:*

$$\cosh(x + iy) = \cosh x \cos y + i \sinh x \sin y, \tag{5.239}$$

$$\sinh(x + iy) = \sinh x \cos y + i \cosh x \sin y, \tag{5.240}$$

$$\operatorname{sech}(x+iy) = 2\frac{\cosh x \cos y - i \sinh x \sin y}{\cosh(2x) + \cos(2y)}, \tag{5.241}$$

$$\operatorname{csch}(x+iy) = 2\frac{\sinh x \cos y - i \cosh x \sin y}{\cosh(2x) - \cos(2y)}, \tag{5.242}$$

$$\tanh(x+iy) = \frac{\sinh(2x) + i\sin(2y)}{\cosh(2x) + \cos(2y)}, \tag{5.243}$$

$$\coth(x+iy) = \frac{\sinh(2x) - i\sin(2y)}{\cosh(2x) - \cos(2y)}, \tag{5.244}$$

namely, for the hyperbolic cosine (Equation 5.239), sine (Equation 5.240), secant (Equation 5.241), cosecant (Equation 5.242), tangent (Equation 5.243), and cotangent (Equation 5.244). Because Equations 5.233 through 5.238 (Equations 5.239 through 5.244) are equivalent, the proofs are similar (Subsection 5.8.3).

5.8.3 Combination of Addition Formulas and the Imaginary Change of Variable

A consequence of the formulas of addition of two variables for circular (Equations 5.47a, 5.47b, 5.48a, and 5.48b) [hyperbolic (Equations 5.49a, 5.49b, 5.50a, and 5.50b)] functions is obtained by setting $u = x$, $v = iy$, and using the imaginary transformations [Equations 5.5a and 5.5b (Equations 5.6a and 5.6b)], which lead to the formulas giving the real and imaginary parts of circular and hyperbolic functions with complex variable. For example, for the circular (Equation 5.245) [hyperbolic (Equation 5.246)] cosine:

$$\cos(x + iy) = \cos x \cos(iy) - \sin x \sin(iy) = \cos x \cosh y - i \sin x \sinh y, \tag{5.245}$$

$$\cosh(x + iy) = \cosh x \cosh(iy) + \sinh x \sinh(iy) = \cosh x \cos y + i \sinh x \sin y; \tag{5.246}$$

these prove Equation 5.245 ≡ Equation 5.233 (Equation 5.246 ≡ Equation 5.239). The proof of Equation 5.234 (Equation 5.240) can be made directly or by transformation of Equation 5.233 ≡ Equation 5.247 ≡ Equation 5.234 (Equation 5.234 ≡ Equation 5.248 ≡ Equation 5.240):

$$\begin{aligned}\sin(x+iy) &= -\cos(x+\pi/2+iy) = -\cos(x+\pi/2)\cosh y + i\sin(x+\pi/2)\sinh y\\ &= \sin x\cosh y + i\cos x\sinh y,\end{aligned} \tag{5.247}$$

$$\begin{aligned}\sinh(x+iy) &= -i\sin(ix-y) = -i\left[\sin(-y)\cosh x + i\cos(-y)\sinh x\right]\\ &= \sinh x\cos y + i\cosh x\sin y\end{aligned} \tag{5.248}$$

completing the real and imaginary parts of the primary functions cosine (sine).

The first pair of secondary functions is the secant (cosecant), which is their inverses, for example:

$$\begin{aligned}\sec(x+iy) &= \frac{1}{\cos(x+iy)} = \frac{1}{\cos x\cosh y - i\sin x\sinh y}\\ &= \frac{\cos x\cosh y + i\sin x\sinh y}{\cos^2 x\cosh^2 y + \sin^2 x\sinh^2 y},\end{aligned} \tag{5.249}$$

$$\begin{aligned}\operatorname{sech}(x+iy) &= \frac{1}{\cosh(x+iy)} = \frac{1}{\cosh x\cos y + i\sinh x\sin y}\\ &= \frac{\cosh x\cos y - i\sinh x\sin y}{\cosh^2 x\cos^2 y + \sinh^2 x\sin^2 y}\end{aligned} \tag{5.250}$$

where the denominators simplify to

$$\begin{aligned}
&\left\{\cos^2 x\cosh^2 y+\sin^2 x\sinh^2 y, \cosh^2 x\cos^2 y+\sinh^2 x\sin^2 y\right\} \\
&=\left\{\cosh^2 y+\sin^2 x\left(\sinh^2 y-\cosh^2 y\right), \cosh^2 x+\sin^2 y\left(\sinh^2 x-\cosh^2 x\right)\right\} \\
&=\left\{\cosh^2 y-\sin^2 x, \cosh^2 x-\sin^2 y\right\} \\
&=\frac{1}{2}\left\{\cosh(2y)+\cos(2x), \cosh(2x)+\cos(2y)\right\};
\end{aligned} \tag{5.251a and 5.251b}$$

substitution of Equation 5.251a (Equation 5.251b) in Equation 5.249 (Equation 5.250) proves Equation 5.235 (Equation 5.241). Similar proofs apply to Equation 5.236 (Equation 5.242):

$$\begin{aligned}
\csc(x+iy) &= -\sec(x+\pi/2+iy) = -2\frac{\cos(x+\pi/2)\cosh y+i\sin(x+\pi/2)\sinh y}{\cosh(2y)+\cos(2x+\pi)} \\
&= 2\frac{\sin x\cosh y-i\cos x\sinh y}{\cosh(2y)-\cos(2x)},
\end{aligned} \tag{5.252}$$

$$\begin{aligned}
\operatorname{csch}(x+iy) &= i\csc(ix-y) = 2i\frac{\sin(-y)\cosh x-i\cos(-y)\sinh x}{\cosh(2x)-\cos(-2y)} \\
&= 2\frac{\sinh x\cos y-i\cosh x\sin y}{\cosh(2x)-\cos(2y)}
\end{aligned} \tag{5.253}$$

which can also be obtained by transformation of Equation 5.235 ≡ Equation 5.252 ≡ Equation 5.236 (Equation 5.241 ≡ Equation 5.253 ≡ Equation 5.242). The remaining formulas to be proved concern the circular (Equation 5.237) [hyperbolic (Equation 5.243)] tangent of complex variable expressed in terms of real variables:

$$\begin{aligned}
\tan(x+iy) &= \frac{\sin(x+iy)}{\cos(x+iy)} = \frac{\sin x\cosh y+i\cos x\sinh y}{\cos x\cosh y-i\sin x\sinh y} \\
&= \frac{(\sin x\cosh y+i\cos x\sinh y)(\cos x\cosh y+i\sin x\sinh y)}{\cos^2 x\cosh^2 y+\sin^2 x\sinh^2 y} \\
&= \frac{\sin x\cos x+i\cosh y\sinh y}{\cosh^2 y-\sin^2 x} = \frac{\sin(2x)+i\sinh(2y)}{\cosh(2y)+\cos(2x)},
\end{aligned} \tag{5.254}$$

$$\begin{aligned}\tanh(x+iy) &= \frac{\sinh(x+iy)}{\cosh(x+iy)} = \frac{\sinh x\cos y + i\cosh x\sin y}{\cosh x\cos y + i\sinh x\sin y} \\ &= \frac{(\sinh x\cos y + i\cosh x\sin y)(\cosh x\cos y - i\sinh x\sin y)}{\cosh^2 x\cos^2 y + \sinh^2 x\sin^2 y} \\ &= \frac{\sinh x\cosh x + i\cos y\sin y}{\cosh^2 x - \sin^2 y} = \frac{\sinh(2x) + i\sin(2y)}{\cosh(2x)+\cos(2y)};\end{aligned} \tag{5.255}$$

in the denominator of Equation 5.254 (Equation 5.255), Equation 5.251a (Equation 5.251b) was used. A similar proof applies to Equation 5.238 (Equation 5.244), which can also be obtained by transformation of Equation 5.237 ≡ Equation 5.256 ≡ Equation 5.238 (Equation 5.243 ≡ Equation 5.257 ≡ Equation 5.244):

$$\begin{aligned}\cot(x+iy) &= -\tan(x+\pi/2+iy) = -\frac{\sin(2x+\pi) + i\sinh(2y)}{\cosh(2y)+\cos(2x+\pi)} \\ &= \frac{\sin(2x) - i\sinh(2y)}{\cosh(2y)-\cos(2x)},\end{aligned} \tag{5.256}$$

$$\begin{aligned}\coth(x+iy) &= i\cot(ix+i^2y) = i\cot(ix-y) = i\frac{\sin(-2y) - i\sinh(2x)}{\cosh(2x)-\cos(-2y)} \\ &= \frac{\sinh(2x) - i\sin(2y)}{\cosh(2x)-\cos(2y)}.\end{aligned} \tag{5.257}$$

These formulas show that the circular (hyperbolic) functions of a complex variable $z \equiv x + iy$ can be calculated by Equations 5.233 through 5.238 (Equations 5.239 through 5.244) in terms of circular (hyperbolic) functions of real variable, plotted in Figure 5.2a through c (Figure 5.2d through f) and represented in Figure 5.1a and c (Figure 5.1b and d).

5.8.4 Modulus and Argument of Complex Functions

From Equation 5.233 (Equation 5.239) follow the arguments of the circular (Equation 5.258) [hyperbolic (Equation 5.259)] cosine:

$$\arg(\cos z) = \arctan\left(-\frac{\sin x\sinh y}{\cos x\cosh y}\right) = -\arctan(\tan x\tanh y) = -\arg(\sec z), \tag{5.258}$$

$$\arg(\cosh z) = \arctan\left(\frac{\sinh x\sin y}{\cosh x\cos y}\right) = \arctan(\tanh x\tan y) = -\arg(\operatorname{sech} z), \tag{5.259}$$

and also their moduli [Equation 5.260 (Equation 5.261)] using Equation 5.251a (Equation 5.251b):

$$\begin{aligned}|\cos z| &= \left|\cos^2 x\cosh^2 y + \sin^2 x\sinh^2 y\right|^{1/2} = \left|\cosh^2 y - \sin^2 x\right|^{1/2} \\ &= \left|\sinh^2 y + \cos^2 x\right|^{1/2} = \left|\frac{1}{2}\left[\cosh(2y) + \cos(2x)\right]\right|^{1/2} = \frac{1}{|\sec z|},\end{aligned} \tag{5.260}$$

$$\begin{aligned}|\cosh z| &= \left|\cosh^2 x\cos^2 y + \sinh^2 x\sin^2 y\right|^{1/2} = \left|\cosh^2 x - \sin^2 y\right|^{1/2} \\ &\left|\sinh^2 x + \cos^2 y\right|^{1/2} = \left|\frac{1}{2}\left[\cosh(2x) + \cos(2y)\right]\right|^{1/2} = \frac{1}{|\operatorname{sech} z|}.\end{aligned} \tag{5.261}$$

Concerning the circular (Equation 5.234) [hyperbolic (Equation 5.240)] sine, the argument is given by Equation 5.262 (Equation 5.263):

$$\arg(\sin z) = \arctan\left(\frac{\cos x\sinh y}{\sin x\cosh y}\right) = \arctan(\cot x\tanh y) = -\operatorname{arc}(\csc z), \tag{5.262}$$

$$\arg(\sinh z) = \arctan\left(\frac{\cosh x\sin y}{\sinh x\cos y}\right) = \arctan(\coth x\tan y) = -\operatorname{arc}(\operatorname{csch} z), \tag{5.263}$$

and the modulus is given by Equation 5.264 (Equation 5.265):

$$\begin{aligned}|\sin z| &= \left|\sin^2 x\cosh^2 y + \cos^2 x\sinh^2 y\right|^{1/2} = \left|\cosh^2 y - \cos^2 x\right|^{1/2} \\ &\left|\sinh^2 y + \sin^2 x\right|^{1/2} = \left|\frac{1}{2}\left[\cosh(2y) - \cos(2x)\right]\right|^{1/2} = \frac{1}{|\csc z|},\end{aligned} \tag{5.264}$$

$$\begin{aligned}|\sinh z| &= \left|\sinh^2 x\cos^2 y + \cosh^2 x\sin^2 y\right|^{1/2} = \left|\cosh^2 x - \cos^2 y\right|^{1/2} \\ &= \left|\sinh^2 x + \sin^2 y\right|^{1/2} = \left|\frac{1}{2}\left[\cosh(2x) - \cos(2y)\right]\right|^{1/2} = \frac{1}{|\operatorname{csch} z|}.\end{aligned} \tag{5.265}$$

From the ratio of Equations 5.264 and 5.260 (Equations 5.265 and 5.261) follows the modulus of the circular (Equation 5.266) [hyperbolic (Equation 5.267)] tangent:

$$|\tan z| = \left|\frac{\sin z}{\cos z}\right| = \left|\frac{\cosh(2y) - \cos(2x)}{\cosh(2y) + \cos(2x)}\right|^{1/2} = \frac{1}{|\cot z|}, \tag{5.266}$$

$$|\tanh z| = \left|\frac{\sinh z}{\cosh z}\right| = \left|\frac{\cosh(2x) - \cos(2y)}{\cosh(2x) + \cos(2y)}\right|^{1/2} = \frac{1}{|\coth z|}, \tag{5.267}$$

and from Equation 5.237 (Equation 5.243) follows the argument of the circular (Equation 5.268) [hyperbolic (Equation 5.269)] tangent:

$$\arg(\tan z) = \operatorname{arc}\tan\left(\frac{\sinh(2y)}{\sin(2x)}\right) = -\operatorname{arc}(\cot z), \tag{5.268}$$

$$\arg(\tanh z) = \operatorname{arc}\tan\left(\frac{\sin(2y)}{\sinh(2x)}\right) = -\operatorname{arc}(\coth z). \tag{5.269}$$

Thus have been obtained, *the modulus and argument of the circular (hyperbolic) cosine and secant [Equations 5.260 and 5.258 (Equations 5.261 and 5.259)], sine and cosecant [Equations 5.264 and 5.262 (Equations 5.265 and 5.263)], and tangent and cotangent [Equations 5.266 and 5.268 (Equations 5.267 and 5.269)].*

5.9 Periods, Symmetries, Values, and Limits

The circular (hyperbolic) functions have a real (imaginary) period and thus do (do not) appear to be periodic as functions of real variable (Subsection 5.9.1). The circular (hyperbolic) functions of complex variable can be calculated (Section 5.8) from those of real variable [Subsection 5.8.1 (Subsection 5.8.2)]. This gives more importance to the properties of circular and hyperbolic functions of real variable such as (1) symmetry and translation formulas reducing the width of the fundamental strip from 2π to $\pi/4$ (Subsection 5.9.1) or to $\pi/12$ (Subsection 5.9.3); (2) values for particular arguments (Subsection 5.9.2); (3) limits and inequalities (Subsection 5.9.4) for real and imaginary arguments; (4) inequalities for the real circular cosine and sine; and (5) modulus and inequalities for complex functions (Subsection 5.9.6).

5.9.1 Period, Symmetry, and Translation Formulas

Bearing in mind that the exponential (Equations 3.27a and 3.27b) has period $2\pi i$, it follows from Equation 5.1a (Equation 5.1b) that *the circular (hyperbolic) functions have period 2π ($2\pi i$):*

$$m \in |Z: \quad \operatorname{circ}(z) = \operatorname{circ}(z + 2\pi m), \qquad \operatorname{hyp}(z) = \operatorname{hyp}(z + m2\pi i), \tag{5.270a–5.270c}$$

that is, they are known from Equation 5.270b (Equation 5.270c) over the whole complex z plane, provided that [Figure 5.3a (Figure 5.3b)] they be specified in a **fundamental vertical (horizontal) strip** *[Equations 5.271a and 5.271b (Equations 5.271c and 5.271d)] with 2π:*

$$0 \le \operatorname{Im}(z) < 2\pi: \quad \operatorname{circ}(z) = \operatorname{circ}(z + m2\pi); \quad 0 \le \operatorname{Re}(z) < 2\pi: \quad \operatorname{hyp}(z) = \operatorname{hyp}(z + i\,rm\,2\pi). \tag{5.271a–5.271d}$$

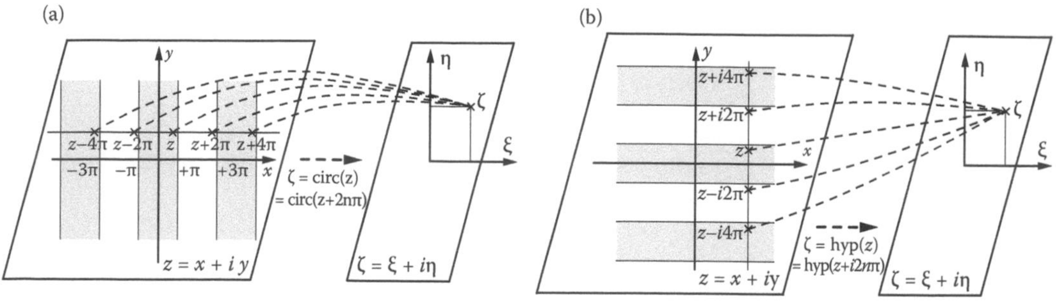

FIGURE 5.3
Circular (hyperbolic) functions are periodic with real (imaginary) period 2π ($2\pi i$), and thus repeat themselves [a (b)] on vertical (horizontal) strip of width 2π. This implies that the infinite set of points differing by $2\pi m$ ($i2\pi m$) with m integer is mapped to the same point. It also implies that the circular functions transform to hyperbolic functions (vice versa) by a rotation of $\pi/2$ ($-\pi/2$) in the positive (negative) direction, that is, counterclockwise (clockwise) in the complex plane.

The horizontal (vertical) fundamental strip of width 2π can be reduced to π, $\pi/2$, and $\pi/4$ by using the **symmetry or translation formulas** *in Table 5.3.* For example, for the circular sine: (1) it is defined in the whole plane (Equation 5.271a ≡ Equation 5.272a) if it is known on the vertical strip (Equations 5.272b and 5.272c) of width 2π:

$$m \in |Z: \quad 0 \le \mathrm{Im}\,(z) < 2\pi: \quad \sin z = \sin\,(z + 2m\pi), \tag{5.272a–5.272c}$$

$$0 \le \mathrm{Im}\,(z) < \pi: \quad \sin\,(z \pm \pi) = \sin z \cos \pi \pm \cos z \sin \pi = -\sin z, \tag{5.273a and 5.273b}$$

TABLE 5.3

Symmetry and Translation Relations for Circular and Hyperbolic Functions

z	$-z$	$\frac{\pi}{2} \pm z$	$\pi \pm z$	$\frac{3\pi}{2} \pm z$	$n\pi \pm z$
$\sin z$	$-\sin z$	$\cos z$	$\mp\sin z$	$-\cos z$	$\pm(-)^n \sin z$
$\cos z$	$\cos z$	$\mp\sin z$	$-\cos z$	$\pm\sin z$	$(-)^n \cos z$
$\tan z$	$-\tan z$	$\mp\cot z$	$\pm\tan z$	$\mp \cot z$	$\pm\tan z$
$\cot z$	$-\cot z$	$\mp\tan z$	$\pm\cot z$	$\mp \tan z$	$\pm\cot z$
$\sec z$	$\sec z$	$\mp\csc z$	$-\sec z$	$\pm\csc z$	$(-)^n \sec z$
$\csc z$	$-\csc z$	$\sec z$	$\mp\csc z$	$-\sec z$	$\pm(-)^n \csc z$
z	$-z$	$i\frac{\pi}{2} \pm z$	$i\pi \pm z$	$i\frac{3\pi}{2} \pm z$	$in\pi \pm z$
$\sinh z$	$-\sinh z$	$i \cosh z$	$\mp\sinh z$	$-i \cosh z$	$\pm(-)^n \sinh z$
$\cosh z$	$\cosh z$	$\pm i \sinh z$	$-\cosh z$	$\mp i \sinh z$	$(-)^n \cosh z$
$\tanh z$	$-\tanh z$	$\pm\coth z$	$\pm\tanh z$	$\pm\coth z$	$\pm\tanh z$
$\coth z$	$-\coth z$	$\pm\tanh z$	$\pm\coth z$	$\pm\tanh z$	$\pm\coth z$
$\mathrm{sech}\, z$	$\mathrm{sech}\, z$	$\mp i\, \mathrm{csch}\, z$	$-\mathrm{sech}\, z$	$\pm i\, \mathrm{csch}\, z$	$(-)^n \mathrm{sech}\, z$
$\mathrm{csch}\, z$	$-\mathrm{csch}\, z$	$-i\, \mathrm{sech}\, z$	$\mp\mathrm{csch}\, z$	$i\, \mathrm{sech}\, z$	$\pm(-)^n \mathrm{csch}\, z$

Note: Circular (hyperbolic) functions are defined in a vertical (horizontal) strip of width 2π in the complex plane [Figure 5.3a (Figure 5.3b)] and repeat themselves periodically in congruent parallel strips. The symmetry and translation formulas reduce the width of the strip from 2π by halving it two times $2\pi \to \pi \to \pi/2 \to \pi/4$ in this table, and even more to $\pi/6$, $\pi/8$, and $\pi/12$ in Table 5.5.

$$0 \le \operatorname{Im}(z) < \frac{\pi}{4}: \quad \sin\left(z \pm \frac{\pi}{2}\right) = \sin z \cos\left(\frac{\pi}{2}\right) \pm \sin\left(\frac{\pi}{2}\right)\cos z = \pm\cos z, \quad \text{(5.274a and 5.274b)}$$

$$0 \le \operatorname{Im}(z) < \frac{\pi}{4}: \quad \sin\left(z \pm \frac{\pi}{4}\right) = \sin z \cos\left(\frac{\pi}{4}\right) \pm \sin\left(\frac{\pi}{4}\right)\cos z = \frac{\sin z \pm \cos z}{\sqrt{2}}; \quad \text{(5.275a and 5.275b)}$$

(2–4) the symmetry or translation formulas (Equations 5.273b, 5.274b, and 5.275b) reduce the strip from 2π in Equation 5.272b to π in Equation 5.273a, $\pi/2$ in Equation 5.274a, and $\pi/4$ in Equation 5.275a. Similar results, for example:

$$\begin{aligned} \sinh\left(i\frac{\pi}{2} \pm z\right) &= \sinh\left[i\left(\frac{\pi}{2} \mp iz\right)\right] = i\sin\left(\frac{\pi}{2} \mp iz\right) \\ &= i\cos(iz) = i\cosh z, \end{aligned} \tag{5.276}$$

$$\begin{aligned} \sinh(i\pi \pm z) &= \sinh\left[i(\pi \mp iz)\right] = i\sin(\pi \mp iz) \\ &= \pm i\sin(iz) = \pm i^2 \sinh z = \mp\sinh z, \end{aligned} \tag{5.277}$$

$$\begin{aligned} \cosh\left(i\frac{3\pi}{2} \pm z\right) &= \cosh\left[i\left(\frac{3\pi}{2} \mp iz\right)\right] = \cos\left(\frac{3\pi}{2} \mp iz\right) \\ &= \mp\sin(iz) = \mp i\sinh z, \end{aligned} \tag{5.278}$$

apply to the other circular and hyperbolic functions in Table 5.3.

5.9.2 Values for Particular Real/Imaginary Arguments

Some of the values of the circular (hyperbolic) functions for particular variables in Table 5.4 were used in Equations 5.273b, 5.274b, and 5.275b. The latter correspond (1) geometrically to Figure 5.1a and c (Figure 5.1b and d); (2) graphically to the plots of circular (hyperbolic) functions of real variable [Figure 5.2a through c (Figure 5.2d through f)]; and (3) from the latter, the circular (hyperbolic) functions of complex variable [Figure 5.3a (Figure 5.3b)] can be calculated by Equations 5.233 through 5.238 (Equations 5.239 through 5.244). This gives more importance to the values of the circular (hyperbolic) functions for particular real (imaginary) values of the variable. The latter can be calculated from projections specifying their values at intervals of $30° = \pi/6$. The angle $\theta = 45°$, for argument $z = \pi/4$ along the bisector of the first quadrant, leads in Figure 5.1c to equal sides $A = B$, so that Pythagoras theorem $A^2 + B^2 = 1$ implies $A = 1/\sqrt{2} = B$. The case $\theta = 30°$, $z = \pi/6$ leads to an equilateral triangle by symmetry on the middle side, so that the smaller side is one-half of the largest $B = 1/2$, and thus, $A = \sqrt{1-2/4} = \sqrt{3}/2$; the values are interchanged for $\theta = 60°$, $z = \pi/3$. Table 5.4 summarizes the values of circular (hyperbolic) functions, for real (imaginary) variable including $\pm\infty$, and multiples of $\theta = 30°, 45°$ ($z = \pi/6, \pi/4$) in the range $0 \le \theta < 360°$ ($0 \le z < 2\pi$). The table specifies any circular (hyperbolic) function of real (imaginary ib) variable listed in the first (last) row, for example, Equation 5.5b implies Equations 5.279a through 5.279c:

TABLE 5.4

Particular Values of Circular and Hyperbolic Functions

z	$\sin z$	$\cos z$	$\tan z$	$\cot z$	$\sec z$	$\csc z$	θ
0	0	1	0	∞	1	∞	0°
$\frac{\pi}{6}$	$\frac{1}{2}$	$\frac{\sqrt{3}}{2}$	$\sqrt{3}$	$\frac{1}{\sqrt{3}}$	$\frac{2}{\sqrt{3}}$	2	30°
$\frac{\pi}{4}$	$\frac{1}{\sqrt{2}}$	$\frac{1}{\sqrt{2}}$	1	1	$\sqrt{2}$	$\sqrt{2}$	45°
$\frac{\pi}{3}$	$\frac{\sqrt{3}}{2}$	$\frac{1}{2}$	$\sqrt{3}$	$\frac{1}{\sqrt{3}}$	2	$\frac{2}{\sqrt{3}}$	60°
$\frac{\pi}{2}$	1	0	∞	0	∞	1	90°
$\frac{2\pi}{3}$	$\frac{\sqrt{3}}{2}$	$-\frac{1}{2}$	$-\sqrt{3}$	$-\frac{1}{\sqrt{3}}$	-2	$\frac{2}{\sqrt{3}}$	120°
$\frac{3\pi}{4}$	$\frac{1}{\sqrt{2}}$	$-\frac{1}{\sqrt{2}}$	-1	-1	$-\sqrt{2}$	$\sqrt{2}$	145°
$\frac{5\pi}{6}$	$\frac{1}{2}$	$-\frac{\sqrt{3}}{2}$	$-\frac{1}{\sqrt{3}}$	$-\sqrt{3}$	$-\frac{2}{\sqrt{3}}$	2	120°
π	0	-1	0	∞	-1	∞	180°
$\frac{7\pi}{4}$	$-\frac{1}{2}$	$-\frac{\sqrt{3}}{2}$	$\frac{1}{\sqrt{3}}$	$\sqrt{3}$	$-\frac{2}{\sqrt{3}}$	-2	210°
$\frac{5\pi}{4}$	$-\frac{1}{\sqrt{2}}$	$-\frac{1}{\sqrt{2}}$	1	1	$-\sqrt{2}$	$-\sqrt{2}$	225°
$\frac{4\pi}{3}$	$-\frac{\sqrt{3}}{2}$	$-\frac{1}{2}$	$\sqrt{3}$	$\frac{1}{\sqrt{3}}$	-2	$-\frac{2}{\sqrt{3}}$	240°
$\frac{3\pi}{2}$	-1	0	∞	0	∞	-1	270°
$\frac{5\pi}{3}$	$-\frac{\sqrt{3}}{2}$	$\frac{1}{2}$	$-\sqrt{3}$	$-\frac{1}{\sqrt{3}}$	2	$-\frac{2}{\sqrt{3}}$	300°
$\frac{7\pi}{4}$	$-\frac{1}{\sqrt{2}}$	$\frac{1}{\sqrt{2}}$	-1	-1	$\sqrt{2}$	$-\sqrt{2}$	315°
$\frac{11\pi}{6}$	$-\frac{1}{2}$	$\frac{\sqrt{3}}{2}$	$-\frac{1}{\sqrt{3}}$	$-\sqrt{3}$	$\frac{2}{\sqrt{3}}$	-2	330°
$-i\infty$	$-i\infty$	$+\infty$	$-i$	$+i$	0	0	–
$+i\infty$	$+i\infty$	$+\infty$	$+i$	$-i$	0	0	–
iz	$i \sinh z$	$\cosh z$	$i \tanh z$	$-i \coth z$	$\operatorname{sech} z$	$-i \operatorname{csch} z$	$-i\theta$

Note: Simplest 16 particular values of the six circular functions with arguments spaced by $\pi/4$ around the circle. They correspond to 16 values of the six hyperbolic functions for imaginary arguments. The values at infinity exist only along the real (imaginary) axis for hyperbolic (circular) functions.

$$\sin\left(\frac{\pi}{6}\right)=\frac{1}{2}, \quad -i\sinh\left(i\frac{\pi}{6}\right)=\frac{1}{2}, \quad \sinh\left(i\frac{\pi}{6}\right)=\frac{i}{2}, \tag{5.279a–5.279c}$$

$$\cosh(\pm\infty)=+\infty, \qquad \cos(\pm i\infty)=\cosh(\pm\infty)=+\infty, \tag{5.280a and 5.280b}$$

and, similarly, Equation 5.5a implies Equation 5.280a and 5.280b.

5.9.3 Reduction of the Width of the Fundamental Strip

The values on the whole complex plane, for example, for the sine can be calculated from those on the fundamental strip (Equations 5.272a through 5.272c) of width 2π. The translation formulas (Equations 5.273a, 5.273b, 5.274a, 5.274b, 5.275a, and 5.275b) halve the width of the strip three times to π, $\pi/2$, and $\pi/4$. Other translations concern $\pi/3$ ($\pi/6$) in Equation 5.281 (Equation 5.282):

$$\sin\left(z\pm\frac{\pi}{3}\right)=\sin z\cos\left(\frac{\pi}{3}\right)\pm\cos z\sin\left(\frac{\pi}{3}\right)=\frac{1}{2}\left(\sin z\pm\sqrt{3}\cos z\right), \tag{5.281}$$

$$\sin\left(z\pm\frac{\pi}{6}\right)=\sin z\cos\left(\frac{\pi}{6}\right)\pm\cos z\sin\left(\frac{\pi}{6}\right)=\frac{1}{2}\left(\pm\cos z+\sqrt{3}\sin z\right). \tag{5.282}$$

Equation 5.282 reduces the width of the strip to $\pi/6$, and it is reduced further to $\pi/8$ ($\pi/12$) in Equation 5.283 (Equation 5.284):

$$\sin\left(z\pm\frac{\pi}{8}\right)=\sin z\cos\left(\frac{\pi}{8}\right)\pm\cos z\sin\left(\frac{\pi}{8}\right)=a\sin z\pm b\cos z, \tag{5.283}$$

$$\sin\left(z+\frac{\pi}{12}\right)=\sin z\cos\left(\frac{\pi}{12}\right)\pm\cos z\sin\left(\frac{\pi}{12}\right)=c\sin z\pm d\cos z. \tag{5.284}$$

In Equation 5.283 (Equation 5.284), (a and b) [(c and d)] appear with the values [Equations 5.285a and 5.285b (Equations 5.286a and 5.286b)]:

$$a,b\equiv\cos,\sin\left(\frac{\pi}{8}\right)=\left|\frac{1}{2}\left[1\pm\cos\left(\frac{\pi}{4}\right)\right]\right|^{1/2}=\left|\frac{1}{2}\left(1\pm\frac{1}{\sqrt{2}}\right)\right|^{1/2}=0.92388,\,0.38268, \tag{5.285a and 5.285b}$$

$$c,d\equiv\cos,\sin\left(\frac{\pi}{12}\right)=\left|\frac{1}{2}\left|1\pm\cos\left(\frac{\pi}{6}\right)\right)\right|^{1/2}=\left|\frac{1}{2}\left(1\pm\frac{\sqrt{3}}{2}\right)\right|^{1/2}=0.96593,\,0.25882. \tag{5.286a and 5.286b}$$

The results (Equations 5.281 through 5.286) are included in the translation formulas by $\pm(\pi/3, \pi/4, \pi/6, \pi/8, \pi/12)$ [$\pm i\,(\pi/3, \pi/4, \pi/6, \pi/8, \pi/12)$] for all six circular (hyperbolic) functions in Table 5.5. These allow the calculation of circular and hyperbolic functions for all x from values in the interval $0\le x<\pi/12$. The values of the circular and hyperbolic functions

TABLE 5.5

Translation Relations for Circular and Hyperbolic Functions

z	$z\pm\frac{\pi}{3}$	$z\pm\frac{\pi}{4}$	$z\pm\frac{\pi}{6}$	$z\pm\frac{\pi}{8}$	$z\pm\frac{\pi}{12}$
$\sin z$	$\frac{\sin z\pm\sqrt{3}\cos z}{2}$	$\frac{\sin z\pm\cos z}{\sqrt{2}}$	$\frac{\sqrt{3}\sin z\pm\cos z}{2}$	$a\sin z\pm b\cos z$	$c\sin z\pm d\cos z$
$\cos z$	$\frac{\sqrt{3}\cos z\mp\sin z}{2}$	$\frac{\cos z\mp\sin z}{\sqrt{2}}$	$\frac{\cos z\mp\sqrt{3}\sin z}{2}$	$a\cos z\mp b\sin z$	$c\cos z\mp d\sin z$
$\tan z$	$\frac{\tan z\pm\sqrt{3}}{\sqrt{3}\mp\tan z}$	$\frac{\tan z\pm 1}{1\mp\tan z}$	$\frac{\sqrt{3}\tan z\pm 1}{1\mp\sqrt{3}\tan z}$	$\frac{a\tan z\pm b}{a\mp b\tan z}$	$\frac{c\tan z\pm d}{c\mp d\tan z}$
$\cot z$	$\frac{\sqrt{3}\cot z\mp 1}{1\pm\sqrt{3}\cot z}$	$\frac{\cot z\mp 1}{1\pm\cot z}$	$\frac{\cot z\mp\sqrt{3}}{\sqrt{3}\pm\cot z}$	$\frac{a\cot z\mp b}{a\pm b\cot z}$	$\frac{c\cot z\mp d}{c\pm d\cot z}$
$\sec z$	$\frac{2}{\sqrt{3}\cos z\mp\sin z}$	$\frac{\sqrt{2}}{\cos z\mp\sin z}$	$\frac{2}{\cos z\mp\sqrt{3}\sin z}$	$\frac{1}{a\cos z\mp b\sin z}$	$\frac{1}{c\cos z\mp d\sin z}$
$\csc z$	$\frac{2}{\sin z\pm\sqrt{3}\cos z}$	$\frac{\sqrt{2}}{\sin z\pm\cos z}$	$\frac{2}{\sqrt{3}\sin z\pm\cos z}$	$\frac{1}{a\sin z\pm b\cos z}$	$\frac{1}{c\sin z\pm d\cos z}$

z	$z \pm i\frac{\pi}{3}$	$z \pm i\frac{\pi}{4}$	$z \pm i\frac{\pi}{6}$	$z \pm i\frac{\pi}{8}$	$z \pm i\frac{\pi}{12}$
$\sinh z$	$\frac{\sinh z \pm i\sqrt{3}\cosh z}{2}$	$\frac{\sinh z \pm i\cosh z}{\sqrt{2}}$	$\frac{\sqrt{3}\sinh z \pm i\cosh z}{2}$	$a\sin z \pm i\,b\cosh z$	$c\sinh z \pm id\cosh z$
$\cosh z$	$\frac{\cosh z \mp i\sqrt{3}\sinh z}{2}$	$\frac{\cosh z \mp i\sinh z}{\sqrt{2}}$	$\frac{\sqrt{3}\cosh z \mp i\sinh z}{2}$	$a\cosh z \mp i\,b\sinh z$	$c\cosh z \mp i\,d\sinh z$
$\tanh z$	$\frac{\tanh z \pm i\sqrt{3}}{1 \mp i\sqrt{3}\tanh z}$	$\frac{\tanh z \pm i}{1 \mp i\tanh z}$	$\frac{\sqrt{3}\tanh z \pm i}{\sqrt{3} \mp i\tanh z}$	$\frac{a\tanh z \pm ib}{a \mp ib\tanh z}$	$\frac{c\tanh z \pm id}{c \mp id\tanh z}$
$\coth z$	$\frac{\coth z \pm i\sqrt{3}}{1 \pm i\sqrt{3}\coth z}$	$\frac{\coth z \mp i}{1 \pm i\coth z}$	$\frac{\sqrt{3}\coth z \mp i}{\sqrt{3} \pm i\coth z}$	$\frac{a\coth z \mp ib}{a \pm ib\coth z}$	$\frac{c\coth z \mp id}{c \pm id\coth z}$
$\operatorname{sech} z$	$\frac{2}{\cosh z \mp i\sqrt{3}\sinh z}$	$\frac{\sqrt{2}}{\cosh z \mp i\sinh z}$	$\frac{2}{\sqrt{3}\cosh z \mp i\sinh z}$	$\frac{1}{a\cosh z \mp ib\sinh z}$	$\frac{1}{c\cosh z \mp id\sinh z}$
$\operatorname{csch} z$	$\frac{2}{\sinh z \pm i\sqrt{3}\cosh z}$	$\frac{\sqrt{2}}{\sinh z \pm i\cosh z}$	$\frac{2}{\sqrt{3}\sinh z \pm i\cosh z}$	$\frac{1}{a\sinh z \pm ib\cosh z}$	$\frac{1}{c\sinh z \pm id\cosh z}$

$$a = \cos\left(\frac{\pi}{8}\right) = \sqrt{\frac{1}{2}+\frac{1}{2\sqrt{2}}} = 0.92388, \quad c \equiv \cos\left(\frac{\pi}{12}\right) = \sqrt{\frac{1}{2}+\frac{\sqrt{3}}{4}} = 0.96593,$$

$$b = \sin\left(\frac{\pi}{8}\right) = \sqrt{\frac{1}{2}-\frac{1}{2\sqrt{2}}} = 0.38268, \quad d \equiv \sin\left(\frac{\pi}{12}\right) = \sqrt{\frac{1}{2}-\frac{\sqrt{3}}{4}} = 0.25882.$$

Note: Translation formulas by $\pi/3$, $\pi/4$, $\pi/6$, $\pi/8$, and $\pi/12$ specify the circular (hyperbolic) functions in the whole complex plane from their values on a strip of width $\pi/12$. The values of the circular and hyperbolic functions for small arguments $|z| \le \pi/12 = 0.26180$ can be calculated accurately from the first few terms of suitable expansions in power series, series of fractions, infinite products, or continued fractions (Table 7.9).

for small variable $x < \pi/12 = 0.26180$ can be calculated with good accuracy from a small number of terms of their general representations by power series, series of fractions, infinite products, or continued fractions (Chapter 7).

5.9.4 Asymptotic Limits, Bounds, and Inequalities

Because the circular (hyperbolic) functions [Figure 5.2a (Figure 5.2b)] have a real (Equation 5.271a) [imaginary (Equation 5.271c)] fundamental period [Equation 5.271b (Equation 5.271d)], they appear (do not appear) as periodic functions of a real variable, and as a consequence, hold *the following* **limits:** *(1) the circular functions of a real variable (Figure 5.3a through c) have no limit as $x \to \pm\infty$; (2) the hyperbolic functions of a real variable (Figure 5.3d through f) have a limit as $x \to \pm\infty$ because (Equations 5.287a, 5.288a, and 5.289a) they are monotonic:*

$$\lim_{x\to\pm\infty}\left\{\cosh(x),\sinh(x)\right\}=\left\{\infty,\pm\infty\right\}=\lim_{y\to\pm\infty}\left\{\cos(iy),-i\sin(iy)\right\}, \qquad (5.287\text{a and }5.287\text{b})$$

$$\lim_{x\to\pm\infty}\left\{\text{sech}(z),\text{csch}(z)\right\}=0=\lim_{y\to\pm\infty}\left\{\sec(iy),i\csc(iy)\right\}, \qquad (5.288\text{a and }5.288\text{b})$$

$$\lim_{x\to\pm\infty}\left\{\tanh(z),\coth(z)\right\}=\pm1=\lim_{y\to\pm\infty}\left\{-i\tan(iy),i\cot(iy)\right\}; \qquad (5.289\text{a and }5.289\text{b})$$

(3) the circular functions have limit at infinity only along the imaginary axis (Equations 5.287b, 5.288b, and 5.289b) as follows from Equations 5.2a, 5.2b and; 5.20a through 5.20d and Equations 5.6a, 5.6b, 5.26a, and 5.26b. The real circular (hyperbolic) functions satisfy the **bounds** *[Equations 5.290a through 5.290d (Equations 5.291a through 5.291d)]:*

$$|\cos x| \le 1 \ge |\sin x|, \qquad |\sec x| \ge 1 \le |\csc x|, \qquad (5.290\text{a–}5.290\text{d})$$

$$\cosh x \ge 1 \ge \text{sech}\, x > 0, \qquad |\tanh x| \le 1 \le |\coth x|. \qquad (5.291\text{a–}5.291\text{d})$$

Concerning **inequalities**, *the real circular sine (tangent) lies below (Equation 5.292d) [above (Equation 5.292b)] the diagonal of the odd quadrants or tangent at the origin, where the relation [Equation 5.292d (5.292b)] holds for all x (Equation 5.292c) (up to $\pi/2$ in Equation 5.292a):*

$$|x|<\frac{\pi}{2}: \quad \frac{\tan x}{x}>1, \quad |x|<\infty: \quad \frac{\sin x}{x}<1; \qquad (5.292\text{a–}5.292\text{d})$$

$$|x|<\infty: \quad \frac{\sinh x}{x}>1>\frac{\tanh x}{x}, \quad 0<x<\frac{\pi}{2}: \quad \frac{2}{\pi}<\frac{\sin x}{x}. \qquad (5.293\text{a–}5.293\text{e})$$

The converse applies to the real hyperbolic sine (tangent) that lies above (below) the diagonal of the odd quadrants or tangent at the origin [Equation 5.293b (Equation 5.293c)] for all x. Besides, the circular sine has the (1) upper bound (Equation 5.292d) in the positive real line because it lies below its tangent at the origin, that is, the straight line of slope unity (Figure 5.2a); and (2) lower bound

(Equation 5.293e) because it lies above the straight line joining the origin to the first maximum $(\pi/2,1)$, hence with slope $2/\pi$ in the range (Equation 5.293d) as shown in Figure I.17.3.

5.9.5 Inequalities for the Real Circular Cosine and Sine

Besides Equations 5.290a, 5.290b, 5.292c, 5.292d, 5.293d and 5.293e, the circular cosine and sine satisfy the inequalities [Equations 5.295a through 5.295c (Equations 5.294d through 5.294f)]:

$$0 \le x \le \pi: \quad \cos x \le \frac{\sin x}{x} \le 1; \quad 0 < x < 1: \quad \pi < \frac{\sin(\pi x)}{x(1-x)} \le 4. \qquad (5.294a\text{–}5.294f)$$

The inequality (Equation 5.294c) corresponds to Equation 5.292d with the value unity attained as $x \to 0$. The inequality (Equation 5.294b) is established considering the function (Equation 5.295a):

$$f(x) = \sin x - x\cos x: \quad f(0) = 0; \quad 0 \le x \le \pi: f'(x) = x \sin x \ge 0; \quad f(x) \ge 0, \qquad (5.295a\text{–}5.295e)$$

that (1) vanishes at the origin (Equation 5.295b); and (2) has positive derivative (Equation 5.295d) in the interval (Equation 5.295c ≡ Equation 5.294a). Thus, the function (Equation 5.295a) is monotonic nondecreasing in the interval (Equation 5.294a), and because it is zero at the origin (Equation 5.295b), it is positive in the rest of the interval (Equation 5.295e), proving Equation 5.294b.

Concerning Equations 5.294d through 5.294f, the lower bound (Equation 5.294e) follows from the infinite product for the sine (Equation 1.148b ≡ Equation 5.296b):

$$0 < x < 1: \quad \frac{\sin(\pi x)}{\pi x} = \sum_{n=1}^{\infty}\left(1 - \frac{x^2}{n^2}\right) \ge 1 - x^2 = (1-x)(1+x) > 1 - x \qquad (5.296a \text{ and } 5.296b)$$

taken in the reduced interval (Equation 5.296a) instead of Equation 5.294a. The remaining inequality (Equation 5.294f) arises from the comparison of two functions. First, the circular sine (Equation 5.297b) vanishes at the ends (Equations 5.297c and 5.297d) of the interval (Equation 5.297a ≡ Equation 5.294d) and has the maximum value unity in the middle (Equation 5.297e):

$$0 < x < 1: \quad h(x) = \sin(\pi x), \quad h(0) = 0 = h(1), \quad [h(x)]_{\max} = h\left(\frac{1}{2}\right) = 1. \qquad (5.297a\text{–}5.297e)$$

Second, the comparison function (Equation 5.298a) (1) also vanishes (Equations 5.298b and 5.298c) at the ends of the interval (Equation 5.297a); (2) its derivative (Equation 5.298d) vanishes in the middle (Equation 5.298e):

$$g(x) = 4x(1-x): \quad g(0) = 0 = g(1), \quad g'(x) = 4(1-2x), \quad g'\left(\frac{1}{2}\right) = 0; \qquad (5.298a\text{–}5.298e)$$

and (3) hence it also has an extremum in the middle (Equation 5.299b) that is a maximum (Equation 5.299c) in the interval and is also equal to unity:

$$0 < x < 1: \quad g''(x) = -8 < 0, \quad \left[g(x)\right]_{max} = g\left(\frac{1}{2}\right) = 1. \qquad (5.299a–5.299c)$$

Both functions are symmetric:

$$g(x) = 4x(1 - x) = g(1 - x), \quad h(x) = \sin(\pi x) = \sin[\pi(1 - x)] = h(1 - x) \qquad (5.300a \text{ and } 5.300b)$$

relative to the middle of the interval.

Thus, if the functions cross inside the interval (Equation 5.294d), their difference (Equation 5.301) vanishes, and it must do so at least at two symmetric points x_1 and $\pi - x_1$ with $0 < x_1 < ½$:

$$j(x) = g(x) - h(x) = 4x(1 - x) - \sin(\pi x), \qquad (5.301)$$

Furthermore, the function (Equation 5.301) vanishes at $x = ½$, so it would have three distinct roots inside the interval (Equation 5.294d) plus two at the ends of the interval, for a total of five roots. This means that the derivative must change sign four times and must have at least three roots. However, the derivative (Equation 5.302a) of Equation 5.301 has at most one root (Equation 5.302b):

$$j'(x) = 4 - 8x - \pi\cos(\pi x) = 0 \quad \Leftrightarrow \quad \cos(\pi x) = \frac{4 - 8x}{\pi} \qquad (5.302a \text{ and } 5.302b)$$

because a straight line through the origin cannot cross cos (πx) more than once in the interval (Equation 5.294d). It follows that the derivative changes sign only once, and thus (Equation 5.301) can have only one root in the interval (Equation 5.294d); that root is $x = ½$. Thus, Equation 5.301 cannot vanish for $0 < x < ½$ $(1/2 < x < 1)$ and its sign can be determined at any point, such as $0 < x = 1/6 < ½$ $(1/2 < x = 2/3 < 1)$ in Equation 5.303a (Equation 5.303b):

$$j\left(\frac{1}{6}\right) = \frac{2}{3}\left(1 - \frac{1}{6}\right) - \frac{1}{2} = \frac{1}{18} > 0, \quad j\left(\frac{2}{3}\right) = \frac{8}{3}\left(1 - \frac{2}{3}\right) - \frac{1}{2} = \frac{7}{18} > 0. \qquad (5.303a \text{ and } 5.303b)$$

Hence, Equation 5.301 is positive (Equation 5.304b) over the whole interval (Equation 5.304a):

$$0 < x < 1: \quad j(x) = 4x(1 - x) - \sin(\pi x) > 0, \qquad (5.304a \text{ and } 5.304b)$$

proving Equation 5.294f.

5.9.6 Bounds and Inequalities for Complex Functions

From Equations 5.260 and 5.264 (Equations 5.261 and 5.265), it follows that *the modulus of the circular (hyperbolic) cosine and sine satisfies the inequalities [Equation 5.305 (Equation 5.306)] that imply [Equation 5.307 (Equation 5.308)]:*

$$\sinh |z| \geq |\sinh y| \leq |\cos z|, \ |\sin z| \leq \cosh y \leq \cosh |z|, \tag{5.305}$$

$$\sinh |z| \geq |\sinh x| \leq |\cosh z|, \ |\sinh z| \leq \cosh x \leq \cosh |z|, \tag{5.306}$$

$$\operatorname{csch} |z| \leq |\operatorname{csch} y| \geq |\sec z|, \ |\csc z| \geq \operatorname{sech} y \geq \operatorname{sech} |z|, \tag{5.307}$$

$$\operatorname{csch} |z| \leq |\operatorname{csch} x| \geq |\operatorname{sech} z|, \ |\operatorname{csch} z| \geq \operatorname{sech} x \geq \operatorname{sech} |z|. \tag{5.308}$$

In Equation 5.307 (Equation 5.308), the following was used:

$$|z| = |x + iy| = |x^2 + y^2|^{1/2} \geq |y|, \ |x|, \tag{5.309}$$

together with the monotonic increasing (decreasing) property of the hyperbolic cosine and sine (cosecant and secant) on the positive real line in Figure 5.2d and e. *The circular and hyperbolic functions of a complex variable satisfy besides the inequalities (Equations 5.305 through 5.308) also the upper bounds for the modulus of the hyperbolic cosine (Equation 5.110c) [sine (Equation 5.110b)] in the unit disk (Equation 5.310a):*

$$|z| < 1: \quad |\sin z|, |\sinh z| < \frac{6}{5}|z|, \quad |\cos z|, |\cosh z| \leq \cosh 1 = \frac{1}{2}\left(e + \frac{1}{e}\right) = 1.54308, \tag{5.310a–5.310c}$$

which are proved next.

The proof of Equation 5.310c is immediate from Equations 5.305 and 5.306:

$$|\cos z| \leq \cosh y \leq \cosh |z| \leq \cosh 1 \geq \cosh |z| \geq \cosh x \geq |\cosh z|. \tag{5.311}$$

The proof of Equation 5.310b uses the power series for the circular (Equation 5.312a ≡ Equation 1.31b) [hyperbolic (Equation 5.6b ≡ Equation 5.312b)] sine:

$$\sin z = z\sum_{n=0}^{\infty} \frac{(-z^2)^n}{(2n+1)!}, \quad \sinh z = -i\sin(iz) = z\sum_{n=0}^{\infty} \frac{z^{2n}}{(2n+1)!}. \tag{5.312a and 5.312b}$$

The modulus of the series cannot (Equations 5.312a and 5.312b) exceed the series of moduli:

$$|\sin z|, |\sinh z| \leq |z| + \frac{|z|^3}{3!} + \sum_{n=2}^{\infty} \frac{|z|^{2n+1}}{(2n+1)!} \leq |z| \left[1 + \frac{|z|^2}{6} + \sum_{n=2}^{\infty} \left(\frac{|z|^2}{6}\right)^n\right], \tag{5.313}$$

where the following was used:

$$n = 2, 3, \ldots: \quad \frac{6^n}{(2n+1)!} = \frac{6}{2n(2n+1)} \frac{6}{(2n-1)(2n-2)} \cdots \frac{6}{5.4} \frac{6}{3.2} < 1. \tag{5.314a and 5.314b}$$

Summing Equation 5.313 as a geometric series (Equations I.21.62a and I.21.62b ≡ Equations 3.148 and 3.148b) leads to

TABLE 5.6

Algebraic Relations for Circular/Hyperbolic Functions

Formula		Cosine	Sine	Tangent	Cotangent	Cosecant	Secant
I: Sum of arguments $z_1 + \ldots + z_N$	N terms	Equations 5.34a and 5.34c/5.35a and 5.35c	Equations 5.34b and 5.34c/5.35b and 5.35c	Equations 5.40 and 5.39a/5.41 and 5.39b	Equations 5.45 and 5.44a/5.46 and 5.44b	–	–
	$N = 2$: double	Equation 5.47a/5.49a	Equation 5.47b/5.49b	Equation 5.48a/5.50a	Equation 5.48b/5.50b	–	–
	$N = 3$: triple	Equation I.3.32a/10.83	Equation I.3.32b/10.84	–	–	–	–
	$N = 4$: quadruple	Equation I.10.24a/10.91	Equation I.10.24b/10.92	–	–	–	–
II: Multiple argument Nz	N times	Equation 5.53/5.57	Equation 5.54/5.58	Equation 5.55/5.59	Equation 5.56b/5.60	–	–
	$N = 2$: duplicating	Equations 5.61a through 5.61c/5.62a through 5.62c	Equation 5.63a/5.63b	Equations 5.64a through 5.64c/5.65a through 5.65c	Equations 5.64b through 5.64d/5.65b through 5.65d	–	–
	$N = 3$: triplication	Equation 5.124a and 5.124b/5.124c and 5.124d	Equation 5.123a/5.123b	Equations 10.89a/10.89b	Equations 10.90a/10.90b	–	–
	$N = 4$: quadruplicating	Equation I.10.25b/10.93a through 10.93c	Equation I.10.25b/10.94a through 10.94c	–	–	–	–
III: Half-argument $z/2$		Equation 5.66a/5.67a	Equation 5.66b/5.67b	Equations 5.68a through 5.68g/5.69a through 5.69g	Equations 5.68b through 5.68h/5.69b through 5.69h	–	–
IV: Product of two functions $\ldots x \ldots$		Equations 5.88a and 5.88c/5.89a and 5.89c	Equations 5.88b and 5.88c/5.89b and 5.89c	–	–	–	–
V: Sum of two functions $\ldots x \ldots$		Equations 5.91a and 5.91b/5.92a and 5.92b	Equations 5.91c and 5.91d/5.92c and 5.92d	–	–	–	–
VI: Difference of squares		Equations 5.97 and 5.98b/5.99a and 5.99b; 5.100a	Equations 5.97 and 5.98a/5.99a and 5.99b; 5.100b	–	–	–	–

VII: Powers of functions	Even powers $2N$	Equation 5.78a/5.80a	Equation 5.79a/5.81a	–	–	–	–
	Odd power $2N + 1$	Equation 5.78b/5.80b	Equation 5.79b/5.81b	–	–	–	–
	Square: 2	Equation 5.82a/5.83a	Equation 5.82b/5.83b	–	–	–	–
	Cube: 3	Equation 5.84a/5.85a	Equation 5.84b/5.85b	–	–	–	–
VIII: Multiple argument: in powers of cosines	Multiple argument: Nz	Equation 5.110/5.110	Equation 5.109/5.109	–	–	–	–
	Even multiple argument: $2Nz$	Equation 5.113/5.113	Equation 5.111/5.111	–	–	–	–
	Odd multiple argument: $(2N + 1)z$	Equation 5.114/5.114	Equation 5.112/5.112	–	–	–	–
	Duplicating: $2z$	Equation 5.124a/5.124c	Equation 5.123a/5.123b	–	–	–	–
	Triplicating: $3z$	Equation 5.126a/5.126b	Equation 5.125a/5.125c	–	–	–	–
IX: Multiple argument: in powers of sines	Multiple argument: Nz	–	–	–	–	–	–
	Even multiple argument: $2Nz$	Equation 5.117/5.121	Equation 5.115/5.119	–	–	–	–
	Odd multiple argument: $(2N + 1)z$	Equation 5.116/5.120	Equation 5.118/5.122	–	–	–	–
	Duplicating: $2z$	Equation 5.124b/5.124d	Equation 5.123a/5.123b	–	–	–	–
	Triplicating: $3z$	Equation 5.126b/5.126d	Equation 5.125b/5.125d	–	–	–	–
X: Fourier series for powers	Even	Equation 5.197	Equation 5.196	–	–	–	–
	Odd	Equation 5.198	Equation 5.195b	–	–	–	–
XI: Definition in terms of exponentials		Equation 5.3a/5.2a	Equation 5.3b/5.2b	Equation 5.21c/5.20c	Equation 5.21d/5.20d	Equation 5.21b/5.20b	Equation 5.21a/5.20a
XII: Imaginary change of variable		Equation 5.5a/5.6a	Equation 5.5b/5.6b	Equation 5.26a/5.26b	Equation 5.27a/5.27b	Equation 5.25a/5.25b	Equation 5.24a/5.24b

(continued)

TABLE 5.6 (Continued)

Algebraic Relations for Circular/Hyperbolic Functions

Formula		Cosine	Sine	Tangent	Cotangent	Cosecant	Secant
XIII: Complex functions	Real and imaginary parts	Equation 5.233/5.239	Equation 5.234/5.240	Equation 5.237/5.243	Equation 5.238/5.244	Equation 5.236/5.242	Equation 5.235/5.241
	Modulus	Equation 5.260/5.261	Equation 5.264/5.265	Equation 5.266/5.267	Equation 5.266/5.267	Equation 5.264/5.265	Equation 5.260/5.261
	Argument	Equation 5.258/5.259	Equation 5.262/5.263	Equation 5.268/5.269	Equation 5.268/5.269	Equation 5.262/5.263	Equation 5.258/5.259
XIV: Limits		Equation 5.287b/5.287a	Equation 5.287b/5.287a	Equation 5.289b/5.289a	Equation 5.289b/5.289a	Equation 5.288b/5.288a	Equation 5.288b/5.288a
XV: Real inequalities		Equations 5.290a; 5.294a through 5.294c/5.291a	Equations 5.290b; 5.292c and 5.292d; 5.293d and 5.293e; 5.294a through 5.294f/5.293a and 5.293b	Equation 5.292a and 5.292b/ 5.291c; 5.293a and 5.293c	–/Equation 5.291d	Equation 5.290d	Equation 5.290c/5.291b
XVI: Complex inequalities		Equations 5.305; 5.310a and 5.310c/5.306; 5.310a and 5.310c	Equations 5.305; 5.310a and 5.310b/5.306; 5.310a and 5.310b	–	–	Equation 5.307/5.308	Equation 5.307/5.308

Note: Algebraic relations between the circular and hyperbolic functions include (1) the addition formulas for the sum of several arguments, for example, two, three, or four; (2) if the arguments in item 1 are equal, the function for multiple arguments, for example, the duplicating, triplicating, and quadruplicating formulas; (3) the inverse of item 2, that is, the formulas for half-arguments; (4) the double summation formulas item 1 lead to the formulas for the product of two functions; (5) from item 4 follow the formulas for the sum of two cosines or sines; (6) from item 2 also follow formulas for the difference of squares of cosines and sines; (7) an extension of item 4 are the formulas for powers of functions in terms of functions of multiple arguments; (8 and 9) the inverse of item 6 is the function of multiple arguments expressed as sum of products of powers of cosines and sines (2), using alternatively only cosines (8) or sines (9); (10) an alternative to item 7 is the expansion of even or odd powers of cosines and sines in Fourier sine or cosine series; (11 and 12) the definition of the circular and hyperbolic functions in terms of exponentials (12) leads to the formulas for the imaginary change of variable (12); (13) combining the latter (12) with the formulas for the sum of two arguments specifies the complex functions, that is, their real and imaginary and modulus and argument parts in terms of the real functions; (14–16) the geometric properties (Figure 5.1), plots (Figure 5.2), and maps (Figure 5.3) lead to limits (14) and inequalities for (15) real and (16) complex functions. Besides these formulas listed in this table, other algebraic results include (17) values for particular arguments (Table 5.4) that lead (18) [(19)] to the translation formulas [Table 5.3 (Table 5.5)], reducing the width of the fundamental strip by $2\pi \to \pi \to \pi/2$ ($\pi/3 \to \pi/4 \to \pi/6 \to \pi/8 \to \pi/12$). The analytic properties of the circular and hyperbolic functions appear in Tables 7.7 through 7.9.

$$\left|\sin z\right| \le \left|z\right| \sum_{n=0}^{\infty} \left(\frac{\left|z\right|}{6}\right)^n = \frac{\left|z\right|}{1-\left|z\right|2/6} \le \frac{\left|z\right|}{1-1/6} = \frac{6}{5}\left|z\right| < \frac{6}{5}, \tag{5.315}$$

that proves Equation 5.310b in the unity disk (Equation 5.310a).

NOTE 5.1 ELEMENTARY REAL AND COMPLEX FUNCTIONS

Although the circular and hyperbolic functions appear to be different as real functions, as complex functions, they correspond to rotations by $\pi/2$ in the plane of the independent and/or dependent variable. As complex functions, the circular and hyperbolic functions are all related to the exponential; for example, the complex exponential is calculated from the real exponential and real circular functions. The circular (hyperbolic) functions are geometrically related to the circle/ellipse (hyperbola), which can serve to illustrate their properties. One of the reasons for the importance of the circular and hyperbolic functions is that they provide the general solutions of linear differential and integral equations with constant coefficients. The elementary transcendental functions can be used to obtain the general integral of linear homogeneous ordinary (partial) differential equations with constant coefficients. The corresponding forced ordinary (partial) differential equations can be solved by methods like variation of constants or using Fourier series (Fourier or Laplace transforms), which are series expansions (parametric integrals) again using elementary functions. Some of the algebraic properties of circular and hyperbolic functions mentioned (Chapter 5) are quite elementary and are summarized in Table 5.6. Other analytic properties, like power series (Chapters I.23, I.25, I.27, and I.29), infinite products, series of fractions, and continued fractions (Chapter 1), are less immediate (Chapter 7), although they are the simplest results of their kind and are summarized in Tables 7.7 through 7.9. Collecting together all the main algebraic (analytic) properties [Table 5.6 (Table 7.7 through 7.9)] of the circular and hyperbolic functions, elementary or not, is the starting point to their subsequent use in connection with generalized functions, differential equations, Fourier series, and integral transforms besides numerical methods.

5.10 Conclusion

The horizontal (vertical) Cartesian coordinate of point on an ellipse (hyperbola) of half-axes a and b (maximum curvature at $x = \pm a$ and asymptotes of slope $\pm b/a$) is proportional [Figure 5.1a (Figure 5.1b)], respectively, to the circular (hyperbolic) cosine and sine, and coincides with it the case of the unit circle (equilateral hyperbola), from which the values of the cosine A and sine B, which are the primary circular (hyperbolic) functions, can be read [Figure 5.1c (Figure 5.1d)]. They are plotted for real variable, together with the secondary circular (hyperbolic) functions, in inverse pairs, namely, the circular (hyperbolic): (1) sine B and cosecant $1/B$ in Figure 5.3a (Figure 5.3d); (2) cosine A and secant $1/A$ in Figure 5.3b (Figure 5.3e); and (3) tangent B/A and cotangent A/B in Figure 5.3c (Figure 5.3f). The secant $1/A$ and cosecant $1/B$ are obtained by inversion, and the tangent B/A and cotangent A/B are their ratios, as indicated in Table 5.1 (Table 5.2) for the circular (hyperbolic) functions. Some of the values of circular (hyperbolic) functions can be read from Figure 5.1a and c (Figure 5.1b and d), the eclipse (hyperbola) in Figure 5.1c (Figure 5.1d), or the plots in

Figure 5.3a through c (Figure 5.3d through f) and are indicated in Table 5.4. Although the circular (hyperbolic) functions appear distinct for real variable in Figure 5.2a through c (Figure 5.2d through f), as functions of a complex variable [Figure 5.3a (Figure 5.3b)], they coincide by a right-hand rotation of the complex plane. This explains the similarities in their properties, for example, with regard to the symmetries and translations in Tables 5.3 and 5.5, which allow the calculation of circular (hyperbolic) function in the whole complex plane from the values in a strip of progressively reducing width $2\pi \to \pi \to \pi/2 \to \pi/3 \to \pi/4 \to \pi/6 \to \pi/8 \to \pi/12$. The algebraic properties of the circular and hyperbolic functions are summarized in Table 5.6, which is complemented by Tables 7.7 through 7.9 for the analytic properties.

6

Membranes, Capillarity, and Torsion

A special case of elasticity with two independent variables that is not plane elasticity (Chapter 4) is a membrane. If the boundary curve is a plane, the membrane remains a plane under tension, unless a transverse force is applied, for example, its own weight; the transverse force deflects the membrane until it is balanced by the transverse component of the tension, so that the deflection is linear (nonlinear) for small (large) slope [Section 6.1 (Section 6.2)]. The surface tension acts as a membrane, since it is associated with pressure or stress differences across an interface between two fluids (Section 6.3); examples are capillary effects (Section 6.4) such as a liquid wetting a wall at an angle or the rise in the free surface of a liquid between walls. A third less obvious analogy is the torsion of a slender straight rod: the cross section is similar to a membrane (Section 6.5), and the moment applied at the ends of the rod causes a relative rotation of successive cross sections (Section 6.6). The cross section of the bar under torsion can be singly or multiply connected; for example, a triangular prism (Section 6.8) has a simply connected cross section, whereas a hollow ellipse (Section 6.7) has a doubly connected cross section. A fourth analogous problem is a potential flow in a rotating vessel (Section 6.9) where the centrifugal energy adds a rotational component that affects the trajectories of fluid particles. Eight more analogies (Notes 6.6 through 6.9) lead to a total of 12.

6.1 Linear and Nonlinear Deflection of a Membrane

A membrane is a thin elastic body that can be assimilated to a surface and is able to support an isotropic tension (Figure 6.1). If the boundary ∂D is a plane curve, the tension maintains the membrane flat on the same plane. If a transverse force per unit area or shear stress is applied, the membrane deflects to a shape such that a transverse component of the tension balances the force; equivalently, the work done by the external force in the deflection is balanced by the elastic energy corresponding to the increase in area under tension (Subsection 6.1.1). This leads to the membrane deflection equation both in the linear (nonlinear) case of small (large) slope [Subsection 6.1.2 (Subsection 6.1.3)]. The corresponding linear and nonlinear membrane equations can be solved for arbitrary radially symmetric loading (Subsection 6.1.4). This specifies, for example, the linear deflection of a homogeneous membrane under constant tension by its own weight in a uniform gravity field for circular (annular) horizontal supports [Subsection 6.1.5 (Subsection 6.1.6)].

6.1.1 Elastic Energy of a Deflected Membrane under Tension

The work done by the force per unit area or shear stress in a deflection is given by Equation 6.1b:

$$\delta W_d = \delta E_d: \quad \delta W_d = \int_D f\,\delta\zeta\,dx\,dy, \quad \delta E_d = \int_D T\delta\left(dS - dx\,dy\right), \qquad (6.1a\text{–}6.1c)$$

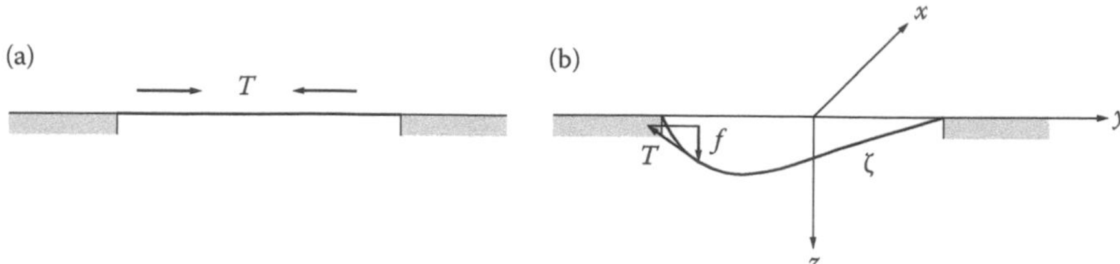

FIGURE 6.1
Membrane is a thin elastic body subject to a tangential tension T. If it is held along a fixed horizontal boundary curve, the tangential tension leads to the minimum area, and the interior of the membrane remains flat (a). If a transverse force per unit area or shear stress f is applied in the interior of the membrane, the latter is deflected (b) until the vertical component of the tangential tension balances the external loading due to the applied shear stress.

and it is balanced (Equation 6.1a) by the elastic energy that equals (Equation 6.1c) the tension times the area change. Let Equation 6.2a be the equation of the deflected membrane:

$$0 = z - \zeta(x,y) \equiv \Phi(x,y,z): \quad 0 = d\Phi = dz - \frac{\partial\zeta}{\partial x}dx_\Phi - \frac{\partial\zeta}{\partial y}dy_\Phi \equiv \vec{M}\cdot d\vec{r}_\Phi, \qquad (6.2a \text{ and } 6.2b)$$

so that an arbitrary displacement along the membrane satisfies Equation 6.2b; vector $\vec{M}$ in Equation 6.2b ($\equiv$ Equation 6.3a) is orthogonal to every displacement $d\vec{r}_\Phi$ on the membrane, so it is a **normal vector**:

$$\vec{M} = \left\{-\frac{\partial\zeta}{\partial x}, -\frac{\partial\zeta}{\partial y}, 1\right\}; \quad \vec{N} \equiv \frac{\vec{M}}{\left|\vec{M}\right|} = \left|1 + (\partial_x\zeta)^2 + (\partial_y\zeta)^2\right|^{-1/2}\left\{-\partial_x\zeta, -\partial_y\zeta, 1\right\}. \qquad (6.3a \text{ and } 6.3b)$$

The **unit normal** vector is obtained (Equation 6.3b) by dividing the normal vector by its modulus. The area of the undeflected membrane is $dxdy$, and when deflected, its area increases (Figure 6.2) by the factor $1/N_z$ in Equation 6.3b relative to its horizontal projection, leading to

$$dS = \frac{dxdy}{N_z} = \left|1 + \left|\nabla\zeta\right|^2\right|^{1/2} dxdy, \qquad (6.4a)$$

where Equation 6.4b is the two-dimensional gradient, and Equation 6.4c is the square of its modulus, that is, the inner product by itself:

$$\nabla\zeta \equiv \vec{e}_x\partial_x\zeta + \vec{e}_y\partial_y\zeta, \quad \left|\nabla\zeta\right|^2 \equiv (\nabla\zeta\cdot\nabla\zeta) = (\partial_x\zeta)^2 + (\partial_y\zeta)^2. \qquad (6.4b \text{ and } 6.4c)$$

The exact nonlinear area change of the deflected membrane is given by Equation 6.5a:

$$dS - dx\,dy = \left\{\left|1 + \left|\nabla\zeta\right|^2\right|^{1/2} - 1\right\}dx\,dy = \frac{1}{2}\left|\nabla\zeta\right|^2 dx\,dy\left\{1 + O\left|\nabla\zeta\right|^2\right\} \qquad (6.5a \text{ and } 6.5b)$$

that simplifies to Equation 6.5b in the linear case of a small slope.

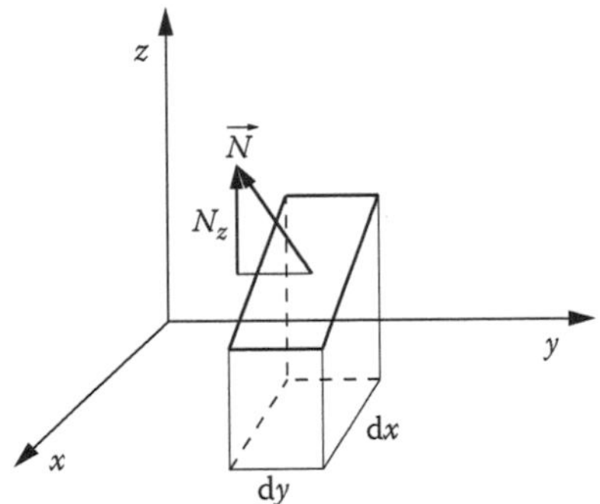

FIGURE 6.2
Infinitesimal area element dS of a surface is a facet projected on a horizontal infinitesimal rectangle dxdy with an orientation specified by the unit normal vector $\vec{N}$. Its vertical component N_z, that is, orthogonal to the (x, y)-plane, corresponds to the direction cos θ, that is, the cosine of the angle between the unit normal to the surface $\vec{N}$ and the unit normal $\vec{e}_z$ to the (x, y) plane: $\vec{N} \cdot e_z = \cos\theta$. The direction cosine specifies the ratio of areas $\mathrm{d}x\,\mathrm{d}y = \cos\theta\,\mathrm{d}S = N_z\,\mathrm{d}S$. This relation fails if $N_z = 0$, that is, the facet has a horizontal unit normal $\theta = \pi/2$ and projects as a line on the (x, y)-plane. In the latter case, (x, z)- or the (y, z)-planes can be used to project the area element of the surface. Since $\vec{N}$ is a unit vector, not more than two of the projections in the (x, y)-, (x, z)-, and (y, z)-planes can vanish.

6.1.2 Balance with the Work of the External Shear Stress

The elastic energy associated with the tension (Equation 6.1c) involves the variation of the area change (Equations 6.5a and 6.5b) that is considered first (Equation 6.6b) in the linear case of a small slope (Equation 6.6a):

$$|\nabla\zeta|^2 << 1: \quad T\delta(\mathrm{d}S - \mathrm{d}x\,\mathrm{d}y) = T\delta\left[\frac{1}{2}(\nabla\zeta\cdot\nabla\zeta)\right] = T\nabla\zeta\cdot\delta(\nabla\zeta) = T(\nabla\zeta\cdot\nabla)(\delta\zeta). \quad \text{(6.6a and 6.6b)}$$

The identity

$$\partial_x (B_x\,\Phi) + \partial_y (B_y\,\Phi) = \Phi\,(\partial_x B_x + \partial_y B_y) + B_x\,\partial_x\,\Phi + B_y\,\partial_y\,\Phi, \quad \text{(6.7a)}$$

is equivalent to

$$\nabla\cdot(\Phi\vec{B}) = \Phi(\nabla\cdot\vec{B}) + \vec{B}\cdot\nabla\Phi; \quad \text{(6.7b)}$$

it is applied to Equation 6.6b with Equations 6.8a and 6.8b leading to Equation 6.8c:

$$\vec{B} \equiv T\nabla\zeta, \Phi \equiv \delta\zeta: \quad \nabla\cdot\left[(\delta\zeta)T\nabla\zeta\right] = (\delta\zeta)\nabla\cdot(T\nabla\zeta) + T(\nabla\zeta\cdot\nabla)(\delta\zeta). \quad \text{(6.8a–6.8c)}$$

Substituting Equations 6.6b and 6.8c in the elastic energy of tension (Equation 6.1c) in the linear case (Equation 6.6a ≡ Equation 6.9b) leads to Equation 6.9c:

$$\begin{aligned} \delta T = 0, \quad |\nabla\zeta|^2 << 1: \quad \delta E_d &= \int_D T\left[(\nabla\zeta\cdot\nabla)(\delta\zeta)\right]\mathrm{d}x\,\mathrm{d}y \\ &= \int_D \left\{-(\delta\zeta)\left[\nabla\cdot(T\delta\zeta)\right] + \nabla\cdot\left[(\delta\zeta)T\nabla\zeta\right]\right\}\mathrm{d}x\,\mathrm{d}y, \end{aligned} \quad \text{(6.9a–6.9c)}$$

where the tension is not varied (Equation 6.9a) and need not be uniform. Using the divergence theorem (Equation I.28.1), the second integral becomes an integral over the boundary (Equation 6.10b):

$$\delta\zeta\big|_{\partial D} = 0: \quad \int_D \nabla\cdot\left[T(\delta\zeta)\nabla\zeta\right]\mathrm{d}x\,\mathrm{d}y = \int_{\partial D} T\frac{\partial\zeta}{\partial n}(\delta\zeta)\,\mathrm{d}s = 0 \qquad \text{(6.10a and 6.10b)}$$

that vanishes because the membrane is fixed at the boundary, where there can be no variation (Equation 6.10a). The remaining term, that is, the first on the right-hand side (r.h.s.) of Equation 6.9c, specifies the elastic energy of tension; it equals (Equation 6.1a) the work of the shear stress (Equation 6.1b) leading to Equation 6.11b:

$$\left|\nabla\zeta\right|^2 << 1: \quad 0 = \int_D \left[f + \nabla\cdot\left(T\nabla\zeta\right)\right]\left(\delta\zeta\right)\mathrm{d}x\,\mathrm{d}y; \qquad \text{(6.11a and 6.11b)}$$

Equation 6.11b is satisfied for arbitrary deflection $\delta\zeta$ if the term in square brackets vanishes:

$$|\nabla\zeta|^2 << 1: \quad -f = \nabla\cdot(T\nabla\zeta) = T\nabla^2\zeta + \nabla T\cdot\nabla\zeta, \qquad \text{(6.12a and 6.12b)}$$

where Equation 6.7b was used in Equation 6.12b in the case of nonuniform tension. For uniform tension (Equation 6.13b), Equation 6.12b simplifies to a Poisson equation (Equation 6.13c):

$$\left|\nabla\zeta\right|^2 << 1; \quad T = \text{const:} \quad -\frac{f}{T} = \nabla^2\zeta. \qquad \text{(6.13a–6.13c)}$$

It has been shown that *the linear deflection of a membrane, that is, with a small slope (Equation 6.12a), is specified by Equation 6.12b where the externally applied transverse force per unit area or shear stress f and the tangential tension T may be nonuniform; if the tension is uniform (Equation 6.13b), the deflection of a membrane is specified by the Poisson equation 6.13c, forced by minus the ratio of the shear stress to the tension.*

6.1.3 Large Slope and Nonlinear Deflection

If the slope is not small somewhere, then the approximation (Equation 6.5b) cannot be made, and the variation is applied to the exact area change (Equation 6.5a):

$$T\delta(\mathrm{d}S - \mathrm{d}x\,\mathrm{d}y) = T\delta\left\{\left|1+(\nabla\zeta\cdot\nabla\zeta)\right|^{1/2} - 1\right\} = T\left|1+\left|\nabla\zeta\right|^2\right|^{-1/2}\nabla\zeta\cdot\nabla\left(\delta\zeta\right). \qquad (6.14)$$

The identity (Equation 6.7b) again applies to the vector (Equation 6.15a), leading to Equation 6.15b:

$$\vec{B} \equiv T\left|1+\left|\nabla\zeta\right|^2\right|^{-1/2}\nabla\zeta: \quad \nabla\cdot\left[(\delta\zeta)\vec{B}\right] = \left(\delta\zeta\right)(\nabla\cdot\vec{B}) + (\vec{B}\cdot\nabla)\left(\delta\zeta\right); \qquad \text{(6.15a and 6.15b)}$$

note that Equations 6.15a and 6.15b reduce to Equations 6.8a and 6.8c in the linear case (Equation 6.6a). In the nonlinear case, Equation 6.14 may be substituted in Equation 6.1c to specify the exact elastic energy of tension:

$$\delta E_d = \int_D \left[(\vec{B}\cdot\nabla)(\delta\zeta)\right] \mathrm{d}x\,\mathrm{d}y = \int_D \left\{-(\delta\zeta)[\nabla\cdot\vec{B}] + \nabla\cdot\left[\vec{B}(\delta\zeta)\right]\right\} \mathrm{d}x\,\mathrm{d}y. \tag{6.16}$$

The divergence theorem (Equation I.28.1) again transforms the second term on the r.h.s. into a boundary integral (Equation 6.17b):

$$\delta\zeta|_{\partial D} = 0: \quad \int_D \nabla\cdot\left[\vec{B}(\delta\zeta)\right] \mathrm{d}x\,\mathrm{d}y = \int_{\partial D} B_n(\delta\zeta)\,\mathrm{d}s = \int_{\partial D} T(\delta\zeta)\left|1+|\nabla\zeta|^2\right|^{-1/2} \frac{\partial\zeta}{\partial n}\,\mathrm{d}s = 0 \tag{6.17a and 6.17b}$$

that vanishes for the same reason as before (Equation 6.10a ≡ Equation 6.17a), that is, the deflection vanishes on the boundary.

The remaining first term on the r.h.s. of Equation 6.16 specifies the elastic energy (Equation 6.1c); equating Equation 6.1a to the work of the shear stress (Equation 6.1b) leads to Equation 6.18a:

$$0 = \int_D \left[f + (\nabla\cdot\vec{B})\right](\delta\zeta)\,\mathrm{d}x\,\mathrm{d}y, \quad -f = \nabla\cdot\vec{B} = \nabla\cdot\left\{T\left|1+(\nabla\zeta\cdot\nabla\zeta)\right|^{-1/2}\nabla\zeta\right\}, \tag{6.18a and 6.18b}$$

and hence for arbitrary displacement to Equation 6.18b. The latter reduces to the Poisson Equation 6.13b in the linear case (Equation 6.13a) with constant tension. *In the nonlinear case of an arbitrary slope, the deflection of the membrane under constant tension (Equation 6.19a) is still forced by the ratio of the shear stress to the tension, but the Laplace operator is replaced by the* **minimal surface operator** *(Equation 6.19b):*

$$T = \text{const}: \quad -\frac{f}{T} = \nabla\cdot\left\{\left|1+(\nabla\zeta\cdot\nabla\zeta)\right|^{-1/2}\nabla\zeta\right\} = \left|1+|\nabla\zeta|^2\right|^{-1/2}\nabla^2\zeta - \frac{1}{2}\left|1+|\nabla\zeta|^2\right|^{-3/2}\left\{(\nabla\zeta\cdot\nabla)\left[(\nabla\zeta\cdot\nabla\zeta)\right]\right\}, \tag{6.19a and 6.19b}$$

or equivalently using the two-dimensional gradient (Equation 6.4b):

$$T = \text{const}: \quad -\frac{f}{T}\left|1+|\nabla\zeta|^2\right|^{3/2} = \left[1+|\nabla\zeta|^2\right]\nabla^2\zeta - \nabla\zeta\cdot\left[(\nabla\zeta\cdot\nabla)\nabla\zeta\right]. \tag{6.20a and 6.20b}$$

In the most general case of nonuniform tension and large slope (Equation 6.18b ≡ Equation 6.21):

$$-f = T\nabla\cdot\left[\left|1+|\nabla\zeta|^2\right|^{-1/2}\nabla\zeta\right] + (\nabla T\cdot\nabla\zeta)\left|1+|\nabla\zeta|^2\right|^{-1/2}, \tag{6.21}$$

the minimal surface operator (Equation 6.19a) appears in the first term on the r.h.s. of Equation 6.21, and the second term on the r.h.s. of Equation 6.21 specifies the effect of the nonuniform tension. All terms appear in

$$-f\left|1+|\nabla\zeta|^2\right|^{3/2}+T\nabla\zeta\cdot\left[(\nabla\zeta\cdot\nabla)\nabla\zeta\right]=\left[1+|\nabla\zeta|^2\right]\nabla\cdot(T\nabla\zeta)$$
$$=\left[1+|\nabla\zeta|^2\right]\left(T\nabla^2\zeta+\nabla T\cdot\nabla\zeta\right). \tag{6.22a and 6.22b}$$

Thus, the shape of a membrane under an externally applied transverse force per unit area or shear stress f (1) is specified by Equation 6.18b ≡ Equation 6.21 ≡ Equation 6.22a ≡ Equation 6.22b) in the general case of nonuniform tangential stress T and a large slope; (2 and 3) it simplifies to Equation 6.19b ≡ Equation 6.20b (Equation 6.12b) for uniform tension (Equation 6.19a ≡ Equation 6.20a) [(small slope (Equation 6.12a ≡ Equation 6.6a))]; and (4) the simplest case combining the small slope (Equation 6.13a) and the uniform tension (Equation 6.13b) is the Poisson Equation 6.13c forced by minus the ratio of the shear stress to the tension.

6.1.4 Linear/Nonlinear Radially Symmetric Deflection

If the applied transverse external force per unit area or shear stress is radially symmetric, that is, in polar coordinates, does not depend on the direction but only on the radius, the same applies to the deflection (Equation 6.23a) in the gradient (Equation 4.46a ≡ Equation I.11.31b) and divergence (Equation 4.30 ≡ Equation I.11.33b) operators [Equation 6.23b (Equation 6.23c)]:

$$\zeta'\equiv\frac{d\zeta}{dr},\quad \nabla=\vec{e}_r\frac{d}{dr},\quad \nabla\cdot=\frac{1}{r}\frac{d}{dr}r; \tag{6.23a–6.23c}$$

this leads to the Laplacian (Equation 6.24a). The divergence (Equation 6.23c) and gradient (Equation 6.23b) operators appear in the deflection equation 6.18b in the general case of large slope and nonuniform tension (Equation 6.24b):

$$\nabla^2=\nabla\cdot\nabla=\frac{1}{r}\frac{d}{dr}r\frac{d}{dr},\quad -f(r)=\frac{1}{r}\frac{d}{dr}\left[\frac{\zeta' rT(r)}{\sqrt{1+\zeta'^2}}\right]. \tag{6.24a and 6.24b}$$

A first integration of Equation 6.24b yields

$$T(r)\frac{\zeta' r}{\sqrt{1+\zeta'^2}}=C_1-\int^r f(\zeta)\,\zeta\,d\zeta\equiv A\left(r;C_1\right), \tag{6.25a}$$

where C_1 is an arbitrary constant. Solving for ζ'

$$\frac{d\zeta}{dr}=\zeta'=\left|\left[\frac{rT(r)}{A\left(r;C_1\right)}\right]^2-1\right|^{-1/2} \tag{6.25b}$$

and integrating again

$$\zeta(r) = C_2 + \int^r \left| \left[\frac{\eta T(\eta)}{A(\eta; C_1)} \right]^2 - 1 \right|^{-1/2} \mathrm{d}\eta \tag{6.25c}$$

introduces another constant of integration C_2. Thus, *the shape of a membrane under radially symmetric shear stress f(r) and with radially nonuniform tension T(r) is given by Equation 6.25c involving the function 6.25a. The two arbitrary constants (C_1, C_2) are determined by boundary or boundedness conditions. The slope (Equation 6.25b) and shape (Equation 6.25c) of the membrane apply for nonuniform tension in the nonlinear case of large slope; if the shear stress and tension are polynomials of the radius, the shape of the membrane (Equation 6.25c) is specified by elliptic integrals (Subsections I.39.7 to I.39.9). In the linear case of small slope (Equation 6.13a ≡ Equation 6.26a) the equilibrium Equation 6.26b is integrated more simply for the slope (Equation 6.26c) and shape (Equation 6.26d) allowing for nonuniform tension:*

$$\zeta'^2 << 1: \quad -f(r) = \frac{1}{r}\frac{\mathrm{d}}{\mathrm{d}r}\left[\zeta' r T(r)\right], \quad rT(r)\zeta'(r) = C_1 - \int^r f(\zeta)\zeta \,\mathrm{d}\zeta, \tag{6.26a–6.26c}$$

$$\zeta(r) = C_2 + \int^r \left[\eta T(\eta)\right]^{-1} \left[C_1 - \int^\eta f(\zeta)\zeta \,\mathrm{d}\zeta\right] \mathrm{d}\eta. \tag{6.26d}$$

Before considering the nonlinear deflection of membranes (Section 6.2), the linear case of a small slope for a circular (annular) membrane deflected by its own weight [Subsection 6.1.5 (Subsection 6.1.6)] is illustrated first.

6.1.5 Heavy Circular Membrane under Tension

In the presence of a gravity field with acceleration g, a membrane with mass density ρ per unit area is subject to a shear stress equal to its weight (Equation 6.27b), and the corresponding linear deflection (Equation 6.13b) under constant tension (Equation 6.27a) is specified by Equation 6.27c:

$$T = \text{const}: \quad f = \rho g, \quad -\frac{\rho g}{T} = \nabla^2 \zeta = \frac{1}{r}\frac{\mathrm{d}}{\mathrm{d}r}\left(r \frac{\mathrm{d}\zeta}{\mathrm{d}r}\right), \tag{6.27a–6.27c}$$

where the Laplacian was written in polar coordinates (Equation 4.111 ≡ Equation I.11.28c) and radial symmetry assumed (Equation 6.24a), for example, for a circular membrane (Figure 6.3a) with radius a held horizontally at the circumference (Equation 6.29c); a first (second) integration of Equation 6.27c yields Equation 6.28b (Equation 6.28c):

$$\rho g = \text{const}: \quad -\frac{\rho g r^2}{2T} = r\frac{\mathrm{d}\zeta}{\mathrm{d}r} + C_1, \quad -\frac{\rho g}{4T} r^2 = \zeta(r) + C_1 \log r + C_2 \tag{6.28a–6.28c}$$

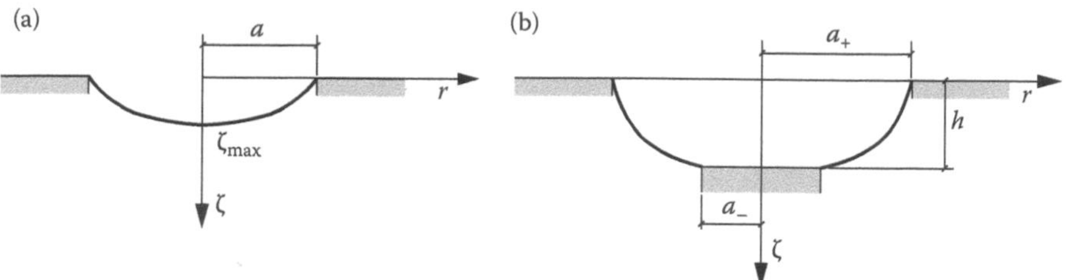

FIGURE 6.3
Axisymmetric deflection of a membrane, which under its own weight can be considered in at least two cases: (a) singly connected with a horizontal circular boundary, leading to a parabolic shape with minimum at the center; and (b) doubly connected, with circular inner and outer boundaries, both horizontal but at different heights, leading to the three subcases in Figure 6.4. Solutions obtained assume a constant weight per unit area, that is, the product of the acceleration of gravity by the mass density per unit area must be constant; this applies, in particular, if both factors are constants.

for a membrane with constant weight per unit area (Equation 6.28a). The constants of integration are determined [Equation 6.29b (Equation 6.29d)] from the conditions of finite (Equation 6.29a) [zero (Equation 6.29c)] displacement at the center (boundary):

$$\zeta(0) < \infty: \quad C_1 = 0; \quad \zeta(a) = 0: \quad C_2 = -\frac{\rho g a^2}{4T}. \tag{6.29a–6.29d}$$

Thus, *the linear deflection (Equation 6.12a) of a circular membrane with radius a held horizontally at the edge (Equation 6.29c) with mass density per unit area ρ in the gravity field g with constant weight per unit area (Equation 6.28a) under a uniform tension (Equation 6.27a) leads to (Figure 6.3a) the shape (Equation 6.30a) of a paraboloid of revolution, whose cross section through the axis is a parabola:*

$$\zeta(r) = \frac{\rho g}{4T}(a^2 - r^2); \quad \zeta_{max} = \zeta(0) = \frac{\rho g a^2}{4T}, \quad \zeta' \equiv \frac{d\zeta}{dr} = -\frac{\rho g r}{2T}, \quad \zeta'(0) = 0; \tag{6.30a–6.30d}$$

the maximum deflection (Equation 6.30b) occurs at the center where the slope (Equation 6.30c) is zero (Equation 6.30d). The slope is maximum in modulus at the boundary (Equation 6.31a), implying that the linear approximation (Equation 6.6a) requires Equation 6.31b:

$$\left|\zeta'_{max}\right| = \left|\zeta'(a)\right| = -\zeta'(a) = \frac{\rho g a}{2T}: \quad 1 \gg \left|\zeta'_{max}\right|^2 = \left(\frac{\rho g a}{2T}\right)^2 = \left(\frac{P}{2\pi a T}\right)^2, \quad P \equiv \pi a^2 \rho g \tag{6.31a–6.31c}$$

that the total weight of the membrane (Equation 6.31c) should be small compared with the tension multiplied by the length of the boundary.

6.1.6 Annular Membrane Hanging from Two Rings

If the membrane is annular (Figure 6.3b) and hangs from two rings of radii $a_+(a_-)$ with the same axis and height difference h:

$$a_- \le r \le a_+: \qquad \zeta(a_+) = 0, \qquad \zeta(a_-) = h; \tag{6.32a–6.32c}$$

the arbitrary constants of integration in the linear solution 6.28c are specified (Equations 6.33a and 6.33b) by the boundary conditions (Equations 6.32b and 6.32c)

$$-\frac{\rho g}{4T}a_+^2 = C_1 \log a_+ + C_2, \quad -\frac{\rho g}{4T}a_-^2 = h + C_1 \log a_- + C_2; \qquad \text{(6.33a and 6.33b)}$$

thus, unlike for the circular membrane (Equations 6.29a and 6.29b), for the case of the annular membrane, both constants of integration in Equations 6.33a and 6.33b are nonzero:

$$C_1 \log\left(\frac{a_+}{a_-}\right) = h - \frac{\rho g}{4T}\left(a_+^2 - a_-^2\right), \qquad \text{(6.34a)}$$

$$C_2 = -\frac{\rho g}{4T}a_+^2 - C_1 \log a_+ = -\frac{\rho g}{4T}a_+^2 - \left[h - \frac{\rho g}{4T}\left(a_+^2 - a_-^2\right)\right]\frac{\log a_+}{\log\left(a_+/a_-\right)} \qquad \text{(6.34b)}$$

because the center (Equations 6.29a and 6.29b) is excluded by Equation 6.32a. Substituting Equations 6.34a and 6.34b in Equation 6.28c, it follows that *the linear displacement (Equation 6.12a) of an annular membrane (Equation 6.32a) with constant weight per unit area (Equation 6.28a) under a uniform tension (Equation 6.27a) in a gravity field g is specified by*

$$\zeta(r) = \frac{\rho g}{4T}\left(a_+^2 - r^2\right) + \left[h - \frac{\rho g}{4T}\left(a_+^2 - a_-^2\right)\right]\frac{\log\left(a_+/r\right)}{\log\left(a_+/a_-\right)} \qquad \text{(6.35)}$$

if (Figure 6.2b) the outer (inner) radius $a_+(a_-)$ is Equation 6.32b (Equation 6.32c) at height 0(h). It can be checked that Equation 6.35 (1) is of the form Equation 6.28c as a function of r and (2, 3) meets the boundary conditions Equations 6.32b and 6.32c. From Equation 6.28b, the condition of horizontal slope (Equation 6.36a) is (Equation 6.36b):

$$\zeta'(b) = 0: \quad -\frac{\rho g b^2}{2T} = C_1, \quad 2b^2 \log\left(\frac{a_+}{a_-}\right) = a_+^2 - a_-^2 - \frac{4hT}{\rho g} \geq 0 \qquad \text{(6.36a–6.36c)}$$

leading by Equation 6.34a to Equation 6.36c; thus, *the membrane has horizontal slope (Equation 6.36a) at a position $r = b$ if the condition 6.36c is met, in which case the maximum deflection is*

$$\zeta_{\max} = \zeta(b) = \frac{\rho g}{4T}a_+^2 + \left[h - \frac{\rho g}{4T}\left(a_+^2 - a_-^2\right)\right]\frac{\log\left(a_+/b\right) + 1/2}{\log\left(a_+/a_-\right)}, \qquad \text{(6.37)}$$

as follows substituting Equation 6.36c in Equation 6.35. Thus, three cases (Equation 6.3a through c) arise for the linear deflection (Equation 6.12a) of the annular membrane (Equation 6.32a) under constant tension (Equation 6.13b) due to uniform weight (Equation 6.28a): (1) if Equation 6.36c holds for $b = a_-$, the membrane (Figure 6.4b) is horizontal at the inner support; (2) if $b < a_-$, the

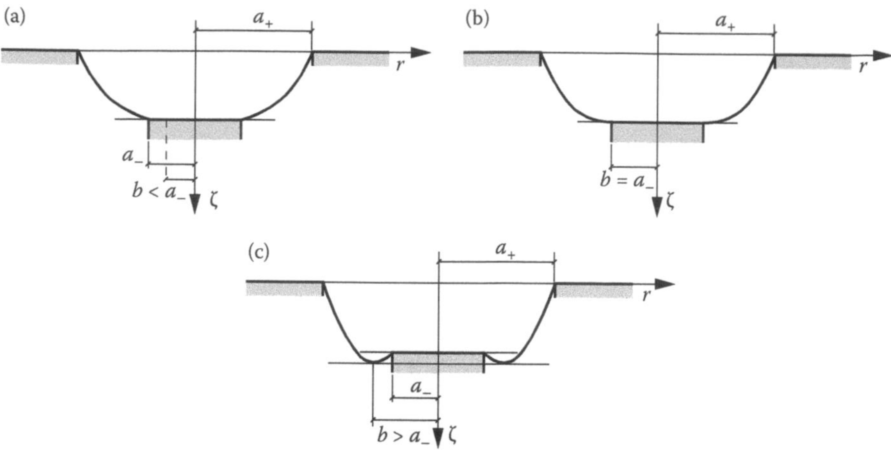

FIGURE 6.4
For a homogeneous elastic membrane deflected by its own weight under uniform gravity supported on horizontal inner and outer circular boundaries at different heights (Figure 6.3b), three cases can arise: (a) the membrane has a negative slope at the lower boundary; (b) the membrane is tangent to the lower boundary; and (c) the membrane has a positive slope at the lower boundary, and "sags" under its own weight to a minimum below the lower boundary. The intermediate case (b) corresponds to a particular combination of tension, weight per unit area, and height difference between the two boundaries. Decreasing the tension, increasing the weight per unit area, or decreasing the height difference between the two boundaries lead to "sagging" that passes from case b to case c; the opposite conditions lead from case b to case a.

membrane has (Figure 6.4a) a positive slope at the inner support because ζ is measured downward and nowhere has a zero slope; and (3) if $a_- < b < a_+$, then the membrane has a zero slope at $r = b$ corresponding (Figure 6.4c) to a maximum deflection, and the slope is negative for $a_+ > r > b$ and positive for $b > r > a_-$.

6.2 Large Deflection of a Membrane by Weight or Pressure

The deflection of a membrane under its own weight was considered for a circular (annular) shape [Subsection 6.1.5 (Subsection 6.1.6)] in the linear case of a small slope. The nonlinear deflection of a membrane, that is, with a large slope (Section 6.2), can be considered with uniform tension (Subsection 6.2.1) for various loads, for example, (1) its own weight that has a fixed vertical direction (Subsection 6.2.2) and (2) a pressure difference across the normal direction (Subsections 6.2.3 through 6.2.7). The latter case is considered for the circular membrane (Subsection 6.2.4) specifying the exact shape (Subsection 6.2.6) and slope (Subsection 6.2.3) and their maximum values in modulus. The maximum load that the membrane can support corresponds to the value unity of the nonlinearity parameter (Subsection 6.2.4); the linear deflection corresponds to the zero-order term in the expansion in powers of the nonlinearity parameter. The first-order nonlinear correction, as well as all corrections of higher order, appear in the exact spherical shape of the membrane that deflects more than the parabolic shape in the linear case (Subsection 6.2.7).

6.2.1 Nonlinear Deflection of a Homogeneous Membrane under Uniform Tension

Consider again (Figure 6.3a) a circular (Equation 6.38a) membrane under a uniform tension (Equation 6.27a), eliminating the restriction of a small slope (Equation 6.31b) so that the deflection equation 6.27c is replaced (Equation 6.24b) by Equation 6.38b:

$$0 \le r \le a: \quad -\frac{T}{r}\frac{\mathrm{d}}{\mathrm{d}r}\left(\frac{\zeta' r}{\sqrt{1+\zeta'^2}}\right) = -f(r). \qquad \text{(6.38a and 6.38b)}$$

The shear stress is considered in two cases: (1) own weight (Subsection 6.2.1) and (2) normal pressure (Subsections 6.2.2 through 6.2.7). The weight has fixed vertical direction and is equal to the acceleration of gravity times the mass per unit horizontal area (Equation 6.39b):

$$\rho = \frac{dm}{dS}: \quad f(r) = g\frac{dm}{dx\,dy} = g\frac{dm}{dS}\frac{dS}{dx\,dy} = \rho g\left|1+\zeta'^2\right|^{1/2}; \qquad \text{(6.39a and 6.39b)}$$

the **mass density** is the mass per unit area (Equation 6.39a), and thus, the shear stress due to the weight is given by Equation 6.39b. This reduces to Equation 6.27b ≡ Equation 6.40b in the linear approximation Equation 6.6a ≡ Equation 6.40a:

$$\zeta'^2 \ll 1: \qquad f = \rho g \leftrightarrow f = \Delta \mathrm{p}. \qquad \text{(6.40a–6.40c)}$$

In the case of a difference of normal pressure (Equation 6.41a), the shear stress is given by Equation 6.41b exactly, even in the nonlinear case of a large slope:

$$\Delta p = p_1 - p_2: \quad f = \Delta p N_z \frac{dxdy}{dS} = \Delta p. \qquad \text{(6.41a and 6.41b)}$$

Equation 6.4a was used in Equations 6.39b and 6.41b. Thus, *the shear stress in the nonlinear case of a strong slope is given by Equation 6.39b (Equation 6.41b) due to own weight (a normal pressure). They are equivalent (Equation 6.40b ≡ Equation 6.40c) only in the linear approximation of a small slope (Equation 6.40a).*

6.2.2 Deflection by Own Weight or by Normal Pressure

In the case of deflection by own weight (Equation 6.39b), the exact nonlinear displacement (Equation 6.38b) is given by

$$T = \text{const:} \quad -\frac{\rho g}{T}\left|1+\zeta'^2\right|^{1/2} = \frac{1}{r}\left[\zeta' r\left|1+\zeta'^2\right|^{-1/2}\right]'. \qquad \text{(6.42a and 6.42b)}$$

Performing the differentiation on the r.h.s. of Equation 6.42b leads to

$$\begin{aligned} r^{-1}\left\{\zeta' r\left|1+\zeta'^2\right|^{-1/2}\right\}' &= r^{-1}\zeta'\left|1+\zeta'^2\right|^{-1/2} + \zeta''\left|1+\zeta'^2\right|^{-1/2} - \zeta''\zeta'^2\left|1+\zeta'^2\right|^{-3/2} \\ &= r^{-1}\zeta'\left|1+\zeta'^2\right|^{-1/2} + \zeta''\left|1+\zeta'^2\right|^{-3/2}. \end{aligned} \qquad (6.43)$$

Substitution of Equation 6.43 in Equation 6.42b leads to *Equation 6.44b for the nonlinear deflection of a circular membrane under constant tension (Equation 6.44a) due to its own weight, where neither the acceleration of gravity nor the mass density per unit area needs to be uniform:*

$$T = \text{const}: \quad -\frac{\rho(r)g(r)}{T}\left(1+\zeta'^2\right)^2 = \zeta'' + \frac{\zeta'}{r}\left(1+\zeta'^2\right); \qquad \text{(6.44a and 6.44b)}$$

$$T = \text{const}: \quad -\frac{\Delta p(r)}{T} = \frac{1}{r}\frac{\mathrm{d}}{\mathrm{d}r}\left(\frac{\zeta' r}{\sqrt{1+\zeta'^2}}\right) = \left|1+\zeta'^2\right|^{-3/2}\left[\zeta'' + \frac{\zeta'}{r}\left(1+\zeta'^2\right)\right]; \qquad \text{(6.45a and 6.45b)}$$

it is replaced by Equations 6.45a and 6.45b in the case of a difference of normal pressure (Equations 6.38b and 6.41b) that need not be uniform. The case of a uniform normal pressure is considered in the sequel (Subsections 6.2.3 through 6.2.7).

6.2.3 Large Deflection by a Difference of Normal Pressure

If both the tension (Equation 6.45a ≡ Equation 6.46a) and the normal pressure difference (Equation 6.46b) are constant, a first integration of Equation 6.45b yields Equation 6.46c:

$$T, \Delta p = \text{const}: \quad C_1 - \frac{r^2 \Delta p}{2T} = \frac{\zeta' r}{\sqrt{1+\zeta'^2}}; \quad \zeta'(0) < \infty: \quad C_1 = 0; \qquad \text{(6.46a–6.46e)}$$

the condition of finite slope on the axis (Equation 6.39b) implies that the constant of integration is zero (Equation 6.46e). Solving Equation 6.46c with Equation 6.46e specifies (Equation 6.47a) the exact slope (Equation 6.47b) where the minus rather than the plus sign was chosen before the square root because the slope is negative with z-axis directed downward (Figure 6.1):

$$\frac{\zeta'}{\sqrt{1+\zeta'^2}} = -\frac{r\Delta p}{2T}: \quad \zeta' = -\frac{r\Delta p}{2T}\left[1-\left(\frac{r\Delta p}{2T}\right)^2\right]^{-1/2} = \frac{\mathrm{d}\zeta}{\mathrm{d}r}. \qquad \text{(6.47a and 6.47b)}$$

Equation 6.47b may be integrated for the exact shape (Equation 6.48b) with the boundary condition 6.41a that the membrane is held horizontally at the edge:

$$\zeta(a) = 0: \quad \zeta(r) = -\int_a^r \frac{\xi\Delta p}{2T}\left[1-\left(\frac{\xi\Delta p}{2T}\right)^2\right]^{-1/2} \mathrm{d}\xi = \frac{2T}{\Delta p}\left\{\left[1-\left(\frac{\xi\Delta p}{2T}\right)^2\right]^{1/2}\right\}_a^r. \qquad \text{(6.48a and 6.48b)}$$

The exact shape (Equation 6.48b) of a membrane with uniform tension (Equation 6.49a) and with a uniform normal pressure difference (Equation 6.49b) can be put in the dimensionless form (Equation 6.49c):

$$T = \text{const}, \Delta p = \text{const}: \quad \frac{\Delta p}{2T}\zeta(r) = \left|1 - \left(\frac{r\Delta p}{2T}\right)^2\right|^{1/2} - \left|1 - \left(\frac{a\Delta p}{2T}\right)^2\right|^{1/2}, \qquad (6.49\text{a–}6.49\text{c})$$

and the exact slope (Equation 6.47b) is also in dimensionless form.

6.2.4 Maximum Load for Nonlinearity Parameter Unity

The **nonlinearity parameter** is defined (Equation 6.50a) as the ratio of total pressure over the area of the membrane to the tension multiplied by the perimeter:

$$q \equiv \frac{a\Delta p}{2T} = \frac{\pi a^2 \Delta p}{2\pi a T}: \quad \frac{q}{a}\zeta(r) = \sqrt{1 - \left(q\frac{r}{a}\right)^2} - \sqrt{1 - q^2}; \qquad (6.50\text{a and } 6.50\text{b})$$

it appears as the only parameter in the exact shape (Equation 6.49c ≡ Equation 6.50b) of the membrane in dimensionless form with the distance r and deflection $\zeta(r)$ normalized to the radius. In the linear case (Equation 6.40a ≡ Equation 6.51a), the nonlinearity parameter (Equation 6.50a) associated with the weight (Equations 6.40b and 6.40c) is (Equation 6.51b) the total weight of the membrane (Equation 6.31c) divided by the product of the tension by the perimeter:

$$\zeta'^2 \ll 1: \quad q = \frac{\rho g a}{2T} = \frac{\pi a^2 \rho g}{2\pi a T} = \frac{P}{2\pi a T}, \quad q^2 \ll 1 \qquad (6.51\text{a–}6.51\text{c})$$

and is small (Equation 6.31b ≡ Equation 6.51c). Returning to the nonlinear deflection of a membrane under normal pressure the deflection (Equation 6.50b) must be real for all radial distances (Equation 6.38a), and thus, *the maximum normal force due to a uniform pressure (Equation 6.49b) on a membrane under a uniform tension (Equation 6.49a) is the product of the latter by the perimeter (Equation 6.52a), corresponding to a value unity (Equation 6.52b) of the nonlinearity parameter (Equation 6.50a):*

$$P = \pi a^2 \Delta p \le 2\pi a T \quad \Leftrightarrow \quad q \le 1 \quad \Leftrightarrow \quad \Delta p \le \frac{2T}{a} \qquad (6.52\text{a–}6.52\text{c})$$

and a maximum constant pressure difference on the two sides (Equation 6.52c). The shape (Equation 6.49c ≡ Equation 6.50b) and the slope (Equation 6.47b ≡ Equation 6.53):

$$\zeta' = -q\frac{r}{a}\left|1 - \left(q\frac{r}{a}\right)^2\right|^{-1/2} \qquad (6.53)$$

depend only on the nonlinearity parameter if the distance and deflection are normalized with regard to the radius. The exact nonlinear shape of the membrane (Equation 6.50b ≡ Equation 6.54a) is

$$\left(\frac{\zeta}{a}+\sqrt{\frac{1}{q^2}-1}\right)^2+\left(\frac{r}{a}\right)^2=\frac{1}{q^2}: \quad z=-a\sqrt{\frac{1}{q^2}-1}, \quad R=\frac{a}{q}=\frac{1}{k}, \quad \tan\alpha=\frac{a}{R} \qquad (6.54a–6.54d)$$

a spherical sector with (1) center on the axis at a position (Equation 6.54b), (2) radius (Equation 6.54c), and (3) aperture α given by Equation 6.54d. It follows that the curvature k is equal to the nonlinearity parameter q in Equation 6.43a divided by a. The spherical shape of a membrane under a uniform normal pressure difference and surface tension corresponds to the spherical shape (Subsection 6.3.5) of a soap bubble (Subsection 6.3.3), that is, a minimal surface (Subsection 6.3.4). The local approximation to the sphere is a paraboloid of revolution, corresponding to the linear deflection of the membrane under its own weight (Subsection 6.1.5) that is equivalent to a normal pressure (Equations 6.40a through 6.40c). The linear deflection corresponds to the lowest order term in the expansion of the deflection (slope) in powers of the nonlinearity parameter [Subsection 6.2.6 (Subsection 6.2.5)]; the remaining terms give the first- and higher-order nonlinear corrections, corresponding to shapes (Subsection 6.2.7) between the paraboloid of revolution and the sphere.

6.2.5 Exact Slope and Maximum Value

The exact slope (Equation 6.53) can be expanded in a binomial series (Equations 3.144a through 3.144c ≡ Equations I.25.37a through I.25.37c) of powers of the nonlinearity parameter:

$$\zeta'=-\sum_{n=0}^{\infty}a_n\left(\frac{qr}{a}\right)^{2n+1}, \quad a_n\equiv(-)^n\binom{-1/2}{n}, \qquad (6.55a \text{ and } 6.55b)$$

with coefficients:

$$\begin{aligned} n\ge 1: \quad a_n &= \frac{(-)^n}{n!}\left(-\frac{1}{2}\right)\left(-\frac{1}{2}-1\right)\ldots\left(-\frac{1}{2}-n+1\right) \\ &= \frac{2^{-n}}{n!}1.3\ldots(2n-1)=2^{-n}\frac{(2n-1)!!}{n!} \end{aligned} \qquad (6.56a \text{ and } 6.56b)$$

involving the double factorial (Equations 1.178a through 1.178c). The first seven coefficients are

$$\{a_{0-6}\}=\left\{1,\frac{1}{2},\frac{3}{8},\frac{5}{16},\frac{35}{128},\frac{63}{256},\frac{231}{1024}\right\}, \qquad (6.57a–6.57g)$$

where Equation 6.56b is used for (Equations 6.57b through 6.57g), and Equation 6.57a follows from the leading term of Equation 6.55a ≡ Equation 6.53. The exact slope is given by

$$\zeta'(r) = -q\frac{r}{a}\left[1+\sum_{n=1}^{\infty}\frac{(2n-1)!!}{n!}\left(\frac{q^2r^2}{2a^2}\right)^n\right]$$
$$= -\frac{r\Delta p}{2T}\left\{1+\sum_{n=1}^{\infty}\frac{(2n-1)!!}{n!}\left[\frac{r^2(\Delta p)^2}{8T^2}\right]^n\right\} \tag{6.58}$$

and to the leading order by

$$\zeta'(r) = -q\frac{r}{a}\left[1+\frac{q^2r^2}{2a^2}+O\left(\left(q\frac{r}{a}\right)^4\right)\right] = -\frac{r\Delta p}{2T}\left[1+\frac{r^2(\Delta p)^2}{8T^2}+O\left(\frac{r\Delta p}{T}\right)^4\right]. \tag{6.59}$$

Using the analogy (Equations 6.40a through 6.40c), the first term on the r.h.s. of Equation 6.59 coincides with the linear approximation (Equation 6.30c); the next term in Equation 6.59 estimates the error of the linear approximation, and all higher order terms appear in Equation 6.58. The maximum slope (Equation 6.53 ≡ Equations 6.55a and 6.56b) is at the boundary

$$-\tan\theta = -\zeta'(a) = \frac{q}{\sqrt{1-q^2}} = q\sum_{n=0}^{\infty}a_n q^{2n} = q\left[1+\sum_{n=0}^{\infty}\frac{(2n-1)!!}{n!}\left(\frac{q^2}{2}\right)^n\right]$$
$$= \frac{a\Delta p}{2T}\left\{1+\sum_{n=0}^{\infty}\frac{(2n-1)!!}{n!}\left[\frac{a^2(\Delta p)^2}{8T^2}\right]^n\right\} = -\frac{a}{z}, \tag{6.60}$$

and equals (Equation 6.54b) the radius divided by the height of the center; using the analogy (Equations 6.40a through 6.40c), the leading term in Equation 6.60 coincides with the linear approximation (Equation 6.31a), and the first-order nonlinear correction is included in

$$-\tan\theta = -\zeta'(a) = \frac{a\Delta p}{2T}\left[1+\frac{a^2(\Delta p)^2}{8T^2}+O\left(\left(\frac{a\Delta p}{T}\right)^4\right)\right]. \tag{6.61}$$

Thus, *the exact (maximum in modulus) nonlinear slope [Equation 6.53 (Equation 6.60)], of a membrane under constant tension (Equation 6.49a), and uniform normal pressure difference (Equation 6.49b) are expanded as a power series [Equation 6.58 (Equation 6.60)] of the nonlinearity parameter, with (1) a zero-order term corresponding (Equations 6.40a through 6.40c) to the linear case [Equation 6.30c (Equation 6.31a)] of a small slope, and (2) first-order nonlinear correction [Equation 6.59 (Equation 6.61)].*

6.2.6 Exact Deflection and Maximum Value

The exact deflection (Equation 6.49c) is given by

$$\frac{\Delta p}{2T}\zeta(r) = \sum_{n=1}^{\infty} b_n \left[\left(\frac{r\Delta p}{2T}\right)^{2n} - \left(\frac{a\Delta p}{2T}\right)^{2n}\right], \qquad b_n \equiv (-)^n \binom{1/2}{n}, \tag{6.62a and 6.62b}$$

with coefficients

$$\begin{aligned} n \geq 2: \quad b_n &= \frac{(-)^n}{n!}\left(\frac{1}{2}\right)\left(-\frac{1}{2}\right)\left(-\frac{3}{2}\right)..\left(\frac{1}{2} - n + 1\right) \\ &= -\frac{2^{-n}}{n!} 1.3.5...(2n-3) = -2^{-n}\frac{(2n-3)!!}{n!}. \end{aligned} \tag{6.63a and 6.63b}$$

The first six coefficients are

$$b_{1-6} = \left\{-\frac{1}{2}, -\frac{1}{8}, -\frac{1}{16}, -\frac{5}{128}, -\frac{7}{256}, -\frac{21}{1024}\right\}, \tag{6.64a–6.64f}$$

where Equation 6.63b is used for Equations 6.64c through 6.64f, and Equations 6.64a and 6.64b follow from Equation 6.62a ≡ Equation 6.49c. The exact shape (Equations 6.62a and 6.63b) of the homogeneous membrane is specified by an expansion in power of the nonlinearity parameter:

$$\begin{aligned} \frac{\zeta(r)}{a} &= \frac{1}{q}\sum_{n=1}^{\infty} b_n q^{2n}\left[\left(\frac{r}{a}\right)^{2n} - 1\right] = -\frac{q}{2}\left(\frac{r^2}{a^2} - 1\right) - \sum_{n=2}^{\infty}\frac{(2n-3)!!}{n!2^n} q^{2n-1}\left[\left(\frac{r}{a}\right)^{2n} - 1\right] \\ &= \frac{\Delta p}{4T}\left(a - \frac{r^2}{a^2}\right) + \sum_{n=2}^{\infty}\frac{(2n-3)!!}{n!2^n}\frac{2T}{a\Delta p}\left[\left(\frac{a\Delta p}{2T}\right)^{2n} - \left(\frac{r\Delta p}{2T}\right)^{2n}\right] \\ &= \frac{\Delta p}{4T}\left(a - \frac{r^2}{a}\right) + \sum_{n=2}^{\infty}\frac{(2n-3)!!}{n!}\frac{2T}{a\Delta p}\left\{\left[\frac{a^2(\Delta p)^2}{8T^2}\right]^n - \left[\frac{r^2(\Delta p)^2}{8T^2}\right]^n\right\}. \end{aligned} \tag{6.65}$$

The leading term in Equation 6.65 coincides via the analogy (Equations 6.40a through 6.40c) with the linear approximation (Equation 6.30a); the first-order nonlinear correction is indicated in Equation 6.66:

$$\begin{aligned} \zeta(r) &= \frac{\Delta p}{4T}(a^2 - r^2) + \frac{(\Delta p)^3}{32T^3}(a^4 - r^4) + O\left(\frac{(\Delta p)^5}{T^5}(a^6 - r^6)\right) \\ &= \frac{\Delta p}{4T}(a^2 - r^2)\left[1 + \frac{(\Delta p)^2}{8T^2}(a^2 + r^2) + O\left(\left(\frac{a\Delta p}{T}\right)^4\right)\right] \end{aligned} \tag{6.66}$$

The maximum deflection (Equation 6.49c) is

$$\frac{\delta}{a} \equiv \frac{\zeta(0)}{a} = \frac{2T}{a\Delta p}\left\{1 - \left|1 - \left(\frac{a\Delta p}{2T}\right)^2\right|^{1/2}\right\} = \frac{1-\sqrt{1-q^2}}{q}$$

$$= -\sum_{n=1}^{\infty} b_n q^{2n-1} = \frac{q}{2} + \sum_{n=2}^{\infty} \frac{(2n-3)!!}{n!2^n} q^{2n-1} \tag{6.67}$$

$$= \frac{a\Delta p}{4T} + \sum_{n=2}^{\infty} \frac{(2n-3)!!}{n!} \frac{2T}{a\Delta p}\left(\frac{a^2(\Delta p)^2}{8T^2}\right)^n.$$

The leading term in Equation 6.67 coincides via the analogy (Equations 6.40a–6.40c) with the linear approximation (Equation 6.30b); the first-order nonlinear correction is included in Equation 6.68:

$$\delta = \zeta(0) = \frac{a^2\Delta p}{4T} + \frac{a^4(\Delta p)^3}{32T^3} + O\left(\frac{a^6(\Delta p)^5}{T^5}\right) = \frac{a^2\Delta p}{4T}\left[1 + \frac{a^2(\Delta p)^2}{8T^2} + O\left(\left(\frac{a\Delta p}{T}\right)^4\right)\right]. \tag{6.68}$$

Thus, *the exact (maximum) nonlinear deflection of a membrane under constant tension (Equation 6.49a) and uniform normal pressure difference (Equation 6.49b) is given by Equation 6.49c (Equation 6.67) as a power series [Equation 6.65 (Equation 6.67)] of the nonlinearity parameter (Equation 6.50a) that (1) coincides at zero order via the analogy (Equations 6.40a through 6.40c) with the linear approximation [Equation 6.30a (Equation 6.30b)] and (2) includes the first-order nonlinear correction [Equation 6.66 (Equation 6.68)].*

6.2.7 Nonlinear Effects on the Deflection and Slope

The maximum deflection normalized to the radius (Equation 6.67) is one-half of the nonlinearity parameter (Equation 6.50a) in the linear case (Equation 6.30b) of small nonlinearity parameter (Equation 6.51c) as seen in Table 6.1 for $q = 0.1$; in the nonlinear case, as the nonlinearity parameter increases toward the maximum value unity in Table 6.1, it is seen that the maximum deflection increases to more than half its value, for example, $\zeta/a = 0.627 > 0.45 = q/2$ for $q = 0.9$. Table 6.1 indicates the following for nine values of the nonlinearity parameter (Equation 6.50a) in steps of 0.1 from 0.1 to 0.9: (1) the maximum deflection normalized to the radius that occurs at the center and (2) the maximum slope that is dimensionless and occurs at the edges and the corresponding angle. In the linear approximation of a small slope, the maximum slope (Equation 6.60) is equal to (Equation 6.50a) the nonlinearity parameter, as seen in Table 6.1 for $q = 0.1$. As the nonlinearity parameter increases, the maximum slope increases faster; for example, it is more than double $|\zeta'(a)| = 2.065 > 1.8 = 2q$ for $q = 0.9$. As the maximum load is approached corresponding to the value unity of the nonlinearity parameter, (1) the maximum slope (Equation 6.60) diverges and (2) the maximum deflection (Equation 6.67) equals the radius, and (3) the center of the sphere (Equation 6.54b) is at zero height, because (1) and (3) are algebraic inverses. The plots of the exact spherical shape of the heavy homogeneous membrane deflected normal pressure difference (Panel 6.1) show that the parabolic shape (Equation 6.30a) in the linear case of the small nonlinearity parameter (Equation 6.31b), as the latter increases, "sags" more, that is, the nonlinear terms enhance the deflection overall, as they do for the maximum deflection and slope in Table 6.1.

TABLE 6.1

Nonlinear Deflection of a Membrane under Normal Pressure

Nonlinearity Parameter	Maximum Deflection	Maximum Slope in Modulus (Angle)
$q \equiv \frac{\rho g a}{2T}$	$\frac{\zeta(0)}{a} = \frac{\delta}{a} = \frac{1-\sqrt{1-q^2}}{q}$	$-\zeta'(a) \equiv \tan\theta = \frac{q}{\sqrt{1-q^2}} = -\frac{a}{z}$
Equation 6.50a	Equation 6.67	Equation 6.60
0.0	0.000	0.000 (0.00°)
0.1	0.050	0.101 (–5.74°)
0.2	0.101	0.204 (–11.54°)
0.3	0.154	0.315 (–17.46°)
0.4	0.209	0.436 (–23.58°)
0.5	0.268	0.577 (–30.00°)
0.6	0.333	0.750 (–36.87°)
0.7	0.408	0.980 (–44.43°)
0.8	0.500	1.333 (–53.13°)
0.9	0.627	2.065 (–64.16°)
1.0	1.000	∞ (–90.00°)

Notes: a, radius of the undeflected membrane; ρ, mass density; T, uniform tangential tension; g, acceleration of gravity; ζ, transverse deflection; r, radial coordinate; $\zeta' \equiv \frac{d\zeta}{dr}$, slope; δ, maximum deflection; θ, maximum angle of inclination; q, nonlinearity parameter; z, height of center of sphere

Maximum slope at the edges and maximum deflection at the center normalized to the radius for the nonlinear deflection of a homogeneous membrane under uniform tension horizontally at the circular edge due to a pressure difference along the normal. The nonlinearity parameter is given nine equally spaced values ranging (1) from small $q = 0.1$ for the linear deflection, that is, with small slope $q^2 \ll 1$; (2) to $q = 0.9$, that is, nonlinear approaching unity, which corresponds to the maximum load that the membrane can support.

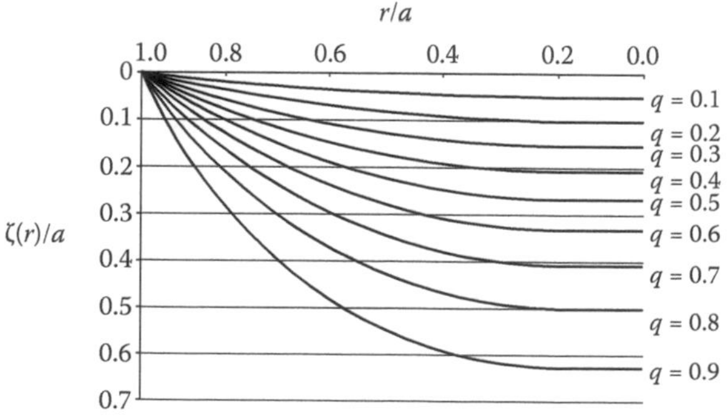

PANEL 6.1

Shape of a membrane under constant tension difference held at a horizontal circular boundary due to a normal pressure is a parabola in the linear case of a small slope $q^2 \ll 1$. As the nonlinearity parameter increases toward the value unity corresponding to the maximum load, the shape is deformed into a spherical sector with a larger slope and deflection than a parabola.

6.3 Boundary Condition with Surface Tension

A physical phenomenon analogous to a membrane (Sections 6.1 and 6.2) is surface tension (Sections 6.3 and 6.4) at the interface between two fluids. Surface tension can be included in the boundary condition at the interface between two fluids (Subsection 6.3.2) in terms of either the area change (Subsection 6.1.1) or the principal curvatures (Subsection 6.3.1). In the absence of pressure or stresses or other surface forces, the shape of the interface is a minimal surface (Subsection 6.3.4). The minimal surfaces were predicted theoretically (Euler 1744; Lagrange 1760) before being found experimentally (Plateau 1849) as soap bubbles (Subsection 6.3.3). The axisymmetric minimal surface (Subsection 6.3.7) is the catenoid that is an anticlastic surface, that is, has principal curvatures with opposite signs (Subsection 6.3.8); the sphere is an extremal (Subsection 6.3.5) but not a minimal surface, because it is a synclastic surface with constant curvature. The spherical shape applies (Subsection 6.3.6) both to a soap bubble (Subsection 6.3.3) [the nonlinear deflection of an elastic membrane (Subsections 6.2.3 through 6.2.7)] under constant tangential tension and normal pressure difference.

6.3.1 Area Change and Principal Curvatures

Consider (Figure 6.5a) a point on a surface where it has a unique normal, that is, a **regular point**, and not an edge or a cusp. Any plane passing through the normal is a **normal plane** and intersects the surface along a **normal section**, which is a curve with a given curvature R. As the normal plane rotates around the normal, the curvature will change, unless the surface is locally equivalent to a sphere. The maximum R_1 and minimum R_2 curvatures are called **principal curvatures**, and it is shown in differential geometry that they lie in orthogonal directions. Consider a line of curvature (Figure 6.5b) resulting from a small deflection $\delta\zeta$ of a straight segment; by small, it means a small (Equation 6.69a) angle so that

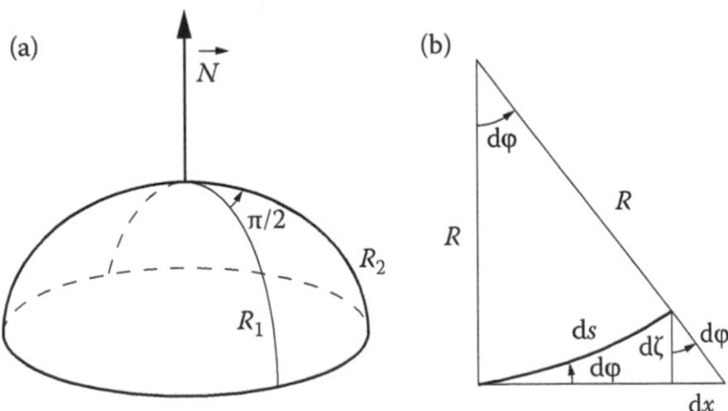

FIGURE 6.5
Point on a surface is a *regular point* if it has a unique unit normal vector $\vec{N}$, that is, it is not a cusp and it does not lie on an edge. Any plane passing through the unit normal vector at a regular point is a *normal plane* and cuts the surface along a *line of curvature*. If the surface is not locally a sphere, the curvature changes as the normal plane rotates around the unit normal; the lines of largest and smallest curvature specify the principal curvatures, and for a regular point, they lie in orthogonal directions, that is, the normal vector and unit tangent vectors to the lines of principal curvature on the surface form a Cartesian frame. The radii of curvature (R_1, R_2) along the lines of principal curvature (a) can be calculated as the radius of curvature R of a plane curve (b). The latter is determined (Figure 6.8b) by the radius of the circle that best approximates the curve in the neighborhood of each point.

(1) the arc length is given by Equation 6.69b as for a circle of radius R, and (2) the change in length is given by Equation 6.69c neglecting infinitesimals of higher than the second order:

$$(\mathrm{d}\varphi)^2 \ll 1: \quad \mathrm{d}\varphi = \frac{\mathrm{d}s}{R}, \quad \mathrm{d}x = \mathrm{d}\zeta \tan(\mathrm{d}\varphi) \sim \mathrm{d}\zeta\, \mathrm{d}\varphi = \mathrm{d}s \frac{\mathrm{d}\zeta}{R}. \qquad (6.69\text{a}–6.69\text{c})$$

The change in area is the product of changes in length in orthogonal directions:

$$\begin{aligned} \mathrm{d}S &= (\mathrm{d}s_1 + \mathrm{d}x_1)(\mathrm{d}s_2 + \mathrm{d}x_2) = \mathrm{d}s_1 \mathrm{d}s_2 \left(1 + \frac{\mathrm{d}\zeta}{R_1}\right)\left(1 + \frac{\mathrm{d}\zeta}{R_2}\right); \\ &= \mathrm{d}s_1\, \mathrm{d}s_2 \left[1 + \left(\frac{1}{R_1} + \frac{1}{R_2}\right)\mathrm{d}\zeta + O\left((\mathrm{d}\zeta)^2\right)\right] \end{aligned} \qquad (6.70)$$

neglecting higher order infinitesimals. Thus, the area increase is given by Equation 6.71a:

$$\mathrm{d}S - \mathrm{d}s_1\, \mathrm{d}s_2 = \left(\frac{1}{R_1} + \frac{1}{R_2}\right)\mathrm{d}\zeta = -\nabla . \left\{\left|1 + |\nabla\zeta|^2\right|^{-1/2} \nabla\zeta\right\}\mathrm{d}\zeta = -\nabla^2\zeta\left[1 + O\left(|\nabla\zeta|^2\right)\right]\mathrm{d}\zeta,$$

(6.71a and 6.71b)

in the nonlinear (Equation 6.71a) [linear (Equation 6.71b)] case, with minus the minimal surface operator (Laplacian) on the r.h.s. of Equation 6.19b ≡ Equation 6.71a (Equation 6.13c ≡ Equation 6.71b).

6.3.2 Boundary Condition with Tension, Pressure, and Surface Forces

The **boundary condition** *at a fluid interface equates (I) the surface tension times area change, expressed by (1) principal curvatures and (2 and 3) Laplacian (Equation 6.72a) [minimal surface operator (Equation 6.72c)] applied to the displacement in the linear (nonlinear) case:*

$$\vec{n}T\left(\frac{1}{R_1} + \frac{1}{R_2}\right) = -\vec{n}T\nabla^2\zeta\left\{1 + O\left(|\nabla\zeta|^2\right)\right\}, \qquad (6.72\text{a and } 6.72\text{b})$$

$$= -\vec{n}T\left|1 + (\nabla\zeta)^2\right|^{-3/2}\left\{\left[1 + |\nabla\zeta|^2\right]\nabla^2\zeta - \nabla\zeta\cdot\left[(\nabla\zeta\cdot\nabla)\nabla\zeta\right]\right\}, \qquad (6.72\text{c})$$

$$= (p_1 - p_2)\vec{n} + f_n\vec{n} + f_s\vec{s} + \nabla T\vec{s} \qquad (6.72\text{d})$$

that acts along the unit normal against (II) the combination (Equations 6.72c and 6.72d) of (Figure 6.6) (1) the pressure difference between the two sides of the interface that acts along the normal, (2) the gradient of the surface tension that acts in the tangential direction, and (3) any other normal f_n or tangential f_s component of the stress vector, which is the surface force per unit area. A zero tangential force leads to a constant surface tension:

$$f_s = 0 \Rightarrow \nabla \mathrm{T} = 0 \Leftrightarrow \mathrm{T} = const. \qquad (6.73)$$

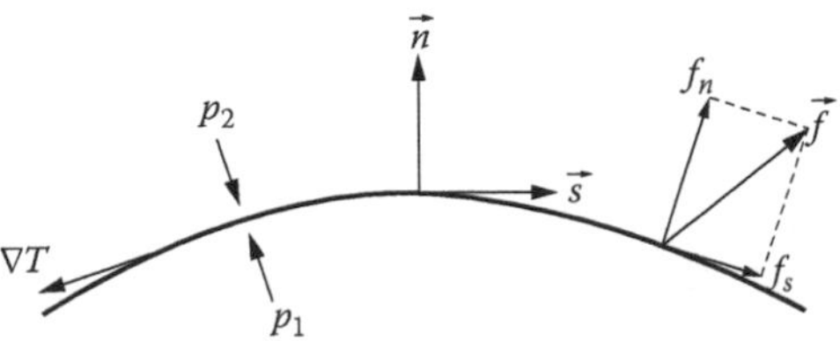

FIGURE 6.6
Boundary condition at the interface between two fluids in the presence of surface tension T involves (1) the pressure p_1 (p_2) on the inner (outer) side along (opposite to) the unit normal; (2) the surface tension and its gradient along the unit tangent $\vec{s}$ to a curve that is a cross section of the interface by a normal plane; and (3) the normal f_n (tangential f_s) components of any other surface forces, that is, forces per unit area. An example of tangential force is the shear stress due to a viscous fluid moving past the interface; in the absence of such a tangential surface force, the surface tension is constant. In the absence of other normal surface forces, it is the pressure difference across the interface that causes its curvature in the presence of surface tension.

The flow of an inviscid (viscous) fluid implies no tangential stress [a tangential viscous stress associated with the rate-of-strain (Note 4.4)]. In Equations 6.72a through 6.72d, the radius of the curvature has the dimensions of length, the tension T, the dimensions of force per unit length, the gradient of the tension ∇T and pressures (p_1, p_2), and the dimensions of force per unit area like (f_s, f_n).

6.3.3 Plateau (1849) Problem and Soap Bubbles

The curvature of the interface is due to the pressure difference alone (Equation 6.74c) in the absence of the other surface forces (Equations 6.74a and 6.74b):

$$f_n = 0 = f_s: \quad T\left(\frac{1}{R_1} + \frac{1}{R_2}\right) = p_1 - p_2. \tag{6.74a–6.74c}$$

If there is no pressure difference (Equation 6.75a) the surface has minimal area (Equation 6.75b) corresponding **(Euler 1744)** *to principal curvatures with the same modulus and opposite signs (Equation 6.75c):*

$$p_1 = p_2: \quad \frac{1}{R_1} + \frac{1}{R_2} = 0 \quad \Leftrightarrow \quad R_1 = -R_2 \tag{6.75a–6.75c}$$

like a saddle (Figure 6.7a). A bounded surface without boundary (Figure 6.7b) under tension is a sphere (Equation 6.76a), for example, a "soap bubble," because it has the minimum area for a given volume. The minimal surface of revolution whose generator is a fixed curve passing through two points outside the axis **(Meusnier 1776)** *is a* **catenoid** *(Figure 6.7c).* The sphere (catenoid) with radius a (radius a for the minimum cross section) is the surface generated by a circle with the same radius (Equation 5.12c ≡ Equation 6.76b) [a **catenary**, that is the curved specified by the hyperbolic cosine (Equation 6.77a)] by rotation (Equation 6.76a) around an axis passing through the center (Figure 6.7b) [an axis parallel to the tangent to the minimum (Figure 5.2e) at a distance a] leading to Equation 6.76c (Equation 6.77b):

$$r^2 \equiv x^2 + y^2 : a^2 = r^2 + z^2 = x^2 + y^2 + z^2 \quad \Leftrightarrow \quad z = \pm \left|a^2 - x^2 - y^2\right|^{1/2} \equiv \zeta_+ (x,y), \tag{6.76a–6.76c}$$

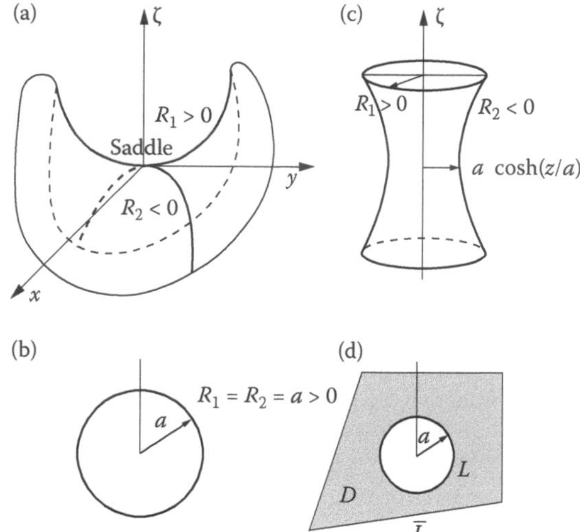

FIGURE 6.7
Surface under tension, such as a soap bubble held at a boundary, takes the shape with a minimum area. If the boundary is a curve with torsion, that is, a curve that does not lie on a plane, the surface with a minimum area supported on it is a minimal surface whose principal curvatures add to zero and thus have opposite signs. Examples of surfaces with principal curvatures with opposite signs include (a) parabolic hyperboloid that has a saddle point that is a minimum (maximum) along one line (the other line) of principal curvature and (b) the catenoid that is the axisymmetric surface obtained rotating the curve cosh or catenary around an axis in the same plane. The surface tension without boundary, like free soap bubbles, leads to a spherical surface (c) that has a minimum area for a given volume. Likewise, a surface under tension held at an outer boundary with any shape, with a flexible string as inner boundary, leads to a circular shape (d), that is, for a given length of string, the maximum internal area and minimum external area.

$$r/a = \cosh(z/a) \quad \Leftrightarrow \quad z = a \arg\cosh\left(\frac{\left|x^2+y^2\right|^{1/2}}{a}\right) \equiv \zeta_-(x,y), \qquad \text{(6.77a and 6.77b)}$$

The sphere is an **extremal surface** (**Lagrange 1760, 1776**) because it has the maximum volume for a given area; the constraint of fixed area introduces a Lagrange multiplier so that the minimal surface equation no longer applies. It will be confirmed in Subsection 6.3.5 that the sphere does not satisfy the minimal surface equation; since the sphere has the same constant curvature in all directions, the two principal curvatures are equal in modulus and sign, and it cannot satisfy the condition (Equation 6.75c) of a minimal surface. In contrast the catenoid defined as the surface of revolution of minimum area generated by a curve of fixed length passing through two points outside the axis of symmetry is subject to no additional constraint; thus the catenoid satisfies the equation of minimal surfaces (Subsection 6.3.6), and its principal curvatures have opposite signs (Subsection 6.3.7) with equal modulus at the minimum radius (Figure 6.7c) corresponding to the saddle point (Figure 6.7a). The catenoid is an example of the **Plateau (1849) problem:** finding the minimal surface supported on a given boundary, for example, the shape of a soap film on a wire boundary. In the case of the catenoid, it is the surface of the soap film supported on two circles orthogonal to a common axis. A minimal surface may be supported on a multiply connected boundary, either in space or in the plane. An example of the latter is *a plane soap film in a doubly connected domain (Figure 6.7d) consisting of a rigid outer boundary and an inner* *boundary made of a flexible string of fixed length. The shape of the hole is circular* because (1) the area of the domain must be minimum, (2) hence the area of the hole must be maximum, and (3) the curve of a given length enclosing the maximum area is the circle.

6.3.4 Necessary Condition for a Minimal Surface (Langrange 1760, 1776)

A **minimal surface** satisfies (**Lagrange 1776**) the equation (6.78b):

$$f = 0: \qquad 0 = [1 + (\nabla\zeta \cdot \nabla\zeta)]\nabla^2\zeta - \nabla\zeta \cdot [(\nabla\zeta \cdot \nabla)\,\nabla\zeta] \equiv M\,\{\zeta\}, \qquad \text{(6.78a and 6.78b)}$$

that follows from Equation 6.20b in the absence of shear stress (Equation 6.78a). The **minimal surface operator** (Equation 6.78b) is written in Cartesian coordinates using the two-dimensional gradient (Equation 6.4b), leading to Equations 6.79a and 6.79b for the second set of terms:

$$(\nabla\zeta\cdot\nabla)\nabla\zeta = \left[(\partial_x\zeta)\partial_x + (\partial_y\zeta)\partial_y\right]\left[(\partial_x\zeta)\vec{e}_x + (\partial_y\zeta)\vec{e}_y\right], \qquad \text{(6.79a)}$$

$$\nabla\zeta \cdot [(\nabla\zeta\cdot\nabla)\nabla\zeta] = (\partial_x\,\zeta)^2\,(\partial_{xx}\,\zeta) + (\partial_y\,\zeta)^2\,(\partial_{yy}\,\zeta) + 2\,(\partial_x\,\zeta)\,(\partial_y\,\zeta)\,(\partial_{xy}\,\zeta); \qquad \text{(6.79b)}$$

the first set of terms in Equation 6.78b is given in Cartesian coordinates by:

$$\lfloor 1 + |\nabla\zeta|^2 \rfloor\,\nabla^2\zeta = [1 + (\partial_x\,\zeta)^2 + (\partial_y\,\zeta)^2]\,(\partial_{xx}\,\zeta + \partial_{yy}\,\zeta). \qquad \text{(6.79c)}$$

The two sides Equation 6.79b ≡ Equation 6.79c of Equation 6.78b lead to:

$$0 = \lfloor 1 + (\partial_y\,\zeta)^2 \rfloor\,(\partial_{xx}\,\zeta) + \lfloor 1 + (\partial_y\,\zeta)^2 \rfloor\,\partial_{xx}\,\zeta - 2(\partial_x\zeta)(\partial_y\zeta)(\partial_{xy}\,\zeta) \equiv \mathrm{M}\,\{\zeta(x,y)\}, \qquad \text{(6.80)}$$

the equation of minimal surfaces (Equation 6.78b) in Cartesian coordinates (Equation 6.80). This condition is necessary but not sufficient since (1) the first variation of the area was considered, that is, a condition that it is stationary, and (2) a minimum requires proof that the second variation is positive. As an example of extremal surface without boundary, the sphere is considered next.

6.3.5 Sphere as an Extremal Surface (Euler 1744)

The sphere cannot be a minimal surface because it has equal principal curvatures and thus cannot satisfy Equation 6.75c. The sphere is an extremal surface in that of all surfaces it has the largest volume for a given area (and the least area for a given volume). The constraint of a given area (volume) introduces a Lagrange multiplier and modifies the equation of minimal surfaces. It is shown next that the sphere (Equation 6.76c) does not satisfy the Equation 6.80 of minimal surfaces, but rather the condition of constant curvature that also involves the unit normal vector. The verification of some of these statements starts (Equation 6.76c) with the partial derivatives of first (Equation 6.81a and 6.81b) and second (Equations 6.81c through 6.81e) order:

$$\left\{\partial_x\zeta, \partial_y\zeta\right\} \equiv \left\{\zeta_x, \zeta_y\right\} = -\frac{\{x,y\}}{\zeta}, \quad \left\{\zeta_{xx}, \zeta_{yy}, \zeta_{xy}\right\} = -\frac{1}{\zeta}\left\{1+\frac{x^2}{\zeta^2},\ 1+\frac{y^2}{\zeta^2},\ \frac{xy}{\zeta^2}\right\}, \qquad \text{(6.81a–6.81e)}$$

where the following was used:

$$\left\{\frac{\partial}{\partial x},\frac{\partial}{\partial y}\right\}\frac{1}{\zeta}\equiv\left\{\partial_x,\partial_y\right\}\frac{1}{\zeta}=-\frac{1}{\zeta^2}\left\{\partial_x\zeta,\partial_y\zeta\right\}=\frac{\{x,y\}}{\zeta^3}. \qquad \text{(6.82a and 6.82b)}$$

From Equations 6.81a through 6.81e follow the terms in Equation 6.80, namely,

$$\zeta^2\left\{1+\zeta_x^2,1+\zeta_y^2\right\}=\left\{\zeta^2+x^2,\zeta^2+y^2\right\}=\left\{a^2-y^2,a^2-x^2\right\}, \qquad \text{(6.83a and 6.83b)}$$

$$-\zeta^3\left\{\zeta_{xx},\zeta_{yy},\zeta_{xy}\right\}=\left\{\zeta^2+x^2,\zeta^2+y^2,xy\right\}=\left\{a^2-y^2,a^2-x^2,xy\right\}. \qquad \text{(6.83c–6.83e)}$$

Substitution of Equations 6.83a through 6.83e in Equation 6.80 proves that the sphere (Equation 6.76a) is not a minimal surface;

$$\begin{aligned}&(1+\zeta_y^2)\zeta_{xx}+(1+\zeta_y^2)\zeta_{xx}-2\zeta_x\zeta_y\zeta_{xy}\\&=-(2/\zeta^5)(a^2-x^2)(a^2-y^2)=2(a^2/\zeta^5)(a^2-x^2-y^2)=-2a^2/\zeta^3,\end{aligned} \qquad \text{(6.83f)}$$

because the r.h.s. of Equation 6.83f is not zero.

The area element or unit normal (Equation 6.4a) for a sphere (Equations 6.76c, 6.81a, and 6.81b) appears in the r.h.s. (Equation 6.84b) of Equation 6.83f:

$$\frac{dS}{dx\,dy}=\frac{1}{N_z}=\left|1+(\zeta_x)^2+(\zeta_y)^2\right|^{1/2}: \qquad \frac{a^2}{\zeta^2}=1+\frac{x^2}{\zeta^2}+\frac{y^2}{\zeta^2}=1+\zeta_x^2+\zeta_y^2=N_z^{-2}. \qquad \text{(6.84a and 6.84b)}$$

Substituting Equation 6.84b in Equation 6.83f it follows that the sphere (Equation 6.76c) satisfies the Equation (6.84d):

$$\lambda\equiv 2/a: \qquad M\left\{\zeta_+(x,y)\right\}=-(2/a)\left|1+\zeta_x^2+\zeta_y^2\right|^{3/2}=-\lambda N_z^{-3}, \qquad \text{(6.84c and 6.84d)}$$

that consists of: (i) the minimal surface operator (Equation 6.80); (ii) plus a term involving the unit normal (Equation 6.84a) and the constant Lagrange multiplier (6.84c). The magnitude of the Lagrange multiplier indicates the severity of the constraint of fixed area for maximum volume (or fixed volume for minimum area); thus the sphere deviates more from a minimal surface for larger Lagrange multiplier, that is, smaller radius. In the absence of boundaries the minimal surface is a plane that satisfies Equation 6.75c trivially with zero principal curvatures, that is, infinite principal radii of curvature; thus an increasing deviation is a larger curvature or a smaller radius of curvature. It has been shown that *the sphere (Equation 6.76c) of radius a satisfies the Equation 6.84c involving the minimal surface operator (Equation 6.80), the unit normal vector (Equation 6.84a) and a Lagrange multiplier (Equation 6.84c). The Equation 6.84d corresponds to a surface with constant curvature, that is the particular case of equal principal curvatures* $R_1 = a = R_2$ *of*

$$M\{\zeta\}=-\left(\frac{1}{R_1}+\frac{1}{R_2}\right)\left|1+(\nabla\zeta\cdot\nabla\zeta)\right|^{3/2}=-2\mathrm{H}\mathrm{N}_z^{-3}, \quad 2\mathrm{H}\equiv\frac{1}{R_1}+\frac{1}{R_2}=k_1+k_2, \qquad \text{(6.84e and 6.84f)}$$

the general relation (Equation 6.84e) between the principal radii of curvature (Equation 6.72c) and the minimal surface operator (Equation 6.78b). It follows that the deviation of an arbitrary surface (Equation 6.84e) from a minimal surface (Equation 6.78b) is proportional to the inverse cube of the unit normal (Equation 6.4a) through the Lagrange multiplier (Equation 6.84f), that coincides with the ***mean curvature*** *H defined equivalently by: (i) the arithmetic mean of the principal curvatures k_1 and k_2; (ii) the sum of the inverses of the principal radii of curvature.* In the case of a sphere (minimal surface) the principal radii of curvature are equal $R_1 = a = R_2$ (are equal in modulus with opposite signs $R_1 + R_2 = 0$), and Equation 6.84e reduces to the condition of constant curvature (Equation 6.84c and 6.84d) [of minimal surface (Equation 6.78b)].

6.3.6 Comparison of Soap Bubbles and Elastic Membranes

For an axisymmetric surface (Equation 6.85a) the gradient (Equation 6.85b) implies that the unit normal takes the form (Equation 6.85c):

$$z = \zeta(r), \quad \nabla\zeta = \vec{e}_r \frac{d\zeta}{dr} \equiv \vec{e}_r \zeta', \quad \frac{1}{N_z} = \left|1 + (\nabla\zeta \cdot \nabla\zeta)\right|^{1/2} = \left|1 + \zeta'^2\right|^{1/2}. \qquad (6.85a\text{–}6.85c)$$

The latter appears in the balance Equation 6.45a ≡ Equation 6.85e for (Equation 6.85d) a membrane with constant tangential tension (Equation 6.46a) and normal pressure difference (Equation 6.46b) whose shape is a sphere:

$$\frac{\Delta p}{T} = \lambda = \text{const}: \qquad \zeta'' + \frac{\zeta'}{r}(1 + \zeta'^2) = -\frac{\Delta p}{T}\left|1 + \zeta'^2\right|^{3/2} = -\lambda N_z^{-3}. \qquad (6.85d\text{–}6.85e)$$

Since both represent a sphere of radius a the Equation 6.84d ≡ Equation 6.85e must coincide, implying that: (i) the l.h.s. of Equation 6.85e must be the axisymmetric minimal surface operator (Equation 6.85f):

$$M\{\zeta(r)\} = \zeta'' + \frac{\zeta'}{r}(1 + \zeta'^2), \qquad a = \frac{\Delta p}{T}, \qquad (6.85f \text{ and } 6.85g)$$

as will be confirmed in the sequel (Equation 6.86b ≡ Equation 6.86d ≡ Equation 6.85f); (ii) the Equation 6.85d shows that the soap bubble deviates most from a minimal surface when the Lagrange multiplier is large, that is for large normal pressure difference and small tangential tension; (iii) the condition (ii) is equivalent to small radius because the Lagrange multipliers coincide (Equation 6.84c ≡ Equation 6.85d) leading to Equation 6.85g (compare with Equation 6.52c). It follows that *the radius of a spherical bubble is the ratio of twice the tangential tension to the normal pressure difference (Equation 6.85g).* The spherical shape is easiest to obtain for a soap bubble because both the tangential tension and normal pressure difference are constant and the deformation of the surface corresponds to a linear stress–strain relation. In the case of an elastic membrane the linear deformation with small slope (Subsection 6.1.5) leads to a parabolic generator (Equation 6.30a) so the shape is a section of a paraboloid of revolution; this is the local approximation to the spherical sector (Equation 6.54a) that is the shape of the nonlinear deflection of an elastic membrane with large slope. This shape as a spherical section may not apply to an inelastic membrane for which large deflections imply a nonlinear stress–strain relation. It has been shown that *the sphere (Equation 6.76c): (i) does not satisfy (Equation 6.83f) the equation of minimal surfaces (Equation 6.80); (ii) it does satisfy the Equation 6.83f ≡ Equation 6.84d adding the constraint of fixed area (Equation 6.84a) through the constant Lagrange multiplier (Equation 6.84c); (iii) the same multiplier (Equation 6.84c ≡ Equation 6.85d) applies to the deflection of a membrane under constant tangential tension and normal pressure difference (Equation 6.52c) whose shape is a sphere of radius (Equation 6.85g) specified by the condition of constant curvature in all directions (Equation 6.45a ≡ Equation 6.83f ≡ Equation 6.84c and 6.84d); (iv) the first term of Equation 6.84d (≡Equation 6.85e) is the minimal surface operator (Equation 6.78b) in Cartesian (Equation 6.80) [axisymmetric (Equation 6.85f ≡ Equation 6.86b ≡ Equation 6.86d) form.* The last statement is confirmed next (Subsection 6.3.6) obtaining the axisymmetric minimal surface operator, and showing that its solution is the catenoid, thus proving that it is a minimal surface, as stated before (Subsection 6.3.3).

6.3.7 Axisymmetric Minimal Surfaces (Meusnier 1776)

The equation (6.78b) of minimal surfaces can be obtained in axisymmetric form (Equation 6.85a) using the gradient (Equation 6.85b) [Laplacian (Equation 6.86a)] in cylindrical (or polar) coordinates [Equation I.11.31b (Equation I.11.28b)] leading to Equation 6.86b:

$$\nabla^2\zeta = \frac{1}{r}\frac{\mathrm{d}}{\mathrm{d}r}\left(r\frac{\mathrm{d}\zeta}{\mathrm{d}r}\right) = \frac{1}{r}\left(\frac{\zeta'}{r}\right)', \quad 0 = (1+\zeta'^2)\frac{\mathrm{d}}{\mathrm{d}r}(\zeta' r) - \zeta''\zeta'^2 \equiv M\{\zeta(r)\}. \quad \text{(6.86a and 6.86b)}$$

In Equation 6.86b the azimuthal derivatives (Equation 6.86c) for an axisymmetric surface were omitted, leaving as only variable the distance from the symmetry axis (Equation 6.76a) in the *equation of axisymmetric (Equation 6.85a) minimal surfaces (Equation 6.86b ≡ Equation 6.86d)*:

$$\frac{\partial}{\partial\varphi} = 0: \quad 0 = (1+\zeta'^2)\left(\zeta'' + \frac{\zeta'}{r}\right) - \zeta''\zeta'^2 = \zeta'' + \frac{\zeta'}{r}(1+\zeta'^2) \equiv M\{\zeta(r)\}; \quad \text{(6.86c and 6.86d)}$$

this confirms that the first term in Equation 6.85e is the axisymmetric minimal surface operator (Equation 6.85f ≡ Equation 6.86b ≡ Equation 6.86d). Rewriting Equation 6.86d ≡ Equation 6.87a allows separation of variables (Equation 6.87b):

$$-\zeta'\left(1+\zeta'^2\right) = r\zeta'' = r\frac{\mathrm{d}\zeta'}{\mathrm{d}r}, \quad -\frac{\mathrm{d}r}{r} = \frac{\mathrm{d}\zeta'}{\zeta'\left(1+\zeta'^2\right)} = \left(\frac{1}{\zeta'} - \frac{\zeta'}{1+\zeta'^2}\right)\mathrm{d}\zeta' \quad \text{(6.87a and 6.87b)}$$

and immediate integration (Equation 6.87c) where a is a constant:

$$-\log\left(\frac{r}{a}\right) = \log\zeta' - \frac{1}{2}\log\left(1+\zeta'^2\right) = \log\left[\zeta'\left|1+\zeta'^2\right|^{-1/2}\right]. \quad \text{(6.87c)}$$

Solving Equation 6.87c ≡ Equation 6.88a for ζ' leads to Equation 6.88b:

$$\zeta'_-\left|1+\zeta_-'^2\right|^{-1/2} = \frac{a}{r}, \quad \zeta'_- = a\left|r^2 - a^2\right|^{-1/2}, \quad \text{(6.88a and 6.88b)}$$

whose primitive (Equation 6.88c) specifies a catenoid:

$$\zeta_-(r) - b = a\,\mathrm{arg\,cosh}\left(\frac{r}{a}\right) = a\,\mathrm{arg\,cosh}\left(\frac{\left|x^2+y^2\right|^{1/2}}{a}\right). \quad \text{(6.88c)}$$

Changing the sign on the l.h.s. of Equation 6.87c is equivalent to the interchange (Equation 6.89a) in Equation 6.88a and leads to Equation 6.89b ≡ Equation 6.89c:

$$(a,r)\to(r,a):\quad \zeta'_+\left|1+\zeta'^2_+\right|^{-1/2}=-\frac{r}{a},\quad \zeta'_+=-r\left|a^2-r^2\right|^{-1/2}, \tag{6.89a–6.89c}$$

whose primitive (Equation 6.89d) specifies a sphere:

$$\zeta_+(r)-b=\left|a^2-r^2\right|^{1/2}=\left|a^2-x^2-y^2\right|^{1/2}. \tag{6.89d}$$

A translation along the axis of symmetry places the center of the sphere (the cross section of minimum radius of the catenoid) at the origin, leading to the coincidence of Equation 6.89d ≡ Equation 6.76c (Equation 6.88c ≡ Equation 6.77b). It shown next that the sphere (catenoid) are surfaces of revolution that: (i) have generators, namely the circle (catenary) with positive (negative) curvature; (ii) since the other principal curvature is the radius of rotation that is positive, the sphere (catenoid) have principal curvatures with the same (opposite) signs, that is, is a synclastic (anticlastic) surface; (iii) a minimal surface must have (Equation 6.75c) principal curvatures with the same modulus and opposite signs, and therefore must be an anticlastic surface; (iv) thus the sphere cannot be a minimal surface; (v) since the catenoid is a minimal surface its principal curvatures must have the same modulus with opposite signs at the saddle point (Figure 6.7a) of minimum radius. This is confirmed next (Subsection 6.3.7).

6.3.8 Synclastic/Anticlastic Surfaces (Gauss 1820)

The sphere (catenoid) are considered next as simple examples of **synclastic (anticlastic) surfaces (Gauss 1820)**, that is surfaces whose principal curvatures (Equations 6.70, 6.71a, and 6.72a) have the same (Equation 6.90a) [opposite (Equation 6.90b)] signs:

$$\text{surface}\begin{cases}\text{synclastic:} & R_1R_2>0, & (6.90a)\\ \text{anticlastic:} & R_1R_2>0. & (6.90b)\end{cases}$$

A synclastic surface looks like a sail in the wind or a pillow; an anticlastic surface looks like a saddle. The one-dimensional form of Equation 6.71a specifies the radius of curvature R and the curvature k of a plane curve:

$$k=\frac{1}{R}=\left\{\zeta'\left|1+\zeta'^2\right|^{-1/2}\right\}'=\zeta''\left|1+\zeta'^2\right|^{-3/2}, \tag{6.90e}$$

where the sign has been reversed by taking the z-axis upward instead of downward. The curvature is calculated next for the generators of the sphere (catenoid) as surfaces of revolution, that is the circle (Equation 6.89d) [catenary (Equation 6.88c)]. *The generator of the sphere, that is the circle (Equation 6.91a) of radius a has slope (Equation 6.91b) leading to (Equations 6.91c and 6.91d) a curvature (Equation 6.91e) that is a positive constant equal to the inverse of the radius:*

$$\zeta_+(r)=-\left|a^2-r^2\right|^{1/2},\quad \zeta'_+=r\left|a^2-r^2\right|^{-1/2},\quad 1+\zeta'^2_+=a^2\left(a^2-r^2\right)^{-1}, \tag{6.91a–6.91c}$$

$$\zeta''_+=\left|a^2-r^2\right|^{-1/2}+r^2\left|a^2-r^2\right|^{-3/2}=a^2\left|a^2-r^2\right|^{-3/2}, \tag{6.91d}$$

$$k_+ = \zeta_+'' \left|1+\zeta_+'^2\right|^{-3/2} = a^2\left|a^2-r^2\right|^{-3/2}\left[a^2\left(a^2-r^2\right)^{-1}\right]^{-3/2} = \frac{1}{a} > 0. \tag{6.91e}$$

The catenary, that is the generator of the catenoid (Equation 6.92a) with minimum radius a, has slope (Equation 6.92b) implying (Equations 6.91c and 6.92d) a curvature (Equation 6.92e) that is negative and not constant:

$$\zeta_-(r) = a \arg\cosh(r/a), \quad \zeta_-' = a\left|r^2-a^2\right|^{-1/2}, \tag{6.92a and 6.92b}$$

$$1+\zeta_-'^2 = r^2(r^2-a^2)^{-1}, \quad \zeta_-'' = -ar\left|r^2-a^2\right|^{-3/2}, \tag{6.92a and 6.92d}$$

$$k_-(r) = \zeta_-''(1+\zeta_-'^2)^{-3/2} = -ar\left|r^2-a^2\right|^{-3/2}\left[r^2(a^2-r^2)^{-1}\right]^{-3/2} = -\frac{a}{r^2} < 0. \tag{6.92c}$$

For all surfaces of revolution the other principal curvature is the radius (Equation 6.92f):

$$R_2^\pm(r) = r: \quad -(R_1^-)_{\min} = \left|R_1^-\right|_{\max} = -\frac{1}{k_-(a)} = \frac{1}{a} = R_2^\pm(a) = R_1^+(a). \tag{6.92f and 6.92g}$$

and therefore (Equation 6.92g): (i) the minimum radius of curvature of the catenoid is minus the radius at the smallest section where the sum of curvatures is zero (Equation 6.75c) corresponding (Figure 6.7c) to a saddle point (Figure 6.7a); (ii) the same value in modulus is the radius of curvature of the sphere in all directions, that could not meet the condition of a minimal surface (Equation 6.75c). The principal curvatures of a surface are the maximum and minimum curvatures of its normal sections at a point and thus are the curvatures of plane curves. The curvature of a plane curve is applied next (Section 2.4) to a phenomenun related to surface tension, namely, the capillary rise of a fluid wetting one (or more) wall(s).

6.4 Wetting Angle and Capillary Rise

The surface tension causes a liquid to attach to a vertical flat wall at an angle (Subsection 6.4.2) and to rise between two vertical plates (Subsection 6.4.3), both cases involving a single curvature (Subsection 6.4.1). In the case of two vertical plates, the capillary rise is different on the inner and outer sides of the walls (Subsection 6.4.4); the exact solutions in terms of elliptic functions take an elementary explicit form in the case of close vertical plates (Subsection 6.4.5). This specifies the capillary rise at the midpoint between (inside and outside of) two vertical plates [Subsection 6.4.6 (Subsection 6.4.7)].

6.4.1 Tangent Circle and Radius of Curvature

For a **cylindrical surface,** (1) one principal curvature is zero in the direction of the **generators** (Figure 6.8a), and (2) the nonzero principal curvature is orthogonal to the generators and corresponds to the **directrix** (Figure 6.8b). The latter is a plane curve whose **radius of**

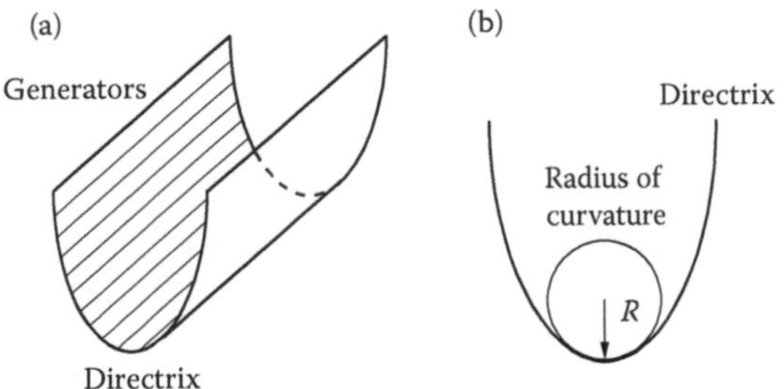

FIGURE 6.8
Cylindrical surface is obtained by translation (a) of a plane curve, the directrix, normal to its plane, so that the generators are straight lines, with zero curvature, and the nonzero principal curvature is associated with the directrix (b).

curvature R is that of the tangent circle that most closely approximates the curve. Since the equation of a circle involves three parameters, namely, the radius R and coordinates of the center, for example, the origin in Equation 6.93a:

$$x^2 + \zeta^2 = R^2, \qquad \zeta(x) \in \zeta^2, \tag{6.93a and 6.93b}$$

it can approximate an arbitrary curve (Equation 6.93b) to second order at a point with a continuous second-order derivative. The variation of the radius of curvature would be of the third order, and thus it is neglected; this implies that the radius of curvature may be taken as a constant when differentiating (Equation 6.94a) twice (Equations 6.94b and 6.94c) the equation (Equation 6.93a) of the circle:

$$\zeta' \equiv \frac{d\zeta}{dx}: \quad 0 = x + \zeta'\zeta, \quad 0 = 1 + \zeta'^2 + \zeta''\zeta. \tag{6.94a–6.94c}$$

Expressing the coordinates (x, ζ) in Equations 6.94b and 6.94c in terms of the derivatives (ζ', ζ'') leads to

$$\zeta = -\frac{1+\zeta'^2}{\zeta''}, \quad x = \zeta'\frac{1+\zeta'^2}{\zeta''^2}; \tag{6.95a and 6.95b}$$

substitution of Equations 6.95a and 6.95b in Equation 6.93a specifies the radius of curvature as a function of the first two derivatives:

$$R^2 = \left(1+\zeta'^2\right)\left(\frac{1+\zeta'^2}{\zeta''}\right)^2 = \frac{\left(1+\zeta'^2\right)^3}{\zeta''^2}. \tag{6.96}$$

From Equation 6.96 *follows the* **curvature** *(Equation 6.97b) and* **radius of curvature** *(Equation 6.97a) of a curve with continuous second-order derivative (Equation 6.93b):*

$$R = \pm\frac{1}{\zeta''}\left|1+\zeta'^2\right|^{3/2}, \quad k = \frac{1}{R} = \pm\zeta''\left|1+\zeta'^2\right|^{3/2}; \tag{6.97a and 6.97b}$$

the upper sign in Equation 6.97a and 6.97b is chosen in order that a minimum $\zeta' = 0 > \zeta''$ (maximum $\zeta' = 0 > \zeta''$), that is a slope ζ' increasing (decreasing) with x corresponds to a positive (Equation 6.98a) [negative (6.98b)] curvature:

$$k = \frac{1}{R} = \zeta'' \left|1 + \zeta'^2\right|^{-3/2} \begin{cases} > 0 & \text{if} \quad \zeta'' > 0 = \zeta' \quad \text{or minimum} \\ < 0 & \text{if} \quad \zeta'' < 0 = \zeta' \quad \text{or maximum.} \end{cases} \tag{6.98a, 6.98b}$$

The one-dimensional form of Equation 6.72b ≡ Equation 6.72c would lead to the curvature with the lower, negative sign in Equation 6.97b, because the ζ-axis was taken downward instead of upward. Reversing the direction of either the x-axis or ζ-axis changes the sign of the curvature; it follows that the translation of a curve from the first to the third (second or fourth) quadrant keeps (reverses) its sign. For example in the Equation 6.76c of the sphere of radius a, either the upper (Equation 6.89d) [or lower (Equation 6.91a)] sign can be taken, corresponding to the upper (lower) half. The lower half (Equation 6.91a) has a minimum (Equation 6.98a) and leads to a positive curvature (Equation 6.91e); if the upper half of the sphere had been chosen (Equation 6.89d) the maximum (Equation 6.98b) would have lead to a negative curvature, that is Equation 6.9le with reversed sign. Thus it the modulus of the curvature that provides an invariant measure of the curvature of a curve; the curvature of the same curve can change in sign by convention (Equations 6.97a and 6.97b), and for the same convention by translation relative to the coordinate axis.

6.4.2 Wetting Angle at a Plane Vertical Wall

In the case of hydrostatic equilibrium of a fluid with constant density ρ in a uniform gravity field g, the pressure difference balancing the surface tension is the weight (Equation 6.99a) of a column of fluid of height ζ leading Equation 6.74c to Equation 6.99b:

$$p_2 - p_1 = \rho g \zeta + \text{const:} \quad \rho g \zeta + T\left(\frac{1}{R_1} + \frac{1}{R_2}\right) = \text{const.} \tag{6.99a and 6.99b}$$

In the case (Figure 6.9a) of a cylindrical surface wetting a vertical plane wall, (1) one curvature is zero (Equation 6.100a), and (2) the other curvature vanishes (Equation 6.100c) far from the wall, since the surface is plane there (Equation 6.100b). Calculating the constant

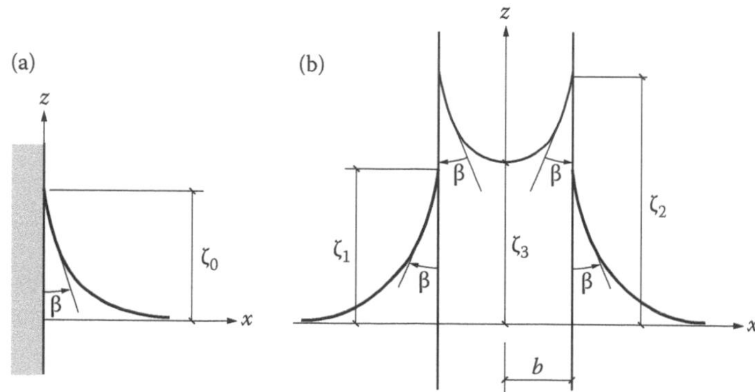

FIGURE 6.9
Surface tension causes a fluid to rise against gravity near a wall, leading (a) to a wetting angle $\beta < \pi/2$, excluding a purely horizontal surface, except asymptotically at a large distance from the wall. In the case of two parallel plates inserted in a fluid (b), the wetting angle β is the same inside and outside, and the surface tension causes a capillary effect raising the fluid inside relative to the fluid outside.

in Equation 6.99b at a large distance, it follows from Equations 6.100a and 6.100c that it is zero leading to Equation 6.100d:

$$R_2 = \infty; \quad \lim_{x\to\infty}\{\zeta, \zeta', \zeta''\} = 0: \quad \lim_{x\to\infty} R_1 = \infty, \frac{\rho g \zeta}{T} = -\frac{1}{R} = \left|1+\zeta'^2\right|^{-3/2} \zeta'', \qquad (6.100a–6.100d)$$

where the radius of curvature Equation 6.98 was used. The first integration of Equation 6.100d is performed multiplying by ζ':

$$0 = \frac{\rho g}{T}\zeta'\zeta - \left(1+\zeta'^2\right)^{-3/2}\zeta'\zeta'' = \frac{d}{dx}\left\{\frac{\rho g}{2T}\zeta^2 + \left(1+\zeta'^2\right)^{-1/2}\right\} \qquad (6.101)$$

and leads to Equation 6.102b where the constant is evaluated at infinity using Equation 6.100b:

$$c^2 \equiv \frac{2T}{\rho g}: \quad \left(\frac{\zeta}{c}\right)^2 + \left|1+\zeta'^2\right|^{-1/2} = \frac{\rho g \zeta^2}{2T} + \left|1+\zeta'^2\right|^{-1/2} = \text{const} = 1; \qquad (6.102a \text{ and } 6.102b)$$

the only parameter in the equation of the free surface (Equation 6.102b) is the **capillary constant** that is equal to the square root of twice the tension divided by the weight and has the dimensions of length since ζ/c is dimensionless. From Equation 6.102b follows:

$$\frac{\rho g}{2T}\left[\zeta(0)\right]^2 - 1 = -\left\{1 + \left[\zeta'(0)\right]^2\right\}^{-1/2} = -\left(1+\cot^2\beta\right)^{-1/2} = -\left(\csc^2\beta\right)^{-1/2} = -\sin\beta, \qquad (6.103)$$

that involves the **wetting angle** β at the wall (Figure 6.9a) given by Equation 6.104a and leading to Equation 6.104b:

$$\zeta'(0) = -\cot\beta: \quad \zeta_0 \equiv \zeta(0) = \sqrt{\frac{2T}{\rho g}}\sqrt{1-\sin\beta} = c\sqrt{1-\sin\beta}. \qquad (6.104a \text{ and } 6.104b)$$

The wetting angle (Equation 6.104a) is related to the **capillary rise** *(Equation 6.104b) through the capillary constant (Equation 6.102b). There is no capillary rise for horizontal wetting:*

$$0 \le \beta \le \frac{\pi}{2}: \quad 0 = \zeta_0\left(\frac{\pi}{2}\right) \le \zeta_0(\beta) \le \zeta(0) = c = \sqrt{\frac{2T}{\rho g}} = \zeta_{max}, \rho g \zeta_{max} = \frac{2T}{\zeta_{max}}, \qquad (6.105a–6.105c)$$

and the maximum capillary rise (Equation 6.105b) corresponds to a column of fluid whose weight is twice the surface tension divided by the height (Equation 6.105c). The factor 2 arises because the fluid wets only one side of the surface.

6.4.3 Capillary Rise between Parallel Vertical Plates

Equation 6.100d also applies to the fluid between and outside two parallel vertical plates (Figure 6.9b), leading to Equation 6.102b where (1) the constant is again unity (Equation 6.102b ≡ Equation 6.106a) outside the walls because then the condition of asymptotically flat horizontal surface (Equation 6.100b) holds, and (2) the condition fails between the parallel vertical plates (Equation 6.106b), so the determination of the constant A is part of the solution of the problem:

$$\frac{\zeta^2}{c^2} + \left|1+\zeta'^2\right|^{-1/2} = \begin{cases} 1 & \text{if} \quad |x| > b, \\ A & \text{if} \quad |x| < b. \end{cases} \quad \begin{matrix} (6.106a) \\ (6.106b) \end{matrix}$$

Thus, (1) the capillary rise is (Equation 6.104b ≡ Equation 6.107a) outside the plates; (2) between the plates is the same (Equation 6.104b) with 1 in Equation 6.106a replaced by A in Equation 6.106b, leading to Equation 6.107b:

$$\{\zeta_1, \zeta_2\} \equiv \{\zeta(\pm b \pm 0),\ \zeta(\pm b \mp 0)\} = c\left\{\sqrt{1-\sin\beta},\ \sqrt{A-\sin\beta}\right\}; \qquad \text{(6.107a and 6.107b)}$$

and (3) by symmetry, the slope must vanish (Equation 6.108a) at the midpoint between plates leading to Equation 6.108c:

$$\zeta'(0) = \lim_{x\to 0}\frac{d\zeta}{dx} = 0: \quad \lim_{x\to 0}\beta = \frac{\pi}{2}, \quad \zeta(0) = c\sqrt{A-1} \equiv \zeta_3. \qquad \text{(6.108a–6.108c)}$$

Solving (Equation 6.106b) for ζ':

$$\frac{d\zeta}{dx} = \zeta' = \sqrt{\frac{1}{\left(A-\zeta^2/c^2\right)^2} - 1} = \frac{\sqrt{1-\left(A-\zeta^2/c^2\right)^2}}{A-\zeta^2/c^2} \qquad (6.109)$$

leads to the integral [Equation 6.110b (Equation 6.111b)] outside (between) the plates, that is, Equation 6.110a (Equation 6.111a):

$$|x| > b: \quad x - b = \int_{\zeta_1}^{\zeta}\left(1-\frac{\zeta^2}{c^2}\right)\left[1-\left(1-\frac{\zeta^2}{c^2}\right)^2\right]^{-1/2} d\zeta, \qquad \text{(6.110a and 6.110b)}$$

$$|x| < b: \quad x = \int_{\zeta_3}^{\zeta}\left(A-\frac{\zeta^2}{c^2}\right)\left[1-\left(A-\frac{\zeta^2}{c^2}\right)^2\right]^{-1/2} d\zeta, \qquad \text{(6.111a and 6.111b)}$$

since $A = 1$ ($A \neq 1$) in Equation 6.106a (Equation 6.106b) and $\zeta = \zeta_1$ ($\zeta = \zeta_3$) at the plates $x = b$ (center $x = 0$).

6.4.4 Exact Solution in Terms of Elliptic Integrals

The change of variable

$$\zeta = c\sqrt{A-\cos\eta}: \quad \cos\eta = A - \frac{\zeta^2}{c^2}, \quad 1-\left(A-\frac{\zeta^2}{c^2}\right)^2 = \sin^2\eta \qquad \text{(6.112a–6.112c)}$$

with differential (Equation 6.113a) transforms Equation 6.111b to an elliptic integral (Equation 6.113b):

$$2\,d\zeta = \frac{c}{\sqrt{A-\cos\eta}}\sin\eta\,d\eta, \quad 2x = c\int_0^{\eta}\left|A-\cos\eta\right|^{-1/2}\cos\eta\,d\eta, \qquad \text{(6.113a and 6.113b)}$$

where (1) the lower limit of integration is the midposition between the vertical plates corresponding to Equation 6.114a in Equation 6.108c, (2) comparing Equation 6.108c with Equation 6.112a follows Equation 6.114b, and (3) the latter leads to Equation 6.114c as the lower limit in the integral (Equation 6.113b) between the plates:

$$\zeta = \zeta_3: \quad \cos\eta_3 = 1, \quad \eta_3 = 0; \quad \zeta = \zeta_2: \quad \cos\eta_2 = \sin\beta, \quad \eta_2 = \frac{\pi}{2} - \beta, \qquad \text{(6.114a–6.114f)}$$

whereas Equations 6.114a through 6.114c correspond to the capillary rise at the midpoint between the plates (Equation 6.108c); the capillary rise on the inner side of the plates is given by Equation 6.113b with $x = b$ leading to the integral Equation 6.115 (1) with upper limit Equation 6.114d, (2) comparing Equation 6.107b with Equation 6.112a leads to Equation 6.114e, and (3) the corresponding value (Equation 6.114f) applies in Equation 6.113b for a distance $x = b$ equal to half the spacing of the vertical plates:

$$2b = c\int_0^{\pi/2-\beta\equiv\eta_2}\left|A-\cos\eta\right|^{-1/2}\cos\eta\,d\eta. \qquad (6.115)$$

The condition 6.115 specifies the constant A in terms of the distance $2b$ between the parallel vertical plates. The same change of variable (Equations 6.112a through 6.112c and 6.113a) with $A = 1$ specifies (Equation 6.116a) the shape of the free surface (Equation 6.116c) outside the plates:

$$\zeta = c\sqrt{1-\cos\xi}, \quad \xi_1 = \frac{\pi}{2} - \beta: \quad 2(x-b) = c\int_{\pi/2-\beta}^{\xi}\left|1-\cos\xi\right|^{-1/2}\cos\xi\,d\xi; \qquad \text{(6.116a–6.116c)}$$

comparison of ζ_1 in Equation 6.107a with Equation 6.116a leads as in Equations 6.114e and 6.114f to the lower limit of integration (Equation 6.116b) in Equation 6.116c. Thus, *the shape of the free surface (Figure 6.9b) is given as follows: (1) implicitly $\zeta(x)$ eliminating η from Equations 6.112a and 6.113b between (Equation 6.111a) the vertical plates, (2) the constant A is specified (Equation 6.115) by the distance between the vertical plates; (3) outside (Equation 6.110a) the vertical plates leading by elimination of ξ between Equations 6.116a and 6.116c to $\zeta(x)$ the implicit shape of the free surface; (4) the capillary constant (Equation 6.102b) appears in Equations 6.112a, 6.113b, 6.116a, 6.116c, and 6.115; (5) together with the wetting angle, it specifies the capillary rise on the outside (Equation 6.107a) [inside (Equation 6.107b)] of the vertical plates; and (6) the capillary rise at the midposition between the plates (Equation 6.108c) depends explicitly only on the capillary constant (Equation 6.102b) and not on the wetting angle except through the constant A in Equation 6.115.*

6.4.5 Capillary Rise between Two Closely Spaced Plates

For very close plates (Equation 6.117a), the constant in Equation 6.115 is large (Equation 6.117b), and Equation 6.115 is approximated by Equation 6.117c:

$$b \to 0, A \to \infty: \quad 2b\sqrt{A} = c\int_0^{\pi/2-\beta} \cos\eta \, d\eta = c[\sin\eta]_0^{\pi/2-\beta} = c\cos\beta. \quad (6.117a–6.117c)$$

The constant (Equation 6.117c ≡ Equation 6.118a)

$$A = \left(\frac{c}{2b}\right)^2 \cos^2\beta \gg 1: \quad \zeta_{20} \sim \zeta_{30} \sim c\sqrt{A} = \frac{c^2}{2b}\cos\beta = \frac{T}{\rho g b}\cos\beta \equiv \zeta_0 \quad (6.118a \text{ and } 6.118b)$$

determines (Equations 6.107b and 6.108c) the capillary rise (Equation 6.118b) that (Equation 6.111b) to the lowest order is the same in the whole space between the walls; there is a difference between the capillary rise on the inner side of the plates (Equation 6.107b) and at the midposition between the plates (Equation 6.108c) to the next order, as follows from

$$\begin{aligned} \Delta\zeta_{23} \equiv \zeta_2 - \zeta_3 &= c\left(\sqrt{A - \sin\beta} - \sqrt{A-1}\right) = c\sqrt{A}\left(\sqrt{1 - \frac{\sin\beta}{A}} - \sqrt{1 - \frac{1}{A}}\right) \\ &= c\sqrt{A}\left[\left(1 - \frac{\sin\beta}{2A}\right) - \left(1 - \frac{1}{2A}\right) + O\left(\frac{1}{A^2}\right)\right] = \frac{c}{2}\frac{1-\sin\beta}{\sqrt{A}} + O\left(\frac{c}{A\sqrt{A}}\right) \quad (6.119) \\ &= b\frac{1-\sin\beta}{\cos\beta} + O\left(\frac{b^3}{c^2}\sec^3\beta\right) = b\frac{1-\sin\beta}{\cos\beta}\left[1 + O\left(\frac{\rho g b^2}{2T}\sec^3\beta\right)\right], \end{aligned}$$

where the lowest order (Equation 6.118b) and the capillarity parameter (Equation 6.102b) were used. *The capillary rise (Equation 6.118b) of a fluid of constant density ρ between parallel closely spaced vertical plates is equal to the following: (1) the product of the surface tension T divided by the weight ρgb of a column of fluid of height equal to half the distance between the walls, (2) the cosine of the wetting angle so that it is the vertical component of the surface tension that balances the weight; and (3) the equivalent height of a column of fluid is half the distance between the walls because there is wetting on both sides. There is no capillary rise for horizontal wetting:*

$$0 \le \beta \le \frac{\pi}{2}: \quad 0 = \zeta_{20}\left(\frac{\pi}{2}\right) \le \zeta_{20}(\beta) \sim \zeta_{30}(\beta) \le \zeta_{20}(0) = \frac{T}{\rho g b}, \quad (6.120a \text{ and } 6.120b)$$

and the maximum, for vertical wetting, corresponds to a tension equal to the weight of a column of the fluid with height equal to the half-distance between walls because there is wetting at two surfaces.

6.4.6 Capillary Rise at the Midpoint between Two Vertical Plates

The capillary rise (Equations 6.118a, 6.118b, 6.120a, and 6.120b) is uniform between the parallel plates in the limit (Equation 6.117b) of zero spacing (Equation 6.117a); to the next order, there is a difference of capillary rise between the inside of the walls and the center (Equation 6.119 ≡ Equation 6.121)

$$\Delta\zeta_{23} \equiv \zeta_2 - \zeta_3 = b\frac{1-\sin\beta}{\cos\beta} = b\frac{1-\sin\beta}{\sqrt{1-\sin^2\beta}} = b\sqrt{\frac{1-\sin\beta}{1+\sin\beta}} = b\frac{\sqrt{1-\sin^2\beta}}{1+\sin\beta} = b\frac{\cos\beta}{1+\sin\beta} \tag{6.121}$$

that varies from zero for horizontal wetting to half the distance between the plates for vertical wetting:

$$0 = \Delta\zeta_{23}\left(\frac{\pi}{2}\right) = \left(\Delta\zeta_{23}\right)_{min} \le \Delta\zeta_{23}(\beta) \le \left(\Delta\zeta_{23}\right)_{max} = \Delta\zeta_{23}(0) = b. \tag{6.122}$$

Thus, the capillary rise (Equations 6.118b and 6.122) is less ζ_3 at the center between the plates than at the inside ζ_2 of the plates:

$$\zeta_2, \zeta_3 = \zeta_0 \pm \frac{1}{2}\Delta\zeta_{23} = \frac{T}{\rho g b}\cos\beta \pm \frac{b}{2}\frac{\cos\beta}{1+\sin\beta} = b\cos\beta\left(\frac{T}{\rho g b^2} \pm \frac{1/2}{1+\sin\beta}\right), \tag{6.123a and 6.123b}$$

with extreme values (Equations 6.124a and 6.124b) for vertical wetting:

$$\zeta_{2\max} = \zeta_2(0) = \frac{T}{\rho g b} + \frac{b}{2}, \quad \zeta_{3\min} = \zeta_3(0) = \frac{T}{\rho g b} - \frac{b}{2}, \quad \zeta_{2\max} - \zeta_{3\min} = b \tag{6.124a–6.124c}$$

differing by half the distance between the plates (Equation 6.124c).

6.4.7 Capillary Rise on the Inside/Outside of Two Vertical Plates

The difference in capillary rise between the inner side of Equation 6.107b (the middle point between Equation 6.108c), the vertical plates, and the outer side of the vertical plates (Equation 6.107a) is given (Equations 6.123a and 6.123b) by Equation 6.125a (Equation 6.125b):

$$\begin{aligned}
\{\Delta\zeta_{21}, \Delta\zeta_{31}\} &\equiv \{\zeta_2 - \zeta_1, \zeta_3 - \zeta_1\} = \zeta_0 - \zeta_1 \pm \frac{1}{2}\Delta\zeta_{23} \\
&= \frac{T}{\rho g b}\cos\beta \pm \frac{b}{2}\frac{\cos\beta}{1+\sin\beta} - \sqrt{\frac{2T}{\rho g}}\sqrt{1-\sin\beta} \\
&= \left(\frac{T}{\rho g b} \pm \frac{b/2}{1+\sin\beta}\right)\sqrt{1-\sin^2\beta} - \sqrt{\frac{2T}{\rho g}}\sqrt{1-\sin\beta} \\
&= \sqrt{\frac{2T}{\rho g}}\sqrt{1-\sin\beta}\left[\sqrt{\frac{T}{2\rho g}}\frac{\sqrt{1+\sin\beta}}{b} \pm \sqrt{\frac{\rho g}{2T}}\frac{b/2}{\sqrt{1+\sin\beta}} - 1\right],
\end{aligned} \tag{6.125a and 6.125b}$$

where Equations 6.102b and 6.107a were used. *The capillary rise (Figure 6.9b) on the inside of the vertical plates is equal to that outside only for horizontal wetting and exceeds it otherwise (Equation 6.105a) with a maximum for vertical wetting:*

$$\begin{aligned} 0 &= \Delta\zeta_{21}\left(\frac{\pi}{2}\right) = \left(\Delta\zeta_{21}\right)_{min} \le \Delta\zeta_{21}(\beta) \le \left(\Delta\zeta_{21}\right)_{max} \\ &= \Delta\zeta_{21}(0) = \frac{T}{\rho g b} + \frac{b}{2} - \sqrt{\frac{2T}{\rho g}} = \left(\sqrt{\frac{T}{\rho g b}} - \sqrt{\frac{b}{2}}\right)^2 > 0. \end{aligned} \tag{6.126}$$

The capillary rise on the outside of the vertical plates is equal to that at the midpoint between the plates (Equation 6.127a) for (Equation 6.125b) the wetting angle (Equation 6.127b):

$$\Delta\zeta_{31}(\beta_*) = 0: \quad 1 + \sqrt{\frac{\rho g}{2T}} \frac{b/2}{\sqrt{1+\sin\beta_*}} = \sqrt{\frac{T}{2\rho g}} \frac{\sqrt{1+\sin\beta_*}}{b}. \tag{6.127a and 6.127b}$$

For a larger wetting angle, the capillary rise is greater at the midpoint between the plates than at the outside of the plates (Equation 6.128a) and vice versa in the opposite case (Equation 6.128c):

$$\zeta_3 \begin{cases} > \zeta_1 & \text{if} \quad \beta > \beta_*. & (6.128a) \\ = \zeta_1 & \text{if} \quad \beta = \beta_*. & (6.128b) \\ > \zeta_1 & \text{if} \quad \beta < \beta_*. & (6.128c) \end{cases}$$

The conclusion (Equation 6.128a) follows noting that if $\beta > \beta_*$, the r.h.s. (l.h.s.) of Equation 6.127b increases (decreases), so the difference is positive, that is, the term in square brackets in Equation 6.125b is positive, implying $\Delta\zeta_{31} > 0$ and hence Equation 6.128a.

6.5 Warping, Stress, and Displacement Functions

An analogy with the linear deflection of a membrane (Sections 6.1 and 6.2) is the torsion of a slender rod that can be described for the cross section alternatively in terms of (1) a warping function (Subsection 6.5.1), (2) a stress function (Subsection 6.5.2), and (3) a displacement function (Subsection 6.5.3).

6.5.1 Torsion Rate and Warping Function (Saint-Venant 1855)

Consider a **slender rod**, that is, of length much larger than the cross section (Figure 6.10a), that is constant but may have an arbitrary shape (Figure 6.10b). The torsion of the rod is

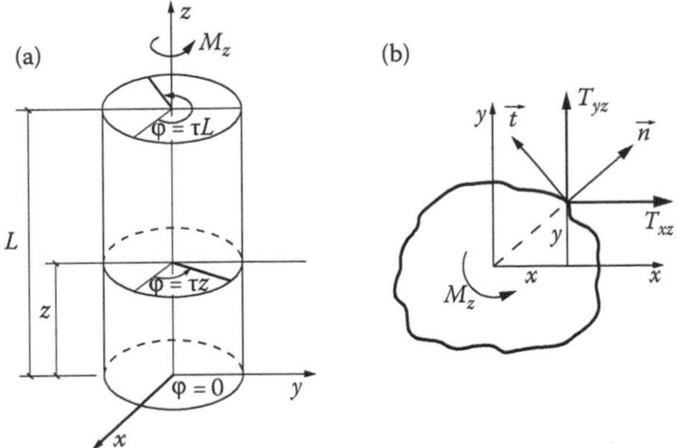

FIGURE 6.10
If a torque M_z is applied at one end of slender rod (a), the transverse cross sections are rotated by an angle φ proportional to the distance along the axis z, that is, the torsion rate $\tau = d\varphi/dz$ is constant. In the plane of a cross section (b), the axial torsion does not cause normal stresses along the unit normal $\vec{n}$ and tangent $\vec{t}$ vector to the boundary, nor shear stresses in the plane of the cross section. There is no axial normal stress either, but only shear stresses between the axial direction and the plane of the cross section. The torsion generally causes the initially plane cross section to become curved, so the expression "plane of the cross section" refers to the initial reference condition without applied axial moment or torque.

specified by the **torsion rate** that is the rate of the rotation along the length of the **angle of torsion** with a fixed horizontal line (Equation 6.129a):

$$\tau \equiv \frac{d\varphi}{dz} = \text{const}; \quad \vec{E}_2 = \vec{e}_z \tau z. \qquad (6.129a \text{ and } 6.129b)$$

It will be shown (Subsection 6.6.3) that the torsion rate is constant, and thus, the rotation vector at a distance z from the base is Equation 6.129b. The corresponding displacement in the plane (x,y) of the cross section is

$$\vec{u} = \vec{E}_2 \wedge \left(\vec{e}_x x + \vec{e}_y y\right) = \vec{e}_z \tau z \wedge \left(\vec{e}_x x + \vec{e}_y y\right) = \vec{e}_y \tau z x - \tau z y \vec{e}_x. \qquad (6.130)$$

Besides the horizontal displacements (Equation 6.130 ≡ Equations 6.131a and 6.131b) due to the rotation:

$$u_x = -\tau z y, \qquad u_y = \tau z x, \qquad u_z = \tau \Phi(xy), \qquad (6.131a\text{–}6.131c)$$

there may be a vertical displacement along the axis specified (Equation 6.131c) by a **warping function**: (1) if the warping function is zero or constant, the cross section of the bar remains flat under warping, for example, for the circular shape (Subsection 6.7.2); and (2) otherwise, the warping function specifies the deviation from the plane undeformed shape of a cross section when the bar undergoes torsion, for example, for the elliptic shape (Subsection 6.7.1).

From the displacements (Equations 6.131a through 6.131c), it follows that (Equations 4.12a through 4.12d): (1) there are no strains in the cross section (Equations 6.132a through 6.132c):

$$S_{xx} = \partial_x u_x = 0 = \partial_y u_y = S_{yy}, \qquad 2S_{xy} = \partial_x u_y + \partial_y u_x = 0, \qquad S_{zz} = \partial_z u_z = 0, \qquad (6.132a\text{–}6.132d)$$

nor in the axial direction (Equation 6.132d); (2) there are only distortions between the axial direction and the cross section (Equations 6.113a and 6.113b):

$$2S_{xz} = \partial_x u_z + \partial_z u_x = \tau\left(\partial_x \Phi - y\right) = \frac{1}{\mu} T_{xz}, \tag{6.133a}$$

$$2S_{yz} = \partial_y u_z + \partial_z u_y = \tau\left(\partial_y \Phi + x\right) = \frac{1}{\mu} T_{yz}; \tag{6.133b}$$

(3) the corresponding stresses by the Hooke law (Equations 4.67a through 4.67c) are shears (Equations 6.134b through 6.134e) and involve (Equations 6.133a and 6.133b) only the shear modulus (Equation 6.134a):

$$\mu \equiv \frac{1}{2}\frac{E}{1+\sigma}: \quad T_{xx} = T_{yy} = T_{zz} = T_{xy} = 0. \tag{6.134a–6.134c}$$

and (4) the latter is related (Equation 4.75b) to the Young modulus E and Poisson ratio σ by Equation 6.134a.

The stress balance (Equations 4.39c and 4.39d) in the absence of external forces (Equation 6.135a) for the z-direction (Equation 6.135b)

$$\vec{f} = 0: \quad 0 = \partial_x T_{xz} + \partial_y T_{yz} = \tau\mu\left(\partial_{xx}\Phi + \partial_{yy}\Phi\right), \quad \nabla^2\Phi = 0, \tag{6.135a–6.135c}$$

shows that the warping function is harmonic, that is, satisfies the Laplace equation 6.135c. The cross section of the rod (Figure 6.10b) has boundary (Equations 4.42a through 4.42c) with arc length (Equation 4.21d), unit tangent (Equation 4.43a), and normal (Equation 4.43b) vectors, and thus, the condition of absence of stresses along the boundary is

$$\partial D: \quad 0 = \mathrm{d}s\left(T_{xz} n_x + T_{yz} n_y\right) = T_{yz}\,\mathrm{d}x - T_{xz}\,\mathrm{d}y = \tau\mu\left[\mathrm{d}x\left(\partial_y \Phi\right) - \mathrm{d}y\left(\partial_x \Phi\right) + \mathrm{d}\left(\frac{x^2+y^2}{2}\right)\right]. \tag{6.136}$$

Thus, *the torsion of a slender rod is specified by a warping function that is harmonic (Equation 6.135c) in the cross section and satisfies the condition 6.136 on its boundary; it determines the nonzero stresses and strains (Equations 6.133a and 6.133b) and displacements (Equations 6.131a through 6.131c).*

6.5.2 Stress Function and Boundary Condition

The least straightforward aspect of the formulation using the warping function is the boundary condition 6.136. An alternative is to use a **stress function**:

$$2\mu S_{xz} = T_{xz} = \mu\tau\partial_y\Theta, \qquad 2\mu S_{yz} = T_{yz} = -\mu\tau\partial_x\Theta, \tag{6.137a and 6.137b}$$

that satisfies the stress balance (Equation 6.135b ≡ Equation 6.138a) in the domain D of the cross section:

$$D: \qquad \partial_x T_{xz} + \partial_y T_{yz} = \mu\tau\left(\partial_x\Theta - \partial_{yx}\Theta\right) = 0 \tag{6.138a}$$

$$\partial D: \qquad 0 = T_{xz}\,\mathrm{d}y - T_{yz}\,\mathrm{d}x = \mu\tau\left(\mathrm{d}y\partial_y\Theta + \mathrm{d}x\,\partial_x\Theta\right) = \mu\tau\mathrm{d}\Theta \tag{6.138b}$$

and is constant on the boundary (Equation 6.136 ≡ Equation 6.138b). The stress Θ and warping Φ functions are related (Equations 6.137a and 6.137b) ≡ Equations 6.133a and 6.133b) by

$$\partial_y \Theta = \partial_x \Phi - y, \qquad -\partial_x \Theta = x + \partial_y \Phi; \qquad \text{(6.139a and 6.139b)}$$

it follows that the stress function satisfies the Poisson equation:

$$0 = \partial_{yx}\Phi - \partial_{xy}\Phi = \partial_y (\partial_y \Theta + y) + \partial_x (\partial_x \Theta + x) = \partial_{xx}\Theta + \partial_{yy}\Theta + 2 = \nabla^2\Theta + 2, \qquad \text{(6.140a)}$$

$$d\Phi = dx\,(\partial_x \Phi) + dy\,(\partial_y \Phi) = dx\,(\partial_y \Theta) - dy\,(\partial_x \Theta) + y dx - x dy \qquad \text{(6.140b)}$$

and specifies the warping function through Equation 6.140b. Thus, *the torsion of a slender rod is specified by a stress function that satisfies a Poisson equation (Equation 6.140a ≡ Equation 6.141b) in the cross section (Equation 6.141a) and is constant (Equation 6.139b ≡ Equation 6.141c) on the boundary:*

$$D: \qquad \nabla^2\Theta = -2; \qquad \Theta|_{\partial D} = \text{const.}; \qquad \text{(6.141a–6.141c)}$$

it determines the nonzero stresses and strains (Equations 6.137a and 6.137b) and, through the warping function (Equation 6.140b), the displacements (Equations 6.131a through 6.131c).

6.5.3 Displacement Function and Complex Torsion Potential

Introducing the **displacement function** (Equation 6.142a):

$$\Psi \equiv \Theta + \frac{x^2 + y^2}{2}: \quad \partial_x \Psi = \partial_x \Theta + x, \quad \partial_y \Psi = \partial_y \Theta + y \qquad \text{(6.142a–6.142c)}$$

it follows from Equations 6.142b and 6.142c that Equations 6.139a and 6.139b become the Cauchy–Riemann conditions 6.143a and 6.143b:

$$\partial_x \Phi = \partial_y \Psi, \qquad \partial_y \Phi = -\partial_x \Psi; \qquad \Phi(x,y) + i\Psi(x,y) = f(x + iy) \in \mathcal{A}(|C), \qquad \text{(6.143a–6.143c)}$$

and these imply the existence of a **complex torsion potential** (Equation 6.143c), that is, a holomorphic function (Equations I.11.10a and I.11.10b). Hence, the real (imaginary) part of the complex potential, that is, the warping Φ (displacement Ψ) function, satisfies [Equation 6.135c (Equation 6.144)] the Laplace equation:

$$0 = \partial_{yx}\Phi - \partial_{xy}\Phi = \partial_{yy}\Psi + \partial_{xx}\Psi = \nabla^2\Psi. \qquad \text{(6.144)}$$

The displacement function 6.142a satisfies the boundary condition (Equation 6.141c ≡ Equation 6.145a):

$$\left[\Psi(x,y) - \frac{x^2 + y^2}{2}\right]_{\partial D} = \text{const}, \qquad \text{(6.145a)}$$

$$\mu S_{xz} = T_{xz} = \mu\tau(\partial_y \Psi - y), \qquad \mu S_{yz} = T_{yz} = \mu\tau(x - \partial_x \Psi) \qquad \text{(6.145b and 6.145c)}$$

and specifies (Equations 6.142b and 6.142c) the stresses and strains (Equations 6.137a and 6.137b) through Equations 6.145b and 6.145c. Thus, *a complex analytic function specifies a complex torsion potential (Equation 6.143c) whose real (imaginary) part is the warping Φ (displacement Ψ) function, specifying the displacements (Equations 6.131a through 6.131c) [the boundary condition (Equation 6.145a)]; both satisfy the Laplace equation [Equation 6.135c (Equation 6.144)] and determine the nonzero stresses and strains [Equations 6.137a and 6.137b (Equations 6.145b and 6.145c)]. They can be determined one from the other using Equation 6.146a (Equation 6.146b)*:

$$d\Phi = (\partial_x\Phi)\,dx + (\partial_y\Phi)\,dy = (\partial_y\Psi)\,dx - (\partial_x\Psi)\,dy, \tag{6.146a}$$

$$d\Psi = (\partial_x\Psi)\,dx + (\partial_y\Psi)\,dy = -(\partial_y\Phi)\,dx + (\partial_x\Phi)\,dy, \tag{6.146b}$$

and from the stress function by Equation 6.140b (Equation 6.142a); conversely, the stress function is determined from the warping (displacement) function by Equation 6.147 (Equation 6.142a):

$$d\Theta = (\partial_x\Theta)dx + (\partial_y\Theta)dy = -(\partial_y\Phi)dx + (\partial_x\Phi)dy - d\left(\frac{x^2+y^2}{2}\right) \tag{6.147}$$

using Equations 6.139a and 6.139b. In Equations 6.146a and 6.146b, Equations 6.143a and 6.143b were used.

6.6 Torsional Stiffness of a Multiply Connected Section

The moment along the axis (elastic energy) is proportional to (to one-half of the square of) the torsion through [Subsection 6.6.1 (Subsection 6.6.2)] the torsional stiffness that can be calculated for a multiply connected cross section. The total elastic energy of deformation of the rod is minimum for a constant torsion angle (Subsection 6.6.3) as assumed initially (Subsection 6.5.1).

6.6.1 Axial Moment and Torsional Stiffness

The nonzero stresses (Equations 6.137a and 6.137b) associated with the torsion cause (Figure 6.10b) an axial moment (Equation 6.148a) expressed in terms of the stress function by Equation 6.148b:

$$\vec{M} = \vec{e}_z \int_D \left(T_{yz}x - T_{xz}y\right) dx\,dy = -\vec{e}_z\mu\tau \int_D \left[x\left(\partial_x\Theta\right) + y\left(\partial_y\Theta\right)\right] dx\,dy; \tag{6.148a and 6.148b}$$

using also Equations 6.139a, 6.139b, 6.142b, and 6.142c, *the* **torsional moment** *(Figure 6.10b) along the axis of a slender rod (Figure 6.10a) is proportional to the torsion rate (Equation 6.149a)*:

$$\begin{aligned} M_z = C\tau:\quad C &\equiv -\mu \int_D \left[x\left(\partial_x\Theta\right) + y\left(\partial_y\Theta\right)\right] dx\,dy \\ &= \mu \int_D \left[x^2 + y^2 + x\left(\partial_y\Phi\right) - y\left(\partial_x\Phi\right)\right] dx\,dy \\ &= \mu \int_D \left[x^2 + y^2 - x\left(\partial_x\Psi\right) - y\left(\partial_y\Psi\right)\right] dx\,dy \end{aligned} \tag{6.149a–6.149d}$$

through the **torsional stiffness** *that can be expressed in terms of (1) the stress function (Equation 6.148b ≡ Equation 6.149b) using Equations 6.137a and 6.137b (2) the warping function (Equation 6.149c) using Equations 6.139a and 6.139b; and (3) the displacement function (Equation 6.149d) using Equations 6.142b and 6.142c.* Since the boundary condition is simpler for the stress function (Equation 6.141c) than for the displacement function (Equation 6.145a) and for the warping function (Equation 6.136), the stress function is used in the sequel; the integrand (Equation 6.149b) in the torsional stiffness takes the form

$$x\,(\partial_x\Theta) + y\,(\partial_y\Theta) = \partial_x\,(x\Theta) + \partial_y\,(y\Theta) - 2\Theta; \tag{6.150}$$

substituting Equation 6.150 in Equation 6.149b yields

$$C = 2\mu\int_D \Theta\,dx\,dy - \int_D\left[\partial_x(x\Theta)+\partial_y(y\Theta)\right]dx\,dy. \tag{6.151}$$

The integrand in the second term on the r.h.s. is the divergence (Equation 6.152b) of the vector (Equation 6.152a):

$$\vec{A} \equiv \Theta\left(\vec{e}_x x+\vec{e}_y y\right), \quad \nabla\cdot\vec{A} = \partial_x(x\Theta)+\partial_y(y\Theta), \tag{6.152a and 6.152b}$$

and the divergence theorem leads to

$$\begin{aligned}\int_D\left[\partial_x(x\Theta)+\partial_y(y\Theta)\right]dx\,dy &= \int_{\partial D}\left(\nabla\cdot\vec{A}\right)dx\,dy = \int_D\left(\vec{n}\cdot\vec{A}\right)ds\\ &= \int_{\partial D}\Theta\left[\left(\vec{e}_x x+\vec{e}_y \vec{y}\right)\cdot\vec{n}\right]ds\end{aligned} \tag{6.153}$$

an integral over the boundary (Equation 6.153) with arc length ds.

Two cases arise. First, if the cross section is simply connected, (1) there is only one outer boundary (Figure 6.11a); (2) the stress function is constant on the boundary (Equation 6.141c); (3) the stress function is defined by Equations 6.137a and 6.137b to within an added constant Θ+ const that changes neither the strains nor the stresses; (4) the constant may be chosen so that the stress function vanishes on the boundary (Equation 6.158a); and (5) thus Equation 6.153 vanishes and the torsional stiffness (Equation 6.151) simplifies to Equation 6.158b. The second case concerns (Figure 6.11b) a multiply connected cross section: (6) the stress function may again be taken to vanish on the outer boundary (Equation 6.159a); (7) the values on the inner boundaries will be constant but generally nonzero and distinct (Equation 6.159b); (8) thus Equation 6.153 simplifies to

$$\int_D\left[\partial_x(x\Theta)+\partial_y(y\Theta)\right]dx\,dy = \sum_{n=1}^{N}\Theta_n\int_{\partial D_n}\left(n_x x+n_x y\right)ds = -2\sum_{n=1}^{N}\Theta_n S_n, \tag{6.154}$$

where the sum is over all the inner boundaries; (9) it involves the integrals

$$-2S_n = \int_{\partial D_n}\left(n_x x+n_x y\right)ds = \int_{\partial D_n}(y\,dx - x\,dy), \tag{6.155}$$

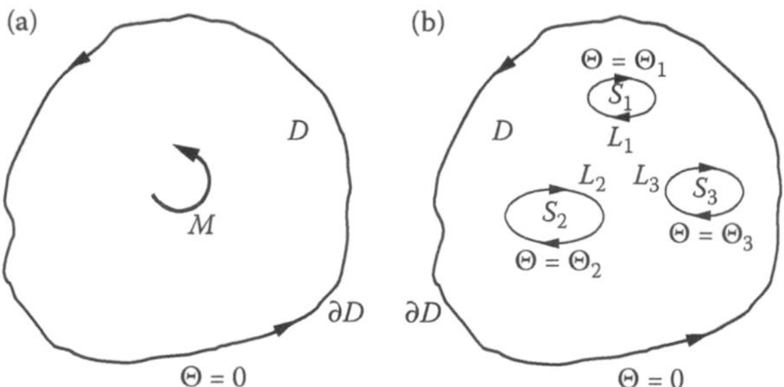

FIGURE 6.11
Torsion of a slender rod (Figure 6.10a) is specified by a stress function that specifies the nonzero strains [stresses (Figure 6.10b)], that is, the distortions (shears) between the axial and transverse directions. The stress function satisfies a Poisson equation in the domain D of the cross section and is constant on the boundary ∂D curve(s). The stress function is defined to within an arbitrary added constant that does not affect the strains or stresses that are obtained by differentiation of the stress function. In the case of a simply connected cross section (Figure 6.11a), that is, a slender massive or nonhollow rod, the stress function can be taken to be zero on the boundary by fixing the value of the arbitrary added constant. In the case of a multiply connected cross section (Figure 6.11b), that is, a hollow slender rod with one or more holes: (a) the arbitrary added constant can be chosen so that the stress function is still zero on the outer boundary; and (b) the stress function is constant on all inner boundaries, generally taking a distinct nonzero value on each of them. The situation is similar for the stream function in an incompressible vortical flow (Sections 2.2 through 2.9).

where the unit normal vector (Equation 4.43b) was used; (10) the area element in Cartesian coordinates is

$$2\,\mathrm{d}S = (\vec{x} \wedge \mathrm{d}\vec{x}) \cdot \vec{e}_3 = \begin{vmatrix} x & y \\ \mathrm{d}x & \mathrm{d}y \end{vmatrix} = x\,\mathrm{d}y - y\,\mathrm{d}x, \tag{6.156}$$

and thus S_n in Equation 6.155 is the area of the hole inside the loop ∂D_n; and (11) substitution of Equation 6.154 in Equation 6.151 leads to Equation 6.159c.

Thus, *the torsional stiffness is specified in terms of the stress function by*

$$C = 2\mu \int_D \Theta\,\mathrm{d}x\,\mathrm{d}y + \mu \int_{\partial D} \Theta\left(x\,\mathrm{d}y - y\,\mathrm{d}x\right). \tag{6.157}$$

For a simply connected cross section (Figure 6.11a), the stress function is constant on the boundary (Equation 6.141c); it is defined within an additive constant that can be put equal to zero on the boundary (Equation 6.158a) so that the torsional stiffness simplifies to the first term in Equation 6.157, namely, Equation 6.158b:

$$\Theta\big|_{\partial D} = 0: \quad C = 2\mu \int_D \Theta(x,y)\,\mathrm{d}x\,\mathrm{d}y. \qquad \text{(6.158a and 6.158b)}$$

$$\Theta\big|_{L_0} = 0, \quad \Theta\big|_{L_n} = \Theta_n: \quad C = 2\mu \int_D \Theta\,\mathrm{d}x\,\mathrm{d}y + 2\mu \sum_{n=1}^{N} \Theta_n S_n. \qquad \text{(6.159a–6.159c)}$$

For a multiply connected region (Figure 6.11b) to the integral of the stress function over the cross section (Equation 6.158b): (1) there is no need to add an integral over the outer boundary L_0 because the stress function can be put equal to zero there (Equation 6.159a); (2) its product by the area S_n in Equation 6.155 must be added for each hole n = 1..., N in which in the boundary L_n the stress function (Equation 6.159b) is a constant Θ_n. In the last term on the r.h.s. of Equations 6.157 and 6.159c, the inner loops are described in the opposite direction to the outer loop (Figure 6.11b). This result (Equation 6.159c) will be confirmed next (Subsection 6.6.2) in an independent calculation of the elastic energy.

6.6.2 Elastic Energy Density in the Cross Section and for the Rod

The elastic energy density (Equation 4.86b ≡ Equation 6.160a) associated with the torsion is given by Equation 6.160b in terms of the stresses (Equation 6.160c) and of (Equations 6.160d and 6.160e) the stress function (Equations 6.137a and 6.137b):

$$\begin{aligned} E_d = S_{xz}T_{xz} + S_{yz}T_{yz} &= \frac{1}{2\mu}\left[\left(T_{xz}\right)^2 + \left(T_{yz}\right)^2\right] \\ &= \frac{\mu\tau^2}{2}\left[\left(\partial_x\Theta\right)^2 + \left(\partial_y\Theta\right)^2\right] = \frac{\mu\tau^2}{2}\left|\nabla\Theta\right|^2. \end{aligned} \qquad (6.160a\text{–}6.160e)$$

The elastic energy density (Equations 6.160a through 6.160e) is proportional to the square of the torsion rate, and when integrated over the cross section (Equation 6.161a):

$$\bar{E}_d = \int_D E_d \, dx \, dy = \frac{1}{2}C\tau^2 = \frac{1}{2}M_z\tau, \quad C \equiv \mu\int_D (\nabla\Theta\cdot\nabla\Theta)\, dx\, dy, \qquad (6.161a\text{–}6.161c)$$

the coefficient is one-half of the torsional stiffness (Equation 6.161c). Thus, it also equals (Equation 6.161b) one-half of the torsion rate times the torsional moment (Equation 6.149a). It has been anticipated that the two expressions Equations 6.161c and 6.159c for the torsional stiffness coincide; this will be proven next using the identity (Equation 6.162b) that follows from Equation 6.7b with Equation 6.162a and replacing Φ by Θ:

$$\vec{B} = \nabla\Theta: \quad (\nabla\Theta\cdot\nabla\Theta) = \nabla\cdot(\Theta\nabla\Theta) - \Theta\,\nabla^2\Theta = \nabla\cdot(\Theta\nabla\Theta) + 2\Theta; \qquad (6.162a \text{ and } 6.162b)$$

in Equation 6.162b, Equation 6.141b was used. Substituting Equation 6.162b in Equation 6.161c yields

$$C - 2\mu\int_D \Theta\, dx\, dy = \mu\int_D \nabla\cdot\left(\Theta\,\nabla\Theta\right) dx\, dy = \mu\int_{\partial D} \Theta\frac{\partial\Theta}{\partial n}\, ds, \qquad (6.163)$$

that coincides with Equation 6.157 ≡ Equation 6.163 noting that (1) the first integral, over the cross section, is the same, (2) the second integral over the boundary is zero (Equation 6.158a) for a simply connected cross section (Figure 6.11a) proving the coincidence of Equation 6.163 ≡ Equation 6.158b; and (3) in the case of a multiply connected cross section (Figure 6.11b), it remains to prove (Equation 6.163 ≡ Equation 6.159c) next that the r.h.s. of Equation 6.163 coincides with the second term on the r.h.s. of Equation 6.159c. The equality of Equation 6.163 ≡ Equation 6.159c using Equation 6.159b ≡ Equation 6.164a leads to Equation 6.164b:

$$\Theta\Big|_{L_n} = \Theta_n: \quad \int_{L_n} \Theta \frac{\partial \Theta}{\partial n} \mathrm{d}s = \Theta_n \int \frac{\partial \Theta}{\partial n} \mathrm{d}s = 2\Theta_n S_n. \qquad \text{(6.164a and 6.164b)}$$

It remains to prove Equation 6.164b.

The latter expression (Equation 6.164b) can be calculated as follows: (1) the stress function Θ is not single-valued in a multiply connected region, but the axial displacement (Equation 6.131c) and hence the warping function are single-valued; (2) thus, the integral of the warping function along the boundary is zero:

$$\begin{aligned} 0 &= \int_{\partial D} \mathrm{d}\Phi = \int_{\partial D} \left[(\partial_x \Phi) \mathrm{d}x + (\partial_y \Phi) \mathrm{d}y \right] \\ &= \int_{\partial D} (y \,\mathrm{d}x - x \,\mathrm{d}y) + \int_{\partial D} \left[(\partial_y \Theta) \mathrm{d}x - (\partial_x \Theta) \mathrm{d}y \right] = -2S_n + \int_{\partial D} \frac{\partial \Theta}{\partial n} \mathrm{d}s, \end{aligned} \qquad (6.165)$$

where Equations 6.139a and 6.139b ≡ Equation 6.140b was used; (3) in Equation 6.165, the area element (Equation 6.155) and the normal (Equation 4.43b) derivative of the stress function (Equation 6.166) were also used:

$$\frac{\partial \Theta}{\partial n} = n_x \partial_x \Theta + n_y \partial_y \Theta = \frac{\mathrm{d}x}{\mathrm{d}s} \partial_y \Theta - \frac{\mathrm{d}y}{\mathrm{d}s} \partial_x \Theta; \qquad (6.166)$$

and (4) in Equations 6.163 and 6.165, the loop integral was taken in the same direction. Summarizing the preceding results in reverse order (Equation 6.165) proves Equation 6.164b that, in turn, shows the coincidence of (Equation 6.163 ≡ Equation 6.159c) the two expressions for the torsional stiffness. QED. Thus, *the elastic energy density (Equations 6.160a through 6.160d) is proportional to the square of the torsion rate, and integrated over the cross section, Equation 6.161a involves as a coefficient one-half of the torsional stiffness (Equation 6.149b ≡ Equation 6.159c ≡ Equation 6.163) in terms of the stress function; the elastic energy of the whole rod (Equation 6.167)*

$$\bar{\bar{E}}_d = \int_0^L E_d \,\mathrm{d}z \,\mathrm{d}x \,\mathrm{d}y = \int_0^L \bar{E}_d \,\mathrm{d}z = \frac{1}{2} \int_0^L \mathrm{C}\tau^2 \,\mathrm{d}z = \frac{1}{2} \int_0^L \mathrm{C} \left(\frac{\mathrm{d}\varphi}{\mathrm{d}z} \right)^2 \mathrm{d}z \qquad (6.167)$$

involves another integration along the axial direction.

6.6.3 Minimum Energy and Constant Torsion Rate

The initial assumption (Equation 6.129a) of constant torsion rate will be reconsidered for a variation (Equation 6.168b ≡ Equation 6.168c) of the total elastic energy of torsion (Equation 6.167) of the rod; the cross section and material properties are not varied, and hence torsional stiffness (Equation 6.168a) is not varied:

$$\delta \mathrm{C} = 0: \quad \delta \bar{\bar{E}}_d = \frac{\mathrm{C}}{2} \int_0^L \delta \left[\left(\frac{\mathrm{d}\varphi}{\mathrm{d}z} \right)^2 \right] \mathrm{d}z = \mathrm{C} \int_0^L \frac{\mathrm{d}\varphi}{\mathrm{d}z} \mathrm{d}(\delta\varphi). \qquad \text{(6.168a–6.168c)}$$

Integrating by parts leads to

$$\left[C\left(\frac{d\varphi}{dz}\right)\delta\varphi\right]_0^L - \int_0^L C\left[\frac{d}{dz}\left(\frac{d\varphi}{dz}\right)\right]\delta\varphi\,\delta z = \delta\bar{\bar{E}}_d = \left[M_z\delta\varphi\right]_0^L, \tag{6.169}$$

where the variation of the elastic energy equals the work done by the axial moment to cause the rotation; collecting together the terms evaluated at the ends of the rod leads to

$$0 = \left[\left(C\tau - M_z\right)\delta\varphi\right]_0^L + \int_0^L C\left(\frac{d\tau}{dz}\right)\delta\varphi\, dz. \tag{6.170}$$

Since the first (second) term on the r.h.s. of Equation 6.170 is a boundary (volume) term at the ends (along the length) of the rod, both must vanish. From the boundary part first on the r.h.s. of Equation 6.170, it follows that the torsional moment equals the product of the torsion rate by the torsional stiffness as had been shown before (Equation 6.171b ≡ Equation 6.149a). Furthermore, the volume part second on the r.h.s. of Equation 6.170 shows that *the condition of stationary (Equation 6.171a) total elastic energy of the rod (Equation 6.167) is that torsion rate is constant along its length (Equation 6.171c ≡ Equation 6.129a):*

$$\delta\bar{\bar{E}}_d = 0: \quad M_z - C\tau = 0 = \frac{d\tau}{dz}; \quad \frac{\delta^2}{\delta\tau^2}\bar{\bar{E}}_d = \int_0^L C\,dz > 0. \tag{6.171a–6.171d}$$

The extremum (Equation 6.171a) of the elastic energy is actually a minimum (Equation 6.171d) because the torsional stiffness is positive. This also confirms that the torsional stiffness is the same in the torsional moment (Equations 6.149a ≡ Equation 6.171b) and elastic energy of the cross section (Equation 6.161a).

6.7 Hollow Elliptical or Thin or Cut Cross Sections

The hollow elliptical cross section (Subsection 6.7.1) includes the particular circular case for which the wall thickness is constant (Subsection 6.7.3); this allows a comparison of the torsional stiffness of massive and hollow cross sections (Subsection 6.7.2), that is, the loss of torsional stiffness due to a hole (Subsection 6.7.4). The comparison of a thin annular section with a thin rectangular rod (Subsection 6.7.5) shows that a short closed section is torsionally stiffer than a thin long section; also there is a further loss of torsional stiffness if a cut is made (Subsection 6.7.7) in a closed thin section (Subsection 6.7.6).

6.7.1 Similar Ellipses as Inner and Outer Boundaries

Consider the torsion of a slender (Figure 6.10a) hollow rod whose cross section (Figure 6.12a) has for inner (outer) boundaries similar ellipses, with the same center and axis, and half-axis (a, b) in the ratio λ:

$$x' \equiv \frac{x}{a}, \quad y' \equiv \frac{y}{b}; \quad 0 < \lambda < 1: \quad \lambda^2 \le \left(\frac{x}{a}\right)^2 + \left(\frac{y}{b}\right)^2 = x'^2 + y'^2 \le 1. \tag{6.172a–6.172d}$$

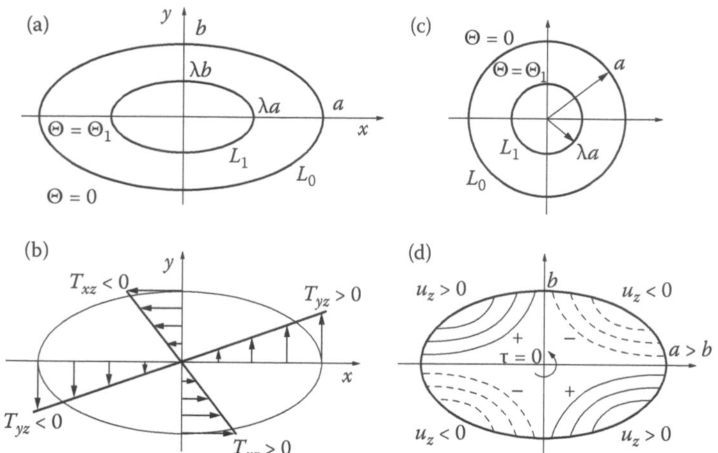

FIGURE 6.12
Example of a doubly connected cross section is a slender hollow rod with elliptical inner and outer boundaries (a) with half-axis in the same ratio. The particular case of equal half-axis leads to a hollow slender rod with circular annular cross section (c). Shrinking the inner boundary to zero leads from case a (case c) to the massive or nonhollow slender rod with an elliptical (circular) cross section. In all cases, the nonzero shear stresses between the axial and transverse directions are linear functions (b) of the coordinates in the plane of the cross section. An initially plane cross section does not remain plane under torsion for an elliptical rod (d), with alternating quadrants being deflected up and down; the only exception are the circular cylindrical rods, for which the cross sections remain plane under torsion.

The stress function (Equation 6.173a) vanishes on the outer boundary (Equation 6.173b):

$$\Theta(x,y) = B\left(1-\frac{x^2}{a^2}-\frac{y^2}{b^2}\right), \quad \Theta\big|_{L_0} = 0, \quad \Theta\big|_{L_1} = B\left(1-\lambda^2\right) \equiv \Theta_1 \qquad \text{(6.173a–6.173c)}$$

and takes the constant value (Equation 6.173c) on the inner boundary. It satisfies the Poisson equation 6.141b in the form Equation 6.174a:

$$-2 = \nabla^2\Theta = \partial_{xx}\Theta + \partial_{yy}\Theta = -2B\left(\frac{1}{a^2}+\frac{1}{b^2}\right), \quad B = \frac{a^2b^2}{a^2+b^2}, \qquad \text{(6.174a and 6.174b)}$$

that determines the constant (Equation 6.174b). Thus, *the stress function (Equations 6.173a and 6.174b) for a straight slender hollow rod with elliptical (Figure 6.12a) inner and outer boundaries (Equations 6.172a through 6.172d) is specified by Equation 6.175a:*

$$\Theta(x,y) = \frac{a^2b^2 - b^2x^2 - a^2y^2}{a^2+b^2}; \quad \left\{T_{xz}, T_{yz}\right\} = \frac{\mu\tau}{a^2+b^2}\left\{-2a^2y, 2b^2x\right\}; \qquad \text{(6.175a–6.175c)}$$

the corresponding (Equations 6.137a and 6.137b) stresses (Equations 6.175b and 6.175c) are linear functions of position, that is, for (Figure 6.12b) positive or counterclockwise torsion, (1) the vertical stresses are positive (negative) or upward (downward) to the right (left), and (2) the horizontal stresses are negative (positive) or to the left (right) above (below).

The displacement function (Equation 6.142a) corresponding to the stress function (Equation 6.175a) is given to within a constant by

$$\Psi(x,y) = \frac{x^2+y^2}{2} - \frac{b^2x^2+a^2y^2}{a^2+b^2} = \frac{a^2-b^2}{a^2+b^2}\frac{x^2-y^2}{2}. \tag{6.176}$$

The latter is the imaginary part of a complex torsion potential:

$$\Phi(x,y)+i\Psi(x,y) = f(x+iy) = i\frac{a^2-b^2}{a^2+b^2}\frac{(x+iy)^2}{2} = \frac{a^2-b^2}{a^2+b^2}\left(-xy+i\frac{x^2-y^2}{2}\right), \tag{6.177}$$

whose real part is the warping function:

$$\Phi(x,y) = \operatorname{Re}\left[f(x+iy)\right] = -\frac{a^2-b^2}{a^2+b^2}xy = \frac{1}{\tau}u_z(x,y). \tag{6.178}$$

A slender rod with hollow elliptical cross section (Equations 6.172a through 6.172d) under torsion has complex torsion potential (Equation 6.177), displacement function (Equation 6.176), and warping function (Equation 6.178). The latter shows that there is no vertical displacement for a circular cross section $a = b$, that is, the cross section of an annular rod (Figure 6.12c) remains plane under torsion. This is not so for an elliptical cross section for which the cross section does not remain plane under torsion (Figure 6.12d). Denoting by $a > b$ the longer half-axis (Equation 6.179a):

$$a > b: \quad u_z(x,y) = -\tau\frac{a^2-b^2}{a^2+b^2}xy\begin{cases} <0 & \text{if} \quad xy>0, \\ =0 & \text{if} \quad x=0 \text{ or } y=0, \\ >0 & \text{if} \quad xy<0, \end{cases} \tag{6.179a–6.179d}$$

the axial displacement is given as follows (1 and 2) negative (positive) or downward (upward) in the odd (even) quadrants [Equation 6.179b (Equation 6.179d)], that is, the cross section is displaced downward (upward) in the first and third (second and fourth) quadrants; (3) the separating lines are the axis of the ellipse, along which the vertical displacement is zero; (4) the nonzero level curves Ψ $(x, y) = const \neq 0$ are hyperbolas $x, y = const \neq 0$ with the axis of the ellipse as asymptotes; and (5) the center of the ellipse is a saddle point (Figure 6.7a).

6.7.2 Torsional Stiffness of a Massive/Hollow Ellipse

The torsional stiffness of a massive elliptical cross section (Equation 6.172d) is specified by setting $\lambda = 0$ in the integral (Equation 6.158b) of the stress function (Equations 6.173a and 6.174b); for general λ, the integral is evaluated by

$$\begin{aligned}\frac{C(\lambda)}{2\mu B} &= \int_{\lambda^2 \le x^2/a^2 + y^2/b^2 \le 1} dx \int dy \left(1 - \frac{x^2}{a^2} - \frac{y^2}{b^2}\right) \\ &= ab \int_{\lambda \le x'^2 + y'^2 \le 1} dx' \int dy' \left(1 - x'^2 - y'^2\right) \\ &= ab \int_0^{2\pi} d\varphi' \int_\lambda^1 \left(1 - r'^2\right) r' \, dr' \\ &= 2\pi ab \left[\frac{r'^2}{2} - \frac{r'^4}{4}\right]_\lambda^1 = \pi ab \left(1 - \lambda^2 - \frac{1-\lambda^4}{2}\right) \\ &= \frac{\pi ab}{2}\left(1 - 2\lambda^2 + \lambda^4\right) = \frac{\pi ab}{2}\left(1-\lambda^2\right)^2,\end{aligned} \tag{6.180}$$

where the change of the variable (Equations 6.172a and 6.172b) was used to transform the ellipse into the unit circle (Equations 6.181b and 6.181c), leading to a simple integration in polar coordinates (Equation 6.181a):

$$dx\, dy = ab\, dx' dy' = ab r' dr' d\varphi': \qquad 0 \le r' \equiv |x'^2 + y'^2|^{1/2} \le 1, \tag{6.181a and 6.181b}$$

$$0 \le \varphi' = \text{arc tan}\left(\frac{y'}{x'}\right) = \text{arc tan}\left(\frac{ay}{bx}\right) < 2\pi. \tag{6.181c}$$

The case $\lambda = 0$ in Equation 6.180 together with Equation 6.174b shows that *the torsional stiffness of a massive rod with elliptical cross section (Figure 6.12d) with half-axis (a, b) is*

$$C_0 \equiv C(0) = \mu\pi Bab = \mu\pi \frac{a^3 b^3}{a^2 + b^2}. \tag{6.182}$$

In the case of a hollow cross section (Equations 6.172a through 6.172d) with a hole shaped like a similar ellipse (Figure 6.12a) with half-axis (λa, λb) and area (Equation 6.183a), the torsional stiffness is reduced (Equation 6.159c) to the value Equation 6.183b:

$$S_1 = \pi\lambda^2 ab: \qquad C_1 = C(\lambda) + 2\mu S_1 \Theta_1 = \mu\pi Bab\left[(1 - 2\lambda^2 + \lambda^4) + 2\lambda^2(1 - \lambda^2)\right], \tag{6.183a and 6.183b}$$

where Equations 6.173c and 6.180 were used:

$$C_1 = \mu\pi ab B\left(1 - \lambda^4\right) = \mu\pi \frac{a^3 b^3}{a^2 + b^2}\left(1 - \lambda^4\right) = C_0\left(1 - \lambda^4\right). \tag{6.184}$$

The torsional stiffness of the rod with the elliptic cross section reduces from Equation 6.182 in the massive case to Equation 6.184 in the hollow case, down to zero $C_1 = 0$ for zero thickness $\lambda = 1$, demonstrating the loss of torsional stiffness due to the hole.

6.7.3 Rod with Annular Cross Section

A particular case is the rod with a circular cross section (Figure 6.12c) with outer radius a and inner radius λa in Equation 6.185a for which the stress function (Equation 6.175a) simplifies to Equations 6.185b and 6.185c:

$$b \equiv a \geq r \geq \lambda a, r \equiv \left|x^2 + y^2\right|^{1/2}: \quad \Theta(x,y) = \frac{a^2 - x^2 - y^2}{2} = \frac{a^2 - r^2}{2} \equiv \Theta(r); \qquad (6.185a–6.185c)$$

it vanishes at the outer boundary (Equation 6.186a) and takes the value (Equation 6.186b) at the inner boundary:

$$\Theta(a) = 0, \quad \Theta(\lambda a) = a^2 \frac{1-\lambda^2}{2} \equiv \Theta_1. \qquad (6.186a and 6.186b)$$

The stresses (Equations 6.175b and 6.175c) have nonzero Cartesian components (Equations 6.187a and 6.187b):

$$\{T_{xz}, T_{yz}\} = \mu\tau\{-y, x\}, \qquad \{T_{zr}, T_{z\varphi}\} = \mu\tau r\{0, 1\}, \qquad (6.187a–6.187d)$$

implying that they are tangential as shown by the polar components (Equations 6.187c and 6.187d). The derivation of Equations 6.187c and 6.187d from Equations 6.187a and 6.187b uses the transformation (Equations 6.188a and 6.188b) from polar (r, φ) to Cartesian coordinates to show that the shear stress vector (Equation 6.188c):

$$\{x, y\} = r\{\cos\varphi, \sin\varphi\}: \quad \{T_{xz}, T_{yz}\} = \mu\tau r\{-\sin\varphi, \cos\varphi\} = \mu\tau r\left\{\cos\left(\varphi + \frac{\pi}{2}\right), \sin\left(\varphi + \frac{\pi}{2}\right)\right\}, \qquad (6.188a–6.188c)$$

lies in the direction $\varphi + \frac{\pi}{2}$, that is, at the positive right angle to the radial direction, which is in the azimuthal direction, implying that (1) the radial component is zero (Equation 6.187c), and (2) the azimuthal component has the direction of the torsion (Equation 6.187d) and modulus (Equation 6.188c).

The same result can be obtained using the relations

$$\begin{bmatrix} T_{zr} \\ T_{z\varphi} \end{bmatrix} = \begin{bmatrix} \cos\varphi & \sin\varphi \\ -\sin\varphi & \cos\varphi \end{bmatrix} \begin{bmatrix} T_{zx} \\ T_{zy} \end{bmatrix} = \begin{bmatrix} \frac{x}{r} & \frac{y}{r} \\ -\frac{y}{r} & \frac{x}{r} \end{bmatrix} \begin{bmatrix} -\mu\tau y \\ \mu\tau x \end{bmatrix} = \frac{\mu\tau}{r} \begin{bmatrix} 0 \\ x^2 + y^2 \end{bmatrix} = \mu\tau \begin{bmatrix} 0 \\ r \end{bmatrix} \qquad (6.188d)$$

that apply (Equations I.16.55a and I.16.55b) to a rotation by an angle φ, that is, to the transformation from Cartesian (x, y) to polar (r, φ) unit base vectors (Equations 4.19a and 4.19b ≡ Equation 6.189a ≡ Equation 6.188d) and its inverse (Equation 6.189b):

$$\begin{bmatrix} \vec{e}_r \\ \vec{e}_\varphi \end{bmatrix} = \begin{bmatrix} \cos\varphi & \sin\varphi \\ -\sin\varphi & \cos\varphi \end{bmatrix} \begin{bmatrix} \vec{e}_x \\ \vec{e}_y \end{bmatrix}, \quad \begin{bmatrix} \vec{e}_x \\ \vec{e}_y \end{bmatrix} = \begin{bmatrix} \cos\varphi & -\sin\varphi \\ \sin\varphi & \cos\varphi \end{bmatrix} \begin{bmatrix} \vec{e}_r \\ \vec{e}_\varphi \end{bmatrix}. \qquad (6.189a \text{ and } 6.189b)$$

The rotation by φ corresponds (Equation 6.190a) to the transformation Equations 6.190b and 6.190c:

$$z' = e^{-i\varphi}z, \; z \equiv x + iy, \; z' \equiv x' + iy': \qquad x' + iy' = (x + iy)(\cos\varphi - i\sin\varphi), \qquad (6.190a\text{–}6.190c)$$

leading to

$$x' = x\cos\varphi + y\sin\varphi, \qquad y' = x\cos\varphi - y'\sin\varphi, \qquad (6.191a \text{ and } 6.191b)$$

which has the same transformation matrix as Equations 6.191a and 6.191b ≡ Equation 6.189a ≡ Equation 6.188d). *The warping function (Equation 6.176) is zero (Equation 6.192b) for the massive or hollow straight rod with a circular cross section (Equation 6.192a) implying by Equation 6.131c that the sections remain plane under torsion (Equation 6.192c):*

$$a = b: \qquad \Psi(x, y) = 0 = u_z(x, y); \qquad f(x + iy) = 0 = \Phi(x, y). \qquad (6.192a\text{–}6.192e)$$

The displacement function (Equation 6.178) [and hence the complex torsion potential (Equation 6.177)] is also zero [Equation 6.192d (Equation 6.192e)] whereas the stress function (Equation 6.185c) is nonzero for a circular or annular cross section (Equations 6.185a and 6.185b).

6.7.4 Effect of a Hole on the Torsional Stiffness

The torsional stiffness of a rod with an annular cross section follows from the integrals (Equations 6.159b and 6.159c) of the stress function (Equation 6.185c) leading to

$$\begin{aligned} C(\lambda) &= \mu\int_0^{2\pi} d\varphi \int_{\lambda a}^{a} \left(a^2 - r^2\right) r\,dr = 2\pi\mu\left[\frac{a^2r^2}{2} - \frac{r^4}{4}\right]_{\lambda a}^{a} \\ &= \frac{\mu\pi a^4}{2}\left[2\left(1-\lambda^2\right) - \left(1-\lambda^4\right)\right] = \frac{\mu\pi a^4}{2}\left(1 - 2\lambda^2 + \lambda^4\right) = \frac{\mu\pi a^4}{2}\left(1-\lambda^2\right)^2. \end{aligned} \qquad (6.193)$$

Thus, *a slender cylindrical rod with massive (Equation 6.194a ≡ Equation 6.194b) circular cross section with radius a has torsional stiffness (Equation 6.194c)*

$$\lambda = 0: \quad 0 \le r \le a; \quad C_0 \equiv C(0) = \frac{\mu\pi a^4}{2}. \qquad (6.194a\text{–}6.194c)$$

An annular cross section with inner radius λa reduces the torsional stiffness (Equation 6.159c) to Equation 6.195b:

$$S_1 = \pi(\lambda a)^2 = \pi\lambda^2 a^2: \quad C_1 = C_0 + 2\mu\Theta_1 S_1 = C_0 + 2\mu\pi\lambda^2 \frac{a^2}{2}(1-\lambda^2)$$

$$= \frac{\mu\pi a^4}{2}\left[(1-2\lambda^2+\lambda^4)+2\lambda^2(1-\lambda^2)\right] = \frac{\mu\pi a^4}{2}(1-\lambda^4) = C_0(1-\lambda^4), \tag{6.195a and 6.195b}$$

where the area of the hole (Equation 6.195a) and the value of the stress function on the boundary (Equation 6.186b) were used. The reduction in torsional stiffness from the massive to the hollow cross section is specified by the factor in curved brackets, that is, the same as (Equation 6.184 ≡ Equation 6.195b) for the elliptic cross section.

For a thickness-to-radius ratio (Equation 6.196a), the loss of torsional stiffness of an annular relative to a massive circular rod is given by Equation 6.196b:

$$\frac{h}{R} = 1-\lambda: \quad \frac{C_1}{C_0} = 1-\lambda^4 = 1-\left(1-\frac{h}{R}\right)^4 = 4\frac{h}{R} - 6\frac{h^2}{R^2} + 4\frac{h^3}{R^3} - \frac{h^4}{R^4}. \tag{6.196a and 6.196b}$$

This is shown in Table 6.2 for (1) nine values of the ratio of thickness to radius spaced 0.1 between 0.1 and 1.0, and (2) a further nine values spaced 0.01 from 0.01 to 0.09, for example, $C_1/C_0 = 0.9375$ (0.3439) for an annulus or circular hole with inner radius a fraction $\lambda = 1/2$ (9/10) of the outer radius, corresponding to a thickness $h/R = 1 - \lambda = 0.5$ (0.1) one-half (one-tenth) of the radius; the torsional stiffness is still 11.5% of that of the massive cross section for a thickness only 3% of the radius. Thus, a thin hollow tube retains some torsional stiffness with a considerable saving in the amount, volume, and weight of the material relative to a massive cylinder. To be more precise, *comparing a massive with a hollow circular or elliptical cross section, (1) the saving in material or mass is Equation 6.196a, (2) the loss of torsional stiffness is Equation 6.196b ≡ Equation 6.196d, and (3) this leads to a gain in stiffness per unit mass given by Equation 6.196c:*

$$\frac{m_1}{m_0} = 1-\lambda^2, \quad \frac{C_1}{C_0} = 1-\lambda^4, \quad \frac{C_1/m_1}{C_0/m_0} = \frac{1-\lambda^4}{1-\lambda^2} = 1+\lambda^2, \tag{6.196c–6.196e}$$

since the ratio of radii (half-axis) of the hollow circular (elliptical) cross section satisfies (Equation 6.172c) the maximum gain in stiffness-to-mass ratio (Equation 6.196e) compared with a massive cross section is for a very thin shell λ → 1 by a factor of two 1 + λ² → 2. The torsional stiffness of a massive (Equation 6.194c) [hollow (Equation 6.195b)] circular cross section is proportional to the fourth power of the radius; since the mass increases with the square of the radius, the torsional stiffness per unit mass increases with the square of the radius. Next the thin closed and open cross sections are compared (Subsections 6.7.5 and 6.7.6) showing a significant loss of torsional stiffness if a cut is made in a closed cross section (Subsection 6.7.7).

6.7.5 Rod with Slender Rectangular Cross Section

For a thin annulus (Equations 6.197a and 6.197b) of thickness Equation 6.197c:

$$\varepsilon^2 \ll 1, \qquad \lambda = 1-\varepsilon, \qquad h \equiv \varepsilon a: \qquad C_1 = C_0\left[1-(1-\varepsilon)^4\right] = 4\varepsilon C_0 = 2\varepsilon\mu a^4 = 2\mu\pi a^3 h, \tag{6.197a–6.197d}$$

TABLE 6.2

Loss of Torsional Stiffness due to a Central Hole and a Cut across the Boundary

Ratio of Thickness to Radius	Loss of Torsional Stiffness		
	Closed Section with/without Hole	Open/Closed Sector	
		Constant Shear Stress	Constant Torsion Rate
$\frac{h}{R}$	Equation 6.196b	Equation 6.215b	Equation 6.214b
0.9	0.99990	0.30000	0.2700
0.8	0.99840	0.26667	0.21333
0.7	0.91900	0.23333	0.16333
0.6	0.97440	0.20000	0.12000
0.5	0.93750	0.16667	0.08333
0.4	0.87040	0.13333	0.05333
0.3	0.75990	0.10000	0.03000
0.2	0.59040	0.06667	0.01333
0.1	0.34390	0.03333	0.00333
0.09	0.31425	0.03000	0.00270
0.08	0.28361	0.02667	0.00213
0.07	0.25195	0.02333	0.00163
0.06	0.21925	0.02000	0.00120
0.05	0.18549	0.01667	0.00083
0.04	0.15065	0.01333	0.00053
0.03	0.11471	0.01000	0.00030
0.02	0.07763	0.00667	0.00013
0.01	0.03940	0.00333	0.00003

Note: Ratio of torsional stiffness for a hollow (massive) cross section with (without) a hole, in the case of an ellipse or circle; also the ratio of torsional stiffness for an open (closed) thin cross section with (without) a cut in the cases of either constant shear stress or constant torsion rate. In all three cases, the ratio of thickness to radius is given two sets nine values in steps of 0.1 (0.01) from 0.1 (0.01) to 0.9 (0.09).

the torsional stiffness (Equation 6.197d ≡ Equation 6.198b) can be expressed:

$$L = 2\pi a: \quad C_1 = 2\mu\pi a^3 h = \frac{\mu}{4\pi^2} L^3 h \qquad \text{(6.198a and 6.198b)}$$

in terms of the perimeter (Equation 6.198a) of the cross section. For a given thickness, the torsional stiffness of a thin cylindrical shell increases with the cube of the radius; the mass is proportional to the radius and thickness. This confirms for a cylindrical shell the same conclusion as for a massive or hollow cylinder: the stiffness-to-mass ratio increases with the square of the radius.

The thin-walled rod with a closed cross section can be compared with an open cross section, for example, a thin long rectangle in Figure 6.13a:

$$0 \le x \equiv s \le L, \qquad 0 \le y \equiv n \le h, \qquad h^2 \ll L^2 \qquad \text{(6.199a–6.199c)}$$

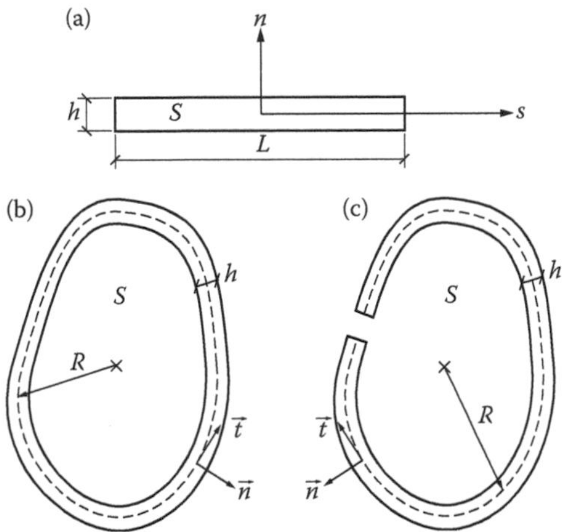

FIGURE 6.13
Simplest torsion problem is for a long, thin rectangular rod (a) neglecting the end effects, so that only the dependence on one transversal coordinate need be considered. The thin rectangular rod (a) may be bent into an open (c) [closed (b)] shell if the ends are separated (joined rigidly). The thin closed shell (b) gives a moderate torsional stiffness against axial torques, with saving on the amount of material by having a large central "hole." There is a considerable loss of torsional stiffness for the same shape, material, and size of the hole if a cut is made in the closed shell (b) replacing it by an open shell (c).

Far from the edges, the stress function varies mainly in the transverse direction (Equation 6.200a), and the Poisson equation 6.141b reduces to Equation 6.200b, whose solution (Equation 6.200c) involves two arbitrary constants:

$$\partial_{yy}\Theta \gg \partial_{xx}\Theta: \quad -2 = \frac{d^2\Theta}{dy^2}, \quad \Theta(x) = -y^2 + Ay + B. \qquad (6.200a–6.200c)$$

The boundary conditions stating that the stress function vanishes at the upper and lower surfaces (Equations 6.201a and 6.201b) specify the constants of integration (Equations 6.201c and 6.201d):

$$0 = \Theta\left(\pm\frac{h}{2}\right) = -\frac{h^2}{4} \pm \frac{Ah}{2} + B: \quad A = 0, \quad B = \frac{h^2}{4}; \quad \Theta(y) = \frac{h^2}{4} - y^2; \qquad (6.201a–6.201e)$$

substitution of Equations 6.201c and 6.201d in Equation 6.200c specifies the stress function (Equation 6.201e). It implies (Equations 6.137a and 6.137b) that there is only one nonzero stress (Equation 6.202a) corresponding to a shear strain with equal modulus and opposite signs (Equation 6.202b) at the two boundary surfaces:

$$2\mu S_{xz}(y) = T_{xz}(y) = -2\mu\tau y, \quad 2\mu S_{xz}\left(\pm\frac{h}{2}\right) = T_{xz}\left(\pm\frac{h}{2}\right) = \mp\mu\tau h. \qquad (6.202a \text{ and } 6.202b)$$

The stress function (Equation 6.201e) leads (Equation 6.158b) to the torsional stiffness:

$$C_2 = 2\mu \int_{-L/2}^{+L/2} dx \int_{-h/2}^{+h/2} \left(\frac{h^2}{4} - y^2 \right) dy = 2\mu L \left[\frac{h^2 y}{4} - \frac{y^3}{3} \right]_{-h/2}^{+h/2} = \frac{\mu L h^3}{3}. \tag{6.203}$$

Comparing (1) a thin-walled rod (Equation 6.203 ≡ Equation 6.204a) with a rectangular cross section of thickness h and length L with a (2) closed hollow circular section (Equation 6.198b) with the same thickness h and perimeter L, the ratio of torsional stiffness is Equation 6.204b:

$$C_2 = \frac{1}{3}\mu L h^3, \quad \frac{C_1}{C_2} = \frac{3}{4\pi^2}\frac{L^2}{h^2} = \left(\frac{L}{L_0}\right)^2; \quad L_0 = \frac{2\pi}{\sqrt{3}} h = 3.62760; \qquad \text{(6.204a and 6.204b)}$$

thus for the length-to-thickness ratio given by Equation 6.204c, the circular and rectangular thin-walled rods have the same torsional stiffness; for larger $L > L_0$ (smaller $L < L_0$), the thin annular rod is stiffer (less stiff) than the thin rectangular rod. For example, for a perimeter 10 times the thickness $L/h = 10$, the circular rod is stiffer than the rectangular one by a factor of $75/\pi^2 = 7.60$. If the length-to-thickness ratio is smaller than Equation 6.204b, the conclusion that the thin rectangular cross section is torsionally stiffer than the closed cylindrical shell is limited by the assumption (Equation 6.199c) implying that $(h/L_0)^2 = 1/(3.2760)^2 = 3/(4\pi^2) = 0.07599 \le (h/L)^2 \ll 1$. If this assumption is not met, for example, for a thick rectangular $h \sim L$ or square $h = L$ cross section, the torsion of a rod can be considered using Fourier series. The torsion of a thin rectangular rod is analogous to the viscous flow between parallel walls (Note 6.8).

6.7.6 Torsion of Rod with Thin Closed Cross Section

The preceding example suggests that there should be a significant loss of torsional stiffness if a cut is made (Figure 6.13c) in a thin closed section (Figure 6.13b). Considering first the closed thin section (Figure 6.13b), the variations are larger along normal dn than along the tangential ds direction (Equation 6.205a), and thus (Equations 6.205a and 6.205b) the main shear stress is tangential (Equation 6.205b), where h is the thickness:

$$\frac{\partial}{\partial n} \gg \frac{\partial}{\partial s}: \quad T_{nz} \ll T_{sz} = \mu\tau\partial_n\Theta = \mu\tau\frac{\partial\Theta}{\partial n} \sim \mu\tau\frac{\Theta}{h}; \qquad \text{(6.205a and 6.205b)}$$

the stress function is constant on the boundaries (Equation 6.206a) implying Equation 6.206b:

$$T_{sz}h \sim \text{const} \equiv \mu\tau\Theta; \qquad (T_{sz})_{\max} \Leftrightarrow h_{\min}. \qquad \text{(6.206a and 6.206b)}$$

Thus, *for a long thin closed section (Figure 6.13b) with variable thickness h, the shear stress (Equation 6.206a) is largest at the thinnest point (Equation 6.206b), that is, the location of first structural failure.*

The torsional stiffness of the long thin closed section of thickness h and perimeter L is specified by Equation 6.159c with one hole of area S, so that the second term on the r.h.s. dominates (Equation 6.207b) the first

$$S \gg Lh: \qquad C_{closed} = 2\mu\Theta S \gg 2\mu\Theta L h, \qquad \text{(6.207a and 6.207b)}$$

if the area of the cross section is small compared with the area of the central hole (Equation 6.207a). The **average radius** (Equation 6.208c) is defined as the ratio of the area of the central hole to twice the perimeter:

$$S = \pi R^2, \quad L = 2\pi R, \quad R = \frac{2S}{L}, \tag{6.208a–6.208c}$$

in agreement with the case of the circle (Equations 6.208a and 6.208b).

The condition Equation 6.207a states that the average radius should be much larger than twice the thickness (Equation 6.209a):

$$R \equiv \frac{2S}{L} \gg 2h: \quad C_{closed} = \mu\Theta RL = \frac{1}{\tau} RLhT_{sz}, \tag{6.209a and 6.209b}$$

in which case the torsional stiffness (Equation 6.207b) simplifies Equation 6.208c to Equation 6.209b, where the shear stress was substituted from Equation 6.206a. The shear strain is of the order of half the torsion rate times the radius (Equation 6.210a) and multiplied by twice the shear modulus specifies the shear stress (Equation 6.210b):

$$S_{sz} \sim \frac{\tau}{2} R, \quad T_{sz} = 2\mu S_{zz} \sim \mu\tau R: \quad C_{closed} \sim \mu R^2 Lh = 2\pi\mu R^3 h = 4\mu S^2 \frac{h}{L}; \tag{6.210a–6.210e}$$

substitution of the shear stress (Equation 6.210b) in Equation 6.209b leads to the torsional stiffness (Equation 6.210c) for the closed section; also Equation 6.208b (Equation 6.208c) was used in Equation 6.210d (Equation 6.210e).

6.7.7 Loss of Torsional Stiffness due to a Cut

The case of a circular cross section with outer radius a and inner radius λa in Equations 6.211a and 6.211b leads to the thickness Equation 6.211c, average radius Equation 6.211d, and perimeter Equation 6.211e:

$$\lambda a < r < a, \quad 0 < \lambda < 1: \quad h = (1-\lambda)a, \quad R = a\frac{1+\lambda}{2}, \quad L = 2\pi R = \pi a(1+\lambda). \tag{6.211a–6.211e}$$

Substituting Equations 6.211c and 6.211d in the approximate torsional stiffness (Equation 6.210d) leads to Equation 6.212a:

$$C_{closed} \sim \frac{\pi\mu a^4}{4}(1+\lambda)^3(1-\lambda); \quad C_1 = \frac{\pi\mu a^4}{2}\left(1-\lambda^4\right); \tag{6.212a and 6.212b}$$

comparison with the exact value (Equation 6.195b ≡ Equation 6.212b) leads to a relative error:

$$\begin{aligned} \varepsilon \equiv \frac{C_1}{C_{closed}} - 1 &= \frac{2}{1-\lambda}\frac{1-\lambda^4}{(1+\lambda)^3} - 1 = \frac{2}{(1+\lambda)^2}\frac{1-\lambda^4}{1-\lambda^2} - 1 = 2\frac{1+\lambda^2}{(1+\lambda)^2} - 1 \\ &= \frac{2+2\lambda^2-(1+\lambda)^2}{(1+\lambda)^2} = \frac{1-2\lambda+\lambda^2}{(1+\lambda)^2} = \left(\frac{1-\lambda}{1+\lambda}\right)^2 = \left(\frac{h}{2R}\right)^2 = \left(\frac{hL}{4S}\right)^2 \end{aligned} \tag{6.213}$$

using Equations 6.208c, 6.211c, and 6.211d. Thus, *the approximate torsional stiffness (Equations 6.210c through 6.210e) of a closed cross section of thickness h, perimeter length L, and average radius R with a hole area S in Equation 6.208c has a relative error of the order of the square of the ratio of the thickness to the diameter (Equation 6.213). This value is comparable to the loss of torsional stiffness (Equation 6.214b) of a thin hollow cross section when a cut is made at a constant torsion rate*; the latter result is proved next.

The torsional stiffness of the open section is taken from the thin rectangle (Equation 6.203 ≡ Equation 6.214a) with the same thickness and length (Equation 6.208b):

$$C_{open} = \frac{1}{3}\mu L h^3 = \frac{2}{3}\pi\mu R h^3; \quad \left\{\frac{C_{open}}{C_{closed}}\right\}_{\tau=\text{const}} = \frac{h^2}{3R^2} = 3\left(\frac{hL}{6S}\right)^2; \qquad \text{(6.214a and 6.214b)}$$

comparison with the closed section (Equation 6.210c) leads to Equation 6.214b for the same torsion rate. For the same torsion rate, the shear stresses are in the ratio Equation 6.215a leading to Equation 6.215b:

$$\left\{\frac{(T_{sz})_{open}}{(T_{sz})_{closed}}\right\}_{\tau=const} = \frac{h}{R}, \quad \left\{\frac{C_{open}}{C_{closed}}\right\}_{T_{sz}=const} = \frac{h}{3R} = \frac{hL}{6S}. \qquad \text{(6.215a and 6.215b)}$$

It has been shown that *if a cut is made (Figure 6.13c) in a thin, closed cross section (Figure 6.13b) of thickness h, length L, interior area S, and average radius (Equation 6.208c ≡ Equation 6.209a), there is a loss of torsional stiffness at constant shear stress (Equation 6.215b) [constant torsion rate (Equation 6.214b)] that is (1) equal to (equal to three times the square of) one-third of the ratio of the thickness to the average radius, and (2) equivalently one-sixth of the ratio of the area of the cross section to the area of the central hole.*

Thus, a cut in a thin-walled cross section reduces significantly the torsional stiffness, more so the larger the central "hole" or the thinner the wall relative to the radius. Table 6.2 lists the following (1) the loss of torsional stiffness of a hollow elliptic (Equation 6.184 ≡ Equation 6.195b) or a circular cylinder relative to a massive cylinder; (2 and 3) the further loss of torsional stiffness of the hollow section if a cut is made across it, in the case of constant shear stress (Equation 6.215b) [constant torsion rate (Equation 6.214b)]; (4 and 5) the loss of torsional stiffness relative to a massive cylinder of a hollow section with a cut is given by the product of items 1 and 2 (item 3) for constant shear stress (torsion angle). The effect of making a cut in a hollow cross section is much more severe at the constant torsion rate than at the constant shear stress. Even in the latter case, the effect of a cut in a hollow cross section is much more severe than the change from a massive to a hollow cross section. Thus, a thin hollow tube retains some of the torsional stiffness of the massive cylinder at a considerable saving of weight and volume of the material, as long as the cross section remains closed. A cut across the boundary causes a significant loss of torsional stiffness; this corresponds to the everyday experience that a tube is twisted much more easily if it is cut parallel to the axis. Another example is that the removal of the roof of a car, from the saloon to the cabriolet, reduces the torsional stiffness; in order to restore some of the torsional stiffness for the open car, structural reinforcement is made in the lower body, for example, adding cross members. As a result, the cabriolet or open top version of a closed or saloon car is usually heavier while having less torsional stiffness, due to the loss of the roof; the latter is the lightest and most efficient way to increase torsional stiffness.

6.8 Torsion of Prisms with Triangular Cross Section (Saint-Venant 1885; Campos and Cunha 2010)

The stress function can be obtained for a triangular prism most simply for an equilateral cross section (Subsection 6.8.1), leading to the warping and displacement functions (Subsection 6.8.2) and also to strains, stress, and stiffness (Subsection 6.8.3); the extension to triangular nonequilateral cross sections (Subsection 6.8.4) is made using a rigid body transformation equivalent to the choice of a torsion center and axis (Subsection 6.8.5); this includes the case of an isosceles triangle (Subsection 6.8.6). In general, it specifies the stresses and strains (torsional stiffness) [Subsection 6.8.7 (Subsection 6.8.8)] for a two-parameter family of triangular cross sections, including nonequilateral and nonisosceles shapes.

6.8.1 Stresses and Strains in a Triangle (Saint-Venant 1855)

Consider a slender rod with a triangular cross section with (Figure 6.14a) vertices at the points $(a, 0)$, $(-b, 0)$, $(c, 0)$, so that the cross section occupies the region:

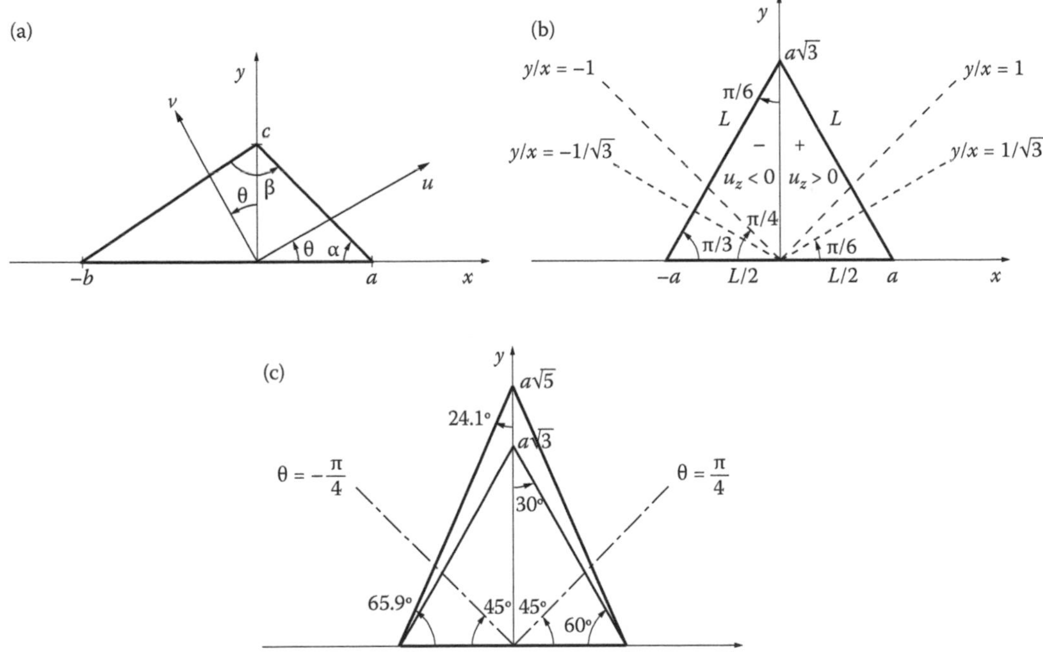

FIGURE 6.14
Stresses and strains associated with the torsion of slender bars with an elliptical (triangular) cross section [Figure 6.12 (Figure 6.14a)] are linear (quadratic) functions of the coordinates in the plane of the cross section. The torsion of a slender bar with an equilateral triangular cross section (b) was one of the original applications of the theory of torsion, in the sense that both were considered in the same time frame. The extension to non-equilateral triangular cross sections has been made recently, more than one-and-a-half centuries later; it is still not complete, since it concerns a two-parameter family of shapes, rather than the general case of independent choice of the three sides of the triangle. Besides the equilateral triangle, it does include the isosceles triangle (c), and other nonequilibrium, nonisosceles shapes.

$$y \geq 0, \quad \frac{y}{c} + \frac{x}{a} \leq 1 \geq \frac{y}{c} - \frac{x}{b}. \tag{6.216a–6.216c}$$

The stress function

$$\begin{aligned}
\Theta(x,y) &= By\left(1 - \frac{y}{c} - \frac{x}{a}\right)\left(1 - \frac{y}{c} + \frac{x}{b}\right), \\
&= By\left[\left(1 - \frac{y}{c}\right)^2 + \left(1 - \frac{y}{c}\right)x\left(\frac{1}{b} - \frac{1}{a}\right) - \frac{x^2}{ab}\right] \\
&= By\left\{1 + x\left(\frac{1}{b} - \frac{1}{a}\right) - \frac{x^2}{ab} - \frac{y}{c}\left[2 + x\left(\frac{1}{b} - \frac{1}{a}\right)\right] + \frac{y^2}{c^2}\right\}
\end{aligned} \tag{6.217a–6.217c}$$

vanishes on the boundary (Equations 6.216a through 6.216c); the stress function Equation 6.217c satisfies the Poisson equation (Equation 6.141b ≡ Equation 6.218):

$$-2 = \partial_{xx}\Theta + \partial_{yy}\Theta = 2By\left(\frac{3}{c^2} - \frac{1}{ab}\right) - \frac{2B}{c}\left[2 + x\left(\frac{1}{b} - \frac{1}{a}\right)\right], \tag{6.218}$$

if the three conditions (Equations 6.219a through 6.219c) are met:

$$\frac{L}{2} \equiv a = b = \frac{c}{\sqrt{3}}, \; B = \frac{c}{2}: \quad \Theta(x,y) = \frac{yc}{2}\left[\left(1 - \frac{y}{c}\right)^2 - \frac{3x^2}{c^2}\right], \tag{6.219a–6.219d}$$

corresponding to (1) an equilateral triangle (Figure 6.14b) with sides of length L and height c and (2) the stress function (Equation 6.217b ≡ Equation 6.219d). The corresponding stresses (Equations 6.137a and 6.137b) ≡ Equations 6.220a and 6.220b) are

$$2\mu\left\{S_{xz}, S_{yz}\right\} = \left\{T_{xz}, T_{yz}\right\} = \frac{\mu\tau c}{2}\left\{1 - 4\frac{y}{c} + 3\frac{y^2 - x^2}{c^2}, \frac{6xy}{c^2}\right\} \tag{6.220a and 6.220b}$$

quadratic functions, in general, except on the diagonals (Equation 6.221a):

$$y = \pm x: \quad 2\mu\left\{S_{xz}, S_{yz}\right\} = \left\{T_{xz}, T_{yz}\right\} = \frac{\mu\tau c}{2}\left\{1 - \frac{4y}{c}, \pm\frac{6y^2}{c^2}\right\}, \tag{6.221a–6.221c}$$

where one of the stresses (Equation 6.221b) is linear (Figure 6.14b). The other stress (Equation 6.220b) vanishes on the axis $x = 0 = y$.

6.8.2 Warping, Displacement, and Stress Functions

The displacement function (Equation 6.142a) associated with the stress function (Equation 6.219d) is

$$\Psi(x,y) = \frac{y^3 - 3x^2y}{2c} + \frac{x^2 - y^2}{2} + \frac{yc}{2}; \tag{6.222}$$

it is the imaginary part of the complex torsion potential:

$$\Phi + i\Psi = f(x+iy) = -\frac{(x+iy)^3}{2c} + i\frac{(x+iy)^2}{2} + (x+iy)\frac{c}{2}; \tag{6.223}$$

$$\Phi(x,y) = x\left(\frac{3y^2 - x^2}{2c} - y + \frac{c}{2}\right) = \frac{1}{\tau}u_z(x,y); \tag{6.224a and 6.224b}$$

the real part of the complex torsion potential (Equation 6.223) is the displacement function (Equation 6.224a) that specifies (Equation 6.131c) the axial displacement (Equation 6.224b); it is generally a cubic, and reduces to a quadratic function (Equation 6.225b) on two straight lines through the origin with a slope (Equation 6.225a) corresponding to an angle 30° orthogonal to the sides of the equilateral triangle (Figure 6.14b):

$$\frac{y}{x} = \pm\frac{1}{\sqrt{3}}: \quad u_z\left(x, \pm\frac{x}{\sqrt{3}}\right) = \tau x\left(\frac{c}{2} \mp \frac{x}{\sqrt{3}}\right). \tag{6.225a and 6.225b}$$

The vertical displacement (Equation 6.224b) is (1 and 2) positive or upward (negative or downward) on the r.h.s. (Equation 6.226a) [l.h.s. (Equation 6.226c)] of the triangle, and (3) it vanishes (Figure 6.14b) on the separating line between items 1 and 2, that is, the y-axis (Equation 6.226b):

$$u_z(x,y)\begin{cases} >0 & \text{if} \quad x>0, & (6.226a)\\ =0 & \text{if} \quad x=0, & (6.226b)\\ <0 & \text{if} \quad x<0. & (6.226c)\end{cases}$$

The preceding result (Equations 6.226a through 6.226c) follows from Equation 6.224b because the term in the curved bracket is always positive (Equation 6.227b):

$$x^2 \le a^2 = \frac{c^2}{3}: \quad 3y^2 - 2cy + c^2 - x^2 = \left(y\sqrt{3} - \frac{c}{\sqrt{3}}\right)^2 + \frac{2}{3}c^2 - x^2 > 0, \tag{6.227a and 6.227b}$$

where Equations 6.219b and 6.227a were used.

6.8.3 Torsional Stiffness of an Equilateral Triangle

The torsional stiffness is given (Equation 6.158b) by

$$C = 4\mu \int_0^a dx \int_0^{c(1-x/a)} dy \Theta(x,y); \tag{6.228}$$

the two-dimensional integration (Equation 6.228) of the stress function (Equation 6.219d) over the equilateral triangle is performed using the dimensionless variables (Equations 6.172a and 6.172b ≡ Equations 6.229a and 6.229b), leading to Equation 6.229c:

$$x' \equiv \frac{x}{a}, \quad y' \equiv \frac{y}{c}: \quad C = 2\mu a c^3 \int_0^1 dx' \int_0^{1-x'} dy' y' \left[\left(1-y'\right)^2 - x'^2 \right]; \tag{6.229a–6.229c}$$

the latter is evaluated (Equations 6.230a through 6.230c) as

$$C = \frac{\mu a c^3}{15} = \frac{\sqrt{3}}{5} \mu a^4 = \frac{\mu c^4}{15\sqrt{3}} = \frac{\sqrt{3}}{80} \mu L^4. \tag{6.230a–6.230d}$$

Thus, *an equilateral triangle (Figure 6.14b) with (Equations 6.219a through 6.219c) side L, half-side a, and height c has torsional stiffness (Equations 6.230a through 6.230d), where μ is the shear modulus (Equation 6.134a); the warping (Equation 6.224a), displacement (Equation 6.222), and stress (Equation 6.219d) functions, and complex torsion potential (Equation 6.223), specify (1) the strains and stresses (Equations 6.220a and 6.220b) that are quadratic, except on the diagonals (Equations 6.221a through 6.221c); (2) the axial displacement (Equation 6.224b) that is a cubic, except on the lines (Equations 6.225a and 6.225b) where it is quadratic; and (3) that the vertical displacement is zero on the y-axis (Equation 6.226b) and upward (downward) to the right (Equation 6.226a) [left (Equation 6.226c)].* The horizontal displacements are specified by Equations 6.131a and 6.131b, as for any cross section. The integral (Equation 6.229c) for the torsional stiffness may be evaluated in two steps. Omitting the primes for brevity, (1) the inner integral:

$$I(x) \equiv \int_0^{1-x} \left[y^3 - 2y^2 + \left(1-x^2\right) y \right] dy = I_1(x) + I_2(x) \tag{6.231}$$

splits into two parts (Equations 6.232a and 6.232b) that are both expressible in powers of $1 - x$:

$$I_1(x) \equiv \int_0^{1-x} \left(y^3 - 2y^2\right) dy = \frac{(1-x)^4}{4} - 2\frac{(1-x)^3}{3}, \tag{6.232a}$$

$$\begin{aligned} I_2(x) &\equiv \left(1-x^2\right) \int_0^{1-x} y\, dy = \left(1-x^2\right) \frac{(1-x)^2}{2} = \frac{(1-x)^3}{2}(1+x) \\ &= \frac{(1-x)^3}{2}\left[2-(1-x)\right] = (1-x)^3 - \frac{(1-x)^4}{2}; \end{aligned} \tag{6.232b}$$

(2) the powers of $1 - x$ facilitate the evaluation of the outer integral after adding (Equations 6.232a and 6.232b):

$$C = 2\mu ac^3 \int_0^1 \left[I_1(x) + I_2(x) \right] dx = 2\mu ac^3 \int_0^1 \left[\frac{(1-x)^3}{3} - \frac{(1-x)^4}{4} \right] dx$$
$$= 2\mu ac^3 \left[-\frac{(1-x)^4}{12} + \frac{(1-x)^5}{20} \right]_0^1 = \mu ac^3 \left(\frac{1}{6} - \frac{1}{10} \right) = \frac{\mu ac^3}{15}; \quad (6.233)$$

this leads to Equation 6.233 ≡ Equation 6.230a.

6.8.4 Nonequilateral Triangle as the Cross Section (Campos and Cunha 2010)

The preceding solution may be extended from equilateral to nonequilateral triangles (1) by starting with the stress function (Equations 6.217a through 6.217c) that vanishes at the boundary of an arbitrary triangle (Equations 6.216a through 6.216c) in a particular position (Figure 6.14a); (2) by using rotation and translation to place the triangle at an arbitrary position; and (3) since the transformation of coordinates (2) is orthonormal, the Laplacian is unchanged, and the same Poisson equation 6.141b is satisfied in the new coordinates. Step (1) of the preceding method is (Figure 6.14a) thus a counterclockwise (Equation 6.189a ≡ Equations I.16.56a and I.16.56b) rotation by an angle θ plus a translation by (x_0, y_0):

$$x = x_0 + u\cos\theta + v\sin\theta, \qquad y = y_0 - u\sin\theta + v\cos\theta; \qquad (6.234a \text{ and } 6.234b)$$

the rotation in Equations 6.234a and 6.234b corresponds to $\theta = \varphi$ in Equations 6.191a and 6.191b since both are a positive or counterclockwise rotation. The *general* **plane rigid body transformation** *(Equations 6.234a and 6.234b) consists of (1) a translation of the origin to (x_0, y_0) and (2) a positive or counterclockwise rotation of the axis by the angle θ.* In the present context, this corresponds to the introduction of a **torsion axis (center)** to be determined next θ (x_0, y_0).

Substituting Equations 6.234a and 6.234b in the stress function (Equation 6.217a) leads to

$$\Theta(u,v) = B\,(y_0 + v\cos\theta - u\sin\theta)\,(\alpha_0 + \alpha_1 u + \alpha_2 v)\,(\beta_0 + \beta_1 u + \beta_2 v), \qquad (6.235)$$

where the constant coefficients are given by

$$\{\alpha_0, \alpha_1, \alpha_2\} \equiv \left\{ 1 - \frac{x_0}{a} - \frac{y_0}{c}, -\frac{\cos\theta}{a} + \frac{\sin\theta}{c}, -\frac{\sin\theta}{a} - \frac{\cos\theta}{c} \right\}, \qquad (6.236a\text{–}6.236c)$$

$$\{\beta_0, \beta_1, \beta_2\} \equiv \left\{ 1 + \frac{x_0}{b} - \frac{y_0}{c}, \frac{\cos\theta}{b} + \frac{\sin\theta}{c}, \frac{\sin\theta}{b} - \frac{\cos\theta}{c} \right\}. \qquad (6.237a\text{–}6.237c)$$

Since (2) the rotation and translation (Equations 6.234a and 6.234b) are orthogonal transformations with unit scale factors:

$$h_u \equiv \left(\frac{\partial x}{\partial u}\right)^2 + \left(\frac{\partial y}{\partial u}\right)^2 = \cos^2\theta + \sin^2\theta = 1 = \left(\frac{\partial x}{\partial v}\right)^2 + \left(\frac{\partial y}{\partial v}\right)^2 \equiv h_v, \qquad (6.238a \text{ and } 6.238b)$$

the Laplacian is the same in both coordinate systems (x, y) and (u, v). Thus, the stress function satisfies a Poisson equation 6.141b that consists (3) of second-order derivates implying that all terms vanish except those involving squares or higher powers of u or v:

$$\begin{aligned} B^{-1}\Theta(u,v) &= y_0\left(\alpha_1\beta_1 u^2 + \alpha_2\beta_2 v^2\right) + v^2\cos\theta\left[\alpha_2\left(\beta_0+\beta_1 u\right)+\beta_2\left(\alpha_0+\alpha_1 u\right)\right] \\ &-u^2\sin\theta\left[\alpha_1\left(\beta_0+\beta_2 v\right)+\beta_1\left(\alpha_0+\alpha_2 v\right)\right] + \alpha_2\beta_2 v^3\cos\theta - \alpha_1\beta_1 u^3\sin\theta + O(u,v,uv); \end{aligned} \tag{6.239}$$

for this reason, the constant term and those linear in u or v and bilinear in u and v were omitted from Equation 6.239.

Substituting Equation 6.239 in the Poisson equation 6.141b leads to

$$\begin{aligned} -B^{-1} &= (2B)^{-1}\left(\partial_{uu}+\partial_{vv}\right)\Theta = y_0\left(\alpha_1\beta_1+\alpha_2\beta_2\right) + 3\alpha_2\beta_2 v\cos\theta - 3\alpha_1\beta_1 u\sin\theta \\ &+\cos\theta\left[\alpha_2\left(\beta_0+\beta_1 u\right)+\beta_2\left(\alpha_0+\alpha_1 u\right)\right] - \sin\theta\left[\alpha_1\left(\beta_0+\beta_2 v\right)+\beta_1\left(\alpha_0+\alpha_2 v\right)\right]; \end{aligned} \tag{6.240}$$

this specifies a system of three equations by equating to zero the constant term (Equation 6.241c) and coefficients [Equation 6.241a (Equation 6.241b)] of u (v):

$$3\alpha_1\beta_1\tan\theta = \alpha_2\beta_1 + \alpha_1\beta_2 = 3\alpha_2\beta_2\cot\theta, \qquad \text{(6.241a and 6.241b)}$$

$$-B^{-1} = y_0\left(\alpha_1\beta_1+\alpha_2\beta_2\right) + \cos\theta\left(\alpha_2\beta_0+\alpha_0\beta_2\right) - \sin\theta\left(\alpha_1\beta_0+\beta_1\alpha_0\right). \tag{6.241c}$$

This completes the three-step method (1–3) that is shown next to allow independent choice of two sides of the triangle; for example, it applies to the torsion of a prism whose cross section is an isosceles triangle.

6.8.5 Choice of Torsion Center and Axis

Equations 6.241a through 6.241c show that (1) the translation appears only in Equation 6.241c and can be omitted (Equations 6.242a and 6.242b) in Equations 6.234a and 6.234b so that there is no need for a "torsion center" that remains at the origin:

$$x_0 = 0 = y_0: \qquad x = u\cos\theta + v\sin\theta, \qquad y = -u\sin\theta + v\cos\theta; \qquad \text{(6.242a–6.242d)}$$

the "torsion axis" alone appears in the rotation (Equations 6.242c and 6.242d) as a result of omitting the translation (Equations 6.242a and 6.242b) from the rigid body transformation (Equations 6.234a and 6.234b). The use of Equations 6.242a and 6.242b ≡ Equations 6.243a and 6.243b simplifies Equation 6.236a (Equation 6.237a) to Equation 6.243c (Equation 6.243d); also the condition 6.241c becomes (Equation 6.212e):

$$x_0 = 0 = y_0: \qquad \alpha_0 = 1 = \beta_0, \qquad B^{-1} = (\alpha_1+\beta_1)\sin\theta - (\alpha_2+\beta_2)\cos\theta; \qquad \text{(6.243a–6.243e)}$$

Substitution of Equations 6.236b, 6.236c, 6.237b, and 6.237c in Equations 6.241a and 6.241b gives two equations:

$$\tan^2\theta = \frac{\alpha_2\beta_2}{\alpha_1\beta_1} = \frac{ab\cos^2\theta - c^2\sin^2\theta + c(b-a)\cos\theta\sin\theta}{ab\sin^2\theta - c^2\cos^2\theta - c(b-a)\cos\theta\sin\theta}, \tag{6.244a}$$

$$3\cot\theta = \frac{\alpha_1\beta_2 + \alpha_2\beta_1}{\alpha_2\beta_2} = \frac{c(b-a)\cos(2\theta) - (ab+c^2)\sin(2\theta)}{ab\cos^2\theta - c^2\sin^2\theta + c(b-a)\cos\theta\sin\theta} \tag{6.244b}$$

that follow from Equation 6.244a ≡ Equation 6.245a (Equation 6.244b ≡ Equation 6.245b):

$$\begin{aligned}\tan^2\theta &= \frac{\left(-\frac{\sin\theta}{a} - \frac{\cos\theta}{c}\right)\left(\frac{\sin\theta}{b} - \frac{\cos\theta}{c}\right)}{\left(-\frac{\cos\theta}{a} + \frac{\sin\theta}{c}\right)\left(\frac{\cos\theta}{b} + \frac{\sin\theta}{c}\right)} \\ &= \frac{\frac{\cos^2\theta}{c^2} - \frac{\sin^2\theta}{ab} + \frac{\cos\theta\sin\theta}{c}\left(\frac{1}{a} - \frac{1}{b}\right)}{\frac{\sin^2\theta}{c^2} - \frac{\cos^2\theta}{ab} - \frac{\cos\theta\sin\theta}{c}\left(\frac{1}{a} - \frac{1}{b}\right)} \\ &= \frac{ab\cos^2\theta - c^2\sin^2\theta + c(b-a)\cos\theta\sin\theta}{ab\sin^2\theta - c^2\cos^2\theta - c(b-a)\cos\theta\sin\theta},\end{aligned} \tag{6.245a}$$

$$\begin{aligned}3\cot\theta &= \frac{\left(-\frac{\cos\theta}{a} + \frac{\sin\theta}{c}\right)\left(\frac{\sin\theta}{b} - \frac{\cos\theta}{c}\right) + \left(-\frac{\sin\theta}{a} - \frac{\cos\theta}{c}\right)\left(\frac{\cos\theta}{b} + \frac{\sin\theta}{c}\right)}{\left(-\frac{\sin\theta}{a} - \frac{\cos\theta}{c}\right)\left(\frac{\sin\theta}{b} - \frac{\cos\theta}{c}\right)} \\ &= \frac{\frac{\cos^2\theta - \sin^2\theta}{c}\left(\frac{1}{a} - \frac{1}{b}\right) - 2\cos\theta\sin\theta\left(\frac{1}{ab} + \frac{1}{c^2}\right)}{\frac{\cos^2\theta}{c^2} - \frac{\sin^2\theta}{ab} + \frac{\cos\theta\sin\theta}{c}\left(\frac{1}{a} - \frac{1}{b}\right)} \\ &= \frac{c(b-a)\cos(2\theta) - \left(ab + c^2\right)\sin(2\theta)}{ab\cos^2\theta - c^2\sin^2\theta + c(b-a)\cos\theta\sin\theta}.\end{aligned} \tag{6.245b}$$

Thus, *given (Figure 6.14a) two sides (a, b): (1) the third side c and the direction θ of the* **torsion axis** *are determined by Equations 6.244a and 6.244b, and (2) the constant B follows from Equation 6.243e completing with (Equations 6.236b, 6.236c, 6.237b, 6.237c, and 6.243a through 6.243d) the determination of the stress function (Equation 6.235). The triangular cross section no longer needs to be equilateral, since two sides can be given independently.*

6.8.6 Isosceles Triangle as the Cross Section

For example, (1) assume that the torsion axis lies along the diagonal of the quadrants (Equation 6.246a) implying (Equations 6.246b through 6.246e):

$$\theta = \pm\frac{\pi}{4}: \quad \tan^2\theta = 1 = \pm\cot\theta, \quad \cos\theta = \frac{1}{\sqrt{2}} = \pm\sin\theta; \qquad (6.246a–6.246e)$$

(2) substitution of Equations 6.246b through 6.246d in Equation 6.244a leads to Equation 6.247a so that (Equation 6.247b) the triangular cross section is isosceles:

$$ab - c^2 \pm c(b - a) = ab - c^2 \mp c(b - a) \Rightarrow a = b; \qquad (6.247a \text{ and } 6.247b)$$

(3) substitution of Equations 6.246b through 6.246e and 6.247b in Equation 6.244b leads to Equation 6.248a and hence to Equation 6.248b:

$$\pm\frac{3}{2} = \mp\frac{c^2 + a^2}{a^2 - c^2}, \quad 5a^2 = c^2, \quad \cot\alpha = \frac{a}{c} = \frac{1}{\sqrt{5}}, \qquad (6.248a–6.248c)$$

that specifies the ratio of the half-base to the height (Equation 6.216c); (4) hence follows the equal internal angles (Equation 6.249d) and the distinct angle (Equation 6.249e):

$$\theta = \pm\frac{\pi}{4}: \quad a = b; \quad \frac{c}{a} = \sqrt{5} = \tan\alpha, \quad \alpha = 65.905°, \quad \beta = \pi - 2\alpha = 48.190°. \qquad (6.249a–6.249e)$$

Thus, *the particular case of the torsion axis along the diagonal of the quadrants (Equation 6.249a) leads to an isosceles triangle (Equation 6.249b), with a ratio of height to half-base (Equation 6.249c) and hence unequal (equal) internal angles [Equation 6.249e (Equation 6.249d). The isosceles triangle is compared (Figure 6.14c) with the equilateral triangle (Figure 6.14b). This specifies (Equation 6.250a) the parameters (Equations 6.236b, 6.236c, 6.237b, 6.237c ≡ Equations 6.250b through 6.250e):*

$$\theta = \frac{\pi}{4}: \quad \alpha_1 = \left(\frac{1}{c} - \frac{1}{a}\right)\frac{1}{\sqrt{2}} = -\beta_2, \quad \beta_1 = \left(\frac{1}{c} + \frac{1}{a}\right)\frac{1}{\sqrt{2}} = -\alpha_2; \qquad (6.250a–6.250e)$$

the coefficient (Equation 6.241c) of the stress function (Equation 6.239) simplifies to Equation 6.251a:

$$\frac{1}{B} = \frac{\alpha_1 + \beta_1 - \alpha_2 - \beta_2}{\sqrt{2}} = (\alpha_1 + \beta_1)\sqrt{2} = \frac{2}{c}, \quad B = \frac{c}{2}, \qquad (6.251a \text{ and } 6.251b)$$

that is related to the height of the triangle (Equation 6.251b ≡ Equation 6.219c) as in the equilateral case.

Substitution of Equations 6.243b through 6.243d, 6.250b through 6.250e, and 6.251b in Equation 6.235 specifies the stress function:

$$\Theta(u, v) = \frac{c}{2}\frac{v - u}{\sqrt{2}} - \frac{(v - u)^2}{2} - \frac{5}{4}\frac{v - u}{\sqrt{2}}\frac{(u + v)^2}{c} + \frac{(v - u)^3}{4c\sqrt{2}}, \qquad (6.252)$$

that follows from simplification of

$$\begin{aligned}\frac{2}{c}\frac{\sqrt{2}}{v-u}\Theta(u,v) &= (1+\alpha_1 u-\beta_1 v)(1+\beta_1 u-\alpha_1 v)\\ &= 1+(\alpha_1+\beta_1)(u-v)+\alpha_1\beta_1(u^2+v^2)-uv\left[(\alpha_1)^2+(\beta_1)^2\right]\\ &= 1+\frac{\sqrt{2}}{c}(u-v)+\frac{u^2+v^2}{2}\left(\frac{1}{c^2}-\frac{1}{a^2}\right)-uv\left(\frac{1}{c^2}+\frac{1}{a^2}\right)\\ &= 1+\frac{\sqrt{2}}{c}(u-v)-2\frac{u^2+v^2}{c^2}-6\frac{uv}{c^2}\\ &= 1+\frac{\sqrt{2}}{c}(u-v)-\frac{5}{2}\left(\frac{u+v}{c}\right)^2+\frac{1}{2}\left(\frac{v-u}{c}\right)^2,\end{aligned} \tag{6.253}$$

where Equation 6.248b was used. The same result follows using the coordinate transformation (Equations 6.234a and 6.234b) with Equation 6.250a ≡ Equation 6.254a and Equations 6.243a and 6.243b ≡ Equations 6.254b and 6.254c leading to Equations 6.254d and 6.254e:

$$\theta=\frac{\pi}{4}; \quad x_0=0=y_0: \quad x=\frac{u+v}{\sqrt{2}}, \quad y=\frac{v-u}{\sqrt{2}}; \tag{6.254a–6.254e}$$

substitution of Equations 6.251b, 6.254d and 6.254e in the stress function (Equation 6.217c) with Equation 6.247b leads to

$$\begin{aligned}\Theta(x,y) &= By\left(1-2\frac{y}{c}-\frac{x^2}{a^2}+\frac{y^2}{c^2}\right)=\frac{c}{2}\frac{v-u}{\sqrt{2}}\left(1-\sqrt{2}\frac{v-u}{c}-5\frac{x^2}{c^2}+\frac{y^2}{c^2}\right)\\ &= \frac{c}{2}\frac{v-u}{\sqrt{2}}\left[1+\sqrt{2}\frac{u-v}{c}-\frac{5}{2}\left(\frac{u+v}{c}\right)^2+\frac{1}{2}\left(\frac{u-v}{c}\right)^2\right]\equiv\Theta(u,v),\end{aligned} \tag{6.255}$$

that coincides with Equation 6.255 ≡ Equation 6.252. From the stress function, as before (Subsection 6.8.2), the torsion and displacement functions, the displacement vector, the strain and stress tensors, and the torsional stiffness follow. The latter two are calculated next for the generic two-parameter family of triangles (Subsection 6.8.6) of which the isosceles triangle (Equations 6.249a through 6.249e) is a particular case.

6.8.7 Stresses/Strains in Aligned/Torsional Axis

Using the torsion axis but not the torsion center (Equations 6.243a and 6.243b), the stress function (Equation 6.235) with parameters Equations 6.236b, 6.236c, 6.237b, and 6.237c) is given by Equation 6.256b:

$$B=\frac{c}{2}: \quad \Theta(u,v)=\frac{c}{2}(v\cos\theta-u\sin\theta)$$
$$\left[1+\left(\frac{u}{c}-\frac{v}{a}\right)\sin\theta-\left(\frac{u}{a}+\frac{v}{c}\right)\cos\theta\right]\left[1+\left(\frac{u}{c}+\frac{v}{b}\right)\sin\theta+\left(\frac{u}{b}-\frac{v}{c}\right)\cos\theta\right]; \tag{6.256a and 6.256b}$$

the parameter Equation 6.243e takes the value Equation 6.256a not only for equilateral (Equation 6.219c) and isosceles (Equation 6.251b) but also for all other shapes (Equation 6.257 ≡ Equation 6.256a ≡ Equation 6.251b ≡ Equation 6.219c):

$$\frac{1}{B} = \sin\theta\left[\frac{2}{c}\sin\theta + \left(\frac{1}{b} - \frac{1}{a}\right)\cos\theta\right] - \cos\theta\left[-\frac{2}{c}\cos\theta + \left(\frac{1}{b} - \frac{1}{a}\right)\sin\theta\right] = \frac{2}{c}. \tag{6.257}$$

The original (x,y) [rotated (u,v) in Equations 6.242c and 6.242d] Cartesian coordinates, with x (u)-axis along one side of the triangle (along the torsion axis), are used in the stress function [Equations 6.217a through 6.217c (Equation 6.256b) that applies to a two-parameter family of triangles (Figure 6.14a) whose sides (a,b,c) satisfy two relations (Equations 6.244a and 6.244b) that also specify the direction θ of the torsion axis. Two particular cases are the equilateral (isosceles) triangle [Equations 6.219a through 6.219c (Equations 6.249a through 6.249e)] for which the stress function simplifies to Equation 6.219d (Equation 6.252 ≡ Equation 6.255).

These two particular cases and all other cases of the two parameter family of triangles (Equations 6.244a and 6.244b) hold the stresses (Equations 6.137a and 6.137b) in (1) the original coordinate system with x-axis along one side of the triangle using (Equation 6.217a)

$$2\mu S_{xz} = T_{xz} = \frac{\mu\tau c}{2}\left\{\left(1 - \frac{y}{c} - \frac{x}{a}\right)\left(1 - \frac{y}{c} + \frac{x}{b}\right) - \frac{y}{c}\left[2 - 2\frac{y}{c} + x\left(\frac{1}{b} - \frac{1}{a}\right)\right]\right\}, \tag{6.258a}$$

$$2\mu S_{yz} = T_{yz} = \frac{\mu\tau c}{2} y\left[\left(1 - \frac{y}{c}\right)\left(\frac{1}{a} - \frac{1}{b}\right) + \frac{2x}{ab}\right]; \tag{6.258b}$$

(2) the coordinate system (Equations 6.242c and 6.242d) rotated by θ to place the u-axis along the torsion axis using (Equation 6.256b)

$$\begin{aligned}
&2\mu\{S_{uz}, S_{vz}\} = \{T_{uz}, T_{vz}\} = \mu\tau B\{\cos\theta, \sin\theta\} \\
&\left[1 + \left(\frac{u}{c} - \frac{v}{a}\right)\sin\theta - \left(\frac{u}{a} + \frac{v}{c}\right)\cos\theta\right]\left[1 + \left(\frac{u}{c} + \frac{v}{b}\right)\sin\theta + \left(\frac{u}{b} - \frac{v}{c}\right)\sin\theta\right] \\
&\qquad + \frac{\mu\tau c}{2}(v\cos\theta - u\sin\theta) \\
&\left\{-\frac{\sin\theta}{a} - \frac{\cos\theta}{c}, -\frac{\sin\theta}{c} + \frac{\cos\theta}{a}\right\}\left[1 + \left(\frac{u}{c} + \frac{v}{b}\right)\sin\theta + \left(\frac{u}{b} - \frac{v}{c}\right)\cos\theta\right] \\
&\qquad + \frac{\mu\tau c}{2}(v\cos\theta - u\sin\theta) \\
&\left\{\frac{\sin\theta}{b} - \frac{\cos\theta}{c}, -\frac{\sin\theta}{c} - \frac{\cos\theta}{b}\right\}\left[1 + \left(\frac{u}{c} - \frac{v}{a}\right)\sin\theta - \left(\frac{u}{b} + \frac{v}{c}\right)\cos\theta\right].
\end{aligned} \tag{6.259a and 6.259b}$$

The torsional stiffness is a scalar and can be calculated in any coordinate system, for example, the original by substituting Equations 6.217a through 6.217c in Equation 6.158b, leading to

$$C = 2\mu\left\{\int_0^a dx\int_0^{c(1-x/a)} dy + \int_{-b}^0 dx\int_0^{c(1+x/b)} dy\right\}\Theta(x,y)$$
$$= \frac{\mu c^3}{40}\left[a\left(1+\frac{a}{3b}\right)+b\left(1+\frac{b}{3a}\right)\right] \quad (6.260)$$

for the two-parameter family of triangles (Equations 6.244a and 6.244b). The two terms in square brackets in Equation 6.260 correspond to interchanging (a,b) as follows from the pair of integrals (Subsection 6.8.7). Thus, *the torsion of a prism with triangular cross section (Figure 6.14a) with two arbitrary sides (Equations 6.216a through 6.216c) and with third side c and torsion axis θ given by Equations 6.244a and 6.244b, such as the equilateral (isosceles) cross section [Equations 6.219a through 6.219c (Equations 6.249a through 6.249e)], is specified in (1) the Cartesian coordinates with x-axis along one side of the triangle by the stress function (Equations 6.217a through 6.217c) and stresses and strains (Equations 6.258b); and (2) the orthonormal coordinate system (u, v) obtained by rotation (Equations 6.242c and 6.242d) of θ to place the u-axis along the torsion axis, the stress function (Equation 6.256b), and stresses and strains (Equations 6.259a and 6.259b). The torsional stiffness is given by Equation 6.260 and simplifies to Equations 6.230a through 6.230d (Equations 6.261a through 6.261c):*

$$C = \frac{\mu c^3 a}{15} = \frac{\mu c^4}{15\sqrt{5}} = \frac{\mu a^4\sqrt{5}}{3}, \quad (6.261a\text{–}6.261c)$$

for the equilateral (Equations 6.219a and 6.219b) [particular isosceles (Equations 6.249b and 6.249c)] triangular cross sections.

6.8.8 Torsional Stiffness for an Irregular Triangular Cross Section

The stress function for an irregular triangular cross section (Equations 6.217a and 6.251b) is unaffected by the change of variable and parameters (Equation 6.262a):

$$\Theta(x,y;a,b,c) = \Theta(-x,\pm y;b,a,\pm c); \quad I(a,b) = \int_0^a dx\int_0^{c(1-x/a)} dx\,\Theta(x,y). \quad (6.262a \text{ and } 6.262b)$$

It follows that the first integral (Equation 6.262b) on the r.h.s. of Equation 6.260 coincides with the second (Equations 6.263a and 6.263b) by exchanging the parameters (*a*, *b*):

$$\xi = -x:\quad \int_{-b}^0 dx\int_0^{c(1+x/b)} dx\,\Theta(x,y;a,b,c)$$
$$= \int_0^b d\xi\int_0^{c(1-\xi/b)} dy\,\Theta(-\xi,y;a,b;c) = \int_0^b dx\int_0^{c(1-x/a)} dy\,\Theta(x,y;b,a,c) = I(b,a).$$
(6.263a and 6.263b)

Thus, the torsional stiffness for the irregular triangular cross section is specified (Equation 6.264a) by the sum of integrals (Equations 6.262b and 6.263b):

$$C = 2\mu\left[I(a,b) + I(b,a)\right]: \quad I(a,b) = B\int_0^a dx \int_0^{c(1-x/a)} dy \left[\left(1-\frac{y}{c}\right)^2 + \left(1-\frac{y}{c}\right)x\left(\frac{1}{b}-\frac{1}{a}\right) - \frac{x^2}{ab}\right]y, \tag{6.264a and 6.264b}$$

where the stress function (6.217b) was substituted (Equation 6.264b). Using the dimensionless coordinates (Equations 6.229a and 6.229b), the integral (Equation 6.264b) simplifies to

$$I(a,b) = \frac{c^3 a}{2}\int_0^1 dx' \int_0^{1-x'} dy' \left[(1-y')^2 + (1-y')x'\left(\frac{a}{b}-1\right) - \frac{a}{b}x'^2\right]y', \tag{6.265a}$$

where the value (Equation 6.257) of the coefficient B was used. Henceforth, the primes will be deleted in Equation 6.265a to lighten the notation when evaluating the integral:

$$I(a,b) = \frac{c^3 a}{2}\int_0^1 dx \int_0^{1-x} dy \left\{y^3 - y^2\left[2 + x\left(\frac{a}{b}-1\right)\right] + y\left[1 + x\left(\frac{a}{b}-1\right) - \frac{a}{b}x^2\right]\right\}. \tag{6.265b}$$

The dy integration in Equation 6.265b leads to powers of $1-x$:

$$I(a,b) = \frac{c^3 a}{2}\int_0^1 \left\{\frac{(1-x)^4}{4} - \frac{(1-x)^3}{3}\left[2 + x\left(\frac{a}{b}-1\right)\right] + \frac{(1-x)^2}{2}\left[1 + x\left(\frac{a}{b}-1\right) - \frac{a}{b}x^2\right]\right\} dx; \tag{6.266}$$

the coefficients in square brackets are also written as a sum of powers of $1-x$:

$$2 + x\left(\frac{a}{b}-1\right) = 2 + \left[1-(1-x)\right]\left(\frac{a}{b}-1\right) = 1 + \frac{a}{b} + (1-x)\left(1-\frac{a}{b}\right), \tag{6.267a}$$

$$\begin{aligned} 1 + x\left(\frac{a}{b}-1\right) - \frac{a}{b}x^2 &= 1 - x + \frac{a}{b}x(1-x) = 1 - x + \frac{a}{b}\left[1-(1-x)\right](1-x) \\ &= \left(1+\frac{a}{b}\right)(1-x) - \frac{a}{b}(1-x)^2. \end{aligned} \tag{6.267b}$$

Substituting Equations 6.267a and 6.267b in Equation 6.266 and collecting powers of $1 - x$, as in the case (Equations 6.232a, 6.232b, and 6.233) of the equilateral triangle, leads to

$$\begin{aligned}
I(a,b) &= \frac{c^3 a}{2}\int_0^1 \left[(1-x)^4\left(\frac{1}{4}+\frac{a}{3b}-\frac{1}{3}-\frac{a}{2b}\right)+(1-x)^3\left(-\frac{1}{3}-\frac{a}{3b}+\frac{1}{2}+\frac{a}{2b}\right)\right]\mathrm{d}x \\
&= \frac{c^3 a}{2}\left[-\frac{(1-x)^5}{5}\left(-\frac{1}{12}-\frac{a}{6b}\right)-\frac{(1-x)^4}{4}\left(\frac{1}{6}+\frac{a}{6b}\right)\right]_0^1 \\
&= \frac{c^3 a}{2}\left[\frac{1}{5}\left(-\frac{1}{12}-\frac{a}{6b}\right)+\frac{1}{4}\left(\frac{1}{6}+\frac{a}{6b}\right)\right] \\
&= \frac{c^3 a}{2}\left[\frac{1}{24}-\frac{1}{60}+\frac{a}{6b}\left(\frac{1}{4}-\frac{1}{5}\right)\right]=\frac{c^3 a}{2}\frac{1}{40}\left(1+\frac{1}{3b}\right).
\end{aligned} \tag{6.268}$$

Substitution of Equation 6.268 in Equation 6.264a specifies (Equation 6.260) the torsional stiffness of a slender prism with the irregular triangular cross section belonging to the two-parameter family (Equations 6.244a and 6.244b).

6.9 Trajectories of Fluid Particles in a Rotating Vessel

Besides the analogies with the transversely loaded membrane (Sections 6.1 and 6.2) and surface tension (Sections 6.3 and 6.4), the torsion of a rod (Sections 6.5 through 6.8) is also analogous to the potential flow (Subsection 6.9.2) in a rotating vessel (Subsection 6.9.1). An example is the trajectory of fluid particles in a rotating elliptic cylinder (Subsection 6.9.3) that differs from rigid body rotation by an angular velocity of drift (Subsection 6.9.4).

6.9.1 Path Function for a Rotating Fluid

Consider (Figure 6.15a) a rotating vessel containing a plane (x, y) inviscid incompressible fluid, so that a constant velocity of rotation (Equation 6.269a) along the x_3- axis (Equation 6.239b) normal to the plane (x, y) of the position vector (Equation 6.239c) is superimposed (Equation 6.269d) to the flow velocity $\vec{v}$:

$$\Omega \equiv \frac{\mathrm{d}\varphi}{\mathrm{d}t} = \text{const}, \quad \vec{\Omega} = \Omega \vec{e}_3, \quad \vec{r} = \vec{e}_x x + \vec{e}_y y: \quad \vec{u} = \vec{v} + \Omega \wedge \vec{r}; \tag{6.269a–6.269d}$$

the sum of the velocity (Equation 6.270a) due to the incompressible flow with stream function Ψ and the velocity due to the uniform rotation of the vessel (Equations 6.269a through 6.269c) is the total velocity (Equation 6.269d) whose Cartesian components are Equations 6.270b and 6.270c:

$$\vec{v} = \nabla \wedge (\vec{e}_3 \Psi): \quad u_x = \partial_y \Psi - \Omega y, \quad u_y = -\partial_x \Psi + \Omega x, \tag{6.270a–6.270c}$$

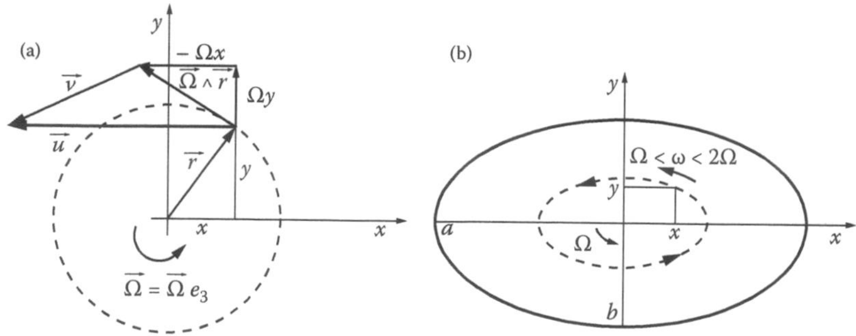

FIGURE 6.15
Problem analogous to (1) the deflection of membranes (Figures 6.1 through 6.4), (2) surface tension and capillarity (Figures 6.5 through 6.9), and (3) torsion of slender beams (Figures 6.10 through 6.14) is (4) the potential flow in a uniformly rotating vessel. Comparing the last two (3 and 4), the torsion is replaced by the angular velocity of rotation and the stress function by the path function (a). In the case of an elliptical beam, the curvature of the cross section under torsion (Figure 6.12d) corresponds to the drift of the fluid in an elliptical rotating vessel (b), that is, the fluid particles rotate at an angular velocity that is less than that of the vessel. The exception is the circular cross section for which (1) the cross section remains plane under torsion, and (2) there is no drift, that is, the fluid particles in a circular cylindrical vessel rotate at the same angular velocity as the vessel, as for a rigid body.

as follows from

$$\vec{u} = \left(\vec{e}_x \partial_x + \vec{e}_y \partial_y\right) \wedge \left(\vec{e}_3 \Psi\right) + \vec{e}_3 \Omega \wedge \left(\vec{e}_x x + \vec{e}_y y\right) = \vec{e}_x \left(\partial_y \Psi - \Omega y\right) + \vec{e}_y \left(-\partial_x \Psi + \Omega x\right). \tag{6.271}$$

There is an analogy between the velocity that is the second term on the r.h.s. of Equations 6.270b and 6.270c [displacement (Equations 6.131b and 6.131c)] vector due to the angular velocity (Equation 6.269a) [torsion rate (Equation 6.129a)].

The stream function exists because the flow is incompressible, that is, has zero dilatation:

$$\dot{D}_2 \equiv \nabla \cdot \vec{u} = \partial_x u_x + \partial_y u_y = \partial_{xy} \Psi - \partial_{yx} \Psi = 0; \tag{6.272}$$

there is generally no velocity potential because the flow is rotational since

$$\vec{\varpi} = \nabla \wedge \vec{u} = \vec{e}_3 \left(\partial_x u_y - \partial_y u_x\right) = \left(-\partial_{xx} \Psi - \partial_{yy} \Psi + 2\Omega\right) \vec{e}_3 = 2\Omega \vec{e}_3 = 2\vec{\Omega}; \tag{6.273}$$

the vorticity (Equation 6.273) is twice the angular velocity of rigid body rotation [compared with elasticity (Equations 4.8a through 4.8c)] because there is no contribution to the vorticity:

$$0 = \nabla \wedge \vec{v} = \vec{e}_3 \left(\partial_x v_y - \partial_y v_x\right) = \vec{e}_3 \left(\partial_{xx} \Psi + \partial_{yy} \Psi\right) = \vec{e}_3 \nabla^2 \Psi, \tag{6.274}$$

from the irrotational component of the flow.

The total velocity (Equations 6.270b and 6.270c) can be calculated (Equations 6.275b and 6.275c) in terms of

$$\Lambda(x, y) = \Psi(x, y) - \Omega \frac{x^2 + y^2}{2}: \quad u_x = \partial_y \Lambda, \quad u_y = -\partial_x \Lambda, \tag{6.275a–6.275c}$$

a **path function** *(Equation 6.275a) that (1) differs from (Equation 6.270a) the stream function for the incompressible (Equation 6.272) flow of the inviscid fluid by minus a centrifugal energy per unit mass associated with rotation; (2) specifies the total velocity (Equations 6.275b and 6.275c ≡ Equations 6.270b and 6.270c); (3) is constant along the trajectories of fluid particles:*

$$\frac{dy}{dx} = \frac{dy/dt}{dx/dt} = \frac{u_y}{u_x} = -\frac{\partial_x \Lambda}{\partial_y \Lambda} = \left(\frac{dy}{dx}\right)_\Lambda, \tag{6.276a}$$

as shown equivalently by Equation 6.276a or 6.276b:

$$d\Lambda = (\partial_x \Lambda)dx + (\partial_x \Lambda)\,dy = -u_y\,dx + u_x\,dy = -\frac{dy}{dt}\,dx + \frac{dx}{dt}\,dy = 0; \tag{6.276b}$$

(4) for example, is constant at the wall where the velocity is tangential; and (5) satisfies (Equation 6.273) in a potential flow (Equation 6.274) a Poisson equation:

$$-2\Omega = \partial_y u_x - \partial_x u_y = \partial_{yy}\Lambda + \partial_{xx}\Lambda = \nabla^2\Lambda, \tag{6.277}$$

forced by minus the double of the angular velocity. Thus, the stream (Equation 6.270a) [path (Equation 6.275a)] function is constant along the streamlines in a rotating frame [along the paths (Equations 6.270b and 6.270c ≡ Equations 6.275b and 6.275c) in a frame at rest].

6.9.2 Analogies among Membrane, Torsion, and Rotation

The preceding results show *the analogy of the torsion of a slender rod (with the incompressible flow of an inviscid fluid in a rotating vessel), with constant cross section and uniform shear modulus (velocity of rotation) as concerns: (1) the torsion rate (Equation 6.129a) [angular velocity (Equation 6.269a)] is constant; (2) the stress Θ (path Λ) function is constant on the boundary (Equation 6.141c) [along the trajectory of the particles (Equation 6.276a ≡ Equation 6.276b)] and satisfies a Poisson equation 6.141b (Equation 6.277) forced by −2 (−2Ω, where Ω is the angular velocity of rotation); (3) the displacement (Equation 6.142a) [stream (Equation 6.275a)] function Ψ, that is, a harmonic function satisfying the Laplace equation 6.144 (Equation 6.274 for an irrotational flow); (4) the displacement (stream) function is the imaginary part of the complex torsion (Equation 6.143c) [velocity (Equation 2.167a)] potential and the real part also satisfies the Laplace equation 6.135c (2.143c) and is the warping function (Equation 6.131c) specifying the displacement normal to [the (Equation 6.278a) potential specifying the irrotational part of the velocity (Equations 6.278b and 6.278c) in] the cross section:*

$$\vec{v} = \nabla\Phi: \quad v_x = \partial_x \Phi = \partial_y \Psi, \quad v_y = \partial_y \Phi = -\partial_x \Psi; \tag{6.278a–6.278c}$$

(5) the derivatives of the stress (path) function specify the shear stress and distortions (Equations 6.137a and 6.137b) between the cross section and the axial direction [the total velocity (Equations 6.275b and 6.275c) in the cross section]; (6) the torsional stiffness (Equation 6.158b) [moment of inertia (Equation 6.279a)] of the cross section; (7) the torsion moment (Equation 6.149a) [angular

momentum (Equation 6.279b)]; and (8) the elastic energy (Equation 6.161a) [kinetic energy or centrifugal energy (Equation 6.279c)] of the cross section:

$$I = \rho \int_D \left(x^2 + y^2\right) \mathrm{d}x \, \mathrm{d}y, \quad L = I\Omega, \quad E = \frac{1}{2} I\Omega^2. \tag{6.279a–6.279c}$$

The relations 6.279a through 6.279c, where ρ denotes the mass density, are proven in the classical mechanics of rigid bodies with a fixed axis of rotation.

The analogy can be extended to the linear deflection of a membrane under a transverse load:

$$D: \quad \nabla^2 \zeta = -\frac{f}{T}, \quad \nabla^2 \Theta = -2, \quad \nabla^2 \Lambda = -2\Omega, \tag{6.280a–6.280c}$$

$$\partial D_0: \quad \zeta\big|_{\partial D_0} = 0, \quad \Theta\big|_{\partial D_0} = 0, \quad \Lambda\big|_{\partial D_0} = 0, \tag{6.281a–6.281c}$$

$$\partial D_n: \quad \zeta\big|_{\partial D_n} = \zeta_n, \quad \Theta\big|_{\partial D_n} = \Theta_n, \quad \Lambda\big|_{\partial D_n} = \Lambda_n, \tag{6.282a–6.282c}$$

where (1) deflection ζ (2) stress function Θ, and (3) path function Λ satisfy a Poisson equation forced by, respectively, (1) the ratio (Equation 6.280a ≡ Equation 6.13c) of the transverse force per unit area or shear stress to the tangential tension; (2) –2 in Equation 6.280b ≡ Equation 6.141b; and (3) –2 times (Equation 6.280c ≡ Equation 6.277) the angular velocity of rotation Ω. In case, 2 (case 3), the stress (path) function is (a) zero [Equation 6.281b ≡ Equation 6.158a (Equation 6.281c ≡ Equation 6.276b)] at the boundary of a simply connected region (Figure 6.11a), and (b) zero [Equation 6.281b ≡ Equation 6.159a (Equation 6.281c ≡ Equation 6.276a)] at the outer boundary and constant [Equation 6.282b ≡ Equation 6.159b (Equation 6.282c ≡ Equation 6.276b)] at the inner boundaries of a multiply connected region (Figure 6.11b). In case 1, the boundary condition depends on how the membrane is fixed; for example, (Equation 6.281a ≡ Equation 6.29c) (Equation 6.282a ≡ Equation 6.32c) was considered for a circular (annular) membrane [Subsection 6.2.1 (Subsection 6.2.2)] in Figure 6.3a (Figure 6.3b).

6.9.3 Trajectories of Fluid Particles in a Rotating Elliptic Cylinder

From the comparison of the stress (Equation 6.280b) [path (Equation 6.280c)] function for an elliptic cylinder [Equation 6.175a (Equation 6.283)]:

$$\Lambda(x, y) = -\Omega \frac{b^2 x^2 + a^2 y^2}{a^2 + b^2}, \tag{6.283}$$

the latter is given by omitting a constant term. It is constant (Equation 6.284a) on the elliptic boundary (Equation 6.284b):

$$\frac{x^2}{a^2} + \frac{y^2}{b^2} = 1: \quad \Lambda(x, y) = -2\Omega \frac{a^2 b^2}{a^2 + b^2}\left(\frac{x^2}{a^2} + \frac{y^2}{b^2}\right) = -2\Omega \frac{a^2 b^2}{a^2 + b^2}. \tag{6.284a and 6.284b}$$

From *the path function (Equation 6.283) follows (Equations 6.275b and 6.275c) the total velocity (Equations 6.285a and 6.285b) of fluid particles in Figure 6.15b in an elliptic cylinder of half-axis (a, b) rotating with angular velocity Ω*:

$$\left\{\frac{dx}{dt},\frac{dy}{dt}\right\}=\left\{u_x,u_y\right\}=\left\{\partial_y\Lambda,-\partial_x\Lambda\right\}=-\frac{2\Omega}{a^2+b^2}\left\{a^2y,-b^2x\right\}. \qquad (6.285a\text{ and }6.285b)$$

The positions of fluid particles in a rotating vessel satisfy a system of coupled first-order ordinary differential equations 6.285a and 6.285b similar to the trajectories of an electron in a magnetic field (Equations I.6.5a and I.6.5b) if $a = b$. Thus, the case $a = b$ corresponds to circular paths and $a \neq b$ corresponds to elliptical paths, as shown next. The system (Equations 6.285a and 6.285b) can be decoupled as a second-order linear ordinary differential equation with constant coefficients:

$$\frac{d^2x}{dt^2}=-\frac{2\Omega}{a^2+b^2}a^2\frac{dy}{dt}=-\left(\frac{2\Omega\, ab}{a^2+b^2}\right)^2 x, \qquad (6.286a)$$

$$\frac{d^2y}{dt^2}=\frac{2\Omega}{a^2+b^2}b^2\frac{dx}{dt}=-\left(\frac{2\Omega\, ab}{a^2+b^2}\right)^2 y. \qquad (6.286b)$$

It is the same differential equation 6.286a ≡ Equation 6.287b (Equation 6.286b ≡ Equation 6.287c) involving a single parameter (Equation 6.287a) with the dimensions of frequency:

$$\omega\equiv\Omega\frac{2ab}{a^2+b^2}: \quad \frac{d^2x}{dt^2}+\omega^2x=0=\frac{d^2y}{dt^2}+\omega^2y; \qquad (6.287a\text{–}6.287c)$$

the general solution of Equation 6.287b is Equation 6.288a, where the constants of integration are determined from the initial conditions (Equations 6.288b and 6.288c):

$$x(t)=C_1\cos(\omega t)+C_2\sin(\omega t): \quad 0=\dot{x}(0)=C_2\omega, \quad ka=x(0)=C_1; \qquad (6.288a\text{–}6.288c)$$

the time $t = 0$ was chosen at the point of zero x-component of the velocity (Equation 6.288b), corresponding (Equation 6.288c) to the amplitude ka in the x-direction. Substituting Equations 6.288b and 6.288c in Equation 6.288a leads to Equation 6.289a:

$$x(t) = ka\cos(\omega t); \qquad y(t) = kb\sin(\omega t); \qquad (6.289a\text{ and }6.289b)$$

also Equation 6.289b follows from Equations 6.285a and 6.285b ≡ Equations 6.290a and 6.290b):

$$\frac{dx}{dt}\equiv\dot{x}=-\omega\frac{a}{b}y, \quad \frac{dy}{dt}\equiv\dot{y}=\omega\frac{b}{a}x, \qquad (6.290a\text{ and }6.290b)$$

using (Equation 6.287a)

$$y = -\frac{b}{\omega a}\frac{dx}{dt} = -\frac{b}{\omega a}\frac{d}{dt}\left[ka\cos(\omega t)\right] = kb\sin(\omega t). \tag{6.291}$$

Thus, *the trajectories of fluid particles are given by Equations 6.289a and 6.289b, with initial conditions 6.288b and 6.288c in an elliptical vessel with half-axis (a, b) rotating with angular velocity Ω, where Equation 6.287a is the angular velocity along the path; it is generally distinct from the angular velocity of the vessel, and the difference is the angular drift velocity* considered next (Subsection 6.9.4).

6.9.4 Drift Velocity Relative to Rigid Body Rotation

Eliminating the time between Equations 6.289a and 6.289b leads to *the paths that are ellipses (Equation 6.292c) with half-axis (ka, kb), where ka specifies the initial position at time t = 0 in Equation 6.292a ≡ Equation 6.288c and Equation 6.292b ≡ Equations 6.288b and 6.289a:*

$$\{x(0), y(0)\} \equiv \{ka, 0\}: \quad \left[\frac{x(t)}{a}\right]^2 + \left[\frac{y(t)}{b}\right]^2 = k^2 \equiv \left(\frac{x_0}{a}\right)^2. \tag{6.292a–6.292c}$$

The angular velocity along the trajectory is Equation 6.287a, implying an **angular drift velocity** *relative to rigid body rotation:*

$$0 < \Omega - \omega = \Omega\left(1 - \frac{2ab}{a^2+b^2}\right) = \Omega\frac{(a-b)^2}{a^2+b^2} < \Omega. \tag{6.293a and 6.293b}$$

There is no angular drift (Equation 6.294b) for a circular cylinder (Equation 6.294a), that is, (1) the fluid particles follow circular trajectories (Equations 6.294d and 6.249e) with the initial radius (Equation 6.294c) as a rigid body:

$$a \equiv b: \quad \omega = \Omega, \quad x_0 = ka, \quad \{x(t), y(t)\} = x_0\{\cos(\Omega t), \sin(\Omega t)\}; \tag{6.294a–6.294e}$$

and (2) the path function (Equation 6.283) depends (Equations 6.295a and 6.295c) only on the radius (Equation 6.295b):

$$a = b; \quad r^2 = x^2 + y^2: \quad \Omega(x, y) = -\Omega\frac{x^2+y^2}{2} = -\frac{\Omega^2 r^2}{2}. \tag{6.295a–6.295c}$$

The angular drift velocity is nonzero for an elliptic cross section and cannot exceed the velocity of rotation (Equation 6.293b), so that the fluid particle is never at rest (Equation 6.293a); the inequalities in Equations 6.293a and 6.293b follow from Equation 6.296c:

$$ab > 0: \qquad 0 < (a-b)^2 = a^2 + b^2 - 2ab < a^2 + b^2 \tag{6.296a–6.296c}$$

using Equations 6.296a and 6.296b. The rigid body (nonrigid body) rotation without (with) drift angular velocity (Equations 6.293a and 6.293b) in an circular (elliptic) vessel corresponds to the torsion of a rod with circular (elliptic) cross section that remains (Subsection 6.7.3) [does not remain (Subsection 6.7.2)] plane.

NOTE 6.1 TORSION AND OTHER DEFORMATIONS OF RODS

Four analogies were demonstrated: (1) the torsion of a slender rod by a torque at the end (Sections 6.5 through 6.8); (2) the surface tension (Sections 6.3 and 6.4); (3) the linear deflection of a membrane by a shear stress (Sections 6.1 and 6.2); and (4) the incompressible motion of an inviscid fluid in a uniformly rotating vessel (Section 6.9). A fifth analogy concerns the unidirectional shear flow of a viscous fluid in two dimensions (Notes 6.7 through 6.9). These five analogies belong to a set of 12 problems satisfied by the Poisson equation (Note 6.6) besides two specified by the biharmonic equation (Note 6.5). Torsion was considered for a straight rod of constant cross section (Sections 6.5 through 6.8). It can be extended to a straight shaft with a cross section varying axially; a bar or beam can be subject also to bending and compression or traction. The three deformations or loads can be combined for straight beams and for beams that are already curved in the undeformed state: (1) torsion due to axial moments; (2) bending due to transverse forces or moments; and (3) tractions or compressions due to axial forces. The analogy between the deflection of a membrane (Sections 6.1 and 6.2) and surface tension (Sections 6.3 and 6.4) applies in the linear case of small slope, and extends to the geometric nonlinearity associated with curvature for large slope. The assumption of constant tension applies to large deflection for a soap bubble, whereas for a membrane it may be modified by a nonlinear stress–strain relation. Both subjects can be approached by the calculus of variations that applies to minimal surfaces and minimization of the elastic energy. The torsion of a rod with multiply-connected cross section is also analogue to the deflection of a membrane supported on a multiply connected boundary.

NOTE 6.2 THREE-DIMENSIONAL ELASTICITY IN CURVILINEAR AND OBLIQUE COORDINATES

The preceding account (Chapters 4 and 6) is the minimum to cover some of the basics of plane elasticity in Cartesian and polar coordinates. The same topics can be considered in a broader physical context, as the mathematical background to support it develops, for example: (1) three-dimensional stresses including principal stresses can be considered using the linear algebra of matrices and eigenvalues and eigenvectors; (2) three-dimensional strains in curvilinear coordinates, orthogonal or oblique, use the metric and covariant derivative; (3) the Hooke law (Section 4.3) in three dimensions can be extended to include thermal and electromagnetic effects as part of the general constitutive relations using the tensor calculus; (4) the fundamental equations of elasticity (Section 4.4) in three dimensions can be extended to anisotropic (nonlinear) materials such as crystals (rubber) using the variational calculus; and (5) the general forms of the invariant differential operators (gradient, divergence, curl, and Laplacian) for vectors (Section 4.5) and tensors also appear in the general form of the theory of elasticity and other rheological models such as plasticity and viscoelasticity.

NOTE 6.3 TWELVE SPECIAL CASES OF ELASTICITY

There are specific cases of elasticity, besides plane elasticity, that (Diagram 6.1) are amenable to special or simpler treatment, for example: (1) the deflection of strings that have no bending stiffness; (2) the bending and torsion (Sections 6.5 through 6.8) of straight rods;

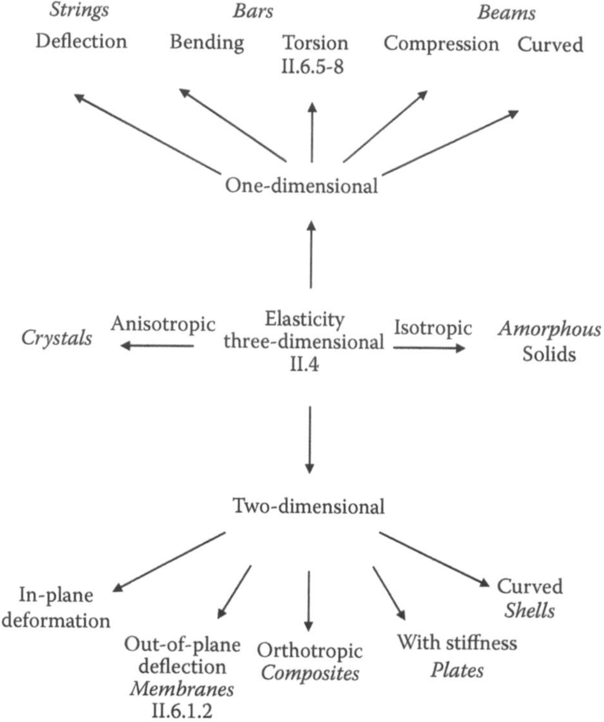

DIAGRAM 6.1
Relation between the 12 special cases of elasticity indicating the book, chapter, and section where the subject is addressed.

(3) the compressive axial loads that can cause elastic instability and buckling of beams; (4) the deflection of membranes (Sections 6.1 and 6.2) that relates to surface tension and capillarity (Sections 6.3 and 6.4); and (5) the bending of plates that, unlike membranes, have stiffness. Other one-dimensional (cases 1 to 3) [two-dimensional (cases 4 and 5)] cases of elasticity include curved beams (shells). There are unsteady extensions to (6) waves in strings and membranes, (7) longitudinal, transversal, and torsional vibrations of bars and beams, (8) bending and in-plane vibrations of plates; and (9) compressive, shear, and surface waves in three-dimensional bodies including crystals. An example of a class of problems is those with cylindrical symmetry such as the following: (1) the stresses in solid and hollow cylinders and cylindrical cavities (Section 4.5); (2) the steady heat conduction (Chapter I.32) in the same geometry (Subsections I.32.5 through I.32.8); and (3) the two may be combined in anisothermal elasticity, for example, to consider thermal stresses due to temperature gradients. Thus, the sequence of topics mentioned, starting with plane elasticity (Chapter 4), serves as a gradual introduction to general elasticity, as outlined in Diagram 6.1. Diagram 6.1 itself is an introduction since only a part of it has been covered so far in this series.

NOTE 6.4 VIBRATIONS AND DISSIPATION IN SOLIDS

Steady elasticity excludes time-dependent phenomena like vibrations and dissipation that apply to elastic bodies in one, two, or three dimensions. The steady (unsteady) elastic problems in one dimension include (I-1) the deformation (vibrations) of strings; (I-2) the deformation (vibrations) of bars that unlike strings have bending stiffness; (I-3) the elastic stability of beams, such as a buckling of a bar under a compressive axial load;

(I-4) the torsion of beams, for example, by torques (Sections 6.5 through 6.8); and (I-5) the coupling of torsion and bending in curved beams. The steady (unsteady) elastic problems, in two dimensions, include (II-1) the deflection (Sections 6.1 and 6.2) (vibrations) of membranes; (II-2) the deformations (vibrations) of plates that unlike membranes have bending stiffness; and (II-3) the deformations and vibration of shells that have stiffness like plates and are also curved in the undeformed state, for example, a spherical shell. The three-dimensional elastic problems include (III-1) deformations, strains, and stresses in bodies subject to forces or torques, and/or in contact, including compressive, shear, and surface waves; and (III-2) dissipative processes like heat conduction or viscosity leading to rheological laws more general than elasticity. The general foundations of elasticity are best formulated using the tensor calculus for stresses, strains, and the fundamental equations, allowing the use of curvilinear coordinates, oblique or orthogonal. The preceding topics go beyond the 23 analogous potential field problems considered in the present chapter that appear in List 6.1.

LIST 6.1

Twenty-Three Analogous Potential Field Problems

A. Deflection of a Membrane under Its Own Weight
- 1) General: Subsection 6.1.3
- 2) Linear for circular membrane—under own weight: Subsection 6.1.5
- 3) Linear annular membrane under own weight: Subsection 6.1.6
- 4) Nonlinear for a circular membrane under own weight: Subsections 6.2.1 and 6.2.2
- 5) Nonlinear for a circular membrane under normal pressure: Subsections 6.2.3 through 6.2.7

B. Minimal Surface due to Surface Tension
- 6) Sphere: Subsections 6.3.3, 6.3.5, and 6.3.6
- 7) Catenoid: Subsections 6.3.3 and 6.3.7

C. Capillary Effects
- 8) Wetting of a wall: Subsection 6.4.2
- 9) Rise between walls: Subsections 6.4.3 through 6.4.7

D. Torsion of a Slender Rod with Constant Cross Subsection
- a) Simply connected
  - 10) α—elliptic: Subsections 6.7.1 and 6.7.2
  - 11) β—circular: Subsections 6.7.3 and 6.7.4
  - 12) χ—thin rectangular: Subsection 6.7.5
  - 13) δ—equilateral triangle: Subsections 6.8.1 and 6.8.2
  - 14) ε—nonequilateral triangle: Subsections 6.8.3 through 6.8.8
- b) Doubly connected
  - 15) α—elliptic: Subsections 6.7.1 and 6.7.2
  - 16) β—circular: Subsections 6.7.3 and 6.7.4
  - 17) γ—open thin walled: Subsections 6.7.4 and 6.7.5
  - 18) δ—closed thin walled: Subsection 6.7.6
  - 19) ε—closed thin walled with a cut: Subsection 6.7.7

E. Incompressible Flow of Inviscid Fluid in Rotating Vessel
- 20) Elliptic cylinder: Subsection 6.9.3
- 21) Circular cylinder: Subsection 6.9.3

F. Unidirectional Shear Flow of a Viscous Fluid
- 22) Between parallel walls: Note 6.8
- 23) In a triangular pipe: Note 6.9

NOTE 6.5 ANALOGOUS BIHARMONIC FIELDS IN FLUID/SOLID MECHANICS

The analogy has been made between plane continuum mechanics in two cases. First is the elastic medium (Section 4.4) at rest with no external forces, for which the stress function is biharmonic (Equation 6.297b ≡ Equation 4.106e) and specifies the stress tensor (Equation 6.297c ≡ Equation 4.58d) in a domain (Equation 6.297a), and the pressure (Equation 6.298b) and tangential stress (Equation 6.298c) on the boundary (Equation 6.298a) with (Equations 6.42a through 6.42c) normal vector $\vec{N}$:

$$D: \quad \nabla^4\Theta = 0, \{T_{nn}, T_{ss}, T_{sn}\} = \{\partial_{ss}\,\Theta, \partial_{nn}\,\Theta, -\partial_{ns}\,\Theta\}, \tag{6.297a–6.297c}$$

$$\partial D: \quad -p = T_{nn}\,N_n + T_{ns}\,N_s, \qquad t = T_{ns}\,N_n + T_{ss}\,N_s. \tag{6.298a–6.298c}$$

The second case is the steady incompressible creeping flow of a viscous fluid (Note 4.11), for which the stream function is biharmonic (Equation 6.299b ≡ Equation 4.361d) and specifies the velocity vector (Equation 6.299c ≡ Equation 4.352b) in the domain (Equation 6.299a); it has normal and tangential derivatives zero at a rigid impermeable boundary (Equations 6.299d and 6.299e ≡ Equations 4.357b and 4.357c):

$$D: \; \nabla^4\Psi = 0, \quad \{v_n, v_s\} = \left\{-\frac{\partial\Psi}{\partial s}, \frac{\partial\Psi}{\partial n}\right\}; \quad \left.\frac{\partial\Psi}{\partial s}\right|_{\partial D} = 0 = \left.\frac{\partial\Psi}{\partial n}\right|_{\partial D}. \tag{6.299a–6.299e}$$

These two biharmonic fields appear in Table 6.3 as Cases 13 and 14. Cases 1–12 are solutions of the Laplace or Poisson equations that satisfy a second-order partial differential equation, so that only one of two boundary conditions (Equations 6.299d and 6.299e) can be imposed. Imposing both boundary conditions (Equations 6.299d and 6.299e ≡ Equations

TABLE 6.3

Twelve Harmonic and Two Biharmonic Forced Two-Dimensional Fields

Case	Volume, Chapter, and Section	Physical Conditions
1	I.12.1, I.18.1; II.2.1	Irrotational, compressible flow
2	I.12.2, I.18.2; II.2.2–II.2.6, II.9.9	Incompressible, rotational, inviscid flow
3	I.12, I.14, I.16, I.28, I.34, I.36, I.38; II.2.7–II.2.9, II.9.1–II.9.8	Potential flow: irrotational and incompressible
4	I.18	Gravity field
5	I.14, I.36	Electrostatic field
6	I.26, I.36	Magnetostatic field
7	I.32	Steady diffusion: heat, mass, electricity
8	II.6.1 and II.6.2	Linear deflection of a membrane
9	II.6.3 and II.6.4	Surface tension
10	II.6.5–II.6.8	Torsion of a prism
11	II.6.9	Incompressible uniformly rotating flow
12	Notes II.6.7–II.6.9	Unidirectional viscous shear flow in a pipe
13	II.4	Plane elasticity
14	Notes II.4.4–II.4.11	Plane creeping viscous flow

Note: List of 12 (2) analogous harmonic (biharmonic) problems, indicating the physical conditions in which they occur, and the volume, chapter, and section where they are considered.

6.300e and 6.300f) to a harmonic field (Equation 6.300b) in a domain (Equation 6.300a) would lead to a zero solution (Equations 6.300g and 6.300h):

$$D:\quad \nabla^2\Psi = 0;\quad \partial D:\quad \frac{\partial \Psi}{\partial n} = 0 = \frac{\partial \Psi}{\partial s} \quad \Rightarrow \quad D:\quad \Psi = 0. \qquad (6.300a\text{–}6.300h)$$

The interpretation is that an inviscid (viscous) fluid has zero normal (total, that is, normal and tangential) velocity at a rigid impermeable boundary. The 12 harmonic unforced (forced) fields that are the solutions of the Laplace (Poisson) equation are considered next (Note 6.6).

NOTE 6.6 TWELVE ANALOGIES BETWEEN HARMONIC AND POTENTIAL FIELDS

To the seven analogies between potential fields (potential, incompressible and irrotational flow, gravity, electrostatics, magnetostatics, and steady heat conduction) covered before in volume I, the present chapter adds four more (membranes, surface tension, torsion of rods, and flow in a rotating vessel), plus reference in the sequel (Notes 6.7 through 6.9), to unidirectional viscous shear creeping flow, making the total 12. The following is an example of each of the potential fields consisting of only four elements: (1) the fundamental equation, that is, the Laplace (Poisson) equation, for unforced (forced) fields; (2) an associated vector field; (3) a boundary condition; and (4) some properties of the medium may be involved in some or all cases 1–3.

The first three basic potential fields are flows. The first is the irrotational (Equation 6.301a) flow (Subsections I.12.1, I.18.1, Sections 2.1, and 2.4) due to line sources or sinks for which (1) the velocity is the gradient (Equation 6.301b ≡ Equation 6.18.1b) of a potential satisfying the Poisson equation (Equation 6.301c ≡ Equation I.18.2b) forced by the dilatation or time rate of change of volume; and (2) the normal component of the velocity at a boundary has a discontinuity equal to (Equation 6.301d ≡ Equation I.18.9b) the output of surface source q:

$$\nabla \wedge \vec{v} = 0:\quad \vec{v} = \nabla\Phi_v,\quad \dot{D}_2 = \nabla \cdot \vec{v} = \nabla^2\Phi_v,\quad \left[v_n\right] \equiv \left[\frac{\partial \Phi_v}{\partial n}\right] = q, \qquad (6.301a\text{–}6.301d)$$

and is zero for a rigid impermeable wall. The second potential field is the inviscid incompressible (Equation 6.302a) flow (Subsections I.12.2, I.18.2, Sections 2.2 through 2.9, and 8.9) for which (1) the velocity is the two-dimensional curl (Equation 6.302b ≡ Equation I.18.10b) of a stream function satisfying the Poisson equation (Equation 6.302c ≡ Equation I.18.11b) forced by minus the vorticity; and (2) the tangential component of the velocity at a boundary (Equation 6.302d ≡ Equation I.18.17b) has a discontinuity equal to the surface circulation γ:

$$\left\{v_x, v_y\right\} = \left\{\partial_y \Psi_v, -\partial_x \Psi_v\right\},\quad \nabla^2\Psi_v = -\varpi,\quad \left[v_s\right] = \frac{\partial \Psi_v}{\partial n} = -\gamma, \qquad (6.302a\text{–}6.302d)$$

and in its absence is continuous. The third potential flow (Chapters I.12, I.14, I.16, I.28, I.34, I.36, I.38, and 2.8) (1) is both irrotational (Equation 6.303a) and incompressible (Equation 6.303b); (2) has a potential (stream function) satisfying the Laplace equation [Equation 6.303a ≡ Equation I.12.22a (Equation 6.303d ≡ Equation I.12.22b)] specifying the components of the velocity vector [Equation 6.303e ≡ Equation I.12.7a (Equation 6.303f ≡ Equation I.12.14)]:

$$\nabla \wedge \vec{v} = 0 = \nabla \cdot \vec{v}:\quad \nabla^2\Phi_v = 0 = \nabla^2\Psi_v,\quad \nabla\Phi_v = \vec{v} = -\nabla \wedge \left(\vec{e}_3 \Psi_v\right); \qquad (6.303a\text{–}6.303f)$$

and (3) the normal (Equation 6.304a ≡ Equation 6.299b) [tangential (Equation 6.304b ≡ Equation 6.299c)] component of the velocity at a boundary has a discontinuity:

$$[v_n] = \left[\frac{\partial \Phi}{\partial n}\right] = \left[-\frac{\partial \Psi}{\partial s}\right] = q, \quad [v_s] = \left[\frac{\partial \Phi}{\partial s}\right] = \left[\frac{\partial \Psi}{\partial n}\right] = -\gamma \qquad \text{(6.304a and 6.304b)}$$

due to the surface flow rate q (circulation γ); for a rigid impermeable wall, the normal (tangential) velocity is (is generally not) zero.

The next three basic potential fields are force fields. The fourth one is the gravity field (Subsections I.18.3 through I.18.8): (1) it is irrotational (Equation 6.305a) that is equal to (Equation 6.305b ≡ Equation I.18.18b) minus the gradient of a gravity potential, that satisfies a Poisson equation (Equation 6.305c ≡ Equation I.18.19b) forced by minus the mass density ρ times the gravitational constant G; and (2) there is a normal discontinuity (Equation 6.305d ≡ Equation I.18.22b) across a surface of mass density μ:

$$\nabla \wedge \vec{g} = 0: \quad \vec{g} = -\nabla \Phi_g, \quad G\rho = -\nabla \cdot \vec{g} = \nabla^2 \Phi_g, \quad [g_n] = \left[-\frac{\partial \Phi_g}{\partial n}\right] = -G\mu. \qquad \text{(6.305a–6.305d)}$$

The fifth potential field (Chapters I.24 and I.36) is the electrostatic field (Equation 6.306a ≡ Equation I.24.4a) that is irrotational (Equation 6.303a) and (1) that is equal to minus the gradient of an electrostatic potential (Equation 6.306b) that satisfies a Poisson equation (Equation 6.306c ≡ Equation I.24.5b) forced by minus the electric charge density per unit volume q divided by the dielectric permittivity of the medium ε; and (2) the normal component of the electrostatic field at a boundary multiplied by the dielectric permeability (Equation 6.306d ≡ Equation I.24.8b) has a discontinuity equal to the surface electric charge σ:

$$\nabla \wedge \vec{E} = 0: \quad \vec{E} = -\nabla \Phi_e, \quad -\frac{q}{\varepsilon} = -\nabla \cdot \vec{E} = \nabla^2 \Phi_e, \quad [\varepsilon E_n] = \left[-\varepsilon \frac{\partial \Phi_e}{\partial n}\right] = \sigma \qquad \text{(6.306a–6.306d)}$$

and is continuous for an insulating boundary. The sixth potential field (Chapters I.26 and I.36) is the magnetostatic field (Equation 6.307a ≡ Equation I.26.4a) is specified by the magnetic induction that is solenoidal and equal to the two-dimensional curl of a field function (Equation 6.307b) satisfying a Poisson equation (Equation 6.307c ≡ Equation I.26.6b) forced by minus the electric current density j divided by the speed of light *in vacuo* c and multiplied by the magnetic permeability μ of the medium; the tangential component of the magnetic induction at a boundary (Equation 6.307d ≡ Equation I.26.9b) has a discontinuity specified by the surface electric current ϑ divided by the speed of light *in vacuo* and magnetic permeability:

$$\nabla \cdot \vec{B} = 0, \quad \vec{B} \wedge (\vec{e}_3 \Psi_m), \quad -\frac{\mu j}{c} = \vec{e}_3 \cdot (\nabla \wedge \vec{B}) = \nabla^2 \Psi_m, \quad [B_s] = \left[\frac{\partial \Psi_m}{\partial n}\right] = \frac{\mu \vartheta}{c} \qquad \text{(6.307a–6.307d)}$$

and is continuous for an insulating boundary.

Besides the three flows (fields), the seventh basic potential field is considered next together with the five additional potential fields. This seventh basic potential field is steady heat conduction (Chapter I.32) for which (1) the heat flux is (Equation 6.308a ≡ Equation

I.32.4b) minus the thermal conductivity k times the temperature gradient; (2) the temperature satisfies a Poisson equation (Equation 6.308b ≡ Equation I.32.5c) forced by minus the density of heat sources w divided by the thermal conductivity; and (3) a convection boundary condition from a fluid at temperature T_0 involves (Equation 6.308c ≡ Equation I.32.7c) the surface thermal conductivity h:

$$\vec{G} = -k\nabla T, \quad -w = -\nabla\cdot\vec{G} = k\nabla^2 T, \quad G_n = -k\frac{\partial T}{\partial n} = h\left(T - T_0\right). \qquad \text{(6.308a–6.308c)}$$

The five additional potential fields are related to deformable solids and fluids. The eighth one is the linear (nonlinear) deflection ζ of a membrane (Sections 6.1 and 6.2) that is fixed at the boundaries and satisfies a Poisson (Equation 6.13c) [minimal surface operator (Equation 6.19b)] equation involving the uniform membrane tension and shear stress. The ninth potential field are the same operators (Equations 6.72a through 6.72c) at the interface (Sections 6.3 and 6.4) between two fluids subject to surface tension as well as the balance of pressures and other stresses on the two sides. The tenth one is the torsion (Sections 6.5 through 6.8) of a slender beam, for example, specified by a stress function that is constant on the boundary (Equation 6.141c), and satisfies a Poisson equation 6.141b and determines the shear stresses and distortion strains (Equations 6.137a and 6.137b). The eleventh are the trajectories of the particles (Sections 6.9) of an inviscid fluid for an incompressible flow in a vessel rotating with angular velocity Ω that correspond to a constant (Equation 6.276a ≡ Equation 6.276b) path function (Equation 6.275a), which satisfies a Poisson equation 6.277 and specifies the velocity (Equations 6.275b and 6.275c). The twelfth potential field is the flow of a viscous fluid in a pipe, that is, a unidirectional shear flow, which is not inviscid (Sections 2.4 and 2.5) but rather viscous (Note 6.7), for example, the viscous flow (1) between two parallel walls (Note 6.8) and (2) in a tube of a triangular cross section (Note 6.9).

NOTE 6.7 STEADY VISCOUS UNIDIRECTIONAL SHEAR FLOW IN A TUBE (POISEUILLE, 1840)

Consider a cylindrical tube with an arbitrary cross section (Figure 6.16a) containing (Equation 6.309a) a unidirectional shear flow with velocity parallel to the axis and depending only on the transverse coordinates in the cross section (Equation 6.310b):

$$D:\quad \vec{v} = \vec{e}_z v\left(x,y\right); \quad \partial D:\quad v\left(x,y\right) = 0; \qquad \text{(6.309a–6.309d)}$$

for a viscous fluid (Figure 6.16b), the velocity is zero (Equation 6.309d) on the boundary (Equation 6.309c). The unidirectional shear flow has zero convective acceleration (Equation 6.310a), and thus the momentum equation 2.7 for an inviscid fluid reduces to Equation 6.310b:

$$\left(\vec{v}\cdot\nabla\right)\vec{v} = \vec{e}_z v\frac{\partial v}{\partial z} = 0: \quad \rho\frac{\partial\vec{v}}{\partial t} = -\nabla p; \qquad \text{(6.310a and 6.310b)}$$

$$\nabla\cdot\vec{v} = \frac{\partial v}{\partial z} = 0: \quad \rho\frac{\partial\vec{v}}{\partial t} = -\nabla p + \eta\nabla^2\vec{v}, \qquad \text{(6.311a and 6.311b)}$$

the unidirectional shear flow (Equation 6.310a) is also incompressible (Equation 6.311a), and the viscous momentum or Navier–Stokes equation 4.345b without external forces $\vec{f} = 0$ involves (Equation 6.311b) only the shear (bar not bulk) viscosity. In the inviscid (Equation 6.310b) [viscous (Equation 6.311b)] case, the momentum equation is linear,

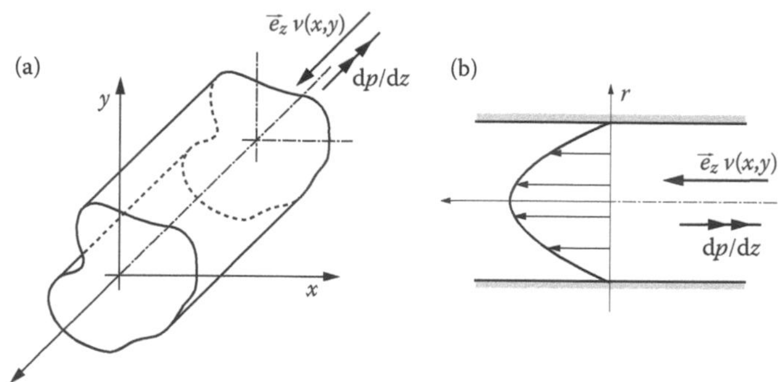

FIGURE 6.16
Fifth analogy is the steady unidirectional shear flow of a viscous fluid in a cylindrical pipe with an arbitrary cross section (a) for which (1) the velocity has the axial direction and varies (b) with the transverse coordinates in the cross section; (2) the velocity vanishes at the walls because the fluid is viscous; and (3) the motion is driven by a constant axial pressure gradient that is proportional to the flow rate. An analogy concerns the velocity (stress function) vanishing at the boundary of the singly connected cross section [Figure 6.16b (Figure 6.11a)] of the flow tube (slender rod under torsion). The flow is in the direction of decreasing pressure, with flow rate increasing with the modulus of the pressure gradient and the square area of the cross section and decreasing with the viscosity.

without restriction to creeping flow or low Reynolds number (Note 4.10), because the convective acceleration vanishes (Equation 6.310a) for a unidirectional shear flow (Equation 6.309a).

In the case of a steady (Equation 6.312a) inviscid (Equation 6.312b) unidirectional (Equation 6.312c ≡ Equation 6.309a) shear flow, the pressure is constant (Equation 6.312d ≡ Equation 6.312e):

$$\frac{\partial \vec{v}}{\partial t} = 0 = \eta, \quad \vec{v} = \vec{e}_z v(x,y): \quad \nabla p = 0 \quad \Leftrightarrow \quad p = \text{const}; \qquad (6.312\text{a}–6.312\text{e})$$

$$\frac{\partial \vec{v}}{\partial t} = 0 \neq \eta, \quad \vec{v} = \vec{e}_z v(x,y): \quad \nabla p = \vec{e}_z \eta \left(\frac{\partial^2 v}{\partial x^2} + \frac{\partial^2 v}{\partial y^2} \right); \qquad (6.313\text{a}–6.313\text{d})$$

in the case of a viscous flow (Equation 6.313b), the pressure gradient is balanced by the viscous stresses (Equation 6.313d ≡ Equation 6.314b) and has only an axial component (Equation 6.314a):

$$\nabla p = \vec{e}_z \frac{\mathrm{d}p}{\mathrm{d}z}, \quad \nabla^2 v = \frac{\partial^2 v}{\partial x^2} + \frac{\partial^2 v}{\partial y^2} = \frac{1}{\eta} \frac{\mathrm{d}p}{\mathrm{d}z}. \qquad (6.314\text{a and } 6.314\text{b})$$

It has been shown that *the steady unidirectional shear flow (Equation 6.309a) of a fluid with shear viscosity (Equation 6.313b) in a pipe is due to an axial pressure gradient (Equation 6.314a) that forces the Poisson equation 6.314b for the velocity; the velocity vanishes (Equation 6.309d) on the boundary (Equation 6.309c). This is analogous to the stress function for torsion of a slender beam*

(Section 6.5) that vanishes on the boundary (Equation 6.141c ≡ Equation 6.315a) and satisfies a Poisson equation 6.141b analogous to Equation 6.315b:

$$\Theta\big|_{\partial D} = 0 = v\big|_{\partial D}: \quad \nabla^2\Theta = -2 = \nabla^2\left(-\frac{2\eta}{dp/dz}v\right). \qquad (6.315a\text{ and }6.315b)$$

As examples, (1) [(2)] the unidirectional viscous shear flow between two flat walls (in a pipe with a triangular cross section) that is [Note 6.8 (Note 6.9)] analogous to [Subsection 6.7.3 (Subsection 6.8.1)] the torsion of a slab (slender prism with an equilateral triangular cross section) will be considered. Another analog problem is the unidirectional viscous shear flow in a solid or hollow tube with an elliptic or circular (nonequilateral triangular) cross section [Subsections 6.7.1 through 6.7.3 (Subsections 6.8.3 through 6.8.7)].

NOTE 6.8 VISCOUS UNIDIRECTIONAL SHEAR FLOW BETWEEN PARALLEL WALLS

The simplest problem of a steady unidirectional viscous shear flow concerns parallel walls (Equation 6.316a):

$$\vec{v} = \vec{e}_y v(x): \quad v\left(\pm\frac{h}{2}\right) = 0; \quad -\frac{h}{2} \le x \le \frac{h}{2}: \quad \eta\frac{d^2v}{dx^2} = \frac{dp}{dy} \qquad (6.316a\text{–}6.316d)$$

with the velocity (1) varying transversely (Equation 6.316a); (2) vanishing at the walls (Equation 6.316b); and (3) satisfying the viscous momentum equation (Equation 6.314b ≡ Equation 6.316d) between the walls (Equation 6.316c). Assuming a constant pressure gradient (Equation 6.317a), the velocity profile (Equation 6.317c ≡ Equation 6.316d) is parabolic (Equation 6.317b):

$$k_1 \equiv -\frac{1}{2\eta}\frac{dp}{dy} \equiv \text{const}: \quad v(x) = k_1\left(\frac{h^2}{4} - x^2\right), \quad \frac{d^2v}{dx^2} = -2k_1 = \frac{1}{\eta}\frac{dp}{dy}, \qquad (6.317a\text{–}6.317c)$$

with vanishing velocity at the walls (Equation 6.316b). The velocity (Equations 6.317a and 6.317c ≡ Equation 6.318b) is maximum (Equation 6.318c) at an equal distance from the walls:

$$\frac{dp}{dz} < 0: \quad v(x) = \frac{1}{2\eta}\frac{dp}{dz}\left(x^2 - \frac{h^2}{4}\right), \quad v_{\max} = v(0) = -\frac{h^2}{8\eta}\frac{dp}{dz}, \qquad (6.318a\text{–}6.318c)$$

and the flow is driven by a negative pressure gradient (Equation 6.318a). The velocity profile resembles the stress function (Equation 6.201e) of a thin rectangular prism that vanishes on the sides (Equations 6.201a and 6.201b). The role of torsional stiffness (Equation 6.203) is played by the volume flow rate (Equation 6.319):

$$\begin{aligned} Q &= \int_{-h/2}^{+h/2} v(x)\,dx = k_1\int_{-h/2}^{+h/2}\left(\frac{h^2}{4} - x^2\right)dx \\ &= k_1\left[\frac{h^2x}{4} - \frac{x^3}{3}\right]_{-h/2}^{+h/2} = \frac{1}{6}k_1h^3 = -\frac{1}{12\eta}\frac{dp}{dy}. \end{aligned} \qquad (6.319)$$

The nonzero viscous stresses (strain rates) are linear (Equation 6.320a) as the stresses (strains) in Equation 6.202a, with the shear or incompressible viscosity (Equation 6.313b) replacing the shear modulus of elasticity (Equation 6.114a):

$$\sigma_{xy}(x) = 2\eta\dot{S}_{xy}(x) = \eta\frac{dv}{dx} = -2k_1\eta x = x\frac{dp}{dy}; \tag{6.320a}$$

the shear stress (Equation 6.320a) has equal modulus and opposite signs on the walls:

$$\sigma_{xy}\left(\pm\frac{h}{2}\right) = 2\eta\dot{S}_{xy}\left(\pm\frac{h}{2}\right) = \pm\frac{h}{2}\frac{dp}{dy}. \tag{6.320b}$$

The Notes 6.8 and 6.9 are examples of *the analogy between torsion of a prism (steady unidirectional shear flow of a viscous incompressible fluid in a tube) with the same cross section: (1) the stress function (longitudinal velocity) vanishes on the boundary (walls); (2) the constant torsion rate corresponds to a constant pressure gradient along the axis; (3) it satisfies a Poisson equation; (4) the integral over the cross section determines the torsional stiffness (volume flow rate); and (5) multiplying by the torsion (mass density) specifies the torsional moment (mass flux).*

NOTE 6.9 VISCOUS FLOW IN A PIPE WITH TRIANGULAR CROSS SECTION

The analogy with the torsion of a prism (Note 6.8) also applies to the steady unidirectional shear flow of a viscous incompressible fluid in a straight tube with an equilateral triangular cross section; like the stress function (Equation 6.219d), the velocity profile (Equation 6.321a) in the cross section is (Equations 6.321b and 6.321c):

$$\begin{aligned}\vec{v} = \vec{e}_z v(x,y); \quad v(x,y) &= -k_2 y\left(y - L\frac{\sqrt{3}}{2} - \sqrt{3}x\right)\left(y - L\frac{\sqrt{3}}{2} + \sqrt{3}x\right)\\ &= -k_2 y\left[\left(y - L\frac{\sqrt{3}}{2}\right)^2 - 3x^2\right],\end{aligned} \tag{6.321a–6.321c}$$

so that the longitudinal velocity (Equation 6.321b) vanishes (Equations 6.322d through 6.322f) the three sides (Figure 6.14b) of the triangle (Equations 6.322a through 6.322c):

$$y = 0 = \frac{y}{\sqrt{3}} - \frac{L}{2} \mp x: \quad v(x,0) = 0 = v\left(x, \frac{\sqrt{3}}{2}L \pm x\sqrt{3}\right). \tag{6.322a–6.322f}$$

The viscous momentum equation 6.314b with shear kinematic viscosity η is satisfied (Equation 6.323a) by the velocity profile (Equation 6.321b) with a constant longitudinal pressure gradient (Equation 6.323b):

$$\frac{1}{\eta}\frac{dp}{dz} = \frac{\partial^2 v}{\partial x^2} + \frac{\partial^2 v}{\partial y^2} \equiv 2\sqrt{3}k_2 L, \quad k_2 = \frac{1}{2\sqrt{3}L\eta}\frac{dp}{dz}. \tag{6.323a and 6.323b}$$

When substituting the unidirectional shear velocity profile (Equation 6.321b) in the Poisson Equation 6.323a, only cubic and quadric terms matter:

$$\begin{aligned}\nabla^2 v &= \left(\frac{\partial^2}{\partial x^2}+\frac{\partial^2}{\partial y^2}\right)\left\{-k_2 y\left[y^2-\sqrt{3}Ly-3x^2+O(1)\right]\right\}\\ &= -k_2\left(6y-6y-2\sqrt{3}L\right)=2\sqrt{3}k_2 L,\end{aligned} \tag{6.324}$$

in agreement with Equation 6.324 ≡ Equations 6.323a and 6.323b.

The velocity profile in the cross section (Equations 6.321c and 6.323b ≡ Equations 6.325a and 6.325b):

$$\frac{dp}{dz}=\text{const:}\quad \vec{v}=-\vec{e}_z\frac{dp}{dz}\frac{y}{2\sqrt{3}L\eta}\left[\left(y-L\frac{\sqrt{3}}{2}\right)^2-3x^2\right] \tag{6.325a and 6.325b}$$

leads to the nonzero rates of strain and viscous stresses:

$$2\eta\dot{S}_{zx}=\sigma_{zy}=\eta\partial_x v_x=\frac{dp}{dz}\frac{\sqrt{3}}{L}xy, \tag{6.326a}$$

$$2\eta\dot{S}_{zy}=\sigma_{zy}=\eta\partial_y v_z=\frac{dp}{dz}\left[\frac{\sqrt{3}}{2L}\left(x^2-y^2\right)+y-L\frac{\sqrt{3}}{8}\right]; \tag{6.326b}$$

these coincide with Equation 6.220b ≡ Equation 6.326a (Equation 6.220a ≡ Equation 6.326b) using the upper (lower) sign in Equation 6.237a, besides Equation 6.237a ≡ Equation 6.219a:

$$c=\frac{\sqrt{3}}{L}L: \qquad \frac{dp}{dz}\quad\leftrightarrow\quad \pm 2\mu\tau. \tag{6.327a and 6.237b}$$

The interchange of Equations 6.326a and 6.326b with Equations 6.220a and 6.220b and the different signs in Equation 6.237b arise from the distinct relation between the elastic (Equations 6.137a and 6.137b) [viscous (Equations 4.332a through 4.332c)] stresses and the stress function (flow velocity).

The volume flow rate is given by

$$Q=2\int_0^{L/2}dx\int_0^{\sqrt{3}(L/2-x)}dy\,v(x,y)=-\frac{3}{160}k_2L^5=-\frac{\sqrt{3}}{320}\frac{L^4}{\eta}\frac{dp}{dz} \tag{6.328}$$

and is analogous to the torsional stiffness. The torsional stiffness is related to the flow rate through the same cross section by (1) suppressing the factor 2μ in Equation 6.158b by division and (2) inserting the factor in Equation 6.315b for the velocity relative to the stress function again by division:

$$Q = \frac{C}{2\mu}\left(-\frac{1}{2\eta}\frac{dp}{dz}\right) = -\frac{C}{4\mu\eta}\frac{dp}{dz}. \tag{6.329}$$

In the case of the equilateral triangular cross section, the torsional stiffness (Equation 6.230d) substituted in Equation 6.329 leads to the flow rate for a steady unidirectional viscous flow (Equation 6.328). Thus, *the steady unidirection shear flow of a viscous incompressible fluid in a straight tube whose cross section (Equations 6.322a through 6.322c) is an equilateral triangle (Figure 6.14b) is specified by the longitudinal (Equation 6.321a) velocity profile (Equations 6.325a and 6.325b) corresponding to rates-of-strain and viscous stresses (Equations 6.326a and 6.326b); the flow rate (Equation 6.328) is (1) directly proportional to the modulus of the negative or downstream pressure gradient that drives the flow and (2) inversely proportional to the shear viscosity that slows down the flow to zero velocity at the wall (3) to the fourth power (square) of the lengthscale (area) of the cross section. For an inviscid fluid in a pipe the flow rate is proportional to the area of the cross section; for a viscous fluid the flow rate is proportional to the square of the area of the cross section, showing that a larger cross section is less affected by the slowing down of the flow by viscosity in the boundary layers near to the walls.* The extension to a nonequilateral cross section can be made as in Subsections 6.8.3 through 6.8.7.

NOTE 6.10 DISTINCTION BETWEEN SOLIDS AND FLUIDS, AND LIQUIDS AND GASES

The harmonic (biharmonic) fields have been considered [Note 6.5 (Notes 6.6 through 6.9)] for several fields (gravity, electric, magnetic) and states of matter (solids, fluids). The usual distinction between the states of matter is that (1) solids have a definite shape that may have finite deformations, and (2) a fluid adapts to the shape of the vessel that contains it, and its flow may lead to infinite displacements with a finite velocity over an infinite time. Both are described in terms of finite quantities: (1) the displacement for a solid and (2) the velocity for a fluid. The fluids are usually separated into (1) liquids that are nearly incompressible, that is, take the shape of the vessel that contains them, without changing volume, thus either filling or overflowing; and (2) gases that can be compressed or expanded and always fill all available space. An incompressible flow (Subsections I.14.6 and II.2.1) corresponds to a Mach number $M = v/c$ or ratio of velocity to sound speed, which is small $M^2 \ll 1$; for example, (1) for sea water with normal salinity, the sound speed is $c = 1550$ m/s, so low Mach number $M \le 0.3$ applies up to a large speed $u \sim 0.3 \times 1550$ m/s $= 465$ m/s $= 1674$ km/h, which is hardly attainable by a ship; and (2) for the air at sea level at normal temperature, $c = 340$ m/s, and the velocity for incompressible flow $v_a \sim 0.3 \times 340$ m/s $= 102$ m/s $= 367$ km/h is routinely exceeded by aircraft. Although these examples explain the difference between gases and liquids, it is not true that a liquid is strictly incompressible: sound waves are compressive perturbations and propagate in gases and liquids, as well as solids, either with small or large amplitude. The distinction between gases and liquids, and also solids, relates to the thermodynamic equation of state; the simplest is for a perfect gas. The simplest equation of state that allows for phase changes such as solidification, vaporization, melting, and condensation in van der Waal's.

NOTE 6.11 PLASMA, STATES OF MATTER, AND INTERACTIONS

The classification of matter into solids, liquids, and gases dates back to the Greek civilization and is connected with everyday experience on the surface of the earth. Most of the matter of the universe exists in the state of plasma that is decomposed into charged particles, namely, negative (electrons), positive (ions), and neutrals; plasma in nature is farther; for example, it constitutes stars and galaxies, fills intergalactic and interplanetary space, and is closest in the high atmosphere or ionosphere of the earth. The plasma may be too rarefied for a continuum description and lead to statistical or quantum effects; in the molten core of the earth, as for a plasma, electromagnetic effects are important. These bring another set of fundamental equations (Maxwell equations) and constitutive relations. The presence of matter by itself creates another field: gravity and gravitation that affects space–time. An example of interactions between various classes of phenomena is the magneto-acoustic-gravity-inertial wave in a fluid combining (1) compressibility and pressure perturbations as in sound, (2) ionization and the magnetic field as in hydromagnetic waves, (3) stratification and the gravity field as in internal waves, and (4) rotation and associated forces as for inertial waves. There are other examples of interactions, such as magnetohydrodynamics (piezomagnetism) combining fluids (solids) with magnetic fields.

NOTE 6.12 MATHEMATICAL METHODS, PHYSICAL MODELS, AND ENGINEERING INVENTIONS

When studying interdisciplinary phenomena, the gradual approach is (1) to address each subject separately on its own, (2) to consider all possible couplings, and (3) to conclude with multiple interactions of all effects. This applies as well to (1) the observation and reproduction of phenomena in nature and (2) the design and operation of engineering devices invented by mankind. In both cases, a reliable physical model (A) and accurate mathematical model (B) are needed. The gradual approach (1–3) applies to all three interlinked activities: engineering design, physical models, and mathematical models. This is the approach adopted in the present study: to develop the physical models and engineering applications as they are enabled by mathematical methods. The potential harmonic (biharmonic) fields [Note 6.6 (Note 6.7)] apply to the basic force fields and simple states of matter; they model the basic phenomena modified by more complex interactions.

6.10 Conclusion

(1) A membrane with a plane boundary curve remains plane due to tension in the absence of a shear stress (Figure 6.1a). (2) A shear stress deflects the membrane until it is balanced by the transverse component of the tension, that is tangential (Figure 6.1b). (3) The deflection of the membrane is associated with an increase in the area (Figure 6.2), which, multiplied by the tension, determines the elastic energy of deformation. (4) An example of applied force is the membrane's own weight that, if small compared with the tension effect, causes a linear deflection with a small slope into a parabolic shape (Figure 6.3a). (5) The shape is modified for an annular membrane supported on two coaxial spaced rings (Figure 6.3b), leading to three cases (Figures 6.4a through 6.4c) including "sagging" (Figure 6.4c). (6) In the nonlinear case of a large slope, the spherical deflection of the membrane exceeds the parabolic shape (Panel 6.1). (7) The maximum deflection and slope (Table 6.1) increase with the nonlinearity parameter that takes the value unity for the largest load that the membrane can support. A second, analogous problem concerns the effect of surface tension on a curved surface (Figure 6.5a), for example, the interface between two fluids at different pressures (Figure 6.6). The surface has two principal curvatures (Figure 6.5a) that (1) are

equal in modulus with opposite signs for a saddle-type surface (Figure 6.7a); (2) are equal and positive for spherical surface, that is, the minimal surface without boundaries (Figure 6.7b), for example, a soap bubble; and (3) have opposite signs for a catenoid (Figure 6.7c), that is, the minimal surface between two coaxial rings generated by the rotation of a curve with a fixed length. The effect of surface tension also applies in the plane, for example, a circular hole (Figure 6.7d) in a soap film held between an outer plane wire frame and an inner flexible string of constant length, leading to a circular shape of the hole to minimize the area of the membrane. A cylindrical surface (Figure 6.8a) has straight lines as generators with zero curvature and the directrix (Figure 6.8b) as the nonzero principal curvature. The latter relates to the angle of wetting of a vertical wall (Figure 6.9a) and to the capillary rise between vertical walls (Figure 6.9b). A third analogous problem is the torsion of a slender beam by an axial torque (Figure 6.10b), causing a relative rotation of the cross sections (Figure 6.10a) without forces on the side surface. The cross section may be simply (multiply) connected [Figure 6.11a (Figure 6.11b)], for example, a massive elliptical (Figure 6.12b and d) [hollow elliptical (Figure 6.12a) or circular (Figure 6.12c) cross section] is simply (doubly) connected. The axial torsion causes shear stresses in the cross section (Figure 6.12d) consistent with zero normal stress on the side surface; they may cause an axial displacement, for example, a plane elliptical cross section under torsion becomes (Figure 6.12b) a surface with a saddle point (Figure 6.7a) at a mountain pass joining two valleys (–) between two hills (+). The torsional stiffness of a closed thin cross section (Figure 6.13b) can be compared (Table 6.2) with (1) the same cross section with a cut (Figure 6.13c) and (2) a long thin rectangular section (Figure 6.13a). Another cross section for a prism under torsion is a triangle: equilateral (Figure 6.14b), isosceles (Figure 6.14c), or irregular (Figure 6.14a). A fourth analogy is with an inviscid fluid in a rotating vessel (Figure 6.15a), superimposing on the incompressible flow a rotational component; in the case of a rotating elliptic cylinder (Figure 6.15b), the fluid particles follow elliptic trajectories with an angular velocity that drifts more away from the velocity of rotation for larger eccentricity, so that rigid body rotation applies only to the circular cylinder. A fifth analogy concerns the steady unidirectional shear flow of a viscous fluid (Figure 6.16b) in a pipe of constant cross section with an arbitrary shape (Figure 6.16a). These complementary five problems involving solutions of the Laplace and/or Poisson equations appear together with the seven basic cases in the set of 12 harmonic fields in Table 6.3; the latter also includes two biharmonic fields, one each for fluid and solid mechanics. The solid mechanics, in the specific case of elasticity, leads to the 12 special cases in Diagram 6.1.

7

Infinite and Cyclometric Representations

Concerning the six circular (hyperbolic) functions, namely, the two primary functions sine and cosine, and the four secondary functions secant, cosecant, tangent, and cotangent, the algebraic (analytic) properties are discussed in Chapter 5 (Chapter 7). The direct functions are periodic and hence have as inverses the 12 cyclometric functions that are many-valued and can be made single-valued by choosing the principal branch in a complex plane with suitable branch cuts (Section 7.2). Concerning infinite representations: (1) the four primary functions are integral, with an infinite number of zeros, and thus can be represented by power series with infinite radius of convergence (Section 7.1), by infinite products (Section 7.7), and by continued fractions (Section 7.8); (2) the eight secondary functions are meromorphic and can be represented by power series with finite radius of convergence (Section 7.1), by series of fractions (Section 7.6), and by continued fractions (Section 7.8); and (3) the cyclometric functions have branch points (branch cuts) that limit the radius of convergence (domain of validity) of their representation by power series (Section 7.4) [continued fractions (Section 7.8)]. In the process of establishing these infinite representations: (1) the Euler and Bernoulli numbers are introduced to specify the coefficients of the power series for secondary functions (Section 7.1); (2 and 3) the zeros and slopes (poles and residues) lead (Section 7.5) to infinite products (series of fractions) for the primary (secondary) functions [Section 7.7 (Section 7.6)]; (4) the derivatives and primitives (Section 7.3) specify the coefficients in the power series for the direct (inverse), that is, primary and secondary (cyclometric) functions [Section 7.1 (Section 7.4)]; (5) the continued fractions (Section 7.8) are obtained by transformation [Subsections 1.8.1 and 1.8.2 (Subsections 1.8.4 and 1.8.5)] of power series (infinite products) or by the Lambert method (Section 1.9) applied to functions of the hypergeometric family (Section 5.9); and (6) the power series, infinite products, and continued fractions can be used to calculate irrational numbers like e (π), that are the base of natural logarithms (Subsection 1.8.5 and Section 3.9) [solve the quadrature of the circle (Subsection 1.8.7 and Section 7.9)].

7.1 Power Series and Euler (1755)/Bernoulli (1713) Numbers

The primary circular and hyperbolic functions, namely, sine and cosine, are integral functions whose MacLaurin series has an infinite radius of convergence and simple coefficients (Subsection 7.1.1). The secondary functions, namely, secant, cosecant, tangent, and cotangent, are meromorphic, and their poles limit the radius of convergence of their power series expansions; since secant and tangent (cosecant and cotangent) are analytic (have a simple pole) at the origin, they have a MacLaurin (Laurent–MacLaurin) series (Subsection 7.1.2) with a radius of convergence $\pi/2$ (π), that is, limited by the nearest pole. The coefficients of the power series for the secondary functions are less simple than for the primary functions and involve the Euler (Bernoulli) numbers for the secant (cosecant, tangent, and cotangent) that can be obtained (1) for the lowest orders by inversion (product) of series

[Subsections 7.1.3 and 7.1.4 (Subsections 7.1.5 and 7.1.6)] and (2) for an arbitrary order from recurrence relations [Subsection 7.1.7 (Subsection 7.1.8)] derived from a generating function. The Bernoulli numbers also appear in other related power series (Subsection 7.1.9).

7.1.1 Derivation of and Series for the Primary Functions

Using the definition of the primary hyperbolic (Equation 5.1a) [circular (Equation 5.1b)] functions, namely, cosine and sine, and the differentiation property (Equation 3.1a) of the exponential leads to

$$\frac{d^n\{\cosh z + \sinh z\}}{dz^n} = \frac{d^n(e^z)}{dz^n} = e^z = \cosh z + \sinh z, \tag{7.1}$$

$$\frac{d^n\{\cos z + i \sin z\}}{dz^n} = \frac{d^n(e^{iz})}{dz^n} = i^n e^{iz} = i^n \cosh z + i^{n+1} \sinh z. \tag{7.2}$$

Equating the even and odd parts in Equation 7.1 leads to the even and odd derivates of the hyperbolic sine and cosine:

$$\cosh^{(2n)} z = \cosh z, \quad \cosh^{2n+1} z = \sinh z, \qquad \text{(7.3a and 7.3b)}$$

$$\sinh^{(2n)} z = \sinh z, \quad \sinh^{(2n+1)} z = \cosh z, \qquad \text{(7.4a and 7.4b)}$$

and to their values at the origin:

$$\cosh^{(2n)}(0) = 1 = \sinh^{(2n+1)}(0), \quad \cosh^{(2n+1)}(0) = 0 = \sinh^{(2n)}(0). \qquad \text{(7.5a–7.5d)}$$

Substitution of Equations 7.5a and 7.5c) (Equations 7.5b and 7.5d) in the MacLaurin series (Equation 1.24) leads to the power series for the hyperbolic cosine [sine (Equations 7.7a and 7.7b)]:

$$|z| < \infty: \quad \cosh z = \sum_{n=0}^{\infty} \frac{z^n}{n!} \cosh^{(n)}(0) = \sum_{n=0}^{\infty} \frac{z^{2n}}{(2n)!} \cosh^{(2n)}(0)$$
$$= \sum_{n=0}^{\infty} \frac{z^{2n}}{(2n)!} = 1 + \frac{z^2}{2} + \frac{z^4}{24} + \frac{z^6}{720} + \cdots. \qquad \text{(7.6a and 7.6b)}$$

$$|z| < \infty: \quad \sinh z = \sum_{n=0}^{\infty} \frac{z^n}{n!} \sinh^{(n)}(0) = \sum_{n=0}^{\infty} \frac{z^{2n+1}}{(2n+1)!} \sinh^{(2n+1)}(y)$$
$$= \sum_{n=0}^{\infty} \frac{z^{2n+1}}{(2n+1)!} = z + \frac{z^3}{6} + \frac{z^5}{120} + \frac{z^7}{5040} + \cdots. \qquad \text{(7.7a and 7.7b)}$$

These series can also be obtained from the series (Equation 1.26b) for the exponential:

$$|z| < \infty: \quad \cosh z + \sinh z = e^z = \sum_{n=0}^{\infty} \frac{z^n}{n!} = \sum_{n=0}^{\infty} \frac{z^{2n}}{(2n)!} + \sum_{n=0}^{\infty} \frac{z^{2n+1}}{(2n+1)!}, \quad (7.8a \text{ and } 7.8b)$$

separating the even (odd) powers that appear in Equation 7.6b (Equation 7.7b).

The series for the exponential with imaginary argument (Equation 1.30) also leads to the series for the circular cosine (Equation 1.31a) [sine (Equation 1.31b)] by (1) separation into even and odds parts for real z and (2) separation into real and imaginary parts for real z, followed by analytic continuation (Subsection I.31.2) from real to complex z. Either method 1 or 2 applied to Equation 7.2 specifies the derivatives of even and odd orders of the circular cosine and sine:

$$\cos^{(2n)} z = (-)^n \cos z, \quad \cos^{(2n+1)} z = (-)^{n+1} \sin z, \quad (7.9a \text{ and } 7.9b)$$

$$\sin^{(2n)} z = (-)^n \sin z, \quad \sin^{(2n+1)} z = (-)^n \cos z, \quad (7.10a \text{ and } 7.10b)$$

and their values at the origin:

$$\cos^{(2n)}(0) = (-)^n = \sin^{(2n+1)}(0), \quad \cos^{(2n+1)}(0) = 0 = \sin^{(2n)}(0). \quad (7.11a\text{–}7.11d)$$

Theses values [Equations 7.11a and 7.11c (Equations 7.11b and 7.11d)] substituted in the MacLaurin series (Equation 1.24) lead to the power series for the circular cosine (Equations 7.12a and 7.12b) [sine (Equations 7.13a and 7.13b)]:

$$|z| < \infty: \quad \cos z = \sum_{n=0}^{\infty} \frac{z^n}{n!} \cos^{(n)}(0) = \sum_{n=0}^{\infty} \frac{z^{2n}}{(2n)!} \cos^{(2n)}(0) = \sum_{n=0}^{\infty} \frac{(-)^n}{(2n)!} z^{2n}, \quad (7.12a \text{ and } 7.12b)$$

$$|z| < \infty: \quad \sin z = \sum_{n=0}^{\infty} \frac{z^n}{n!} \sin^{(n)}(0) = \sum_{n=0}^{\infty} \frac{z^{2n+1}}{(2n+1)!} \sin^{(2n+1)}(0) = \sum_{n=0}^{\infty} \frac{(-)^n}{(2n+1)!} z^{2n+1} \quad (7.13a \text{ and } 7.13b)$$

in agreement with Equation 1.31a ≡ Equations 7.12a and 7.12b (Equation 1.31b ≡ Equations 7.13a and 7.13b). The MacLaurin series for the circular (hyperbolic) cosine [Equations 7.12a and 7.12b (Equations 7.6a and 7.6b)] and sine [Equations 7.13a and 7.13b (Equations 7.7a and 7.7b)] differ only in the presence (absence) of alternating signs, as follows from the imaginary transformation [Equations 5.5a and 5.6a (Equations 5.5b and 5.6b)]; the alternating (fixed) sign in the power series for a real variable allows the circular cosine and cosine to be periodic (causes the hyperbolic cosine and sine to be monotonic) functions [Figure 5.2a and 5.2b (Figure 5.2d and 5.2e)]. *The primary circular (hyperbolic) functions are integral functions whose power series have an infinite radius of convergence, that is, converge absolutely* $|z| < \infty$ *and totally for* $|z| \le M < \infty$*, namely, the series for the circular (hyperbolic) cosine [Equation 7.14b ≡ Equations 7.12a and 7.12b ≡ Equation 1.34a (Equation 7.14c ≡ Equations 7.6a and 7.6b)] and sine*

TABLE 7.1

Derivatives of All Orders and Their Values at the Origin for the Primary Circular and Hyperbolic Functions, That Is, the Cosine and Sine

Function	Circular		Hyperbolic	
$f(z)$	$\cos z$	$\sin z$	$\cosh z$	$\sinh z$
$f'(z)$	$-\sin z$	$\cos z$	$\sinh z$	$\cosh z$
$f^{(2n)}(z)$	$(-)^n \cos z$	$(-)^n \sin z$	$\cosh z$	$\sinh z$
$f^{(2n+1)}(0)$	$(-)^{n+1} \sin z$	$(-)^n \cos z$	$\sinh z$	$\cosh z$
$f^{(2n)}(0)$	$(-)^n$	0	1	0
$f^{(2n+1)}(0)$	0	$(-)^n$	0	1

[Equation 7.15b ≡ Equations 7.13a and 7.13b ≡ Equation 1.34c (Equation 7.15c ≡ Equations 7.7a and 7.7b)]:

$$\left|z\right| < \infty: \quad \cos z = \sum_{n=0}^{\infty} (-)^n \frac{z^{2n}}{(2n)!}, \qquad \cosh z = \sum_{n=0}^{\infty} \frac{z^{2n}}{(2n)!}, \tag{7.14a–7.14c}$$

$$\left|z\right| < \infty: \quad \sin z \sum_{n=0}^{\infty} (-)^n \frac{z^{2n+1}}{(2n+1)!} \qquad \sinh z = \sum_{n=0}^{\infty} \frac{z^{2n+1}}{(2n+1)!}. \tag{7.15a–7.15c}$$

The coefficients of these MacLaurin series [Equations 7.14a through 7.14c (Equations 7.15a through 7.15c)] involve the derivatives of all orders of the circular and hyperbolic cosine (Equations 7.3a, 7.3b, 7.9a, and 7.9b) [sine (Equations 7.4a, 7.4b, 7.10a, and 7.10b)] at the origin [Equations 7.5a, 7.5c, 7.11a, and 7.11c (Equations 7.5b, 7.5d, 7.11b, and 7.11d)] that are listed in Table 7.1.

7.1.2 Power Series for the Secondary Functions

The primary functions are integral and have power series (Subsection 7.1.1) with an infinite radius of convergence; the secondary functions are meromorphic, and hence their power series has a radius of convergence limited by the nearest pole. For example (Figure 7.1), the secant is analytic at the origin and has the nearest poles at $\pm \pi/2$; hence, its MacLaurin series has a radius of convergence $\pi/2$ and consists of even powers because the function is even:

$$|z| < \frac{\pi}{2}: \quad \sec z \equiv \sum_{n=0}^{\infty} (-)^n \frac{z^{2n}}{(2n)!} E_{2n}, \quad \operatorname{sech} z \equiv \sum_{n=0}^{\infty} \frac{z^{2n}}{(2n)!} E_{2n}, \tag{7.16a–7.16c}$$

where the coefficients of Equation 7.16b define the **Euler numbers** E_{2n}. Similarly the cotangent has (Figure 7.1) poles at the origin and nearest at $\pm \pi$ and is an odd function; thus, its Laurent–MacLaurin series has a radius of convergence equal to π and proceeds in odd powers, starting with a pole of residue unity at the origin:

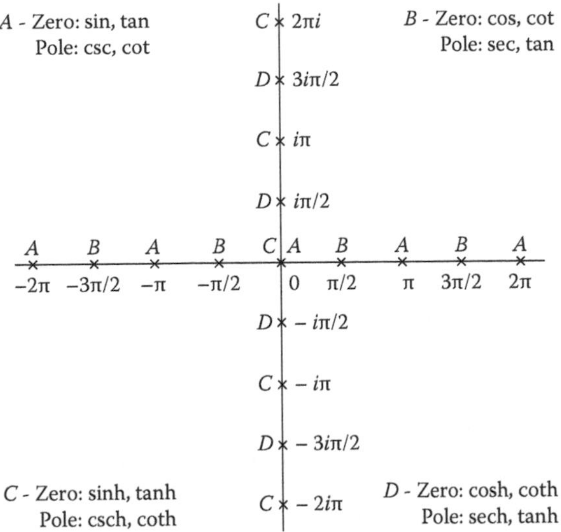

FIGURE 7.1
Zeros of the circular (hyperbolic) functions sine, cosine, tangent, and cotangent lie on the real (imaginary) axis and are equally spaced and interlaced. Poles of the circular (hyperbolic) secondary functions secant, cosecant, tangent, and cotangent arise from algebraic inversion and thus are all simple like the zeros and have corresponding locations, that is, are also equally spaced and interlaced along the real (imaginary) axis.

$$|z| < \pi: \quad \cot z \equiv \frac{1}{z}\sum_{n=0}^{\infty}(-)^n \frac{(2z)^{2n}}{(2n)!} B_{2n}, \quad \coth z \equiv \frac{1}{z}\sum_{n=0}^{\infty} \frac{(2z)^{2n}}{(2n)!} B_{2n}, \qquad (7.17a\text{–}7.17c)$$

where the coefficients of Equation 7.17b define the **Bernoulli numbers** B_{2n}. The other two secondary functions, cosecant (Equation 7.18) and tangent (Equation 7.19), are specified by

$$\operatorname{csch} z = \frac{2-1}{\sinh z} = \frac{2\cosh^2\left(\frac{z}{2}\right) - 2\sinh^2\left(\frac{z}{2}\right) - 1}{2\cosh\left(\frac{z}{2}\right)\sinh\left(\frac{z}{2}\right)} = \frac{\cosh\left(\frac{z}{2}\right)}{\sinh\left(\frac{z}{2}\right)} - \frac{\cosh z}{\sinh z} = \coth\left(\frac{z}{2}\right) - \coth z, \qquad (7.18)$$

$$\tanh z = \frac{(4-2)\sinh^2 z}{2\cosh z \sinh z} = \frac{4\sinh^2 z + 2 - 2\cosh^2 z}{2\sinh z \cosh z} = 2\frac{\cosh(2z)}{\sinh(2z)} - \frac{\cosh z}{\sinh z} = 2\coth(2z) - \coth z, \qquad (7.19)$$

as linear combinations of the tangent

$$\operatorname{csch}(z) = \coth\left(\frac{z}{2}\right) - \coth(z), \quad \tanh(z) = 2\coth(2z) - \coth(z), \qquad (7.20a \text{ and } 7.20b)$$

and thus do not introduce any new numbers. Substituting Equations 7.17a through 7.17c in Equations 7.20a and 7.20b leads to the power series for the remaining secondary functions:

$$|z| < \pi: \quad \operatorname{csch} z = \frac{1}{z}\sum_{n=0}^{\infty} \frac{2-2^{2n}}{(2n)!} B_{2n} z^{2n}, \quad \csc z = \frac{1}{z}\sum_{n=0}^{\infty} (-)^n \frac{2-2^{2n}}{(2n)!} B_{2n} z^{2n} \tag{7.21a–7.21c}$$

$$|z| < \frac{\pi}{2}: \quad \tanh z = \frac{1}{z}\sum_{n=1}^{\infty} \frac{2^{2n}-1}{(2n)!} B_{2n} (2z)^{2n}, \quad \tan z = \frac{1}{z}\sum_{n=1}^{\infty} (-)^n \frac{1-2^{2n}}{(2n)!} B_{2n} (2z)^{2n}. \tag{7.22a–7.22c}$$

The power series for the secondary circular (hyperbolic) functions, namely, the secant [Equation 7.16b (Equation 7.16c)], cotangent [Equation 7.17b (Equation 7.17c)], cosecant [Equation 7.21c (Equation 7.21b)], and tangent [Equation 7.22c (Equation 7.22b)] (1) differ only in the presence (absence) of alternating signs in agreement with the imaginary change of variables (Equations 5.24a and 5.24b/Equations 5.27a and 5.27b/Equations 5.25a and 5.25b/ Equations 5.26a and 5.26b, respectively); and (2) have the poles at positions rotated by $\pi/2$ around the origin (Figure 7.1), so that the nearest pole is at the same distance from the origin, and the radius of convergence is the same (Equation 7.16a/Equation 7.17a/Equation 7.21a/Equation 7.22a, respectively). Recalling the theorems on regular (singular) power series [Section I.23.7 (Section I.25.6)], it follows that *the MacLaurin (Laurent–MacLaurin) series for the circular and hyperbolic secant (Equations 7.16a through 7.16c) and tangent (Equations 7.22a through 7.22c) [cotangent (Equations 7.17a through 7.17c) and cosecant (Equations 7.21a through 7.21c)] converge absolutely for* $|z| < \pi/2$ $(0 < |z| < \pi)$ *and totally for* $|z| \le \pi/2 - \varepsilon$ $(\delta \le z \le |\pi| - \varepsilon)$ *with* $0 < \varepsilon < \pi/2$ $(0 < \delta < \pi - \varepsilon < \pi)$.

7.1.3 Coefficients for Algebraic Inversion of Power Series

The reversion of the power series for $w = f(z)$ specifies (Subsection I.25.8) the power series for the inverse function $z = f^{-1}(w)$. Next a distinct case is considered, namely, the power series for the algebraic inverse function $1/w = 1/f(z) = g(z)$. This process is used to obtain the power series for the secant (cosecant) from those of the cosine (sine); this also specifies the first few Euler (Bernoulli) numbers. Starting with the power series for the circular cosine (Equation 7.14b ≡ Equation 7.23b), that is, an integral function (Equation 7.23a), the first three coefficients are (Equation 7.23c)

$$|z| < \infty: \quad \cos z = a_0 + a_1 z^2 + a_2 z^4 + O(z^6): \quad \{a_0, a_1, a_2\} = \left\{1, -\frac{1}{2}, \frac{1}{24}\right\}. \tag{7.23a–7.23c}$$

The algebraic inverse function is the circular secant that is analytic in a circle of radius up to the nearest pole (Equation 7.24a); the first three terms of the power series (Equation 7.24b):

$$|z| < \frac{\pi}{2}: \quad \sec z = b_0 + b_1 z^2 + b_2 z^4 + O(z^6), \tag{7.24a and 7.24b}$$

must satisfy

$$\begin{aligned} 1 = \cos z \sec z &= \left[a_0 + a_1 z^2 + a_2 z^4 + O(z^6)\right]\left[b_0 + b_1 z^2 + b_2 z^4 + O(z^6)\right] \\ &= a_0 b_0 + \left(a_0 b_1 + a_1 b_0\right) z^2 + \left(a_0 b_2 + a_1 b_1 + a_2 b_0\right) z^4 + O(z^6). \end{aligned} \tag{7.25}$$

Equating the coefficients of powers of z on the left-hand side (l.h.s.) and the right-hand side (r.h.s.) of Equation 7.25 leads to three identities:

$$a_0 b_0 = 1, \qquad a_0 b_1 + a_1 b_0 = 0 = a_0 b_2 + a_1 b_1 + a_2 b_0, \tag{7.26a–7.26c}$$

that specify one set of coefficients in terms of the other:

$$b_0 = \frac{1}{a_0}, \quad b_1 = -b_0 \frac{a_1}{a_0} = -\frac{a_1}{(a_0)^2}, \quad b_2 = -b_0 \frac{a_2}{a_0} - b_1 \frac{a_1}{a_0} = -\frac{a_2}{(a_0)^2} + \frac{(a_1)^2}{(a_0)^3}; \tag{7.27a–7.27c}$$

the **formulas for the algebraic inversion of series** *(Equations 7.27a through 7.27c) specify the first three coefficients of the series of the algebraic inverse function (Equation 7.24b) in terms of the coefficients of the original function (Equation 7.23b). The formulas are symmetric with the two sets of coefficients* (a_0, a_1, a_2) *and* (b_0, b_1, b_2).

7.1.4 Power Series for the Secant and Cosecant

The coefficients (Equation 7.23c) of the power series for the circular cosine (Equation 7.23b) lead by Equations 7.27a through 7.27c to the coefficients (Equations 7.28b through 7.28d) of the power series for the circular secant (Equations 7.28a and 7.28e):

$$|z| < \frac{\pi}{2}: \quad \{b_0, b_1, b_2\} = \left\{1, \frac{1}{2}, \frac{5}{24}\right\}, \quad \sec z = 1 + \frac{z^2}{2} + \frac{5z^4}{24} + O(z^6). \tag{7.28a–7.28e}$$

Comparing Equation 7.28e with the first three terms of the power series for the circular secant (Equations 7.16a and 7.16b ≡ Equations 7.29a and 7.29b) specifies the first three even Euler numbers:

$$|z| < \frac{\pi}{2}: \quad \operatorname{sech} z = E_0 - E_2 \frac{z^2}{2} + E_4 \frac{z^4}{24} + O(z^6); \quad \{E_0, E_2, E_4\} = \{1, -1, 5\}, \tag{7.29a–7.29e}$$

in agreement with Table 7.2. The circular sine is an integral function specified by the power series (Equations 7.15a and 7.15b ≡ Equations 7.30a and 7.30b) with first three coefficients (Equations 7.30c through 7.30e):

TABLE 7.2

First 10 Euler (Bernoulli) Numbers That Appear as Coefficients in the Power Series for the Circular and Hyperbolic Secondary Function(s) Secant (Cosecant, Tangent, and Cotangent)

n	0	2	4	6	8	10	12
E_n	1	−1	5	−61	1385	−50 521	27 02765
B_n	1	$\frac{1}{6}$	$-\frac{1}{30}$	$\frac{1}{42}$	$-\frac{1}{30}$	$\frac{5}{66}$	$-\frac{691}{2730}$

n	14	16	18	20
E_n	−1993 60981	1 93915 12145	−240 48796 17541	37037 11882 37525
B_n	$\frac{7}{6}$	$-\frac{3617}{510}$	$\frac{43867}{798}$	$-\frac{174611}{330}$

$$|z| < \infty: \quad \sin z = a_0 z + a_1 z^3 + a_2 z^5 + O(z^7); \quad \{a_0, a_1, a_2\} = \left\{1, -\frac{1}{6}, \frac{1}{120}\right\}. \tag{7.30a–7.30e}$$

The corresponding first three coefficients (Equations 7.27a through 7.27c) are Equations 7.31b through 7.31d for the algebraic inverse function, that is, the circular cosecant (Equations 7.31a and 7.31e):

$$|z| < \pi: \quad \{b_0, b_1, b_2\} = \left\{1, \frac{1}{6}, \frac{7}{360}\right\}, \quad \csc z = \frac{1}{z} + \frac{z}{6} + \frac{7z^3}{360} + O(z^5). \tag{7.31a–7.31e}$$

The cosecant has a simple pole at the origin, where the sine has a zero (Equation 7.30b) and hence the Laurent–MacLaurin series (Equation 7.31e). The even powers $(1,z^2,z^4)$ in Equations 7.23b and 7.24b are replaced by odd powers $z,z^3,z^5(1/z,z,z^3)$ in Equation 7.30b (Equation 7.31e) leading to the same relations (Equations 7.26a through 7.26c ≡ Equations 7.27a through 7.27c) among the coefficients of the series of two algebraically inverse functions. Comparing Equation 7.31e with the first three terms of the power series for the circular cosecant (Equations 7.21a and 7.21c ≡ Equations 7.32a and 7.32b) specifies the first three even Bernoulli numbers (Equations 7.32c through 7.32e):

$$|z| < \pi: \quad \csc z = \frac{B_0}{z} + B_2 z - \frac{7}{12} B_4 z^3 + O(z^5); \quad \{B_0, B_2, B_4\} = \left\{1, \frac{1}{6}, -\frac{1}{30}\right\}, \tag{7.32a–7.32e}$$

in agreement with Table 7.2.

7.1.5 Product of Series and the Tangent and Cotangent

The power series for the secant (Equation 7.28e) [cosecant (Equation 7.31e)] were obtained (Subsection 7.1.4) by algebraic inversion of the power series for the cosine (Equation 7.14b) [sine (Equation 7.15b)]. Multiplying by the power series for sine (Equation 7.15b) [cosine (Equation 7.14b)] leads to the first three terms of the power series for the circular tangent (Equations 7.33a and 7.33b) [cotangent (Equations 7.34a and 7.34b)]:

$$|z| < \frac{\pi}{2}: \quad \tan z = \sin z \sec z = \left[z - \frac{z^3}{6} + \frac{z^5}{120} + O(z^7)\right]\left[1 + \frac{z^2}{2} + \frac{5z^4}{24} + O(z^6)\right]$$

$$= z + z^3\left(\frac{1}{2} - \frac{1}{6}\right) + z^5\left(\frac{5}{24} - \frac{1}{12} + \frac{1}{120}\right) + O(z^6) \tag{7.33a and 7.33b}$$

$$= z + \frac{z^3}{3} + \frac{2z^5}{15} + O(z^7),$$

$$0 < |z| < \pi: \quad \cot z = \cos z \csc z = \left[1 - \frac{z^2}{2} + \frac{z^4}{24} + O(z^6)\right]\left[\frac{1}{z} + \frac{z}{6} + \frac{7z^3}{360} + O(z^5)\right]$$

$$= \frac{1}{z} + z\left(-\frac{1}{2} + \frac{1}{6}\right) + z^3\left(\frac{1}{24} - \frac{1}{12} + \frac{7}{360}\right) + O(z^5) \tag{7.34a and 7.34b}$$

$$= \frac{1}{z} - \frac{z}{3} - \frac{z^3}{45} + O(z^5).$$

Comparing Equation 7.33b (Equation 7.34b) with the first three terms of the power series for the circular tangent (Equations 7.22a and 7.22c ≡ Equations 7.35a and 7.35b) [cotangent (Equations 7.17a and 7.17b ≡ Equations 7.36a and 7.36b)]:

$$|z| < \frac{\pi}{2}: \quad \tan z = 6B_2 z - 10B_4 z^3 + \frac{84}{15}B_6 z^5 + O(z^7), \tag{7.35a and 7.35b}$$

$$0 < |z| < \pi: \quad \cot z = \frac{B_0}{z} - 2B_2 z + \frac{2}{3}B_4 z^3 + O(z^5), \quad B_6 = \frac{1}{42}, \tag{7.36a–7.36c}$$

confirms the first three even Bernoulli numbers (Equations 7.32c through 7.32e) and adds a fourth (Equation 7.36c), in agreement with Table 7.2.

7.1.6 Power Series for the Secondary Hyperbolic Functions

The first three terms of the power series for the secondary circular (hyperbolic) functions, namely, secant [Equations 7.28a and 7.28e ≡ Equations 7.29a and 7.29b (Equations 7.16a and 7.16c ≡ Equations 7.37a–7.37c)], cosecant [Equations 7.31a and 7.31e ≡ Equations 7.32a and 7.32b (Equations 7.21a and 7.21b ≡ Equations 7.38a–7.38c)], tangent [Equations 7.33a and 7.33b ≡ Equations 7.35a and 7.35b (Equations 7.22a–7.22c ≡ Equations 7.39a–7.39c)], and cotangent [Equations 7.34a and 7.34b ≡ Equations 7.36a and 7.36b (Equations 7.17a and 7.17c ≡ Equations 7.40a–7.40c)]:

$$|z| < \frac{\pi}{2}: \quad \operatorname{sech} z = E_0 + E_2\frac{z^2}{2} + E_4\frac{z^4}{24} + O(z^5) = 1 - \frac{z^2}{2} + \frac{5z^4}{24} + O(z^6), \tag{7.37a–7.37c}$$

$$0 < |z| < \pi: \quad \operatorname{csch} z = \frac{B_0}{z} - B_2 z - \frac{7}{12}B_4 z^3 + O(z^5) = \frac{1}{z} - \frac{z}{6} + \frac{7z^3}{360} + O(z^5), \tag{7.38a–7.38c}$$

$$|z| < \pi: \quad \tanh z = 6B_2 z + 10B_4 z^3 + \frac{84}{15}B_6 z^5 + O(z^7) = z - \frac{z^3}{3} + \frac{2z^5}{15} + O(z^7), \tag{7.39a–7.39c}$$

$$0 < |z| < \pi: \quad \coth z = \frac{B_0}{z} + 2B_2 z + \frac{2}{3}B_4 z^3 + O)(z^5) = \frac{1}{z} + \frac{z}{3} - \frac{z^3}{45} + O(z^5), \tag{7.40a–7.40c}$$

follow from each other using (1) the Euler (Equations 7.29c through 7.29e) and Bernoulli (Equations 7.32c through 7.32e and 7.36c) numbers and (2) the imaginary change of variable (Equations 5.24a, 5.24b, 5.25a, 5.25b, 5.26a, 5.26b, 5.27a, and 5.27b), respectively.

7.1.7 Generation of Euler (1755) Numbers

The power series for the primary functions (Equations 7.14a through 7.14c and 7.15a through 7.15c) have elementary coefficients because their nth derivatives are simple (as shown in Table 7.1). The secondary functions, for example, the circular secant (Equation 7.16b) [cotangent (Equation 7.17b)] have power series with nonelementary coefficients because their derivates became increasingly intricate beyond the first order, for example [Equation 7.41 (Equation 7.42)]:

$$\frac{d(\sec z)}{dz} = \frac{d}{dz}\left(\frac{1}{\cos z}\right) = \frac{\sin z}{\cos^2 z} = \sec z \tan z, \tag{7.41}$$

$$\frac{d(\cot z)}{dz} = \frac{d}{dz}\left(\frac{\cos z}{\sin z}\right) = -1 - \frac{\cos^2 z}{\sin^2 z} = -\frac{1}{\sin^2 z} = -\csc^2 z; \tag{7.42}$$

thus, in order to calculate the coefficients, that is, the Euler (Bernoulli) numbers, it is preferable to regard Equation 7.16c (Equation 7.17c) as **generating functions** from which recurrence formulas can be deduced. The Euler numbers are defined by Equation 7.16c that can be rewritten:

$$|z| < \frac{\pi}{2}: \quad 1 = \sum_{n=0}^{\infty} \frac{E_{2n}}{(2n)!} z^{2n} \cosh z = \sum_{n,m=0}^{\infty} \frac{E_{2n}}{(2n)!} \frac{z^{2n+2m}}{(2m)!} = \sum_{k=0}^{\infty} z^{2k} \sum_{n=0}^{k} \frac{E_{2n}}{(2n)!(2k-2n)!}, \tag{7.43}$$

where the Cauchy rule (Equation I.21.25a) for the product of series (Equation 7.6b) was used. Equating coefficients of powers of z in Equation 7.43 leads to the recurrence formula:

$$E_0 = 1, \quad \sum_{n=0}^{k} \frac{E_{2n}}{(2n)!(2k-2n)!} = 0, \tag{7.44a and 7.44b}$$

that allows successive Euler numbers to be calculated recursively, starting with $E_0 = 1$, and setting $k = 1,2,\ldots,$ in Equation 7.44b. Thus, *the Euler numbers E_{2n} have the generating function Equation 7.16c and satisfy the recurrence formulas 7.44a and 7.44b; the first 10 Euler numbers are indicated in Table 7.2.* The recurrence formula (Equations 7.44a and 7.44b) for $k = 1, 2, 3$

$$0 = \frac{E_0 + E_2}{2} = \frac{E_0 + E_4}{4!} + \frac{E_2}{2!2!} = \frac{E_0 + E_6}{6!} + \frac{E_2 + E_4}{2!4!} \tag{7.45a–7.45c}$$

specifies the three Euler numbers beyond (Equation 7.44a):

$$E_2 = -E_0 = -1, \quad E_4 = -E_0 - \frac{4!}{2!2!} E_2 = -1 - 6E_2 = 5, \tag{7.46a and 7.46b}$$

$$E_6 = -E_0 - \frac{6!}{2!4!}\left(E_2 + E_4\right) = -1 - 15\left(E_2 + E_4\right) = -61, \tag{7.46c}$$

in agreement with Equations 7.29c through 7.29e and Table 7.2.

7.1.8 Generation of Bernoulli (1713) Numbers

The Bernoulli numbers have the generating function (Equation 7.17c):

$$|z| < \pi: \quad \sum_{n=0}^{\infty} B_{2n} \frac{z^{2n}}{(2n)!} = \frac{z}{2}\coth\left(\frac{z}{2}\right) = \frac{z}{2}\frac{1+e^{-z}}{1-e^{-z}} = \frac{z}{1-e^{-z}} - \frac{z}{2}, \tag{7.47a and 7.47b}$$

that leads to the identity:

$$|z| < \pi: \quad 1 + \frac{e^{-z} - 1}{2} = \frac{1 - e^{-z}}{z} \sum_{n=0}^{\infty} \frac{B_{2n}}{(2n)!} z^{2n}; \tag{7.48a and 7.48b}$$

equating powers of z in Equations 7.48a and 7.48b ≡ Equations 7.49a and 7.49b:

$$\begin{aligned} k = m + 2n: \quad 1 + \frac{1}{2}\sum_{k=1}^{\infty} \frac{(-)^k}{k!} z^k &= \sum_{n,m=0}^{\infty} \frac{(-)^m}{(m+1)!} z^m \frac{B_{2n}}{(2n)!} z^{2n} \\ &= \sum_{k=0}^{\infty} z^k \sum_{n=0}^{\leq k/2} \frac{(-)^{k-2n}}{(k-2n+1)!} \frac{B_{2n}}{(2n)!}, \end{aligned} \tag{7.49a and 7.49b}$$

and replacing k by $2k$ leads to the recurrence formula:

$$B_0 = 1, \quad \frac{1}{2} = \sum_{n=0}^{k} \frac{(2k)!}{(2k-2n+1)!} \frac{B_{2n}}{(2n)!}. \tag{7.50a and 7.50b}$$

Thus, *the Bernoulli numbers B_{2n} have the generating function (Equations 7.47a and 7.47b ≡ Equations 7.48a and 7.48b) and satisfy the recurrence formulas (Equations 7.50a and 7.50b). The first 10 Bernoulli numbers appear in Table 7.2.* The recurrence formula 7.50b with $k = 1, 2, 3$:

$$\frac{1}{2} = \frac{B_0}{3} + B_2 = \frac{B_0}{5} + 2B_2 + B_4 = \frac{B_0}{7} + 3B_2 + 5B_4 + B_6 \tag{7.51a–7.51c}$$

specifies the first three Bernoulli numbers beyond (Equation 7.50a):

$$B_2 = \frac{1}{2} - \frac{1}{3} = \frac{1}{6}, \quad B_4 = \frac{1}{2} - \frac{B_0}{5} - 2B_2 = \frac{1}{2} - \frac{1}{5} - \frac{1}{3} = -\frac{1}{30}, \tag{7.52a and 7.52b}$$

$$B_6 = \frac{1}{2} - \frac{B_0}{7} - 3B_2 - 5B_4 = \frac{1}{2} - \frac{1}{7} - \frac{1}{2} + \frac{1}{6} = \frac{1}{42}, \tag{7.52c}$$

in agreement with Equations 7.32c through 7.32e and 7.36a and Table 7.2.

7.1.9 Some Power Series Involving Bernoulli Numbers

The Bernoulli numbers appear in other power series that can be obtained from the preceding equations by integration, for example, from [Equations 7.22a and 7.22c) ≡ Equations 7.53a and 7.53b (Equations 7.17a and 7.17b ≡ Equations 7.54a and 7.54b)]:

$$|z| < \frac{\pi}{2}: \quad \frac{d}{dz}\left[\log(\cos z)\right] = -\frac{\sin z}{\cos z} = -\tan z = \sum_{n=1}^{\infty} \frac{(-)^n (2^{2n} - 1) 2^{2n}}{(2n)!} B_{2n} z^{2n-1}, \tag{7.53a and 7.53b}$$

$$|z| < \pi: \quad \frac{d}{dz}\left[\log(\sin z)\right] = \frac{\cos z}{\sin z} = \cot z = \frac{1}{z} + \sum_{n=1}^{\infty} \frac{(-)^n 2^{2n}}{(2n)!} B_{2n} z^{2n-1}; \tag{7.54a and 7.54b}$$

integrating Equations 7.53a and 7.53b between 0 and z leads to (Equations 7.55a and 7.55b)

$$|z| < \frac{\pi}{2}: \quad \log(\cos z) = \sum_{n=1}^{\infty} \frac{(-)^n 2^{2n-1} (2^{2n} - 1)}{(2n)!\,n} B_{2n} z^{2n}, \tag{7.55a and 7.55b}$$

$$|z| < \frac{\pi}{2}: \quad \log(\cosh z) = \sum_{n=1}^{\infty} \frac{2^{n-1} (2^{2n} - 1)}{(2n)!\,n} B_{2n} z^{2n}, \tag{7.56a and 7.56b}$$

and the analogous (Equations 5.5a and 5.6a) result (Equations 7.56a and 7.56b). Rewriting Equation 7.54b in the form Equation 7.57b

$$\lim_{z \to 0} \frac{\sin z}{z} = 1: \quad \sum_{n=1}^{\infty} \frac{(-)^n 2^{2n}}{(2n)!} B_{2n} z^{2n-1} = \frac{d}{dz}\left[\log(\sin z) - \log z\right] = \frac{d}{dz}\left[\log\left(\frac{\sin z}{z}\right)\right], \tag{7.57a and 7.57b}$$

and using Equation 7.57a in the integration from 0 to z leads to

$$|z| < \pi: \quad \log\left(\frac{\sin z}{z}\right) = \sum_{n=1}^{\infty} \frac{(-)^n 2^{2n-1}}{(2n)!n} B_{2n} z^{2n}, \tag{7.58a and 7.58b}$$

$$|z| < \pi: \quad \log\left(\frac{\sinh z}{z}\right) = \sum_{n=1}^{\infty} \frac{2^{2n-1}}{(2n)!n} B_{2n} z^{2n}. \tag{7.59a and 7.59b}$$

From Equations 7.55a, 7.55b, 7.58a, and 7.58b follow Equations 7.60a and 7.60b:

$$\begin{aligned} |z| < \pi: \quad \log\left(\frac{\tan z}{z}\right) &= \log\left(\frac{\sin z}{z}\right) - \log(\cos z) \\ &= -\log(z \cot z) = \sum_{n=1}^{\infty} \frac{(-)^n 2^{2n}(1-2^{2n-1})}{(2n)!n} B_{2n} z^{2n}, \end{aligned} \tag{7.60a and 7.60b}$$

$$|z| < \frac{\pi}{2}: \quad \log\left(\frac{\tanh z}{z}\right) = -\log(z \coth z) = \sum_{n=1}^{\infty} \frac{2^{2n}(1-2^{2n-1})}{(2n)!n} B_{2n} z^{2n}. \tag{7.61a and 7.61b}$$

Thus, *the Bernoulli numbers (Equations 7.50a and 7.52a through 7.52c ≡ Equations 7.32c through 7.32e and 7.36c) appear (Table 7.2) in the power series for the logarithms of circular (hyperbolic) cosine [Equations 7.55a and 7.55b (Equations 7.56a and 7.56b)], sine [Equations 7.58a and 7.58b (Equations 7.59a and 7.59b)], and tangent and cotangent [Equations 7.60a and 7.60b (Equations 7.61a and 7.61b)].*

7.2 Branch Points and Branch Cuts for Cyclometric Functions

Since the circular (hyperbolic) functions are periodic with period $2\pi(2\pi i)$, the inverse or cyclometric functions (Subsection 7.2.1) are many-valued like the logarithm (Subsection 7.2.2) and become single-valued choosing the principal branch in the complex plane with suitable branch cuts [Subsection 7.2.4 (Subsection 7.2.3)].

7.2.1 Cyclometric Functions Expressed as Logarithms

The direct functions are combinations of exponentials, and thus the cyclometric functions should be expressible in terms of logarithms. The argument of the logarithm specifies the branch cuts to be made in the complex plane to render the cyclometric function single-valued by taking its principal branch; the cyclometric functions are generally many-valued because they are inverses of circular (hyperbolic) functions that are periodic with period 2π $(2\pi i)$. The process of construction of cyclometric functions has already been considered in connection with branch points and branch cuts (Chapter I.7) for the arc sin

(Equations 7.67a and 7.67b ≡ Equations I.7.23a and I.7.26b), arc csc (Equations 7.71a and 7.71b ≡ Equations I.7.27a and I.7.28), and arg cosh (Equations 7.66a and 7.66b ≡ Equations I.7.32a and I.7.36b). As a further example, the arc tan is considered:

$$w = \arctan z: \quad iz = i \tan w = \frac{e^{iw} - e^{-iw}}{e^{iw} + e^{-iw}} = \frac{e^{2iw} - 1}{e^{2iw} + 1}; \qquad \text{(7.62a and 7.62b)}$$

the solution of Equation 7.62b ≡ Equation 7.63a is

$$e^{2iw} = \frac{1 + iz}{1 - iz} = \frac{i - z}{i + z}, \quad w = \frac{i}{2} \log\left(\frac{i + z}{i - z}\right), \qquad \text{(7.63a and 7.63b)}$$

showing that (1) the arc tan is the logarithm (Equation 7.63b ≡ Equation 7.73b) of a rational function; (2) the principal branch of arc tan corresponds to choosing $n = 0$ in Equation 7.64b:

$$n \in |N: \quad \arctan z = \frac{i}{2} \log\left(\frac{i + z}{i - z}\right) + i2\pi n; \qquad \text{(7.64a and 7.64b)}$$

(3) the arc tan is a many-valued function (Equation 7.64b) that becomes single-valued in the complex z-plane by choosing the principal branch $n = 0$ and making branch cuts such that $(i + z)/(i - z)$ is not a negative real number; (4) the last condition excludes imaginary $z = iy$ with modulus larger than unity $|y| > 1$, as indicated in Figure 7.2b and in Equation 7.73a; and (5) the branch points $z \equiv \pm i$ are included in the branch cut because $\tan(\pm i) = \infty$ is singular.

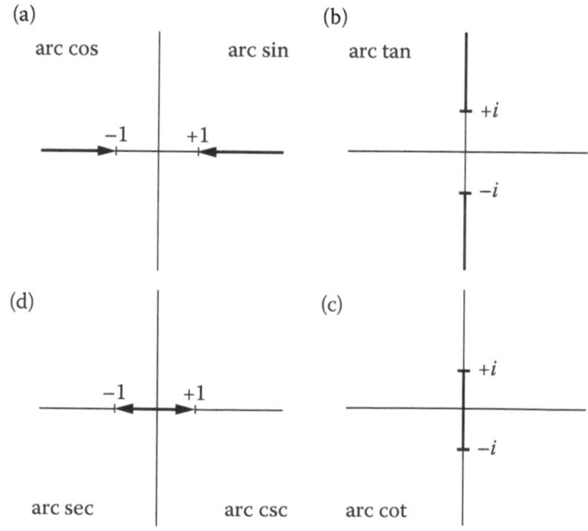

FIGURE 7.2
Exponentials with imaginary argument (circular functions) have period 2π, and thus their inverses, namely, the logarithm and circular cyclometric functions, are many-valued. They become single-valued by choosing the principal branch in the complex plane with branch cuts for (a) inverse circular sine and cosine; (b) inverse circular tangent; (c) inverse circular secant and cosecant; and (d) inverse circular cotangent.

7.2.2 Principal Branches in Complex-Cut Plane

The representations of the remaining cyclometric functions as logarithms are included in

$$z \neq)-\infty,-1(\cup)+1,+\infty(: \quad \arccos z = -i\log\left\{z + i\sqrt{1-z^2}\right\}, \tag{7.65a and 7.65b}$$

$$z \neq)-\infty,1(: \quad \arg\cosh z = \log\left\{z + \sqrt{z^2-1}\right\}, \tag{7.66a and 7.66b}$$

$$z \neq)-\infty,-1(\cup)+1,+\infty(: \quad \arcsin z = -i\log\left\{iz + \sqrt{1-z^2}\right\}, \tag{7.67a and 7.67b}$$

$$z \neq)-i\infty,-i(\cup)+i,+i\infty(: \quad \arg\sinh z = \log\left\{z + \sqrt{z^2+1}\right\}, \tag{7.68a and 7.68b}$$

$$z \notin)-1,+1(: \quad \operatorname{arc}\sec z = -i\log\left\{\frac{1}{z} + i\sqrt{1-\frac{1}{z^2}}\right\}, \tag{7.69a and 7.69b}$$

$$z \notin)-\infty,0)\cup)+1,+\infty(: \quad \arg\operatorname{sech} z = \log\left\{\frac{1}{z} + \sqrt{\frac{1}{z^2}-1}\right\}, \tag{7.70a and 7.70b}$$

$$z \notin)-1,+1(: \quad \operatorname{arc}\csc z = -i\log\left\{\frac{i}{z} + \sqrt{1-\frac{1}{z^2}}\right\}, \tag{7.71a and 7.71b}$$

$$z \notin)-i,+i(: \quad \arg\operatorname{csch} z = \log\left\{\frac{1}{z} + \sqrt{\frac{1}{z^2}+1}\right\}, \tag{7.72a and 7.72b}$$

$$z \notin)-i\infty,-i)\cup(+i,+i\infty(: \quad \arctan z = \frac{i}{2}\log\left(\frac{i+z}{i-z}\right), \tag{7.73a and 7.73b}$$

$$z \notin)-\infty,-1)\cup(+1,+\infty(: \quad \arg\tanh z = \frac{1}{2}\log\left(\frac{1+z}{1-z}\right), \tag{7.74a and 7.74b}$$

$$z \notin (-i,+i): \quad \operatorname{arc}\cot z = \frac{i}{2}\log\left(\frac{z-i}{z+i}\right), \tag{7.75a and 7.75b}$$

$$z \notin (-1,+1): \quad \arg\coth z = \frac{1}{2}\log\left(\frac{z+1}{z-1}\right). \tag{7.76a and 7.76b}$$

Thus, *the inverse circular and hyperbolic, that is, cyclometric functions, are single-valued if identified with the principal branch of the logarithm of the algebraic expressions (Equations 7.65b through 7.76b), implying that branch cuts (Equations 7.65a through 7.76a) be made in the complex z-plane, joining the branch points between themselves or to infinity, as shown in Figures 7.2a through d and 7.3a through f. The branch cuts include (do not include) the branch points that are excluded (allowed) values for the functions, in the case of the tangent and cotangent (sine, cosine, secant, and cosecant) in Figures 7.2b and c and 7.3e and f (Figures 7.2a and b and 7.3a through d); this is indicated by the blunt (sharp) end at the branch point of the thick line representing the branch cut.*

7.2.3 Branch Cuts Connecting Branch Points

The starting point for the proofs is the algebraic relation between the hyperbolic (or circular) functions and the exponential (Section 5.1):

$$w = \arg\cosh, \sinh z: \quad z = \cosh, \sinh w = \frac{e^{w} \pm e^{-w}}{2}, \tag{7.77a and 7.77b}$$

$$w = \arg\operatorname{sech}, \operatorname{csch} z: \quad z = \operatorname{sech}, \operatorname{csch} w = \frac{2}{e^{w} \pm e^{-w}}, \tag{7.78a and 7.78b}$$

$$w = \arg\tanh z: \quad z = \tanh w = \frac{e^{w} - e^{-w}}{e^{w} + e^{-w}} = \frac{e^{2w} - 1}{e^{2w} + 1}, \tag{7.79a and 7.79b}$$

$$w = \arg\tanh z: \quad z = \coth w = \frac{e^{w} + e^{-w}}{e^{w} - e^{-w}} = \frac{e^{2w} + 1}{e^{2w} - 1}. \tag{7.80a and 7.80b}$$

The algebraic relations 7.77b ≡ 7.81b, 7.78b ≡ 7.82b, 7.79b, and 7.80b are solved for w, respectively, in Equations 7.81c, 7.82c, 7.83b and 7.83d:

$$w = \arg\cosh, \sinh z: \quad e^{2w} - 2ze^{w} \pm 1 = 0 \quad \Rightarrow \quad e^{w} = z \pm \sqrt{z^2 \mp 1}, \tag{7.81a–7.81c}$$

$$w = \arg\operatorname{sech}, \operatorname{csch} z: \quad e^{2w} - \frac{2}{z}e^{w} \pm 1 = 0 \quad \Rightarrow \quad e^{w} = \frac{1}{z} \pm \sqrt{\frac{1}{z^2} \mp 1}, \tag{7.82a–7.82c}$$

$$w = \arg\tanh z: \quad e^{2w} = \frac{1+z}{1-z}; \quad w = \arg\coth z: \quad e^{2w} = \frac{z+1}{z-1}. \tag{7.83a–7.83d}$$

Applying logarithms to Equation 7.83b (Equation 7.83d) yields Equation 7.74b (Equation 7.76b); also applying logarithms to Equation 7.81c (Equation 7.82c) and choosing the + sign before the radical lead to Equations 7.66b and 7.68b (Equations 7.70b and 7.72b). The branch cuts exclude negative real values of the argument of the logarithms; for Equations 7.83b and 7.83d, this implies (Equations 7.84a through 7.84e):

$$x > 0: -x \neq \frac{1+z}{1-z}, \frac{z+1}{z-1}, z \neq \frac{x+1}{x-1}, \frac{x-1}{x+1}: \quad \operatorname{Im}(z) \neq 0 \text{ or } |\operatorname{Re}(z)| >, < 1, \tag{7.84a–7.84h}$$

which means that the branch cut(s) for arg tanh (arg coth) lies on the real axis (Equation 7.84f) outside (inside) the unit interval [Figure 7.3f (Figure 7.3e)] including the branch points $z = \pm 1$ where the function is singular arg tanh $(\pm 1) = \pm\infty =$ arg coth (± 1). For Equation 7.81c (Equation 7.82c), the argument of the logarithm never vanishes as can be seen from Equation 7.85a (Equation 7.85b):

$$z^2 \neq z^2 \mp 1, \frac{1}{z^2} \neq \frac{1}{z^2} \mp 1; \quad x > 0: \quad -x \neq z^2 \mp 1, \frac{1}{z^2} \mp 1, \qquad (7.85a\text{–}7.85c)$$

The branch cuts are thus specified by nonnegative real values of the arguments of the radicals (Equations 7.85c through 7.85e), leading to the following: (1) for arg cosh (Equation 7.81c) the radical (Equation 7.85d) must have both factors nonnegative (Equation 7.85e) excluding real z with $z < 1$ as indicated in Equation 7.66a and Figure 7.3b:

$$\sqrt{z^2 - 1} = \sqrt{z-1}\sqrt{z+1}: \quad -x \neq z \pm 1; \quad \sqrt{\frac{1}{z^2} - 1} = \sqrt{\frac{1}{z} - 1}\sqrt{\frac{1}{z} + 1}: \quad -x \neq \frac{1}{z} \pm 1, \qquad (7.85d\text{–}7.85g)$$

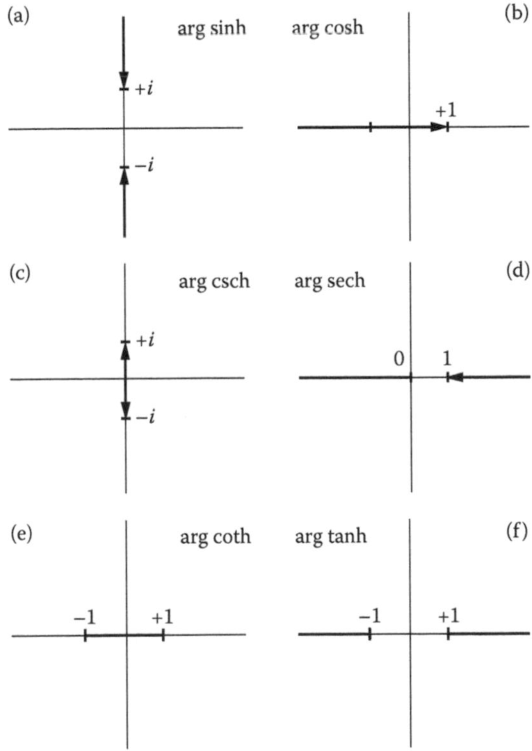

FIGURE 7.3
Exponentials with real argument (hyperbolic functions) have period $2\pi i$ and thus their inverses, namely, the logarithm (hyperbolic cyclometric) functions, are many-valued. They become single-valued choosing the principal branch in the complex plane with branch cuts for (a) inverse hyperbolic sine; (b) inverse hyperbolic cosine; (c) inverse hyperbolic cosecant; (d) inverse hyperbolic secant; (e) inverse hyperbolic cotangent; and (f) inverse hyperbolic tangent. In Figures 7.2 and 7.3, branch cuts include (exclude) the branch points if the function is singular (regular) there, corresponding to a closed (open) integral at one end represented by a slash (arrow).

(2) for arc sinh then $z^2 + 1$ must be nonnegative excluding imaginary z with $|z| > 1$, as indicated in Equation 7.68a and Figure 7.3a; (3) for arc sech (Equation 7.82c) the radical (Equation 7.85f) must have both factors nonnegative (Equation 7.85g) excluding real z outside the unit interval (Equation 7.70a) as shown in Figure 7.3d; (4) for arc csch then $1/z^2 + 1$ must be nonnegative excluding imaginary z with $|z| < 1$ as indicated in Equation 7.72a and Figure 7.3c. In all cases (1 through 4), the branch cuts do not include the branch points since the functions are finite there, that is, arg cosh $(1) = 0$, arg sinh $(\pm i) = \log(\pm i) = \pm i\pi/2$, arg sech $(1) = 0$, and arg csch $(\pm i) = \log(\mp i) = \mp i\pi/2$; the only exception is arg sech $(0) = \pm\infty$, so that the origin is included in the lower semi-infinite branch cut for arg sech in Equation 7.70a and in Figure 7.3d. This completes the proof of the expressions of the cyclometric hyperbolic functions as logarithms and the associated branch cuts in Equations 7.66a, 7.66b, 7.68a, 7.68b, 7.70a, 7.70b, 7.72a, 7.72b, 7.74ab, 7.76a, and 7.76b and in Figure 7.3a through f.

7.2.4 Branch Cuts Including/Excluding Branch Points

The branch cuts exclude (include) the branch points if the cyclometric functions are regular (singular) there for the hyperbolic (circular) case [Subsection 7.2.3 (Subsection 7.2.4)]. Of the cyclometric circular functions, the arc tan has already been considered (Equations 7.62 through 7.64), so that five remain:

$$w = \operatorname{arc}\cos z, \quad z = \cos w = \frac{e^{iw} + e^{-iw}}{2}, \qquad \text{(7.86a and 7.86b)}$$

$$w = \operatorname{arc}\sin z, \quad z = \sin w = \frac{e^{iw} + e^{-iw}}{2i}, \qquad \text{(7.87a and 7.87b)}$$

$$w = \operatorname{arc}\sec z, \quad z = \sec w = \frac{2}{e^{iw} + e^{-iw}}, \qquad \text{(7.88a and 7.88b)}$$

$$w = \operatorname{arc}\csc z, \quad z = \csc w = \frac{2i}{e^{iw} - e^{-iw}}, \qquad \text{(7.89a and 7.89b)}$$

$$w = \operatorname{arc}\cot z, \quad z = \cot w = i\frac{e^{iw} + e^{-iw}}{e^{iw} - e^{-iw}} = i\frac{e^{2iw} + 1}{e^{2iw} - 1}. \qquad \text{(7.90a and 7.90b)}$$

The algebraic relations 7.86b ≡ 7.91b, 7.87b ≡ 7.92b, 7.88b ≡ 7.93b, 7.89b ≡ 7.94b, and 7.90b ≡ 7.95b are solved for w, respectively, in Equations 7.91c, 7.92c, 7.93c, 7.94c, and 7.95c:

$$w = \operatorname{arc}\cos z: \quad e^{2iw} - 2ze^{iw} + 1 = 0, \quad e^{iw} = z \pm i\sqrt{1 - z^2}, \qquad \text{(7.91a–7.91c)}$$

$$w = \operatorname{arc}\sin z: \quad e^{2iw} - 2ize^{iw} - 1 = 0, \quad e^{iw} = iz \pm \sqrt{1 - z^2}, \qquad \text{(7.92a–7.92c)}$$

$$w = \operatorname{arc}\sec z: \quad e^{2iw} - \frac{2}{z}e^{iw} + 1 = 0, \quad e^{iw} = \frac{1}{z} \pm i\sqrt{1 - \frac{1}{z^2}}, \qquad \text{(7.93a–7.93c)}$$

$$w = \text{arc csc } z: \quad e^{2iw} - \frac{2i}{z} e^{iw} - 1 = 0, \quad e^{iw} = \frac{i}{z} \pm \sqrt{1 - \frac{1}{z^2}}, \tag{7.94a–7.94c}$$

$$w = \text{arc cot } z: \quad e^{2iw} = \frac{z+i}{z-i}, \quad w = \frac{i}{2} \log\left(\frac{z-i}{z+i}\right). \tag{7.95a–7.95c}$$

Applying logarithms to Equations 7.91c, 7.92c, 7.93c, and 7.94c and choosing the plus sign before the radical lead, respectively, to Equations 7.65b, 7.67b, 7.69b, and 7.71b; also Equation 7.95c ≡ Equation 7.75b. Concerning the arc cot (Equation 7.95c ≡ Equation 7.75b, the branch cut excludes negative real values of the argument (Equations 7.96a and 7.96b) that imply that z cannot be (Equation 7.96c) imaginary with modulus less than unity (Equations 7.96d and 7.96e):

$$x > 0: \quad -x \neq \frac{z-i}{z+i}, \quad z \neq i\frac{1-x}{1+x}, \quad \text{Re}(z) \neq 0 \ or \ |z| < 1, \tag{7.96a–7.96c}$$

as indicated in Equation 7.75a and Figure 7.2c; the branch points $z = \pm i$ are included in the branch cut because the function (Equation 7.95a ≡ Equation 7.95b) is singular there. Concerning the arc cos, sin, sec, and csc in Equations 7.65b, 7.67b, 7.69b, and 7.71b, the arguments of the logarithm are never zero as follows from

$$z^2 - 1 \neq z^2, \quad 1 - z^2 \neq (iz)^2 = -z^2, \quad \frac{1}{z^2} \neq \frac{1}{z^2} - 1, \quad 1 - \frac{1}{z^2} \neq \left(\frac{i}{z}\right)^2 = -\frac{1}{z^2}. \tag{7.97a–7.97d}$$

Thus, the branch cuts correspond to negative real values of the radicals: (1 and 2) for the arc cos (sin) in Equation 7.65b (Equation 7.67b), nonnegative real $1 - z^2$ excludes real z larger in modulus than unity, as stated in Equation 7.65a ≡ Equation 7.67a and in Figure 7.2a; (3 and 4) for the arc sec (csc) in Equation 7.69b (Equation 7.71b), nonnegative real $1 - 1/z^2$ excludes real z smaller than unity in modulus, as indicated in Equation 7.69a ≡ Equation 7.71a and in Figure 7.2d. In all cases 1 through 4, the functions are finite at the branch points $z = \pm 1$, that is, arc cos $(\pm 1) = 0$, $\pi =$ arc sec (± 1), and arc sin $(\pm 1) = \pm\pi/2 =$ arc csc (± 1), so they are excluded from the branch cuts. This completes the proof of the expressions of the cyclometric circular as logarithms and the associated branch cuts in Equations 7.65a, 7.65b, 7.67a, 7.67b, 7.69a, 7.69b, 7.71a, 7.71b, 7.73a, 7.73b, 7.75a, and 7.75b and in Figure 7.2a through d.

7.3 Derivates/Primitives of Direct/Inverse Functions

The derivatives of the direct functions, that is, primary and secondary circular and hyperbolic functions, follow (Subsection 7.3.1) from those of the cosine and sine; the primitives require in some cases (Subsection 7.3.3) intermediate algebraic transformations. The derivatives of cyclometric functions (Subsection 7.3.2) use the derivatives of the direct function; the primitives of the cyclometric functions may require (Subsection 7.3.4) integration by parts or use of the logarithmic representation. The 24 derivates and primitives of all six direct and inverse circular and hyperbolic functions are listed in Table 7.3. The primitive of the inverse circular secant leads to the Guderman function (Subsection 7.3.5).

TABLE 7.3

48 Derivatives and Primitives of the 24 Circular and Hyperbolic Direct and Inverse Functions

Primitive	Direct Function	Derivative		Inverse Function (Cyclometric)	Primitive
$\int w(z)\,dz$	$w(z)$	$\frac{dw}{dz}$	$\frac{dz}{dw}$	$z(w)$	$\int z(w)\,dw$
$-\cos z$	$\sin z$	$\cos z$	$\frac{1}{\sqrt{1-w^2}}$	$\text{arc sin } w$	$w \text{ arc sin } w + \sqrt{1-w^2}$
$\cosh z$	$\sinh z$	$\cosh z$	$\frac{1}{\sqrt{1+w^2}}$	$\text{arg sinh } w$	$w \text{ arc sinh } w + \sqrt{1+w^2}$
$\sin z$	$\cos z$	$-\sin z$	$-\frac{1}{\sqrt{1-w^2}}$	$\text{arc cos } w$	$w \text{ arc cos } w - \sqrt{1-w^2}$
$\sinh z$	$\cosh z$	$\sinh z$	$\frac{1}{\sqrt{w^2-1}}$	$\text{arg cosh } w$	$w \text{ arg cosh } w - \sqrt{w^2-1}$
$-\log(\cos z)$	$\tan z$	$\sec^2 z$	$\frac{1}{1+w^2}$	$\text{arc tan } w$	$w \text{ arc tan } w - \frac{1}{2}\log(1+w^2)$
$\log(\cosh z)$	$\tanh z$	$\text{sech}^2 z$	$\frac{1}{1-w^2}$	$\text{arg tanh } w$	$w \text{ arg tanh } w + \frac{1}{2}\log(1-w^2)$
$\log(\sin z)$	$\cot z$	$-\csc^2 z$	$-\frac{1}{1+w^2}$	$\text{arc cot } w$	$w \text{ arc cot } w + \frac{1}{2}\log(1+w^2)$
$\log(\sinh z)$	$\coth z$	$-\text{csch}^2 z$	$\frac{1}{1-w^2}$	$\text{arg coth } w$	$w \text{ arg coth } w + \frac{1}{2}\log(w^2-1)$
$\log(\sec z+\tan z)$ (4)	$\sec z$	$\sec z \tan z$	$\frac{1}{w\sqrt{w^2-1}}$	$\text{arc sec } w$	$w \text{ arc sec } w \mp w\log\left(w+\sqrt{w^2-1}\right)$ (2)
$\text{arc tan}(\sinh z)$	$\text{sech } z$	$-\text{sech } z \tanh z$	$\mp\frac{1}{w\sqrt{1-w^2}}$ (1)	$\text{arg sech } w$	$w \text{ arg sech } w \pm \text{arc sin } w$ (1)
$\log(\csc z - \cot z)$(5)	$\csc z$	$-\csc z \cot z$	$-\frac{1}{w\sqrt{w^2-1}}$	$\text{arc csc } w$	$w \text{ arc csc } w \pm \log\left(w+\sqrt{w^2-1}\right)$ (3)
$\log\left[\tanh\left(\frac{z}{2}\right)\right]$	$\text{csch } z$	$-\text{csch } z \coth z$	$\mp\frac{1}{w\sqrt{1+w^2}}$ (1)	$\text{arg csch } w$	$w \text{ arg csch } w \pm \text{arg sinh } w$ (1)

Note: Some of the $6 \times 2 \times 2 \times 2 = 48$ derivatives and primitives have multiple expressions. (1) Signal: – for $\text{Re}(w) > 0$, + for $\text{Re}(w) < 0$; (2) signal: – for $0 < \text{arc sec } z < \pi/2$; + for $\pi/2 < \text{arc sec } z < \pi$; (3) signal: + for $0 < \text{arc csc } z < \pi/2$, – for $-\pi/2 < \text{arc csc } z < 0$; (4) see also Equation 7.139; (5) see also Equations 7.136a and 7.138b.

7.3.1 Derivatives of Direct Circular and Hyperbolic Functions

The direct circular (hyperbolic) functions have derivatives

$$(\sin z)' = \cos z, \quad (\sinh z)' = \cosh z, \tag{7.98a and 7.98b}$$

$$(\cos z)' = -\sin z, \quad (\cosh z)' = \sinh z, \tag{7.99a and 7.99b}$$

$$(\tan z)' = \sec^2 z, \quad (\tanh z)' = \text{sech}^2 z, \tag{7.100a and 7.100b}$$

$$(\cot z)' = -\csc^2 z, \quad (\coth z)' = -\text{csch}^2 z, \tag{7.101a and 7.101b}$$

$$(\csc z)' = -\csc z \cot z, \quad (\text{csch } z)' = -\text{csch } z \coth z, \tag{7.102a and 7.102b}$$

$$(\sec z)' = \sec z \tan z, \quad (\text{sech } z)' = -\text{sech } z \tanh z, \tag{7.103a and 7.103b}$$

namely, for sine (Equations 7.98a and 7.98b), cosine (Equations 7.99a and 7.99b), tangent (Equations 7.100a and 7.100b), cotangent (Equations 7.101a and 7.101b), cosecant (Equations 7.102a and 7.102b), and secant (Equations 7.103a and 7.103b). The proofs are given as follows: (1) from Equations 7.3b, 7.4b, 7.9b, and 7.10b for the primary functions circular and hyperbolic sine and cosine with $n = 0$ follow, respectively, Equations 7.99b, 7.98b, 7.99a, and 7.98a; (2) from Equation 7.41 (Equation 7.42) for the circular secant (cotangent) in Equation 7.103a (Equation 7.101a) and likewise for the hyperbolic secant (cotangent) in Equation 7.103b ≡ Equation 7.104 (Equation 7.101b ≡ Equation 7.105):

$$\frac{d(\text{sech } z)}{dz} = \frac{d}{dz}\left(\frac{1}{\cosh z}\right) = -\frac{\sinh z}{\cosh^2 z} = -\text{sech } z \tanh z; \tag{7.104}$$

$$\frac{d(\coth z)}{dz} = \frac{d}{dz}\left(\frac{\cosh z}{\sinh z}\right) = \frac{\sinh^2 z - \cosh^2 z}{\sinh^2 z} = -\frac{1}{\sinh^2 z} = -\text{csch}^2 z; \tag{7.105}$$

(3) the derivatives of the circular (hyperbolic) tangent are [Equation 7.106 ≡ Equation 7.100a (Equation 7.107 ≡ Equation 7.100b)]

$$\frac{d(\tan z)}{dz} = \frac{d}{dz}\left(\frac{\sin z}{\cos z}\right) = \frac{\cos^2 z + \sin^2 z}{\cos^2 z} = \frac{1}{\cos^2 z} = \sec^2 z, \tag{7.106}$$

$$\frac{d(\tanh z)}{dz} = \frac{d}{dz}\left(\frac{\sinh z}{\cosh z}\right) = \frac{\cosh^2 z - \sinh^2 z}{\cosh^2 z} = \frac{1}{\cosh^2 z} = \text{sech}^2 z; \tag{7.107}$$

and (4) the derivatives of the circular (hyperbolic) cosecant are [Equation 7.108 ≡ Equation 7.102a (Equation 7.109) ≡ Equation 7.102b)]

$$\frac{d(\csc z)}{dz} = \frac{d}{dz}\left(\frac{1}{\sin z}\right) = -\frac{\cos z}{\sin^2 z} = -\csc z \cot z, \tag{7.108}$$

$$\frac{d(\operatorname{csch} z)}{dz} = \frac{d}{dz}\left(\frac{1}{\sinh z}\right) = -\frac{\cosh z}{\sinh^2 z} = -\operatorname{csch} z \coth z. \tag{7.109}$$

This completes the derivatives of the secondary circular and hyperbolic functions (Subsection 7.3.1) to be followed by the derivatives of the cyclometric functions (Subsection 7.3.2).

7.3.2 Derivatives of Cyclometric Functions

The cyclometric functions are the inverse circular and hyperbolic functions, and their derivatives (Equations 7.98a through 7.103b) follow from

$$w = \operatorname{arc}\sin z: \quad (\operatorname{arc}\sin z)' = \frac{dw}{dz} = \left(\frac{dz}{dw}\right)^{-1} = \frac{1}{(\sin w)'} = \frac{1}{\cos w} = \frac{1}{\sqrt{1-\sin^2 w}} = \frac{1}{\sqrt{1-z^2}}, \tag{7.110a and 7.110b}$$

$$w = \operatorname{arc}\cos z: \quad (\operatorname{arc}\cos z)' = \frac{1}{(\cos w)'} = -\frac{1}{\sin w} = -\frac{1}{\sqrt{1-\cos^2 w}} = -\frac{1}{\sqrt{1-z^2}}, \tag{7.111a and 7.111b}$$

$$w = \operatorname{arc}\tan z: \quad (\operatorname{arc}\tan z)' = \frac{1}{(\tan w)'} = \frac{1}{\sec^2 w} = \frac{1}{1+\tan^2 w} = \frac{1}{1+z^2}, \tag{7.112a and 7.112b}$$

$$w = \operatorname{arc}\cot z: \quad (\operatorname{arc}\cot z)' = \frac{1}{(\cot w)'} = -\frac{1}{\csc^2 w} = -\frac{1}{1+\cot^2 w} = -\frac{1}{1+z^2}, \tag{7.113a and 7.113b}$$

$$w = \operatorname{arc}\sec z: \quad (\operatorname{arc}\sec z)' = \frac{1}{(\sec w)'} = \frac{1}{\sec w \tan w} = \frac{1}{z\sqrt{z^2-1}}, \tag{7.114a and 7.114b}$$

$$w = \operatorname{arc}\csc z: \quad (\operatorname{arc}\csc z)' = \frac{1}{(\csc w)'} = -\frac{1}{\csc w \cot w} = -\frac{1}{z\sqrt{z^2-1}}, \tag{7.115a and 7.115b}$$

for the circular cyclometric functions and from

$$w = \operatorname{arg}\sinh z: \quad (\operatorname{arg}\sinh z)' = \frac{1}{(\sinh w)'} = \frac{1}{\cosh w} = \frac{1}{\sqrt{1+\sinh^2 w}} = \frac{1}{\sqrt{1+z^2}}, \tag{7.116a and 7.116b}$$

$$w = \arg\cosh z: \quad (\arg\cosh z)' = \frac{1}{(\cosh w)'} = \frac{1}{\sinh w} = \frac{1}{\sqrt{\cosh^2 w - 1}} = \frac{1}{\sqrt{z^2 - 1}}, \tag{7.117a and 7.117b}$$

$$w = \arg\tanh z: \quad (\arg\tanh z)' = \frac{1}{(\tanh w)'} = \frac{1}{\operatorname{sech}^2 w} = \frac{1}{1 - \tanh^2 w} = \frac{1}{1 - z^2}, \tag{7.118a and 7.118b}$$

$$w = \arg\coth z: \quad (\arg\coth z)' = \frac{1}{(\coth w)'} = -\frac{1}{\operatorname{csch}^2 w} = -\frac{1}{\coth^2 w - 1} = \frac{1}{1 - z^2}, \tag{7.119a and 7.119b}$$

$$w = \arg\operatorname{sech} z: \quad (\arg\operatorname{sech} z)' = \frac{1}{(\operatorname{sech} w)'} = -\frac{1}{\operatorname{sech} w \tanh w} = \mp\frac{1}{z\sqrt{1 - z^2}}, \tag{7.120a and 7.120b}$$

$$w = \arg\operatorname{csch} z: \quad (\arg\operatorname{csch} z)' = \frac{1}{(\operatorname{csch} w)'} = -\frac{1}{\operatorname{csch} w \coth w} = \mp\frac{1}{z\sqrt{1 + z^2}}, \tag{7.121a and 7.121b}$$

for the hyperbolic cyclometric functions.

Thus, *the derivatives of the cyclometric functions are given by*

$$(\arcsin z)' = \frac{1}{\sqrt{1 - z^2}}, \quad (\arg\sinh z)' = \frac{1}{\sqrt{1 + z^2}}, \tag{7.122a and 7.122b}$$

$$(\arccos z)' = -\frac{1}{\sqrt{1 - z^2}}, \quad (\arg\cosh z)' = \frac{1}{\sqrt{z^2 - 1}}, \tag{7.123a and 7.123b}$$

$$(\arctan z)' = \frac{1}{1 + z^2}, \quad (\arg\tanh z)' = \frac{1}{1 - z^2}, \tag{7.124a and 7.124b}$$

$$(\operatorname{arccot} z)' = -\frac{1}{1 + z^2}, \quad (\arg\cot z)' = \frac{1}{1 - z^2}, \tag{7.125a and 7.125b}$$

$$(\operatorname{arc\,sec} z)' = \frac{1}{z\sqrt{z^2 - 1}}; \quad \operatorname{Re}(z) >< 0: \quad (\arg\operatorname{sech} z)' = \mp\frac{1}{z\sqrt{1 - z^2}}, \tag{7.126a–7.126c}$$

$$(\operatorname{arc\,csc} z)' = -\frac{1}{z\sqrt{z^2 - 1}}; \quad \operatorname{Re}(z) >< 0: \quad (\arg\operatorname{csch} z)' = \mp\frac{1}{z\sqrt{1 + z^2}}, \tag{7.127a–7.127c}$$

namely, for the arc (arg) of circular (hyperbolic) sine Equation 7.122a ≡ Equations 7.110a and 7.110b (Equation 7.122b ≡ Equations 7.116a and 7.116b), cosine Equation 7.123a ≡ Equations 7.111a and 7.111b (Equation 7.123b ≡ Equations 7.117a and 7.117b), tangent Equation 7.124a ≡ Equation 7.112a (Equation 7.124b ≡ Equations 7.118a and 7.118b), cotangent Equation 7.125a ≡ Equations 7.113a and 7.113b (Equation 7.125b ≡ Equations 7.119a and 7.119b), secant Equation 7.126a ≡ Equations 7.114a and 7.114b (Equation 7.126b ≡ Equations 7.120a and 7.120b), and cosecant Equation 7.127a ≡ Equations 7.115a and 7.115b (Equation 7.127b ≡ Equations 7.121a and 7.121b). In the derivatives of the inverse hyperbolic secant (Equation 7.126c) [cosecant (Equation 7.127c)], the sign is −/+ in the right-hand/left-hand z-plane, for example, for $z \equiv x$ real, the slope in modulus decreases toward infinity in Figure 5.2e (Figure 5.2d).

7.3.3 Primitives of Primary and Secondary Functions

Replacing derivatives in Equations 7.98 through 7.103 by primitives taken as indefinite integrals leads, respectively, to

$$\int \sin z \, dz = -\cos z, \quad \int \sinh z \, dz = \cosh z, \tag{7.128a and 7.128b}$$

$$\int \cos z \, dz = \sin z, \quad \int \cosh z \, dz = \sinh z, \tag{7.129a and 7.129b}$$

$$\int \tan z \, dz = \int \frac{\sin z}{\cos z} dz = -\int \frac{(\cos z)'}{\cos z} dz = -\log(\cos z), \tag{7.130a}$$

$$\int \tanh z \, dz = \int \frac{\sinh z}{\cosh z} dz = \int \frac{(\cosh z)'}{\cos z} dz = \log(\cosh z), \tag{7.130b}$$

$$\int \cot z \, dz = \int \frac{\cos z}{\sin z} dz = \int \frac{(\sin z)'}{\sin z} dz = \log(\sin z), \tag{7.131a}$$

$$\int \coth z \, dz = \int \frac{\cosh z}{\sinh z} dz = \int \frac{(\sinh z)'}{\sinh z} dz = \log(\sinh z), \tag{7.131b}$$

$$\begin{aligned} \int \csc z \, dz &= \int \frac{dz}{\sin z} = \int \frac{dz}{2\sin(z/2)\cos(z/2)} = \int \frac{\sec^2(z/2)}{\tan(z/2)} \frac{dz}{2} \\ &= \int \frac{[\tan(z/2)]'}{\tan(z/2)} d\left(\frac{z}{2}\right) = \log\left[\tan\left(\frac{z}{2}\right)\right], \end{aligned} \tag{7.132a}$$

$$\int \operatorname{csch} z\, dz = \int \frac{dz}{\sinh z} = \int \frac{dz}{2\sinh(z/2)\cosh(z/2)} = \frac{1}{2}\int \frac{\operatorname{sech}^2(z/2)}{\tanh(z/2)}dz$$
$$= \int \frac{[\tanh(z/2)]'}{\tanh(z/2)} d\left(\frac{z}{2}\right) = \log\left[\tanh\left(\frac{z}{2}\right)\right], \tag{7.132b}$$

$$\int \sec z\, dz = \int \frac{\sec z \tan z + \sec^2 z}{\tan z + \sec z} dz = \int \frac{(\tan z + \sec z)'}{\tan z + \sec z} dz = \log(\tan z + \sec z), \tag{7.133a}$$

$$\int \operatorname{sech} z\, dz = \int \frac{dz}{\cosh z} = \int \frac{1}{\cosh z}\frac{\cosh^2 z}{1+\sinh^2 z} dz = \int \frac{d(\sinh z)}{1+\sinh^2 z} = \operatorname{arc\,tan}(\sinh z), \tag{7.133b}$$

using various algebraic properties (Chapter 5).

The primitives of the circular (hyperbolic) functions are (1) Equation 7.128a (Equation 7.128b) and Equation 7.129a (Equation 7.129b) for the primary functions sine and cosine, respectively; (2) Equation 7.130a ≡ Equation 7.134a (Equation 7.130b ≡ Equation 7.134b), Equation 7.131a ≡ Equation 7.135a (Equation 7.131b ≡ Equation 7.135b), Equation 7.132a ≡ Equation 7.136a (Equation 7.132b ≡ Equation 7.136b), and Equation 7.133a ≡ Equation 7.137a (Equation 7.133b ≡ Equation 7.137b) for the secondary functions tangent, cotangent, cosecant, and secant, respectively:

$$\int \tan z\, dz = -\log(\cos z), \quad \int \tanh z\, dz = \log(\cosh z), \tag{7.134a and 7.134b}$$

$$\int \cot z\, dz = \log(\sin z), \quad \int \coth z\, dz = \log(\sinh z), \tag{7.135a and 7.135b}$$

$$\int \csc z\, dz = \log\left[\tan\left(\frac{z}{2}\right)\right], \quad \int \operatorname{csch} z\, dz = \log\left[\tanh\left(\frac{z}{2}\right)\right], \tag{7.136a and 7.136b}$$

$$\int \sec z\, dz = \log(\tan z + \sec z), \quad \int \operatorname{sech} z\, dz = \operatorname{arc\,tan}(\sinh z), \tag{7.137a and 7.137b}$$

$$\int \csc z\, dz = \log(\csc z - \cot z) = \frac{1}{2}\log\left(\frac{1-\cos z}{1+\cos z}\right), \tag{7.138a and 7.138b}$$

$$\int \sec z\, dz = \log \tan\left(\frac{z}{2} + \frac{\pi}{4}\right). \tag{7.139}$$

The last two primitives for the circular cosecant (Equations 7.138a and 7.138b ≡ Equation 7.140) [cosecant (Equation 7.139 ≡ Equation 7.141)] follow from

$$\begin{aligned}\int \csc z\, dz &= \int \frac{\csc^2 z - \csc z \cot z}{\csc z - \cot z}\, dz = \int \frac{(\csc z - \cot z)'}{\csc z - \cot z}\, dz \\ &= \log(\csc z - \cot z) = \log\left(\frac{1-\cos z}{\sin z}\right) = \log\left(\frac{1-\cos z}{\sqrt{1-\cos^2 z}}\right) \\ &= \log\left(\frac{\sqrt{1-\cos z}}{\sqrt{1+\cos z}}\right) = \frac{1}{2}\log\left(\frac{1-\cos z}{1+\cos z}\right),\end{aligned} \tag{7.140}$$

$$\begin{aligned}\int \sec z\, dz &= \log(\sec z + \tan z) = \log\left(\frac{1+\sin z}{\cos z}\right) \\ &= \log\left[\frac{\cos^2(z/2) + \sin^2(z/2) + 2\sin(z/2)\cos(z/2)}{\cos^2(z/2) - \sin^2(z/2)}\right] \\ &= \log\left\{\frac{\left[\cos(z/2) + \sin(z/2)\right]^2}{\cos^2(z/2) - \sin^2(z/2)}\right\} = \log\left[\frac{\cos(z/2) + \sin(z/2)}{\cos(z/2) - \sin(z/2)}\right] \\ &= \log\left[\frac{1+\tan(z/2)}{1-\tan(z/2)}\right] = \log\left[\frac{\tan(\pi/4) + \tan(z/2)}{1-\tan(\pi/4)\tan(z/2)}\right] = \log\tan\left(\frac{z}{2}+\frac{\pi}{4}\right),\end{aligned} \tag{7.141}$$

as an alternative to Equation 7.136a (Equation 7.137a).

7.3.4 Primitives of Cyclometric Functions

The primitives of cyclometric functions use integration by parts:

$$\int \arcsin z\, dz = \int d(z \arcsin z) - \int \frac{z}{\sqrt{1-z^2}}\, dz = \arcsin z + \sqrt{1-z^2}, \tag{7.142}$$

$$\int \arccos z\, dz = \int d(z \arccos z) + \int \frac{z}{\sqrt{1-z^2}}\, dz = \arccos z - \sqrt{1-z^2}, \tag{7.143}$$

$$\int \arctan z\, dz = \int d(z \arctan z) - \int \frac{z}{1+z^2}\, dz = \arctan z - \frac{1}{2}\log(1+z^2), \tag{7.144}$$

$$\int \operatorname{arccot} z\, dz = \int d(z \operatorname{arccot} z) + \int \frac{z}{1+z^2}\, dz = \operatorname{arccot} z + \frac{1}{2}\log(1+z^2), \tag{7.145}$$

$$\int \operatorname{arcsec} z\, dz = \int d(z \operatorname{arcsec} z) - \int \frac{dz}{\sqrt{z^2-1}} = z \operatorname{arcsec} z \mp \log\left(z + \sqrt{z^2-1}\right), \tag{7.146}$$

$$\int \text{arc csc}\, z\, dz = \int d\left(z\, \text{arc csc}\, z\right) + \int \frac{dz}{\sqrt{z^2-1}} = z\, \text{arc csc}\, z \pm \log\left(z + \sqrt{z^2-1}\right); \tag{7.147}$$

in Equation 7.146 (Equation 7.147), the following was used:

$$\frac{d\left[\log\left(z+\sqrt{z^2-1}\right)\right]}{dz} = \frac{1}{z+\sqrt{z^2-1}}\left(1+\frac{z}{\sqrt{z^2-1}}\right) = \frac{1}{\sqrt{z^2-1}}, \tag{7.148}$$

and the upper sign applies for $0 < \text{arc sec z} < \pi/2$ ($0 < \text{arc csc z} < \pi/2$), and the lower sign applies for $\pi/2 < \text{arc sec z} < \pi$ ($-\pi/2 < \text{arc csc z} < 0$).

The primitives of the inverse hyperbolic functions are also obtained via integration by parts:

$$\int \text{arc sinh}\, z\, dz = \int d\left(z \arg \sinh z\right) - \int \frac{z}{\sqrt{1+z^2}} dz = z \arg \sinh z - \sqrt{1+z^2}, \tag{7.149}$$

$$\int \arg \cosh z\, dz = \int d\left(z \arg \cosh z\right) - \int \frac{z}{\sqrt{z^2-1}} dz = z \arg \sinh z - \sqrt{z^2-1}, \tag{7.150}$$

$$\int \arg \tanh z\, dz = \int d\left(z \arg \tanh z\right) - \int \frac{z}{1-z^2} dz = z \arg \tan z + \frac{1}{2}\log\left(1-z^2\right), \tag{7.151}$$

$$\int \arg \coth z\, dz = \int d\left(z \arg \coth z\right) + \int \frac{z}{z^2-1} = z \arg \coth z + \frac{1}{2}\log(z^2-1), \tag{7.152}$$

$$\begin{aligned}
\int {}^{\text{sech}}_{\text{csch}}\, z\, dz &= \int \log\left(z^{-1} + \sqrt{z^{-2} \mp 1}\right) dz \\
&= \int d\left[z \log\left(z^{-1} + \sqrt{z^{-2} \mp 1}\right)\right] + \int \frac{z^{-1} + z^{-2}/\sqrt{z^{-2} \mp 1}}{z^{-1} + \sqrt{z^{-2} \mp 1}} \\
&= z \log\left(z^{-1} + \sqrt{z^{-2} \mp 1}\right) + \int \frac{1 + 1/\sqrt{1 \mp z^2}}{1 + \sqrt{1 \mp z^2}} dz;
\end{aligned} \tag{7.153}$$

in Equation 7.153, the logarithmic representation of the inverse hyperbolic secant (Equation 7.70b) [cosecant (Equation 7.72b)] was used leading to Equation 7.154 (Equation 7.155):

$$\int \arg \text{sech}\, z\, dz = z \log\left(\frac{1}{z} + \sqrt{\frac{1}{z^2} - 1}\right) + \int \frac{dz}{\sqrt{1-z^2}} = z \arg \text{sech}\, z \pm \text{arc} \sin z, \tag{7.154}$$

$$\int \arg\operatorname{csch} z\,dz = z\log\left(\frac{1}{z}+\sqrt{\frac{1}{z^2}+1}\right)+\int\frac{dz}{\sqrt{1+z^2}} = z\arg\operatorname{csch} z \pm \arg\sinh z, \qquad (7.155)$$

together with the derivative of the inverse circular (hyperbolic) sine [Equation 7.122a (Equation 7.122b)]. This completes *the primitives of the cyclometric functions, namely, the inverse circular (hyperbolic) sine [Equation 7.142 (Equation 7.149)], cosine [Equation 7.143 (Equation 7.150)], tangent [Equation 7.144 (Equation 7.151)], cotangent [Equation 7.145 (Equation 7.152)], secant [Equation 7.146 (Equation 7.154)], and cosecant [Equation 7.147 (Equation 7.155)].*

7.3.5 Gudermann Function and Inverse (Cayley)

The Gudermann function is related to the equilateral hyperbola (Figure 5.1d) through the following geometric construction (Figure 7.4): (1) consider the equilateral hyperbola, that is, with asymptotes along the diagonal of the quadrants, cutting the real axis at a:

$$x^2 - y^2 = a^2: \quad x = a\cosh u, \quad y = a\sinh u; \qquad (7.156a\text{–}7.156c)$$

(2) draw a circle of radius a with centre at the origin, that is, tangent to the hyperbola at the vertex; (3) from the horizontal projection A of an arbitrary point P on the hyperbola, draw a tangent to the circle at B, so that the angle Equation 7.157a leads to Equation 7.157b:

$$\theta = A\,O\,B: \quad a\sec\theta = x = a\cosh u. \qquad (7.157a \text{ and } 7.157b)$$

The one-to-one relation (Equation 7.158c) between the circular secant (Figure 5.2b) on the finite interval (Equation 7.158a) and the hyperbolic cosine (Figure 5.2e) on the real line (Equation 7.158b):

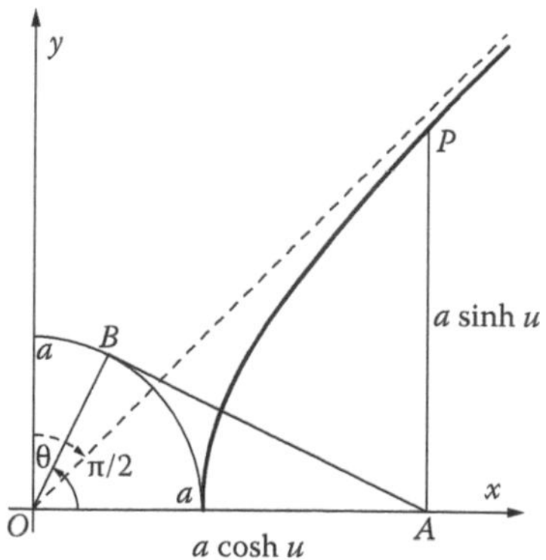

FIGURE 7.4
Consider the equilateral hyperbola, whose asymptotes are the diagonals of quadrants (Figure 5.1d) cutting the real axis at $x = a$. An arbitrary point P on the hyperbola has horizontal projection A from which the tangent is taken touching at B the circle of radius a and center at the origin. The angle θ of OB with the real axis is related to the hyperbolic coordinate u by the inverse Gudermann function $\theta(u)$. Thus, the inverse Gudermann function $u(\theta)$ specifies the height on the Mercator map (1569) of the point of latitude θ on the earth. The simplest type of earth projection [Example I.10.18 (Example I.10.19)] maps the sphere onto a tangent cylinder (cone).

$$-\frac{\pi}{2} < \theta < \frac{\pi}{2}, \quad -\infty < u < +\infty: \quad \sec\theta = \cosh u, \tag{7.158a–7.158c}$$

determines the **Gudermann function** *(Equation 7.159a):*

$$\theta = \text{arc sec}\,(\cosh u) = gd\,(u), \quad u = \text{arg cosh}\,(\sec\theta) = gd^{-1}\,(\theta), \tag{7.159a and 7.159b}$$

and its inverse (Equation 7.159b).

The Gudermann identity (Equation 7.158c) implies

$$\sinh u = \sqrt{\cosh^2 u - 1} = \sqrt{\sec^2\theta - 1} = \tan\theta, \tag{7.160}$$

with the choice of + sign before the radical. From Equations 7.158c and 7.160 also follows

$$\begin{aligned} \tan\left(\frac{\theta}{2}\right) &= \frac{\sin(\theta/2)\cos(\theta/2)}{\cos^2(\theta/2)} = \frac{\sin\theta}{1+\cos\theta} = \frac{\tan\theta}{1+\sec\theta} \\ &= \frac{\sinh u}{1+\cosh u} = \frac{\sinh(u/2)\cosh(u/2)}{\cosh^2(u/2)} = \tanh\left(\frac{u}{2}\right). \end{aligned} \tag{7.161}$$

Thus, *the Gudermann function (Equation 7.159a) and its inverse (Equation 7.159b) satisfy five equivalent identities, namely, Equations 7.158c and 7.160 ≡ Equation 7.162a (Equation 7.161 ≡ Equation 7.162b) and also Equation 7.163 (Equation 7.164):*

$$\sinh u = \tan\theta, \quad \tanh\left(\frac{u}{2}\right) = \tan\left(\frac{\theta}{2}\right), \tag{7.162a and 7.162b}$$

$$e^u = \cosh u + \sinh u = \sec\theta + \tan\theta, \tag{7.163}$$

$$gd^{-1}(\theta) \equiv u = \log(\sec\theta + \tan\theta) = \log\left[\tan\left(\frac{\theta}{2} + \frac{\pi}{4}\right)\right] = \int \sec\theta \, d\theta; \tag{7.164}$$

Equation 7.164 shows that the inverse Gudermann function (Equation 7.159b) is the primitive (Equation 7.141) of the circular secant. The Gudermann function appears in the *Mercator map* (1569) of the earth; simpler mappings of the earth are obtained projecting from the center onto [Example I.10.18 (Example I.10.19)] a cylinder (cone) tangent to the equator (a parallel).

7.4 Power Series for Cyclometric Functions

In much the same way as the poles of the direct secondary functions limit the radius of convergence of their power series, the branch points or branch cuts of cyclometric functions limit the region of validity of their representations in power series. The derivatives of the cyclometric functions are rational functions (Subsection 7.3.2) that can be expanded in uniformly convergent power series and integrated term-by-term, for example, for the inverse circular sine and tangent (Subsection 7.4.1). These two ascending power series lead (Subsection 7.4.4) through algebraic transformations (Subsection 7.4.3) to a total of 16 ascending or descending power series for all cyclometric functions (Subsection 7.4.2), with conditions of convergence indicated in Table 7.4.

7.4.1 Power Series for the Inverse Sine and Tangent

In order to obtain power series for cyclometric functions, five steps are needed, illustrated in the case of the arc sin: (1) since a cyclometric function is the logarithm of an algebraic function, its derivative is an algebraic function, for example, Equation 7.110b; (2) the algebraic function can be expanded in power series, with a radius of convergence determined by the closest branch point; for example, the binomial series (Equations I.25.37a through I.25.37c ≡ Equations 3.144a through 3.144c) applied to Equation 7.110b yields

$$|z| < 1: \quad \frac{d(\arc\sin z)}{dz} = \frac{1}{\sqrt{1-z^2}} = \sum_{n=0}^{\infty} \binom{-1/2}{n} (-z^2)^n = 1 + \sum_{n=1}^{\infty} a_n z^{2n}, \qquad (7.165a \text{ and } 7.165b)$$

in which double factorials (Equations 1.178a through 1.178c) can be used:

$$\begin{aligned} a_n &\equiv (-)^n \binom{-1/2}{n} = \frac{(-)^n}{n!}\left(-\frac{1}{2}\right)\left(-\frac{1}{2}-1\right)\cdots\left(-\frac{1}{2}-n+1\right) \\ &= \frac{1}{n!}\frac{1}{2}\frac{3}{2}\cdots\left(n-\frac{1}{2}\right) = \frac{2n-1}{2n}\frac{2n-3}{2n-2}\cdots\frac{3}{4}\frac{1}{2} = \frac{(2n-1)!!}{(2n)!!}; \end{aligned} \qquad (7.166c)$$

(3) the branch points of the cyclometric function are singularities of the derivative 7.110b, that is, lie at $z = \pm 1$, thus determining the open region of convergence (Equation 7.165a) of the series for the derivate; (4) the latter is uniformly convergent in a closed subregion $|z| \le 1 - \varepsilon$ with $0 < \varepsilon < 1$, where it can be integrated term-by-term (Subsection I.21.6), leading to the power series for the cyclometric function, whose convergence properties can be investigated using the combined test (Subsection I.39.1); and (5) integrating Equation 7.165b between 0 and z leads to the ascending power series for the inverse circular sine (Equation 7.167b):

$$\begin{aligned} |z| < 1: \quad \arc\sin z &= z + \sum_{n=1}^{\infty} a_n \frac{z^{2n+1}}{2n+1} = z + \frac{(2n-1)!!}{(2n)!!}\frac{z^{2n+1}}{2n+1} \\ &= z + \sum_{n=1}^{\infty} \frac{1}{n!}\frac{1}{2}\frac{3}{2}\cdots\left(n-\frac{1}{2}\right)\frac{z^{2n+1}}{2n+1}, \end{aligned} \qquad (7.167a\text{–}7.167c)$$

valid in the unit disk (Equation 7.167a), where it is absolutely convergent.

The derivative of the inverse circular tangent (Equation 7.112b) can be expanded (Equation 7.168b) in an arithmetic series (Equation I.21.62c ≡ Equations 3.147a through 3.147c):

TABLE 7.4

Convergence of Power Series for 12 Cyclometric Functions, That Is, the Six Inverse Circular Plus Hyperbolic Functions

Function	Series	D.	O.	C.C.	A.C.	U.C.	T.C.	Case
arc sin z	Equation 7.172b	$\|z\| > 1$	–	–	$\|z\| \le 1$	$\|z\| \le 1-\varepsilon$	$\|z\| \le 1-\varepsilon$	I
arg sinh z	Equation 7.173b	$\|z\| > 1$	–	–	$\|z\| \le 1$	$\|z\| \le 1-\varepsilon$	$\|z\| \le 1-\varepsilon$	I
arg csch z	Equation 7.177b	$\|z\| < 1$	–	–	$\|z\| \ge 1$	$\|z\| \ge 1+\delta$	$\|z\| \ge 1+\delta$	II
arc csc z	Equation 7.176b	$\|z\| < 1$	–	–	$\|z\| \ge 1$	$\|z\| \ge 1+\delta$	$\|z\| \ge 1+\delta$	II
arc cos z	Equation 7.174b	$\|z\| > 1$	–	–	$\|z\| \le 1$	$\|z\| \le 1-\varepsilon$	$\|z\| \le 1-\varepsilon$	I
arg cosh z	Equation 7.175b	$\|z\| < 1$	–	–	$\|z\| \le 1$	$\|z\| \le 1-\varepsilon$	$\|z\| \le 1-\varepsilon$	I
arg sech z	Equation 7.179b	$\|z\| > 1$	–	–	$\|z\| \ge 1$	$\|z\| \ge 1+\delta$	$\|z\| \ge 1+\delta$	II
arc sec z	Equation 7.178b	$\|z\| < 1$	–	–	$\|z\| \ge 1$	$\|z\| \ge 1+\delta$	$\|z\| \ge 1+\delta$	II
arc tan z	Equations 7.180a and 7.180b	$\|z\| > 1$ or $z = \pm i$	–	$\|z\| = 1$ and $z \ne \pm i$	$\|z\| < 1$	$\|z\| < 1$ and $\|z \pm i\| > \varepsilon$	$\|z\| \le 1-\varepsilon$	III
	Equations 7.180c and 7.180d	$\|z\| < 1$ or $z = \pm i$	–	$\|z\| = 1$ and $z \ne \pm i$	$\|z\| > 1$	$\|z\| > 1$ and $\|z \pm i\| > \varepsilon$	$\|z\| \ge 1+\delta$	IV
arg tanh z	Equations 7.181a and 7.181b	$\|z\| > 1$ or $z = \pm 1$	–	$\|z\| = 1$ and $z \ne \pm 1$	$\|z\| < 1$	$\|z\| < 1$ and $\|z \pm 1\| > \varepsilon$	$\|z\| \le 1-\varepsilon$	V
	Equations 7.181c and 7.181d	$\|z\| < 1$ or $z = \pm 1$	–	$\|z\| = 1$ and $z \ne \pm 1$	$\|z\| > 1$	$\|z\| > 1$ and $\|z \pm 1\| > \varepsilon$	$\|z\| \ge 1+\delta$	VI
arg coth z	Equations 7.182a and 7.182b	$\|z\| < 1$ or $z = \pm 1$	–	$\|z\| = 1$ and $z \ne \pm 1$	$\|z\| > 1$	$\|z\| > 1$ and $\|z \pm 1\| > \varepsilon$	$\|z\| \ge 1+\delta$	VI
	Equations 7.182c and 7.182d	$\|z\| > 1$ or $z = \pm 1$	–	$\|z\| = 1$ and $z \ne \pm 1$	$\|z\| < 1$	$\|z\| < 1$ and $\|z \pm 1\| > \varepsilon$	$\|z\| \le 1-\varepsilon$	V
arc cot z	Equations 7.183a and 7.183b	$\|z\| < 1$ or $z = \pm i$	–	$\|z\| = 1$ and $z \ne \pm i$	$\|z\| > 1$	$\|z\| > 1$ and $\|z \pm i\| > \varepsilon$	$\|z\| \ge 1+\delta$	IV
	Equations 7.183c and 7.183d	$\|z\| > 1$ or $z = \pm i$	–	$\|z\| = 1$ and $z \ne \pm i$	$\|z\| < 1$	$\|z\| < 1$ and $\|z \pm i\| > \varepsilon$	$\|z\| \le 1-\varepsilon$	III

Note: Radius of convergence is limited by the nearest point on the branch cuts in Figures 7.2 and 7.3. D.—divergent; O. —oscillatory; C.C. —conditionally convergent; A.C.—absolutely convergent; U.C.—uniformly convergent; T.C.—totally convergent (see Chapters I.21 and I.29).

$$|z| < 1: \quad \frac{d(\arctan z)}{dz} = \frac{1}{1+z^2} = \sum_{n=0}^{\infty} (-)^n z^{2n}, \tag{7.168a and 7.168b}$$

with a radius of convergence unity (Equation 7.112a ≡ Equation 7.168a); in a closed subradius $|z| \leq 1 - \varepsilon$ with $0 < \varepsilon < 1$, the series is uniformly convergent and can be integrated term-by-term from 0 to z leading to Equation 7.169b:

$$|z| < 1: \quad \arctan z = \sum_{n=0}^{\infty} (-)^n \int_0^z z^{2n}\, dz = \sum_{n=0}^{\infty} (-)^n \frac{z^{2n+1}}{2n+1}. \tag{7.169a and 7.169b}$$

The series Equation 7.169a ≡ Equation 7.170a has a general term Equation 7.170b and a ratio of successive terms Equation 7.170c:

$$\arctan z = \sum_{n=0}^{\infty} g_n, \quad g_n = (-)^n \frac{z^{2n+1}}{2n+1}, \tag{7.170a and 7.170b}$$

$$\frac{g_{n+1}}{g_n} = -z^2 \frac{1+1/(2n)}{1+3/(2n)} = -z^2\left[1 + \frac{1}{n}\left(\frac{1}{2} - \frac{3}{2}\right) + O\left(\frac{1}{n^2}\right)\right] = -z^2\left[1 - \frac{1}{n} + O\left(\frac{1}{n^2}\right)\right]; \tag{7.170c}$$

using the combined convergence test (Subsection I.29.1), this specifies the convergence at all points of the complex z-plane (Table 7.4). Likewise for the series Equation 7.167c ≡ Equation 7.171a, with a general term Equation 7.171b and a ratio of successive terms Equation 7.171c:

$$\arcsin z = \sum_{n=0}^{\infty} f_n, \quad f_n = \binom{-1/2}{n} \frac{z^{2n+1}}{2n+1} = \left(-\frac{1}{2}\right)\left(-\frac{3}{2}\right)\ldots\left(-\frac{1}{2} - n + 1\right)\frac{1}{n!}\frac{z^{2n+1}}{2n+1}, \tag{7.171a and 7.171b}$$

$$\frac{f_{n+1}}{f_n} = z^2 \frac{2n+1}{2n+3}\frac{-n-1/2}{n+1} = -z^2 \frac{1+1/(2n)}{1+3/(2n)}\frac{1+1/(2n)}{1+1/n}$$
$$= -z^2\left[1 + \frac{1}{n}\left(\frac{1}{2} - \frac{3}{2} + \frac{1}{2} - 1\right) + O\left(\frac{1}{n^2}\right)\right] = -z^2\left[1 - \frac{3}{2n} + O\left(\frac{1}{n^2}\right)\right]; \tag{7.171c}$$

the combined convergence test (Equation I.29.1) specifies the convergence in the whole complex z-plane (Table 7.4).

7.4.2 Ascending and Descending Power Series for Cyclometric Functions

From the ascending power series for the inverse circular sine (Equations 7.167a and 7.167b) and tangent (Equations 7.169a and 7.169b) follow 14 additional related ascending and descending power series for other cyclometric functions:

$$|z| < 1: \quad \text{arc}\sin z = z + \sum_{n=1}^{\infty} \frac{(2n-1)!!}{(2n)!!} \frac{z^{2n+1}}{2n+1}; \tag{7.172a and 7.172b}$$

$$|z| < 1: \quad \text{arg}\sinh z = z + \sum_{n=1}^{\infty} (-)^n \frac{(2n-1)!!}{(2n)!!} \frac{z^{2n+1}}{2n+1}, \tag{7.173a and 7.173b}$$

$$|z| < 1: \quad \text{arc}\cos z = \frac{\pi}{2} - z - \sum_{n=1}^{\infty} \frac{(2n-1)!!}{(2n)!!} \frac{z^{2n+1}}{2n+1}, \tag{7.174a and 7.174b}$$

$$|z| < 1: \quad \text{arg}\cosh z = \pm i\frac{\pi}{2} \mp iz \mp i\sum_{n=1}^{\infty} \frac{(2n-1)!!}{(2n)!!} \frac{z^{2n+1}}{2n+1}, \tag{7.175a and 7.175b}$$

$$|z| > 1: \quad \text{arc}\csc z = \frac{1}{z} + \sum_{n=1}^{\infty} \frac{(2n-1)!!}{(2n)!!} \frac{z^{-2n-1}}{2n+1}, \tag{7.176a and 7.176b}$$

$$|z| > 1: \quad \text{arg}\,\text{csch}\, z = \frac{1}{z} + \sum_{n=1}^{\infty} (-)^n \frac{(2n-1)!!}{(2n)!!} \frac{z^{-2n-1}}{2n+1}, \tag{7.177a and 7.177b}$$

$$|z| > 1: \quad \text{arc}\sec z = \frac{\pi}{2} - \frac{1}{z} - \sum_{n=1}^{\infty} \frac{(2n-1)!!}{(2n)!!} \frac{z^{-2n-1}}{2n+1}, \tag{7.178a and 7.178b}$$

$$|z| > 1: \quad \text{arg}\,\text{sech}\, z = \pm i\frac{\pi}{2} \mp iz \mp i\sum_{n=1}^{\infty} \frac{(2n-1)!!}{(2n)!!} \frac{z^{-2n-1}}{2n+1}, \tag{7.179a and 7.179b}$$

$$\text{arc}\tan z = \begin{cases} \displaystyle\sum_{n=0}^{\infty} (-)^n \frac{z^{2n+1}}{2n+1} & \textit{for} \quad |z| < 1, \quad \text{(7.180a and 7.180b)} \\[2em] \displaystyle\frac{\pi}{2} - \sum_{n=0}^{\infty} (-)^n \frac{z^{-2n-1}}{2n+1} & \textit{for} \quad |z| > 1, \quad \text{(7.180c and 7.180d)} \end{cases}$$

$$\arg\tanh z = \begin{cases} \sum_{n=0}^{\infty} \frac{z^{2n+1}}{2n+1} & \text{for} \quad |z| < 1, \qquad \text{(7.181a and 7.181b)} \\ -i\frac{\pi}{2} + \sum_{n=0}^{\infty} \frac{z^{-2n-1}}{2n+1} & \text{for} \quad |z| > 1, \qquad \text{(7.181c and 7.181d)} \end{cases}$$

$$\arg\coth z = \begin{cases} \sum_{n=0}^{\infty} \frac{z^{-2n-1}}{2n+1} & \text{for} \quad |z| > 1, \qquad \text{(7.182a and 7.182b)} \\ i\frac{\pi}{2} + \sum_{n=0}^{\infty} \frac{z^{2n+1}}{2n+1} & \text{for} \quad |z| < 1, \qquad \text{(7.182c and 7.182d)} \end{cases}$$

$$\operatorname{arc}\cot z = \begin{cases} \sum_{n=0}^{\infty} (-)^n \frac{z^{-2n-1}}{2n+1} & \text{for} \quad |z| > 1, \qquad \text{(7.183a and 7.183b)} \\ \frac{\pi}{2} - \sum_{n=0}^{\infty} (-)^n \frac{z^{2n+1}}{2n+1} & \text{for} \quad |z| < 1, \qquad \text{(7.183c and 7.183d)} \end{cases}$$

These results are proved next (Subsection 7.4.3) using relations among the circular and hyperbolic cyclometric functions.

7.4.3 Relations among Inverse Circular and Hyperbolic Functions

The series expansions Equations 7.172a through 7.183b are related by three sets of conditions: (1) change the variable from z to $1/z$:

$$\operatorname{arc}\{\sec, \csc, \cot\}(z) = \operatorname{arc}\{\cos, \sin, \tan\}\left(\frac{1}{z}\right), \qquad \text{(7.184a–7.184c)}$$

$$\arg\{\operatorname{sech}, \operatorname{csch}, \coth\}(z) = \arg\{\cosh, \sinh, \tanh\}\left(\frac{1}{z}\right); \qquad \text{(7.185a–7.185c)}$$

(2) using the imaginary change of variable:

$$\arg\{\cosh, \operatorname{sech}\}(z) = \pm i \operatorname{arc}\{\cos, \sec\}(z), \qquad \text{(7.186a and 7.186b)}$$

$$\arg\{\sinh, \tanh\}(z) = -i \operatorname{arc}\{\sin, \tan\}(iz); \tag{7.187a and 7.187b}$$

$$\arg\{\operatorname{csch}, \coth\}(z) = i \operatorname{arc}\{\csc, \cot\}(iz); \tag{7.188a and 7.188b}$$

(3) the algebraic identities

$$\frac{\pi}{2} = \arg\cos z + \operatorname{arc}\sin z = \operatorname{arc}\sec z + \operatorname{arc}\csc z = \operatorname{arc}\tan z + \arg\cot z. \tag{7.189a–7.189c}$$

$$\begin{aligned} i\frac{\pi}{2} &= \pm\arg\cosh z + \arg\sinh(iz) \\ &= \pm\arg\operatorname{sech} z - \arg\operatorname{csch}(iz) = \arg\tanh(iz) - \arg\coth(iz); \end{aligned} \tag{7.190a–7.190c}$$

(4) the change of sign of the argument:

$$\operatorname{arc}\{\sin,\csc,\tan,\cot\}(-z) = -\operatorname{arc}\{\sin,\csc,\tan,\cot\}(z), \tag{7.191a–7.191d}$$

$$\operatorname{arc}\{\cos,\sec\}(-z) = \pi - \operatorname{arc}\{\cos,\sec\}(z), \tag{7.191e and 7.191f}$$

$$\arg\{\sinh,\operatorname{csch},\tan,\coth\}(-z) = -\arg\{\sinh,\operatorname{csch},\tanh,\tanh\}(z), \tag{7.192a–7.192d}$$

$$\arg\{\cosh,\operatorname{sech}\}(-z) = \pm\pi i - \arg\{\cosh,\operatorname{sech}\}(z). \tag{7.192e and 7.192f}$$

These formulas for the cyclometric functions follow from the corresponding circular and hyperbolic functions, as shown next.

The preceding identities are proved as follows: (1) from Equation 7.193a and 7.193b follows Equation 7.193c:

$$z = \sec w, \quad \frac{1}{z} = \cos w: \quad \operatorname{arc}\sec z = w = \operatorname{arc}\cos\left(\frac{1}{z}\right), \tag{7.193a–7.193c}$$

proving Equation 7.193c ≡ Equation 7.184a, and likewise for the pairs of algebraic inverse functions (sin, csc) and (tan, cot) in Equations 7.184b and 7.184c, and (cosh, sech), (sinh, csch), and (tanh, coth) in Equations 7.185a through 7.185c; (2) from Equation 7.193d (Equation 7.193f) follows Equation 7.193e ≡ Equation 7.191a (Equation 7.193g ≡ Equation 1.191e):

$$\sin(-w) = z = -\sin w, \qquad \operatorname{arc}\sin(-z) = w = -\operatorname{arc}\sin z, \tag{7.193d and 7.193e}$$

$$\cos(\pi - w) = z = -\cos w, \qquad \operatorname{arc}\cos z = \pi - w = \pi - \operatorname{arc}\cos(-z), \tag{7.193f and 7.193g}$$

and likewise for Equations 7.191b, 7.191c, and 7.191d (7.191f); (3) the imaginary transformation Equation 5.5a ≡ Equation 7.193h/Equation 5.6b ≡ Equation 7.193j/Equation 5.25b ≡ Equation 5.193m lead respectively to Equation 7.193i ≡ Equation 7.186a/Equation 7.193k ≡ Equation 1.187a/Equation 7.193n ≡ Equation 7.188a:

$$\cos w = z = \cosh(\pm iw): \qquad \text{arc} \cos z = w = \mp i \arg \cosh z \qquad \text{(7.193h and 7.193i)}$$

$$\sin(iw) = iz = i \sinh w: \qquad \text{arc} \sinh z = w = -i \arg \sinh(iz), \qquad \text{(7.193j and 7.193k)}$$

$$\csc(iw) = -iz = -i \operatorname{csch} w: \qquad \text{arc} \csc z = w = -i \arg \operatorname{csch}(-iz) = i \arg \operatorname{csch}(iz), \qquad \text{(7.193m and 7.193n)}$$

and likewise Equation 7.186b/Equation 7.187b/Equation 7.188b follow from Equation 5.24b/Equation 5.26b/Equation 5.27b; (4) using the imaginary change of variable (Equation 7.187a/7.188a/7.186a) leads to Equation 7.193p ≡ Equation 7.192a/Equation 7.193q ≡ Equation 7.192b/Equation 7.193r ≡ Equation 7.192e:

$$\arg \sinh(-z) = -i \,\text{arc} \sin(-iz) = i \,\text{arc} \sin(iz) = i^2 \arg \sinh z = -\arg \sinh z, \qquad \text{(7.193p)}$$

$$\arg \operatorname{csch}(-z) = i \,\text{arc} \csc(-iz) = -i \,\text{arc} \csc(iz) = (-i)^2 \arg \operatorname{csch} z = -\arg \operatorname{csch} z, \qquad \text{(7.193q)}$$

$$\begin{aligned} \arg \cosh(-z) &= \pm i \,\text{arc} \cos(-z) = \pm i\pi \mp i \,\text{arc} \cos(z) \\ &= \pm i\pi + (\mp i)^2 \arg \cosh z = \pm i\pi - \arg \cosh z, \end{aligned} \qquad \text{(7.193r)}$$

and likewise for Equations 7.192c, 7.192d, and 7.192f; (5) from Equations 7.193g and 7.193t follows Equation 7.193u ≡ Equation 7.189a, 7.190a, and a similar proof applies to Equation 7.189b/Equation 7.189c leading to 7.193w ≡ Equation 7.190b/Equation 7.193x ≡ Equation 7.190c.

$$\cos w = z = \sin\left(\frac{\pi}{2} - w\right); \quad \text{arc} \sin z = \frac{\pi}{2} - w = \frac{\pi}{2} - \text{arc} \cos z, \qquad \text{(7.193c–7.193u)}$$

$$\begin{aligned} \frac{\pi}{2} &= \text{arc} \cos z + \text{arc} \sin(-i^2 z) = \mp i \arg \cosh z + i \arg \sinh(-iz) \\ &= \mp i \arg \cosh z - i \arg \sinh(iz), \end{aligned} \qquad \text{(7.193v)}$$

$$\begin{aligned} \frac{\pi}{2} &= \text{arc} \sec z + \text{arc} \csc(-i^2 z) = \mp i \arg \operatorname{sech} z - i \arg \operatorname{csch}(-iz) \\ &= \mp i \arg \operatorname{sech} z + i \arg \operatorname{csch}(iz), \end{aligned} \qquad \text{(7.193w)}$$

$$\begin{aligned} \frac{\pi}{2} &= \text{arc} \tan(-i^2 z) + \text{arc} \cot(-i^2 z) = i \arg \tanh(-iz) - i \arg \coth(-iz) \\ &= -i \arg \tanh(iz) + i \arg \coth(iz), \end{aligned} \qquad \text{(7.193x)}$$

The preceding properties (Subsection 7.4.3) are used next (Subsection 7.4.4) to prove all 16 power series for cyclometric functions (Subsection 7.4.2) from only two of them (Subsection 7.4.1).

7.4.4 Relations between Cyclometric Power Series

The ascending power series for the inverse circular sine (Equations 7.167a and 7.167b ≡ Equations 7.172a and 7.172b) [tangent (Equations 7.169a and 7.169b ≡ Equations 7.180a and 7.180b)] lead to 14 more ascending and descending power series as follows: (1) using the inverse variable (Equation 7.184a) in Equations 7.167a and 7.167b ≡ Equations 7.172a and 7.172b leads to Equations 7.176a and 7.176b; (2 and 3) the imaginary change of variable Equation 7.187a (Equation 7.188a) applied to Equations 7.172a and 7.172b (Equations 7.176a and 7.176b) leads to Equations 7.173a and 7.173b (Equations 7.177a and 7.177b); (4 and 5) the property Equation 7.189a (Equation 7.189b) applied to Equations 7.172a and 7.172b (Equations 7.176a and 7.176b) yields Equations 7.174a and 7.174b (Equations 7.178a and 7.178b); (6 and 7) the imaginary change of variable Equation 7.186a (Equation 7.186b) applied to Equations 7.174a and 7.174b (Equations 7.178a and 7.178b) leads to Equations 7.175a and 7.175b (Equations 7.179a and 7.179b); (8) the inverse variable Equation 7.184c applied to Equations 7.180a and 7.180b ≡ Equations 7.169a and 7.169b leads to Equations 7.183a and 7.183b; (9 and 10) the imaginary transformation Equation 7.187b (Equation 7.188b) applied to Equations 7.180a and 7.180b (Equations 7.183a and 7.183b) yields Equations 7.181a and 7.181b (Equations 7.182a and 7.182b); (11 and 12) the second series representation for the circular tangent (Equations 7.180c and 7.180d) [cotangent (Equations 7.183c and 7.183d)] follows from Equations 7.183a and 7.183b (Equations 7.180a and 7.180b) using Equation 7.189c; and (13 and 14) the second series representation for the hyperbolic tangent (Equations 7.181c and 7.181d) [cotangent (Equations 7.182c and 7.182d)] follows from Equations 7.180c and 7.180d (Equations 7.183c and 7.183d) using Equation 7.187b (Equation 7.188b). In all cases, only those transformations (Equations 7.184 through 7.189) for which the region of convergence (Equations 7.172a through 7.183a) excludes the branch cuts (Figures 7.2 and 7.3) for the cyclometric functions (Equations 7.65a through 7.76a) have been selected. Thus, *the cyclometric functions are specified by power series, namely, (1 and 2) ascending for the inverse circular (hyperbolic) sine [Equations 7.172a and 7.172b ≡ Equations 7.167a and 7.167b (Equations 7.173a and 7.173b)]; (3 and 4) descending for the inverse circular (hyperbolic) cosecant [Equations 7.176a and 7.176b (Equations 7.177a and 7.177b)]; (5 and 6) ascending for the inverse circular (hyperbolic) cosine [Equations 7.174a and 7.174b (Equations 7.175a and 7.175b); (7 and 8) descending for the inverse circular (hyperbolic) secant [Equations 7.178a and 7.178b (Equations 7.179a and 7.179b)]; (9 and 10) ascending/descending for the inverse circular (hyperbolic) tangent [Equations 7.180a and 7.180b/Equations 7.180c and 7.180d (Equations 7.181a and 7.181b/Equations 7.181c and 7.181d)]; and (9 and 10) descending/ascending for the inverse circular (hyperbolic) cotangent [Equations 7.183a and 7.183b/Equations 7.183c and 7.183d (Equations 7.182a and 7.182b/Equations 7.182c and 7.182d)]. The convergence of these series (Equations 7.172 and 7.183) at all points of the complex z-plane is indicated in Table 7.4;* this leads to six cases using the combined convergence test (Subsection I.29.1) as for the inverse circular sine (tangent) in Equations 7.170a through 7.170c (Equations 7.171a through 7.171c).

7.5 Slopes at Zeros and Residues at Singularities

The zeros and slopes are considered for the circular (hyperbolic) functions [Subsection 7.5.1 (Subsection 7.5.2)]. The residues at singularities (Subsection 7.5.3) concern the poles (essential singularity) of the secondary (primary) functions in the finite plane (at infinity).

7.5.1 Zeros and Slopes of Primary and Secondary Functions

The zeros and slopes of the circular functions (Table 7.5) are

$$m \in |Z: \quad \sin(m\pi) = 0, \quad \sin'(m\pi) = \cos(m\pi) = (-)^m, \tag{7.194a–7.194c}$$

$$\cos\left(m\pi + \frac{\pi}{2}\right) = 0, \quad \cos'\left(m\pi + \frac{\pi}{2}\right) = -\sin\left(m\pi + \frac{\pi}{2}\right) = (-)^{m+1}, \tag{7.195a and 7.195b}$$

$$\sec(\pm i\,\infty) = 0, \quad \sec'(\pm i\,\infty) = \sec(\pm i\,\infty)\tan(\pm i\,\infty) = 0 \tag{7.196a and 7.196b}$$

$$\csc(\pm i\infty) = 0, \quad \csc'(\pm i\infty) = -\csc(\pm i\infty)\cot(\pm i\infty) = 0, \tag{7.197a and 7.197b}$$

$$\tan(m\pi) = 0, \quad \tan'(m\pi) = \sec^2(m\pi) = 1, \tag{7.198a and 7.198b}$$

$$\cot\left(m\pi + \frac{\pi}{2}\right) = 0, \quad \cot'\left(m\pi + \frac{\pi}{2}\right) = -\csc^2\left(m\pi + \frac{\pi}{2}\right) = -1, \tag{7.199a and 7.199b}$$

that is, *(1) primary (secondary) circular functions sine and cosine (tangent and cotangent) have an infinite number of interlaced zeros at equal distances (Figure 7.1) with alternating (equal) slopes [Figure 5.2a and b (Figure 5.2c)]; (2) the secondary function circular secant (cosecant) has no zeros*

TABLE 7.5

Zeros of 12 Circular and Hyperbolic Functions and Corresponding Slopes

Function	Zero	Slope	Pole	Residue	Essential Singularity
$f(z)$	$f(a_m) = 0$	$A_n \equiv f'(a_m)$	$f(b_n) = \infty$	$B_n \equiv f_{(1)}(b_n)$	$z = b_n$
$\sin z$	$m\pi$	$(-)^m$	–	1	∞
$\cos z$	$m\pi+\pi/2$	$(-)^{m+1}$	–	0	∞
$\sec z$	$\pm i\,\infty$	0	$m+\pi/2$	$(-)^{m+1}$	–
$\csc z$	$\pm i\,\infty$	0	$m\pi$	$(-)^m$	–
$\cot z$	$m\pi+\pi/2$	-1	$m\pi$	$+1$	
$\tan z$	$m\pi$	$+1$	$m\pi+\pi/2$	-1	–
$\tanh z$	$im\pi$	$+1$	$im\pi+i\pi/2$	$+1$	–
$\coth z$	$im\pi+i\pi/2$	$+1$	$im\pi$	$+1$	–
$\text{csch}\, z$	$\pm\infty$	0	$im\pi$	$(-)^m$	–
$\text{sech}\, z$	$\pm\infty$	0	$im\pi+i\pi/2$	$i\,(-)^{m+1}$	–
$\cosh z$	$im\pi+i\pi/2$	$i(-)^m$	–	0	∞
$\sinh z$	$im\pi$	$(-)^m$	–	1	∞

Note: Singularities of the same functions, either poles or essential singularities, and the corresponding residues. Zeros and poles are all simple. Functions either have an infinite number of zeros or just one at infinity along the imaginary (real) axis for the circular (hyperbolic) secant and cosecant. Secondary functions, that is, the secant, cosecant, tangent, and cotangent, are meromorphic and thus have an infinite number of poles without accumulation point in the finite plane. Primary functions, that is, the cosine and sine, are integral and thus have no poles in the finite plane and an essential singularity at infinity. $m = 0, \pm1, \pm2, \in |Z$.

[Figure 5.2h (Figure 5.2a)] in the finite plane and vanishes with zero slope at infinity along the imaginary axis [Equations 7.200a and 7.200b (Equations 7.201a and 7.201b)]:

$$\sec(\pm i\infty) = \text{sech}\,(\pm\infty) = 0, \quad \sec'(\pm i\infty) = i\,\text{sech}(\pm\infty)\tanh(\pm\infty) = \pm i\,\text{sech}(\pm\infty) = 0, \tag{7.200a and 7.200b}$$

$$\csc(\pm i\infty) = -i\,\text{csch}(\pm\infty) = 0, \quad \csc'(\pm i\infty) = i(-i)^2\,\text{csch}(\pm\infty)\coth(\pm\infty) = \pm\,\text{csch}(\pm\infty) = 0, \tag{7.201a and 7.201b}$$

since they correspond to the hyperbolic secant (cosecant) at infinity along the real axis [Figure 5.2e (Figure 5.2d)].

7.5.2 Zeros and Slopes of Circular and Hyperbolic Functions

The imaginary change of variable relates the zeros and slopes of the circular (Equations 7.194 through 7.199) and hyperbolic (Equations 7.202 through 7.207) functions in Table 7.5:

$$m \in |Z: \quad \sinh(im\pi) = 0, \quad \sinh'(im\pi) = \cosh(im\pi) = \cos(m\pi) = (-)^m, \tag{7.202a–7.202c}$$

$$\cosh\left(im\pi + i\frac{\pi}{2}\right) = 0, \quad \cosh'\left(im\pi + i\frac{\pi}{2}\right) = \sinh\left(im\pi + i\frac{\pi}{2}\right) = i\sin\left(m\pi + \frac{\pi}{2}\right) = (-)^m i, \tag{7.203a and 7.203b}$$

$$\text{sech}(\pm\infty) = 0, \quad \text{sech}'(\pm\infty) = \text{sech}(\pm\infty)\coth(\pm\infty) = \pm\text{sech}(\pm\infty) = 0, \tag{7.204a and 7.204b}$$

$$\text{csch}(\pm\infty) = 0, \quad \text{csch}'(\pm\infty) = -\text{csch}\,(\pm\infty)\coth(\pm\infty) = \mp\text{csch}(\pm\infty) = 0, \tag{7.205a and 7.205b}$$

$$\tanh(im\pi) = 0, \quad \tanh'(im\pi) = \text{sech}^2(im\pi) = \sec^2(m\pi) = 1, \tag{7.206a and 7.206b}$$

$$\coth\left(im\pi + i\frac{\pi}{2}\right) = 0, \quad \coth'\left(im\pi + i\frac{\pi}{2}\right) = -\text{csch}^2\left(im\pi + i\frac{\pi}{2}\right) = -\left[-i\csc\left(m\pi + \frac{\pi}{2}\right)\right]^2 = 1, \tag{7.207a and 7.207b}$$

showing that *(1) the primary (secondary) hyperbolic functions sine and cosine (tangent and cotangent) have interlaced zeros along the imaginary axis (Figure 7.1) with alternating (fixed) slopes, and the only real zero is at the origin for the hyperbolic sine (tangent) in Figure 5.2d (Figure 5.2f); (2) the secondary function hyperbolic secant (cosecant) has no zeros in the finite complex plane and vanishes only at infinity along the real axis [Figure 5.2e (Figure 5.2d)].*

7.5.3 Residues at Poles and Essential Singularities

The primary circular and hyperbolic functions, namely, sine (cosine), are integral functions, and hence their only singularity is an essential singularity at infinity, with residue specified by the coefficient of the O(z) term (Equation 7.208a), so that the residue is zero (unity) for sine (Equations 7.208b and 7.208c) [cosine (Equations 7.209a through 7.209c)]:

$$f(z) = \ldots + A_{-1}\, z + \ldots: \quad \sin z = z + \ldots, \quad \sinh z = z + \ldots, \quad A_{-1} = 1, \qquad (7.208\text{a–}7.208\text{d})$$

$$\cos z = 1 - \frac{z^2}{2} + \cdots, \quad \cosh z = 1 + \frac{z^2}{2} + \cdots, \quad A_{-1} = 0, \qquad (7.209\text{a–}7.209\text{c})$$

and both circular and hyperbolic.

The secondary functions are meromorphic, that is, they have a finite number of poles and no essential singularity, so that there is no accumulation point in the finite plane. All the poles (Figure 7.1) of the secondary circular functions, namely, cosecant and secant (cotangent and tangent), are equally spaced along the real axis and are interlaced, with alternating (fixed) residues:

$$m \in |Z: \quad \csc(m\pi) = \infty, \quad \csc_{(1)}(m\pi) = (-)^m, \qquad (7.210\text{a–}7.210\text{c})$$

$$\sec\left(m\pi + \frac{\pi}{2}\right) = \infty, \quad \sec_{(1)}\left(m\pi + \frac{\pi}{2}\right) = (-)^{m+1}, \qquad (7.211\text{a and }7.211\text{b})$$

$$\tan\left(m\pi + \frac{\pi}{2}\right) = \infty, \quad \tan_{(1)}\left(m\pi + \frac{\pi}{2}\right) = -1, \qquad (7.212\text{a and }7.212\text{b})$$

$$\cot(m\pi) = \infty, \qquad \cot_{(1)}(m\pi) = 1; \qquad (7.213\text{a and }7.213\text{b})$$

the resides are calculated by Equation 7.110c ≡ Equations 1.38, Equation 7.211b ≡ Equation 1.52, Equation 7.212b ≡ Equation 1.79, and Equation 7.213b ≡ Equation 1.55.

The secondary hyperbolic functions are also meromorphic, with the cosecant and cotangent (secant and tangent) having equally spaced interlaced simple poles along the imaginary axis (Figure 7.1) with alternating (fixed) residues:

$$m \in |Z: \quad \operatorname{csch}(im\pi) = \infty, \quad \operatorname{csch}_{(1)}(im\pi) = (-)^m, \qquad (7.214\text{a–}7.214\text{c})$$

$$\operatorname{sech}\left(im\pi + i\frac{\pi}{2}\right) = \infty, \quad \operatorname{sech}_{(1)}\left(m\pi + i\frac{\pi}{2}\right) = (-)^{m+1} i, \qquad (7.215\text{a and }7.215\text{b})$$

$$\tanh\left(im\pi + i\frac{\pi}{2}\right) = \infty, \quad \tanh_{(1)}\left(im\pi + i\frac{\pi}{2}\right) = +1; \tag{7.216a and 7.216b}$$

$$\coth(im\pi) = \infty, \quad \coth_{(1)}(im\pi) = +1; \tag{7.217a and 7.217b}$$

this follows from Equations 7.214b and 7.214c ≡ Equation 7.218, Equations 7.215a and 7.215b ≡ Equation 7.219, Equations 7.216a and 7.216b ≡ Equation 7.220, and Equations 7.217a and 7.217b ≡ Equation 7.221:

$$\operatorname{csch}_{(1)}\left(im\pi\right) = \lim_{z\to im\pi} \frac{z - im\pi}{\sinh z} = \lim_{z\to im\pi} \frac{1}{\cosh z} = \frac{1}{\cosh\left(im\pi\right)} = \sec\left(m\pi\right) = (-)^m, \tag{7.218}$$

$$\begin{aligned} \operatorname{sech}_{(1)}\left(im\pi + i\pi/2\right) &= \lim_{z\to im\pi+i\pi/2} \frac{z - im\pi - i\pi/2}{\cosh z} = \frac{1}{\sinh\left(im\pi + i\pi/2\right)} \\ &= \frac{1}{i\sin\left(m\pi + \pi/2\right)} = (-i)\,(-)^m = (-)^{m+1}\,i, \end{aligned} \tag{7.219}$$

$$\begin{aligned} \tanh_{(1)}\left(im\pi + i\pi/2\right) &= \lim_{z\to im\pi+i\pi/2} \frac{z - im\pi - i\pi/2}{\coth z} = -\frac{1}{\operatorname{csch}^2\left(im\pi + i\pi/2\right)} \\ &= -\frac{1}{\left[-i\csc\left(m\pi + \pi/2\right)\right]} = \sin^2\left(m\pi + \pi/2\right) = 1. \end{aligned} \tag{7.220}$$

$$\begin{aligned} \coth_{(1)}\left(im\pi\right) &= \lim_{z\to im\pi} \frac{z - im\pi}{\tanh z} = \lim_{z\to im\pi} \frac{1}{\operatorname{sech}^2 z} = \cosh^2\left(im\pi\right) \\ &= \cos^2\left(m\pi\right) = 1, \end{aligned} \tag{7.221a–7.221c}$$

where the L'Hospital rule (Equation I.19.41 ≡ Equation 1.55) was used.

7.6 Series of Fractions for Meromorphic Functions

The *primary functions*, namely, sine and cosine, have an infinite number of zeros, all simple and mutually interlaced (Figure 7.1); they coincide with the simple poles of the secondary functions obtained by inversion, namely, cosecant and secant. The *secondary functions* defined by ratios of primary functions, namely, tangent and cotangent, have an infinite number of both zeros and poles, all simple and mutually interlaced. Table 7.5 indicates (Figure 7.1) for all circular and hyperbolic functions the zeros a_n (poles b_n) and the respective slopes A_n (residues B_n). The secondary functions are meromorphic, that is, they are analytic except at an infinite number of poles, and also meet the condition of boundedness in a sequence of circles, leading to expansions in series of fractions (Sections 1.2 and 1.3):

$$m \in |Z: \quad z \neq m\pi: \quad \cot z = \frac{1}{z} - 2z\sum_{n=1}^{\infty}\frac{1}{n^2\pi^2 - z^2}, \quad \csc z = \frac{1}{z} - 2z\sum_{n=1}^{\infty}\frac{(-)^n}{n^2\pi^2 - z^2}, \tag{7.222a–7.222d}$$

$$z \neq im\pi: \quad \coth z = \frac{1}{z} + 2z\sum_{n=1}^{\infty}\frac{1}{n^2\pi^2 + z^2}, \quad \operatorname{csch} z = \frac{1}{z} + 2z\sum_{n=1}^{\infty}\frac{(-)^n}{n^2\pi^2 + z^2}, \tag{7.223a–7.223c}$$

$$z \neq im\pi + \pi/2: \quad \tan z = 8z\sum_{n=1}^{\infty}\frac{1}{(2n-1)^2\pi^2 - z^2}, \quad \sec z = 1 + 4\pi\sum_{n=1}^{\infty}\frac{(-)^n(2n-1)}{(2n-1)^2\pi^2 + z^2}, \tag{7.224a–7.224c}$$

$$z \neq im\pi + i\pi/2: \quad \tan z = 8z\sum_{n=1}^{\infty}\frac{1}{(2n-1)^2\pi^2 + z^2}, \quad \operatorname{sech} z = 1 + 4\pi\sum_{n=1}^{\infty}\frac{(-)^n(2n-1)}{(2n-1)^2\pi^2 + z^2}. \tag{7.225a–7.225c}$$

The series of fractions for the circular (hyperbolic) cotangent [Equations 7.222c (Equation 7.223b)], cosecant [Equation 7.222d (Equation 7.223c)], secant [Equation 7.224c (Equation 7.225c)], and tangent [Equation 7.224b (Equation 7.225b)] converge absolutely outside the poles $|z - b_n| > 0$ *and converge totally outside a neighborhood* $|z - b_n| \geq \varepsilon_n > 0$. The series of fractions for the secondary circular functions coincide and were obtained before (Sections 1.2 and 1.3) for the cotangent (Equation 7.222c ≡ Equation 1.58), cosecant (Equation 7.224d ≡ Equation 1.85), tangent (Equation 7.224b ≡ Equation 1.78), and secant (Equation 7.224c ≡ Equation 1.54). The imaginary change of variable (Equations 5.27b, 5.25b, 5.26b, and 5.24b; Section 5.2) relates the series of fractions for the secondary hyperbolic (circular) functions cotangent (Equation 7.223b), cosecant (Equation 7.223c), tangent (Equation 7.225b), and secant (Equation 7.225c), respectively. The latter series of fractions could also be obtained via the Mittag–Leffler theorem (Sections 1.2 and 1.3).

7.7 Relation with Factorization in Infinite Products

The primary functions (that is, sine and cosine) are analytic for finite z, have an infinite number of zeros, and meet the condition of having bounded logarithmic derivative on a sequence of circles, as required by the Mittag–Leffer theorem (Sections 1.4 and 1.5). This leads to *the infinite products for the circular (hyperbolic) sine [Equation 7.226b (Equation 7.227b)] and cosine [Equation 7.226c (Equation 7.227a)]:*

$$|z| < \infty: \quad \sin z = z \prod_{n=1}^{\infty} \left(1 - \frac{z^2}{n^2 \pi^2}\right), \quad \cos z = \prod_{n=0}^{\infty} \left[1 - \frac{4z^2}{(2n+1)^2 \pi^2}\right], \tag{7.226a–7.226c}$$

$$|z| < \infty: \quad \sinh z = z \prod_{n=1}^{\infty} \left(1 + \frac{z^2}{n^2 \pi^2}\right), \quad \cosh z = \prod_{n=0}^{\infty} \left[1 + \frac{4z^2}{(2n+1)^2 \pi^2}\right], \tag{7.227a–7.227c}$$

which are valid in the whole complex plane. The two formulas for the primary circular functions, namely, sine and cosine, coincide with Equation 7.226b ≡ Equation 1.75 and Equation 7.226c ≡ Equation 1.73, respectively; the infinite products for the primary hyperbolic functions can be obtained by the same method (Sections 1.4 and 1.5) or by using the imaginary change of variable [Equation 5.6a (Equation 5.6b)] for hyperbolic sine (Equation 7.227b) [cosine (Equation 7.227c)]. Although the circular (hyperbolic) tangent (cotangent) also has an infinite number of zeros, it does not meet the conditions of the Mittag–Leffer theorem (Subsection 1.5.5), so that the latter does not supply representations as infinite products.

7.8 Continued Fractions for Direct/Inverse Functions

The elementary transcendental functions have no exact rational representation in finite terms, leading to four infinite representations, namely, (1) infinite products (Section 7.7) for the primary functions cosine and sine; (2) series of fractions (Section 7.6) for the secondary functions secant, cosecant, tangent, and cotangent; (3) power series for all direct, that is, primary and secondary functions (Section 7.1) and also for all inverse or cyclometric functions (Section 7.4); and (4) continued fractions (Section 7.8) for all direct and inverse functions. The continued fractions are obtained by three methods: (1 and 2) transformation [Subsection 1.8.1 (Subsection 1.8.3)] of infinite series (products) and (3) the Lambert method applied to functions of the hypergeometric family (Sections 1.9 and 5.9). The continued fractions can be extended (1) between circular and hyperbolic functions using the imaginary change of variable (Section 5.1) and (2) between algebraically inverse functions using the inversion relation (Subsection 1.6.5). As examples of these methods, 24 continued fractions are considered: (1) eight (four) for the sine and cosecant (also cosine and secant) [Subsection 7.8.1 (Subsection 7.8.2)] by transformations of infinite products (series); (2) four for tangent and cotangent using the hypergeometric function (Subsection 7.8.3); and (3) eight for the cyclometric functions (Subsection 7.8.4) using infinite series and the Gaussian hypergeometric function. The Gaussian hypergeometric function is also used to obtain 16 more (a total of 40) continued fractions; (4) four for the ratio of cyclometric and rational functions (Subsection 7.8.5); and (5 and 6) four (two) for the tangent and cotangent of a multiple angle as a function of the tangent (the exponential of the inverse tangent) [Subsection 7.8.8 (Subsection 7.8.9)] using continued fractions for bilinear combinations (six ratios) of binomial functions [Subsection 7.8.6 (Subsection 7.8.7)].

7.8.1 Continued Fractions for the Sine and Cosecant

The transformation of the infinite product for the circular sine (Equation 1.148a) into a continued fraction for the circular sine (Equations 1.151a and 1.151b ≡ Equations 7.228a and

7.228b) [cosecant (Equations 1.152a and 1.152b ≡ Equations 7.230a and 7.230b)] also leads via the imaginary change of variable [Equation 5.6b (Equation 5.25b)] to the continued fraction for the hyperbolic sine (Equations 7.229a and 7.229b) [cosecant (Equations 7.231a and 7.231b)]:

$$|z| < \infty: \quad \frac{\sin(\pi z)}{\pi z} = \frac{1-z}{1-}\,\frac{z}{1+z+}\,\frac{1+z}{1-z+}\,\frac{2(2-z)}{z+}\,\frac{2(2+z)}{1-z+}\cdots \frac{n(n-z)}{z+}\,\frac{n(n+z)}{1-z+}\cdots, \tag{7.228a and 7.228b}$$

$$|z| < \infty: \quad \frac{\sinh(\pi z)}{\pi z} = \frac{\sin(i\pi z)}{i\pi z} = \frac{1-iz}{1-}\,\frac{iz}{1+iz+}\,\frac{1+iz}{1-iz+}\,\frac{2(2-iz)}{iz+}\,\frac{2(2+iz)}{1-iz+}\cdots\frac{n(n-iz)}{iz+}\,\frac{n(n+iz)}{1-iz+}\cdots$$
$$= \frac{i+z}{i+}\,\frac{z}{i-z}\,\frac{i-z}{i+z+}\,\frac{2(z+2i)}{z+}\,\frac{2(z-2i)}{i+z+}\cdots\frac{n(z+ni)}{z+}\,\frac{n(z-ni)}{i+z+}\cdots \tag{7.229a and 7.229b}$$

$$z \notin Z|: \quad \pi z\csc(\pi z) = \frac{1}{1-z} - \frac{z}{1+z+}\,\frac{1+z}{1-z+}\,\frac{2(2-z)}{z+}\,\frac{2(2+z)}{1+z+}\cdots\frac{n(n-z)}{z+}\,\frac{n(n+z)}{1+z+}\cdots, \tag{7.230a and 7.230b}$$

$$z \notin Z|: \quad \pi z\csc h(\pi z) = i\pi z\csc(i\pi z) = \frac{1}{1-iz} - \frac{iz}{1+iz+}\,\frac{1+iz}{1-iz+}\,\frac{2(2-iz)}{iz+}\,\frac{2(2+iz)}{1-iz}\cdots\frac{n(n-iz)}{iz+}\,\frac{n(n+iz)}{1-iz+}\cdots$$
$$= \frac{1}{1-iz} + \frac{z}{i-z+}\,\frac{i-z}{i+z+}\,\frac{2(z+2i)}{z+}\,\frac{2(z-2i)}{i+z+}\cdots\frac{n(z+ni)}{z+}\,\frac{n(z-ni)}{i+z+}. \tag{7.231a and 7.231b}$$

The continued fraction for the circular (hyperbolic) (1) sine [Equation 7.228b (Equation 7.229b)] converges absolutely for $|z| < \infty$ and totally for $|z| \le M < \infty$; and (2) cosecant [Equation 7.230b (Equation 7.231b)] converges absolutely except at the poles $z \neq m\pi$ and totally excluding a neighborhood $|z - m\pi| \ge \varepsilon > 0$.

7.8.2 Continued Fractions for the Cosine and Secant

The series (infinite products) listed in Sections 7.1, 7.4, and 7.6 (Section 7.7) can be transformed into continued fractions by using Equation 1.128b (Equation 1.133). For example, the series Equation 7.12b ≡ Equation 7.232a (Equation 7.13b ≡ Equation 7.233a) for the circular cosine (sine) has a general term Equation 7.232b (Equation 7.233b) and a ratio of successive terms Equation 7.232c (Equation 7.233c):

$$\cos z = \sum_{n=0}^{\infty} f_{n+1}, \quad f_{n+1} = \frac{(-)^n}{(2n)!} z^{2n}, \quad \frac{f_{n+1}}{f_n} = -\frac{z^2}{2n(2n-1)}, \tag{7.232a–7.232c}$$

$$\sin z = \sum_{n=0}^{\infty} f_{n+1}, \quad f_{n+1} = \frac{(-)^n}{(2n+1)!} z^{2n+1}, \quad \frac{f_{n+1}}{f_n} = -\frac{z^2}{2n(2n+1)}; \tag{7.233a–7.233c}$$

substitution of Equation 7.232c (Equation 7.233c) in Equation 1.128a leads to the continued fraction for the circular cosine (Equation 7.231) [sine (Equation 7.231)]:

$$\cos z = \frac{1}{1-}\,\frac{-z^2/2}{1-z^2/2-}\,\frac{-z^2/4.3}{1-z^2/4.3}\,\frac{-z^2/6.5}{1-z^2/6.5-} \cdots \frac{-z^2/\left[2n(2n-1)\right]}{1-z^2/\left[2n(2n-1)\right]-} \cdots \tag{7.234}$$

$$\sin z = \frac{z}{1-}\,\frac{-z^2/2.3}{1-z^2/2.3-}\,\frac{-z^2/4.5}{1-z^2/4.5-} \cdots \frac{-z^2/\left[2n(2n+1)\right]}{1-z^2/\left[2n(2n+1)\right]-} \cdots, \tag{7.235}$$

that simplifies to Equation 7.236b (Equation 7.238b). Using also the imaginary change of variable (Equations 5.6a, 5.6b, 5.25b, and 5.24b) and the algebraic inversion of continued fractions (Equations 1.96 and 1.108) leads to

$$|z| < \infty: \quad \cos z = \frac{1}{1+}\,\frac{z^2}{2.1-z^2+}\,\frac{2.1z^2}{4.3-z^2+}\,\frac{4.3z^2}{6.5-z^2+} \cdots \frac{2n(2n-1)z^2}{(2n+2)(2n+1)-z^2+} \cdots, \tag{7.236a and 7.236b}$$

$$|z| < \infty: \quad \cosh z = \frac{1}{1-}\,\frac{z^2}{2.1+z^2-}\,\frac{2.1z^2}{4.3+z^2-}\,\frac{4.3z^2}{6.5+z^2-}\,\frac{2n(2n-1)z^2}{(2n+2)(2n+1)+z^2-} \cdots, \tag{7.237a and 7.237b}$$

$$|z| < \infty: \quad \sin z = \frac{z}{1+}\,\frac{z^2}{2.3-z^2+}\,\frac{2.3z^2}{4.5-z^2+}\,\frac{4.5z^2}{6.7-z^2+} \cdots \frac{2n(2n+1)z^2}{(2n+2)(2n+3)-z^2+} \cdots, \tag{7.238a and 7.238b}$$

$$|z| < \infty: \quad \sinh z = \frac{z}{1-}\,\frac{z^2}{2.3+z^2-}\,\frac{2.3z^2}{4.5+z^2-}\,\frac{4.5z^2}{6.7+z^2-} \cdots \frac{2n(2n+1)z^2}{(2n+2)(2n+3)+z^2-} \cdots, \tag{7.239a and 7.239b}$$

$$m \in |Z; \quad z \neq m\pi: \quad \csc z = \frac{1}{z} + \frac{z^2}{2.3 - z^2 +} \frac{z^2}{4.5 - z^2 +} \frac{4.5}{6.7 - z^2 +} \cdots \frac{2n(2n+1)z^2}{(2n+2)(2n+3) - z^2 +} \cdots, \quad (7.240a \text{ and } 7.240b)$$

$$z \neq im\pi; \quad \operatorname{csch} z = \frac{1}{z} - \frac{z^2}{2.3 + z^2 -} \frac{2.3z^2}{4.5 + z^2 -} \frac{4.5z^2}{6.7 + z^2 -} \cdots \frac{2n(2n+1)z^2}{(2n+2)(2n+3) + z^2 -} \cdots, \quad (7.241a \text{ and } 7.241b)$$

$$z \neq m\pi + \frac{\pi}{2}: \quad \sec z = 1 + \frac{z^2}{2.1 - z^2 +} \frac{2.1z^2}{4.3 - z^2 +} \frac{4.3z^2}{6.5 - z^2 +} \cdots \frac{2n(2n-1)z^2}{(2n+2)(2n+1) - z^2 +} \cdots, \quad (7.242a \text{ and } 7.242b)$$

$$z \neq im\pi + i\frac{\pi}{2}: \quad \operatorname{sech} z = 1 - \frac{z^2}{2.1 + z^2 -} \frac{2.1z^2}{4.3 + z^2 -} \frac{4.3z^2}{6.5 + z^2 -} \cdots \frac{2n(2n-1)z^2}{(2n+2)(2n+1) + z^2 -} \cdots \quad (7.243a \text{ and } 7.243b)$$

Thus, *the continued fractions for the circular (hyperbolic) (1) cosine [Equation 7.236b (Equation 7.237b)] and sine [Equation 7.238b (Equation 7.239b)] converge absolutely in the finite z-plane* $|z| < \infty$ *and totally in the finite disk* $|z| \leq M < \infty$ *(2) cosecant [Equation 7.240b (Equation 7.241b)] converge absolutely outside the poles [Equations 7.241a and 7.241b (Equation 7.241a)] and totally excluding a neighborhood* $|z - m\pi| \geq \varepsilon > 0$ $(|z - im\pi| \geq \varepsilon > 0)$*; and (3) secant [Equation 7.242b (Equation 7.243b)] converge absolutely outside the poles [Equation 7.242a (Equation 7.243a)] and totally excluding a neighborhood* $|z - m\pi - \pi/2| \geq \varepsilon > 0$ $(|z - im\pi - i\pi/2| \geq \varepsilon > 0)$.

7.8.3 Continued Fractions for the Tangent and Cotangent

The Lambert method (Section 9.1) applied to the hypogeometric function (Equation 1.159) leads to the continued fraction (Equations 1.181a and 1.181b ≡ Equations 7.177b and 7.177c) for the circular tangent; the algebraic inversion (Equations 1.196 and 1.108) specifies the continued fraction for the circular cotangent (Equations 7.146a and 7.146b), and the imaginary change of variable (Equations 5.26b and 5.27b) leads to the corresponding hyperbolic functions (Equations 7.145a, 7.145b, 7.147a, and 7.147b).

$$m \in |Z; \quad z \neq m\pi + \frac{\pi}{2}: \quad \tan z = \frac{z}{1-} \frac{z^2}{3-} \frac{z^2}{5-} \frac{z^2}{7-} \frac{z^2}{9-} \cdots \frac{z^2}{2n+1-} \cdots \quad (7.244a\text{–}7.244c)$$

$$z \neq im\pi + i\frac{\pi}{2}: \quad \tanh z = \frac{z}{1+} \frac{z^2}{3+} \frac{z^2}{5+} \frac{z^2}{7+} \frac{z^2}{9+} \cdots \frac{z^2}{2n+1+} \cdots, \quad (7.245a \text{ and } 7.245b)$$

$$z \neq im\pi: \quad \cot z = \frac{1}{z} - \frac{z^2}{3-}\,\frac{z^2}{5-}\,\frac{z^2}{7-}\,\frac{z^2}{9-}\cdots\frac{z^2}{2n+1-}\cdots, \qquad \text{(7.246a and 7.246b)}$$

$$z \neq im\pi: \quad \coth z = \frac{1}{z} + \frac{z^2}{3+}\,\frac{z^2}{5+}\,\frac{z^2}{7+}\,\frac{z^2}{9+}\cdots\frac{z^2}{2n+1+}\cdots. \qquad \text{(7.247a and 7.247b)}$$

The inversion (Equations 1.95 and 1.108) of the continued fraction Equation 7.244c ≡ Equation 7.248a leads to Equation 7.248b ≡ Equation 7.246b:

$$\frac{\tan z}{z} = \frac{1}{1-}\,\frac{z^2}{3-}\,\frac{z^2}{5-}\,\frac{z^2}{9-}\cdots\frac{z^2}{2n+1-}\cdots \qquad \text{(7.248a)}$$

$$z \cot z = 1 - \frac{z^2}{3-}\,\frac{z^2}{5-}\,\frac{z^2}{7-}\,\frac{z^2}{9-}\cdots\frac{z^2}{2n+1-}\cdots \qquad \text{(7.248b)}$$

As $z \to 0$ the r.h.s. of Equations 7.236a, 7.23b, 7.242b and 7.243b also tend to unity, the r.h.s. of Equations 7.238b, 7.239b, 7.244c and 7.245c tend to zero, and the r.h.s. of Equations 7.240b, 7.241b, 7.246b and 7.247b have a simple pole at the origin with residue unity. *The continued fractions for the circular (hyperbolic) (1) tangent [Equation 7.244c (Equation 7.245b)] converge absolutely except at the poles [Equation 7.244b (Equation 7.245a)] and converge totally excluding a neighbourhood* $|z - m\pi - \pi/2| \geq \varepsilon > 0$ ($|z - im\pi - i\pi/2| \geq \varepsilon > 0$)*; and (2) cotangent [Equation 7.246b (Equation 7.247b)] converge absolutely excluding the poles [Equation 7.246a (Equation 7.247a)] and totally excluding a neighborhood of the poles* $|z - m\pi| \geq \varepsilon > 0$.

7.8.4 Continued Fractions for Cyclometric Functions

The cyclometric functions are the solution of trigonometric inversion problems that specify the angles (arguments) corresponding to given values of circular (hyperbolic) functions. As an alternative to trigonometric inversion using power series (Section 7.4), these may be transformed into continued fractions (Subsection 1.8.1), leading to expressions valid in the whole complex plane except for branch cuts. For example, the arc tan z has power series Equation 7.180b ≡ Equation 7.249a with a general term Equation 7.249b and a ratio of successive terms Equation 7.249c:

$$\operatorname{arc\,tan} z = \sum_{n=1}^{\infty} f_n, \quad f_n = \frac{(-)^{n-1}}{2n-1} z^{2n-1}, \quad \frac{f_{n+1}}{f_n} = -z^2\,\frac{2n-1}{2n+1}, \qquad \text{(7.249a–7.249c)}$$

leading (Equation 1.128a) to the continued fraction (Equation 7.249d):

$$\operatorname{arc\,tan} z = \frac{z}{1-}\,\frac{-z^2/3}{1-z^2/3-}\,\frac{-3z^2/5}{1-3z^2/5-}\,\frac{-5z^2/7}{1-5z^2/7-}\cdots\frac{-(2n-1)z^2/(2n+1)}{1-(2n-1)z^2/(2n+1)+}\cdots; \qquad \text{(7.249d)}$$

this can be simplified to Equations 7.250a and 7.250b:

$$z \notin)-i\infty,-i) \cup (+i,+i\infty(: \quad \arctan z = \frac{z}{1+}\frac{z^2}{3-z^2+}\frac{3^2 z^2}{5-3z^2+}\frac{5^2 z^2}{7-5z^2+}\cdots\frac{(2n-1)^2 z^2}{2n+1-(2n-1)z^2+}\cdots \tag{7.250a and 7.250b}$$

$$z \notin)-\infty,-i) \cup (+1,\infty(: \quad \arg\tanh z = \frac{z}{1-}\frac{z^2}{3+z^2-}\frac{3^2 z^2}{5+3z^2-}\frac{5^2 z^2}{7+5z^2-}\cdots, \frac{\left(2n-1\right)^2 z^2}{2n+1+\left(2n-1\right)z^2-}\cdots, \tag{7.251a and 7.251b}$$

$$z \notin \left(-i,+i\right): \quad \operatorname{arc}\cot z = \frac{1}{z}+\frac{z^2}{3-z^2+}\frac{3^2 z^2}{5-3z^2+}\frac{5^2 z^2}{7-5z^2+} \cdots\frac{\left(2n-1\right)^2 z^2}{2n+1-\left(2n-1\right)z^2+}\cdots, \tag{7.252a and 7.252b}$$

$$z \notin \left(-1,+1\right): \quad \arg\coth z = \frac{1}{z}-\frac{z^2}{3+z^2-}\frac{3^2 z^2}{5+3z^2-}\frac{5^2 z^2}{7+5z^2-} \cdots\frac{\left(2n-1\right)^2 z^2}{2n+1+\left(2n-1\right)z^2-}\cdots; \tag{7.253a and 7.253b}$$

also Equations 7.252a and 7.252b follow from Equations 7.250a and 7.250b by algebraic inversion of continued fractions (Equations 1.95 and 1.108), and Equations 7.251a and 7.251b (Equations 7.253a and 7.253b) follow from Equations 7.250a and 7.250b (Equations 7.252a and 7.252b) by the imaginary change of variable Equation 5.187b (Equation 5.188b). All the continued fractions (Equations 7.228b through 7.231b) are valid except at the branch cuts (Equations 7.250a through 7.253a), respectively, in Figures 7.2b, 7.3f, 7.2c, and 7.3e.

The series Equation 7.180b ≡ Equation 7.254 for the inverse circular tangent is a particular case of the Gaussian hypergeometric series (Equation 3.101b):

$$\begin{aligned} \arctan z &= \sum_{n=0}^{\infty}\frac{(-)^n}{2n+1}z^{2n+1} = z\sum_{n=0}^{\infty}\frac{(-z^2)^n}{2n+1} \\ &= z\sum_{n=0}^{\infty}\frac{(-z^2)^n}{n!}\frac{(1)_n(1/2)_n}{(3/2)_n} = zF\left(\frac{1}{2},1;\frac{3}{2};-z^2\right), \end{aligned} \tag{7.254}$$

where Equation 3.121b was used; the imaginary change of variable Equation 5.187b leads from Equation 7.254 ≡ Equation 7.255a to Equation 7.255b

$$\arc\tan z = zF\left(\frac{1}{2},1;\frac{3}{2};-z^2\right), \quad \arg\tanh z = zF\left(\frac{1}{2},\frac{3}{2};1;z^2\right). \qquad \text{(7.255a and 7.255b)}$$

The logarithmic representation of the inverse hyperbolic tangent (Equations 7.74a and 7.74b) agrees with relations 3.121a and 7.256 with the Gaussian hypergeometric function:

$$\arg\tanh z = \frac{1}{2}\log\left(\frac{1+z}{1-z}\right) = zF\left(\frac{1}{2},1;\frac{3}{2};z^2\right). \qquad (7.256)$$

This specifies (Equation 3.122e) its continued fraction (Equations 7.258a and 7.258b) and yields

$$z \notin)-i\infty,-i) \cup (i,+\infty(: \quad \arc\tan z = \frac{z}{1+}\frac{z^2}{3+}\frac{4z^2}{5+}\frac{9z^2}{7+}\frac{16z^2}{9+}\cdots, = \frac{n^2 z}{2n+1+}\cdots, \qquad \text{(7.257a and 7.257b)}$$

$$z \notin)-\infty,-1) \cup (1,+\infty(: \quad \arg\tanh z = \frac{z}{1-}\frac{z^2}{3-}\frac{4z^2}{5-}\frac{9z^2}{7-}\frac{16z^2}{9-} - \frac{n^2z^2}{2n+1-}\cdots, \qquad \text{(7.258a and 7.258b)}$$

$$z \notin (-1,+1): \quad \arg\coth z = \frac{1}{z} - \frac{z}{3-}\frac{4z^2}{5-}\frac{9z^2}{7-}\frac{16z^2}{9-} \cdots \frac{n^2z^2}{2n+1-}\cdots, \qquad \text{(7.259a and 7.259b)}$$

$$z \notin (-i,+i): \quad \arc\cot z = \frac{1}{z} + \frac{z}{3+}\frac{4z^2}{5+}\frac{9z^2}{7+}\frac{16z^2}{9+}\cdots\frac{n^2z^2}{2n+1+}\cdots, \qquad \text{(7.260a and 7.260b)}$$

where (1) algebraic inversion of continued fractions (Equations 1.96 and 1.108) leads from Equations 7.258a and 7.258b to Equations 7.259a and 7.259b; and (2) the imaginary change of variable 7.187b (Equation 7.188b) applied to Equations 7.258a and 7.258b (Equations 7.259a and 7.259b) yields Equations 7.257a and 7.257b (Equations 7.260a and 7.160b). Thus, *the circular (hyperbolic) cyclometric functions have continued fractions for the (1) inverse tangent [Equations 7.250b and 7.257b (Equations 7.251b and 7.258b)] valid outside the branch cut [Equation 7.250a ≡ Equation 7.257a (Equation 7.251a ≡ Equation 7.258a)] in Figure 7.2b (Figure 7.3f); and (2) inverse cotangent [Equations 7.252b and 7.260b (Equations 7.253b and 7.259b)] valid outside the branch cut [Equation 7.252a ≡ Equation 7.260a (Equation 7.253a ≡ Equation 7.259a)] in Figure 7.2c (Figure 7.3e).*

7.8.5 Continued Fractions for the Ratio of Cyclometric and Rational Functions

The series Equations 7.167a and 7.167c ≡ Equations 7.261a and 7.261b for the inverse circular sine are also a particular case of the Gaussian hypergeometric series (Equation 3.101b).

$$|z| < 1: \quad \arc\sin z = z + \sum_{n=1}^{\infty} \frac{z^{2n+1}}{2n+1}\frac{1}{n!}\frac{1}{2}\frac{3}{2}\cdots\left(n-\frac{3}{2}\right)\left(n-\frac{1}{2}\right)$$

$$= z + z\sum_{n=1}^{\infty}\frac{(z^2)^n}{n!}\frac{(1/2)_n(1/2)_n}{(3/2)_n} = z\left(F\frac{1}{2},\frac{1}{2};\frac{3}{2};z\right), \qquad \text{(7.261a and 7.261b)}$$

where Equation 3.121b was used; the imaginary change of variable Equation 7.187a applied to Equation 7.261b ≡ Equation 7.262d leads to Equation 7.262e:

$$(a,b,c)=\left(\frac{1}{2},\frac{1}{2};\frac{3}{2}\right): \quad \arcsin z = zF\left(\frac{1}{2},\frac{1}{2};\frac{3}{2};z^2\right), \quad \arg\sinh z = zF\left(\frac{1}{2},\frac{1}{2};\frac{3}{2};-z^2\right). \tag{7.262a–7.262e}$$

The Gaussian hypergeometric function Equation 7.261b has parameters Equations 7.262a through 7.262c, and the contiguous hypergeometric function with parameters Equations 7.263a through 7.263c is (Equation 3.143b) the binomial Equation 7.263d:

$$(a,b;c)=\left(-\frac{1}{2},\frac{1}{2};\frac{1}{2}\right): \quad F\left(-\frac{1}{2},\frac{1}{2};\frac{1}{2};z^2\right)=\sqrt{1-z^2}. \tag{7.263a–7.263d}$$

The ratio of Equations 7.262d and 7.263d is the ratio of two contiguous hypergeometric functions (Equation 7.264a) that has (Equation 3.115d) the continued fraction Equation 7.264b:

$$\begin{aligned}\frac{\arcsin z}{\sqrt{1-z^2}} &= z\frac{F\left(1/2,1/2;3/2;z^2\right)}{F\left(1/2,-1/2;1/2;z^2\right)}\\ &= \frac{z}{1+}\,\frac{\alpha_1 z^2}{1+}\,\frac{\alpha_2 z^2}{1+}\cdots\frac{\alpha_{2n+1} z^2}{1+}\,\frac{\alpha_{2n+2} z^2}{1+}\cdots;\end{aligned} \tag{7.264a and 7.264b}$$

the symmetry of the Gaussian hypergeometric function for the upper parameters (Equation 3.111a) was used when passing from Equations 7.263d to 7.264a; the coefficients (Equations 3.113a and 3.113b) of the continued fraction Equation 7.264b for Equations 7.265a through 7.265c are given by Equations 7.265d and 7.265e:

$$(a,b;c)=\left(\frac{1}{2},-\frac{1}{2};\frac{1}{2}\right): \quad \left\{\alpha_{2n+1},\alpha_{2n+2}\right\}=-\frac{2+2n}{3+4n}\left\{\frac{1+2n}{1+4n},\frac{1+2n}{5+4n}\right\}. \tag{7.265a–7.265e}$$

Substitution of Equations 7.265d and 7.265e in Equation 7.264b leads to the continued fraction:

$$\begin{aligned}\frac{\arcsin z}{\sqrt{1-z^2}} &= \frac{z}{1+}\,\frac{-2z^2/3}{1+}\,\frac{-2z^2/3.5}{1+}\,\frac{-3.4z^2/5.7}{1+}\,\frac{-3.4z^2/7.9}{1+}\\ &\quad\cdots\frac{-(2+2n)(1+2n)z^2/(1+4n)(3+4n)}{1+}\,\frac{-(2+2n)(1+2n)z^2/(3+4n)(5+4n)}{1+}\cdots\\ &= \frac{z}{1-}\,\frac{1.2z^2}{3-}\,\frac{1.2z^2}{5-}\,\frac{3.4z^2}{7-}\,\frac{3.4z^2}{9-}\cdots\frac{(2n-1)2nz^2}{2n-1-}\,\frac{(2n-1)2nz^2}{2n+1-}\cdots,\end{aligned} \tag{7.266}$$

that coincides with Equations 7.267a and 7.267b.

Thus, the continued fractions follow:

$$z \notin \left)-\infty,-1\right(\cup\left)+1,+\infty\right(: \quad \frac{\arcsin z}{\sqrt{1-z^2}} = \frac{z}{1-}\,\frac{1.2z^2}{3-}\,\frac{1.2z^2}{5-}\,\frac{3.4z^2}{7-}\,\frac{3.4z^2}{9-} \cdots \frac{(2n-1)2nz^2}{2n-1-}\,\frac{(2n-1)2nz^2}{2n+1-}\cdots, \tag{7.267a and 7.267b}$$

$$z \notin \left)-i\infty,-i\right(\cup\left)+i,+i\infty\right(: \quad \frac{\operatorname{arc\,sinh} z}{\sqrt{1-z^2}} = \frac{z}{1+}\,\frac{1.2z^2}{3+}\,\frac{1.2z^2}{5+}\,\frac{3.4z^2}{7+}\,\frac{3.4z^2}{9+} \cdots \frac{(2n-1)2nz^2}{2n-1+}\,\frac{(2n-1)2nz^2}{2n+1+}\cdots, \tag{7.268a and 7.268b}$$

$$z \notin \left)-1,+1\right(: \quad \frac{\operatorname{arc\,csc} z}{\sqrt{z^2-1}} = \frac{1}{z^2-}\,\frac{1.2}{3z^2-}\,\frac{1.2}{5z^2-}\,\frac{3.4}{7z^2-}\,\frac{3.4}{9z^2-} \cdots \frac{(2n-1)2n}{(2n-1)z^2-}\,\frac{(2n-1)2n}{(2n+1)z^2-}\cdots, \tag{7.269a and 7.269b}$$

$$z \notin \left)-i,+i\right(: \quad \frac{\operatorname{arg\,csch} z}{\sqrt{z^2+1}} = \frac{1}{z^2+}\,\frac{1.2}{3z^2+}\,\frac{1.2}{5z^2+}\,\frac{3.4}{7z^2+}\,\frac{3.4}{9z^2+} \cdots \frac{(2n-1)2n}{(2n-1)z^2+}\,\frac{(2n-1)2n}{(2n+1)z^2+}\cdots, \tag{7.270a and 7.270b}$$

where Equations 7.268a and 7.268b (Equations 7.270a and 7.270b) follow from Equations 7.267a and 7.267b (Equations 7.268a and 7.268b) by the imaginary change of variable Equation 7.187a (Equation 7.188a). It remains to prove Equations 7.269a and 7.269b, that follow from Equations 7.267a and 7.267b) as shown next.

The change of variable Equation 7.271a together with the property Equation 7.184b leads to Equation 7.271b:

$$z \to \frac{1}{z}: \quad \frac{\arcsin(1/z)}{\sqrt{1-1/z^2}} = \frac{z}{\sqrt{z^2-1}} \operatorname{arc\,csc}(z); \tag{7.271a and 7.271b}$$

substitution of Equation 7.267b in Equation 7.271b leads to the continued fraction:

$$\begin{aligned} \frac{\operatorname{arc\,csc}(z)}{\sqrt{z^2-1}} &= \frac{1/z^2}{1-}\,\frac{1.2/z^2}{3-}\,\frac{1.2/z^2}{5-}\,\frac{3.4/z^2}{9-}\,\frac{3.4z^2}{9-} \cdots \frac{(2n-1)2n/z^2}{2n-1-}\,\frac{(2n-1)2n/z^2}{2n+1-} \\ &= \frac{1}{z^2-}\,\frac{1.2}{3z^2-}\,\frac{1.2}{5z^2-}\,\frac{3.4}{7z^2-}\,\frac{3.4}{9z^2-} \cdots \frac{(2n-1)2n}{(2n-1)z^2-}\,\frac{(2n-1)2n}{(2n+1)z^2-}, \end{aligned} \tag{7.272}$$

that coincides with Equation 7.272 ≡ Equation 7.269b. The functions in the numerators and denominators of Equations 7.267b through 7.270b have the same branch cut (Equations 7.267a through 7.270a) in Figures 7.2a, 7.3a, 7.2d, and 7.3c. Thus, *the ratio of the inverse circular (hyperbolic) cyclometric functions to the square roots with the same branch cuts has the continued fractions (1) for the inverse sine [Equation 7.267b (Equation 7.268b)] valid outside the branch cut [Equation 7.267a (Equation 7.268a)] in Figure 7.2a (Figure 7.3a); and (2) for the inverse cosecant [Equation 7.269b (Equation 7.270b)] valid outside the branch cut [Equation 7.269a (Equation 7.270a)] in Figure 7.2d (Figure 7.3c).* Two more pairs of continued fractions are obtained next [Subsection 7.8.7 (Subsection 7.8.6)].

7.8.6 Continued Fractions for Bilinear Combinations of Binomials

The binomial series (Equation 3.144a through 3.144c) with Equations 7.273a and 7.273b lead to Equation 7.273c:

$$b=1, z\to\pm z:\quad (1\pm z)^{\nu}=\sum_{n=0}^{\infty}\frac{(\pm z)^{n}}{n!}\nu(\nu-1)\cdots(\nu-n+1). \qquad (7.273\text{a–}7.273\text{c})$$

Adding (subtracting) the two series (Equation 7.273c) leads to Equation 7.274 (Equation 7.275):

$$(1+z)^{\nu}+(1-z)^{\nu}=2\sum_{n=0}^{\infty}\frac{z^{2n}}{(2n)!}\nu(\nu-1)\cdots(\nu-2n+1), \qquad (7.274)$$

$$(1+z)^{\nu}-(1-z)^{\nu}=2\sum_{n=0}^{\infty}\frac{z^{2n+1}}{(2n+1)!}\nu(\nu-1)\cdots(\nu-2n). \qquad (7.275\text{b})$$

Both series [Equation 7.274 (Equation 7.275)] are particular cases [Equation 7.276 (Equation 7.277)] of the Gaussian hypergeometric series Equation 3.101b:

$$\begin{aligned}(1+z)^{\nu}+(1-z)^{\nu}&=2\sum_{n=0}^{\infty}z^{2n}\frac{\nu(\nu-1)\cdots(\nu-2n+1)}{1.2\cdots 2n}\\ &=2\sum_{n=0}^{\infty}z^{2n}\frac{(\nu/2)(\nu/2-1)(\nu/2-n+1)}{1.2\cdots n}\frac{[(\nu-1)/2]\cdots[(\nu-1)/2-n+1]}{(1/2)(3/2)\cdots(1/2+n-1)}\\ &=2\sum_{n=0}^{\infty}\frac{(z^2)^n}{n!}\frac{(-\nu/2)_n(1/2-\nu/2)_n}{(1/2)_n}=2F\left(\frac{1-\nu}{2},-\frac{\nu}{2};\frac{1}{2};z^2\right),\end{aligned} \qquad (7.276)$$

$$\begin{aligned}(1+z)^{\nu}-(1-z)^{\nu}&=2\nu z\sum_{n=0}^{\infty}z^{2n}\frac{(\nu-1)\cdots(\nu-2n)}{2\cdots(2n+1)}\\ &=2\nu z\sum_{n=0}^{\infty}z^{2n}\frac{[(\nu-1)/2]\cdots[(\nu-1)/2-n+1]}{1.2\cdots n}\frac{(\nu/2-1)\cdots(\nu/2-n)}{(3/2)(5/2)\cdots(1/2+n)}\\ &=2\nu z\sum_{n=0}^{\infty}\frac{(z^2)^n}{n!}\frac{(1/2-\nu/2)_n(1-\nu/2)_n}{(3/2)_n}=2\nu zF\left(\frac{1-\nu}{2},1-\frac{\nu}{2};\frac{3}{2};z^2\right).\end{aligned} \qquad (7.277)$$

The two hypergeometric functions Equation 7.276 ≡ Equation 7.278a (Equation 7.277 ≡ Equation 7.278b):

$$\left(1+z\right)^{\nu} \pm \left(1-z\right)^{\nu} = 2F\left(\frac{1-\nu}{2}, -\frac{\nu}{2}; \frac{1}{2}; z^2\right), 2\nu z F\left(\frac{1-\nu}{2}, 1-\frac{\nu}{2}; \frac{3}{2}; z^2\right), \qquad \text{(7.278a and 7.278b)}$$

are contiguous in parameters (*b*, *c*), and thus their ratio:

$$\nu z \frac{\left(1+z\right)^{\nu} + \left(1-z\right)^{\nu}}{\left(1+z\right)^{\nu} - \left(1-z\right)^{\nu}} = \frac{F\left(1/2-\nu/2, -\nu/2; 1/2; z^2\right)}{F\left(1/2-\nu/2, 1-\nu/2; 3/2; z^2\right)}$$
$$= 1 + \frac{\alpha_1 z^2}{1+} \frac{\alpha_2 z^2}{1+} \cdots \frac{\alpha_{2n+1} z^2}{1+} \frac{\alpha_{2n+2} z^2}{1+} \cdots, \qquad \text{(7.279)}$$

is specified by the continued fraction Equation 3.115c.

The coefficients (Equations 3.113a and 3.113b) of the continued fraction (Equation 7.279) are given for Equations 7.280a through 7.280c by Equations 7.280d and 7.280e:

$$\left(a,b;c\right) = \left(\frac{1}{2} - \frac{\nu}{2}, -\frac{\nu}{2}; \frac{1}{2}\right): \quad \left\{\alpha_{2n+1} \alpha_{2n+2}\right\} = -\frac{1}{4n+3} \left\{\frac{\left(2n+1\right)^2 - \nu^2}{4n+1}, \frac{\left(2n+2\right)^2 - \nu^2}{4n+5}\right\}. \qquad \text{(7.280a–7.280c)}$$

The coefficients in Equations 7.280b and 7.280c have the common form in Equation 7.281a and substituted in Equation 7.279 lead to Equation 7.281b:

$$\alpha_n = \frac{1}{2n-1} \frac{\nu^2 - n}{2n+1}: \quad \nu z \frac{\left(1+z\right)^2 + \left(1-z\right)^{\nu}}{\left(1+z\right)^{\nu} - \left(1-z\right)^{\nu}}$$
$$= 1 + \frac{\left(\nu^2-1\right) z^2 / 1.3}{1+} \frac{\left(\nu^2-4\right) z^2 / 3.5}{1+} \frac{\left(\nu^2-9\right) z^2 / 5.7}{1+}$$
$$\cdots \frac{\left(\nu^2 - n^2\right) / \left[(2n-1)(2n+1)\right]}{1+} \cdots \qquad \text{(7.281a and 7.281b)}$$

that simplifies to Equation 7.282b:

$$z \notin \left(-1,+1\right): \quad \nu z \frac{\left(1+z\right)^2 + \left(1-z\right)^{\nu}}{\left(1+z\right)^{\nu} - \left(1-z\right)^{\nu}} = 1 + \frac{\left(\nu^2-1\right)}{3+} \frac{\left(\nu^2-4\right) z^2}{5+} \frac{\left(\nu^2-9\right) z^2}{7+} \cdots \frac{\left(\nu^2 - n^2\right) z^2}{2n+1+} \cdots, \qquad \text{(7.282a and 7.282b)}$$

$z \notin)-\infty,-1) \cup (+1,+\infty($:

$$\frac{(1+z)^{2}-(1-z)^{\nu}}{(1+z)^{\nu}+(1-z)^{\nu}} = \frac{\nu z}{1+}\,\frac{(\nu^2-1)z^2}{3+}\,\frac{(\nu^2-4)z^2}{5+}\,\frac{(\nu^2-9)z^2}{7+}\cdots\frac{(\nu^2-n^2)z^2}{2n+1+}\cdots,$$

(7.283a and 7.283b)

the continued fraction (Equation 7.283b) follows from Equation 7.281b by algebraic inversion (Equations 1.95 and 1.108). *The continued fractions for the ratio of sums and differences of binomials [Equation 7.282b (Equation 7.283b)] are valid in the whole complex plane outside the branch cut [Equation 7.282a (Equation 7.283a)] in Figure 7.3e (Figure 7.3f).*

7.8.7 Continued Fractions for Ratios of Binomials (Laguerre 1877)

The change of variable Equation 7.284a transforms the continued fraction Equation 7.283b to Equation 7.284b:

$$z \to \frac{1}{z}: \quad \frac{(z+1)^{\nu}-(z-1)^{\nu}}{(z+1)^{\nu}+(z-1)^{\nu}} = \frac{\nu/z}{1+}\,\frac{(\nu^2-1)/z^2}{3+}\,\frac{(\nu^2-4)/z^2}{5+}\,\frac{(\nu^2-9)/z^2}{7+}\cdots\frac{(\nu^2-n^2)/z^2}{2n+1+}\cdots$$

(7.284a and 7.284b)

that simplifies to Equation 7.285b:

$$z \notin (-1,+1): \quad \frac{(z+1)^{\nu}-(z-1)^{\nu}}{(z+1)^{\nu}+(z-1)^{\nu}} = \frac{\nu}{z+}\,\frac{\nu^2-1}{3z+}\,\frac{\nu^2-4}{5z+}\,\frac{\nu^2-9}{7z+}\cdots\frac{\nu^2-n^2}{(2n+1)z+}\cdots \qquad \text{(7.285a and 7.285b)}$$

$$z \notin)-\infty,-1) \cup (+1,+\infty(: \quad \nu\frac{(z+1)^{\nu}+(z-1)^{\nu}}{(z+1)^{\nu}-(z-1)^{\nu}} = z + \frac{\nu^2-1}{3z+}\,\frac{\nu^2-4}{5z+}\,\frac{\nu^2-9}{7z+}\cdots\frac{\nu^2-n^2}{2n+1+}\cdots$$

(7.286a and 7.286b)

where (1) the inversion Equation 7.284a changes the branch cut from Equation 7.283a in Figure 7.3f to Equation 7.285a in Figure 7.3e; (2) the algebraic inversion of continued fractions (Equations 1.95 and 1.108) leads from Equations 7.285a and 7.285b to Equations 7.286a and 7.286b. Thus, *the bilinear combinations of rational functions have continued fractions [Equations 7.282b and 7.285b (Equations 7.283b and 7.286b)] valid in the complex z-plane excluding the branch cuts [Equation 7.282a ≡ Equation 7.285a (Equation 7.283a ≡ Equation 7.286a)] in Figure 7.3e (Figure 7.3f).*

Subtracting ν from Equation 7.286b leads to the continued fraction:

$$z \notin)-\infty,-1) \cup (+1,+\infty(: \quad \nu\frac{(z+1)^{\nu}+(z-1)^{\nu}}{(z+1)^{\nu}-(z-1)^{\nu}} - \nu = \frac{2\nu(z-1)^{\nu}}{(z+1)^{\nu}-(z-1)^{\nu}}$$

$$= z - \nu + \frac{\nu^2-1}{3z+}\,\frac{\nu^2-4}{5z+}\,\frac{\nu^2-9}{7z+}\cdots\frac{\nu^2-n^2}{(2n+1)z+}\cdots \qquad \text{(7.287a and 7.287b)}$$

whose algebraic inverse (Equations 1.95 and 1.108) is

$$\frac{1}{2\nu}\left[\left(\frac{z+1}{z-1}\right)^{\nu}-1\right]=\frac{1}{z-\nu+}\frac{\nu^2-1}{3z+}\frac{\nu^2-4}{5z+}\frac{\nu^2-9}{7z+}\cdots\frac{\nu^2-n^2}{(2n+1)z+}\cdots \tag{7.288}$$

This simplifies to the continued fraction **(Laguerre 1877)**:

$$z\notin(-1,+1):\quad \left(\frac{z+1}{z-1}\right)^{\nu}=1+\frac{2\nu}{z-\nu+}\frac{\nu^2-1}{3z+}\frac{\nu^2-4}{5z+}\frac{\nu^2-9}{7z+}\cdots\frac{\nu^2-n^2}{(2n+1)z+}\cdots \tag{7.289a and 7.289b}$$

The continued fractions for the ratio of binomial functions [Equation 7.289b (Equation 7.287b)] hold in the complex z-plane with the branch cuts [Equation 7.289a (Equation 7.287a)] in Figure 7.3e (Figure 7.3f).

7.8.8 Continued Fraction for the Tangent of Multiple Argument (Euler 1813)

The change of variable Equation 7.290a maps the branch cut Equation 7.283a ≡ Equation 7.290b to Equation 7.290c:

$$z=i\tan w:\quad z\notin\,)-\infty,-1\cup(+1,+\infty(\quad\Leftrightarrow\quad \operatorname{Im}(\tan w)\notin\,)-\infty,-1)\cup(+1,+\infty(; \tag{7.290a–7.290c}$$

the condition Equation 7.290c is met on the strip Equations 7.290e ≡ Equations 7.290f of the w-plane (Equation 7.290d) as follows (Equation 5.237) from Equation 7.290g:

$$w=x+iy,\ |\operatorname{Re}(w)|\le\frac{\pi}{4};\quad \cos(2x)\ge 0:$$

$$|\operatorname{Im}(\tan w)|=\left|\frac{\sinh(2y)}{\cosh(2y)+\cos(2x)}\right|\le\left|\frac{\sinh(2y)}{\cosh(2y)}\right|=|\tanh(2y)|\le 1. \tag{7.290d–7.290g}$$

The function (Equation 7.283b) becomes:

$$\frac{(1+z)^{\nu}-(1-z)^{\nu}}{(1+z)^{\nu}+(1-z)^{\nu}}=\frac{(1+i\tan w)^{\nu}-(1-i\tan w)^{\nu}}{(1+i\tan w)^{\nu}+(1-i\tan w)^{\nu}}$$

$$=\frac{(\cos w+i\sin w)^{\nu}-(\cos w-i\sin w)^{\nu}}{(\cos w+i\sin w)^{\nu}+(\cos w-i\sin w)^{\nu}}=\frac{e^{i\nu w}-e^{-i\nu w}}{e^{i\nu w}+e^{-i\nu w}}=i\tan(\nu w). \tag{7.291}$$

Substituting Equations 7.290a through 7.290c and 7.291 in Equation 7.283b leads to

$$-\frac{\pi}{4} \le \mathrm{Re}(z) \le +\frac{\pi}{4}; \quad m \in |Z; \quad \nu z \ne m\pi + \frac{\pi}{2}:$$

$$\tan(\nu z) = \frac{\nu \tan z}{1-}\,\frac{(\nu^2-1)\tan^2 z}{3-}\,\frac{(\nu^2-4)\tan^2 z}{5-}\,\frac{(\nu^2-9)\tan^2 z}{7-}\cdots\frac{(\nu^2-n^2)\tan^2 z}{2n+1-}\cdots, \tag{7.292a–7.292d}$$

$$-\frac{\pi}{4} \le \mathrm{Im}(z) \le +\frac{\pi}{4}; \quad \nu z \notin im\pi + \frac{\pi}{2}:$$

$$\tanh(\nu z) = \frac{\nu \tanh z}{1+}\,\frac{(\nu^2-1)\tanh^2 z}{3+}\,\frac{(\nu^2-4)\tanh^2 z}{5+}\,\frac{(\nu^2-9)\tanh^2 z}{7-}\cdots\frac{(\nu^2-n^2)\tanh^2 z}{2n+1+}\cdots, \tag{7.293a–7.293c}$$

$$-\frac{\pi}{4} \le \mathrm{Re}(z) \le +\frac{\pi}{4}; \quad \nu z \notin m\pi:$$

$$\cot(\nu z) = \frac{\cot z}{\nu} - \frac{(\nu^2-1)\tan^2 z}{3-}\,\frac{(\nu^2-4)\tan^2 z}{5-}\,\frac{(\nu^2-9)\tan^2 z}{7-}\cdots\frac{(\nu^2-n^2)\tan^2 z}{2n+1-}\cdots \tag{7.294a–7.294c}$$

$$-\frac{\pi}{4} \le \mathrm{Im}(z) \le +\frac{\pi}{4}; \quad \nu z \ne im\pi:$$

$$\coth(\nu z) = \frac{\coth z}{\nu} + \frac{(\nu^2-1)\tanh^2 z}{3+}\,\frac{(\nu^2-4)\tanh^2 z}{5+}\,\frac{(\nu^2-9)\tanh^2 z}{7+}\cdots\frac{(\nu^2-n^2)\tanh^2 z}{2n+1+}\cdots \tag{7.295a–7.295c}$$

where (1) algebraic inversion leads from Equation 7.292d to Equation 7.294c; and (2) the imaginary change of variable Equation 5.26b (Equation 5.27b) leads from Equation 7.292d (Equation 7.294c) to Equation 7.293c (Equation 7.295c). *The continued fractions for the circular (hyperbolic) tangent [Equation 7.292d (Equation 7.293c)] and cotangent [Equation 7.294c (Equation 7.295c)] of multiple argument* **(Euler 1813)** *hold in a strip of the fundamental region determined by their period (Equations 7.292a through 7.295a) outside their poles (Equations 7.292b, 7.292c, and 7.293b through 7.295b).*

7.8.9 Continued Fraction for Exponentials of Cyclometric Functions (Laguerre 1879)

The function (Equation 7.289b) with complex exponent is by definition

$$\left(\frac{z+1}{z-1}\right)^{\nu} = \exp\left[\nu \log\left(\frac{z+1}{z-1}\right)\right] = \exp(2\nu \arg\coth z), \tag{7.296}$$

using Equation 7.76b. Recalling the identity 7.185c and using Equation 7.289b lead to the continued fraction:

$$z \notin (-1,+1): \quad \exp(2\,\nu \arg\coth z) = \exp\left[2\,\nu \arg\tanh\left(\frac{1}{z}\right)\right]$$

$$= 1 + \frac{2\nu}{z-\nu+}\,\frac{\nu^2-1}{3z+}\,\frac{\nu^2-9}{7z+}\cdots\frac{\nu^2-n^2}{(2n+1)z+}\cdots, \qquad (7.297a \text{ and } 7.297b)$$

$$z \notin (-i,+i): \quad \exp(2\,\nu \operatorname{arc}\cot z) = \exp\left[2\,\nu \operatorname{arc}\tan\left(\frac{1}{z}\right)\right]$$

$$= 1 + \frac{2\nu}{z-\nu+}\,\frac{\nu^2+1}{3z+}\,\frac{\nu^2+4}{5z+}\,\frac{\nu^2+9}{7z+}\cdots\frac{\nu^2+n^2}{(2n+1)\,z+}\cdots. \qquad (7.298a \text{ and } 7.298b)$$

The continued fraction Equation 7.298b is related to Equation 7.297b using the imaginary changes of variable Equations 7.299a and 7.299b, together with Equations 7.75b and 7.184c in Equation 7.299c:

$$z \to iz, \quad \nu \to i\nu: \quad \left(\frac{iz+i}{iz-i}\right)^{i\nu} \equiv \exp\left[i\nu \log\left(\frac{iz+1}{iz-1}\right)\right] = \exp\left[i\nu \log\left(\frac{z-i}{z+i}\right)\right]$$

$$= \exp(2\nu \operatorname{arc}\cot z) = \exp\left[2\nu \operatorname{arc}\tan\left(\frac{1}{z}\right)\right]. \qquad (7.299a\text{–}7.299c)$$

The same imaginary changes of variable Equation 7.299a and exponent Equation 7.299b transforms Equations 7.289a and 7.289b to

$$z \in (-i,+i): \quad \left(\frac{iz+1}{iz-1}\right)^{i\nu} = 1 + \frac{2i\nu}{iz-i\nu}\,\frac{-\nu^2-1}{3iz+}\,\frac{-\nu^2-4}{5iz+}\,\frac{-\nu^2-9}{7iz+}\cdots\frac{-\nu^2-n^2}{(2n+1)iz+}\cdots$$

$$= 1 + \frac{\nu^2}{z-\nu}\,\frac{\nu^2+1}{3z+}\,\frac{\nu^2+4}{5z+}\,\frac{\nu^2+9}{7z+}\cdots\frac{\nu^2+n^2}{(2n+1)z+}\cdots$$

(7.300a and 7.300b)

equating Equation 7.299c ≡ Equation 7.300b proves Equation 7.298b valid outside the branch cut Equation 7.300a ≡ Equation 7.298a. *The continued fractions* **(Laguerre 1879)** *for the exponential of hyperbolic (circular) cyclometric function inverse cotangent [Equation 7.297b (Equation 7.298b)] hold in the whole complex plane excluding the branch cuts [Equation 7.297a (Equation 7.298a)] in Figure 7.3f (Figure 7.2c).*

7.9 Gregory, Leibnitz, Brouncker, and Wallis Quadratures

A classical inverse trigonometric problem is the **quadrature of the circle:** *to find the side L of a square, such that its area L^2 equals that of a circle πR^2 of radius R. The solution of $L^2 = \pi R^2$ reduces to $L = R\sqrt{\pi}$, the determination of the number π,* for which several of the preceding formulas can be used, for example, (1) setting Equation 7.301a (Equation 7.302a) in the power series Equation 7.180b (Equation 7.172b) for the inverse circular tangent (sine) leads to an expression for Equation 7.301b (Equation 7.302b), that is, the **Gregory (Leibnitz) quadrature:**

$$z = 1: \quad \frac{\pi}{4} = \operatorname{arc\,tan} 1 = \sum_{n=0}^{\infty} \frac{(-)^n}{2n+1} = 1 - \frac{1}{3} + \frac{1}{5} - \frac{1}{7} + \frac{1}{9} - \cdots, \tag{7.301a and 7.301b}$$

$$z = \frac{1}{2}: \quad \frac{\pi}{6} = \frac{1}{2} + \sum_{n=0}^{\infty} \frac{2^{-2n-1}}{2n+1} = \frac{(2n-1)!!}{(2n)!!}$$
$$= \frac{1}{2} + \frac{1}{2.3.2^3} + \frac{1.3}{2.4.5.2^5} + \frac{1.3.5}{2.4.6.7.2^7} + \frac{1.3.5.7}{2.4.6.8.9.2^9} - \cdots; \tag{7.302a and 7.302b}$$

(2) setting Equation 7.303a (Equation 7.304a) in the continued fractions Equation 7.250b (Equation 7.257b) for the inverse circular tangent leads to Equation 7.303b (Equation 7.304b), the **Brouncker's quadrature:**

$$z = 1: \quad \frac{\pi}{4} = \operatorname{arc\,tan}(1) = \frac{1}{1+}\,\frac{1}{2+}\,\frac{9}{2+}\,\frac{25}{2+}\,\frac{49}{2+}\,\frac{81}{2+}\,\frac{121}{2+}\,\frac{169}{2+}\,\frac{225}{2+}\cdots, \tag{7.303a and 7.303b}$$

$$z = 1: \quad \frac{\pi}{4} = \operatorname{arc\,tan}(1) = \frac{1}{1+}\,\frac{1}{3+}\,\frac{4}{5+}\,\frac{9}{7+}\,\frac{16}{9+}\,\frac{25}{11+}\,\frac{36}{13+}\,\frac{49}{15+}\,\frac{64}{17+}\cdots; \tag{7.304a and 7.304b}$$

(3) setting Equation 7.305a in the infinite product for the sine Equation 7.226b:

$$z = \frac{\pi}{4}: \quad 1 = \sin\left(\frac{\pi}{2}\right) = \frac{\pi}{2}\prod_{n=1}^{\infty}\left(1 - \frac{1}{4n^2}\right), \quad \pi = 2\prod_{n=1}^{\infty}\frac{4n^2}{4n^2-1}, \tag{7.305a–7.305c}$$

leads (Equation 7.305b ≡ Equation 7.305c ≡ Equation 7.306) to **Wallis (1656) quadrature:**

$$\frac{\pi}{2} = \prod_{n=1}^{\infty} \frac{2n}{2n-1}\cdot\frac{2n}{2n+1} = \frac{2}{1}\,\frac{2}{3}\,\frac{4}{3}\,\frac{4}{5}\,\frac{6}{5}\,\frac{6}{7}\,\frac{8}{7}\,\frac{8}{9}\,\frac{10}{9}\,\frac{10}{11}\cdots. \tag{7.306}$$

The convergent, nonterminating continued fractions (Equations 7.303b and 7.304b) prove that the number π is irrational; it can also be computed using the series Equations 7.301b and 7.302b or the infinite product Equation 7.306.

The number π is given with 100 decimals **(Mansell 1929)** *by*

$$\pi = 3.14159\ 26535\ 89793\ 23846\ 26433\ 83279\ 50288\ 41971\ 69399\ 37510$$
$$58209\ 74844\ 59230\ 78164\ 06286\ 20899\ 86280\ 34825\ 34211\ 70679. \tag{7.307}$$

The series Equation 7.302b converges totally on the circle of convergence $|z| = 1$ at all points by the combined convergence test (Subsection I.29.1) except (Table 7.4) for the neighborhood of $z = \pm i$; since Equation 7.301a is included in this region, the series Equation 7.301b is uniformly and absolutely convergent. The series Equation 7.301b and product Equation 7.306 give alternate values above and below π, but are slowly converging, for example, 500 terms are needed for a three-digit accuracy. The continued fractions Equations 7.303b and 7.304b also give alternative values above and below π because they are of the first kind (Subsection 1.7.1), but since they are not simple, the convergence is slow, unlike in the case of the calculation of e (Section 3.2). The infinite product Equation 7.306 also alternates between lower and higher values, but convergence is again slow as for a preceding infinite product (Subsection 1.8.6). The most practical formula is the series Equation 7.302b, which is computed in Table 7.6 with

$$\begin{aligned}\pi = 3 &+ \frac{1}{2^3} + \frac{6.3}{4.5.2^6} + \frac{3.5}{4.7.2^8} + \frac{3.5.7}{4.8.9.2^{10}} + \frac{3.5.7.9}{4.8.10.11.2^{12}} \\ &+ \frac{3.5.7.9.11}{4.8.10.12.13.2^{14}} + \frac{3.5.7.9.11.13}{4.8.10.12.14.15.2^{16}} + \cdots + \frac{3.5\cdots 19.12}{4.8.10.12.\cdots 20.22.23.2^{24}},\end{aligned} \tag{7.308}$$

up to 12 terms S_n, retaining the digits π_n that have not changed since the preceding iteration; this yields nine accurate digits using 12 terms.

NOTE 7.1 PROPERTIES OF CIRCULAR, HYPERBOLIC, AND CYCLOMETRIC FUNCTIONS

The properties of circular and hyperbolic (inverse or cyclometric) functions are extensions of the properties of the exponential (Sections 3.1 through 3.4 and Tables 4.1 and 4.2) [logarithm (Sections 3.5 through 3.9 and Tables 3.3 and 3.4)]. The simpler algebraic properties of circular and hyperbolic functions were considered first (Chapter 5 and Tables 5.1 through 5.6); the analytical properties of circular and hyperbolic functions were considered together with those of cyclometric functions (Chapter 7 and Tables 7.1 through 7.9),

TABLE 7.6

Quadrature of Circle or Calculation of π

n	1	2	3	4	5
S_n	3.00000 0000	3.12500 0000	3.13906 25000	3.14115 5134	3.14151 1172
π_n	–	3	3.1	3.14	3.14
n	6	7	8	9	
S_n	3.14157 6715	3.14158 9425	3.14159 1982	3.14159 2511	
π_n	3.1416	3.1416	3.14159	3.14159	
n	10	11	12		
S_n	3.14159 2623	3.14159 2647	3.14159 2652		
π_n	3.14159 3	3.14159 26	3.14159 265		

Note: First 12 terms (Equation 7.308) of the power series Equation 7.302b give estimates increasing monotonically toward the irrational number π with all 12 terms yielding 9 accurate digits.

TABLE 7.7

Properties of Circular and Hyperbolic Functions

Function	Relation with Exponential	Imaginary Change of Variable	Zeros and Slopes	Singularities and Residues	Plot	Fundamental Period
cos	Equation 5.3a	Equation 5.6a	Equations 7.195a and 7.195b	Equations 7.209a and 7.209c	Figure 5.2b	2π
sin	Equation 5.3b	Equation 5.6b	Equations 7.194a through 7.194c	Equations 7.208b and 7.208d	Figure 5.2a	2π
sec	Equation 5.21a	Equation 5.24b	Equations 7.196a and 7.196b	Equations 7.211a and 7.211b	Figure 5.2b	2π
csc	Equation 5.21b	Equation 5.25b	Equations 7.197a and 7.197b	Equations 7.210a and 7.210b	Figure 5.2a	2π
tan	Equation 5.21c	Equation 5.26b	Equations 7.198a and 7.198b	Equations 7.212a and 7.212b	Figure 5.2c	π
cot	Equation 5.21d	Equation 5.27b	Equations 7.199a and 7.199b	Equations 7.213a and 7.213b	Figure 5.2c	π
cosh	Equation 5.2a	Equation 5.5a	Equations 7.203a and 7.203b	Equations 7.209b and 7.209c	Figure 5.2e	$2\pi i$
sinh	Equation 5.2b	Equation 5.5b	Equations 7.202a through 7.202c	Equations 7.208c and 7.208d	Figure 5.2d	$2\pi i$
sech	Equation 5.20a	Equation 5.24a	Equations 7.204a and 7.204b	Equations 7.215a and 7.215b	Figure 5.2e	$2\pi i$
csch	Equation 5.20b	Equation 5.25a	Equations 7.205a and 7.205b	Equations 7.214a and 7.214b	Figure 5.2d	$2\pi i$
tanh	Equation 5.20c	Equation 5.26a	Equations 7.206a and 7.206b	Equations 7.216a and 7.216b	Figure 5.2f	πi
coth	Equation 5.20d	Equation 5.27a	Equations 7.207a and 7.207b	Equations 7.217a and 7.217b	Figure 5.2f	πi

Note: Analytic properties of the direct circular and hyperbolic primary (cosine and sine) and secondary (secant, cosecant, tangent, and cotangent) functions include (1) their definition from the exponential; (2) imaginary change of variable between circular and hyperbolic functions; (3) location of zeros and slopes, that is, values of the derivatives there; (4) location of singularities (poles or essential singularities) and residues, that is, coefficient of first inverse power in Laurent series; (5) plots for real variable; and (6) fundamental period, that is, real (imaginary) for circular (hyperbolic) functions.

TABLE 7.8

Properties of Inverse or Cyclometric Functions

Function	Relation with Logarithm	Imaginary Change of Variable	Inverse Change of Variable	Branch Cut(s)	Plot	Identity	Change of Sign
arc cos	Equation 7.65b	Equation 7.186a	Equation 7.184a	Equation 7.65a	Figure 7.2a	Equation 7.189a	Equation 7.191e
arc sin	Equation 7.67b	Equation 7.187a	Equation 7.184b	Equation 7.67a	Figure 7.2a	Equation 7.189a	Equation 7.191a
arc sec	Equation 7.69b	Equation 7.186b	Equation 7.184a	Equation 7.69a	Figure 7.2d	Equation 7.189b	Equation 7.191f
arc csc	Equation 7.71b	Equation 7.188a	Equation 7.184b	Equation 7.71a	Figure 7.2d	Equation 7.189b	Equation 7.191b
arc tan	Equation 7.73b	Equation 7.187b	Equation 7.184c	Equation 7.73a	Figure 7.2b	Equation 7.189c	Equation 7.191c
arc cot	Equation 7.75b	Equation 7.188b	Equation 7.184c	Equation 7.75a	Figure 7.2c	Equation 7.189c	Equation 7.191d
arg cosh	Equation 7.66b	Equation 7.186a	Equation 7.185a	Equation 7.66a	Figure 7.3b	Equation 7.190a	Equation 7.192e
arg sinh	Equation 7.68b	Equation 7.187a	Equation 7.185b	Equation 7.68a	Figure 7.3a	Equation 7.190a	Equation 7.192a
arg sech	Equation 7.70b	Equation 7.186b	Equation 7.185a	Equation 7.70a	Figure 7.3d	Equation 7.190b	Equation 7.192f
arg csch	Equation 7.72b	Equation 7.188a	Equation 7.185b	Equation 7.72a	Figure 7.3c	Equation 7.190b	Equation 7.192b
arg tanh	Equation 7.74b	Equation 7.187b	Equation 7.185c	Equation 7.74a	Figure 7.3f	Equation 7.190c	Equation 7.192c
arg coth	Equation 7.76b	Equation 7.188b	Equation 7.185c	Equation 7.76a	Figure 7.3e	Equation 7.190c	Equation 7.192d

Note: Analytic properties of inverse or cyclometric circular and hyperbolic primary (cosine and sine) and secondary (secant, cosecant, tangent, and cotangent) functions include (1) their definition from the logarithm; (2) imaginary change of variable between cyclometric circular and hyperbolic functions; (3) algebraic inverse change of variable that relates algebraically inverse functions, that is, cosine/sine/tangent, respectively, with secant/cosecant/cotangent; (4 and 5) branch cuts that render the functions single-valued as principal branches and join the branch points between them or to infinity and may include (exclude) the branch point if the function is singular (regular) there, as shown in Figure 7.2 (Figure 7.3) for cyclometric circular (hyperbolic) functions; and (6) algebraic identities between pairs of circular (hyperbolic) cyclometric functions cosine/sine, secant/cosecant, and tangent/cotangent that add to give the value $\pi/2$ $(i\pi/2)$; (7) relations between cyclometric functions of variable $\pm z$.

providing perhaps the simplest set of examples of the methods of analysis, for example, the infinite representations in Table 7.9.

NOTE 7.2 PROPERTIES OF ELEMENTARY AND HIGHER TRANSCENDENTAL FUNCTIONS

The present volume has been concerned mostly with the properties of the elementary transcendental functions, namely, (1) the exponential and its inverse, that is, the logarithm (Chapter 3); (2) the direct circular and hyperbolic functions (Chapter 5), that is, the primary (sine and cosine) and secondary, namely, the inverses (cosecant and secant) and ratios (tangent and cotangent); and (3) the inverse circular and hyperbolic functions (Chapter 7), that is, the cyclometric functions arc circ and arg hyp. The

TABLE 7.9

Infinite Representations for Elementary Transcendental Functions

Function	Power Series	Series of Fractions	Continued Fraction	Infinite Product	Number
exp	Equation 1.26b ≡ 3.1	–	Equations 1.137b, 1.138, 1.139, 1.140b, 1.141b, 3.5a ≡ 3.136a, 3.5b ≡ 3.138b, 3.137	–	1 + 0 + 8 + 0 = 9
log	Equations 3.47a, 3.47b, 3.48a, 3.48b, 3.49a, 3.49b, 3.50a, 3.50b, 3.64a through 3.64c, 3.66a through 3.66c, 3.67a through 3.67c	–	Equations 3.58a and 3.58b ≡ Equations 3.100b and 3.100c, 3.59a, 3.59b, 3.60a, 3.60b, 3.61a, 3.61b, 3.122e, 3.123a, and 3.123b	–	8 + 0 + 6 + 0 = 14
cos	Equation 1.31a ≡ Equations 7.12a and 7.12b	–	Equations 7.236a and 7.236b	Equation 1.73 ≡ Equations 7.226a and 7.226c	1 + 0 + 1 + 1 = 3
cosh	Equations 7.6a and 7.6b	–	Equations 7.237a and 7.237b	Equations 7.227a and 7.227c	1 + 0 + 1 + 1 = 3
sin	Equation 1.31b ≡ Equations 7.13a and 7.13b	–	Equations 1.151a and 1.151b ≡ Equations 7.228a and 7.228b, 7.238a, and 7.238b	Equation 1.75 ≡ Equations 7.226a and 7.226b	1 + 0 + 2 + 1 = 4
sinh	Equations 7.7a and 7.7b	–	Equations 7.229a, 7.229b, 7.239a, and 7.239b	Equation 7.227a and 7.227b	1 + 0 + 2 + 1 = 4
sec	Equations 7.16a and 7.16b ≡ Equations 7.28e ≡ Equations 7.29a and 7.29b	Equation 1.54 ≡ Equations 7.224a and 7.224c	Equations 7.242a and 7.242b	–	1 + 1 + 1 + 0 = 3
sech	Equations 7.16b and 7.16c ≡ Equations 7.37a through 7.37c	Equations 7.225a and 7.225c	Equations 7.243a and 7.243b	–	1 + 1 + 1 + 0 = 3
csc	Equations 7.21a and 7.21c ≡ Equation 7.31e ≡ Equations 7.32a and 7.32b	Equation 1.85 ≡ Equations 7.222b and 7.222d	Equations 1.152a and 1.152b ≡ Equations 7.230a and 7.230b, 7.240a and 7.240b	–	1 + 1 + 2 + 0 = 4
csch	Equations 7.21a and 7.21b ≡ Equations 7.38a through 7.38c	Equations 7.223a and 7.223c	Equations 7.231a, 7.231b, 7.241a, and 7.241b	–	1 + 1 + 2 + 0 = 4
tan	Equations 7.22a and 7.22c ≡ Equations 7.33a and 7.33b ≡ Equations 7.35a and 7.35b	Equation 1.78 ≡ Equations 7.224a 7.224b	Equations 1.181a and 1.181b ≡ Equations 7.244a through 7.244c, 7.292a through 7.292d	–	1 + 1 + 2 + 0 = 4

tanh	Equations 7.22a and 7.22b ≡ Equations 7.39a through 7.39c	Equations 7.225a and 7.225b	Equations 7.245a, 7.245b, 7.293a through 7.293c	–	1 + 2 + 2 + 0 = 4
cot	Equations 7.17a and 7.17b ≡ Equations 7.34a and 7.34b ≡ Equations 7.36a and 7.36b	Equation 1.58 ≡ Equations 7.222b and 7.222c	Equations 7.246a and 7.246b, 7.294a through 7.294c	–	1 + 1 + 2 + 0 = 4
coth	Equations 7.17a and 7.17c ≡ Equations 7.40a through 7.40c	Equations 7.223a and 7.223b	Equations 7.247a, 7.247b, 7.295a, and 7.295b	–	1 + 1 + 2 + 0 = 4
arc cos	Equations 7.174a and 7.174b	–	+	–	1 + 0 + 0 + 0 = 1
arg cosh	Equations 7.175a and 7.175b	–	+	–	1 + 0 + 0 + 0 = 1
arg sinh	Equations 7.173a and 7.173b	–	Equations 7.268a and 7.268b	–	1 + 1 + 0 + 0 = 2
arc sin	Equations 7.172a and 7.172b	–	Equations 7.267a and 7.267b	–	1 + 0 + 1 + 0 = 2
arc sec	Equations 7.178a and 7.178b	–	+	–	1 + 0 + 0 + 0 = 1
arg sech	Equations 7.179a and 7.179b	–	+	–	1 + 0 + 0 + 0 = 1
arc csc	Equations 7.176a and 7.176b	–	Equations 7.269a and 7.269b	–	1 + 0 + 1 + 0 = 2
arg csch	Equations 7.177a and 7.177b	–	Equations 7.270a and 7.270b	–	1 + 0 + 1 + 0 = 2
arc tan	Equations 7.180a through 7.180d	–	Equations 7.250a, 7.250b, 7.257a, 7.257b, 7.298a, and 7.298b	–	2 + 0 + 3 + 0 = 5
arc tanh	Equations 7.181a through 7.181d	–	Equations 7.251a, 7.251b, 7.258a, 7.258b, 7.297a, and 7.297b	–	2 + 0 + 3 + 0 = 5
arc cot	Equations 7.183a through 7.183d	–	Equations 7.252a, 7.252b, 7.260a, 7.260b, 7.298a, and 7.298b	–	2 + 0 + 3 + 0 = 5
arg coth	Equations 7.182a through 7.182d	–	Equations 7.253a, 7.253b, 7.259a, 7.259b, 7.297a, and 7.297b	–	2 + 0 + 3 + 0 = 5
Generalized logarithm	Equations 3.96a and 3.96b	–	Equations 3.99a through 3.99c	–	1 + 0 + 1 + 0 = 2
Compound logarithm	Equations 7.55a, 7.55b, 7.56a, 7.56b, 7.58a, 7.58b, 7.59a, 7.59b, 7.60a, 7.60b, 7.61a, and 7.61b	–	–	–	6 + 0 + 0 + 0 = 6

(continued)

TABLE 7.9 (Continued)

Infinite Representations for Elementary Transcendental Functions

Function	Power Series	Series of Fractions	Continued Fraction	Infinite Product	Number
Binomials	Equations 3.144a through 3.144c, 3.146a through 3.146c	–	Equations 3.151, 3.152, 3.153b, 7.282a, 7.282b, 7.283a, 7.283b, 7.285a, 7.285b, 7.286a, 7.286b, 7.287a, 7.287b, 7.288, 7.289a, 7.289b	–	4 + 0 + 10 + 0 = 14
Hypogeometric	Equation 1.159	–	Equations 1.173a through 1.173d	–	1 + 0 + 1 + 0 = 2
Confluent Hypergeometric	Equations 3.124a and 3.124b	–	Equations 3.131a through 3.131d, 3.132d, 3.132e, and 3.134a through 3.134d	–	1 + 0 + 2 + 0 = 3
Gaussian hypergeometric	Equations 3.101a and 3.101b	–	Equations 3.115a through 3.115d, 3.113a, 3.113b, 3.117a through 3.117d, 3.116d, and 3.116e	–	1 + 0 + 2 + 0 = 3
Generalized hypergeometric	Equation 3.155	–	+	–	1 + 0 + 0 + 0 = 1
Total	53	8	65	4	130

Note: 130 basic infinite representations for the elementary transcendental functions include (1) 53 power series; (2) 8 series of fractions; (3) 65 continued fractions; and (4) 4 infinite products. The table includes all the elementary transcendental functions (exponential, logarithm, circular, hyperbolic, and cyclometric) plus a few other functions (rational and hypergeometric) used to prove some properties of the former. – does not exist; + the continued fraction can be obtained from the series using Equations 1.128a through 1.128c; * the continued fraction can also be obtained from the infinite product using Equation 1.132 ≡ Equation 1.133.

properties of the elementary transcendental functions have analogs for the higher transcendental functions (List I.39.1), namely, (4) the auxiliary functions like the digamma function (Subsection I.29.5.2); (5) the special functions like the hypergeometric family (Subsection I.29.9, Example 30.20 and 40.20; Sections 1.9, 3.9, and 7.8); and (6) the specific functions like the elliptic functions of Weierstrass and Jacobi (Subsection I.39.9). The common properties include (1) power series expansions (Chapters I.23 and I.25) with a radius of convergence (Chapters I.21 and I.29) limited by the nearest singularity (Sections 7.1 and 7.4); (2) series of fractions (Sections 1.2 and 1.3) in the case (Section 7.6) of meromorphic functions; (3) infinite products (Sections 1.4 and 1.5) for functions with an infinite number of zeros (Section 7.7); (4) continued fractions (Sections 1.5 through 1.9 and 3.9) that may be the most accurate for a given level of complexity (Section 7.8); (5) differentiation formulas (Section 7.3 and Tables 7.1 and 7.3); (6) particular values of the function derivatives and residues at poles (Table 7.5); (7 and 8) addition or multiplication formulas for the arguments (Sections 5.4 through 5.8); (9) for multivalued functions (Chapter I.7) and selection of the principal branch in the complex cut plane (Section 7.2); and (10) representations as parametric integrals (Subsections I.13.8 and I.13.9). These properties 1 through 10 and other properties usually take simpler forms and are easier to prove, for elementary than for higher transcendental functions, so that the former are the more straightforward illustrations, for example, of infinite representations in Table 7.9, including both the direct (inverse or cyclometric) circular and hyperbolic functions in Table 7.7 (Table 7.8).

7.10 Summary

The zeros and poles of the circular (hyperbolic) functions (Figure 7.1) lie on the real (imaginary) axis as implied in Tables 7.1 and 7.5 of zeros (singularities) and slopes (residues). The branch cuts in the z-plane for the cyclometric functions (Figures 7.2 and 7.3) limit the radius of convergence of their power series (Table 7.4); one of the series is used to calculate the irrational number π in Table 7.6. The power series, together with the series of fractions, continued fractions, and infinite products, appear in the infinite representations in Table 7.9 both for direct (inverse) functions [Table 7.7 (Table 7.8)]. The latter include the power series for the secondary functions that involve the Euler and Bernoulli numbers in Table 7.2 as coefficients. Among the derivatives and primitives in Table 7.3 appears the Gudermann function that is related to the Mercator map in Figure 7.4.

8

Confined and Unsteady Flows

The basics of plane steady incompressible irrotational flow (Chapters I.12 and I.14) have been applied to a number of problems: (1) flows past bodies (Chapter I.28); (2) sources, sinks, and vortices in corners, channels, slits, and wells (Chapter I.36); (3) the aerodynamics of airfoils and wings (Chapter I.34); and (4) jets with free boundaries (Chapter I.38). Some consideration was given to compressible (Sections I.12.4, I.12.5, 1.1, and 1.2) and rotational (Sections 1.3 through 1.9) flows, and confined and unsteady flows are addressed next (Chapter 8). A flow is confined if it is limited to a compact region, such as a cavity, and free if it extends to infinity in all directions, for example, an airfoil in a free stream; an intermediate case is a duct, where a flow is partially confined by walls. The flow due to a rigid body moving uniformly in a fluid is steady in a reference frame moving with the body and unsteady in any other reference frame relative to which the body is moving; for example, a pulsating body will cause an unsteady flow. The examples of confined and/or unsteady flows start with a cylinder moving in a large cylindrical cavity (Sections 8.1); the flow is confined by the cavity, and the cylinder moves relative to the cavity. The case of two cylinders not contained in one another and in relative motion leads to a flow that includes the particular cases where the cylinders move along (across) the line of centers [Section 8.2 (Example 10.10)]. The case of cylinders with the same radius and velocity is equivalent to a cylinder moving perpendicular (parallel) to a wall at an equal distance from the two cylinders, since each cylinder is the image of the other on the wall. The flow past two cylinders is specified exactly by an infinite number of images and approximately by the first few images if the distance between the centers is much larger than the radii. A third case (Section 8.3) is two cylinders at rest with a noncoincident parallel axis. This problem can be solved using coaxial coordinates that also apply to cylindrical log standing on a wall. If the cylinder recesses into the wall, it becomes a circular mound, and if it is then removed, it creates a ditch; by symmetry this is equivalent to a biconvex or biconcave cylinder. Another type of obstacle on a wall is a backward (forward) ramp or step (Section 8.4); the symmetry with regard to a line parallel to the wall leads to a parallel-sided duct with a ramp or step-like expansion (contraction). A parallel-sided channel may contain a thick semi-infinite plate (Section 8.5) with a sharp or blunt front end; the latter can also exist in a free stream, and in the limit of zero thickness becomes the edge of a flat plate. The flat plate is the simplest airfoil (Section I.34.1) and serves to study two aspects: (1) the increase in lift of an airfoil due to the use of a flap (Section 8.6) and (2) the lift loss due to flow separation (Section 8.7). The flap may have a slot separating it from the wing, which helps delay flow separation until a larger angle-of-attack. A set of parallel instead of tandem airfoils forms a cascade that can change the speed and direction of an incident stream (Section 8.8). The separated flow, either from the trailing edge of an airfoil or from farther upstream, forms a wake; its simplest representation, as a vortex sheet, is unstable to small amplitude sinusoidal disturbances (Section 8.9). The vortex sheet may separate two jets with different velocities and/or mass densities.

8.1 Cylinder Moving in Large Cavity

When measuring the forces on a body in a wind tunnel, there are wall effects (Subsection 8.1.4), that is, a correction is needed in order to obtain the forces in a free stream. The example considered next is a cylinder moving in a large cylindrical cavity (Subsection 8.1.1) showing the wall effect on the velocity (Subsection 8.1.2) and pressure and lift (Subsection 8.1.3).

8.1.1 Inner Moving and Outer Fixed Boundary Conditions

The complex potential (Equation 8.1a) is taken as the superposition of (1) a uniform flow (Equation 2.171a) with (2) a dipole (Equation I.12.43a) and (3) a vortex (Equation 2.192a):

$$f(z) = Az - \frac{P_1}{2\pi z} - \frac{i\Gamma}{2\pi}\log z, \quad v^*(z) = \frac{df}{dz} = A + \frac{P_1}{2\pi z^2} - \frac{i\Gamma}{2\pi z}, \qquad (8.1a\text{ and }8.1b)$$

corresponding to the complex conjugate velocity (Equation 8.1b); for the latter to represent (Figure 8.1) a cylinder of radius a with circulation Γ moving at velocity U in a potential flow, in a cylindrical cavity of much larger radius b, the radial velocity on two cylinders must satisfy

$$v_r(a, \varphi) = -U\cos\varphi, \quad v_r(b, \varphi) = 0, \qquad (8.2a\text{ and }8.2b)$$

that is, (1) a reference frame fixed to the cylinder is chosen in order to have a steady flow, and then the flow velocity is $-U$ in the x-direction and has a normal or radial component

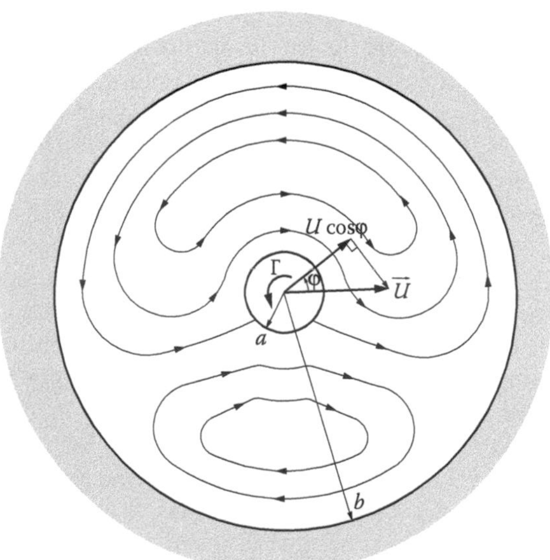

FIGURE 8.1
Potential flow due to a cylinder moving uniformly in a large cylindrical cavity satisfies (1) the Laplace equation in the domain between the cylinders; (2) zero normal velocity at the rigid cylindrical cavity; and (3) normal velocity on the moving cylinder opposite to the radial projection of the velocity of translation. The potential flow is represented by a dipole in a uniform flow; the free stream velocity and dipole moment depend on the radii of the cylinder and cavity and on the velocity of the former.

(Equation 8.2a) at the surface of the cylinder; and (2) the normal velocity is zero at the surface of the large cylindrical cavity (Equation 8.2b). The complex conjugate velocity (Equation 8.1b) specifies (Equation 2.190b)

$$v_r - iv_\varphi = e^{i\varphi} v^*\left(re^{i\varphi}\right) = Ae^{i\varphi} - \frac{i\Gamma}{2\pi r} + \frac{P_1}{2\pi r^2} e^{-i\varphi}, \tag{8.3}$$

the radial velocity (Equation 8.4a):

$$v_r(r,\phi) = \left(A + \frac{P_1}{2\pi r^2}\right)\cos\varphi; \quad A + \frac{P_1}{2\pi a^2} = -U, \quad A = -\frac{P_1}{2\pi b^2}; \tag{8.4a–8.4c}$$

it meets the conditions 8.2a and 8.2b for the choice (Equations 8.4b and 8.4c) of constants, leading to (Equations 8.5a and 8.5b)

$$A = U\frac{a^2}{b^2 - a^2}, \quad P_1 = -2\pi U\frac{a^2 b^2}{b^2 - a^2}. \tag{8.5a and 8.5b}$$

The limit of a large radius of the cavity relative to the cylinder (Equation 8.6a):

$$b^2 \gg a^2: \quad A = UO\left(\frac{a^2}{b^2}\right), \quad P_1 = -2\pi U a^2, \quad v^*(z) \sim -U\frac{a^2}{z^2} - \frac{i\Gamma}{2\pi z} \tag{8.6a–8.6d}$$

corresponds to (1) a dipole of moment (Equation 8.6c ≡ Equation I.28.89a) as for a cylinder of radius a in an unbounded free stream; and (2) a negligible free stream velocity (Equation 8.6b) leaving unaffected the circulation in Equation 8.1b ≡ Equation 8.6d. Thus, the **wall effect** is of the order of the square of the ratio of the radii of the cylinder and the cavity (Equation 8.6b).

8.1.2 Potential, Stream Function, and Stagnation Points

The two constants specify the free stream velocity (Equation 8.5a) and dipole moment (Equation 8.5b) in the complex potential (Equation 8.1a ≡ Equation 8.7):

$$f(z) = \frac{Ua^2}{b^2 - a^2}\left(z + \frac{b^2}{z}\right) - \frac{i\Gamma}{2\pi}\log z; \tag{8.7}$$

it leads to the scalar potential Equation 8.8a and stream function Equation 8.8b:

$$\Phi(r,\varphi) = \mathrm{Re}\left[f\left(re^{i\varphi}\right)\right] = \frac{Ua^2}{b^2 - a^2}\left(r + \frac{b^2}{r}\right)\cos\varphi + \frac{\Gamma\varphi}{2\pi}, \tag{8.8a}$$

$$\Psi(r,\varphi) = \mathrm{Im}\left[f\left(re^{i\varphi}\right)\right] = \frac{Ua^2}{b^2 - a^2}\left(r - \frac{b^2}{r}\right)\sin\varphi - \frac{\Gamma}{2\pi}\log r, \tag{8.8b}$$

confirming that the outer cylinder or cylindrical cavity is the streamline $\Psi(b, \varphi) - (\Gamma/2\pi) \log b \equiv \text{const}$. The complex conjugate velocity corresponding to the complex potential Equation 8.7 is

$$v^*(z) = \frac{df}{dz} = U \frac{a^2}{b^2 - a^2}\left(1 - \frac{b^2}{z^2}\right) - \frac{i\Gamma}{2\pi z}. \tag{8.9}$$

It vanishes (Equation 8.10a)

$$v^*(z_\pm) = 0: \quad z^2 - i\frac{\Gamma}{2\pi U}\frac{b^2 - a^2}{a^2} z - b^2 = 0, \tag{8.10a and 8.10b}$$

at (Equation 8.10b) the stagnation points:

$$z_\pm = ic \pm \sqrt{b^2 - c^2}, \quad c \equiv \frac{\Gamma}{4\pi U}\frac{b^2 - a^2}{a^2}. \tag{8.11a and 8.11b}$$

The **critical circulation** is defined (Equation 8.12c) as that for which the stagnation points (Equation 8.11a) coincide (Equation 8.12a) at the top (Equation 8.12b) of the cavity:

$$c = b: \quad z_\pm = ic = ib, \quad \Gamma = \frac{4\pi U a^2 b}{b^2 - a^2} \equiv \Gamma_*. \tag{8.12a–8.12c}$$

If the circulation is less (Equation 8.13a) than the critical value (Equation 8.12c), then (Equation 8.13b) the stagnation points (Equation 8.11a ≡ Equation 8.13c) are distinct and lie on the outer cavity (Equation 8.13d):

$$\Gamma < \Gamma_*: \quad c < b, \quad z_+ = ic + |b^2 - c^2|^{1/2} \neq z_- = ic - |b^2 - c^2|^{1/2}, \; |z_\pm| = b^2, \tag{8.13a–8.13d}$$

$$\Gamma > \Gamma_*: \quad c > b, \quad z_\pm = ic \pm i|b^2 - c^2|^{1/2}, \; |z_\pm| > c > b; \tag{8.14a–8.14d}$$

if the circulation exceeds (Equation 8.14a) the critical value (Equation 8.12c), then (Equation 8.14b) the stagnation points (Equation 8.11a ≡ Equation 8.14c) lie outside the cavity, that is, there are no stagnation points in the flow region:

$$z_\pm = \begin{cases} b\left(i\dfrac{\Gamma}{\Gamma_*} \pm \left|1 - \dfrac{\Gamma^2}{\Gamma_*^2}\right|^{1/2}\right) & \text{if} \quad \Gamma < \Gamma_*, \\ ib & \text{if} \quad \Gamma = \Gamma_* = \dfrac{4\pi U a^2 b}{b^2 - a^2}, \\ b\left(i\dfrac{\Gamma}{\Gamma_*} \pm \left|\dfrac{\Gamma^2}{\Gamma_*^2} - 1\right|^{1/2}\right) & \text{if} \quad \Gamma > \Gamma_*. \end{cases} \tag{8.15a–8.15c}$$

In Equations 8.15a through 8.15c, Equations 8.11b and 8.12c ≡ Equation 8.16a was used:

$$\frac{c}{b} = \frac{\Gamma}{4\pi U}\frac{b^2 - a^2}{a^2 b} = \frac{\Gamma}{\Gamma_*}, \quad z_\pm = b\left(i\frac{\Gamma}{\Gamma_*} \pm \sqrt{1 - \frac{\Gamma^2}{\Gamma_*^2}}\right), \qquad \text{(8.16a and 8.16b)}$$

which implies Equation 8.16b. The positions of stagnation points may be compared for a cylinder in a cavity (Equations 8.15a through 8.15c) [in a free stream (Equations I.28.124a and I.28.124b)].

8.1.3 Cartesian and Polar Components of the Velocity

From Equation 8.9 also follows the flow velocity (Equations 2.190a and 2.190b) in Cartesian:

$$v_x(r,\varphi) = \operatorname{Re}\left[v^*\left(re^{i\varphi}\right)\right] = U\frac{a^2}{b^2 - a^2}\left[1 - \frac{b^2}{r^2}\cos(2\varphi)\right] - \frac{\Gamma}{2\pi r}\sin\varphi, \qquad \text{(8.17a)}$$

$$v_y(r,\varphi) = -\operatorname{Im}\left[v^*\left(re^{i\varphi}\right)\right] = -U\frac{a^2}{b^2 - a^2}\frac{b^2}{r^2}\sin(2\varphi) + \frac{\Gamma}{2\pi r}\cos\varphi, \qquad \text{(8.17b)}$$

and polar:

$$v_r(r,\varphi) = \operatorname{Re}\left[e^{i\varphi}v^*\left(re^{i\varphi}\right)\right] = U\frac{a^2}{b^2 - a^2}\left(1 - \frac{b^2}{r^2}\right)\cos\varphi, \qquad \text{(8.18a)}$$

$$v_\varphi(r,\varphi) = -\operatorname{Im}\left[e^{i\varphi}v^*\left(re^{i\varphi}\right)\right] = -U\frac{a^2}{b^2 - a^2}\left(1 + \frac{b^2}{r^2}\right)\sin\varphi + \frac{\Gamma}{2\pi r}, \qquad \text{(8.18b)}$$

components. *The potential flow (Figure 8.1) past a cylinder of radius a moving with velocity U in a cylindrical cavity of radius b is specified by the complex potential (Equation 8.7) [conjugate velocity (Equation 8.9)] leading to the scalar potential (Equation 8.8a) and stream function (Equation 8.8b) [the Cartesian (Equations 8.17a and 8.17b) and polar (Equations 8.18a and 8.18b) components of the velocity]. The stagnation points (Equation 8.11a) coincide at the top of the cavity (Equation 8.12b ≡ Equation 8.15b) for the critical value (Equation 8.12c ≡ Equation 8.16a) of the circulation; the stagnation points separate (Equation 8.13c) and still lie on the cavity (Equation 8.13d) for smaller circulation (Equation 8.15a); for larger circulation (Equation 8.15c), there are no stagnation points in the flow region.*

8.1.4 Forces on the Fixed Cylindrical Cavity

The velocity (Equations 8.18a and 8.18b) is tangential on the outer cylindrical cavity:

$$v_r(b,\varphi) = 0, \quad v_\varphi(b,\varphi) = -2U\frac{a^2}{b^2 - a^2}\sin\varphi + \frac{\Gamma}{2\pi b}. \qquad \text{(8.19a and 8.19b)}$$

The Bernoulli equation (Equation 2.12b ≡ Equation I.14.27c) specifies the corresponding pressure distribution:

$$p(b,\varphi)-p_0=-\frac{1}{2}\rho\left[v_\varphi(b,\varphi)\right]^2=-\frac{\rho\Gamma^2}{8\pi^2b^2}-2\rho\left(\frac{Ua^2\sin\varphi}{b^2-a^2}\right)^2+\frac{\rho\Gamma U}{\pi b}\frac{a^2}{b^2-a^2}\sin\varphi, \quad (8.20)$$

where p_0 is the stagnation pressure, and ρ is the mass density; the total force on the cylindrical cavity due to the pressure distribution (Equation 8.20) corresponds to zero drag (Equation 8.21a):

$$F_x^+(a,b)=-b\int_0^{2\pi}p(b,\varphi)\cos\varphi\,d\varphi=0, \quad (8.21a)$$

$$F_y^+(a,b)=-b\int_0^{2\pi}p(b,\varphi)\sin\varphi\,d\varphi=-\rho\Gamma U\frac{a^2}{b^2-a^2}, \quad (8.21b)$$

and to the lift (Equation 8.21b) proportional to the circulation; the lift vanishes $F_y^+\to 0$ if the cavity recedes to infinity $b\to\infty$. When substituting Equation 8.20 in Equations 8.21a and 8.21b, the elementary integrals are used:

$$\int_0^{2\pi}\left\{\cos\varphi,\sin\varphi,\sin^2\varphi\cos\varphi,\sin^3\varphi\right\}d\varphi$$

$$=\left[-\sin\varphi,\cos\varphi,\frac{\sin^3\varphi}{3},-\cos\varphi+\frac{\sin^3\varphi}{3}\right]_0^{2\pi}=0, \quad (8.22a–8.22e)$$

$$\int_0^{2\pi}\sin^2\varphi\,d\varphi=\frac{1}{2}\int_0^{2\pi}\left[1-\cos(2\varphi)\right]d\varphi=\left[\frac{\varphi}{2}-\frac{\sin(2\varphi)}{4}\right]_0^{2\pi}=\pi, \quad (8.23)$$

where (1) the periodic integrands lead to zero integrals (Equations 8.22a through 8.22e) and (2) the nonzero integral Equation 8.23 arises from the constant nonperiodic term.

8.1.5 Forces on the Moving Cylinder

The velocity (Equations 8.18a and 8.18b) on the inner cylinder:

$$v_r(a,\varphi)=-U\cos\varphi,\quad v_\varphi(a,\varphi)=-U\frac{b^2+a^2}{b^2-a^2}\sin\varphi+\frac{\Gamma}{2\pi a} \quad (8.24a \text{ and } 8.24b)$$

(1) confirms (Equation 8.24a ≡ Equation 8.2a) from the radial component that it moves with velocity $-U$ along the real axis; (2) the azimuthal component (Equation 8.24b) leads in the limit (Equation 8.6a) to a uniform flow $-U\sin\varphi$; and (3) the circulation Γ affects only the azimuthal component (Equation 8.24b) of the velocity. The total velocity specifies the pressure distribution on the inner cylinder:

$$p(a,\varphi) = p_0 - \frac{\rho\Gamma^2}{8\pi^2 a^2} + \frac{\rho\Gamma U}{2\pi a}\frac{b^2+a^2}{b^2-a^2}\sin\varphi - \frac{\rho}{2}\left(\frac{U}{b^2-a^2}\right)^2\left[a^4+b^4-2a^2b^2\cos(2\varphi)\right]. \tag{8.25}$$

The passage from Equations 8.24a and 8.24b to Equation 8.25 involves the Bernoulli equation (Equation 2.12b ≡ Equation I.14.27c) in the form:

$$\begin{aligned} p(a,\varphi) - p_0 &= -\frac{1}{2}\rho\left\{\left[v_r(a,\varphi)\right]^2 + \left[v_\varphi(a,\varphi)\right]^2\right\} \\ &= -\frac{1}{2}\rho U^2\left[\cos^2\varphi + \left(\frac{b^2+a^2}{b^2-a^2}\right)^2\sin^2\varphi\right] \\ &\quad + \frac{\rho\Gamma U}{2\pi a}\frac{b^2+a^2}{b^2-a^2}\sin\varphi - \frac{\rho\Gamma^2}{8\pi^2 a^2}, \end{aligned} \tag{8.26}$$

where the term in the square brackets simplifies to

$$\begin{aligned} \cos^2\varphi + \left(\frac{b^2+a^2}{b^2-a^2}\right)^2\sin^2\varphi &= \frac{\left(b^2-a^2\right)^2\left[1+\cos(2\varphi)\right] + \left(b^2+a^2\right)^2\left[1-\cos(2\varphi)\right]}{2\left(b^2-a^2\right)^2} \\ &= \frac{\left(b^2-a^2\right)^2 + \left(b^2+a^2\right)^2 + \left[\left(b^2-a^2\right)^2 - \left(b^2+a^2\right)^2\right]\cos(2\varphi)}{2\left(b^2-a^2\right)^2} \\ &= \frac{a^4+b^4-2a^2b^2\cos(2\varphi)}{\left(b^2-a^2\right)^2}. \end{aligned} \tag{8.27}$$

The total force on the moving cylinder due to the pressure distribution (Equation 8.25) is given (Equations 8.28a and 8.28b) as in Equations 8.21a and 8.21b replacing b by a:

$$F_y^-(a,b) = -a\int_0^{2\pi} p(a,\varphi)\cos\varphi \, d\varphi = 0, \tag{8.28a}$$

$$F_y^-(a,b) = -a\int_0^{2\pi} p(a,\varphi)\sin\varphi \, d\varphi = -\frac{1}{2}\rho U\Gamma\frac{b^2+a^2}{b^2-a^2}. \tag{8.28b}$$

The passage from Equation 8.25 to Equations 8.28a and 8.28b involves, besides the elementary integrals (Equations 8.22a through 8.22e and 8.23), also

$$\int_0^{2\pi} \cos(2\varphi)\{\cos\varphi, \sin\varphi\} d\varphi = \frac{1}{2}\int_0^{2\pi} \{\cos(3\varphi) + \cos\varphi, \sin(3\varphi) - \sin\varphi\} d\varphi$$
$$= \left[\frac{\sin(3\varphi)}{6} + \frac{\sin\varphi}{2}, -\frac{\cos(3\varphi)}{6} + \frac{\cos\varphi}{2}\right]_0^{2\pi} = 0. \quad \text{(8.29a and 8.29b)}$$

These are zero (Equations 8.29a and 8.29b) like Equations 8.22a through 8.22e because the integrands are periodic. The drag is zero on the fixed (Equation 8.21a) [moving (Equation 8.28a)] cylinder because the flow is symmetric relative to the y-axis through their center perpendicular to the motion; the circulation causes an asymmetry relative to the x-axis in the direction of motion and produces a lift on the fixed (Equation 8.21b) [moving (Equation 8.28b)] cylinder.

8.1.6 Wall Effect on the Lift on a Body

The lift on the moving cylinder (Equation 8.28b) always exceeds that on the fixed cavity (Equation 8.21b):

$$\Gamma < 0: \quad F_y^-(a,b) = -\frac{1}{2}\rho U\Gamma \frac{b^2 + a^2}{b^2 - a^2} = \frac{b^2 + a^2}{2a^2} F_y^+(a,b) > F_y^+(a,b). \quad \text{(8.30a and 8.30b)}$$

As the cavity recedes to infinity (Equation 8.31a), the lift (Equation 8.21b ≡ Equation 8.31b) on the outer cylinder becomes negligible:

$$b^2 \gg a^2: \quad F_y^+(a,b) \sim -\rho\Gamma U \frac{a^2}{b^2}, \quad F_y^-(a,b) = -\frac{\rho}{2} U\Gamma, \quad \text{(8.31a–8.31c)}$$

and the lift (Equation 8.28b ≡ Equation 8.31c) on the inner cylinder (1) is equal to one-half of that for a cylinder at rest (Equation I.28.29b); (2) is of the same sign; and (3) has a correction of order $2a^2/b^2$ for the wall effect relative to Equation I.28.29b=Equation 2.333a ≡ Equation 8.32a. The passage from Equations 8.28b to 8.31c to the first order in Equation 8.31a is

$$-\frac{2}{\rho U\Gamma} F_y^-(a,b) = \frac{b^2 + a^2}{b^2 - a^2} = \frac{1 + b^2/a^2}{1 - b^2/a^2}$$
$$= \left(1 + \frac{a^2}{b^2}\right)\left[1 + \frac{a^2}{b^2} + O\left(\frac{a^4}{b^4}\right)\right] = 1 + 2\frac{a^2}{b^2} + O\left(\frac{a^4}{b^4}\right). \quad \text{(8.31d)}$$

Concerning (1) as the cavity recedes to infinity (Equation 8.31a), the component of the tangential velocity on the cylinder (Equation 8.24b) unrelated to the circulation is $-U \sin\varphi$; this is one-half of the tangential velocity (Equation I.28.94b) for a cylinder at rest, without circulation; the circulation adds (Equation I.28.118b) the last term on the r.h.s. of Equation 8.24b. Thus, the velocity is doubled for a cylinder in a reference frame at rest relative to a reference frame moving with the cylinder with zero velocity at infinity.

The velocity appears linearly in the term of the pressure responsible for the lift, namely, the third on the r.h.s. of Equation 8.25. Thus, the lift on the inner cylinder, in a reference frame where it is at rest, is the double of Equation 8.31c as concerns the first term; it is assumed that the wall correction, that is, the second term on the r.h.s. of Equation 8.31c, is the same in a frame at rest and in motion, leading to the total lift Equation 8.32b:

$$F_{y0} = -\rho U \Gamma, \quad F_y(a,b) = -\rho U \Gamma - \rho U \Gamma \frac{a^2}{b^2} = F_{0y}\left(1 + \frac{a^2}{b^2}\right). \quad \text{(8.32a and 8.32b)}$$

The potential flow (Figure 8.1) past a cylinder of radius a moving with velocity U in a cylindrical cavity of radius b leads to a pressure distribution [Equation 8.20 (Equation 8.25)] on the outer (inner) fixed (moving) cylinder, corresponding to the zero drag [Equation 8.21a (Equation 8.28a)] and lift [Equation 8.21b (Equation 8.28b)] in the ratio (Equations 8.30a and 8.30b). The total lift on the inner cylinder in a reference frame at rest (Equation 8.32b) shows that the correction to the lift (Equation 8.32a) due to the wall effect is a^2/b^2, where a (b) is the radius of the inner (outer) cylinder. The "wall effect" is due to the "blockage" of the flow in the cavity by the inner cylinder: (1) for an incompressible flow, the volume flow rate must be constant; (2) a relative "blockage" scaling like the radii of the inner to the outer cylinder a/b leads to a relative velocity increase a/b; (3) by the Bernoulli law, the pressure scales on the square of the velocity; and (4) since lift scales on the pressure, the relative wall effect scales as $(a/b)^2$, as shown in Equation 8.32b. The potential flow for two cylinders, neither contained in the other, is considered next.

8.2 Two Cylinders in Relative Motion

Instead of one cylinder being inside the other (Figure 8.1), two cylinders external to each other are considered next (Figure 8.2), with possibly distinct radii *a* (*b*) and uniform velocities *U* (*V*) with arbitrary direction. Let *O* be the point of intersection of the trajectories of

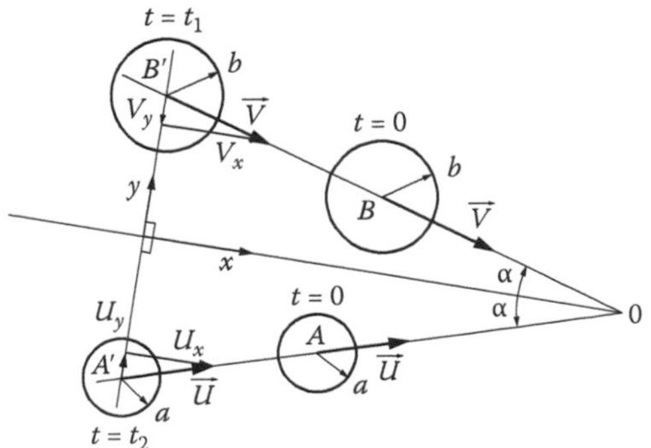

FIGURE 8.2
Potential flow past two cylinders can be considered in three cases: (1) the cylinders move along the line of centers (Figure 8.4a); (2) the cylinders move orthogonal to the line of centers (Figure 10.1a); and (3) in the general case of motion in oblique paths is a combination of cases 1 and 2.

the centers of the cylinders; it does not follow that the cylinders collide near O because their centers may pass there at different times. Suppose that at an initial time $t = 0$ the centers of the cylinders are at A (B). At some earlier $t_1 < 0$ or later $t_1 > 0$ time the cylinders will be at point A' (B') the same distance D from O. Then their velocity can be decomposed into components along (orthogonal to) the line of centers; since the potential flow is linear, it can be obtained by superimposing the two cases that are solved by similar methods [Section 8.2 (Example 10.10)]. The exact velocity potential for two cylinders involves an infinite set of pairs of images (Subsection 8.2.1). Only the first images appear in the complex potential that takes a simple form if the distance between the cylinders is large compared with their radii (Subsections 8.2.2 and 8.2.3); it also specifies the kinetic energy (Subsection 8.2.4) and added mass (Subsection 8.2.5). The particular case of two cylinders with the same radius moving toward or away from each other with the same velocity is equivalent to images on a wall at equal distance from the two cylinders (Subsection 8.2.6); it leads to the trajectory, that is, the position as a function of time including the wall effect (Subsection 8.2.7).

8.2.1 Infinite Set of Pairs of Images in Two Cylinders

In the case (Figure 8.2) of two cylinders of radii (a, b) with centers at a distance $c > a + b$ that may not be large compared with (a, b), the circle theorem (Sections I.24.7 and 2.5.3) leads (Figure 8.3) to an infinite set of pairs of images (Table 8.1), as follows: (1) the first cylinder, with radius a and center at the origin (Equation 8.33a), is represented by a dipole with moment Equation 8.33b ≡ Equation 8.6c:

$$z_0 = 0, \quad P_0 = -2\pi U a^2; \quad z_1 = c - \frac{b^2}{c}: \quad P_1 = -P_0\left(\frac{b}{c}\right)^2; \tag{8.33a–8.33d}$$

(2) the image on the second cylinder is located at Equation 8.33c and is a dipole with strength Equation 8.33d; (3) the next image on the first cylinder is located at Equation 8.34b and is a dipole with moment Equation 8.34c:

$$n = 1, 2, \ldots: \quad z_{2n} = \frac{a^2}{z_{2n-1}}, \quad P_{2n} = -P_{2n-1}\left(\frac{a}{z_{2n-1}}\right)^2, \tag{8.34a–8.34c}$$

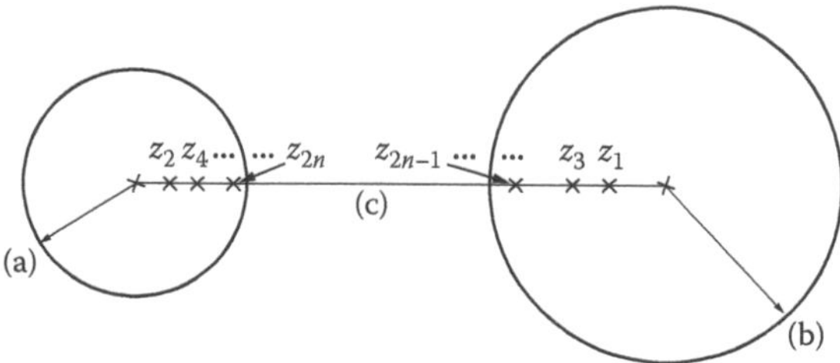

FIGURE 8.3
Potential flow past two cylinders, generally with distinct radii and velocities, is represented as follows: (1) to zero order by the dipoles at the center of each cylinder; (2) the interaction appears to the first order as the image of each dipole in the other cylinder; and (3) the higher orders lead to two infinite sets of images on each cylinder, with accumulation point at the surface of each cylinder closest to the other cylinder. The higher-order approximations are important for close cylinders; the first-order correction is sufficient if the distance between the centers of the cylinders is large compared with the radii.

TABLE 8.1

Images on Two Cylinders

Cylinders with Centers at Distance c			
First Cylinder: Radius a		**Second Cylinder: Radius b**	
Location of Image	**Dipole moment**	**Location of Image**	**Dipole Moment**
$z_0 = 0$	$P_0 = -2\pi Ua^2$	$z_1 = c - \frac{b^2}{c}$	$P_1 = -P_0\left(\frac{b}{c}\right)^2$
$z_2 = \frac{a^2}{z_1}$	$P_2 = -P_1\left(\frac{a}{z_1}\right)^2$	$z_3 = c - \frac{b^2}{c - z_2}$	$P_3 = -P_2\left(\frac{b}{c - z_2}\right)^2$
$z_4 = \frac{a^2}{z_3}$	$P_4 = -P_3\left(\frac{a}{z_3}\right)^2$	$z_5 = c - \frac{b^2}{c - z_4}$	$P_5 = -P_4\left(\frac{b}{c - z_4}\right)^2$
$z_6 = \frac{a^2}{z_5}$	$P_6 = -P_5\left(\frac{a}{z_5}\right)^2$	$z_7 = c - \frac{b^2}{c - z_6}$	$P_7 = -P_6\left(\frac{b}{c - z_6}\right)^2$
$z_{2n} = \frac{a^2}{z_{2n-1}}$	$P_{2n} = -P_{2n-1}\left(\frac{a}{z_{2n-1}}\right)^2$	$z_{2n+1} = c - \frac{b^2}{c - z_{2n}}$	$P_{2n+1} = -P_{2n}\left(\frac{b}{c - z_{2n}}\right)^2$

Notes: Potential flow past two close cylinders consists of the following: (1) the potential flow due to one cylinder represented by a dipole; (2) the effect of the second cylinder is represented to lowest order by the image dipole; and (3) the higher orders correspond to alternating images on each cylinder (Figure 8.3). The table indicates the locations and moments of the dipoles of all orders in each cylinder.

and likewise for all images of even order (Equation 8.34c); and (4) proceeding in the same way, the odd images (Equation 8.35a) are located at Equation 8.35b and correspond to a dipole with moment Equation 8.35c:

$$n = 1, 2, \ldots.: \quad z_{2n+1} = c - \frac{b^2}{c - z_{2n}}, \quad P_{2n+1} = -P_{2n}\left(\frac{b}{c - z_{2n}}\right)^2. \qquad (8.35\text{a}–8.35\text{c})$$

The images (Figure 8.3) in the second (Equation 8.35b) [first (Equation 8.34b)] cylinder as $n \to \infty$ have an accumulating point on the surface closest to the first (Equation 8.36a) [second (Equation 8.36b)] cylinder:

$$\lim_{n\to\infty} z_{2n+1} = c - b, \quad \lim_{n\to\infty} z_{2n} = a; \qquad (8.36\text{a and } 8.36\text{b})$$

the successive dipole moments decay like Equation 8.37:

$$\begin{aligned} P_{2n} &= P_0 \prod_{m=1}^{n} \frac{P_{2m}}{P_{2m-1}} \frac{P_{2m-1}}{P_{2m-2}} = -2\pi Ua^2 \prod_{m=1}^{n} \left(\frac{a}{z_{2m-1}}\right)^2 \left(\frac{b}{c - z_{2m-2}}\right)^2 \\ &= -2\pi Ua^{2+2n} b^{2n} c^{-4n} \prod_{m=1}^{n} \left(\frac{c}{z_{2m-1}}\right)^2 \left(1 - \frac{z_{2m-2}}{c}\right)^{-2} \end{aligned} \qquad (8.37)$$

where Equations 8.34b and 8.35b were used; the terms in the product Equation 8.37 have (Equations 8.36a and 8.36b) the limit Equation 8.38:

$$\lim_{m\to\infty}\left(\frac{c}{z_{2m-1}}\right)^2\left(1-\frac{z_{2m-2}}{c}\right)^{-2}=\left(\frac{c}{c-b}\frac{c}{c-a}\right)^2, \tag{8.38}$$

that is, of order unity if the radii are small relative to the distance between the centers. Thus, *the complex potential due to two cylinders of radii a (b) with centers at z = 0 (z = c) is given by Equation 8.39a*:

$$f_1(z)=-\frac{1}{2\pi}\sum_{n=0}^{\infty}\frac{P_n}{z-z_n},\quad P_{2n}\sim O\left(a^2\left(\frac{ab}{a^2}\right)^{2n}\right), \tag{8.39a and 8.39b}$$

in terms of the locations (Equations 8.33a, 8.34b, and 8.35b) and dipole moments (Equations 8.33b, 8.34c, and 8.35c) of the images (Figure 8.3 and Table 8.1). The images in the first (second) cylinder have accumulation points [Equation 8.36b (Equation 8.36a)], and the dipole moments (Equation 8.37) scale Equation 8.38 as Equation 8.39b. The potential due to the first cylinder (Equations 8.39a and 8.39b) transforms to that of the second cylinder via the substitutions:

$$(f_1, z, a, b, c)\leftrightarrow(f_2, z-c, b, a, c). \tag{8.40}$$

In both cases, the series like Equation 8.39a converges more rapidly (Equation 8.39b) for the distance between the centers larger than the radii; only the first few terms are needed if $c^2 \gg a^2, b^2$ in the **method of images.** A similar conclusion is reached by an alternative **method of perturbation potentials** used next (Subsection 8.2.2).

8.2.2 Two Cylinders Moving along the Line of Centers

The velocity potential must satisfy (1) the Laplace equation 8.41a; and (2) a boundary condition like Equation 8.2a on each cylinder using polar coordinates (r, φ) $[(s, \phi)]$ with center at the cylinder of radius a (b) in Figure 8.4a moving at velocity U (V) in Equation 8.41b (Equation 8.41c):

$$\nabla^2\Phi=0:\quad \left.\frac{\partial\Phi}{\partial r}\right|_{r=a}=-U\cos\varphi,\quad \left.\frac{\partial\Phi}{\partial s}\right|_{s=b}=-V\cos\phi, \tag{8.41a–8.41c}$$

using an external fixed reference frame. Since the problem is linear, the total potential is expressible as a linear combination (Equation 8.42a) of

$$\Phi=U\Phi_1+V\Phi_2,\quad \left.\frac{\partial\Phi_{1,2}}{\partial r}\right|_{r=a}=\{-\cos\varphi,0\},\quad \left.\frac{\partial\Phi_{1,2}}{\partial s}\right|_{s=b}=\{0,-\cos\phi\}, \tag{8.42a–8.42e}$$

the unit potentials due to the first cylinder moving with unit velocity with the other at rest (Equations 8.42b and 8.42c) [vice versa (Equations 8.42d and 8.42e)]. In the absence of the second cylinder, the unit potential of the first cylinder would be Equation 8.43a:

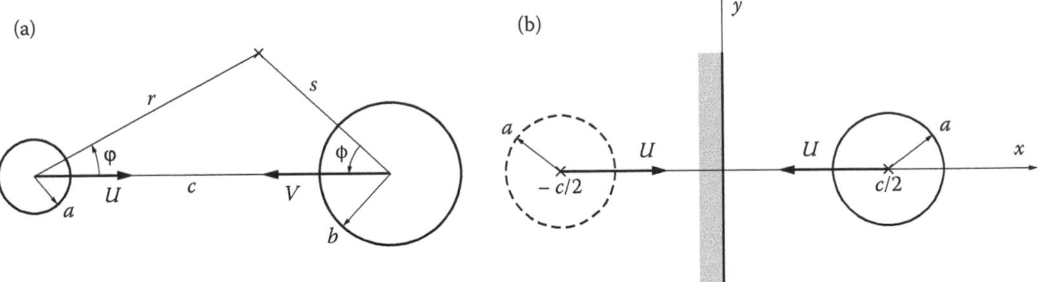

FIGURE 8.4
Potential flow associated with two cylinders with distinct radii and velocities moving along the line of centers (a). In the particular case of two cylinders with the same radius moving with the same velocity toward or away from each other, one cylinder corresponds to the image of the other on a static wall at equal distance between the two; this specifies the potential flow due to a cylinder moving perpendicularly toward or away from a plane rigid infinite wall (b).

$$\Phi_{10} = \frac{a^2}{r}\cos\varphi, \quad r = \left|c - se^{i\phi}\right|, \quad c = r\cos\varphi + s\cos\phi, \tag{8.43a–8.43c}$$

where the distance r (s) and angles (φ, ϕ) from the center of the two cylinders satisfy the geometric relations Equations 8.43b and 8.43c. The potential of the first cylinder in isolation (Equation 8.31a) satisfies the boundary condition 8.42b on the first but not (Equation 8.42d) on the second cylinder. In order to meet the latter, a perturbation potential Φ_{12} must be added to Equation 8.43a. The perturbation potential is found as follows: (1) the distance of an arbitrary point from the center of the first cylinder is the third side (Equation 8.43b) of a triangle with sides (c, s) making an angle ϕ:

$$\begin{aligned} r^2 &= \left|c - se^{i\phi}\right|^2 = \left|c - s\cos\phi - i\sin\phi\right|^2 \\ &= \left(c - s\cos\phi\right)^2 + \left(s\sin\phi\right)^2 = c^2 + s^2 - 2cs\cos\phi \\ &= \left(c - se^{i\phi}\right)\left(c - se^{-i\phi}\right) = c^2 + s^2 - cs\left(e^{i\phi} + e^{-i\phi}\right); \end{aligned} \tag{8.44}$$

(2) the asymptotic form of (Equation 8.44) leads to

$$r^{-2} = \left[c^2 + s^2 - 2cs\cos\phi\right]^{-1} = c^{-2}\left[1 + 2\frac{s}{c}\cos\phi + O\left(\frac{s^2}{c^2}\right)\right]; \tag{8.45}$$

(3) together with Equation 8.43c, this specifies the free potential of the first cylinder (Equation 8.43a) near the second cylinder:

$$\begin{aligned} \Phi_{10} &= \frac{a^2}{r^2}(c - s\cos\phi) = \frac{a^2}{c}\left[1 - \frac{s}{c}\cos\phi\right]\left[1 + 2\frac{s}{c}\cos\phi + O\left(\frac{s^2}{c^2}\right)\right] \\ &= \frac{a^2}{c}\left[1 + \frac{s}{c}\cos\phi + O\left(\frac{s^2}{c^2}\right)\right]; \end{aligned} \tag{8.46}$$

(4) thus, the free potential of the first cylinder causes at the second cylinder a normal velocity (Equation 8.47a):

$$\left.\frac{\partial\Phi_{10}}{\partial s}\right|_{s=b}=\frac{a^2}{c^2}\cos\phi=-\left.\frac{\partial\Phi_{12}}{\partial s}\right|_{s=b};\quad \Phi_{12}=\frac{a^2}{c^2}\frac{b^2}{s}\cos\phi; \qquad \text{(8.47a and 8.47b)}$$

and (5) in order to meet the boundary condition 8.42d on the second cylinder, a perturbation potential Equation 8.47b is introduced satisfying Equation 8.47a. Thus, the total potential of the first cylinder (Equation 8.48a) consists of the potential of the first cylinder in isolation (Equation 8.43a) plus the perturbation potential (Equation 8.47b) due to the second cylinder:

$$\Phi_1=\Phi_{10}+\Phi_{12}=\frac{a^2}{r}\cos\varphi+\frac{a^2b^2}{c^2s}\cos\phi; \qquad \text{(8.48a)}$$

$$\Phi_1\big|_{r=a}=a\cos\varphi\left[1+O\left(\frac{ab^2}{c^3}\right)\right]; \qquad \text{(8.48b)}$$

the total potential of the first cylinder takes the value Equation 8.48b on the first cylinder, where the term neglected specifies the order of accuracy of the method of the perturbation potential after one iteration. A higher accuracy requires more iterations. The total potential of the second cylinder (Equation 8.49b) can be obtained by an interchange of variables (Equation 8.49a):

$$(a,r,\varphi)\leftrightarrow(b,s,\phi):\quad \Phi_2=\frac{b^2}{s}\cos\phi+\frac{a^2b^2}{c^2r}\cos\varphi; \qquad \text{(8.49a and 8.49b)}$$

$$\Phi_2\big|_{s=b}=b\cos\varphi\left[1+O\left(\frac{ba^2}{c^3}\right)\right]; \qquad \text{(8.49c)}$$

the value of the second potential on the second cylinder is (Equation 8.49c) to within the accuracy of the first iteration of the method of the perturbation potential.

8.2.3 Values of Two Unit Potentials on Each Cylinder

In order to calculate the kinetic energy and added mass (Subsection 8.2.4), the values of the two potentials on the two cylinders are needed: (1) the values of first (second) potential on the corresponding first (second) cylinder have already been obtained [Equation 8.48b (Equation 8.49c)]; (2) the values of each potential on the other cylinder are obtained next; (3) the interchange of variables (Equation 8.49a) can be used to obtain one case from the other; and (4) thus, it suffices to consider one case, for example, the value of the potential of the second cylinder on the first cylinder. The method to do so is similar to the determination of the perturbation potential (Equations 8.45, 8.46, 8.47a, and 8.47b) made before: (1) the relation 8.50 inverse to Equation 8.45 is

$$s^{-2} = \left[c^2 + r^2 - 2c\,r\cos\varphi\right]^{-1} = c^{-2}\left[1 + 2\frac{r}{c}\cos\varphi + O\left(\frac{r^2}{c^2}\right)\right], \tag{8.50}$$

which is similar to Equation 8.45 with the change of variables Equation 8.49a; (2) it is used together with Equation 8.43c in the potential of the second cylinder in isolation:

$$\begin{aligned}\Phi_{20} &= \frac{b^2}{s}\cos\phi = \frac{b^2}{s^2}\left(c - r\cos\varphi\right)\\ &= \frac{b^2}{c}\left(1 - \frac{r}{c}\cos\varphi\right)\left[1 + 2\frac{r}{c}\cos\varphi + O\left(\frac{r^2}{c^2}\right)\right]\\ &= \frac{b^2}{c}\left[1 - \frac{r}{c}\cos\varphi + O\left(\frac{r^2}{c^2}\right)\right];\end{aligned} \tag{8.51}$$

(3) this may be added to the perturbation potential in the total potential Equation 8.49b of the second cylinder:

$$\frac{a^2b^2}{c^2r}\cos\varphi = \Phi_{21} = \Phi_2 - \Phi_{20} = \Phi_2 - \frac{b^2}{c} - \frac{b^2r}{c^2}\cos\varphi + O\left(\frac{b^2r^2}{c^3}\right); \tag{8.52}$$

(4) thus (Equation 8.52) the value of the potential of the second cylinder at the first cylinder (Equation 8.53a) is specified:

$$\Phi_2\big|_{r=a} = \frac{b^2}{c} + 2\frac{ab^2}{c^2}\cos\varphi; \qquad \Phi_1\big|_{s=b} = \frac{a^2}{c} + 2\frac{b\,a^2}{c^2}\cos\phi, \tag{8.53a and 8.53b}$$

and (5) the value of the potential of the first cylinder on the second cylinder (Equation 8.53b) is specified from Equation 8.53a via the interchange of variables Equation 8.49a. Thus, *the potential flow due to two cylinders of radii a (b) moving with velocity U (V) along the line of centers (Figure 8.4a) at a distance c has a total potential (Equation 8.42a) that is a linear combination of the unit potentials due to the first (Equation 8.48a) [second (Equation 8.49b)] cylinder; the unit potential of the first (second) cylinder takes the value Equation 8.48b (Equation 8.53a) on the first cylinder and Equation 8.53b (Equation 8.49c) on the second cylinder. The consideration of Equations 8.48b and 8.49c shows that the first iteration of the method of perturbation potentials is valid for Equations 8.69a and 8.69b).*

8.2.4 Kinetic Energy of a Potential Flow

The kinetic energy of a flow in a two-dimensional domain D with area element dS involves (Equation 8.54b) the mass density and modulus of the velocity:

$$\vec{v} = \nabla\Phi: \quad E_v \equiv \frac{1}{2}\int_D \rho|\vec{v}|^2\,dS = \frac{1}{2}\int_D \rho(\nabla\Phi.\nabla\Phi)\,dS; \tag{8.54a–8.54c}$$

in the case of an irrotational flow, for which the velocity is the gradient of a potential (Equation 8.54a) and the kinetic energy, Equation 8.54b becomes Equation 8.54c. Using the identity Equation 6.7b with Equation 8.55a leads to Equation 8.55b:

$$\vec{B} = \nabla\Phi: \quad \nabla.(\Phi\nabla\Phi) - \nabla\Phi.\nabla\Phi = \Phi\nabla.(\nabla\Phi) = \Phi\nabla^2\Phi; \qquad \text{(8.55a and 8.55b)}$$

substituting Equation 8.55b in Equation 8.52c specifies an alternate form for the kinetic energy of a potential flow:

$$2E_v = \int_D \rho[\nabla.(\Phi\nabla\Phi) - \Phi\nabla^2\Phi]dS. \qquad (8.56)$$

If the flow is also incompressible (Equation 8.57a), the potential satisfies the Laplace equation 8.57b and the identity Equation 8.55b simplifies to Equation 8.57c:

$$\rho = const: \quad \nabla^2\Phi = 0 \quad \Rightarrow \quad \nabla.(\Phi\nabla\Phi) = \nabla\Phi.\nabla\Phi = \left|\nabla\Phi\right|^2; \qquad \text{(8.57a–8.57c)}$$

in the latter case, the kinetic energy is given (Equations 8.56, 8.57a, and 8.57b) by an integral (Equation 8.58a) over the domain:

$$\frac{2}{\rho}E_v = \int_D \nabla.(\Phi\nabla\Phi)ds = \int_{\partial D} \Phi\frac{\partial\Phi}{\partial n}d\ell, \qquad \text{(8.58a and 8.58b)}$$

which by the divergence theorem (Equations I.28.1a and I.28.1b) becomes an integral (Equation 8.58b) over the boundary ∂D with arc length $d\ell$.

Thus, *the kinetic energy of a two-dimensional flow in a domain D of area element dS is specified by Equation 8.54b in terms of the mass density ρ and velocity $\vec{v}$. In the case of an irrotational flow (Equation 8.54a), it is expressed by Equation 8.54c ≡ Equation 8.56 in terms of the scalar real velocity potential Φ. If the flow is also incompressible (Equation 8.57a), hence for a potential flow Equation 8.57b, the kinetic energy Equation 8.58a can be expressed as an integral Equation 8.58b over the boundary ∂D with arc length $d\ell$*. In the present case:

$$\frac{2}{\rho}E_v = \left(-\int_{r=a} - \int_{s=b} + \int_{|z|\to\infty}\right)\Phi\frac{\partial\Phi}{\partial n}d\ell, \qquad (8.59)$$

the boundary consists of (1 and 2) the surface of the two cylinders, where the unit normal points inward, hence the minus signs in the first two terms on the r.h.s. of Equation 8.59; (3) the "surface at infinity" $|z| = R \to \infty$ where the unit normal points outward, hence the positive sign in the last term on the r.h.s. of Equation 8.59. Concerning the last term on the r.h.s. of Equation 8.59: (1) the potential of a dipole decays (Equation 8.60b) like the inverse of the radius (Equation 8.60a); (2) the normal derivative, that is, the radial velocity, decays like the inverse square of the radius (Equation 8.60c); and (3) hence the integral over "surface at infinity" (Equation 8.60d) taken as a circle of radius R:

$$|z| = R: \quad \Phi \sim \frac{1}{R}, \quad \frac{\partial \Phi}{\partial n} \sim \frac{1}{R^2}, \quad \int_{|z|=R} \Phi \frac{\partial \Phi}{\partial n} d\ell \sim \int \frac{R d\varphi}{R^3} \sim O\left(\frac{1}{R^2}\right) \quad (8.60a–8.60d)$$

vanishes as $R \to \infty$ and makes no contribution to the kinetic energy of the flow.

8.2.5 Added Mass of Two Interacting Cylinders

Substituting Equation 8.42a in Equation 8.59 specifies the kinetic energy of the flow past the two cylinders:

$$\frac{2}{\rho} E_v = -\left(\int_{r=a} + \int_{s=b}\right)(U\,\Phi_1 + V\,\Phi_2)\left(U\frac{\partial \Phi}{\partial n} + V\frac{\partial \Phi}{\partial n}\right) d\ell; \quad (8.61)$$

it is a bilinear function of the velocities of the two cylinders:

$$\frac{2}{\rho} E_v = I_1 U^2 + I_2 V^2 + 2 I_3 UV, \quad (8.62)$$

with three coefficients. The coefficient of the square of the velocity of the first (Equation 8.63a) [second (Equation 8.63b)] cylinder:

$$I_1 \equiv -\left(\int_{r=a} + \int_{s=b}\right)\Phi_1 \frac{\partial \Phi_1}{\partial n} d\ell = -\int_{r=a} \Phi_1 \frac{\partial \Phi_1}{\partial r} d\ell, \quad (8.63a)$$

$$I_2 \equiv -\left(\int_{r=a} + \int_{s=b}\right)\Phi_2 \frac{\partial \Phi_2}{\partial n} d\ell = -\int_{s=b} \Phi_2 \frac{\partial \Phi_2}{\partial s} d\ell, \quad (8.63b)$$

involves by [Equation 8.42d (Equation 8.42c)] only the potential of the first (second) cylinder. The product of the velocities of the two cylinders has the coefficient

$$\begin{aligned} 2I_3 &= -\left(\int_{r=a} + \int_{s=b}\right)\left(\Phi_1 \frac{\partial \Phi_1}{\partial n} + \Phi_2 \frac{\partial \Phi_1}{\partial n}\right) d\ell \\ &= -\int_{r=a} \Phi_2 \frac{\partial \Phi_1}{\partial r} d\ell - \int_{s=b} \Phi_1 \frac{\partial \Phi_2}{\partial s} d\ell, \end{aligned} \quad (8.64)$$

where Equations 8.42c and 8.42d were used. Both potentials satisfy the Laplace equation (Equations 8.65a and 8.65b)

$$\nabla^2 \Phi_1 = 0 = \nabla^2 \Phi_2: \quad \nabla.(\Phi_1 \nabla \Phi_2 - \Phi_2 \nabla \Phi_1) = \Phi_1 \nabla^2 \Phi_2 - \Phi_2 \nabla^2 \Phi_1 = 0, \quad (8.65a–8.65c)$$

leading to the **second Green identity** (Equation 8.65c); then the divergence theorem

$$0 = \int_D \nabla.(\Phi_1 \nabla \Phi_2 - \Phi_2 \nabla \Phi_1) dS = \int_{\partial D} \left[\Phi_1 \frac{\partial \Phi_2}{\partial n} - \Phi_2 \frac{\partial \Phi_1}{\partial n} \right] d\ell \tag{8.66}$$

shows (Equation 8.66) that

$$\int_{r=a} \Phi_1 \frac{\partial \Phi_1}{\partial r} d\ell = -I_3 = \int_{s=b} \Phi_1 \frac{\partial \Phi_2}{\partial s} d\ell \tag{8.67a and 8.67b}$$

the two terms of Equation 8.64 are equal.

The expression 8.62 for the kinetic energy of the potential flow (Equation 8.42a) past two cylinders can be applied to the exact potentials (Equations 8.39a and 8.40) [approximate (Equations 8.48a and 8.49b) potentials when the distance between the centers is arbitrary (much larger than the radii (Equations 8.69a and 8.69b)]. Taking the simple case as an example, the identity Equations 8.67a and 8.67b can be confirmed (Equation 8.68c ≡ Equation 8.68d) calculating all integrals (Equations 8.68a and 8.68b) from Equations 8.48b, 8.53b, 8.53a, 8.49c, 8.42b, and 8.42e:

$$I_1 = a \int_{r=a} \Phi_1 \cos\varphi \, d\varphi = a^2 \int_0^{2\pi} \cos^2\varphi \, d\varphi = \pi a^2, \tag{8.68a}$$

$$I_2 = b \int_{s=b} \Phi_2 \cos\phi \, d\phi = b^2 \int_0^{2\pi} \cos^2\phi \, d\phi = \pi b^2, \tag{8.68b}$$

$$I_3 = a \int_{r=a} \Phi_2 \cos\varphi \, d\varphi = 2\frac{a^2 b^2}{c^2} \int_0^{2\pi} \cos^2\varphi \, d\varphi = 2\pi \frac{a^2 b^2}{c^2}, \tag{8.68c}$$

$$I_3 = b \int_{s=b} \Phi_1 \cos\phi \, d\phi = 2\frac{a^2 b^2}{c^2} \int_0^{2\pi} \cos^2\phi \, d\phi = 2\pi \frac{a^2 b^2}{c^2}. \tag{8.68d}$$

The integrals Equations 8.68a and 8.68b interchange (a, b) in agreement with Equation 8.49a; the integrals Equation 8.68c ≡ Equation 8.68d coincide confirming Equations 8.67a and 8.67b. The **added area** for each cylinder (Equations 8.68a and 8.68b) corresponds to the area of displaced fluid, as had been found before for a single cylinder (Equation I.28.104b). Substituting Equations 8.68a through 8.68d in Equation 8.62 specifies

$$ab^2, a^2 b << c^3: \quad E_v = \rho \frac{\pi}{2} \left(U^2 a^2 + V^2 b^2 + 4 U V \frac{a^2 b^2}{c^2} \right) \tag{8.69a–8.69c}$$

the kinetic energy (Equation 8.69c) of the potential flow of a fluid of density ρ due to cylinders of radii $a(b)$ moving with velocities $U(V)$ along the line of centers at a distance c much larger than the radii (Equations 8.69a and 8.69b); it consists of the displaced mass of each cylinder plus an interaction term. The total potential (Equation 8.70) follows from (Equations 8.42a, 8.48a, and 8.49b):

$$\Phi = \frac{a^2}{r}\left(U + \frac{b^2}{c^2}V\right)\cos\varphi + \frac{b^2}{s}\left(V + \frac{a^2}{c^2}U\right)\cos\phi, \tag{8.70}$$

with the same approximation (Equations 8.69a and 8.69b).

8.2.6 Flow due to a Cylinder Moving Perpendicular to a Wall

In the case of cylinders of equal radii (Equation 8.71a) moving with the same velocity (Equation 8.71c) toward (away) from each other, they are images on a wall at rest at equal distance x from both (Figure 8.4b), simplifying (1) the total potential Equation 8.70 to Equation 8.71b:

$$a = b: \quad \Phi = U(\Phi_1 + \Phi_2) = Ua^2\left(1 + \frac{a^2}{c^2}\right)\left(\frac{\cos\varphi}{r} + \frac{\cos\phi}{s}\right), \qquad \text{(8.71a and 8.71b)}$$

$$U = V: \quad E = \frac{1}{2}E_v = \frac{1}{4}\rho U^2(I_1 + I_2 + 2I_3) = \frac{\pi}{2}\rho U^2 a^2\left(1 + \frac{2a^2}{c^2}\right), \qquad \text{(8.71c and 8.71d)}$$

and (2) the kinetic energy (Equation 8.69c) halved in Equation 8.71d) corresponding to the flow filling the half-space on one side of the wall. A cylinder of mass m_0 at a distance Equation 8.72a from a wall has a total energy Equation 8.72b:

$$x = \frac{c}{2}: \quad const \equiv \bar{E} = \frac{m_0}{2}U^2 + E = \left[\frac{m_0}{2} + \frac{\pi}{2}\rho a^2\left(1 + \frac{a^2}{2x^2}\right)\right]U^2, \qquad \text{(8.72a and 8.72b)}$$

that is, the sum of the kinetic energies corresponding to the following (Equation 8.73a): (1) the cylinder of mass m_0 moving at velocity U; (2) the **added mass** *of entrained fluid (Equation 8.73b) where Equation 8.73c is the* **displaced mass** *of one cylinder; (3) the term in curved brackets is the wall effect; and (4) it reduces to unity if the distance from the wall is large compared with the radius $2x^2 \gg a^2$:*

$$\bar{E} = \frac{1}{2}(m_0 + \bar{m})U^2, \quad \bar{m} \equiv m\left(1 + \frac{a^2}{2x^2}\right), \quad m = \rho\pi a^2. \qquad \text{(8.73a–8.73c)}$$

The total energy must be constant, and thus the velocity of the cylinder is maximum (minimum) at infinity $x = \infty$ (in contact with the wall $x = a$):

$$\left|\frac{2\bar{E}}{m_0 + m}\right|^{1/2} \equiv U_\infty \ge U(x) \equiv \frac{dx}{dt} \ge U_0 \equiv \left|\frac{2\bar{E}}{m_0 + 3m/2}\right|; \tag{8.74}$$

using Equation 8.73c, the velocity at infinity (at the wall) is given by Equation 8.75a (Equation 8.75b):

$$U_\infty \equiv \left|\frac{2\bar{E}}{m_0+\rho\pi a^2}\right|^{1/2} > U_0 \equiv \left|\frac{2\bar{E}}{m_0+3\rho\pi a^2/2}\right|^{1/2}. \tag{8.75a and 8.75b}$$

Solving Equation 8.72b for the velocity leads to

$$\left(\frac{\mathrm{d}x}{\mathrm{d}t}\right)^2 = U^2 = \frac{2\bar{E}}{m_0+m\left(1+\dfrac{a^2}{2x^2}\right)} = \frac{2\bar{E}}{m_0+m}\,\frac{m_0+m}{m_0+m+\dfrac{ma^2}{2x^2}} = \frac{U_\infty^2}{1+\dfrac{m}{m_0+m}\dfrac{a^2}{2x^2}}, \tag{8.76}$$

where the velocity at infinity (Equation 8.50) was introduced.

8.2.7 Trajectory of a Cylinder with Wall Effect

The equation of motion (Equation 8.76 ≡ Equation 8.77a):

$$U_\infty^2 = \left(1+\frac{1}{\mu}\frac{a^2}{x^2}\right)\left(\frac{\mathrm{d}x}{\mathrm{d}t}\right)^2, \quad \mu \equiv 2\frac{m_0+m}{m} = 2\left(1+\frac{\rho\pi a^2}{m}\right), \tag{8.77a and 8.77b}$$

involves: (1) the radius of the cylinder; (2) the velocity at infinity (Equation 8.75a); and (3) the dimensionless **mass ratio parameter** *(Equation 8.77b). The trajectory (Equation 8.78) with time measured from the contact with the wall is specified by*

$$U_\infty t = \frac{a}{\sqrt{\mu}}\left[\sqrt{1+\mu\frac{x^2}{a^2}} - \sqrt{1+\mu} + \arg\operatorname{csch}\left(\sqrt{\mu}\right) - \arg\operatorname{csch}\left(\sqrt{\mu}\,\frac{x}{a}\right)\right]. \tag{8.78}$$

The latter follows using the change of variable:

$$\xi = \frac{a}{\sqrt{\mu}}\operatorname{csch}\eta: \ \mathrm{d}\xi = -\frac{a}{\sqrt{\mu}}\operatorname{csch}\eta\coth\eta\,\mathrm{d}\eta, \quad \left|1+\frac{1}{\mu}\frac{a^2}{\xi^2}\right|^{1/2} = \left|1+\sinh^2\eta\right|^{1/2} = \cosh\eta \tag{8.79a–8.79c}$$

in the integral

$$\begin{aligned} U_\infty t = \int_a^x \left|1+\frac{1}{\mu}\frac{a^2}{\xi^2}\right|^{1/2}\mathrm{d}\xi &= -\frac{a}{\sqrt{\mu}}\int_{\arg\operatorname{csch}\left(\sqrt{\mu}\right)}^{\arg\operatorname{csch}\left(x\sqrt{\mu}/a\right)} \cosh\eta\operatorname{csch}\eta\coth\eta\,\mathrm{d}\eta \\ &= -\frac{a}{\sqrt{\mu}}\left[\eta-\coth\eta\right]_{\arg\operatorname{csch}\left(\sqrt{\mu}\right)}^{\arg\operatorname{csch}\left(x\sqrt{\mu}/a\right)}, \end{aligned} \tag{8.80}$$

where (1) the primitive is

$$\cosh\eta\operatorname{csch}\eta\coth\eta\,\mathrm{d}\eta = \coth^2\eta\,\mathrm{d}\eta = \left(1+\operatorname{csch}^2\eta\right)\mathrm{d}\eta = \mathrm{d}\left(\eta-\coth\eta\right); \tag{8.81}$$

(2) use of

$$\eta = \arg\operatorname{csch}\left(\sqrt{\mu}\,\frac{x}{a}\right), \quad \operatorname{csch}\eta = \sqrt{\mu}\,\frac{x}{a}, \quad \coth^2\eta = 1 + \operatorname{csch}^2\eta = 1 + \mu\frac{x^2}{a^2} \qquad \text{(8.82a and 8.82b)}$$

in Equation 8.80 leads to Equation 8.78. The trajectory, that is, the position as a function of time $x(t)$, is given by Equation 8.53 in inverse form as a time function of position $t(x)$, starting $t = 0$ at the wall $x = a$. The preceding result applies if the cylinder moves toward or away from the wall; in the latter case, it will never come into contact with the wall. The case of two cylinders moving perpendicular to the line of centers (Example 10.10) includes in particular a cylinder moving parallel to a wall. The potential flow past two cylinders at rest (Example 10.11) includes, in the case of equal radii, the images on a wall; the electrostatic analog concerns a conducting/insulating cylinder in the presence of a conducting/insulating wall.

8.3 Eccentric, Biconcave/Biconvex, and Semi-Recessed Cylinders

Besides a cylinder moving in a cylindrical cavity (Section 8.1) and two cylinders external to each other in relative motion (Section 8.2), another case is two static eccentric cylinders one inside the other (Section 8.3). The corresponding complex potential (Subsection 8.3.4) can be obtained in terms of coaxial coordinates (Subsection 8.3.1); the latter are conformal coordinates (Subsection I.33.3.1), that is, plane orthogonal curvilinear coordinates specified by a holomorphic function. The holomorphic function is the complex potential of a vortex with unit circulation near a wall (Section I.16.2; Subsection 2.7.4), leading to (1 and 2) the transformations between Cartesian and coaxial coordinates in complex (real) form [Subsection 8.3.1 (Subsection 8.3.2)]; and (3) the coordinate curves in coaxial coordinates (that is, equipotentials and streamlines) that are orthogonal families of circles (Subsection 8.3.3). The coaxial coordinates can also be used to describe the potential flow (1) between two eccentric cylinders (Subsection 8.3.4) including the force between them (Subsection 8.3.5); (2) past a cylindrical log on a wall (Subsection 8.3.8) including the pressure distribution (Subsection 8.3.9) and the wall effect on the velocity (Subsection 8.3.10); (3) by recessing (removing) the cylinder, past a cylindrical mound (ditch) in a wall (Subsection 8.3.6); and (4) past a biconcave (biconvex) cylinder by symmetry removing the wall (Subsection 8.3.7).

8.3.1 Vortex Images and Conformal Coaxial Coordinates

A vortex with circulation $\Gamma = 2\pi$ located at $(c, 0)$ has an opposite image with circulation $-\Gamma = -2\pi$ at $(-c, 0)$ if the y-axis is taken as a wall (Figure 8.5a). The corresponding complex potential (Equation 2.285b) specifies the **coaxial coordinates** (Equation 8.83b):

$$z \equiv x + iy: \quad \xi + i\eta \equiv \zeta = -i\log(z-c) + i\log(z+c) = i\log\left(\frac{z+c}{z-c}\right), \qquad \text{(8.83a and 8.83b)}$$

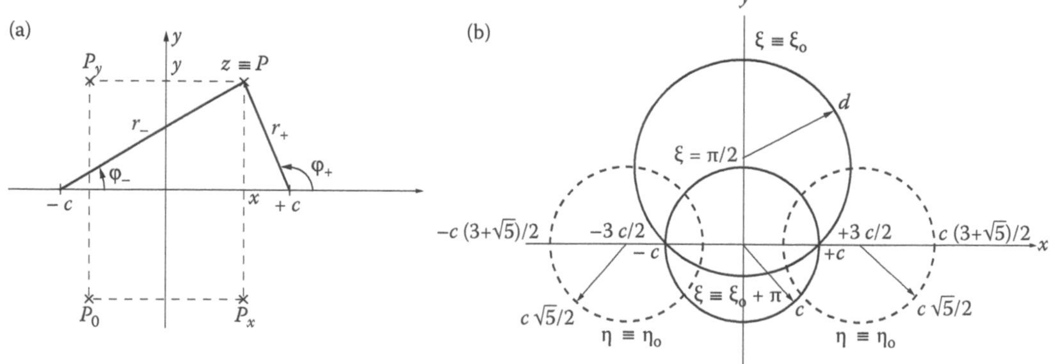

FIGURE 8.5
Coaxial coordinates are a conformal coordinate system (a) corresponding to (1) a vortex near a wall with its image; and (2) equivalently a pair of opposite vortices. The coaxial coordinate curves (b) are (1) circles with center on the imaginary axis; and (2) pairs of circles with centers on the real axis symmetric relative to the imaginary axis. The coaxial coordinates can be used to specify the potential flow associated with (1) eccentric cylinders with parallel axis (Figure 8.6a), including the particular case of a cylinder near a wall (Figure 8.6b); (2) a cylindrical mound or ditch on a wall (Figure 8.7a through f), including as limit case a cylindrical log on a wall (Figure 8.7j); and (3) biconcave (biconvex) cylinders [Figure 8.7g (Figure 8.7i)] leading as intermediate case to the circular cylinder (Figure 8.7h).

that is, the coaxial coordinate curves coincide with the equipotentials $\xi = \text{const}$ (streamlines $\eta = \text{const}$) of the flow. The arc length in the $z(\zeta)$-plane [Equation 8.84a (Equation 8.84b)]:

$$(\mathrm{d}s)^2 = (\mathrm{d}x)^2 + (\mathrm{d}y)^2 = |\mathrm{d}z|^2 = \left|\frac{\mathrm{d}z}{\mathrm{d}\zeta}\mathrm{d}\zeta\right|^2 = h^2\,|\mathrm{d}\zeta|^2 = h^2[(\mathrm{d}\xi)^2 + (\mathrm{d}\eta)^2] \quad \text{(8.84a and 8.84b)}$$

shows that (1) there is no cross term $\mathrm{d}\xi\,\mathrm{d}\eta$, and thus the coaxial coordinates are orthogonal, like all conformal coordinates (Subsection I.33.3.1); (2) the scale factor is the same along the two coordinate curves because it is specified by the modulus of the derivative of a holomorphic function that is independent of the direction:

$$\begin{aligned} h &\equiv \left|\frac{\mathrm{d}z}{\mathrm{d}\zeta}\right| = \left|\frac{\mathrm{d}\zeta}{\mathrm{d}z}\right|^{-1} = \left|i\left(\frac{1}{z+c} - \frac{1}{z-c}\right)\right|^{-1} = \left|\frac{z^2-c^2}{-2ic}\right| = \frac{1}{2c}\left|(x+iy)^2 - c^2\right| \\ &= \frac{1}{2c}\left|x^2+y^2-c^2+2ixy\right| = \frac{1}{2c}\left|(x^2+y^2-c^2)^2 - 4x^2y^2\right|^{1/2}. \end{aligned} \quad (8.85)$$

Introducing (Figure 8.5a) the distances $r_+(r_-)$ from the focal points $+c(-c)$ and the angles $\varphi_+(\varphi_-)$ with the real axis of the lines joining an arbitrary point (Equation 8.83a) to the focal points

$$z \mp c = r_\pm \exp(i\,\varphi_\pm): \quad \xi + i\eta = i\log\left|\frac{z+c}{z-c}\right| - \arg\left(\frac{z+c}{z-c}\right) \quad \text{(8.86a and 8.86b)}$$

TABLE 8.2

Coaxial Coordinates in the Four Quadrants

Quadrant	First	Second	Third	Fourth
(x,y)	$x > 0 < y$	$x < 0 < y$	$x < 0 > y$	$y < 0 < x$
Point	z	$-z^*$	$-z$	z^*
Cartesian coordinates	(x,y)	$(-x,y)$	$(x,-y)$	$(-x,-y)$
Correspondence	*Original*	*Reflection on the y-axis*	*Reflection on the origin*	*Reflection on the x-axis*
Coaxial coordinates	$\eta > 0$	$\eta < 0$	$\eta < 0$	$\eta > 0$
	$\xi > 0$	$\xi > 0$	$\xi < 0$	$\xi < 0$
Point	ζ	ζ^*	$-\zeta$	$-\zeta^*$
Coaxial coordinate	(ξ,η)	$(\xi,-\eta)$	$(-\xi,-\eta)$	$(-\xi,\eta)$

Notes: Coaxial coordinates (ξ,η) have distinct pairs of signs in the four quadrants; they have the same modulus $(|\xi|,|\eta|)$ at a point z in the first quadrant and (1) its image z^* on the x-axis that lies in the fourth quadrant; (2) its image $-z$ on the origin that lies in the third quadrant; and (3) its image $-z^*$ on the imaginary axis that lies in the second quadrant. The correspondence $(z,-z^*,-z,z^*) \leftrightarrow (\zeta,\zeta^*,-\zeta,-\zeta^*)$ arises because of the imaginary factor in Equation 8.83b.

leads (Equation 8.83b ≡ Equation 8.86b to the two coordinates:

$$\eta = \log\left|\frac{z+c}{z-c}\right| = \log\left(\frac{r_+}{r_-}\right), \quad \xi = \arg(z-c) - \arg(z+c) = \varphi_+ - \varphi_-. \tag{8.87a and 8.87b}$$

In particular, on the imaginary (real) axis, $r_- = r_+(\varphi_+ - \varphi_- = 0, \pi)$ implying $\eta = 0$ $(\xi = 0)$. Also the coaxial coordinates have the same modulus and distinct signs (Table 8.2) at the points in the four quadrants (P,P_x,P_y,P_0) that are (Figure 8.5a ≡ Figure I.1.1) symmetric relative to the origin (P_0), real (P_x), and imaginary (P_y) axes.

8.3.2 Transformations between Cartesian and Coaxial Coordinates

The transformation from complex Cartesian z to complex coaxial ζ coordinates (Equation 8.83b) can be inverted using (Equation 5.21d):

$$z = -c\frac{1+e^{-i\zeta}}{1-e^{-i\zeta}} = -c\frac{e^{i\zeta/2}+e^{-i\zeta/2}}{e^{i\zeta/2}-e^{-i\zeta/2}} = ic\cot\left(\frac{\zeta}{2}\right). \tag{8.88}$$

From Equations 8.87a and 8.87b follows the transformation from real Cartesian (x, y) to real coaxial (ξ,η) coordinates:

$$e^{2\eta} = \left|\frac{z+c}{z-c}\right|^2 = \left|\frac{x+c+iy}{x-c+iy}\right|^2 = \frac{(x+c)^2+y^2}{(x-c)^2+y^2}, \tag{8.89a}$$

$$\cot\xi = \cot(\varphi_+ - \varphi_-) = \frac{\cot(\varphi_+)\cot(\varphi_-)+1}{\cot(\varphi_-)-\cot(\varphi_+)} = \frac{\dfrac{x-c}{y}\dfrac{x+c}{y}+1}{\dfrac{x+c}{y}-\dfrac{x-c}{y}} = \frac{x^2+y^2-c^2}{2cy}. \tag{8.89b}$$

The identity

$$2\sin\left(\frac{\zeta}{2}\right)\sin\left(\frac{\zeta^*}{2}\right) = \cos\left(\frac{\zeta-\zeta^*}{2}\right) - \cos\left(\frac{\zeta+\zeta^*}{2}\right)$$
$$= \cos(i\eta) - \cos(\xi) = \cosh\eta - \cos\xi, \tag{8.90}$$

may be used in Equation 8.88 leading to

$$\{2x, 2iy\} = z \pm z^* = ic\left[\cot(\zeta/2) \mp \cot(\zeta^*/2)\right]$$
$$= ic\left[\frac{\cos(\zeta/2)}{\sin(\zeta/2)} \mp \frac{\cos(\zeta^*/2)}{\sin(\zeta^*/2)}\right] = ic\frac{\sin(\zeta^*/2)\cos(\zeta/2) \mp \sin(\zeta/2)\cos(\zeta^*/2)}{\sin(\zeta/2)\sin(\zeta^*/2)} \tag{8.91}$$
$$= 2ic\frac{\sin\left[(\zeta^* \mp \zeta)/2\right]}{2\sin(\zeta/2)\sin(\zeta^*/2)} = 2ic\frac{\{\sin(-i\eta), \sin(\xi)\}}{\cosh\eta - \cos\xi} = 2ic\frac{\{-i\sinh\eta, \sin\xi\}}{\cosh\eta - \cos\xi}.$$

The same result can be obtained from (Equations 5.238 and 8.88):

$$x + iy = z = ic\cot\left(\frac{\xi + i\eta}{2}\right) = ic\frac{\sin\xi - i\sinh\eta}{\cosh\eta - \cos\xi}. \tag{8.92}$$

The denominator in Equation 8.92 never vanishes because $\cosh\eta > 1 > |\cos\xi|$; the case $\cosh\xi = 1 = \cos\eta$, that is, $\xi = 0 = \eta$, when the numerator also vanishes implies $\zeta = 0$ in Equation 8.83b that $z + c = z - c$, which is impossible for noncoincident foci $c \neq 0$. These *(Equation 8.91 ≡ Equation 8.92) specify the transformation from real coaxial (η, ξ) to Cartesian (x, y) coordinates:*

$$\{x,y\} = c(\cosh\eta - \cos\xi)^{-1}\{\sinh\eta, \sin\xi\} \tag{8.93a and 8.93b}$$

showing that (1) the focal points ($\pm c$,0) correspond by Equation 8.89a or 8.87a to $\eta = \infty$; (2) on Equation 8.93b the real axis $y = 0$ except at the focal points $\xi = 0$; and (3) $y > 0$ for $\xi < \pi$ and $y < 0$ for $\xi > \pi$ in Equation 8.93b so that the arc circle $0 < \xi \equiv \xi_0 < \pi$ above the real axis corresponds to the arc of circle $\xi = \xi_0 + \pi$ below the real axis (Figure 8.5b).

8.3.3 Scale Factors and Coordinate Curves

The scale factor of coaxial coordinates is given by (1) Equation 8.85 in Cartesian coordinates and (2) Equation 8.88 in coaxial coordinates:

$$h = \left|\frac{dz}{d\zeta}\right| = \left|\frac{ic}{2}\sec^2\left(\frac{\zeta}{2}\right)\right| = \frac{c}{2}\left|\sin\left(\frac{\xi + i\eta}{2}\right)\right|^{-2} = \frac{c}{\cosh\eta - \cos\xi}, \tag{8.94}$$

where Equation 5.264 was used. *(1) The coaxial coordinates being conformal have the same scale factor along both coordinate curves; (2) the scale factor is given by Equation 8.85 (Equation 8.94) in terms of the Cartesian (coaxial) coordinates; and (3) this leads (Equations 8.85a through 8.85c) to the* **arc length** *in coaxial coordinates:*

$$(ds)^2 = |dz|^2 = h^2 |d\zeta|^2 = \frac{c^2\left[(d\xi)^2 + (d\eta)^2\right]}{\left[\cosh\eta - \cos\xi\right]^2}. \tag{8.95}$$

The coordinate curves $\xi = \text{const}$ follow from (Equation 8.89b):

$$\begin{aligned} 0 &= x^2 + y^2 - c^2 - 2cy\cot\xi = x^2 + \left(y - c\cot\xi\right)^2 - c^2(1 + \cot^2\xi) \\ &= x^2 + (y - c\cot\xi)^2 - c^2\csc^2\xi \end{aligned} \tag{8.96}$$

using Equation 5.23b. The coordinate curve $\xi = \text{const}$ follows from Equation 8.89a in the form

$$\begin{aligned} 0 &= \left[(x-c)^2 + y^2\right]e^{\eta} - e^{-\eta}\left[(x+c)^2 + y^2\right] \\ &= (x^2 + y^2 + c^2)(e^{\eta} - e^{-\eta}) - 2cx(e^{\eta} + e^{-\eta}) \\ &= 2(x^2 + y^2 + c^2)\sinh\eta - 4cx\cosh\eta, \end{aligned} \tag{8.97a}$$

which is equivalent to

$$\begin{aligned} 0 &= x^2 + y^2 + c^2 - 2cx\coth\eta = (x - c\coth\eta)^2 + y^2 - c^2\ (\coth^2\eta - 1) \\ &= (x - c\coth\eta)^2 + y^2 - c^2\operatorname{csch}^2\eta, \end{aligned} \tag{8.97b}$$

where was used Equation 5.22b. Thus, Equations 8.89a and 8.89b are transformed to Equation 8.96 ≡ Equation 8.98a (Equation 8.97b ≡ Equation 8.98b):

$$\left(x - c\coth\eta\right)^2 + y^2 = c^2\operatorname{csch}^2\eta, \quad x^2 + \left(y - c\cot\xi\right)^2 = c^2\csc^2\xi, \tag{8.98a and 8.98b}$$

The latter show that *the coaxial coordinates (Equations 8.83a and 8.83b) are conformal coordinates with scale factor Equation 8.85 ≡ Equation 8.94 and inverse Equation 8.88; the coordinate curves η = const (ξ = const) are (Figure 8.5b) circles with the center on the real (imaginary) axis at (c coth η, 0)[(0, c cot ξ)], with radii c csch η(c csc ξ). The points ±c are inverse with regard to the circles η = const with the center on the real axis because (Equation I.35.51) the product of their distances from the center is the square of the radius:*

$$(c\coth\eta - c)(c\coth\eta + c) = c^2(\coth^2\eta - 1) = c^2\operatorname{csch}^2\eta. \tag{8.99}$$

In particular, (1) the circles with center on the real axis at $x = \pm 3c/2$ correspond to $\coth\eta = \pm 3/2$ and hence have radii $\operatorname{csch}\eta = \sqrt{\coth^2\eta - 1} = \sqrt{5}/2$; and (2) the circle with center at the origin $y = 0$ corresponds to $\xi = \pi/2$ and has a radius $c\csc\xi = c$.

8.3.4 Eccentric Cylinder with Circulation in a Cylindrical Cavity

Using coaxial coordinates (Equation 8.83b) in the complex potential Equation 8.100a:

$$\Phi + i\Psi = f = -\frac{\Gamma\zeta}{2\pi}: \quad \Phi = -\frac{\Gamma\xi}{2\pi}, \quad \Psi = -\frac{\Gamma\eta}{2\pi}, \tag{8.100a–8.100c}$$

it follows that (1) the circles η = const in Equation 8.98a with (Figure 8.5b) inverse points $(\pm c, 0)$ are (Equation 8.100c) the streamlines Ψ = const (Figure 8.6a); (2) along a streamline, the potential Equation 8.100b changes by $-\Gamma$, so Γ is the circulation. Thus, *the complex potential of a cylinder $\eta = \eta_1 \equiv$ const with circulation Γ in a cylindrical tunnel $\eta = \eta_2 =$ const $< \eta_1$ is specified by Equations 8.100a and 8.88 ≡ Equation 8.101a in inverse form:*

$$z = -ic\cot\left(\frac{\pi}{\Gamma}f\right), \quad f = \frac{\Gamma}{\pi}\operatorname{arc}\cot\left(i\frac{z}{c}\right), \tag{8.101a and 8.101b}$$

leading to the direct potential Equation 8.101b and complex conjugate velocity:

$$v^*(z) = \frac{df}{dz} = -\frac{i\Gamma}{\pi c}\frac{1}{1+(iz/c)^2} = i\frac{\Gamma}{\pi}\frac{c}{z^2-c^2} = \frac{i\Gamma}{2\pi}\left(\frac{1}{z-c} - \frac{1}{z+c}\right), \tag{8.102}$$

corresponding to opposite image vortices with circulations $\pm\Gamma$ at the inverse points $z = \pm c$. The Blasius theorem (Equation I.28.20a ≡ Equation 2.205b) specifies through the residue (Equations I.15.33a and I.15.33b ≡ Equation 1.17) at the double pole at $z = c$ in the tunnel:

$$\begin{aligned} F^* \equiv F_x - iF_y &= i\frac{\rho}{2}\int^{(c+)}\left(\frac{df}{dz}\right)^2 dz = -i\frac{\rho\Gamma^2c^2}{2\pi^2}\int^{(c+)}\frac{dz}{(z^2-c^2)^2} \\ &= -i\frac{\rho\Gamma^2c^2}{2\pi^2}2\pi i\lim_{z\to c}\frac{d}{dz}\left[(z+c)^{-2}\right] = -\frac{2}{(2c)^3}\frac{\rho\Gamma^2c^2}{\pi} \end{aligned} \tag{8.103}$$

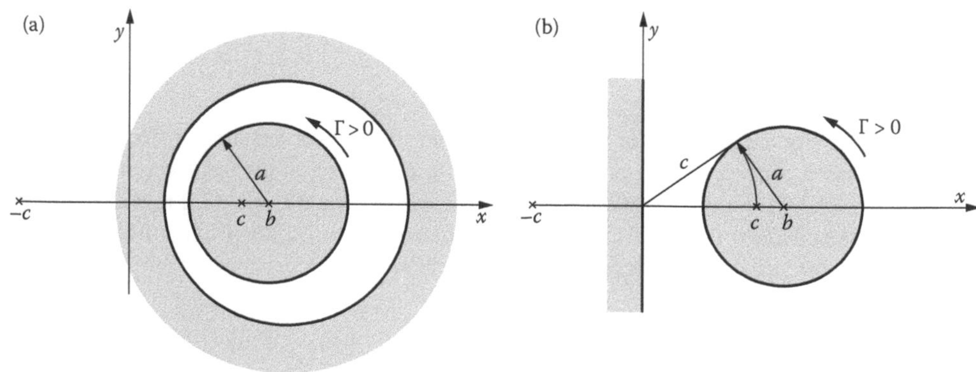

FIGURE 8.6
Potential flow between two eccentric cylinders with parallel but not coincident axis and radii such that one lies inside the other (Figure 8.6a) can be represented using coaxial coordinates (Figure 8.5). The particular case when the outer cylinder has an infinite radius corresponds to a cylinder near a wall (Figure 8.6b).

the force between two eccentric cylinders:

$$F_x = -\frac{\rho\Gamma^2}{4c\pi}, \quad F_y = 0, \tag{8.104a and 8.104b}$$

which tends to decrease the distance between their axis. The particular case of a cylinder near a wall is considered next (Subsection 8.3.5).

8.3.5 Force of Attraction of a Cylinder towards a Tunnel/Wall

In $\eta_2 \to 0$, the radius (center) tends to infinity $c \operatorname{csch} \eta_2 \to \infty$ ($c \coth \eta_2 \to \infty$), and the outer cylindrical cavity or tunnel becomes a plane $x = 0$ (Figure 8.6b). Denoting by b the distance of the center of the cylinder from the wall, since the reciprocal points are at $\pm c$, the product of their distances from the center equals the square of the radius (Equation 8.105a):

$$a^2 = (b+c)(b-c) = b^2 - c^2; \quad c = \left|b^2 - a^2\right|^{-1/2}; \tag{8.105a and 8.105b}$$

this implies (Equation 8.105b) that the distance of the reciprocal points from the wall $|c|$ equals the length of the tangent to the circle taken from the origin (Figure 8.6b). Substituting Equation 8.105a ≡ Equation 8.105b in Equation 8.104a leads to

$$F_x = -\frac{\rho\Gamma^2}{4\pi}\frac{1}{\sqrt{b^2 - a^2}}, \quad F_y = 0 = M, \tag{8.106a–8.106c}$$

the force of attraction (Equations 8.106a and 8.106b) of a cylinder with radius a and circulation Γ toward a wall at a distance b in a potential flow (Figure 8.6b) of mass density ρ; the pitching moment (Equation 8.106c) is zero, both for the cylindrical tunnel (Equations 8.104a and 8.104b) and flat wall (Equations 8.106a and 8.106b) because the Blasius theorem (Equation I.28.20b ≡ Equation 2.205c) implies:

$$\begin{aligned} M &= -\operatorname{Re}\left\{\frac{\rho}{2}\int\left(\frac{df}{dz}\right)^2 z\,dz\right\} = \operatorname{Re}\left\{\frac{\rho\Gamma^2c^2}{2\pi^2}\int^{(c+)}\frac{z}{(z^2-c^2)^2}dz\right\} \\ &= \operatorname{Re}\left\{-\frac{\rho\Gamma^2c^2}{4\pi^2}\int^{(c+)} d\left(\frac{1}{z^2-c^2}\right)\right\} = 0, \end{aligned} \tag{8.107}$$

since the primitive in Equation 8.107 is single-valued and returns to the same value after describing a contraclockwise circuit around $z = c$. There is no lift on Equation 8.104b on the inner cylinder because the two vortices in Equation 8.102 have opposite circulations. There is no pitching moment Equation 8.107 because there are neither spiral sources nor dipoles (Equation I.28.31b). The force of attraction of the inner cylinder toward the outer cylindrical cavity (Equation 8.104a) can be explained as follows: (1) the vortices create a circulatory flow around the inner cylinder; (2) since flow is incompressible, the velocity is larger in the narrower gap from the outer cylinder; (3) by the Bernoulli theorem (Equation 2.12b), the pressure is lower on the side of the inner cylinder closer to the outer cylindrical cavity; and (4) as a result, the former is attracted to the latter.

8.3.6 Flow Parallel to a Wall with a Cylindrical Mound or Ditch

Using coaxial coordinates (Equation 8.88 ≡ Equation 8.108b) in the complex potential (Equation 8.108a):

$$f = i\frac{2cU}{k}\cot\left(\frac{\zeta}{k}\right), \quad z = ic\cot\left(\frac{\zeta}{2}\right), \qquad \text{(8.108a and 8.108b)}$$

for $\zeta = \xi$ real (Equation 8.83b), z is imaginary in Equation 8.108b, and f is imaginary (Equation 8.108a), that is, it equals i times the stream function $f = i\Psi$; the streamline $\Psi = 0$ corresponds to $\cot(\xi/k) = 0$, that is, $\xi = k\pi/2$. This suggests considering the complex potential Equation 8.108a for Equation 8.109a, showing that (1) it is real, that is, coincides with the scalar potential and (2) thus Equation 8.109a corresponds to the streamline Equation 8.109c:

$$\xi = k\frac{\pi}{2}: \quad f\left(k\frac{\pi}{2} + i\eta\right) = \frac{2cU}{k}\tanh\left(\frac{\eta}{k}\right) \equiv \Phi\left(k\frac{\pi}{2}, \eta\right), \quad \Psi\left(k\frac{\pi}{2}, \eta\right) = 0; \qquad \text{(8.109a–8.109c)}$$

in Equation 8.109b, Equation 5.26b was used:

$$f\left(k\frac{\pi}{2} + i\eta\right) = i\frac{2cU}{k}\cot\left(\frac{\pi}{2} + i\frac{\eta}{k}\right) = -i\frac{2cU}{k}\tan\left(i\frac{\eta}{k}\right) = \frac{2cU}{k}\tanh\left(\frac{\eta}{k}\right). \qquad \text{(8.110)}$$

Thus, *Equations 8.108a and 8.108b are the complex potential of the flow past a wall* $x = 0$ *with a circular arc (Figure 8.5a) corresponding (Equation 8.98b) to coaxial coordinate* $\xi = k\pi/2$ *(Figure 8.7a through f); since f is real (Equation 8.108a and 8.108b) for real z, the reflection on the real axis (Subsection I.32.2.2) leads to compound cylinders (Figures 8.7g through i). The following particular cases are included (Table 8.3) of circular arc (Equation 8.98b ≡ Equation 8.111a) with radius Equation 8.111b and center at Equation 8.111c:*

$$(x - x_0)^2 + (y - y_0)^2 = a^2: \quad a = c\csc\left(\frac{k\pi}{2}\right), \quad \{x_0, y_0\} = \left\{0, c\cot\left(\frac{k\pi}{2}\right)\right\}, \qquad \text{(8.111a–8.111c)}$$

namely, (1) a semicircular mound $k = 1$ *(ditch* $k = 3$*) in Figure 8.7b (Figure 8.7e) leading to a circular cylinder in Figure 8.7h; (2) a concave mound (ditch) in Figure 8.7a (Figure 8.7f) for* $k < 1$ *(*$k > 3$*) leading to a biconcave cylinder (Figure 8.7g); (3 and 4) a convex mound (ditch) in Figure 8.7c (Figure 8.7d) for* $1 < k < 2(3 > k > 2)$ *leading to a biconvex cylinder in Figure 8.7i; and (5) the highest circular mound (Figure 8.7a) still touching the wall is a cylindrical log (Figure 8.7j).*

The complex conjugatet velocity:

$$v^*(\zeta) \equiv \frac{df}{dz} = \frac{df/d\zeta}{dz/d\zeta} = \frac{4U}{k^2}\sin^2\left(\frac{\zeta}{2}\right)\csc^2\left(\frac{\zeta}{k}\right) \qquad \text{(8.112)}$$

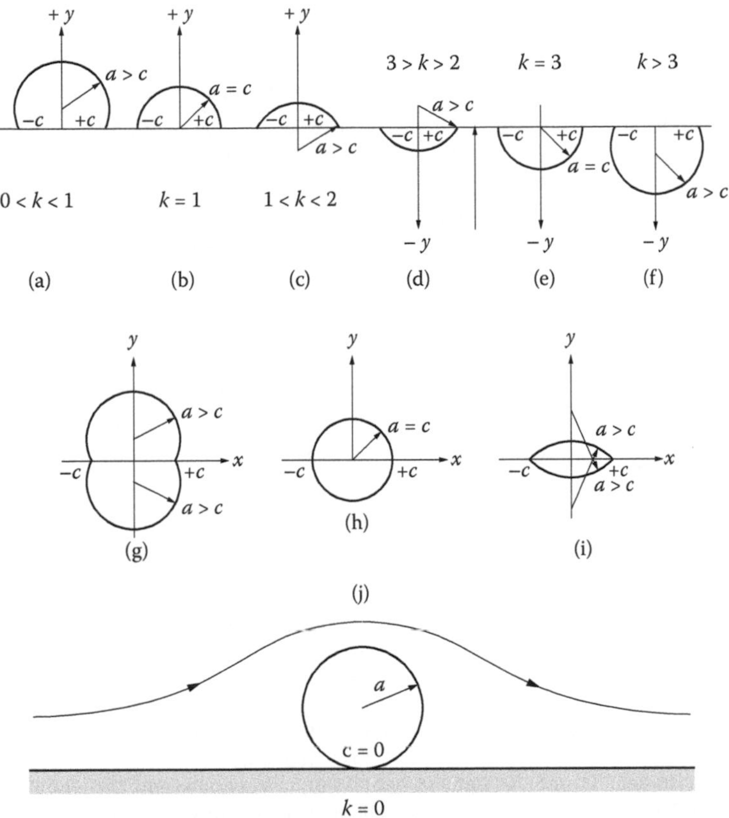

FIGURE 8.7
Coaxial coordinates (Figure 8.5) can also be used to specify the potential flow past a wall with (1) a cylindrical ditch, whose cross section can be a half-circle (e) or smaller (d) or larger (f); and (2) a cylindrical mound, whose cross section can be a half-circle (b) or larger (a) or smaller (c). The ditch (1) and mound (2) are images of each other on the wall and lead to a circular cylinder (h) as an intermediate case between cylinders with biconcave (g) and biconvex (i) cross sections. The limiting case of the mound (2) is a cylindrical log on a wall (j).

confirms that at infinity $z \to \infty$ in Equation 8.88, then $\zeta \to 0$ and

$$\lim_{z\to\infty} v^* = \lim_{\zeta\to 0} \frac{4U}{k^2}\sin^2\left(\frac{\zeta}{2}\right)\csc^2\left(\frac{\zeta}{k}\right) = \lim_{\zeta\to\infty} \frac{4U}{k^2}\left(\frac{\zeta}{2}\right)^2\left(\frac{k}{\zeta}\right)^2 = U, \tag{8.113}$$

so that U in Equation 8.108a is the free stream velocity. Using Equation 8.90, the modulus of the velocity (Equation 8.112) is given by

$$\left(\frac{k^2|v^*|}{4U}\right)^2 = \left(\frac{k^2}{4U}\right)^2 v^* v = \left[\frac{\sin(\zeta/2)\sin(\zeta^*/2)}{\sin(\zeta/k)\sin(\zeta^*/k)}\right]^2$$

$$= \left[\frac{\cosh\eta - \cos\xi}{\cosh(2\eta/k) - \cos(2\xi/k)}\right]^2 = \left[\frac{e^{\eta} + e^{-\eta} - e^{i\xi} - e^{-i\xi}}{e^{2\eta/k} + e^{-2\eta/k} - e^{2i\xi/k} - e^{-2i\xi/k}}\right]^2; \tag{8.114}$$

TABLE 8.3

Potential Flows in Coaxial Coordinates

Parameter	$k=0$	$0<k<1$	$k=1$	$1<k<2$	$k=2$	$2<k<3$	$k=3$	$k>3$
Unsymmetric	Cylindrical log	Concave mound	Circular mound	Convex mound	Flat wall	Convex ditch	Circular ditch	Concave ditch
Figure	8.7j	8.7a	8.7b	8.7c	–	8.7d	8.7	8.7f
Symmetric	Double cylinder log	Biconcave airfoil	Circular airfoil	Biconvex airfoil	Flat wall	Biconvex airfoil	Circular airfoil	Biconcave airfoil
Figure	–	8.7g	8.7h	8.7i	–	8.7i	8.7h	8.7g
Example	$k=0$	$k=\frac{1}{2}$	$k=1$	$k=\frac{3}{2}$	$k=2$	$k=\frac{5}{2}$	$k=3$	$k=\frac{7}{2}$
$\xi =$	0	$\frac{\pi}{4}$	$\frac{\pi}{2}$	$\frac{3\pi}{4}$	π	$\frac{5\pi}{4}$	$\frac{3\pi}{2}$	$\frac{7\pi}{4}$
Center at $x = 0, y = c\lvert\cot\xi\rvert$	a	c	0	c	∞	c	0	c
Radius $a = c\lvert\csc\xi\rvert$	a	$c\sqrt{2}$	c	$c\sqrt{2}$	∞	$c\sqrt{2}$	c	$c\sqrt{2}$

Notes: Potential flows described by coaxial coordinates include the following: (1) the eccentric cylinders (Figure 8.6a), that is, cylinders with circulation in a noncoaxial cylindrical cavity or tunnel: (2) a particular case is a cylinder near a wall but not touching it (Figure 8.6b); (3) a distinct case is a cylindrical log resting on a wall with a free stream parallel to the wall (Figure 8.7j); (4) the cylinder may be "sunk" into the wall to form a mound (Figure 8.7a through c); (5) removing the cylinder leaves a ditch in the wall (Figure 8.7d through f); (6 and 7) symmetry on the real axis leads to the compound concave (convex) cylinder or biconcave (biconvex) airfoil [Figure 8.7g (Figure 8.7i)]; (8) simplest case is a semicircular mound (Figure 8.7b) or ditch (Figure 8.7c) corresponding to a circular cylinder in a free stream (Figure 8.7h); and (9) trivial case of a uniform stream parallel to a wall is also included. The cylindrical log on a wall (Figure 8.8b) has a wall effect not present for the same cylinder in a free stream (Figure 8.8a).

thus at the focal points $z = \pm c$ that are at the edges (Figure 8.5b) and where $\eta \to \infty$ by Equation 8.87a, the velocity is

$$\lim_{\eta\to\pm c}\left|v(z)\right| = \lim_{\eta\to\infty}\frac{4U}{k^2}\exp\left[2\eta\left(1-\frac{2}{k}\right)\right] = \begin{cases}\infty & \text{if} \quad k>2,\\ U & \text{if} \quad k=2,\\ 0 & \text{if} \quad k<2,\end{cases} \tag{8.115a–8.115c}$$

zero (infinite) for $k < 2 (k > 2)$ corresponding to the mound (ditch) in Figure 8.7a through c (Figure 8.7d through f) in Equation 8.115c (Equation 8.115a) that has an acute (obtuse) corner; the intermediate case (Equation 8.115b ≡ Equation 8.116a) is

$$k=2: \quad f_1 = icU\cot\left(\frac{\zeta}{2}\right) = Uz, \quad \frac{df_1}{dz} = U, \tag{8.116a–8.116c}$$

corresponding to the uniform flow Equation 8.116b with velocity Equation 8.116c over a flat wall. Thus, *the complex potential Equations 8.108a and 8.108b corresponds to the complex conjugate velocity Equation 8.112 where U is the free stream velocity at infinity (Equation 8.113). The modulus of the velocity (Equation 8.114) shows that the velocity is zero (Equation 8.115c) [infinite (Equation 8.115a)] at the edge of a ditch (Figures 8.7d through 8.7f) [mound (Figures 8.7a through 8.7c)], with the uniform flow over a flat wall as the intermediate case (Equations 8.116a through 8.116c).*

8.3.7 Biconvex or Biconcave Airfoil in a Stream

The potential flow past a biconcave (biconvex) airfoil [Figure 8.7g (Figure 8.7i)] is obtained adding Equation 8.116b to Equation 8.108a, leading to

$$f_2 = f + f_1 = icU\left[\frac{2}{k}\cot\left(\frac{\zeta}{k}\right) + \cot\left(\frac{\zeta}{2}\right)\right]. \tag{8.117}$$

The flow with angle-of-attack α and velocity U past the **biconcave (biconvex) airfoil** *has complex potential:*

$$f_3 = i\frac{2Uc}{k}\left[\cos\alpha\cot\left(\frac{\zeta}{k}\right) - i\sin\alpha\csc\left(\frac{\zeta}{k}\right)\right] \tag{8.118}$$

and complex conjugate velocity:

$$v_3^*(\zeta) \equiv \frac{\mathrm{d}f_3}{\mathrm{d}z} = \frac{\mathrm{d}f_3/\mathrm{d}\zeta}{\mathrm{d}z/\mathrm{d}\zeta} = \frac{4U}{k^2}\sin^2\left(\frac{\zeta}{2}\right)\left[\cos\alpha\csc\left(\frac{\zeta}{k}\right) - i\sin\alpha\cot\left(\frac{\zeta}{k}\right)\right]\csc\left(\frac{\zeta}{k}\right). \tag{8.119}$$

This follows from the following (1) on the airfoil (Equation 8.120a), the complex potential Equation 8.118 is real (Equation 8.120b) corresponding to the streamline (Equation 8.120c):

$$\xi = \frac{k\pi}{2}:\quad f\left(\frac{k\pi}{2} + i\eta\right) = \Phi\left(\frac{k\pi}{2},\eta\right),\quad \Psi\left(\frac{k\pi}{2},\eta\right) = 0; \tag{8.120a–8.120c}$$

(2) at infinity, the limit of the conjugate velocity (Equation 8.119) is

$$\lim_{z\to\infty} v^* = \lim_{\zeta\to\infty} v_3^*(\zeta) = U(\cos\alpha - i\sin\alpha) = U\,e^{-i\alpha}, \tag{8.121}$$

corresponding to a uniform stream of velocity U and angle-of-attack α. The latter result (Equation 8.121) follows from Equation 8.119 taking the equivalent (Equation 8.88) limits $z \to \infty$ or $\zeta \to 0$:

$$\lim_{z\to\infty} v_3^*(z) = \lim_{\zeta\to 0}\frac{4U}{k^2}\left(\frac{\zeta}{2}\right)^2(\cos\alpha - i\sin\alpha)\left(\frac{k}{\zeta}\right)^2 = U(\cos\alpha - i\sin\alpha) = U\,e^{-i\alpha}. \tag{8.122}$$

The former result (1) follows from the complex potential Equation 8.118 that is given on the airfoil (Equation 8.120a) by

$$\begin{aligned} f_3\left(\frac{k\pi}{2} + i\eta\right) &= i\frac{2Uc}{k}\left[\cos\alpha\cot\left(\frac{\pi}{2} + i\frac{\eta}{k}\right) - i\sin\alpha\csc\left(\frac{\pi}{2} + i\frac{\eta}{k}\right)\right] \\ &= -i\frac{2Uc}{k}\left[\cos\alpha\tan\left(i\frac{\eta}{k}\right) + i\sin\alpha\sec\left(i\frac{\eta}{k}\right)\right] \\ &= \frac{2Uc}{k}\left[\cos\alpha\tanh\left(\frac{\eta}{k}\right) + \sin\alpha\,\mathrm{sech}\left(\frac{\eta}{k}\right)\right] \equiv \Phi\left(\frac{k\pi}{2},\eta\right) \end{aligned} \tag{8.123}$$

using Equations 5.26b and 5.24b; since Equation 8.123 is real, it coincides with the real potential Equation 8.120b and the airfoil is the streamline Equation 8.120c.

8.3.8 Flow Past a Cylindrical Log on a Wall

The cylindrical log on a wall (Figure 8.7j) corresponds to the coincidence $c \to 0$ of the focal points (Figure 8.5b) and for the mound (Equation 8.124a) leads by Equation 8.98b ≡ Equation 8.111a or 8.111b to the radius Equation 8.124b:

$$\xi = \frac{k\pi}{2}: \quad a = \lim_{c\to 0} c\csc\xi = \lim_{k\to 0} c\csc\left(\frac{k\pi}{2}\right) = \frac{2c}{k\pi}, \qquad (8.124\text{a and }8.124\text{b})$$

where a finite position (Equation 8.108b) [complex potential (Equation 8.108a)] for $c \to 0$ implies $\zeta \to 0$ $(k \to 0)$. The arc of the circle below the real axis corresponds to Equation 8.124a, and it reduces to a point $c \to 0$ if $k \to 0$ as indicated in Equation 8.124b. For finite z, the following approximations (Equations 8.125b and 8.125c) apply in Equations 8.108a and 8.108b in the limit Equation 8.125a:

$$c \to 0: \quad \zeta = \frac{2ic}{z} = \frac{ik\pi a}{z}, \quad f = iU\pi a\cot\left(\frac{i\pi a}{z}\right) = U\pi a\coth\left(\frac{\pi a}{z}\right) \qquad (8.125\text{a–}8.125\text{c})$$

using Equation 5.27b. Thus, *the complex potential (Equation 8.125c ≡ Equation 8.126a) [conjugate velocity (Equation 8.126b)] of a uniform stream of velocity U past a* **cylindrical log** *of radius a on a wall (Figure 8.7j) follows:*

$$f(z) = U\pi a\coth\left(\frac{\pi a}{z}\right), \quad v^*(z) = \frac{df}{dz} = U\left(\frac{\pi a}{z}\right)^2 \operatorname{csch}^2\left(\frac{\pi a}{z}\right). \qquad (8.126\text{a and }8.126\text{b})$$

The complex potential (Equation 8.126a≡Equation 8.127b):

$$r \equiv \left|x^2 + y^2\right|^{1/2}: \quad f(x+iy) = U\pi a\coth\left(\frac{\pi a}{x+iy}\right) = U\pi a\coth\left(\pi a\frac{x-iy}{r^2}\right) = \Phi + i\Psi$$

(8.127a and 8.127b)

is real specifying the streamline [Equation 8.128c (Equation 8.129c)] consisting of two parts: (1) the wall (Equation 8.128a) where the complex potential is real (Equation 8.128b):

$$y = 0: \quad f(x) = U\pi a\coth\left(\frac{\pi a x}{r^2}\right) = \Phi(x,0), \quad \Psi(x,0) = 0; \qquad (8.128\text{a–}8.128\text{c})$$

$$\frac{\pi a y}{r^2} = \frac{\pi}{2}: \quad f\left(x + i\frac{r^2}{2a}\right) = U\pi a\tanh\left(\frac{\pi a x}{r^2}\right) = \Phi\left(x, \frac{r^2}{2a}\right), \quad \Psi\left(x, \frac{r^2}{2a}\right) = 0;$$

(8.129a–8.129c)

and (2) the curve (Equation 8.129a) where the complex potential is also real:

$$f\left(x+i\frac{r^2}{2a}\right)=U\pi a\coth\left(\frac{\pi ax}{r^2}-i\frac{\pi}{2}\right)=iU\pi a\cot\left(i\frac{\pi ax}{r^2}+\frac{\pi}{2}\right)$$
$$=-iU\pi a\tan\left(i\frac{\pi ax}{r^2}\right)=U\pi a\tanh\left(\frac{\pi ax}{r^2}\right)=\Phi\left(x,\frac{2ay}{r^2}\right), \quad (8.130)$$

as follows using Equations 5.26b and 5.27b. The curve (Equation 8.129a ≡ Equation 8.131) corresponds to the cylinder of radius a and center at (0, a) on the imaginary axis and tangent to the real axis:

$$0=r^2-2ay=x^2+y^2-2ay=x^2+(y-a)^2-a^2. \quad (8.131)$$

At infinity $z\to\infty$, the velocity is U as follows [Equation 8.132a (Equation 8.132b)]:

$$\lim_{z\to\infty}f(z)=U\pi a\frac{z}{\pi a}=Uz, \quad \lim_{z\to\infty}v^*(z)=U\left(\frac{\pi a}{z}\right)^2\left(\frac{\pi a}{z}\right)^{-2}=U \quad \text{(8.132a and 8.132b)}$$

from Equation 8.126a (Equation 8.126b).

8.3.9 Velocity and Pressure Distribution on a Cylindrical Log

The velocity (Equation 8.126b) is (1) horizontal (Equation 8.133b) and thus tangential (Equation 8.133c) at the wall (Equation 8.133a):

$$y=0: \quad v_y(x,0)=0, \quad v_x(x,0)=U\left(\frac{\pi a}{x}\right)^2\operatorname{csch}^2\left(\frac{\pi a}{x}\right); \quad \text{(8.133a–8.133c)}$$

(2) since the log (Equation 8.134a) is a streamline (Equation 8.129c), the radial velocity is zero (Equation 8.134b):

$$r^2=2ay: \quad v_r(\varphi)=0; \quad -v_\varphi(\varphi)=\left|v^*\left(x+i\frac{r^2}{2a}\right)\right|=U\left(\frac{\pi a}{r}\right)^2\left|\operatorname{csch}^2\left(\pi a\frac{x-iy}{r^2}\right)\right|; \quad \text{(8.134a–8.134c)}$$

and (3) the tangential velocity on the log is counterclockwise (Equation 8.134c) and its value is the modulus (Equation 8.135) of Equation 8.126b at Equation 8.134a:

$$|v^*|=U\left(\frac{\pi a}{r}\right)^2\left|\operatorname{csch}^2\left(\frac{\pi ax}{r^2}-i\frac{\pi}{2}\right)\right|=U\left(\frac{\pi a}{r}\right)^2\operatorname{sech}^2\left(\frac{\pi ax}{r^2}\right)=U\frac{\pi^2 a}{2y}\operatorname{sech}^2\left(\frac{\pi x}{2y}\right). \quad (8.135)$$

In Equation 8.135, the following was used:

$$\operatorname{csch}\left(\frac{\pi a x}{r^2} - i\frac{\pi}{2}\right) = i \csc\left(i\frac{\pi a x}{r^2} + \frac{\pi}{2}\right) = i \sec\left(i\frac{\pi a x}{r^2}\right) = i \operatorname{sech}\left(\frac{\pi a x}{r^2}\right), \tag{8.136}$$

from Equations 5.24b and 5.25b.

An alternative of the derivation (Equations 8.134a through 8.134c and 8.135) is to obtain (Equation 8.137b) the complex conjugate velocity (Equation 8.126b) on the cylindrical log (Equation 8.137a):

$$\begin{aligned}
&z = ia + ae^{i\varphi}: \quad v_r(\varphi) - i v_\varphi(\varphi) \equiv e^{i\varphi} v^*(ia + ae^{i\varphi}) \\
&= e^{i\varphi} U \left(\frac{\pi}{i + e^{i\varphi}}\right)^2 \operatorname{csch}^2\left(\frac{\pi}{i + e^{i\varphi}}\right) = \frac{\pi^2 U}{e^{i\varphi} - e^{-i\varphi} + 2i} \operatorname{csch}^2\left(\frac{\pi}{\cos\varphi + i(1 + \sin\varphi)}\right) \\
&= \frac{\pi^2 U}{2i(1 + \sin\varphi)} \operatorname{csch}^2\left(\pi \frac{\cos\varphi - i(1 + \sin\varphi)}{\cos^2\varphi + (1 + \sin\varphi)^2}\right) = -i\frac{\pi^2}{2}\frac{U}{1 + \sin\varphi} \operatorname{csch}^2\left(\frac{\pi}{2}\frac{\cos\varphi}{1 + \sin\varphi} - i\frac{\pi}{2}\right) \\
&= i\frac{\pi^2}{2}\frac{U}{1 + \sin\varphi} \operatorname{sech}^2\left(\frac{\pi}{2}\frac{\cos\varphi}{1 + \sin\varphi}\right)
\end{aligned} \tag{8.137a and 8.137b}$$

using Equation 8.136. Thus, *a cylindrical log (Equation 8.131 ≡ Equation 8.137a) of radius a standing on an infinite wall in a free stream of velocity U parallel to the wall (Figure 8.7j) has a tangential surface velocity:*

$$v_\varphi(\varphi) = -U\frac{\pi^2 a}{2y} \operatorname{sech}^2\left(\frac{\pi x}{2y}\right) = -\frac{\pi^2}{2}\frac{U}{1 + \sin\varphi} \operatorname{sech}^2\left(\frac{\pi}{2}\frac{\cos\varphi}{1 + \sin\varphi}\right); \tag{8.138a and 8.138b}$$

this leads by the Bernoulli law (Equation 2.12b) to the pressure distribution:

$$p(\varphi) = p_0 - \frac{1}{2}\left[v_\varphi(\varphi)\right]^2 = p_0 - \frac{\rho\pi^4}{8}\frac{U^2}{(1 + \sin\varphi)^2} \operatorname{sech}^4\left(\frac{\pi}{2}\frac{\cos\varphi}{1 + \sin\varphi}\right), \tag{8.139}$$

where p_0 is the stagnation pressure. The consistency of Equation 8.138a ≡ Equations 8.135 and 8.134c and Equation 8.138b ≡ Equation 8.137b follows (Equation 8.137a ≡ Equation 8.140a) from Equations 8.140b and 8.140c:

$$x + iy = z = ia + ae^{i\varphi}: \quad x = a\cos\varphi, \quad y = a(1 + \sin\varphi). \tag{8.140a–8.140c}$$

The tangential velocity (Equations 8.138a and 8.138b) demonstrates the wall effect in accelerating the flow past the cylinder (Subsection 8.3.10).

8.3.10 Wall Effect in Accelerating the Flow Past Cylindrical Log

The velocity (Equation 8.126b) on the cylindrical log (Equations 8.140a through 8.140c) is tangential (Equations 8.138a and 8.138b) and thus (1) vertical upward (downward) at the upstream (downstream) side point with

$$\begin{aligned}
-v_\varphi(\pi) &= -v_\varphi(0) = v_y(-a,a) = -v_y(a,a) = \left|v^*(\mp a + i a)\right| \\
&= U\frac{\pi^2}{2}\left|\sinh^2\left[\frac{\pi}{2}(\mp 1 - i)\right]\right|^{-1} = U\frac{\pi^2}{2}\left|\sin\left[i\frac{\pi}{2}(\mp 1 - i)\right]\right|^{-2} \\
&= U\frac{\pi^2}{2}\left|\cos\left[\left(\mp i\frac{\pi}{2}\right)\right]\right|^{-2} = U\frac{\pi^2}{2}\left|\cosh\left[\left(\frac{\pi}{2}\right)\right]\right|^{-2} \\
&= U\frac{\pi}{2}\operatorname{sech}^2\left(\frac{\pi}{2}\right) = 0.78380U
\end{aligned} \tag{8.141}$$

modulus smaller than the free stream velocity; (2) zero at the lowest point where the slope is zero $y/x \to 0$:

$$\begin{aligned}
v_\varphi\left(-\frac{\pi}{2}\right) &= \left|v^*(0)\right| = \lim_{\substack{y/x\to 0\\ x\to 0}} U\frac{\pi^2 a^2}{x^2\left(1+y^2/x^2\right)}\left|\operatorname{csch}^2\left(\frac{\pi a}{x}\frac{1-iy/x}{1+y^2/x^2}\right)\right| \\
&= \lim_{x\to 0} U\left(\frac{\pi a}{x}\right)^2\left|\operatorname{csch}^2\left(\frac{\pi a}{x}\right)\right| = \lim_{x\to 0} U\left(\frac{2\pi a}{x}\right)^2\exp\left(-\frac{2\pi a}{x}\right) = 0;
\end{aligned} \tag{8.142}$$

and (3) horizontal at the highest point:

$$\begin{aligned}
v_\varphi\left(\frac{\pi}{2}\right) &= v_x(0,2a) = \left|v^*(0,2ia)\right| = U\frac{\pi^2}{4}\left|\operatorname{csch}^2\left(-i\frac{\pi}{2}\right)\right| \\
&= U\frac{\pi^2}{4}\left|\sin\left(\frac{\pi}{2}\right)\right|^{-2} = U\frac{\pi^2}{4} = 2.46740\,U,
\end{aligned} \tag{8.143}$$

where it is larger than the free stream velocity. The maximum velocity on a cylindrical log on a wall (Equation 8.143) is larger than the maximum velocity $2U$ on a cylinder in a free stream (Equations I.28.94a and I.28.94b ≡ Equation 2.226d by a factor $\pi^2/8 = 1.23370$; thus, *the horizontal wall on which the cylindrical log is supported (Figure 8.8b) accelerates the potential flow past the highest point relative to a free stream by 23.3%. At the stagnation points for the cylinder in a free stream (Figure 8.8a), the wall effect on the cylindrical log causes a velocity of 78.4% of the free stream velocity. The acceleration of the incompressible flow is due to the convergence of the streamlines as the obstacle is approached; there is the reverse deceleration as the obstacle is left behind.* This is another example of wall effect (Subsections 8.1.4 and 8.3.6) in a potential flow. The rounded top of the cylinder leads to a finite velocity, in contrast to a concave

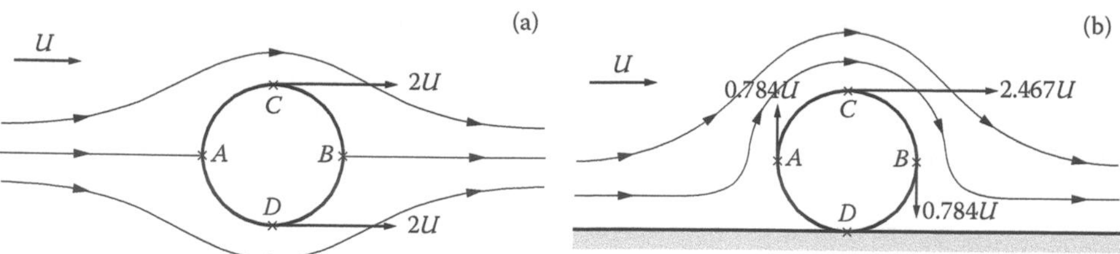

FIGURE 8.8
Cylinder in a free stream of velocity U has (a) two stagnation points A and B on the line through the axis parallel to the free stream direction; and maximum velocity $2U$ at the points C and D on the line through the axis perpendicular to the free stream direction. If the cylinder is placed on an infinite wall (b) as a cylindrical log, the wall effect causes the following: (1) the contact point at the bottom D to be a stagnation point; (2) the velocity to be 78.4% of the free stream velocity at the side points A and B; and (3) at the top C the velocity is 23.3% higher. This demonstrates the wall effect of acceleration of the flow over a smooth obstacle.

(convex) corner with a zero (infinite) velocity at the edge; the latter is demonstrated next for the flow past a ramp or step in a wall (Section 8.4).

8.4 Ramp/Step in a Wall and Thick Pointed/Blunt Plate

The real axis may be mapped (Figure 8.9a) into a polygon by the Schwartz–Christoffel transformation (Chapter I.33) with the edges corresponding to the critical points (Subsection 8.4.1). There are two critical points (Subsection 8.4.2) in the case [Figure 8.9b (Figure 8.9c)] of the uniform flow parallel past a ramp (step) in a wall [Subsection 8.4.2 (Subsection 8.4.3)]. The reflection on the lower wall leads from the ramp (step) in a wall [Subsection 8.4.4

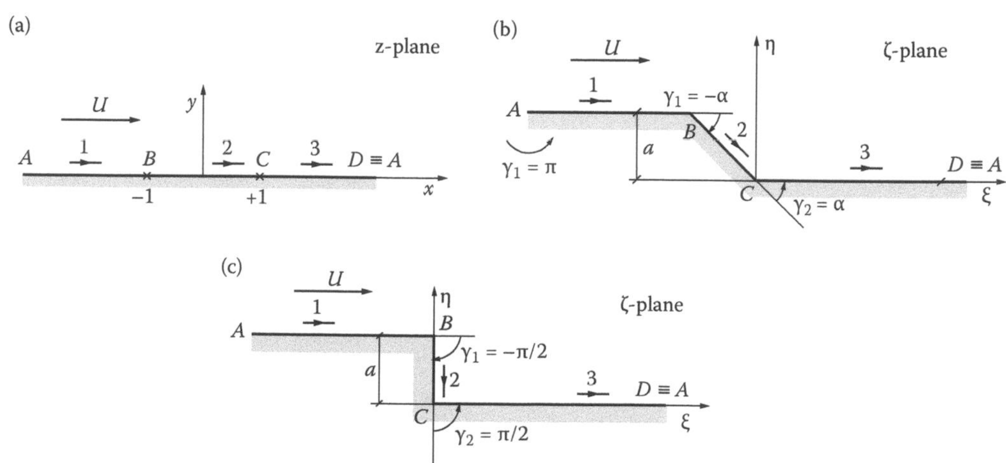

FIGURE 8.9
Exterior conformal polygonal transformation can be used to map the real axis (a) into a wall with a ramp (b); if the angle of the ramp is 90°, it becomes a rectangular step (c). The upper-half z-plane is mapped into the flow region. The critical point B (C) lies at the salient (re-entrant) edge, where the corner angle is larger (smaller) than 180°, implying that the velocity is infinite (zero).

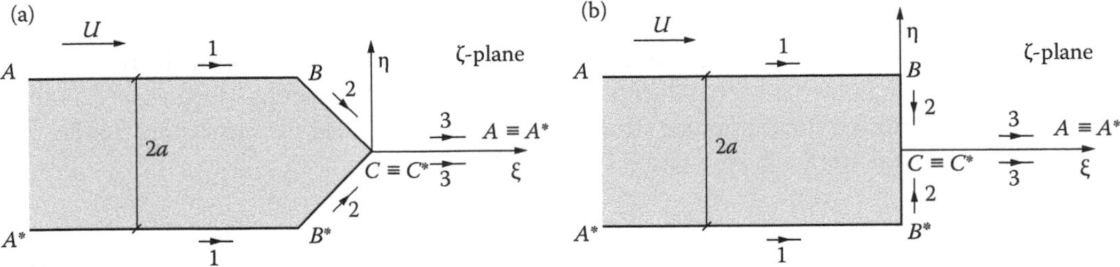

FIGURE 8.10
Reflection of the wall with ramp (step) in Figure 8.9b (Figure 8.9c) leads to the thick semi-infinite plate with sharp (blunt) nose in a (b). The critical points correspond to (1) the outer edges (B, B^*) where the velocity is infinite; and (2) the midpoint $C \equiv C^*$ that is a stagnation point. The free stream goes past the trailing edge for $U > 0$ and is incident upon the leading edge for $U < 0$.

(Subsection 8.4.5)] to a thick semi-infinite plate [Figure 8.10a (Figure 8.10b)] with a pointed (blunt) end (Subsection 8.4.6) in a free stream. The pressure force on the blunt end specifies the base drag of a body in a stream (Subsection 8.4.7).

8.4.1 Polygonal Interior Transformation (Schwartz 1890; Christoffel 1868)

The potential flow past a ramp (step) in Figure 8.9b (Figure 8.9c) can be obtained from the uniform flow over a flat wall in the z-plane (Figure 8.9a) by mapping the real z-axis into a polygonal boundary. This uses one of the four polygonal or Schwartz–Christoffel (1868) transformations (Chapter I.33), namely, the interior polygonal transformation (Equation I.33.25 ≡ Equation 8.144b):

$$x_1 < x_2 < \ldots < x_n : \quad \frac{d\zeta}{dz} = A \prod_{n=1}^{N} (z - x_n)^{-\gamma_n/\pi}. \qquad \text{(8.144a and 8.144b)}$$

If $z = x$ follows the real axis from $-\infty$ to $+\infty$, for $x < x_1$ in Equation 8.144a, then $\arg(d\zeta/dz)$ is constant in Equation 8.144b. As x crosses the critical point $x = x_1$, the first factor in Equation 8.144b changes sign:

$$(-)^{-\gamma_1/\pi} = (e^{i\pi})^{-\gamma_1/\pi} = e^{-i\gamma_1}, \tag{8.145}$$

and thus displacement in the ζ-plane turns in the positive or counterclockwise direction by an angle γ_1. As x increases, the argument of Equation 8.144b or direction in the ζ-plane is constant, corresponding to a straight line, except when it crosses x_n, when there is a turn with external angle γ_n. Thus, *the* **interior polygonal transformation** *(Equation 8.144b) maps the real axis of the z-plane to a polygonal line in the ζ-plane with external angles γ_n at the critical points $\zeta_n = \zeta(x_n)$. The upper complex z-plane is mapped to the interior of the polygon* because (Chapter I.37) the function $\zeta(z)$ is differentiable except at the critical points and thus continuous, so that the l.h.s. in the z-plane corresponds to the l.h.s. in the ζ-plane. The polygon in the ζ-plane has finite (infinite) sides if x_n is mapped to a finite distance from the origin $|\zeta_n| = |\zeta(x_n)| < \infty$ [to infinity $\zeta_n = \zeta(x_n) = \infty$].

8.4.2 Backward Ramp and Incomplete Beta Function

The real axis of the z-plane in Figure 8.9a is mapped to a backward ramp in the ζ-plane in Figure 8.9b by a Schwartz–Christoffel external polygonal mapping (Equation 8.144b ≡ Equation 8.146e) with critical points in the z-plane (Equation 8.144a ≡ Equations 8.146a and 8.146b) and external angles (8.146c and 8.146d) in the ζ-plane:

$$z_2 = 1 = -z_1, \quad \gamma_2 = \alpha = -\gamma_1: \quad \frac{d\zeta}{dz} = A\prod_{n=1}^{2}(z - z_n)^{-\gamma_n/\pi} = A\left(\frac{z+1}{z-1}\right)^{\alpha/\pi}. \tag{8.146a–8.146e}$$

The change of variable (Equation 8.147a) corresponds (Subsections I.35.5.2 and I.38.9.4) to a self-inverse function (Equation 8.147b); its derivative (Equation 8.147c) leads to Equation 8.147d:

$$u \equiv \frac{z+1}{z-1}: \quad z = \frac{u+1}{u-1}, \quad \frac{dz}{du} = -\frac{2}{(u-1)^2}, \quad d\zeta = \frac{d\zeta}{dz}\frac{dz}{du}du = -2A\frac{u^{\alpha/\pi}}{(1-u)^2}du; \tag{8.147a–8.147d}$$

the latter can be integrated:

$$\zeta = -2A\int_0^u v^{\alpha/\pi}(1-v)^{-2}\,dv + C = -2AB\left(1 + \frac{\alpha}{\pi}, -1; \frac{z+1}{z-1}\right) + C \tag{8.148}$$

in terms of an **incomplete Beta function** (Equation 8.149c):

$$\mathrm{Re}(a) > 0,\, u < 1: \quad B(a,b;u) \equiv \int_0^u v^{a-1}(1-v)^{b-1}dv. \tag{8.149a–8.149c}$$

(1) The incomplete Beta function is specified by the integral Equation 8.149c with lower limit zero requiring Equation 8.149a; (2) the condition 8.149b leaves b unrestricted; (3) the complete Beta function corresponds to $u = 1$ in which case $\mathrm{Re}(b) > 0$ in Equation 8.149c; (4) both the complete and incomplete Beta functions are particular cases of the Gaussian hypergeometric function (Equations 3.101a and 3.101b ≡ Equation I.29.74); and (5) the latter provides analytic continuation of Equations 8.149a–8.149c for other real or complex values of the variable z and parameters (a, b). In Equation 8.148, two arbitrary constants (A, C) appear.

8.4.3 Potential Flow Past a Backward Ramp

The uniform flow with velocity V in the z-plane (Equation 8.150a) corresponds (Equation 8.145b) in the ζ-plane to the complex conjugate velocity (Equation 8.150b):

$$\frac{df}{dz} = V: \quad v^* = \frac{df}{d\zeta} = \frac{df/dz}{d\zeta/dz} = \frac{V}{A}\left(\frac{z-1}{z+1}\right)^{\alpha/\pi} \tag{8.150a and 8.150b}$$

showing that (1) the flow at infinity is uniform with velocity Equation 8.151a:

$$\lim_{z\to\infty} v^* = \frac{V}{A} \equiv U; \quad \lim_{z\to -1} v^* = \infty, \quad \lim_{z\to 1} v^* = 0, \tag{8.151a–8.151c}$$

and (2 and 3) at the convex (concave) corner at point B (C), the velocity [Equation 8.151b (Equation 8.151c)] is infinite (zero), as expected for a corner flow (Section I.14.8). From the complex conjugate velocity (Equations 8.150a and 8.150b), it follows that (1) the velocity is horizontal (Equations 8.152c and 8.152d) before and after the ramp (Equations 8.152a and 8.152b):

$$z = x \in |R; \quad |x| > 1: \quad v^*(z) = U\left|\frac{x-1}{x+1}\right|^{\alpha/\pi} = v_x, \quad v_y = 0; \tag{8.152a–8.152d}$$

$$|x| < 1: \quad v^*(z) = U\left(e^{i\pi}\,\frac{x-1}{x+1}\right)^{\alpha/\pi} = U\,e^{i\alpha}\left|\frac{x-1}{x+1}\right|^{\alpha/\pi}, \tag{8.153a and 8.153b}$$

and (2) the velocity makes an angle $-\alpha$ on the ramp (Equation 8.153a) and hence is tangential in the downward direction (Equation 8.153b) for positive U; for negative U the direction of the flow is reversed. Hence, *the potential flow with velocity U at infinity past a backward ramp (Figure 8.9a) with angle α has complex conjugate velocity (Equation 8.150b) leading to a tangential velocity on the ramp (Equations 8.153a and 8.153b) and on the walls (Equations 8.152a through 8.152d).* The constants of integration (A, C) in Equation 8.148 are determined for the ramp by (1) placing the lower corner at the origin (Equation 8.154a):

$$0 = \zeta(1), \qquad -a\,e^{-i\alpha} = \zeta(-1) \tag{8.154a and 8.154b}$$

and (2) specifying the height a of the step by the position of the outer corner (Equation 8.154b).

8.4.4 Edge and Corner Flows near a Backward Step

In the case (Equation 8.156a) of an orthogonal step (Figure 8.9c), the polygonal transformation (Equation 8.146e) simplifies to Equation 8.155a:

$$\frac{d\zeta}{dz} = A\sqrt{\frac{z+1}{z-1}} = A\frac{z+1}{\sqrt{z^2-1}}, \quad \zeta(z) = A\left[\sqrt{z^2-1} + \arg\cosh z\right] + B, \tag{8.155a and 8.155b}$$

whose primitive Equation 7.123b is Equation 8.155b; the condition Equation 8.154a (Equation 8.154b) leads to Equation 8.156b (Equation 8.156c):

$$\alpha = \frac{\pi}{2}: \quad 0 = \zeta(1) = B, \quad ia = \zeta(-1) = A\arg\cosh(-1) = iA\arg\cos(-1) = i\pi A \tag{8.156a–8.156c}$$

using Equation 7.186a. Substituting Equations 8.156b and 8.156c in Equations 8.155a and 8.155b specifies *(Equation 8.157a and 8.157b) the mapping of the upper half complex z-plane (Figure 8.9a) into a backward step in the ζ-plane (Figure 8.9c):*

$$\zeta = \frac{a}{\pi}\left[\sqrt{z^2-1} + \arg\cosh z\right], \quad \frac{d\zeta}{dz} = \frac{a}{\pi}\sqrt{\frac{z+1}{z-1}} \qquad \text{(8.157a and 8.157b)}$$

and its derivative Equation 8.157b. A uniform flow of velocity V in the z-plane (Equation 8.158a) becomes (Equation 8.157b) in the ζ-plane the flow Equation 8.158b with velocity U at infinity (Equation 8.158c):

$$f(z) = Vz, \quad v^* \equiv \frac{df}{d\zeta} = \frac{df/dz}{d\zeta/dz} = U\sqrt{\frac{z-1}{z+1}}, \quad U \equiv \frac{\pi}{a}V = \lim_{z\to\infty} v^*(z) \qquad \text{(8.158a–8.158c)}$$

in the case of the step (Equation 8.156a). Both in the cases of the ramp (step), the velocity [Equation 8.150b (Equation 8.158b)] vanishes (is infinite) at the inner (Equation 8.151c) [outer (Equation 8.151a)] corner, that is, the point $C(B)$ in Figure 8.9a (Figure 8.9c). The complex conjugate velocity is given in the ζ-plane by eliminating z between Equation 8.157a and 8.158b leading to $v^*(\zeta)$. The parameter z may be replaced by the parameter t in Equation 8.159a:

$$z \equiv \cosh t: \quad \zeta = \frac{a}{\pi}(t + \sinh t), \quad f = Vz = \frac{Ua}{\pi}\cosh t, \quad v^* = U\sqrt{\frac{\cosh t - 1}{\cosh t + 1}} \qquad \text{(8.159a–8.159d)}$$

that appears in (1) the Schwartz–Christoffell conformal mapping (Equation 8.159b); and (2 and 3) the complex potential (Equation 8.159c) [conjugate velocity (Equation 8.159d)].

8.4.5 Velocity Field near an Orthogonal Step

The Schwartz–Christoffel transformation (Equation 8.157a) shows that the real axis of the z-plane (Equation 8.160a) consists of three segments separated by the two critical points. The first segment, after the second critical point (Equation 8.160b), leads in Equation 8.157a to a positive real value (Equation 8.160c) corresponding to the horizontal wall along the positive real axis (Equations 8.160d and 8.160e):

$$z = x \in |R; x > 1: \quad \zeta(z) = \frac{a}{\pi}\left[\left|x^2-1\right|^{1/2} + \arg\cosh x\right] \equiv \zeta_+(x) > 0, \quad \operatorname{Im}(\zeta) = 0. \qquad \text{(8.160a–8.160e)}$$

The second segment before the first critical point (Equation 8.160a) leads to the value (Equation 8.160b) of Equation 8.157a:

$$z = x \in |R; x < -1: \quad \zeta(z) - \frac{a}{\pi}\left|x^2-1\right|^{1/2} = \frac{a}{\pi}\arg\cosh\left(-|x|\right)$$
$$= \frac{a}{\pi}\left[-\arg\cosh|x| + i\pi\right] = \zeta_-(x) + ia \qquad \text{(8.161a and 8.161b)}$$

that has (1) the negative real part (Equation 8.161c) and (2) the constant imaginary part equal to the height of the step (Equation 8.161d):

$$\zeta_-(x) = \frac{a}{\pi}\left[\left|x^2-1\right|^{1/2} - \arg\cosh|x|\right] < 0, \quad \operatorname{Im}(\zeta) = a; \qquad \text{(8.161c and 8.161d)}$$

hence it corresponds to the horizontal wall before the step. The third segment between the two critical points (Equation 8.162a) leads in Equation 8.157a to an imaginary value (Equation 8.162b):

$$-1 < x < +1: \quad \zeta = i\frac{a}{\pi}\left[\left|1-x^2\right|^{1/2} + \arg\cos x\right] = i\zeta_0(x), \qquad \text{(8.162a and 8.162b)}$$

$$\operatorname{Re}(\zeta) = 0, \quad 0 = \zeta_0(1) \le \zeta_0(x) \le \zeta_0(-1) = a \qquad \text{(8.162c and 8.162d)}$$

that varies in a range (Equation 8.162d) equal to the height of the step. In Equation 8.162b, Equation 8.163a was used Equation 7.186a ≡ Equation 8.163a and Equation 7.192e ≡ Equation 8.163b, choosing the upper sign in the former:

$$\arg\cosh x = i\arg\cos x; \quad \arg\cosh(-x) + \arg\cosh(x) = i\pi; \qquad \text{(8.163a and 8.163b)}$$

an alternate proof of Equation 8.163b follows from (Equations 5.5a, 5.9a, and 5.49a):

$$\begin{aligned}
&\cosh\left[\arg\cosh(-x) + \arg\cosh(x)\right] \\
&= \cosh\left[\arg\cosh(-x)\right]\cosh\left[\arg\cosh(x)\right] + \sinh\left[\arg\cosh(-x)\right]\sinh\left[\arg\cosh(x)\right] \\
&= (-x)x + \left|x^2-1\right|^{1/2}\left|(-x)^2-1\right|^{1/2} = -x^2 + x^2 - 1 = -1 = \cos(\pi) = \cosh(i\pi).
\end{aligned} \qquad \text{(8.164)}$$

The velocity is horizontal (Equations 8.152a through 8.152d) [vertical (Equations 8.153a and 8.153b) with Equation 8.156a) for Equations 8.160a through 8.160e and 8.161a through 8.161d (Equations 8.162a through 8.162d), hence tangent to the wall (step) in both cases.

8.4.6 Thick Semi-Infinite Plate in a Uniform Stream

The conformal mapping (Equation 8.157a) is specified by an analytic function that is real for real $z \equiv x > 1$, and thus (Subsection I.32.2.2) the reflection principle applies; it follows that *the complex conjugate velocity of a uniform stream U past a semi-infinite plate with thickness 2a is specified by Equation 8.150b (Equation 8.158b) for a pointed plate of nose angle 2α (blunt plate with square corners) in Figure 8.10a (Figure 8.10b).* The edges of the thick plate (Equation 8.165a) correspond in Equation 8.159a to Equation 8.165b and hence Equation 8.159b to Equation 8.165c:

$$z = -1: \quad t = \pm\arg\cosh(-1) = \mp i\arg\cos(-1) = \pm i\pi, \qquad \text{(8.165a and 8.165b)}$$

$$\zeta(\pm i\pi) = \frac{a}{\pi}\left[\pm i\pi + \sinh(\pm i\pi)\right] = \pm ia \pm i\frac{a}{\pi}\sin\pi = \pm ia; \qquad \text{(8.165c)}$$

thus the velocity (Equation 8.158b ≡ Equation 8.159d) on the face of the plate (Equations 8.166a and 8.166b) is Equation 8.166c:

$$t = i\theta, -\pi \le \theta \le +\pi: \quad v^*(i\theta) = U\sqrt{\frac{\cosh(i\theta)-1}{\cosh(i\theta)+1}} = \pm iU\sqrt{\frac{1-\cos\theta}{1+\cos\theta}} = -iv_y(\theta) \qquad \text{(8.166a–8.166c)}$$

that is (1) vertical (Equations 8.167a and 8.167b):

$$v_x(\theta) = 0, \quad v_y(\theta) = U \operatorname{sgn}(\theta) \left| \frac{1-\cos\theta}{1+\cos\theta} \right|^{1/2}, \qquad \text{(8.167a and 8.167b)}$$

$$v_y(\theta > 0) < 0 < v_y(\theta < 0), \quad v_y(0) = 0, \qquad \text{(8.168a–8.168c)}$$

(2) with downward (upward) direction in the upper (lower) part [Equation 8.168a (Equation 8.168b) for $U > 0$, that is, a uniform stream past the trailing edge of a thick semi-infinite plate (Figure 8.10b); (3) the reverse of (2) for $U < 0$, that is, a uniform stream incident upon the leading edge of a thick semi-infinite plate; and (4) in both cases, there is a stagnation point (Equation 8.168c) at the midposition on the dividing streamline. In the case of $U > 0$, the flow past the trailing edge of a thick semi-infinite plate, the pressure at the aft causes a force with direction opposite to the flow, corresponding to the **base drag** of the plate; and for $U < 0$, the flow incident on the leading edge of a thick semi-infinite plate, the pressure force is the same, and since it has the direction of the velocity, it corresponds to a **thrust force**. These forces are calculated next (Subsection 8.4.7).

8.4.7 Base Drag of a Thick Blunt Semi-Infinite Plate

The pressure distribution on the plate leads by the Bernoulli equation (Equation I.14.27c ≡ Equation 8.12b) to force per unit width of the plate:

$$F_x = -\int_{-a}^{+a} p \, d\eta = -\int_{-a}^{+a} \left(p_0 - \frac{\rho}{2} |v|^2 \right) d\eta = -2 p_0 a + \frac{\rho}{2} \int_{-\pi}^{+\pi} \left[v_y(\theta) \right]^2 \frac{d\eta}{d\theta} d\theta, \qquad \text{(8.169)}$$

where p_0 is the stagnation pressure, and ρ is the mass density; using Equations 8.159b and 8.166a on the face of the plate

$$i\eta = \zeta = \frac{a}{\pi} \left[i\theta + \sinh(i\theta) \right] = i \frac{a}{\pi} (\theta + \sin\theta), \quad \frac{d\eta}{d\theta} = \frac{a}{\pi} (1 + \cos\theta) \qquad \text{(8.170a and 8.170b)}$$

and substituting Equations 8.167b and 8.170b in Equation 8.169 leads to Equation 8.171a:

$$F_x + 2 p_0 a = \frac{\rho}{2} U^2 \frac{a}{\pi} \int_{-\pi}^{\pi} (1 - \cos\theta) d\theta = \rho U^2 a; \quad p_0 = p_\infty + \frac{1}{2} \rho U^2; \qquad \text{(8.171a and 8.171b)}$$

the Bernoulli equation (Equation I.14.27c ≡ Equation 1.12b) leads to Equation 8.171b denoting by p_∞ the pressure at infinity where the velocity is U. This specifies **the drag (thrust) coefficient** *Equation 8.172a exerted by a potential flow with velocity U, mass density ρ, and stagnation pressure p_0 along (against) a semi-infinite plate of thickness $2a$:*

$$-F_x = \frac{\rho}{2} U^2 (2a) C_D = \rho U^2 C_D a; \quad C_D = \frac{2 p_0}{\rho U^2} - 1 = \frac{2 p_\infty}{\rho U^2}. \qquad \text{(8.172a and 8.172b)}$$

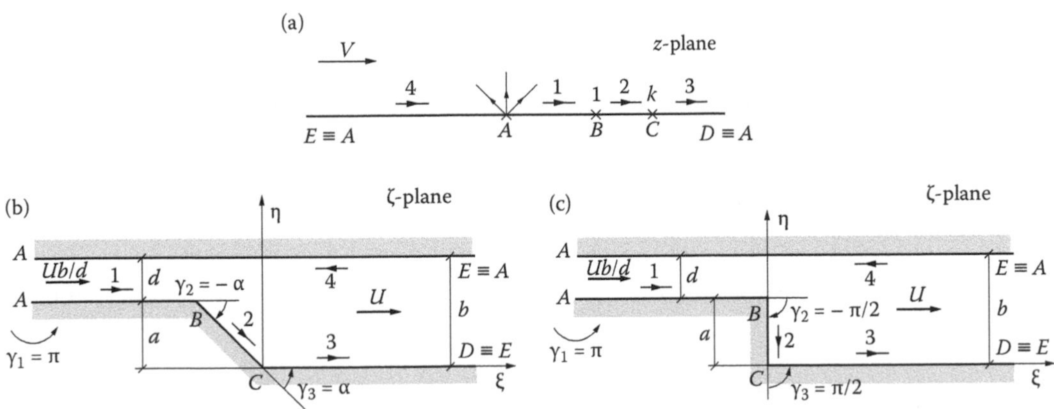

FIGURE 8.11
To the wall with a ramp (step) in Figure 8.9b (Figure 8.9c) a parallel wall may be added leading to a parallel-sided channel with a ramp (step) on one side in b (c). This corresponds in the exterior polygonal transformation to the mapping from the real axis (Figure 8.9a) with a third critical point (a) mapped to infinity; the point at infinity corresponds to the channel far upstream (downstream) where the flow is uniform with distinct velocities related to the width of the channel to conserve the volume flux in the incompressible flow. The upper-half z-plane (Figure 8.11a) is mapped into the interior of the parallel-sided channel with ramp (step) on one side [Figure 8.11b (Figure 8.11c)].

The base drag coefficient Equation 8.172b is determined by the pressure at infinity (Equation 8.171b) and corresponds to slowing down the flow from the horizontal velocity U to zero at the front of the blunt thick plate. For a closed body, the integral of the constant stagnation pressure on the surface is zero, so it does not affect the lift, drag, or pitching moment (Equations I.28.31a through I.28.31c ≡ Equations 2.205a through 2.205c), and this is also the case for a thin semi-infinite plate; for a thick plate, the effects of the stagnation pressure on the upper and lower sides balance, but there is no "forward end" to compensate the pressure on the "aft end." The value (Equation 8.172b) of the drag/thrust coefficient of a thick semi-infinite plate in a free stream (Figure 8.10b) will be confirmed (Subsection 8.5.2) as a particular case of when it lies between two parallel walls (Figure 8.12b); the latter is equivalent to an abrupt change of the cross section in a channel (Figure 8.11c) and is considered next (Subsection 8.5.1). Another problem solvable via the Schwartz–Christoffell transformation is a plate oblique (orthogonal) to a wall (Example 10.12) like a ramp (step) without a back material; by symmetry, it leads to the potential flow incident on a straight (bent) lamina.

8.5 Channel with Contraction/Expansion and a Thick Plate

An extension of the preceding problem adds a parallel wall, leading to a gradual (an abrupt) increase in the cross section [(Subsections 8.5.1 and 8.5.2 (Subsections 8.5.3 and 8.5.4)]. The symmetry principle then leads to a thick semi-infinite plate with a sharp (blunt) nose in a channel [Subsection 8.5.5 (Subsection 8.5.6)]. The latter specifies the base drag of a thick semi-infinite plate in a parallel-sided channel (Subsection 8.5.7). Compared with the same plate in a free stream (Subsection 8.4.7), this specifies the reduction in the base drag due to the wall effect (Subsection 8.5.8).

8.5.1 Gradual Change of Cross Section of a Channel

A parallel-sided channel of width d has a gradual change of cross section to b in Figure 8.11b. The upper-half z-plane is mapped into the channel (Figure 8.11a) by the interior polygonal transformation (Equations 8.144a and 8.144b ≡ Equation 8.173c) similar to Equations 8.146a through 8.146e, but with three critical points at positions (Equation 8.173a) in the z-plane corresponding to the external angles (Equation 8.173b) in the ζ–plane:

$$x_{1-3} = \{0, 1, k\}, \quad \gamma_{1-3} = \{\pi, -\alpha, \alpha\}, \quad \frac{d\zeta}{dz} = A\prod_{n=1}^{3}(z - x_n)^{-\gamma_n/\pi} = \frac{A}{z}\left(\frac{z-1}{z-k}\right)^{\alpha/\pi}. \quad \text{(8.173a–8.173c)}$$

The geometry of the change of the cross section of the channel (Figure 8.11a) is specified by the positions of three points B, C, and E; since the Schwartz–Christoffell mapping (Equation 8.146e) involves only one multiplying constant A and an integration constant B, the position of the third critical point $z_3 = k$ has been left variable. In this way, there are three parameters in the conformal mapping to meet three constraints. The change of variable Equation 8.174a leads (Equations 8.174b and 8.174c):

$$u = \frac{z-1}{z-k}: \quad z = \frac{ku-1}{u-1}, \quad \frac{dz}{du} = \frac{1-k}{(u-1)^2}, \quad d\zeta = \frac{d\zeta}{dz}\frac{dz}{du}du = A\frac{1-k}{u-1}\frac{u^{\pi/\alpha}}{ku-1}du \quad \text{(8.174a–8.174d)}$$

to the transformation (Equation 8.174d ≡ Equation 8.175):

$$\zeta = A(1-k)\int_0^u v^{\alpha/\pi}\left[(v-1)(kv-1)\right]^{-1} dv + B; \quad \text{(8.175)}$$

this is no longer reducible to an incomplete Eulerian Beta function (Equation 8.149c).

8.5.2 Potential Flow in a Channel with a Ramp

The origin in the z-plane is the point at infinity in the ζ–plane where a flow source is located for which half the flow rate (Equation 8.174a) equals the flow rate in the channel; the corresponding complex potential (Equation 2.192a) is Equation 8.176b:

$$\frac{Q}{2} = Ub: \quad f = \frac{Q}{2\pi}\log z = \frac{Ub}{\pi}\log z, \quad \frac{df}{dz} = \frac{Ub}{\pi z}, \quad \text{(8.176a–8.176c)}$$

and the complex conjugate velocity is Equation 8.176c. From Equations 8.173c and 8.176c follows the complex conjugate velocity in the channel (Equation 8.177a):

$$v^* = \frac{df}{d\zeta} = \frac{df/dz}{d\zeta/dz} = \frac{Ub}{\pi A}\left(\frac{z-k}{z-1}\right)^{\alpha/\pi}; \quad U = \lim_{z\to\infty}\frac{df}{d\zeta} = \frac{Q}{2\pi A} = \frac{Ub}{\pi A}; \quad \text{(8.177a and 8.177b)}$$

the velocity at infinity downstream is Equation 8.177b, specifying the constant Equation 8.178a:

$$A = \frac{b}{\pi}: \quad V \equiv U\frac{b}{d} = \lim_{z \to 0}\frac{df}{d\zeta} = \frac{Q}{2\pi A}k^{\alpha/\pi} = Uk^{\alpha/\pi}, \quad k = \left(\frac{b}{d}\right)^{\pi/\alpha}; \quad (8.178a–8.178c)$$

the velocity at the other end of the channel, that is, infinity upstream, is Equation 8.178b, specifying the second constant (Equation 8.178c). This determines the two constants (A, k) from the geometry of the channel flow. Substituting Equations 8.178a and 8.178c in Equation 8.173c (Equation 8.177a) leads to the Schwartz–Christoffel transformation (Equation 8.179a) [complex conjugate velocity (Equation 8.179b)] for the parallel-sided channel with ramp:

$$\frac{d\zeta}{dz} = \frac{b}{\pi z}\left[\frac{z-1}{z-(b/d)^{\pi/\alpha}}\right]^{\alpha/\pi}, \quad v^* = \frac{df}{d\zeta} = U\left[\frac{z-(b/d)^{\pi/\alpha}}{z-1}\right]^{\alpha/\pi}. \quad (8.179a \text{ and } 8.179b)$$

From the complex conjugate velocity (Equation 8.179b), it follows that (1) the segment of the real axis (Equation 8.180a) outside (Equations 8.180b and 8.180c) the points B and C corresponds to a horizontal velocity (Equation 8.180d) on the walls:

$$z = x \in |R; \quad x < 1 \quad \text{or} \quad x > k: \quad v^*(x) = U\left|\frac{x-k}{x-1}\right|^{\alpha/\pi} = v_x(x), \quad v_y(x) = 0; \quad (8.180a–8.180d)$$

$$1 < x < k: \quad v^*(x) = U\left(e^{i\pi}\frac{x-k}{x-1}\right)^{\alpha/\pi} = Ue^{i\alpha}\left|\frac{x-k}{x-1}\right|^{\alpha/\pi} = |v^*(x)|e^{i\alpha}; \quad (8.181a \text{ and } 8.181b)$$

(2) the segment (Equation 8.181a) of the real axis (Equation 8.180a) between the points B and C corresponds to a velocity (Equation 8.181b) making an angle $-\alpha$, that is, tangent to the ramp in the downward direction.

8.5.3 Abrupt Change of Cross Section of a Parallel-Sided Channel

In the particular case (Figure 8.11c) of the abrupt change of the cross section at an orthogonal step (Equation 8.182a), from Equations 8.178b and 8.178c follow Equations 8.182b and 8.182c leading to Equations 8.182d and 8.182e:

$$\alpha = \frac{\pi}{2}: \quad k = \frac{b^2}{d^2}, \quad V = U\sqrt{k} = U\frac{b}{d}, \quad \frac{d\zeta}{dz} = \frac{b}{\pi z}\sqrt{\frac{z-1}{z-(b/d)^2}}, \quad v^* = U\sqrt{\frac{z-(b/d)^2}{z-1}}. \quad (8.182a–8.182e)$$

The preceding results *specify (1) the mapping [Equation 8.179a (Equation 8.182d)] of the upper-half z-plane (Figure 8.11a) to a parallel-sided channel with a gradual (abrupt) change of the cross section [Figure 8.11b (Figure 8.11c)] from d to b through a ramp of angle α [an orthogonal step (Equation 8.182a)]; and (2) the complex conjugate velocity of the potential flow in the channel [Equation 8.179b (Equation 8.182e)] with velocity [Equations 8.178b and 8.179c (Equation 8.177b)] at infinity upstream at the side of cross section d, where U is the velocity in the downstream side of cross section b.* The velocity vanishes (is infinite) at the inner $z = k$ (outer $z = 1$) corner corresponding to the point C (B) in Figure 8.11b (Figure 8.11c). The interior polygonal mapping (Equation 8.182d) for the orthogonal step (Figure 8.11c) in channel, corresponding (Figure 8.12b) to the blunt-nosed thick plate (Equation 8.182a ≡ Equation 8.183a) is integrated via the change of variable (Equation 8.183b):

$$\alpha = \frac{\pi}{2}: \quad z = \frac{b^2/d^2 - t^2}{1-t^2}, \quad \frac{1}{t^2} = \frac{z-1}{z - b^2/d^2}, \tag{8.183a–8.183c}$$

that implies

$$\frac{\mathrm{d}z}{z} = \mathrm{d}\,(\log z) = \mathrm{d}\left[\log\left(\frac{b^2}{d^2} - t^2\right)\right] - \mathrm{d}\left[\log(1-t^2)\right] = 2t\left(\frac{1}{1-t^2} - \frac{1}{b^2/d^2 - t^2}\right)\mathrm{d}t; \tag{8.184}$$

substituting Equation 8.184 in Equation 8.182d leads to

$$\begin{aligned}\frac{\mathrm{d}\zeta}{\mathrm{d}t} &= \frac{\mathrm{d}\zeta}{\mathrm{d}z}\frac{\mathrm{d}z}{\mathrm{d}t} = \frac{b}{\pi z}\frac{1}{t}\frac{\mathrm{d}z}{\mathrm{d}t} = 2\frac{b}{\pi}\left(\frac{1}{1-t^2} - \frac{1}{b^2/d^2 - t^2}\right)\\ &= \frac{b}{\pi}\left(\frac{1}{1-t} + \frac{1}{1+t}\right) - \frac{d}{\pi}\left(\frac{1}{b/d+t} + \frac{1}{b/d-t}\right),\end{aligned} \tag{8.185a and 8.185b}$$

which may be integrated

$$\zeta(t) = \frac{b}{\pi}\log\left(\frac{1+t}{1-t}\right) - \frac{d}{\pi}\log\left(\frac{b/d+t}{b/d-t}\right); \tag{8.186}$$

the constant of integration is zero because the point C in Figure 8.11c corresponds to (1) the origin in the ζ–plane (Equation 8.187a); (2) by Figure 8.11b to Equation 8.187b; and (3) by Equation 8.183c to Equation 8.187c:

$$\zeta = 0, \quad z = k = \left(\frac{b}{d}\right)^2, \quad t = 0. \tag{8.187a–8.187c}$$

The substitution of Equation 8.183c in Equation 8.186 specifies the Schwartz–Christoffel mapping $\zeta(z)$ into the channel with ramp (Figure 8.11c) that is analyzed next (Subsection 8.5.4).

8.5.4 Interaction of Potential Flow in Convex and Concave Corners

The Schwartz–Christoffel transformation (Equations 8.183c and 8.186) leads to four segments of the real axis of the z-plane (Equation 8.188a) separated by the three critical points. The first segment to the right of the third critical point (Equation 8.188b) leads (Equation 8.183c) to t in the modulus less than unity (Equation 8.188c); hence both logarithms in Equation 8.186 have real arguments larger than unity, and since $b > d$, the whole expression is real positive (Equation 8.188d), corresponding to the lower wall along the positive real axis (Equation 8.188e):

$$z = x \in |R; \quad x > \frac{b^2}{d^2}: \quad |t| < 1 < \frac{b}{d}, \quad \zeta = \frac{b}{\pi}\log\left|\frac{1+t}{1-t}\right| - \frac{d}{\pi}\log\left|\frac{b/d+t}{b/d-t}\right| \equiv \bar{\zeta}(t), \quad \bar{\zeta}(t<1) > 0.$$

(8.188a–8.188e)

The second segment before the first critical point (Equation 8.189a) leads (Equation 8.183c) to t having a modulus in the range Equation 8.189b, and thus the first (second) logarithm on the r.h.s. of Equation 8.186 has negative (positive) real argument, leading to Equation 8.189c:

$$x < 0: \quad 1 < |t| < \frac{b}{d}: \quad \zeta = \frac{b}{\pi}\log\left(-\frac{t+1}{t-1}\right) - \frac{d}{\pi}\log\left(\frac{b/d+t}{b/d-t}\right) = \bar{\zeta}(t) + ib, \quad \text{Im}(\zeta) = b;$$

(8.189a–8.189d)

the constant imaginary part Equation 8.189d corresponds to the height of the upper wall. The third segment between the first two critical points (Equation 8.190a) leads (Equation 8.183c) to t in the modulus exceeding Equation 8.190b, and thus both logarithms in Equation 8.186 have negative real arguments:

$$0 < x < 1: \quad |t| > \frac{b}{d} > 1, \quad \zeta = \frac{b}{\pi}\log\left(-\frac{t+1}{t-1}\right) - \frac{d}{\pi}\log\left(-\frac{b/d+t}{b/d-t}\right) = \bar{\zeta}(t) + i(b-d); \quad (8.190a–8.190c)$$

this implies (Equation 8.190c) from the imaginary (Equation 8.190d) [real (Equation 8.190e)] part that it corresponds to the lower wall at height Equation 8.190d (along the negative real axis Equation 8.190e):

$$\text{Im}(\zeta) = b - d = a, \quad \text{Re}(\zeta) = \bar{\zeta}\left(t > \frac{b}{d}\right) < 0. \qquad (8.190d and 8.190e)$$

The fourth segment between the last two critical points (Equation 8.191a) leads to a negative value (Equation 8.191b) in Equation 8.183c), hence (Equation 8.191c) to imaginary t, so that Equation 8.186 is purely imaginary (Equation 8.191d):

$$1 < x < \frac{b^2}{d^2}: \quad \frac{1}{t^2} < 0, \quad t = i|t|, \qquad (8.191a–8.191c)$$

$$\zeta = \frac{b}{\pi}\log\left(\frac{1+i|t|}{1-i|t|}\right) - \frac{d}{\pi}\log\left(\frac{b/d+i|t|}{b/d-i|t|}\right) = i\frac{2}{\pi}\left[b\arctan|t| - d\arctan\left(\frac{d}{b}|t|\right)\right] = i\tilde{\zeta}(t); \qquad (8.191d)$$

this corresponds to the vertical step along the imaginary axis (Equation 8.191e) with height (Equation 8.191f):

$$\text{Re}(\zeta) = 0, \quad 0 = \tilde{\zeta}(0) \le \text{Im}(\zeta) = \tilde{\zeta}(t) \le \tilde{\zeta}(+\infty) = b - d = a. \quad (8.191e and 8.191f)$$

The velocity is horizontal (Equations 8.180a and 8.180b) [vertical (Equations 8.181a and 8.18b) with Equation 8.183a] for the first three segments (Equations 8.188a through 8.188e, Equations 8.189a through 8.189d and Equations 8.190a through 8.190e) [the last or fourth segment (Equation 8.191a through 8.191e)] and thus tangential in all cases. Extending t in Equation 8.191f from $0 \le t < \infty$ to $-\infty < t < +\infty$ would lead from the step $0 \le \zeta \le a$ of height a to a semi-infinite plate $-a \le \zeta \le +a$ of thickness $2a$, which is considered next (Subsection 8.5.5).

8.5.5 Thick Semi-Infinite Plate in a Channel

Since the conformal mapping (Equation 8.179a) is real ζ for real $z \equiv x > k \ge 1$, the reflection principle (Subsection I.31.2.2) applies, and *Equation 8.179b (Equation 8.182e) is the complex conjugate velocity of the potential flow past a semi-infinite plate of thickness 2a with a sharp nose of angle 2α (blunt nose α = π/2) in Figure 8.12a (Figure 8.12b), placed in a parallel-sided channel of width 2b*. Substitution of Equation 8.183b in Equation 8.176b specifies

$$f(t) = U\frac{b}{\pi}\log z = U\frac{b}{\pi}\log\left(\frac{b^2/d^2 - t^2}{1 - t^2}\right) \tag{8.192}$$

the complex potential Equation 8.192 in the ζ–plane (Equation 8.186) with t as a parameter for (Figure 8.12b) a plate of thickness $2a = 2(b - d)$ in a channel of width $2b$, symmetrically placed to leave a gap of width d on each side. From Equation 8.192 follows

$$\frac{df}{dt} = \frac{2Ubt}{\pi}\left(\frac{1}{1-t^2} - \frac{1}{b^2/d^2 - t^2}\right); \tag{8.193}$$

comparing with Equation 8.185a shows that the potential in the t-plane

$$v^* = \frac{df}{d\zeta} = \frac{df/dt}{d\zeta/dt} = Ut, \tag{8.194}$$

corresponds to a uniform flow with velocity U. Thus, *the complex potential (conjugate velocity) of the potential flow with downstream velocity U in a channel of width 2b past the trailing edge*

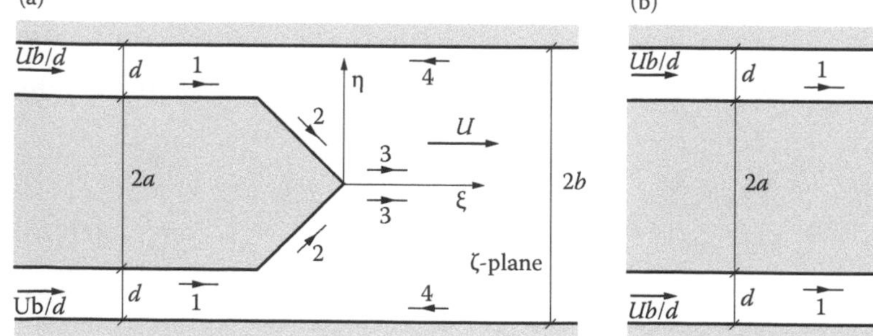

FIGURE 8.12
Image of the parallel-sided channel with ramp (step) on one side [Figure 8.11b (Figure 8.11c)]; on that side with a larger cross section leads to the thick semi-infinite plate with a sharp (blunt) edge [Figure 8.12a (Figure 8.12b)] in a parallel-sided channel; this includes the thick semi-infinite plate in a free stream [Figure 8.10a (Figure 8.10b)] when the wall effect is removed by placing the walls at infinity.

of a thick blunt semi-infinite plate of width 2a, placed symmetrically so as to leave a gap of d = b − a on each side (Figure 8.12b), is specified eliminating the parameter t between Equations 8.186 and 8.192 (Equation 8.194).

8.5.6 Velocity and Pressure on a Thick Plate in a Channel

The latter result (Equation 8.194) facilitates the calculation (Equation 8.169) of the drag (thrust) force (Equation 8.195) per unit length of plate for flow parallel (against) the plate:

$$\bar{F}_x + 2p_0\, a = \frac{\rho}{2}\int_{-a}^{+a} |v|^2\, d\eta = \frac{1}{2}\rho U^2 \int_{-i\infty}^{+i\infty} |t|^2 \frac{d\eta}{dt}\, dt; \tag{8.195}$$

the limits of integration in Equation 8.195 correspond to the following: (1) the edges of the blunt semi-infinite plate (Equation 8.196a); (2) point B in Figure 8.11c approached from the end side (Equation 8.196b); (3) from Equation 8.183c follows Equation 8.196c; and (4) the latter corresponds to Equation 8.196d:

$$\zeta = i\eta = \pm ia = \pm i|b-d|, \quad z \to 1+0, \quad \frac{1}{t^2} \to -0, \quad t \to \pm i\infty, \qquad (8.196\text{a–}8.196\text{d})$$

that is, the limits of integration in Equation 8.195. On the face of the plate, parameter t is (Equation 8.191c) imaginary (Equation 8.197a) leading by Equation 8.185a to Equation 8.197b:

$$t = i\psi: \quad i\, d\eta = d\zeta = \frac{d\zeta}{dt}dt = \frac{2b}{\pi}\left(\frac{1}{1+\psi^2} - \frac{1}{b^2/d^2+\psi^2}\right) i\, d\psi; \qquad (8.197\text{a and } 8.197\text{b})$$

substituting Equations 8.197a and 8.197b in Equation 8.195 leads to the integral Equation 8.198a:

$$\bar{F}_x + 2p_0\, a = \frac{\rho U^2 b}{\pi}\int_{-\infty}^{\infty} h(\psi)\, d\psi, \quad h(\psi) = \psi^2\left(\frac{1}{1+\psi^2} - \frac{1}{b^2/d^2+\psi^2}\right) \qquad (8.198\text{a and } 8.198\text{b})$$

with integrand Equation 8.198b.

8.5.7 Evaluation of Drag Integral by Residues

The integral Equation 8.198a may be evaluated (Section I.17.4) by the following: (1) closing the path of integration in the upper (lower) complex-ψ plane, since the integrand satisfies (Equation 8.199b) in all directions (Equation 8.199a) the asymptotic decay condition (Equation I.17.19):

$$\begin{aligned} 0 \le \arg(\psi) < 2\pi: \quad & \lim_{\psi\to\infty} \psi\, h(\psi) \\ &= \lim_{\psi\to\infty} \psi^3 \frac{b^2/d^2 - 1}{b^2/d^2 + (1+b^2/d^2)\psi^2 + \psi^4} = \lim_{\psi\to\infty} O\left(\frac{1}{\psi}\right) = 0; \end{aligned} \qquad (8.199\text{a and } 8.199\text{b})$$

(2) the integrand Equation 8.198b has (Equation I.15.24b ≡ Equation 1.18) simple poles at $\psi = \pm i$ ($\pm ib/d$) with residues Equation 8.200a (Equation 8.200b):

$$h_1(\pm i) = \lim_{\psi \to \pm i} (\psi \mp i)\, h(\psi) = \lim_{\psi \to \pm i} \psi^2 \frac{\psi \mp i}{\psi^2 + 1} = \lim_{\psi \to \pm i} \frac{\psi^2}{\psi \pm i} = \frac{i^2}{\pm 2i} = \pm \frac{i}{2}, \tag{8.200a}$$

$$\begin{aligned} h_1\left(\pm i\frac{b}{d}\right) &= \lim_{\psi \to \pm ib/d} \left(\psi \mp i\frac{b}{d}\right) h(\psi) = -\lim_{\psi \to \pm ib/d} \psi^2 \frac{\psi \mp ib/d}{\psi^2 + b^2/d^2} \\ &= -\lim_{\psi \to \pm ib/d} \frac{\psi^2}{\psi \pm ib/d} = -\frac{(ib/d)^2}{\pm 2ib/d} = \mp \frac{ib}{2d}; \end{aligned} \tag{8.200b}$$

(3) the integral equals (Equation I.15.45) the sum of the residues at the poles in the upper (lower) complex ψ-plane multiplied by $+2\pi i(-2\pi i)$ since the real axis is closed by a semi-circle of large radius $R \to \infty$ enclosing the poles taken in the positive (negative) or counter-clockwise (clockwise) direction:

$$\int_{-\infty}^{\infty} h(\psi)\, d\psi = \pm 2\pi i \left[h_{(1)}(\pm i) + h_{(1)}\left(\pm i\frac{b}{d}\right) \right] = \pm 2\pi i \left(\pm\frac{i}{2} \mp \frac{ib}{2d} \right) = -\pi \left(1 - \frac{b}{d}\right); \tag{8.201}$$

and (4) this completes the evaluation of the integral (Equations 8.198a and 8.198b) by residues.

8.5.8 Wall Effect on the Drag of a Blunt Thick Semi-Infinite Plate

Substituting Equation 8.201 in Equation 8.198a:

$$\bar{F}_x + 2\,p_0\,a = -\rho U^2 b \left(1 - \frac{b}{d}\right) = \rho U^2 \frac{b}{d}(b - d) = \rho U^2 a \frac{b}{d} = \rho U^2 a \left(1 + \frac{a}{d}\right), \tag{8.202}$$

leads to *(Equation 8.202 ≡ Equation 8.203a) the base drag (thrust) of the flow along (against) a thick semi-infinite plate with width 2a at equal distance d from the walls of a channel of width 2b with a stream of velocity U, mass density ρ, and stagnation pressure* p_0:

$$-\bar{F}_x = \rho U^2\, \bar{C}_D\, a; \quad \bar{C}_D = \frac{2p_0}{\rho U^2} - 1 - \frac{a}{d} = \frac{2p_\infty}{\rho U^2} - \frac{a}{d} = C_D - \frac{a}{d}; \tag{8.203a and 8.203b}$$

the drag coefficient Equation 8.203b is smaller than for a flat plate in a free stream (Equation 8.172b), with an extra term equal to the ratio of the half-width of the plane a to the gap d between it and the wall. The wall effect on the base drag of a thick blunt semi-infinite plate can be explained as follows: (1) relative to the same plate in a free stream (Figure 8.10b), the presence of the walls of the channel (Figure 8.12b) accelerates the flow to conserve the volume flux of the incompressible fluid; (2) a larger velocity corresponds by the Bernoulli Equation 8.12 to a lower pressure and hence a reduction of pressure force and base drag. Thus, the base drag measured in a wind tunnel is reduced by the blockage effect of the model and must be corrected to yield the correct larger base drag in free flight. This effect of flow acceleration through the gap between the plate and the wall disappears for $a/d = 0$, that is, in two cases: (1) a thin plate $a = 0 \neq d$ that does not disturb the flow in the channel; and (2) a thick plate $a \neq 0$ in an unbounded stream $d = \infty$ for which there is no wall effect. The thin semi-infinite plate in a parallel-sided duct has for electrostatic analog the triple plate condenser (Example 10.13). Other geometries include a reservoir with exit channel (a duct corner) leading [Example 10.14 (Example 10.15)] by symmetry to Y-junction (T-junction); an alternative to the latter is a channel with (Example 10.16) side branch(es).

8.6 Flat Plate with Partially Separated Flow

The limit of zero thickness in Figure 8.10a and b (Figure 8.12a and b) leads to the semi-infinite plate in a free stream (Section I.14.9.3) [in a channel (Example 1.13) like a triple plate condenser]. The finite flat plate is the simplest airfoil (Section I.34.1) and is considered next with partially separated flow (Subsection 8.6.1). The procedure is similar to the Sedov method (Section 8.7) in that it involves the choice of a separation function (Subsection 8.6.3) meeting the Kutta condition (Subsection 8.6.4) ensuring a finite velocity at the trailing edge of the airfoil (Subsection 8.6.2). The associated circulation (Subsection 8.6.5) specifies the loss of lift due to flow separation (Subsection 8.6.6); alternatively it specifies the lift coefficient and lift slope of an airfoil with partially separated flow (Subsection 8.6.7).

8.6.1 Airfoil with Circulation and Flow Separation

Consider a flat plate of chord c in a uniform stream of velocity U with an angle-of-attack α in Figure 8.13a. The total velocity consists (Equation 8.204b) of the free stream velocity plus a perturbation:

$$0 \leq s \equiv \frac{x}{c} \leq 1: \quad v^*(z) = U\,e^{-i\alpha} + \tilde{v}^*(z), \quad \tilde{v}_x(x > c, y = -0) = 0 = \tilde{v}_x(x > sc, y + 0) \tag{8.204a–8.204d}$$

whose (1) horizontal component vanishes in the wake of the lower surface (Equation 8.204c) and on the upper surface (Equation 8.204d) after the separation point at $x = sc$,

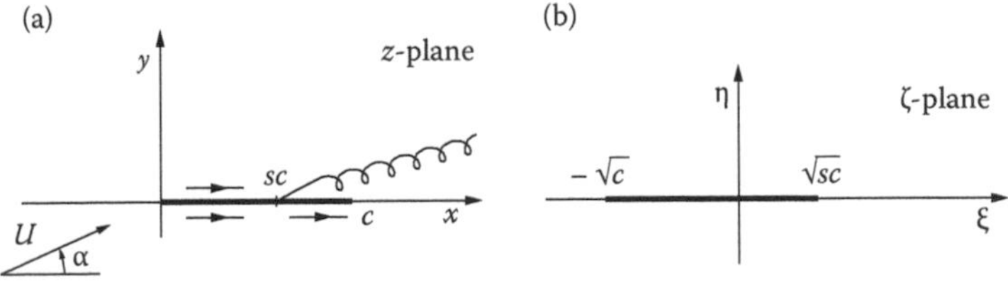

FIGURE 8.13
Simple representation of an airfoil in an incident stream with angle-of-attack and attached flow on the underside and detached flow on the upper side near the trailing edge (a) is the mapping from an unsymmetric strip (b); the asymmetry corresponds to the fraction of the upper surface of the flat plate covered by the detached flow.

where Equation 8.204a is the fraction of the chord of the upper surface behind the leading edge covered by the attached flow; (2) the vertical component cancels the free stream velocity on the whole of the lower surface (Equation 8.205a) and on the upper surface (Equation 8.205b) up to the separation point:

$$\tilde{v}_y\left(0<x<c, y=-0\right)=-U\sin\alpha=\tilde{v}_y\left(0<x<sc, y=+0\right). \quad \text{(8.205a and 8.205b)}$$

The circulation around the airfoil is specified by Equation 8.206a:

$$\Gamma=\int_0^c \tilde{v}_x\left(x,-0\right)\mathrm{d}x-\int_0^{sc}\tilde{v}_x\left(x,+0\right)\mathrm{d}x, \quad v^*\left(z\right)-Ue^{-i\alpha}=-\frac{i\Gamma}{2\pi z}+O\left(\frac{1}{z^2}\right) \quad \text{(8.206a and 8.206b)}$$

and determines the perturbation velocity (Equation 8.140b) far from the airfoil, where Γ is the circulation.

8.6.2 Corner Flows and the Thin Semi-Infinite Plate

The simplest Schwartz–Christoffell exterior polygonal transformation (Equation 8.144b) has Equation 8.207c only one critical point at the origin of the z-plane (Equation 8.207a) corresponding to an arbitrary external angle γ and the internal angle β in the ζ-plane (Equation 8.207b):

$$x_1=0, \quad \gamma_1=\gamma=\pi-\beta: \quad \frac{\mathrm{d}\zeta}{\mathrm{d}z}=A\,z^{-\gamma/\pi}=A\,z^{-1+\beta/\pi}; \quad \text{(8.207a–8.207c)}$$

integration of Equation 8.207c leads to Equation 8.208a where the constant is determined by placing the critical point at the origin of the ζ-plane (Figure 8.14a and b):

$$\zeta=\frac{A\pi}{\beta}z^{\beta/\pi}+B, \quad 0=\zeta_1\equiv\zeta\left(z_1\right)=\zeta(0)=B: \quad \zeta=\frac{A\pi}{\beta}z^{\beta/\pi}. \quad \text{(8.208a–8.208c)}$$

The Schwartz–Christoffell exterior polygonal transformation (Equations 8.208a and 8.208b $\equiv$ Equation 8.208c) implies that a uniform flow in the ζ-plane (Equation 8.209a) becomes a flow in a corner (Equation 8.209b) with internal angle β:

$$U\equiv\left(\frac{A\pi}{\beta}\right)^{-\pi/\beta}V: \quad f=Vz=V\left(\frac{\zeta\beta}{A\pi}\right)^{\pi/\beta}=U\zeta^{\pi/\beta}, \quad \frac{df}{d\zeta}=\frac{U\pi}{\beta}\zeta^{\pi/\beta-1}. \quad \text{(8.209a–8.209c)}$$

Thus, *the potential flow in a corner with internal angle β has complex potential (Equation 8.209b $\equiv$ Equation I.14.75b) [conjugate velocity (Equation 8.209c $\equiv$ Equation I.14.75c)]. The simplest particular case (Equation 8.210a) is the uniform flow with velocity U over a flat infinite wall [Equation 8.210b (Equation 8.210e)]:*

$$\beta=\pi: \quad f=U\zeta, \quad \frac{df}{d\zeta}=U; \quad \beta=2\pi: \quad f=U\sqrt{\zeta}, \quad \frac{df}{d\zeta}=\frac{U}{2\sqrt{\zeta}}. \quad \text{(8.210a–8.210f)}$$

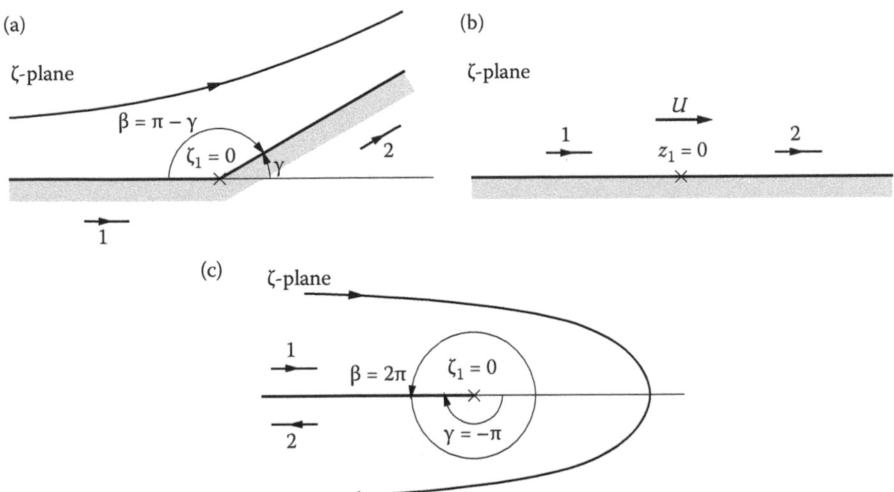

FIGURE 8.14
Thick semi-infinite plate in a free stream (between parallel walls) is obtained [Figure 8.10a (Figure 8.12a)] using an exterior polygonal transformation with two (three) critical points [Figure 8.9a (Figure 8.11a)]. The simplest case of a single critical point (b) leads to the corner flow (a) that includes the thin semi-infinite plate (c).

The case of a semi-infinite plate (Figure 8.14c) corresponds to the external angle (Equation 8.210d) and complex potential (Equation 8.210e) [conjugate velocity (Equation 8.210f)]. Other values of the external angle β may be considered (Sections I.14.8 and I.14.9). The case of a semi-infinite plate (Equations 8.211a through 8.211d) applies close to any sharp edge, for example, near either edge of a finite plate (Subsection 8.6.2).

8.6.3 Calculation of the Perturbation Velocity

The problem of finding the perturbation velocity (Equation 8.204b) that meets all the conditions (Equations 8.204c, 8.204d, 8.205a, and 8.205b) can be solved by using (Chapters I.33 and I.37) the conformal transformation (Equation 8.211a) that maps (Figure 8.13b) the lower (upper) side of the airfoil to (Equation 8.211c) (Equation 8.211d for the "wetted" region before flow separation):

$$\xi + i\eta = \zeta = \sqrt{z}; \quad z^2 = \zeta: \quad -\sqrt{c} < \operatorname{Re}(\zeta) < 0, \quad 0 < \operatorname{Re}(\zeta) < \sqrt{sc}; \qquad (8.211a\text{–}8.211d)$$

the different signs, that is, positive (Equation 8.211d) [negative (Equation 8.211c)] along the real axis of the ζ-plane, correspond to just above (Equations 8.212a and 8.212b) [below (8.212c and 8.212d)] the real axis of the z-plane:

$$z \equiv x + i0 = xe^{i0}: \quad \sqrt{z} = \sqrt{x}; \quad z = x - i0 = xe^{i2\pi}: \quad \sqrt{z} = e^{i\pi}\sqrt{z} = -\sqrt{x}. \qquad (8.212a\text{–}8.212d)$$

Using Equation 8.211a ≡ Equation 8.211b leads to the complex conjugate velocity:

$$\tilde{v}_x - i\,\tilde{v}_y = \tilde{v}^* = \frac{df}{dz} = \frac{df}{d\zeta}\frac{d\zeta}{dz} = \frac{1}{2\sqrt{z}}\frac{df}{d\zeta} = \frac{1}{2\zeta}\frac{df}{d\zeta} = \frac{df/d\zeta}{dz/d\zeta}, \qquad (8.213)$$

that [Equations 8.204c and 8.204d (Equations 8.205a and 8.205b)] satisfies [Equations 8.214a and 8.214b (Equation 8.214c)]:

$$\tilde{v}_\xi\left(-\infty<\xi<-\sqrt{c},\eta=0\right)=0=\tilde{v}_\xi\left(\sqrt{sc}<\xi<\infty,\eta=0\right), \quad \text{(8.214a and 8.214b)}$$

$$\tilde{v}_\eta\left(-\sqrt{c}<\xi<\sqrt{sc},\eta=0\right)=i2\zeta\tilde{v}^*=2\xi\tilde{v}_y=-2\alpha U\xi\sin\alpha. \quad \text{(8.214c)}$$

Since $df/d\zeta$ is analytic in the upper-half ζ-plane and decays like $O(1/\zeta)$, the second Cauchy theorem (I.15.8) reduces to an integral along the real axis:

$$\frac{1}{g(\zeta)}\in\mathcal{A}\left(\operatorname{Im}(\zeta)>0\right): \quad \frac{df}{d\zeta}\frac{2\pi i}{g(\zeta)}=\int_{-\infty}^{+\infty}\frac{df}{dt}\frac{1}{g(t)}\frac{dt}{t-\zeta}=⨍_{-\infty}^{+\infty}\frac{\tilde{v}^*(t)}{g(t)}\frac{dt}{t-\zeta}+\pi i\frac{\tilde{v}^*(\zeta)}{g(\zeta)},$$

(8.215a and 8.215b)

where (1) $g(\zeta)$ is any function whose inverse is analytic in the upper-half complex ζ–plane (Equation 8.215a) and (2) the integral equals its principal value (Sections I.17.8 through I.17.9) plus πi times the residue at a pole on the boundary (Equation 8.215b). From Equation 8.215b, it follows that

$$\frac{g(\zeta)}{\pi i}⨍_{-\infty}^{+\infty}\frac{\tilde{v}^*(t)}{g(t)}\frac{dt}{t-\zeta}=2\frac{df}{d\zeta}-\tilde{v}^*(\zeta)=\tilde{v}^*(\zeta)=\frac{df}{d\zeta} \quad \text{(8.216)}$$

and substituting Equation 8.214c leads to

$$\frac{df}{d\zeta}=-\frac{i}{\pi}g(\zeta)⨍_{-\infty}^{+\infty}\frac{-i\tilde{v}_\eta(t)}{g(t)}\frac{dt}{t-\zeta}=\frac{2U\sin\alpha}{\pi}g(\zeta)\int_{-\sqrt{c}}^{\sqrt{sc}}\frac{t}{g(t)}\frac{dt}{t-\zeta}. \quad \text{(8.217)}$$

Thus, *the complex conjugate perturbation velocity is given by Equation 8.217 where the* **Sedov function** *(Equation 8.215a) is chosen to lead to a complex conjugate velocity that is regular (singular) at the edges of the airfoil if the Kutta condition imposing a finite velocity is met (not met); examples of both cases are given in the sequel (Subsections 8.6.4 and 8.8.1 through 8.8.7).*

8.6.4 Separation Function Meeting the Kutta Condition

The choice of Sedov function in Equation 8.217 is the **separation function** is chosen to satisfy the Kutta condition (Subsection I.34.1.4) at the leading edge and separation point (Equation 8.216a) by suppressing the inverse square root singularity (Equation 8.210f ≡ Equation 8.213) of the velocity in (Equation 8.215b):

$$g_1(\zeta)=\left[\left(\zeta+\sqrt{c}\right)\left(\sqrt{sc}-\zeta\right)\right]^{1/2}=i\left[\left(\zeta+\sqrt{c}\right)\left(\zeta-\sqrt{sc}\right)\right]^{1/2}. \quad \text{(8.218a and 8.218b)}$$

The separation function is real on the segment (Equations 8.211c and 8.211d) of the ξ-axis of the ζ-plane corresponding to the "wetted" part of the airfoil, whose midposition (length) is Equation 8.219a (Equation 8.219b):

$$\bar{\xi} = \frac{\sqrt{sc} - \sqrt{c}}{2}, \quad \ell = \frac{\sqrt{c} + \sqrt{sc}}{2}; \quad \{\zeta, t\} = \ell\{\cos\theta, \cos\phi\} + \bar{\xi}; \quad (8.219\text{a}–8.219\text{d})$$

this suggests the change of variable (Equations 8.219c and 8.219d ≡ Equations 8.220a and 8.220b):

$$\{\zeta, t\} = \frac{\sqrt{c} + \sqrt{sc}}{2}\{\cos\theta, \cos\phi\} + \frac{\sqrt{sc} - \sqrt{c}}{2}, \quad (8.220\text{a and } 8.220\text{b})$$

that leads to

$$\phi = 0, \pi: \quad t = \sqrt{sc}, -\sqrt{c}; \quad \frac{dt}{t - \zeta} = -\frac{\sin\phi\, d\phi}{\cos\phi - \cos\theta}; \quad (8.221\text{a}–8.221\text{c})$$

the separation function 8.218b satisfies

$$\begin{aligned}\left[g_1(t)\right]^2 &= -\left(t + \sqrt{c}\right)\left(t - \sqrt{sc}\right) = \frac{1}{4}\left(\sqrt{sc} + \sqrt{c}\right)^2 (1 + \cos\phi)(1 - \cos\phi) \\ &= \frac{1}{4}\left(\sqrt{sc} + \sqrt{c}\right)^2 \sin^2\phi = \left(\frac{\sqrt{sc} + \sqrt{c}}{2}\sin\phi\right)^2,\end{aligned} \quad (8.222\text{a})$$

implying

$$\frac{t}{g_1(t)} = \frac{2t}{\left(\sqrt{sc} + \sqrt{c}\right)\sin\phi} = \cot\phi + \frac{\sqrt{sc} - \sqrt{c}}{\sqrt{sc} + \sqrt{c}}\csc\phi. \quad (8.222\text{b})$$

Substitution of Equations 8.221a through 8.221c and 8.222b in Equation 8.217 leads to

$$\frac{\pi}{2\sin\alpha U}\frac{df}{d\zeta}\frac{1}{g_1(\zeta)} = \fint_0^{\pi}\left(\cos\phi + \frac{\sqrt{sc} - \sqrt{c}}{\sqrt{sc} + \sqrt{c}}\right)\frac{d\phi}{\cos\phi - \cos\theta} = I_1 + \frac{\sqrt{sc} - \sqrt{c}}{\sqrt{sc} + \sqrt{c}}I_0 \quad (8.223)$$

involving the principal value of the integrals (Equation I.34.191 ≡ Equation 8.224a) in the particular cases (Equations 8.224b and 8.224c):

$$I_n \equiv \fint_0^{\pi}\frac{\cos(n\phi)}{\cos\phi - \cos\theta}d\phi = \pi\frac{\sin(n\theta)}{\sin\theta}: \quad I_0 = 0, \quad I_1 = \pi. \quad (8.224\text{a}–8.224\text{c})$$

The integral (Equation 8.224b) is evaluated directly from the primitive (Equations I.34.184a and I.34.184b); the integral (Equation 8.224c) follows from (Equation 8.224a):

$$I_1 = \fint_0^{\pi}\frac{\cos\phi}{\cos\phi - \cos\theta}d\phi = \fint_0^{\pi}\left(1 + \frac{\cos\theta}{\cos\phi - \cos\theta}\right)d\phi = \fint_0^{\pi}d\phi + I_0\cos\theta = \pi. \quad (8.225)$$

Only Equations 8.224b and 8.224c ≡ Equation 8.225 appear in Equation 8.223. The remaining integrals (Equation 8.221a) for other values of $n = 2,3,\ldots$ appear in the calculation of the downwash velocity of a lifting-line representing a wing (Subsection I.34.7.4). Substitution of Equations 8.224b and 8.224c) in Equation 8.223 leads to

$$\frac{df}{d\zeta} = 2U \sin\alpha\, g_1(\zeta) = i2U \sin\alpha\left[\left(\zeta+\sqrt{c}\right)\left(\zeta-\sqrt{sc}\right)\right]^{1/2} \tag{8.226}$$

using the separation function (Equation 8.218b). From Equation 8.226, it follows that the velocity is tangential (Equation 8.227c):

$$-\sqrt{c} < \xi < \sqrt{sc}: \quad \tilde{v}_x(\xi) = 2U \sin\alpha\left[\left(\xi+\sqrt{c}\right)\left(\sqrt{sc}-\xi\right)\right]^{1/2} \quad \tilde{v}_y(\xi) = 0 \qquad \text{(8.227a–8.227c)}$$

and equals Equation 8.227b on the wetted part (Equation 8.227a) of the airfoil.

8.6.5 Circulation and Lift with Partially Separated Flow

The circulation around the airfoil (Equation 8.206a) is specified (Equations 8.227a and 8.227b) by

$$x = \xi^2: \quad -\Gamma = \int_{-\sqrt{c}}^{\sqrt{sc}} \tilde{v}_x(\xi)\, d\xi = 2U \sin\alpha \int_{-\sqrt{c}}^{\sqrt{sc}} \left[\left(\xi+\sqrt{c}\right)\left(\sqrt{sc}-\xi\right)\right]^{1/2} d\xi \qquad \text{(8.228a–8.228c)}$$

involving (Equation 8.228b ≡ Equation 8.229a) the integral (Equation 8.229b):

$$\Gamma = -2UI \sin\alpha, \quad I \equiv \int_{-\sqrt{c}}^{\sqrt{sc}} \left[\left(\xi+\sqrt{c}\right)\left(\sqrt{sc}-\xi\right)\right]^{1/2} d\xi. \qquad \text{(8.229a and 8.229b)}$$

The integral (Equation 8.229b) in the circulation (Equation 8.229a) is evaluated by a method similar to Example I.40.19 via two changes of variable: (1) the first change of variable (Equation 8.230a) shifts the origin to the midposition (Equation 8.219a) and eliminates the cross term in the square root in the integrand of Equation 8.229b:

$$\eta \equiv \xi - \bar{\xi} = \xi + \frac{\sqrt{c}}{2} - \frac{\sqrt{sc}}{2}: \quad I \equiv \int_{-\left(\sqrt{c}+\sqrt{sc}\right)/2}^{\left(\sqrt{c}+\sqrt{sc}\right)/2} \left[\left(\frac{\sqrt{c}+\sqrt{sc}}{2}\right)^2 - \eta^2\right]^{1/2} d\eta; \qquad \text{(8.230a and 8.230b)}$$

and (2) then the trigonometric change of variable (Equation 8.230c) eliminates the square root in Equation 8.230b ≡ Equation 8.230d:

$$2\eta = \left(\sqrt{c}+\sqrt{sc}\right)\sin\psi: \quad I = \left[\left(1+\sqrt{s}\right)\frac{\sqrt{c}}{2}\right]^2 \int_{-\pi/2}^{+\pi/2} \cos^2\psi\, d\psi = \frac{\pi}{8} c\left(1+\sqrt{s}\right)^2.$$

(8.230c and 8.230d)

Substituting Equation 8.230d in Equation 8.229a determines the circulation (Equation 8.231a):

$$\Gamma = -\frac{\pi}{4} Uc \sin\alpha \left(1+\sqrt{s}\right)^2, \quad L = -\rho U \Gamma = \frac{\pi}{4} \rho U^2 c \sin\alpha \left(1+\sqrt{s}\right)^2 \qquad \text{(8.231a and 8.231b)}$$

and hence the lift (Equation 8.231b) using the Kutta–Joukowski theorem (Equation I.28.29b ≡ Equation 2.233a). For fully attached flow (Equation 8.232a):

$$s = 1: \quad \Gamma = -\pi U c \sin\alpha, \quad L = -\rho U \Gamma = \pi \rho U^2 c \sin\alpha, \qquad \text{(8.232a–8.232c)}$$

the circulation (Equation 8.231a) [lift (Equation 8.231b)] simplifies to Equation 8.232b ≡ Equation I.34.12c (Equation 8.232c ≡ Equation I.34.17).

8.6.6 Lift Loss due to Flow Separation

The lift coefficient (Equation 8.233a) of an airfoil is given (Equation 8.232c ≡ Equation 8.233a) by Equation 8.234b for separated flow:

$$L = \frac{1}{2} \rho U^2 a C_L: \quad C_L(\alpha; s) = \frac{\pi \sin\alpha}{2} \left(1+\sqrt{s}\right)^2, \quad C_L(\alpha; 1) = 2\pi \sin\alpha \qquad \text{(8.233a–8.233c)}$$

compared with the value (Equation 8.233c) in the absence of separated flow, that is, for attached flow. The value of the lift coefficient (Equation 8.233c ≡ Equation I.34.18a) for attached flow applies to a flat plate with attached flow and simplifies at small angle-of-attack $\sin\alpha \backsim \alpha$ for $\alpha^2 \ll 1$. It applies up to the angle-of-attack for the onset of flow separation (Equation 8.234a) corresponding (Figure 8.15) to the maximum lift coefficient (Equation 8.234b):

$$0 \le \alpha \le \alpha_s: \quad C_{L\max} \equiv C_L(\alpha_s; 1) = 2\pi \sin\alpha_s. \qquad \text{(8.234a and 8.234b)}$$

Beyond the angle-of-attack for flow separation (Equation 8.235a), the lift coefficient diminishes (Equation 8.235b):

$$\alpha \ge \alpha_s: \quad 2\pi \sin\alpha_s = C_{L\max} \ge C_L(\alpha; s) = \frac{\pi \sin\alpha}{2} \left(1+\sqrt{s}\right)^2; \qquad \text{(8.235a and 8.235b)}$$

thus, the region of separated flow must extend at least to a fraction (Equation 8.236a) of the (Equation 8.204a) airfoil chord:

$$s \le \left(2\sqrt{\frac{\sin\alpha_s}{\sin\alpha}} - 1\right)^2; \quad s = 1: \quad \alpha \le \alpha_s, \qquad \text{(8.236a–8.236c)}$$

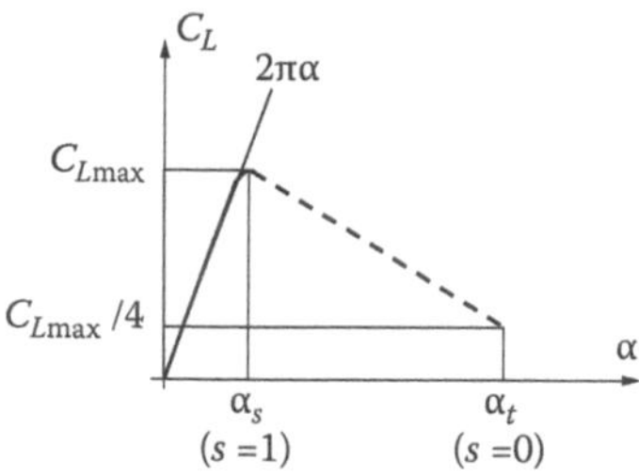

FIGURE 8.15
Lift of a flat plate airfoil increases with the angle-of-attack α with a slope 2π as long as (c) the maximum angle-of-attack for flow separation is not reached $\alpha < \alpha_s$; beyond that angle-of-attack $\alpha < \alpha_s$, a region of separated flow appears on the upper surface of the airfoil near the trailing edge. As the angle-of-attack is further increased, the region of separated flow increases by starting closer to the leading edge $\alpha_s < \alpha < \alpha_t$, causing a progressive reduction in the lift coefficient. The lift coefficient is reduced to one-quarter of the maximum for the angle-of-attack α_t corresponding to totally separated flow on the upper surface, when only the attached flow on the lower surface contributes to the lift.

in particular, if the flow is attached over the whole of the upper surface of the airfoil up to the trailing edge (Equation 8.236b) for angle-of-attack (Equation 8.231c) not exceeding the separation angle-of-attack.

8.6.7 Lift of an Airfoil with Partially Separated Flow

Thus, *the lift coefficient is given by Equation 8.233b ≡ Equation 8.237a (Equation 8.233a ≡ Equation 8.237b) for nonseparated (separated) flow for angle-of-attack below (above) the maximum angle-of-attack α_s for fully attached flow:*

$$C_L(\alpha) = \begin{cases} 2\pi\sin\alpha & \text{if} \quad 0 \le \alpha \le \alpha_s, \\ \dfrac{\pi}{2}\sin\alpha\left(1+\sqrt{s}\right)^2 & \text{if} \quad \alpha_s \le \alpha \le \alpha_t. \end{cases} \qquad (8.237a) \; (8.237b)$$

The ratio of lift coefficients for separated (Equation 8.233a) and attached (Equation 8.233b) flow at the same angle-of-attack is

$$D(s) \equiv \frac{C_L(s)}{C_{L0}(1)} = \left(\frac{1+\sqrt{s}}{2}\right)^2 = \frac{C_{L\alpha}(\alpha;s)}{C_{L\alpha}(\alpha;1)} \qquad (8.238a \text{ and } 8.238b)$$

and applies also to the lift slope (Equations 8.240a and 8.240b); it depends only on the fraction (Equation 8.204a) of the chord of the upper surface covered by flow separation. It is assumed that the flow remains attached over the whole of the lower surface of the airfoil; for totally separated flow over the whole of the upper surface of the airfoil (Equation 8.239a), the lift is one quarter (Equation 8.239b) of the lift with fully attached flow on both the upper and lower surfaces:

$$s=0: \quad C_L(\alpha;0) = \frac{\pi\sin\alpha}{2} = \frac{1}{4}C_L(\alpha;1); \quad \sin\alpha_t = 4\sin\alpha_s; \qquad (8.239a\text{–}8.239c)$$

since in both cases the flow remains attached to the lower surface of the airfoil, the latter contributes one quarter of the lift, that is, one-third of the contribution of the upper surface, where the velocity of the perturbed flow is increased relative to the free stream. The condition 8.239a of fully separated flow over the whole of the upper surface of the airfoil corresponds (Equation 8.236a) to a sine of the angle-of-attack (Equation 8.239c) equal to four times the sine of the angle-of-attack for maximum lift (Figure 8.15).

The lift coefficient Equation 8.238a (Equation 8.238b) corresponds to the lift slope of the flat plate airfoil without (Equation 8.240a) [with (Equation 8.240b)] flow separation on the upper surface:

$$C_{L\alpha} \equiv \frac{\partial C_L}{\partial \alpha} = \begin{cases} 2\pi\cos\alpha & \text{if} \quad 0 \le \alpha \le \alpha_s, \quad (8.240a) \\ \dfrac{\pi}{2}\left(1+\sqrt{s}\right)^2 \cos\alpha & \text{if} \quad \alpha_s \le \alpha \le \alpha_t. \quad (8.240b) \end{cases}$$

Table 8.4 shows the evolution of the lift slope (Equation 8.240b) and lift loss (Equations 8.238a and 8.238b) as the region of separated flow extends over the upper surface of the airfoil, in the upstream direction starting from the trailing edge; the lift loss is already significant if the region of flow separation approaches half the chord. The detached flow is represented by a region of zero velocity; since the flow is potential, this implies that the pressure equals the stagnation pressure. If the flow were attached, the Bernoulli law would lead to a pressure lower than the stagnation pressure. The larger pressure on the upper surface beyond the separation point leads to a lift loss. The latter has been calculated in the context of a potential flow. Thus, the preceding theoretical values of the lift loss due to flow separation are at best orders of magnitude, since no viscous or turbulent flow effects have been modeled. They suggest that flow separation on the wing is a flight condition to be avoided, as it can reduce

TABLE 8.4

Separated versus Attached Flow

Separation Point $s \equiv \frac{x}{c}$	**Lift Slope** $\frac{dC_L}{d\alpha} = \frac{\pi}{2}\left(1+\sqrt{s}\right)^2$	**Ratio of Lift Coefficients** $\frac{C_L(s)}{C_L(1)} = \frac{\left(1+\sqrt{s}\right)^2}{4}$
1.000	6.28	1.000
0.900	5.96	0.949
0.800	5.64	0.897
0.700	5.30	0.843
0.600	4.95	0.787
0.500	4.58	0.729
0.400	4.19	0.666
0.300	3.76	0.599
0.200	3.29	0.524
0.100	2.72	0.433
0.000	1.57	0.250

Notes: Lift of an airfoil is reduced by flow separation. It is assumed that the flow remains attached to the whole of the lower surface of the flat plate airfoil and to the fraction s of the upper surface closer to the leading edge (Figure 8.13). As the separation point moves forward, the region of separated flow near the trailing edge increases. The lift slope reduces, and thus the lift coefficient with separated flow becomes a smaller fraction of the lift coefficient with attached flow (Figure 8.15).

lift significantly, possibly making steady flight impossible and leading to a stall. The flow separation spreads through the propagation of viscous instability waves. The preceding analysis is limited by the neglect of the viscous boundary layer that forms around the airfoil in a stream. After the onset of flow separation, the boundary layer becomes unsteady and turbulent. All these phenomena have been replaced in the present model with a simpler representation, as a region of detached flow near the trailing edge of the upper surface of the airfoil where the velocity is zero and the pressure equals the stagnation pressure, with potential flow over the rest of the upper surface plus the whole of the lower surface. Another form of detached flow occurs for a multielement airfoil, with slotted leading-edge slats and/or trailing-edge flaps, which are considered next (Section 8.7).

8.7 Airfoil with Plain/Slotted Flap/Slat

The lift of an airfoil may be increased by using a leading (trailing) edge **slat (flap)** in Figure 8.16a (Figure 8.16b); large deflections lead to flow separation, which can be delayed by having a slot between the airfoil and the **high-lift device**. A **slotted flap (slat)** produces [Figure 8.16c (Figure 8.16d)] more lift than a nonslotted or **plain flap/slat**. The use of **double slotted** or **triple slotted** flaps or slats further increases lift at the expense of more complexity and weight of the actuation mechanism. These high-lift devices also increase drag; the resulting additional fuel consumption is of lesser importance for short periods of flight, such as take-off or landing, than higher lift and lower speed on approach to land (lift-off at take-off), allowing the use of shorter runways. The extra drag may serve to decelerate the aircraft more rapidly, for example, on approach to land, allowing a steeper and shorter descent path. The effect of a plain (Subsection 8.7.3) or slotted (Subsection 8.7.4)

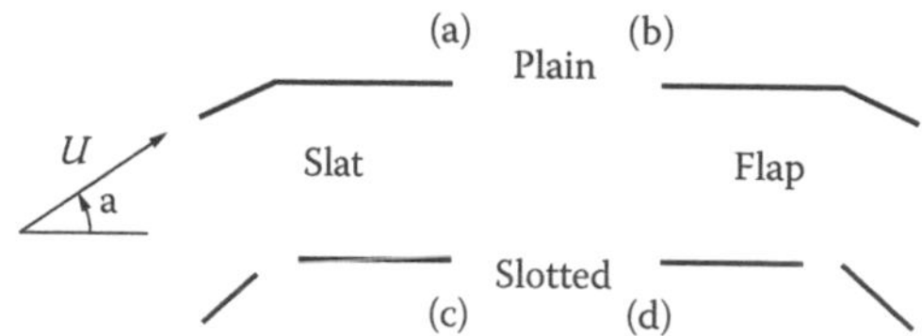

FIGURE 8.16
Lift of an airfoil may be increased by fitting a leading-edge slat (a) [trailing-edge flap (b)] that causes a greater deflection of the incident stream. This effect is increased by adding a slot between the slat (c) [flap (d)] and the leading (trailing) edge. The slot causes additional deflection of the free stream and may reduce or eliminate the region of separated flow (Figure 8.14a) by "energizing" the boundary layer (Figure 8.19), that is, (1) the lower (higher) speed of the perturbed flow on the lower (upper) side of the airfoil corresponds to a higher (lower) pressure, leading to a lift on the airfoil, enhanced by a plain flap or slat; (2) adding a slot, the pressure difference causes an upflow through the slot, accelerating the flow above the airfoil, thus delaying or preventing flow separation; and (3) this allows a larger maximum angle-of-attack without flow separation and a larger maximum lift coefficient (Figure 8.15). The lift on the airfoil is thus (1) increased by the addition of a plain flap or slat and (2) further increased in the case of a slotted flap or slat. The increase in lift leads to an increase in induced drag to the square and hence the need for higher thrust to keep steady flight; the additional fuel consumption may not be relevant for a short fraction of the flight time close to landing or takeoff. More important is that the increase in lift allows steady flight at lower speed reducing the takeoff roll and landing distance; this makes possible the use of shorter runways. The possibility to change the drag also serves to modulate the airspeed as an alternative or supplement to using the throttle to change engine thrust.

flap on lift is calculated considering close airfoils (Subsection 8.7.2) and using the Sedov theorem (Subsection 8.7.1); although the slot decreases the lifting area, it can prevent flow separation (Subsection 8.7.5) on the upper surface of the airfoil (Section 8.6), thus leading to an increase in lift (Subsection 8.7.6).

8.7.1 Sedov (1965) Theorem and Multiple Airfoils

Consider as before (Equation 8.204b) a uniform stream incident on a flat plate airfoil in which chord c was placed symmetrically relative to the origin on the real axis (Equation 8.241a):

$$-\frac{c}{2} < x < +\frac{c}{2}: \quad \tilde{v}_x(z) - i\tilde{v}_y(z) = \frac{1}{2\pi i}\int^{(z+)}\left[\tilde{v}_x(\zeta) - i\tilde{v}_y(\zeta)\right]\frac{d\zeta}{\zeta - z}d\zeta, \quad \text{(8.241a and 8.241b)}$$

and let the perturbation velocity be given by the Cauchy integral (Equation 8.241b ≡ Equation I.15.8) around a loop L in Figure 8.17. The loop L around z may be deformed into a loop at infinity L_∞ for which the integral (Equation 8.241b) is zero, and a loop L_0 around the airfoil, which on the real axis (Equation 8.242a) outside the airfoil (Equation 8.242b) leads by equating real and imaginary parts in Equation 8.241b to Equations 8.242c and 8.242d:

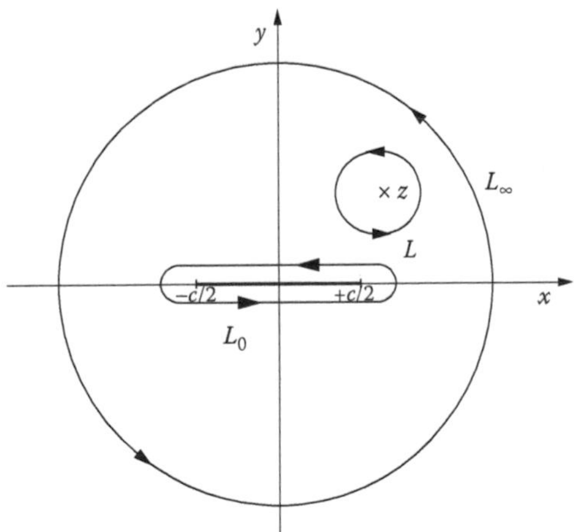

FIGURE 8.17
Flow with circulation around an airfoil in a free stream can be specified: (1) from the flow around a cylinder using a Joukowski conformal transformation (Figure 8.20) to map the circular cross section into an airfoil shape; (2) by using the Sedov method imposing a zero normal velocity on the airfoil and using the airfoil and circle at infinity as the boundaries to calculate the flow in the interior via the Cauchy integral satisfied by an analytic function. The application of the Kutta condition enforcing a finite velocity at the trailing edge is made: (1) in the Joukowski method specifying the circulation around the airfoil; and (2) in the Sedov method choosing an airfoil function that suppresses the inverse square root singularity of the velocity at the edge. The two methods lead to the same: (a) circulation around the airfoil; (b) lift on the airfoil; (c) finite tangential velocity at the trailing edge; and (d) infinite normal velocity at the leading edge associated with the suction force. The suction force along the flat plate airfoil balances the horizontal component of the lift; the lift force is orthogonal to the free stream direction and thus has components along (normal to) the airfoil, corresponding to minus the suction force at the leading edge (the net upward force at the center of lift).

$$z \equiv x, |x| > \frac{c}{2}: \quad 2\pi\{\tilde{v}_x(x,0), \tilde{v}_y(x,0)\} = \int_{L_0} (\zeta - x)^{-1}[-\tilde{v}_y(\zeta), \tilde{v}_x(\zeta)]\, d\zeta. \quad (8.242a–8.242d)$$

Thus, *the horizontal (vertical) velocity outside the airfoil is specified [Equation 8.242c (Equation 8.242d)] by the vertical (horizontal) velocity on the loop around the airfoil.* The vertical perturbation velocity must cancel the free stream velocity on the airfoil (Equation 8.243a) implying that the integral (Equation 8.242c) is zero (Equation 8.243b), that is, the perturbation velocity is vertical (Equation 8.243c) on the real axis outside the airfoil:

$$\tilde{v}_y\left(|x| < \frac{c}{2}, \pm 0\right) = -U \sin\alpha, \tilde{v}_x\left(|x| < \frac{c}{2}, 0\right) = 0, \quad \tilde{v}^*\left(|x| > \frac{c}{2}, 0\right) = -i\tilde{v}_y\left(|x| > \frac{c}{2}, 0\right). \quad (8.243a–8.243c)$$

Since the complex conjugate perturbation velocity is analytic outside the real axis, it can, by the reflection principle (Subsection I.31.2.2), be extended analytically:

$$\tilde{v}_x(x,y) - i\,\tilde{v}_y(x,y) = \tilde{v}^*(x + i\,y, 0) = -i\,\tilde{v}_y(x + i y, 0), \quad (8.244)$$

leading by reflection on the real axis to

$$\begin{aligned}\tilde{v}_x(x,-y) - i\tilde{v}_y(x,-y) &= -i\tilde{v}_y(x - iy, 0) = -i\tilde{v}_y^*(x + iy, 0) = [i\tilde{v}_y(x + iy, 0)]^* \\ &= \left[-\tilde{v}^*(x + iy, 0)\right]^* = -\tilde{v}(x + iy, 0) = -\tilde{v}_x(x,y) - i\tilde{v}_y(x,y);\end{aligned} \quad (8.245)$$

the latter (Equation 8.245) proves

$$\tilde{v}_x(x,-y) = -\tilde{v}_x(x,y), \quad \tilde{v}_y(x,-y) = \tilde{v}_y(x,y), \quad (8.246a \text{ and } 8.246b)$$

showing that *the perturbation velocity parallel (Equation 8.246a) [orthogonal (Equation 8.246b)] to the airfoil is skew-symmetric (symmetric) relative to the straight line passing through the airfoil, that is has opposite (equal) directions above and below.*

In particular, Equation 8.246a leads on the airfoil to Equation 8.247a:

$$\tilde{v}_x(x,+0) = -\tilde{v}_x(x,-0), \quad \tilde{v}_x(x,0) = 0, \quad (8.247a \text{ and } 8.247b)$$

which implies Equation 8.247b; the latter (Equation 8.247b) confirms Equation 8.243b on the airfoil and holds also outside the airfoil (Equation 8.243c). Thus, Equation 8.247b holds on the whole real axis. The perturbation velocity is given by the Cauchy integral:

$$2\pi i \frac{\tilde{v}^*(z)}{g(z)} = \int_{L \equiv L_\infty - L_0} \frac{\tilde{v}^*(\zeta)}{g(\zeta)} \frac{d\zeta}{\zeta - z}, \quad (8.248)$$

where (1) the function $g(z)$ has an analytic inverse (Equation 8.215a), except for suitable singularities to satisfy the Kutta condition at the edges of airfoils; and (2) the loops L_∞ and L_0 are described in opposite directions, that is, for L_∞ (L_0) the same as (opposite to) L.

Deforming the loop L into L_∞ and L_0 in Figure 8.17 and using Equation 8.244 on the latter lead to

$$2\pi i\frac{\tilde{v}^*(z)}{g(z)} = \int_{L_\infty} \frac{\tilde{v}^*(\zeta)}{g(\zeta)}\frac{d\zeta}{\zeta - z} + i\int_{L_0} \frac{\tilde{v}_y(\zeta)}{g(\zeta)}\frac{d\zeta}{\zeta - z}, \tag{8.249}$$

where the sign of the second integral on the r.h.s. of Equation 8.249 was reversed so that both loops are taken in the same positive or counterclockwise direction. Thus, *the* **theorem of Sedov (1965)** *has been obtained, expressing the complex conjugate velocity around airfoils as an integral of the normal velocity on the airfoils (Equation 8.249), added to an integral at infinity; both integrals involve an* **Sedov function** *$g(\zeta)$ with analytic inverse (Equation 8.215a) except at singularities chosen to meet the Kutta condition at the edges of airfoils.* The Sedov method applies to single and multiple airfoils and is applied next to an airfoil (Section 8.8) with a slotted flap (Subsection 8.7.2).

8.7.2 Flat Plate with a Single-Slotted Flap

Consider (Figure 8.18) a flat plate of chord c, with a flap of chord a and a slot of width b. The incident stream of velocity U and angle-of-attack α impinges on the fixed part $0 \le x \le c - a$ of the airfoil, and if the flap has a **deflection angle** δ, the angle-of-attack relative to the free stream is $\alpha + \delta$. The total velocity of the potential flow past the airfoil with single-slotted flap is given by Equation 8.204b ≡ Equation 8.250b for small angle-of-attack (Equation 8.250a) of the free stream:

$$\alpha^2 \ll 1: \quad v^*(z) = U(1 - i\alpha) + \tilde{v}^*(z); \tag{8.250a and 8.250b}$$

the perturbation velocity on the airfoil (Equation 8.251a) [flap (Equation 8.251b)] cancels the normal velocity:

$$v_y(0 < x < c - a, 0) = -\alpha U, \qquad v_y(c - a + b < x < c + b, 0) = -(\alpha + \delta)U. \tag{8.251a and 8.251b}$$

The perturbation velocity is given in the whole flow region by Equation 8.252b:

$$\Gamma = -\pi U c\alpha: \quad \tilde{v}^* + \frac{i\Gamma}{2\pi z} = \tilde{v}^* - i\frac{Uc\alpha}{2z} = \frac{g(z)}{2\pi}\int_{L_f} \frac{\tilde{v}_y(\zeta)}{g(\zeta)}\frac{d\zeta}{\zeta - z}, \tag{8.252a and 8.252b}$$

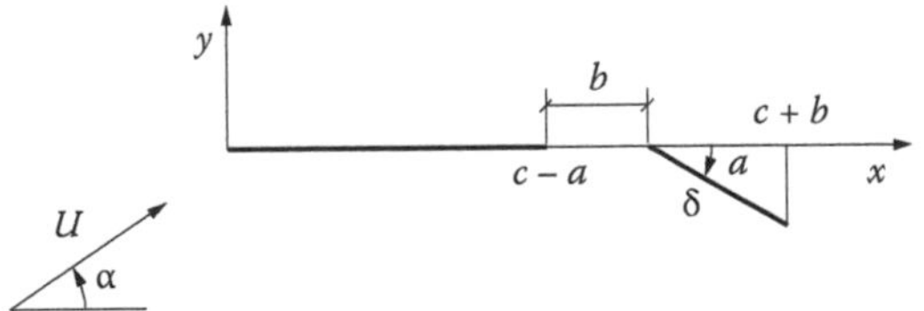

FIGURE 8.18
Airfoil with slotted flap (slat) in Figure 8.16c (Figure 8.16d) is a doubly connected region. The simplest representation consists (Figure 8.18) of a large (small) flat plate represent the airfoil (flap) with the slot in between. The geometrical parameters are (1–3) the chord of the airfoil c and the fraction a/c (b/c) occupied by the flap (slot); (4 and 5) the angle-of-attack of the free stream α and the deflection angle δ of the flap relative to the airfoil. These parameters determine the lift increase relative to the (1) flapless airfoil due to the addition of (2) a plain flap or (3) a slotted flap.

where (1) the circulation around the airfoil is given by Equation 8.252a ≡ Equation 8.232a for small angle-of-attack (Equation 8.250a) of the incident stream and attached flow $s = 1$ leading to the second term on the r.h.s. of Equation 8.252b; and (2) the last term on the r.h.s. of Equation 8.252b concerns the effect of the slotted flap corresponding to the second term on the r.h.s. of Equation 8.249 with the integral along a loop L_f enclosing the flap. The **slotted airfoil function** is chosen to satisfy the Kutta condition (Subsections I.34.1.4 and 8.6.3) suppressing the inverse square root singularity at (Equation 8.210f) the trailing edge $\zeta = c - a(\zeta = c + b)$ of the airfoil (flap):

$$\frac{1}{g_2(\zeta)} = \left[\frac{(\zeta - c + a)(\zeta - c - b)}{\zeta(\zeta - c + a - b)}\right]^{1/2}; \tag{8.253}$$

the leading edge $\zeta = 0$ ($\zeta = c - a + b$) of the airfoil (flap) also appears in Equation 8.253. In the far field, the airfoil function (Equation 8.253) reduces to unity (Equation 8.254a), and the complex conjugate velocity (Equation 8.252b) simplifies to Equation 8.254b:

$$\lim_{z\to\infty} g_2(z) = 1: \quad \tilde{v}^* = -\frac{i\Gamma}{2\pi z} + \int_{L_f} \tilde{v}_y(\zeta)\frac{d\zeta}{g_2(\zeta)} = i\frac{Uc\alpha}{2z} - \frac{U\delta}{2\pi z}2\int_0^a \frac{dx}{g_2(x)}, \tag{8.254a and 8.254b}$$

where on the r.h.s of Equation 8.254b (1) the first term corresponds to the circulation (Equation 8.252a) around the airfoil; and (2) the second term involves an integral around the flap, with tangential velocity (Equation 8.251b) differing from that on the airfoil (Equation 8.251a) by $-U\delta$. The tangential perturbation velocity on the airfoil and flap vanishes (Equation 8.247b) and does not affect the circulation (Equation 8.252a) and the integral in Equation 8.254b. The lift (Equation 8.255) is specified (Equation I.28.29b = Equation 2.233a) by the circulation:

$$\begin{aligned} F_y &= -\rho U\Gamma = -\rho U\int^{(0+)} \tilde{v}^*(z)\,dz = \frac{\rho U^2}{2}\left[-ic\alpha - \frac{2\delta}{\pi}\int_0^a \frac{dx}{g_2(x)}\right]\int^{(0+)}\frac{dz}{z} \\ &= \pi\rho U^2 c\alpha - i2\rho U^2\delta\int_0^a \frac{dx}{g_2(x)} = \frac{1}{2}\rho U^2 C_{L1}\,c; \end{aligned} \tag{8.255}$$

it follows that the lift coefficient is given by Equation 8.256b:

$$C_L = 2\pi\alpha: \quad C_{L1} = \frac{2F_y}{\rho U^2 c} = 2\pi\alpha - i\frac{4\delta}{c}\int_0^a \frac{dx}{g_2(x)}, \tag{8.256a and 8.256b}$$

which adds the lift coefficient of the flat plate (Equation 8.256a) that is the first term on the r.h.s. of Equation 8.256b; and the second term on the r.h.s. of Equation 8.256b is the effect of the slotted flap, involving the airfoil function (Equation 8.253). The ratio of lift coefficients with (Equation 8.256b) [without (Equation 8.256a)] slotted flap is given by Equation 8.257b:

$$x = \xi^2 c: \quad \frac{C_{L1}}{C_L} - 1 = \frac{2\delta}{\pi\alpha c}\int_0^a \frac{dx}{ig(x)} = \frac{4\delta}{\pi\alpha}\int_0^{\sqrt{a/c}} \frac{\xi}{ig_2(\xi^2 c)}d\xi; \tag{8.257a and 8.257b}$$

the airfoil function (Equation 8.253) of the variable (Equation 8.257a)

$$\frac{1}{ig_2\left(\xi^2 c\right)} = \left[\frac{\left(\xi^2 c - c + a\right)\left(\xi^2 c - c - b\right)}{-\xi^2 c\left(\xi^2 c - c + a - b\right)}\right]^{1/2} = \frac{1}{\xi}\left[\frac{\left(1 - a/c - \xi^2\right)\left(1 + b/c - \xi^2\right)}{1 - a/c + b/c - \xi^2}\right]^{1/2} \tag{8.258}$$

appears in the integrand in Equation 8.257b.

8.7.3 Lift and Drag of an Airfoil with a Plain Flap

Substituting Equation 8.258 in Equation 8.257b specifies *the change in the lift coefficient of an airfoil of chord c due to a slotted flap of chord a with gap b, in terms of the angle-of-attack α of the free stream and flap deflection δ:*

$$\frac{C_{L1}}{C_L} = 1 + \frac{4\delta}{\pi\alpha}\int_0^{\sqrt{a/c}}\left[\frac{\left(1 + b/c - \xi^2\right)\left(1 - a/c - \xi^2\right)}{1 + b/c - a/c - \xi^2}\right]^{1/2} d\xi. \tag{8.259}$$

This simplifies to Equation 8.260b for a plain flap (Equation 8.260a), that is, without slot (Figure 8.16b):

$$b = 0: \quad \frac{C_{L2}}{C_L} = 1 + \frac{4\delta}{\pi\alpha}\int_0^{\sqrt{a/c}}\sqrt{1 - \xi^2}\, d\xi = 1 + \frac{4\delta}{\pi\alpha} I_0\left(\sqrt{\frac{a}{c}}\right) \tag{8.260a and 8.260b}$$

involving the integral Equation 8.261b, which is evaluated via the change of variable (Equation 8.261a):

$$\xi \equiv \sin\theta: \quad I_0(\eta) \equiv \int_0^{\eta}\sqrt{1 - \xi^2}\, d\xi = \int_0^{\arcsin\eta}\cos^2\theta\, d\theta$$
$$= \int_0^{\arcsin\eta}\frac{1 + \cos(2\theta)}{2}\, d\theta = \left[\frac{\theta}{2} + \frac{\sin(2\theta)}{4}\right]_0^{\arg\sin\eta}. \tag{8.261a and 8.261b}$$

Bearing in mind

$$\theta = \arcsin\eta: \quad \frac{\sin(2\theta)}{2} = \sin\theta\cos\theta = \sin\theta\sqrt{1 - \sin^2\theta} = \eta\sqrt{1 - \eta^2}, \tag{8.262a and 8.262b}$$

the integral Equation 8.261b is evaluated by Equation 8.263

$$I_0(\eta) = \int_0^{\eta}\sqrt{1 - \xi^2}\, d\xi = \frac{1}{2}\left(\arcsin\eta + \eta\sqrt{1 - \eta^2}\right). \tag{8.263}$$

Substitution of Equation 8.263 in Equation 8.260b specifies

$$b=0: \quad \frac{C_{L2}}{C_L} = 1 + \frac{2\delta}{\pi\alpha}\left[\arcsin\left(\sqrt{\frac{a}{c}}\right) + \sqrt{\frac{a}{c}}\sqrt{1-\frac{a}{c}}\right] = \sqrt{\frac{C_{D2i}}{C_{Di}}} \quad \text{(8.264a–8.264c)}$$

the ratio of lift coefficients of two airfoils same angle-of-attack α, one without flap C_L and the other with a flap of chord a deflected by an angle δ; the ratio of induced drag coefficients is the square (Equation I.34.208 ≡ Equation 8.264c). The effect on lift and induced drag depends only on (1) the ratio of the chord of the flap a to the chord of the airfoil c and (2) the ratio of the flap deflection angle δ to the angle-of-attack α of the free stream. The angle-of-attack of the free stream acts as an "airfoil deflection angle"; a positive (negative) relative deflection of the flap increases (decreases) lift. Table 8.5 indicates the ratio of lift coefficients (Equation 8.264b) for a plain flap (Equation 8.264a) compared with a flat plate (Equation 8.256a), for all combinations of eight (ten) values of the ratio of flap chord to airfoil chord a/c (flap deflection δ to angle-of-attack α), starting at 5% (0.5) up to 40% (5.0) in steps of 5% (0.5). The effect on lift is significant for flap chord ≥10%–15% of the airfoil

TABLE 8.5

Ratio of Lift Coefficients for Flat Plate with and without Plain (Slotted) Flap

	$\frac{a}{c}$ = 0.05	0.10	0.15	0.20	0.25	0.30	0.35	0.40
$\frac{\delta}{\alpha}$ = 0.5	1.141 (1.170)	1.198 (1.258)	1.240 (1.333)	1.275 (1.402)	1.304 (1.467)	1.330 (1.530)	1.353 (1.590)	1.374 (1.649)
1.0	1.282 (1.314)	1.396 (1.466)	1.481 (1.592)	1.550 (1.705)	1.609 (1.810)	1.661 (1.910)	1.707 (2.006)	1.748 (2.098)
1.5	1.423 (1.459)	1.594 (1.674)	1.721 (1.850)	1.825 (2.007)	1.913 (2.152)	1.991 (2.290)	2.060 (2.420)	2.122 (2.546)
2.0	1.564 (1.603)	1.792 (1.881)	1.961 (2.108)	2.100 (2.310)	2.218 (2.495)	2.321 (2.669)	2.413 (2.835)	2.496 (2.995)
2.5	1.706 (1.749)	1.990 (2.089)	2.201 (2.366)	2.375 (2.612)	2.522 (2.837)	2.652 (3.050)	2.767 (3.251)	2.869 (3.443)
3.0	1.847 (1.893)	2.187 (2.296)	2.441 (2.624)	2.649 (2.914)	2.827 (3.180)	2.982 (3.429)	3.120 (3.666)	3.243 (3.892)
3.5	1.988 (2.039)	2.385 (2.504)	2.682 (2.883)	2.924 (3.216)	3.131 (3.522)	3.313 (3.810)	3.473 (4.081)	3.617 (4.340)
4.0	2.129 (2.182)	2.583 (2.712)	2.922 (3.141)	3.199 (3.519)	3.436 (3.866)	3.643 (4.189)	3.827 (4.497)	3.991 (4.789)
4.5	2.270 (2.327)	2.781 (2.920)	3.162 (3.399)	3.474 (3.821)	3.740 (4.209)	3.973 (4.569)	4.180 (4.911)	4.365 (5.238)
5.0	2.412 (2.472)	2.979 (3.128)	3.403 (3.658)	3.749 (4.124)	4.044 (4.545)	4.304 (4.950)	4.533 (5.326)	4.739 (5.687)

Notes: Lift coefficient of a flat plate airfoil is increased by the addition of a flap at the trailing edge. The effect is larger for (1) larger flap chord as a fraction of the airfoil chord (columns); and (2) larger ratio of the flap deflection angle to the angle-of-attack of the airfoil (lines). The lift coefficient of the plain flap (Figure 8.16b) is further increased by using (Figure 8.16d) a slotted flap (values in brackets). The effect of the size of the slot is less important than reducing or eliminating the region of separated flow that otherwise could exist (Figure 8.19). Values in brackets show the increase in the lift coefficient under the assumption that the flow was separated only in the upper surface up to the flap hinge, and that the change to a slotted flap eliminates the flow separation leading to fully attached flow both on the upper and lower surfaces of the airfoil and flap. *c*: airfoil chord (including flap); α: angle-of-attack of airfoil; *a*: flap chord; δ: flap deflection angle.

chord and deflections with more than double of the angle attack $\delta \geq 2\alpha$; the induced drag increases more rapidly than the lift, that is, like the square (Equation 8.264c). It is assumed in Equation 8.264b and in Table 8.5 that there is no flow separation; otherwise, a significant lift loss occurs (Section 8.6); it is more difficult to preserve fully attached flow for larger deflection of a flap covering a bigger fraction of the chord. The slotted flap (Figure 8.16d) can delay flow separation and is considered next (Subsections 8.7.4 and 8.7.5). Even for double- or triple-slotted flaps, it is difficult to achieve absolute lift coefficients more than about $C_L \backsim 3$; thus, the larger ratios of lift coefficients in Table 8.5 for larger angles of deflection of wide-chord flaps apply only if the initial lift coefficient without flap deflection C_L is small enough so that $C_{L2} \leq 3.0$.

8.7.4 Effect of the Slot on the Lifting Area

The chord of a leading-edge slat (trailing-edge flap) may be up to $a/c \leq 0.15$ ($a/c \leq 0.4$) of the airfoil chord, so the approximation $a^2 << c^2$ is quite (is not so) accurate. The slot width is always small compared with the airfoil chord $b/c \leq 0.05$, so even for a triple-slotted flap $b/c \leq 0.15$, the approximation (Equation 8.265a) can be used in the ratio (Equation 8.259) of lift coefficients leading to Equation 8.265b:

$$b^2 << c^2: \quad \frac{C_{L3}}{C_L} = 1 + \frac{4\delta}{\alpha\pi}\int_0^{\sqrt{a/c}} \left[1-\xi^2 - \frac{ab/c^2}{1-\xi^2}\right]^{1/2} d\xi, \qquad (8.265\text{a and } 8.265\text{b})$$

as follows from

$$\begin{aligned}
\frac{(1-\xi^2+b/c)(1-\xi^2-a/c)}{1-\xi^2-a/c+b/c} &= \left(1+\frac{b/c}{1-\xi^2}\right)\frac{1-\xi^2}{1+\dfrac{b/c}{1-\xi^2-a/c}} \\
&= (1-\xi^2)\left[1+\frac{b}{c}\left(\frac{1}{1-\xi^2}-\frac{1}{1-\xi^2-a/c}\right)\right] = (1-\xi^2)\left[1+\frac{b}{c}\frac{1}{1-\xi^2}\left(1-\frac{1-\xi^2}{1-\xi^2-a/c}\right)\right] \qquad (8.266) \\
&= (1-\xi^2)\left[1+\frac{b}{c}\frac{1}{1-\xi^2}\left(1-\frac{1}{1-\dfrac{a/c}{1-\xi^2}}\right)\right] = (1-\xi^2)\left[1-\frac{ab}{c^2}\frac{1}{(1-\xi^2)^2}\right] = 1-\xi^2-\frac{ab/c^2}{1-\xi^2}
\end{aligned}$$

using the additional approximation (Equation 8.267a). The latter approximation (Equation 8.267a) simplifies Equation 8.265b to Equation 8.267b:

$$\begin{aligned}
ba^2 << c^3: \quad \frac{C_{L4}}{C_L} &= 1+\frac{4\delta}{\alpha\pi}\int_0^{\sqrt{a/c}} \sqrt{1-\xi^2}\left[1-\frac{ab}{c^2}(1-\xi^2)^{-2}\right]^{1/2} \\
&= 1+\frac{4\delta}{\alpha\pi}\int_0^{\sqrt{a/c}}\left[\sqrt{1-\xi^2}-\frac{ab}{2c^2}(1-\xi^2)^{-3/2}\right]d\xi,
\end{aligned} \qquad (8.267\text{a and } 8.267\text{b})$$

where (1) the first term on the r.h.s. of Equation 8.267b coincides with the second term on the r.h.s. of Equation 8.260b that is evaluated by (Equation 8.263):

$$\frac{C_{L4}}{C_L} = 1 + \frac{2\delta}{\alpha\pi} I_0\left(\sqrt{\frac{a}{c}}\right) - \frac{2\delta}{\alpha\pi}\frac{ab}{c^2} I_1\left(\sqrt{\frac{a}{c}}\right); \tag{8.268}$$

and (2) the second term on the r.h.s. of Equation 8.267b ≡ Equation 8.268 involves the integral (Equation 8.269)

$$I_1(\eta) \equiv \int_0^\eta \left(1-\xi^2\right)^{-3/2} d\xi = \frac{\eta}{\sqrt{1-\eta^2}} = \frac{1}{\sqrt{1/\eta^2 - 1}}. \tag{8.269}$$

Its evaluation uses the same change of variable (Equation 8.261a) as for (Equation 8.261b):

$$I_1(\eta) = \int_0^{\arcsin\eta} \sec^2\theta \, d\theta = [\tan\theta]_0^{\arcsin\eta} = \left.\frac{\sin\theta}{\sqrt{1-\sin\theta}}\right|_{\theta=\arcsin\eta} = \frac{\eta}{\sqrt{1-\eta^2}}. \tag{8.270}$$

Substitution of Equations 8.263 and 8.269 in Equation 8.268 leads to

$$\frac{C_{L4}}{C_L} = 1 + \frac{2\delta}{\alpha\pi}\left[\arcsin\left(\sqrt{\frac{a}{c}}\right) + \sqrt{\frac{a}{c}}\sqrt{1-\frac{a}{c}} - \frac{ab}{c^2}\frac{\sqrt{a/c}}{\sqrt{1-a/c}}\right], \tag{8.271}$$

valid with the approximations (Equation 8.265a and 8.267a). Comparing the lift coefficients for the slotted (Equation 8.271) [plain (Equation 8.264b)] flap leads to

$$\frac{C_{L4}-C_L}{C_{L2}-C_L} = \frac{C_{L4}/C_L - 1}{C_{L2}/C_L - 1} = 1 - \frac{ab/c^2}{\sqrt{1-a/c}}\left(\sqrt{\frac{c}{a}}\arcsin\sqrt{\frac{a}{c}} + \sqrt{1-\frac{a}{c}}\right)^{-1}. \tag{8.272}$$

For a small flap, whose chord is small relative to the airfoil chord (Equation 8.273a), the approximations [Equation 8.273b (Equation 8.273c)] hold [Equations 7.172a and 7.172b (Equations 3.144a through 3.144c)]:

$$a^2 \ll c^2: \quad \arcsin\sqrt{\frac{a}{c}} = \sqrt{\frac{a}{c}}\left(1+\frac{a}{6c}\right), \quad \sqrt{1-\frac{a}{c}} = 1 - \frac{a}{2c}. \tag{8.273a–8.273c}$$

Substitution of Equations 8.273b and 8.273c in Equation 8.264b (Equation 8.271) leads to the lift coefficient for a flat plate airfoil with a small flap without (Equations 8.274a through 8.274c) [with (Equations 8.275a through 8.276c)] gap:

$$b = 0, \quad a^2 \ll c^2: \quad \frac{C_{L2}}{C_L} = 1 + \frac{4\delta}{\pi\alpha}\sqrt{\frac{a}{c}}\left(1-\frac{a}{6c}\right), \tag{8.274a–8.274c}$$

$$a^2, b^2 \ll c^2: \quad \frac{C_{L4}}{C_L} = 1 + \frac{4\delta}{\pi\alpha}\sqrt{\frac{a}{c}}\left(1 - \frac{a}{6c} - \frac{ab}{2c^2}\right). \tag{8.275a–8.275c}$$

The conditions (Equations 8.265a and 8.273a ≡ Equations 8.275a and 8.275b) make Equation 8.267a redundant and suggest that the difference between Equations 8.274c and 8.275c is small, as shown by the ratio:

$$a^2 << c^2: \quad \frac{C_{L4} - C_L}{C_{L2} - C_L} = \frac{C_{L4}/C_L - 1}{C_{L2}/C_L - 1} = \frac{1 - \frac{a}{6c} - \frac{ab}{2c^2}}{1 - \frac{a}{6c}} = 1 - \frac{ab}{2c^2} + O\left(\frac{a^2 b}{c^3}\right), \quad (8.276a–8.276c)$$

valid with approximation Equation 8.267a. The result (Equation 8.276b) also follows by substituting Equations 8.273b and 8.273c in Equation 8.272. Thus, *for a flat plate airfoil with a flap of large [small (Equation 8.273a)] chord a, if a slot of width b is used, the lift coefficient changes from Equation 8.264b (Equation 8.274c) to Equation 8.271 (Equation 8.275c). The relative change of lift coefficient [Equation 8.272 (Equation 8.276b) is a reduction due to the loss of lifting area; the reduction equals one-half of the product of the slot width and flap chord normalized to the chord of the airfoil. The approximations [Equations 8.265a and 8.267a (Equations 8.276a and 8.277b)] imply that the term on the r.h.s. of Equation 8.272 (Equation 8.276b) subtracting from unity is the product of two small factors; it specifies the small lift loss relative (Equation 8.264c) to the airfoil with a plain or unslotted flap [Figure 8.16b (Figure 8.16d)] due to the reduction in the chord. This effect is smaller than the lift gain due to the slot delaying flow separation (Section 8.6) as shown next (Subsection 8.7.5).*

8.7.5 Effect of the Slot in Delaying/Preventing Flow Separation

The slot can have an additional effect: it causes an upflow from the high pressure region below the airfoil to the low pressure above, delaying flow separation; the flow separation at the position (Equation 8.277a) of the slot causes a lift loss (Equation 8.238a) if the slot does not exist. Preventing flow separation thus causes the inverse (Equation 8.277b) lift gain:

$$0 < s < 1: \quad \frac{C_{L5}}{C_{L4}} = \frac{C_L(1)}{C_L(s)} = \left(\frac{2}{1 + \sqrt{s}}\right)^2. \quad (8.277a \text{ and } 8.277b)$$

In particular, if flow separation would occur at the slot position (Equation 8.278a) in the absence of the slot, and is prevented by the presence of the slot, there is a lift gain (Equation 8.278b):

$$s = 1 - \frac{a}{c}: \quad \frac{C_{L5}}{C_{L4}} = \left(\frac{2}{1 + \sqrt{1 - a/c}}\right)^2 = \left[\frac{2}{2 - a/(2c) + O(a^2/c^2)}\right]^2$$
$$= \left[1 - \frac{a}{4c} + O\left(\frac{a^2}{c^2}\right)\right]^{-2} = 1 + \frac{a}{2c} + O\left(\frac{a^2}{c^2}\right). \quad (8.278a \text{ and } 8.278b)$$

The lift gain due to the prevention of flow separation is a first-order correction (Equation 8.278b), whereas the lift loss due to slot reducing the airfoil chord and lifting area is a second-order correction (Equation 8.276b). From Equations 8.271 and 8.277b (Equation 8.275c) follows *the lift coefficient of an airfoil with a large (Equation 8.279e) [small (Equation 8.279f)] slotted flap preventing the flow separation that would otherwise occur on the upper surface:*

$$\alpha^2 << 1 >> (\alpha+\delta)^2, \quad \frac{b^2}{c^2} << 1 >> \frac{a^2 b}{c^3}; \tag{8.279a–8.279d}$$

$$\begin{aligned} C_{L5} = C_L \frac{C_{L4}}{C_L}\frac{C_{L5}}{C_{L4}} &= 2\pi\alpha\left(\frac{2}{1+\sqrt{s}}\right)^2 \left\{1+\frac{2\delta}{\pi\alpha}\left[\arcsin\left(\sqrt{\frac{a}{c}}\right)+\sqrt{\frac{a}{c}}\sqrt{1-\frac{a}{c}}-\frac{ab/c^2}{\sqrt{c/a-1}}\right]\right\}, \\ &= 2\pi\alpha\left(\frac{2}{1+\sqrt{s}}\right)^2 1+\frac{4\delta}{\pi\alpha}\sqrt{\frac{a}{c}}\left[1-\frac{a}{6c}-\frac{ab}{2c^2}+O\left(\frac{a^2}{c^2}\right)\right], \end{aligned} \tag{8.279e and 8.279f}$$

where (1 and 2) the airfoil (flap) chord is c(a); (3 and 4) the slot width (fraction of the chord of the upper surface with separated flow) is b(s); and (5 and 6) the angle-of-attack of the airfoil (flap deflection angle) is α(δ). In Equation 8.279e, the following were assumed: (1 and 2) small angle of attack (Equation 8.279a) in Equations 8.250a and 8.252b when (Equation 8.251b) the flap deflection is added (Equation 8.279b); (3 and 4) small slot width relative to the chord (Equation 8.265a ≡ Equation 8.279c) and small cross product (Equation 8.267a ≡ Equation 8.279d) involving also the ratio of flap a to airfoil c. In Equation 8.279f, small flap chord (Equation 8.273a) in addition was assumed. In the particular case (Equation 8.278a ≡ Equation 8.280a) that the slot prevents flow separation that would otherwise occur at the slot position, the lift coefficient [Equation 8.279e (Equation 8.279f)] for a large (small) flap simplifies on replacing the factor in curved brackets on the r.h.s. by Equation 8.278b, leading to Equation 8.280b (Equation 8.280c):

$$\begin{aligned} s = 1-\frac{a}{c}: \quad C_{L5} &= 2\pi\alpha\left(1+\frac{a}{2c}\right)\left[1+\frac{2\delta}{\pi\alpha}\left(\arcsin\left(\sqrt{\frac{a}{c}}\right)+\sqrt{\frac{a}{c}}\sqrt{1-\frac{a}{c}}-\frac{ab/c^2}{\sqrt{c/a-1}}\right)\right], \\ &= 2\alpha\pi\left(1+\frac{a}{2c}\right)1+\frac{4\delta}{\alpha\pi}\sqrt{\frac{a}{c}}\left[1-\frac{a}{6c}-\frac{ab}{2c^2}+O\left(\frac{a^2}{c^2}\right)\right]. \end{aligned} \tag{8.280a–8.280c}$$

This is equivalent (Equation 8.226a) to the relative lift change (Equation 8.281b):

$$a^2 << c^2: \quad \frac{C_{L5}-C_L}{C_L} = \frac{a}{2c}+\frac{4\delta}{\alpha\pi}\sqrt{\frac{a}{c}}\left[1+\frac{a}{3c}\left(1-\frac{3b}{2c}\right)\right], \tag{8.281a and 8.281b}$$

where the following was used:

$$\frac{C_{L5} - C_L}{C_L} = \frac{C_{L5}}{2\pi\alpha} - 1 = \left(1 + \frac{a}{2c}\right)\left[1 + \frac{4\delta}{\alpha\pi}\sqrt{\frac{a}{c}}\left(1 - \frac{a}{6c} - \frac{ab}{2c^2}\right)\right] - 1$$

$$= \frac{a}{2c} + \frac{4\delta}{\alpha\pi}\sqrt{\frac{a}{c}}\left(1 - \frac{a}{6c} - \frac{ab}{2c^2}\right) + \frac{a}{2c}\frac{4\delta}{\alpha\pi}\sqrt{\frac{a}{c}} + O\left(\frac{a^2}{c^2}\right) \tag{8.282}$$

$$= \frac{a}{2c} + \frac{4\delta}{\alpha\pi}\sqrt{\frac{a}{c}}\left(1 + \frac{a}{3c} - \frac{ab}{2c^2}\right),$$

valid with the approximation (Equation 8.281a).

8.7.6 Comparison of Plain, Flapped, and Slotted Airfoils

Starting with the plain airfoil (Figure 8.19a), (1) adding a slotless flap (Figure 8.19b) increases the relative lift more for a flap occupying a larger fraction of the chord; (2) the lift also increases with the ratio of flap deflection to the angle of attack, up to the limit flow separation (Figure 8.19c); (3) a higher order effect is the product of (1) and (2); and (4) the slot width appears as an even higher order effect multiplying (3) by the slot width divided by the chord. The ratio of lift coefficients of a flat plate with a plain (a slotted) flap without (with) brackets in Table 8.5 shows the improvement in lift of the slotted relative to the plain flap, which increases with the ratio of flap to airfoil chord and flap deflection to airfoil angle-of-attack; the small effect of the slot width can be omitted to avoid a third parameter with limited relevance. The lift coefficient for the flat plate airfoil with a plain flap was calculated from Equation 8.264b; the increase in the lift of the airfoil with flap due to the slot was calculated by Equation 8.280b in Table 8.5 assuming (Figure 8.19d) that the slot prevented flow separation from occurring at its location (Equation 8.278a). Thus, the effect of the slot in increasing the lift is more significant for a larger flap chord relative to the airfoil chord, since it prevents flow separation over a larger region; the presence of the slot also increases the effectiveness of the flap in producing higher lift for larger deflection relative to the angle-of-attack of the airfoil. The slot can prevent flow separation further upstream; for example, if it prevents flow separation at a larger distance upstream, the increase in lift is specified by Table 8.4. Thus, the effects of the slot on lift are to create an upflow through the slot and delay flow separation, and allow a larger angle-of-attack for the airfoil and a larger angle of deflection of the slotted relative to the plain flap. The slot delays flow separation by "energizing" the boundary layer, that is, the passage between the lower (upper) side of the airfoil at high (low) pressure causes an upflow through the slot and accelerates the boundary layer delaying flow separation. Thus, (1) the flap increases lift by deflecting the flow downward, and (2) the slot contributes to the lift by creating an upflow through the gap that delays flow separation until larger flap deflections and/or larger angle-of-attack. In the preceding analysis, it was assumed that the airfoil chord is unchanged by the deflection of the plain or slotted flap. If the flap moves back as it is deployed, there is also an additional lift due to the increased chord. A further lift gain may be obtained using double or triple slotted airfoils, at the expense of greater weight and complexity of the flap (or slat) extension and retraction mechanism.

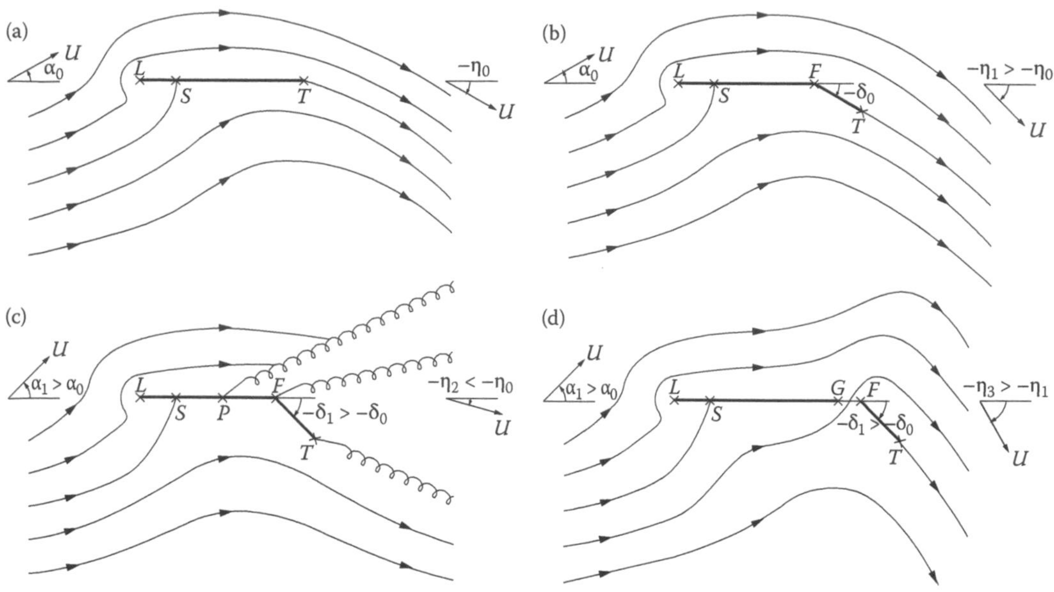

FIGURE 8.19
Flat plate airfoil in a free stream of velocity U and angle-of-attack α_0 has (a) stagnation point S at the underside near the leading edge L; at the trailing edge T, if the Kutta condition is applied, the stream velocity is turned through the downwash angle $-\eta_0$. Adding a trailing-edge flap F (b) with deflection δ_0 gives a larger downwash angle $-\eta_1 > -\eta_0$ and increases the lift. If the angle-of-attack $\alpha_1 > \alpha_0$ or flap deflection $\delta_1 > \delta_0$ is increased too much (c), there is flow separation at point P on the upper surface and a reduction in downwash angle $-\eta_2 < -\eta_0$ and lift loss. A slotted flap energizes the flow by causing an upstream through the gap GF (d), avoiding flow separation and allowing a larger downwash angle $-\eta_3 > -\eta_1$ and lift. In the region of separated flow, there may exist a recirculation bubble, and transition to turbulence may occur; since the potential flow does not apply in these conditions, the region of separated flow is indicated in (c) by the same curly line as in Figure 8.13a. The angles-of-attack and downwash angles in (a) through (c) are exaggerated to make the physical effects more visible; the relations between the angles remain valid, even if for most aircraft the actual values are smaller. The main purpose is to explain that the increase in lift from (a) to (b) to (d) and also the flow separation (c) are associated with an increase in drag. Reducing drag is important to minimize free consumption and emissions during long duration cruise flight when the flap is retracted (a). For short periods of flight, like takeoff and landing, the increase in the lift due to the flap (b) and slotted flap (d) is more important that allows (1) lower stalling speed, that is, minimum velocity for straight and level steady flight; (2) lower takeoff speed; (3) lower speed on approach to land; and (4) the use of shorter runways at airports.

8.8 Flow Past Tandem Airfoils and Cascades

The Sedov theorem (Equation 8.249) specifies the complex conjugate velocity for a single airfoil or for multiple airfoils. The single airfoil is considered in three cases: (1) circulation without mean flow (Subsections 8.8.1 through 8.8.4); (2) mean flow without circulation (Subsections 8.8.5 and 8.8.6); and (3) mean flow with circulation specified by the Kutta condition (Subsections 8.8.7 and 8.8.8). The cases of multiple airfoils considered are (4) the airfoil with slotted flap (Subsections 8.7.4 through 8.7.6); (5 and 6) tandem airfoils with circulation in the absence of mean flow (Subsection 8.8.9) or in the presence of mean flow applying the Kutta condition (Subsection 8.8.10); and (7) a cascade of parallel airfoils

(Subsections 8.8.11 through 8.8.14). A comparison is made between the Sedov method (Joukowski transformation) that is [Subsection 8.8.1 (Subsection 8.8.3)] a particular case of the Cauchy theorem (Subsections I.15.3 and I.37.8) [one (Subsection 8.8.2) of the four Schwartz–Christoffell mappings (Subsections I.33.6 through I.33.9)]. The comparison of the Sedov method (Joukowski conformal mapping) concerns a single airfoil: (1) with circulation and without mean flow [Subsection 8.8.1 (Subsection 8.8.4)]; (2) without circulation in a uniform stream with angle-of-attack [Subsection 8.8.5 (Subsection 8.8.6)]; and (3) in case 2 with the circulation specified by the Kutta condition of finite velocity at the trailing edge Subsection 8.8.7 (Subsection 8.8.9)]. The Sedov method applies to multiply connected region, such as (1) two tandem airfoils with circulation (Subsections 8.8.9 and 8.8.10) and (2) a cascade consisting of an infinity of parallel airfoils (Subsections 8.8.11 and 8.8.12). The Kutta condition imposing a finite velocity at the edges specifies the circulation and lift; if the Kutta condition is not applied at an edge, the inverse square root singularity of the velocity leads to a suction force.

8.8.1 Single Airfoil with Circulation

In the absence of mean flow, the perturbation velocity cancels the integral around (Equation 8.283a) the airfoil L_0 in the Sedov formula 8.249, and the remaining integral around L_∞ implies that there is a flow only if $\tilde{v}^* \sim O(1/\zeta)$, which corresponds (Equation 8.283b) to an airfoil with circulation and without mean flow:

$$-\frac{c}{2} \le x \le \frac{c}{2}: \quad v^*(\zeta) = -\frac{i\Gamma}{2\pi\zeta}; \quad g_3(\zeta) = \left(\zeta^2 - \frac{c^2}{4}\right)^{-1/2}. \tag{8.283a–8.283c}$$

The complex conjugate velocity must have (Equation 8.210f) inverse square root singularities at the sharp edges (Section I.14.9.2) of the airfoil at $\zeta = \pm c/2$; these conditions are enforced by choosing the **airfoil function** (Equation 8.283c). Substitution of Equations 8.283a and 8.283b in the first term on the r.h.s. of Equation 8.249 leads to

$$\begin{aligned} v^*(z) &= \frac{\left(z^2 - c^2/4\right)^{-1/2}}{2\pi i} \lim_{R\to\infty} \int_{|\zeta|=R} \frac{\left(\zeta^2 - c^2/4\right)^{1/2}}{\zeta - z} \left(-\frac{i\Gamma}{2\pi\zeta}\right) d\zeta \\ &= \frac{\left(z^2 - c^2/4\right)^{-1/2}}{2\pi i} \lim_{R\to\infty} \int_{|\zeta|=R} \left(1 - \frac{c^2}{4\zeta^2}\right)^{1/2} \left(-\frac{i\Gamma}{2\pi\zeta}\right) d\zeta \\ &= -\frac{i\Gamma}{2\pi} \frac{\left(z^2 - c^2/4\right)^{-1/2}}{2\pi i} \int^{(0+)} \frac{d\zeta}{\zeta} = -\frac{i\Gamma}{2\pi} \frac{1}{\sqrt{z^2 - c^2/4}}. \end{aligned} \tag{8.284}$$

In Equation 8.284, the property Equations 2.230a and 2.230b ≡ Equations I.28.24a through I.28.24c was used. Thus, *Equation 8.284 is the complex conjugate velocity of a potential flow with circulation Γ around a flat plate airfoil (Equation 8.283a) chord c.* The same result Equation 8.284 ≡ Equation I.36.68b can be obtained using a Schwartz–Christoffell or Joukowski conformal mapping, instead of Sedov's method. A succinct account of the latter approach follows (Subsection 8.8.3) using the Schwartz–Christoffel exterior disk transformation (Subsection 8.8.2).

8.8.2 Schwartz–Christoffell Exterior Disk Transformation

The Schwartz–Christoffell polygonal interior (exterior disk) transformation [Section I.33.6 (Section I.33.9)] is specified by [Equation 8.144b ≡ Equation I.33.25 (Equation 8.285b ≡ Equation I.33.41)]

$$0 \le \varphi_1 < \varphi_2 < \ldots < \varphi_n: \quad \frac{d\zeta}{dz} = A \prod_{n=1}^{N} \left(1 - \frac{z_n}{z}\right)^{\gamma_n/\pi}, \qquad \text{(8.285a and 8.285b)}$$

where (1) A is a constant allowing rotations and homotheties; and (2) the critical points z_n are ordered along [Figure 8.20a (Figure 8.20c)] the real axis (Equation 8.144a) [along a circle (Equation 8.286a) of radius a]:

$$z_n = ae^{i\varphi_n}; \quad z = ae^{i\varphi}: \quad \left(1 - \frac{z_n}{z}\right)^{\gamma_n/\pi} = \left[1 - e^{i(\varphi_n - \varphi)}\right]^{\gamma_n/\pi}. \qquad \text{(8.286a–8.286c)}$$

As z travels along the real axis (circle of radius a) from $-\infty$ to $+\infty$ (in the positive or counterclockwise direction starting on the real axis), if it crosses a critical point, the corresponding factor in Equation 8.144b (Equation 8.285b) changes sign leading [Equation 8.145 (Equation 8.287a)] to (Figure 8.20b) an external angle γ_n:

$$(-)^{\gamma_n/\pi} = \left(e^{-i\pi}\right)^{\gamma_n/\pi} = e^{-i\gamma_n}; \quad \sum_{n=0}^{N} \gamma_n = 2\pi. \qquad \text{(8.287a and 8.287b)}$$

If the external angles add to 2π, they form a closed polygon (Equation 8.287b). Thus, *the Schwartz–Christoffel interior polygonal (Equation 8.144b) [disk exterior (Equation 8.285b)] transformation maps [Figure 8.20a (Figure 8.20c)] the real axis (circle of radius a with center at the origin) to a polygon (Figure 8.20b) with vertices ζ_n at the critical points $\zeta_n = \zeta(z_n)$ and external angles γ_n. The upper-half complex z-plane (exterior of the circle) is mapped to the interior (exterior) of the polygon; the lower-half complex z-plane (interior of the circle) is mapped to the interior (exterior) of the image polygon with external angles $-\gamma_n$. The exterior of both the original and image polygons is not mapped into (is mapped into twice). The polygons are closed if the sum of the external angles (Equation 8.287b) is 2π.*

The Schwartz–Christoffell exterior disk transformation (Equation 8.144b) involves powers of linear functions of z and thus (1) is single-valued for integral powers and (2) can be made single-valued for nonintegral rational, irrational, or complex powers using the principal branch (Section 3.7). The Schwartz–Christoffell exterior disk transformation (Equation 8.285b) involves powers of linear functions of $1/z$ and scales asymptotically for large z as

$$\frac{d\zeta}{dz} = A\left[1 - \frac{1}{\pi z}\left(\sum_{n=1}^{N} \gamma_n z_n\right) + O\left(\frac{1}{z^2}\right)\right]. \qquad \text{(8.288)}$$

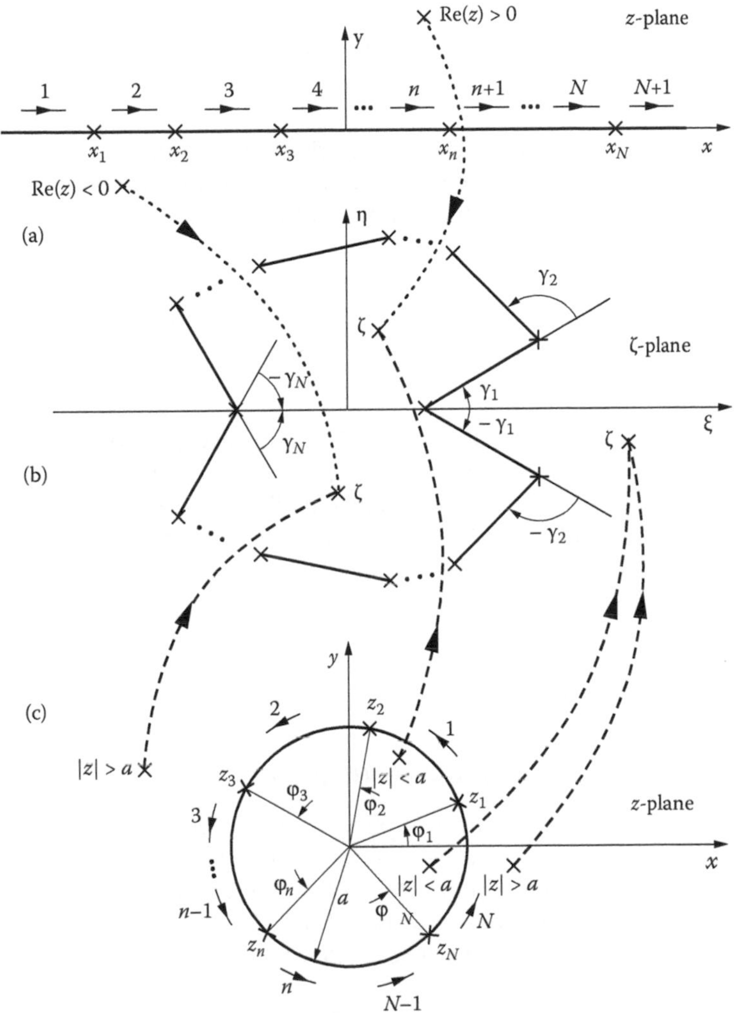

FIGURE 8.20
Polygon is specified in the ζ-plane (b) by the position of its vertices ζ_n and the external angles γ_n; the polygon is closed iff the external angles add to 2π. The Schwartz–Christoffel interior polygonal (exterior disk) transformation transforms the real axis (a) [circle with center at the origin (c)] to the polygon with the critical points mapped to the vertices. The upper z half-plane (exterior of the disk) is mapped to the interior (exterior) of the polygon. The lower z half-plane (interior of the disk) is mapped to the polygon that has opposite angles $-\gamma_n$ at the vertices. The points exterior to both the original and image polygon are not mapped into (are mapped into twice).

The primitive of Equation 8.288 is

$$\zeta(z) = B + Az - \frac{1}{\pi}\left(\sum_{n=1}^{N} \gamma_n z_n\right)\log z + O\left(\frac{1}{z}\right); \tag{8.289}$$

the logarithmic term shows (Section 3.5) that the Schwartz–Christoffell exterior disk transformation is many-valued, unless the coefficient vanishes (Equation 8.290a):

$$\sum_{n=1}^{N} \gamma_n z_n = 0: \quad \zeta(z) = B + Az + O\left(\frac{1}{z}\right). \tag{8.290a and 8.290b}$$

The condition of unicity (Equation 8.290a) simplifies the asymptotic form of the Schwartz–Christoffel exterior disk transformation (Equation 8.289) to a linear function (Equation 8.290b), so that (1) there is no deformation at infinity (Equation 8.291a) if the constant is chosen to be unity:

$$1 = \sum_{n \to \infty}^{N} \frac{d\zeta}{dz} = A; \quad B = 0: \quad \zeta = z + O\left(\frac{1}{z^2}\right); \tag{8.291a–8.291c}$$

and (2) eliminating the translation (Equation 8.291b) preserves the origin (Equation 8.291c). Thus, *the Schwartz–Christoffel exterior disk transformation (Equations 8.285a and 8.285b) is unique iff the condition 8.290a is met leading to Equation 8.290b; there is no deformation at infinity (the origin is preserved) if Equation 8.291a (Equation 8.291b) is also met leading to Equation 8.291c. The Schwartz–Christoffell interior polygonal (Equation 8.144b) [exterior disk (Equation 8.285b)] transformation can have only one or more critical points (Equation 8.144a) [must have at least two critical points (Equation 8.285a) to meet the unicity condition (Equation 8.290a)]. The simplest case of Schwartz–Christoffell interior polygon (exterior disk) transformation with one (two) critical point corresponds to the corner flow (Subsection 8.6.2) [flow past a flat plate airfoil (Subsection 8.8.3)].*

8.8.3 Mapping of a Circle into a Flat Plate (Joukowski 1910)

The z-plane is now chosen as the physical plane and the ζ-plane as the plane of the circle (Equation 8.291a) so that the simplest Schwartz–Christoffell exterior disk transformation (Equation 8.285b) with two critical points (z_1,z_2) and angles (γ_1,γ_2) must satisfy (1) the unicity condition (Equation 8.290a) leading to Equation 8.292a and (2) the condition that the polygon is closed (Equation 8.287b) leading to Equation 8.292b:

$$\gamma_1\zeta_1 + \gamma_2\zeta_2 = 0, \qquad \gamma_1 + \gamma_1 = 2\pi; \qquad \gamma_1 = \gamma_1 = \pi, \qquad -\zeta_1 = \zeta_2 = a, \tag{8.292a–8.292d}$$

in the case of a flat plate (Figure 8.21a), the two angles are equal (Equation 8.292c) and the critical points must lie (Equation 8.292d) diametrically opposite (Figure 8.21b). The real axis of the ζ-plane may be chosen through the critical points, and the Schwartz–Christoffell exterior disk transformation (Equation 8.285b) with two critical points (Equation 8.292d) and angles (Equation 8.292c) leads to

$$\frac{dz}{d\zeta} = A\left(1 - \frac{\zeta_1}{\zeta}\right)^{\gamma_1/\pi}\left(1 - \frac{\zeta_2}{\zeta}\right)^{\gamma_1/\pi} = A\left(1 - \frac{a}{\zeta}\right)\left(1 + \frac{a}{\zeta}\right) = A\left(1 - \frac{a^2}{\zeta^2}\right). \tag{8.293}$$

The integration of Equation 8.293 leads to Equation 8.294a:

$$z = A\left(\zeta + \frac{a^2}{\zeta}\right) + B; \quad B = 0, \quad 1 = \lim_{\zeta \to \infty} \frac{z}{\zeta} = A: \quad z = \zeta + \frac{a^2}{\zeta}. \tag{8.294a–8.294d}$$

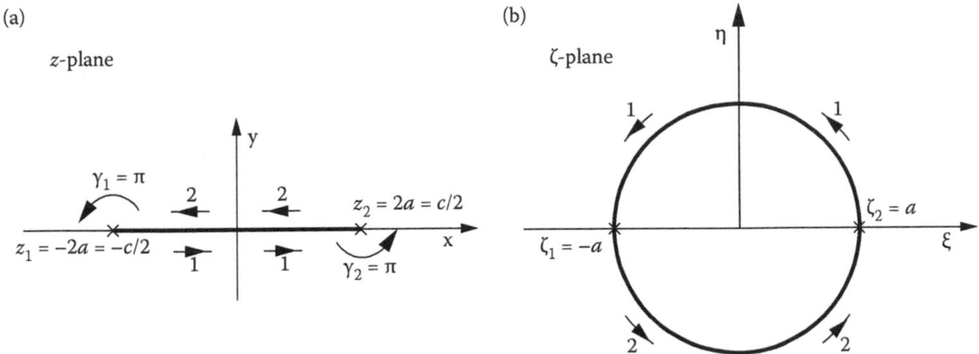

FIGURE 8.21
Flat plate (Figure 8.20a) is the simplest conformal mapping of the type Schwartz–Christoffell exterior disk transformation, since it involves the minimum number of critical points to ensure unicity; the location of the two critical points on the real axis symmetrically to the imaginary axis (Figure 8.20b) leads to the Joukowski transformation of the circle onto a flat plate.

The constant of integration corresponding to a translation can be omitted (Equation 8.291b); the constant A is specified by the condition (Equation 8.294c ≡ Equation 8.291a) that there is no deformation at infinity, leading to Equation 8.294d. It can be confirmed that when ζ lies on a circle of radius a in Equation 8.295a, then z lies on a segment of the real axis (Equation 8.295b):

$$\zeta = ae^{i\varphi}: \quad -\frac{c}{2} \equiv -2a \le z = a\left(e^{i\varphi} + e^{-i\varphi}\right) = 2a\cos\varphi \le 2a \equiv +\frac{c}{2}, \qquad \text{(8.295a and 8.295b)}$$

corresponding to airfoil with chord $c = 4a$. The conformal mapping (Equation 8.294d ≡ Equation I.34.32a) is the **Joukowski (1910) transformation** *(Equation 8.296b) for an airfoil with chord (Equation 8.296a) and may be inverted by Equation 8.296c:*

$$c = 4a: \quad z = \zeta + \frac{c^2}{16\zeta}, \quad \zeta^2 - \zeta z + \frac{c^2}{16} = 0, \quad 2\zeta = z + \sqrt{z^2 - c^2/4}, \qquad \text{(8.296a–8.296d)}$$

where the plus sign was used in Equation 8.296d to meet the condition (Equation 8.294c) of no deformation at infinity.

8.8.4 Flat Plate with Circulation and Suction Forces

The complex conjugate velocity (Equation 8.283b) due to a vortex circulation Γ in the ζ- plane corresponds to the complex potential (Equation 8.297a) in the z-plane:

$$f = -\frac{i\Gamma}{2\pi}\log\zeta = -\frac{i\Gamma}{2\pi}\log\left(\frac{z + \sqrt{z^2 - c^2/4}}{2}\right), \quad v^* = \frac{df}{dz} = -\frac{i\Gamma}{2\pi}\frac{1}{\sqrt{z^2 - c^2/4}} \qquad \text{(8.297a and 8.297b)}$$

and leads (Equation 7.148) to the velocity (Equation 8.297b ≡ Equation 8.284) in the z-plane of airfoil. The velocity (Equation 8.297b) has (Equation 8.210f) an inverse square root singularity (Subsection I.14.9.3) at both the leading (Equation 8.297a) [trailing (Equation 8.297b)] edges of the airfoil:

$$v^*(z) = \frac{C_\pm}{\sqrt{z \mp c/2}} + O(1), \quad C_\pm = -i\frac{\Gamma}{2\pi}\frac{1}{\sqrt{\pm c}}, \quad F_\pm = -\rho\pi(C_\pm)^2 = \pm\frac{\rho\Gamma^2}{4\pi c}, \qquad \text{(8.298a–8.298d)}$$

leading (Equation 8.298c) to **suction forces** *(Equation 8.298d ≡ Equation I.34.16b) that are equal and opposite at the two edges. The suction force (Equation 8.298c ≡ Equation 8.299c) is associated with the inverse square root singularity of the velocity (Equation 8.210f ≡ Equation 8.299a) by the Blasius theorem (Equation 2.205b ≡ Equation 8.298b) on forces:*

$$v^* \sim \frac{C}{\sqrt{z}}, \quad F^* = i\frac{\rho}{2}\int^{(0+)} v^{*2}\,dz = \frac{i}{2}\rho C^2\int^{(0+)}\frac{dz}{z} = -\rho\pi C^2; \qquad \text{(8.299a–8.299c)}$$

in Equation 8.299a, the singularity was placed at the origin and it is enclosed (Equation 8.299b) by the loop integral (Equation 2.230b ≡ Equation I.28.24b). This approach provides an alternative proof of Equation 8.284 and its implications (Equation 8.298a–8.298d) using a conformal mapping instead of the Sedov method.

8.8.5 Airfoil in a Uniform Stream

If there is no circulation around the airfoil, the perturbation potential is due to a dipole (Section I.28.6), and the complex conjugate velocity scales as Equation 8.300a implying that there is no contribution from the integral at infinity, that is, the first on the r.h.s. of Equation 8.249; in the presence of a free stream of velocity U and angle-attack α, the vertical velocity (Equation 8.243a) with the same airfoil function (Equation 8.283c) specifies the complex conjugate velocity (Equation 8.300b):

$$\begin{aligned} v^* \sim O\left(\frac{1}{\zeta^2}\right)&: \quad -\frac{2\pi}{U\sin\alpha}\sqrt{z^2 - c^2/4}\,\tilde{v}^*(z) = \int_{L_0}(\zeta - z)^{-1}\left(\zeta^2 - c^2/4\right)^{1/2} d\zeta \\ &= \int_{L_\infty}(\zeta - z)^{-1}\left(\zeta^2 - c^2/4\right)^{1/2} d\zeta - 2\pi i \lim_{\zeta\to z}\sqrt{\zeta^2 - c^2/4} \\ &= \lim_{R\to\infty}\int_{|\zeta|=R}\left[1 + \frac{z}{\zeta} + O\left(\frac{1}{\zeta^2}\right)\right] d\zeta - 2\pi i\sqrt{z^2 - c^2/4} \\ &= 2\pi i\left(z - \sqrt{z^2 - c^2/4}\right), \end{aligned} \qquad \text{(8.300a and 8.300b)}$$

where (1) in Equation 8.300b, the loop was stretched from the airfoil L_0 to infinity L_∞, crossing the pole at $\zeta = z$, so that (Equations I.15.24a and I.15.24b) the residue times $2\pi i$ must be subtracted leading to the second term on the r.h.s.; and (2) in the first term on the r.h.s., Equations 2.230a and 2.230b ≡ Equations I.28.24a through I.28.24c was used. The perturbation velocity (Equation 8.300b) must be added to the free stream velocity to specify (Equation 8.204b) the total velocity (Equation 8.301):

$$v^*(z) = Ue^{-i\alpha} + \tilde{v}^*(z) = U\cos\alpha - iU\sin\alpha\left(1 + \frac{z - \sqrt{z^2 - c^2/4}}{\sqrt{z^2 - c^2/4}}\right)$$
$$= U\left\{\cos\alpha - \frac{i\,z\sin\alpha}{\sqrt{z^2 - c^2/4}}\right\}. \tag{8.301}$$

Thus, *Equation 8.301 is the complex conjugate velocity of an airfoil of chord c at angle-of-attack α in a free stream of velocity U.* The same result is obtained next via the Joukowski (Subsection 8.8.6) conformal mapping.

8.8.6 Comparison of the Sedov Method and Joukowski Mapping

A flat plate that does not disturb the uniform flow of velocity U in the z-plane without angle-of-attack (Equation 8.302a) parallel to the x-axis has (Equation 8.296b) complex potential (Equation 8.302b):

$$\alpha = 0:\quad f = Uz = U\left(\zeta + \frac{c^2}{16\zeta}\right);\quad f - U\zeta = \frac{Uc^2}{16\zeta} = -\frac{P_1}{2\pi\zeta},\quad P_1 = -\frac{\pi}{8}c^2U, \tag{8.302a–8.302d}$$

and thus the potential flow past a cylinder of radius c in the ζ-plane is specified by Equation 8.302b that consists (Equation 8.302c) of a uniform stream plus a dipole with moment (Equation 8.302d); the velocity of the dipole (Equation 8.302c) decays like Equation 8.300a. The dipole moment (Equation 8.302d) for the flat plate airfoil of chord c agrees with the dipole moment (Equation 8.6c) of a cylinder of radius a bearing in mind Equation 8.296a The Joukowski transformation (Equation 8.296d) has the property Equation 8.303a implying Equation 8.303b:

$$\left[z + \sqrt{z^2 - c^2/4}\right]\left[z - \sqrt{z^2 - c^2/4}\right] = z^2 - \left(z^2 - c^2/4\right) = c^2/4, \tag{8.303a}$$

$$\frac{c^2/4}{z \pm \sqrt{z^2 - c^2/4}} = z \mp \sqrt{z^2 - c^2/4}. \tag{8.303b}$$

The potential flow with free stream velocity U and angle-of-attack α past a cylinder of radius c/4 has complex potential (Equation 8.304b) obtained by a rotation (Equation 8.304a) from Equation 8.302b:

$$\zeta \to \zeta e^{-i\alpha}:\quad f = U\left(\zeta e^{-i\alpha} + \frac{c^2}{16\zeta}e^{i\alpha}\right) = \frac{U}{2}\left[e^{-i\alpha}\left(z + \sqrt{z^2 - c^2/4}\right) + e^{i\alpha}\frac{c^2}{4}\frac{e^{i\alpha}}{z + \sqrt{z^2 - c^2/4}}\right]; \tag{8.304a–8.304c}$$

the Joukowski transformation (Equation 8.296d) maps (Equations 8.303b and 8.304b ≡ Equation 8.305a) to the potential flow (Equation 8.305b) with angle-of-attack α past an airfoil of chord c:

$$f = \frac{U}{2}\left[e^{-i\alpha}\left(z + \sqrt{z^2 - c^2/4}\right) + e^{i\alpha}\left(z - \sqrt{z^2 - c^2/4}\right)\right]$$
$$= U\left(z\cos\alpha - i\sin\alpha\sqrt{z^2 - c^2/4}\right); \tag{8.305a and 8.305b}$$

the corresponding complex conjugate velocity:

$$\frac{df}{dz} = U\left(\cos\alpha - \frac{iz\sin\alpha}{\sqrt{z^2 - c^2/4}}\right) \equiv v^*(z) \tag{8.306}$$

coincides with Equation 8.306 ≡ Equation 8.301. The velocity Equation 8.306 has inverse square root singularities at the edges of the airfoil (Equation 8.307a) leading (Equation 8.307b)

$$v^*(z) = \frac{C^{\pm}}{\sqrt{z \mp c/2}} + O(1), \quad C_{\pm} = -iU\sin\alpha\frac{\sqrt{\pm c}}{2} \qquad \text{(8.307a and 8.307b)}$$

to opposite (Equation 8.308) suction forces (Equation 8.299c):

$$F_{\pm} = -\rho\pi(C^{\pm})^2 = \pm\frac{1}{4}\rho\pi U^2 c\sin^2\alpha. \tag{8.308}$$

The suction forces coincide (Equation 8.309a) for the same airfoil chord in a free stream of velocity U and angle-of-attack in the pressure of circulation if (Equation 8.308 ≡ Equation 8.298d) the circulation is given by Equation 8.309b:

$$F_{\pm} = F^{\pm}: \qquad \Gamma = -\pi cU\sin\alpha. \qquad \text{(8.309a and 8.309b)}$$

The minus sign was chosen in Equation 8.309b in agreement with Equation 8.232b ≡ Equation I.34.17). The latter is the circulation specified next by the Kutta condition (Subsection 8.8.7).

8.8.7 Kutta (1902a, 1902b) Condition for an Airfoil in a Stream

The Kutta condition eliminates the inverse square root singularity (Equation 8.210f) at the trailing edge of the airfoil $z = c$ by choosing the **circulation function** Equation 8.310a:

$$g_4(\zeta) = \sqrt{\frac{\zeta - c/2}{\zeta + c/2}}; \quad v^*(\zeta) \sim O\left(\frac{1}{\zeta}\right). \qquad \text{(8.310a and 8.310b)}$$

This leads to the presence of circulation (Equation 8.310b ≡ Equation 8.283b); there is no contribution from infinity L_∞ due to higher order terms (Equation 8.300a), and the perturbation velocity is specified by the second term in Equation 8.249 with Equation 8.243a, leading to

$$\begin{aligned} -\frac{2\pi}{U\sin\alpha}\sqrt{\frac{2z+c}{2z-c}}\,\tilde{v}^*(z) &= \int_{L_0}\sqrt{\frac{2\zeta+c}{2\zeta-c}}\frac{d\zeta}{\zeta - z} \\ &= \int_{L_\infty}\left[1 + \frac{c}{2\zeta} + O\left(\frac{1}{\zeta^2}\right)\right]\frac{d\zeta}{\zeta} - 2\pi i\sqrt{\frac{2z+c}{2z-c}} \\ &= 2\pi i\left(1 - \sqrt{\frac{2z+c}{2z-c}}\right), \end{aligned} \tag{8.311}$$

where (1) the loop was extended from the airfoil to infinity, crossing a pole, so that $2\pi i$ times the residue is subtracted (Equations I.15.24a and I.15.24b); (2) the asymptotic approximation of the inverse circulation function (Equation 8.310a) is

$$\frac{1}{g_2(\zeta)} = \sqrt{\frac{2\zeta+c}{2\zeta-c}} = \sqrt{\frac{1+c/(2\zeta)}{1-c/(2\zeta)}} = \left\{\left(1+\frac{c}{2\zeta}\right)\left[1+\frac{c}{2\zeta}+O\left(\frac{c^2}{2\zeta}\right)\right]\right\}^{1/2}$$
$$= \left[1+\frac{c}{\zeta}+O\left(\frac{c^2}{\zeta^2}\right)\right]^{1/2} = 1+\frac{c}{2\zeta}+O\left(\frac{c^2}{\zeta^2}\right); \tag{8.312}$$

and (3) only the first term on the r.h.s. of Equation 8.295 contributes (Equations 2.230a and 2.230b ≡ Equations I.28.24a through I.28.24c) to the integral in Equation 8.311 with residue unity. The perturbation (Equation 8.311) and free stream velocities add in the total velocity (Equation 8.204b) specified by

$$v^*(z) = U\,e^{-i\alpha} - iU\sin\alpha\left(\sqrt{\frac{2z-c}{2z+c}}-1\right) = U\left(\cos\alpha - i\,\sin\alpha\sqrt{\frac{2z-c}{2z+c}}\right). \tag{8.313}$$

Thus, *Equation 8.313 is the complex conjugate velocity of an airfoil of chord c at angle-of-attack α in a free stream of velocity U, with the Kutta condition ensuring a finite velocity at the trailing edge (Equation 8.314a):*

$$v^*\left(\frac{c}{2}\right) = U\cos\alpha = v_x; \quad \Gamma = -\pi Uc\sin\alpha, \quad L = -\rho U\Gamma = \rho U^2\pi c\sin\alpha. \tag{8.314a–8.314c}$$

The velocity (Equation 8.314a) is tangential and corresponds to (1) the circulation (Equation 8.314b) from Equation 8.309b, in agreement with Equation 8.232b; and (2) the lift (Equation 8.314c) from Equation I.28.29b ≡ Equation 2.233c. The result (Equation 8.314b) for the flat plate airfoil obtained via the Sedov theorem can, as before (Subsections 8.8.1, 8.8.5, and 8.8.7), also be obtained using the Joukowski transformation (Subsections 8.8.2 through 8.8.4, 8.8.6, and 8.8.8, respectively), proving also Equations 8.315a and 8.315c in the process. The Sedov method (Joukowski transformation) is [Subsection 8.8.1 (Subsection 8.8.3)] a particular form of the Cauchy theorem (Sections 15.3 and 37.8) [conformal mapping of one (Subsection 8.8.2) of the four Schwartz–Christoffell types (Sections I.33.6 through I.33.9)].

8.8.8 Balance of Lift and Suction Forces

A vortex with circulation Γ has (Equation 8.283b), for streamlines in the ζ-plane, any circle and can be added to Equation 8.304b ≡ Equations 8.305a and 8.305b ≡ Equation 8.306); thus, *the potential flow of a cylinder of radius a (a flat plate airfoil of chord c = 4a) in a uniform stream of velocity U and angle-of-attack α, in the presence of circulation Γ, is specified by the complex potentials in the ζ-plane (Equation 8.315a) [z-plane (Equation 8.315b)]:*

$$f(\zeta) = U\left(\zeta\,e^{i\alpha} + \frac{c^2}{16\,\zeta}e^{-i\alpha}\right) - \frac{i\Gamma}{2\pi}\log\zeta, \tag{8.315a}$$

$$f(z) = U\left(z\cos\alpha - i\sin\alpha\sqrt{z^2 - c^2/4}\right) - \frac{i\Gamma}{2\pi}\log\left(z + \sqrt{z^2 - c^2/4}\right), \tag{8.315b}$$

that are related by the Joukowski conformal mapping (Equation 8.296a through 8.296d). The corresponding complex conjugate velocity is Equation 8.316a (Equation 8.316b):

$$\frac{df}{d\zeta} = U\left[e^{-i\alpha} - \left(\frac{a}{4\zeta}\right)^2 e^{i\alpha}\right] - \frac{i\Gamma}{2\pi\zeta}, \tag{8.316a}$$

$$\frac{df}{dz} = U\cos\alpha - \frac{i}{\sqrt{z^2 - c^2/4}}\left(Uz\sin\alpha + \frac{\Gamma}{2\pi}\right) \tag{8.316b}$$

for the cylinder (airfoil).

The inverse square root singularity of the velocity (Equation 8.316b) at the trailing edge (Equation 8.317a) of the airfoil may be eliminated by imposing the **Kutta condition** *that specifies the circulation (Equation 8.317b):*

$$z = \frac{c}{2}: \quad \Gamma = -\lim_{z\to\frac{c}{2}} 2\pi Uz\sin\alpha = -\pi Uc\sin\alpha, \tag{8.317a and 8.317b}$$

in agreement with Equation 8.317b = Equation 8.309b ≡ Equation 8.232b ≡ Equation I.34.12c. The circulation (Equation 8.317b) simplifies the complex conjugate velocity (Equation 8.316b) to

$$\frac{df}{dz} - U\cos\alpha = -iU\sin\alpha\frac{z - c/2}{\sqrt{z^2 - c^2/4}} = -iU\sin\alpha\sqrt{\frac{z - c/2}{z + c/2}}. \tag{8.318}$$

From Equation 8.318 ≡ Equation 8.313, it follows that the velocity is finite (Equation 8.314a) at the trailing edge. The inverse square root singularity of the velocity (Equation 8.318) remains at the leading edge (Equation 8.319a) implying (Equation 8.319b):

$$v^*(z) = \frac{C}{\sqrt{z + c/2}} + O(1), \quad C \equiv -iU\sin\alpha\sqrt{-c}; \tag{8.319a and 8.319b}$$

the suction force (Equation 8.320a):

$$F_s = -\rho\pi C^2 = -\pi\rho U^2 c\sin^2\alpha = -L\sin\alpha = \mp 4F^{\pm} \tag{8.320a–8.320c}$$

(1) that is equal to the lift (Equation 8.314c) projected transversely to the flow direction (Equation 8.320b) and (2) is four times (Equation 8.320c) the suction force compared with the case (Equation 8.308) when the Kutta condition is not applied.

The circulation (Equation 8.314b ≡ Equation 8.232b ≡ Equation I.34.12c) and lift (Equation 8.314c ≡ Equation 8.232c ≡ Equation I.34.17) agree as derived by the Sedov method (Joukowski transformation) in Subsection I.34.1 (Subsection 8.8.3). *The lift force (Equation 8.321b) is orthogonal to the flow direction (Equation 8.321a):*

$$\vec{U} = U\left(\vec{e}_x \cos\alpha + \vec{e}_y \sin\alpha\right):$$

$$\vec{F}_\ell = L\left[\vec{e}_x \cos\left(\alpha + \frac{\pi}{2}\right) + \vec{e}_x \sin\left(\alpha + \pi/2\right)\right] = L\left(-\vec{e}_x \sin\alpha + \vec{e}_y \cos\alpha\right), \quad \text{(8.321a and 8.321b)}$$

so that (1) the vertical force is Equation 8.322a and (2) the horizontal force (Equation 8.322b) is balanced (Equation 8.322c) by the suction force (Equation 8.300c):

$$F_{\ell y} = L \cos\alpha, \qquad F_{\ell x} = -L \sin\alpha = F_s. \quad \text{(8.322a–8.322c)}$$

The suction force on the upper (lower) side of the airfoil near the leading edge is positive (Equation 8.320a) [negative (Equation I.34.26)] with the same modulus (Equation 8.322a). The suction force turns the airflow around the leading edge by pushing it below $F_s < 0$ and pulling it above $-F_s > 0$. This corresponds to the Coanda effect turning a jet around an edge (Section I.38.7). The Coanda effect applies at the leading edge (Equations 8.322a through 8.322c) and also at the trailing edge (Equation 8.308) if the Kutta condition is not applied.

8.8.9 Circulation around Two Tandem Airfoils

The potential flow past the flat plate airfoil can be considered (1) in the ζ-plane of the cylinder (Section I.34.1); it is the appropriate starting point to consider elliptic cylinders (Section I.34.2) and Joukowski airfoils (Sections I.34.3 and I.34.4) and other wing sections (Sections I.34.5 and I.34.6) and (2) in the z-plane of the flat plate (Subsections 8.8.1 and 8.8.2) that provides a more straightforward comparison with the Sedov method. Approach 1 via the Joukowski conformal mapping applies to a simply connected region (Chapter I.37), whereas the Sedov method (Subsection 8.7.1) applies as well to multiply connected regions, for example, the airfoil with a flap (Subsections 8.7.2 through 8.7.6) is a doubly connected region. Next, the following are considered: (1) another doubly connected region, namely, two airfoils in tandem, without (with) the Kutta condition [Subsection 8.8.9 (Subsection 8.8.10)] and (2) an infinitely multiply connected region, namely, a cascade of airfoils (Subsections 8.8.11 and 8.8.14). Consider (Figure 8.22) two (Equation 8.323a) of tandem airfoils (Equation 8.323b) separated along the real axis (Equation 8.323):

$$n = 1{,}2: \qquad a_\pm \le x \le b_\pm, \qquad -\infty < a_- < b_- < a_+ < b_+ < +\infty. \quad \text{(8.323a–8.323c)}$$

In the absence of mean flow, the complex conjugate perturbation velocity (Equation 8.324a) involves the sum of circulations (Equation 8.324b) [dipole moments (Equation 8.324c)]:

$$2\pi v^*(\zeta) = -\frac{i\Gamma}{\zeta} - \frac{P}{\zeta^2} + O\left(\frac{1}{\zeta^3}\right), \quad \{\Gamma, P\} = \sum_{n=1}^{2}\{\Gamma_n, P_n\}. \quad \text{(8.324a–8.324c)}$$

The **tandem airfoil function** (Equation 8.325b) is chosen to enforce inverse square root singularities at all leading and trailing edges, and its asymptotic scaling is given by Equation 8.325c involving the constant Equation 8.325a:

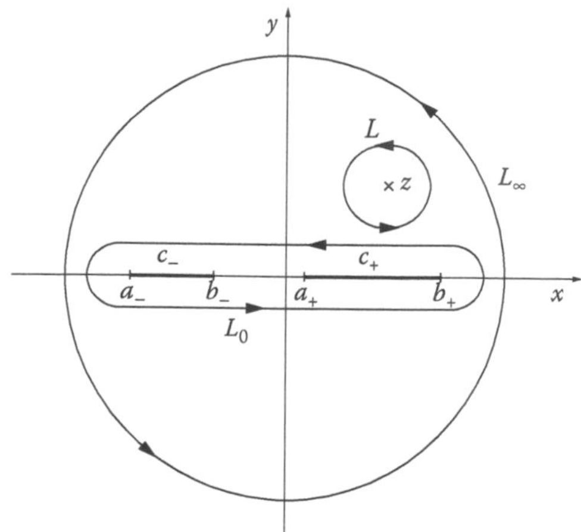

FIGURE 8.22
Another case of a doubly connected region, besides the airfoil with slotted slat (Figure 8.16c) or flap (Figures 8.16d and 8.18), is a tandem airfoil (Figure 8.22) consisting of two aligned airfoils with chords generally different but of comparable magnitude instead of a relatively small flap or slat. The Sedov method extends from one (Figure 8.17) to two (Figure 8.21) or any number, even infinite (Figure 8.22) of airfoils in any positions, using loops around the airfoils and infinity. In the case of two tandem airfoils with circulation in a potential flow, the lift is the same as for a single airfoil adding the chords because there is no wake effect (Figures 8.24 and 8.25). For the multiple airfoils, like the tandem case (Figure 8.22), the Kutta condition is applied at all the trailing edges and there is a suction force at each leading edge.

$$2A \equiv a_+ + a_- + b_+ + b_-: \quad \frac{1}{g_5(\zeta)} = \left[(\zeta - a_-)(\zeta - b_-)(\zeta - a_+)(\zeta - b_+)\right]^{1/2} = \zeta^2\left[1 - \frac{A}{\zeta} + O\left(\frac{1}{\zeta^2}\right)\right] \tag{8.325a–8.325c}$$

as follows from

$$\begin{aligned}\frac{1}{g_5(\zeta)} &= \zeta^2\left[\left(1-\frac{a_-}{\zeta}\right)\left(1-\frac{b_-}{\zeta}\right)\left(1-\frac{a_+}{\zeta}\right)\left(1-\frac{b_+}{\zeta}\right)\right]^{1/2} \\ &= \zeta^2\left[1 - \frac{a_+ + a_- + b_+ + b_-}{\zeta} + O\left(\frac{1}{\zeta^2}\right)\right]^{1/2} = \zeta^2\left[1 - \frac{A}{\zeta} + O\left(\frac{1}{\zeta^2}\right)\right]\end{aligned} \tag{8.326}$$

using Equation 8.325a.

The integral L_0 around the airfoils makes no contribution, and the remaining integral at infinity in Equation 8.249 is evaluated:

$$\begin{aligned}\frac{2\pi i}{g_5(z)}v^*(z) &= \frac{1}{2\pi}\int_{L_\infty}\frac{1}{g_5(\zeta)}\left[-i\frac{\Gamma}{\zeta}-\frac{P}{\zeta^2}+O\left(\frac{1}{\zeta^3}\right)\right]\frac{d\zeta}{\zeta-z}\\ &= \lim_{R\to\infty}\frac{1}{2\pi}\int_{|\zeta|=R}\left[1-\frac{A}{\zeta}+O\left(\frac{1}{\zeta^2}\right)\right]\left[-i\Gamma-\frac{P}{\zeta}+O\left(\frac{1}{\zeta^2}\right)\right]\left[1+\frac{z}{\zeta}+O\left(\frac{z^2}{\zeta^2}\right)\right]d\zeta\\ &= \frac{1}{2\pi}\lim_{R\to\infty}\int_{|\zeta|=R}\left[-i\Gamma+\frac{i\Gamma A-P-i\Gamma z}{\zeta}+O\left(\frac{1}{\zeta^2}\right)\right]d\zeta\\ &= \frac{1}{2\pi}2\pi i\left[i\Gamma(A-z)-P\right]=-\Gamma(A-z)-iP\end{aligned} \tag{8.327}$$

using the asymptotic form (Equation 8.325c ≡ Equation 8.326) of the inverse tandem airfoil function (Equation 8.325b).

Thus, *the complex conjugate velocity of the potential flow around two tandem airfoils (Figure 8.22) is given by*

$$2\pi v^*(z)=\frac{i\Gamma(A-z)-P}{\sqrt{(z-a_+)(z-a_-)(z-b_+)(z-b_-)}} \tag{8.328}$$

involving the total circulation (Equation 8.324b) and the dipole moment (Equation 8.324c) augmented by Equation 8.325a. In the case of nonzero circulation, its value can be chosen to apply the Kutta condition at the trailing-edge or leading-edge one of the airfoils. Without restricting the value of the circulation, the velocity has inverse square root singularities at the leading (Equations 8.329a and 8.330) [trailing (Equations 8.329b and 8.331)] edges of both airfoils:

$$v^*(z)=\frac{C_\pm^-}{\sqrt{z-a_\pm}}+O(1),\quad v^*(z)=\frac{C_\pm^+}{\sqrt{z-b_\pm}}+O(1), \tag{8.329a and 8.329b}$$

$$C_\pm^-\equiv\frac{i\Gamma}{2\pi}\lim_{z\to a_\pm}\left(A-a_\pm+iP\right)\sqrt{z-a_\pm}\,g_5(z)=\frac{i\Gamma}{4\pi}\frac{b_++b_-+a_\mp-a_\pm+2iP}{\sqrt{(a_\pm-a_\mp)(a_\pm-b_+)(a_\pm-b_-)}}, \tag{8.330}$$

$$C_\pm^+\equiv\frac{i\Gamma}{2\pi}\lim_{z\to b_\pm}\left(A-b_\pm+iP\right)\left(z-b_\pm\right)g_5(z)=\frac{i\Gamma}{4\pi}\frac{a_++a_-+b_\mp-b_\pm+2iP}{\sqrt{(b_\pm-b_\mp)(b_\pm-b_-)(b_\pm-a_+)}}. \tag{8.331}$$

In the absence of a dipole term [Equation 8.332a (Equation 8.333a)], the corresponding suction forces are [Equation 8.332b (Equation 8.333b)]

$$P=0:\quad F_\pm^-=-\rho\pi\left(C_\pm^-\right)^2=\frac{\rho\Gamma^2}{16\pi}\frac{\left(b_++b_-+a_\mp-a_\pm\right)^2}{(a_\pm-a_\mp)(a_\pm-b_+)(a_\pm-b_-)}, \tag{8.332a and 8.332b}$$

$$P=0:\quad F_\pm^+=-\rho\pi\left(C_\pm^+\right)^2=\frac{\rho\Gamma^2}{16\pi}\frac{\left(a_++a_-+b_\mp-b_\pm\right)^2}{(b_\pm-b_\mp)(b_\pm-a_-)(b_\pm-a_+)} \tag{8.333a and 8.333b}$$

and transform into each other interchanging the edges ($a_\pm$, $b_\pm$).

8.8.10 Lift Generation by Two Tandem Airfoils

The Kutta condition can be applied at the trailing edges of both airfoils by choosing, instead of Equation 8.325b, the **tandem circulation function** (Equation 8.334b):

$$2B = b_- - a_- + b_+ - a_+ = c_+ + c_-: \quad \frac{1}{g_6(\zeta)} = \sqrt{\frac{\zeta - a_-}{\zeta - b_+}\frac{\zeta - a_+}{\zeta - b_-}} = 1 + \frac{B}{\zeta} + O\left(\frac{1}{\zeta^2}\right), \quad (8.334a\text{–}8.334c)$$

whose inverse has an asymptotic scaling (Equation 8.335) involving the sum of the chords (Equation 8.334a) of the forward c_- (rear c_+) airfoils as follows from

$$\frac{1}{g_6(\zeta)} = \left(\frac{1 - a_-/\zeta}{1 - b_-/\zeta}\frac{1 - a_+/\zeta}{1 - b_+/\zeta}\right)^{1/2} = \left[1 + \frac{b_- + b_+ - a_- - a_+}{\zeta} + O\left(\frac{1}{\zeta^2}\right)\right]^{1/2} = 1 + \frac{B}{\zeta} + O\left(\frac{1}{\zeta^2}\right), \quad (8.335)$$

where Equation 8.334a was used.

The circulation Γ implies an asymptotic velocity (Equation 8.283b) and no contribution from infinity L_∞ from higher order terms in Equation 8.249. The first term on the r.h.s. of Equation 8.249 involves an integration over both airfoils with velocity Equation 8.243a leading to

$$\begin{aligned} -\frac{2\pi}{U\sin\alpha}\frac{\tilde{v}^*(z)}{g_6(z)} &= \sum_{n=1}^{2}\int_{L_0}\frac{1}{g_6(\zeta)}\frac{d\zeta}{\zeta - z} \\ &= -\frac{2\pi i}{g_6(z)} + \int_{L_\infty}\left[1 + \frac{B}{\zeta} + O\left(\frac{1}{\zeta^2}\right)\right]\frac{d\zeta}{\zeta} = 2\pi i\left[1 - \frac{1}{g_6(z)}\right], \end{aligned} \quad (8.336)$$

where the loop L_0 was stretched back to infinity L_∞ crossing a pole. Thus, *the perturbation velocity Equation 8.336 is given exactly by Equation 8.337a:*

$$\tilde{v}^*(z) = iU\sin\alpha\left[1 - \sqrt{\frac{z - b_-}{z - a_-}\frac{z - b_+}{z - a_+}}\right] = iU\frac{B}{z}\sin\alpha + O\left(\frac{1}{z^2}\right), \quad (8.337a \text{ and } 8.337b)$$

leading (Equation 8.334c) to the asymptotic form Equation 8.337b. The total velocity (Equation 8.204a) is given exactly (asymptotically) by Equation 8.338a) (Equation 8.338b):

$$\begin{aligned} v^*(z) = Ue^{-i\alpha} + \tilde{v}^*(z) &= U\cos\alpha - iU\sin\alpha\sqrt{\frac{z - b_-}{z - a_-}\frac{z - b_+}{z - b_+}} \\ &= U\cos\alpha - iU\sin\alpha\left[1 - \frac{B}{z} + O\left(\frac{1}{z^2}\right)\right] \end{aligned} \quad (8.338a \text{ and } 8.338b)$$

for two airfoils (Equation 8.323a through 8.323c) in tandem (Figure 8.21) at angle-of-attack α in a uniform stream of velocity U; the circulation (Equation 8.319b) is specified by the Kutta condition as 2π times the coefficient of $-i/z$ in the velocity Equation 8.338b.

$$c = c_+ + c_-: \quad \Gamma = -2\pi B U \sin\alpha = -\pi\left(c_+ + c_-\right)\sin\alpha = -\pi c U \sin\alpha, \qquad \text{(8.339a and 8.339b)}$$

$$L = -\rho U \Gamma = 2\pi\rho U^2 B \sin\alpha = \pi\rho U^2\left(c_+ + c_-\right)\sin\alpha = \pi\rho U^2 c \sin\alpha. \qquad \text{(8.339c)}$$

The circulation (Equation 8.339b) [lift (Equation 8.339c)] for the tandem airfoils (Figure 8.22) is specified by the sum of the chords, that is, the same as [Equation 8.339b ≡ Equation 8.314b (Equation 8.339c ≡ Equation 8.314c)] for a single airfoil with the same total chord (Equation 8.339a). The total velocity Equation 8.338b equals Equation 8.314a at the trailing edges, as expected when the Kutta condition is applied. At the leading edges, the perturbation (Equation 8.337b) [total (Equation 8.338b)] velocities both have an inverse square root singularity (Equation 8.340a) leading (Equation 8.340b) to the suction forces (Equation 8.340c):

$$v^*(z) \sim \frac{C^\pm}{\sqrt{z - a_\pm}}, \quad C^\pm = -iU\sin\alpha\sqrt{\frac{\left(a_\pm - b_-\right)\left(a_\pm - b_+\right)}{a_\mp - a_\pm}}, \qquad \text{(8.340a and 8.340b)}$$

$$F^\pm = -\rho\pi\left(C^\pm\right)^2 = \rho\pi\frac{\left(a_\pm - b_+\right)\left(a_\pm - b_-\right)}{a_\mp - a_\pm}U^2\sin^2\alpha. \qquad \text{(8.340c)}$$

The final example of the Sedov method is an application to an infinitely multiply connected region, namely, a cascade of airfoils (Subsections 8.8.11 through 8.8.14).

8.8.11 Infinite Cascade of Flat Plate Airfoils

The total flow velocity (Equation 8.204b) including the perturbation specified by the Sedov theorem (Equation 8.249) is given by

$$v^*(z) - U e^{-i\alpha} = \frac{g(z)}{2\pi i}\sum_{m=-\infty}^{+\infty}\int_{L_m}\frac{v^*(\zeta)}{g(\zeta)}\,\mathrm{d}\left[\log(\zeta - z)\right], \qquad \text{(8.341)}$$

where for a cascade of airfoils (Figure 8.23) the integration is performed summing over the infinity of loops L_m with $m = 0, \pm 1, \pm 2, \ldots$ surrounding all the airfoils of the cascade. The transformation (Equation 8.342a) has a differential (Equation 8.342b):

$$\log(\zeta - z) \to \log\left[\sinh\left(\pi\frac{\zeta - z}{b}\right)\right] - \pi\frac{\zeta - z}{b}, \quad \mathrm{d}\left[\log(\zeta - z)\right] \to \frac{\pi}{b}\left[\coth\left(\pi\frac{\zeta - z}{b}\right) - 1\right]\mathrm{d}\zeta, \qquad \text{(8.342a and 8.342b)}$$

which, when substituted in Equation 8.341, gives the total velocity

$$v^*(z) - U e^{-i\alpha} = \frac{g(z)}{2ib}\int_{L_0}\frac{v^*(\zeta)}{g(\zeta)}\left[\coth\left(\pi\frac{\zeta - z}{b}\right) - 1\right]\mathrm{d}\zeta. \qquad \text{(8.343)}$$

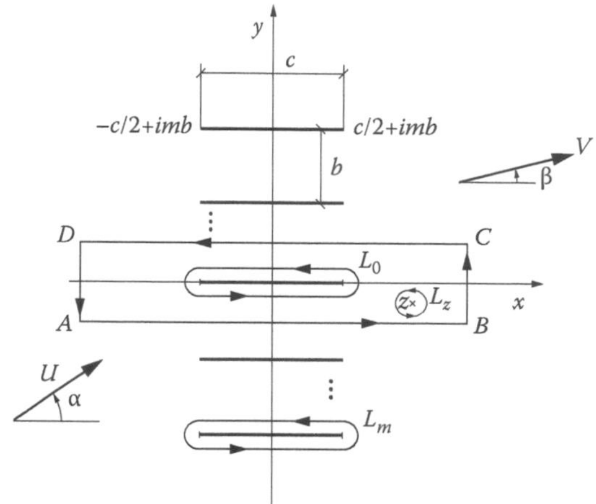

FIGURE 8.23
Sedov method applies to a single airfoil (Figure 8.17), two airfoils (Figure 8.22), or any number of them, even infinite; an example is a linear cascade consisting of an infinite number of flat plates with equal chords and equally spaced in the normal direction. The annular cascade used in turbomachinery has a finite number of airfoils with the same chord and angular spacing. A cascade changes the magnitude and direction of the flow velocity between the incident flow far upstream and the wake flow far downstream. Increasing the chord of the airfoils and decreasing the spacing between them tend to "straighten" more the flow into the direction of the airfoils and reduce the magnitude of the downstream relative to the upstream velocity. A larger reduction in flow angle and velocity magnitude corresponds to a larger lift on the cascade.

The integration in Equation 8.343 is performed along only one loop L_0 surrounding the airfoil on the real axis because (1) the transformation Equation 8.242a maps the airfoil on the real axis to all parallel airfoils of the cascade at a distance $\pm b$, $\pm 2b, \ldots$, as shown next (Subsection 8.8.11); and (2) the inverse square root singularities are preserved at the leading and trailing edges of all airfoils, as shown in the sequel (Subsection 8.8.12). The transformation (Equation 8.342a ≡ Equation 8.344c) maps a cascade (Equation 8.344a) of flat plate airfoils (Equation 8.344b) with chord c and (Equation 8.344c) spacing b:

$$m \in |Z, \quad |x| \le \frac{c}{2}, \quad \zeta - z = x + imb:$$

$$w = \frac{b}{\pi}\log\left[\sinh\left(\pi\frac{\zeta - z}{b}\right)\right] - (\zeta - z) = \frac{b}{\pi}\log\left[\sinh\left(\pi\frac{x}{b} + im\pi\right)\right] - x - imb \tag{8.344a–8.344d}$$

to the w-plane at Equation 8.344d and leads to Equation 8.345 using Equations 5.5b and 5.6b):

$$\begin{aligned} w + x + imb &= \frac{b}{\pi}\log\left[-i\sin\left(i\pi\frac{x}{b} - m\pi\right)\right] = \frac{b}{\pi}\log\left[-(-)^m i\sin\left(i\pi\frac{x}{b}\right)\right] \\ &= \frac{b}{\pi}\log\left[(-)^m \sinh\left(\pi\frac{x}{b}\right)\right] = \frac{b}{\pi}\log\left|\sinh\left(\pi\frac{x}{b}\right)\right| + \{0, ib\}, \end{aligned} \tag{8.345}$$

in which (1) if the term in square brackets is real and positive, then w is real; and (2) if the term in square brackets is negative, then $i\,b$ is added. The transformation Equation 8.345 leads (Equation 8.344b ≡ Equation 8.346a) to a segment of the real axis (Equation 8.346b):

$$|x| \le \frac{c}{2}: \quad |\mathrm{Re}(w)| \le \frac{c}{2} + \frac{b}{\pi} \log \left| \sinh\left(\pi \frac{x}{b}\right) \right|, \quad \mathrm{Im}(w) = -mb + \{0, b\} \qquad (8.346\text{a}–8.346\text{c})$$

repeated at all heights spaced b since the cascade has an infinite number of airfoils (Equations 8.344a and 8.346c). The choice of the transformation (Equations 8.342a and 8.342b) takes into account the period $2\pi i$ of the hyperbolic sine (Equation 5.270c) that leads to

$$m \in |Z: \quad \sinh\left(\pi \frac{\zeta - z}{b}\right) = \sinh\left(\pi \frac{\zeta - z}{b} + im\pi\right) = \sinh\left[\frac{\pi}{b}(\zeta - z + imb)\right], \qquad (8.347\text{a and } 8.347\text{b})$$

showing that the function Equation 8.347b takes the same value at all airfoils (Equation 8.347a).

8.8.12 Separation, Circulation, Airfoil, and Cascade Functions

The Sedov function (Equation 8.249) has been applied in seven cases: (1) as separation function (Equations 8.218a and 8.218b) for an airfoil with partial flow separation on the upper surface to enforce inverse square root singularities of the velocity at the separation point and trailing edge; (2) as a slotted airfoil function (Equation 8.253) to satisfy the Kutta conditions by eliminating the singularities of the velocity at the trailing edge of the airfoil and flap; (3) the separation function (case 1) is of similar form to the airfoil function (Equation 8.283c) that enforces inverse square root singularities for the complex conjugate velocity (Equation 8.210f) at the leading and trailing edges; (4) the slotted airfoil function (case 2) is the product of two circulation functions (Equation 8.310a), each enforcing one Kutta condition by eliminating a singularity of the velocity at a trailing edge; (5 and 6) the tandem airfoil (circulation) function [Equation 8.325b (Equation 8.334b)] is the product of two airfoil (case 3) [circulation (case 4)] functions, one for each airfoil; and (7) as in case 5, the cascade function (Equation 8.348c) is the product of circulation functions (Equation 8.310a), one for each of the infinite number of airfoils of the cascade, with edges at Equations 8.348a and 8.348b:

$$m \in |Z; \quad \zeta_m^{\pm} = \pm c/2 + imb: \quad g_7(\zeta) = \prod_{m=-\infty}^{+\infty} \sqrt{\frac{\zeta - \zeta_m^+}{\zeta - \zeta_m^-}} = \left[\prod_{m=-\infty}^{+\infty} \left(\frac{\zeta - c/2 - imb}{\zeta + c/2 + imb}\right)\right]^{1/2}, \qquad (8.348\text{a}–8.348\text{c})$$

applying the Kutta condition by suppressing the singularities of the velocity at the trailing edges ζ_n^+ but not at the leading edges ζ_n^-.

The infinite product in the argument of the square root in the cascade function (Equation 8.348c) can be expressed (Equations 7.227a and 7.227b) in terms of the hyperbolic sine:

$$\begin{aligned} [g_7(\zeta)]^2 &= \frac{\zeta - c/2}{\zeta + c/2} \prod_{n=1}^{+\infty} \frac{(\zeta - c/2 + inb)(\zeta - c/2 - inb)}{(\zeta + c/2 + inb)(\zeta + c/2 - inb)} \\ &= \frac{\zeta - c/2}{\zeta + c/2} \prod_{n=1}^{+\infty} \frac{(\zeta - c/2)^2 + n^2b^2}{(\zeta + c/2)^2 + n^2b^2} \\ &= \frac{\zeta - c/2}{\zeta + c/2} \prod_{m=1}^{+\infty} \frac{1 + (\zeta - c/2)^2/n^2b^2}{1 + (\zeta + c/2)^2/n^2b^2} = \frac{\sinh[\pi(\zeta - c/2)/b]}{\sinh[\pi(\zeta + c/2)/b]}. \end{aligned} \qquad (8.349)$$

It can be confirmed that the **cascade function**

$$g_7(\zeta) = \left\{ \frac{\sin\left[\pi\left(z - c/2\right)/b\right]}{\sin\left[\pi\left(z + c/2\right)/b\right]} \right\}^{1/2} \tag{8.350}$$

satisfies the condition that its inverse:

$$k \in Z|: \quad \frac{1}{g_7(\zeta)} = \frac{D_0}{\sqrt{\zeta - c/2 - ikb}} + O(1) \tag{8.351a and 8.351b}$$

has an inverse square root singularity (Equation 8.351b) at the trailing edge of every airfoil (Equation 8.351a):

$$D_0 = \sqrt{\frac{b}{\pi}\sinh\left(\frac{\pi c}{b}\right)} = \sqrt{c}\prod_{\substack{m=-\infty \\ m\neq 0}}^{+\infty}\left(1 - \frac{ic}{mb}\right) \tag{8.352a and 8.352b}$$

with coefficient Equation 8.352a ≡ Equation 8.352b.

The coefficient can be calculated [Equation 8.352a ≡ Equation 8.353a (Equation 8.352b ≡ Equation 8.353b)] from Equation 8.350 (Equation 8.347c):

$$\begin{aligned} D_0 &= \lim_{\zeta\to c/2}\frac{\sqrt{\zeta - c/2}}{g_7(\zeta)} = \lim_{\zeta\to c/2}\left[\frac{\zeta - c/2}{\sinh\left[\pi\left(\zeta - c/2\right)/b\right]}\right]^{1/2}\left\{\sinh\left[\pi\left(\zeta + c/2\right)/b\right]\right\}^{1/2} \\ &= \left[\frac{b}{\pi}\sinh\left(\frac{\pi c}{b}\right)\right]^{1/2}, \end{aligned} \tag{8.353a}$$

$$D_k = \lim_{\zeta\to\zeta_k^+}\frac{\sqrt{\zeta - \zeta_k^+}}{g_7(\zeta)} = \sqrt{\zeta_k^+ - \zeta_k^-}\prod_{\substack{m=-\infty \\ m\neq k}}^{+\infty}\sqrt{\frac{\zeta_k^+ - \zeta_m^-}{\zeta_k^+ - \zeta_m^+}} = \sqrt{c}\prod_{\substack{m=-\infty \\ m\neq k}}^{+\infty}\sqrt{\frac{c + i(k-m)b}{i(k-m)b}}. \tag{8.353b}$$

The coefficient Equation 8.353a was calculated for the airfoil on the real axis, and it is equal for all other airfoils (Equation 8.353b) as follows from

$$\begin{aligned} s = k - m: \quad D_k &= \sqrt{c}\prod_{\substack{s=-\infty \\ s\neq 0}}^{+\infty}\sqrt{\frac{c + isb}{isb}} = \sqrt{c}\prod_{n=1}^{\infty}\sqrt{1 - \frac{ic}{nb}}\sqrt{1 + \frac{ic}{nb}} \\ &= \sqrt{c}\prod_{n=1}^{\infty}\sqrt{1 + \frac{c^2}{n^2 b^2}} = \sqrt{c}\sqrt{\frac{b}{\pi c}\sinh\left(\frac{\pi c}{b}\right)} = \sqrt{\frac{b}{\pi}\sinh\left(\frac{\pi c}{b}\right)} = D_0, \end{aligned} \tag{8.354}$$

where again the infinite product (Equations 7.227a and 7.227b) for the hyperbolic sine was used.

8.8.13 Velocity Field Past a Cascade of Airfoils

The path of integration is the loop around the airfoil, so Equation 8.243a is used in Equation 8.343 leading to Equation 8.355a):

$$v^*(z) = Ue^{-i\alpha} - \frac{U}{2b}\sin\alpha I, \quad \mathrm{I} \equiv g_7(z)\int_{L_0}\left[\coth\left[\left(\pi\frac{\zeta - z}{b}\right)\right] - 1\right]\frac{d\zeta}{g_7(\zeta)}, \qquad \text{(8.355a and 8.355b)}$$

in terms of the integral (Equation 8.355b) involving the cascade function (Equation 8.350). The integral is evaluated by deformation of the loop around z into an infinite rectangle ABCDA in Figure 8.23, thus crossing the pole of the integrand at $\zeta = z$ and implying that the residue times $2\pi i$ must be subtracted.

$$\begin{aligned}\eta = \pi\frac{\zeta - z}{b}: \quad & -2\pi i g_7(z)\lim_{\zeta\to z}\frac{\zeta - z}{g_7(\zeta)}\left[\coth\left(\pi\frac{\zeta - z}{b}\right) - 1\right]\\ &= -2\pi i\lim_{\zeta\to z}(\zeta - z)\left(\frac{\cosh\eta}{\sinh\eta} - 1\right)\\ &= -2\pi i\lim_{\eta\to 0}\frac{b}{\pi}\frac{\eta}{\sinh\eta}(\cosh\eta - \sinh\eta)\\ &= -2ib\lim_{\eta\to 0}\frac{\eta}{\sinh\eta}e^{-\eta} = -2ib.\end{aligned} \qquad \text{(8.356a and 8.356b)}$$

The term in square brackets in Equations 8.343, 8.356b, and 8.355b can also be written (Equation 8.357) in terms of the variable Equation 8.356a:

$$\coth\eta - 1 = \frac{\cosh\eta}{\sinh\eta} - 1 = \operatorname{csch}\eta(\cosh\eta - \sinh\eta) = e^{-\eta}\operatorname{csch}\eta. \qquad (8.357)$$

The term Equation 8.356b appears in Equation 8.355b first on the r.h.s. of

$$\begin{aligned} I &= -2\pi i\lim_{\zeta\to z}(\zeta - z)\operatorname{csch}\left(\pi\frac{\zeta - z}{b}\right)\exp\left(-\pi\frac{\zeta - z}{b}\right)\frac{g_7(z)}{g_7(\zeta)}\\ &+ g_7(z)\int_{ABCDA}\left[\coth\left[\left(\pi\frac{\zeta - z}{b}\right)\right] - 1\right]\left\{\frac{\sinh\left[\pi(\zeta + c/2)/b\right]}{\sinh\left[\pi(\zeta - c/2)/b\right]}\right\}^{1/2} d\zeta\\ &= -2bi + i2bg_7(z)\exp\left(-\frac{\pi c}{2b}\right).\end{aligned} \qquad (8.358)$$

The second term on the r.h.s. of Equation 8.358 is the integral along the rectangle ABCDA in Figure 8.23 where (1) the integrand is equal on the sides (AB,CD) parallel to the real axis, which are taken in opposite directions and thus cancel; (2) the limit Equation 8.359a implies that the integral along the vertical upstream side CB vanishes:

$$\lim_{\zeta\to\pm\infty}\cot\left[\pi\frac{\zeta-z}{b}\right]=\pm 1;\quad \lim_{\zeta\to-\infty}\sinh\left[\left(\pi\frac{\zeta\pm c/2}{b}\right)\right]=-\frac{1}{2}\exp\left[-\pi\frac{\zeta\pm c/2}{b}\right],\qquad (8.359\text{a}–8.359\text{c})$$

and (3) the limit Equation 8.359b implies that the remaining integral along the vertical downstream side AD is given by

$$-ibg_7(z)\lim_{\zeta\to-\infty}\left[\coth\left(\pi\frac{\zeta-z}{b}\right)-1\right]\left\{\frac{\sinh\left(\pi\left(\zeta+c/2\right)/b\right)}{\sinh\left(\pi\left(\zeta-c/2\right)/b\right)}\right\}^{1/2}$$
$$=2ibg_7(z)\lim_{\zeta\to-\infty}\left\{\frac{\exp\left(-\pi\left(\zeta+c/2\right)/b\right)}{\exp\left(-\pi\left(\zeta-c/2\right)/b\right)}\right\}^{1/2}=2ibg_7(z)\exp\left(-\frac{\pi c}{2b}\right),\qquad (8.360)$$

which specifies the second term on the r.h.s. of Equation 8.358. Substitution of Equation 8.358 into Equation 8.355a specifies the flow velocity past a cascade of airfoils:

$$v^*(z)=U\left(\cos\alpha-i\sin\alpha\right)+\frac{U}{2b}\sin\alpha 2ib\left[1-e^{-\pi c/2b}g_7(z)\right]=U\cos\alpha-iUe^{-\pi c/2b}g_7(z),\qquad (8.361)$$

showing its effect on the change in magnitude and direction (Subsection 8.8.14). The vertical component of the free stream velocity $-iU\sin\alpha$ is canceled by the vertical component $iU\sin\alpha$ of the perturbation velocity Equation 8.205a for (1) the cascade of parallel flat plate airfoils (Equation 8.361); (2) the tandem airfoils (Equations 8.338a and 8.338b); and (3 and 4) the single airfoil with (Equation 8.313) [without (Equation 8.301)] the Kutta condition.

8.8.14 Change of Flow Velocity and Direction across a Cascade

Substitution of the cascade function Equation 8.350 in Equation 8.361 leads to

$$v^*(z)=U\cos\alpha-iU\exp\left(-\frac{\pi c}{2b}\right)\sin\alpha\left\{\frac{\sinh\left[\pi\left(z-c/2\right)/b\right]}{\sinh\left[\pi\left(z+c/2\right)/b\right]}\right\}^{1/2}\qquad (8.362)$$

the complex conjugate velocity of the potential flow past a cascade of flat plate airfoils with chord c and spacing b in Figure 8.23 in an incident stream of velocity U and angle-of-attack α. At the trailing edge of the airfoils, the velocity is horizontal (Equation 8.314a), as should be expected from the Kutta condition; again, as should be expected at a sharp edge without the Kutta condition, the velocity has an inverse square root singularity (Equations 8.363b and 8.363c) at the leading edges (Equation 8.363a):

$$m\subset|Z:v^*(z)-\frac{C_0}{\sqrt{z+c/2-imb}},\quad C_0=-iU\sin\alpha\, e^{-\pi c/2b}\sqrt{-\frac{b}{\pi}\sinh\left(\frac{\pi c}{b}\right)}.\qquad (8.363\text{a and }8.363\text{b})$$

The corresponding suction force (Equation 8.364a) is related to the lift (Equation 8.314c) of a single isolated airfoil by Equation 8.364b:

$$F_0 = -\rho\pi\left(C_0\right)^2 = -\rho U^2 b \sin^2\alpha\, e^{-\pi c/b}\sinh\left(\frac{\pi c}{b}\right) = -\frac{L}{2\pi^*}\frac{b}{c}\sin\alpha\left(1 - e^{-2\pi c/2b}\right). \tag{8.364a and 8.364b}$$

Far downstream (Equation 8.365a), the velocity has (Equation 8.362) modulus (angle) given by Equation 8.365b (Equation 8.365c):

$$\lim_{z\to\infty} v^*(z) = U\left[\cos\alpha - i\sin\alpha\exp\left(-\frac{\pi c}{2b}\right)\right] = Ve^{-i\beta}, \tag{8.365a}$$

$$V = U\left|\cos^2\alpha + e^{-\pi c/b}\sin^2\alpha\right|^{1/2}, \quad \tan\beta = e^{-\pi c/2b}\tan\alpha. \tag{8.365b and 8.365c}$$

These values are (1) close to the incident stream for Equations 8.366b and 8.366c for short widely spaced airfoils (Equation 8.366a):

$$\left(\frac{\pi c}{b}\right)^2 << 1: \quad \frac{V}{U} = 1 - \frac{\pi c}{2b}\sin^2\alpha, \quad \tan\beta = \left(1 - \frac{\pi c}{2b}\right)\tan\alpha; \tag{8.366a–8.366c}$$

$$\frac{\pi c}{b} >> 1: \quad V = U\cos\alpha\left\{1 + \frac{1}{2}e^{-\pi c/b}\tan^2\alpha\right\}, \quad \beta = e^{-\pi c/b}\tan\alpha, \tag{8.367a–8.367c}$$

and (2) close to the horizontal velocity (Equations 8.367b and 8.367c) for (Equation 8.367a) long closely spaced airfoils.

In the passage from Equation 8.362 to Equations 8.363a and 8.363b, the limit (Equations 8.368a and 8.368b) was used:

$$\eta = \pi\frac{z + c/2}{b}: \quad \lim_{z\to -c/2}\left\{\frac{z + c/2}{\sinh\left[\pi\left(z + c/2\right)/b\right]}\right\}^{1/2} = \sqrt{\frac{b}{\pi}}\lim_{\eta\to 0}\left(\frac{\eta}{\sinh\eta}\right)^{1/2} = \sqrt{\frac{b}{\pi}}. \tag{8.368a and 8.368b}$$

The derivation of Equation 8.366b (Equation 8.367b) from Equation 8.365b is given by Equation 8.369 (Equation 8.370):

$$\begin{aligned}\frac{V}{U} &= \left|\cos^2\alpha + \left[1 - \frac{\pi c}{b} + O\left(\frac{\pi c}{b}\right)^2\right]\sin^2\alpha\right|^{1/2} \\ &= \left|1 - \frac{\pi c}{b}\sin^2\alpha + O\left(\frac{\pi c}{b}\right)^2\right|^{1/2} = 1 - \frac{\pi c}{2b}\sin^2\alpha + O\left(\left(\frac{\pi c}{b}\right)^2\right),\end{aligned} \tag{8.369}$$

$$\frac{V}{U} = \cos\alpha \left|1 + e^{-\pi c/b}\tan^2\alpha\right|^{1/2} = \cos\alpha\left[1 + \frac{1}{2}e^{-\pi c/b}\tan^2\alpha + O\left(e^{-2\pi c/b}\right)\right]. \quad (8.370)$$

The passage from Equations 8.365c to 8.366c is immediate and Equation 8.367c is simplified for small angle $\tan\beta \backsim \beta$. Table 8.6 shows (1) the ratio of flow velocities far downstream and upstream of the cascade (Equation 8.365b) and (2) the angle of the flow far downstream (Equation 8.365c) for all combinations of (a) six values of the angle-of-attack of the incident stream spaced 15° from 0° to 75°; and (b) seven values of the ratio of the chord c to the spacing of the airfoils b from 0 to ∞ mainly in the range 0.1 – 1. When the airfoil chord is comparable to or larger than the spacing $c > b$, the cascade is quite effective at "straightening" the incident flow, even for large angle-of-attack, into a flow parallel to the cascade far downstream, with an associated reduction of velocity close to the value of the projection parallel to the cascade. The loss of vertical momentum equates to a lift on the cascade and is the driving principle of turbo machinery. A smaller airfoil spacing relative to the chord and a larger angle-of-attack of the incident stream lead (Table 8.6) to the following: (1) a larger deflection of the flow through the cascade; (2) a greater speed loss from far upstream to far downstream; and (3) both 1 and 2 increase the lift on the cascade, up to the onset of flow separation on the airfoils (Section 8.6). The separated flow, either from the trailing edge or from further upstream of one or multiple airfoils or a cascade, forms a wake; its

TABLE 8.6

Effect of a Cascade in a Free Stream

	$\frac{c}{b} = 0.0$	0.1	0.2	0.3	0.4	05	1	∞
$\alpha = 0°$	$\frac{V}{U} = 1$	1	1	1	1	1	1	1
	$\beta = 0°$	0°	0°	0°	0°	0°	0°	0°
15°	1	0.9909	0.9843	0.9793	0.9757	0.9731	0.9674	0.9659
	15°	12.90°	11.07°	9.495°	8.135°	6.965°	3.188°	0.000°
30°	1	0.9657	0.9399	0.9206	0.9062	0.8955	0.8722	0.8660
	30°	26.26°	22.87°	19.82°	17.12°	14.75°	6.844°	0.000°
45°	1	0.9302	0.8756	0.8336	0.8014	0.7771	0.7222	0.7071
	45°	40.52°	36.14°	31.97°	28.08°	24.51°	11.74°	0.000°
60°	1	0.8932	0.8063	0.7364	0.6808	0.6371	0.5314	0.5000
	60°	55.96°	51.68°	47.23°	42.74°	38.30°	19.80°	0.000°
75°	1	0.8651	0.7515	0.6562	0.5767	0.5108	0.3276	0.2588
	75°	72.59°	69.85°	66.77°	63.33°	59.56°	37.80°	0.000°

Notes: Infinite cascade of identical parallel flat plate airfoils at equal distances changes (Figure 8.23) the magnitude and direction of an incident stream: (1) the change in the magnitude of the velocity is indicated by the ratio of the modulus of the velocities far upstream U and downstream V of the cascade; (2) the change in the direction of the flow is indicated by the difference between the angle-of-attack α of the incident stream and the flow angle β far downstream of the cascade. The two effects depend on two parameters: (a) the absolute value of the angle-of-attack (lines); and (b) the ratio of the chord to the spacing of the airfoils (columns). The effect (b) varies between two extremes: (α) a cascade with long closed spaced airfoils straightens the flow horizontally and reduces the velocity significantly; (β) a cascade with short fairly spaced airfoils deflects less the flow and gives a smaller reduction in velocity. The change in flow direction and reduction of flow velocity across the cascade specifies the lift; this is the principle of operation of turbomachinery that uses an annular cascade in a hollow cylindrical duct. c: chord of airfoils; U: velocity far upstream; α: angle-of-attack far upstream; b: spacing of airfoils; V: velocity far downstream; β: angle-of-attack far downstream.

simplest representation, as a vortex sheet, is unstable to small amplitude sinusoidal disturbances, as shown next (Section 8.9).

8.9 Instability of a Plane Vortex Sheet (Helmholtz 1868; Kelvin 1871)

The simplest representation of the separated flow from an airfoil (Section 8.6) or the wakes from the trailing edges of airfoils and flaps (Section 8.7) and cascades (Section 8.8) is a plane vortex sheet. Broadly speaking, irrotational (rotational) flows tend to be stable (unstable), as shown by the simplest cases of a uniform flow (vortex sheet). The general plane vortex sheet is an interface (Figure 8.24) without surface tension between fluids with distinct densities and velocities on the two sides (Subsection 8.9.1). The boundary conditions (Subsection 8.9.2) of continuity of displacement and pressure across the vortex sheet show that small perturbations generally grow with time, leading to linear instability (Subsection 8.9.3). The Kelvin–Helmhotz or plane vortex sheet instability is due to the accumulation of circulation over time (Subsection 8.9.4) and is affected by the velocities and mass densities of the two streams and by the acceleration of gravity (Subsection 8.9.5); this leads to the consideration of convected reference frames and the Doppler shifted frequency (Subsection 8.9.6).

8.9.1 Sinusoidal Perturbation of a Plane Vortex Sheet

The shape of the perturbed vortex sheet in three dimensions is assumed to be (Equation 8.371a) a sinusoid in the along (across) flow or $x-$ ($z-$) directions with wavenumber $k_x(k_z)$:

$$\zeta(x,z,t) = A(t)\exp\{i(k_x x + k_z z)\}, \quad k = \left|(k_x)^2 + (k_z)^2\right|^{1/2}, \qquad \text{(8.371a and 8.371b)}$$

where Equation 8.371b is the total wavenumber, and the amplitude is a function of time; its growth (decay) with time implies instability (stability). The flow below (above) the vortex

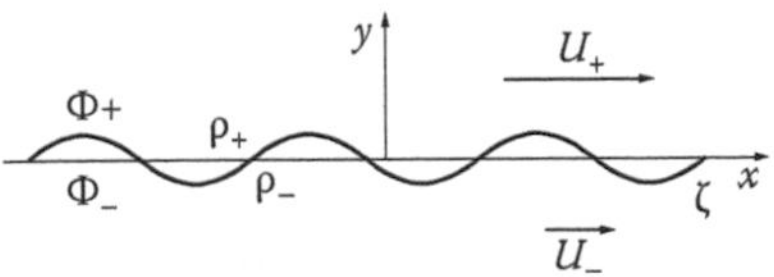

FIGURE 8.24
Potential flow does not consider the vortex sheet issued from the trailing edge of an isolated airfoil (Figure 8.17) or all airfoils in a cascade (Figure 8.22). In the case of tandem airfoils (Figure 8.21), the wake of the aft airfoil is free as before, but the wake of the forward airfoil impinges on the aft airfoil. This modifies the flow around the aft airfoil, so that the total lift is no longer the same as for a single airfoil with the total chord. The vortex sheet separates two jets with distinct tangential velocity and/or distinct mass densities on the two sides. The Laplace equation applies to the potential flow on either side of the vortex sheet; the change of tangential velocity across the vortex sheet leads to an infinite vorticity, so the flow is not irrotational. The effect of the vortex sheet of negligible thickness is accounted for by the continuity of the displacement and pressure on the two sides. It follows that a small wave-like disturbance of a flat vortex sheet always grows with time implying that it is unstable, that is, leads to a turbulent shear layer of finite thickness. The instability is due to the vorticity concentrating more at the points where it is higher due to the initial disturbance.

sheet has a potential $\Phi_-(\Phi_+)$ with the same dependence on the coordinates (x, z) in the plane of the undisturbed vortex sheet $y = 0$, that is, Equation 8.372a:

$$\Phi_\pm(x,y,z,t) = B_\pm(t) F_\pm(y) \exp\{i(k_x x + k_z z)\}; \quad \lim_{y\to\infty} F_\pm(y) = 0. \qquad \text{(8.372a and 8.372b)}$$

Furthermore, the potential must tend to a constant far from the vortex sheet, and the constant may be taken to be zero (Equation 8.372b) since its value does not affect the velocity that is obtained by the differentiation of the potential. Outside the vortex sheet, the flow is potential, so both potentials satisfy Laplace equation (Equations 8.373a and 8.373b) on both sides:

$$0 = \nabla^2 \Phi_\pm = (\partial_{xx} + \partial_{yy} + \partial_{zz})\Phi_\pm = B_\pm(t)\exp[i(k_x x + k_z z)]\left\{\frac{d^2 F_\pm}{dy^2} - k^2 F_\pm\right\}. \qquad \text{(8.373a and 8.373b)}$$

The term in curly brackets vanishes for $F_\pm \backsim \exp(\pm ky)$, and the solution that satisfies the boundary condition (Equation 8.372b) at infinity is Equation 8.374a:

$$F_\pm(y) = \exp(\mp ky): \quad \Phi_\pm(x,y,z,t) = B_\pm(t)\exp[\mp ky + i(k_x x + k_z z)], \qquad \text{(8.374a and 8.374b)}$$

leading to the potentials Equation 8.374b. The amplitudes of the displacement (Equation 8.371a) and potentials (Equation 8.374b) are determined from the boundary conditions at the vortex sheet.

8.9.2 Continuity of the Pressure and Displacement across the Vortex Sheet

The stability analysis will be made in the **linear approximation** of small amplitudes, implying that (1) the squares and products of amplitudes can be neglected and (2) the boundary conditions can be applied at the undisturbed position of the vortex sheet $y = 0$ instead of at the exact position $y = \zeta$ in Equation 8.375a. The normal velocity at the interface (Equation 8.375a) is the material derivative (Equation 2.6b) of the displacement (Equation 8.375b) taken above d^+ (below d^-) of the vortex sheet:

$$y = \zeta: \quad \partial_y \Phi_\pm = v_y^\pm = \frac{d^\pm \zeta}{dt} = \partial_t \zeta + (v_x + U_\pm)\partial_x \zeta + v_y \partial_y \zeta + v_z \partial_z \zeta, \qquad \text{(8.375a and 8.375b)}$$

where $U_+(U_-)$ is the mean flow velocity above (below) the vortex sheet, and the notations Equation 8.376b are used. The linear approximation simplifies Equation 8.375b to Equation 8.376c at the mean position (Equation 8.376a) of the vortex sheet:

$$y = 0; \quad \{\partial_x, \partial_y, \partial_t\}\zeta \equiv \left\{\frac{\partial \zeta}{\partial x}, \frac{\partial \zeta}{\partial y}, \frac{\partial \zeta}{\partial t}\right\}: \quad \partial_y \Phi_\pm = v_y^\pm = \partial_t \zeta + U_\pm \partial_x \zeta. \qquad \text{(8.376a–8.376c)}$$

The pressure is specified (Equation I.14.25c ≡ Equation 2.11) by the unsteady Bernoulli equation 8.377b including uniform gravity, that is, linearized (Equation 8.377c):

$$\frac{p_0^\pm - p}{\rho_\pm} - g\zeta - \frac{\partial \Phi}{\partial t} = \frac{(v_x + U_\pm)^2 + (v_y)^2 + (v_z)^2}{2} = \frac{1}{2}(U_\pm)^2 + U_\pm \partial_x \Phi. \qquad \text{(8.377a–8.377c)}$$

In Equations 8.377a and 8.377b the acceleration of gravity g is constant as well as the mass densities ρ_+ (ρ_-) and stagnation pressures $p_0^+ (p_0^-)$ above (below) the interface. There is pressure balance at the vortex sheet for the mean flow (Equation 8.377d):

$$0 = \left[p_0 - \frac{1}{2}\rho U^2 \right] \equiv p_0^+ - p_0^- - \frac{1}{2}\rho_+ U_+^2 + \frac{1}{2}\rho_- U_-^2. \tag{8.377d}$$

Concerning the linearized pertubations the displacement (Equation 8.378a) and pressure (Equation 8.378b) are continuous across the vortex sheet leading (Equation 8.377c) to the boundary condition Equation 8.378c:

$$\begin{aligned} [\zeta] = 0 = [p]: \quad 0 &= \left[\rho \left(\partial_t \Phi + U\, \partial_x \Phi + g\zeta \right) \right] \\ &= \rho_+ \partial_t \Phi_+ - \rho_- \partial_t \Phi_- + \rho_+ U_+ \partial_x \Phi_+ - \rho_- U_- \partial_x \Phi_- + g(\rho_+ - \rho_-)\zeta. \end{aligned} \tag{8.378a–8.378c}$$

Substitution of the displacement (Equation 8.371a) and potentials (Equation 8.374b) in the boundary conditions of continuity of displacement (Equation 8.376c) [pressure (Equation 8.378c)] leads to Equations 8.379a and 8.379b (Equation 8.379c):

$$\frac{\mathrm{d}A}{\mathrm{d}t} + i\, k_x U_\pm A = \mp k\, B_\pm, \tag{8.379a and 8.379b}$$

$$\frac{\mathrm{d}(\rho_+ B_+ - \rho_- B_-)}{\mathrm{d}t} + i\, k_x (\rho_+ U_+ B_+ - \rho_- U_- B_-) = g(\rho_- - \rho_+) A; \tag{8.379c}$$

the system of three coupled first-order ordinary differential equations 8.379a through 8.379c) specifies the dependence of the amplitudes on time.

8.9.3 Growth of the Amplitude of Pertubations with Time

The system (Equations 8.379a through 8.379c) of three coupled first-order ordinary differential equations is of order 2 rather than 3, as follows eliminating for the amplitude of the displacement of the vortex sheet:

$$\begin{aligned} (\rho_+ + \rho_-)\frac{\mathrm{d}^2 A}{\mathrm{d}t^2} + ik_x(\rho_+ U_+ + \rho_- U_-)\frac{\mathrm{d}A}{\mathrm{d}t} &= -k\frac{\mathrm{d}(\rho_+ B_+ - \rho_- B_-)}{\mathrm{d}t} \\ &= ik_x k(\rho_+ U_+ B_+ - \rho_- U_- B_-) - kg(\rho_- - \rho_+)A \\ &= kg(\rho_+ - \rho_-)A - ik_x \rho_+ U_+ \left(\frac{\mathrm{d}A}{\mathrm{d}t} + ik_x U_+ A \right) - ik_x \rho_- U_- \left(\frac{\mathrm{d}A}{\mathrm{d}t} + ik_x U_- A \right), \end{aligned} \tag{8.380}$$

leading to a second-order ordinary differential equation:

$$\begin{aligned} &(\rho_+ + \rho_-)\frac{\mathrm{d}^2 A}{\mathrm{d}t^2} + 2ik_x(\rho_+ U_+ + \rho_- U_-)\frac{\mathrm{d}A}{\mathrm{d}t} \\ &- \left[(k_x)^2 (\rho_+ U_+^2 + \rho_- U_-^2) + k\, g(\rho_+ - \rho_-) \right] A = 0, \end{aligned} \tag{8.381}$$

that specifies the time dependence of the amplitude of the displacement of the vortex sheet. Since the ordinary differential equation 8.381 is linear with constant coefficients, it has solutions sinusoidal in time (Equation 8.382a) with frequency ω related to the wavenumber k_x by the **dispersion relation** (Equation 8.382b):

$$A(t) \sim e^{-i\omega t}: \quad 0 = \omega^2 (\rho_+ + \rho_-) - 2k_x\omega(\rho_+U_+ + \rho_-U_-) + (k_x)^2 (\rho_+U_+^2 + \rho_-U_-^2) + kg(\rho_+ - \rho_-) \equiv (\omega - \omega_+)(\omega - \omega_-). \quad \text{(8.382a and 8.382b)}$$

The discriminant of the binomial (Equation 8.382b) is always negative (Equation 8.383b):

$$\rho_+ \geq \rho_-: a^2 \equiv -\frac{D}{4} = (\rho_+ + \rho_-) \left[(k_x)^2 (\rho_+U_+^2 + \rho_-U_-^2) + kg(\rho_+ - \rho_-)\right] - \left[k_x(\rho_+U_+ + \rho_-U_-)\right]^2 = (k_x)^2 \rho_-\rho_+ (U_+ - U_-)^2 + k\,g(\rho_+^2 - \rho_-^2) > 0,$$

(8.383a and 8.383b)

designating the + side as that whose mass density is not lower (Equation 8.383a); in Equation 8.383b, the following simplification was used:

$$(\rho_+ + \rho_-)(\rho_+U_+^2 + \rho_-U_-^2) - (\rho_+U_+ + \rho_-U_-)^2 = \rho_+\rho_-(U_+^2 + U_-^2 - 2U_+U_-) = \rho_+\rho_-(U_+ - U_-)^2. \quad \text{(8.384)}$$

It follows from Equation 8.383b that the dispersion relation 8.382b always has a pair of complex conjugate roots (Equation 8.385a):

$$\omega_\pm = \omega_0 \pm ib: \quad \omega_0 \equiv k_x \frac{\rho_+ U_+ + \rho_- U_-}{\rho_+ + \rho_-}, \quad b = \frac{\sqrt{-D}}{2(\rho_+ + \rho_-)} = \frac{a}{\rho_+ + \rho_-}. \quad \text{(8.385a–8.385c)}$$

The two solutions (Equation 8.382a) corresponding (Equation 8.386) to the complex frequencies (Equation 8.385a):

$$\exp(-i\,\omega_\pm t) = \exp(-i\omega_0\,t)\exp(\pm\,b\,t) \quad \text{(8.386)}$$

(1) have the same **frequency** of oscillation with time (Equation 8.385b), corresponding to the real part, and (2) include always an exponentially growing mode, corresponding to one of the imaginary parts of Equation 8.385a, where Equations 8.383b and 8.385c specify the **exponential amplification factors**. There is amplification (Equation 8.383b) if the jets on the two sides of the shear layer have different (1) velocities, (2) mass densities, or (3) both.

8.9.4 Accumulation of Circulation and Density Change as Instability Mechanisms

The displacement of the vortex sheet (Equation 8.371a) has time dependence (Equation 8.382a) specified by a linear combination of Equation 8.386, namely, Equation 8.387a:

$$\zeta(x,z,t) = \exp\left[i\left(k_x x + k_z z - \omega_0 t\right)\right]\left(C_+ e^{bt} + C_- e^{-bt}\right), \tag{8.387a}$$

$$\Phi_\pm(x,y,z,t) = \left(\pm\frac{i}{k}\right)\exp(\mp ky)\exp\left[i\left(k_x x + k_z z - \omega_0 t\right)\right]$$
$$\times\left[C_+\left(\omega_0 + ib - k_x U_\pm\right)e^{bt} + C_-\left(\omega_0 - ib - k_x U_\pm\right)e^{-bt}\right], \tag{8.387b and 8.387c}$$

with the arbitrary constants $C_\pm$ appearing (Equations 8.374a and 8.374b) also in the potentials (Equations 8.387b and 8.387c); the latter are specified by Equation 8.385a, with the relations (Equations 8.379a and 8.379b):

$$\mp k B_\pm = \left[-i\left(\omega_0 + ib\right) + ik_x U_\pm\right]C_+ e^{bt} + \left[-i\left(\omega_0 - ib\right) + ik_x U_\pm\right]C_- e^{-bt}. \tag{8.388}$$

The potentials (Equations 8.387b and 8.387c) are out-of-phase by $\pm\pi/2$ relative to the displacement since $\pm i = \exp(\pm i\pi/2)$; the circulation, which equals the difference of potentials across the vortex sheet:

$$\Gamma(x,z,t) = \Phi_+(x,z,0,t) - \Phi_-(x,z,0,t) = \exp\left[i\left(k_x x + k_z z\right)\right]\left[B_+(t) - B_-(t)\right], \tag{8.389}$$

is given by Equations 8.387b and 8.387c by Equation 8.390:

$$\Gamma(x,z,t) = i\frac{k_x}{k}\exp\left[i\left(k_x x + k_z z - \omega_0 t\right)\right]\left(U_- - U_+\right)\left(C_+ e^{bt} + C_- e^{-bt}\right). \tag{8.390}$$

Thus, when the velocity of the streams on each side of the vortex sheet is different $U_+ \neq U_-$, the circulation accumulates for one of the modes, leading to instability. The preceding account concerns the **Kelvin–Helmhotz instability** *as linear perturbation of a plane vortex sheet (Figure 8.24) between jets with velocities* $U_\pm$ *and mass densities* $\rho_\pm$*; it preserves the horizontal wavenumbers for the normal displacement (Equation 8.387a), and the potentials (Equations 8.387b and 8.387c) decay toward infinity on both sides (Equation 8.374a). The motion is sinusoidal in time with frequency (Equation 8.385b) and always one of the modes grows exponentially with time at a rate Equation 8.385c if the two streams have (Equation 8.383b) different velocities and/or mass densities. This growth also affects the circulation (Equation 8.390) if the two streams have different velocities.*

8.9.5 Effects of Velocity, Mass Density, and Gravity

The frequency (Equation 8.385b) and exponential growth rate (Equations 8.383b and 8.385c) simplify, in particular, the following cases:

$$\rho_+ = \rho_-, U_- = 0 \neq U_+ \equiv U: \quad \omega_0 = k_x\frac{U}{2} = b; \tag{8.391a–8.391e}$$

$$\rho_+ = \rho_-, U_\pm = \pm\frac{U}{2}: \quad \omega_0 = 0, \quad b = k_x\frac{U}{2}; \tag{8.392a–8.392e}$$

$$\rho_+ > \rho_-, U_\pm = U: \quad \omega_0 = k_x U, \quad b = \sqrt{kg}\,\frac{\sqrt{\rho_+^2 - \rho_-^2}}{\rho_+ + \rho_-}, \tag{8.393a–8.393d}$$

namely, *for cases 1 and 2, identical mass densities with one fluid at rest (Equations 8.391a through 8.391c) [two fluids (Figure 8.25a) with opposite velocities (Equations 8.392a through 8.392c)] leading to the same amplification (Equation 8.391e ≡ Equation 8.392e) and nonzero (zero) frequency [Equation 8.391d (Equation 8.392d)]; for case 3, same velocities (Figure 8.25b) but different mass densities (Equations 8.393a and 8.393b), where the frequency (Equation 8.393c) is independent of the mass densities, and the amplification depends (Equation 8.393d ≡ Equation 8.394c) on the mass densities:*

$$\rho_+ > \rho_-, U_\pm = U: \quad b = \sqrt{kg}\sqrt{\frac{\rho_+ - \rho_-}{\rho_+ + \rho_-}} = \sqrt{kg}\left[1 - \frac{\rho_-}{\rho_+} + O\left(\left(\frac{\rho_-}{\rho_+}\right)^2\right)\right] \tag{8.394a–8.394d}$$

and simplifies to Equation 8.394d when they are very dissimilar. The approximation from Equations 8.394c to 8.394d follows by

$$\begin{aligned}\sqrt{\frac{\rho_+ - \rho_-}{\rho_+ + \rho_-}} &= \sqrt{\frac{1-\rho_-/\rho_+}{1+\rho_-/\rho_+}} = \left\{\left(1 - \frac{\rho_-}{\rho_+}\right)\left[1 - \frac{\rho_-}{\rho_+} + O\left(\left(\frac{\rho_-}{\rho_+}\right)^2\right)\right]\right\}^{1/2} \\ &= \left[1 - 2\frac{\rho_-}{\rho_+} + O\left(\left(\frac{\rho_-}{\rho_+}\right)^2\right)\right]^{1/2} = 1 - \frac{\rho_-}{\rho_+} + O\left(\left(\frac{\rho_-}{\rho_+}\right)^2\right),\end{aligned} \tag{8.395}$$

completing the three subcases that are reconsidered next.

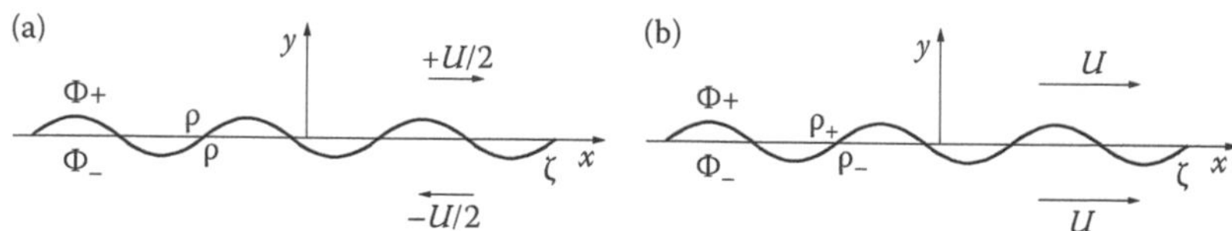

FIGURE 8.25
Vortex sheet separates (Figure 8.24) two tangential jets that may have distinct velocities and/or mass densities. The two simplest cases are (1) two jets with the same velocity and distinct mass densities (Figure 8.24b) and (2) two jets with the same mass density and different tangential velocities (Figure 8.24a). In case 1 of jets with equal tangential velocities, choosing a reference frame moving at their common velocity puts the whole fluid at rest; this leads to the interface between two fluids at rest, for example, air over water, or two immiscible liquids. In case 2 of two jets with distinct velocities, the moving reference frame may be chosen (2-a) at the velocity of one jet, leading to the interface between a fluid at rest on that side and a jet with the sum of velocities on the other side of the vortex sheet; and (2-b) at the arithmetic mean of the velocities of the two jets, so that the flows of the two sides of the vortex sheet move in opposite directions, with the same velocity in modulus, equal to the difference between the original velocities.

8.9.6 Convected Reference Frame and Doppler (1893) Shifted Frequency

The three subcases are comprehensive in the sense that (1) if the velocities coincide (Figure 8.25b), a vortex sheet exists only for distinct mass densities on each side (Equations 8.393a through 8.393d) corresponding to case 3; and (2) if the mass densities are the same, the vortex sheet exists only for distinct tangential velocities (Figure 8.25a). In the latter situation, a convected frame may be considered, namely, (2-a) moving with the velocity of one of the jets (Equation 8.396a) so that it is at rest (Equation 8.396b) and there is a nonzero velocity only on the other side of the vortex sheet (Equation 8.396c), as (Equations 8.391a through 8.391e) in case 1:

$$V \equiv U - U_-: \quad V_- = 0 \neq V_+ = U_+ - U_-; \qquad (8.396a–8.396c)$$

$$V \equiv U - \frac{1}{2}\left(U_+ + U_-\right): \quad V_\pm = U_\pm - \frac{1}{2}\left(U_+ + U_-\right) = \pm\frac{1}{2}\left(U_+ - U_-\right), \qquad (8.397a–8.397c)$$

and (2-b) moving with the average velocity of the two jets (Equation 8.397a), the velocities on the two sides of the vortex sheet are equal with opposite signs (Equations 8.397b and 8.397c) corresponding (Equations 8.392a through 8.392e) to case 2. The general case of distinct tangential velocities and mass densities on the two sides of the vortex sheet leads to the frequency (Equation 8.385b) and amplification factor (Equations 8.383b and 8.385c ≡ Equation 8.398):

$$b = \frac{a}{\rho_- + \rho_+} = \left[\left(k_x\right)^2 \frac{\rho_- \rho_+ \left(U_- - U_+\right)^2}{\left(\rho_+ + \rho_-\right)^2} + kg\frac{\rho_+ - \rho_-}{\rho_+ + \rho_-}\right]^{1/2} \qquad (8.398)$$

that follow from the dispersion relation (Equation 8.382b). The dispersion relation (Equation 8.382b) can be written in the form Equation 8.399b:

$$\begin{aligned}\omega_\pm \equiv \omega - k_x U_\pm: \quad 0 &= \rho_+\left(\omega - k_x U_+\right)^2 + \rho_-\left(\omega - k_x U_-\right)^2 + kg\left(\rho_+ - \rho_-\right) \\ &= \rho_+\left(\omega_+\right)^2 + \rho_-\left(\omega_-\right)^2 + kg\left(\rho_+ - \rho_-\right),\end{aligned} \qquad (8.399a \text{ and } 8.399b)$$

involving the **Doppler (1893) shifted frequencies** (Equation 8.399a), that is, the frequencies measured by an observer moving at the local flow velocity on each side of the vortex sheet. The general dispersion relation (Equations 8.399a and 8.399b) for the vortex sheet instability implies that (1) if the fluids have the same density $\rho_+ = \rho_-$, the sum of squares of the Doppler shifted frequencies is zero; (2) if gravity is neglected, the square of each Doppler shifted frequency is "weighted" by the mass density; (3) the gravity effect is proportional to the difference of mass densities; and (4 and 5) the gravity (Doppler shift) effect is isotropic (anisotropic) since it involves the modulus of the wavenumber (Equation 8.382b) [the wavenumber k_x in (Equation 8.399a) the flow direction].

NOTE 8.1 POTENTIAL AND OTHER TYPES OF FLOWS

The present section has addressed some cases other than the steady potential flows (Chapters I.12, I.14, I.16, I.28, I.34, I.36, and I.38), namely, some unsteady and rotational flows; it has not included viscosity (Notes 4.4 through 4.11) nor other dissipative processes like diffusion of heat and mass; also the brief consideration of separated flows has not included turbulence. Thus, it is no more than a faint hint of the multiple aspects of fluid mechanics that lie beyond potential flows. The potential flow can, in some cases, signal its own limitations; for example, (1) if the pressure in Bernoulli's equation approaches zero, such as when the flow is accelerated near a body, then cavitation will occur; and (2) if the pressure gradient is negative and large in a modulus in a flow region, such as behind a bluff body, then flow separation may occur. The potential flow does not apply to either of the preceding cases because (1) cavitation is a two-phase flow involving gas bubbles in a liquid, in which the gas phase has much higher compressibility than the liquid; and (2) the separated flow is vortical if deterministic and random if turbulent, and can interact strongly with viscosity and momentum transport. Thus, the potential flow provides a basic understanding of fluids, and its limitations point to the need for more general approaches to fluid mechanics.

The main subject of volumes I and II is the incompressible, irrotational, steady flow in two dimensions (Chapters I.12, I.14, I.16, I.28, I.34, I.36, and I.38). There is some mention of compressible (Sections I.14.6, I.14.7, and 2.1), rotational (Chapter 2), and unsteady flow (Section 8.9); three-dimensional potential flow is a generalization not including viscous (Notes 4.4 through 4.11) and turbulent flows. The irrotational, incompressible, steady two-dimensional flow supplies solutions to a large number of problems with reasonable economy of effort, providing in the process an illustration of the methods of analysis, in particular, the theory of functions of a complex variable. It is no more than an introduction to the more realistic vortical, compressible, unsteady, viscous, and turbulent flows; it must be admitted though that the latter are less tractable, and as more of these effects are combined, the number of solutions available decreases as the effort to obtain them increases. The properties of the simpler potential flows are also the baseline that serves for comparison of other effects. On balance, potential flows are a good introduction to fluid mechanics, but no more than a start.

NOTE 8.2 LIST OF POTENTIAL FLOW AND FIELD PROBLEMS

Classification 8.1 (pages 645–656) lists 150 potential flow or field problems considered in the first volume and the present volume in the context of (1) steady irrotational incompressible flow (Chapters I.12, I.14, I.16, I.28, I.34, I.36, I.38, II.2, and II.8); (2) steady irrotational compressible flow (Sections I.14.1 through I.14.7 and II.2.1); (3) steady rotational incompressible flow (Sections II.2.2 through II.2.9); (4) unsteady incompressible irrotational or rotational flow (Section II.8.9); (5) gravity field (Chapter I.18); (6) electrostatics (Chapters I.24 and I.36); (7) magnetostatics (Chapters I.20 and I.36); and (8) steady heat conduction (Chapter I.32). It groups together under (9) plane elasticity (Chapter II.4) and also (10 and 11) deflection of membranes (Sections II.6.1 and II.6.2) and surface tension and capillarity (Sections II.6.3 and II.6.4); (12) torsion of beams (Sections II.6.5 through II.6.8); (13) rotating flow (Section II.6.9); and (14) viscous flow (Notes II.4.4 through II.4.11, II.6.8, and II.6.9). Classification 8.1 includes (a) the text; (b) examples; (c) tables; (d) diagrams; (e) figures; (f) panels; and (g) a list of 50 conformal mappings (Chapters I.33.35, I.33.37, I.33.39, II.2, and II.8) including 25 Schwartz–Christoffell transformations.

The eight potential flow problems in the present chapter in List 8.1 (pages 658–659) are included, together with relevant examples from the rest of volume II, in Classification 8.1.

CLASSIFICATION 8.1

Classification of Potential Fields and Problems

Potential Fields	Topics	Problems	Potential Flow	Gravity Field	Electro Statics	Magneto Statics	Steady Heat Conduction	Plane Elasticity	C.M. (S.C.)
I General 8	a. Fundamental equations	I.32.3, II.2.4, 4.6; C.I.38.1; T.I.26.1; N.I.26.1, N.II.4.4–11; F.I.12.1, 14.1; F.II.4.1–9, 6.1–4, 6.10–11	I.14.1–6	I.18.3–5	I.24.1–3	I.26.1–4	I.32.1–3	II.4.1–4; II.6.1–2, 5–8	–
	b. Conditions of validity	I.32.3	I.14.1–6	I.18.3	I.24.1–2	I.26.1–2	I.32.1	II.4.3	–
	c. Poisson/ biharmonic equation	I.37.8–9; II.2.4, 4.6.1–8	I.12.1–3	I.18.3	I.24.2	I.26.2	I.32.1	II.4.4	–
	d. Boundary conditions	I.38.1.1; D.I.38.1	I.18.1–2; F.I.18.1–2	I.18.3; F.I.18.5, C.I.38.1	I.24.2	I.26.2; F.I.26.4	I.32.2; T.I.32.1–2		–
	e. Force	I.28.2–3, 28.9; E.I.30.13–16; F.I.28.2–3, 6–9, 11; F.II.4.5–6, 4.13–16 6.1–2, 6.9–10	I.28.2–3	I.18.5, I.28.3; F.I.18.3–4	I.24.3, I.28.3	I.26.3, I.28.3	–	II.4.2, 4	–
	f. Energy	I.28.1; II.4.3.7–8; N.I.28.2	I.28.1; E.I.30.12	I.18.3, I.28.1	I.24.2, I.28.1	I.26.2, I.28.1	I.32.1	II.4.3.7–8, 6.1.1–2	–
	g. Irrotational (R) or solenoidal (S)	I.26.3, 32.2; C.I.38.1; T.I.26.1; D.I.40.2	C.I.38.1	C.I.38.1	C.I.38.1	C.I.38.1	C.I.38.1	II.D.6.1	–
II Basic 14	A Corners 8	1.Uniform flow	I.14.8.1; F.I.14.5–6	×	×	×	I.32.4.1	×	–
		2. Rectangular corner	I.14.8.2; F.I.14.7–8	×	×	×	I.32.4.1	×	1(1)
		3. Semi-infinite plate	I.14.9.1; F.I.14.10b	×	×	×	I.32.4.1	×	1(1)

(continued)

CLASSIFICATION 8.1 (Continued)

Classification of Potential Fields and Problems

Potential Fields	Topics	Problems	Potential Flow	Gravity Field	Electro Statics	Magneto Statics	Steady Heat Conduction	Plane Elasticity	C.M. (S.C.)
II Basic 14	A Corners 8	4. Submultiple of 2π or π	I.14.8.2; F.I.14.9–10	×		×	I.32.4.1	×	1(1)
		5. Arbitrary corner $0 < \alpha < 2\pi$	I.14.8–9, 36.1; E.I.20.8; T.I.14.1–2; F.I.14.5–10, 36.1a, 36.4a,b	×	×	×	I.32.4.1	×	1(1)
		6. Incidence on a wedge	I.14.9.2; F.I.14.11	×	×	×	×	II.4.7, II.4.6.9–10; F.II.4.11–13	2
		7. Rounded corner	I.36.1.1; F.I.36.2	×	×	×	×	×	3
		8. Corner with indentation	I.36.1.2; F.I.36.5a,b	×	I.36.1.3; F.I.36.2	I.36.1.3; F.I.36.3	×	×	4
	B Singularities 6	9. Source/sink	I.12.4; E.I.20.12; F.I.12.2a	×	I.24.3	–	×	II.4.5, 8.9; F.II.4.10,16	–
		10. Vortex	I.12.5; E.I.20.12; F.I.12.2b	×	–	I.26.3,5	×	II.4.5, 8.9; F.II.4.10,16	–
		11. Monopole/spiral	I.12.6; E.I.20.12; F.I.12c,d; F.I.28.8	×	–	–	×	II.4.5, 8.9; F.II.4.10,16	–
		12. Dipole	I.12.7; F.I.12.3–4	×	I.24.4	I.26.5	×	×	–
		13. Quadrupole	I.12.8; F.I.12.5–6	×	I.24.4	I.26.5	×	×	–
		14. Multipole	I.12.9; E.I.20.3; F.I.12.7	×	I.24.4	I.26.5	×	×	–

III Distributions 9	C Finite 6	15. Slab	×	I.18.6–7; F.I.18.5	×	×	×	×	–
		16. Crossed plates	×	E.I.20.19; F.I.20.4c	×	×	×	×	–
		17. Two parallel plates	×	E.I.20.18; F.I.20.4a	×	×	×	×	–
		18. Stack of parallel plates	×	E.I.20.18; F.I.20.4b	×	×	×	×	–
		19. Plane potential	×	×	I.38.1.2–3; E.I.40.15.1; F.I.38.1a; F.I.40.9a	×	×	×	–
		20. Cylinder potential	×	×	I.38.1.2, 4; E.I.40.15.2; F.I.38.1b, F.I.40.9b	×	×	×	–
	D Infinite 3	21. Monopole: irrotational	×	I.18.8; T.I.18.1	×	×	×	II.4.7.5–6	–
		22. Dipole: irrotational	×	E.I.20.20	×	–	×	×	–
		23. Dipole: solenoidal	×	–	–	I.26.9; F.I.26.4	–	×	–
IV Images 16	E On a plane 4	24. Source/sink	I.16.1; F.I.16.1–2	×	I.24.4; F.I.24.1	–	×	×	5
		25. Vortex	I.16.2; F.I.16.3	×	–	I.26.5	–	×	5
		26. Monopole	I.16.3; II.2.7.4; F.II.2.10–12	×	–	–	–	×	5
		27. Multipole	I.16.4	×	I.24.4	I.26.5	×	×	5
IV Images 16	F Rectangular corner 3	28. Source/sink	I.16.5–7; F.I.16.4–5	×	I.24.5; F.I.24.2	–	×	×	1
		29. Vortex	I.16.5–7; F.I.16.4–5	×	–	I.26.6	–	×	1
		30. Monopole	I.16.5–7; E.I.20.11			×	×	×	1

(continued)

CLASSIFICATION 8.1 (Continued)

Classification of Potential Fields and Problems

Potential Fields	Topics	Problems	Potential Flow	Gravity Field	Electro Statics	Magneto Statics	Steady Heat Conduction	Plane Elasticity	C.M. (S.C.)
IV Images 16	G Acute corner 5	31. Source/sink	I.16.8; I.36.2.1; F.I.36.4a	–	I.24.5; I.36.2.2; F.I.24.2; F.I.36.4a,c	–	I.32.4.2	II.4.7.2; F.II.4.11	1
		32. Vortex	I.16.9; I.36.2.1; F.I.36.4b	–	–	I.26.6; I.36.2.4; F.I.36.4b,c	–	II.4.7.3; F.II.4.12	1
		33. Monopole	I.16.9; I.36.2.1	–	I.24.5	–	–		1
		34. Monopole with indentation	I.36.2.3; E.I.40.11; F.I.36.5a,b	×	I.36.2.4; E.I.40.10.1–4; F.I.36.5a,c; F.I.40.4	I.36.2.4; F.I.36.5b,c	×	×	1
		35. Multipole	E.I.40.11	×	×	×	×	×	1
	H Interfaces 4	36. Plane	×	×	I.24.9; F.I.24.6	×	×	II.4.7.4–5; F.II.4.13	–
		37. Cylindrical	×	×	×	I.26.8; F.I.26.2–3	×	II.4.5.1–4,8; F.II.4.10a–d,f	–
		38. Multiple layers	×	×	×	×	I.32.8; F.I.32.8	II.4.5.5–7; F.II.4.10e	–
		39. Inhomogeneous layer	×	×	×		I.32.9; F.I.32.9	×	–
V One cylinder 14	I Isolated 2	40. Cylinder in uniform flow	I.28.6;F.I.28.4a	×	I.24.6; F.I.24.3	I.26.7.1; F.I.26.1	I.32.5–7; F.I.32.3 –7	II.4.5,9; F.I.4.10,16	6
		41. Cylinder with circulation	I.28.7; F.I.28.4b–d	–	–	×	–	II.4.5.8; F.II.4.10f	6

V One cylinder 14	J Images on cylinder 4	42. Source/sink	I.28.8; F.I.28.13	×	I.24.7–8; F.I.24.4–5	×	×	×	6
		43. Vortex	I.28.8; F.I.28.13	×	–	I.26.7.2	×	×	6
		44. Dipole	I.28.9; F.I.28.14	×	×	×	×	×	6
		45. Two trailing vortices	II.2.8–9; T.II.3.4; F.II.2.16–20	–	–	×	–	×	7
	K Cylinder with wall 6	46. Depression on wall	II.8.3.6; F.II.8.7d–f	×	×	×	×	×	8
		47. Bump on wall	II.8.3.6; F.II.8.7a–c	×	×	×	×	×	8
		48. On wall: cylindrical log	II.8.3.8–10; F.II.8.7j, 8	×	×	×	×	×	8
		49. Separate from wall	E.II.11; 1–5; F.II.10.2	×	E.II.11.6.9; F.II.10.3	E.II.11.9–10; F.II.10.4	×	×	9
		50. Moving perpendicular to	II.8.2.6–7; F.II.8.4b	×	×	×	×	×	9
		51. Moving parallel to a wall	E.II.10.4; F.II.10.1b						
	L Noncircular 2	52. Elliptic	I.34.2; F.I.34,2a	×	×	×	×	×	10
		53. Biconvex/ Biconcave	II.8.3.6; F.II.8.7g,i	×	×	×	×	×	11
VI Two cylinders 7	M Internal 4	54. Hollow cylinder and cylindrical shell	×	×	×	×	I.32.6–7; F.I.32.5–7	II.4.5.3–4; F.II.4.10c,d	–
		55. Concentric	×	×	×	×	I.32.8; F.I.32.8	×	–
		56. Eccentric	II.8.3.4–5; F.II.8.5–6a	×	×	×	×	×	12
		57. Moving in a cavity	II.8.1; F.II.8.1	×	×	×	×	×	13

(*continued*)

CLASSIFICATION 8.1 (Continued)

Classification of Potential Fields and Problems

Potential Fields	Topics	Problems	Potential Flow	Gravity Field	Electro Statics	Magneto Statics	Steady Heat Conduction	Plane Elasticity	C.M. (S.C.)
VI Two cylinders 7	N External 3	58. General	II.8.2.1; T.II.8.1; F.I.8.2–3	×	×	×	×	×	9
		59. Moving toward each other	II.8.2.2–5; F.II.8.4a	×	×	×	×	×	9
		60. Moving in parallel	E.II.10.1–3; F.II.10.1a	×	×	×	×	×	9
VII Bodies 29	O Flat plates 17	61. Flat plate with circulation	I.36.4.1; II.8.8.1; F.I.36.9a; F.II.8.17, 21	I.18.6–7; F.I.18.5	I.36.4.3; F.I.36.9a,d	E.36.4.3; F.I.36.9a,d	×	×	10(2)
		62. Flat plate in a stream	I.34.1; F.I.34.1; II.8.8.2–8	×	×	×	×	×	–
		63. With flap	I.8.7.1-3; T.II.8.5; F.II.8.16a,b, 19a–c	×	×	×	×	×	–
		64. With slotted flap	II.8.7.4–6; T.II.8.5; F.II.8.16c,d, 18, 19d	×	×	×	×	×	–
		65. Tandem aligned with circulation	II.8.8.9; F.II.8.22	×	×	×	×	×	–
		66. Tandem in a stream	II.8.8.10; FII.8.22	×	×	×	×	×	–
		67. Cascade	II.8.8.11–14; T.II.8.6; F.II.8.23	×	×	×	×	×	11
VII Bodies 29	O Flat plates 17	68. Thick sharp plate	II.8.4.6; F.II.8.10a	×	×	×	×	×	12(3)
		69. Thick blunt plate	II.8.4.7; F.II.8.10b	×	×	×	×	×	12(3)
		70. Thick sharp plate in channel	II.8.5.5; F.II.8.12a	×	×	×	×	×	13(4)

		71. Thick blunt plate in a channel	II.8.5.6–8; F.II.8.12b	×	×	×	×	×	13(4)
		72. Thin plate in a channel		×	E.II.10.13; F.II.10.6	×	×	×	14(5)
		73. With one oblique plate	E.II.10.18.1–8; F.II.10.14–15	×	×	×	×	×	15–16(6–7)
		74. With two oblique plates	E.II.10.18.9–11; F.II.10.16–18	×	×	×	×	×	17–18(8–9)
		75. T-junction	E.II.10.18.2, 7; F.II.10.15	×	×	×	×	×	19 (6–7)
		76. Cruciform	E.II.10.18.10; F.II.10.18	×	×	×	×	×	20 (8–9)
		77. Crossed plates	II.10.18.9; F.II.10.17	×	×	×	×	×	20 (8–9)
	P Airfoils 7	78. Circular arc	I.34.3.2; F.I.34.2d,3a	×	×	×	×	×	21(10)
		79. Symmetric Joukowski	I.34.3.1; F.I.34.2c	–	–	–	–	–	21(10)
		80. Cambered Joukowski	I.34.4; E.I.40.6; F.I.34.3b; E.I.40.2	–	–	–	–	–	21(10)
		81. Karman–Trefftz	I.34.5.1; F.I.34.4a–d	–	–	–	–	–	23
VII Bodies 29	P Airfoils 7	82. Von Mises	I.34.5.2; F.I.34.4e	–	–	–	–	–	22
		83. Carafoli	I.34.5.3; F.I.34.4f	–	–	–	–	–	21–23
		84. Generic	I.34.6; D.I.34.1; F.I.34.5	–	–	–	–	–	24
	Q Aerodynamics 5	85. Bent lamina	E.II.10.14; F.II.10.4a,c		×	×	×	×	25(11)
		86. Wing	I.34.7–9; F.I.34.6–11	–	–	–	–	–	–

(*continued*)

CLASSIFICATION 8.1 (Continued)

Classification of Potential Fields and Problems

Potential Fields	Topics	Problems	Potential Flow	Gravity Field	Electro Statics	Magneto Statics	Steady Heat Conduction	Plane Elasticity	C.M. (S.C.)
VII Bodies 29	Q Aerodynamics 5	87. Rankine fairings	I.28.4; F.I.28.5	×	×	×	×	×	–
		88. Valley between mountains	I.28.5.3; F.I.28.10a	×	×	×	×	×	–
		89. Rankine oval	I.28.5.1–2; F.I.28.7		×	×	×	×	–
VIII Jets 12	R Jets 7	90. Normal to a plate	I.38.4; F.I.38.4	×	–	–	–	–	26
		91. Oblique to a plate	I.38.5; T.I.38.1; F.I.38.6	×	–	–	–	–	27
		92. Reverse flow behind arrow	I.38.6; T.38.2; F.I.38.7b	×	–	–	–	–	28
		93. Colliding/ splitting jets	I.38.9.1; F.I.38.10	×	–	–	–	–	29
		94. Surfing on a stream	I.38.5; F.I.38.6	–	–	–	–	–	30
		95. Deflection by vortex	E.II.10.17; F.II.10.13	–	–	–	–	–	31
		96. With internal source	E.I.40.10; F.I.40.10	×	–	–	–	–	32
VIII Jets 12	S With wall 5	97. Orthogonal to a wall	I.38.9.2–4; F.I.38.11	×	–	–	–	–	32
		98. Deflection by source (fluidic effect)	I.38.8; F.I.38.9	–	–	–	–	–	33
		99. Attachment to a wall (Coanda effect)	I.38.7; F.I.38.8	–	–	–	–	–	34
		100. Slit in a reservoir	I.38.2; F.I.38.3a,b	–	–	–	–	–	35
		101. Tube in a reservoir	I.38.3; F.I.38.3a,c	–	–	–	–	–	36

IV Ducts 25	T Between walls 5	102. Source/sink between walls	I.36.6; E.I.40.12; FI.36.13,14a; 14a; F.I.40.6a,7a	×	I.36.7.2; F.I.36.14a,b	×	–	×	37(12)
		103. Source/sink on a wall	I.36.8.1–2; F.I.38.15	×	I.36.8.1–2; F.I.36.15	×	–	×	37(12)
		104. Vortex between walls	I.36.7; E.I.40.12; F.I.36.14c; F.I.40.6b, 7a	×	–	I.36.7.3; F.I.36.14c,d	–	×	37(12)
		105. Spiral between walls	E.I.40.13; T.I.40.3; F.I.40.7	×	–	–	–	×	37(12)
		106. Dipole between walls	E.I.40.14; T.I.40.4; F.I.40.8	×	×	×	×	×	37(12)
IV Ducts 25	T In a well 4	107. Source/sink at a corner	I.36.8.1–2; F.I.36.15	×	I.36.8.1–2; F.I.36.15	–	×	×	38(13)
		108. Source/sink in a well	I.36.8.3; F.I.36.16a	×	I.36.9.2–4, I.40.10.6–7; F.I.36.16a,b; E.I.40.5	–	×	×	38(13)
		109. Vortex in a well: vortex street	I.36.9.1; F.I.36.16c,d	×	–	I.36.9.2–3; F.I.36.16c,d	×	×	38(13)
		110. Staggered double vortex street	I.36.9.5; F.I.36.17	–	–	×	–	–	38(13)
	U Open 6	111. Convergent plates	I.36.5.1; F.I.36.11	×	×	×	×	×	39(14)
		112. Parallel-sided channel	I.36.5.2; F.I.36.12a,b	×	I.36.5.4; F.I.36.12c	×	×	×	40(15)
		113. Slit in a wall	I.36.4.2; F.I.36.8b,10	×	I.36.4.3; F.I.36.9c,d,10	I.36.4.3; F.I.36.9c,d,10	×	×	41(16)
		114. Plate on wall	E.II.10.12; F.II.10.5a,b	×	×	×	×	×	42(17)
		115. Exit along wall	E.II.14.1; F.II.10.7a,b	×	×	×	×	×	43(18)

(*continued*)

CLASSIFICATION 8.1 (Continued)

Classification of Potential Fields and Problems

Potential Fields	Topics	Problems	Potential Flow	Gravity Field	Electro Statics	Magneto Statics	Steady Heat Conduction	Plane Elasticity	C.M. (S.C.)
IV Ducts 25	V Nonuniform 5	116. Exit from reservoir	E.II.15.2; F.II.10.9a,b	×	×	×	×	×	43(18)
		117. Potential flow	I.14.7.1; F.I.14.3a	×	–	–	–	–	–
		118. Rotational flow	I.14.7.2; F.I.14.3b	×	–	–	–	–	–
		119. Ramp/step in a wall	II.8.4.1–3; F.II.8.9a,b	×	×	×	×	×	44(19)
		120. Duct with contraction/ expansion	II.8.4.4–5; F.II.8.9a,c	×	×	×	×	×	45(20)
		121. Throated duct	I.28.5.3; F.I.28.10b	×	×	×	×	×	–
IV Ducts 25	W Junctions 5	122. One side branch	E.II.10.16.1–3; F.II.10.12a	×	×	×	×	×	46(21)
		123. Two side branches	E.II.10.16.4; F.II.10.12b	×	×	×	×	×	46(21)
		124. Corner	E.II.10.15; F.II.10.9a,b, 10a	×	×	×	×	×	47(22)
		125. T- or Y- junction	E.II.10.15; F.II.10.9a,c, 10b	×	×	×	×	×	48(23)
		126. Triple plate	E.II.10.15; F.II.10.5	×	×	×	×	×	49(24)
X Nonpotential flows 13	X Compressible 5	127. Pitot tube	I.14.6; I.36.5.3; F.I.14.2a; F.I.36.12b	×	–	–	–	–	50(25)
		128. Venturi tube	I.14.7; F.I.14.2b	×	–	–	–	–	–
		129. Source/sink	II.2.1.9; F.II.2.1b	×	–	–	–	–	–
		130. Vortex	II.2.1.8; F.II.2.1a	×	–	–	–	–	–
		131. Monopole	II.2.1.10; F.II.2.1b	×	×	×	–	–	–

X Nonpotential flows 13	Y Vortical 6	132. General	II.2.3–6; F.II.2.2–7	×	–	×	–	×	–
		133. Nonpotential vortex	II.2.2; E.II.10.3–4; T.II.2.1–2; P.II.1–3	×	–	–	–	×	–
		134. Unidirectional shear flow	II.2.4; F.II.2.3	×	–	–	×	×	–
		135. Cylinder in shear flow	II.2.6; F.II.2.4–7	×	–	–	×	×	–
		136. Rotating	II.6.9; F.II.6.14	×	–	–	–	×	–
		137. Viscous	N.II.4.4–11; N.II.6.8–9; F.II.15	×	–	–	–	×	–
	Z Unsteady 2	138. Airfoil with separation	II.8.2; T.II.8.2; F.II.8.12–13	×	–	–	–	–	–
		139. Vortex sheet	II.8.9; F.II.8.11–12	×	–	–	–	–	–
XI Paths 11	Trajectories 11	140. General	II.2.9.1–2; T.II.2.4	×	×	×	–	–	–
		141. Near wall	I.16.1–2; II.2.7.4; F.II.2.11–12	×	×	×	–	–	–
		142. Vortex/source/sink past semi-infinite plate	I.36.3; T.I.36.1–3; F.I.36.6–7	×	×	×	–	–	–
		143. In a corner	I.16.6–7; F.I.16.4–5	×	×	×	–	–	–
		144. Vortex pair	II.2.7.3; F.II.2.9	×	×	×	–	–	–
		145. Two sources and/or sinks	II.2.7.2; F.II.2.8	×	×	×	–	–	–

(*continued*)

CLASSIFICATION 8.1 (Continued)

Classification of Potential Fields and Problems

Potential Fields	Topics	Problems	Potential Flow	Gravity Field	Electro Statics	Magneto Statics	Steady Heat Conduction	Plane Elasticity	C.M. (S.C.)
XI Paths 11	Trajectories 11	146. Source/sink near a cylinder	I.28.8.4; F.I.28.13	×	×	×	–	–	–
		147. Vortex near cylinder	I.28.8.3; II.2.9.3; F.I.28.13	×	×	×	–	–	–
		148. In a duct	II.2.7.5–9; F.II.2.13–15	×	×	×	–	–	–
		149. Pair inside/ outside cylinder	II.2.9.4–6; F.II.2.20	×	×	×	–	–	–
		150. Multiple monopoles	II.2.7.1	×	×	×	–	–	–

Note: Recollection lists in tabular form 150 problems related to potential field starting with general properties and ending with the paths of singularities. The potential flow is taken as the reference case, since it corresponds to the majority of explicit solutions. The analog problems for (1) gravity, (2) electrostatic and (3) magnetostatic fields, and for (4) heat conduction and (5) plane elasticity are mentioned; the analogy is closer in cases 1–3 than in cases 4 and 5. A small fraction of the analog problems were treated explicitly; of the majority that were not treated explicitly, the relevant cases are indicated by "×." For the cases treated explicitly, they are indicated with the location of the text (in the format volume, chapter, section, and subsection) and also related tables, notes, diagrams, figures, examples, and lists. Among the 150 problems used are 50 conformal mappings of which 25 are of Schwartz–Christoffell type. A further 10 topics appearing in volumes I and II are listed at the end. The volume, chapter, section, subsection, classification (C), diagram (D), example (E), figure (F), list (L), note (N), panel (P), or table (T) where the problem is explicitly discussed is indicated; "×" means that the problem is relevant but has not been explicitly discussed; for example, it is analogous to another problem that has been treated in detail; – means that it is not possible or not too relevant in this context.

Related potential field problems—*In space domain*–151. Deflection of membrane: II.6.1–2, T.II.6.1, P.I.6.1, F.II.6.1–4; 152. Surface tension and capillarity: II.6.3–4, F.II.6.5–9; 153. Torsion of rods: II.6.5–8, T.II.6.2, F.II.6.10–14; 154. Fluid in a rotating vessel: II.6.9, F.II.6.15; 155. Inviscid shear flow: II.2.2–6, T.II.2.1–2, F.II.2.2–7, P.II.2.1–3; 156. Viscous shear flow: N.II.4.4–11, N.II.6.7–9, F.II.6.16. *In time domain* – 157. Stability of vortex sheet: II.8.9, F.II.8.24–25; 158. Stability of equilibrium position: I.2, E.I.10.3–5, T.I.2.1, F.I.2.1–2; 159. Motion of pendulum or ship: I.8, E.I.10.13–14, F.I.8.1; 160. Associations of mechanical or electrical impedances: I.4, E.I.10.8, T.I.4.1, F.I.4.1–3, F.I.10.2; 161. Trajectories of electron in electromagnetic field: I.6, E.I.10.10–12, T.I.6.1–2, F.I.6.1–9, F.I.10.3; 162. Refraction of light in a lens: I.22, E.I.30.4–5, F.I.22.1–6, F.I.30.1.

Symbols for Classification II.8.1: Text (Volume. Chapter. Section. Subsection); C-Classification; D-Diagram: E-Example; F-Figure; L-List; N-Note; P-Panel: T-Table.

Last column for classification II.8.1: Conformal mapping (C.M.) [Schwartz–Christoffell transformation (S.C.)] numbers 1 to 50 (1 to 25).

The latter lists 150 potential flow problems from volumes I and II. The text is located by the sequence volume, chapter, section, and subsection. All these 150 potential flow problems are two-dimensional; there are three-dimensional analogs as well as extensions to higher dimensions. Nearly all of the 150 problems have been considered for a (1) potential flow, which is taken as the baseline. Many of the problems have analogs for (2) gravity, (3) electrostatic and (4) magnetostatic fields, (5) steady heat conduction, and (6) plane elasticity. The foundations of these six major types of potential fields are discussed, as well as a typical sample of a few problems. Most potential field problems have analogs in several of areas 1–6; this is indicated by a cross, and the details are not repeated. The cases when the analogy does not apply are indicated by "–"; for example, the source/sink (vortex) in a potential flow has as analog the electric charge (current) in the electrostatic (magnetostatic) field, but not the other way around: the electrostatic (magnetostatic) field is irrotational (solenoidal) and thus has no analog for a vortex (source/sink).

All conformal mappings obtained from the same formula giving different values to the parameters are considered as nondistinct, that is, the "same" transformation, and counted as one; for example, (1) the corner flow is one Schwartz–Christoffell mapping, with one critical point, and applies to the uniform flow, the rectangular, acute, or obtuse corner, and corner angles submultiples or not submultiples of π and 2π; (2) the Joukowski transformation is one mapping with different parameters for the flat plate, elliptic cylinder, circular arc, thick symmetric, and thick cambered airfoils; and (3) the von Mises and von Karmann–Trefftz airfoils are two distinct mappings, but the Carafoli airfoils are choices of parameters in these and the Joukowski airfoils. The (a) 150 two-dimensional flow problems, including (b) 50 distinct conformal mappings, of which (c) 25 are distinct Schwartz–Christoffell transformations, form three sets (a, b, c) of examples. The volumes I and II of the present series address: (1) the 150 potential field problems in the Classification 8.1; (2) twelve additional related problems, six each in the space and time domains, indicated in the footnote to Classification 8.1.

8.10 Conclusion

The potential flow due to a cylinder moving uniformly in a large cylindrical cavity (Figure 8.1) involves different boundary conditions on the inner moving (outer static) cylinder. In the case of two dissimilar cylinders one outside the other, the relative motion (Figure 8.2) can be decomposed into motion along [Figure 8.4a (Figure 10.1a)] (across) the line of centers. In the case of close cylinders (Figure 8.3), the potential flow is represented by an infinite set of pairs of images (Table 8.1). Only the first images are needed for cylinders whose distance is large compared with the radius. The case of two identical cylinders moving with the same velocity away (Figure 8.4b) or toward each other is equivalent to images on a wall at equal distances. The images of a line vortex on a wall (Figure 8.5a) lead to equipotential and streamlines that specify coaxial coordinates (Figure 8.5b and Table 8.2). These orthogonal curvilinear conformal coordinates can be used (Table 8.3) to describe the following: (1) the potential flow of a cylinder with circulation placed eccentrically in a cylindrical tunnel (Figure 8.6a); (2) the case of a tunnel with an infinite radius leads to a cylinder with circulation near a wall (Figure 8.6b) (3) past a wall with a cylindrical mound or ditch (Figure 8.7a through f); (4) a cylinder with biconcave or biconvex cross section made of circular arcs (Figure 8.7g through i); and (5) a cylindrical log on a plane wall (Figure 8.7j). The cylindrical log on a wall (Figure 8.8b) shows the wall effect of flow acceleration over the upper rounded part, when compared with the same cylinder in a free stream (Figure 8.8a).

Another "obstacle" on a wall is [Figure 8.9b (Figure 8.9c)] a backward ramp (step) that can be obtained from a polygonal mapping (Figure 8.9a) with two critical points at the edges (corners); by symmetry, it leads [Figure 8.10a (Figure 8.10b)] to a thick semi-infinite plate with a sharp (blunt) nose in a free stream. Adding a parallel wall [Figure 8.11b (Figure 8.11c)] to the backward ramp (step) leads to gradual (abrupt) increase in the cross section of a channel; it can be obtained from a polygonal mapping with three critical points (Figure 8.11a) and leads by symmetry [Figure 8.12a (Figure 8.12b)] to a thick semi-infinite plate with sharp (blunt) nose placed at equal distance from parallel walls. The polygonal mappings for the thick plate in a free stream (between parallel walls) have [Figures 8.10 and 8.12 (Figures 8.11 and 8.12)] two (three) critical points. The simplest case of one critical point (Figure 8.14b) leads to the corner flow (Figure 8.14a). A particular case is the thin semi-infinite plate (Figure 8.14c); the latter is also the particular case of zero thickness of the thick semi-infinite plate (Figure 8.10). The thickness is taken to be zero in the simplest model of an airfoil, that is, flat plate (Figure 8.13b), that leads to separated flow (Figure 8.13a) beyond the angle-of-attack for the maximum lift coefficient (Figure 8.15); the airfoil with separated flow has a lift loss (Table 8.4) relative to an airfoil with attached flow; conversely flow separation, may be delayed or prevented, thus increasing the lift of an airfoil by using on the leading (trailing) edge a slat (flap); as shown in Figure 8.16a and 8.16c (Figure 8.16b and d), it may be either plain [Figure 8.16a (Figure 8.16b)] or slotted [Figure 8.16c (Figure 8.16d)]; the deflection of the flap (and/or slat) increases the lift (Table 8.5) and the slot delays separation until higher angle-of-attack (Figure 8.19). Thus the lift of an airfoil (Figure 18.9a) is increased using a plain flap (Figure 8.19b); the flow separation and lift loss at higher angles-of-attack (Figure 8.19c) may be prevented or delayed until larger angle-of-attack by using a slotted flap (Figure 8.19d).

The Schwartz–Christoffell interior polygonal (exterior circle) transformation (Figure 8.20) includes as simplest cases the corner flow (Figure 8.13) [flat plate airfoil (Figure 8.21). The latter can be considered via the Joukowski transformation (Figure 8.21) or by the Sedov method (Figure 8.17). The Sedov theorem can be used to consider the flow past (1) a single (Figures 8.17) airfoil and (2) multiple airfoils, for example, for a slotted airfoil (Figure 8.18), two airfoils in tandem (Figure 8.22) or a cascade (Figure 8.23). Depending on the chord of the airfoils and spacing between them, the cascade may change more or less the magnitude and direction of the velocity far downstream compared with the incident flow (Table 8.6). The simplest wake of airfoils, such as in a cascade (Figure 8.23), with flaps (Figure 8.18) or with separated flow (Figure 8.13a), is a plane vortex sheet (Figure 8.24) between jets with distinct velocities and mass densities; the vortex sheet is unstable relative to small perturbations, which amplify the displacement over time, due to the accumulation of circulation when the velocities (mass densities) on the two sides are different [Figure 8.25a (Figure 8.25b)]. The preceding flows are related to a wider class of potential fields (Classification 8.1).

LIST 8.1

TWENTY-EIGHT CONFINED AND UNSTEADY FLOWS

A. Moving cylinder(s):

1. In a cylindrical cavity: Section 8.1
2. Two close cylinders: Subsection 8.2.1
3. Moving along the line of centers: Subsections 8.2.2 through 8.2.5
4. Moving perpendicular to the line of centers: Example 10.10
5. Moving perpendicular to a wall: Subsections 8.2.6 and 8.2.7
6. Moving parallel to a wall: Example 10.10

B. Static cylinder(s):

7. Static eccentric cylinders: Subsections 8.3.4 and 8.3.5
8. Cylindrical mound/ditch on a wall: Subsection 8.3.6
9. Cylindrical log on a wall: Subsections 8.3.8 through 8.3.10
10. Biconvex/biconcave cylinder in a stream: Subsection 8.3.7

C. Ramp (s)/step(s) on wall(s):

11. Oblique ramp/rectangular step on a wall: Subsections 8.4.2 and 8.4.3
12. Rectangular step on a wall: Subsections 8.4.4 and 8.4.5
13. Gradual change of cross section of a channel: Subsections 8.5.1 and 8.5.2
14. Sharp change of cross section of a channel: Subsections 8.5.3 and 8.5.4

D. Thick sharp/blunt-nosed semi-infinite plate:

15. Sharp-nosed thick plate in a free stream: Subsection 8.4.6
16. Blunt square plate in a free stream: Subsection 8.4.7
17. Sharp-nosed thick plate in a parallel-sided channel: Subsection 8.5.5
18. Blunt thick plate in a parallel sided channel: Subsections 8.5.6 and 8.5.7

E. Single airfoil:

19. With circulation: Subsections 8.8.3 and 8.8.4
20. In a free stream: Subsections 8.8.5 and 8.8.6
21. With circulation in a free stream: Subsections 8.8.7 and 8.8.8

F. Multiple airfoils:

22. Tandem with circulation: Subsection 8.8.9
23. Tandem in a free stream: Subsection 8.8.10
24. Airfoil with plain slat: Subsections 8.7.2 and 8.7.3
25. Airfoil with slotted flap: Subsections 8.7.4 through 8.7.6
26. Cascade of parallel airfoils: Subsections 8.8.11 and 8.8.14

G. Vortical and separated flows:

27. Airfoil with separated flow: Section 8.6
28. Instability of vortex sheet: Section 8.9

9

Infinite Processes and Summability

The questions of convergence arise for infinite processes such as (1) series including power series (Chapters I.23, I.25 and I.27 and Section 1.1) or series of fractions (Sections 1.2 and 1.3) for functions analytic except for poles or essential singularities; (2) infinite products for functions with an infinite number of zeros that meet the Mittag-Leffler (Weierstrass) theorems [Sections 1.4 and 1.5 (Section 9.2)]; (3) nonterminating continued fractions (Sections 1.5 through 1.9) including recurrent and nonrecurrent (Section 9.3); and (4) improper integrals of the first (second) kind that is (Section I.17.1) with one or two infinite limits (an integral singular at a point within or at one end of the path of integration). A combined test of convergence has been derived (Chapter I.29) that applies at all points of the complex plane for a wide class of series of functions. The combined convergence test can also be applied to other infinite processes by relating them to series, namely, (1) an improper integral of the first or second kind of a monotonic function has lower and upper bounds that are specified by series and thus can be used to establish convergence (Section 9.1); (2) to each infinite product a series is associated that converges, diverges, and oscillates in the same conditions (Section 9.2); and (3) the convergence of a continued fraction can be established by considering some associated series (Section 9.3).

A recurrent series or infinite product consisting beyond a certain order of a set of repeated terms generally does not converge; in contrast, recurrent continued fractions with positive real numbers converge to surd numbers, that is, numbers of the form $a+\sqrt{b/c}$ with a, b, c positive integers (Section 9.3). The convergent infinite processes are those that have a limit, for example, the sum of an infinite series. There are two types of convergence stronger than ordinary convergence (Chapter I.21). The absolute convergence or convergence of the series of moduli allows (Section I.21.3) derangement of the order of the terms of the original series without changing its sum; this is not the case for a conditionally convergent series, which is a convergent series that is not absolutely convergent. For a conditionally convergent series, the terms can be deranged (Section 9.4) in such a way that (1) it converges to any number x given *a priori*; (2) it oscillates in any interval (a,b) given *priori*; and (3) it diverges. A second, distinct, and independent form of convergence is uniform convergence of a series of functions in a domain of the variable, which allows (Sections I.21.5 and I.21.6) taking limits, differentiation, and integration term-by-term. The combination of absolute and uniform convergence is total convergence (Section I.21.7). A convergent series has a finite sum, for example, a series whose general term is a rational function that is analytic except for poles can be summed using the residues at the poles (Section 9.5).

The usual definition of convergence to class C0 is a particular case of class Cn, which applies to series and other infinite processes convertible to series, such as infinite products. If a series converges to class Cn, then it also converges to class Cm for all $m > n$ and has the same sum; however, some series that do not converge to class Cn may converge to class $Cn + 1$, that is, in some cases, it is possible to sum to class Ck series that are oscillating or divergent to class C0. A divergent series that is not summable to any order is by definition (Section 9.6) unbounded; thus, its sum tends to infinity. There are distinct "kinds of infinity"; for example, (1) the number of elements of the set of positive integers is infinite but denumerable; (2) the number of real numbers in a finite interval is infinite and not denumerable; and (3) the number of possible values of a discontinuous function in a finite interval is infinite nondenumerable like item 2 both for the independent and dependent variables. The three examples correspond, respectively, to (1) enumerable, (2) continuum, and (3) discontinuous (Section 9.7). These are "transfinite numbers" of two kinds: (1) the "cardinal transfinite numbers" specify the number of elements of a set with an infinite number of elements, such as the integers, rational, irrational, real, complex numbers, and ordered n-tuples made of those numbers; and (2) the "ordinal transfinite numbers" arise from imposing in a set with an infinite number of elements an order relation that establishes a precedence between them. The basic operations of the algebra of numbers, such as the sum and product, can be extended (Section 9.8) to the transfinite cardinal and ordinal numbers; although the theory of transfinite numbers succeeds in classifying distinct kinds of infinity, beyond a certain point, it leads to contradictions or antimonies (Section 9.9) that restrict or question further developments.

9.1 Bounds for Integrals of Monotonic Functions

Since a complex integral (series) consists of two real integrals (series), namely, the real and imaginary parts, only real expressions need be considered as concerns proper or improper integrals (Section I.17.1). Next are obtained bounds for improper integrals of the first (second) kind of a real monotonic function [Subsection 9.1.1 (Subsection 9.1.2)].

9.1.1 Monotonic Nonincreasing Functions and Improper Integrals of the First Kind

A real function $g(x)$ [$f(x)$] is **monotonic decreasing** $\mathcal{M}^-$ in Equation 9.1a (**monotonic nonincreasing** $\mathcal{M}_0^-$ in Equation 9.1b) if its value decreases (decreases or remains constant) as the argument increases:

$$\mathcal{M}^- \equiv \{g(x): y > x \Rightarrow g(y) < g(x)\} \subset \mathcal{M}_0^- \equiv \{f(x): y > x \Rightarrow f(y) \le f(x)\}. \quad \text{(9.1a and 9.1b)}$$

Consider next improper integrals of the first kind, that is, with an infinite range of integration; the bilateral integral over the real line (Equation 9.2) is the sum of integrals over the positive and negative half-real lines, that is, the sum of two unilateral integrals:

$$\int_{-\infty}^{+\infty} f(x)\,dx = \left(\int_{-\infty}^{0} + \int_{0}^{\infty}\right) f(x)\,dx = \int_{0}^{\infty} \{f(x) + f(-x)\}\,dx; \quad (9.2)$$

any unilateral integral that is with one limit of integration + ∞ (−∞) can be reduced to an integral between 0 and ∞(0 and −∞):

$$\int_{a}^{\infty} f(x)\,dx = \left(\int_{a}^{0} + \int_{0}^{\infty}\right) f(x) = \left(\int_{0}^{\infty} - \int_{0}^{a}\right) f(x)\,dx, \quad \text{(9.3a and 9.3b)}$$

plus a proper integral along the interval $(0, a)$ that raises no convergence issues. In all cases (Equations 9.2, 9.3a, and 9.3b), only the convergence of a unilateral improper integral of the first kind has to be considered, with integration along the positive real half-axis:

$$\int_0^\infty f(x)\,\mathrm{d}x = \sum_{n=0}^\infty \int_n^{n+1} f(x)\,\mathrm{d}x; \tag{9.4}$$

an integrand that is (Equation 9.1) monotonic nonincreasing

$$\sum_{n=0}^\infty f(n)\int_n^{n+1} \mathrm{d}x \geq \int_0^\infty f(x)\,\mathrm{d}x \geq \sum_{n=0}^\infty f(n+1)\int_n^{n+1} \mathrm{d}x \tag{9.5}$$

leads to the upper and lower bounds:

$$f(x) \in \mathcal{M}_0^-(0,+\infty(: \quad \sum_{n=0}^\infty f(n) \geq \int_0^\infty f(x)\,\mathrm{d}x \geq \sum_{n=1}^\infty f(n). \tag{9.6a–9.6c}$$

It has been shown that *the convergence of a bilateral (Equation 9.2) [unilateral (Equations 9.3a and 9.3b)] improper integral of the first kind is reducible to that (Equation 9.4) of a unilateral improper integral between 0 and ∞ only (plus a proper integral); if the integrand is monotonic nonincreasing (Equation 9.1b ≡ Equation 9.6a), the latter integral has upper (Equation 9.6b) and lower (Equation 9.6c) bounds given by the series.*

9.1.2 Monotonic Nondecreasing Functions and Improper Integrals of the Second Kind

The convergence of an improper integral of the third kind, that is, a mixed integral with singularity at b:

$$-\infty > b > a,\, f(b) = \infty: \quad \int_{\pm\infty}^a f(x)\,\mathrm{d}x = \int_{\pm\infty}^b f(x)\,\mathrm{d}x + \int_b^a f(x)\,\mathrm{d}x, \tag{9.7a–9.7c}$$

is reduced to the study of convergence of improper integrals of the first and second kinds. Since the former has already been considered, the latter is considered next:

$$y = \frac{1}{x-b},\, x = b + \frac{1}{y},\, \mathrm{d}x = -\frac{\mathrm{d}y}{y^2}: \quad \int_b^a f(x)\,\mathrm{d}x = \int_{1/(a-b)}^{+\infty} f\left(b + \frac{1}{y}\right) y^{-2}\,\mathrm{d}y, \tag{9.8a–9.8d}$$

showing that *an improper integral of the second kind transforms (Equation 9.8d) to the first kind using an inversion (Equations 9.8a through 9.8c) in the variable of integration.* If the function $g(x)$ [$f(x)$] is **monotonic increasing** $\mathcal{M}^+$ in (Equation 9.9a) [**monotonic nondecreasing** $\mathcal{M}_0^+$ in (Equation 9.9b)]:

$$\mathcal{M}^+ = \{g(x): y > x \Rightarrow g(y) > g(x)\} \subset \mathcal{M}_0^+ \equiv \{f(x): y > x \Rightarrow f(y) \geq f(x)\}, \tag{9.9a and 9.9b}$$

then $f(b + 1/y)$ is monotonic decreasing (Equation 9.1a) [monotonic nonincreasing (Equation 9.1b)], that is, belongs to $\mathcal{M}^-(\mathcal{M}_0^-)$. In this case, Equation 9.8d becomes (Equation 9.10a) an improper integral of the first kind between 1 and ∞:

$$f\left(b+\frac{1}{y}\right) \in \mathcal{M}_0^-(a,b): \quad \int_b^a f(x)\,\mathrm{d}x = \left(\int_{1/(a-b)}^1 + \int_1^\infty\right) f\left(b+\frac{1}{y}\right)\frac{\mathrm{d}y}{y^{-2}}; \qquad \text{(9.10a and 9.10b)}$$

thus the bounds (Equations 9.6b and 9.6c) apply:

$$f(x) \in \mathcal{M}_0^+(b,a): \quad \sum_{n=1}^\infty n^{-2} f\left(b+\frac{1}{n}\right) \ge \int_1^\infty f\left(b+\frac{1}{y}\right) y^{-2}\,\mathrm{d}y \ge \sum_{n=2}^\infty n^{-2} f\left(b+\frac{1}{n}\right). \qquad \text{(9.11a–9.11c)}$$

It has been shown that *an improper integral of the third kind can be decomposed into (Equations 9.7a through 9.7c) the sum of improper integrals of the first and second kinds; the improper integrals of the first (Equation 9.4) [second (Equations 9.8a through 9.8d)] kind, of a nonincreasing (Equation 9.1b ≡ Equation 9.6a) [nondecreasing (Equation 9.9b ≡ Equation 9.11a)] function, has upper and lower bounds given by Equations 9.6b and 9.6c (Equations 9.11b and 9.11c)*. As an example of Equation 9.5, consider the factorial notation $n!$ that is an increasing function for $n > 0$, so that $1/n!$ is a decreasing function, and

$$e = \sum_{n=0}^\infty \frac{1}{n!} \ge \int_0^\infty \frac{\mathrm{d}x}{x!} \ge \sum_{n=1}^\infty \frac{1}{n!} = e - 1, \qquad (9.12)$$

where e is the irrational number (Equation 3.7).

9.2 Genus of an Infinite Canonical Product (Weierstrass 1876)

The classification of infinite products according to their convergence properties (Subsection 9.2.1) is similar to that of infinite series (Section I.21.1); this allows associating to each infinite product an infinite series with the same converge properties (Subsection 9.2.2). The consideration of the zeros and essential singularities of integral functions (Subsection 9.2.3) leads to the relations with meromorphic functions (Subsection 9.2.4). The canonical infinite products for functions with an infinite number of zeros extend (Subsections 9.2.5 and 9.2.6) the Mittag-Leffler (1876, 1884) theorem (Sections 1.4 and 1.5). The Mittag-Leffler infinite products are the case of genus unity of a Weierstrass canonical infinite product (Subsection 9.2.7).

9.2.1 Convergent, Divergent, and Oscillatory Infinite Products

The infinite product of functions is defined by Equation 9.13b:

$$g_n(z) \neq 0: \quad \prod_{n=1}^\infty g_n(z) \equiv \lim_{N\to\infty}\left\{\prod_{n=1}^N g_n(z)\right\} \qquad \text{(9.13a and 9.13b)}$$

and is nontrivial if no factor is zero (Equation 9.13a). As for an infinite series (Section I.21.1), an infinite product

$$\prod_{n=1}^{\infty} g_n(z) \begin{cases} = G < \infty & \text{C.: convergent,} \quad (9.14a) \\ = \pm\infty & \text{D.: divergent,} \quad (9.14b) \\ = A, B < \infty & \text{O.: oscillating} \quad (9.14c) \end{cases}$$

is (1) convergent if Equation 9.13b tends to a finite limit (Equation 9.14a); (2) divergent if Equation 9.13b is unbounded (Equation 9.14b); and (3) oscillating in the remaining case (Equation 9.14b), that is, if Equation 9.13b is bounded but tends to no limit or has at least two distinct accumulation points A and B. *If the general term $g_n(z)$ of an infinite product tends to a number g smaller (larger) than unity $g < 1$ ($g > 1$), the infinite product (Equation 9.13b) converges to zero $G = 0$ in Equation 9.15a [diverges (Equation 9.15b)]:*

$$\lim_{n\to\infty} g_n(z) = g \begin{cases} > 1 \\ \\ < 1 \end{cases} \Rightarrow \begin{rcases} \infty \\ \\ 0 \end{rcases} = \prod_{n=1}^{\infty} g_n(z). \qquad (9.15\text{a and } 9.15\text{b})$$

It follows that *a necessary but not sufficient condition that an infinite product be convergent to a value other than zero is that the general term tends to unity:*

$$\prod_{n=1}^{\infty} g_n(z) = G(z) \neq 0 \quad \Rightarrow \quad \lim_{n\to\infty} g_n(z) = 1. \qquad (9.16)$$

The condition is necessary because, if it is not met, there is either divergence (Equation 9.15a) or convergence to zero (Equation 9.15b). It is not sufficient because the convergence properties of an infinite product coincide with those of the series of logarithms:

$$\prod_{n=1}^{\infty} g_n(z) = G(z) \quad \Leftrightarrow \quad \sum_{n=1}^{\infty} \log\{g_n(z)\} = \log\{G(z)\}, \qquad (9.17\text{a and } 9.17\text{b})$$

and the condition $g_n \to 1$, $\log(g_n) \to 0$ as $n \to \infty$ is necessary but not sufficient (Subsection I.29.2.1) for the convergence of the series. The property (Equations 9.17a and 9.17b) can be used to prove Equation 9.15b (Equation 9.15a); if $g_n \to g < 1$ ($g > 1$) as $n \to \infty$, then $\log(g_n) \to \log g < 0$ (>0), and the series diverges to $\log G = -\infty$ ($+\infty$), that is, $G \to 0$ (∞).

9.2.2 Series with Same Convergence Properties as an Infinite Product

The property Equation 9.16 suggests that the general term of the infinite product (Equation 9.3b) be rewritten in the form Equation 9.18a leading

$$g_n(z) = 1 + h_n(z): \quad \prod_{n=1}^{\infty} \{1 + h_n(z)\} \quad \Leftrightarrow \quad \sum_{n=1}^{\infty} \log\{1 + h_n(z)\} \quad \Leftrightarrow \quad \sum_{n=1}^{\infty} h_n(z) \qquad (9.18\text{a–}9.18\text{d})$$

to the infinite product (Equation 9.18b), that has the same convergence properties (Equation 9.17b) as the series of logarithms (Equation 9.18c), which in turn behaves like the series Equation 9.18d. To prove the latter point, note that if the last series (Equation 9.18d) converges (Equation 9.19a), then (1) the general term must tend to zero (Equation 9.19b); (2) this implies that beyond a certain order, it cannot exceed one-half in modulus (Equation 9.19c):

$$\sum_{n=1}^{\infty} h_n(z)C. \quad \Rightarrow \quad \lim_{n\to\infty} h_n(z) = 0 \quad \Rightarrow \quad \exists_{N\in|N} \forall_{n\in|N}: n > N \quad \Rightarrow \quad |\mathrm{h}_n(z)| < \frac{1}{2}; \quad (9.19a\text{–}9.19c)$$

(3) using the logarithmic series (Equations 3.47a and 3.47b) follows

$$\left|\frac{\log(1+h_n)-1}{h_n}\right| \le \left|-\frac{h_n}{2}+\frac{h_n^2}{3}-\frac{h_n^3}{4}\right| \le \left|\frac{h_n}{2}\right| + \left|\frac{h_n}{3}\right|^2 + \left|\frac{h_n}{4}\right|^3 + \cdots$$
$$\le 2^{-2}+2^{-3}+2^{-4}+\cdots = \frac{2^{-2}}{1-2^{-1}} = \frac{1}{2}; \quad (9.19d)$$

(4) this implies that the series of logarithms is bounded, above and below, by the series of general term:

$$\exists_{N\in N} \forall_{n>N}: \quad \frac{1}{2}|h_n(z)| \le \log|1+h_n(z)| \le \frac{3}{2}|h_n(z)|; \quad (9.20)$$

and (5) hence the convergence of the series Equation 9.18d implies that of the series Equation 9.18c and that of the infinite product Equation 9.18a. If the series Equation 9.18d diverges, then (1) either h_n does not tend to zero, and the infinite product Equation 9.18b diverges because $1 + h_n$ does not tend to unity (Equation 9.16); (2) $h_n \to 0$ as $n \to \infty$, so that Equation 9.20 holds, and the divergence of the series Equation 9.18d implies that of the series Equation 9.18c and infinite product Equation 9.18b. Since the series Equations 9.18b and 9.18c and product Equation 9.18a converge and diverge together, the same applies to the remaining case, that is oscillation.

It has been proved that *the infinite product (Equation 9.16 ≡ Equation 9.18b) with general term Equation 9.18a, the series of logarithms Equation 9.18b, and the series Equation 9.18d diverge, oscillate, and converge together (Equations 9.14a through 9.14c); this applies also to conditional, absolute, uniform, and total convergence.* Thus, the convergence of infinite products can be investigated (Example 10.19) considering the corresponding series and applying the known convergence tests for the latter (Chapter I.21), such as the combined convergence test that holds at all points of the complex plane (Chapter I.29). For example, the infinite products for the sine (cosine), either circular [Equation 7.226b (Equation 7.226c)] or hyperbolic [Equation 7.227b (Equation 7.227c)], are of the form [Equation 9.21a (Equation 9.21b)]:

$$\sin z \sim z \prod_{n=1}^{\infty}\left[1+O\left(\frac{1}{n^2}\right)\right] \sim \sinh z, \cos z \sim \prod_{n=1}^{\infty}\left[1+O\left(\frac{1}{n^2}\right)\right] \sim \cosh z; \quad (9.21a \text{ and } 9.21b)$$

their general term (Equation 9.22b) implies that their convergence is the same as that of the series Equation 9.22c, that is (Section I.29.4.2), totally convergent in the finite complex plane (Equation 9.22a):

$$|z| \le \mathcal{M} < \infty: \quad g_n \sim 1 + O\left(\frac{1}{n^2}\right), \quad h_n \sim 1 + O\left(\frac{1}{n^2}\right) \quad \text{T.C.} \qquad (9.22a\text{–}9.22c)$$

Thus, it has been proved that *the infinite products for the circular (hyperbolic) cosine [Equation 7.226c (Equation 7.227c)] and sine [Equation 7.226b (Equation 7.227b)] are (1) absolutely convergent for* $|z| < \infty$ *in agreement with Equation 7.226a ≡ Equation 7.227a and (2) also uniformly convergent for* $|z| \le M < \infty$.

9.2.3 Function without Zeros or with a Single Essential Singularity

If the function $f(z)$ is analytic and has no zeros (Equation 9.23b):

$$f(z) \in \mathcal{A}(|z| < \infty), \quad f'(z) \ne 0: \quad g(z) \equiv \frac{f'(z)}{f(z)} = \frac{\mathrm{d}}{\mathrm{d}z} = \frac{\mathrm{d}\{\log(f(z))\}}{\mathrm{d}z}, \qquad (9.23a \text{ and } 9.23b)$$

then its logarithmic derivative (Equation 9.23c) is also analytic (Equation 9.24a) and can be integrated (Equation 9.24b):

$$g(z) \in \mathcal{A}(|z| < \infty): \quad f(z) = \exp\{g(z)\}. \qquad (9.24a \text{ and } 9.24b)$$

Thus, *a nonzero analytic function Equations 9.23a and 9.23b is the exponential of an analytic function Equations 9.24a and 9.24b. Generalizing, it follows that an analytic function Equation 9.25a with (Equations 9.25b and 9.25c) a finite number N of zeros* a_n *of orders* α_n*:*

$$f(z) \in \mathcal{A}(|z| < \infty); \quad n = 1, \cdots, N: \quad f^{(\alpha_n)}(a_n) \ne 0 = f^{(\alpha_n - 1)}(a_n) \qquad (9.25a\text{–}9.25c)$$

is given by Equation 9.26b:

$$g(z) \in \mathcal{A}(|z| < \infty): \quad f(z) = \exp\{g(z)\} \prod_{n=1}^{N} (z - a_n)^{\alpha_n}, \qquad (9.26a \text{ and } 9.26b)$$

where Equation 9.26a is an analytic function. In the limit $N \to \infty$ of an infinite number of zeros, the function $g(z)$ must be chosen so as to assure the convergence of the infinite product (Subsection 9.2.1). If a function has a finite number of poles, then it is a rational function, and it will have zeros, unless the numerator reduces to a constant.

9.2.4 Relation between Integral and Meromorphic Functions

A more interesting case is a function $F(z)$ analytic and without zeros, except in the neighborhood of a single essential singularity $z = a$, leading to the two Laurent series:

$$F(z) = \sum_{n=1}^{\infty} A_{-n}(z-a)^{-n}, \quad \frac{F'(z)}{F(z)} = \sum_{m=2}^{\infty} a_{-m}(z-a)^{-m} \qquad \text{(9.27a and 9.27b)}$$

because both $F'(z)$ and $1/F(z)$ also have an essential singularity at $z = a$ (Subsection I.39.2.2). Integrating Equation 9.27b:

$$\log\{F(z)\} = \sum_{m=2}^{\infty} \frac{a_{-m}}{1-m}(z-a)^{1-m} = G\left(\frac{1}{z-a}\right) \qquad (9.28)$$

leads to an integral (Equation 9.29a) function (Equation 9.29b) of variable $1/(z - a)$:

$$G(1/(z-a)) \in \mathcal{J}: \quad F(z) = \exp\left\{G\left(\frac{1}{z-a}\right)\right\}. \qquad \text{(9.29a and 9.29b)}$$

Hence, *a function* $F(z)$ *analytic and without zeros, except for an essential singularity (Equation 9.27a) at* $z = a$*, is the exponential (Equation 9.29b) of an integral function (Equation 9.29a) of* $\zeta \equiv 1/(z - a)$*, that is, the exponential of an analytic function for finite* ζ *with an essential singularity at* $\zeta = \infty$ *or* $z = a$. The poles of a meromorphic function $f(z)$ can be eliminated by multiplication by an integral function $g(z)$ that has zeros at the same points a_n, with the same order α_n. This leads $f(z)g(z) = h(z)$ to an integral function and confirms the result (Subsection 1.3.1) that

$$\forall_{F(z)\in\mathcal{J}} \quad \exists_{g(z),\in\mathcal{J}}: \quad f(z) = \frac{g(z)}{h(z)} \qquad \text{(9.30a–9.30c)}$$

a meromorphic function (Equation 9.30a) is the ratio of two integral functions (Equation 9.30b). The meromorphic function has an infinite number of poles and no essential singularity; the integral functions have no poles and only one essential singularity at infinity.

9.2.5 Canonical Infinite Product (Weierstrass 1876)

Infinite product representations have been obtained for functions with an infinite number of zeros, subject to the restriction (Section 1.4) that the logarithmic derivative be bounded on a sequence of circles tending to infinity. This restriction will be lifted next, and it will be shown that the exponential factors, some of which already appeared for example in Equation 1.67, may be needed to assure convergence. Consider an analytic function in the finite complex plane with an infinite number (Equation 9.31d ≡ Equation 9.25c) of zeros a_n of order α_n including the origin (Equation 9.31b) of order α:

$$f \in \mathcal{A} \equiv (|z| < \infty): \quad f^{(\alpha)}(0) \neq 0 = f^{(\alpha-1)}(0); \quad n = 1, \cdots, \infty: \quad f^{(\alpha_N)}(a_n) \neq 0 = f^{(\alpha_n-1)}(a_n); \qquad \text{(9.31a–9.31d)}$$

it has the representation (Equation 9.32c):

$$g, g_n \in \mathcal{A}(|z| < \infty): \quad f(z) = z^{\alpha} e^{g(z)} \prod_{n=1}^{\infty} \left(1 - \frac{z}{a_n}\right)^{\alpha_n} \exp\{g_n(z)\}, \qquad \text{(9.32a–9.32c)}$$

where the functions Equations 9.32a and 9.32b are analytic and can be chosen so as to ensure the convergence of the infinite product (Equation 9.32c). A function analytic in the finite complex plane can be a polynomial in which case (1) it has a finite number of zeros; (2) their orders add to the degree of the polynomial; and (3) the latter is also the order of the pole at infinity. A function analytic in the finite plane that is not a polynomial must have an essential singularity at infinity, and hence is an integral function. An integral function may have (1) no zeros, for example, the exponential; (2) a finite number of zeros, for example, the product of the exponential and a polynomial; and (3) an infinite number of zeros, for example, the circular and hyperbolic cosine and sine. In case 3, the infinite number of zeros cannot have an accumulation point in the finite plane; if it had it would be an essential singularity in the finite plane, and the function would be polymorphic rather than integral. Thus, (1) the function analytic in the finite plane (Equation 9.31a) and with an infinite number of zeros (Equations 9.31c and 9.31d) must be an integral function (Equation 9.33a); and (2) if the zeros are ordered by modulus (Equation 9.33b), they must tend to infinity (Equation 9.33c):

$$f(z) \in \mathcal{J}; \quad |a_1| \le |a_2| \le \cdots \le |a_n| \le \cdots: \quad \lim_{n\to\infty} |a_n| = \infty. \tag{9.33a–9.33c}$$

Let R be a positive real number, and separate the roots a_n in two sets, namely, those whose modulus is smaller (larger) than R:

$$|a_1| \le |a_2| \le \cdots \le |a_s| < R < |a_{s+1}| \le \cdots \le |a_n| \le \cdots. \tag{9.33d}$$

The roots inside the circle of radius R and center at the origin correspond to the first s factors of the infinite product (Equation 9.32b) and specify an integral function (Equation 9.34a):

$$h(z) = \prod_{n=1}^{s} \left(1 - \frac{z}{a_s}\right) \exp\left[g_s(z)\right]; \tag{9.34a}$$

$$f(z) = z^{\alpha} e^{s(z)} h(z) \prod_{n=s+1}^{\infty} \left(1 - \frac{z}{a_n}\right) \exp\left[g_n(z)\right]; \tag{9.34b}$$

thus the convergence of the infinite product (Equation 9.32c ≡ Equation 9.34b) is determined by the roots outside the radius R. The infinite product (Equation 9.32c ≡ Equation 9.34b) has the same convergence properties (Equation 9.18b) as the series Equation 9.18c of general term (Equation 9.35):

$$f_n(z) \equiv \log\left\{\left(1 - \frac{z}{a_n}\right)^{\alpha_n} \exp\left(g_n(z)\right)\right\} = \alpha_n \log\left(1 - \frac{z}{a_n}\right) + g_n(z) = g_n(z) - \alpha_n \sum_{m=1}^{\infty} \frac{(z/a_n)^m}{m}, \tag{9.35}$$

where the logarithmic series Equations 3.47a and 3.47b was used.

9.2.6 Total Convergence of Canonical Product

Choosing the analytic function $g_n(z)$ to be the polynomial of degree β_n, which cancels the first β_n terms in Equation 9.35:

$$g_n(z) \equiv \alpha_n \left[\frac{z}{a_n} + \frac{1}{2}\left(\frac{z}{a_n}\right) + \cdots + \frac{1}{\beta_n}\left(\frac{z}{a_n}\right)^{\beta_n} \right], \tag{9.36}$$

(1) the expansion (Equation 9.35) starts with the power $\beta_n + 1$:

$$f_n(z) = -\frac{\alpha_n}{1+\beta_n}\left(\frac{z}{a_n}\right)^{\beta_n+1}\left(1 + \frac{1+\beta_n}{2+\beta_n}\frac{z}{a_n} + \cdots\right); \tag{9.37}$$

(2) thus Equation 9.37 has an upper bound in the modulus (Equation 9.38b):

$$|z| \le R \le |a_n|: \quad |f_n(z)| \le \frac{\alpha_n}{1+\beta_n}\left(\frac{R}{|a_n|}\right)^{\beta_n+1}\sum_{m=0}^{\infty}\left(\frac{|z|}{a_n}\right)^m \le \frac{\alpha_n}{1+\beta_n}\left(\frac{R}{|a_n|}\right)^{1+\beta_n}\left(1-\frac{R}{|a_n|}\right)^{-1}, \tag{9.38a and 9.38b}$$

where the geometric series (Equation 3.148b ≡ Equations I.21.62a and I.21.62b) in the circle (Equation 9.38a) was used; (3) choosing Equation 9.39a leads from Equation 9.38b to Equation 9.39b:

$$\beta_n = n-1: \quad |f_n(z)| \le \alpha_n\left(1-\frac{R}{|a_n|}\right)^{-1}\left(\frac{R}{|a_n|}\right)^n; \tag{9.39a and 9.39b}$$

(4) taking the upper bounds (Equations 9.40a and 9.40b) from Equation 9.39b follows (Equation 9.40c):

$$|\alpha_n| \le \alpha, \quad \frac{R}{|a_n|} \le \varepsilon < 1: \quad \sum_{n=s+1}^{\infty}|f_n(z)| \le \frac{\alpha}{1-\varepsilon}\sum_{n=s+1}^{\infty}\varepsilon^n = \frac{\alpha\varepsilon^{s+1}}{1-\varepsilon}\sum_{n=0}^{\infty}\varepsilon^n = \alpha\varepsilon^{s+1}, \tag{9.40a–9.40c}$$

where the geometric series was used again; (5) since the series of moduli with general term Equation 9.35 has an upper bound (Equation 9.40c) independent of z, it is absolutely and uniformly convergent in the circle (Equation 9.38a) of radius R; and (6) this proves the total convergence of the canonical infinite product (Equation 9.32c ≡ Equation 9.43b) in the finite complex z-plane (Equation 9.43a).

9.2.7 Genus of Canonical Infinite Product

Step 5 of the preceding derivation can be proved in an alternate way using the Cauchy necessary and sufficient condition for the convergence of a series (Subsection I.29.2.2). The latter states that, for any δ however small, there is an order s beyond which the sum of the terms of the series does not exceed δ:

$$\forall_{\delta>0} \quad \exists_{s\in|N}: \quad \sum_{n=s+1}^{\infty} |f_n| < \delta. \tag{9.41}$$

Comparing Equation 9.41 with Equation 9.40c, it follows (Equation 9.42a) that it is sufficient to choose Equation 9.42b for the order:

$$\alpha\, \varepsilon^s < \delta \quad \Rightarrow \quad s > \frac{\log \delta - \log \alpha}{\log \varepsilon}. \tag{9.42a and 9.42b}$$

Since the condition 9.41 is met (Equation 9.42b) for the series of moduli, and independently of z in Equation 9.38a, the convergence is absolute and uniform; besides, the value of the radius of convergence R is arbitrary (Equation 9.43a). The points $z = a_n$ that have to be excluded from Equation 9.33d do not affect the convergence in Equation 9.32c. The preceding result generalizes the Mittag-Leffler theorem on infinite products (Sections 1.4 and 1.5) by dispensing with the condition of boundedness on a sequence of circles (Figure 1.1) that is not essential. Thus, has been proved: the **theorem on canonical products (Weierstrass 1876)** *an integral function (Equation 9.31a ≡ Equation 9.33a) with a zero of order* α *at the origin (Equation 9.31b), and an infinite number (Equation 9.31c) of zeros (Equation 9.31d) of orders* α_n *at* a_n*, tending to infinity (Equations 9.33b and 9.33c), can be represented in the finite complex plane (Equation 9.43a) by a totally convergent infinite product (Equation 9.43b):*

$$|z| \le R < \infty: \quad f(z) = z^{\alpha} e^{g(z)} \prod_{n=1}^{\infty} \frac{z}{a_n} \exp\left\{\alpha_n \left[\frac{z}{a_n} + \left(\frac{z}{a_n}\right)^2 + \cdots + \frac{1}{\beta_n}\left(\frac{z}{a_n}\right)^{\beta_n}\right]\right\}, \tag{9.43a and 9.43b}$$

where $g(z)$ *is an integral function (Equation 9.26a), and the positive integer numbers* β_n *are chosen so that the series of general term (Equation 9.37) can be totally convergent*. If all $\beta_n \equiv m$ are equal, then Equations 9.43a and 9.43b are a **canonical product of genus m**, for example, $m = 1$ in Equation 1.67 showing that Mittag-Leffler's theorem leads to a canonical product of genus unity; thus, *an integral function, with an infinite number of zeros* a_n *of order* α_n*, such that* $a_n \to \infty$ *as* $n \to \infty$*, and with bounded logarithmic derivative in a sequence of circles* $|z| = R_n$ *containing the first* n *zeros, is of genus 1*. The simplest case is a function of genus 0 (Equation 9.44a):

$$\beta_n = 0: \quad f(z) = z^{\alpha} e^{g(z)} \prod_{n=1}^{\infty} \left(1 - \frac{z}{a_n}\right)^{\alpha_n}, \tag{9.44a and 9.44b}$$

which is the only case not needing (Equation 9.44b) exponential factors to ensure the convergence of the infinite product. The genus of the canonical infinite product of an integral function with an infinite number of zeros (Subsection 9.2.5) bears no relation to the genus of a Riemann surface that serves as the topological representation of a many-valued function with several branch-cuts (Section I.9.9).

9.3 Convergent and Periodic Continued Fractions

As for infinite products (Subsection 9.2.2), the equivalent infinite series with the same convergence properties (Subsection 9.3.1) can be associated to a complex nonterminating continued fraction. This leads to two alternative convergence criteria (Subsection 9.3.2) for a real nonterminating continued fraction of the first kind whose numerators and denominators are all positive (Subsection 1.7.1); the particular case of numerators all unity leads to the simple continued fraction (Subsection 1.7.3) for which there is one simpler convergence criterion (Subsection 9.3.3). In the particular case of a simple recurrent nonterminating continued fraction, whose terms are repeated cyclically beyond a certain order (Subsection 9.3.4), the reduction to an equivalent terminating continued fraction specifies the limit. For a simple positive periodic continued fraction, whose terms are positive integers and repeat cyclically from the beginning, the limit is generally a root of a binomial (Subsection 9.3.5), as illustrated numerically in a specific case (Subsection 9.3.6).

9.3.1 Series with Same Convergence Properties as a Continued Fraction

The convergence (Equation 9.45) of the general continued fraction (Equation 1.96)

$$f = a_1 + \frac{b_2}{a_2 +}\frac{b_3}{a_3 +}\frac{b_4}{a_4 +}\cdots\frac{b_n}{a_n +}\cdots = \lim_{n\to\infty} X_n \equiv \lim_{n\to\infty}\frac{p_n}{q_n} \tag{9.45}$$

may be considered comparing with an equivalent series whose sum of the first n terms equals the nth convergent (Equation 9.46a) leading to Equation 9.46b:

$$\sum_{k=1}^{n} f_k = X_n, \quad f_n = \sum_{k=1}^{n} f_k - \sum_{k=1}^{n-1} f_k = X_n - X_{n-1}. \tag{9.46a and 9.46b}$$

Using Equation 1.107a follows the series

$$f = \sum_{n=1}^{\infty} f_n = \sum_{n=1}^{\infty} (-)^n \frac{b_2 \cdots b_n}{q_n q_{n-1}}, \tag{9.47}$$

whose ratio of successive terms, for application of the combined test (Section I.29.1), is

$$\frac{f_n(z)}{f_{n+1}(z)} = -\frac{1}{b_{n+1}(z)}\frac{q_{n+1}(z)}{q_{n-1}(z)}. \tag{9.48}$$

It has been proved that *the general complex continued fraction (Equation 9.45) has the same convergence properties as the series Equation 9.47, with ratio of succeeding terms Equation 9.48, where q_n satisfies the recurrence relations (Equation 1.102b) with initial values (Equations 1.99b and 1.99d)*. The combined convergence test is readily applied to Equation 9.47 if q_n is known, most simply for $q_n = 1$. Criteria of convergence that do not depend on q_n may be obtained for continued fractions of the first kind (Subsection 1.7.2), for which all terms

(Equations 9.49a and 9.49b) are positive and the even (odd) continuants form a decreasing (increasing) sequence, which exceeds (Equation 9.49c) [lies below (Equation 9.49d)] the limit:

$$a_n, b_n > 0: \quad \lim_{n\to\infty} X_{2n} = A \geq B = \lim_{n\to\infty} X_{2n-1}. \tag{9.49a–9.49d}$$

The result (Equations 9.49a through 9.49d) shows that *a continued fraction (Equation 9.45) of the first kind (Equations 9.49a and 9.49b) cannot diverge; it oscillates in the range (B, A) if the even and odd convergences have distinct limits, and it converges if the limits coincide A = B*. The preceding theorem excludes divergence but does not prove convergence because oscillation is also possible. Next the oscillatory case is excluded leading to a convergence theorem.

9.3.2 Convergence of Continued Fractions of the First Kind

In order to obtain an explicit convergence test for continued fraction of the first kind, excluding the oscillatory case, the recurrence formula 1.102b is used, bearing in mind that Equations 9.49a and 9.49b imply Equation 9.50a and hence also Equation 9.50b:

$$q_{n-1} = a_{n-1}\, q_{n-2} + a_{n-2}\, q_{n-3} > a_{n-1}\, q_{n-2}, \tag{9.50a}$$

$$q_n = a_n\, q_{n-1} + b_n\, q_{n-2} > (a_n\, a_{n-1} + b_n) q_{n-2}; \tag{9.50b}$$

continuing by iteration leads to

$$q_n\, q_{n-1} > (a_n\, a_{n-1} + b_n) \cdots (a_2\, a_3 + b_3) q_1\, q_2 = (a_n\, a_{n-1} + b_n) \cdots (a_2\, a_3 + b_3) a_2 \tag{9.51}$$

using Equations 1.99b and 1.99d. The series Equation 9.47 that is equivalent (Equation 9.51) to the continued fraction Equation 9.45 has a general term:

$$(-)^n f_n < \frac{b_2 \cdots b_n}{a_2 (a_2\, a_3 + b_3) \cdots (a_n\, a_{n-1} + b_n)} = \frac{b_2 / a_2}{(1 + a_2\, a_3 / b_3) \cdots (1 + a_n\, a_{n-1} / b_n)} \tag{9.52}$$

from which convergence conditions can be deduced. It is shown next that either of condition 9.53c or 9.53d:

$$a_n, b_n > 0: \quad \lim_{n\to\infty} \frac{a_{n-1}\, a_n}{b_n} > 0 \; or \; \sum_{n=1}^{\infty} \frac{a_{n-1}\, a_n}{b_n} D. \quad \Rightarrow \quad a_1 + \frac{a_2}{b_2 +} \cdots \frac{a_n}{b_n +} \cdots C. \tag{9.53a–9.53e}$$

implies the convergence (Equation 9.53c) of the continued fraction (Equation 9.45) of the first kind (Equations 9.53a and 9.53b). Note that Equation 9.53c implies Equation 9.53d, but the inverse is not true.

9.3.3 Convergence of Positive Simple Continued Fractions

The proof concerns two cases. First condition 9.53c implies that Equation 9.52 is an alternating series of decreasing terms, hence convergent (Equation I.29.56). Second if the series Equation 9.53d diverges, so does the infinite product:

$$\infty = \lim_{n\to\infty} \prod_{k=1}^{n} \left(1 + \frac{a_{k-1}\, a_k}{b_k}\right) = \lim_{n\to\infty} \frac{q_n\, q_{n-1}}{b_1 \cdots b_n}; \tag{9.54}$$

it follows that Equation 9.47 is an alternating series, whose general term tends to zero $f_n \to 0$ as $n \to \infty$, and hence converges (Equation I.29.56). It has been proved that *if either of the two conditions 9.53c and 9.53d is met, then the continued fraction Equation 9.45 of the first kind (Equations 9.53a and 9.53b) converges (Equation 9.53e).* For example, the simple continued fraction Equation 9.55 ≡ Equation 1.142d (Equation 9.56 ≡ Equation 7.304b) for the irrational number e (π):

$$e = 1 + \frac{1}{0+}\frac{1}{1+}\frac{2}{2+}\frac{3}{3+}\cdots\frac{n}{n+}\cdots \tag{9.55}$$

$$\frac{\pi}{4} = \frac{1}{1+}\frac{1}{3+}\frac{4}{5+}\frac{9}{7+}\frac{16}{9+}\frac{25}{11+}\cdots\frac{n^2}{2n+1}\frac{(n+1)^2}{2n+3}\cdots \tag{9.56}$$

has numerators and denominators given by Equations 9.57a through 9.57c (Equations 9.58a through 9.58c):

$$a_n = n, \quad a_{n-1} = n-1, \quad b_n = n: \quad \sum \frac{a_n\, a_{n-1}}{b_n} = \sum O(n)\, D., \tag{9.57a–9.57d}$$

$$a_n = 2n+1, \quad a_{n-1} = 2n-1, \quad b_n = n^2; \quad \lim_{n\to\infty} \frac{a_n\, a_{n-1}}{b_n} = 4; \tag{9.58a–9.58d}$$

the criterion [Equation 9.53d (Equation 9.53c)] proves [Equation 9.57d (Equation 9.58d)] the convergence of the simple continued fraction [Equation 9.55 (Equation 9.56)]. A simple continued fraction (Subsection 1.7.3) has all numerators equal to unity (Equation 9.59a ≡ Equation 1.118a), and if it is also of the first kind (Equation 9.53b ≡ Equation 9.59b), the condition 9.53d that simplifies to Equation 9.59c ensures convergence (Equation 9.59d):

$$b_n = 1, \quad a_n > 0: \quad \sum a_n\, a_{n-1}\, D. \quad \Rightarrow \quad 1 + \frac{1}{a_1+}\cdots\frac{1}{a_n+}C. \tag{9.59a–9.59d}$$

Thus, *a simple (Equation 9.59a) positive (Equation 9.59b) continued fraction (Equation 9.45) whose denominators lead to the divergent series (Equation 9.59c) is convergent (Equation 9.59d).* A special case of simple positive continued fraction is recurrence, that is, repetition of the same set of terms beyond a certain order; in this case, the nonterminating recurrent continued fraction can be reduced to a terminating continued fraction; this not only proves convergence but also specifies the limit (Subsection 9.3.4).

9.3.4 Recurrent/Periodic Simple Positive Continued Fractions

A simple continued fraction has all numerators unity (Equation 9.59a), and it is **periodic** if the denominators consist of the same set of numbers repeated sequentially (Equation 9.60a):

$$x = a_1 + \frac{1}{a_1 +}\cdots\frac{1}{a_n +}\frac{1}{a_1 +}\cdots\frac{1}{a_n +}\frac{1}{a_1 +}\cdots\frac{1}{a_n +}\cdots$$
$$\equiv \underline{a_1} + \frac{1}{a_2 +}\cdots\frac{1}{a_3 +}\cdots\frac{1}{a_{n-1} +}\cdots\frac{1}{\underline{\underline{a_n}}}; \quad \text{(9.60a and 9.60b)}$$

the notation (Equation 9.60b) uses one (two) underbar(s) indicating the beginning (end) of a **cycle**. A simple continued fraction is **recurrent** if it becomes periodic after a certain order (Equations 9.61a and 9.61b):

$$y = c_1 + \frac{1}{c_2 +}\cdots\frac{1}{c_s +}\frac{1}{a_1 +}\cdots\frac{1}{a_n +}\frac{1}{a_1 +}\cdots\frac{1}{a_n +}\cdots$$
$$\equiv c_1 + \frac{1}{c_2 +}\cdots\frac{1}{c_s +}\frac{1}{\underline{a_1} +}\cdots\frac{1}{\underline{\underline{a_n}}} = c_1 + \frac{1}{c_2 +}\cdots\frac{1}{c_g + x}. \quad \text{(9.61a–9.61c)}$$

The calculation of a nonterminating recurrent simple continued fraction (Equations 9.61a and 9.61b) reduces (Equation 9.61c) to that of a terminating continued fraction with a periodic continued fraction (Equations 9.60a and 9.60b) as the last term. Thus, it is sufficient to indicate how to calculate a simple periodic continued fraction.

A simple periodic continued fraction (Equation 9.60a ≡ Equation 9.60b) satisfies the identity Equation 9.62a:

$$x = a_1 + \frac{1}{a_2 +}\cdots\frac{1}{a_n + x}; \quad \frac{p'}{q'} = a_1 + \frac{1}{a_2 +}\cdots\frac{1}{a_{n-1}}, \quad \frac{p}{q} = a_1 + \frac{1}{a_2 +}\cdots\frac{1}{a_n}. \quad \text{(9.62a–9.62c)}$$

The last (last but one) continuant of the cycle Equation 9.60b is Equation 9.62a (Equation 9.62b). The terminating continued fraction (Equation 9.62a) has a last continuant (Equation 9.63g) related to the two preceding continuants (Equations 9.62b and 9.62c) by Equations 1.102a and 1.102b whose coefficients are Equations 9.63a through 9.63f:

$$b_n = 1,\ a_n = x;\ (p_{n-1}, q_{n-1}) = (p, q);\ (p_{n-2}, q_{n-2}) \equiv (p', q');$$
$$x = \frac{p_n}{q_n} = \frac{a_n\, p_{n-1} + b_n\, q_{n-2}}{a_n\, q_{n-1} + b_n\, q_{n-2}} = \frac{p\, x + p'}{q\, x + q'}. \quad \text{(9.63a–9.63g)}$$

Thus, the simple periodic continued fraction (Equation 9.60a ≡ Equation 9.60b) if it converges must tend to a value x that is a root of Equation 9.63g ≡ Equation 9.64a), that is, a binomial:

$$q\,x^2+(q'-p)x-p'=0;\quad x_\pm=\frac{p-q'}{2q}\pm\frac{1}{2q}\left|(p-q')^2+4\,p'\,q\right|; \qquad \text{(9.64a and 9.64b)}$$

the binomial Equation 9.64a has two roots (Equation 9.64b) and the continued fraction can converge only to one of them if they are real and distinct. If all terms in the periodic continued fraction (Equation 9.60a) are real and positive (Equation 9.65a), then (1) the numerators and denominators of the continuants are also all positive (Equations 9.65b through 9.65e); (2) of the two roots (Equation 9.59b), only one is positive (Equation 9.65g):

$$a_n>0:\quad p>0<q,\quad p'>0<q',\quad x_-<0<x_+\equiv x; \qquad \text{(9.65a–9.65g)}$$

and (3) the value of the continued fraction (Equation 9.64b) must also be real and positive so it can converge only to the positive root (Equation 9.65g).

9.3.5 Surd Number and Recurrent Continued Fractions

It has been proved that *a simple periodic continued fraction (Equation 9.60a) with positive real terms (Equation 9.65a ≡ Equation 9.66a) converges to (Equation 9.66b) the positive root (Equation 9.65g ≡ Equation 9.64b) of the binomial (Equation 9.64a):*

$$a_n>0:\quad a_1+\frac{1}{a_2+}\cdots\frac{1}{\underline{\underline{a_n}}}=\frac{p-q'}{2q}+\frac{1}{2q}\left|(p-q')^2+4\,p'\,q\right|^{1/2}\equiv x \qquad \text{(9.66a and 9.66b)}$$

that involves the last two continuants (Equations 9.62b and 9.62c) of the cycle (Equation 9.60b). If all terms are positive integers (Equation 9.67a), the simple periodic continued fraction converges to a **surd number** *(Equation 9.67b) that may be irrational, with all terms integers (Equations 9.67d and 9.67e) and two positive (Equations 9.67f and 9.67g):*

$$a_n\in|N:\quad \underline{a_1}+\frac{1}{a_2+}\cdots\frac{1}{\underline{\underline{a_n}}}=\frac{A+\sqrt{B}}{C}\equiv x,\quad A,B,C\in|N,\quad B>0<C; \qquad \text{(9.67a–9.67g)}$$

conversely, any irrational surd number can be represented by a recurrent simple continued fraction whose terms are all integers. The statement (Equations 9.67a through 9.67g) follows immediately from Equations 9.66a and 9.66b and the converse is proved next. A surd number (Equation 9.68d) whose elements are positive integers (Equations 9.68a through 9.68c) is one of the roots of the binomial (Equation 9.68e):

$$A,B,C\in|N:\quad x=\frac{A+\sqrt{B}}{C},0=\left(x-\frac{A+\sqrt{B}}{C}\right)\left(x-\frac{A-\sqrt{B}}{C}\right)=\left(x-\frac{A}{C}\right)^2-\frac{B}{C^2}. \qquad \text{(9.68a–9.68e)}$$

The binomial (Equation 9.68e ≡ Equation 9.69a) has coefficients (Equations 9.69b and 9.69c) that are ratios of integers:

$$0=\left(x-\frac{A}{C}\right)^2-\frac{B}{C^2}=x^2-a\,x-b,\quad \{a,b\}=\left\{\frac{2\,A}{C},\frac{B-A^2}{C^2}\right\}. \tag{9.69a–9.69c}$$

The binomial (Equation 9.69a) like a double recurrence formula of order 2 (Equation 1.90b) can be transformed (Equations 1.92a, 1.92b, 1.93a, and 1.93b) into a continued fraction:

$$\begin{aligned} x &= a+\frac{b}{x}=a+\frac{1}{x/b}=a+\frac{1}{a/b+1/x}=a+\frac{1}{a/b+}\,\frac{1}{x} \\ &= a+\frac{1}{a/b+}\,\frac{1}{a+b/x}=a+\frac{1}{a/b+}\,\frac{1}{a+}\,\frac{b}{x}=a+\frac{1}{a/b+}\,\frac{1}{a+}\,\frac{1}{x/b}; \end{aligned} \tag{9.70}$$

since there are only two coefficients, Equation 9.70 leads to a periodic continued fraction:

$$x=a+\frac{1}{a/b+}\,\frac{1}{a+}\,\frac{1}{a/b+}\,\frac{1}{a+}\,\frac{1}{a/b+}\cdots=\underline{a}+\frac{1}{\underline{\underline{a/b}}}. \tag{9.71a}$$

$$=a+\frac{b}{a+}\,\frac{b}{a+}\,\frac{b}{a+}\cdots=a+\frac{b}{\underline{a+}}\,\frac{b}{\underline{\underline{a+}}}. \tag{9.71b}$$

Thus, *the binomial (Equation 9.69a) has the simple periodic continued fraction (Equation 9.71a ≡ Equation 9.71b) as a root; using Equations 9.69b and 9.69c, the simple periodic continued fraction represents the surd number (Equation 9.67b ≡ Equation 9.72a ≡ Equation 9.72b ≡ Equation 9.72c):*

$$\frac{A+\sqrt{B}}{C}=\frac{2A}{C}+\frac{1}{2A/(B-A^2)+}\,\frac{1}{2A/C+}\,\frac{1}{2A/(B-A^2)+}\,\frac{1}{2A/C+}\cdots=\frac{2A}{\underline{C+}}\,\frac{2A}{\underline{\underline{B-A^2}}}, \tag{9.72a}$$

$$=\frac{2\,A}{C}+\frac{(B-A^2)/C^2}{2\,A/C+}\,\frac{(B-A^2)/C^2}{2\,A/C+}\,\frac{(B-A^2)/C^2}{2\,A/C+}=\frac{2\,A}{C}+\frac{(B-A^2)/C^2}{\underline{2\,A/C}}\,\frac{(B-A^2)/C^2}{\underline{\underline{2\,A/C}}}, \tag{9.72b}$$

$$=\frac{2A}{C}+\frac{B-A^2}{2AC+}\,\frac{B-A^2}{2AC+}\,\frac{B-A^2}{2AC+}\cdots=\frac{2A}{C}+\frac{B-A^2}{\underline{2AC+}}\,\frac{B-A^2}{\underline{\underline{2AC}}}. \tag{9.72c}$$

The preceding results are illustrated next by two numerical examples.

9.3.6 Numerical Examples of Recurrent/Periodic Continued Fractions

The periodic simple continued fraction:

$$x=1+\frac{1}{2+}\frac{1}{1+}\frac{1}{1+}\frac{1}{2+}\frac{1}{1+}\frac{1}{1+}\frac{1}{2+}\frac{1}{1+}\cdots=\underline{1}+\frac{1}{2+}\frac{1}{\underline{1}} \tag{9.73}$$

has last two continuants (Equations 9.74a and 9.74b):

$$\frac{p'}{q'}=1+\frac{1}{2}=\frac{3}{2},\quad \frac{p}{q}=1+\frac{1}{2+}\frac{1}{1+}=1+\frac{1}{3}=\frac{4}{3}; \tag{9.74a and 9.74b}$$

this leads to Equations 9.75a through 9.75d to Equation 9.64b ≡ Equation 9.66b to the limit Equation 9.75e:

$$q'=2,\quad p'=3=q,\, p=4:\quad x_+=\frac{2}{6}+\frac{\sqrt{40}}{6}=\frac{1+\sqrt{10}}{3}. \tag{9.75a–9.75e}$$

Thus, the periodic continued fraction (Equation 9.73) has the value (Equation 9.75e ≡ Equation 9.76):

$$1+\frac{1}{2+}\frac{1}{1+}\frac{1}{2+}\frac{1}{1+}\frac{1}{1+}\frac{1}{2+}\frac{1}{1+}\cdots=\underline{1}+\frac{1}{2+}\frac{1}{\underline{1}}=\frac{1+\sqrt{10}}{3}; \tag{9.76}$$

$$3+\frac{1}{2+}\frac{1}{1+}\frac{1}{1+}\frac{1}{2+}\frac{1}{1+}\frac{1}{1+}\frac{1}{2+}\frac{1}{1+}1\cdots=3+\frac{1}{2+}\frac{1}{\underline{1}+}\frac{1}{2+}\frac{1}{\underline{1}}=\frac{20-\sqrt{10}}{5}; \tag{9.77}$$

the recurrent simple continued fraction Equation 9.77 has value:

$$\begin{aligned} y&=3+\frac{1}{2+}\frac{1}{\underline{1}+}\frac{1}{2+}\frac{1}{\underline{1}+}=3+\frac{1}{2+}\frac{1}{x}=3+\frac{1}{2+}\frac{1}{\left(1+\sqrt{10}\right)/3}\\ &=3+\frac{1}{2+}\frac{3}{1+\sqrt{10}}=3+\frac{1+\sqrt{10}}{5+2\sqrt{10}}=\frac{16+7\sqrt{10}}{5+2\sqrt{10}}\\ &=\frac{16+7\sqrt{10}}{5+2\sqrt{10}}\frac{5-2\sqrt{10}}{5-2\sqrt{10}}=\frac{80-14.10+3\sqrt{10}}{25-4.10}=\frac{60-3\sqrt{10}}{15}=4-\sqrt{\frac{2}{5}}, \end{aligned} \tag{9.78}$$

proving Equation 9.78 ≡ Equation 9.77) and writing the result as a surd number (Equation 9.77).

It is possible to reverse the process and, starting from the surd number, obtain the corresponding periodic continued fraction (Subsection 1.7.4). The method is to choose at each stage the integer part, that is, the largest integer smaller than the value. Starting with the value Equation 9.75e leads to

$$
\begin{aligned}
\frac{1+\sqrt{10}}{3} &= 1+\frac{-2+\sqrt{10}}{3} = 1+\frac{1}{3/\left(-2+\sqrt{10}\right)} \\
&= 1+\frac{1}{2+\left(7-2\sqrt{10}\right)/\left(-2+\sqrt{10}\right)} = \frac{1}{2+}\frac{7-2\sqrt{10}}{-2+\sqrt{10}} \\
&= 1+\frac{1}{2+}\frac{1}{\left(-2+\sqrt{10}\right)/\left(7-2\sqrt{10}\right)} = 1+\frac{1}{2+}\frac{1}{1+\left(-9+3\sqrt{10}\right)/\left(7-2\sqrt{10}\right)} \\
&= 1+\frac{1}{2+}\frac{1}{1+}\frac{-9+3\sqrt{10}}{7-2\sqrt{10}} = 1+\frac{1}{2+}\frac{1}{1+}\frac{1}{\left(7-2\sqrt{10}\right)/\left(-9+3\sqrt{10}\right)} \\
&= 1+\frac{1}{2+}\frac{1}{1+}\frac{1}{1+\left(16-5\sqrt{10}\right)/\left(-9+3\sqrt{10}\right)} = 1+\frac{1}{2+}\frac{1}{1+}\frac{1}{1+}\frac{16-5\sqrt{10}}{-9+3\sqrt{10}} = \underline{1}+\frac{1}{2+}\frac{1}{\underline{\underline{1}}},
\end{aligned}
\tag{9.79}
$$

where (1) the first fraction has an integer part Equation 9.80a; (2) the second fraction has an integer part Equation 9.80b; (3) the third fraction has an integer part Equation 9.80c; (4) the fourth fraction has an integer part Equation 9.80d:

$$
Ip\left(\frac{1+\sqrt{10}}{3}\right) = Ip\,(1.4) = 1, \quad Ip\left(\frac{3}{-2+\sqrt{10}}\right) = Ip\,(2.6) = 2, \tag{9.80a and 9.80b}
$$

$$
Ip\left(\frac{-2+\sqrt{10}}{7-2\sqrt{10}}\right) = Ip\,(1.7) = 1, \quad Ip\left(\frac{7-2\sqrt{10}}{-9+3\sqrt{10}}\right) = Ip\,(1.4) = 1; \tag{9.80c and 9.80d}
$$

and (5) the fourth fraction has the same integer part (Equation 9.80d) as the first (Equation 9.80a), and proceeding similarly, it follows that the cycle (1, 2, 1) repeats itself leading to the simple periodic continued fraction Equation 9.76 ≡ Equation 9.75e.

9.4 Derangement of Conditionally Convergent Series

A convergent series (product) has a finite sum (logarithm of the sum) that can be calculated with reordering of the terms (factors) if the convergence is absolute (Section I.21.3). If the convergence is conditional (Equation 9.81a), that is, the series converges (Equation 9.81b) but the series of moduli does not (Equation 9.81c), then the sum can be altered by rearrangement of the terms, as shown before in an example (Section I.21.3). It is possible to prove the far stronger result:

$$
\sum_{n=1}^{\infty} f_n \ \text{C.C.} \ \Leftrightarrow \ \sum_{n=1}^{\infty} f_n \ \ C. \ \wedge \ \sum_{n=1}^{\infty} |f_n| \ \ D. \tag{9.81a–9.81c}
$$

if the series (Equation 9.81a) is conditionally convergent, that is, it converges (Equation 9.81b) but the series of moduli diverges (Equation 9.81c), then by derangement, that is, recording of the terms, it can be made to (a) diverge to + ∞ or – ∞; (b) converge to any value f given a priori; and (c) oscillate between any pair of lower μ and upper ν bounds. All three cases 1–3 are proved (Subsection 9.4.2) by noting that a conditionally convergent series is the difference of two divergent series (Subsection 9.4.1).

9.4.1 Conditionally Convergent Series as Difference of Two Divergent Series

To prove the theorem, the series Equation 9.81a is separated without reordering into positive p_n and negative $-q_n$ terms:

$$p_n, q_n > 0: \quad \sum_{n=1}^{\infty} f_n = p_2 + \ldots + p_{m_1} - q_1 - \ldots - q_{k_1} \qquad \text{(9.82a–9.82c)}$$
$$+ p_{m_1+1} + \ldots + p_{m_2} - q_{k_1+1} - \ldots - q_{k_2} + \ldots;$$

the series of moduli of Equation 9.81c coincides with Equation 9.82c with + signs in all terms:

$$\sum_{n=1}^{\infty} |f_n| = p_1 + \ldots + p_{m_1} + q_1 + \ldots + q_{k_1} + p_{m_1+1} + \ldots + p_{m_2} + q_{k_1+1} \ldots + q_{k_2} + \ldots. \qquad (9.83)$$

Since the series Equation 9.81c ≡ Equation 9.83 diverges, at least one of the series p_n or q_n diverges; if one was divergent and the other convergent, then Equation 9.81c would diverge, which contradicts the assumption that Equation 9.82c ≡ Equation 9.81b converges. Thus, both p_n and q_n must diverge:

$$p_1 + p_2 + \ldots + p_n + \ldots = \infty = q_1 + q_2 + \ldots + q_m + \ldots. \qquad \text{(9.84a and 9.84b)}$$

This proves that *if the series Equation 9.81a is conditionally convergent (Equations 9.81b and 9.81c), then the subseries (Equations 9.82a through 9.82c) of positive p_n (negative q_n) terms diverges (Equations 9.84a and 9.84b). This means that a conditionally convergent series consists of terms extracted from two divergent series, one with positive and the other with negative terms;* for example, the conditionally convergent series (Equation 3.47b ≡ Equation 9.85a):

$$\log 2 = 1 - \frac{1}{2} + \frac{1}{3} - \frac{1}{4} + \frac{1}{5} - \ldots, \quad 1 + \frac{1}{3} + \frac{1}{5} + \ldots = \infty = \frac{1}{2} + \frac{1}{4} + \frac{1}{6} + \ldots \qquad \text{(9.85a–9.85c)}$$

consists of terms taken alternatively from two divergent arithmetic (Equation I.21.7) series (Equations 9.85b and 9.85c) with opposite signs. The essence of the proof of the theorem is to show that deranging the series, by increasing the proportions of positive and negative terms, it is possible to (a) cause divergence to +∞ or −∞, (c) or oscillation in any interval (μ, ν) or (b) convergence to any value μ = ν ≡ *f*.

9.4.2 Divergence, Oscillation, or Convergence by Derangement of Conditionally Convergent Series

In order to rearrange the series (Equation 9.82c) so that it diverges to $+\infty$, choose a number $X > 0$ and take a sequence of integers $n_1 < n_2 < n_3 < \ldots$ such that

$$p_1 + \ldots p_{n_1} - q_1, p_{n_1+1} + \ldots + p_{n_2} - q_2, \ldots > X; \tag{9.86}$$

this is always possible because $\sum p_n \to \infty$ and $q_n \to 0$. The rearranged series satisfies

$$p_1 + \ldots p_{n_1} - q_1 + p_{n_1+1} + \ldots + p_{n_2} - q_2 + p_{n_2+1} + \ldots + p_{n_k} - q_k > k\,X, \tag{9.87}$$

and letting $k \to \infty$ leads to a divergent series, tending to $+\infty$. A series diverging to $-\infty$ could be obtained exchanging the roles of p_n, q_n, thus proving statement a in the theorem. To prove statement c, choose i_1, j_1 such that

$$p_1 + \ldots + p_{i_1-1} < \mu_1, \quad p_1 + \ldots + p_{i_1} \geq \mu_1, \tag{9.88a and 9.88b}$$

$$p_1 + \ldots + p_{i_1} - q_1 - \ldots - q_{j_1-1} > \nu_1, \quad p_1 + \ldots + p_{i_1} - q_1 - \ldots - q_{j_1} \leq \nu_1; \tag{9.89a and 9.89b}$$

this is always possible for any μ_1, ν_1 because p_n, $q_n > 0$ and $\sum p_n, \sum q_n \to \infty$, so that (1) enough p_n are summed to exceed μ_1; and (2) then enough q_n are subtracted until the total sum falls below $\nu_1 < \mu_1$. Proceeding to find $i_1 < i_2 < i_3 \ldots$ and $j_1 < j_2 < j_3 \ldots$ in the same way leads to sequences of lower μ_n (upper ν_n) bounds, which can be chosen so that

$$\mu_1 < \mu_2 < \ldots < \mu_n < \ldots < \mu < \nu < \ldots < \nu_n < \ldots < \nu_1. \tag{9.90}$$

The increasing (decreasing) sequences must tend to a value that is a lower μ(upper ν) bound for the deranged series. The deranged series

$$\begin{aligned}\sum_{n=1}^{\infty} g_n = {} & p_1 + \ldots + p_{i_1} - q_1 - \ldots - q_{j_1} + p_{i_1+1} \\ & + \ldots + p_{i_2} - q_{j_1+1} - \ldots - q_{j_2} - p_{i_2+1} + \ldots + p_{i_3} - \ldots\end{aligned} \tag{9.91}$$

has partial sums $r_1, s_1, r_2, s_2, \ldots$, after i_1, $i_1 + j_1$, $i_2 + j_1$, $i_2 + j_2$, … terms, such that

$$\lim_{n\to\infty} |r_n - s_n| \leq \lim_{n\to\infty} p_{i_n} = 0 = \lim_{n\to\infty} q_{j_n} = \lim_{n\to\infty} |s_n - \nu_n|. \tag{9.92}$$

Hence, $r_n \to \mu$, $s_n \to \nu$ as $n \to \infty$, and the points μ, ν are accumulation points for the sum of the series. Also any partial sum that is not $r_n(s_n)$ has fewer $p_n(q_n)$ terms, and hence lies between them. It follows that any accumulation point σ other than μ, ν must lie in between $\mu < \sigma < \nu$, and thus the series oscillates between μ and ν, proving statement c. If the upper and lower bounds are chosen to coincide $\mu = \nu \equiv f$, then a derangement of the series converging to f follows, thus proving proposition b.

9.5 Summation of Series of Rational Functions

Given a series, it can be summed without (with) derangement if it is conditionally (absolutely) convergent; the summation can be affected by residues if the general term $f(n)$ is a rational function $f(z)$, with a finite number of poles a_k of residues b_k. This is shown next in general (Subsection 9.5.1), followed by some examples (Subsection 9.5.2).

9.5.1 Rational Function with Simple Poles as the General Term

Consider a rational function (Figure 9.1) with M simple poles (Equation 9.93a) with residues (Equation 9.93b):

$$m = 1,\ldots, M: \quad f_{(1)}(a_k) \equiv \lim_{z\to a_k} (z-a_k) f(z) \equiv b_k. \qquad \text{(9.93a and 9.93b)}$$

The functions [Equation 9.84a (Equation 9.95a)] have simple poles: (1) at the same points with residues [Equation 9.94d (Equation 9.95d)] since the poles a_k of the function are not [Equation 9.94c (Equation 9.95c)] poles of the circular cotangent (Equation 9.94b) [cosecant (Equation 9.95b)]; and (2) also at the poles $m\pi$ with m integer [Equation 9.94b (Equation 9.95b)] that are not poles of the function, the residues are given [Equation 7.213b (Equation 7.210c)] by Equation 9.94e (Equation 9.95e):

$$F(z) \equiv \pi f(z) \cot(\pi z); m \in |Z,\ a_k \neq m\pi:\ F_{(1)}(a_k) = \pi b_k \cot(a_k\pi);\ F_{(1)}(n) = f(n), \qquad \text{(9.94a–9.94e)}$$

$$G(z) \equiv \pi f(z) \csc(\pi z): m \in |Z, a_k \neq m\pi:\ G_{(1)}(a_k) = \pi b_k \csc(a_k\pi), G_{(1)}(n) = (-)^n f(n). \qquad \text{(9.95a–9.95e)}$$

The conditions (Equations 9.94b and 9.94c ≡ Equations 9.95b and 9.95c) ensure that all poles are simple, that is, there are no double poles; the residues of the circular cotangent (cosecant) at their simple poles are given by Equation 1.55 (Equation 1.38) and Equation 9.94e (Equation 9.95e) is used. Applying the theorem of residues (Section I.15.9) to the square L_n with vertices at $(N + 1/2)\ (\pm 1 \pm i)$ that (Figure 9.1) contains the first $2N + 1$ poles $\pm n\pi$ with $n = 0, 1, \ldots, N$ of the cotangent (cosecant) leads from Equations 9.94a through 9.94e (Equations 9.95a through 9.95e) to Equation 9.96 (Equation 9.97):

$$(2\pi i)^{-1} \oint_{L_N} f(z)\, \pi \cot(\pi z)\, \mathrm{d}z = \sum_{n=-N}^{+N} f(n) + \pi \sum_{k=1}^{M} b_k \cot(\pi a_k), \qquad (9.96)$$

$$(2\pi i)^{-1} \oint_{L_N} f(z)\, \pi \csc(\pi z)\, \mathrm{d}z = \sum_{n=-N}^{+N} (-)^n f(n) + \pi \sum_{k=1}^{M} b_k \csc(\pi a_k); \qquad (9.97)$$

since the functions $\pi\cot(\pi z)$, $\pi\csc(\pi z)$ are bounded on the square, a sufficient condition for the loop integrals to vanish (Equation 9.98b) is that the function $z\, f(z)$ vanishes at infinity in all directions (Equation I.17.19 ≡ Equation 9.98a):

$$\lim_{z\to\infty} z\, f(z) = 0 \Rightarrow \quad \lim_{z\to\infty} \oint_{L_N} f(z)\pi\{\cot, \csc(\pi z)\}\mathrm{d}z = 0. \qquad \text{(9.98a and 9.98b)}$$

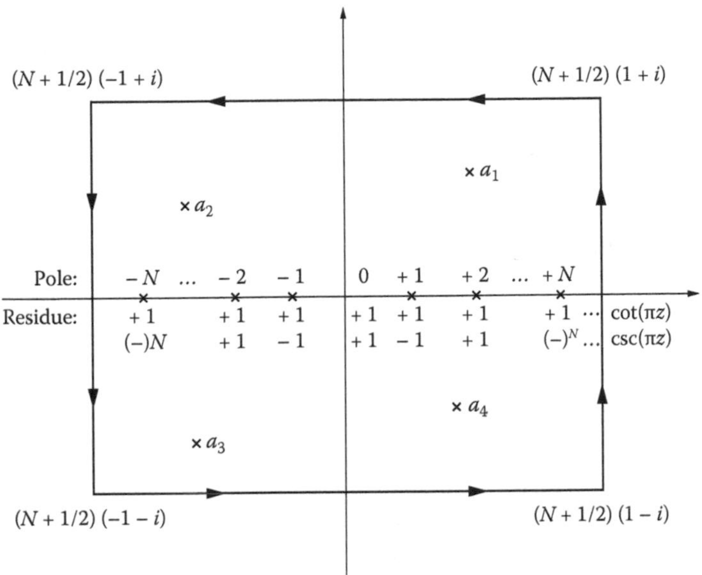

FIGURE 9.1
In order to sum the series of general term $f(n)[(-)^n f(n)]$, where $f(z)$ is a function analytic for all z, except for simple poles a_k, with residues b_k for $k = 1,\dots, M$, which decays at infinity faster than $1/z$, the cotangent (cosecant) is used as an auxiliary function; the functions $\cot(\pi z)[\sec(\pi z)]$ have simple poles at $z = \pm n$, with residues $+1[(-)^n]$, and are bounded on the square of vertices $(N+1/2)(\pm 1 \pm i)$ that contains $2N + 1$ poles.

Taking the limit $N \to \infty$, the l.h.s. of Equation 9.96 (Equation 9.97) vanishes, and the r.h.s. proves that the series converge, in agreement with $f(n) = o(1/n)$ in Equation 9.98a; this also evaluates the sum using the residues b_k at the poles a_k of $f(z)$.

9.5.2 Sum of Series in Terms of Cotangents and Cosecants

Thus, Equation 9.96 (Equation 9.97) leads with $N \to \infty$ and zero on the l.h.s. to the **theorem of summation of series by residues:** *if the function $f(z)$ is analytic except at M simple poles a_k with residues b_k in Equations 9.93a and 9.93b and is $o(1/z)$ at infinity in all directions (Equation 9.98a), then the nonalternating (Equation 9.99a) [alternating (Equation 9.99b)] series converge to the sum indicated:*

$$\sum_{n=-\infty}^{+\infty} f(n) = -\pi \sum_{k=1}^{M} b_k \cot(\pi a_k), \quad \sum_{n=-\infty}^{+\infty} (-)^n f(n) = -\pi \sum_{k=1}^{M} b_k \csc(\pi a_k); \quad \text{(9.99a and 9.99b)}$$

the bilateral series (Equations 9.100a and 9.100b) summed for n from $-\infty$ to $+\infty$ can be transformed to unilateral series summed for n from 0 or 1 to ∞:

$$\sum_{n=-\infty}^{+\infty} f(n) = f(0) + \sum_{n=1}^{+\infty} \{f(n) + f(-n)\} = -f(0) + \sum_{n=0}^{+\infty} \{f(n) + f(-n)\}. \quad (9.100)$$

Several instances of the summation of series by the method of residues are given in Example 10.1. As examples, Equation 9.99a (Equation 9.99b) with Equations 9.101a and

9.101b is used to deduce the series of fractions for the cotangent (cosecant); in this case, the function Equation 9.101c with a simple pole for $k = 1$ at Equation 9.101a is chosen with residue Equation 9.101b:

$$a_k\pi = z, \quad b_k\pi = -1: \quad f(n) = \frac{b_1}{n - a_1} = \frac{1}{\pi}\frac{1}{n - z/\pi} = \frac{1}{z - n\pi}; \qquad (9.101a\text{–}9.101c)$$

substituting in Equation 9.99a (Equation 9.99b) leads to the series of fractions for the cotangent (Equation 9.102) [cosecant (Equation 9.103)]:

$$\cot(z) = \sum_{n=-\infty}^{+\infty} \frac{1}{z - n\pi} = \frac{1}{z} + 2z\sum_{n=1}^{\infty} \frac{1}{z^2 - n^2\pi^2}, \qquad (9.102)$$

$$\csc(z) = \sum_{n=-\infty}^{+\infty} \frac{(-)^n}{z - n\pi} = \frac{1}{z} + 2z\sum_{n=1}^{\infty} \frac{(-)^n}{z^2 - n^2\pi^2}; \qquad (9.103)$$

these coincide with Equation 1.58 ≡ Equation 9.102 (Equation 1.85 ≡ Equation 9.103) that was obtained from the Mittag-Leffler theorem (Sections 1.2 through 1.5).

9.6 Extension of Convergence to Summability (Euler 1755; Césaro 1890)

The usual definition (Section I.21.1) of convergent series is the case C0 of the Césaro (1890) sum of class Ck when $k = 0$. It can be proved (Subsection 9.6.5) that if a series converges Ck, then it converges Ck' for all $k' \geq k$; the proof uses ordinary and arithmetic sums (Subsection 9.6.4). The converse need not be true, that is, a series may be oscillating or divergent to the class C0 (Ck) and become convergent to the class C1 (Ck' for some $k' \geq k$); this is shown by induction stating that the proof of convergence to class C0 implies (Subsection 9.6.1) convergence to class C1 with the same sum. Thus, the interest of Césaro summability lies in being able to sum some series that are oscillating (divergent) according to the ordinary criterion C0 of convergence [Subsection 9.6.2 (Subsection 9.6.3)].

9.6.1 Ordinary Convergence and Summability to Class C1

Given a series of general term f_n, the usual sum of the first n terms is the **sum of class zero**:

$$\text{C0:} \quad f = \sum_{n=1}^{\infty} f_n = \lim_{n\to\infty} S_n^{(0)}, \quad S_n^{(0)} \equiv f_1 + \cdots + f_n; \qquad (9.104a \text{ and } 9.104b)$$

the **Césaro sum of class one (C1)** is defined by

$$\text{C1:} \quad S_n^{(1)} \equiv \frac{S_1^{(0)} + \cdots + S_n^{(0)}}{n} = \sum_{k=1}^{n} \frac{n-k+1}{n} f_k = \sum_{k=1}^{n}\left(1 - \frac{k-1}{n}\right) f_k. \qquad (9.105)$$

The series converges to class zero (unity) if the limit of Equation 9.104b (Equation 9.105) as $n \to \infty$ exists and is finite. First must be proved a **theorem of consistency**: *if the series Equation 9.104a converges C0 to f (Equation 9.106a), then it also converges C1 to (Equation 9.105) the same sum (Equation 9.106b):*

$$\lim_{n\to\infty} S_n^{(0)} = f \quad \Rightarrow \quad \lim_{n\to\infty} S_n^{(1)} = f. \qquad \text{(9.106a and 9.106b)}$$

The relation in Equations 9.106a and 9.106b is an implication, not an equivalence; there are series that do not converge to order 0, but do converge to order 1, and this is why the concept of convergence C1 is useful. Before giving examples, the theorem (Equations 9.106a and 9.106b) is proved.

In order to prove Equations 9.106a and 9.106b, it is assumed first that Equations 9.104a and 9.104b converge, so that the Cauchy necessary and sufficient condition (Subsection I.29.2.2) holds. This states that if the series Equation 9.104a converges, for any real positive ε however small, there is an order n such that the sum of any number m of terms following n cannot exceed ε in modulus:

$$\forall_{\varepsilon>0} \exists_{n\in N} \forall_{m\in N}: \ \varepsilon > \left| \sum_{k=n+1}^{n+m} f_k \right| = \left| S_{n+m}^{(0)} - \sum_{k=1}^{n} f_k \right|. \qquad (9.107)$$

The sum C1 of the first $n + m$ terms is given (Equation 9.105) by

$$\begin{aligned} S_{n+m}^{(1)} = \sum_{k=1}^{n+m} \left(1 - \frac{k-1}{n+m}\right) f_k &= f_1 + \cdots + \left(1 - \frac{n-1}{n+m}\right) f_n \\ &+ \left(1 - \frac{n}{n+m}\right) f_{n+1} + \cdots + \left(1 - \frac{n+m-1}{n+m}\right) f_{n+m}. \end{aligned} \qquad (9.108)$$

The second set of terms has an upper bound:

$$\begin{aligned} &\left| \left(1 - \frac{n}{n+m}\right) f_{n+1} + \cdots + \left(1 - \frac{n+m-1}{n+m}\right) f_{n+m} \right| \\ &< \left(1 - \frac{n}{n+m}\right) \left| f_{n+2} + \ldots + f_{n+m} \right| < \left(1 - \frac{n}{n+m}\right) \varepsilon, \end{aligned} \qquad (9.109)$$

implying Equation 9.110a:

$$\left| S_{n+m}^{(1)} - \sum_{k=1}^{n} \left(1 - \frac{k-1}{n+m}\right) f_k \right| < \left(1 - \frac{n}{n+m}\right) \varepsilon; \quad \lim_{m\to\infty} \left| S_{n+m}^{(1)} - \sum_{k=0}^{n} f_k \right| < \varepsilon; \qquad \text{(9.110a and 9.110b)}$$

letting $m \to \infty$ in Equation 9.110a leads to Equation 9.110b. From Equations 9.110b and 9.107 follows

$$\lim_{m\to\infty}\left|S_{n+m}^{(1)}-S_{n+m}^{(0)}\right| = \lim_{m\to\infty}\left|\left(S_{n+m}^{(1)}-\sum_{k=1}^{n}f_k\right)-\left(S_{n+m}^{(0)}-\sum_{k=1}^{n}f_k\right)\right|$$
$$\le \lim_{m\to\infty}\left\{\left|S_{n+m}^{(1)}-\sum_{k=1}^{n}f_k\right|+\left|S_{n+m}^{(0)}-\sum_{k=1}^{n}f_k\right|\right\} < 2\varepsilon \tag{9.111}$$

From Equation 9.111, it follows that $S_\infty^{(1)} = S_\infty^{(0)}$, as stated in Equations 9.106a and 9.106b.

9.6.2 C1-Sum of a C0-Oscillating Series (Euler 1755)

The **Euler (1755) series** (Equation 9.112) consisting of the alternating values plus and minus one

$$1-1+1-1+1-1+1-1+\cdots = \frac{1}{2} \tag{9.112}$$

has sum unity and zero in alternation, so that the arithmetic mean 1/2 might crudely be taken as the sum. This can be justified (**Euler 1755**) using the geometric series (Equation 3.148b ≡ Equation I.21.62b):

$$\sum_{n=0}^{\infty}(-)^n = \lim_{x\to 1-0}\sum_{n=0}^{\infty}(-x)^n = \lim_{x\to 1-0}\frac{1}{1+x} = \frac{1}{2} \tag{9.113}$$

bearing in mind (Section I.21.9 and Table I.21.2) that (1) it converges (diverges) for $x < 1$ $(x > 1)$, so that the limit must be taken from below $x \to 1-0$; (2) the series converges uniformly for $|z| \le 1$ and $|z+1| \ge \varepsilon > 0$, that is, on the circle of convergence $|z| = 1$ excluding a neighborhood of the point $z = -1$, which is a simple pole of the sum; and (3) thus the limit $z \equiv x \to 1-0$ is permissible. The result (Equation 9.112) can be proved rigorously by a summation to class C1:

$$S_{2n-1}^{(0)} = 1, \quad S_{2n}^{(0)} = 0: \quad S_{2n-1}^{(1)} = \frac{n}{2n-1} \to \frac{1}{2} = S_{2n}^{(1)}; \tag{9.114a–9.114d}$$

(1) the sums to class C0 as for ordinary convergence (Equation 9.104a) are unity (zero) for odd (Equation 9.114a) [even (Equation 9.114b)] order, so the series Equation 9.112 is oscillating to class C0; and (2) the sums class C1 are (Equation 9.105) one-half for even order (Equation 9.114d) and tend to the same value for odd order (Equation 9.114c) as $n \to \infty$, so the series converges class C1 to one-half. This agrees with the Euler limit (Equation 9.113) and rough reasoning (Equation 9.112). Thus, *the series Equation 9.112, which is oscillatory to class C0, converges to 1/2 to class unity C1.*

9.6.3 C2-Sum of a C0-Divergent Series

The summation by arithmetic means (Equation 9.105) or C1 process has been used to sum an oscillating series (Equation 9.112), and by applying the method iteratively, it can sum divergent series as well. Starting with Equations I.25.40a through I.25.40c the binomial expansion (Equation 9.115) with

$$(1-x)^{-2} = \sum_{n=0}^{\infty} \binom{n+1}{1} x^n = \sum_{n=0}^{\infty} (n+1)x^n \tag{9.115}$$

and taking as in Equation 9.113 the limit $x \to -1$ lead to

$$1-2+3-4+5-6+7-8+9-10+\cdots = \frac{1}{4}, \tag{9.116}$$

as the sum of a series that is divergent oscillating to the class C0. In order to justify this result, the process of summation by arithmetic means is applied twice, that is, to class 2 as shown in Table 9.1, namely, *the series Equation 9.116 (1) is divergent (Equations 9.117a through 9.117c) to class C0; (2) is oscillating (Equations 9.118a through 9.118d) to class C1; and (3) converges (Equations 9.119a through 9.119c) to 1/4 to class C2:*

$$S_{2n-1}^{(0)} = +n, \quad S_{2n}^{(0)} = -n, \quad \lim_{n\to\infty} \left| S_n^{(0)} \right| = \infty, \tag{9.117a–9.117c}$$

$$S_{2n-1}^{(1)} = \frac{n}{2n-1}, \quad S_{2n}^{(1)} = 0, \quad \lim_{n\to\infty} \left| S_{2n-1}^{(1)} \right| = \frac{1}{2} \neq 0 = S_{2n}^{(1)}; \tag{9.118a–9.118d}$$

$$(2n-1)\, S_{2n-1}^{(2)} = \sum_{k=1}^{n} \frac{k}{2k-1} = 2n\, S_{2n}^{(1)}, \quad \lim_{n\to\infty} S_{2n-1}^{(2)} = \lim_{n\to\infty} \left[\frac{1}{2} + O\left(\frac{1}{n}\right) \right] \frac{n}{2n+0(1)} = \frac{1}{4}; \tag{9.119a–9.119c}$$

in Equation 9.119c, it was taken into account that the terms of the sums Equations 9.119a and 9.119b tend to 1/2 as $k \to n \to \infty$, and there are n terms divided by $2n - 1$ or $2n \to 2n$. The convergence of the C0-oscillating (Equation 9.112) [C0-divergent (Equation 9.116)] series is established in one iteration (Equations 9.114a through 9.114d) [two iterations (Equations 9.117a through 9.117c, 9.118a through 9.118d, and 9.119a through 9.119c)] at class C1 (C2).

TABLE 9.1

Series That Is Divergent C0, Oscillating C1, and Convergent C2

n	1	2	3	4	5	6	7	8	$2n-1$	$2n$	$n \to \infty$
f_n	+1	−2	+3	−4	5	−6	+7	−8	$2n-1$	$-2n$	$\pm\infty$
$S_n^{(0)}$	+1	−1	+2	−2	3	−3	+4	−4	$+n$	$-n$	$\pm\infty$
$S_n^{(1)}$	1	0	$\frac{2}{3}$	0	$\frac{3}{5}$	0	$\frac{4}{7}$	0	$\frac{n}{2n-1}$	0	1/2,0
$S_n^{(2)}$	1	$\frac{1}{2}$	$\frac{5}{9}$	$\frac{5}{12}$	$\frac{34}{75}$	$\frac{34}{90}$	$\frac{298}{735}$	$\frac{298}{840}$	$\sim \frac{n/2}{2n-1}$	$\sim \frac{n/2}{2n}$	$\frac{1}{4}$

Note: Summation to increasing order can lead from a divergent series at class C0 to an oscillating series at class C1 and a convergent series at class C2.

9.6.4 Ordinary and Arithmetic Sums of All Orders

The preceding example uses a generalization of Equation 9.104b (Equation 9.105) that is the definition of **ordinary (arithmetic) sum of order k** of a series [Equations 9.120a and 9.120b (Equations 9.121a and 9.121b)] by

$$P_n^{(0)} \equiv f_1 + \ldots + f_n, \quad P_n^{(k)} \equiv P_0^{(k-1)} + \ldots + P_n^{(k-1)}, \qquad \text{(9.120a and 9.120b)}$$

$$S_n^{(0)} \equiv f_1 + \ldots + f_n, \quad S_n^{(k)} \equiv \frac{S_0^{(k-1)} + \ldots + S_n^{(k-1)}}{n}. \qquad \text{(9.121a and 9.121b)}$$

Also the series converges **(Césaro, 1890)** class Ck if the limit of Equations 9.121a and 9.121b as $n \to \infty$ exists (Equation 9.122a):

$$Ck: \quad \lim_{n\to\infty} S_n^{(k)} = g \quad \Rightarrow \quad \forall_{p>k}: \quad \lim_{n\to\infty} S_k^{(p)} = g, \qquad \text{(9.122a and 9.122b)}$$

and as an extension of Equations 9.106a and 9.106b, it follows by induction that *if a series Equation 9.122a converges to the class Ck, then it converges to all the classes Cp, with p = k + 1, k + 2,... with the same sum (Equation 9.122b).* The converse is not true; for example, Equation 9.116 converges C2 but not C1, and Equation 9.112 converges C1 but not C0. Next the following theorem will be proved—*the ordinary (arithmetic) sum of order k is related to the terms of the series by Equation 9.123a (Equation 9.123b):*

$$P_n^{(k)} = \sum_{p=0}^{n} \binom{k+p}{k} f_{n-p}, \quad S_n^{(k)} = \binom{k+n}{k}^{-1} P_n^{(k)}; \qquad \text{(9.123a and 9.123b)}$$

it follows that *(Equations 9.121b and 9.122a) the Césaro limit can be evaluated directly to the class Ck from*

$$S_n^{(k)} = \binom{k+n}{k}^{-1} \sum_{p=0}^{n} \binom{k+p}{k} f_{n-p} = \sum_{p=0}^{n} f_{n-p} \frac{(k+p)!}{p!} \frac{n!}{(k+n)!} = \sum_{p=0}^{n} f_{n-p} \frac{(p+1)\cdots(p+k)}{(n+1)\cdots(n+k)}, \qquad (9.124)$$

which follows from Equation 9.124.

9.6.5 Consistency of Convergence to All Classes *Ck*

It is sufficient to prove only Equations 9.123a and 9.123b since Equation 9.124 follows from them. Concerning Equation 9.123b, note that it holds: (1) at order $k = 0$ since Equation 9.120a ≡ Equation 9.121a, then $P_n^{(0)} = S_n^{(0)}$ that agrees (Equation 9.123b) with Equation 9.105, that is, the sequence $P_1^{(0)}, \ldots, P_n^{(0)}$ is compared with the unit sequence 1,…,1; (2) at order k, the comparison of $P_1^{(k)}, \ldots, P_n^{(k)}$ with the sum of order k of the unit sequence, that is, to $\binom{n+k}{k}$, yields Equation 9.123b in the form

$$\sum_{n=1}^{\infty} P_n^{(k-1)} x^n = \sum_{n=1}^{\infty} \left(P_n^{(k)} - P_{n-1}^{(k)}\right) x^n = (1-x) \sum_{n=1}^{\infty} P_n^{(k)} x^n; \qquad (9.125)$$

the result (Equation 9.125) also follows (Equations 9.123a and 9.123b) from

$$\begin{aligned}\binom{n+k}{k}-\binom{n-1+k}{k}&=\frac{1}{k!}\left[\frac{(k+n)!}{n!}-\frac{(k+n-1)!}{(n-1)!}\right]\\&=\frac{1}{k!}\frac{(k+n-1)!}{(n-1)!}\left(\frac{k+n}{n}-1\right)=\frac{1}{k!}\frac{(k+n-1)!}{(n-1)!}\frac{k}{n}\\&=\frac{(k+n-1)!}{n!(k-1)!}=\binom{n+k-1}{k-1}.\end{aligned}\tag{9.126}$$

Applying Equation 9.125 iteratively from k to 0 leads to

$$\sum_{n=1}^{\infty}P_n^{(k)}x^n=(1-x)^{-k}\sum_{n=1}^{\infty}P_n^{(0)}x^n=(1-x)^{-k-1}\sum_{n=1}^{\infty}f_n\,x^n. \qquad \text{(9.127a and 9.127b)}$$

The series for the inverse power (Equation 9.128 ≡ Equations I.25.40a through I.25.40c) is the particular case $\nu = -1$ of the binomial series (Equations 3.144a through 3.144c ≡ Equations I.25.37a through I.25.37c):

$$\begin{aligned}(1-x)^{-k-1}&=\sum_{q=0}^{\infty}\binom{-k-1}{q}(-x)^q=\sum_{q=0}^{\infty}\frac{(-)^q}{q!}(-k-1)\cdots(-k-q)x^q\\&=\sum_{q=0}^{\infty}\frac{x^q}{q!}(k+1)\cdots(k+q)=\sum_{q=0}^{\infty}\binom{k+q}{q}x^q.\end{aligned}\tag{9.128}$$

Substituting Equation 9.128 in Equation 9.127b leads to the identity (Equation 9.123b)

$$n=p+q:\quad \sum_{n=1}^{\infty}P_n^{(k)}x^n=\sum_{q=0}^{\infty}\binom{q+k}{k}x^q\sum_{p=1}^{\infty}f_px^p=\sum_{n=1}^{\infty}x^n\sum_{q=1}^{\infty}\binom{q+k}{k}f_{n-q}, \qquad \text{(9.129a–9.129c)}$$

where the Cauchy (1821) rule for the product of series (Equations I.21.25a through I.21.25c) was used replacing the sum by lines and columns in Equation 9.129b by a sum by diagonals in Equations 9.129a and 9.129c. From Equations 9.129b and 9.129c follows Equation 9.123a by equating the coefficients of equal powers of x^n.

9.7 Cardinals of Enumerable, Continuum, and Discontinuous (Cantor 1874, 1878)

When a series cannot be summed to any order (Section 9.6), it either oscillates or diverges; if it diverges, it is unbounded and tends to "infinity." The stereographic projection of the sphere

(Section I.9.2) suggests that the complex plane has one "point at infinity" corresponding to the $+\infty$ of the real line. The number of positive integers and of integers is infinite, and the number of rational, irrational, and real numbers in the unit interval is also infinite. This raises the question of whether all these "infinities" are "equivalent" in some sense or can be ranked in some "order of magnitude." The concept of cardinal allows a distinction to be made between the infinities corresponding to (1) an enumerable set (Subsection 9.7.1); (2) the continuum (Subsection 9.7.3); and (3) the discontinuous functions (Subsection 9.7.4). These three transfinite cardinal numbers are related to the properties of numbers and sets (Subsection 9.7.3).

9.7.1 Cardinals of Finite and Enumerable Sets

Two sets (A, B) have the same **cardinal** denoted by # iff (≡ if and only if) a one-to-one relation can (Equation 9.130a) be established between their elements:

$$\# A = \# B: \quad \Leftrightarrow \quad \forall_{a\in A}\, \exists^1_{b\in B}\colon\, b = f(a) \wedge \forall_{b\notin B}\, \exists^1_{a\notin A}\, b = f(a) \wedge \forall_{a_1,a_2\notin A}\colon\, f(a_1) \neq f(a_2), \tag{9.130a–9.130d}$$

namely, the relation is (1) injective (Equation I.9.1 and Figure I.9.1a), that is, assigns one element of B to each element of A in Equation 9.130b; (2) surjective (Equation I.9.3 and Figure I.9.1c), that is, all elements of B are images of one element of A in Equation 9.130c; and (3) bijective (Equation I.9.5 and Figure I.9.1e), that is, distinct elements of A correspond to distinct elements of B in Equation 9.130d. Thus, two finite sets with the same cardinal have the same number of elements. If a set is infinite, the number of its elements cannot be determined; two infinite sets with the same cardinal are said to be **comparable**. An **enumerable set** is a set whose elements (Equation 9.131b) can be put into a one-to-one correspondence or **isomorphism** with the positive integers:

$$A \text{ is an enumerable set:} \quad \# A \equiv |N \equiv \bar{b} \quad \Leftrightarrow \quad \forall_{a\in A}\, \exists^1_{n\in|N}\colon \quad f(a) = n, \tag{9.131a and 9.131b}$$

and hence its cardinal (Equation 9.125a) equals the cardinal of the positive integers, denoted by $\bar{b}$. Thus, the **cardinal of the enumerable** $\bar{b}$ *is, in particular, the cardinal of (1) the positive integers* $|N$ *(Equation 9.132a); (2) the nonnegative integers* $|N_0$ *(Equation 9.132b); (3) the integers* $|Z$ *(Equation 9.132c); (4) the rational numbers* $|L$ *(Equation 9.132d); and (5) any ordered sequence of rational numbers* $|L^n$ *(Equation 9.132e):*

$$n \in |N: \quad \bar{b} = \#|N = \#|N_0 = \#|Z = \#|L = \#|L^n \tag{9.132a–9.132e}$$

because all these sets are enumerable.

To prove this, it is sufficient to note that (1 and 2) the positive (Equation 9.133b) and nonnegative (Equation 9.133a) integers are put in a one-to-one relation by adding unit to the latter:

$$|N_0 \equiv \{0,1,2,3,\ldots\}, \quad |N = \{1,2,3,\ldots\} = |N_0 + \{1\}; \tag{9.133a and 9.133b}$$

(3) ordering the integers as in Equation 9.134:

$$|Z \equiv \{0,+1,-1,+2,-2,+3,-3,\ldots.\}, \tag{9.134}$$

they can be put into a one-to-one relation with the positive integers; (4) the rational numbers are the ratio of two integers and can be ordered by numerators smaller than denominators:

$$|L \equiv x/y: \quad x, y \in |Z = \left\{\pm\frac{1}{2}, \pm\frac{1}{3}, \pm\frac{2}{3}, \pm\frac{1}{4}, \pm\frac{3}{4}, \cdots\right\} \tag{9.135}$$

leading to a one-to-one relation with the positive integers; and (5) an ordered pair of rational numbers can also be ordered:

$$|L^2 \equiv \left\{\left(\pm\frac{1}{2}, \pm\frac{1}{2}\right), \left(\pm\frac{1}{2}, \pm\frac{1}{3}\right), \left(\pm\frac{1}{3}, \pm\frac{1}{3}\right), \left(\pm\frac{1}{2}, \pm\frac{2}{3}\right), \left(\pm\frac{1}{3}, \pm\frac{2}{3}\right), \cdots\right\}, \tag{9.136}$$

and likewise for an ordered sequence of rational numbers because *an enumerable sequence of enumerable sets is enumerable and has the cardinal of the enumerable.* The latter statement can be proved by considering the elements $a_{m,n}$ of sets a_n, and writing them in diagonal form (Equation 9.176), they can then be put in one-to-one correspondence with the integers.

9.7.2 Dense Sets and Rational and Irrational Numbers (Cantor 1874)

The existence of a cardinal larger than that of the enumerable (Equations 9.131a, 9.131b, and 9.132a through 9.132e) is proved by **Cantor (1874) first theorem**: *the aggregate that consists of the continuum of numbers in a given interval is not enumerable.* As an example, consider the interval of the real line (a, b). *The set of rational numbers is* **dense** *on any interval of the real line, since any subinterval has at least one rational number:*

$$\forall_{a<c<d<b\notin|R}: \quad \exists_{x,y\in|Z}: \quad c < \frac{x}{y} < d. \tag{9.137}$$

The Cantor first theorem can be restated: *a set dense in the unit interval has in every subinterval, no matter how small, at least one accumulation point not belonging to the set.* Thus, it is possible to define at least one sequence of rational numbers in the interval whose limit is not a rational number:

$$n \in |N; \quad \exists_{(x_n, y_n)\in|Z}: \quad \lim_{n\to\infty} \frac{x_n}{y_n} \notin |L. \tag{9.138}$$

The accumulation point specifies a **Dedekind section** (Section I.1.2) and hence is an **irrational number**. It follows that *any closed unit interval of the real line, which includes all its accumulation points, consists of rational and irrational numbers.* Before proceeding, the Cantor first theorem is proved.

Let (a, b) be an interval of the real line and $\{p_n\}$ enumerable set of numbers in that interval. If the set is not dense, there is a finite interval without a number of the set, and the theorem is proved. If the set is dense, consider any subinterval (α_0, β_0) such that $a < \alpha_0 < \beta_0 < b$, no matter how small $|\beta_0 - \alpha_0| < \varepsilon$, and let p_{n1}, p_{n2} be the first two points that lie in the interval. Take the largest (smallest) of the two as β_1 (α_1) and consider the interval (α_1, β_1); all numbers p_n with $n > n_2 > n_1$ lie outside the interval. Next consider the first numbers (p_{n3}, p_{n4}) inside the interval (α_1, β_1) and use them to define a smaller interval:

$$(\alpha_n, \beta_n) \equiv \left(\inf(p_{2n-1}, p_{2n}), \sup(p_{2n-1}, p_{2n})\right). \tag{9.139}$$

The intervals (α_n, β_n) are each contained in all the preceding and have an accumulation point on the left and right:

$$\alpha_1 < \alpha_2 < \ldots \alpha_n < \ldots \rightarrow \alpha \leq \beta \leftarrow \ldots < \beta_n < \ldots < \beta_2 < \beta_1. \tag{9.140}$$

If the accumulation points do not coincide $\alpha \neq \beta$, the interval (α, β) has no elements of the set, so $\{p_n\}$ is not dense, contrary to the hypothesis. Thus, the accumulation points must coincide $\alpha = \beta$, and since they do not belong to the sequence, the theorem is proved.

9.7.3 Cardinal of the Continuum and of a Bounded Domain (Cantor 1878)

The first Cantor theorem shows that given an enumerable set, every interval of the real line has points not belonging to the set. Thus, an interval of the real line is not denumerable and cannot have the cardinal of the denumerable. The **cardinal of the continuum** *is defined (Equation 9.141a) as the cardinal of the unit interval of the real line:*

$$n \in |N: \quad \bar{c} \equiv \#(0,1) = \#(a,b) = \#|I = \#|R = \#|R^n \tag{9.141a–9.141f}$$

and equals (1) the cardinal of any finite interval (Equation 9.141c); (2) the cardinal of the irrational numbers (Equation 9.141d); (3) the cardinal of the real numbers (Equation 9.141e); and (4) the cardinal of the set of n ordered real numbers (Equation 9.141f). To prove item 1, it is sufficient to note that the function 9.142a establishes a one-to-one relation between any interval (Equation 9.142b) of the real line and the unit interval (Equation 9.142c):

$$a, b \in |R: \quad y \equiv \frac{x-a}{b-a}, \quad a \leq y \leq b \quad \Rightarrow \quad 0 \leq x \leq 1. \tag{9.142a–9.142c}$$

The rational numbers are enumerable (Equation 9.132d), and thus *the cardinal of the irrational numbers equals the cardinal of the real numbers (Equations 9.141c and 9.141d) because subtracting an enumerable set (Equation 9.143b) does not change (Equation 9.143c) the cardinal of the continuum (Equation 9.143a):*

$$\#|R = \bar{c}, \quad \#|L = \bar{b}, \quad \#|I = (|R - |L) = \bar{c} - \bar{b} = \bar{c}. \tag{9.143a–9.143c}$$

The proof of the latter theorem is as follows: (1) if A is a continuum (Equation 9.144a) and B_1 an enumerable set (Equation 9.144b), then $A_1 = A - B_1$ is (Equation 9.144d) nonenumerable (otherwise, $A = A_1 + B_1$ is the union of enumerable sets and is also enumerable); and (2) let B_2 be another enumerable set (Equation 9.144c) so that $A_2 = A_1 - B_2 = A - B_1 - B_2$ is still nonenumerable:

$$\#A = \bar{c}, \ \#B_1 = \bar{b} = \#B_2: \quad \#A_1 = \#(A - B_1) \neq \#\bar{b} \neq \#A_2 = \#(A_2 - B_1 - B_2); \tag{9.144a–9.144c}$$

$$\#(B_1+B_1)=\bar{b}:\ \bar{b}=\#(B_1+B_2)=\#(A-A_2)=\#A-\#A_2=\#A-\#(A_1-B_2)=\#A-\#A_1+\bar{b}, \quad (9.145a\text{ and }9.145b)$$

it follows that A and A_1 have the same cardinal (Equation 9.145b), namely, that of the continuum, although they differ by an enumerable set B_1.

The last statement in Equation 9.141e is a particular case of the **Cantor (1878) second theorem:** *the cardinal of any hyperparallelipiped (Equation 9.146b) of any dimension (Equation 9.146a), possibly denumerably infinite, has the cardinal of the continuum (Equation 9.146c):*

$$n\in|N:\quad X\equiv\left\{(x_1,\ldots x_n,\ldots):a_n<x_n<b_n;\quad (a_n,b_n)\in|R\right\}\quad\Rightarrow\quad \#X=\bar{c}. \quad (9.146a\text{–}9.146c)$$

The hyperparalleliped is specified by a set of n ordered real numbers, each in an interval. Thus, (1) for $n = 2$ ($n = 4$), it includes [Section I.1.3 (Section I.1.9)] the finite complex numbers (quaternions); (2) for finite n, the points in an n-dimensional space like a vector space (Subsection 5.7.5); and (3) for infinite n, the points in a space of infinite dimension, like a function space (Section 5.7). Since there is a one-to-one mapping (Equations 9.142a through 9.142c) between any interval of the real axis and the unit interval, the hyperparalleliped (Equation 9.146a) may be replaced by the unit hypercube (Equation 9.147c) choosing the unit interval (Equations 9.147a and 9.147b) along all axis:

$$a_n=0,\ b_n=1:\quad Y=\left\{(x_1,\cdots,x_1,\cdots):\ 0<x_n<1\right\}:\quad x_n\in\left|I\right.\quad\Rightarrow\quad \#Y=\bar{c}; \quad (9.147a\text{–}9.147e)$$

also the irrational numbers (Equation 9.147d) have the same cardinal as the real numbers (Equation 9.143c), and thus only the irrational numbers need be considered (9.147e). An irrational number in the unit interval has (Subsection 1.7.4) a unique representation as a nonterminating continued fraction:

$$x=\frac{1}{a_2+}\cdots\frac{1}{a_n+}\cdots; \quad (9.148)$$

this coincides with Equation 1.124c where $a_1 = 0$ since the integer part of a number in the unit interval is zero, that is, zero is the largest integer smaller than any number in the unit interval. The continuants are obtained truncating the continued fraction on the right, leading to rational numbers. If instead the continued fraction is truncated on the left, it specifies uniquely a sequence of irrational numbers:

$$x_1=\frac{1}{a_1+}\frac{1}{a_{p+1}+}\frac{1}{a_{2p+1}+}\ldots,\quad x_2=\frac{1}{a_2+}\frac{1}{a_{p+2}+}\frac{1}{a_{2p+2}+}\ldots,$$
$$\ldots x_p=\frac{1}{a_p+}\frac{1}{a_{2p}+}\frac{1}{a_{3p}+}\ldots, \quad (9.149a\text{–}9.149c)$$

that all lie in the unit interval; conversely the enumerable sequence (Equations 9.149a–9.149c) specifies a unique irrational number (Equation 9.148) in the unit interval. Thus, a one-to-one relation has been established between the unit interval and the unit hypercube, proving that they have the same cardinal for any denumerable dimension n.

9.7.4 Continuous Functions and Cardinal of the Discontinuous

A continuous function is specified by its values at the rational numbers of the domain because they form a dense set, and the values at other (irrational) numbers are obtained by continuity; at each point, the range of values of the function is the continuum. Thus, a continuous function is specified by the continuum of values in the range, corresponding to an enumerable set of points in the domain. By the Cantor second theorem, *the cardinal of a continuous function (Equation 9.150a) is the cardinal of the continuum (Equation 9.150b):*

$$y = f(x) \in C(a,b): \quad \#f = \bar{c}; \quad y = g(x) \in \mathcal{U} - C(a,b): \#g = \bar{d}, \qquad (9.150a–9.150d)$$

whereas the cardinal of a discontinuous function (Equation 9.150c) is (Equation 9.150d) the **cardinal of the discontinuous $\bar{d}$**. The latter cannot be the cardinal of the continuum because the values of a discontinuous function are totally independent at neighboring points, and thus both the range and the domain have the cardinal of the continuum. Since the rational (real) points of the domain are enumerable (nonenumerable), they can (cannot) be ordered when considering the product by the continuum of the range. Thus, *three cardinal numbers have been introduced: the cardinal of the enumerable (Equations 9.132a through 9.132e), the cardinal of the continuum (Equations 9.141a through 9.141f), and the cardinal of the discontinuous (Equations 9.150c and 9.150d):*

$$\#|N \equiv \bar{b} < \#|R^n \equiv \bar{c} < \#\mathcal{U} - C \equiv \bar{d}, \qquad (9.151)$$

each larger than the preceding (Equation 9.151) in the sense that (1) there is no one-to-one relation between sets with different cardinals; and (2) the cardinal of one set does not change if a subset with a lower cardinal is subtracted:

$$\bar{d} = \bar{d} - \bar{c} = \bar{d} - \bar{c} - \bar{b}, \quad \bar{c} = \bar{c} - \bar{b}. \qquad (9.152a \text{ and } 9.152b)$$

Although the theory of cardinals answers some questions, others are left open: a set with a cardinal between that of the denumerable and continuum has not yet been found, nor it has been proved that its existence would lead to any contradiction. Rather than proceed with the search for larger cardinal numbers, the question addressed next is whether they can be obtained as an extension of ordinary numbers. This leads to the two theories of Cantor on transfinite numbers, namely, the ordinal and cardinal numbers (Section 9.8).

9.8 Transfinite Cardinal and Ordinal Numbers (Cantor 1883a,b; Hardy 1903)

Like the Peano postulates for the positive integers (Section I.1.1), the Cantor postulates specify the transfinite cardinal numbers (Subsection 9.8.1), that can be partially ordered

(Subsection 9.8.2). The simple ordering or isomorphism (normal ordering) leads [Subsection 9.8.4 (Subsection 9.8.5)] to the ordinal (aleph) numbers. The ordinal and cardinal numbers coincide if finite and are distinct if infinite, namely, the sum and product of cardinal (ordinal) numbers have distinct properties, for example, they are (are not) commutative.

9.8.1 Postulates (Cantor 1883a,b) for and Construction (Hardy 1903) of Transfinite Numbers

The integers can be defined by the Peano postulates (Section I.1.1), that establish a **principle of generation**, whereby from the first integer unity, all others are obtained via a successor function that adds unity. The **Cantor postulates (1883a,b)** specify the two principles of generation of **transfinite numbers:** ($P1$) after any number, a successor number, immediately following it, is obtained by adding unity; and ($P2$) after any endless succession of numbers, a new number is formed that succeeds all numbers in the sequence and has no number immediately preceding it. Thus, the Cantor postulates differ from the Peano postulates in that a number may be generated ($P2$) without using the successor function ($P1$). The first principle specifies the finite cardinal numbers, that is, the **first class**; the second principle extends these to the transfinite cardinal numbers of the **second class.** The cardinal of the enumerable $\bar{b}$ coincides with the first cardinal number of second class, which follows all integers and specifies the **first aleph number** $\chi_0 \equiv \bar{b}$. The numbers of the second class may be formed by the **Hardy method (1903):** (1) consider the sequence of integers truncated at left in succession, for example, omitting the first (Equation 9.153a), second (Equation 9.153b), third (Equation 9.153c) positive integer, and so on:

$$\left\{(1) \equiv \underline{1},2,3,4,5,...;\quad (2) \equiv 2,\underline{3},4,5,6,...;\quad (3) \equiv 3,4,\underline{5},6,7,\cdots;\cdots\right\}; \qquad (9.153a–9.153c)$$

(2) the first transfinite cardinal number of the second kind, which lies beyond all integers, may be obtained from the diagonal sequence (Equation 9.154a) that picks the first element of Equation 9.153a, the second of Equation 9.153b, the third of Equation 9.153c, and so on:

$$\left\{(\bar{b}) \equiv \underline{1},3,5,7,9,\cdots;\quad (\bar{b}+1) \equiv 3,\underline{5},7,9,11\cdots;\quad (\bar{b}+2) \equiv 5,7,\underline{9},11,13,\cdots\right\}; \qquad (9.154a–9.154c)$$

the same process (Equations 9.153a through 9.153c) is applied to Equation 9.154a, omitting the first (Equation 9.154b), second (Equation 9.154c) element, and so on, leading to all transfinite members of the form $\bar{b}+n$; (3) after all the successors $\bar{b}+n$ of the first transfinite cardinal number of the second kind, the principle P_2 introduces another number of the second kind $\bar{b}+\bar{b}=2\bar{b}$, which is obtained from the diagonal sequence of Equations 9.154a through 9.154c, namely

$$\left\{(2\bar{b}) \equiv (\underline{1},5,9,13,17,\cdots),\quad (2\bar{b}+1) \equiv (5,\underline{9},13,17,21),\cdots\right\}; \qquad (9.155a \text{ and } 9.155b)$$

(4) the diagonal of Equations 9.155a and 9.155b specifies $3\bar{b}$ in Equation 9.156a:

$$\left\{(3b) \equiv (1,9,17,25,33,...),...\right\},\quad (4\bar{b}) \equiv \left\{1,17,33,49,55,...\right\}, \qquad (9.156a \text{ and } 9.156b)$$

and so Equation 9.156b for $4\bar{b}$,…, $n\bar{b}$; (5) after all $n\bar{b}$ a new number $\bar{b}\ x\ \bar{b} = \bar{b}^2$ is introduced by P_2 corresponding to the diagonal:

$$(\bar{b}^2) \equiv (\bar{b}, 2\bar{b}, 3\bar{b}, 4\bar{b}, 5\bar{b}, \ldots) \equiv (1, 5, 17, 49, \ldots); \tag{9.157}$$

(6) after $\bar{b}^2$ follows $\bar{b}^2 + 1, \bar{b}^2 + 2, \ldots$, then $\bar{b}^2 + \bar{b}$, …, then $\bar{b}^2 + 2\bar{b}, \ldots$, and then

$$a_n \in |N: \quad a_n \bar{b}^n + a_{n-1} \bar{b}^{n-1} + \ldots + a_1 \bar{b} + a_0 \to \bar{b}^{\bar{b}} \to \bar{b}^{\bar{b}^{\bar{b}}}, \qquad \text{(9.158a and 9.158b)}$$

after that follows (7) the power (Equation 9.158a); (8) after all powers (Equation 9.158b), the first transfinite cardinal number of the **third kind** could be introduced, but this would require a third postulate (P3). The Hardy method leads to sequences b_n, each consisting of a denumerable set of numbers $b_{n,m}$ with the property that each succeeding sequence contains larger numbers at the same positions:

$$\forall_{n \in |N} \exists_{s \in |N}: \quad m > s \quad \Rightarrow \quad b_{n+1,m} > b_{n,m}, \tag{9.159}$$

and thus all sequences are distinct. *The set of all Hardy sequences is in one-to-one correspondence with the set of all cardinals of numbers first and second kind and specifies the* **second aleph number** χ_1 *as its cardinal.* This last statement relates to the theory of ordinal numbers (Subsections 9.8.4 through 9.8.6), which is the Cantor second approach (1895, 1897) to transfinite numbers.

9.8.2 Partial Ordering of Cardinal Numbers (Cantor 1895, 1897)

An **aggregate** is defined as a collection of elements drawn from the **universe** of available objects; although this rather general and vague definition will give rise to objections later (Section 9.9), it is taken as the starting point for the present section. Two aggregates are **equivalent** *(symbol ~)* if a one-to-one relation can be established between them, and then they have the same **cardinal number** (Equations 9.130a through 9.130d). A partial order relation can be introduced into the cardinal numbers by defining the cardinal $\# A$ of the set A to be larger (Equation 9.160a) than the cardinal $\# B$ of the set B if (1) a subset A_1 of A is equivalent (Equation 9.160b) to B; and (2) no subset B_1 of B is (Equation 9.160c) equivalent to A.

$$\#A \geq \#B \quad \Leftrightarrow \quad \exists_{A_1 \subset A}: \#A_1 = \#B \quad \wedge \quad \forall_{B_1 \subset B}: \#B_1 \neq \#A. \qquad \text{(9.160a–9.160c)}$$

This is a **partial order relation** because the following possibilities exist given two sets A, B: (1) the case (Equation 9.160a) that the cardinal of A is larger than the cardinal of B; (2) the reverse case that the cardinal of B is larger than the cardinal of A if a subset B_1 of B is equivalent to A, but no subset A_1 of A is equivalent to B; and (3) the conjunction of items 1 and 2 so that

$$\#A \geq \#B \quad \wedge \quad \#B \geq \#A \quad \Rightarrow \quad \#A = \#B \tag{9.161}$$

because Equations 9.162a and 9.162b imply Equation 9.162c and hence Equation 9.162d:

$$A \supseteq A_1 - B \wedge B \supseteq B_1 \sim A: \#B = \#A_1 \leq \#A = \#B_1 \leq \#B \quad \Rightarrow \quad \#A = \#B; \qquad \text{(9.162a–9.162d)}$$

and (4) there is no subset A_1 of A equivalent to B, and no subset B_1 of B is equivalent to A. The possibility of item 4 means that the order relation is partial or nonstrict: given two sets it does not follow that

$$\tilde{A} > \tilde{B} \quad \vee \quad \tilde{A} = \tilde{B} \quad \vee \quad \tilde{B} > \tilde{A}, \tag{9.163a–9.163c}$$

as would be the case for a **strict order relation** (Equations 9.163a through 9.163c). The **transitive property**

$$\#A > \#B \quad \wedge \quad \#B > \#C \quad \Rightarrow \quad \#A > \#C \tag{9.164}$$

holds for the partial ordering of cardinal numbers (Equations 9.160a through 9.160c).

9.8.3 Sum, Product, and Power of Cardinal Numbers

Given two sets A, B without a common element (Equation 9.165a), the cardinal of the union is the **sum** *of the cardinals (Equation 9.165b):*

$$A \cap B = \phi: \quad \#A + \#B = \#(A \cup B) = \#B + \#A, \quad \#A + (\#B + \#C) = (\#A + \#B) + \#C, \tag{9.165a–9.165d}$$

and it has the commutative (Equation 9.165c) and associative (Equation 9.165d) properties. Given any two sets A, B, the **product** *is the set formed by combining each element of the first set with all elements of the second set, and the cardinal is the product of cardinals (Equation 9.166a) and has the commutative (Equation 9.166b) and associative (Equation 9.166c) properties:*

$$(\#A)(\#B) = (\#B)(\#A), \quad (\#A)[(\#B)(\#C)] = \left[(\#A)(\#B)\right](\#C); \tag{9.166a–9.166c}$$

also the product has the distributive property relative to the sum:

$$(\#A + \#B)(\#C) = (\#A)(\#C) + (\#B)(\#C). \tag{9.167}$$

The iterated product of cardinal numbers leads to the **power** of a cardinal number.

9.8.4 Simple Ordering and Isomorphism

A set is **simply ordered** if an **order function** can be defined for any pair of elements such that (1) it is zero if they coincide (Equation 9.168b); and (2 and 3) it is +1(−1) if the first follows (precedes) the second [Equation 9.168a (Equation 9.168c)]:

$$f \text{ order function in } A: \quad \forall_{a,b \in A}: \quad f(a,b) = \begin{cases} +1 & \text{if} \quad b < a & (9.168\text{a}) \\ 0 & \text{if} \quad b = a & (9.168\text{b}) \\ -1 & \text{if} \quad b > a. & (9.168\text{c}) \end{cases}$$

Two ordered aggregates are **similar** if there is an **isomorphism**, that is, a one-to-one relation between them that preserves the order:

$$f \text{ isomorphism} \quad A \to B: \quad \forall_{a_1,a_2 \in A} \; \exists^1_{b_1,b_2 \in B}: \; f(a_1,a_2) = f(b_1,b_2). \tag{9.169}$$

The **sum** *of two simply ordered sets (Equations 9.170a and 9.170b) is a set (Equation 9.170c) that preserves the orders of elements in A and B, and such that every element of A precedes every element of B:*

$$A = \{a_1, a_2, \ldots\}, \quad B \equiv \{b_1, b_2, \ldots\}, \quad C \equiv A + B = \{a_1, a_2, \ldots; b_1, b_2, \ldots\} \neq B + A, \tag{9.170a–9.170d}$$

and it is noncommutative (Equation 9.170d). The **product** *of two ordinal numbers or simply ordered sets (Equations 9.170a and 9.170b) is the set obtained combining each element of the first set with each element of the second set:*

$$A\, x\, B = \{a_1b_1, a_1b_2, \ldots, a_2b_1, a_2b_2, \ldots, a_3b_1, a_3b_2 \ldots\} \neq B\, x\, A \tag{9.171a and 9.171b}$$

and is noncommutative (Equation 9.171b). The cardinal (ordinal) numbers have distinct properties, for example, the sum [Equations 9.165a through 9.165d (Equations 9.170a and 9.170d)] and product [Equation 9.166a (Equations 9.171a and 9.171b)] are (are not) commutative.

9.8.5 Normal Ordering and Ordinal Numbers

An aggregate M is **normally ordered** if (1) there is an element m of lower rank than all the others; and (2) if M_1 is a part of M, and if M contains elements m_1, m_2, ... of higher rank than all the elements in M_1, then there exists one element m' of M that immediately follows M_1, and there are no elements of M between m' and M_1:

$$\forall_{m' \in M} \quad \exists^1_{m \in M}: \; m' < m_i; \quad M_1 \subset M, \quad \forall_{m_i \in M - M_1} \quad \forall_{m_j \in M_1} \quad \exists_{m' \in M}: \; m_j < m' \leq m_i. \tag{9.172a and 9.172b}$$

The order type of a normally ordered aggregate specifies an **ordinal number**. The set of all order types of an aggregate is its **cardinal number**. If a normally ordered aggregate M has an element similar to M' but M' has no element similar to M, then the ordinal number $\tilde{M}$ is larger than the ordinal number $\tilde{M}'$. Thus, the ordinal numbers are strictly ordered since Equations 9.163a through 9.163c hold for ordinal numbers. The reasons are that (1) of the four possibilities in Subsection 9.8.2, the fourth can be excluded; and (2) only the first three possibilities exist and they are mutually exclusive. This follows from (1) if M is isomorphic to a subset of M' and M' is isomorphic to a subset of M, then M is isomorphic to M' and they are similar, that is, they have the same ordinal; (2) if a subset of M' is isomorphic to M, but no subset of M' is isomorphic to M, then $\tilde{M} > M'$; (3) in the opposite case, $M' > \tilde{M}$, and all cases 1–3 are mutually exclusive; and (4) if there is no element of M isomorphic to M' and no element of M' isomorphic to M then a contradiction follows. A part M_1 of M must be isomorphic to a part M_1' of M', for example, the lowest element. Adding an element of $M - M_1$ to M_1 and an element of $M' - M_1'$ to M_1', the isomorphism is maintained. Proceeding in this way until the elements of M (or M') are exhausted leads to an isomorphism between $M(M')$ and a part of $M(M')$, or both if $M = M'$.

9.8.6 Second Class of Ordinal Numbers

Every finite ordered aggregate (Equations 9.173a and 9.173b) is normally ordered; its order type or ordinal number is the number of its elements and coincides (Equation 9.173c) with the cardinal number:

$$n \in |N: \quad A \equiv \{a_1, \ldots, a_n\}, \quad \#A = \tilde{A}. \tag{9.173a–9.173c}$$

The aggregates that have transfinite cardinal number can be normally ordered in an infinite number of distinct ways, and thus an infinite set of ordinal numbers corresponds to the same transfinite cardinal number. The cardinal number of the enumerable $\bar{b} = \chi_0$ is the first aleph number χ_0, and the ordinal number of all order types (Equation 9.174a) with cardinal number χ_0 is (Equation 9.174b) the **ordinal number of the second class** $\chi_1 = Z(\chi_0)$:

$$B \equiv \left\{A: \quad \#A = \bar{b} \equiv \chi_0\right\}, \quad \tilde{B} \equiv \chi_1. \tag{9.174a and 9.174b}$$

From Equations 9.174a and 9.174b, it follows that the first aleph number is unchanged by addition, multiplication, or power (Equation 9.175d) of an integer, or by taking itself as an exponent (Equation 9.175d):

$$n \in |N: \quad \chi_0 = \chi_0 + n = n\,\chi_0 = (\chi_0)^n = (\chi_0)^{\chi_0}. \tag{9.175a–9.175d}$$

The last relation (Equation 9.175d) can be proved as follows: (1) the first aleph number χ_0 can be identified with a series $a_{11},\ldots,a_{1n}$ and its power with the infinite matrix $a_{n,m}$; (2) ordering the elements by diagonals as in Cauchy's rule for the product of series (Equations I.21.25a through I.21.25c):

$$n, m \,\varepsilon\, |N: \quad \left\{a_{n,m}\right\} = \left\{a_{11}; a_{12}, a_{21}, a_{13}; a_{22}, a_{31}; \ldots\right\} \quad \leftrightarrow \quad \{1, \ldots, k\} \equiv |N \tag{9.176}$$

leads to a one-to-one mapping to the integers, hence χ_0 again. The preceding proof is the same as the one used before (Subsection 9.7.1) to show that an enumerable set whose elements are enumerable sets is itself enumerable.

9.8.7 First and Second Aleph Numbers

It is next shown that *the* **second aleph number** *or cardinal of the aggregate of all ordinal numbers of the second class {α} is larger than the cardinal of the enumerable that is the cardinal of all ordinal numbers of the first class:*

$$\chi_1 = \#\{\alpha\} > \chi_0 = \#|N = \bar{b}. \tag{9.177}$$

The aggregate of all numbers of the second class contains a subaggregate whose cardinal is χ_0:

$$\#\left\{\bar{b},\bar{b}_{+1},\bar{b}_{+2},\cdots\right\}=\#\,|\,N=\bar{b}. \tag{9.178}$$

Also a subset of an enumerable aggregate is (1) either finite or (2) infinite, and in the latter case can be put into a one-to-one relation with the integers, and hence its cardinal is χ_0. Thus, an infinite subset of a set with the cardinal of the enumerable has also the cardinal of the enumerable. Thus, it is sufficient to prove that χ_1 does not coincide with χ_0. If $\chi_0=\chi_1$ then the ordinal numbers of the second kind could be arranged in a sequence of cardinal χ_0:

$$\alpha_1,\alpha_2,\cdots \quad \rightarrow \quad \alpha_{p1}<\alpha_{p2}<\ldots<\alpha, \tag{9.179}$$

which could be put in an ascending order and would have an upper limit (Equation 9.179). Thus, α would be larger than every number α_n of the second class. However, this is impossible since α_n contains every number of the second class. Thus, $\chi_1=\chi_0$ leads to a contradiction and Equation 9.178 must hold. Proceeding in this way, it follows that

$$1,2,3,\ldots.,n,\bar{b},\bar{b}+1,\ldots.,\bar{b}^2,\ldots.,\bar{b}^{\bar{b}},a,\ldots \tag{9.180a}$$

$$1,2,3,\ldots.,n,\chi_0,\chi_1,\ldots.,\chi_n,\ldots\chi_{\bar{b}},\chi_{\bar{a}},\ldots, \tag{9.180b}$$

there is a correspondence between (1) the ordered aggregate (Equation 9.180a) containing every ordinal number of every class and (2) the cardinal numbers (Equation 9.180b) of each class of ordinal numbers. The finite elements are coincident positive integers ≡ ordinal ≡ cardinal numbers; the transfinite ordinal numbers correspond to the **aleph numbers**.

9.9 Three Antinomies and the Axiom of Selection (Burali-Forti 1897; Russell 1903; Zermelo 1908)

Cantor made the conjecture that following the cardinal of the enumerable as first aleph number $\bar{b}=\chi_0$, the cardinal of the continuum would be the second aleph number $\bar{c}=\chi_1$. Others have made the conjecture that the cardinal of the continuum exceeds all aleph numbers $\bar{c}>\chi_n$. In the mean time, several objections or possible contradictions have been raised, in connection with the concept of aggregate. Three of these **antimonies** are mentioned: (1) the "admissible" aggregates (Subsection 9.9.1); (2) the self-contradictory aggregates (Subsection 9.9.2); and (3) the axiom of selection (Subsection 9.9.3). This leaves open a number of questions concerning cardinals larger than that of the enumerable (Subsection 9.9.4).

9.9.1 Admissible Aggregates (Burali-Forti 1897; Bernstein 1905)

The first antimony or apparent contradiction concerns (Subsection 9.8.7) the aggregate W of all ordinal numbers (Equation 9.180a). It is a normally ordered aggregate (Subsection

9.8.5) because (1) it has a lowest element 1; (2) any part W_1 of W such that W contains higher order elements is a part of W_1 that also contains other elements of W; and (3) since W_2 is normally ordered, it has one element m' that immediately follows W_1, and m' belongs to W. Thus, W is a normally ordered aggregate and has an ordinal number $\tilde{W}$ and a cardinal number # W. The ordinal number $\tilde{W}$ must be in the aggregate and can only be the largest element of W. However, there can be no largest element of W because every ordinal number $\tilde{W}$ has a successor $\tilde{W}+1$. Thus, a contradiction has been attained.

The contradiction was known to Cantor (1895), who communicated it to Hilbert (1896) and Dedekind (1899); meanwhile, it was published by Burali-Forti (1897). The contradiction may be avoided by restricting the "admissible aggregates" so as to exclude Equation 9.180a; this is possible if W lacks some property that most or all other aggregates of interest possess. This leads (Bernstein 1905) to the **successor property:** *an* **admissible aggregate** *is such that a new aggregate can be formed by adding another element of higher rank. The aggregate of all ordinal numbers (Equation 9.180a) by definition does not have this property, whereas all normally ordered aggregates (Equations 9.172a and 9.172b) do and are admissible.*

9.9.2 Self-Contradictory Statement (Russell 1903)

The **aggregate of first (second) kind $k_1(k_2)$** is defined as an aggregate that does not (does) contain itself as an element [Equation 9.181a (Equation 9.181b)]:

$$k_1 \notin k_1, \quad k_2 \in k_2. \qquad \text{(9.181a and 9.181b)}$$

The aggregate of all aggregates of the first kind is self-contradictory because it consists of all aggregates k_1 but does not contain k_1. The contradiction does not arise for the aggregate of all aggregates of the second kind k_2, since k_2 is included. Thus, the aggregate of the second kind is the only choice. The antinomy can be taken further. The construction of the aggregate of the second kind supposes that it is already known as an element. Is this like locking a safe and leaving the key inside? It looks like a "chicken and egg problem": the element precedes the aggregate and the aggregate precedes the element. In any case, the definition of aggregate as a collection of elements drawn from the available universe (Subsection 9.8.2) needs to be restricted to exclude self-contradictory aggregates. The use of the word "collection" rather than "set" of elements is made to keep the latter concept free from antimonies in the theory of sets.

It is equally possible to construct self-contradictory statements in ordinary language, for example, the **paradox of the barber:** there is in a village a barber who shaves all the men who do not shave themselves and only those. This must be false, because (a) if the barber shaves himself, he does shave one man who shaves himself and not only those who do not shave themselves (2); (b) if the barber does not shave himself, then there is one man in the village who does not shave himself and whom the barber does not shave (1). Thus, it is not possible to satisfy 1 and 2 simultaneously.

A self-contradictory statement can be constructed deliberately, as in the **tale of the condemned:** a man is condemned to death by hanging; he is asked to make a single statement. If it is true (false) he will be hanged from the tree of truth (falsehood). He says, "I will be hanged from the tree of falsehood." Thus, he cannot be hanged because he cannot be (1) hanged from the tree of falsehood because he would have told the truth; or (2) hanged from the tree of the truth because he would have told a lie. Thus, he was saved by constructing a self-contradictory statement.

9.9.3 Axiom of Selection (Zermelo 1908)

The **axiom of selection** (Zermelo 1908) states that an aggregate A consisting (Equation 9.182a) of separate parts $A_1, A_2, \ldots$, each of which contains at least one element (Equation 9.182b), has at least one part A_0 that (Equation 9.182c) shares exactly one element with each of the A_n:

$$A \equiv \underset{n}{U} A_n, \quad A_n \neq \phi: \quad \exists_{A_0 \subset A} \quad \forall_{n \in |N}: \quad \#(A_0 \cap A_n) = 1. \qquad (9.182a\text{–}9.182c)$$

The axiom of selection is implicit not only in theorems in the theory of transfinite numbers but also in other results in mathematics obtained before. In some cases, a **criterion of selection** can be defined that makes recourse to the axiom unnecessary; for example, the proof of the Osgood-Vitali theorem (Subsection I.37.2.1) involves selecting a diagonal sequence from a matrix of a set of sequences, as in the Hardy method (Subsection 9.8.1) of construction of transfinite cardinal numbers. An example of the opposite situation is the aggregate of sets of real numbers (Equation 9.183a) whose union is the interval (Equation 9.183b):

$$S_n \equiv \left\{ n: \frac{1}{n+1} \leq |x| < \frac{1}{n} \right\}, \quad \underset{n}{\cup} S_n = (-1, +1(\qquad (9.183a \text{ and } 9.183b)$$

closed at left and open at right (Equation 9.177b ≡ Equation 9.177c):

$$S_n \equiv \left\{ x: \frac{1}{n+1} \leq x < \frac{1}{n} \;\; v \;\; -\frac{1}{n} < x \leq -\frac{1}{n+1} \right\}. \qquad (9.183c)$$

A sequence of points $p_n \in s_n$ converges to zero $p_n \to 0$ as $n \to \infty$ if such a set can be formed; however, the existence of such a sequence is implied by the axiom of selection. Some other condition, like a suitable norm, must be invoked if the axiom of selection is not used.

9.9.4 Cardinals Larger than That of the Enumerable

The following three statements are equivalent: (1) the axiom of selection; (2) any two aggregates are **comparable** *in the sense that one is equivalent to a part of the other; and (3) any aggregate can be normally ordered. Statement 2 would imply that the continuum can be normally ordered without necessarily showing how it could be done. If any of statements 1–3 is false, then no aggregate exists whose cardinal exceeds all aleph numbers. This would exclude the cardinal of the continuum* $\bar{c}$ *being larger than all aleph numbers. The cardinal of the enumerable can be exceeded, in several ways: (1) the cardinal of all subsets of N-dimensional continuum is Equation 9.184a because all points belonging (not belonging) to a set can be equal to 1 (0); (2) the cardinal of all closed sets of the n-dimensional continuum is the cardinal of the continuum (Equation 9.184b) because a closed set consists of the denumerable set of rational numbers plus all its limit points, that is, the irrational numbers; (3) the cardinal of all discontinuous functions is Equation 9.184c because at each point of the domain (a continuum) there is an independent range of values that is also a continuum:*

$$\#\left(A \subset |R^n\right) = 2^{\bar{b}}, \quad \#(a,b) = \bar{c}, \quad \#\mathcal{U} - C = \bar{d} = \bar{c}^{\bar{c}}; \qquad (9.184a\text{–}9.184c)$$

and (4) the cardinal of a single continuous (discontinuous) function is $\bar{b}+\bar{c}(\bar{c}+\bar{c})$, *hence* $\bar{c}$ *in both cases.* Whereas the theory of transfinite numbers (1) addresses convincingly the cardinal of the enumerable (Section 9.8), (2) it is not definitive about the cardinal of the continuum (Section 9.9); (3) hence, it leaves largely open cardinals larger than those of the continuum.

NOTE 9.1 BOUNDARIES OF METAMATHEMATICS AND PHILOSOPHY

The controversies, contradictions, or antimonies concerning transfinite numbers should not be surprising, bearing in mind that science is relative and not absolute. The **scientific proof** of a fact A is based on showing that it follows logically from another fact B taken as "primitive" or "known" or "obvious" without proof; science cannot prove something out of nothing. In this sense, the **scientific method** can add more links to the chain of knowledge but cannot provide either (1) a start or (2) an end. There is no (1) start since something cannot be proved out of nothing; there is (2) no end since there is no assurance of ever having exhausted knowledge. The theory of transfinite number starts with the ambitious aim of clarifying the meaning of infinity. It succeeds in proving a number of properties of transfinite ordinal and cardinal numbers, of which only a few were presented (Sections 9.7 through 9.9). The properties selected are sufficient to lead to the antimonies that concern not so much the ordinal and cardinal numbers but rather the concept of "aggregate." The concept of "aggregate" is taken as known or "primitive"; to define the concept of "aggregate" would require adding another chain to link, that is, some concept(s) would have to be taken as "obvious" or "primitive" to allow this definition. However, the definition of the concept of aggregate may transcend the mathematical method and cross the boundary into metamathematics or philosophy, which are beyond the scope of the present work.

The present account on transfinite numbers (Sections 9.7 through 9.9) has had three aims. First is to complement the bare minimum of the theory of numbers used at the start of the course to (Chapter I.1) introduce the complex numbers; in this respect, it is noted that the proof of the Cantor second theorem (Subsection 9.7.3), that is, the cardinal of a bounded domain in an n-dimensional continuum is the cardinal of the continuum, uses the properties of continued fractions (Subsection 1.7.4) and could not logically be presented earlier. A second objective is to complement the theory of infinite processes, which is relevant to the convergence of improper integrals (Chapter I.17), series (Chapters I.23, I.25, and I.27), series of fractions, infinite products, and continued fractions (Chapters 1 and 9). All these infinite processes raise convergence issues (Chapters I.21 and I.29 and Sections 9.1 through 9.5) and relate to summability (Section 9.6). The third objective is to present some useful concepts like the cardinals of the denumerable, continuum, and discontinuous (Section 9.7) and their relation with transfinite ordinal and cardinal numbers (Section 9.8). When the application of the scientific method to the elucidation of the nature of infinity leads to as many questions as answers (Section 9.9), the subject may be considered beyond the realm of mathematics and is out of the scope of the present work.

NOTE 9.2 PROPERTIES OF TRANSCENDENTAL FUNCTIONS

The properties of complex functions (Volume I; Chapters 1 and 9) are well illustrated (Chapters 3, 5, and 7) by the elementary transcendental (i.e., nonrational) functions, of which the simplest is the exponential; the exponential can be defined as the function that equals its derivative everywhere and takes the value unity at the origin. From this simple definition follow all the properties not only of the exponential but also the trigonometric and hyperbolic functions, and their inverses, namely, the cyclometric functions and the logarithm. All these apparently distinct real functions of a real variable form a single family of complex functions with one complex variable and one fundamental period. The elementary functions illustrate several types of infinite representations and their convergence properties: (1) power series for all the functions; (2) series of fractions for the meromorphic functions, that is, the secant, cosecant, tangent, and cotangent, which have an infinite number of poles with accumulation point only at infinity; (3) infinite products for the sine and cosine, which have an infinite number of zeros; and (4) continued fractions whose truncations supply approximations by rational functions, which may be the most accurate for a given level of complexity. In order to realize to what extent the elementary transcendental functions are just an introduction, it is sufficient to consider the 35 functions in Table I.39.1, of which the elementary functions are the first 6 and the auxiliary, special, and extended functions are the remaining 29. On one hand, it will be only gradually, as the subject unfolds, that the full reach of these functions will be appreciated; on the other hand, this is an opportunity to preview, at an early stage, one of the main lines of development.

NOTE 9.3 TOPICS ON FUNCTIONS OF A COMPLEX VARIABLE

A constant feature of the present series of volumes is the link among mathematical methods, physical models, and practical applications. The present volume completes the theory of functions of a complex variable that occupied volume I. The complex elementary functions apply to a wide variety of problems from potential fields to vibrations and waves. With the same aim of closing the account on the theory of complex functions and plane potential fields, Examples 10.1 through 10.20 include the following: (1) Examples 10.1 through 10.11 and 10.19 and 10.20 are related to the present volume; and (2) the remaining examples relate to major topics started in volume I, namely, Examples 10.12 through 10.16 and 10.18 using the Schwartz–Christoffel conformal mappings (Chapter I.33) and Example 10.17 using the hodograph method (Chapter I.38). The conclusion of the part of the course on the theory of complex functions (plane potential fields) suggests listing in Classification 8.1 (Classification 9.1) at the end of Chapter 8 (Chapter 9) the main concepts and results. Classification 9.1 recalls the 225 topics that have been addressed in complex functions and may be used again in subsequent volumes. It complements on the mathematical side Classification 8.1 of 150 potential flow problems (Chapter 8) on the physical side. Classification 8.1 (Classification 9.1) in Chapter 8 (Chapter 9) each serves as a review, reminder, and cross-check of subjects already addressed in volumes I and II on the mathematical (physical) side that are closely related. The two classifications 8.1 (9.1) together serve (1) as a structured index of the theory (applications) and (2) as a checklist of concepts (examples) to be retained.

9.10 Conclusion

The infinite processes leading to convergence issues include improper integrals (Section 9.1), infinite products (Section 9.2), continued fractions (Section 9.3), and series (Section 9.4). In the case of convergence, a single finite limit exists and can in principle be determined, provided that a method of calculation be found (e.g., Section 9.5). Some cases of nonconvergence, such as oscillation or divergence, can be brought back to convergence by improved summation methods (Section 9.6) that are consistent, that is, do not change convergent cases. The cases of divergence that cannot be tamed by improved methods of summation lead as a limit to "infinity." The concept of infinity can be analyzed to some extent but using transfinite numbers (Sections 9.7 through 9.9).

CLASSIFICATION 9.1

Methods of Analysis and Complex Functions

Subject	Concept	Topic	Introduction	Examples	Applications
I	A	1. Positive integers ≡ natural	I.1.1	x	x
Algebra	Numbers	2. Nonnegative integers	I.1.1	x	x
20	11	3. Integers	I.1.1	x	x
		4. Rational	I.1.1	x	x
		5. Irrational	I.1.2	x	x
		6. Real	I.1.2	x	x
		7. Complex	I.1.3–I.1.8; F. I.1.1	x	I.2; F.I.2.2
		8. Quaternions	I.1.9	x	x
		9. Hypercomplex	N.I.1.1	x	x
		10. Cardinal (transfinite)	II.9.7–9	x	x
		11. Ordinal (transfinite)	II.9.8–9	x	x
	B	12. Sum	I.3.1–2; F.I.3.1–2	x	x
	Operations	13. Subtraction	I.3.1–2; F.I.3.1–2	x	x
	9	14. Product	I.3.3–4; F.I.3.3–4	x	x
		15. Division	I.3.3–4; F.I.3.3–4	x	x
		16. Power	I.5.1–2; F.I.5.1	I.3,8–9, I.5.8–9	3.4.6–3.4.9
		17. Root	I.5.3–4; F.I.5.2–4	E.I.10.9	I.39.3–6
		18. Real/imaginary part	I.1.3, I.3.6–7; I.5.5	E.I.10.2	x
		19. Modulus and argument	I.1.4; F.I.3.5	x	x
		20. Conjugate	I.1.4, I.1.9, I.1.3.6; F.I.1.1	x	x

(continued)

CLASSIFICATION 9.1 (Continued)

Methods of Analysis and Complex Functions

Subject	Concept	Topic	Introduction	Examples	Applications
II Functions 24	C Classes 9	21. Single-valued	I.9.1; N.I.37.4; D.I.37.1; F.I.9.1a	x	x
		22. Injective	I.9.1; N.I.37.4; D.I.37.1; F.I.9.1d	x	x
		23. Surjective	I.9.1; N.I.37.4; D.I.37.1; F.I.9.1c	x	x
		24. Bijective	I.9.1; N.I.37.4; D.I.37.1; F.I.9.1e, F.I.37.8c	x	x
		25. Multivalued, branch-cut/point	I.7,9,37; N.I.37.4; D.I.37.1; F.I.9.1b; F.I.37.8a	x	T5.2
		26. Multivalent	I.37.6; N.I.37.4; D.I.37.1; F.I.37.8b, F.I.9.1b	x	T5.2
		27. Univalent	I.37.4–5; N.I.37.4; D.I.37.1; F.I.37.8c	x	x
		28. Many-valued manyvalent	N.I.37.4; D.I.37.1; F.I.37.8d	x	x
		29. Branch-points/cuts	I.7,9; F.I.7.1–5; F.I.9.3–5	x	x
	D Points 15	30. Regular	I.27.3, I.39.1; C.I.39.1	x	x
		31. Ordinary	I.27.3, I.39.1; C.I.39.1	x	x
		32. Simple zero	I.31.5–6	x	x
		33. Multiple zero	I.39.1–2; C.I.39.1	x	x
		34. Extraordinary	I.27.3, I.39.1; C.I.39.1	x	x
		35. Special	I.27.3, I.39.1; C.I.39.1	x	x
		36. Algebraic branch-point	I.7.2.3, I.39.1; C.I.39.1	x	x
		37. Logarithmic branch-point	I.7.2.3, I.39.1; C.I.39.1	x	x
		38. Simple pole	I.15.7, I.19.9.1, I.27.4, I.31.5, I.39.1–2; T.I.271. C.I.39.1	x	x
		39. Multiple pole	I.15.8, I.19.9.2, I.27.4, I.31.6; I.39.1–2; T.I.27.1; C.I.39.1	x	x
		40. Essential singularity	I.27.4, I.39.1–6; T.I.27.1; C.I.39.1	x	x
		41. Regular singularity	I.39.1; C.I.39.1	x	x
		42. Irregular singularity	I.39.1; C.I.39.1	x	x
		43. At infinity	I.27.5; T.I.27.1	x	x
		44. Ignorable singularity	I.19.6	x	x

(continued)

III Calculus 21	E Derivates 12	45. Limit	I.11.1, I.19.8	x	x
		46. Continuity	I.11.2; F.I.11.1	x	x
		47. Uniformity	I.11.1; I.13.4	x	x
		48. Infinitesimals	I.19.7–9; T.I.19.1	x	x
		49. Derivate	I.13.3–5	x	x
		50. Holomorphic	I.11.3-5; F.I.11.2–4	x	x
		51. Caudy–Riemann conditions	I.11.3–5	x	x
		52. Chain rule	1.13.7	x	x
		53. Parametric differentiation	I.13.8–9	E.I.20.6–7	x
		54. Of parametric integrals	I.13.4, 6	I.13.4, 6	x
		55. Of integrals at end-points	I.13.9; F.I.13.1h	E.I.20.7	x
		56. L'Hôspital rule	I.19.8	x	x
	F Integrals 9	57. Primitive	I.13.1	x	x
		58. Riemann	I.13.2; F.I.13.1a	x	x
		59. Contour	I.13.5; F.I.13.1a–c	x	x
		60. Loop	I.13.5; F.I.13.1d–g	x	x
		61. Proper	I.17.1–3; F.I.17.1–2	E.I.20.13, 17	x
		62. Improper: first kind	I.17.1.4–6; F.I.17.3	E.I.20.14–15, 17	3.88
		63. Improper: second kind	I.17, 1.7; 3F.I.17.2	E.I.20.16,17	x
		64. Improper: third kind	I.17.1.7–9	E.I.20.17	x
		65. Principal value (Cauchy)	I.17.8–9; F.I.17.5	I.18.6–8	x

(*continued*)

CLASSIFICATION 9.1 (Continued)

Methods of Analysis and Complex Functions

Subject	Concept	Topic	Introduction	Examples	Applications
IV Sequences 15	G Power Series 7	66. Stirling–MacLaurin	I.23.7, 9; D.I.25.1	E.I.30.3, 8	x
		67. Taylor	I.23.7; D.I.25.1	x	x
		68. Lagrange–Burmann	I.23.6,8; N.I.25.1; N.II.1.2; F.I.25.1	E.I.30.6	x
		69. Laurent–MacLaurin	I.25.6; D.I.25.1	E.I.30.9	x
		70. Laurent	I.25.6, I.27.3–9; T.I.27.1; F.I.25.2	x	x
		71. Teixeira	II.25.5; N.II.1.2; D.I.25.1	x	x
		72. Darboux	I.23.9	x	x
	H Continued Fractions 8	73. Series of fractions	II.1.2–3; F.I.1.1–2	E.II.10.1	1.79, II.7.6
		74. Infinite products	I.1.4–5	x	1.79, II.7.7
		75. Canonical infinite products	II.9.2	x	x
		76. Continued fractions	II.1.6–9, II.3.9, II.7.8	E.II.10.2	1.79, II.7.8
		77. Continued fractions of the first kind	II.1.7, II.9.7	x	x
		78. Simple continued fractions	II.1.7, II.9.3	x	x
		79. Periodic continued fractions	II.9.3.4–6	x	x
		80. Recurrent continued fractions	II.9.3.4–6	x	x
V Families 29	I Properties 17	81. Bounded	I.19.1–2	x	x
		82. Monotonic	II.9.1	x	x
		83. Continuous	I.11.2; F.I.11.1	x	x
		84. Sectionally continuous	I.27.1	x	x
		85. Uniformly continuous	I.11.2, I.13.4, I.19.5	x	x
		86. Bounded oscillation	I.27.9; II.5.7.5; F.I.27.2–5	x	x
		87. Holomorphic	I.11.3–5; F.I.11.2–4	x	x
		88. Differentiable	I.11.3–5, I.27.1–2; F.I.11.2–4	x	x
		89. Sectionally differentiable	I.27.1	x	x
		90. Smooth	I.27.1–2	x	x
		91. Analytic	I.27.1–2	x	x

		92. Monogenic	I.31.1	x	x
		93. Lacunary	I.31.1; E.I.30.6	x	x
		94. Automorphic	I.37.6	x	x
		95. Harmonic	I.13.1–5; I.27.1	x	x
		96. Biharmonic	II.4.6.1–3, 8	x	x
		97. Multiharmonic	II.4.6.4–7	x	x
	J Classification 12	98. Constant	I.27.6, I.9.3–6; T.I.27.2	x	x
		99. Polynomial	I.27.7, I.31.4–7; T.I.27.2	E.I.40.3	E.II.10.2
		100. Rational	I.27.7, I.31.8–9; T.I.27.2	x	x
		101. Integral	I.27.9, I.39.3–6; T.I.27.2	x	x
		102. Rational-integral	I.27.9; T.I.27.2	x	x
		103. Meromorphic	I.27.8–9; II.1.2–3, II.9.2	x	x
		104. Polymorphic	I.27.9	x	x
		105. Elementary	I.39.8; II.3, 5, 7; L.I.39.1	x	x
		106. Auxiliary	I.29.5; L.I.39.1	x	x
		107. Specific	I.39.7, 9; L.I.39.1	x	x
		108. Special	I.29.9; II.1.9, II.3.9, II.7.8	E.I.40.20	x
		109. Extended	N.I.23.2–3	x	x
VI Convergence 25	K Particular Series 7	110. Arithmetic	I.11.1	x	x
		111. Geometric	I.11.8–9, I.23.4, I.25.2; T.I.21.2	II.7.1.4-6; E.I.30.2-3; E.II.10.5-6	x
		112. Harmonic	I.29.4, E.I.30.18	x	x
		113. Binomial	I.25.9, II.3.9.10	x	x
		114. Hypergeometric	I.29.9; II.1.9, II.3.9, II.7.8; E.I.30.20; N.II.3.1	x	x
		115. Operations on	I.21.1–7, I.25.8.8, II.7.1.3; T.I.21.1	II.7.1.4–6; E.I.30.7; E.II.10.5–6	x
		116. Double	I.21.4	x	x

(continued)

CLASSIFICATION 9.1 (Continued)

Methods of Analysis and Complex Functions

Subject	Concept	Topic	Introduction	Examples	Applications
VI Convergence 25	L Types 7	117. Divergent	I.21.1; II.9.6; D.I.21.1	E.I.30.1; T.I.30.1–4	x
		118. Oscillatory	I.21.1; II.9.6; D.I.21.1	E.I.30.1; T.I.30.1–4	x
		119. Conditionally convergent	I.21.3; II.9.4; D.I.21.1	x	x
		120. Absolutely convergent	I.21.3; D.I.21.1	x	x
		121. Uniformly convergent	I.21.5–6; D.I.21.1; F.I.21.1–3	x	x
		122. Totally convergent	I.21.7; T.I.29.1; F.I.29.1	x	x
		123. Summable *Ck*	II.9.6	x	x
	M Tests 11	124. Combined	I.29.1	x	x
		125. Comparison	I.29.3–5	x	x
		126. Comparison with integral	I.29.4; II.9.1	x	x
		127. Cauchy	I.29.2	x	x
		128. Weierstrass-*M*	I.21.7	x	x
		129. Weierstrass-*K*	I.29.7	x	x
		130. D'Alembert ratio	I.29.3	x	x
		131. Cauchy ratio	E.I.30.17	x	x
		132. Gauss	I.29.5.3	x	x
		133. Abel	I.29.6	E.I.30.19	
		134. Abel-Dirichlet	I.29.6	x	x

VII Processes 18	N Evaluation of Integrals 9	135. By the primitive	I.13.1	x	x
		136. By parts	I.13.7	x	x
		137. By partial fractions	I.31.8–9	E.I.40.4	x
		138. By residues	I.15.7–9, I.19.9	E.I.20.9	x
		139. Along the unit circle	I.17.2–3; F.I.17.1	E.I.20.13	x
		140. Along the real axis	I.17.4; F.I.17.2	E.I.20.14	x
		141. With oscillating factor	I.17.5; F.I.17.2–3	E.I.20.15	x
		142. Around a branch-cut	I.17.7; F.I.17.4	E.I.20.16	x
		143. Localization lemma	I.17.6	x	x
	P Decompositions 9	144. Factorization of polynomials	I.31.6–7	x	x
		145. Rational partial fractions	I.31.8–9	E.I.40.4	x
		146. Remainder of series	I.23.9	x	x
		147. Radius of series	I.39.8	x	x
		148. Exponent of series	I.39.8	x	x
		149. Landau radius	I.39.6; E.I.40.18; F.I.40.11	x	x
		150. Julia rays	I.39.5–6; F.I.39.5–6	x	x
		151. Exponential order	I.39.5.3–4	x	x
		152. Summation of series by residues	II.9.5; F.II.9.1	E.II.10.1	x
VIII Extensions 11	Q Continuation 6	153. Continuation	I.31.1–2; F.I.31.1–2	x	x
		154. Reflection	I.31.3; F.I.31.3–4	x	x
		155. Analytic	I.23, 25, 27, I.31.1–2	x	x
		156. Plemelj	I.31.3; F.I.31.5	x	x
		157. Schwartz	I.37.8.2, I.37.9.1; F.I.37.7a	E.I.40.15.2; F.I.40.9b	I.38.1.4; F.I.38.1b
		158. Poisson	I.37.9; F.I.37.7b	E.I.40.15.1; F.I.40.9a	I.38.1.3; F.I.38.1a
	R Relations 5	159. Reflection	I.1.6; F.I.1.1	x	x
		160. Inversion	I.3.5; F.I.3.3, F.I.35.3	x	x
		161. Homothety	I.35.2; F.I.35.1	x	x
		162. Rotation	I.35.1–2; F.I.35.1–2	x	x
		163. Translation	I.31, I.35.1	x	x

(continued)

CLASSIFICATION 9.1 (Continued)

Methods of Analysis and Complex Functions

Subject	Concept	Topic	Introduction	Examples	Applications
IX Conformal Mapping 14	S Basic 10	164. General	I.33.1–5; D.I.33.1; T.I.33.1; F.I.33.1–4	E.I.40.5; T.I.40.1–2	x
		165. Isometric	I.35.1	x	x
		166. Linear	I.35.2–3	x	x
		167. Bilinear or mobius	I.35.4–9; T.I.35.1–2; F.I.35.5–6	E.I.40.7–9	x
		168. Self-inverse	I.35.5.2; N.38.1–2	II.8.4.2	I.38.9.4
		169. Reciprocal points	I.24.7–8, I.26.8, I.35.4–9; F.I.24.4, 26.2, 35.4	x	x
		170. Simple	I.37.2–3, 5, 7	x	x
		171. Monogenic	I.31.1; F.I.31.1–2	x	x
		172. Normal (family)	I.37.2.2	x	x
		173. Automorphism (group)	I.37.6	F5.7F	x
	T Schwartz–Christoffel 4	174. Half-plane to polygon (interior)	I.33.6, II.8.4.1; T.I.33.2; F.I.33.5–6	II.36.1,5,6, II.8.4.5; E.II.13–16,18; F.I.36.1,11,13, F.II.8.9–14, 10.5–12,10.14–17	I.36.1,5,6, II.8.4,5; E.II.10.13–16,18; T.I.33.2; F.I.33.5–6
		175. Half-plane to polygon (exterior)	I.33.7; T.I.33.2	x	x
		176. Circle to polygon (interior)	I.33.9; II.8.8.2; T.I.33.2; F.I.33.7a	x	x
		177. Circle to polygon (exterior)	I.33.9; II.8.8.2; T.I.33.2; F.I.33.7b	I.34.1–4; E.II.10.18; F.II.34.13, F.II.10.18	x
X Geometry 17	U Curves 10	178. General	I.13.3	x	x
		179. Regular	I.28.1	x	x
		180. Rectifiable	I.13.3; F.I.13.1a	x	x
		181. Cusp	I.6.6; E.I.20.9, 34.2–6, 39.4; F.I.6.3, 20.3c, d, 34.2c, 34.3b, 39.4	x	x
		182. Angular point	I.15.5; E.I.20.9, 33.5.9; N.I.37.3; IT.I.37.1; F.I.15.2a, b, 20.3a, b, 33.5–7, 34.4a–c	x	x
		183. Double point	I.6.7, 12.7, 28.5, 28.8; N.I.37.3; T.I.37.1; F.I.6.6, 12.4, 28.4d, 28.10, 28.12d	x	x

		184. Multipole point	I.12.8–9; N.I.37.3; T.I.37.1; F.I.12.5, 7	x	x
		185. Orthogonal curves	I.11.8–9; F.I.11.4	x	x
		186. Orthogonal coordinates	I.36.4.4; II.8.3.1–3; F.I.36.10, F.II.8.5	x	x
		187. Conformal coordinates	I.33.3.1	I.36.4.4; F.I.36.10	II.8.3.1–3, F.II.8.5
	V Domains 7	188. Simply connected	I.15.1; F.I.15.1b–d	I.23.3; F.I.23.1	x
		189. Doubly connected	I.15.2; F.I.15.1a	I.25.1; F.I.25.1	x
		190. Multiply connected	I.15.6–9; F.I.15.3	I.37.6; F.I.37.4	x
		191. Boundary	I.37.7; F.I.37.5	x	x
		192. Riemann p-surface	I.7; F.I.7.1–5	x	x
		193. Riemann sphere	I.9.1–2; F.I.9.2	x	x
		194. Surface of genus g	I.9.2–9;F.I.9.3–5	x	x
XI Invariant 10	W Operators 7	195. Gradient	I.11.7.9; II.4.4	x	x
		196. Divergence	I.11.7.9; II.4.4	x	x
		197. Curl	I.11.7.9; II.4.4	x	x
		198. Scalar Laplacian	I.11.6, 9; II.4.4	x	x
		199. Vector Laplacian	II.4.4	x	x
		200. Biharmonic	II.4.6.1–4	x	x
		201. Multiharmonic	I.II.4.6.4–7	x	x
	X Identities 3	202. Divergence theorem	I.14.1–2, 26.1, I.28.1; F.I.28.1	x	x
		203. First Green identity	II.8.2.4	x	x
		204. Second Green identity	II.8.2.4	x	x

(continued)

CLASSIFICATION 9.1 (Continued)

Methods of Analysis and Complex Functions

Subject	Concept	Topic	Introduction	Examples	Applications
XII Functions 21	Y Elementary 9	205. Exponential	II.3.1–4; F.II.3.1–2	x	
		206. Logarithm	II.3.5–9; F.II.3.3–4	x	x
		207. Circular	I.3.7,5.5,7.8,37.8, II.5.7: E.I.10.7, E.II.10.9: F.II.5.1a,c,5.2a-c,5.3a	x	x
		208. Hyperbolic	I.7.8,37.8, II.5.7; E.I.10.7,E.II.10.9; F.II.5.1b,d,5.2d-f,5.3b–g	x	x
		209. Cyclometric	II.7.5-9	x	x
		210. Self-reciprocal	E.I.30.6	x	x
		211. Self-inverse	I.35.5.2, 38.9.4; N.I.38.1–2	x	x
		212. Osgood	I.31.1.2	x	x
		213. Dirichlet	N.I.13.1	x	x
	Z Higher 12	214. Digamma	I.29.5.2	x	x
		215. Euler number	II.7.1.7; T.II.17.1	x	x
		216. Bernoulli number	II.7.1.8; T.II.17.1	x	x
		217. Weierstrass elliptic	I.39.9.1	x	x
		218. Jacobi elliptic	I.39.9.1–2; F.I.39.11	x	x
		219. Hyperelliptic	I.39.7, F.I.39.8–9	x	x
		220. Schwartz triangular	I.39.4.1; F.I.39.1–2; T.I.39.2	x	x
		221. Modular	I.37.5.3, I.39.3–7	x	x
		222. Hypogeometric	II.1.9	x	x
		223. Hypergeometric confluent	II.3.9.7–10	x	x
		224. Hypergeometric Gaussian	I.29.9; II.3.9.2–6	x	x
		225. Hypergeometric generalized	E.I.40.20; N.II.3.1	x	x

Note: 225 main topics of complex analysis addressed in volumes I and II are listed in a similar format to the 150 potential field (and related) problems (Classification 8.1) solved by using complex function theory.
Text: volume, chapter, section, subsection (C: classification; D: diagrams; E: examples; F: figures; L: lists; N: notes; T: tables).

10

Twenty Examples

10.1 Examples 10.1 through 10.20

EXAMPLE 10.1: SUMMATION OF SERIES OF FRACTIONS IN TERMS OF TRIGONOMETRIC FUNCTIONS

Sum the following series of rational functions:

$$\sum_{n=-\infty}^{+\infty} \frac{1}{n^2-(a+b)n+ab} = \frac{\pi}{a-b}\left[\cot(\pi b)-\cot(\pi a)\right]$$
$$= \frac{1}{ab} + 2\sum_{n=1}^{\infty} \frac{n^2+ab}{n^4-(a^2+b^2)n^2+a^2b^2}, \tag{10.1}$$

$$\sum_{n=-\infty}^{+\infty} (-)^n \frac{n^2-(c+d)n+cd}{n^4-(a^2+b^2)n^2+a^2b^2} = \frac{\pi}{a^2-b^2}\left[\left(b+\frac{cd}{b}\right)\csc(\pi b)-\left(a+\frac{cd}{a}\right)\csc(\pi a)\right]$$
$$= \frac{cd}{a^2b^2} + 2\sum_{n=1}^{\infty} (-)^n \frac{n^2+cd}{n^4-(a^2+b^2)n^2+a^2b^2}, \tag{10.2}$$

$$\sum_{n=-\infty}^{\infty} \frac{n^2}{n^4-a^4} = \frac{\pi}{2a}\left[\coth(\pi a)-\cot(\pi a)\right] = 2\sum_{n=1}^{\infty} \frac{n^2}{n^2-a^2}, \tag{10.3}$$

$$\sum_{n=-\infty}^{\infty} \frac{(-)^n}{n^4-a^4} = -\frac{\pi}{2a^3}\left[\operatorname{csch}(\pi a)+\csc(\pi a)\right] = -\frac{1}{a^4} + 2\sum_{n=1}^{\infty} \frac{(-)^n}{n^4-a^4}, \tag{10.4}$$

$$\sum_{n=-\infty}^{\infty} (-)^n \frac{n^2-c^2}{n^4-a^4} = \frac{\pi}{4a^3}\left[(a^2+c^2)\operatorname{csch}(\pi a)-(a^2-c^2)\csc(\pi a)\right],$$
$$= \frac{c^2}{a^4} + 2\sum_{n=1}^{\infty} (-)^n \frac{n^2-c^2}{n^4-a^4}, \tag{10.5}$$

where a, b, c, and d are constants, and a, b are not integers.

The series are summed by decomposing the rational coefficients into partial fractions (Subsections I.31.8 and I.31.9) and then using known series of fractions (Sections 1.2 and

1.3). For example, Equation 10.1 has the partial fraction decomposition arising from the simple poles at (a, b):

$$\sum_{n=-\infty}^{+\infty} \frac{1}{n^2-(n+b)n+ab} = \sum_{n=-\infty}^{+\infty} \frac{1}{(n-a)(n-b)} = \frac{1}{a-b}\sum_{n=-\infty}^{+\infty}\left(\frac{1}{n-a}-\frac{1}{n-b}\right); \tag{10.6}$$

substitution of the series of fractions for the cotangent (Equation 1.57 ≡ Equation 10.7):

$$\pi\cot(\pi z) = \sum_{n=-\infty}^{\infty} \frac{1}{z-n} \tag{10.7}$$

leads to Equation 10.1. In a similar way, Equation 10.2 leads to the partial fraction decomposition arising from the simple poles at $(\pm a, \pm b)$:

$$\begin{aligned}
A &\equiv \sum_{n=-\infty}^{+\infty} (-)^n \frac{n^2-(c+d)n+cd}{n^4-(a^2-b^2)n^2+a^2b^2} = \sum_{n=-\infty}^{+\infty} (-)^n \frac{(n-c)(n-d)}{(n^2-a^2)(n^2-b^2)} \\
&= \sum_{n=-\infty}^{+\infty} (-)^n \frac{(n-c)(n-d)}{(n-a)(n+a)(n-b)(n+b)} \\
&= \sum_{n=-\infty}^{+\infty} \frac{(-)^n}{a^2-b^2}\left\{\frac{1}{2a}\left[\frac{(a-c)(a-d)}{n-a}-\frac{(a+c)(a+d)}{n+a}\right]-\frac{1}{2b}\left[\frac{(b-c)(b-d)}{n-b}-\frac{(b+c)(b+d)}{n+b}\right]\right\};
\end{aligned} \tag{10.8}$$

the series of fractions for the cosecant (Equation 1.84 ≡ Equation 10.9):

$$\pi\csc(\pi z) = \sum_{n=-\infty}^{+\infty} \frac{(-)^n}{z-n} \tag{10.9}$$

may be substituted in Equation 10.8, yielding

$$\begin{aligned}
A &= \frac{\pi}{a^2-b^2}\left\{\left[-\frac{(a-c)(a-d)+(a+c)(a+d)}{2a}\right]\csc(\pi a)\right. \\
&\quad \left.+\left[\frac{(b-c)(b-d)+(b+c)(b+d)}{2b}\right]\csc(\pi b)\right\} \\
&= \frac{\pi}{a^2-b^2}\left[\left(b+\frac{cd}{b}\right)\csc(\pi b)-\left(a+\frac{cd}{a}\right)\csc(\pi a)\right],
\end{aligned} \tag{10.10}$$

which proves Equation 10.2. In the case Equation 10.3, the simple poles are at $(\pm a, \pm ia)$, leading to the partial fraction decomposition:

$$\begin{aligned}
\sum_{n=-\infty}^{+\infty} \frac{n^2}{n^4-a^4} &= \frac{1}{2}\sum_{n=-\infty}^{+\infty}\left(\frac{1}{n^2-a^2}+\frac{1}{n^2+a^2}\right) \\
&= \frac{1}{4a}\sum_{n=-\infty}^{+\infty}\left(\frac{1}{n-a}-\frac{1}{n+a}-\frac{i}{n-ia}+\frac{i}{n+ia}\right) \\
&= -\frac{\pi}{2a}\left[\cot(\pi a)-i\cot(i\pi a)\right] = \frac{\pi}{2a}\left[\coth(\pi a)-\cot(\pi a)\right].
\end{aligned} \tag{10.11}$$

where Equations 10.7 and 5.27b are used to simplify Equation 10.11 to Equation 10.3. In the case Equation 10.4, the four simple poles are again at ($\pm a$, $\pm ia$), leading to a similar partial fraction decomposition:

$$\sum_{n=-\infty}^{+\infty} \frac{(-)^n}{n^4 - a^4} = \frac{1}{2a^2} \sum_{n=-\infty}^{+\infty} (-)^n \left(\frac{1}{n^2 - a^2} - \frac{1}{n^2 + a^2} \right)$$

$$= \frac{1}{4a^3} \sum_{n=-\infty}^{+\infty} (-)^n \left(\frac{1}{n-a} - \frac{1}{n+a} + \frac{i}{n-ia} - \frac{i}{n+ia} \right) \tag{10.12}$$

$$= -\frac{\pi}{2a^3} \left[\csc(\pi a) + i\csc(i\pi a) \right] = -\frac{\pi}{2a^3} \left[\csc(\pi a) + \operatorname{csch}(\pi a) \right],$$

with the alternating signs leading to cosecants (Equation 5.25b) in Equation 10.9 instead of cotangents in Equation 10.7. The poles are again at ($\pm a$, $\pm ia$) for the partial fraction decomposition of Equation 10.5:

$$\sum_{n=-\infty}^{+\infty} (-)^n \frac{n^2 - c^2}{n^4 - a^4} = \sum_{n=-\infty}^{+\infty} (-)^n \left[\frac{1}{2} \left(\frac{1}{n^2 - a^2} + \frac{1}{n^2 + a^2} \right) - \frac{c^2}{2a^2} \left(\frac{1}{n^2 - a^2} - \frac{1}{n^2 + a^2} \right) \right]$$

$$= \sum_{n=-\infty}^{+\infty} (-)^n \left[\frac{1}{4a} \left(\frac{1}{n-a} - \frac{1}{n+a} - \frac{i}{n-ia} + \frac{i}{n+ia} \right) - \frac{c^2}{4a^2} \left(\frac{1}{n-a} - \frac{1}{n+a} + \frac{i}{n-ia} - \frac{i}{n+ia} \right) \right]$$

$$= \frac{1}{4a^3} \sum_{n=-\infty}^{+\infty} (-)^n \left\{ (a^2 - c^2) \left[\frac{1}{n-a} - \frac{1}{n+a} \right] - (a^2 + c^2) \left[\frac{i}{n-ia} - \frac{i}{n+ia} \right] \right\}$$

$$= \frac{1}{2a^3} \sum_{n=-\infty}^{+\infty} \left[-(a^2 - c^2)\csc(\pi a) + i(a^2 + c^2)\csc(i\pi a) \right]$$

$$= \frac{\pi}{2a^3} \left[(a^2 + c^2)\operatorname{csch}(\pi a) - (a^2 - c^2)\csc(\pi a) \right] \tag{10.13}$$

where Equation 10.9 is substituted leading to Equation 10.5.

A particular case $c = a$ of Equation 10.13 is

$$\sum_{n=-\infty}^{+\infty} \frac{(-)^n}{n^2 + a^2} = \frac{\pi}{a} \cosh(\pi a) = \frac{1}{a^2} + 2\sum_{n=1}^{\infty} \frac{(-)^n}{n^2 + a^2}. \tag{10.14}$$

The sums from $n = -\infty$ to $n = +\infty$ not only in Equation 10.14 but also in Equations 10.1 through 10.5 can be replaced by sums from $n = 1$ to $n = +\infty$:

$$\sum_{n=-\infty}^{+\infty} \frac{1}{n^2 - (a+b)n + ab} = \frac{1}{ab} + \sum_{n=1}^{+\infty} \left[\frac{1}{n^2 + ab - (a+b)n} + \frac{1}{n^2 + ab + (a+b)n} \right]$$

$$= \frac{1}{ab} + 2\sum_{n=1}^{+\infty} \frac{n^2 + ab}{(n^2 + ab)^2 - (a+b)^2 n^2} = \frac{1}{ab} + 2\sum_{n=1}^{+\infty} \frac{n^2 + ab}{n^4 - (a^2 + b^2)n^2 + a^2 b^2}, \tag{10.15}$$

$$\sum_{n=-\infty}^{+\infty} (-)^n \frac{n^2-(c+d)n+cd}{n^4-(a^2+b^2)^2 n^2+a^2b^2} = \frac{cd}{a^2b^2} + 2\sum_{n=1}^{+\infty} (-)^n \frac{n^2+cd}{n^4-(a+b)^2 n^2+a^2b^2}, \tag{10.16}$$

$$\sum_{n=-\infty}^{+\infty} \frac{n^2}{n^4-a^4} = 2\sum_{n=-\infty}^{+\infty} \frac{n^2}{n^4-a^4}, \tag{10.17}$$

$$\sum_{n=-\infty}^{+\infty} \frac{(-)^n}{n^4-a^4} = -\frac{1}{a^4} + 2\sum_{n=1}^{+\infty} \frac{(-)^n}{n^4-a^4}, \tag{10.18}$$

$$\sum_{n=-\infty}^{+\infty} (-)^n \frac{n^2-c^2}{n^4-a^4} = \frac{c^2}{a^4} + 2\sum_{n=1}^{+\infty} (-)^n \frac{n^2-c^2}{n^4-a^4}. \tag{10.19}$$

Thus, the series from $n = 1$ to $n = \infty$ in Equations 10.15 through 10.19 are also summed, respectively, by Equations 10.1 through 10.5.

EXAMPLE 10.2: TRANSFORMATION OF SERIES INTO CONTINUED FRACTIONS

Transform the following polynomials into terminating continued fractions:

$$\sum_{n=1}^{N} a_n z^n = \frac{a_1 z}{1-} \frac{a_2 z}{a_1+a_2 z-} \frac{a_1 a_3 z}{a_2+a_3 z-} \cdots \frac{a_{n-2} a_n z}{a_{n-1}+a_n z}, \tag{10.20}$$

$$\sum_{n=1}^{N} \frac{z^n}{b_n} = \frac{z}{b_1-} \frac{b_1^2 z}{b_1 z+b_2-} \frac{b_2^2 z}{b_2 z+b_3-} \cdots \frac{b_{n-1}^2 z}{b_{n-1} z+b_n}, \tag{10.21}$$

$$\sum_{n=1}^{N} \frac{a_1 \ldots a_n}{b_1 \ldots b_n} z^n = \frac{a_1 z}{b_1-} \frac{b_1 a_2 z}{b_2+a_2 z-} \frac{b_2 a_3 z}{b_3+a_3 z-} \cdots \frac{b_{n-1} a_n z}{b_n+a_n z}; \tag{10.22}$$

hence obtain the following terminating continued fractions:

$$|z|<1: \quad \frac{1}{1-z} = 1+\frac{z}{1-} \frac{z}{1+z-} \frac{z}{1+z-} \frac{z}{1+z-} \cdots \frac{z}{1+z}, \tag{10.23a and 10.23b}$$

$|z|<1, \quad \alpha \neq -1,-2,\ldots:$

$$(1+z)^\alpha = 1+\frac{\alpha z}{1-} \frac{(\alpha-1)z}{2+(\alpha-1)z-} \frac{2(\alpha-2)z}{3+(\alpha-2)z-} \cdots \frac{n(\alpha-n)}{n+1+(\alpha-n)z}; \tag{10.24a–10.24c}$$

setting Equations 10.26a and 10.26b in Equation 10.25c, obtain Equation 10.26c:

$$\alpha = \frac{1}{2}, z = -\frac{1}{4}: \quad \frac{\sqrt{3}}{2} = 1 - \frac{1}{8-}\,\frac{8}{2.8+1-}\,\frac{2.3.8}{3.8+3-}\cdots\frac{8n(2n-1)}{8(n+1)+2n-1} \tag{10.25a–10.25c}$$

and conclude that $\sqrt{3}$ is an irrational number.

The general terms of the series Equation 10.20/Equation 10.21/Equation 10.22 ≡ Equation 10.26a are, respectively, Equations 10.26b/10.26c/10.26d):

$$\sum_{n=1}^{N} f_n(z): \qquad f_n \equiv \left\{ a_n z^n, \frac{z^n}{b_n}, \frac{a_1 \ldots a_n}{b_1 \ldots b_n} z^n \right\}, \tag{10.26a–10.26d}$$

leading to the ratio of coefficients:

$$\frac{f_n}{f_{n-1}} = \left\{ \frac{a_n}{a_{n-1}} z, \frac{b_{n-1}}{b_n} z, \frac{a_n}{b_n} z \right\}. \tag{10.27a–10.27c}$$

Substitution of Equations 10.27a/10.27b/10.27c in Equation 1.128a transforms the polynomials Equations 10.27a and 10.27b/10.27c/10.27d, respectively, into terminating continued fractions (Equations 10.28/10.29/10.30):

$$\begin{aligned} \sum_{n=1}^{N} a_n z^n &= \frac{a_1 z}{1-}\,\frac{a_2 z/a_1}{a_2 z/a_1 + 1-}\,\frac{a_3 z/a_2}{a_3 z/a_2 + 1-}\cdots\frac{a_n z/a_{n-1}}{a_n z/a_{n-1} + 1} \\ &= \frac{a_1 z}{1-}\,\frac{a_2 z}{a_1 + a_2 z-}\,\frac{a_1 a_3 z}{a_2 + a_3 z-}\cdots\frac{a_{n-2} a_n z}{a_{n-1} + a_n z}, \end{aligned} \tag{10.28a and 10.28b}$$

$$\begin{aligned} \sum_{n=1}^{N} \frac{z^n}{b_n} &= \frac{z/b_1}{1-}\,\frac{b_1 z/b_2}{b_1 z / b_2 + 1-}\,\frac{b_2 z/b_3}{b_2 z/b_3 + 1-}\cdots\frac{b_{n-1} z/b_n}{b_{n-1} z/b_{n+1}} \\ &= \frac{z}{b_1}\,\frac{b_1^2 z}{b_2 + b_1 z-}\,\frac{b_2^2 z}{b_3 + b_2 z-}\cdots\frac{b_{n-1}^2 z}{b_n + b_{n-1} z}, \end{aligned} \tag{10.29a and 10.29b}$$

$$\begin{aligned} \sum_{n=1}^{N} \frac{a_1 \ldots a_n}{b_1 \ldots b_n} z^n &= \frac{a_1 z/b_1}{1-}\,\frac{a_2 z/b_2}{a_2 z/b_2 + 1-}\,\frac{a_3 z/b_3}{a_3 z/b_3 + 1-}\cdots\frac{a_n z/b_n}{a_n z/b_n + 1} \\ &= \frac{a_1 z}{b_1-}\,\frac{b_1 a_2 z}{b_2 + a_2 z-}\,\frac{b_2 a_3 z}{b_3 + a_3 z-}\cdots\frac{b_{n-1} a_n z}{b_n + a_n z} \end{aligned} \tag{10.30a and 10.30b}$$

proving, respectively, Equation 10.28b ≡ Equation 10.20/10.29b ≡ Equation 10.21/10.30b ≡ Equation 10.22.

The geometric series Equation 3.148b ≡ Equation I.11.62a ≡ Equation 10.31b corresponds to Equation 10.20 with coefficients Equation 10.31a and leads (Equation 10.20) to the continued fraction Equation 10.31c ≡ Equation 10.23b:

$$a_n = 1: \qquad \frac{1}{1-z} = \sum_{n=0}^{\infty} z^n = 1 + \sum_{n=1}^{\infty} z^n = 1 + \frac{z}{1-}\,\frac{z}{1+z-}\,\frac{z}{1+z-}\cdots\frac{z}{1+z}. \tag{10.31a–10.31c}$$

The binomial series (Equations 3.144a and 3.144b ≡ Equations I.25.37a through I.25.37c ≡ Equation 10.32a)

$$(1+z)^\alpha = \sum_{n=0}^{\infty} \binom{\alpha}{n} z^n = 1 + \sum_{n=1}^{\infty} a_n z^n \tag{10.32a}$$

has coefficients (Equation 10.32b) satisfying the recurrence formula (Equation 10.32c):

$$a_n \equiv \binom{\alpha}{n} = \frac{\alpha(\alpha-1)\ldots(\alpha-n+1)}{n!}, \qquad \frac{a_n}{a_{n-1}} = \frac{\alpha-n+1}{n}; \tag{10.32b and 10.32c}$$

substitution of Equation 10.32c in Equation 10.28a leads to the continued fraction

$$\begin{aligned}(1+z)^\alpha &= 1 + \frac{\alpha z}{1-}\,\frac{(\alpha-1)z/2}{1+(\alpha-1)z/2-}\,\frac{(\alpha-2)z/3}{1+(\alpha-2)z/3-}\cdots\frac{(\alpha-n+1)z/n}{1+(\alpha-n+1)z/n}\\ &= 1 + \frac{\alpha z}{1-}\,\frac{(\alpha-1)z}{2+(a-1)z-}\,\frac{2(\alpha-2)z}{3+(\alpha-2)z-}\cdots\frac{(n-1)(\alpha-n+1)z}{n+(\alpha-n+1)z},\end{aligned} \tag{10.33}$$

that proves Equation 10.24c ≡ Equation 10.33a and is an alternative continued fraction to Equation 3.151. The continued fraction Equation 10.24c does not hold for α a negative integer (Equation 10.24b) and does not supply an alternative to Equation 10.23b. The radius of convergence (Equation 10.24a) of the continued fraction Equation 10.24c is the same as that of the binomial series (Equations 3.145a through 3.145d ≡ Equations I.25.37a through I.25.37c). The radius of convergence of the continued fraction Equation 10.23b is (Equation 10.23a) the same as that of the geometric series (Equation 3.148b ≡ Equation I.21.62).

The particular case (Equations 10.25a and 10.25b) of Equation 10.33 yields

$$\begin{aligned}\frac{\sqrt{3}}{2} &= 1 + \frac{-1/8}{1-}\,\frac{1/8}{2+1/8-}\,\frac{3/4}{3+3/8-}\cdots\frac{(n-1)(n-3/2)/4}{n+(n-3/2)/4}\\ &= 1 - \frac{1}{8-}\,\frac{8}{2.8+1-}\,\frac{2.3.8}{3.8+3-}\cdots\frac{8(n-1)(2n-3)}{8n+2n-3};\end{aligned} \tag{10.34}$$

the corresponding nonterminating nonperiodic continued fraction Equation 10.25c ≡ Equation 10.34 proves that $\sqrt{3}$ is an irrational number. The first three continuants (Equations 10.35b through 10.35d) of the continued fraction Equation 10.34 are

$$\sqrt{3} = 1.73205: \qquad \left(\sqrt{3}\right)_1 = 2\left(1 - \frac{1}{8}\right) = \frac{7}{4} = 1.75000, \tag{10.35a and 10.35b}$$

$$\left(\sqrt{3}\right)_2 = 2\left(1-\frac{1}{8-8/17}\right) = 2-\frac{1}{4-4/17} = 2-\frac{17}{64} = \frac{111}{64} = 1.734375, \tag{10.35c}$$

$$\begin{aligned}\left(\sqrt{3}\right)_3 &= 2\left[1-\frac{1}{8-8/(17-48/27)}\right] = 2-\frac{1}{4-108/(17.27-48)}\\ &= 2-\frac{411}{4.411-108} = 2-\frac{411}{1536} = \frac{2661}{1536} = 1.73242;\end{aligned} \tag{10.35d}$$

they give, respectively, two (Equation 10.35b), three (Equation 10.35a), and four (Equation 10.35d) accurate digits of the irrational number (Equation 10.35a).

EXAMPLE 10.3: VORTEX WITH GAUSSIAN TANGENTIAL VELOCITY

Consider a vortex with the tangential velocity

$$v_\varphi(r) = Ar\exp\left(-\frac{r^2}{a^2}\right) \tag{10.36}$$

and determine the angular velocity Ω and vorticity ϖ. Relate the pressure p at radius r to the pressure at infinity p_∞. Indicate the extrema of v_φ, Ω, ϖ, and p, and any constraints on the vortex core radius.

The derivative of the tangential velocity (Equation 10.37a) vanishes at Equation 10.37b, that is, the location of the maximum value (Equation 10.37c):

$$\frac{dv_\varphi}{dr} = Ae^{-r^2/a^2}\left(1-\frac{2r^2}{a^2}\right), \quad r_1 = \frac{a}{\sqrt{2}}, \quad v_{\varphi\max} = v_\varphi\left(\frac{a}{\sqrt{2}}\right) = \frac{Aa}{\sqrt{2e}}. \tag{10.37a–10.37c}$$

The angular velocity (Equation 10.38a) decays monotonically in modulus with the radius from the value A at the center (Equation 10.38b) to zero at infinity (Equation 10.38c):

$$\Omega(r) = \frac{v_\varphi(r)}{r} = A\exp\left(-\frac{r^2}{a^2}\right), \quad |A| = |\Omega|_{\max} = |\Omega(0)| \geq |\Omega(r)| > 0 = \Omega(\infty). \tag{10.38a and 10.38b}$$

The vorticity is given by

$$\varpi = \frac{1}{r}\frac{d}{dr}(v_\varphi r) = \frac{A}{r}\frac{d}{dr}\left(r^2 e^{-r^2/a^2}\right) = 2Ae^{-r^2/a^2}\left(1-\frac{r^2}{a^2}\right), \tag{10.39a}$$

and its derivative (Equation 10.39b)

$$\frac{d\varpi}{dr} = -\frac{4Ar}{a^2}e^{-r^2/a^2}\left(2-\frac{r^2}{a^2}\right), \qquad r_2 = a\sqrt{2} = 2r_1 \tag{10.39b and 10.39c}$$

vanishes at the radius (Equation 10.39c).

Thus, (1) the vorticity is equal to twice the angular velocity at the center (Equation 10.40a), where the rotation is rigid body (Equation 10.40b):

$$\varpi(0) = 2A = 2\Omega(0), \quad v_\varphi(r) = Ar + O(r^3); \quad \varpi(a) = 0 = \varpi(\infty), \tag{10.40a–10.40d}$$

(2) the vorticity vanishes at the core radius (Equation 10.40c) and also at infinity (Equation 10.40d) like angular velocity (Equation 10.38b); (3) it changes sign across the vortex core radius, that is, for a counterclockwise or positive vortex (Equation 10.41a), the vorticity decays from a maximum (Equation 10.38b) at the center through zero at the vortex core radius (Equation 10.40d) down to a minimum (Equation 10.41b) at the radius (Equation 10.39c) and then increases back to zero at infinity (Equation 10.40d):

$$A>0: \quad \Omega_{\max}=\Omega(0)=2A\geq\Omega(r)\geq\Omega(r_2)=-\frac{2A}{e^2}=\Omega_{\min}; \quad r_2=a\sqrt{2}>a>\frac{a}{\sqrt{2}}=r_1; \tag{10.41a–10.41c}$$

(4) for a clockwise vortex, the signs are reversed in Equations 10.41a and 10.41b, interchanging the maxima and minima; and (5) the extrema of the velocity (Equation 10.37c) [vorticity (Equation 10.41b)] occur (Equation 10.41c) inside (Equation 10.37b) [outside (Equation 10.39c)] the vortex core radius $r_1 < a < r_2$.

The pressure at radius r is related to the pressure at infinity by Equation 2.80 ≡ Equation 10.42:

$$\begin{aligned} p(r)-p_\infty &= \frac{\rho}{2}\int_\infty^r \frac{1}{\xi}\left[v_\varphi(\xi)\right]d\xi = \frac{\rho}{2}A^2\int_\infty^r \xi e^{-2\xi^2/a^2}\,d\zeta \\ &= -\frac{\rho}{4}A^2a^2\left[e^{-2\xi^2/a^2}\right]_\infty^r = -\frac{\rho}{2}A^2a^2e^{-2r^2/a^2}. \end{aligned} \tag{10.42}$$

Thus, the pressure (Equation 10.42 ≡ Equation 10.43) is maximum (minimum) at infinity (the center):

$$p_{\max}=p_\infty=p(\infty)\geq p(r)=p_\infty-\frac{\rho}{4}A^2a^2e^{-2r^2/a^2}\geq p_\infty-\frac{\rho}{4}A^2a^2=p(0)=p_{\min}\geq 0, \tag{10.43}$$

implying that the vortex core radius cannot exceed Equation 10.44a:

$$a\leq\frac{2}{A}\sqrt{\frac{p_\infty}{\rho}}; \qquad \{r_1,r_2\}=a\left\{\frac{1}{\sqrt{2}},\sqrt{2}\right\}\leq\frac{2}{A}\left\{\sqrt{\frac{p_\infty}{2\rho}},\sqrt{\frac{2p_\infty}{\rho}}\right\}; \tag{10.44a–10.44c}$$

the upper bound for the vortex core radius implies the upper bounds [Equation 10.44b (Equation 10.44c)] for the location [Equation 10.37b (Equation 10.39c)] of maximum velocity (vorticity).

EXAMPLE 10.4: VORTEX WITH POWER LAW TANGENTIAL VELOCITIES

Consider a vortex with the tangential velocity:

$$A,B,n,m,a>0: \qquad v_\varphi(r)=\begin{cases} A\left(\dfrac{r}{a}\right)^n & \text{if} \quad 0\leq r\leq a, \quad (10.45a) \\ B\left(\dfrac{a}{r}\right)^m & \text{if} \quad a\leq r<\infty, \quad (10.45b) \end{cases}$$

where all parameters are constant. Determine the angular velocity Ω and vorticity ϖ; if the velocity is continuous, determine the jump of the vorticity at the core radius. Relate the pressure inside and outside the core radius to the pressure at infinity and obtain any constraint arising from the condition of nonnegative pressure.

The continuity of the velocity at the vortex core radius requires Equation 10.46a and thus the angular velocity is also continuous (Equation 10.46b):

$$A = v_\varphi(a-0) = v_\varphi(a+0) = B, \quad \Omega(a-0) = \frac{A}{a} = \Omega(a+0), \qquad \text{(10.46a and 10.46b)}$$

as confirmed by its expression by Equation 10.46c (Equation 10.46d) inside (outside) the vortex core radius:

$$\Omega(r) = \begin{cases} Aa^{-n}r^{n-1} & \text{if} \quad 0 \le r \le a, \\ Aa^{n}r^{-m-1} & \text{if} \quad a \le r < \infty. \end{cases} \qquad \text{(10.46c and 10.46d)}$$

The vorticity is given by

$$\varpi = \frac{A}{r}\frac{d}{dr}\left[r\left\{\left(\frac{r}{a}\right)^n, \left(\frac{a}{r}\right)^m\right\}\right] = A\left\{(n+1)a^{-n}r^{n-1}, (1-m)a^m r^{-1-m}\right\}$$

(10.47a and 10.47b)

inside (Equation 10.47a ≡ Equation 10.48a) [outside (Equation 10.47b ≡ Equation 10.48b)] the core radius:

$$\varpi(r) = \begin{cases} (1+n)A\,a^{-n}r^{n-1} & \text{if} \quad 0 \le r < a, \qquad (10.48a) \\ (1-m)A\,a^{m}r^{-m-1} & \text{if} \quad a < r < \infty; \qquad (10.48b) \end{cases}$$

it is discontinuous at the core radius, with value Equation 10.49a (Equation 10.49b) on the inner (outer) side:

$$\Omega(a-0) = A\frac{n+1}{a}, \quad \Omega(a+0) = A\frac{1-m}{a}, \quad \Omega \equiv \Omega(a-0) - \Omega(a+0) = A\frac{n+m}{a},$$

(10.49a–10.49c)

with jump (Equation 10.49c) across the core radius.

The pressure outside the core radius is related to the pressure at infinity by Equation 2.80 ≡ Equation 10.50:

$$\begin{aligned} p(r \ge a) - p_\infty &= \frac{\rho}{2}\int_\infty^r \frac{1}{\xi}\left[v_\varphi(\xi \ge a)\right]^2 d\xi \\ &= \frac{\rho}{2}A^2 a^{2m}\int_\infty^r \xi^{-2m-1}\,d\xi = -\frac{\rho A^2}{4m}\frac{a^{2m}}{r^{2m}}; \end{aligned} \qquad (10.50)$$

thus the pressure reduces monotonically from infinity to the core radius:

$$p_{\max} = p_\infty = p(\infty) \geq p(r) = p_\infty - \frac{\rho A^2}{4m}\left(\frac{a}{r}\right)^{2m} \geq p_\infty - \frac{\rho A^2}{4m} = p(a). \tag{10.51}$$

The pressure inside the core radius is given by

$$p(r \leq a) = p(a) + \frac{\rho}{2}\int_a^r \frac{1}{\xi}\left[v_\varphi(\xi \leq a)\right]^2 \mathrm{d}\xi = p_\infty - \frac{\rho A^2}{4m} + \frac{\rho}{2}A^2 a^{-2n}\int_a^r \xi^{2n-1}\,\mathrm{d}\xi, \tag{10.52a}$$

leading to

$$p(r \leq a) = p_\infty - \frac{\rho A^2}{4m} - \frac{\rho A^2}{4n}\left[1 - \left(\frac{r}{a}\right)^{2n}\right]. \tag{10.52b}$$

The minimum pressure is at the vortex center (Equation 10.53a):

$$p_{\min} = p(0) = p_\infty - \frac{\rho A^2}{4}\left(\frac{1}{m} + \frac{1}{n}\right) \geq 0, \qquad A \leq 2\sqrt{\frac{p_\infty}{\rho}}\sqrt{\frac{mn}{m+n}}, \tag{10.53a and 10.53b}$$

and it limits the velocity at the core radius (Equation 10.46a) to the value Equation 10.53b.

EXAMPLE 10.5: SUMMATION OF RATIONAL-EXPONENTIAL SERIES

Show that the insertion of the terms:

$$f(n) \equiv \sum_{m=0}^{M} a_m \frac{n!}{(n-m)!}, \qquad \sum_{n=0}^{\infty} f(n)\frac{z^n}{n!} = e^z \sum_{m=0}^{M} a_m z^m \tag{10.54a and 10.54b}$$

in the exponential series Equation 3.1 leads to the **rational-exponential series** Equation 10.54b. As an example, establish the following expansions:

$$\sum_{n=0}^{\infty} \frac{n+1}{(n-1)!} = z(z+2)e^z, \qquad \sum_{n=1}^{\infty} \frac{n^2-1}{(n-1)!} z^n = z^2(z+3)e^z, \tag{10.55a and 10.55b}$$

$$\sum_{n=0}^{\infty} \frac{(n+1)^2}{n!} z^n = (z^2+3z+1)e^z, \qquad \sum_{n=0}^{\infty} \frac{n+a}{n!} z^n = (z+a)e^z, \tag{10.55c and 10.55d}$$

$$\sum_{n=0}^{\infty} \frac{(n-a)(n-b)}{n!} z^n = \left\{z^2 + (1-a-b)z + ab\right\} e^z \tag{10.55e}$$

by suitable choice of a_m or $f(n)$.

The proof of Equations 10.54a and 10.54b follows from Equation 10.56b:

$$k = n - m: \quad \sum_{n=0}^{\infty} \frac{z^n}{n!} f(n) = \sum_{n=0}^{\infty} \sum_{m=0}^{M} \frac{z^n}{(n-m)!} a_m$$
$$= \sum_{k=0}^{\infty} \frac{z^k}{k!} \sum_{m=0}^{M} z^m a_m = e^z \sum_{m=0}^{M} a_m z^m, \quad \text{(10.56a and 10.56b)}$$

where a change of dummy variable of summation (Equation 10.56a) was made and the exponential series Equation 3.1 was used. The derivation of Equations 10.55a through 10.55e is made in five steps: (1) identification of the functions Equation 10.54b appearing in the coefficients of the series Equations 10.55a through 10.55e, respectively:

$$f(n) = \left\{n(n+1), n(n^2-1), (n+1)^2, n+a, (n-a)(n-b)\right\}$$
$$= \left\{n^2 + n, n^3 - n, n^2 + 2n + 1, n + a, n^2 - (a+b)n + ab\right\}; \quad \text{(10.57a–10.57e)}$$

(2) comparison with the function Equation 10.54a shows that the order Equation 10.58a is the highest power in Equations 10.57a through 10.57e:

$$M = 3: \quad f(n) = a_0 + a_1 n + a_2 n(n-1) + a_3 n(n-1)(n-2)$$
$$= a_0 + (a_1 - a_2 + 2a_3)n + (a_2 - 3a_3)n^2 + a_3 n^3; \quad \text{(10.58a–10.58c)}$$

(3) comparison of Equation 10.58c with Equations 10.57a through 10.57e specifies the respective coefficients in Equation 10.54b:

$$\{a_3, a_2, a_1, a_0\} = \{0,1,2,0;\ 1,3,0,0;\ 0,1,3,1;\ 0,0,1,a;\ 0,1,1-a-b, ab\}; \quad \text{(10.59a–10.59e)}$$

(4) the coefficients Equations 10.59a through 10.59e specify the polynomials multiplying e^z in Equations 10.55a through 10.55e:

$$e^z \sum_{m=0}^{3} a_n z^m = e^z \left\{z^2 + 2z, z^3 + 3z^2, z^2 + 3z + 1, z + a, z^2 + (1-a-b)z + ab\right\};$$
$$\text{(10.60a–10.60e)}$$

and (5) it is possible to cross-check the final result; for example, Equation 10.59b ≡ Equations 10.61a through 10.61d:

$$a_3 = 1, \quad a_2 = 3, \quad a_1 = 0 = a_0: \qquad \sum_{m=0}^{3} a_m z^m = z^2(z+3) \quad \text{(10.61a–10.61e)}$$

corresponds to Equation 10.61f:

$$f(n) = \sum_{m=0}^{3} a_m \frac{n!}{(n-m)!} = \frac{n!}{(n-3)!} + 3\frac{n!}{(n-2)!} = n(n-1)(n-2+3), \quad \text{(10.61f)}$$

$$\frac{f(n)}{n!} = \frac{n(n-1)(n+1)}{n!} = \frac{n(n^2-1)}{n!} = \frac{n^3-n}{n!} = \frac{n^2-1}{(n-1)!}, \qquad \text{(10.61g and 10.61h)}$$

which implies Equation 10.61g (Equation 10.61h) in agreement with Equation 10.57b (Equation 10.55b).

EXAMPLE 10.6: SUMMATION OF GENERALIZED LOGARITHMIC SERIES

Establish also the **generalized logarithmic series:**

$$|z|<1: \quad -\sum_{n=1}^{\infty} \frac{z^n}{n+m} = z^{-m}\log(1-z) + \sum_{k=1}^{m} \frac{z^{k-m}}{k} \qquad \text{(10.62a and 10.62b)}$$

and use it to sum the following series:

$$|z|<1: \quad -\sum_{n=m+1}^{\infty} \frac{z^n}{n-m} = z^m \log(1-z), \qquad \text{(10.63a and 10.63b)}$$

$$|z|<1: \quad -\sum_{n=1}^{\infty} \frac{z^n}{n+2} = \frac{1}{z^2}\log(1-z) + \frac{1}{z} + \frac{1}{2}, \qquad \text{(10.64a and 10.64b)}$$

$$|z|<1: \quad \sum_{n=1}^{\infty} \frac{(-)^n}{n+3} z^n = \frac{1}{z^3}\log(1+z) - \frac{1}{z^2} + \frac{1}{2z} - \frac{1}{3}, \qquad \text{(10.65a and 10.65b)}$$

$$|z|<1: \quad -\sum_{n=1}^{\infty} \frac{n}{n+1}\frac{z^n}{n+2} = \left(\frac{2}{z^2} - \frac{1}{z}\right)\log(1-z) + \frac{2}{z}, \qquad \text{(10.66a and 10.66b)}$$

$$|z|<1: \quad \sum_{n=1}^{\infty} (-)^n z^n \frac{n+1}{n} = -\log(1+z) + \frac{1}{1+z} - 1, \qquad \text{(10.67a and 10.67b)}$$

$$|z|<1: \quad \sum_{n=1}^{\infty} (-)^n \frac{n^2}{n+1}\frac{z^n}{n+2} = \frac{1}{1+z} + \left(\frac{4}{z^2} + \frac{1}{z}\right)\log(1+z) - \frac{4}{z}, \qquad \text{(10.68a and 10.68b)}$$

as particular instances of Equations 10.62a and 10.62b or cases reducible to combinations of Equations 10.62a and 10.62b.

The proof of Equation 10.62b follows from Equation 10.69b:

$$k = n + m: \quad \sum_{n=1}^{\infty} \frac{z^n}{n+m} = \sum_{k=1+m}^{\infty} \frac{z^{k-m}}{k} = z^{-m} \sum_{k=1}^{\infty} \frac{z^k}{k} - \sum_{k=1}^{m} \frac{z^{k-m}}{k}$$

$$= -z^{-m} \log(1-z) - \sum_{k=1}^{m} \frac{z^{k-m}}{k}, \qquad \text{(10.69a and 10.69b)}$$

where the change of dummy variable of summation (Equation 10.69a) was made and substituted the logarithmic series (Equation 3.48b) valid in the unit disk (Equation 3.48a ≡ Equation 10.62a). The proof of Equation 10.63b is similar:

$$k = n - m: \quad \sum_{n=m+1}^{\infty} \frac{z^n}{n-m} = z^m \sum_{k=1}^{\infty} \frac{z^k}{k} = -z^m \log(1-z). \qquad (10.70)$$

From Equations 10.62a and 10.62b, the following are obtained: Equations 10.64a and 10.64b setting $m = 2$ and Equations 10.65a and 10.65b setting $m = 3$ and changing the sign of $-z$. The case Equations 10.66a and 10.66b involves a partial fraction decomposition (Equation 10.70a) before using Equations 10.62a and 10.62b:

$$\frac{n}{(n+1)(n+2)} = \frac{2}{n+2} - \frac{1}{n+1}, \qquad (10.70a)$$

before using Equation 10.62b:

$$-\sum_{n=1}^{\infty} \frac{n}{n+1} \frac{z^n}{n+2} = \sum_{n=1}^{\infty} \frac{z^n}{n+1} - 2\sum_{n=1}^{\infty} \frac{z^n}{n+2}$$

$$= -\frac{1}{z} \log(1-z) - 1 + 2\left[\frac{1}{z^2} \log(1-z) + \frac{1}{z} + \frac{1}{2}\right] \qquad (10.70b)$$

$$= \left(\frac{2}{z^2} - \frac{1}{z}\right) \log(1-z) + \frac{2}{z},$$

to prove Equation 10.70b ≡ Equation 10.66b.

The case Equations 10.67a and 10.67b includes a geometric (Equations 3.147a and 3.147b ≡ Equations I.21.62a and I.21.62c) and a logarithmic (Equations 3.47a and 3.47b) series:

$$\sum_{n=1}^{\infty} (-)^n z^n + \sum_{n=1}^{\infty} (-)^n \frac{z^n}{n} = \frac{1}{1+z} - 1 - \log(1+z). \qquad (10.71)$$

The case Equations 10.68a and 10.68b involves a partial fraction deposition Equation 10.72a:

$$\frac{n^2}{(n-1)(n+1)} = \frac{n^2}{n^2+3n+2} = 1 - \frac{3n+2}{n^2+3n+2} = 1 + \frac{1}{n+1} - \frac{4}{n+2}, \qquad (10.72a)$$

which leads to the following: (1) a geometric series (Equations 3.147a and 3.147b ≡ Equations I.21.62a and I.21.62c) in the first term on the r.h.s. of Equation 10.72b; and (2) the second (third) terms on the r.h.s. follow from Equation 10.62b with $m = 1(m = 2)$ using the variable $-z$:

$$\sum_{n=1}^{\infty}(-)^n \frac{n^2}{n+1}\frac{z^n}{n+2} = \sum_{n=1}^{\infty}(-z)^n + \sum_{n=1}^{\infty}\frac{(-z)^n}{n+1} - 4\sum_{n=1}^{\infty}\frac{(-z)^n}{n+2}$$

$$= \frac{1}{1+z} - 1 + \frac{1}{z}\log(1+z) - 1 + 4\left[\frac{1}{z^2}\log(1+z) - \frac{1}{z} + \frac{1}{2}\right] \tag{10.72b}$$

$$= \frac{1}{1+z} + \left(\frac{4}{z^2} + \frac{1}{z}\right)\log(1+z) - \frac{4}{z},$$

proving Equation 10.68b ≡ Equation 10.72b.

EXAMPLE 10.7: LOGARITHMS OF BASE e, 10, π, AND i

Obtain the constant coefficients in Table 10.1 relating the logarithms of base e, 10, π, and i.

The logarithms of base 10, π, and i can be calculated from natural logarithms using (Equation 3.85a):

$$\{\log_{10} z, \log_\pi z, \log_i z\} = \{k_1, k_2, k_3\} \log_e z \tag{10.73a–10.73c}$$

with the coefficients

$$k_1 = \log_{10} e = \frac{1}{\log_e 10} = 0.43429, \quad k_2 = \log_\pi e = \frac{1}{\log_e \pi} = 0.87357, \tag{10.74a and 10.74b}$$

$$k_3 = \log_i e = \frac{1}{\log_e i} = \frac{1}{\log|i| + i\arg(i)} = \frac{1}{i\pi/2} = -i\frac{2}{\pi} = -i0.63662, \tag{10.74c}$$

$$k_4 \equiv \log_i 10 = \frac{1}{\log_{10} i} = (\log_e 10)(\log_i e) = -\frac{2i}{k_1 \pi} = -i1.46587, \tag{10.74d}$$

TABLE 10.1

Logarithms with Different Basis

	$\log_e z=$	$\log_{10} z=$	$\log_\pi z=$	$\log_i z=$
$\log_e z\ \times$	1	$k_1 = 0.43429$	$k_2 = 0.87357$	$k_3 = -i0.63662$
$\log_{10} z\ \times$	$1/k_1 = 2.3026$	1	$k_5 = 2.0115$	$k_4 = -i1.4659$
$\log_\pi z\ \times$	$1/k_2 = 1.1447$	$1/k_5 = 0.49715$	1	$k_6 = -\,i0.72876$
$\log_i z\ \times$	$1/k_3 = i1.5708$	$1/k_4 = i0.68219$	$1/k_6 = i1.3722$	1

Notes: Most common bases for logarithms are (1) the decimal logarithms used in tables, since for a real number $10^N \le x < 10^{N+1}$ the integer part of the logarithm is immediate Ip $[\log(x)] = N$, and only the decimal part Dp $[\log(x)] = \log x - N$ needs to be tabulated for numbers in the range $1 < x < 10$; (2) the natural or Naperian logarithms use as base the irrational number e and lead to the simplest analytical formulas. Using any other base, for example, π or i, involves the constant coefficients listed in the table. Examples: $\log_{10} z = 0.43429 \log_e z$; $\log_i z = -i0.63662 \log_e z$; $\log_{10} z = 0.49715 \log_\pi z$; $\log_i z = -i1.4659 \log_{10} z$; and $\log_\pi z = i1.3722 \log_i z$.

In addition to Equations 10.74a through 10.74d, the coefficients

$$k_5 = \log_\pi 10 = \frac{1}{\log_{10}\pi} = \frac{\log_e 10}{\log_e \pi} = \frac{k_2}{k_1} = 2.0115, \tag{10.74e}$$

$$k_6 = \log_i \pi = \frac{1}{\log_\pi i} = \frac{\log_e \pi}{\log_e i} = \frac{k_2}{i\pi/2} = -\frac{2i}{\pi}k_2 = -i0.72876 \tag{10.74f}$$

also appear in Table 10.1 whose elements are: (1) unity along the diagonal; (2) algebraic inverse (k_n, $1/k_n$) with $n = 1,2,\ldots,6$ symmetrically placed relative to the diagonal. An instance of the use of one of the preceding formulas (Equations 10.73c and 10.74c) or of Table 10.1 is the solution (Equation 10.75b) of Equation 10.75a:

$$0.5 = i^a: \quad a = \log_i 0.5 = (\log_i e)(\log_e 0.5) = -k_3 \log_e 2 = i\frac{2}{\pi}\log_e 2 = i0.44127,$$

(10.75a and 10.75b)

which can be checked

$$\begin{aligned} i^a = i^{i0.44127} &\equiv \exp\left[\log(i^{0.44127})\right] = \exp(i0.44127 \log i) \\ &= \exp\left[i0.44127\left(\log|i| + i\frac{\pi}{2}\right)\right] = \exp\left(-0.44127 \times \frac{\pi}{2}\right) = 0.50000 \end{aligned} \tag{10.75c}$$

by taking the inverse operations in Equation 10.75c.

EXAMPLE 10.8: COMPATIBILITY RELATION FOR THE STRAIN TENSOR IN POLAR COORDINATES

Obtain the compatibility relation (Subsection 4.1.5) for the strain tensor in Cartesian coordinates (Equation 4.92b) using instead polar coordinates.

The derivation is made in six steps: (1) the comparison of the rotation (Equation 4.29 ≡ Equation 10.76a) and shear strain (Equation 4.28 ≡ Equation 10.76b):

$$2E_2 = \frac{1}{r}\left[\partial_r(u_\varphi r) - \partial_\varphi u_r\right] = \partial_r u_\varphi + \frac{1}{r}(u_\varphi - \partial_\varphi u_r), \tag{10.76a}$$

$$2S_{r\varphi} = \partial_r u_\varphi + \frac{1}{r}(\partial_\varphi u_r - u_\varphi) = 2\partial_r u_\varphi - 2E_2 \qquad \text{(10.76b and 10.76c)}$$

leads to the preliminary identity (Equation 10.76c); (2) the partial derivatives of the two polar components of the displacement are given (Equations 4.27a through 4.27d) by

$$\partial_r u_r = S_{rr}, \quad \partial_\varphi u_\varphi = S_{\varphi\varphi} r - u_r, \quad \partial_r u_\varphi = S_{r\varphi} + E_2, \qquad \text{(10.77a–10.77c)}$$

$$\partial_\varphi u_r = 2S_{r\varphi} r + u_\varphi - r\partial_r u_\varphi = 2S_{r\varphi} r + u_\varphi - r(S_{r\varphi} + E_2) = S_{r\varphi} r + u_\varphi - E_2 r \tag{10.77d}$$

using Equation 10.76c in Equation 10.77c and then Equations 10.76b and 10.77c in Equation 10.77d; (3) the differential of the radial (Equation 10.78a) [azimuthal (Equation

10.78b)] displacement follows from Equations 10.77a and 10.77d (Equations 10.77b and 10.77c):

$$\mathrm{d}u_r = (\partial_r u_r)\mathrm{d}r + (\partial_\varphi u_r)\mathrm{d}\varphi = S_{rr}\mathrm{d}r + (S_{r\varphi}r + u_\varphi - E_2 r)\mathrm{d}\varphi, \tag{10.78a}$$

$$\mathrm{d}u_\varphi = (\partial_r u_\varphi)\mathrm{d}r + (\partial_\varphi u_\varphi)\mathrm{d}\varphi = (S_{r\varphi} + E_2)\mathrm{d}r + (S_{\varphi\varphi}r - u_r)\mathrm{d}\varphi; \tag{10.78b}$$

(4) since Equation 10.78a (Equation 10.78b) is an exact differential, the second-order cross-derivatives must be equal, leading to

$$\begin{aligned}0 &= \partial_{\varphi r} u_r - \partial_{r\varphi} u_r = \partial_\varphi S_{rr} - \partial_r (S_{r\varphi} r + u_\varphi - rE_2) = \partial_\varphi S_{rr} - \partial_r (S_{r\varphi} r) - \partial_r u_\varphi + \partial_r (rE_2) \\ &= \partial_\varphi S_{rr} - \partial_r (S_{r\varphi} r) - S_{r\varphi} - E_2 + \partial_r (rE_2) = \partial_\varphi S_{rr} - 2S_{r\varphi} - r\partial_r S_{r\varphi} + r\partial_r E_2,\end{aligned} \tag{10.79a}$$

$$\begin{aligned}0 &= \partial_{\varphi r} u_\varphi - \partial_{r\varphi} u_\varphi = \partial_\varphi (S_{r\varphi} + E_2) - \partial_r (rS_{\varphi\varphi} - u_r) \\ &= \partial_\varphi S_{r\varphi} - S_{\varphi\varphi} - r\partial_r S_{\varphi\varphi} + S_{rr} + \partial_\varphi E_2,\end{aligned} \tag{10.79b}$$

where Equation 10.76c was used to eliminate all derivatives of the displacement vector, so that only the strain tensor and derivatives of the rotation appear in Equations 10.79a and 10.79b; (5) the differential of the rotation is specified in terms of the derivatives in Equations 10.79a and 10.79b by

$$\begin{aligned}\mathrm{d}E_2 &= (\partial_r E_2)\mathrm{d}r + (\partial_\varphi E_2)\mathrm{d}\varphi \\ &= (-\partial_\varphi S_{rr} + 2S_{r\varphi} + r\partial_r S_{r\varphi})\frac{\mathrm{d}r}{r} + (-S_{rr} - \partial_\varphi S_{r\varphi} + S_{\varphi\varphi} + r\partial_r S_{\varphi\varphi})\mathrm{d}\varphi,\end{aligned} \tag{10.79c}$$

where only the three components of the strain tensor appear; and (6) since the rotation (Equation 10.79c) is an exact differential, the second-order cross-derivatives must be equal:

$$0 = \partial_{\varphi r} E_2 - \partial_{r\varphi} E_2 = -r^{-1}\partial_{\varphi\varphi} S_{rr} + 2r^{-1}\partial_\varphi S_{r\varphi} + \partial_{\varphi r} S_{r\varphi} + \partial_r S_{rr} + \partial_{r\varphi} S_{r\varphi} - \partial_r S_{\varphi\varphi} - r\partial_{rr} S_{\varphi\varphi} - \partial_r S_{\varphi\varphi}. \tag{10.80}$$

The relation Equation 10.80 ≡ Equation 10.81 involves only the strain tensor and is

$$\partial_{\varphi\varphi} S_{rr} + r^2 \partial_{rr} S_{\varphi\varphi} = 2r\partial_{r\varphi} S_{r\varphi} - 2r\partial_r S_{\varphi\varphi} + r\partial_r S_{rr} + 2\partial_\varphi S_{r\varphi} \tag{10.81}$$

the **compatibility relation** *for the strain tensor in polar coordinates, ensuring that its three components* ($S_{rr}, S_{\varphi\varphi}, S_{r\varphi}$) *correspond to (1) a unique rotation (Equation 10.79c) and (2) a unique pair of polar displacements (Equations 10.78a and 10.78b). The compatibility relation for the stress tensor (Equation 10.82) in polar coordinates*

$$\begin{aligned}\partial_{\varphi\varphi} T_{rr} + r^2 \partial_{rr} T_{\varphi\varphi} &= \sigma\left(\partial_{\varphi\varphi} T_{\varphi\varphi} + r^2 \partial_{rr} T_{rr}\right) + 2(1+\sigma)\left[r\partial_{r\varphi} T_{r\varphi} + \partial_\varphi T_{r\varphi}\right] \\ &\quad + (1+2\sigma)r\partial_r T_{rr} - (2+\sigma)r\partial_r T_{\varphi\varphi}\end{aligned} \tag{10.82}$$

follows from Equation 10.81 using the inverse Hooke law (Equations 4.78a through 4.78c).

EXAMPLE 10.9: TRIPLE/QUADRUPLE FORMULAS FOR CIRCULAR/HYPERBOLIC FUNCTIONS

Obtain (1) the triple and quadruple addition formulas for the hyperbolic cosine and sine, and the corresponding triplicating/quadruplicating formulas; (2) the triple addition and triplicating formulas for the circular and hyperbolic tangent and cotangent; and (3) the cube and fourth powers of the circular and hyperbolic cosine and sine.

E10.9.1: Triple Addition and Triplicating Formulas

The triple addition formulas for the circular sine and cosine (Equations I.3.32a and I.3.32b) have as analogues the triple addition formulas for the hyperbolic cosine (Equation10.83) [sine (Equation 10.84)] that follow from Equations 5.35a and 5.35c (Equations 5.35b an 5.35c):

$$\begin{aligned}\cosh(u+v+w) &= \cosh u \cosh v \cosh w + \cosh u \sinh v \sinh w \\ &\quad + \sinh u \cosh v \sinh w + \sinh u \sinh v \cosh w,\end{aligned} \tag{10.83}$$

$$\begin{aligned}\sinh(u+v+w) &= \cosh u \cosh v \sinh w + \cosh u \sinh v \cosh w \\ &\quad + \sinh u \cosh v \cosh w + \sinh u \sinh v \sinh w.\end{aligned} \tag{10.84}$$

The ratios specify the triple addition formulas for the hyperbolic (Equation 10.85) [circular (Equation 10.86)] tangent:

$$\tanh(u+v+w) = \frac{\tanh u + \tanh v + \tanh w + \tanh u \tanh v \tanh w}{1 + \tanh u \tanh v + \tanh u \tanh w + \tanh u \tanh w}, \tag{10.85}$$

$$\tan(u+v+w) = \frac{\tan u + \tan v + \tan w - \tan u \tan v \tan w}{1 - \tan u \tan v - \tan u \tan w - \tan v \tan w}. \tag{10.86}$$

and for the hyperbolic (Equation 10.87) [circular (Equation 10.88)] cotangent:

$$\coth(u+v+w) = \frac{\coth u + \coth v + \coth w + \coth u \coth v \coth w}{1 + \coth u \coth v + \coth u \coth w + \coth u \coth w}, \tag{10.87}$$

$$\cot(u+v+w) = \frac{\cot u + \cot v + \cot w - \cot u \cot v \cot w}{1 - \cot u \cot v - \cot u \cot w - \cot u \cot w}. \tag{10.88}$$

The triplicating formulas for the hyperbolic cosine (sine) follow from Equation 10.83 (Equation 10.84) with equal arguments:

$$\cosh(3z) = \cosh^3 z + 3 \cosh z \sinh^2 z = 4 \cosh^3 z - 3 \cosh z, \tag{10.89a}$$

$$\sinh(3z) = 3 \cosh^2 z \sinh z + \sinh^3 z = 3 \sinh z + 4 \sinh^3 z; \tag{10.89b}$$

they have analogues in the triplicating formulas for the circular cosine (Equation I.3.33b) [sine (Equation I.3.33c)]. Their ratio of Equation 10.85 (Equation 10.86) specifies

$$\tanh(3z) = \frac{3 \tanh z + \tanh^3 z}{1 + 3 \tanh^2 z}, \qquad \tan(3z) = \frac{3 \tan z - \tan^3 z}{1 - 3 \tan^2 z}$$

(10.90a and 10.90b)

the triplicating formula for the hyperbolic (Equation 10.90a) [circular (Equation 10.90b)] tangent, and also

$$\coth(3z) = \frac{3\coth z + \coth^3 z}{1 + 3\coth^2 z}, \qquad \cot(3z) = \frac{3\cot z - \cot^3 z}{1 - 3\cot^2 z} \tag{10.90c and 10.90d}$$

for the hyperbolic (Equation 10.90c) [circular (Equation 10.90d)] cotangent.

E10.9.2: Quadruple Addition and Quadruplicating Formulas

The quadruple addition formulas for the hyperbolic cosine (Equation 10.91) [sine (Equation 10.92)] follow from Equations 5.35a and 5.35c (Equations 5.35b and 5.35c):

$$\begin{aligned}\cosh(u+v+w+z) &= \cosh u \cosh v \cosh w \cosh z + \sinh u \sinh v \sinh w \sinh z \\ &+ \cosh u \cosh v \sinh w \sinh z + \cosh u \sinh v \cosh w \sinh z \\ &+ \cosh u \sinh v \sinh w \cosh z + \sinh u \cosh v \cosh w \sinh z \\ &+ \sinh u \cosh v \sinh w \cosh z + \sinh u \sinh v \cosh w \cosh z,\end{aligned} \tag{10.91}$$

$$\begin{aligned}\sinh(u+v+w+z) &= \sinh u \cosh v \cosh w \cosh z + \cosh u \sinh v \cosh w \cosh z \\ &+ \cosh u \cosh v \sinh w \cosh z + \cosh u \cosh v \cosh w \sinh z \\ &+ \sinh u \sinh v \sinh w \cosh z + \sinh u \sinh v \cosh w \sinh z \\ &+ \sinh u \cosh v \sinh w \sinh z + \cosh u \sinh v \sinh w \sinh z;\end{aligned} \tag{10.92}$$

the analogue quadruple addition formulas for the circular cosine (Equation I.10.24a) [sine (Equation I.10.24b)] were obtained in Example I.2.7; the latter also contains the quadruplicating formulas for the circular cosine (Equation I.10.25a) [sine (Equation I.10.25b)] corresponding to

$$\begin{aligned}\cosh(4z) &= \cosh^4 z + \sinh^4 z + 6\cosh^2 z \sinh^2 z \\ &= 8\cosh^4 z - 8\cosh^2 z + 1 = 8\sinh^4 z + 8\sinh^2 z + 1,\end{aligned} \tag{10.93a–10.93c}$$

$$\begin{aligned}\sinh(4z) &= 4\sinh z \cosh^3 z + 4\sinh^3 z \cosh z \\ &= 4\sinh z \cosh z(\cosh^2 z + \sinh^2 z) = 2\sinh(2z)\cosh(2z) \\ &= 4\sinh z \cosh z + 8\cosh z \sinh^3 z = -4\sinh z \cosh z + 8\sinh z \cosh^3 z\end{aligned} \tag{10.94a–10.94c}$$

the quadruplicating formulas for the hyperbolic cosine (Equations 10.93a through 10.93c) [sine (Equations 10.94a through 10.94c)].

E10.9.3: Cube and Fourth-Power Formulas

From the preceding follow the formulas for the **cubes and fourth powers** of cosine and sine:

$$\cos^3 z = \frac{\cos(3z) + 3\cos(z)}{4}, \qquad \sin^3 z = \frac{3\sin(z) - \sin(3z)}{4}, \tag{10.95a and 10.95b}$$

$$\cos^4 z = \frac{\cos(4z) + 4\cos(2z) + 3}{8}, \qquad \sin^4 z \equiv \frac{\cos(4z) + 4\cos(2z) - 5}{8}, \qquad \text{(10.96a and 10.96b)}$$

$$\cosh^3 z = \frac{\cosh(3z) + 3\cosh(z)}{4}, \qquad \sinh^3 z = \frac{\sinh(3z) - 3\sinh(z)}{4}, \qquad \text{(10.97a and 10.97b)}$$

$$\cosh^4 z = \frac{\cosh(4z) + 4\cosh(2z) + 3}{8}, \qquad \sinh^4 z = \frac{\cosh(4z) - 4\cosh(2z) + 3}{8}; \qquad \text{(10.98a and 10.98b)}$$

for example, from Equation 10.93b (Equation 10.93c), using Equation 5.83a (Equation 5.83b) follows Equation 10.99a ≡ Equation 10.98a (Equation 10.99b ≡ Equation 10.98b):

$$\cosh(4z) = 8\cosh^4 z - 4\cosh(2z) - 3 = 8\sinh^4 z + 4\cosh(2z) - 3. \qquad \text{(10.99a and 10.99b)}$$

The cubes (fourth) powers of the circular cosine [Equation 10.95a (Equation 10.96a)] and sine [Equation 10.95b (Equation 10.96b)] and hyperbolic cosine [Equation 10.97a (Equation 10.98a)] and sine [Equation 10.97b (Equation 10.98b)] are the particular cases N = 1 of Equations 5.78a, 5.78b, 5.79a, 5.79b, 5.80a, 5.80b, 5.81a, and 5.81b). Table 5.6 include all the algebraic relations between circular and hyperbolic functions.

EXAMPLE 10.10: POTENTIAL FLOW PAST TWO CYLINDERS MOVING ORTHOGONALLY TO THE LINE OF CENTERS

Consider the potential flow past two cylinders with distinct radii moving at different velocities orthogonal to the line of centers; consider also the particular case of cylinders with the same radius and velocity, that is, two image cylinders moving parallel to a wall.

E10.10.1: Free, Perturbation, and Total Potentials of Each Cylinder

The polar coordinates $(r,\varphi)[(s,\phi)]$ are chosen with center at the cylinder of radius a (b), and the angle is measured from the velocity U (V), that is, perpendicular (parallel) to the line of centers in Figure 10.1a (Figure 8.4a) in Example 10.10 (Section 8.2). The same method of perturbation potentials (Subsection 8.2.2) applies if the distance between the centers c is large compared with the radii of both cylinders. The total potential can be decomposed into a linear combination (Equation 8.43a) of unit potentials satisfying the boundary conditions (Equations 8.43b through 8.43e). In the absence of the second cylinder, the potential due to the first cylinder would be Equation 10.100b ≡ Equation 8.44a:

$$r\cos\varphi = s\cos\phi: \qquad \Phi_{10}(r,\varphi) = \frac{a^2}{r}\cos\varphi = \frac{a^2}{r^2}s\cos\phi, \qquad \text{(10.100a–10.100c)}$$

where geometric relation Equation 10.100a can be used to obtain Equation 10.100c (Figure 10.1a). The latter specifies the unit potential of the first cylinder near the second cylinder (Equation 10.101a):

$$\Phi_{10}\Big|_{s=b} = \frac{a^2}{c^2}b\cos\phi, \qquad \frac{\partial\Phi_{10}}{\partial s}\bigg|_{s=b} = \frac{a^2}{c^2}\cos\phi = -\frac{\partial\Phi_{12}}{\partial s}\bigg|_{s=b}, \qquad \text{(10.101a–10.101c)}$$

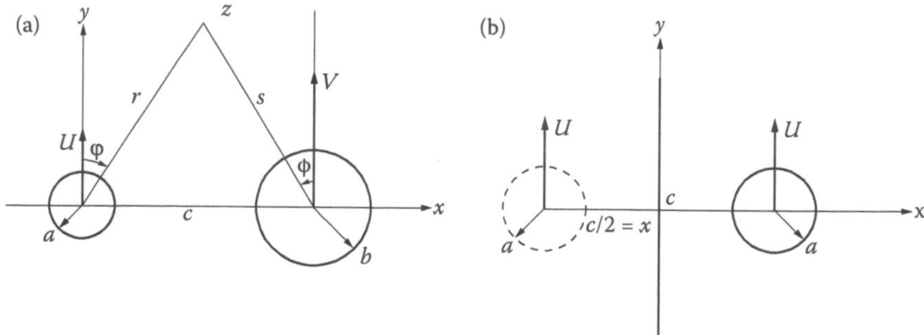

FIGURE 10.1
Potential flow past two cylinders with distinct velocities and radii moving in a direction orthogonal to (along) the line of centers [Figure 10.1a (Figure 8.2a)] can be solved to the lowest order in the iteration using a perturbation method [Example 10.10 (Section 8.2)]: the particular case of two cylinders with equal radii moving at the same velocity orthogonal to (along) the line of centers corresponds to taking one cylinder as the image of the other [Figure 10.1b (Figure 8.4b)] on a wall at equal distance from the two centers; thus it specifies the effect of a wall on a cylinder moving in the parallel (orthogonal) direction.

implying a velocity (Equation 10.101b) normal to the surface; the latter is canceled (Equation 10.101c) by the perturbation potential (Equation 10.102a):

$$\Phi_{12} = \frac{a^2}{c^2}\frac{b^2}{s}\cos\phi; \qquad \Phi_1 = \Phi_{10} + \Phi_{12} = \frac{a^2}{r}\cos\varphi + \frac{a^2b^2}{c^2 s}\cos\phi, \tag{10.102a and 10.102b}$$

the total unit potential of the first cylinder (Equation 10.102b), and the sum of the free space (Equation 10.100b) and perturbation (Equation 10.102a) potentials, and satisfies the rigid wall boundary conditions on the two cylinders to $O(a^2b^2/c^4)$. The unit potential of the second cylinder is obtained (Equation 10.103b) by an interchange of variables (Equation 10.103a) in Equation 10.102b:

$$(a, r, \varphi, U) \leftrightarrow (b, s, \phi, V): \qquad \Phi_2 = \Phi_{20} + \Phi_{21} = \frac{b^2}{s}\cos\phi + \frac{a^2b^2}{c^2 r}\cos\varphi; \tag{10.103a and 10.103b}$$

it consists of

$$\Phi_{20} = \frac{b^2}{s}\cos\phi, \qquad \Phi_{21} = \frac{a^2b^2}{c^2 r}\cos\varphi \tag{10.104a and 10.104b}$$

the free potential of the second cylinder (Equation 10.104a) plus the perturbation potential (Equation 10.104b) due to the first cylinder.

E10.10.2: Values of the Unit Potentials on Each Cylinder

The geometric relation Equation 10.100a may be used to express the unit potential of the first (Equation 10.102b) [second (Equation 10.103b)] cylinder involving only the angle φ relative to the center of the first cylinder in Equation 10.105a (Equation 10.105b):

$$\Phi_1 = a^2 r\left(\frac{1}{r^2} + \frac{h^2}{c^2 s^2}\right)\cos\varphi, \qquad \Phi_2 = b^2 r\left(\frac{1}{s^2} + \frac{a^2}{c^2 r^2}\right)\cos\varphi; \tag{10.105a and 10.105b}$$

this specifies the values on the first cylinder of the unit potential of the first (second) cylinder [Equation 10.106a (Equation 10.106b)]:

$$\Phi_1\Big|_{r=a} = a\cos\varphi\left[1+O\left(\frac{a^2b^2}{c^4}\right)\right], \qquad \Phi_2\Big|_{r=a} = 2\frac{b^2}{c^2}a\cos\varphi. \qquad \text{(10.106a and 10.106b)}$$

The geometric relation Equation 10.100a can also be used to specify the total unit potential of the first (Equation 10.102b) [second (Equation 10.103b)] cylinder involving only the angle ϕ relative to the center of the second cylinder [Equation 10.107a (Equation 10.107b)]:

$$\Phi_1 = a^2 s\left(\frac{1}{r^2}+\frac{b^2}{c^2s^2}\right)\cos\phi, \qquad \Phi_2 = b^2 s\left(\frac{1}{s^2}+\frac{a^2}{c^2r^2}\right)\cos\phi; \qquad \text{(10.107a and 10.107b)}$$

this specifies the values on the second cylinder of the potential of the first (Equation 10.108a) [second (Equation 10.108b)] cylinder:

$$\Phi_1\Big|_{s=b} = 2\frac{a^2}{c^2}b\cos\phi, \qquad \Phi_2\Big|_{s=b} = b\left[1+O\left(\frac{a^2b^2}{c^4}\right)\right]\cos\phi. \qquad \text{(10.108a and 10.108b)}$$

The three pairs of expressions for the unit potentials (Equations 10.102b, 10.103b, 10.105a, 10.105b, 10.107a, and 10.107b) and their values on the two cylinders (Equations 10.106a, 10.106b, 10.108a, and 10.108b) all satisfy the interchange of variables (Equation 10.103a). Thus, *two cylinders of radii a (b) moving with velocity U (V) orthogonal to the line of centers (Figure 10.1a) at a distance c have a potential (Equation 10.109a ≡ Equation 8.43a) that is a linear combination of unit potentials of the first (Equation 10.102b ≡ Equation 10.105a ≡ Equation 10.107a) [second (Equation 10.103b ≡ Equation 10.105b ≡ Equation 10.107b)] cylinder; the latter take the values Equations 10.106a and 10.108a (Equations 10.106b and 10.108b) on the two cylinders, within the accuracy (Equation 10.106a ≡ Equation 10.108b ≡ Equation 10.111a) of the first iteration of the perturbation method. This accuracy is distinct from the case of cylinders moving in the line of centers (Equations 8.70a and 8.70b) also in the first iteration of the perturbation method.*

E10.10.3: Kinetic Energy and Added Mass

The kinetic energy of fluid is given by Equation 10.109a ≡ Equation 8.63:

$$\Phi = \Phi_1 U + \Phi_2 V: \qquad E_v = \frac{\rho}{2}(I_1U^2 + I_2V^2 + 2I_3UV) \qquad \text{(10.109a and 10.109b)}$$

in terms (Equations 8.64a, 8.64b, and 8.65) of the integrals

$$I_1 = -\int_{r=a}\Phi_1\frac{\partial\Phi_1}{\partial n}\mathrm{d}s = a\int_0^{2\pi}\Phi_1\cos\varphi\,\mathrm{d}\varphi = a^2\int_0^{2\pi}\cos^2\varphi\,\mathrm{d}\varphi = \pi a^2, \qquad \text{(10.110a)}$$

$$I_2 = -\int_{s=b}\Phi_2\frac{\partial\Phi_2}{\partial n}\mathrm{d}s = b\int_0^{2\pi}\Phi_2\cos\phi\,\mathrm{d}\phi = b^2\int_0^{2\pi}\cos^2\phi\,\mathrm{d}\phi = \pi b^2, \qquad \text{(10.110b)}$$

$$I_3 = -\int_{r=a} \Phi_2 \frac{\partial \Phi_1}{\partial n} ds = a \int_0^{2\pi} \Phi_2 \cos\varphi \, d\varphi = 2\frac{a^2 b^2}{c^2} \int_0^{2\pi} \cos^2 \varphi \, d\varphi = 2\pi a^2 \frac{b^2}{c^2},$$
$$= -\int_{s=b} \Phi_1 \frac{\partial \Phi_2}{\partial n} ds = b \int_0^{2\pi} \Phi_1 \cos\phi \, d\phi = 2\frac{a^2 b^2}{c^2} \int_0^{2\pi} \cos^2 \phi \, d\phi = 2\pi a^2 \frac{b^2}{c^2}, \tag{10.110c}$$

where Equations 8.43b, 8.43e, 10.106a, 10.106b, 10.108a, and 10.108b were used, and Equations 10.110a through 10.110c satisfy the symmetry relations (Equation 10.103a). Substitution of Equations 10.110a through 10.110c in Equation 10.109b specifies

$$a^2 b^2 << c^4: \qquad E_v = \frac{\rho\pi}{2}\left(U^2 a^2 + V^2 b^2 + 4UV \frac{a^2 b^2}{c^2} \right) \tag{10.111a and 10.111b}$$

the kinetic energy of the potential flow due to two cylinders of radii a (b) moving with velocities U (V) orthogonal to the line centers (Figure 10.1a) at a distance c; it involves the displaced mass of each cylinder plus an interaction term.

E10.10.4: Cylinder Moving Parallel to a Wall

The case of two cylinders with the same radius (Equation 10.112a) and velocity (Equation 10.112b) corresponds to images (Figure 10.1b) on a plane at equal distance (Equation 10.112c) from the two centers; the kinetic energy on the half-space on one side of the wall is Equation 10.112d:

$$a = b, V = U, x = \frac{c}{2}: \qquad E = \frac{1}{2} E_v = \frac{\rho\pi}{2} U^2 a^2 \left(1 + \frac{a^2}{2x^2} \right); \tag{10.112a–10.112d}$$

the kinetic energy Equation 10.112d ≡ Equation 10.113a corresponds to a displaced mass (Equation 10.113c) plus (Equation 10.113b) an added mass:

$$E_v = \frac{1}{2} \bar{m} U^2, \qquad \bar{m} = m_0 \left(1 + \frac{a^2}{2x^2} \right), \qquad m_0 = \rho\pi a^2. \tag{10.113a–10.113c}$$

If the cylinder has mass m, the total kinetic energy including entrained fluid is

$$\bar{E} = \frac{1}{2}(m + \bar{m}) U^2 = \frac{1}{2}\left[m + \rho\pi a^2 \left(1 + \frac{a^2}{2x^2} \right) \right] U^2. \tag{10.114}$$

The total energy is conserved (Equation 10.114), and thus the velocity of the cylinder of radius a at a distance x from a wall is given by

$$U_\infty \equiv \left| \frac{2\bar{E}}{m_0 + \pi\rho a^2} \right|^{1/2} \equiv U(\infty) \geq U(x) \geq U(a) = \left| \frac{2\bar{E}}{m_0 + 3\rho\pi a^2/2} \right|^{1/2} \equiv U_0 \tag{10.115a and 10.115b}$$

and varies from a maximum (Equation 10.115a) at infinity to a minimum (Equation 10.115b) close to the wall. Since the cylinder moves parallel to the wall, its trajectory (Equation 10.116a) is

$$\frac{dy}{dt} = U(x), \qquad y(t) = y(0) + tU(x) \tag{10.116a and 10.116b}$$

a straight line taken at constant velocity (Equation 10.116b). This contrast with the nonuniform motion (Equation 8.79) of a cylinder perpendicular to a wall.

EXAMPLE 10.11: CYLINDER IN A FLOW/FIELD WITH A WALL

Consider a cylinder of radius a at distance b from a wall in a uniform stream of velocity U parallel to the wall. Representing the cylinder by a dipole at its axis, and the wall effect by an image dipole, determine the highest order in the parameter a/b for which the boundary conditions are met on the wall and on the cylinder. Consider both the cases without and with circulation around the cylinder. Consider also the analogous problems of a conducting or insulating cylinder near a conducting or insulating wall in the presence of a uniform electro(magneto)static field.

E10.11.1: Potential Flow Past a Cylinder near a Wall

The complex potential (Equation 10.117) consists of (1) a uniform stream with velocity U; (2) a dipole with real moment P_1 at $z = ib$; and (3) an image dipole with the same moment at $z = -ib$:

$$f(z) = Uz - \frac{P_1}{2\pi}\left(\frac{1}{z-ib} + \frac{1}{z+ib}\right) = Uz - \frac{P_1}{\pi}\frac{z}{z^2+b^2}. \tag{10.117}$$

The boundary condition at the cylinder requires that it be a streamline. The complex potential (Equation 10.117) on the cylinder (Equation 10.118a) is given by Equation 10.118b:

$$\begin{aligned} z = ib + ae^{i\varphi}: \qquad & f(ib+ae^{i\varphi}) - ibU - Uae^{i\varphi} = -\frac{P_1}{\pi}\frac{ib+ae^{i\varphi}}{2ibae^{i\varphi}+a^2e^{2i\varphi}} \\ &= -\frac{P_1}{2\pi a}e^{-i\varphi}\left(1 - i\frac{a}{b}e^{i\varphi}\right)\left(1 - i\frac{a}{2b}e^{i\varphi}\right)^{-1} \\ &= -\frac{P_1}{2\pi a}e^{-i\varphi}\left(1 - i\frac{a}{b}e^{i\varphi}\right)\left[1 + i\frac{a}{2b}e^{i\varphi} - \frac{a^2}{4b^2}e^{2i\varphi} + O\left(\frac{a^3}{b^3}e^{3i\varphi}\right)\right] \\ &= -\frac{P_1}{2\pi a}e^{-i\varphi} + \frac{iP_1}{4\pi b} - \frac{P_1 a}{8\pi b^2}e^{i\varphi} + O\left(P_1\frac{a^2}{b^3}e^{2i\varphi}\right). \end{aligned} \tag{10.118a and 10.118b}$$

The stream function (or potential) cannot be made constant to order $(a/b)^3$ since the term $e^{2i\varphi}$ on the r.h.s. of Equation 10.118b has no matching $e^{-2i\varphi}$ term on the l.h.s. of Equation 10.118b. Thus, the boundary conditions on the surface of the cylinder could be met to order $(a/b)^3$ only by considering a quadrupole term. Representing the cylinder by a dipole implies that the boundary condition on the surface of the cylinder can be met only to order Equation 10.121a, as shown by the stream function on the cylinder (Equation 10.119):

$$\Psi(ib+ae^{i\varphi}) = \operatorname{Im}\left[f(ib+ae^{i\varphi})\right] = bU + \frac{P_1}{4\pi b} + \left(Ua - \frac{P_1 a}{8\pi b^2} + \frac{P_1}{2\pi a}\right)\sin\varphi. \tag{10.119}$$

The stream function becomes a constant on the cylinder (Equation 10.120a) by choosing the dipole moment (Equation 10.120b) to cancel the term in curved brackets on the r.h.s. of Equation 10.119:

$$\Psi(ib + ae^{i\varphi}) = bU + \frac{P_1}{4\pi b} = const: \qquad -2\pi U a^2 = P_1\left(1 - \frac{a^2}{4b^2}\right). \qquad \text{(10.120a and 10.120b)}$$

To the order of approximation (Equation 10.121a), the dipole moment (Equation 10.120b) consists (Equation 10.121b) of

$$a^3 << b^3: \qquad P_1 = -2\pi U a^2\left(1 + \frac{a^2}{4b^2}\right) = -2\pi U a^2 - \frac{\pi}{2} U \frac{a^4}{b^2} \qquad \text{(10.121a–10.121c)}$$

two terms: (1) the first on the r.h.s. of Equation 10.121b is the dipole moment (Equation 8.6c ≡ Equation I.28.89a) for a cylinder of radius a in free stream with velocity U; and (2) the second on the r.h.s. of Equation 10.121b is the lowest order correction for the wall effect. The correction is $O(a^2/b^2)$ for the cylinder of radius a both in a cylindrical cavity of radius b (at a distance b from an infinite wall) with [Section 8.1 (Example 10.10.1)] a coefficient:

$$P_1 = -\frac{2\pi a^2 U}{1 - a^2/b^2} = -2\pi a^2 U\left[1 + \frac{a^2}{b^2} + O\left(\frac{a^2}{b^4}\right)\right], \qquad \text{(10.121d)}$$

equal to 1 (1/4) in Equation 8.5b ≡ Equation 10.121d (Equation 10.121b).

E10.11.2: Velocity in the Flow and on the Cylinder

The potential on the cylinder is given (Equation 10.118b) by

$$\begin{aligned} \Phi(ib + e^{i\varphi}) &= \operatorname{Re}\left[f(ib + ae^{i\varphi})\right] = \left[Ua - \frac{P_1}{2\pi a}\left(1 + \frac{a^2}{4b^2}\right)\right]\cos\varphi \\ &= Ua\left[1 + \left(1 + \frac{a^2}{4b^2}\right)^2\right]\cos\varphi = 2Ua\left(1 + \frac{a^2}{4b^2}\right)\cos\varphi, \end{aligned} \qquad \text{(10.122)}$$

where the dipole moment (Equation 10.121b) was substituted. *The velocity is tangential (Equation 10.123a) on the cylinder (Equation 2.186c):*

$$v_r(ib + ae^{i\varphi}) = 0, \quad v_\varphi(ib + ae^{i\varphi}) = \frac{1}{a}\frac{\partial\Phi}{\partial\varphi} = -2U\left(1 + \frac{a^2}{4b^2}\right)\sin\varphi \qquad \text{(10.123a and 10.123b)}$$

and consists (Equation 10.123b) of (1) the tangential velocity (Equation 8.226d ≡ Equation I.28.94b) on a cylinder of radius a in a free stream of velocity U, that is, the first term on the r.h.s. of Equation 10.123b; and (2) the second term on the r.h.s. of Equation 10.123b is the wall correction to lowest order (a^2/b^2) with (3) the coefficient is:

$$\Gamma = 0: \quad v_\varphi(a, \varphi) = -U\frac{1 + a^2/b^2}{1 - a^2/b^2}\sin\varphi = -U\left[1 + 2\frac{a^2}{b^2} + O\left(\frac{a^2}{b^2}\right)\right]\sin\varphi$$

(10.123c and 10.123d)

equal to 1/4 (2) for a cylinder at a distance b from an infinite wall (Equation 9.123b) [in a cylindrical cavity (Equation 8.24b ≡ Equation 10.123d) of radius b without circulation (Equation

10.123c)]. The tangential velocity has leading factor $-2U \sin \varphi$ ($-U \sin \varphi$) in Equation 10.123b (Equation 10.125e) because the cylinder is at rest (moving with velocity U) relative to the infinite wall at distance b (cylindrical cavity of radius b).

The dipole moment (Equation 10.121b) appears (Equation 10.117) in the complex potential (Equation 10.124) [conjugate velocity (Equation 10.125)]:

$$f(z) = Uz\left[1 + \frac{2a^2}{z^2 + b^2}\left(1 + \frac{a^2}{4b^2}\right)\right], \tag{10.124}$$

$$v^*(z) = \frac{\mathrm{d}f}{\mathrm{d}z} = U\left[1 + 2a^2\frac{b^2 - z^2}{(z^2 + b^2)^2}\left(1 + \frac{a^2}{4b^2}\right)\right], \tag{10.125}$$

where the wall correction applies only to the dipole term, since the free stream is parallel to the wall and unaffected by it. The velocity is horizontal at the wall so the boundary condition is met:

$$v^*(x) = U\left[1 + 2a^2\frac{b^2 - x^2}{(x^2 + b^2)^2}\left(1 + \frac{a^2}{4b^2}\right)\right] = v_x(x,0), \quad v_y(x,0) = 0. \tag{10.126a and 10.126b}$$

The velocity is also horizontal (Equation 10.127b) on the line perpendicular to the wall and passing through the axis of the cylinder:

$$v^*(iy) = U\left[1 + 2a^2\frac{b^2 + y^2}{(b^2 - y^2)^2}\left(1 + \frac{a^2}{4b^2}\right)\right] = v_x(0,y) > U, \quad v_y(x,0) = 0, \tag{10.127a and 10.127b}$$

and it exceeds the free stream velocity (Equation 10.127a) because the flow is accelerated around the cylinder to preserve the flow rate of the incompressible fluid. Thus, *a cylinder of radius a in a uniform stream of velocity U parallel to an infinite rigid impermeable wall at a distance b > a from the center (Figure 10.2a) is represented to order Equation 10.121a by a dipole of moment Equation 10.121b, and has a complex potential (Equation 10.124) [conjugate velocity (Equation 10.125)], corresponding on the cylinder to the real potential Equation 10.122 and constant stream function Equation 10.120a (tangential velocity Equations 10.123a and 10.123b). The flow velocity is horizontal both on the wall (Equations 10.126a and 10.126b) and on the line perpendicular to the wall passing through the center (Equations 10.127a and 10.127b).*

E10.11.3: Cylinder with Circulation near a Wall

In the presence of circulation Γ around the cylinder (Figure 10.2b), a vortex with circulation Γ at ib and the image vortex with opposite circulation $-\Gamma$ at $-ib$ must be added (Equation 10.128) to the complex potential (Equation 10.117):

$$\bar{f}(z) - f(z) = -\frac{i\Gamma}{2\pi}\log(z - ib) + \frac{i\Gamma}{2\pi}\log(z + ib) = \frac{i\Gamma}{2\pi}\log\left(\frac{z + ib}{z - ib}\right). \tag{10.128}$$

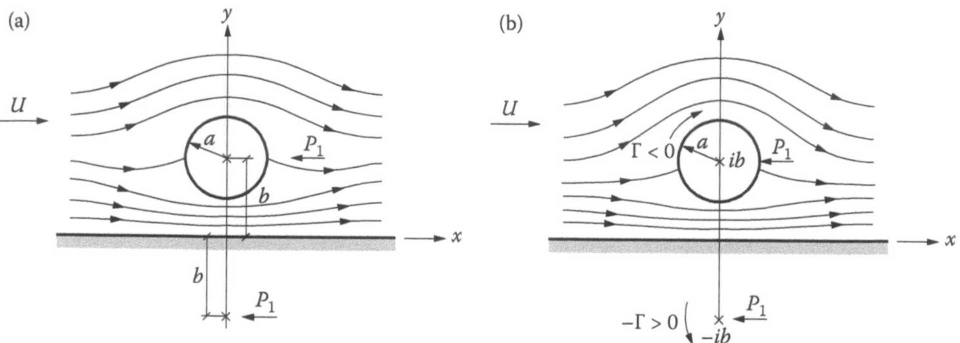

FIGURE 10.2
Potential flow past a cylinder of radius a with center at a distance b from an infinite wall, both rigid and impermeable, can be represented by a dipole and its image (Example 10.11) as an alternative to the perturbation method (Example 10.10). In both cases, the accuracy is limited by the boundary condition on the surface of the cylinder to $O(a^3/b^3)$ for the dipole method and to $O(a^4/b^4)$ for the perturbation method to the lowest order. The $O(a^3/b^3)$ accuracy of the dipole method allows a wall effect of order $O(a^2/b^3)$ on the dipole moment since the velocity field of a dipole decays on the inverse square of the distance (a); in the presence of circulation around the cylinder (b), there is an image vortex with opposite circulation, specifying a wall effect of order $O(a/b)$ in the dipole moment because the velocity field of vortex decays with the inverse of the distance.

The term (Equation 10.128) must be added to the potential (Equation 10.118b) on the cylinder, leading to

$$\begin{aligned}\bar{f}(ib+ae^{i\varphi})-f(ib+ae^{i\varphi}) &= \frac{i\Gamma}{2\pi}\log\left(\frac{2ib+ae^{i\varphi}}{ae^{i\varphi}}\right)\\ &= \frac{i\Gamma}{2\pi}\left[-\log(ae^{i\varphi})+\log(2ib)+\log\left(1-i\frac{a}{2b}e^{i\varphi}\right)\right]\\ &= \frac{i\Gamma}{2\pi}\left[-\log a-i\varphi+\log(2b)+i\frac{\pi}{2}-i\frac{a}{2b}e^{i\varphi}+O\left(\frac{a^2}{b^2}e^{2i\varphi}\right)\right]\\ &= -\frac{\Gamma}{4}+\frac{\Gamma\varphi}{2\pi}+\frac{i\Gamma}{2\pi}\log\left(\frac{2b}{a}\right)+\frac{\Gamma a}{4\pi b}e^{i\varphi}+O\left(\Gamma\frac{a^2}{b^2}e^{2i\varphi}\right).\end{aligned} \tag{10.129}$$

Substitution of Equation 10.118b in Equation 10.129 specifies the complex potential (Equation 10.130b) on the cylinder with circulation and wall effect to order Equation 10.130a:

$$\begin{aligned}a^2 \ll b^2: \quad \bar{f}(ib+ae^{i\varphi}) &= ibU-\frac{\Gamma}{4}+\frac{iP_1}{4\pi b}+\frac{i\Gamma}{2\pi}\log\left(\frac{2b}{a}\right)\\ &\quad+\left(Ua-\frac{P_1 a}{8\pi b^2}+\frac{\Gamma a}{4\pi b}\right)e^{i\varphi}-\frac{P_1}{2\pi a}e^{-i\varphi}+\frac{\Gamma\varphi}{2\pi};\end{aligned} \tag{10.130a and 10.130b}$$

considering the term $e^{\pm 2i\varphi}$ would require adding a quadrupole.

The representation of the cylinder as a dipole implies that the boundary condition on the cylinder can be met to order $a^2/b^2(a/b)$ for the dipole moment (circulation of the vortex), as shown by stream function on the cylinder:

$$\bar{\Psi}(ib+ae^{i\varphi}) = \mathrm{Im}\left[\bar{f}(ib+ae^{i\varphi})\right] = bU + \frac{P_1}{4\pi b} + \frac{\Gamma}{2\pi}\log\left(\frac{2b}{a}\right) + \left[Ua + \frac{P_1}{2\pi a}\left(1-\frac{a^2}{4b^2}\right) + \frac{\Gamma a}{4\pi b}\right]\sin\varphi. \tag{10.131}$$

The stream function becomes a constant (Equation 10.132b) to order Equation 10.132a on the cylinder if the dipole moment is chosen (Equation 10.133b) to cancel the term in square brackets in Equation 10.131:

$$a^2 \ll b^2: \quad \bar{\Psi}(ib+ae^{i\varphi}) = bU + \frac{P_1}{4\pi b} + \frac{\Gamma}{2\pi}\log\left(\frac{2b}{a}\right) = const, \tag{10.132a and 10.132b}$$

$$\Gamma \sim P_1\frac{a}{b}: \quad P_1 = \left(1-\frac{a^2}{4b^2}\right)^{-1}\left(-2\pi a^2 U - \frac{\Gamma a^2}{2b}\right); \tag{10.133a and 10.133b}$$

comparing the terms neglected in Equation 10.118b and in Equation 10.129 that apply both to Equation 10.130b, the approximation to lowest powers of a/b is consistent if Equation 10.133a holds. In this case (Equation 10.132a ≡ Equation 10.134a and Equation 10.133a ≡ Equation 10.134b), the dipole moment (Equation 10.132d) is given by

$$\frac{a^3}{b^3} \ll 1 \sim \frac{P_1 a}{\Gamma b}: \quad P_1 = -2\pi U a^2\left(1+\frac{a^2}{4b^2}\right) - \frac{\Gamma a^2}{2b}; \tag{10.134a–10.134c}$$

the dipole moment (Equation 10.134c) adds to (Equation 10.121b) the effect of circulation. Thus, *a cylinder of radius a in a free stream of velocity U parallel to a wall a distance b from the center in the presence of circulation Γ can be represented by a dipole with moment Equation 10.134c with the approximations Equation 10.134a and 10.134b that imply that (1) the dipole moment is valid to $a^2/b^2(a^3/b^3)$ in the presence (absence) of circulation; (2) in the presence of circulation the latter scales on the dipole moment as $\Gamma b \sim P_1 a$. If the circulation is much larger (smaller) it is dominant (can be neglected)*

E10.11.4: Effect of Circulation on Velocity and Lift

The potential on the cylinder is given (Equation 10.130b) by

$$\begin{aligned}
\bar{\Phi}(ib+ae^{i\varphi}) = \mathrm{Re}\left[\bar{f}(ib+ae^{i\varphi})\right] &= -\frac{\Gamma}{4} + \frac{\Gamma\varphi}{2\pi} + \left[Ua - \frac{P_1}{2\pi a}\left(1+\frac{a^2}{4b^2}\right) + \frac{\Gamma a}{4\pi b}\right]\cos\varphi \\
&= -\frac{\Gamma}{4} + \frac{\Gamma\varphi}{2\pi} + \left\{Ua + \frac{\Gamma a}{4\pi b} + \left(1+\frac{a^2}{4b^2}\right)\left[Ua\left(1+\frac{a^2}{4b^2}\right) + \frac{\Gamma a}{4\pi b}\right]\right\}\cos\varphi \\
&= -\frac{\Gamma}{4} + \frac{\Gamma\varphi}{2\pi} + \left\{Ua\left[1+\left(1+\frac{a^2}{4b^2}\right)^2\right] + \frac{\Gamma a}{4\pi b}\left(2+\frac{a^2}{4b^2}\right)\right\}\cos\varphi \\
&= -\frac{\Gamma}{4} + \frac{\Gamma\varphi}{2\pi} + \left[2Ua\left(1+\frac{a^2}{4b^2}\right) + \frac{\Gamma a}{2\pi b}\right]\cos\varphi,
\end{aligned} \tag{10.135}$$

where the dipole moment (Equation 10.134c) was substituted and the approximations (Equations 10.133a and 10.134a) were made. The velocity is tangential (Equation 10.136a) on the cylinder:

$$v_r(ib+ae^{i\varphi})=0; \quad v_\varphi(ib+ae^{i\varphi})=\frac{1}{a}\frac{\partial\bar{\Phi}}{\partial\varphi}=-2U\sin\varphi+\frac{\Gamma}{2\pi a}-\left(\frac{\Gamma}{2\pi b}+\frac{Ua^2}{2b^2}\right)\sin\varphi.$$

(10.136a and 10.136b)

The tangential velocity (Equation 10.136b) consists of the following: (1) the first two terms on the r.h.s. of Equation 10.136b corresponding (Equation I.28.118b) to a cylinder of radius a with circulation Γ in a free stream with velocity U; and (2) the last term on the r.h.s. of Equation 10.132c is the effect of the wall at a distance b from the center of the cylinder that appears to first (second) order in a/b in the circulation (dipole moment) and is most significant $|\sin\varphi| = 1$ for the points on the cylinder closest $\varphi = -\pi/2$ and farthest $\varphi = \pi/2$ from the wall, that lie on the line perpendicular to the wall passing through the center of the cylinder. From Equation 10.136b or 10.135 it follows that the wall effect is equivalent to a change in free stream velocity (Equation 10.137a) implying that the same effect to the square appears [Equation 10.137b (Equation 10.137c)] in the lift force (Equation 8.233a ≡ Equation I.34.17) [lift coefficient (Equation 2.233b ≡ Equation I.34.135b)]:

$$\bar{U}=U\left(1+\frac{a^2}{4b^2}\right)+\frac{\Gamma}{4\pi b}: \qquad \frac{\bar{L}}{L}=\frac{\bar{C}_L}{C_L}=\left(\frac{\bar{U}}{U}\right)^2=\left(1+\frac{a^2}{4b^2}+\frac{\Gamma}{4\pi Ub}\right)^2;$$

(10.137a and 10.137b)

to the level of approximation (Equation 10.134a), the wall effect (Equation 10.137b) on the lift is specified by

$$\frac{\bar{L}}{L}=\frac{\bar{C}_L}{C_L}=1+\frac{a^2}{2b^2}+\frac{\Gamma}{2\pi Ub}+\frac{\Gamma a^2}{8\pi Ub^3}=\left(1+\frac{a^2}{2b^2}\right)+\frac{\Gamma}{2\pi Ub}\left(1+\frac{a^2}{4b^2}\right). \tag{10.137c}$$

Placing the cylinder of radius a with the centre at a distance b from a wall increases the lift because the flows is accelerated in the gap, this being equivalent to a larger free stream velocity; a negative that is counterclockwise circulation, that produces lift, decreases the wall effect, because it opposes the acceleration of the flow in the gap between the cylinder and the wall.

E10.11.5: Exact, Near-, and Far-Field Flow

The dipole moment (Equation 10.134c) appears (Equations 10.117 and 10.128) in the complex potential

$$f(z)=Uz+2\frac{a^2z}{z^2+b^2}\left[U\left(1+\frac{a^2}{4b^2}\right)+\frac{\Gamma}{4\pi b}\right]+\frac{i\Gamma}{2\pi}\log\left(\frac{z+ib}{z-ib}\right); \tag{10.138}$$

it involves (Equation 10.138 ≡ Equation 10.139) the equivalent free stream velocity (Equation 10.137a):

$$\bar{f}(z)=Uz+\bar{U}\frac{a^2z}{z^2+b^2}+\frac{i\Gamma}{2\pi}\log\left(\frac{z+ib}{z-ib}\right), \tag{10.139}$$

$$\bar{v}^*(z) = \frac{df}{dz} = U + \bar{U}a^2 \frac{b^2 - z^2}{(b^2 + z^2)^2} + \frac{\Gamma}{\pi} \frac{b}{z^2 + b^2} \tag{10.140}$$

and leads to the complex conjugate velocity (Equation 10.140).

The velocity is tangential to the wall proving that the boundary condition is met:

$$\bar{v}^*(x) = U + \bar{U}\frac{a^2(b^2 - x^2)}{(b^2 + x^2)^2} + \frac{\Gamma}{\pi}\frac{b}{b^2 + x^2} = \bar{v}_x(x,0), \qquad \bar{v}_y(x,0) = 0.$$

(10.141a and 10.141b)

The velocity is also horizontal on a line perpendicular to the wall and passing through the axis of the cylinder:

$$\bar{v}^*(iy) = U + \bar{U}\frac{a^2(b^2 + y^2)}{(b^2 - y^2)^2} + \frac{\Gamma}{\pi}\frac{b}{b^2 - y^2} = \bar{v}_x(0,y), \qquad \bar{v}_y(0,y) = 0.$$

(10.142a and 10.142b)

In particular, at the point on the wall closest to the cylinder, the velocity is given by

$$\bar{v}_y(0,0) = 0, \qquad \bar{v}_y(0,0) = U + \bar{U}\frac{a^2}{b^2} + \frac{\Gamma}{\pi b} = U\left(1 + \frac{a^2}{b^2}\right) + \frac{\Gamma}{\pi b},$$

(10.143a and 10.143b)

where Equation 10.137a was used with the approximations Equations 10.134a and 10.134b. *The velocity is increased in the gap between the cylinder and the wall, in proportion to the square of the radius of the cylinder to the distance of the center from the wall, with factor unity, instead of 1/4 for the dipole moment (Equation 10.121b) and tangential velocity on the cylinder (Equation 10.123b) without circulation. A negative (positive), that is clockwise (counterclockwise) circulation, that decreases (increases) the velocity below the cylinder, decreases (increases) the equivalent free stream velocity (Equation 10.137a) and has a similar effect at the point on the wall closest to the cylinder (Equation 10.143b).*

In the far-field (Equation 10.144a), the complex conjugate velocity (Equation 10.140) reduces (Equation 10.144b):

$$|z|^2 >> b^2 > a^2: \qquad \bar{v}^*(z) = U - \bar{U}\frac{a^2}{z^2} + \frac{\Gamma b}{\pi z^2} = U + \frac{\bar{P}_1}{2\pi z^2},$$

(10.144a and 10.144b)

corresponding to a dipole at the origin with moment:

$$\begin{aligned}\bar{P}_1 &= -2\pi\bar{U}a^2 + 2\Gamma b = -2\pi a^2\left[U\left(1 + \frac{a^2}{4b^2}\right) + \frac{\Gamma}{4\pi b}\right] + 2\Gamma b \\ &= -2\pi U a^2\left(1 + \frac{a^2}{4b^2}\right) + 2\Gamma b\left(1 - \frac{a^2}{4b^2}\right).\end{aligned} \tag{10.145}$$

The dipole moment (Equation 10.145) has four terms: (1) the first is the dipole moment (Equation 8.6c ≡ Equation I.28.89a) of a cylinder of radius a in a free stream of velocity U

due to the equal image on the wall; (2) the second is due to the vortex with circulation Γ and its image with opposite circulation at a distance $2b$; (3 and 4) the third (fourth) term is a wall effect on the dipole (vortex), scaling like a^2/b^2 with a coefficient 1/4 (1/2). Thus, *a cylinder with radius a and circulation Γ in a free stream of velocity U parallel to a wall at a distance b from the center (Figure 10.2b) is represented to the order of approximation Equations 10.134a and 10.134b by a dipole with moment Equation 10.134c, corresponding to the complex potential Equation 10.138 ≡ Equation 10.139 involving the equivalent free steam velocity Equation 10.137a. The complex conjugate velocity (Equation 10.140) shows that the velocity is (1) tangential (Equations 10.136a and 10.136b), on the cylinder, (2 and 3) horizontal on the wall (Equations 10.142a and 10.142b) and on the line perpendicular to the wall passing through the center of the cylinder (Equations 10.143a and 10.143b) and (4) at infinity (Equations 10.144a and 10.144b) corresponds to a dipole of moment Equation 10.145 including the images on the wall of the dipole (vortex) representing the cylinder (the circulation around the cylinder).*

E10.11.6: Electrostatic Field of Insulating Cylinder near an Insulating Wall

The potential flow past a cylinder near a wall, both rigid impermeable surfaces (Figure 10.2a), is analogous (case I) to the electrostatic field of an insulating cylinder near an insulating wall (Figure 10.3a). In all cases, the expression near a wall is meant to be under the influence of a wall but at a sufficient distance for Equation 10.121a to hold, implying that the radius is small relative to distance of the axis from the wall, for

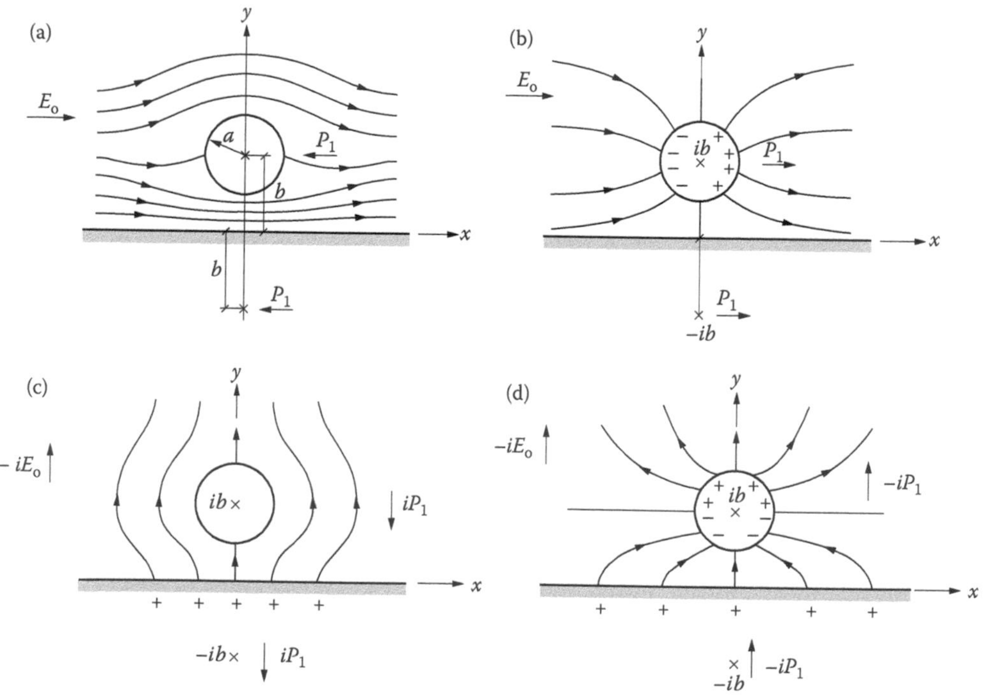

FIGURE 10.3
Electrostatic analogues of the potential flow past a cylinder in a uniform stream parallel to an infinite wall (Figures 10.1b and 10.2a) are a conducting or insulating cylinder near a conducting or insulating wall, leading to four combinations (a–d). The external electric field is parallel (orthogonal) to the insulating (conducting) wall [a, b (c, d)] ensuring that the wall is a field line (equipotential); the dipole moment is parallel (antiparallel) to the external electric field for a conducting (insulating) cylinder [b, d (a, c)] ensuring that the cylinder is a field line (equipotential).

example, $a/b \sim 0.5$ leads to $(a/b)^3 \sim 0.125 \ll 1$, so that the cylinder could come at most to within one radius of the wall $b - a \sim a$. The complex electrostatic potential (Equation 10.146a) is given by (Equation 10.117) replacing the free stream velocity U by the external electric field E_0:

$$f_e^{++}(z) = -E_0 z + \frac{P_{++}}{\pi\varepsilon}\frac{z}{z^2+b^2}; \qquad P_{++} = -2\pi a^2 E_0 \varepsilon\left(1+\frac{a^2}{4b^2}\right); \tag{10.146a and 10.146b}$$

the condition that the surface of the insulating cylinder is a field line leads (Equation 10.120b) to order Equation 10.121a to the dipole moment (Equation 10.146b). Substituting Equation 10.146b in Equation 10.146a specifies

$$a^3 \ll b^3: \qquad f_e^{++}(z) = -E_0\left[z + \frac{2a^2 z}{z^2+b^2}\left(1+\frac{a^2}{4b^2}\right)\right] \tag{10.147a and 10.147b}$$

the electrostatic potential of an insulating cylinder of radius a, in a uniform external electric field E_0, at a distance b from an insulating wall (Equation 10.147b) but (Figure 10.3a) not too close to it (Equation 10.147a).

The corresponding complex conjugate electric field is

$$E_e^{++}(z) = -\frac{df_e^{++}}{dz} = E_0\left[1+\frac{2a^2(b^2-z^2)}{(b^2+z^2)^2}\left(1+\frac{a^2}{4b^2}\right)\right]. \tag{10.148}$$

Since the electric field is tangent to the cylinder (Equations 10.149a and 10.149b) [the wall (Equations 10.150a and 10.150b)]:

$$\sigma^{++}(x) = \varepsilon E_r^{++}(ib + ae^{i\varphi}) = 0, \quad E_\varphi^{++}(ib+ae^{i\varphi}) = -2E_0\left(1+\frac{a^2}{4b^2}\right)\sin\varphi, \tag{10.149a and 10.149b}$$

$$\sigma^{++}(x) = \varepsilon E_y^{++}(x,0) = 0, \qquad E_x^{++}(x,0) = E_0\left[1+\frac{2a^2(b^2-x^2)}{(b^2+x^2)^2}\left(1+\frac{a^2}{4b^2}\right)\right], \tag{10.150a and 10.150b}$$

there are no induced electric charges on the surfaces for any finite value of the dielectric permittivity ε of the medium. Just as the flow is accelerated between the cylinder and the wall, the electric field is enhanced, for example:

$$E_y^{++}(0,y) = 0, \qquad E_x^{++}(y) = E_0\left[1+\frac{2a^2(b^2+y^2)}{(b^2-y^2)^2}\left(1+\frac{a^2}{4b^2}\right)\right] \tag{10.151a and 10.151b}$$

on the line perpendicular to the wall and passing through the axis of the cylinder.

E10.11.7: Conducting Cylinder and Insulating Wall

In this case II (Figure 10.3b), since the wall remains insulating, Equation 10.146a still holds (Equation 10.152a) because (1) the external electric field is parallel to the wall hence horizontal; and (2) the image on the wall is identical:

$$f_e^{+-}(z) = -E_0 z + \frac{P_{+-}}{\pi\varepsilon}\frac{z}{z^2+b^2}; \qquad P_{+-} = 2\pi a^2 E_0 \varepsilon\left(1-\frac{a^2}{4b^2}\right). \qquad \text{(10.152a and 10.152b)}$$

The difference between the insulating (Equation 10.146a) [conducting (Equation 10.152a)] cylinder is that the dipole moment has an opposite sign, that is, antiparallel (Equation 10.146b) [parallel (Equation 10.152b)] to the external electric field. The presence of an insulating wall increases (Equation 10.146b) [decreases (Equations 10.152b)] the modulus of the dipole moment representing the insulating (conducting) cylinder. The complex potential (Equations 10.118a and 10.118b) is the same, but for the insulating (conducting) cylinder, the surface is a field line (Equation 10.119) [an equipotential (Equation 10.153c) from Equation 10.118b with the substitutions Equations 10.153a and 10.153b that arise from the comparison of Equation 10.152a ≡ Equation 10.117]:

$$(U, P_1) \to \left(-E_0, -\frac{P_{+-}}{\varepsilon}\right): \quad \Phi_e^{+-}(ib + ae^{i\varphi}) = \operatorname{Re}\left[f_e^{+-}(ib + ae^{i\varphi})\right]$$

$$= \operatorname{Re}\left(-ibE_0 - E_0 a e^{i\varphi} + \frac{P_{+-}}{2\pi\varepsilon a}e^{-i\varphi} - \frac{iP_{+-}}{4\pi\varepsilon b} + \frac{P_{+-}a}{8\pi\varepsilon b^2}e^{i\varphi}\right)$$

$$= -\left(E_0 a - \frac{P_{+-}}{2\pi\varepsilon a} - \frac{P_{+-}a}{8\pi\varepsilon b^2}\right)\cos\varphi, \qquad \text{(10.153a–10.153c)}$$

and the field function (Equation 10.120a) [potential (Equation 10.154a)] is constant:

$$\Phi_e^{+-}(ib + a^{i\varphi}) = 0: \qquad P_{+-} = 2\pi a^2 E_0 \varepsilon\left(1+\frac{a^2}{4b^2}\right)^{-1}, \qquad \text{(10.154a and 10.154b)}$$

if the dipole moment is given by Equation 10.120b (Equation 10.154b) leading to Equation 10.121b (Equation 10.152b).

The dipole moment for the insulating (Equation 10.146b) [conducting (Equation 10.152b)] cylinder differs in the leading term with opposite sign because the dipole is antiparallel (parallel) to the external electric field; combined with the same sign for the wall effect, this leads to a dipole moment that is increased (decreased) in modulus for the insulating (conducting) cylinder. This leads (Equations 10.152a and 10.152b) to the complex electrostatic potential (Equation 10.155a) [conjugate electric field (Equation 10.155b)] of a conducting cylinder of radius a at a distance b from an insulating wall in a uniform external electric field (Figure 10.3b):

$$f_e^{+-}(z) = -E_0 z\left[1 - \frac{2a^2}{z^2+b^2}\left(1-\frac{a^2}{4b^2}\right)\right], \qquad \text{(10.155a)}$$

$$E_-^{+-}(z) = -\frac{df^{+-}}{dz} = E_0\left[1 - \frac{2a^2(b^2-z^2)}{(b^2+z^2)^2}\left(1-\frac{a^2}{4b^2}\right)\right]. \qquad \text{(10.155b)}$$

The electric field is horizontal (1) on the line orthogonal to the wall passing through the axis of the cylinder (Equations 10.156a and 10.156b):

$$E_y^{+-}(0,y)=0, \qquad E_x^{+-}(0,y)=E_0\left[1-\frac{2a^2(b^2+y^2)}{(b^2-y^2)^2}\left(1-\frac{a^2}{4b^2}\right)\right]; \qquad \text{(10.156a and 10.156b)}$$

$$\sigma^{+-}(x)=\varepsilon E_y^{+-}(x,0)=0, \qquad E_x^{+-}(x,0)=E_0\left[1-\frac{2a^2(b^2-x^2)}{(b^2+x^2)}\left(1-\frac{a^2}{4b^2}\right)\right]; \qquad \text{(10.157a and 10.157b)}$$

(2) on the insulating wall (Equation 10.157b), without induced surface electric charges (Equation 10.157a). There is an induced electric change distribution on the surface of the cylinder specified by the normal component of the electric field:

$$E_\varphi^{+-}(ib+ae^{i\varphi})=0, \quad \sigma^{+-}(\varphi)=\varepsilon E_r^{+-}(ib+ae^{i\varphi})=2\varepsilon E_0\left(1-\frac{a^2}{4b^2}\right)\cos\varphi=\frac{P_{+-}}{\pi a^2}\cos\varphi. \qquad \text{(10.158a and 10.158b)}$$

The normal component of the electric field on the cylinder (Equation 10.158b) is obtained from Equations 10.153c, 10.152b and 2.186b:

$$\begin{aligned}\Psi_e^{+-}\left(ib+ae^{i\varphi}\right)&=\operatorname{Im}\left[f_e^{+-}\left(ib+ae^{i\varphi}\right)\right]=-E_0b-\left[E_0a+\frac{P_{+-}}{2\pi\varepsilon a}\left(1-\frac{a^2}{4b^2}\right)\right]\sin\varphi\\&=-E_0b-E_0a\left[1+\left(1-\frac{a^2}{4b^2}\right)^2\right]\sin\varphi-=-E_0b-2E_0a\left(1-\frac{a^2}{4b^2}\right)\sin\varphi,\end{aligned} \qquad \text{(10.158c)}$$

$$E_r^{+-}\left(ib+ae^{i\varphi}\right)=-\frac{1}{a}\frac{\partial\Psi_e^{+-}}{\partial\varphi}=2E_0\left(1-\frac{a^2}{4b^2}\right)\cos\varphi. \qquad \text{(10.158d)}$$

The total electric charge on the conducting cylinder is zero (Equation 10.159a):

$$a\int_0^{2\pi}\sigma^{+-}(\varphi)\{1,a\cos\varphi,a\sin\varphi\}\,d\varphi=\{0,P_{+-},0\}, \qquad \text{(10.159a–10.159c)}$$

and the dipole moment is horizontal (Equation 10.159c) and coincides with Equation 10.154b ≡ Equation 10.159b.

E10.11.8: Insulating Cylinder near a Conducting Wall

In case III of a conducting wall (Figure 10.3c), the electrostatic potential (Equation 10.147b) is modified (Equation 10.160) because (1) the external electric field is orthogonal to the conducting wall and hence vertical; (2) the dipole for an insulating cylinder is antiparallel and hence also vertical; and (3) the image is opposite conjugate and hence identical:

$$f_e^{-+}(z) = iE_0 z + \frac{1}{2\pi\varepsilon}\left[\frac{iP_{-+}}{z-ib} - \frac{(iP_{-+})^*}{z+ib}\right]$$
$$= iE_0 z + \frac{iP_{-+}}{2\pi\varepsilon}\left(\frac{1}{z-ib} + \frac{1}{z+ib}\right) = i\left(E_0 z + \frac{P_{-+}}{\pi\varepsilon}\frac{z}{z^2+b^2}\right). \quad (10.160)$$

Since the cylinder is insulating, it must be a field line; using Equation 10.118b with the substitution Equations 10.161a and 10.161b that arise from the comparison of Equation 10.160 ≡ Equation 10.117 leads to the field function Equation 10.161c:

$$(U, P_1) \to \left(iE_0, -i\frac{P_{-+}}{\varepsilon}\right): \quad \Psi_e^{-+}(ib + ae^{i\varphi}) = \operatorname{Im}\left[f_e^{-+}(ib + ae^{i\varphi})\right]$$
$$= \operatorname{Im}\left(-E_0 b + iE_0 a e^{i\varphi} + \frac{iP_{-+}}{2\pi\varepsilon a}e^{-i\varphi} - \frac{P_{-+}}{4\pi\varepsilon b} + \frac{iP_{-+}a}{8\pi\varepsilon b^2}e^{i\varphi}\right)$$
$$= \left(E_0 a + \frac{P_{-+}}{2\pi\varepsilon a} + \frac{P_{-+}a}{8\pi\varepsilon b^2}\right)\cos\varphi. \quad (10.161a–10.161c)$$

It is constant (Equation 10.162a) if the dipole moment is given by Equation 10.162b:

$$\Psi_e^{-+}(ib + ae^{i\varphi}) = 0, \qquad P_{-+} = -2\pi a^2 E_0 \varepsilon\left(1 + \frac{a^2}{4b^2}\right)^{-1} = -2\pi a^2 E_0 \varepsilon\left(1 - \frac{a^2}{4b^2}\right).$$
(10.162a and 10.162b)

The dipole moment is (1) opposite to the external electric field because the cylinder is insulating and (2) weakened by the wall effect because the wall is a conductor.

Substituting Equation 10.162b in Equation 10.160 specifies *the electrostatic potential (Equation 10.163a) [conjugate electric field (Equation 10.163b)] of an insulating cylinder of radius a at a distance b from a conducting wall in the presence of a constant external electric field (Figure 10.3c):*

$$f_e^{-+}(z) = iE_0 z\left[1 - \frac{2a^2}{z^2+b^2}\left(1 - \frac{a^2}{4b^2}\right)\right], \quad (10.163a)$$

$$E_{-+}^*(z) = -\frac{df_e^{-+}}{dz} = -iE_0\left[1 - \frac{2a^2(b^2 - z^2)}{b^2+z^2}\left(1 - \frac{a^2}{4b^2}\right)\right]. \quad (10.163b)$$

The electric field is vertical (1) on the line normal to the wall passing through the axis of the cylinder (Equations 10.164a and 10.164b):

$$E_x^{-+}(0,y) = 0, \qquad E_y^{-+}(0,y) = E_0\left[1 - \frac{2a^2(b^2+y^2)}{(b^2-y^2)^2}\left(1 - \frac{a^2}{4b^2}\right)\right];$$
(10.164a and 10.164b)

$$E_x^{-+}(x,0)=0, \qquad \sigma^{-+}(x)=\varepsilon E_y^{-+}(x,0)=E_0\varepsilon\left[1-\frac{2a^2(b^2-x^2)}{(b^2+x^2)}\left(1-\frac{a^2}{4b^2}\right)\right]; \tag{10.165a and 10.165b}$$

and (2) on the conducting wall (Equation 10.165a) with induced electric charge distribution (Equation 10.165b). The electric field is tangent to the cylinder (Equation 10.166b):

$$\sigma^{-+}(x)=\varepsilon E_r^{-+}(ib+ae^{i\varphi})=0, \qquad E_\varphi^{-+}(ib+ae^{i\varphi})=-2E_0\left(1-\frac{a^2}{4b^2}\right)\cos\varphi, \tag{10.166a and 10.166b}$$

which is an insulator without surface electric changes (Equation 10.166a). The tangential component of the electric field on the cylinder (Equation 10.166b ≡ Equation 10.166d) follows (Equation 10.161c) from the potential (Equations 10.166c and 10.162b):

$$\begin{aligned}\Phi_e^{-+}\left(ib+ae^{i\varphi}\right)&=\operatorname{Re}\left[f_e^{-+}(ib+ae^{i\varphi})\right]\\&=-E_0b-\frac{P_{-+}}{4\pi\varepsilon b}-\left[E_0a-\frac{P_{-+}}{2\pi\varepsilon a}\left(1-\frac{a^2}{4b^2}\right)\right]\sin\varphi\\&=-E_0b\left(1-\frac{a^2}{2b^2}\right)-E_0a\left[1+\left(1-\frac{a^2}{4b^2}\right)^2\right]\sin\varphi\\&=-E_0\left(b-\frac{a^2}{2b}\right)-2E_0a\left(1-\frac{a^2}{4b^2}\right)\sin\varphi,\end{aligned} \tag{10.166c}$$

$$E_\varphi^{+-}\left(ib+ae^{i\varphi}\right)=\frac{1}{a}\frac{\partial\Phi_e^{+-}}{\partial\varphi}=-2E_0a\left(1-\frac{a^2}{4b^2}\right)\cos\varphi, \tag{10.166d}$$

using Equation 2.186c.

E10.11.9: Conducting Cylinder near a Conducting Wall

In case IV (Figure 10.3d), the electrostatic potential (Equation 10.147b) is modified (Equation 10.167a) because (1) the wall is conducting so the external electric field is vertical; (2) the cylinder is also conducting so the dipole moment is parallel to the external electric field; and (3) the image has opposite complex conjugate moment and hence equal:

$$f_e^{--}(z)=iE_0z+i\frac{P_{--}}{\pi\varepsilon}\frac{z}{z^2+b^2}; \qquad P_{--}=2\pi a^2\varepsilon\left(1+\frac{a^2}{4b^2}\right); \tag{10.167a and 10.167b}$$

the dipole moment (Equation 10.167b) is aligned with the electric field and increased by the conducting wall. This is a consequence of the conducting cylinder being an equipotential, with potential (Equation 10.168c) given by the real part of Equation 10.118b with the substitutions Equations 10.168a and 10.168b that arise from the comparison of Equation 10.117 ≡ Equation 10.167a:

$$(U,P_1)\to\left(iE_0,-i\frac{P_{--}}{\varepsilon}\right):\quad \Phi_e^{--}(ib+ae^{i\varphi})=\operatorname{Re}\left[f_e^{--}(ib+ae^{i\varphi})\right]$$

$$=\operatorname{Re}\left(-E_0b+iE_0ae^{i\varphi}+\frac{iP_{--}}{2\pi\varepsilon a}e^{-i\varphi}+\frac{P_{--}a}{4\pi\varepsilon b}+\frac{iP_{--}}{8\pi\varepsilon b^2}e^{i\varphi}\right)$$

$$=-E_0b+\frac{P_{--}}{4\pi\varepsilon b}-\left(E_0a-\frac{P_{--}}{2\pi\varepsilon a}+\frac{P_{--}}{8\pi\varepsilon b^2}\right)\sin\varphi.$$

(10.168a–10.168c)

The potential is constant (Equation 10.169a) if the dipole moment satisfies Equation 10.169b:

$$\Phi_e^{--}(ib+ae^{i\varphi})=-E_0b+\frac{P_{--}}{4\pi\varepsilon b}=\text{const},\qquad 2\pi a^2E_0=P_{--}\left(1-\frac{a^2}{4b^2}\right),\tag{10.169a and 10.169b}$$

leading to Equation 10.167b.

This specifies (Equations 10.167a and 10.167b) *the electrostatic potential (Equation 10.170c) of a conducting cylinder of radius a at a distance b from a conducting wall in the presence of a uniform external electric field (Figure 10.3d):*

$$f_e^{--}(z)=iE_0z\left[1+\frac{2a^2}{z^2+b^2}\left(1+\frac{a^2}{4b^2}\right)\right],\tag{10.170a}$$

$$E_{--}^{*}(z)=-iE_0\left[1+\frac{2a^2(b^2-z^2)}{(b^2+z^2)^2}\left(1+\frac{a^2}{4b^2}\right)\right].\tag{10.170b}$$

The electric field is vertical (1) on the line normal to the wall passing through the axis of the cylinder (Equations 10.171a and 10.171b):

$$E_x^{--}(0,y)=0,\qquad E_y^{--}(0,y)=E_0\left[1+\frac{2a^2(b^2+y^2)}{(b^2-y^2)^2}\left(1+\frac{a^2}{4b^2}\right)\right];\tag{10.171a and 10.171b}$$

$$E_x^{--}(0,y)=0,\qquad \sigma^{--}(x)=\varepsilon E_y^{--}(x,0)=E_0\varepsilon\left[1+\frac{2a^2(b^2-x^2)}{(b^2+x^2)^2}\left(1+\frac{a^2}{4b^2}\right)\right];$$

(10.172a and 10.172b)

and (2) on the conducting wall (Equation 10.172a) with introduced electric charge distribution (Equation 10.172b). The electric field is radial on the conducting cylinder (Equation 10.173a):

$$E_\varphi^{--}(ib+a^{i\varphi})=0,\quad \sigma^{--}(\varphi)=\varepsilon E_r^{--}(ib+a^{i\varphi})=2E_0\varepsilon\left(1+\frac{a^2}{4b^2}\right)\sin\varphi=\frac{P_{--}}{\pi a^2}\sin\varphi$$

(10.173a and 10.173b)

and specifies the surface electric charge distribution (Equation 10.173b). The total electric charge on the cylinder is zero (Equation 10.174a):

$$a\int_0^{2\pi} \sigma^{--}(\varphi)\{1, a\cos\varphi, a\sin\varphi\}\, d\varphi = \{0, 0, P_{--}\}, \qquad (10.173c–10.174e)$$

and the dipole moment is vertical (Equation 10.174b) and coincides with Equation 10.174c ≡ Equation 10.167b. The normal component of the electric field on the cylinder (Equation 10.173b ≡ 10.173d) is obtained (Equation 10.168c) from the field function (Equation 10.173c):

$$\begin{aligned}\Psi_e^{--}(ib + ae^{i\varphi}) &= \operatorname{Im}\left[f_e^{--}(ib + ae^{i\varphi})\right]\\ &= \left[E_0 a + \frac{P_{--}}{2\pi\varepsilon a}\left(1 + \frac{a^2}{4b^2}\right)\right]\cos\varphi \\ &= E_0 a\left[1 + \left(1 + \frac{a^2}{4b^2}\right)^2\right]\cos\varphi = 2E_0 a\left(1 + \frac{a^2}{4b^2}\right)\cos\varphi,\end{aligned} \qquad (10.174a)$$

$$E_r^{--}(ib + ae^{i\varphi}) = -\frac{1}{a}\frac{\partial \Psi_e^{--}}{\partial \varphi} = 2E_0 a\left(1 + \frac{a^2}{4b^2}\right)\sin\varphi, \qquad (10.174b)$$

using Equations 10.167b and 2.176c.

E10.11.10: Magnetostatic Field of a Cylinder near a Wall

In the case of the magnetostatic field, there are the same four combinations of the conducting or insulating cylinder near a conducting or insulating wall. The differences between the magnetostatic (electrostatic) field [Examples E11.10.6 through E11.10.9 (Example E11.10.10)] start with [Chapter I.24 (Chapter I.26)] *the Maxwell equations 10.175a and 10.175b (Equations 10.175c and 10.175d):*

$$\nabla \wedge \vec{E} = 0, \qquad \nabla . \vec{E} = \frac{q}{\varepsilon}; \qquad \nabla . \vec{H} = 0, \qquad \nabla \wedge \vec{H} = \frac{\vec{j}}{c}, \qquad (10.175a–10.175d)$$

where $\vec{E}(\vec{H})$ is the electric (magnetic) field, $q(\vec{j})$ is the electric charge (current) per unit volume, $\varepsilon(\mu)$ is the dielectric permittivity (magnetic permeability) of the medium, and c is the speed of light in vacuo. From Equation 10.175b (Equation 10.175d), it follows that the electric charge (current) density per unit area $\sigma(\vartheta)$ on a conductor is specified by the normal (tangential) component of the electric (Equation 10.176a ≡ Equation I.24.8b) [magnetic (Equation 10.176b ≡ Equation I.26.9b)] field:

$$\frac{\sigma}{\varepsilon} = E_n, \qquad \frac{\vartheta}{c} = H_s: \qquad \frac{\sigma}{\varepsilon} \leftrightarrow \frac{\vartheta}{c}, \qquad E_n \leftrightarrow H_s, \qquad (10.176a–10.176d)$$

implying the transformations (Equations 10.176c and 10.176d). The complex conjugate electric (magnetic) field due to electric charge e (current J) per unit length is [Equation 10.175a ≡ Equation I.26.20b (Equation 10.177b ≡ Equation I.26.21b)]

$$E^* = \frac{q}{2\pi\varepsilon z}, \qquad H^* = -\frac{iJ}{2\pi cz}: \qquad \frac{q}{\varepsilon} \leftrightarrow \frac{J}{c}, \qquad E^* \leftrightarrow iH^*, \tag{10.177a–10.177d}$$

leading to the transformation (Equation 10.177c ≡ Equation 10.176c) and hence also Equation 10.177d.

These transformations (Example E10.11.10) applied to the preceding results (Examples E10.11.6 through E10.11.9) lead to three sets of results. First the dipole moment is given by Equation 10.146b ≡ Equation 10.178a, Equation 10.152b ≡ Equation 10.178b, Equation 10.162b ≡ Equation 10.178c, and Equation 10.167b ≡ Equation 10.178d:

$$\{P_{++}, P_{+-}, P_{-+}, P_{--}\} = 2\pi a^2 c H_0 \left\{-1 - \frac{a^2}{4b^2}, 1 - \frac{a^2}{4b^2}, -1 + \frac{a^2}{4b^2}, 1 + \frac{a^2}{4b^2}\right\}; \tag{10.178a–10.178d}$$

thus *the dipole moment is parallel +1 (antiparallel −1) to the external magnetic field for a conducting P_{x-} (insulating P_{x+}) cylinder and is enhanced (reduced) when the wall and cylinder are both insulating (P_{++}) or both conducting (P_{--}) [one is conducting and the other insulating (P_{+-}, P_{-+})]. Second the magnetic field (Equation 10.177d) is given by Equation 10.148 ≡ Equation 10.179a, Equation 10.155b ≡ Equation 10.179b, Equation 10.163b ≡ Equation 10.179c, and Equation 10.170b ≡ Equation 10.179d:*

$$\left\{H^*_{++}, H^*_{+-}, H^*_{-+}, H^*_{--}\right\}(z) = H_0\left\{-i[1 + h_+(z)], -i[1 - h_-(z)], -[1 - h_-(z)], -[1 + h_+(z)]\right\}; \tag{10.179a–10.179d}$$

in the case of an insulating H^*_{x+} (conducting H^*_{x-}) cylinder near an insulating H^*_{+x} (conducting H^*_{-x}) wall, Equations 10.180a and 10.180b were used:

$$h_\pm(z) = \frac{2a^2(b^2 - z^2)}{(b^2 + z^2)^2} g_\pm, \qquad g_\pm = 1 \pm \frac{a^2}{4b}. \tag{10.180a and 10.180b}$$

Third the induced electric currents are given by Equations 10.149a and 10.150a ≡ Equations 10.181a and 10.182a, Equations 10.157a and 10.158b ≡ Equations 10.181b and 10.182b, Equations 10.165b and 10.166a ≡ Equations 10.181c and 10.182c, and Equations 10.172b and 10.173b ≡ Equations 10.181d and 10.182d:

$$\vartheta^{++}, \vartheta^{+-}, \vartheta^{-+}, \vartheta^{--}(x) = \frac{2H_0}{c}\{0, 0, 1 - h_-(x), 1 + h_+(x)\}, \tag{10.181a–10.181d}$$

$$\vartheta^{++}, \vartheta^{+-}, \vartheta^{-+}, \vartheta^{--}(\varphi) = \frac{2H_0}{c}\{0, g_- \cos\varphi, 0, g_+ \sin\varphi\}; \tag{10.182a–10.182d}$$

the surface induced electric currents apply only to a conductor if the magnetic field is zero on the opposite side. The induced electric currents on the conducting cylinder for an insulating (Equation 10.182b) [conducting (Equation 10.182d)] wall lead to a zero total current [Equation 10.159a (Equation 10.174a)] and a horizontal (Equation 10.159c) [vertical (Equation 10.174b)] moment equal to the moment of the dipole [Equation 10.159b (Equation 10.174c)]. The induced electric current or charges in the conducting wall with insulating (Equation 10.181c) [conducting (Equation 10.181d)] cylinder are considered next.

E10.11.11: Total Charge and Dipole Moment Induced on the Wall

The electric charges induced on the conducting wall in case III (case IV) of a conducting (Equation 10.172b) [insulating (Equation 10.165b)] cylinder consist (Equation 10.183a)

$$\sigma^{-\mp}(x) = E_0\varepsilon\left[1 \pm 2a^2\left(1 \pm \frac{a^2}{4b^2}\right)q(x)\right], \qquad q(x) = \frac{x^2 - b^2}{(x^2 + b^2)^2} \qquad \text{(10.183a and 10.183b)}$$

of the following: (1) a constant term first on the r.h.s. of Equation 10.183a and (2) the second term on the r.h.s. of Equation 10.183a with dependence on the position on the wall specified by the function Equation 10.183b. The total charge (dipole moment) has a contribution from the second term (Equation 10.183b) of Equation 10.183a that is proportional to Equation 10.184a (Equation 10.184b) and hence is zero:

$$M_0 = \int_{-\infty}^{+\infty} q(x)\,\mathrm{d}x = 0, \qquad M_1 = \int_{-\infty}^{+\infty} xq(x)\,\mathrm{d}x = 0. \qquad \text{(10.184a and 10.184b)}$$

The dipole moment involves the integration of an odd function $-xq(-x) = -xq(x)$ over the real line and thus vanishes. The total charge is specified by the integral Equation 10.184a that can be evaluated (Equation 10.184c) by closing (Subsection I.15.8) the real axis by a large semicircle in the upper (lower) complex plane as $+2\pi i(-2\pi i)$ times the residue at the double pole at $+ib(-ib)$.

$$\int_{-\infty}^{+\infty} q(x)\,\mathrm{d}x = \int_{-\infty}^{+\infty} \frac{x^2 - b^2}{(x^2 + b^2)^2}\,\mathrm{d}x = \pm 2\pi i \lim_{x\to\pm ib} \frac{\mathrm{d}}{\mathrm{d}x}\left[(x \mp ib)^2 \frac{x^2 - b^2}{(x^2 + b^2)^2}\right]$$

$$= \pm 2\pi i \lim_{x\to\pm ib}\left[\frac{\mathrm{d}}{\mathrm{d}x}\frac{x^2 - b^2}{(x \pm ib)^2}\right] = \pm 2\pi i \lim_{x\to\pm ib} \frac{2b(b \pm ix)}{(x \pm ib)^3} = 0; \qquad \text{(10.184c)}$$

besides the total electric charge (Equation 10.184a) and dipole moment (Equation 10.184b) of the charge distribution (Equation 10.183b), the moments of all orders are considered next (Example E10.11.12).

E10.11.12: Smooth Distribution with All Moments Zero or Nonexistent

The distribution of electric charge (Equation 10.183a) leads by Equation 10.183b to

$$M_n = \int_{-\infty}^{+\infty} q(x) x^n\,\mathrm{d}x = \lim_{\substack{a\to-\infty \\ b\to+\infty}} \int_a^b x^n \frac{x^2 - b^2}{(x^2 + b^2)^2}\,\mathrm{d}x \qquad \text{(10.185)}$$

the moments (Equation 10.185) of order n with the following particular properties: (1) they do not exist for $n = 2,3,\ldots$ because the integral Equation 10.185 diverges, for example, there is no variance M_2; (2) since for $n = 0$ the integral is zero (Equation 10.184a ≡ Equation 10.184c), the root mean square (r.m.s.) value would be infinite; (3) the first moment (Equation 10.184b) is zero if the limit in Equation 10.185 is taken symmetrically $-a = b \to \infty$ because the integrand is an odd function; (4) in this case, a mean value cannot be defined because it is the ratio of the moments of first and zero orders, which are both zero; (5) if the general definition of improper integral of the first kind (Subsection I.17.1) is followed, then the limits ($a \to -\infty, b \to \infty$) are independent, and the integral (Equation 10.185) for the first moment does not exist; and (6) in this case, the only moment that exists is of order zero (Equation 10.184a) and is zero. The surface electric change or current distribution (Equation 10.183b) is even (Equation 10.186a), vanishes at the points (Equation 10.186b), and decays as Equation 10.186c at infinity:

$$q(x) = q(-x), \qquad q(\pm b) = 0, \qquad q(x) \sim \frac{1}{x^2} + O\left(\frac{1}{x^3}\right); \qquad (10.186a\text{–}10.186c)$$

the derivative (Equation 10.187c) shows that the minimum is at the origin (Equation 10.187b) and the maximum at Equation 10.187c:

$$\frac{dq}{dx} = \frac{2x(3b^2 - x^2)}{(x^2 + b^2)^3}: \qquad q_{\min} = q(0) = -\frac{1}{b^2}, \qquad q_{\max} = q\left(\pm b\sqrt{3}\right) = \frac{1}{8b^2}. \qquad (10.187a\text{–}10.187c)$$

The function Equation 10.183b is illustrated in Figure 10.4 in the dimensionless form:

$$Q = b^2 q, X \equiv \frac{x}{b}: \quad Q = \frac{X^2 - 1}{(X^2 + 1)^2}, \quad \frac{dQ}{dX} = \frac{2(3 - X^2)}{(X^2 + 1)^2}, Q_{\min} = Q(0) = -1, Q_{\max} = Q\left(\sqrt{3}\right) = \frac{1}{8}. \qquad (10.188a\text{–}10.188f)$$

The property of having all moments zero or nonexistent applies not only to the smooth distribution (Equation 10.183b) in Figure 10., but also to any distribution with the following three properties: (1) it has zero integral (Equation 10.184a) on the real line so that the moment of order

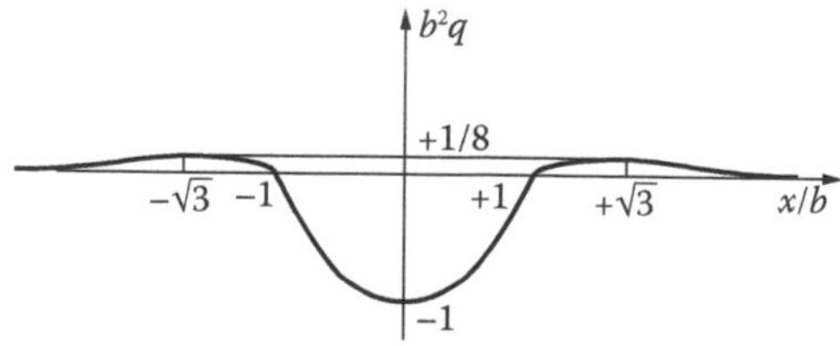

FIGURE 10.4
Surface electric charge density induced on an infinite conducting wall by a uniform vertical electric field in the presence of a conducting (insulating) cylinder [Figure 10.3d (Figure 10.3c)] in an example of a distribution for which all moments either vanish or do not exist: (1) the moment of order zero vanishes because the areas above and below the axis are equal; (2) the first moment is zero because the distribution is an even function of x; (3) the moments of order two or higher do not exist because the decay like the inverse square at infinity leads to divergent integrals; (4) the mean value does not exist because it is an indeterminate ratio of zero quantities (items 1 and 2); and (5) the variance, root mean square (r.m.s.) value, skewness, excess, and all other higher order centroids are infinite because the moment of zero order is zero and those of orders two and higher diverge.

zero vanishes; (2) it is an even function (Equation 10.186a) so that the first moment (Equation 10.184b) also vanishes; and (3) it decays at infinity like an inverse square (Equation 10.186c) so that the moments (Equation 10.185) of orders two or more do not exist.

EXAMPLE 10.12: POTENTIAL FLOW PAST A PLATE OBLIQUE TO A WALL

Consider the potential flow due to a stream parallel to a wall with an inclined plate and, in particular, the case of an orthogonal plate.

E10.12.1: Mapping into a Plane with a Wall with an Oblique Plate

The hodograph method (Subsection I.38.6) can be used to consider two related problems of jets with free boundaries: (1) a jet parallel to an infinite wall with an inclined plate (Figure I.38.7a) and (2) a jet incident from the front or rear of an arrow (Figure I.38.7b). In the present example, the geometry 1 (geometry 2) is the same in Figure 10.5b (Figure 10.5c), but the flow occupies all space, and there is (are) no free streamline(s) issuing from the edge(s) and no "vacuous" or "recirculation" region behind the obstacle. The wall with inclined plate (Figure 10.5b) is mapped from the real axis (Figure 10.5a) by the Schwartz–Christoffel interior polygonal transformation (Equation I.33.25 ≡ Equation 8.145b ≡ Equation 10.189c) with three critical points Equation 10.189a with angles Equation 10.189b:

$$x_{1-3} = \{-a, 0, a\}, \quad \gamma_{1-3} = \{\alpha, -\pi, \pi - \alpha\}:$$

$$\frac{d\zeta}{dz} = A\prod_{n=1}^{3}\left(z - x_n\right)^{-\gamma_n/\pi} = A(z+a)^{-\alpha/\pi} z(z-a)^{\alpha/\pi - 1}. \tag{10.189a–10.189c}$$

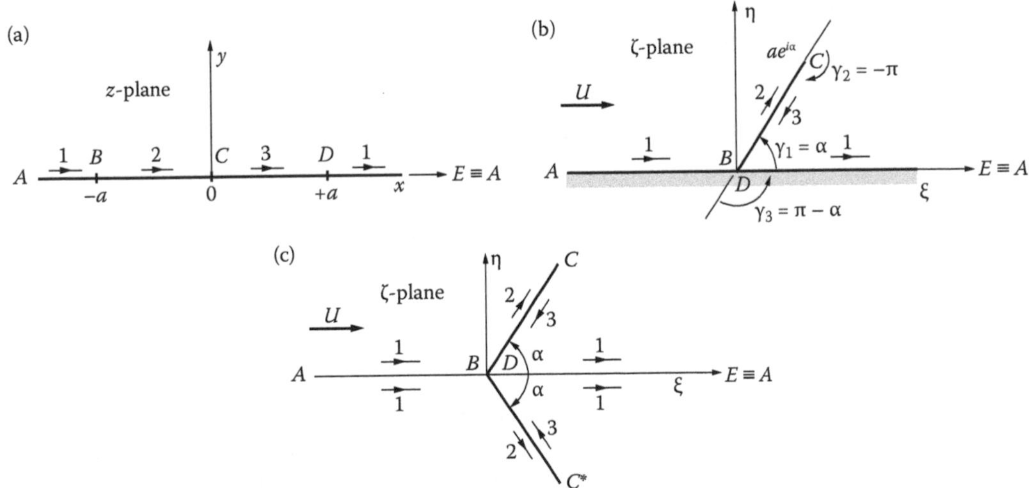

FIGURE 10.5
Schwartz–Christoffel internal polygonal mapping of the real axis with three critical points (a) into a flat plate oblique to a wall (b) specifies (1) the perturbation of a uniform stream parallel to a wall due to an inclined plate (b); (2) by the reflection principle (Figure I.31.4) the potential flow past an arrow in an incident stream (c). In the present case, the fluid fills all space, and the horizontal force on the two sides of the plate (arrow) balances [Figure 10.4b (Figure 10.4c)] so that there is no net drag. There is drag for the same geometry of a plate inclined to the wall (an arrow or bent lamina) in the case of a jet [Figure I.38.7a (Figure I.38.7b)], with free surfaces issuing from the edge(s); horizontal force is due to the pressure on the face on the upstream side and no such effect on the face on the downstream side.

The mapping (Equation 10.189c) involves three parameters, namely, (a, A) and an integration constant B in Equation 10.192c; there are three constraints, namely, the position of edge C and two base points B, D of the plate in Figure 10.5b and c. The critical points $x_3 = a = -x_1$ were chosen symmetrically relative to the origin, with variable distance a, so that there are three parameters to meet three constraints. The flow in the z-plane is uniform (Equation 10.190a) with velocity U, leading to the complex conjugate velocity in the ζ-plane (Equation 10.190b):

$$f = Uz, \qquad \frac{df}{d\zeta} = \frac{df/dz}{d\zeta/dz} = \frac{U}{A}\left(1 - \frac{a}{z}\right)\left(\frac{z+a}{z-a}\right)^{\alpha/\pi}, \qquad \text{(10.190a and 10.190b)}$$

that is, singular (vanishes) at the tip (base) of the plate $z = 0(z = \pm a)$. The constant A is determined (Equation 10.191b) from the condition (Equation 10.190a) that the velocity is U at infinity:

$$U = \lim_{z\to\infty} \frac{df}{d\zeta} = \frac{U}{A}, \qquad A = 1: \qquad v^*(\zeta) = U\left(1 - \frac{a}{z}\right)\left(\frac{z+a}{z-a}\right)^{\alpha/\pi}, \qquad \text{(10.191a–10.191c)}$$

thus specifying the complex conjugate velocity (Equation 10.191c).

E10.12.2: Complex Potential of Flow Past a Plate Orthogonal to a Wall

In the particular case (Equation 10.192a) of the plate orthogonal to the wall, the conformal mapping (Equation 10.189c) with Equation 10.191b simplifies to Equation 10.192b:

$$\alpha = \frac{\pi}{2}: \qquad \frac{d\zeta}{dz} = \frac{z}{\sqrt{z^2 - a^2}}, \qquad \zeta(z) = \sqrt{z^2 - a^2} + B, \qquad \text{(10.192a–10.192c)}$$

whose primitive is Equation 10.192c. The constant B is determined from the position of the base of the plate (Equation 10.193a):

$$0 = \zeta(\pm a) = B; \qquad \zeta(0) = \sqrt{-a^2} = ia: \qquad \zeta = \sqrt{z^2 - a^2}; \qquad \text{(10.193a–10.193c)}$$

it can be checked that the choice of the other constant (Equation 10.191b) leads to the correct position of the tip of the plate (Equation 10.193b). The conformal mapping (Equation 10.193c ≡ Equations 10.192c and 10.193a) together with the complex potential in the z-plane (Equation 10.190a) specifies the complex potential in ζ-plane (Equation 10.194a):

$$f = Uz = U\sqrt{\zeta^2 + a^2}; \quad v^*(\zeta) \equiv \frac{df}{d\zeta} = \frac{U\zeta}{\sqrt{\zeta^2 + a^2}}, \quad \lim_{\zeta\to\infty} \frac{df}{d\zeta} = U; \qquad \text{(10.194a–10.194c)}$$

the corresponding complex conjugate velocity (Equation 10.194b) is uniform at infinity (Equation 10.194c) in agreement with Equation 10.191a.

E10.12.3: Velocity Distribution on a Plate Orthogonal to a Wall

The velocity (Equation 10.194b) is (1) horizontal along the wall (Equations 10.195a and 10.195b) or the real-ζ or ξ-axis; (2) vertical along the plate (Equations 10.196a and 10,196b) with opposite signs on the two sides; and (3) horizontal on the straight line extending from the plate into the flow (Equations 10.197a and 10.197b):

$$v_\xi(\xi,0) = U\xi\left|\xi^2 + a^2\right|^{-1/2}, \qquad v_\eta(\xi,0) = 0; \qquad \text{(10.195a and 10.195b)}$$

$$v_\xi(0, 0 \le \eta < a) = 0, \qquad v_\eta(0, 0 \le \eta < a) = -U\eta\left|a^2 - \eta^2\right|^{-1/2} \operatorname{sgn}(\xi);$$
(10.196a and 10.196b)

$$v_\xi(0, \eta > a) = U\eta\left|\eta^2 - a^2\right|^{-1/2}, \qquad v_\eta(0, \eta > a) = 0. \qquad \text{(10.197a and 10.197b)}$$

In Equation 10.192b, the sign function was used defined [Equation 10.198a (Equation 10.198c)] to be –1 (+1) for negative (positive) argument and to vanish (Equation 10.198b) for zero argument:

$$\operatorname{sgn}(\xi) = \begin{cases} -1 & \text{if} \quad \xi < 0 & \text{(10.198a)} \\ 0 & \text{if} \quad \xi = 0 & \text{(10.198b)} \\ +1 & \text{if} \quad \xi > 0. & \text{(10.198c)} \end{cases}$$

It implies that (1 and 2) the velocity is upward (downward) in the upstream (Equation 10.199a) [downstream (Equation 10.199c)] side of the plate; (3 and 4) it is zero (infinite) at the base (Equation 10.199b) [tip (Equation 10.199d)] of the plate:

$$v_\xi(0, 0 \le \eta < a) = -U\eta\left|a^2 - \eta^2\right|^{-1/2} \times \begin{cases} -1 & \text{if} \quad \xi < 0 & \text{(10.199a)} \\ 0 & \text{if} \quad \xi = 0 = \eta & \text{(10.199b)} \\ +1 & \text{if} \quad \xi > 0 & \text{(10.199c)} \\ \infty & \text{if} \quad \xi > 0, \eta = \text{a}. & \text{(10.199d)} \end{cases}$$

The velocity (Equations 10.194a and 10.194b) specifies through the Bernoulli law (Equation 2.12b ≡ Equation I.14.27c) the pressure distribution on the plate (Equation 10.200) where p_0 is the stagnation pressure:

$$p_0 - p(\eta) = \frac{\rho}{2}\left[v_\xi(0, 0 < \eta < a)\right]^2 = \frac{\rho}{2}U^2\frac{\eta^2}{a^2 - \eta^2}; \qquad \text{(10.200)}$$

$$0 = C_p(0) \le C_p(\eta) \equiv 2\frac{p_0 - p(\eta)}{\rho U^2} = \frac{\eta^2}{a^2 - \eta^2} = \frac{1}{a^2/\eta^2 - 1} \le C_p(a) = \infty; \qquad \text{(10.201)}$$

the corresponding pressure coefficient (Equation 10.201) varies from zero at the base of the plate to infinity at the tip where the velocity is singular.

E10.12.4: Pressure and Force on a Plate Perpendicular to a Wall

The pressure distribution (Equation 10.200) leads to a horizontal force on the upstream side of the plate orthogonal to the wall specified by Equation 10.202b:

$$s = \frac{\eta}{a}: \quad F_x = \int_0^a p(\eta)\,d\eta = \int_0^a \left(p_0 - \frac{\rho}{2}U^2 \frac{\eta^2}{a^2 - \eta^2} \right) d\eta$$
$$= p_0 a - \frac{\rho}{2}U^2 \int_0^a \left(\frac{a^2}{a^2 - \eta^2} - 1 \right) d\eta$$
$$= p_0 a + \frac{\rho}{2}U^2 a \left(1 - \int_0^1 \frac{ds}{1 - s^2} \right) ds \qquad \text{(10.202a and 10.202b)}$$
$$= p_0 a + \frac{\rho}{2}U^2 a (1 - \operatorname{arc\,tanh} 1),$$

where the change of variable (Equation 10.202a) was used to perform the integration (Equation 7.118b). Thus, the horizontal force on the upstream side of the plate is given by Equation 10.203a:

$$F_x = p_0 a + \frac{\rho}{2}U^2 a (1 - \operatorname{arc\,tanh} 1), \qquad (10.203)$$

$$\bar{C}_p = 2\frac{F_x - p_0 a}{\rho U^2 a} = 1 - \operatorname{arc\,tanh} 1 = 0.11863 \qquad (10.204)$$

corresponding to a total pressure coefficient (Equation 10.204); since the force on the downstream side of the plate is equal and opposite, there is no net drag. There would be drag in the case of a free jet (Equation I.38.64b) due to the presence (absence) of flow pressure on the upstream (downstream) side of the plate (Subsection I.38.4).

E10.12.5: Suction Force at the Tip of Lamina Orthogonal to a Stream

Both Equations 10.196b and 10.197a show that the velocity (Equation 10.194b) has (Equations 10.205a and 10.205b) an inverse square root singularity (Equation 8.211f ≡ Equation I.14.83c) at the tip of the plate:

$$v^*(\zeta) = \frac{C}{\sqrt{\zeta - ia}}, \quad C \equiv U\sqrt{\frac{ia}{2}}: \quad F_s^* = -\rho\pi C^2 = \frac{i}{2}a\rho\pi U^2 \qquad \text{(10.205a–10.205c)}$$

corresponding (Equation I.34.15a ≡ Equation 8.300a) to a suction force (Equation I.34.16b ≡ Equation 8.300c) that is (Equation 10.205c) vertical downward, to turn the flow around the edge. By the reflection principle (Subsection I.31.2.2), the present solution applies both to the plate to a wall (Figure 10.5b) and to a bent lamina symmetrically placed in a free stream (Figure 10.5c). Thus, *a plate of length a at an angle α with a wall (Figure 10.5b) [a bent lamina of length 2a and angle 2α (Figure 10.5c)] in a free stream of velocity U has complex conjugate velocity (Equation 10.191c) in terms of the variable (Equations 10.189c and 10.191b) in the conformal z-plane (Figure 10.5a). The case (Equation 10.192a) of a* *plate (straight lamina) perpendicular to the wall and free stream (to the free stream) leads to the complex potential (Equation 10.194a) and conjugate velocity (Equation 10.194b) in the physical ζ-plane (Equation 10.193c) of the plate. The complex conjugate velocity (Equation 10.194b) leads to the velocities along the wall (Equations 10.195a and 10.195b), along the plate (lamina) (Equations 10.196a and 10.196b ≡ Equations 10.199a through 10.199d), and on the straight line extending from it into the flow (Equations 10.197a and 10.197b). The pressure distribution on the plate (Equation 10.200) specifies the pressure coefficient (Equation 10.201) that vanishes at the stagnation point at the base of the plate (center of the lamina). The inverse square root singularity (Equations 10.205a and 10.205b) at the tip (tips) of the plate (lamina) leads (lead) to the suction force (Equation 10.205c), which turns the flow around the edge of the plate. The overall pressure coefficient (Equation 10.204) corresponds to a horizontal force (Equation 10.203) that is equal and opposite on both sides of the plate so that there is no drag.* There is a clear distinction between the potential flow filling all space (Example E10.12) [leaving a cavity (Section I.38.6)] behind an orthogonal/inclined plate on a wall [Figure 10.5b (Figure I.38.7a)] or around a bent lamina [Figure 10.5c (Figure I.38.7b)]; both can be solved using a Schwartz–Christoffel conformal mapping, but the former (latter) does not (does) involve the hodograph method (Chapter I.38).

EXAMPLE 10.13: ELECTROSTATIC FIELD IN TRIPLE PLATE CONDENSER

Consider the electrostatic potential of a triple plate condenser consisting of a semi-infinite plate at equal or unequal distance from two infinite plates, with a potential difference between them.

E10.13.1: Mapping into a Channel with a Thin Plate

The potential flow past a thick plate in a channel (Subsection 8.5.2) in the case of a thin semi-infinite plate $b = d$ in Equation 8.180b reduces to a uniform stream $df/d\zeta = U$ because it is obtained by reflection (Figure 8.12) of two parallel-sided ducts. Thus, the Schwartz–Christoffel transformation is applied directly next, without using the Schwartz reflection principle. Considering a thin semi-infinite plate at unequal distance from two infinite plates (Figure 10.6b), there are three critical points (Equation 10.206a)

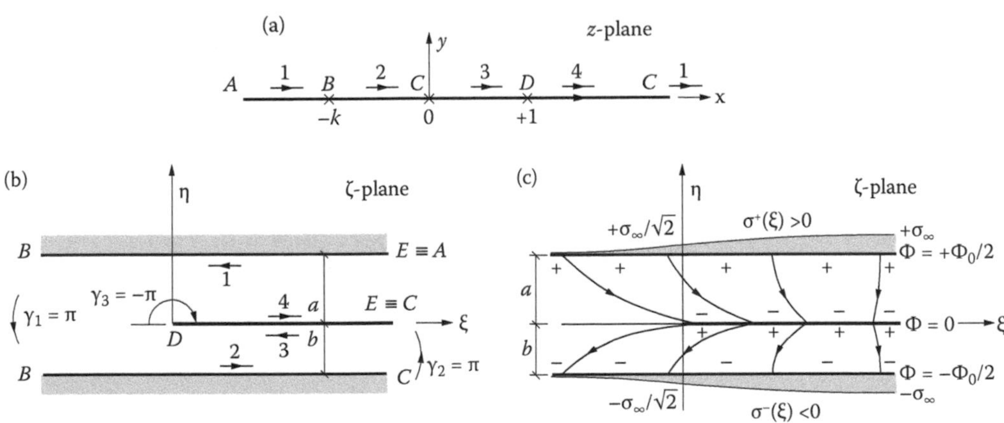

FIGURE 10.6
Schwartz–Christoffel interior polygonal mapping from the real axis with three critical points (a) onto the triple plate condenser (b) specifies the electric charge distribution in the two infinite plates and in the middle semi-infinite plate, in particular, when the latter is at equal distance from the former (c). The edge of the middle plate is a singularity separating the middle plate (its extension) on the positive (negative) real axis that is an equipotential (field line).

with angles (Equation 10.206b) leading to the Schwartz–Christoffel interior polygonal transformation (Equation I.33.25 ≡ Equation 8.145b ≡ Equation 10.206c):

$$x_{1-3} = \{-k, 0, +1\}, \quad \gamma_{1-3} = \{\pi, \pi, -\pi\}: \qquad \frac{d\zeta}{dz} = A\prod_{n=1}^{3}(z - x_n)^{-\gamma_n/\pi} = \frac{A}{z}\frac{z-1}{z+k}. \tag{10.206a–10.206c}$$

There are three parameters in the mapping (Equation 10.206c), namely, (a, k) and an integration constant B in Equation 10.207b; these are determined (Figure 10.6a and b) by the distance between the plates (a, b) and the position of the edge of the middle plate. The critical point $z_1 = -k$ was left variable to have three parameters to meet three constraints. The rational function (Equation 10.206c ≡ Equation 10.207a) is integrated by (Subsection I.31.8) decomposition into partial fractions (Equation 10.207b):

$$\frac{d\zeta}{dz} = A\left(\frac{1+1/k}{z+k} - \frac{1}{kz}\right), \qquad \zeta = A\left[\left(1+\frac{1}{k}\right)\log(z+k) - \frac{1}{k}\log z\right] + B. \tag{10.207a and 10.207b}$$

The critical points $z_1 = -k$ $(z_2 = 0)$ are mapped by Equation 10.207b to infinity $\zeta = \infty$, that is, correspond to the points B (C, E) in Figure 10.6a and b that lie at infinity on the walls (middle plate); the critical point $z_2 = 1$ corresponds to the edge of the semi-infinite plate or point D in Figure 10.6a at the origin, leading to a relation between the arbitrary constants (Equation 10.208a):

$$0 = \zeta(1) = A\left(1+\frac{1}{k}\right)\log(1+k) + B, \tag{10.208a}$$

$$\zeta(z) = A\left(1+\frac{1}{k}\right)\log\left(\frac{z^{k/(1+k)} + z^{-1/(1+k)}k}{1+k}\right); \tag{10.208b}$$

substitution of Equation 10.208a in Equation 10.207b leads to

$$\begin{aligned}
\zeta(z) &= A\left[\left(1+\frac{1}{k}\right)\left[\log(z+k) - \log(1+k)\right] - \frac{1}{k}\log z\right] \\
&= A\frac{k+1}{k}\left[\log\left(\frac{z+k}{1+k}\right) - \frac{1}{k+1}\log z\right] \\
&= A\left(1+\frac{1}{k}\right)\log\left(z^{-1/(1+k)}\frac{z+k}{1+k}\right) \\
&= A\left(1+\frac{1}{k}\right)\log\left(\frac{z^{1-1/(1+k)} + z^{-1/(1+k)}k}{1+k}\right),
\end{aligned} \tag{10.209}$$

which proves Equation 10.208b ≡ Equation 8.209.

E10.13.2: Middle Plate at Unequal/Equal Distance

The position of the two infinite plates (Equation 10.210a) determines the two remaining constants in Equation 10.210b:

$$a,-b = \operatorname{Im}\left[\zeta(-k\mp 0)\right] = \lim_{z\to k\mp 0}\operatorname{Im}[\zeta(z)] = \lim_{\varepsilon\to 0}\operatorname{Im}[\zeta(-k\mp\varepsilon)]. \qquad \text{(10.210a and 10.210b)}$$

The limit in Equation 10.210b involves in Equation 10.208b the expression:

$$\begin{aligned} z\to -k\mp\varepsilon:\quad & z^{k/(1+k)} + z^{-1/(1+k)}k \to (-k\mp\varepsilon)^{k/(1+k)} + (-k\mp\varepsilon)^{-1/(1+k)}k \\ &= (-k)^{k/(1+k)}\left[\left(1\pm\frac{\varepsilon}{k}\right)^{k/(1+k)} - \left(1\pm\frac{\varepsilon}{k}\right)^{-1/(1+k)}\right] \\ &= k^{k/(1+k)}e^{i\pi k/(1+k)}\left[\pm\frac{\varepsilon}{k}\left(\frac{k}{1+k}+\frac{1}{1+k}\right)+O\left(\frac{\varepsilon^2}{k^2}\right)\right] \\ &= k^{k/(1+k)}e^{i\pi k/(1+k)}\left[\pm\frac{\varepsilon}{k}+O\left(\frac{\varepsilon^2}{k^2}\right)\right]. \end{aligned} \qquad \text{(10.210c and 10.210d)}$$

Substitution of Equation 10.210c in Equation 10.210b and 10.208b leads to:

$$\begin{aligned} a,-b &= \operatorname{Im}\left\{A\left(1+\frac{1}{k}\right)\lim_{\varepsilon\to 0}\log\left|\frac{k^{1/(1+k)}}{1+k}e^{i\pi k/(1+k)}\left(\pm\frac{\varepsilon}{k}\right)\right|\right\} \\ &= A\left(1+\frac{1}{k}\right)\left\{\frac{\pi k}{1+k},\frac{\pi k}{1+k}-\pi\right\} = A\left\{1,+\frac{1}{k}\right\}. \end{aligned}$$

(10.210e and 10.210f)

Solving Equations 10.210e and 10.210f ≡ Equations 10.211a and 10.211b specifies the parameters (k, A) in terms of the distance a (b) of the upper (lower) infinite plate from the middle semi-infinite plate (Equations 10.211a and 10.211d):

$$a = A\pi,\quad b=\frac{A\pi}{k}:\quad A=\frac{a}{\pi},\quad k=\frac{A\pi}{b}=\frac{a}{b}. \qquad \text{(10.211a–10.211d)}$$

The appear in the Schwartz-Christoffell mapping (Equation 10.208b ≡ Equation 10.211f):

$$\zeta(z) = \frac{a+b}{\pi}\log\left[\frac{bz^{a/(a+b)}+az^{-b/(a+b)}}{b+a}\right]. \qquad \text{(10.211f)}$$

The simplest case of Equations 10.210a and 10.210b ≡ Equations 10.211c and 10.211d is when the semi-infinite plate is at equal distance from the infinite plates (Equation 10.212b), corresponding to Equation 10.212a:

$$k=1:\quad a=b=\pi A,\quad B=-2A\log 2,\quad \zeta(z)=\frac{2a}{\pi}\log\left(\frac{\sqrt{z}+1/\sqrt{z}}{2}\right),$$

(10.212a–10.212d)

The Equation 10.212d can also be obtained substituting Equation 10.213c in Equation 10.211f, and has the alternative forms:

$$a=b: \quad \zeta(z)=\frac{2a}{\pi}\log\left(\frac{(1+z)/\sqrt{z}}{2}\right)=\frac{a}{\pi}\log\left[2\log(1+z)-\log z-2\log 2\right]. \tag{10.213a–10.213c}$$

This completes the calculation of the *Schwartz-Christoffel transformation from the real line (Figure 10.6a) to a triple plate condenser (Figure 10.6d) with the middle semi-infinite plate at unequal (Equation 10.211f) [equal (Equation 10.212d ≡ 10.213a–c)] distance from the parallel infinite plates.*

E10.13.3: Potential on the Middle Plate of the Condenser

The complex potential is taken as a monopole or logarithmic singularity at the edge of the semi-infinite plate (Equation 10.214a):

$$f=i\frac{\Phi_0}{2\pi}\log z=-\frac{\Phi_0}{\pi}arc\cos\left[\exp\left(\frac{\pi\zeta}{2a}\right)\right], \tag{10.214a and 10.214b}$$

leading to Equation 10.214b by elimination of z between Equations 10.212d and 10.214a, that is, Equation 10.214b ≡ Equation 10.215:

$$\exp\left(\frac{\pi\zeta}{2a}\right)=\frac{\sqrt{z}+1/\sqrt{z}}{2}=\frac{e^{-i\pi f/\Phi_0}+e^{i\pi f/\Phi_0}}{2}=\cos\left(\frac{\pi f}{\Phi_0}\right). \tag{10.215}$$

The choice of sign in Equation 10.215 ≡ Equation 10.214b determines whether the electric field is upward or downward between the plates; for $\Phi_0>0$, the minus sign will imply a downward electric field, as shown in the sequel. Using Equation 10.215 in the form

$$\zeta=\xi: \quad \exp\left(\frac{\pi\xi}{2a}\right)=\cos\left(-\frac{\pi f}{\Phi_0}\right)=\cosh\left(-i\frac{\pi f}{\Phi_0}\right) \tag{10.216a and 10.216b}$$

specifies the complex potential (Equation 10.214b ≡ Equations 10.217a and 10.217b) on the positive real-ζ or ξ-axis:

$$\begin{aligned}\Phi+i\Psi=f(\xi>0,\eta=0)&=-\frac{\Phi_0}{\pi}\arg\cos\left[\exp\left(\frac{\pi|\xi|}{2a}\right)\right]\\&=-i\frac{\Phi_0}{\pi}\arg\cosh\left[\exp\left(\frac{\pi|\xi|}{2a}\right)\right];\end{aligned} \tag{10.217a and 10.217b}$$

using Equation 7.187b shows that it is imaginary, implying that the middle plate is the equipotential (Equation 10.218b) at zero potential (Equation 10.218a):

$$\Phi(\xi > 0, \eta = 0) = 0, \quad \Psi(\xi > 0, \eta = 0) = -\frac{\Phi_0}{\pi} \arg\cosh\left[\exp\left(\frac{\pi|\xi|}{2a}\right)\right]. \tag{10.218a and 10.218b}$$

The complex potential (Equation 10.214b) on the negative real axis (Equation 10.219a):

$$f(\xi < 0, \eta = 0) = -\frac{\Phi_0}{\pi} \arccos\left[\exp\left(-\frac{\pi|\xi|}{2a}\right)\right] = \Phi(\xi < 0, \eta = 0), \quad \Psi(\xi < 0, \eta = 0) = 0 \tag{10.219a and 10.219b}$$

is real showing that it is a field line (Equation 10.219b). Thus, the middle semi-infinite plate is the zero equipotential and is extended across the singularity at the edge by the zero field line. The difference between Equations 10.217a, 10.217b, 10.218a, and 10.218b (Equations 10.219a and 10.219b) is that the exponential is larger (smaller) than unity and thus arg cos is imaginary (real), leading to the zero equipotential (field line).

E10.13.4: Potential Difference between the Plates

The complex potential (Equation 10.214b) is given on the infinite plates by

$$\zeta = \xi \pm ia: \qquad f = -\frac{\Phi_0}{\pi} \arccos\left[\exp\left(\frac{\pi\xi}{2a} \pm i\frac{\pi}{2}\right)\right]. \tag{10.220a and 10.220b}$$

Using the identity Equation 10.220b ≡ Equation 10.221:

$$\begin{aligned} \exp\left(\frac{\pi\xi}{2a}\right) &= e^{\mp i\pi/2} \exp\left(\frac{\pi\xi}{2a} \pm i\frac{\pi}{2}\right) = \mp i \exp\left(\frac{\pi\zeta}{2a}\right) = \mp i \cos\left(-\frac{\pi f}{\Phi_0}\right) \\ &= -i\sin\left(-\frac{\pi f}{\Phi_0} \pm \frac{\pi}{2}\right) = i\sin\left(\frac{\pi f}{\Phi_0} \mp \frac{\pi}{2}\right) = \sinh\left(i\frac{\pi f}{\Phi_0} \mp i\frac{\pi}{2}\right) \end{aligned} \tag{10.221}$$

leads (Equation 5.5b) by inversion to

$$i\frac{\pi f}{\Phi_0} \mp i\frac{\pi}{2} = \arg\sinh\left[\exp\left(\frac{\pi\xi}{2a}\right)\right]. \tag{10.222}$$

Inverting Equation 10.222, it follows that the complex potential on the infinite plates is given by

$$f(\xi \pm ia) = \pm\frac{\Phi_0}{2} - i\frac{\Phi_0}{\pi} \arg\sinh\left[\exp\left(\frac{\pi\xi}{2a}\right)\right]. \tag{10.223}$$

The real part of the complex potential (Equation 10.223) is the first term on the r.h.s., showing the following: (1) the scalar potential is constant and equal to $+\Phi_0/2$ ($-\Phi_0/2$) on the upper (lower) infinite plate, and the potential difference between them is Φ_0; (2) for $\Phi_0 > 0$, this leads to a positive (negative) upper (lower) infinite plate; (3) this implies a downward electric field between the plates (Figure 10.6a), that is, from the higher to the lower potential; and (4) it follows that the induced electric charges are positive (negative) on the upper (lower) infinite plate where the field lines start (end), as will be confirmed next.

E10.13.5: Electric Field and Charge Distributions

The complex conjugate electric field corresponding to the electrostatic potential (Equation 10.214b) is

$$\begin{aligned} E^*(\zeta) &= -\frac{df}{d\zeta} = \frac{\Phi_0}{2a}\exp\left(\frac{\pi\zeta}{2a}\right)\left[1-\exp\left(\frac{\pi\zeta}{a}\right)\right]^{-1/2} \\ &= \frac{\Phi_0}{2a}\exp\left(\frac{\pi\zeta}{4a}\right)\left[\exp\left(-\frac{\pi\zeta}{2a}\right)-\exp\left(\frac{\pi\zeta}{2a}\right)\right]^{-1/2} \\ &= \frac{\Phi_0}{2a}\exp\left(\frac{\pi\zeta}{4a}\right)\left[-2\sinh\left(\frac{\pi\zeta}{2a}\right)\right]^{-1/2} \\ &= \frac{i}{\sqrt{2}}\frac{\Phi_0}{2a}\exp\left(\frac{\pi\zeta}{4a}\right)\left[\operatorname{csch}\left(\frac{\pi\zeta}{2a}\right)\right]^{-1/2}. \end{aligned} \tag{10.224}$$

The complex conjugate electric field (Equation 10.224) at the middle plate (Equation 10.225)

$$E^*(\xi > 0, \eta = 0) = \frac{i}{\sqrt{2}}\frac{\Phi_0}{2a}\exp\left(\frac{\pi|\xi|}{4a}\right)\left[\operatorname{csch}\left(\frac{\pi|\xi|}{2a}\right)\right]^{-1/2} = E_\xi - iE_\eta \tag{10.225}$$

shows the following: (1) the electric field is normal to the middle plate (Equation 10.226a), in agreement with the latter being an equipotential (Equation 10.218a):

$$E_\xi(\xi > 0, \eta = 0) = 0, \tag{10.226a}$$

$$E_\eta(\xi > 0, \eta = 0) = -\frac{\Phi_0}{2a\sqrt{2}}\exp\left(\frac{\pi|\xi|}{4a}\right)\left[\operatorname{csch}\left(\frac{\pi|\xi|}{2a}\right)\right]^{-1/2} = \varepsilon\,\operatorname{sgn}(\eta)\sigma_0(\xi); \tag{10.226b}$$

(2) the electric field normal to the middle plate (Equation 10.226b) specifies the induced surface electric charge distribution in a medium of dielectric permittivity ε; (3) the induced electric charges are (Equations 10.198a through 10.198c) negative (positive) on the upper (lower) side of the middle plate for $\Phi_0 > 0$ in Figure 10.6c; (4) this agrees with the electric field pointing downward, that is, from the upper infinite plate at higher positive potential $\Phi_0/2 > 0$ to the lower infinite plate at lower negative potential $-\Phi_0/2 < 0$ in Equation 10.223; and (5) the signs are reversed for $\Phi_0 < 0$. The electric charge distribution has an integrable singularity at the edge of the middle plate (Equation 10.227a)

corresponding to the inverse square root singularity (Equations 10.227b and 10.227c) of the electric field (Equation 10.224):

$$\zeta \to 0: \quad 1-\exp\left(\frac{\pi\zeta}{2a}\right) \sim -\frac{\pi\zeta}{2a}, \quad E^* \sim \frac{\Phi_0}{2a}\left(-\frac{\pi\zeta}{2a}\right)^{-1/2} = \mp i\frac{\Phi_0}{\sqrt{2\pi a\zeta}}. \tag{10.227a–10.227c}$$

The integrable singularity of the electric charge distribution on the middle semi-infinite plate is equivalent to an added length (Subsection I.36.5.4). Across the singularity at the edge of the middle plate, the electric field (Equation 10.224) is horizontal:

$$E^*(\xi < 0, \eta = 0) = \frac{\Phi_0}{2a}\exp\left(-\frac{\pi|\xi|}{4a}\right)\left|1-\exp\left(-\frac{\pi|\xi|}{2a}\right)\right|^{-1/2} = E_\xi(\xi < 0, \eta = 0), \tag{10.228a}$$

$$E_\eta(x < 0, \eta = 0) = 0, \tag{10.228b}$$

in agreement with the negative real axis being a field line (Equation 10.219b). The difference between the middle plate along the positive real axis (Equations 10.226a and 10.226b) [its prolongation to the negative real axis (Equations 10.228a and 10.228b)] is that the exponential is greater (smaller) than unity changing the complex conjugate electric field from imaginary to real, that is, from vertical to horizontal.

E10.13.6: Middle Semi-Infinite and Upper/Lower Infinite Plates

The electric field (Equation 10.224) on the two infinite plates (Equation 10.229a) involves Equation 10.227b:

$$\begin{aligned}\zeta = \xi \pm ia: \quad -2\sinh\left(\frac{\pi\zeta}{2a}\right) &= -2\sinh\left(\frac{\pi\xi}{2a} \pm i\frac{\pi}{2}\right)\\ &= 2i\sin\left(i\frac{\pi\xi}{2a} \mp \frac{\pi}{2}\right) = \mp 2i\cos\left(i\frac{\pi\xi}{2a}\right)\\ &= 2e^{\mp i\pi/2}\cosh\left(\frac{\pi\xi}{2a}\right),\end{aligned} \tag{10.229a and 10.229b}$$

where Equation 5.6b was used. Substitution of Equations 10.229a and 10.229b in Equation 10.224 specifies the electric field on the two infinite plates:

$$\begin{aligned}E^*(\xi \pm ia) &= \frac{\Phi_0}{2a}\exp\left(\frac{\pi\xi}{4a}\right)e^{\pm i\pi/4}\left[2e^{\mp i\pi/2}\cosh\left(\frac{\pi\xi}{2a}\right)\right]^{-1/2}\\ &= \frac{\Phi_0}{2a}\frac{e^{\pm i\pi/2}}{\sqrt{2}}\exp\left(\frac{\pi\xi}{4a}\right)\left[\operatorname{sech}\left(\frac{\pi\xi}{2a}\right)\right]^{1/2}\\ &= \pm\frac{i}{\sqrt{2}}\frac{\Phi_0}{2a}\exp\left(\frac{\pi\xi}{4a}\right)\left[\operatorname{sech}\left(\frac{\pi\xi}{2a}\right)\right]^{1/2} = E_\xi - iE_\eta^\pm.\end{aligned} \tag{10.230}$$

The result (Equation 10.230) can be checked directly from Equation 10.224.

$$
\begin{aligned}
E^*(\xi \pm ia) &= \frac{\Phi_0}{2a}\exp\left(\frac{\pi\xi}{2a} \pm i\frac{\pi}{2}\right)\left[1-\exp\left(\frac{\pi\xi}{a} \pm i\pi\right)\right]^{-1/2} \\
&= \pm i\frac{\Phi_0}{2a}\exp\left(\frac{\pi\xi}{2a}\right)\left[1+\exp\left(\frac{\pi\xi}{a}\right)\right]^{-1/2} \\
&= \pm i\frac{\Phi_0}{2a}\exp\left(\frac{\pi\xi}{4a}\right)\left[\exp\left(-\frac{\pi\xi}{2a}\right)+\exp\left(\frac{\pi\xi}{2a}\right)\right]^{-1/2} \\
&= \pm i\frac{\Phi_0}{2a}\exp\left(\frac{\pi\xi}{4a}\right)\left[2\cosh\left(\frac{\pi\xi}{2a}\right)\right]^{-1/2} \\
&= \pm\frac{i}{\sqrt{2}}\frac{\Phi_0}{2a}\exp\left(\frac{\pi\xi}{4a}\right)\left[\operatorname{sech}\left(\frac{\pi\xi}{2a}\right)\right]^{-1/2}.
\end{aligned}
\tag{10.231}
$$

Thus, the electric field is normal (Equation 10.232a) to the two infinite plates specifying the induced surface electric charge distributions (Equation 10.231b):

$$
E_\xi(\xi, \pm a) = 0: \qquad \sigma^\pm(\xi) = -\varepsilon E_\eta(\xi, \pm a) = \pm\frac{\varepsilon}{\sqrt{2}}\frac{\Phi_0}{2a}\exp\left(\frac{\pi\xi}{4a}\right)\left[\operatorname{sech}\left(\frac{\pi\xi}{2a}\right)\right]^{1/2}.
$$

(10.232a and 10.232b)

The inner normal is downward at the upper infinite plate, and hence the density of electric charges is $\sigma^+ = -\varepsilon E_\eta^+$ in Equation 10.232b; the electric charge is opposite at the lower semi-infinite plate $\sigma^- = -\sigma^+ = \varepsilon E_\eta^+ = -\varepsilon E_\eta^-$ corresponding to the same electric field with opposite normal. The electric charge distribution is (Figure 10.6c) positive (negative) on upper (lower) infinite plate $\sigma^+ > 0 > \sigma^-$ in Equation 10.231b because it is at higher $\Phi_0/2$ (lower $-\Phi_0/2$) potential (Equation 10.223) for $\Phi_0 > 0$.

E10.13.7: Near-Field and Far-Field Electric Charges

Inside the condenser between the two infinite plates (Equation 10.233a), Equation 10.233b holds:

$$
|\xi| \le a: \quad \left|\exp\left(\frac{\pi\zeta}{2a}\right)\right| = \exp\left(\frac{\pi\xi}{2a}\right)\left|\exp\left(\frac{i\pi\eta}{2a}\right)\right| = \exp\left(\frac{\pi\xi}{2a}\right) = \exp\left[\frac{\pi|\xi|}{2a}\operatorname{sgn}(\xi)\right].
$$

(10.233a and 10.233b)

Substituting Equation 10.233b in Equation 10.224, it follows that the far electric field on the upper and lower infinite plate is (1) zero (Equation 10.234a) on the side opposite of the middle plate:

$$
\lim_{\xi\to-\infty} E^*(\xi \pm ia) = 0; \quad \lim_{\xi\to+\infty} E^*(\xi \pm ia) = -i\frac{\Phi_0}{2a} = -iE_\eta(\xi, \pm a) = -i\sigma_\infty;
$$

(10.234a and 10.234b)

and (2) constant and vertical upward on the side of the middle plate (Equation 10.234b). Since the normal to the upper (lower) infinite plate is downward (upward) in Equation 10.237a, the induced electric charge has the opposite (same) sign (Equation 10.237b) as the electric field:

$$\vec{n}_{\pm} = (0, \mp 1): \qquad \sigma^{\pm}(\xi) = \varepsilon \vec{n}_{\pm} . \vec{E}(\xi, \pm a) = \mp \varepsilon E_{\eta}(\xi; \pm a). \tag{10.235 and 10.236}$$

Using the far-field limits (Equations 10.234a and 10.234b), it follows that the surface electric charge on the plates (Equation 10.237b) taken in modulus (1) vanishes far on the opposite side of the middle plate; and (2) takes a constant value σ_{∞} far on the side of the middle plate (10.238a):

$$0 = \sigma^{\pm}(-\infty) \le \left|\sigma^{\pm}(\xi)\right| \le \frac{\Phi_0 \varepsilon}{2a} = \sigma^{\pm}(+\infty) \equiv \sigma_{\infty}; \quad \sigma^{\pm}(0) = \pm \frac{\varepsilon \Phi_0}{2a\sqrt{2}} = \pm \frac{\sigma_{\infty}}{\sqrt{2}}; \tag{10.237 and 10.238}$$

the electric charge density at the point on the upper and lower infinite plates (Equation 10.232b) closest to the edge of the middle plate (Equation 10.238b) is $1/\sqrt{2}$ of the far-field value.

E10.13.8: Asymptotic Electric Charges and Two-Plate Condenser

The asymptotic electric charge density (Equation 10.238a) can be obtained very simply noting that in the triple condenser far opposite to the middle plate, the electrostatic potential depends only on the distance between the upper and lower infinite plates (Equation 10.239a); also it satisfies the Laplace equation (Equation 10.239b) whose solution (Equation 10.239c) involves two arbitrary constants:

$$\lim_{\xi \to -\infty} \Phi(\xi, \eta) = \bar{\Phi}(\eta), \qquad \frac{d^2 \bar{\Phi}}{d\eta^2} = 0, \qquad \bar{\Phi}(\eta) = C_0 + C_1 \eta. \tag{10.239a–10.239c}$$

The constants are specified (Equations 10.240c and 10.240d) by the potential at the two infinite plates (Equations 10.240a and 10.240b), leading to Equations 10.240e and 10.240f:

$$\Phi(\pm a) = \pm \frac{\Phi_0}{2}: \qquad C_0 \pm C_1 a = \pm \frac{\Phi_0}{2}, \qquad C_0 = 0, \qquad C_1 = \frac{\Phi_0}{2a}. \tag{10.240a–10.240f}$$

Substituting Equations 10.240e and 10.240f in Equation 10.239c specifies the potential (Equation 10.241a) that meets the boundary conditions (Equations 10.240a and 10.240b) on the infinite plates and vanishes at the middle position between them (Equation 10.241b):

$$\bar{\Phi}(\eta) = \frac{\Phi_0 \eta}{2a}: \qquad \bar{\Phi}(0) = 0; \qquad \bar{E}_{\eta} = -\frac{d\bar{\Phi}}{d\eta} = -\frac{\Phi_0}{2a}, \qquad \bar{\sigma} = \mp \varepsilon \bar{E}_y = \pm \frac{\Phi_0 \varepsilon}{2a} = \pm \sigma_{\infty}; \tag{10.241a–10.241d}$$

the corresponding electric field (Equation 10.241c) leads (Equation 10.237b) to the surface electric charges (Equation 10.241d) in agreement with Equation 10.238a. Thus, *a triple plate condenser, with two infinite plates at distance 2a and potential* $+\Phi_0/2(-\Phi_0/2)$ *for the upper (lower) plate, with a semi-infinite plate at equal distance (Figure 10.6b) and at zero potential, has complex potential (Equation 10.214b); this is the particular case (Equation 10.213) of a semi-infinite plate at unequal distances* $b \neq a$ *from the infinite plates, when the complex potential* $f(\zeta)$ *is given eliminating z between Equation 10.214a and Equation 10.211f. In the case of middle semi-infinite plate at equal distance from the two infinite plates, the electric field is given by Equation 10.224; it specifies the induced electric charge distributions on the semi-infinite (Equation 10.226b) [infinite (Equation 10.232b)] plates.*

EXAMPLE 10.14: RESERVOIR WITH PARALLEL-SIDED EXIT CHANNEL

Consider the potential flow into a parallel-sided duct that is a straight exit from a reservoir with an oblique or orthogonal wall.

E10.14.1: Straight Exit Duct from a Reservoir with Oblique Walls

The present problem is distinct from (1) the flow out of convergent (parallel) sided channel (Section I.36.5), for which there is no reservoir and the duct opens in free space; and (2) the jet out of a slit (re-entrant tube) in a reservoir [Section I.38.3 (Section I.38.4)] that has two free surfaces and is solved by the hodograph method. In the present problem, (1) the duct is oblique or orthogonal to the wall of the reservoir and (2) the fluid fills all space in the reservoir and discharges into the duct that has no termination. The mapping to the upper complex z-plane (Figure 10.7a) into the reservoir with exit duct (Figure 10.7b) is performed by a Schwartz–Christoffel interior polygonal transforma-

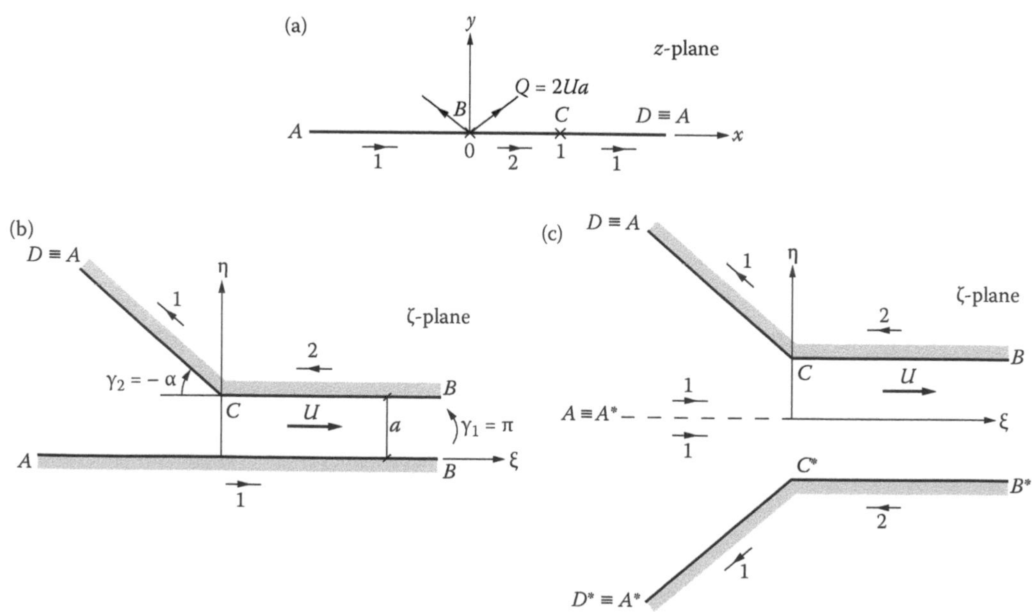

FIGURE 10.7
Schwartz–Christoffel interior polygonal mapping maps the real axis with two critical points (a) into a reservoir with one oblique wall and a straight exit channel (b). The reflection on the real axis (Figure I.31.4) leads to a converging channel joined to a parallel-sided channel (c). This is distinct from a convergent (parallel) sided duct discharging in free space [Figure I.36.11b (Figure I.36.12a)].

tion (Equation I.33.25 ≡ Equation 8.145b ≡ Equation 10.242c) with two critical points Equation 10.242a with angles Equation 10.242b:

$$x_{1-2} = \{0,1\}, \quad \gamma_{1-2} = \{\pi, -\alpha\}: \qquad \frac{d\zeta}{dz} = A \prod_{n=1}^{2} (z - z_n)^{-\gamma_n/\pi} = \frac{A}{z}(z-1)^{\alpha/\pi}. \tag{10.242a–10.242c}$$

The two parameters A in the conformal mapping (Equation 10.242c) plus the integration constant B are sufficient to specify the location of the corner and infinite wall, and thus the position of the critical points (Equation 10.242a) can be chosen at will, so simple values have been taken. The flow with velocity U at the far end of the exit duct of width a corresponds to a sink (Equation 10.243a) whose half flow rate $-Ua$ comes from the far downstream side of the channel, located at the point B that corresponds to $z = 0$ in Equation 10.243b:

$$Q = -2Ua: \qquad f(z) = \frac{Q}{2\pi}\log z = -\frac{Ua}{\pi}\log z; \qquad \frac{df}{dz} = -\frac{Ua}{\pi z}; \tag{10.243a–10.243c}$$

the complex conjugate velocity in the z-plane (Equation 10.243c) leads by Equation 10.242c to the complex conjugate velocity (Equation 10.244) in the ζ-plane:

$$\frac{df}{d\zeta} = \frac{df/dz}{d\zeta/dz} = -\frac{Ua}{\pi A}(z-1)^{-\alpha/\pi}; \tag{10.244}$$

the constant A is determined (Equation 10.245b) by the velocity at the far end of the exit channel (Equation 10.245a):

$$U = \lim_{z\to 0}\frac{df}{d\zeta} = -\frac{Ua}{\pi A}(-)^{-\alpha/\pi}, \qquad A = -\frac{a}{\pi}(-)^{\alpha/\pi}; \tag{10.245a and 10.245b}$$

this leads in Equation 10.242c (Equation 10.244) to the Schwartz–Christoffel transformation (Equation 10.246a) [complex conjugate velocity (Equation 10.246b) in the physical ζ-plane]:

$$\frac{d\zeta}{dz} = -\frac{a}{\pi z}(1-z)^{\alpha/\pi}, \qquad v^* = \frac{df}{d\zeta} = U(1-z)^{-\alpha/\pi}. \tag{10.246a and 10.246b}$$

The velocity (Equation 10.246b) is singular at $z = 1$ that corresponds to the concave corner C between the reservoir and the exit duct, and is $v^* = U$ at $z = 0$ that corresponds to the far downstream end of the duct.

E10.14.2: Transition from a Convergent to a Parallel-Sided Duct

In the case (Equation 10.247a) of a reservoir with orthogonal walls (Figure 10.8a), the Schwartz–Christoffel interior polygonal transformation (Equation 10.246a) simplifies to Equation 10.247b:

$$\alpha = \frac{\pi}{2}: \qquad \frac{d\zeta}{dz} = -\frac{a}{\pi z}\sqrt{1-z} = -\frac{a}{\pi z}\frac{1-z}{\sqrt{1-z}} = \frac{a/\pi}{\sqrt{1-z}} - \frac{a/\pi}{z\sqrt{1-z}}. \tag{10.247a and 10.247b}$$

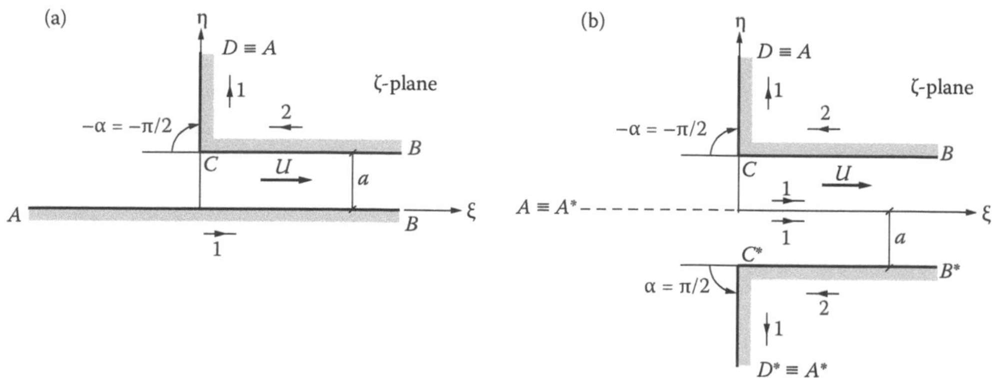

FIGURE 10.8
Particular case of Figure 10.7b (Figure 10.7c) with orthogonal sides is a straight parallel-sided exit duct from a semi-infinite (infinite) reservoir extending to one side (both sides) [a (b)]. In both cases, the fluid occupies all space, in contrast with the jet with free surfaces existing a reservoir through a slit in the wall (Figures I.38.2a and I.38.3a) [a long re-entrant tube (Figures I.38.2b and I.38.3b)].

Using the change of variable (Equation 10.248a), the second term on the r.h.s. of Equation 10.247b is integrated (Equation 10.248b) by an inverse hyperbolic function (Equation 7.126c):

$$z = t^2: \qquad \int \frac{dz}{z\sqrt{1-z}} = 2\int \frac{dt}{t\sqrt{1-t^2}} = -2 \arg \operatorname{sech} t = -2 \arg \operatorname{sech} \sqrt{z}; \tag{10.248a and 10.248b}$$

the integration of the first term on the r.h.s. (Equation 10.247b) is immediate leading to Equation 10.249a:

$$\zeta(z) = -\frac{2a}{\pi}\left[\sqrt{1-z} - \arg \operatorname{sech}\left(\sqrt{z}\right)\right] + B; \qquad ia = \zeta(1) = B; \tag{10.249a and 10.249b}$$

the constant of integration is determined by the position of the corner (Equation 10.249b). Substitution of Equations 10.249b and 10.243b in Equation 10.249a specifies the complex potential $f(\zeta)$ in inverse form $\zeta(f)$, that is,

$$\zeta(f) = ia + \frac{2a}{\pi} \operatorname{arc\,sech}\left[\exp\left(-\frac{\pi f}{2Ua}\right)\right] - \frac{2a}{\pi}\left[1 - \exp\left(-\frac{\pi f}{Ua}\right)\right]^{1/2}; \tag{10.250}$$

the complex conjugate velocity is given (Equations 10.246b and 10.247a) by

$$v^* = \frac{df}{d\zeta} = \frac{U}{\sqrt{1-z}} = U\left[1 - \exp\left(-\frac{\pi f}{Ua}\right)\right]^{-1/2} \tag{10.251a and 10.251b}$$

as a function of z in Equation 10.251a (f in Equation 10.251b) that is related to ζ by Equations 10.249a and 10.249b (Equation 10.250). By the reflection principle, these results also apply (Figure 10.7a and b) to the transition from a symmetric straight convergent duct to a parallel-sided duct. Thus, *the potential flow into a parallel-sided duct from a duct with convergent walls*

(Figure 10.7c) [a reservoir with oblique walls (Figure 10.7b)] making an angle 2α(α) has complex conjugate velocity (Equation 10.246b), where U is the velocity at the far end of the exit duct, and z is the conformal coordinate related to the physical coordinate ζ by Equations 10.249a and 10.249b. In the case in Figure 10.8b (Figure 10.8a) of orthogonal walls (Equation 10.247a), the complex potential is given by Equation 10.250 in inverse form; the complex conjugate velocity is given by Equation 10.251b in terms of the potential, that is, $v^*(\zeta)$ is obtained eliminating f between Equations 10.250 and 10.251b. The case of a convergent (parallel-sided) symmetric channel with an open end was considered in Subsection I.36.5.1 (Subsection I.36.5.2), corresponding in Figure I.36.11b (Figure I.36.12a) to the left (right) hand side of Figure 10.7c with an open end instead of a duct. The reservoir with a straight exit channel orthogonal to the walls with fluid filling all space can be compared with the free jet [Subsection I.38.3 (Subsection I.38.4)] issuing from a slit (a re-entrant tube in the wall) of a [Figures I.38.2a (Figure I.38.2b) and I.38.3a (Figure I.38.3c)] reservoir that differs in having a free surface.

EXAMPLE 10.15: POTENTIAL FLOW IN A Y-JUNCTION OF DUCTS

Consider the potential flow in an angled junction as convergent into a parallel-sided channel including the particular cases of parallel-sided and orthogonal channels.

E10.15.1: Junction between Convergent and Parallel Ducts

The flow in a bent duct (Figure 10.9b) is mapped to the upper half of the complex plane (Figure 10.9a) by the Schwartz–Christoffel interior polygonal transformation (Equation

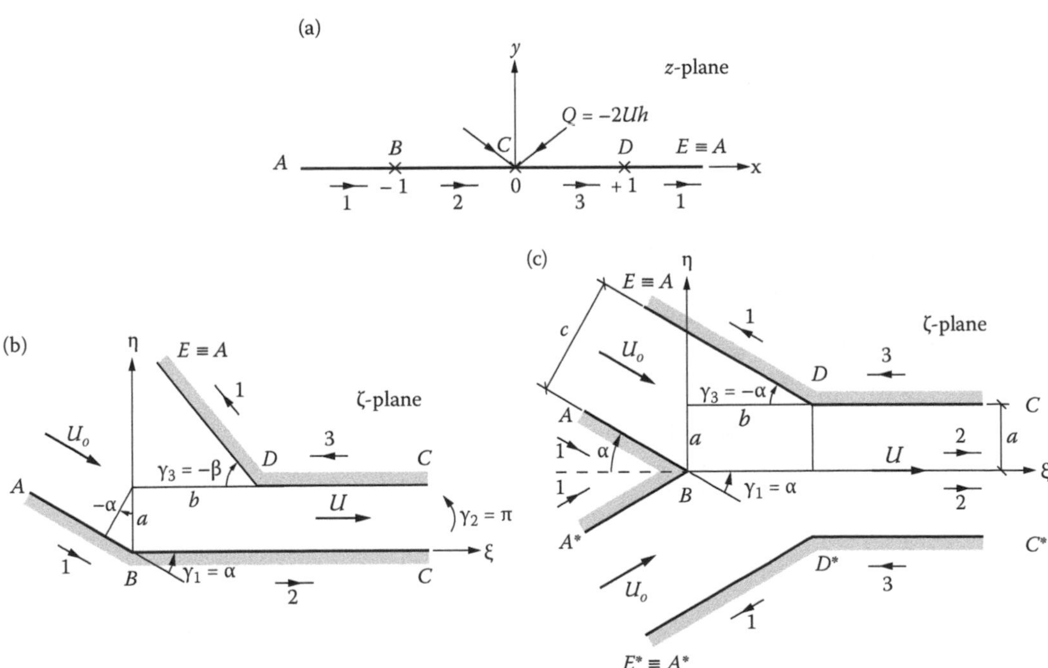

FIGURE 10.9
Schwartz–Christoffel interior polygonal mapping from the real axis (a) into a junction with a parallel-sided downstream channel allows for a convergent (parallel-sided) upstream channel [b (c)]. The reflection principle (Figure I.31.4) leads to a Y-junction of channels (c). Reversing the flow direction corresponds to the junction of a parallel-sided channel with (1) a single divergent or parallel-sided channel (b); and (2) a y-bifurcation into two divergent or parallel-sided jets (c). Since the fluid occupies all space between the walls, this is distinct from the collision, splitting, or merging of jets with free surfaces (Figure I.38.10) or a jet with free surfaces hitting a wall (Figure I.38.11).

I.33.25 ≡ Equation 8.145b ≡ Equation 10.252c) with three critical points Equation 10.252a with angles Equation 10.252b:

$$z_{1-3} = \{-1, 0, +1\},\ \gamma_{1-3} = \{\alpha, \pi, -\beta\}: \quad \frac{d\zeta}{dz} = A \prod_{n=1}^{3} (z - z_n)^{-\gamma_n/\alpha} = \frac{A}{z} \frac{(z-1)^{\beta/\pi}}{(z+1)^{\alpha/\pi}}. \tag{10.252a–10.252c}$$

The two parameters in the conformal mapping, namely, A in Equation 10.252c and the constant of integration B in Equation 10.260a, are sufficient to specify the location of the two corners, so the critical points can be placed at simple fixed positions (Equation 10.252a). If the angles (α, β) of the two sides are distinct ($\beta \neq \alpha$), the upstream channel is convergent (Figure 10.9b) and the velocity U_+ is not uniform, not even far from the bend. If the angles are the same (Equation 10.253a), the conformal mapping (Equation 10.252c) simplifies to Equation 10.253b:

$$\beta = \alpha: \qquad \frac{d\zeta}{dz} = \frac{A}{z}\left(\frac{z-1}{z+1}\right)^{\alpha/\pi}; \tag{10.253a and 10.253b}$$

in this case, the upstream duct is parallel-sided with width c, like the downstream duct with width a, and the velocities far upstream U_0 and far downstream U are related by (Equation 10.254c) the conservation of the volume flux. If the upper corner is located at Equation 10.254a, the width of the upstream duct is Equation 10.254b:

$$\zeta_0 = b + ia: \qquad c = a\cos\alpha + b\sin\alpha, \quad U_0 = \frac{Ua}{c} = \frac{U}{\cos\alpha + (b/a)\sin\alpha} \tag{10.254a–10.254c}$$

and specifies the upstream velocity (Equation 10.254c).

E10.15.2: Potential Flow in a Bent Parallel-Sided Duct

For the convergent (Equations 10.252a through 10.252c) [parallel-sided (Equations 10.254a and 10.254b)] upstream duct, the reflection principle (Subsection I.31.2.1) applies, implying that the flow is the same in a duct bend (Figure 10.9b) or in a Y-junction (Figure 10.9c). In both cases, the downstream duct is parallel sided, and the flow rate U far downstream corresponds to a sink (Equation 10.243a) at the point C far downstream corresponding to the origin in the z-plane (Equations 10.243b and 10.243c). Thus, the complex conjugate velocity is given in the ζ-plane by Equation 10.243c for convergent (Equation 10.255a) upstream duct (Equation 10.255b):

$$\beta \geq \alpha: \quad \frac{df}{d\zeta} = \frac{df/dz}{d\zeta/dz} = -\frac{Ua}{\pi A}\frac{(z+1)^{\alpha/\pi}}{(z-1)^{\beta/\pi}}; \qquad \beta = \alpha: \quad \frac{df}{d\zeta} = -\frac{Ua}{\pi A}\left(\frac{z+1}{z-1}\right)^{\pi/\alpha}; \tag{10.255a–10.255d}$$

it simplifies to Equation 10.255d for the parallel-sided entry channel (Equation 10.255c). The constant A is determined in both cases (Equation 10.256b) by the condition that the velocity is U at the downstream end (Equation 10.256a):

$$U = \lim_{z\to 0}\frac{df}{d\zeta} = -\frac{Ua}{\pi A}(-)^{\beta/\pi}, \qquad A = -\frac{a}{\pi}(-)^{\beta/\pi}; \tag{10.256a and 10.256b}$$

substitution of Equation 10.256b in Equation 10.252c (Equation 10.255b) leads to the Schwartz–Christoffel transformation (Equation 10.257b) [complex conjugate velocity (Equation 10.257c)] in the case (Equation 10.245a) of the convergent upstream duct:

$$\beta \geq \alpha: \qquad \frac{d\zeta}{dz} = -\frac{a}{\pi z}\frac{(1-z)^{\beta/\pi}}{(1+z)^{\alpha/\pi}}, \qquad v^* = \frac{df}{d\zeta} = U\frac{(1+z)^{\alpha/\pi}}{(1-z)^{\beta/\pi}}; \tag{10.257a–10.275c}$$

in the particular case (Equation 10.258a) of a parallel-sided upstream duct, the Schwartz–Christoffel transformation (Equation 10.257b) [complex conjugate velocity (Equation 10.257c)] simplifies to Equation 10.258b (Equation 10.258c):

$$\beta = \alpha: \qquad \frac{d\zeta}{dz} = -\frac{a}{\pi z}\left(\frac{1-z}{1+z}\right)^{\alpha/\pi}, \qquad v^* = \frac{df}{d\zeta} = U\left(\frac{1+z}{1-z}\right)^{\alpha/\pi}. \tag{10.258a–10.258c}$$

In all cases, the velocity is zero (infinite) for $z = -1$ ($z = 1$) corresponding to the concave (convex) corner at point B (D). It has been assumed that the exit or downstream channel is parallel sided. At the junction, the angle β of the upper side of the upstream channel cannot be less than the angle α of the lower side ($\beta \geq \alpha$), leading to two cases: (1) convergent upstream channel (Equations 10.257a through 10.257c) and (2) upstream channel parallel-sided (Equations 10.258a through 10.258c) like the downstream channel (Equations 10.254a through 10.254c).

E10.15.3: Potential Flow in a Right-Angle Corner

In the case (Figure I.10.10a) of a right-angle corner (Equation 10.259a) and both ducts parallel sided (Equation 10.259b), the Schwartz–Christoffel interior polygonal mapping (Equation 10.258b) simplifies to Equation 10.259c:

$$\alpha = \frac{\pi}{2} = \beta: \qquad \frac{d\zeta}{dz} = -\frac{a}{\pi z}\sqrt{\frac{1-z}{1+z}} = -\frac{a}{\pi z}\frac{1-z}{\sqrt{1-z^2}} = \frac{a}{\pi}\left(\frac{1}{\sqrt{1-z^2}} - \frac{1}{z\sqrt{1-z^2}}\right). \tag{10.259a–10.259c}$$

The primitive (Equations 7.122a and 7.126c) of Equation 10.259c is Equation 10.260a:

$$\zeta(z) = \frac{a}{\pi}\left\{\arcsin z - \operatorname{arg\,sech} z\right\} + B; \quad b + ia = \zeta(1) = \frac{a}{2} + B, \quad B = b - \frac{a}{2} + ia; \tag{10.260a–10.260c}$$

the constant of integration is determined (Equation 10.260c) from the position of the inner corner (Equation 10.260b) at $z = 1$ as the point D. The Schwartz–Christoffel mapping (Equations 10.260a and 10.260c) [complex conjugate velocity (Equations 10.258c and 10.259a)] is given by Equation 10.261 (Equation 10.262):

$$\begin{aligned}\zeta - b + \frac{a}{2} - ia &= \frac{a}{\pi}\left[\arcsin z - \operatorname{arg\,sech} z\right]\\ &= \frac{a}{\pi}\left\{\arcsin\left[\exp\left(-\frac{\pi f}{Ua}\right)\right] - \operatorname{arg\,sech}\left[\exp\left(-\frac{\pi f}{Ua}\right)\right]\right\},\end{aligned} \tag{10.261}$$

$$v^* = \frac{df}{d\zeta} = U\sqrt{\frac{1+z}{1-z}} = U\left[\frac{1+\exp(-\pi f/Ua)}{1-\exp(-\pi f/Ua)}\right]^{1/2}$$
$$= U\left[\frac{\exp(\pi f/2Ua)+\exp(-\pi f/2Ua)}{\exp(\pi f/2Ua)-\exp(-\pi f/2Ua)}\right]^{1/2} \tag{10.262}$$
$$= U\left[\coth\left(\frac{\pi f}{2Ua}\right)\right]^{1/2},$$

where Equation 10.243b was used.

E10.15.4: Symmetric Junction of Two Parallel-Sided Channels

Applying the reflection principle (Subsection I.31.1.2) on the real axis, it follows that the preceding results apply to two parallel-sided channels merging in a symmetric junction (Figure 10.9c). Thus, *the potential flow in a sharp bend (Figure 10.9b) [symmetric junction (Figure 10.9c)] of one channel (two channels) has complex conjugate velocity [Equation 10.257c (Equation 10.258c)] in terms of the conformal variable [Equation 10.257b (Equation 10.258b)], where U is the velocity far downstream of the exit channel of width a, and the upstream channel is convergent (Equation 10.257a) [parallel sided (Equation 10.258a)]. In the case of the parallel-sided duct, there is no restriction on the angle α of the junction or the width (Equation 10.254b) of the entry channel, where the velocity (Equation 10.254c) can be chosen by adjusting the parameter b. In the case (Equations 10.259a through 10.259c) of a right-angle corner (junction) in Figure 10.10a (Figure 10.10b), the complex potential f(ζ) is given by Equation 10.261 in inverse form ζ(f); the complex conjugate velocity (Equation 10.262) is obtained eliminating f from Equation 10.261.* The inverse form [Equation 10.261 (Equation 10.262)] may be used to plot the

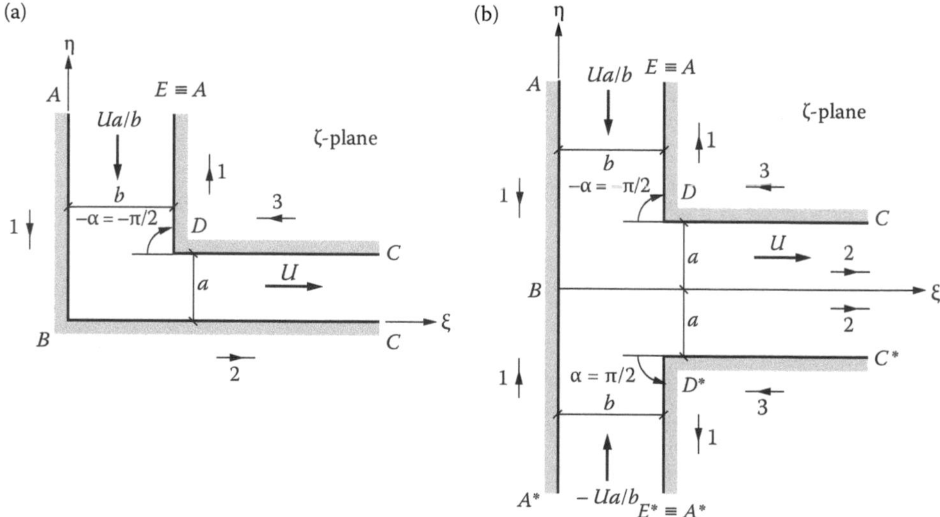

FIGURE 10.10
Particular case of the single junction (Y-junction) of one tube (two tubes) to a single parallel-sided tube [Figure 10.9b (Figure 10.9c)] is the orthogonal case of a right-angle bend (junction) where all channels are parallel sided [a (b)] and may have different widths upstream and downstream. Reversing the flow still leads to a right-angle junction in (a), but changes from a right-angle junction to a bifurcation in (b).

streamlines (equipotentials) and the velocity tangent (normal) to them. For example, the streamline $\Psi = 0$ corresponds to $f = \Phi + i\Psi = \Phi$ real; it follows that $\zeta - b + a/2 - ia$ in Equation 10.261 is real (imaginary plus $a/2$) for $\Phi > 0(\Phi < 0)$, so this part of the streamline $\Psi = 0$ is $\text{Im}(\zeta) = a[\text{Re}(\zeta) = b]$ corresponding to the upper horizontal (outer vertical) wall *DC* (*DA*); $\Phi = 0 = \Psi$ corresponds to the edge $\zeta = b + ia$ at *D*. For $\Psi = 0, f = \Phi$, the velocity (Equation 10.262) is real (imaginary) for $\Phi > 0(\Phi < 0)$, that is, horizontal (vertical) along *DC* (*DA*). The merging of two channels (Figures 10.9c and 10.10b) is distinct from a duct with a side channel (Example 10.16), for which there is no axis of symmetry. Both cases of symmetric merging of ducts (Example 10.15) and duct with side channel (Example 10.16) are distinct from the merging of free jets (Subsection I.38.9). The duct (free jet) problems have fixed boundaries known *a priori* (constant pressure boundaries whose shape is part of the solution of the problem) and thus are solved in the physical (hodograph) plane of the coordinates (of the velocity). Thus, there are three cases: (1) free jets in unbounded space (Figure I.38.11); (2) free jets with boundary surfaces (Figure I.38.12); and (3) ducted jets (Figures 10.9 and 10.10).

EXAMPLE 10.16 POTENTIAL FLOW IN A DUCT WITH A SIDE CHANNEL

Consider the potential flow in a parallel-sided channel with a side branch and a continuing main channel (Figure 10.11a). From the width of the duct h, main h_1 (side h_2) channels, and the velocity U far upstream in the duct, determine the velocities $U_1(U_2)$ far downstream in the main (side) channel.

The present problem is distinct from the splitting/merging of free jets (Subsection I.38.9) that have free surfaces and uses the hodograph method. In the present problem, the flow occupies all space between the walls, and the solution also uses the hodograph plane plus three conformal mappings. Thus, the solution involves the use of four planes: (1) the physical plane of the channels (E10.16.4) in Figure 10.11a; (2) the conformal z-plane where the flow occupies the upper half (E10.16.3) in Figure 10.11b; (3) the hodograph

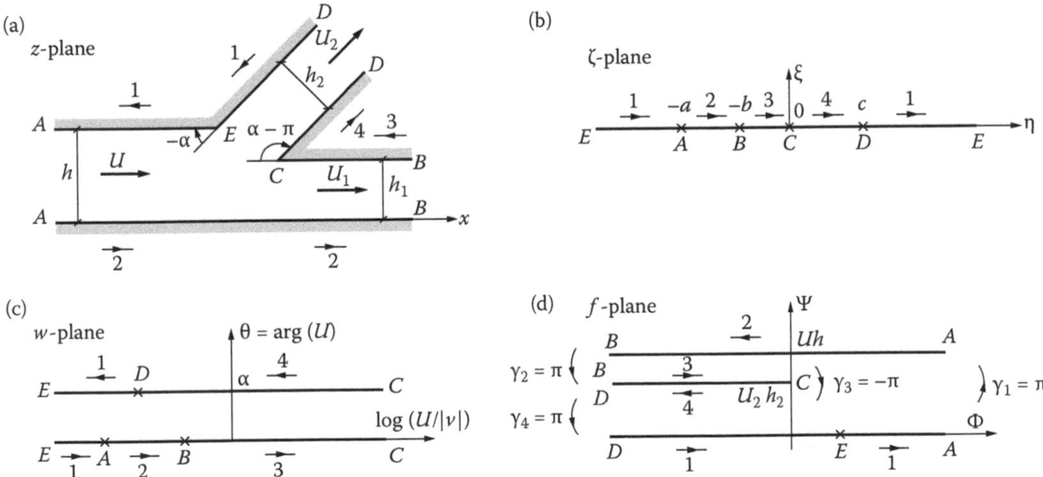

FIGURE 10.11
Potential flow in a tube bifurcating into a "main channel" straight ahead and a "side branch" at an arbitrary angle in the physical z-plane (a) can be obtained via a succession of mappings: (1) starting with a Schwartz–Christoffel interior polygonal mapping with four critical points on the real axis of the conformal ζ-plane (b); (2) mapping the upper-half conformal ζ-plane into a strip of the hodograph w-plane (c) for which the imaginary (real) axis is the direction of the velocity (the logarithm of the far-field velocity divided by the modulus of the local velocity); (3) the strip in the hodograph w-plane is mapped to a triple plate condenser (Figure 10.6b) in the complex potential or f-plane whose real (imaginary) axis is the potential (stream function).

plane (E10.16.1) of the logarithm of the velocity in Figure 10.11c where the flow occupies a strip; and (4) the plane (E10.16.2) of the complex potential in Figure 10.11d where the flow fits inside a triple plate condenser (Example 10.13 and Figure 10.6b).

E10.16.1: Flow Region as a Strip in the Hodograph Plane

The hodograph variable is defined to within a multiplying constant (Equations I.38.29 and I.38.48) as Equation 10.263b as the logarithm of the ratio of the free stream velocity U to the complex conjugate velocity (Equation 10.263a):

$$v^*(z) \equiv \frac{df}{dz} = |v| e^{-i\theta}: \qquad w \equiv \log\left(\frac{U}{v^*}\right) = \log\left(\frac{U}{|v|} e^{i\theta}\right) = i\theta + \log\left(\frac{U}{|v|}\right); \qquad \text{(10.263a and 10.263b)}$$

its imaginary part is the angle of the velocity. Since $\theta = 0$ ($\theta = \alpha$) in (Figure 10.11a) the duct and main (side) channel [Equation 2.264b (Equation 10.264c)]:

$$0 \le \theta \equiv \arg(w) \le \alpha: \qquad \theta = \begin{cases} 0 & \text{for} \quad AB, BC, AE, \\ \alpha & \text{for} \quad CD, ED, \end{cases} \qquad \text{(10.264a–10.264c)}$$

the flow region is (Figure 10.11c) the horizontal strip (Equation 10.264a) in the hodograph plane. The strip (Equation 10.264a) in the w-plane is mapped into the upper complex ζ-plane (Figure 10.11b) by (compare with Equations 2.298a and 2.298b) the transformation

$$\zeta = -\exp\left(-\frac{\pi w}{\alpha}\right), \quad w = -\frac{\alpha}{\pi}\log(-\zeta) = -\frac{\alpha}{\pi}(\log\zeta - i\pi) = i\alpha - \frac{\alpha}{\pi}\log\zeta$$

(10.265a and 10.265b)

because Equation 10.265a is unchanged (Equation 10.266b) adding 2α to Equation 10.265a:

$$w \equiv u + iv: \qquad -\zeta = \exp\left(-\frac{\pi w}{\alpha}\right) = \exp\left(-\frac{\pi u}{\alpha}\right)\exp\left(-\frac{i\pi v}{\alpha}\right)$$
$$= \exp\left(-\frac{\pi u}{\alpha}\right)\exp\left[-\frac{i\pi}{\alpha}(v + 2\alpha)\right].$$

(10.266a and 10.266b)

The f-plane of the potential flow is mapped to the upper-half z-plane by Equations 10.263b and 10.265b leading to

$$\frac{df}{dz} = v^* = Ue^{-w} = U\exp\left[\frac{\alpha}{\pi}(\log\zeta - i\pi)\right] = Ue^{-i\alpha}\zeta^{\alpha/\pi}. \qquad (10.267)$$

The far end of the duct, main, and side channels

$$\zeta = \{-a, -b, c\}, \quad v^* = \{U, U_1, U_2\, e^{-i\alpha}\}, \qquad \text{(10.268a and 10.268b)}$$

$$w = \left\{0, \log\left(\frac{U}{U_1}\right), i\alpha + \log\left(\frac{U}{U_2}\right)\right\}, \quad \{a,b,c\} = \left\{1, \left(\frac{U_1}{U}\right)^{\pi/\alpha}, \left(\frac{U_2}{U}\right)^{\pi/\alpha}\right\}$$

(10.268c and 10.268d)

correspond, respectively, to the following: (1) the coordinates (Equation 10.268a) in the conformal or ζ-plane (Figure 10.11b); (2) the complex conjugate velocities (Equation 10.268b) in the physical z-plane (Figure 10.10a); and (3) the values (Equation 10.268c) of the hodograph variable (Equation 10.263b) in the hodograph or w-plane (Figure 10.11c); the substitution of Equations 10.268a and 10.268c in Equation 10.265a yields Equation 10.268d as follows from

$$b = \exp\left[-\frac{\pi}{\alpha}\log\left(\frac{U}{U_1}\right)\right] = \left(\frac{U_1}{U}\right)^{\pi/\alpha}, \tag{10.269a}$$

$$c = -\exp\left\{-\frac{\pi}{\alpha}\left[i\alpha + \log\left(\frac{U}{U_2}\right)\right]\right\} = -e^{-i\pi}\exp\left[-\frac{\pi}{\alpha}\log\left(\frac{U}{U_2}\right)\right] = \left(\frac{U_2}{U}\right)^{\pi/\alpha}. \tag{10.269b}$$

The positions of three critical points are left as parameters (Equations 10.268a through 10.268d) to match the width of the main channel and positions of the two corners that specify the geometry of the problem.

E10.16.2: Flow Region as a Triple Plate Condenser in the Plane of the Complex Potential

The stream function is defined to within an arbitrary constant (Equation 10.270a) and thus may be taken as zero in the upper wall (Equation 10.270b):

$$f = \Phi + i\Psi: \qquad \Psi = \begin{cases} 0 & \text{in} \quad AED \\ U_2 h_2 = U h - U_1 h_1 & \text{in} \quad CD, CB \\ U h = U_1 h_1 + U_2 h_2 & \text{in} \quad AB \end{cases} \tag{10.270a–10.270d}$$

and the flow rate in the main duct (side channel) specifies the values [Equation 10.270d (Equation 10.270c)]; the conservation of the volume flux (Equation 10.280a) then specifies the flow rate $U_1 h_1$ in the main branch. The flow region is mapped (Figure 10.11d) into a triple plate condenser (like Figure 10.6b) in the plane of the complex potential (Equation 10.270a). A Schwartz–Christoffel transformation with four critical points Equation 10.271a and angles Equation 10.271b maps from the plane of the complex potential to the upper half conformal ζ-plane in Equation 10.271c:

$$\zeta_{1-4} = \{-a, -b, 0, c\}, \quad \gamma_{1-4} = \{\pi, \pi, -\pi, \pi\}:$$

$$\frac{df}{d\zeta} = A\prod_{n=1}^{4}(\zeta - \zeta_n)^{-\alpha/\pi} = \frac{A\zeta}{(\zeta + a)(\zeta + b)(\zeta - c)}.$$

(10.271a–10.271c)

A decomposition of Equation 10.271c into partial fractions (Equation 10.272)

$$\frac{df}{d\zeta} = A\left[\frac{c}{(c+a)(c+b)(\zeta-c)} + \frac{a}{(b-a)(a+c)(\zeta+a)} + \frac{b}{(a-b)(b+c)(\zeta+b)}\right], \quad (10.272)$$

$$B=0: \qquad f(\zeta) = \frac{A}{a-b}\left[\frac{b}{b+c}\log\left(\frac{\zeta+b}{\zeta-c}\right) - \frac{a}{a+c}\log\left(\frac{\zeta+a}{\zeta-c}\right)\right] \quad (10.273\text{a and }10.273\text{b})$$

leads to the primitive (Equation 10.273b), where the constant of integration is zero (Equation 10.273a) because (1) on DE all of $\zeta + b, \zeta - c, \zeta + a$ are positive, and hence $\mathrm{Im}(B) = \mathrm{Im}(f) = \Psi = 0$ on account of Equation 10.270b; and (2) the logarithms in Equation 10.273b are imaginary at $\zeta = 0$, so that $\mathrm{Re}(B) = \mathrm{Re}(f) = \Phi = 0$ since $\Phi = 0$ at point C in Figure 10.11d. In the integration from Equation 10.272, Equation 10.273b is used:

$$\begin{aligned} f(\zeta) - B - \frac{A}{a-b}\left[\frac{b}{b+c}\log(\zeta+b) - \frac{a}{a+c}\log(\zeta+a)\right] &= A\frac{c}{(c+a)(c+b)}\log(\zeta-c) \\ &= \frac{A}{a-b}\left(\frac{a}{a+c} - \frac{b}{b+c}\right)\log(\zeta-c); \end{aligned} \quad (10.274)$$

the partial fraction decomposition on the r.h.s. of Equation 10.272 allows the grouping of terms in Equation 10.274 ≡ Equation 10.273b.

E10.16.3: Relation between Velocities in Three Ducts

The constant A can be determined (1) from Equation 10.270c at point C, that is, $\zeta = 0$ in Equation 10.273b leading to Equation 10.275a; and (2) from Equation 10.270d on AB, where $\zeta + a > 0 > \zeta + b, \zeta - c$, so that Equation 10.273b leads to Equation 10.275b; the difference of Equations 10.275a and 10.275b yields Equation 10.275c:

$$\begin{aligned} \{U_2h_2, Uh, U_1h_1 = Uh - U_2h_2\} &= \Psi(\zeta) \equiv \mathrm{Im}[f(\zeta)] \\ &= \frac{\pi A}{a-b}\left\{\frac{b}{b+c} - \frac{a}{a+c},\quad -\frac{a}{a+c},\quad -\frac{b}{b+c}\right\}. \end{aligned} \quad (10.275\text{a–}10.275\text{c})$$

Substitution of Equations 10.275b and 10.275c eliminates the constant A in Equation 10.273b leading to

$$\pi f(\zeta) = Uh\log\left(\frac{\zeta+a}{\zeta-c}\right) - U_1h_1\log\left(\frac{\zeta+b}{\zeta-c}\right); \quad (10.276)$$

$$\pi\frac{df}{d\zeta} = \frac{Uh}{\zeta+a} - \frac{U_1h_1}{\zeta+b} - \frac{U_2h_2}{\zeta-c}; \quad (10.277)$$

from Equation 10.276 follows Equation 10.277 by differentiation using the conservation of the volume flux (Equation 10.280a). Point C is a stagnation point, where the streamline that separates the flow from the main duct into the main and side branches hits the wall. Thus, at point C, Equations 10.278a and 10.278b hold, implying Equation 10.278c:

$$\zeta = 0, \quad \frac{df}{dz} = 0: \qquad \lim_{\zeta\to 0} \frac{df}{d\zeta} = \lim_{\zeta\to 0} \frac{df}{dz}\frac{dz}{d\zeta} = 0, \qquad \frac{dz}{d\zeta} \sim \zeta^{1-\alpha/\pi} \tag{10.278a–10.278d}$$

since at point C the Schwartz–Christoffel transformation (Equation 8.145b) with a single critical point at the origin and angle $\gamma = \alpha - \pi$ yields Equation 10.278d; for $\alpha < \pi$ this implies that $dz/d\zeta \to 0$ as $\zeta \to 0$. Substituting Equations 10.278a and 10.278c in Equation 10.277 yields Equation 10.279a:

$$0 = \frac{Uh}{a} - \frac{U_1 h_1}{b} + \frac{U_2 h_2}{c} = Uh - U_1 h_1 \left(\frac{U_1}{U}\right)^{-\pi/\alpha} + U_2 h_2 \left(\frac{U_2}{U}\right)^{-\pi/\alpha}, \tag{10.279a and 10.279b}$$

where substitution of Equation 10.268d leads to Equation 10.279b. Thus, *the widths and far stream velocities of the duct* (h, U), *main* (h_1, U_1), *and side* (h_2, U_2) *channels are related by Equation 10.270c ≡ Equation 10.270d ≡ Equation 10.280a and Equation 10.279b ≡ Equation 10.280b:*

$$Uh = U_1 h_1 + U_2 h_2, \qquad h = h_1 \left(\frac{U_1}{U}\right)^{1-\pi/\alpha} - h_2 \left(\frac{U_2}{U}\right)^{1-\pi/\alpha}, \tag{10.280a and 10.280b}$$

where α *is the angle of the side branch; any two quantities can be determined from the other five in* ($U, U_1, U_2, h, h_1, h_2, \alpha$) *using Equations 10.280a and 10.280b. By the principle of reflection, the results also apply to a duct with one unsymmetric (two symmetric) side branch(es) in Figure 10.11a (Figure 10.12a).*

E10.16.4: Duct with a Contraction and an Orthogonal Side Branch

The case of an orthogonal side branch (Equation 10.281a) also allows (Figure 10.12b) a symmetric reflection and simplifies Equation 10.280b to Equation 10.281b:

$$\alpha = \frac{\pi}{2}: \qquad \frac{h}{U} - \frac{h_1}{U_1} + \frac{h_2}{U_2} = 0 = hU - h_1 U_1 - h_2 U_2, \tag{10.281a–10.281c}$$

which can be compared with Equation 10.280a ≡ Equation 10.281c. Eliminating U_2 between Equations 10.281b and 10.281c:

$$\begin{aligned}(h_2)^2 &= \frac{h_2}{U_2}(hU - h_1 U_1) = (hU - h_1 U_1)\left(\frac{h_1}{U_1} - \frac{h}{U}\right) \\ &= -h_1^2 - h^2 + h_1 h \left(\frac{U}{U_1} + \frac{U_1}{U}\right)\end{aligned} \tag{10.282}$$

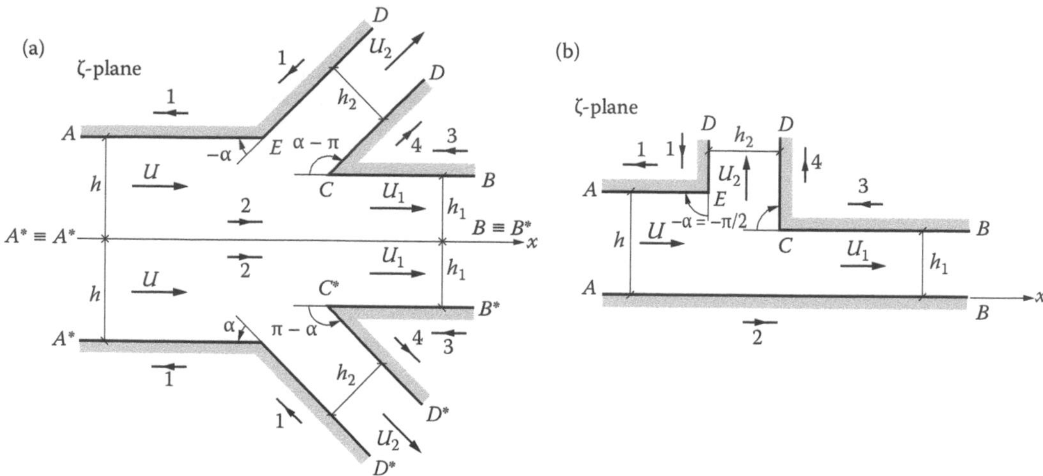

FIGURE 10.12
Applying the reflection principle (Figure I.31.4) to the channel with side branch (Figure 10.11a) leads to the channel with two symmetric side branches (a). This also applies to the particular case of the channel with orthogonal (b) instead of oblique (Figure 10.11a) side branch. The flow direction may be reversed, the angle of the oblique side branch may be acute or obtuse, and the widths of each of the three parallel-sided ducts can be different. The velocity far upstream in the main duct determines the velocities far downstream in the main and side channels. In all cases, the fluid occupies all space between the walls, as distinct from the case of splitting or merging jets with free surfaces (Figure I.38.10) for which (1) there are no rigid boundary surfaces and (2) the shape of the free surfaces is part of the solution of the problem.

leads to a quadratic equation (Equation 10.282 ≡ Equation 10.283b) involving the parameter Equation 10.283a:

$$2k \equiv \frac{h_1}{h} + \frac{h}{h_1} + \frac{(h_2)^2}{h_1 h}: \qquad \left(\frac{U_1}{U}\right)^2 - 2k\left(\frac{U_1}{U}\right) + 1 = 0; \qquad \text{(10.283a and 10.283b)}$$

the roots (Equation 10.284b) of Equation 10.283b are always positive real on account of Equation 10.284a:

$$2k \geq \frac{h_1}{h} + \frac{h}{h_1} \geq 2: \qquad \frac{U_1}{U} = k \pm \sqrt{k^2 - 1} \qquad \text{(10.284a and 10.284b)}$$

and specify the velocity in the main channel; the velocity in the side channel follows substituting Equation 10.284b in Equation 10.281c:

$$\frac{U_2}{U} = \frac{h}{h_2} - \frac{h_1}{h_2}\frac{U_1}{U} = \frac{h}{h_2} - \frac{h_1}{h_2}\left[k \pm \sqrt{k^2 - 1}\right]. \qquad \text{(10.284c)}$$

Thus, *a duct of width h with a flow of velocity U, with an orthogonal side branch of width* h_2 *and a straight contraction or expansion to a width* h_1 *in Figure 10.11a, has velocity [Equation*

10.284b (Equation 10.284c)] in the main (side) branch where K is given by Equation 10.283a. Two particular cases:

$$h_1 = h: \qquad k = 1 + \frac{1}{2}\left(\frac{h_2}{h}\right)^2, \quad \frac{U_1}{U} = 1 + \frac{1}{2}\left(\frac{h_2}{h}\right)^2 \pm \frac{h_2}{h}\left|1 + \left(\frac{h_2}{2h}\right)^2\right|^{1/2},$$

$$\frac{U_2}{U} = -\frac{h_2}{2h} \mp \left|1 + \left(\frac{h_2}{2h}\right)^2\right|^{1/2}, \tag{10.285a–10.285d}$$

$$h_1 = h_2: \qquad k = \frac{h_1}{h} + \frac{h}{2h_1}, \quad \frac{U_1}{U} = \frac{h_1}{h} + \frac{h}{2h_1} \pm \left|\left(\frac{h_1}{h}\right)^2 + \left(\frac{h}{2h_1}\right)^2\right|^{1/2},$$

$$\frac{U_2}{U} = \frac{h}{2h_1} - \frac{h_1}{h} \mp \left|\left(\frac{h_1}{h}\right)^2 + \left(\frac{h}{2h_1}\right)^2\right|^{1/2} \tag{10.286a–10.286d}$$

are (1) the main branch with width equal to the duct (Equations 10.285a through 10.285d) or to (2) the side branch (Equations 10.286a through 10.286d). The common particular case (Equation 10.287d) are all ducts with the same width (Equations 10.287b and 10.287c), besides being orthogonal (Equation 10.287a); there are two solutions both with positive velocity in the main channel:

$$\alpha = \frac{\pi}{2}, h_1 = h = h_2: \quad \frac{U_1}{U} = \frac{3-\sqrt{5}}{2} = 0.38197 < 0.61803 = \frac{-1+\sqrt{5}}{2} = \frac{U_2}{U}, \tag{10.287a–10.287c}$$

$$k = \frac{3}{2}: \quad \frac{U_1}{U} = \frac{3+\sqrt{5}}{2} = 2.61803 > -1.61803 = -\frac{1+\sqrt{5}}{2} = \frac{U_2}{U}; \tag{10.287d and 10.287e}$$

the lower (higher) velocity in the main channel corresponds [Equation 10.287e (Equation 10.287c)] to an outflow toward (inflow from) the side channel. The oblique corner (branch), whether asymmetric [Figure 10.9b (Figure 10.11a)] or symmetric [Figure 10.9c (Figure 10.12a)], differs in that there is no (there is an) upstream duct; the symmetric orthogonal corner (branch) in Figure 10.10b (Figure 10.12b) also differs in that (1) the velocities in the main duct are in opposite directions (the same direction); and (2) the main duct and branch must have the same (can have different) widths. The restriction to the same width of the vertical ducts (Figure 10.9b) results from the reflection principle; it can be removed by not using the reflection principle and using instead a Schwartz–Christoffel transformation with more critical points.

EXAMPLE 10.17: DEFLECTION OF A JET BY AN INTERNAL VORTEX

Consider the deflection of a jet of finite width by a vortex in its interior.

The present problem involves a free jet and thus uses the hodograph method (Chapter I.38); the vortex in the jet is a singularity and creates a stagnation point. Both zero (infinite) velocities at the vortex (stagnation point) are singularities of the hodograph variable (Equation 10.263b), so the hodograph function (Section I.38.8) is used. A source in a jet affects the width of the jet due to its flow rate (Example I.40.16) but causes no deflection. In contrast, it is shown next (Example 10.17) that a vortex deflects the jet. An alternative method to deflect a jet is the Coanda effect near the edge of a semi-infinite plate (Subsection I.38.7). The jet is mapped to the half complex plane (E10.17.1) where the hodograph function (E10.17.2) is used to specify the shape (E10.17.4) and the angle of deflection (E10.17.3) of the jet, hence follows the range of possible jet deflection angles (E10.17.6) that is related to the ratio of the circulation of the vortex to the flow rate of the jet (E10.17.5).

E10.17.1: Source, Vortex, and Stagnation Points in the Upper-Half Plane

The jet in the physical plane (Figure 10.13a) is mapped to the ζ-plane of the conformal variable, with the incident far stream at the origin in Figure 10.13b; the flow is due to a source (Equation 10.288b) for which half the flow rate (Equation 10.288a) goes into the jet of velocity U and width $2b$ with conjugate velocity (Equation 10.288c):

$$Q = 4Ub: \qquad f_0(\zeta) = \frac{Q}{2\pi}\log\zeta = \frac{2Ub}{\pi}\log\zeta, \qquad \frac{df_0}{d\zeta} = \frac{2Ub}{\pi\zeta}. \qquad (10.288a\text{–}10.288c)$$

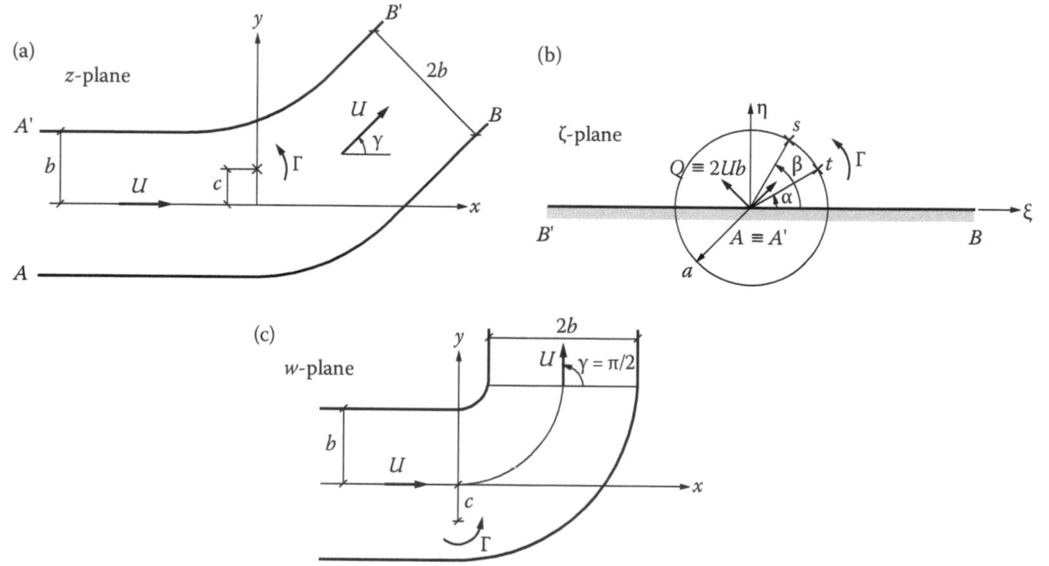

FIGURE 10.13
Free jet can be turned by a vortex in its interior (a) as shown in the physical z-plane with the angle of deflection γ depending on (1) the distance c of the vortex from the jet axis compared with the half-width b of the jet; (2) the circulation Γ of the vortex compared with the flow rate $Q = 2Ub$ in the jet with uniform velocity U far upstream and downstream. The vortex at point t causes the appearance of a stagnation point s in the jet in the physical z-plane, which is mapped on the hodograph ζ-plane in different directions (b), respectively, α and β, at the same distance from the origin. The origin in the hodograph ζ-plane is a source with flow rate $Q = 2Ub$ corresponding to jet of width $2b$ and uniform velocity U far upstream and downstream. A particular case is the turning of a jet in the orthogonal direction by a vortex in its interior (c) in contrast with an orthogonal bend in a duct (Figure 10.10a). For the duct bend, the walls are fixed *a priori*, whereas (1) for the jet with an internal vortex, the width of the jet is fixed but the change of direction is to be determined (Figure 10.1a); (2) for a jet with an internal source/sink it is the asymptotic width rather than direction of the jet that changes (Figure I.40.10); and (3) another method of jet deflection, using a source instead of a vortex, relies on the Coanda effect near the edge of a semi-infinite plate (Figure I.38.9).

The position of the vortex in the conformal or ζ-plane is Equation 10.289a:

$$t \equiv ae^{i\alpha}: \qquad \frac{df}{d\zeta} = \frac{2Ub}{\pi\zeta} - i\frac{\Gamma}{2\pi}\left[\frac{1}{\zeta - t} - \frac{1}{\zeta - t^*}\right], \qquad \text{(10.289a and 10.289b)}$$

and the flow velocity (Equation 10.289b) is due to (1) the source, (2) the vortex at Equation 10.289a, and (3) its image at point t^* conjugate to Equation 10.289a. The flow has a stagnation point s and its image s^* in Equation 10.290:

$$\frac{\pi}{2Ub}\frac{df}{d\zeta} = \frac{(\zeta - s)(\zeta - s^*)}{\zeta(\zeta - t)\zeta - t^*}; \qquad \text{(10.290)}$$

comparison of Equation 10.289b with Equation 10.290 leads to Equation 10.291b:

$$\varepsilon \equiv \frac{\Gamma}{4Ub}: \qquad (\zeta - s)(\zeta - s^*) = \zeta(\zeta - t)(\zeta - t^*)\frac{\pi}{2Ub}\left[\frac{2Ub}{\pi\zeta} - \frac{i\Gamma}{2\pi}\left(\frac{1}{\zeta - t} - \frac{1}{\zeta - t^*}\right)\right]$$

$$= \zeta(\zeta - t)(\zeta - t^*)\left[\frac{1}{\zeta} - i\varepsilon\left(\frac{1}{\zeta - t} - \frac{1}{\zeta - t^*}\right)\right]$$

$$= (\zeta - t)(\zeta - t^*) + i\varepsilon\zeta(t^* - t)$$

(10.291a and 10.291b)

involving the dimensionless vortex strength parameter (Equation 10.291a); the latter compares the circulation of the vortex with the flow rate of the source (Equation 10.288a), that is, twice the flow rate in the jet. The condition that Equation 10.291b be satisfied for arbitrary ζ leads to two equations:

$$0 = s^*s - t^*t = |s|^2 - |t|^2, 0 = s + s^* - (t + t^*) + i\varepsilon(t^* - t) = 2\,\text{Re}(s - t) + 2\varepsilon\,\text{Im}(t). \qquad \text{(10.292a and 10.292b)}$$

The first (Equation 10.292a) implies that the vortex (Equation 10.289a) and stagnation point (Equation 10.293a) are at the same distance from the origin in the hodograph plane (Equation 10.293b):

$$s = ae^{i\beta}: \quad |s| = a = |t|, \quad \cos\beta = \cos\alpha - \varepsilon\sin\alpha; \qquad \text{(10.293a–10.293c)}$$

the second (Equation 10.292b) implies that they are in different directions (Equation 10.293c) since $\varepsilon \neq 0$ implies $\alpha \neq \beta$.

E10.17.2: Construction or the Hodograph Function

The **hodograph function** (Equation I.38.139) is defined (Equation 10.294a) as the derivative of the hodograph variable (Equation 10.263b):

$$\frac{dw}{d\zeta} = d\frac{\left[\log(Ud\zeta/df)\right]}{d\zeta}: \qquad \frac{dw}{d\zeta} = -\frac{1}{\zeta - s} + \frac{1}{\zeta - t} + \frac{1}{\zeta - s^*} - \frac{1}{\zeta - t^*}; \qquad \text{(10.294a and 10.294b)}$$

it has poles (Subsection I.38.2.2) with residue +1(−1) at the vortex (stagnation point) and also at their images with reversed sign in Equation 10.294b. The singularities of the hodograph function result from the following: (1) the complex conjugate velocity having a simple zero (Equation 10.295a) [pole (Equation 10.295b)] at a stagnation point (monopole like a source/sink or vortex):

$$v^*(\zeta) \sim O(\zeta - s), \quad O\left(\frac{1}{\zeta - t}\right): \qquad w(\zeta) = \log\left[\frac{U}{v^*(\zeta)}\right] \sim -\log(\zeta - s), \log(\zeta - t);$$

(10.295a–10.295d)

(2) it follows that the hodograph variable (Equation 10.263b) has a logarithmic singularity [Equation 10.295c (Equation 10.295d)] with coefficient −1 (+1); and (3) the hodograph function (Equation 10.294a) that is the derivative of the hodograph variable has a pole with residue −1 (+1) at the stagnation point (Equation 10.296a) [monopole (Equation 10.296b)] as appears on the first (second) term on the r.h.s. of Equation 10.294b:

$$\frac{dw}{d\zeta} \sim -\frac{1}{\zeta - s}, \frac{1}{\zeta - t}; \qquad t^* = ae^{-i\alpha} = \frac{a^2}{t}, \qquad \zeta^* = ae^{-i\beta} = \frac{a^2}{\zeta};$$

(10.296a–10.296d)

the image of the vortex (Equation 10.288a) [stagnation point (Equation 10.293a)] on the real axis involves its inverse [Equation 10.296c (Equation 10.296d)]; thus the signs in Equation 10.296a (Equation 10.296b) are reversed for the images, as appears in Equation 10.294b. The hodograph function (Equation 10.294b) is imaginary on (Equation 10.297) the real-ζ or ξ-axis:

$$\begin{aligned} \zeta = \xi \in |R: \quad \frac{dw}{dt} &= \frac{1}{\xi - t} - \frac{1}{\xi - t^*} + \frac{1}{\xi - s^*} - \frac{1}{\xi - s} \\ &= \frac{t - t^*}{\xi^2 - (t + t^*)\xi^2 + t^* t} - \frac{s - s^*}{\xi^2 - (s + s^*)\xi + s^* s} \\ &= 2i\left[\frac{\operatorname{Im}(t)}{\xi^2 - 2\xi \operatorname{Re}(t) + |t|^2} - \frac{\operatorname{Im}(s)}{\xi^2 - 2\xi \operatorname{Re}(s) + |s|^2}\right] \end{aligned} \tag{10.297}$$

and is obtained next by integration of Equation 10.294b.

E10.17.3: Determination of the Angle of Deflection of the Jet

The hodograph function (Equation 10.294b) is integrated:

$$\log\left(U \frac{dz}{df}\right) = w = \log\left(\frac{\zeta - t}{\zeta - t^*} \frac{\zeta - s^*}{\zeta - s}\right) + A. \tag{10.298}$$

The constant A is determined (Equation 10.299b) by the condition (Equation 10.299a) that the angle of the jet is the defection angle γ far downstream at B in Figure 10.13a corresponding to $\zeta \to \infty$ in Figure 10.13b:

$$\lim_{\zeta\to\infty} \frac{df}{dz} = Ue^{-i\gamma}: \qquad i\gamma = \lim_{\zeta\to\infty} \log\left(U\frac{dz}{df}\right) = A. \qquad \text{(10.299a and 10.299b)}$$

Substituting Equation 10.299b in Equation 10.298 yields the inverse of the complex conjugate velocity:

$$\frac{1}{v^*} = \frac{dz}{df} = \frac{e^{i\gamma}}{U}\frac{\zeta - t}{\zeta - t^*}\frac{\zeta - s^*}{\zeta - s}. \qquad (10.300)$$

The condition (Equation 10.301a) that the velocity (Equation 10.300) is horizontal and equal to U far upstream at point A in Figure 10.13a corresponding to the origin $\zeta = 0$ in Figure 10.13b leads to Equation 10.301b:

$$\frac{1}{U} = \lim_{\zeta\to 0}\frac{dz}{df} = \frac{e^{i\gamma}}{U}\frac{t}{t^*}\frac{s^*}{s} = \frac{1}{U}\exp\left[i\left(\gamma + 2\alpha - 2\beta\right)\right], \qquad e^{i(\gamma+2\alpha-2\beta)} = 1.$$

(10.301a and 10.301b)

The latter (Equation 10.301b ≡ Equation 10.302a) can be used to eliminate the angle β of the stagnation point (Equation 10.293a) from Equation 10.293b leading to Equation 10.302b:

$$\beta = \alpha + \frac{\gamma}{2}: \qquad \cos\alpha - \varepsilon\sin\alpha = \cos\left(\alpha + \frac{\gamma}{2}\right) = \cos\alpha\cos\left(\frac{\gamma}{2}\right) - \sin\alpha\sin\left(\frac{\gamma}{2}\right);$$

(10.302a and 10.302b)

this can be solved to express

$$\tan\alpha = \frac{1-\cos(\gamma/2)}{\varepsilon - \sin(\gamma/2)} \qquad (10.303)$$

the angle α of the vortex (Equation 10.289a) in terms of the deflection angle γ of the jet (Equation 10.299a) and dimensionless parameter (Equation 10.291a).

E10.17.4: Effect of the Vortex on the Shape of Jet

The product of Equations 10.300 and 10.290 specifies the shape of the jet

$$\frac{\pi}{2b}\frac{dz}{d\zeta} = \frac{\pi}{2b}\frac{dz}{df}\frac{df}{d\zeta} = \frac{e^{i\gamma}}{\zeta}\left(\frac{\zeta - s^*}{\zeta - t^*}\right)^2 \equiv g(\zeta) \qquad (10.304)$$

after integration. The integration of Equation 10.304 is preceded by the partial fraction decomposition (Equation 10.305a):

$$g(\zeta) \equiv \frac{\pi}{2b}\frac{dz}{d\zeta} = \frac{A_0}{\zeta} + \frac{A_1}{\zeta - t^*} + \frac{A_2}{(\zeta - t^*)^2} = \frac{1}{\zeta} + \frac{e^{i\gamma} - 1}{\zeta - t^*} + t^*\left(\frac{1 - e^{i\gamma/2}}{\zeta - t^*}\right)^2, \tag{10.305a and 10.305b}$$

where (1 and 2) the first (second) coefficient A_0 (A_1) is the residue [Equation 10.306a (Equation 10.306b)] at the simple (Equation I.15.24b ≡ Equation 1.18) [double (Equation I.15.33b ≡ Equation 1.16)] pole at $\zeta = 0$ ($\zeta = t^*$):

$$A_0 = \lim_{\zeta \to 0} \zeta g(\zeta) = e^{i\gamma}\left(\frac{s^*}{t^*}\right)^2 = e^{i(\gamma - 2\beta + 2\alpha)} = 1, \tag{10.306a}$$

$$\begin{aligned} A_1 &= \lim_{\zeta \to t^*} \frac{d}{d\zeta}\left[(\zeta - t^*)^2 g(\zeta)\right] = e^{i\gamma} \lim_{\zeta \to t^*} \frac{d}{d\zeta}\left[\frac{(\zeta - s^*)^2}{\zeta}\right] \\ &= e^{i\gamma} \lim_{\zeta \to t^*} \frac{d}{d\zeta}\left(\zeta - 2s^* + \frac{s^{*2}}{\zeta}\right) = e^{i\gamma} \lim_{\zeta \to t^*}\left(1 - \frac{s^{*2}}{\zeta^2}\right) \\ &= e^{i\gamma}\left[1 - \left(\frac{s^*}{t^*}\right)^2\right] = e^{i\gamma} - e^{i(\gamma + 2\alpha - 2\beta)} = e^{i\gamma} - 1 \end{aligned} \tag{10.306b}$$

using Equation 10.301b; (3) the third coefficient can be calculated by the extended rule (Equation I.31.85b) that in the present case takes the form Equation 10.305a ≡ Equation 10.307:

$$\begin{aligned} A_2 &= \lim_{\zeta \to t^*} (\zeta - t^*)^2 g(\zeta) = e^{i\gamma} \lim_{\zeta \to t^*} \frac{(\zeta - s^*)^2}{\zeta} \\ &= e^{i\gamma} \frac{(t^* - s^*)^2}{t^*} = e^{i\gamma} t^* \left(1 - \frac{s^*}{t^*}\right)^2 \\ &= t^*\left[e^{i\gamma/2} - e^{i(\gamma/2 - \beta + \alpha)}\right]^2 = t^*(e^{i\gamma/2} - 1)^2, \end{aligned} \tag{10.307}$$

where Equation 10.301b was used again. The integration of Equation 10.305b is immediate:

$$\frac{\pi}{2b}[z(\zeta) - z_0] = \log\zeta + (e^{i\gamma} - 1)\log(\zeta - ae^{-i\alpha}) - ae^{-i\alpha}\frac{(1 - e^{i\gamma/2})^2}{\zeta - ae^{-i\alpha}}, \tag{10.308}$$

where Equation 10.289a was used.

E10.17.5: Circulation of the Vortex versus Flow Rate of the Jet

The constant z_0 in Equation 10.308 is eliminated by taking the difference (Equation 10.309c) of (1) the position (Equation 10.309a) corresponding to the vortex (Equation 10.289a); and (2) the half width (Equation 10.309b) of the jet in the section far upstream where the source $\zeta = 0$ lies:

$$c = \operatorname{Im}[z(t)], \quad b = \lim_{\zeta\to 0} \operatorname{Im}\{z(\zeta)\}: \qquad \frac{c}{b} = 1 + \frac{c-b}{b} = 1 + \frac{2}{\pi}\operatorname{Im}\left[z(ae^{i\alpha}) - \lim_{\zeta\to 0} z(\zeta)\right]. \tag{10.309a–10.309c}$$

Substitution of Equation 10.308 in Equation 10.309c yields the following, noting that $\zeta \to -0$ at point A in the ζ-plane, corresponding in the z-plane to $\zeta \to -\infty$ along the real axis or arg $(z) = \pi$:

$$\begin{aligned}
\frac{c}{b} &= 1 + \frac{2}{\pi}\operatorname{Im}\left\{ i(\alpha-\pi) + (e^{i\gamma}-1)\log\left(\frac{ae^{i\alpha} - ae^{-i\alpha}}{-ae^{-i\alpha}}\right)\right.\\
&\quad \left. -ae^{-i\alpha}(1-e^{i\gamma/2})^2\left(\frac{1}{ae^{i\alpha}-ae^{-i\alpha}} - \frac{1}{-ae^{-i\alpha}}\right)\right\}\\
&= -1 + \frac{2\alpha}{\pi} + \frac{2}{\pi}\operatorname{Im}\left\{(e^{i\gamma}-1)\left[\log(-2i\sin\alpha) + i\alpha\right] - \left(1-e^{i\gamma/2}\right)^2 \frac{e^{i\alpha}}{e^{i\alpha}-e^{-i\alpha}}\right\}\\
&= -1 + \frac{2\alpha}{\pi} + \frac{2}{\pi}\operatorname{Im}\left\{(e^{i\gamma}-1)\left[\log(2\sin\alpha) - i\frac{\pi}{2} + i\alpha\right]\right.\\
&\quad \left. + \frac{i}{2}\csc\alpha\, e^{i(\alpha+\gamma/2)}(e^{-i\gamma/4} - e^{i\gamma/4})^2\right\}\\
&= -1 + \frac{2\alpha}{\pi} + \frac{2}{\pi}\sin\gamma\log(2\sin\alpha) + (\cos\gamma - 1)\left(\frac{2\alpha}{\pi} - 1\right)\\
&\quad + \frac{1}{2}\csc\alpha\cos\left(\alpha + \frac{\gamma}{2}\right)\left[-2i\sin\left(\frac{\gamma}{4}\right)\right]^2\Bigg\},
\end{aligned} \tag{10.310}$$

which simplifies to

$$\frac{c}{b} = \left(\frac{2\alpha}{\pi} - 1\right)\cos\gamma + \frac{2}{\pi}\left\{\sin\gamma\log(2\sin\alpha) - 2\csc\alpha\cos\left(\alpha+\frac{\gamma}{2}\right)\sin^2\left(\frac{\gamma}{4}\right)\right\}. \tag{10.311}$$

Thus, *a jet with velocity U and width $2b$ is deflected (Figure 10.13a) by a vortex of circulation Γ placed at a distance c from the axis by the angle γ appearing in Equations 10.303 and 10.311, between which the dummy parameter α (that is, the angle of the vortex position in the plane where the jet occupies the upper half) can be eliminated. The vortex strength parameter (Equation 10.291a ≡ Equation 10.312a) compares twice the circulation Γ vortex with the flow rate $Q/2$ in the jet (Equation 10.288a):*

$$\varepsilon = \frac{\Gamma}{4Ub} = \frac{\Gamma}{Q} = \pi\frac{v_0}{U}; \qquad v_0 = \frac{\Gamma}{4\pi b}; \tag{10.312a and 10.312b}$$

an alternative interpretation of the vortex strength parameter (Equation 10.312a) is π times the ratio to the free stream velocity of the velocity due to the vortex of circulation Γ at a distance equal to the jet width (Equation 10.312b).

E10.17.6: Range of Angles of Possible Jet Deflection

The angle of the vortex position (Equation 10.312a) must be positive so that (1) the distance from the axis (Equation 10.309a) is real and (2) the vortex is mapped into the upper half z-plane in Figure 10.10b. A positive vortex angle (Equation 10.313a) implies Equation 10.313b by Equation 10.303:

$$\alpha > 0: \qquad \varepsilon > \sin\left(\frac{\gamma}{2}\right), \qquad \gamma \le 2\arcsin\varepsilon \equiv \gamma_{max}, \qquad |c| < b, \tag{10.313a–10.313d}$$

so that *(1) if the vortex strength parameter (Equation 10.312a) is less than unity, there is a maximum deflection angle (Equation 10.313c); (2) otherwise Equation 10.313c is imaginary and places no restriction on the deflection angle. When the vortex α and jet deflection angle γ are substituted in Equation 10.311, the condition (Equation 10.313d) is required (3) for the vortex to lie within the jet. Conditions 1 and 3 limit the range of possible deflection angles of the jet.* The reference case is taken to be a right-angle deflection of the jet by the vortex (Figure 10.13c), whose effect is comparable to the right-hand corner (Figure 10.10b) or the orthogonal side branch (Figure 10.12b). The baseline case (Equation 10.314a) assumes also a unit vortex strength parameter (Equation 10.314b ≡ Equation 10.312a ≡ Equation 10.291a):

$$\gamma = \frac{\pi}{2}, \quad \varepsilon = \frac{\Gamma}{4Ub} = 1: \qquad \alpha = \frac{\pi}{4}, \qquad \frac{c}{b} = \frac{2}{\pi}\log\left(\sqrt{2}\right) = \frac{1}{\pi}\log 2 = 0.22064 \tag{10.314a–10.314d}$$

leading to (1) the vortex angle (Equation 10.314c) in (Equation 10.303) the ζ-plane; and (2) a distance (Equation 10.314d) of the vortex (Equation 10.311) from the jet axis in the physical z-plane about one-quarter of the distance to the jet boundary.

E10.17.7: Range of Vortex Strength Parameters for a Given Deflection

Two parameters are varied from the baseline case (Equations 10.314a through 10.314d), that is, the angle of deflection (vortex strength) in Table 10.2 (Table 10.3). The angle of deflection of Equation 10.314a in Table 10.2 coincides with the maximum angle of deflection (Equation 10.313c) for *arc* sin $\varepsilon_0 = \pi/4$ or $\varepsilon_0 = 1/\sqrt{2} = 0.707$. Thus, (1) the right-angle deflection cannot be achieved with a smaller vortex strength parameter, for example, $\varepsilon = 0.5$, hence the N.P. ("not possible") in Table 10.2; (2) for $\varepsilon = 0.75$, a right-angle deflection is possible with the vortex inside the jet, hence satisfying the condition Equation 10.313d; (3) for the vortex strength $\varepsilon = 1.5$, the right-angle deflection is possible with the vortex within the jet, changing sides to negative c; and (4) a larger vortex strength $\varepsilon = 3$ achieves a right-angle deflection with the vortex outside the jet on the opposite side and hence the vortex position is indicated in brackets. Thus, there are both lower (Equation 10.315b) and upper (Equation 10.315c) limits of the vortex strength parameter for a right-angle deflection:

$$\gamma = \frac{\pi}{2}: \qquad \varepsilon_{min} = \sin\left(\frac{\gamma}{2}\right) = \frac{1}{\sqrt{2}} = 0.707 \le \varepsilon \equiv \frac{\Gamma}{2Ub} \le 2.033 = \varepsilon_{max}; \tag{10.315a–10315c}$$

TABLE 10.2

Orthogonal Deflection of a Jet by a Vortex

γ	90°	90°	90°	90°	90°	90°	90°	90°
$\varepsilon \equiv \dfrac{\Gamma}{4Ub}$	0.5	0.707	0.75	0.9	1	1.5	2.033	3
α	– 54.74°	90°	81.67°	56.63°	45°	20.27°	12.46°	7.30°
c/b	N.P.	0.573	0.547	0.372	0.573	– 0.459	– 1.000	(– 1.774)
γ_{max}	60.0°	90°	106.3°	128.3°	180°	N.L.	N.L.	N.L.

Note: Vortex in a jet causes a change in its direction (Figure 10.13) that depends on (1) the vortex strength parameter ε comparing the circulation of the vortex Γ with twice the flow rate of the jet of velocity U and width $2b$; and (2) the distance c of the vortex from the jet axis divided by the half-width b of the jet. The right-angle deflection $\gamma = 90°$ of the jet is considered for eight values of the vortex strength parameter from $\varepsilon = 0.5$ to $\varepsilon = 3.0$ indicating (1) the corresponding position of the vortex in the physical plane c/b; (2) the direction α in the hodograph plane; and (3) the maximum deflection possible for the value chosen for the vortex strength parameter. For example, (1) the lowest value of the vortex strength parameter $\varepsilon = 0.5$ leads to a maximum angle-of-deflection $\gamma_{max} = 60°$, so orthogonal deflection $\gamma = 90°$ is not possible (N.P.); (2) the lowest value of the vortex strength parameter for which an orthogonal deflection is possible is $\varepsilon = 1/\sqrt{2}$; (3) for the largest value of the vortex strength parameter in the table $\varepsilon = 3.0$, there is no limit (N.L.) on the maximum deflection, but the vortex would lie outside the jet, and hence its distance from the axis is put in brackets to indicate that this case is not feasible; (4) the limit of feasibility, with the vortex at the edge of the jet $c/b = -1$, occurs for a vortex strength parameter not exceeding $\varepsilon \le 2.033$; (5) the maximum deflection becomes limited to less than $\gamma_{max} < 180°$ if the vortex strength parameter is less than unity $\varepsilon < 1$; and (6) an orthogonal deflection with the vortex within the jet is possible for values of the vortex strength parameter in the range $1/\sqrt{2} < \varepsilon < 1$. U: jet velocity in the far stream; γ: jet deflection angle in the far stream; $2b$: jet width; c: distance of vortex from jet axis; α: angular position of vortex in the upper half ζ-plane; N.P.: not possible; and N.L.: no limit.

the upper limit is determined by interpolation and is also included in Table 10.2 together with the lower limit.

The vortex strength parameter is kept at unity in Table 10.3 and the deflection angle varied from $\pi/4$ to π. Although all deflections are "possible," the constraint (Equation 10.313d) that the vortex must lie within the jet is not met for very small deflections. Thus, for a vortex strength parameter unity (Equation 10.316a), the vortex lies within the jet (Equation 10.302b) for deflection angles exceeding the minimum value (Equation 10.316c):

$$\Gamma = 4Ub: \quad |c| < b \Rightarrow \gamma_{min} = 60.28° \le \gamma \le 180° = \gamma_{max}; \qquad (10.316a\text{–}10.316d)$$

TABLE 10.3

Deflection of a Jet by a Vortex with Unit Strength Parameter

$\varepsilon = \dfrac{\Gamma}{4Ub}$	1	1	1	1	1	1	1	1	1
γ	45°	60°	60.28°	75°	81.60°	90°	120°	150°	180°
α	7.03°	15°	15.19°	27.84°	37.12°	45°	75°	87.37°	90°
c/b	(–1.630)	(–1.013)	–1.000	– 0.338	0.000	0.221	0.672	0.696	0.637

Note: Deflection of a jet by a vortex is considered (Table 10.2) again, this time with a fixed vortex strength parameter equal to unity $\varepsilon = 1$; nine jet deflection angles are considered between 45° and 180°. The first two angles $\gamma = 45°$ and $\gamma = 60°$ are not feasible since the vortex would have to lie outside the jet, at the distance from the axis normalized to the jet half-width indicated in brackets. The smallest possible angle of deflection of the jet with vortex at the edge $c = -b$ is $\gamma = 60.28°$. All larger angles of deflection of the jet are possible, with the vortex crossing from the lower to the upper side of the jet for an angle $\gamma = 81.60°$; the latter is the angle of deflection of the jet for a vortex on an axis with a vortex strength parameter equal to unity.

an inversion of the jet $\gamma = \pi$ is possible for $\alpha = \pi/2$ in Equation 10.303 and vortex at position $c/b = 2/\pi = 0.637$ in Equation 10.311. In the case of a vortex on the axis of the jet (Equation 10.317a), Equation 10.311 simplifies to Equation 10.317b:

$$c = 0: \quad \left(\frac{\pi}{2} - \alpha\right)\cos\gamma = \sin\gamma \log\left(2\sin\alpha\right) - 2\csc\alpha\cos\left(\alpha + \frac{\gamma}{2}\right)\sin^2\left(\frac{\gamma}{4}\right); \tag{10.317a and 10.317b}$$

for each value of the vortex strength parameter (Equation 10.312a), the solution of the coupled system (Equations 10.311 and 10.317b) specifies two angles: (1) the angle of deflection of the jet γ in the physical plane and (2) the angular position of the vortex α in the hodograph plane. For example, in the case of a vortex strength parameter equal to unity and vortex on the jet axis, Table 10.3 shows that (1) the deflection of the jet $\gamma = 81.60°$ is close to a right angle and (2) the vortex lies in the direction $\alpha = 37.12°$ in the hodograph plane. A jet of velocity U and width $2b$ has been considered in two cases: (1) in Example 10.17, the vortex does not change the width of the jet, but changes its direction as an effect of order $O(\Gamma/4bU)$ of the ratio of circulation to the flow rate; and (2) a source (Example I.40.16) does not change the direction of the jet, but changes its width as an effect of order $O(Q/2bU)$, where Q is the flow rate of the source. A small value of the parameter $\varepsilon \equiv Q/2bU$ can have a significant fluidic effect on the jet if the source/sink is placed near the edge of a semi-infinite plate (Subsection I.38.8).

EXAMPLE 10.18: POTENTIAL FLOW PAST A FLAT PLATE WITH ONE (TWO) FLAP(S) AT ANY POSITION

Consider a flat plate with one (two) flap(s) of arbitrary length and angle at any position along the chord [Figure 10.14b (Figure 10.16b)] in a uniform stream without or with circulation. In the latter case, apply the Kutta condition at the trailing edge and determine the lift. Consider in particular the case of flap perpendicular to the airfoil [Figure 10.15a (Figure 10.18a)] as well as intermediate cases.

E10.18.1: Mapping of the Real Line to a Flat Plate with a Flap

Unlike before (Sections 8.6 and 8.7), the flap or slat in the present problem is neither at the leading nor at the trailing edge, but rather at any position along the chord. The flat plate airfoil lies on the real axis in the interval $-b \le x \le d$; it has a flap of length c inclined at an angle β at any position along its chord taken as the origin (Figure 10.14b). The conformal mapping to the exterior of the airfoil with flap from the upper half z-plane

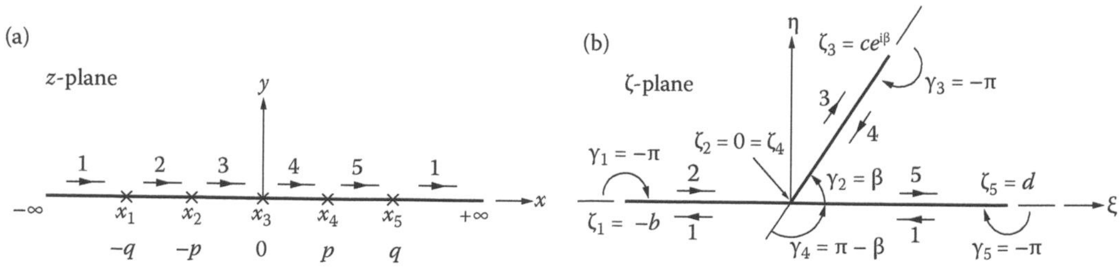

FIGURE 10.14
Schwartz–Christoffel interior polygonal mapping of the real axis with five critical points (a) onto a flat plate airfoil with (b) a flap at any position with any length and angle of inclination can be used to specify the potential flow past the latter in a uniform stream; a particular case is the flat plate airfoil with an orthogonal flap at any position with any length (Figure 10.15a).

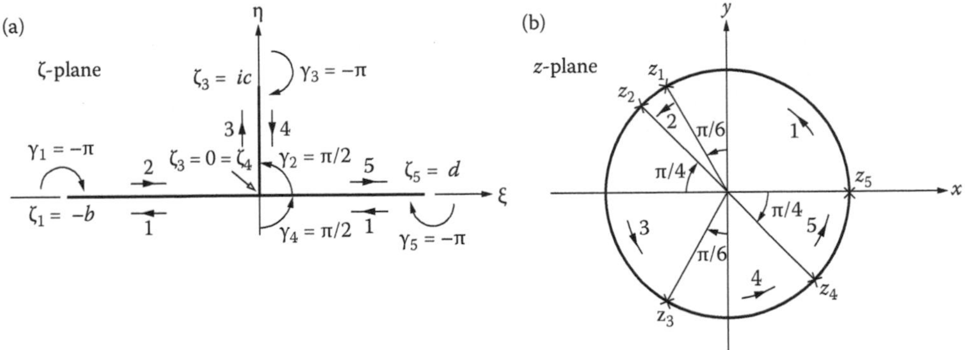

FIGURE 10.15
Same flat plate airfoil with an oblique (orthogonal) flap [Figure 10.14b (Figure 10.15a)] of arbitrary length at an arbitrary position along the chord other than the edge (b) can be represented in a uniform stream (in the presence of circulation) by a Schwartz–Christoffel interior polygonal (exterior disk) transformation [Figure 10.14a (Figure 10.15b)] with five critical points on the real axis (a circle with center at the origin).

(Figure 10.14a) is performed by a Schwartz–Christoffel interior polygonal transformation (Equation I.33.25 ≡ Equation 8.145b ≡ Equation 10.318c) with five critical points (Equation 10.318b) with angles (Equation 10.318a):

$$\gamma_{1-5} = \{-\pi, \beta, -\pi, \pi-\beta, -\pi\}; \quad \zeta_{1-5} = \{-b, 0, ce^{i\beta}, 0, d\}:$$

$$\frac{d\zeta}{dz} = A\prod_{n=1}^{5}(z-x_n)^{-\gamma_n/\pi} = A(z-x_1)(z-x_3)(z-x_5)(z-x_2)^{-\beta/\pi}(z-x_4)^{\beta/\pi-1}. \tag{10.318a–10.318c}$$

The uniform flow with velocity U in the z-plane (Equation 10.319a) corresponds in the ζ-plane to the complex conjugate velocity (Equation 10.319b):

$$f = Uz: \qquad v^* = \frac{df}{d\zeta} = \frac{df}{dz}\frac{dz}{d\zeta} = \frac{U}{A}\left(\frac{z-x_2}{z-x_4}\right)^{\beta/\pi}\frac{z-x_4}{(z-x_1)(z-x_3)(z-x_5)}. \tag{10.319a and 10.319b}$$

Thus, *the upper-half complex z-plane (Figure 10.14a) is mapped into the exterior of a flat plate with a flap at an angle β at any position along its chord (Figure 10.14b) by the interior polygonal Schwartz–Christoffel transformation (Equation 10.318c); the latter leads to the complex conjugate velocity (Equation 10.319b) in a free stream of velocity U. The velocity is singular (vanishes) at the three edges (junctions) of the two plates.* The geometry of the problem is fixed by specifying the position of the three edges (ζ_1, ζ_3, ζ_5) and the coincidence $\zeta_2 = \zeta_4$ of the point of attachment of the flap to the plate at the origin, that is, a total of five conditions; the Schwartz–Christoffel interior polygonal transformation (Equation 10.318c) has seven parameters, namely, the parameter A, the constant of integration B, and the positions of the five critical points. Thus, two conditions can be imposed, for example, the middle critical point at the origin with another two as symmetric pairs. Further restrictions, such as all four points as two symmetric pairs, may restrict the geometry.

E10.18.2: Case of Flap Orthogonal to the Middle of the Plate

In the particular case (Figure 10.15a) of a flap orthogonal to the plate (Equation 10.320a) with the critical points placed symmetrically relative to the origin, Equations 10.320b through 10.320d simplify the Schwartz–Christoffel interior polygonal transformation (Equation 10.318c) to Equation 10.320e:

$$\beta = \frac{\pi}{2}: \quad x_3 = 0, \quad x_4 = p = -x_2, \quad x_5 = q = -x_1:$$

$$\frac{d\zeta}{dz} = Az\frac{z^2 - q^2}{\sqrt{z^2 - p^2}} = Az\left[\sqrt{z^2 - p^2} + \frac{p^2 - q^2}{\sqrt{z^2 - p^2}}\right]. \qquad (10.320a\text{–}10.320e)$$

The integration of Equation 10.320e is elementary:

$$\zeta = \frac{A}{3}(z^2 - p^2)^{3/2} + A(p^2 - q^2)\sqrt{z^2 - p^2} + B. \qquad (10.321)$$

The constants $(A, B; p, q)$ can be determined from the position of the junction (Equations 10.322a and 10.322b) and of the edges (Equations 10.322c through 10.322e):

$$0 = \zeta(\pm p) = B, \qquad ic = \zeta(0) = iAp\left(\frac{2p^2}{3} - q^2\right), \qquad (10.322a\text{–}10.322c)$$

$$-b = \zeta(-q) = -\frac{2A}{3}(q^2 - p^2)^{3/2} = \zeta(q) = d; \qquad (10.322d \text{ and } 10.322e)$$

from Equations 10.322d and 10.322e, it follows that the plate is symmetrically placed relative to the origin with the orthogonal flap in the middle. Further restrictions on the geometry arise as more critical points are fixed.

E10.18.3: Subcase of Flap Length Equal to the Half-Chord

The parameter p can also be chosen (Equation 10.323a) to determine (A, b) in Equations 10.323b and 10.323c:

$$p = 1: \qquad c = A\left(\frac{2}{3} - q^2\right), \qquad \frac{3b}{2} = A(q^2 - 1)^{3/2}; \qquad (10.323a\text{–}10.323c)$$

eliminating A between Equations 10.323b and 10.323c specifies Equation 10.324a ≡ Equation 10.324b:

$$\frac{(q^2 - 1)^{3/2}}{2/3 - q^2} = \frac{3b}{2c}, \qquad 4c^2(q^2 - 1)^3 = 9b^2\left(\frac{2}{3} - q^2\right)^2, \qquad (10.324a \text{ and } 10.324b)$$

so that q is the root of a polynomial of sixth degree:

$$c^2q^6 + 3\left(\frac{3}{4}b^2 - c^2\right)q^4 + 3(b^2 + c^2)q^2 - c^2 - b^2 = 0. \qquad (10.325)$$

If the length of the flap equals half the chord of the airfoil (Equation 10.326a), the roots of Equation 10.325 are independent of the size (Equation 10.326b) of the airfoil with flap:

$$b = c: \quad 4q^6 - 3q^4 + 24q^2 - 8 = 0. \tag{10.326a and 10.326b}$$

In both cases (Equations 10.325 and 10.326b), the polynomial is a cubic in q^2, so $\pm q$ is the solution as follows from symmetry in Figure 10.14b. Thus, *the real line (Figure 10.14a) is mapped to a flat plate with an orthogonal flap in the middle (Figure 10.15a with $d = b$) by the transformation*

$$\zeta = \frac{A}{3}(z^2 - 1)^{3/2} + A(1 - q^2)\sqrt{z^2 - 1}, \tag{10.327}$$

where q is a root of Equation 10.325, and A is determined equivalently by Equation 10.323b or 10.323c; if the length of the flap equals half the chord (Equation 10.326a), then q is a root of Equation 10.326b. The airfoil with orthogonal flap in the middle in a uniform stream of velocity U has the complex potential (Equation 10.319a) and complex conjugate velocity (Equation 10.328):

$$\frac{df}{d\zeta} = \frac{df}{dz}\frac{dz}{d\zeta} = \frac{U}{Az}\frac{\sqrt{z^2 - 1}}{z^2 - q^2}. \tag{10.328}$$

Thus, Equation 10.327 (Equation 10.328) corresponds to Equation 10.321 (Equation 10.319b) with the simplifications (Equations 10.320a through 10.320d, 10.322a, and 10.323a). There is no net force on the plate with flap because (Subsection I.28.2) a potential flow without sources/sinks (circulation) causes no drag (lift). There will be lift around the flat plate with flap in the presence of circulation; this suggests the mappings from a circle (Figure 10.15b).

E10.18.4: Mapping of a Circle to a Flat Plate with Flap

The mapping from (Figure 10.15b) a circle of radius a to a flat plate with an oblique flap at any position along its chord is specified by (Subsection I.33.9) the exterior disk Schwartz–Christoffel transformation (Equation I.33.41 ≡ Equation 8.286b ≡ Equation 8.329b) with the reverse angles (Equation 10.329a) of Equation 10.318a at the five critical points:

$$\gamma_{1-5} = \{\pi, -\beta, \pi, \beta - \pi, \pi\}:$$

$$\frac{d\zeta}{dz} = A\prod_{n=1}^{5}\left(1 - \frac{z_n}{z}\right)^{\gamma_n/\pi} = A\left(1 - \frac{z_1}{z}\right)\left(1 - \frac{z_3}{z}\right)\left(1 - \frac{z_5}{z}\right)\left(1 - \frac{z_2}{z}\right)^{-\beta/\pi}\left(1 - \frac{z_4}{z}\right)^{\beta/\pi - 1}. \tag{10.329a and 10.329b}$$

There are two restrictions: (1) the sum of the angles (Equation 10.329a) at the critical points is 2π in Equation I.33.45a ≡ Equation 8.288b ≡ Equation 10.330a because the polygon is closed:

$$\sum_{n=1}^{5}\gamma_n = 2\pi, \qquad 0 = \sum_{n=1}^{5} z_n\gamma_n = \pi(z_1 + z_3 + z_5 - z_4) + \beta(z_4 - z_2); \tag{10.330a and 10.330b}$$

and (2) the position of the critical points around the circle must satisfy the condition Equation I.33.45b ≡ Equation 8.291a ≡ Equation 10.330b for the transformation to be single valued.

E10.18.5: Circulation around Flat Plate with Flap

The potential flow with velocity U and angle-of-attack α past a cylinder of radius a with circulation Γ is specified by the complex potential (Equation I.34.4b ≡ Equation 8.316a and Equation 8.279a ≡ Equation 10.331):

$$f(z) = U\left(e^{-i\alpha}z + \frac{a^2}{z}e^{i\alpha}\right) - i\frac{\Gamma}{2\pi}\log z, \tag{10.331}$$

$$\frac{df}{dz} = U\left(e^{-i\alpha} - \frac{a^2}{z^2}e^{i\alpha}\right) - \frac{i\Gamma}{2\pi z}; \tag{10.332}$$

corresponding to the complex conjugate velocity is Equation 10.332. Using Equation 10.329b leads to *the complex conjugate velocity of the potential flow with velocity U and angle-of-attack α past a flat plate with oblique flap (Figure 10.14a) in the presence of circulation Γ:*

$$\frac{df}{d\zeta} = \frac{df/dz}{d\zeta/dz} = \frac{1}{A}\left[U(z^2e^{-i\alpha} - a^2e^{i\alpha}) - i\frac{\Gamma z}{2\pi}\right]\left(\frac{z-z_2}{z-z_4}\right)^{\beta/\pi}\frac{z-z_4}{(z-z_1)(z-z_3)(z-z_5)}; \tag{10.333}$$

the velocity vanishes at the junction of the flap with the plate and is singular at the edges, for example, at the trailing edge:

$$\frac{df}{d\zeta} = \frac{C}{z-z_5} + O(1): \quad C \equiv \frac{1}{A}\left\{U\left[(z_5)^2e^{-i\alpha} - a^2e^{i\alpha}\right] - i\frac{\Gamma z_5}{2\pi}\right\}\left(\frac{z_5-z_2}{z_5-z_4}\right)^{\beta/\pi}\frac{z_5-z_4}{(z_5-z_1)(z_5-z_3)},$$
(10.334a and 10.334b)

the Kutta condition suppresses the singularity at the trailing edge (Equation 10.334a) by setting Equation 10.335a:

$$C = 0: \qquad -\frac{L}{\rho U} = \Gamma = i2\pi U\left(\frac{a^2}{z_5}e^{i\alpha} - z_5e^{-i\alpha}\right), \tag{10.335a and 10.335b}$$

which specifies the circulation (Equation 10.335b) and hence the lift (Equation I.28.29b ≡ Equation 2.233a).

E10.18.6: Lift on a Flat Plate with a Flap in a Stream

In the case of a flap orthogonal to the flat plate (Equation 10.336a), the condition Equation 10.330b simplifies to Equation 10.336b:

$$\beta = \frac{\pi}{2}: \qquad 0 = 2(z_1 + z_3 + z_5) - (z_2 + z_4); \tag{10.336a and 10.336b}$$

this can be met by choosing the positions (Equation 10.337) of the critical points around the circle:

$$\beta_{1-5} = \left\{\frac{7\pi}{6}, \frac{3\pi}{4}, -\frac{7\pi}{6}, -\frac{\pi}{4}, 0\right\}, \tag{10.337}$$

$$z_{1-5} = a\exp(i\beta_{1-5}) = a\left\{\frac{-1+i\sqrt{3}}{2}, \frac{-1+i}{\sqrt{2}}, \frac{-1-i\sqrt{3}}{2}, \frac{1-i}{\sqrt{2}}, 1\right\}, \tag{10.338}$$

as shown in Figure 10.15b. The choice of critical points (Equation 10.338) restricts the geometry of the airfoil with the orthogonal flap. Substitution of Equation 10.338 in Equation 10.335b specifies the circulation

$$\frac{L}{\rho U} = -\Gamma = -i2\pi Ua(e^{i\alpha} - e^{-i\alpha}) = -i2\pi Ua2i\sin\alpha = 4\pi Ua\sin\alpha \tag{10.339}$$

and lift in agreement with Equations I.34.17 and I.34.12b ≡ Equations 8.318b and 8.297a. The exterior disk Schwartz–Christoffel transformation (Equation 10.329b) in the case (Equations 10.329a and 10.338)

$$\begin{aligned}\frac{d\zeta}{dz} &= A\left(1-\frac{a}{z}e^{i7\pi/6}\right)\left(1-\frac{a}{z}e^{-i7\pi/6}\right)\left(1-\frac{a}{z}\right)\left[\left(1-\frac{a}{z}e^{i3\pi/4}\right)\left(1-\frac{a}{z}e^{-i\pi/4}\right)\right]^{-1/2}\\ &= A\left(1+\frac{a}{z}+\frac{a^2}{z^2}\right)\left(1-\frac{a}{z}\right)\left(1+i\frac{a^2}{z^2}\right)^{-1/2}\end{aligned} \tag{10.340}$$

simplifies to

$$\frac{d\zeta}{dz} = A\left(1-\frac{a^3}{z^3}\right)\left(1+i\frac{a^2}{z^2}\right)^{-1/2} = A\frac{z-a^3/z^2}{\sqrt{z^2+ia^2}}; \tag{10.341}$$

this is integrated next.

E10.18.7: Explicit Mapping for Orthogonal Flap

The change of variable (Equation 10.342a) leads to Equation 10.342b:

$$z^2 = iu^2: \qquad \frac{d\zeta}{du} = \frac{d\zeta}{dz}\frac{dz}{du} = \sqrt{i}\frac{d\zeta}{dz} = A\left(\frac{\sqrt{i}u}{\sqrt{u^2+a^2}} + \frac{ia^3}{u^2\sqrt{u^2+a^2}}\right); \tag{10.342a and 10.342b}$$

the last term on the r.h.s. of Equation 10.342b can be integrated (Equation 10.343b) via a change of variable (Equation 10.343a):

$$u = a\sinh v: \qquad \int \frac{a^2\,du}{u^2\sqrt{a^2+u^2}} = \int \operatorname{csch}^2 v\,dv = -\coth v = -\frac{\cosh v}{\sinh v} = -\frac{\sqrt{u^2+a^2}}{u}. \tag{10.343a and 10.343b}$$

The integration of the first term on the r.h.s. of Equation 10.342b is elementary, and the last is specified by Equation 10.343b leading to

$$\zeta(u) = A\left(\sqrt{i}\sqrt{u^2+a^2} - i\frac{a}{u}\sqrt{u^2+a^2}\right) + B = A\sqrt{i}\sqrt{u^2+a^2}\left(1-\sqrt{i}\frac{a}{u}\right) + B, \qquad (10.344)$$

where B is an arbitrary constant. Substituting Equation 10.342a in Equation 10.344 leads to

$$\zeta(z) = A\left(1 - i\frac{a}{z}\right)\sqrt{z^2+ia^2} + B. \qquad (10.345)$$

Thus, *the exterior disk Schwartz–Christoffel mapping of a circle (Figure 10.15b) to a flat plate with inclined flap (Figure 10.14b) is specified by Equation 10.329b subject to the constraint Equation 10.330b on the position of the critical points; in the case of a flap orthogonal to the flat plate (Equation 10.336a), the condition Equation 10.330b simplifies to Equation 10.336b, which may be met by a particular choice of the positions (Equation 10.338) of the critical points around the circle of radius a. The mapping then simplifies to Equation 10.345, where the location of the edges (Equations 10.346a through 10.346c) and junction (Equation 10.346d):*

$$\zeta(a) = d, \zeta(ae^{-i7\pi/6}) = ic, \zeta(ae^{i7\pi/6}) = -b, \zeta(ae^{-i3\pi/4}) = \zeta(ae^{ii\pi/4}) \qquad (10.346a\text{–}10.346d)$$

specify the four constants (A, B, b, d), leaving a as a free scale parameter.

E10.18.8: Mapping of the Real Line onto a Flat Plate with Two Flaps

Next a flat plate with two oblique flaps with distinct junction points is considered, lengths and angles on opposite sides (Figure 10.16b) in a uniform stream with or without circulation; in the case with circulation, the Kutta condition applied at the trailing edge determines the lift. Also considered is the particular case when the two flaps have the same junction point (Figure 10.17a) and are aligned corresponding to crossed plates (Figure 10.17b); furthermore, the subcase when the two plates are orthogonal (Figure 10.18a) is also considered. The flat plate airfoil lies on the real axis along the segment $-b \le x \le d$; it has one flap each in the upper

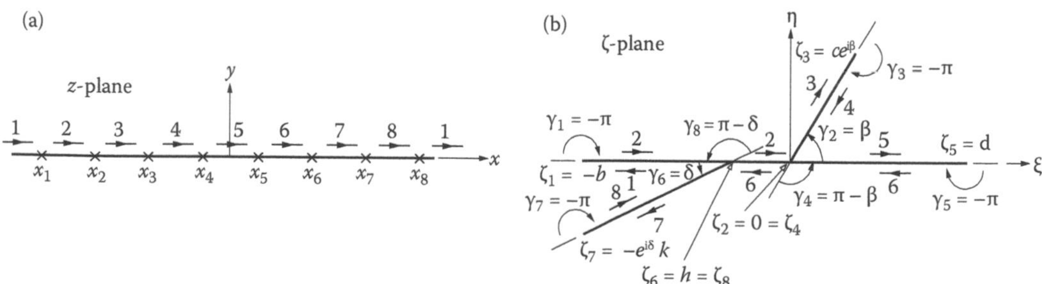

FIGURE 10.16
Schwartz–Christoffel interior polygonal transformation with eight critical points on the real axis (a) can be used to map onto a flat plate airfoil (b) with two flaps, with generally distinct arbitrary lengths and angles relative to the airfoil. The flaps may be at arbitrary positions along the chord of the flat plate airfoil. Two particular cases are as follows: (1) both flaps at either the leading edge or the trailing edge; (2) one flap each at the leading and trailing edges (b). Case 1 (case 2) involves coincidences of critical points reducing their number from 8 to 7 (6) and adding the respective angles. The two flaps are on opposite sides of the flat plate airfoil; if they were on the same side of the flat plate airfoil, this would keep the number of critical points but change their sequence and that of the angles.

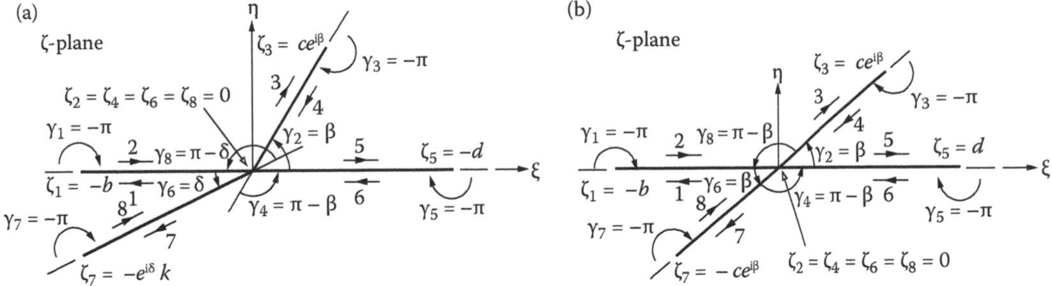

FIGURE 10.17
Flat plate airfoil with two flaps at distinct arbitrary positions along the chord and with arbitrary lengths and inclinations (Figure 10.16b) includes as progressively simpler particular cases (1) the two flaps with the same attachment point on opposite sides of the flat plate airfoil (a) but still distinct lengths and inclinations; (2) the two flaps with the same attachment point and inclination but distinct lengths corresponding to "two crossed plates" (b); (3) the flaps with the same attachment point and both orthogonal to the flat plate, but with distinct length corresponding to "two orthogonal plates"; (4) if the plates have the same length, a "cruciform airfoil" results (Figure 10.18a); and (5) if the length of the flaps is equal to half the chord of the airfoil, the two are interchangeable.

(lower) side with length $c(k)$ at angle $\beta(\delta)$, with junction at the origin $\zeta = 0$ ($\zeta = h$) in Figure 10.16b. The conformal mapping to the exterior of the airfoil with two flaps from the upper-half z-plane is performed (Figure 10.16a) by an interior polygonal Schwartz–Christoffel transformation (Equation I.33.25 ≡ Equation 8.145b ≡ Equation 10.347c) with eight critical points (Equation 10.347b) with angles (Equation 10.347a):

$$\gamma_{1-8} = \left\{-\pi, \beta, -\pi, \pi-\beta, -\pi, \delta, -\pi, \pi-\delta\right\}; \quad z_{1-8} = \left\{-b, 0, ce^{i\beta}, 0, d, h, h-ke^{-i\delta}, h\right\}:$$

$$\frac{d\zeta}{dz} = \prod_{n=1}^{8} (z-x_n)^{-\gamma_n/\pi} = (z-x_1)(z-x_3)(z-x_5)(z-x_7)$$

$$(z-x_2)^{-\beta/\pi}(z-x_4)^{\beta/\pi-1}(z-x_6)^{-\delta/\pi}(z-x_8)^{\delta/\pi-1}. \qquad (10.347a\text{–}10.347c)$$

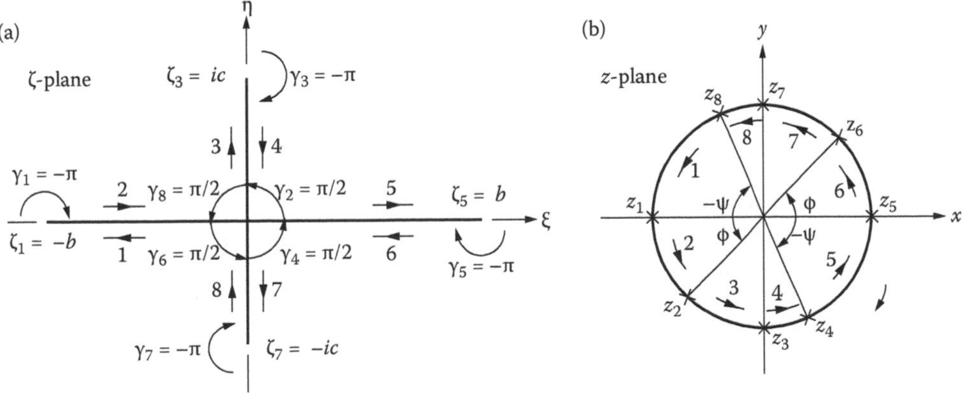

FIGURE 10.18
Same flat plate airfoil with two flaps on opposite sides with distinct attachment points, lengths, and inclinations (Figure 10.16a), as well as all particular cases (Figures 10.16b, 10.17a, 10.17b, and 10.18a), can be considered in a uniform stream (with circulation) using the Schwartz–Christoffel interior polygonal (exterior disk) transformation with eight critical points [Figure 10.16b (Figure 10.18b)] on the real axis (on a circle with center at the origin).

The interior polygonal transformation (Equation 10.347c) has 10 parameters, namely, the eight positions of the critical points plus the parameter A and integration constant B; the constraints are the positions of the four edges and coincidence of the pairs of points at the two junctions, that, is six constraints. This allows a choice of four critical points; if the positions of the two junctions are fixed, then only two critical points can be chosen.

E10.18.9: Potential Flow Past a Plate with Two Flaps

A uniform flow with velocity U in the z-plane (Equation 10.319a) leads (Equation 10.347c) to a complex conjugate velocity in the ζ-plane:

$$\frac{df}{d\zeta} = \frac{df}{dz}\frac{dz}{d\zeta} = \frac{U}{A}\frac{(z-x_2)^{\beta/\pi}(z-x_4)^{1-\beta/\pi}(z-x_6)^{\delta/\pi}(z-x_8)^{1-\delta/\pi}}{(z-x_1)(z-x_3)(z-x_5)(z-x_7)}, \qquad (10.348)$$

showing that the velocity is singular (zero) at the edges (junctions). The particular case when the two flaps have the same junction point (Figure 10.17a) corresponds to the crossing of two bent plates. If in addition the flaps are aligned (Equation 10.349a) like two unbent crossed plates (Figure 10.17b), the transformation (Equation 10.347c) simplifies to Equation 10.349b:

$$\delta = \beta: \quad \frac{d\zeta}{dz} = A\left[(z-x_2)(z-x_6)\right]^{-\beta/\pi}\left[(z-x_4)(z-x_8)\right]^{\beta/\pi-1} (z-x_1)(z-x_3)(z-x_5)(z-x_7); \qquad \text{(10.349a and 10.349b)}$$

a further subcase (Equation 10.350a) is that (Figure 10.18a) of orthogonal flat plates that simplifies the interior polygonal transformation Equation 10.349b to Equation 10.350b:

$$\beta = \frac{\pi}{2}: \quad \frac{d\zeta}{dz} = (z-x_1)(z-x_3)(z-x_5)(z-x_7)\left[(z-x_2)(z-x_4)(z-x_6)(z-x_8)\right]^{-1/2}$$
$$= \prod_{n=1}^{4}(z-x_{2n-1})(z-x_{2n})^{-1/2} = \prod_{n=1}^{4}\frac{z-x_{2n-1}}{\sqrt{z-x_{2n}}}. \qquad \text{(10.350a and 10.350b)}$$

Thus, *the real axis is mapped (Figure 10.16a) onto a flat plate with two flaps (Figure 10.16b) by the polygonal exterior Schwartz–Christoffel transformation (Equation 10.347c), corresponding to the complex conjugate velocity (Equation 10.348) in a free stream of velocity U; successive particular cases are a common junction point (Figure 10.17a), then (Equations 10.349a and 10.349b) for two crossed unbent plates (Figure 10.17b), and (Equations 10.350a and 10.350b) for orthogonal plates (Figure 10.18a).*

E10.18.10: Flow with Circulation Past a Cruciform Airfoil

The *cruciform airfoil* (Figure 10.18a) can also be obtained by an exterior disk Schwartz–Christoffel transformation (Equation I.33.41 ≡ Equation 8.317a ≡ Equation 10.351b) with reversed angles (Equation 10.351a) relative to Equations 10.347a, 10.349a, and 10.350a at the eight critical points around the circle of radius a:

$$\gamma_{1-8} = \left\{\pi, -\frac{\pi}{2}, \pi, -\frac{\pi}{2}, \pi, -\frac{\pi}{2}, \pi, -\frac{\pi}{2}\right\}, \quad \frac{d\zeta}{dz} = \prod_{n=1}^{8}\left(1-\frac{z_n}{z}\right)^{\gamma_n/\pi}$$
$$= A\prod_{n=1}^{4}\left(1-\frac{z_{2n-1}}{z}\right)\left(1-\frac{z_{2n}}{z}\right)^{-1/2} = \frac{A}{z^2}\prod_{n=1}^{4}\frac{z-z_{2n-1}}{\sqrt{z-z_{2n}}}, \qquad \text{(10.351a and 10.351b)}$$

The potential flow (Equation 10.332) with velocity U and angle-of-attack α and circulation Γ around the cruciform airfoil (Equation 10.351b) has complex conjugate velocity (Equation 10.352):

$$v^* = \frac{df}{d\zeta} = \left[U(z^2 e^{-i\alpha} - a^2 e^{i\alpha}) - i\frac{\Gamma z}{2\pi}\right]\prod_{n=1}^{4} \frac{\sqrt{z - z_{2n}}}{z - z_{2n-1}}. \tag{10.352}$$

showing that it is zero at the junction of the two plates and singular at the four tips. The velocity at the trailing edge

$$v^* = \frac{C}{z - z_5} + O(1): \quad C \equiv \left\{U\left[(z_5)^2 e^{-i\alpha} - a^2 e^{i\alpha}\right] - i\frac{\Gamma z_5}{2\pi}\right\}\prod_{n=1}^{4}\sqrt{z - z_{2n}}\prod_{\substack{n=1\\ n\neq 3}}^{4}\frac{1}{z - z_{2n-1}}$$

(10.353a and 10.353b)

is finite if the condition Equation 10.335b is met specifying the circulation and lift (Equation 10.339).

E10.18.11: Mapping the Circle into a Cruciform Airfoil

The cruciform airfoil corresponds to $z_5 = a$, which meets the condition Equation I.33.45a ≡ Equation 10.330b ≡ Equation 10.354:

$$0 = \sum_{n=1}^{8} \gamma_n z_n = \frac{\pi}{2}[2(z_1 + z_3 + z_5 + z_7) - (z_2 + z_4 + z_6 + z_8)], \tag{10.354}$$

$$z_{1-8} = a\{-1, -e^{i\phi}, -i, e^{-i\psi}, 1, e^{i\phi}, i, -e^{-i\psi}\}, \tag{10.355}$$

for example, by choosing four pairs of diametrically opposite critical points around the circle of radius a in Equation 10.355 with four on the coordinate axis (Figure 10.18b). Substituting Equation 10.355 in Equation 10.351b leads to the transformation

$$\begin{aligned}\frac{d\zeta}{dz} &= \frac{A}{z^2}\frac{(z - z_1)(z - z_3)(z - z_5)(z - z_7)}{\sqrt{(z - z_2)(z - z_4)(z - z_6)(z - z_8)}}\\ &= \frac{A}{z^2}\frac{(z + 1)(z^2 + i)(z + 1)(z - i)}{\sqrt{(z + ae^{i\varphi})(z - ae^{-i\psi})(z - ae^{i\varphi})(z + ae^{-i\psi})}}\\ &= \frac{A}{z^2}\frac{(z^2 - 1)(z^2 + 1)}{\sqrt{(z^2 - a^2 e^{i2\varphi})(z^2 - a^2 e^{-i2\psi})}}.\end{aligned} \tag{10.356}$$

The integration of Equation 10.356 can be performed in terms of hyperelliptic functions (Subsection I.39.7). The constant of integration B and $(A, a; \phi, \psi; b, c, d)$ are related to the lengths (Equations 10.357a through 10.357d) of the two plates of the cruciform airfoil:

$$\zeta(a) = d, \zeta(-a) = -b, \zeta(\pm ia) = \pm ic, \zeta(\pm ae^{i\varphi}) = 0 = \zeta(\pm ae^{-i\psi}) \tag{10.357a–10.357h}$$

and by the location (Equations 10.357e through 10.357h) of their junction; thus there are eight constraints to determine eight parameters for a fixed particular geometry of the cruciform airfoil. The function Equation 10.356 is even, so its primitive is odd, to within an arbitrary constant of integration; if the latter is zero, the eight conditions (Equations 10.357a through 10.357h) reduce to four; the remaining four constants (A, a; ϕ, ψ) are sufficient to satisfy Equations 10.357a through 10.357h for the cruciform airfoil (Figure 10.18a). In the general case of the airfoil with two flaps at different angles and with distinct junctions (Figure 10.16b), there are also eight conditions (Equation 10.347b); a choice of the locations of critical points more general than (Equation 10.355) can be made is still satisfying (Equation 10.330b). Choosing z_5 = a to meet Equation 10.335 and satisfy Equation 10.339 leaves six angles in (z_1,...z_6) free; together with the constants (A, B) in the transformation, this adds to eight parameters to meet the eight conditions (Equation 10.347b).

EXAMPLE 10.19: REDUCTION OF CONVERGENCE FROM INFINITE PRODUCTS TO SERIES

Consider the infinite products:

$$|z| < \infty: \quad \prod_{n=1}^{\infty} \left\{ \frac{n^2 + z^2}{n^2 + a^2} \right\}, \qquad \mathrm{Re}(z) < -1: \quad \prod_{n=1}^{\infty} (1 + n^z), \tag{10.358a–10.358d}$$

$$|z| < |a|: \quad \prod_{n=1}^{\infty} \left\{ 1 + \left(\frac{z}{a} \right)^n \right\}, \qquad |z| \le 1: \quad \prod_{n=1}^{\infty} \left(1 + \frac{z^n}{n^2} \right), \tag{10.359a–10.359d}$$

$$\prod_{n=1}^{\infty} \frac{an^2 + bnz + cz^2}{\alpha n^2 + \beta nz + \gamma z^2} \tag{10.360}$$

and establish their convergence properties.

The convergence of an infinite product (Equation 10.361b) is the same (Equations 9.18a through 9.18d) as that of the infinite series (Equation 10.361c) related by Equation 10.361a:

$$v_n \equiv u_n - 1: \quad \prod_{n=1}^{\infty} u_n = \prod_{n=1}^{\infty} (1 + v_n) \Leftrightarrow \sum_{n=1}^{\infty} (u_n - 1) = \sum_{n=1}^{\infty} v_n. \tag{10.361a–10.361c}$$

Thus, the general terms (Equations 10.362a through 10.362d) of the infinite products (Equations 10.358a through 10.358d and 10.359a through 10.359d ≡ Equation 10.361b)

$$u_n \equiv \left\{ \frac{n^2 + z^2}{n^2 + a^2}, 1 + n^z, 1 + \left(\frac{z}{a} \right)^n, 1 + \frac{z^n}{n^2} \right\}, \tag{10.362a–10.362d}$$

$$v_n \equiv \left\{ \frac{z^2 - a^2}{n^2 + a^2}, n^z, \left(\frac{z}{a} \right)^n, \frac{z^n}{n^2} \right\} \tag{10.363a–10.363d}$$

may be replaced (Equation 10.361a) by the general terms (Equations 10.363a through 10.363d) in the series Equation 10.361c. Concerning the series: (1) in Equation 10.363a, the general term is $O(n^{-2})$ leading to absolute convergence (Equation I.29.40d) for finite z in Equation 10.358a; (2) in Equation 10.363b, the general term corresponds to the harmonic series (Equation I.29.40d) that converges for Equation 10.358c; (3) in Equation 10.363c, the general term corresponds to the geometric series (Equations I.21.62a and I.21.62b ≡ Equations 3.148a and 3.148b) that converge for Equation 10.359a; and (4) in Equation 10.363d, the general term (Equation 10.364a) has a ratio of successive terms Equation 10.364b:

$$u_n = \frac{z^n}{n^2}: \qquad \frac{u_{n+1}}{u_n} = z\left(\frac{n}{n+1}\right)^2 = z\left(1+\frac{1}{n}\right)^{-2} = z\left[1-\frac{1}{2n}+O\left(\frac{1}{n^2}\right)\right], \tag{10.364a and 10.364b}$$

so the combined convergence test (Subsection I.29.1.1) may be applied leading to convergence for Equation 10.359c. The resulting convergence properties are indicated in Table 10.4 for the infinite products (Equations 10.358a through 10.358d and 10.359a through 10.359d).

The last infinite product (Equation 10.360) is considered similarly in four stages: (1) replacing the infinite product (Equation 10.361b ≡ Equation 10.360) with Equation 10.365 by the series Equation 10.361c with Equation 10.361a ≡ Equation 10.366:

$$u_n = \frac{an^2+bnz+cz^2}{\alpha n^2+\beta nz+\gamma z^2}, \tag{10.365}$$

$$v_n \equiv u_n - 1 = \frac{(a-\alpha)n^2+(b-\beta)nz+(c-\gamma)z^2}{\alpha n^2+\beta nz+\gamma z^2}; \tag{10.366}$$

(2) the asymptotic form of Equation 10.366 for large n is

$$\begin{aligned} v_n &= \frac{\left(\frac{a}{\alpha}-1\right)+\frac{z}{n}\frac{b-\beta}{\alpha}+\frac{z^2}{n^2}\frac{c-\gamma}{\alpha}}{1+\frac{z}{n}\frac{\beta}{\alpha}+\frac{z^2}{n^2}\frac{\gamma}{\alpha}} \\ &= \left(\frac{a}{\alpha}-1\right)\left[1+\frac{z}{n}\left(\frac{b-\beta}{a-\alpha}-\frac{\beta}{\alpha}\right)+O\left(\frac{z^2}{n^2}\right)\right] \\ &= \left(\frac{a}{\alpha}-1\right)\left[1+\frac{z}{\alpha n}\frac{b\alpha-a\beta}{a-\alpha}+O\left(\frac{z^2}{n^2}\right)\right]; \end{aligned} \tag{10.367}$$

TABLE 10.4

Convergence of Infinite Products

Case	Equation 10.358b	Equation 10.358d	Equation 10.359b	Equation 10.359d
D.	$z = \infty$	$\text{Re}(z) > -1$ or $z = -1$	$\|z\| > a$ or $z = a$	$\|z\| > 1$
O.	–	$\text{Re}(z) = -1 \neq z$	$\|z\| = a \neq z$	–
C.C.	–	–	–	–
A.C.	$\|z\| \le \infty$	$\text{Re}(z) < -1$	$\|z\| < a$	$\|z\| \le 1$
U.C.	$\|z\| \le M < \infty$	$\text{Re}(z) < -1$ and $\|z+1\| \ge \varepsilon > 0$	$\|z\| < a$ and $\|z-a\| \le \varepsilon > 0$	$\|z\| \le 1-\delta$ with $0 < \delta \le 1$
T.C.	$\|z\| \le M < \infty$	$\text{Re}(z) \le -1-\delta$ with $\delta > 0$	$\|z\| \le a-\delta$ with $\delta > 0$	$\|z\| \le 1-\delta$ with $0 < \delta \le 1$

Note: D.: divergent; O.: oscillatory; C.C., A.C., U.C., and T.C.: conditionally, absolutely, uniformly, and totally convergent (Chapter I.21).

(3) the necessary condition for convergence (Equation I.29.16a ≡ Equation 10.368a) implies that convergence is possible only if Equation 10.368b is met:

$$\lim_{n\to\infty} v_n = 0: \qquad a = \alpha, \qquad v_n = \frac{(b-\beta)nz + (c-\gamma)z^2}{\alpha n^2 + \beta nz + \gamma z^2}; \qquad (10.368a–10.368c)$$

(4) the condition Equation 10.368b simplifies Equation 10.366 to Equation 10.368c, which has the asymptotic form (Equation 10.369) for large n:

$$v_n = \frac{\dfrac{b-\beta}{\alpha}\dfrac{z}{n} + \dfrac{c-\gamma}{\alpha}\dfrac{z^2}{n^2}}{1 + \dfrac{\beta}{\alpha}\dfrac{z}{n} + \dfrac{\gamma}{\alpha}\dfrac{z^2}{n^2}} = \frac{b-\beta}{\alpha}\frac{z}{n} + O\left(\frac{1}{n^2}\right); \qquad (10.369)$$

(5) the ratio of coefficients of Equation 10.369 is given by

$$\frac{v_{n+1}}{v_n} = \frac{n+1}{n} + O\left(\frac{1}{n^2}\right) = 1 - \frac{1}{n} + O\left(\frac{1}{n^2}\right), \qquad (10.370)$$

implying that the series diverges unless Equation 10.371a is met:

$$b = \beta: \qquad v_n = \frac{(c-\gamma)z^2}{\alpha n^2 + \beta nz + \gamma z^2} = O\left(\frac{1}{n^2}\right); \qquad (10.371a\text{ and }10.371b)$$

and (6) together Equations 10.368b and 10.371a lead from Equation 10.366 to 10.371b, ensuring total convergence. Thus, the infinite product (Equation 10.360) converges totally if it is the form Equation 10.372b:

$$c \neq \gamma: \qquad \prod_{n=1}^{\infty} \frac{an^2 + bnz + cz^2}{an^2 + bnz + \gamma z^2}\ \text{T.C.}, \qquad (10.372a\text{ and }10.372b)$$

in agreement with Equations 10.368b and 10.371a; the case $c = \gamma$ is excluded (Equation 10.372a) because it would lead to a series of constant unit terms that is divergent. The convergence of the four infinite products (Equations 10.358b, 10.358d, 10.359b, and 10.359d) at all points of the z-plane can be established by applying the combined convergence test (Subsection I.29.1.1) to the corresponding series (Equations 10.363a through 10.363d) leading to the results in Table 10.4.

EXAMPLE 10.20: ACCUMULATION OF BRANCH-CUTS AND POWER SERIES FOR SOME MANY-VALUED FUNCTIONS

Show that the function Equation 10.373b is a particular case of Equation 10.373c where q is a positive integer (Equation 10.373a):

$$q \in |N: \qquad f_2(z) = \log\left[\frac{z}{\pi}\sin\left(\frac{\pi}{z}\right)\right], \qquad f_q(z) = \sum_{n=1}^{\infty} \log\left[1 - (nz)^{-q}\right].$$

(10.373a–10.373c)

Consider the branch-cuts in the z-plane such that these functions become single valued, and obtain power series representation in a suitable region excluding the branch-cuts.

E10.20.1: Accumulation Point of Extrema and Zeros

The function (Equation 10.374c) of a real variable (Equation 10.374a) has maxima (Equation 10.375a) and minima (Equation 10.375b) that occur at the points (Equations 10.374b and 10.347d) accumulating toward the origin (Equation 10.375c):

$$x \in| R; \qquad m \in| Z: \qquad f(x) = \sin\left(\frac{1}{x}\right), \qquad x_m = \frac{1}{m\pi + \pi/2}, \qquad (10.374a\text{–}10.374d)$$

$$f\left(x_{2m}\right) = 1 = f_{\max}, \qquad f\left(x_{2m+1}\right) = -1 = f_{\min}, \qquad \lim_{m\to\pm\infty} x_m = 0. \qquad (10.375a\text{–}10.375c)$$

It follows that the function Equation 10.374c is of unbounded (Equation 10.376b) fluctuation or oscillation (Equations I.27.79a and I.27.79b ≡ Equations 5.165a and 5.165b) in any interval containing the origin (Equation 10.376a):

$$0 < \left|x_{2m}\right| < x_0: \quad F\left(\sin\left(\frac{1}{x}\right); 0, x_0\right) \ge \sum_{n=m}^{\infty} \left|f\left(x_{2n}\right) - f\left(x_{2n+1}\right)\right| = \sum_{n=m}^{\infty} 2 = \infty.$$

(10.376a and 10.376b)

The function Equation 10.374c ≡ Equation 10.377c of real (Equation 10.374a) or complex (Equation 10.377a) variable has zeros (Equation 10.377e) that occur at the points (Equation 10.377d) also accumulating toward the origin (Equations 10.377b and 10.377f):

$$z \in| C; \quad n \in| Z: \quad f(z) = \sin\left(\frac{1}{z}\right): \quad z_n = \frac{1}{n\pi}, \quad f\left(z_n\right) = 0, \quad \lim_{n\to\pm\infty} z_n = 0.$$

(10.377a–10.377f)

The accumulation point at the origin is an essential singularity (Subsection 1.1.39) of the function Equation 10.377c as follows from (Equations 7.15a and 7.15b) the Laurent series expansion:

$$f(z) = \sin\left(\frac{1}{z}\right) = \sum_{n=0}^{\infty} \frac{(-)^n}{(2n+1)!} z^{-2n-1} = \frac{1}{z} + O\left(\frac{1}{z^3}\right); \qquad (10.378)$$

the function Equation 10.379 also has an essential singularity at infinity, since this is not changed by multiplication by any power:

$$g(z) \equiv zf(z) = z\sin\left(\frac{1}{z}\right) = \sum_{n=0}^{\infty} \frac{(-)^n}{(2n+1)!} z^{-2n} = 1 + O\left(\frac{1}{z^2}\right); \qquad (10.379)$$

the function Equation 10.378 has a zero at infinity and becomes finite at infinity (Equation 10.379) multiplying by z. The factor z in Equation 10.379 implies that the extrema of the function instead of being unity (Equations 10.375a and 10.375b) decay as $O(n^{-1})$, that is, an arithmetic series; since the arithmetic series is divergent (Subsection I.21.1), the function Equation 10.379 is still of unbounded fluctuation or oscillation in any interval of the real line containing the origin (Subsections I.27.9.4 and I.27.9.5). A function with an infinite number of zeros becomes a function with an infinite number of branch-points on application of the logarithm; for example, the logarithm of Equation 10.378 (Equation 10.379) has an infinite number of branch-points at Equations 10.377d including (excluding) the point at infinity. The logarithm of Equation 10.379 is considered next.

E10.20.2: Function with an Infinite Number of Branch-Points

The function Equation 10.380b vanishes at two sequences (Equation 10.380c) of points (Equation 10.380a) accumulating toward the origin (Equation 10.380d):

$$n \in |N_0: \qquad g_1(z) = \sin\left(\frac{\pi}{z}\right) = 0 \quad \Rightarrow \quad z_n^{\pm} = \pm\frac{1}{n}, \quad \lim_{n\to\infty} z_n^{\pm} = 0. \tag{10.380a–10.380d}$$

One of the zeros lies at infinity $z_0^{\pm} = 1/0 = \infty$ and is removed multiplying the function Equation 10.380b by z/π in Equation 10.381b:

$$n \in |N: \quad g_2(z) = \sin\left(\frac{z}{\pi}\right) g_1(z) = \frac{z}{\pi}\sin\left(\frac{\pi}{z}\right) = 0 \quad \Rightarrow \quad z_n^{\pm} = \pm\frac{1}{n}, \quad \lim_{n\to\infty} z_n^{+} = 0. \tag{10.381a–10.381d}$$

The zeros of the function Equation 10.381b become the branch-points of its logarithm (Equation 10.373b). Thus, the points (Equations 10.381a and 10.381c) are the branch-points of the function Equation 10.373b and accumulate toward the origin (Equation 10.381d). The logarithm is a many-valued function (Section 3.4) that increases (decreases) by $2\pi i$ describing a counterclockwise (clockwise) loop around a branch-point. It becomes single-valued if all such loops are prevented (Figure 10.19) by joining successive pairs of

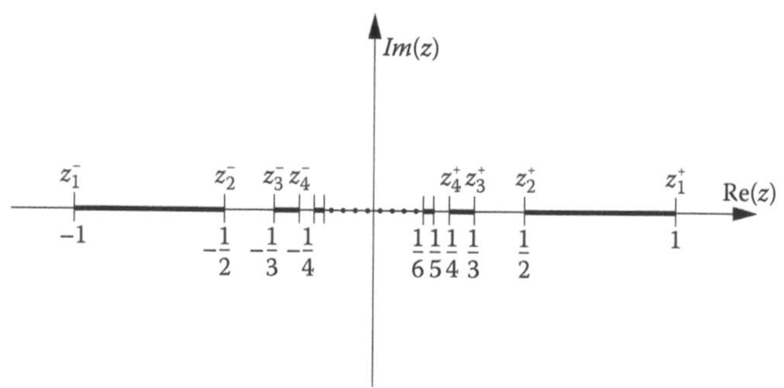

FIGURE 10.19
Function with an infinite sequence of zeros or poles is transformed by application of the logarithm into a many-valued function with an infinite number of branch-points. The points may have an accumulation point that must be an essential singularity in the case of an accumulation of zeros or poles. In the case of an accumulation of branch-points, the many-valued function can be made single-valued by choosing the principal branch and preventing loops around any of the branch-points that would lead to other branches of the function. This is ensured by connecting successive pairs of branch-points by branch-cuts. This leads to a sequence of branch-cuts with length progressively reducing to zero at the accumulation point. One example is the double sequence of branch-points at $z_+ = 1/n$ ($z_- = -1/n$) along the positive (negative) real axis accumulation point at the origin.

branch-points by branch-cuts (Equation 10.382a) whose length (Equation 10.382b) tends to zero (Equation 10.282c) as the origin is approached.

$$\pm z \notin (z_{2n}, z_{2n-1}) = \left(\frac{1}{2n}, \frac{1}{2n-1} \right); L_n = |z_{2n} - z_{2n-1}| = \left| \frac{1}{2n} - \frac{1}{2n-1} \right| = \frac{1}{2n(2n-1)}, \quad \lim_{n\to\infty} L_n = 0. \tag{10.382a–10.382c}$$

This is an example of a many-valued function (Equation 10.373b) with a double infinity of branch-points (Equation 10.381c) with a common accumulation point (Equation 10.381d). It becomes single-valued choosing the principal branch in the complex z-plane with (Figure 10.19) an infinite number of branch-cuts (Equation 10.382a), of decreasing length (Equation 10.382b), accumulating toward the origin (Equation 10.382c) with the branch-points (Equations 10.381a, 10.381c, and 10.381d).

E10.20.3: Descending Power Series Outside the Branch-Cuts

The infinite product for the circular sine (Equation 1.75 ≡ Equation 10.383b) has in each factor one of the zeros (Equations 10.381a and 10.381c):

$$|z| \le M < \infty: \qquad g_2(z) = \frac{z}{\pi} \sin\left(\frac{\pi}{z}\right) = \prod_{n=1}^{\infty} \left(1 - \frac{1}{n^2 z^2}\right); \tag{10.383a and 10.383b}$$

since the infinite product (Equation 10.383b) is uniformly convergent in the finite z-plane (Equation 10.383a), application of logarithms leads to the series

$$\log\left[\frac{z}{\pi}\sin\left(\frac{\pi}{z}\right)\right] = \log\left[g_2(z)\right] = \sum_{n=1}^{\infty} \log\left(1 - \frac{1}{n^2 z^2}\right) = f_2(z), \tag{10.384}$$

that proves that Equation 10.373b is the particular case $q = 2$ of Equation 10.373c. Using the series for the logarithm (Equations 3.48a through 3.48c) in Equation 10.384 leads to the double series Equation 10.385b:

$$|z| > 1: \qquad f_2(z) = -\sum_{n=1}^{\infty}\sum_{m=1}^{\infty} \frac{(nz)^{-2m}}{m} = -\sum_{m=1}^{\infty} \frac{z^{-2m}}{m} \zeta(2m); \tag{10.385a–10.385c}$$

one of the series (Equation 10.385c) was summed using (**Euler 1737; Riemann 1859**) the **zeta-function** (Equation 10.386b):

$$2m > 1: \qquad \zeta(2m) \equiv \sum_{n=1}^{\infty} n^{-2m}. \tag{10.386a and 10.386b}$$

The series defining the zeta-function (Equation 10.386b) is absolutely convergent (Equation I.29.40d) by the Gauss test for Equation 10.386a, that is, met for m a positive integer. The series for the logarithm is absolutely convergent if $|(nz)|^{-2} < 1$ or $|z| > n^{-1}$ implying $|z| > 1$ for the smallest $n = 1$ in Equation 10.385a, that is, outside all branch-cuts in Figure 10.19. It can be confirmed by the D'Alembert ratio test (Equation I.29.31b) that the series (Equation 10.385c ≡ Equation 10.387a), with general term Equation 10.387b, is absolutely convergent (Equation 10.387d) outside the unit disk (Equation 10.385a):

$$f_2(z) \equiv \sum_{m=1}^{\infty} h_m, \quad h_m \equiv -\frac{z^{-2m}}{m}\zeta(2m), \quad \zeta(2m+2) < \zeta(2m) \quad (10.387\text{a}–10.387\text{c})$$

$$\lim_{m\to\infty}\left|\frac{h_{m+1}}{h_m}\right| = \lim_{m\to\infty}\frac{m}{m+1}\frac{1}{\left|z^2\right|}\frac{\zeta(2m+2)}{\zeta(2m)} < \frac{1}{\left|z\right|^2} < 1 \quad (10.387\text{d})$$

noting that the zeta-function (Equation 10.386b) decreases (Equation 10.387c) for increasing m. It has been shown that *the function (Equation 10.373b) has a descending power series (Equation 10.385c), involving the zeta-function (Equation 10.386b) in the coefficients; the series converges absolutely outside the unit disk (Equation 10.385a), thus excluding all the branch-points (Equation 10.381c) and cuts (Equation 10.382a) in Figure 10.19.*

E10.20.4: Multiple Sequences of Branch-Points with the Same Accumulation Point

The function Equation 10.373c contains Equation 10.373b as the particular case $q = 2$ with two sequences of branch-points (Equation 10.381c) both accumulating toward the origin (Equation 10.381d). For $q = 1$ only one sequence of branch-points z_n^+ would exist. For q any positive integer (Equation 10.388a), the branch-points of the function Equation 10.388b form q infinite sequences (Equations 10.388a and 10.388b):

$$p = 0,\cdots,q-1: \qquad z_{n,p} = \frac{1}{n}\sqrt[q]{1} = \frac{1}{n}e^{i2\pi p/q}, \quad \arg(z_{n,p}) = \frac{2\pi p}{q}, \quad \lim_{n\to\infty} z_{n,p} = 0, \quad (10.388\text{a}–10.388\text{d})$$

which accumulate toward the origin (Equation 10.388d) in q equally spaced directions (Equation 10.388c), for example, the diagonals of the quadrants for $q = 4$ in Figure 10.20. The many-valued function (Equation 10.373c) can be made single-valued using similar branch-cuts (Equations 10.389a and 10.389b) in each direction with the same decreasing length (Equation 10.382b ≡ Equations 10.390a and 10.390b) as before.

$$p = 0,\cdots,q-1: \qquad z \notin (z_{2n,p}, z_{2n-1,p}) = \left(\frac{e^{i2\pi p/q}}{2n} - \frac{e^{i2\pi p/q}}{2n-1}\right), \quad (10.389\text{a and } 10.389\text{b})$$

$$n = 1,\cdots,\infty: \qquad L_n = (z_{2n,p} - z_{2n-1,p}) = \left|\frac{1}{2n} - \frac{1}{2n-1}\right| = \frac{1}{2n(2n-1)}. \quad (10.390\text{a and } 10.390\text{b})$$

Thus, *Equation 10.373c is a many-valued function with q infinite sequences of branch-points (Equations 10.388a and 10.388b) accumulating toward the origin (Equation 10.388d) in q equally spaced directions (Equation 10.388c), for example, $q = 4$ in Figure 10.20. The function can be made single-valued choosing the principal branch in the complex z-plane with q sequences*

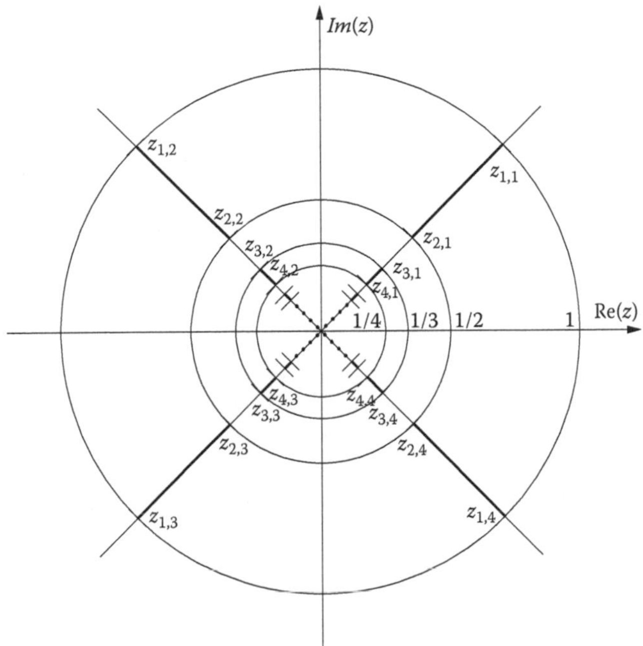

FIGURE 10.20
Another analogous (Figure 10.19) example is a many-valued function with four sequences of branch-points along the diagonals of the quadrants all at the same distance $1/n$ and accumulating toward the origin. This many-valued function can be made single-valued by choosing the principal branch in the complex z-plane with the same branch-cuts accumulating toward the origin in the appropriate directions, for example, the positive and negative real axes (diagonals of quadrants) in Figure 10.19 (Figure 10.20). In all cases, the function is analytic outside the unit disk allowing the expansion in a descending power.

of branch-cuts (Equations 10.389a and 10.389b) of decreasing length (Equations 10.390a and 10.390b).

E10.20.5: Series Converging Absolutely Outside the Unit Disk

As in the particular case (Equation 10.373b), the function Equation 10.373c can be expanded in a descending power series that converges absolutely outside the unit disk, that is, excluding the region where the branch-cuts are located. The method is similar substituting the logarithmic series (Equations 3.48a through 3.48c) in Equation 10.373c leading to a double series (Equation 10.391b):

$$|z| > 1: \qquad f_q(z) = -\sum_{n=1}^{\infty}\sum_{m=1}^{\infty}\frac{1}{m}(nz)^{-mq} = -\sum_{m=1}^{\infty}\frac{z^{-mq}}{m}\zeta(mq) \qquad (10.391a\text{–}10.391c)$$

using again the zeta-function (Equations 10.386a and 10.386b). The logarithmic series (Equations 3.48a through 3.48c) and the series for the zeta-function (Equations 10.386a and 10.386b) are both absolutely convergent in the unit disk (Equation 10.391a), thus allowing derangement of terms (Sections I.21 and 9.3) and summation of the double series (Equation 10.391b) in any order, in particular, Equation 10.391c. The D'Alembert ratio test confirms that the series (Equation 10.391b ≡ Equation 10.392a) with general term Equation 10.392b is absolutely convergent (Equation 10.392d):

$$f_q(z) = \sum_{m=1}^{\infty} h_m, \qquad h_m = -\frac{z^{-mq}}{m}\zeta(mq), \qquad \zeta(mq) > \zeta(mq+2) \tag{10.392a–10.392c}$$

$$\lim_{n\to\infty}\left|\frac{h_{m+1}}{h_m}\right| = \lim_{m\to\infty}\frac{\mathrm{m}}{\mathrm{n}+1}\frac{\zeta(mq+2)}{\zeta(mq)}\left|\mathrm{z}\right|^{-q} \le \left|\mathrm{z}\right|^{-q} < 1, \tag{10.392d}$$

where the decreasing property (Equation 10.392c) of the zeta-function (Equations 10.386a and 10.386b) was used. It has been proved that *the function Equation 10.373c can be expanded in a descending power series (Equation 10.391c), whose coefficients involve zeta-functions (Equations 10.386a and 10.386b); the series is absolutely convergent (Equations 10.392a through 10.392d) outside the unit disk (Equation 10.391a), that is, excluding all the branch-points (Equations 10.389a and 10.389b) and branch-cuts (Equations 10.390a and 10.390b) that are illustrated in the Figure 10.19 (Figure 10.20) in the particular case q = 2 (q = 4).*

NOTE 10.1 CONVERGENT SERIES WITH DIVERGENT RATIO OF SUCCESSIVE TERMS (CÉSARO)

The proof of the absolute convergence [Equations 10.387a through 10.387d (Equations 10.392a through 10.392d)] of the series [Equations 10.375a and 10.375c (Equations 10.391a and 10.391c)] used the D'Alembert ratio test (Subsection I.29.3.2) that specifies a sufficient condition, namely, if the ratio of successive terms in modulus tends to less than unity, the series is absolutely convergent (Equation I.29.31b). This sufficient condition is not necessary, as can be shown by counter example(s) of convergent series whose ratio of successive terms does not tend to less than unity and may even diverge. To construct such counter example to the ratio test, it is sufficient to mix two series with distinct rates of convergence, for example, alternating the terms of two geometric (Equation 10.393b) [harmonic (Equation 10.394b)] series:

$$1 > |b| > |a|: \qquad f = a + b + a^2 + b^2 + \cdots = \sum_{m=1}^{\infty}(a^n + b^n), \tag{10.393a and 10.393b}$$

$$\mathrm{Re}(\beta) > \mathrm{Re}(\alpha) > 1: \qquad g = \frac{1}{1^\beta} + \frac{1}{1^\alpha} + \frac{1}{2^\beta} + \frac{1}{2^\alpha} + \cdots = \sum_{m=1}^{\infty}\left(\frac{1}{n^\beta} + \frac{1}{n^\alpha}\right) \tag{10.394a and 10.394b}$$

with distinct arguments (Equation 10.393a) [exponents (Equation 10.394a)]. Both series (Equations 10.393b and 10.394b) are absolutely convergent because the series of moduli can be summed exactly (Equations 10.395 and 10.396, respectively):

$$\sum_{n=1}^{\infty}\left(|a|^n + |b|^n\right) = \frac{|a|}{1-|a|} + \frac{|b|}{1-|b|}, \tag{10.395}$$

$$\sum_{n=1}^{\infty}\left(\left|n^{-\beta}\right|+\left|n^{-\alpha}\right|\right)=\sum_{n=1}^{\infty}\left[n^{-\mathrm{Re}(\beta)}+n^{-\mathrm{Re}(\alpha)}\right]=\zeta(\mathrm{Re}(\beta))+\zeta(\mathrm{Re}(\alpha)), \tag{10.396}$$

where the sum of the geometric series (Equations I.21.62a and I.21.62b ≡ Equations 3.148a and 3.148b) [the zeta-function (Equations 10.386a and 10.386b)] was used. Besides

$$\begin{aligned}\left|n^{\alpha}\right| &\equiv \left|\exp\left[\log(n^{\alpha})\right]\right| = \left|\exp(\alpha\log n)\right| \\ &= \exp\left|[\mathrm{Re}(\alpha)\log n]\right|\left|\exp[i\,\mathrm{Im}(\alpha)\log n]\right| \\ &= \exp[\mathrm{Re}(\alpha)\log n] = n^{\mathrm{Re}(\alpha)}\end{aligned} \tag{10.397}$$

in Equation 10.396, Equation 10.397 was used in agreement with Equation 3.76b.

NOTE 10.2 MIXING TERMS FROM TWO CONVERGENT SERIES

The series Equation 10.393b ≡ Equation 10.398a (Equation 10.394b ≡ Equation 10.398b) have general terms Equations 10.398c and 10.398e (Equations 10.398d and 10.398f):

$$\{f,g\}=\sum_{n=1}^{\infty}\{f_n,g_n\}: \qquad \{f_{2n-1},g_{n-1}\}=\{a^n,n^{-\alpha}\}, \qquad \{f_{2n},g_{2n}\}=\{b^n,n^{-\beta}\}. \tag{10.398a–10.398f}$$

The ratio of successive terms diverges (tends to zero) if the even terms are divided by the odd terms (Equations 10.399a and 10.399b):

$$\lim_{n\to\infty}\left|\frac{f_{2n}}{f_{2n-1}}\right|=\lim_{n\to\infty}\left(\frac{|b|}{|a|}\right)^{n}=\infty=\lim_{n\to\infty}|n|^{\beta-\alpha}=\lim_{n\to\infty}\left|\frac{g_{2n}}{g_{2n-1}}\right|, \tag{10.399a and 10.399b}$$

$$\lim_{n\to\infty}\left|\frac{f_{2n-1}}{f_{2n}}\right|=\lim_{n\to\infty}\left(\frac{|a|}{|b|}\right)^{n}=0=\lim_{n\to\infty}|n|^{\alpha-\beta}=\lim_{n\to\infty}\left|\frac{g_{2n-1}}{g_{2n}}\right| \tag{10.400a and 10.400b}$$

[vice versa (Equations 10.400a and 10.400b)]. The series [Equation 10.393b (Equation 10.394b)] are the "simplest" mixed series using all terms of each geometric (harmonic) series. Using only part of the terms of each series would lead to other cases, for example, alternating terms from the two series [Equations 10.401a and 10.401b (Equations 10.402a and 10.402b)]:

$$1>|b|>|a|: \qquad h=a+b^2+a^3+b^4+\cdots=\sum_{n=1}^{\infty}(a^{2n-1}+b^{2n}), \tag{10.401a and 10.401b}$$

$$\mathrm{Re}(\beta)>\mathrm{Re}(\alpha)>1: \quad j=\frac{1}{1^{\alpha}}+\frac{1}{1^{\beta}}+\frac{1}{3^{\alpha}}+\frac{1}{4^{\beta}}+\cdots=\sum_{n=1}^{\infty}\left[\left(a_{n-1}\right)^{-\alpha}+(2n)^{-\beta}\right] \tag{10.402a and 10.402b}$$

would lead again to divergent or zero ratio of successive terms, although both series are absolutely convergent, since the series of moduli:

$$\sum_{n=1}^{\infty}\left(|a|^{2n-1}+|b|^{2n}\right)=\frac{|a|}{1-|a|^2}+\frac{1}{1-|b|^2}, \tag{10.403}$$

$$\begin{aligned}\sum_{n=1}^{\infty}\left\{\left|(2n-1)^{-\alpha}\right|+\left|(2n)^{-\beta}\right|\right\} &\le \sum_{n=1}^{\infty}\left\{\left|n^{-\alpha}\right|+\left|n^{-\beta}\right|\right\}\\ &=\sum_{n=1}^{\infty}[n^{-\mathrm{Re}(\alpha)}+n^{-\mathrm{Re}(\beta)}]=\zeta(\mathrm{Re}(\alpha))+\zeta(\mathrm{Re}(\beta))\end{aligned} \tag{10.404}$$

can be summed exactly (Equation 10.403) [have upper bound (Equation 10.404)] as geometric sums [zeta-functions (Equations 10.386a and 10.386b)].

NOTE 10.3 APPLICATIONS OF THE FOUR SCHWARTZ–CHRISTOFFEL TRANSFORMATIONS

The Schwartz–Christoffel transformations (Chapters I.33, I.34, I.36, I.39, I.40, II.2, 9, and 10) are probably the most useful class of conformal mappings due to their ability to map the upper or lower complex plane, or the interior or exterior of a circle, into the exterior or interior of arbitrary polygon, allowing for infinite sides. The ability to handle complex geometries via the Schwartz–Christoffel mapping is limited mostly by the possibility to integrate the transformation in elementary terms; the latter is less straightforward as the number of critical points increases. One of the mappings in Example 1.18 had eight critical points; this is the largest number of critical points of the 25 Schwartz–Christoffel transformations used in volume II and part 4 of volume I. As shown in Classification 8.1, the 25 Schwartz–Christoffel transformations are spread between (1) the first volume in Chapter 33, Sections 34.1 through 34.4, 36.4 through 36.9, 38.3 through 38.6, 39.4, and Examples 40.11 through 40.17; and (2) the present volume. As one of the most powerful methods of complex analysis, the Schwartz–Christoffel transformations are a fitting conclusion to the subject, before proceeding to generalized functions or distributions in the next volume. The generalized functions can represent singularities like multipoles in a space of any dimension. Thus, they provide an extension of potential theory, that is, solution of the Laplace and Poisson equations, from the plane using holomorphic complex functions to the three-dimensional space using generalized functions.

10.2 Conclusion

The potential flow due to two cylinders with different radii and velocities moving perpendicular to the line of centers can be approximated by lowest order perturbations if the mutual distance is much larger than the radii (Figure 10.1a); the case of two cylinders with the same radius moving at the same velocity is equivalent (Figure 10.1b) to images on a wall at equal distance from the centers and parallel to the velocity. The images on the cylinder and wall apply to the following: (1) the potential flow past a cylinder in the presence of a wall, both rigid and impermeable without (with) circulation [Figure 10.2a (Figure 10.2b)]; (2) the former case is analogous to the electrostatic field of an insulating cylinder and wall (Figure 10.3a); and (3) there are three other possibilities, namely, cylinder and wall both conducting (Figure 10.3d), conducting cylinder and insulating wall (Figure 10.3b), and vice versa (Figure 10.3c). The induced change distribution on the wall (Figure 10.4) has all moments zero or infinite. The flow past an oblique plate on a wall (Figure 10.5b) by reflection is equivalent (Figure 10.5a) to the flow past a symmetric bent lamina (Figure 10.5c). Likewise, the flow in a parallel-sided duct out of a reservoir with oblique walls (Figure 10.7b) corresponds (Figure 10.7a) by reflection to a convergent channel changing to a parallel-sided one (Figure 10.7c); in particular, the exit duct may be orthogonal to the wall in an unsymmetric (symmetric) [Figure 10.8a (Figure 10.8b)] configuration. Yet another case is an oblique (orthogonal) corner [Figure 10.9b (Figure 10.10a)] in a parallel-sided duct that corresponds (Figure 10.9a) by symmetry to [Figure 10.9c (Figure 10.10b)] a Y-junction (T-junction). The flow in a duct with main and side branches (Figure 10.11a) includes the case of an orthogonal branch (Figure 10.12b) and also two symmetric side branches (Figure 10.12a); its solution involves mapping the flow region (1) to the upper-half z-plane (Figure 10.11b); (2) to a horizontal strip on the hodograph plane (Figure 10.11c) of the velocity; and (3) on the plane of the complex potential (Figure 10.11d) to a region looking like a triple plate condenser (Figure 10.6a through c). A jet can be deflected without walls by a vortex (Figure 10.13b) leading to an oblique (orthogonal) bend [Figure 10.13a (Figure 10.13c)]. The upper complex z-plane [Figure 10.14a (Figure 10.16a)] can be mapped into a flat plate with one oblique flap (two oblique flaps) at any position along the chord [Figure 10.14b (Figure 10.16b)]. This specifies the corresponding potential flow in a uniform stream without circulation; in the presence of circulation, the mapping should be made from a circle [Figure 10.15b (Figure 10.18b)]. A particular case is the flap(s) orthogonal to the middle of the plate [Figure 10.15a (Figure 10.18a)]. An intermediate case is two oblique flaps with the same junction point (Figure 10.17a) and the same direction (Figure 10.17b); the latter corresponds to two crossed plates (Figure 10.17b) and includes the cruciform airfoil (Figure 10.18a). A many-valued function may have several sequences of branch-cuts accumulating to a point, for example, two (four) along the real axis (diagonals of quadrants) accumulating toward the origin [Figure 10.19 (Figure 10.20)]; since the function is analytic outside a disk containing all the branch-cuts, an expansion in descending power series is possible.

Bibliography

The bibliography of Book 2 mostly complements and, in some cases, adds or supersedes that of Book 1, which remains relevant. The bibliography on general mathematics (Section 1 of Book 1) is not repeated and is complemented by theoretical physics (Section 1 of Book 2). The bibliography on real functions and complex analysis (Sections 2 and 3 of Book 1, respectively) is reproduced with the addition of some extra subjects such as continued fractions (Sections 2 and 3 of Book 2). The bibliography on hydrodynamics and aerodynamics (Sections 4 and 5 of Book 1), although entirely relevant, is not repeated; instead, a bibliography on elasticity and structures is given (Sections 4 and 5 of Book 2) to extend the coverage from fluids to solids. Likewise, the bibliography on electricity and magnetism (Section 6 of Book 1) gives place to bibliography on thermodynamics (Section 6 of Book 2). In Books 1 and 2, the bibliography includes different approaches to the subject; those that have influenced most the contents of the present book are indicated by one, two, or three asterisks.

1. THEORETICAL PHYSICS

Brown, L. M., Pais, A., and Pippard, B. *Twentieth Century Physics*. Institute of Physics and American Institute of Physics, Washington, DC, 1995, 3 vols.

**Courant, R. and Hilbert, D. *Methods of Mathematical Physics*. Springer, New York, 1937, Academic Press, New York, 1953, 2 vols.

Duhamel, J. M. C. *Des méthodes des sciences du raisonnement*. Gauthier-Villars, Paris, 1870–1875, 5 vols.

Feynman, R. P., Leighton, R. B., and Sanos, M. *The Feynman Lectures on Physics*. Addison-Wesley, Reading, MA, 1963, 7th edition 1977, 3 vols.

*Flugge, S. (editor). *Handbuch der Physik*. Springer Verlag, Berlin, 1956, 3rd edition, circa 50 vols.

Hassani, S. *Mathematical Physics*. Springer, New York, 1998.

Hylleraas, E. A. *Mathematical and Theoretical Physics*. Wiley, New York, 1970, 2 vols.

**Jeffreys, H. and Swirles, B. *Methods of Mathematical Physics*. Cambridge University Press, Cambridge, 1st edition 1946, 3rd edition 1956.

***Landau, L. D. and Lifshitz, E. F. *Cours de Physique Theorique*. Editions Mir. Pergamon, New York, 9 vols.

Mathieu, E. *Cours de Physique—Mathématique*. Gauthier-Villars, Paris, 1978.

**Morse, P. M. and Feshbach, H. *Methods of Theoretical Physics*. McGraw-Hill, New York, 1953, 2 vols.

*Murdin, P. (editor). *Encyclopedia of Astronomy and Astrophysics*. Kluwer, The Netherlands, 1997, Institute of Physics Publishing, London, 2001, 4 vols.

Reed, M. and Simon, B. *Methods of Modern Mathematical Physics*. McGraw-Hill, New York, 1970, 4 vols.

*Sommerfeld, A. *Lectures on Theoretical Physics*. Academic Press, New York, 1964, 6 vols.

*Thomson, J. J. and Tait, P. G. *Treatise of Natural Philosophy*. Cambridge University Press, Cambridge, 1879, reprinted University Michigan Library, Dearborn, MI, 2001.

2. REAL FUNCTIONS

*Apostol, T. M. *Mathematical Analysis*. Addison-Wesley, Reading, MA, 1963.
*Appell, P. *Cours d'analyse*. Gauthier-Villars, Paris, 1920.
Appel, P. *Functions algebriques*. Gauthier-Villars, Paris, 1929–1930, reprinted Chelsea, New York, 1976, 2 vols.
Banach, S. *Théorie des operations linéaires*. Lvov, 1932, reprinted Chelsea, New York, 1978.
Berman, G. *Probémes d'analyse mathematique*. Editions Mir, 1976.
Chatterji, S. D. *Cours d'analyse*. Presses Polytechniques et Universitaires Romandes, 1975, 4 vols.
*Chilov, G. *Analyse mathematique*. Editions Mir, 1975.
*Courant, R. *Differential and Integral Calculus*. Interscience, 1st edition 1934, 2nd edition 1937, 2 vols.
Ferreira, J. C. *Análise Matemática*. Fundação Calouste Gulbenkian, Lisbon, 1985.
Franklin, P. *Methods of Advanced Calculus*. McGraw-Hill, New York, 1944.
Gauss, C. F. *Untersuchungen uber Hohere Arithmetik*. Springer, Berlin, 1889, reprinted Chelsea, New York, 1965.
Hardy, G. H. *A Course of Pure Mathematics*. Cambridge University Press, Cambridge, 1st edition 1908, 10th edition 1957.
Hermite, C. *Cours d'analyse*. Gaulthier-Villars, Paris, 1973.
Hildebrand, F. *Advanced Calculus for Engineers*. Prentice-Hall, New York, 1949.
**Hobson, W. *Functions of Real Variable*. Cambridge University Press, Cambridge, 1st edition, 1907, 2nd edition, 1921–1926, reprinted Dover, New York, 2 vols.
*Jordan, C. *Cours d'analyse*. Gaulthier-Villars, Paris, 1909–1915, 3 vols.
*Kinchin, A. Ya. *Continued Fractions*, Dover, New York, 1964.
Klein, F. *Vorlesungen weber Entwicklung der Mathematik*. Springer, Berlin, 1926, reprinted Chelsea, New York, 1970, 2 vols.
Knopp, K. *Problem Book in the Theory of Functions*. Reprinted Dover, New York, 1952, 2 vols.
**Kolmogorov, A. N. and Fomine, S. *Théorie des fonctions et analyse fonctionelle*. Editions Mir, 1937.
Landau, E. *Differential and Integral Calculus*. Groningen, 1934, reprinted Chelsea, New York, 1950.
Landau, E. *Foundations of Analysis*. Springer, Berlin, 1929, reprinted Chelsea, New York, 1951.
Lebedev, N. N., Skalskaya, I. P., and Veyland, Y. S. *Worked Problems in Applied Mathematics*, Dover, New York, 1965.
Littewood, D. E. *Theory of Functions*. Cambridge University Press, Cambridge, 1926.
*Montel, P. *Familes normales*. Gauthier-Villars, Paris, 1927, reprinted Chelsea, New York, 1974.
Olds, C. D. *Continued Fractions*. Random House, Yale University, London, 1963.
Osgood, W. F. *Advanced Calculus*. MacMillan, New York, 1928.
*Picard, E. *Traité d'analyse*. Gauthier-Villars, Paris, 1903–1908, 3 vols.
Picard, E. and Simart, G. *Functions algébriques*. Gauthier-Villars, Paris 1897–1906, reprinted Chelsea, New York, 1971, 2 vols.
Pólya, G. and Szego, G. *Aufgaben und Lehrsatze aus der Analysis*. Springer, Berlin, 1925, 4 Aufgabe 1970.
Randolph, J. F. *Basic Real and Abstract Analysis*. Academic Press, New York, 1968.
Rothe, R., Ollendorf, F., and Pohlausen, K. *Theory of Functions as Applied to Engineering Problems*. MIT Press, Cambridge, MA, 1933, reprinted Dover, New York, 1961.
Rudin, W. *Principles of Mathematical Analysis*. McGraw-Hill, New York, 1953.
Schwartz, L. *Analyse*. Hermann, Paris, 1979.
Schwartz, L. *Analyse Hilbertienne*. Hermann, Paris, 1979.
Sturm, J. C. F. *Course d'annalyse*. Gauthier-Villars, Paris, 1880.
*Teixeira, F. G. *Curso de análise infinitesimal*. Tipografia Ocidental, Porto, Portugal, 1891–1903, 3 vols.
*Tichmarch, E. C. *The Theory of Functions*. Oxford University Press, New York, 1st edition 1932, 2nd edition 1939.
Vallée de Poussin, C. J. *de la Cours d'analyse infinitesimale*. Louvain, 1914, 2 vols.
Weyl, H., Landau, E., and Riemann, B. *Der Kontinuum und andere Monographien*. Springer, Berlin, 1917, reprinted Chelsea, New York, 1960.

3. COMPLEX ANALYSIS

Achieser, N. I. *Theory of Approximation*. Frederick Ungar, New York, 1956, reprinted Dover, New York, 1992.

Ahlfors, L. A. *Complex Analysis*. McGraw-Hill, New York, 1953, 3rd edition 1979.

Ávila, G. *Variáveis complexas*. Livros Técnicos e Científicos, Rio de Janeiro, 2000.

Bieberbach, L. *Conformal Mapping*, 1949. Chelsea, New York, 1953.

Brezinski, C. *Continued Functions and Padé Approximants*. Springer, Berlin, 1991.

**Bromwich, J. T. I. A. *Introduction to the Theory of Infinite Series*. MacMillan, 1908, 2nd edition 1926.

**Caratheodory, C. *Theory of Functions (Real and Complex)*. Verlag Birkhauser, Basel, 1950, Chelsea, New York, 1854, 1+2 vols.

Carrier, G. F., Krook, M., and Pearson, C. E. *Functions of a Complex Variable*. McGraw-Hill, New York, 1966.

Cartan, H. *Functions analytiques d'une ou plusieurs variables complexes*. Hermann, Paris, 1985.

*Chabat, B. *Analyse complexe*. Nauka, Moscow, 1985, Editions Mir 1990, 2 vols.

**Chrystal, G. *Textbook of Algebra*. A&C Black, London, 1904, Chelsea, New York, 1964, 2 vols.

*Churchill, R. V., Brown, J. W., and Verbey, R. E. *Complex Variables and Applications*. McGraw-Hill, 1st edition 1949, 3rd edition 1974.

*Copson, E. T. *An Introduction to the Theory of Functions of Complex Variable*. Oxford University Press, London, 1935.

Courant, R. *Dirichlet's Principle, Conformal Mapping and Minimal Surfaces*. Interscience, New York, 1950.

Davis, H. T. *The Summation of Series*. Principia Press, San Antonio, TX, 1962.

Dettman, J. W. *Applied Complex Variables*. MacMillan, New York, 1965, 5th edition 1970, reprinted Dover, New York, 1984.

Egrafov, M. S. *Recueil de problemes sur la theorie des functions analytiques*. Editions Mir 1974.

Flanigan, F. J. *Complex Variables*. Allyn & Becoon, Boston, MA, 1972, reprinted Dover, New York, 1983.

Ford, W. B. *Divergent Series and Asymptotic Expansions*. Michigan, 1916, reprinted Chelsea, New York, 1960.

*Forsyth, A. R. *Theory of Functions of a Complex Variable*. Cambridge University Press, Cambridge, 1st edition 1893, 3rd edition 1918, Dover, New York, 1965, 2 vols.

Goffman, C. and Pedrick, G. *A First Course in Functional Analysis*. Prentice-Hall, Englewood Cliffs, NJ, 1965.

***Goursat, E. *Cours d'analyse mathematique*. Gauthier-Villars, Paris, 1911, 3 vols., Dover, New York, 1959, 5 vols.

Hardy, G. H. *Divergent Series*. Oxford University Press, New York, 1949, reprinted Chelsea, New York, 1991.

Hille, E. *Analytic Function Theory*. Chelsea, New York, 1976, 2 vols.

Hormander, L. *Complex Analysis in Several Variables*. North-Holland, Amsterdam, 1966.

Jolley, L. B. W. *Summation of Series*. Chapman & Hall, London, 1925, reprinted Dover, New York, 1961.

Jones, W. B. and Thron, W. J. *Continued Fractions: Analytic Theory and Application*. Addison-Wesley, Reading, MA, 1980, Cambridge University Press, Cambridge, 1984.

Knopp, K. *Allgemeine und Aufgabenzammlung der Funktionentheorie*. Tubingen, 1921–1947, 2 vols., Dover, New York, 1945, 4 vols.

*Knopp, K. *Elemente des Funktionentheorie*. Dover, New York, 1952.

**Knopp, K. *Infinite Sequences and Series*. Dover, New York, 1956.

*Knopp, K. *Theory and Applications of Infinite Series*. Tubingen, 1921, 4th edition 1947, reprinted Haffner, Royal Oak, MI, 1971.

Krantz, S. G. *Handbook of Complex Variables*. Birkhauser, Basel, 1999.

Lagrange, J. L. *Theorie des functions analytiques*. Imprimerie de la Republique, Paris, 1794.

Laurent, H. *Théorie des residus*. Gauthier-Villars, Paris, 1866.

*Lavrentiev, M. and Chabat, B. *Méthodes de théorie des fonctions d'une variable complexe*. Editions Mir 1977.
Lusternik, L. and Sobolev, V. *Précis d'analyse fonctionel*. Nauka, Moscow, 1982, Editions Mir 1989.
*Markushevich, A. I. *Theory of Functions of a Complex Variable*. Reprinted Chelsea, New York, 1965, 3 vols.
Michel, A. N. and Herbert, C. J. *Applied Algebra and Functional Analysis*. Prentice-Hall, Englewood Cliffs, NJ, 1981, reprinted Dover, New York, 1993.
Miller, K. S. *Advanced Complex Calculus*. Harper & Row, New York, 1960, Dover, New York, 1970.
Nehari, Z. *Conformal Mapping*. McGraw-Hill, New York, 1952, reprinted Dover, New York, 1975.
Nevanlinna, R. *Le théoreme de Borel-Picard et la théorie des functions méromorphes*. Gauthier-Villars, Paris, 1926.
Nevanlinna, R. *Analytic Functions*. Springer, Berlin, 1973.
Nevanlinna, R. and Paatero, V. *Complex Analysis*. Birkhauser, Basel, 1969, reprinted Chelsea, New York, 1982.
Osgood, W. F. *Lehrbuch der funktionentheorie*. MIT Press, Cambridge, MA, 1929–1932, reprinted Chelsea, New York, 1965.
**Perron, O. *Die Lehre von den Kettenbrüchen*. Teubner, Wiesbaden, Germany, 1918, reprinted University of Michigan Libraries.
Polya, G. and Szego, G. *Isoperimetric Inequalities in Mathematical Physics*. Princeton University Press, Princeton, NJ, 1960.
*Riesz, F. and Sz. Nagy, B. *Functional Analysis*. Frederick Ungar, New York, 1955, reprinted Dover, New York, 1990.
Rivlin, T. J. *Approximation of Functions*. Blaisdell, Waltham, MA, 1969, reprinted Dover, New York, 1981.
Schinzinger, R. and Laura, P. A. A. *Conformal Mapping*. Elsevier, Amsterdam, 1991.
*Sobolev, S. L. *Functional Analysis*. 1950. American Mathematical Society, Providence, RI, 1963.
Sveschinikov, A. and Tikhonov, A. *The Theory of Functions of a Complex Variable*. Mir Publishers, Moscow, 1978.
Timan, A. F. *Approximation of Functions of a Real Variable*. 1960. Clarendon, Oxford, 1963, Dover, New York, 1994.
Tutschke, W. and Vasudeva, H. L. *An Introduction to Complex Analysis*. Chapman & Hall, Boca Raton, FL, 2004.
*Valiron, G. *Théorie des fonctions*. Masson, France, 1st edition 1942, 3rd editions 1966.
Wall, H. S. *Analytic Theory of Continued Functions*. Van Nostrand, New York, 1948.
***Whittaker, E. T. and Watson, G. N. *Course of Modern Analysis*. Cambridge University Press, Cambridge, 1st edition 1902, 4th edition 1927.

4. ELASTICITY AND PLASTICITY

Abir, D. (editor). *Contributions to Mechanics*. Pergamon, New York, 1969.
Antmann, S. S. *Non-linear Problems of Elasticity*. Springer, New York, 1995, 2nd edition 2004.
Chandrasekharaiah, D. S. and Debnath, L. *Continuum Mechanics*. Academic Press, San Diego, CA, 1984.
Chung, T. J. *Applied Continuum Mechanics*. Cambridge University Press, Cambridge, 1996.
Duvaut, G. *Mécanique des millieux continus*. Masson, France, 1990.
Eirich, F. R. *Rheology Theory and Applications*. Academic Press, New York, 1956, 4 vols.
Findley, W. N., Lai, J. S., and Onaran, K. *Creep and Relaxation of Non-linear Viscoelastic Materials*. North-Holland, Amsterdam, 1976, reprinted Dover, New York, 1989.
Flugge, W. *Tensor Analysis and Continuum Mechanics*. Springer, Berlin, 1972.

*Hetnarski, R. B. and Ignaczak, J. *The Mathematical Theory of Elasticity*. CRC Press, Boca Raton, FL, 2001.
*Hill, H. *Theory of Plasticity*. Oxford University Press, London, 1950.
Lai, W., Rubin, D., and Krempl, E. *Continuum Mechanics*. Pergamon, New York, 1993.
*Lamé, G. *Leçons sur la theorie mathematique de l'elasticité des corps solides*. Gauthier-Villars, Paris, 1860.
***Love, A. E. H. *Treatise on the Mathematical Theory of Elasticity*. Cambridge University Press, Cambridge, 1904, 4th edition 1926, reprinted Dover, New York, 1944.
Oliveira, E. R. A. *Teoria de Elasticidade*. IST Press, Lisbon, 1999.
Parton, V. and Morozov, E. M. *Elastic-Plastic Fracture Mechanics*. Editions Mir, 1978.
Prager, W. *Introduction to the Mechanics of Continua*. Reprinted Dover, New York, 1972.
Prager, W. and Hodge, P. G. *Perfectly Plastic Solids*. Wiley, New York, 1951, Dover, New York, 1968.
*Prescott, J. *Applied Elasticity*. Longmans, Green & Co., London, 1924, Dover, New York, 1961.
Rabotnov, Y. N. *Hereditary Solid Mechanics*. Nauka, Moscow, 1977, Mir Publishers, Moscow, 1980.
Roy, M. *Mécanique*. Dunod, Paris, 1965, 2 vols.
Saad, M. H. *Elasticity: Theory, Applications and Numerics*. Elsevier, Amsterdam, 2009.
Sedov, L. I. *Foundations of the Non-linear Mechanics of Continua*. Pergamon, New York, 1966.
Sedov, L. I. *Similitude et Dimensions en Mecanique*. Editions Mir, 1972.
Sedov, L. I. *Mécanique des millieux continus*. Editions Mir, 1973, 2 vols.
Segel, E. A. *Continuum Mechanics*. MacMillan 1977, reprinted Dover, New York, 1987.
*Sokolnikoff, I. S. *Mathematical Theory of Elasticity*. McGraw-Hill, New York, 1956.
*Southwell, R. V. *An Introduction to the Theory of Elasticity*. Oxford University Press, London, 1936, 2nd edition 1941.
Timoshenko, S. P. *Strength of Materials*. 1930. Van Nostrand, New York, 1966.
*Timoshenko, S. P. and Goodier, J. N. *Theory of Elasticity*. 1934. McGraw-Hill, New York, 1970.
Timoshenko, S. P. and Gere, J. M. *Theory of Elasticity Stability*. 1936. McGraw-Hill, New York, 1961.
Timoshenko, S. P. and Woinowsky-Krieger, S. *Theory of Plates and Shells*. 1940. McGraw-Hill, New York, 1959.
Yao, M., Zhong, W., and Lim, C. W. *Sympletic Elasticity*. World Scientific, Singapore, 2009.

5. STRUCTURES

Amabili, M. *Non-linear Vibration and Stability of Shells and Plates*. Cambridge University Press, Cambridge, 2008.
*Asaro, R. J. and Lubarda, V. A. *Mechanics of Solids and Materials*. Cambridge University Press, Cambridge, 2006.
Bauchau, O. A. and Craig, J. J. *Structural Analysis*. Springer, 2009.
Bazart, Z. P. and Cedolin, L. *Stability of Structures*. Dover, New York, 1991.
Bower, A. F. *Applied Mechanics of Solids*. CRC Press, Boca Raton, FL, 2010.
Chakraverty, S. *Vibration Mechanics of Solids*. CRC Press, Boca Raton, FL, 2010.
Clough, R. W. and Benzien, J. *Dynamics of Structures*. McGraw-Hill, New York, 1982.
Gatti, P. L. and Ferrari, V. *Applied Structural and Mechanical Vibration*. E & F N Spon, London, 1999.
Hetnarski, R. B. *Thermal Stresses*. North-Holland, Amsterdam, 1991, 2 vols.
Hodge, P. G. and Goodier, J. N. *Elasticity and Plasticity*. Wiley, New York, 1958.
Kraus, H. *Thin Elastic Shells*. Wiley, New York, 1967.
Krenk, S. *Non-linear Modelling and Analysis of Solids and Structures*. Cambridge University Press, Cambridge, 2009.
Lekhnitski, S. G. *Theory of Elasticity of an Anisotropic Body*. Mir Publishers, Moscow, 1977.
Lin, T. H. *Theory of Inelastic Structures*. Wiley, New York, 1968.
Mansfield, E. H. *The Bending and Stretching of Plates*. Oxford University Press, New York, 1964.

Meirovitch, L. *Elements of Vibration Analysis and Vibrations*. McGraw-Hill, New York, 1975.
*Muskhelishvili, W. I. *Plane Theory of Elasticity, Torsion and Bending*. Nauka, Moscow, 5th edition 1966, Noordhoff, Groningen, 1953.
Norris, C. H., Nilbas, J. B., and Utku, S. *Elementary Structural Analysis*. McGraw-Hill, New York, 1948, 3rd edition 1976.
Palazotto, A. N. and Dennis, S. T. *Non-linear Analysis of Shell Structures*. American Institute of Aeronautics and Astronautics, Washington, DC, 1992.
Parton, V. and Perline, P. *Théorie mathematique de l'elasticité*. Nauka, Moscow, 1981, Editions Mir 1984.
Pilkey, W. D. *Stress Concentration Factors*. Wiley, Hoboken, NJ, 2nd edition 1997.
*Rekach, V. G. *Manuel of Theory of Elasticity*. Editions Mir 1979.
Roark, R. J. and Young, W. C. *Formulas for Stress and Strain*. McGraw-Hill, New York, 1975.
Rodrigues, J. and Martins, P. *Tecnologia Mecânica*. Escolar Editora, Lisbon, 2005, 2 vols.
Sih, G. C., Michopoulos, J. G., and Chou, S. C. *Hygrothermoelasticity*. Martinus-Nijhoff, The Netherlands, 1986.
Vorhees, H. R. *Compilation of Stress–Relaxation Data for Engineering Alloys*. American Society for Testing and Materials, Philadelphia, PA, 1982.
Warburton, G. B. *The Dynamical Behaviour of Structures*. Oxford University Press, 1964, 2nd edition 1976.
Whitney, J. M. *Structural Analysis of Laminated Anisotropic Plates*. Technomic Press, Lancaster, PA, 1987.

6. THERMODYNAMICS AND HEAT

Azevedo, E. G. *Thermodynamics Aplicada*. Escolar Editora, Lisbon, 1995, 2nd edition 2000.
Barrère, M. and Prud'Homme, R. *Aerothermochimie des ecoulements homogénes*. Gauthier-Villars, Paris, 1970.
Bazarov, I. *Thermodynamique*. Nauka, Moscow, 1983, Mir Publishers, Moscow, 1989.
Bird, R. B., Stewart, W. E., and Lightfoot, E. N. *Transport Phenomena*. Wiley, New York, 1960.
Burgers, J. M. *The Non-linear Diffusion Equation*. Reidel, Boston, MA, 1964.
**Callen, H. B. *Thermodynamics*. Wiley, New York, 1960.
Candel, S. M. *Mécanique des fluides*. Dunod, Paris, 1995.
Carnot, L. *Reflexion sur le puissance motrice du feu et sur les machines propes à developer cette puissance*. Gauthier-Villars, Paris, 1878.
*Carslaw, H. S. and Jaeger, H. S. *Heat Conduction in Solids*. Clarendon, Oxford, 1st edition 1946, 2nd edition 1953.
*Chandrasekhar, S. *Radiative Transfer*. Oxford University Press, London, 1950, reprinted Dover, New York, 1960.
Davies, J. T. and Rideal, E. K. *Interfacial Phenomena*. Academic Press, London, 1961.
Denbigh, K. *Chemical Equilibrium*. Cambridge University Press, Cambridge, 1957.
Eckert, E. R. G. and Drake, R. M. *Heat and Mass Transfer*. McGraw-Hill, New York, 1959.
*Fermi, E. *Thermodynamics*. Prentice-Hall, New York, 1973, reprinted Dover, New York, 1956.
Frank-Kamenetskii, D. A. *Diffusion and Heat Exchange in Chemical Kinetics*. Princeton University Press, Princeton, NJ, 1947.
Giles, R. *Thermodynamics*. Pergamon, New York, 1964.
Groot, S. R. and Mazur, P. *Non-equilibrium Thermodynamics*. North-Holland, Amsterdam, 1980.
Groot, S. R. *Thermodynamics of Irreversible Processes*. North-Holland, Amsterdam, 1986.
Guemez, J., Fiolhais, C. and Fiolhais, M. *Termodinamica do equilíbrio*. Fundação Calouste Gulbenkian, Lisbon, 1998.
*Guggenheim, E. A. *Thermodynamics*. North-Holland, Amsterdam, 1949, 3rd edition 1957.

Hahn, H. G. *Methode des finiten Elemente in der Festigkeitsleher*. Akademische Verlag Gesellschaft, Frankfurt, 1975.
Isachenko, V. P., Osipova, V. A., and Sukomel, A. S. *Heat Transfer*. Mir Publishers, Moscow, 1975.
Kestin, J. *Thermodynamics*. Blaisdell, Waltham, MA, 1966.
Kuo, K. K. *Principles of Combustion*. Wiley, Hoboken, NJ, 1986.
Lamé, G. *Leçons sur la Theorie Analytique de la Chaleur*. Gauthier-Villars, Paris, 1861.
Leontiev, A. *Theorie des échanges de chaleur et masse*. Nauka, Moscow, 1979, Editions Mir 1985.
MacAdams, W. H. *Heat Transmission*. McGraw-Will, New York, 1954.
Meyer, E. and Schiffner, E. *Technische Thermodynamik*. VC17 Verlagsgesellschaft, 1986.
Ozisik, M. N. *Boundary-value Problems in Heat Conduction*. International Textbook Company, Scranton, PA, 1968, reprinted Dover, New York, 1989.
Pauling, L. *The Nature of the Chemical Bond*. Cornell University Press, Ithaca, NY, 1939, 3rd edition, 1960.
Philips, N. V. *Problems of Low Temperature Physics and Thermodynamics*. Pergamon, New York, 1962.
*Pipard, A. B. *Thermodynamics*. Cambridge University Press, Cambridge, 1966.
*Planck, M. *Thermodynamics*. Berlin, 1905, 5th edition 1917.
*Planck, M. *Heat Radiation*. Berlin, 1912, 2nd edition Blakiston, 1914, reprinted Dover, New York, 1959.
Poinsont, T. and Veynante, D. *Theoretical and Numerical Combustion*. R.T. Edwards, Philadelphia, PA, 2001.
Prigogine, I. *Thermodynamics of Irreversible Processes*. Wiley Interscience, New York, 1955, 3rd edition 1967.
Roberts, J. K. *Heat and Thermodynamics*. Blackie, London, 1928, 3rd edition 1940.
Roy, M. *Thermodynamique Macroscopique*. Dunod, Paris, 1964.
Shah, R. K. and Sekulic, D. P. *Heat Exchanger Design*. Wiley, Hoboken, NJ, 2003.
*Tisza, L. *Generalized Thermodynamics*. MIT Press, Cambridge, MA, 1966.
*Truesdell, C. A. *Rational Thermodynamics*. McGraw-Hill, New York, 1969.
Zemansky, M. W and Dittman, R. H. *Heat and Thermodynamics*. McGraw-Hill, New York, 1937, 6th edition 1979.

References

(The authors mentioned in the text in chronological order, indicating the date, name, and publications)

1569	Mercator, G. K. Chronologia hoc est Temporum. *Atlas Gerardi Mercatonis*. Amsterdam, 1609.
1614	Napier, J. *Marifi Logarithmorum cannonis discriptio.*
1624	Briggs, H. *Tables of Decimal Logarithms.*
1656	Wallis, J. *Arithmetica Infinitorum* (*Opera Mathematica*, 1). Oxford.
1678	Hooke, R. *De Potentia Restitutiva*. London.
1713	Bernoulli, J. *Ars conjectandi*. Basel.
1737	Euler, L. De Fractiones continuis. *Commentationes Acadamiae Scientarum Imperialis Petropolitana*, 9, 160–168.
1738	Bernoulli, D. *Hydrodinamica*. Argentorati.
1739	Euler, L. Des fractiones continuis obsrvationes. *Commentationes Acadamiae Scientarum Imperialis Petropolitana*, 9.
1744	Euler, L. *Methodus inveniendi lineas curvas maximi minimive proprieate gaudentes, sive solutio problematis isoperimetrici latíssimo sensu accepti*. Lausanne.
1748	Euler, L. *Introductio in Analysis Infinitorum*, Volume 1, Chapter 18. St. Petersburg.
1752	Euler, L. Principles géneraux du mouvement dês fluides. *Histoire de l'Académie de Berlin* 1755, in: *Rational Fluid Mechanics 1687–1765*, edited by C.A. Truesdell.

1755 Euler, L. *Institutiones Calculi Differentialis* (*Opera Omnia* 10). St. Petersburg.

1759 Euler, L. De Principis motus fluidorum. *Novi Commentari Academia Scientarium Imperialis Petropolitanae* 14, 1; *Opera Omnia*. Basel Akademie, Basel.

1760 Lagrange, J. L. *Miscellania Turinensia* 2, 173–195 (*Ouevres* 1, 335–362).

1770 Lambert, J. H. *Beiträge zum Gebrauch der Mathematik und deren Anwendung*. Berlin.

1776 Lagrange, J. L. Sur l'usage des fractions continues dans le calcul integral. *Noveaux Mémoires de L'Académie Royalle des Sciences et Belles Lettres de Berlin (Oeuvres, 6).*

1776 Lagrange, J. L. Essai sur une nouvelle methode pour determiner les máxima et mínima des formulas integrales indéfenies. *Miscellania Turinensia* 4, 163–187 (*Ouevres* 2, 37–63).

1776 Meusnier, J. B. Mémoire sur la courbure des surfaces. *Académie dês Sciences de Paris, Mémoires des Savants Étrangers* 10, 477–510 (1785).

1785 Euler, L. *Opuscula Analytica II.*

1807 Young, T. A course of lectures on natural philosophy and the mechanical arts. London.

1812 Gauss, C. F. Disquitiones generalis circam seriem infinitam... *Commentationes Societones Regia Scientarum Göttensis Recentiores* (*Werke*, 1, 85).

1813 Euler, L. Commentatio in fractionem continuam qua illustris La Grange potestates binomials expressit. *Mémpoires de l'Académie imperiale des Aciences de St. Petersburg*, 6.

1818 Fourier, J. B. J. *Theorie analytique de la chaleur*. Paris. Reprinted Dover, New York, 1955.

1820 Gauss, C. F. *Allgemeine Flachentheorie*, translation Engelmann 1900.

1821 Cauchy, A.L. *Analyse Algebrigue*. Paris.

1822 Navier, C. L. M. H. Mémoires sur les lois du mouvement des fluides. *Memoires de l'Academie des Sciences de Paris* 6, 389.

1827 Jacobi, C.G.J. Ueber eine besondere gattung algebraischer Funktionen, die aus Entwicklung der Funktion $(1 - 2xz + z^2)^{-1/2}$ entstehen. *Journal fur Mathematik* 2 (Werke, 6).

1828 Green, G. *Essay on electricity and magnetism*, Nottingham. *Mathematical Papers*, p. 3. Cambridge University Press, Cambridge, 1871.

1829a Poisson, S. D. Mémoire sur l'équilibre et le mouvement des corps elastiques. *Memóires de l'Académies des Sciences de Paris* 8.

1829b Poisson, S. D. Mémoire sur les equations generales de l'équilibre et mouvement des corps élastiques et des fluides. *Journal de l'Ecole Polytechnique* 8, 1.

1836 Kummer, E.E. Ueber die hypergeometrische Reihe. *Journal fur reine und angewandte Mathematik* 15, 39–82 and 127–172.

1840 Poiseuille, J. L. M. Recherches experimentales sur le mouvement des liquides dans les tubes de très petits diamètres. *Comptes Rendues de l'Academie des Sciences*, 9.

1843 Saint-Venant, B. *Comptes Rendus de l'Academie des Sciences* 13, 1240.

1845 Stokes, G. G. On the theories of the internal friction of fluids in motion. *Transactions of the Cambridge Philosophical Society* 7, 287 (*Papers* 1, 75).

1846 Poiseuille, J. L. M. *Mémoires des Savants Etrangers*, 11.

1849 Kelvin. On the vis-viva of a liquid in motion. *Cambridge and Dublin Mathematical Journal* (*Papers* 1, 107).

1852 Lamé, G. *Leçons sur la théorie mathématique de l'elasticité des corps solides*. Paris.

1855 Saint-Venant, B. Sur la theorie de la torsion. *Mémoires des Savants Étrangers* 14.

1858 Rouché, F. Mémoire sur le development des functions en series ordonées suivant les reduites d'une fraction continue. *Journal de l'École Polytechnique*, 37.

1859 Riemann, B. *Berliner Monatsberichte*, 671–680 (*Gesammelte Werke*, 136–144).

1859 Chebychev, P. L. Sur les questions de minima, qui se rattachent à la representation approximative des functions. *Mémoires de l'Academie Scientifique de St. Petersburg* 7, 199–291 (*Ouevres* 1, 271–378).

1868 Christoffel, E. B. *Annali di Matematica Pura Applicata* 1, 89–103 (1868); 4, 1–9 (1871).

1868 Helmholtz, H.L. von Üeber discontinuirliche Flussigkeitsbewegungen. *Monatshefte der Königlichen Preussische Akademie der Wissenchaften Berlin* 23, 215-218 (Wissenchaften Abhandlungen 1, 146-157).

1871 Kelvin (J.J. Thomson, Lord) Hydrokinetic solution and observations. *Philosophical Magazine* 42, 362–377 (*Papers* 4, 69–85).

1874 Cantor, G. *Crelle Journal* 77, 260.

1876 Mittag-Leffler, G. En method att analystik framställa en funktion at rational karacte. *Öfensigt Konglik Vetenskap Akademic Förhandliger* 33, 3–16.

1876 Weierstrass, K.W.T. Theorie der eindeutingen Funktionnen. *Berlinen Abhandlungen* 11 (*Werke* 2, 77).

1877 Laguere, E. Sur le developement en fraction continue de exp arc tg $(1/x)$. *Bulletin de la Societé Mathematique de France* (*Ouevres* 1).

1878 Cantor, G. *Crelle Journal* 84, 242.

1879 Laguerre, E. Sur la function $\left(\frac{x+1}{x-1}\right)^{\omega}$. *Bulletin de la Societé Mathematique de France* (*Ouevres*, 1).

1880 Kelvin (J.J. Thomson, Lord) On the vibrations of a columnas vortex. *Philosophical Magazine* 5, 155 (*Papers* 4, 152).

1881 Routh, E. J. Some applications of conjugate functions. *Proceedings of the London Mathematical Society* 73.

1883 Reynolds, O. An experimental investigation of the circumstances which determine whether the motion of water shall be direct or sinuous, and of the law of resistance of parallel channels. *Philosophical Transactions of the Royal Society* 124, 935 (Papers 2, 51).

1883a Cantor, G. *Grundlagen einer allgemeinen Mannigfaltigkeitslehre*. Leipzig.

1883b Cantor, G. *Mathematik Annalen* 31, 545.

1884 Mittag-Leffler, G. Sur une fonction analytique des fonctions monogènes uniformes d'une variable independante. *Acta Mathematica* 4, 1–79.

1890 Schwartz, H. A. *Abhandlungen*. Teubner, Berlin, 1890. Reprinted Chelsea, New York, 1972.

1890 Césaro, E. *Bulletin de Sciences Mathematiques* 14, 114.

1892 Flamant, *Comptes Rendus de l'Académie des Sciences*, 114.

1892 Padé, H. Sur la representation approchée d'une function par des fractions rationelles. *Annales de l'École Normale Supérieure*, 9.

1893 Doppler, C. Ueber das farbige Licht der Dopplelsterne und einiger anderer Gestirne des Himmels. *Abhandlungen Bohmischen Gesammelte Wissenschaften* 2, 467 (*Ostwald Klassiker der Exakten Wissenschaften* 161, 1–34).

1895 Cantor, G. Beiträge zur Begründung der Transfiniten Mengenlehre. *Mathematik Annalen* 44.

1897 Cantor, G. Beiträge zur Begründung der transfiniten Mengenlehre. *Mathematik Annalen* 46.

1897 Burali-Forti. *Rendiconti del Circolo Matematico di Palermo* 9, 164.

1898 Lévy, M. *Comptes Rendus de l'Academie des Sciences de Paris* 127, 10.

1899 Padé, H. Mémoire sur les developments en fractions continues de la function exponentielle pouvant server d'introduction à la theorie des fractions continues algebriques. *Annales de l'École Normale Superieure* 16.

1900a Michell, J. H. *Proceedings of the London Mathematical Society* 31, 144–146.

1900b Michell, J. H. *Proceedings of the London Mathematical Society* 32, 35.

1902a Kutta, W. M. Aufstriebskräfte in strömender Flussigkeiten *Illustrierte. Aeronautishe Miheilunger*, 138.

1902b Kutta, W. M. *Sitzberachtung der konigliche Bayerischen Acedemie der Wissenschaften*.

1903 Hardy, C. H. A theorem concerning infinite cardinal numbers. *Quarterly Journal of Mathematics* 35.

1903 Russell, B. *Principles of Mathematics*, 3 vols. Cambridge University Press, Cambridge.

1905 Bernstein. *Mathematik Annalen* 60, 187.

1908 Zermelo. *Mathematik Annalen* 65, 110.

1910 Blasius, H. Funktiontheoretische methoden in der Hydrodynamik. *Zeitschrift for Mathematik und Physik* 58.

1910 Joukowski, N. Y. *Zeitschrift fur Flugtechnik und Motorluftschiffahrt* 1, 281 (1910); 3, 81 (1912).

1911	Karman, T. von Ueber der Mechanismus der widerstands, den ein bewegter Korper in ein Flussigueit erfarht, *Göttinger Nachrichten, Mathematik und Physik Klasse* 12, 509–542.
1912	Karman, T. von Flussiskeits und Wiederstand. *Physikaliche Narichten der Göttinger Gesselschaft fur Wissenschaften* 13, 547–556.
1913	Föppl, L. Wirbelbewegungen himten ein Kreiszylinder. *Sitzbereich der Konigkliche Bayerischen Akademie der Wissenschaften* 1, 7–18.
1929	Mansell, W. E. *Tables of Logarithms to 110 Decimals*. Vol. 8, Royal Society Mathematical Tables, 14 vols. Cambridge University Press, Cambridge, 1964.
1930	Galerkin, B. G. On the analysis of stresses and strains in on elastic isotropic body. *Proceedings of the Academy of Sciences of the USSR* A14, 353–358.
1931	Galerkin, B. G. On the general solution of the elasticity problem in three dimensions using stress and displacement functions. *Proceedings of the Academy of Sciences of the USSR* A10, 281–286.
1965	Sedov, L. I. *Two-dimensional Problems in Hydrodynamics and Aerodynamics*. Wiley, New York.
1997	Hallock, J. N. and Burnham, D. C. Decay characteristics of wake vortex from jet transport aircraft. *American Institute of Aeronautics and Astronautics* paper 87-0060, 35th Aerospace Science Meeting Exhibition, Reno, USA.
2010	Campos, L. M. B. C. and Cunha, F. S. R. P. On the torsion of a prism with non-equilateral triangular cross-section. *International Journal of Engineering Sciences* 48, 718–725.

Index

Note: *Italicized* page references denote notes, figures, tables, diagrams, panels, classifications, lists, and examples.

* See the page in *Complex Analysis with Applications to Flows and Fields*, L.M.B.C. Campos, ISBN 978-1-4200-7118-4.

D

I

J

K

L

M

N

O

P

W

Y

Z

Milton Keynes UK
Ingram Content Group UK Ltd.
UKHW051858071024
449327UK00025B/2005